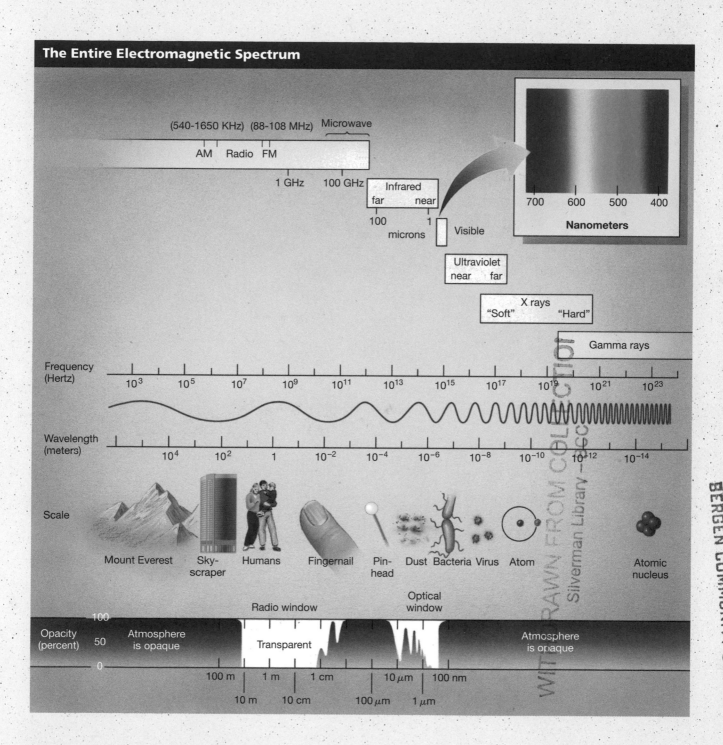

The Entire Electromagnetic Spectrum

Astronomy Today

ABOUT THE AUTHORS

Eric Chaisson

Eric holds a doctorate in astrophysics from Harvard University, where he spent ten years on the Faculty of Arts and Sciences. For five years, Eric was a Senior Scientist and Director of Educational Programs at the Space Telescope Science Institute and Adjunct Professor of Physics at Johns Hopkins University. He then joined Tufts University, where he is now Professor of Physics, Professor of Education, and Director of the Wright Center for Innovative Science Education. He has written nine books on astronomy, which have received such literary awards as the Phi Beta Kappa Prize, two American Institute of Physics Awards, and Harvard's Smith-Weld Prize for Literary Merit. He has published more than 100 scientific papers in professional journals, and has also received Harvard's Bok Prize for original contributions to astrophysics.

Steve McMillan

Steve holds a bachelor's and master's degree in Mathematics from Cambridge University and a doctorate in Astronomy from Harvard University. He held post-doctoral positions at the University of Illinois and Northwestern University, where he continued his research in theoretical astrophysics, star clusters, and numerical modeling. Steve is currently Distinguished Professor of Physics at Drexel University and a frequent visiting researcher at Princeton's Institute for Advanced Study and the University of Tokyo. He has published more than 50 scientific papers in professional journals.

Astronomy Today

Fifth Edition

Eric Chaisson
Tufts University

Steve McMillan
Drexel University

PEARSON

Prentice
Hall

Upper Saddle River, New Jersey 07458

Library of Congress Cataloging-in-Publication Data

Chaisson, Eric.
    Astronomy today / Eric Chaisson, Steve McMillan.—5th ed.
      p. cm.
    Includes index.
    ISBN 0-13-144596-0 (case)—ISBN 0-13-117683-8 (pbk.: v. 1)—ISBN 0-13-117684-6 (pbk.: v. 2)
    1. Astronomy. I. McMillan, S. (Stephen)   II. Title.
    QB43.3.C48 2005
    520--dc22   2004011767

Senior Editor: Erik Fahlgren
Editor in Chief, Science: John Challice
Production Editor: Shari Toron
Electronic Production Specialist/
    Electronic Page Makeup: Joanne Del Ben
Assistant Manager, Formatting: Allyson Graesser
Executive Managing Editor: Kathleen Schiaparelli
Editorial Assistant: Andrew Sobel
Executive Marketing Manager: Mark Pfaltzgraff
Editor in Chief, Development: Carol Trueheart
Development Editor: Erin Mulligan
Vice President of Production
    and Manufacturing: David W. Riccardi
Director of Creative Services: Paul Belfanti
Creative Director: Carole Anson
Art Director: John Christiana
Interior and Cover Designer: Susan Anderson
Managing Editor, Audio & Visual Assets
    and Production: Patty Burns
Manager, Art Production Technologies: Matt Haas
Cover image: Multiwavelength composite image of M81
Credit: NASA/JPL-Caltech/K. Gordon (University of Arizona) & S. Willner (Harvard-Smithsonian Center for Astrophysics)
Physlets® is a registered trademark of Wolfgang Christian.

AV Editor: Connie Long
Art Studio: ArtWorks
Project Coordinator: Daniel Missildine
Illustrator: Mark Landis
Manufacturing Manager: Trudy Pisciotti
Manufacturing Buyer: Alan Fischer
Copy Editor: Brian Baker
Proofreader: Mike Rossa
Image Resource Center Director: Beth Brenzel
Image Coordinator: Debbie Latronica
Photo Researcher: Sheila Norman
Site Supervisor, Central Scanning Services: Joe Conti
Central Scanning Services: Greg Harrison, Corrin Skidds,
    Robert Uibelhoer, Ron Walko
Assistant Managing Editor, Science Media: Nicole Jackson
Media Editor: Michael J. Richards
Associate Editor: Christian Botting
Assistant Managing Editor,
    Science Supplements: Becca Richter

© 2005, 2002, 1999, 1996, 1993 by Pearson Education, Inc.
Pearson Prentice Hall
Pearson Education, Inc.
Upper Saddle River, NJ 07458

Printed in the United States of America

10 9 8 7 6 5 4 3 2

ISBN 0-13-144596-0

Pearson Education LTD., *London*
Pearson Education Australia PTY, Limited, *Sydney*
Pearson Education *Singapore*, Pte. Ltd.
Pearson Education North Asia Ltd., *Hong Kong*
Pearson Education Canada, Ltd., *Toronto*
Pearson Educación de Mexico, S.A. de C.V.
Pearson Education—*Japan*, Tokyo
Pearson Education *Malaysia*, Pte. Ltd.

BRIEF CONTENTS

CONTENTS

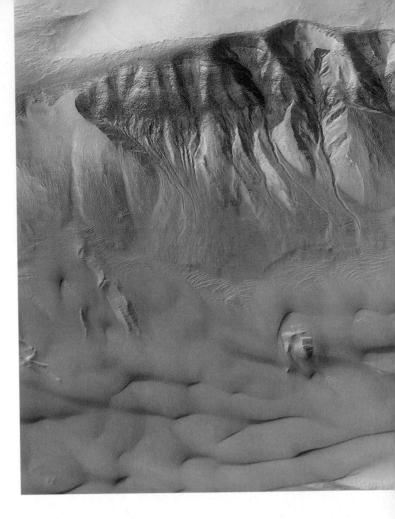

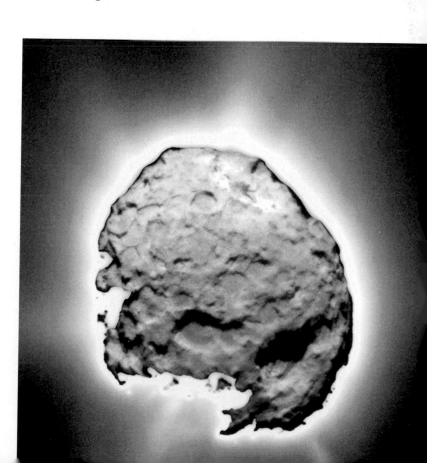

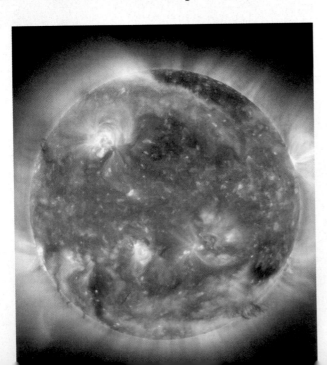

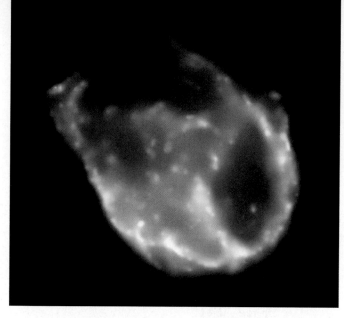

PREFACE

Astronomy is a science that thrives on new discoveries. Fueled by new technologies and novel theoretical insights, the study of the cosmos continues to change our understanding of the universe. We are pleased to have the opportunity to present in this book a representative sample of the known facts, evolving ideas, and frontier discoveries in astronomy today.

Astronomy Today has been written for students who have taken no previous college science courses and who will likely not major in physics or astronomy. It is intended for use in a one- or two-semester, non-technical astronomy course. We present a broad view of astronomy, straightforwardly descriptive and without complex mathematics. The absence of sophisticated mathematics, however, in no way prevents discussion of important concepts. Rather, we rely on qualitative reasoning as well as analogies with objects and phenomena familiar to the student to explain the complexities of the subject without oversimplification. We have tried to communicate the excitement we feel about astronomy and to awaken students to the marvelous universe around us.

Many of you—teachers and students alike—have given us helpful feedback and constructive criticism on earlier editions. From these, we have learned to communicate better both the fundamentals and the excitement of astronomy. Many improvements inspired by your comments have been incorporated into this new edition.

Focus of the Fifth Edition

From the first edition, we have tried to meet the challenge of writing a book that is both accurate and approachable. To the student, astronomy sometimes seems like a long list of unfamiliar terms to be memorized and repeated. You will indeed be introduced to many new terms and concepts in this course, but we hope you will also learn and remember how science is done, how the universe works, and how things are connected. In the fifth edition, we have taken particular care to try to show how astronomers know what they know, and to highlight both the scientific principles underlying their work and the process used in discovery.

New and Revised Material

Astronomy is a rapidly evolving field, and the three years since the publication of the fourth edition of *Astronomy Today* have seen many new discoveries covering the entire spectrum of astronomical research. Almost every chapter in the fifth edition has been substantially updated with new information. Several chapters have also seen significant internal reorganization in order to streamline the overall presentation, strengthen our focus on the process of science, and reflect new understanding and emphases in contemporary astronomy. Among the many changes are:

- Expanded coverage throughout of the scientific method and how astronomers "know what they know."

- New part-opening essays to establish historical context for each section of the text.

- Updated material in Chapter 5 on adaptive optics, Keck, Subaru, Gemini, and the VLT; additional material on infrared and optical interferometry; new coverage of the *Chandra* and *Spitzer* missions.

- An introduction to solar-system formation in Chapter 6, to better frame the discussion of planetary properties that follows.

- New material in Chapter 7 on the Ozone Hole and Global Warming.

- Expanded coverage in Chapters 6 and 10 of the most recent missions to Mars.

- Updates in Chapter 10 on Martian oppositions, gullies, oceans, and ice.

- Final update on the *Galileo/GEM* mission in Chapter 11.

- Coverage of *Stardust*, new Kuiper belt objects, and Pluto's status as a planet in Chapter 14.

- Updated discussion of solar system formation in Chapter 15; expanded coverage of competing theories, planet migration, planetesimal ejection, plutinos, and the angular momentum problem.

- New sections in Chapter 15 on extrasolar planets, with updated material on the latest observations and their implications for the condensation theory of solar system formation.

- Reorganization of presentation in Chapter 16, and an update on neutrino oscillations.

- New information on star names and revised coverage of key concepts in Chapter 17.

- Consistent and up-to-date stellar properties in Examples throughout Part 3.

- Updated information in Chapter 19 on brown dwarfs; new material on competitive accretion and collisions in star formation.

- New coverage in Chapter 20 of the end-states of stellar and binary evolution; more examples of familiar stars in specific evolutionary stages.

- Updated coverage of pulsars and gamma-ray bursts in Chapter 22.

- Reorganized and expanded material in Chapter 22 on Special and General Relativity and their historical development.

- Latest results in Chapter 23 on Sgr A* and the Galaxy's central black hole.

- Reorganization of Chapters 24 and 25, updating all coverage, emphasizing the connection between normal and active galaxies, and expanding the discussion of black holes in galactic nuclei.

- Updated discussion in Chapter 24 of the measurement of Hubble's constant.

- Expanded and substantially revised coverage in Chapter 25 of galaxy collisions, hierarchical merging and galaxy evolution; revised discussion of active galaxy evolution.

- Consistent distances and times in Chapters 24–27, assuming a flat universe with dark matter and dark energy as determined by the WMAP satellite; incorporation of results from recent sky surveys.

- Extensive revision of Chapters 26 and 27 to include the most recent observations of cosmic acceleration and discussion of "dark energy."

- Revised discussions of the cosmological constant and the age of the universe; results from the CBI and *WMAP* experiments suggesting a flat universe.

- Updated coverage of Europa, Mars, interstellar organic molecules, extrasolar planets, and SETI in Chapter 28.

- Expanded Glossary which now includes many additional terms used in the text, but not identified explicitly as keywords.

- New detailed Seasonal Star Charts, courtesy of *Astronomy* Magazine.

The Illustration Program

Visualization plays an important role in both the teaching and the practice of astronomy, and we continue to place strong emphasis on this aspect of our book. We have tried to combine aesthetic beauty with scientific accuracy in the artist's conceptions that adorn the text, and we have sought to present the best and latest imagery of a wide range of cosmic objects. Each illustration has been carefully crafted to enhance student learning; each is pedagogically sound and tied tightly to the nearby discussion of important scientific facts and ideas.

Full Spectrum Coverage and Spectrum Icons

Astronomers exploit the full range of the electromagnetic spectrum to gather information about the cosmos. Throughout this book, images taken at radio, infrared, ultraviolet, X-ray, or gamma-ray wavelengths are used to supplement visible-light images. As it is sometimes difficult (even for a professional) to tell at a glance which images are visible-light photographs and which are false-color images created with other wavelengths, each photo in the text is provided with an icon that identifies the wavelength of electromagnetic radiation used to capture the image and reinforces the connection between wavelength and radiation properties.

Compound Art It is rare that ▶ a single image, be it a photograph or an artist's conception, can capture all aspects of a complex subject. Wherever possible, multiple-part figures are used in an attempt to convey the greatest amount of information in the most vivid way:

- Visible images are often presented along with their counterparts captured at other wavelengths.

- Interpretive line drawings are often superimposed on or juxtaposed with real astronomical photographs, helping students to really "see" what the photographs reveal.

- Breakouts—often multiple ones— are used to zoom in from wide-field shots to closeups so that detailed images can be understood in their larger context.

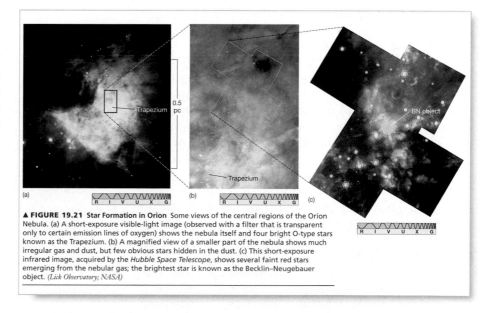

▲ **FIGURE 19.21 Star Formation in Orion** Some views of the central regions of the Orion Nebula. (a) A short-exposure visible-light image (observed with a filter that is transparent only to certain emission lines of oxygen) shows the nebula itself and four bright O-type stars known as the Trapezium. (b) A magnified view of a smaller part of the nebula shows much irregular gas and dust, but few obvious stars hidden in the dust. (c) This short-exposure infrared image, acquired by the *Hubble Space Telescope*, shows several faint red stars emerging from the nebular gas; the brightest star is known as the Becklin–Neugebauer object. (*Lick Observatory; NASA*)

▲ **Explanatory Captions** Students often review a chapter by "looking at the pictures." For this reason, the captions in this book are often a bit longer and more detailed than those in other texts.

H–R Diagrams and Acetate Overlays   All of the ▶
book's H–R diagrams are drawn in a uniform format, using
real data. In addition, a unique set of transparent acetate
overlays dramatically demonstrates to students how the
H–R diagram helps us to organize our information about
the stars and track their evolutionary histories.

Other Pedagogical Features

As with many other parts of our text, instructors have
helped guide us toward what is most helpful for effective
student learning. With their assistance, we have revised
both our in-chapter and end-of-chapter pedagogical apparatus to increase its utility to students.

Learning Goals.   Studies indicate that beginning students have trouble prioritizing textual material. For this
reason, a few (typically 5 or 6) well-defined Learning
Goals are provided at the start of each chapter. These help
students structure their reading of the chapter and then
test their mastery of key facts and concepts. The Goals are
numbered and cross-referenced to key sections in the body
of each chapter. This in-text highlighting of the most important aspects of the chapter also helps students review.
The Goals are organized and phrased in such a way as to
make them objectively testable, affording students a means
of gauging their own progress.

∞ **Concept Links.**   In astronomy, as in many scientific
disciplines, almost every topic seems to have some bearing
on almost every other. In particular, the connection between the astronomical material and the physical principles set forth early in the text is crucial. Practically
everything in Chapters 6–28 of this text rests on the foundation laid in the first five chapters. For example, it is important that students, when they encounter the discussion
of high-redshift objects in Chapter 25, recall not only what
they just learned about Hubble's law in Chapter 24 but also
refresh their memories, if necessary, about the inverse-
square law (Chapter 17), stellar spectra (Chapter 4), and
the Doppler shift (Chapter 3). Similarly, the discussions of
the mass of binary-star components (Chapter 17) and of
galactic rotation (Chapter 23) both depend on the discussion of Kepler's and Newton's laws in Chapter 2.
Throughout, discussions of new astronomical objects and
concepts rely heavily on comparison with topics introduced earlier in the text.

We remind you of these links so you can recall the
principles on which later discussions rest and, if necessary,
review them. To this end, we have inserted "Concept
Links" throughout the text—symbols that mark key intellectual bridges between material in different chapters. The
links, denoted by the symbol ∞ together with a section
reference, signal that the topic under discussion is related
in some significant way to ideas developed earlier, and provide direction to material to review before proceeding.

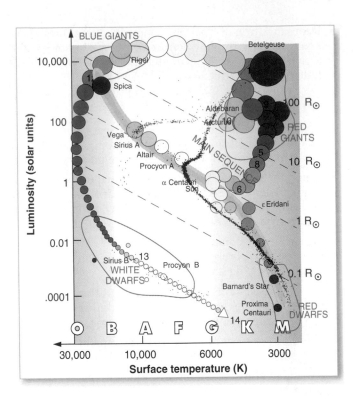

Key Terms.   Like all subjects, astronomy has its own
specialized vocabulary. To aid learning, the most important astronomical terms are boldfaced at their first appearance in the text. Each boldfaced Key Term is also
incorporated in the appropriate chapter summary, together with the page number where it was defined. In addition,
an expanded alphabetical glossary, defining each Key Term
and locating its first use in the text, appears at the end of
the book.

CONCEPT CHECK

✔ In what ways might observations of extrasolar
planets help us understand our own solar
system?

▲ **Concept Checks.**   We incorporate into each chapter
a number of "Concept Checks"—key questions that require the reader to reconsider some of the material just
presented or attempt to place it into a broader context. Answers to these in-chapter questions are provided at the
back of the book.

● **End of Chapter Questions and Problems.**   Many
elements of the end-of-chapter material have seen substantial reorganization:

● Each chapter now incorporates 20 Conceptual Self-Test
Questions, equally divided between "true/false" and
multiple choice formats, allowing students to assess
their understanding of the chapter material. Answers to

all these questions appear at the end of the book. Each chapter also has 20 Review and Discussion Questions, which may be used for in-class review or for assignment. As with the Self-Test Questions, the material needed to answer Review Questions may be found within the chapter. The Discussion Questions explore particular topics more deeply, often asking for opinions, not just facts. As with all discussions, these questions usually have no single "correct" answer.

The end of chapter material includes 15 Problems, based on the chapter contents and entailing some numerical calculation. In many cases the problems are tied directly to quantitative statements made (but not worked out in detail) in the text. The solutions to the Problems are not contained verbatim within the chapter, but the information necessary to solve them has been presented in the text. Answers to odd-numbered Problems appear at the end of the book.

▼ **Discovery Boxes**   Exploring a wide variety of interesting supplementary topics, these features have been expanded and provide the reader with insight into how scientific knowledge evolves, and emphasizing our theme of the process of science.

DISCOVERY 25-1

The Sloan Digital Sky Survey

Many of the photographs used in this book—not to mention most of the headline-grabbing imagery found in the popular media—come from large, high-profile, and usually very expensive, instruments such as NASA's *Hubble Space Telescope* and the European Southern Observatory's *Very Large Telescope* in Chile. ∞ (Secs. 5.3, 5.4) Their spectacular views of deep space have revolutionized our view of the universe. Yet a less well known, considerably cheaper, but no less ambitious, project currently underway may, in the long run, have every bit as great an impact on astronomy and our understanding of the cosmos.

The Sloan Digital Sky Survey (SDSS) is a five year project designed to systematically map out a quarter of the entire sky on a scale and at a level of precision never before attempted. By the time the project is completed in 2005, it will have catalogued more than 100 million celestial objects, recording their apparent brightnesses at 5 different colors (wavelength ranges) spread across the optical and near-infrared part of the spectrum. In addition, spectroscopic follow-up observations will determine redshifts and hence distances to 1 million galaxies and 100,000 quasars. These data will be used to construct even more detailed redshift surveys than those described in the text, and to probe the structure of the universe on very large scales. The sensitivity of the survey is such that it can detect bright galaxies like our own out to distances of more than 1 billion parsecs. Very bright objects, such as quasars and young starburst galaxies, are detectable almost throughout the entire observable universe.

The first figure shows the Sloan Survey telescope, a special-purpose 2.5-m instrument sited in Apache Point Ob-

servatory, near Sunspot, New Mexico. This reflecting telescope (whose box-like structure protects it from the wind) is not space-based, does not employ active or adaptive optics, and cannot probe as deep (*i.e.* far) into space as larger instruments. How can it possibly compete with these other systems? The answer is that, unlike most other large telescopes in current use, where hundreds or even thousands of observers share the instrument and compete for its time, the SDSS telescope was designed specifically for the purpose of the survey. It has a wide field of view and is dedicated to the task, carrying out observations of the sky on *every* clear night during the 5-year duration of the survey project.

The use of a single instrument night after night, combined with tight quality controls on which nights' data are actually incorporated into the survey (nights with poor seeing or other problematic conditions are discarded and the observations repeated) mean that the end-product is a database of exceptionally high quality and uniformity spanning an enormous volume of space—a monumental achievement and an indispensible tool for cosmology. The survey field of view covers much of the sky away from the Galactic plane in the north, together with three broad "wedges" in the south.

Archiving images and spectra on millions of galaxies produces a lot of data. The full survey will consist of roughly 15 *trillion* bytes of information—comparable to the entire Library of Congress! As of mid 2004, data for roughly one third of the survey area, comprising 88 million objects (360,000 with measured spectra), has been released to the public. The second figure shows the distant galaxy NGC 5792 and a bright red star much closer to us, in fact in our own Milky Way Galaxy, just one of hundreds of thousands of images that will make up the full dataset. Among recent highlights, SDSS has detected the largest known structure in the universe, observed the most distant known galaxies and quasars, and has been instrumental in pinning down the key observational parameters describing our universe (see Chapter 26).

SDSS will impact astronomy in areas as diverse as the large-scale structure of the universe, the origin and evolution of galaxies, the nature of dark matter, the structure of the Milky Way, and the properties and distribution of interstellar matter. Its uniform, accurate, and detailed database is likely to be used by generations of scientists for decades to come.

More Precisely Boxes ▶

These provide more quantitative treatments of subjects discussed qualitatively in the text. Removing these more challenging topics from the main flow of the narrative and placing them within a separate modular element of the chapter design (so that they can be covered in class, assigned as supplementary material, or simply left as optional reading for those students who find them of interest) will allow instructors greater flexibility in setting the level of their coverage.

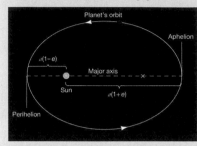

MORE PRECISELY 2-1

Some Properties of Planetary Orbits

Two numbers—semimajor axis and eccentricity—are all that are needed to describe the size and shape of a planet's orbital path. From them, we can derive many other useful quantities. Two of the most important are the planet's *perihelion* (its point of closest approach to the Sun) and its *aphelion* (point of greatest distance from the Sun). From the definitions presented in the text, it follows that if the planet's orbit has semimajor axis a and eccentricity e, the planet's perihelion is at a distance $a(1 - e)$ from the Sun, while its aphelion is at $a(1 + e)$. These points and distances are illustrated in the following figure:

EXAMPLE 1: We can locate the other focus of the ellipse in the diagram, and hence determine the eccentricity from the definition in the text, quite simply. The second focus is placed symmetrically along the major axis, at the point marked with an "X." With a ruler, measure (1) the length of the major axis and (2) the distance between the two foci. Dividing the second distance by the first, you should find an eccentricity of 3.4 cm/6.8 cm = 0.5. Alternatively, we could use the formula given for the perihelion. Measure the perihelion distance to be $a(1 - e) = 1.7$ cm. Dividing this by $a = 3.4$ cm, we obtain $1 - e = 0.5$, so once again, $e = 0.5$.

EXAMPLE 2: A (hypothetical) planet with a semimajor axis of 400 million km and an eccentricity of 0.5 (i.e., with an orbit as shown in the diagram) would range between $400 \times (1 - 0.5) = 200$ million km and $400 \times (1 + 0.5) = 600$ million km from the Sun over the course of one complete orbit. With $e = 0.9$, the range in distances would be 40 to 760 million km, and so on.

No planet has an orbital eccentricity as large as 0.5—the planet with the most eccentric orbit is Pluto, with $e = 0.249$. (See Table 2.1.) However, many meteoroids and all comets (see Chapter 14) have eccentricities considerably greater than that. In fact, most comets visible from Earth have eccentricities very close to $e = 1$. Their highly elongated orbits approach within a few astronomical units of the Sun at perihelion, yet these tiny frozen worlds spend most of their time far beyond the orbit of Pluto.

▼ Interactive eBook

The *Astronomy Today, Fifth Edition* interactive **eBook** is located in the WebCT, BlackBoard, and OneKey courses and has been redesigned for easier and clearer navigation. It contains a full electronic version of the text, with key term hyperlinks and imbedded media elements at point of use. The eBook features:

● **New!** Tutorials: Written by Philip Langill (University of Calgary). These animated, interactive Flash™ files, denoted by an icon in the text, allow students to explore the ideas and concepts from the text in depth. Students are engaged in the thought process as they answer questions and change parameters in these exploratory activities.

● **New!** Physlet® Illustrations for Astronomy: Written by Chuck Niederriter and Steve Mellema (both of Gustavus Adolphus College); Physlets by Wolfgang Christian (Davidson College). Through animation, these brief Java applets, denoted by an icon in the text, further illustrate concepts from the text. Each Physlet is followed by a series of questions that encourage students to think critically about the concept at hand.

● 61 narrated videos and animations imbedded within the text, at point of use. These help to bring text figures and concepts to life.

● All bold key terms in the text are hyperlinked to a glossary definition and an audio pronunciation.

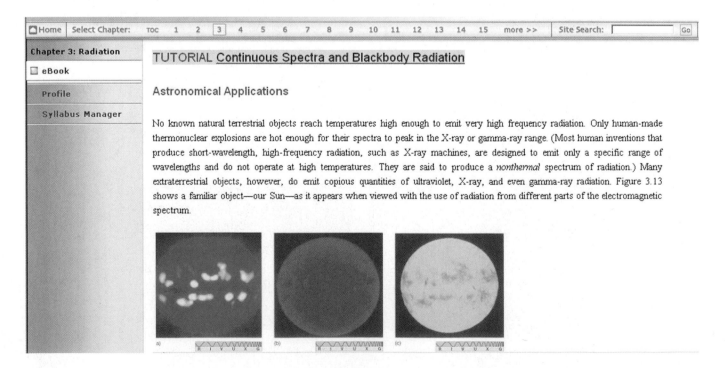

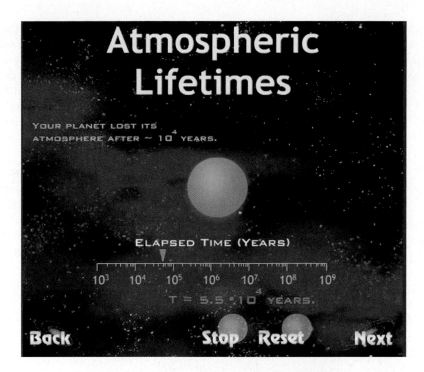

▲ Student Accelerator CD-ROM

The Student Accelerator CD-ROM that is pack-aged with *Astronomy Today, Fifth Edition* contains the Tutorials, Physlet® Illustrations, animations, and videos from the eBook. The CD accelerates the performance of the eBook when students download the high-bandwidth media, so that students are not restricted by slow connections. It can also be used apart from the eBook if a student doesn't have a live Internet connection or just wants to view the media elements.

Companion Website (http://astro.prenhall.com/chaisson)

The text-specific Companion Website for *Astronomy Today, Fifth Edition* organizes material from a variety of sources on the web on a chapter-by-chapter basis, is updated regularly, and provides interactive exercises for each chapter. It includes:

- Annotated images, videos, and animations that are regularly updated to reflect the most recent astronomical discoveries.

- Interactive multiple-choice quizzes with hints and instant feedback.

- Algorithmically generated versions of the end-of-chapter problems from the text.

- Links to associated websites that are regularly updated for currency and relevancy.

Supplementary Material for the Instructor

Course and Homework Management Tools

Prentice Hall's OneKey offers the best teaching and learning resources all in one place. OneKey for *Astronomy Today, Fifth Edition* is all you need to plan and administer your course, and is all your students need for anytime anywhere access to your course materials. Conveniently organized by textbook chapter, these compiled resources help you save time and help your students reinforce and apply what they have learned in class.

WebCT and Blackboard are comprehensive and flexible Web-based educational platforms. With a local installation of either system, Prentice Hall provides content designed especially for *Astronomy Today, Fifth Edition* to create a complete course suite, tightly integrated with the system's course management tools.

Along with all of the material from the Companion Website, our course management cartridges for OneKey, WebCT, and Blackboard include additional materials:

- The interactive eBook
- Gradable follow-up questions to the Flash-based Tutorials and Physlet® Illustrations for Astronomy
- Gradable multiple-choice and true/false self study quizzes
- Gradable Labeling Exercises
- Test bank questions, converted from our *TestGen Test Item File*
- Resources from the Instructor's Resource Center on CD

Online Homework

WebAssign (www.webassign.net) Create assignments from WebAssign's database of problems and questions from *Astronomy Today, Fifth Edition,* or write and customize your own. The instructor has complete control over homework, including due date, content, feedback, and question formats.

Comets. Only Prentice Hall provides you with *Comets,* your annual media update. This unique service is provided at the beginning of each academic year and available free to qualified adopters of Chaisson/McMillan. *Comets* is a library of new videos and slides for you to use in class. The slide kit contains 28 new slides from sources such as NASA, JPL, STScI, GSFC, *HST* Comet LINEAR Investigation Team, APL, JPL, ESA, *Hubble* Heritage Team, IPAC, European Southern Observatory, SDSS/Astrophysical Research Consortium, and the U.S. Department of Defense. Custom animations prepared by the Wright Center for Science Visualization and many other videos of new discoveries and animations from various sources, including NASA, STScI, APL/NRL, ESA, Stanford Lockheed Institute for Space Research, and JPL are provided in both CD and DVD formats. A newsletter provides a cross reference between all the materials in the *Comets* kit and corresponding chapters of both Chaisson/McMillan texts, as well as annotations describing the subject and source of each slide and video in the kit.

Test Item File.   An extensive file of approximately 3200 test questions, newly compiled and revised for the fifth edition by J. Wayne Wooten (Pensacola Junior College), is offered free to qualified adopters. This is available in both printed and electronic formats (see desciption of IRC on CD-ROM). The fifth edition *Test Item File* has been thoroughly revised and includes many new multiple choice and essay questions for added conceptual emphasis. ISBN 0-13-144689-4

Instructor's Resource Manual. By Steven Murrell (Henry Ford Community College) and Leo Connolly (California State University at San Bernardino). This manual provides an overview of each chapter, pedagogical tips, useful analogies, suggestions for classroom demonstrations, writing questions, selected readings, and answers to the end-of-chapter Review and Discussion questions and Problems. ISBN 0-13-144688-6

Instructor's Resource Center on CD-ROM.   This package contains all text illustrations in jpeg, PowerPoint, and pdf formats, as well as the animations, videos, Interactive Tutorials, and Physlet® Illustrations from the eBook CD-ROM. The CDs also contain TestGenerator, an easy to use, fully networkable program for creating tests ranging from short quizzes to long exams. Questions from the *Test Item File* are supplied, and professors can use the Question Editor to modify existing questions or create new questions. This CD-ROM set also contains electronic versions of the *Instructor's Resource Manual* and instructor notes to SkyChart III Projects. Free to qualified adopters. ISBN 0-13-117609-9

Acetates and Slides.   Approximately 275 images from the text are available as a package of color acetates or 35mm slides, and are available free to qualified adopters. ISBN 0-13-144685-1 (Slide set) ISBN 0-13-144687-8 (Transparency pack)

Instructor's Resource Box   This easy to use lecture organizer helps you integrate the entire Chaisson/McMillan supplement package. Free to qualified adopters, each toolbox is organized by chapter and contains:

- The transparency acetates and slides
- The *Test Item File*
- The *Instructor's Resource Manual*
- The Instructor's Resource CDs

ISBN 0-13-147915-6

Learner-Centered Astronomy Teaching:
Strategies for ASTRO 101 by Timothy F. Slater,
University of Arizona and Jeffrey P. Adams, *Montana State University*

Strategies for ASTRO 101 is a guide for instructors of the introductory astronomy course for non-science majors. Written by two leaders in astronomy education research, this book details various techniques instructors can use to increase students' understanding and retention of astronomy topics, with an emphasis on making the lecture a forum for active student participation. Drawing from the large body of recent research to discover how students learn, this guide describes the application of multiple classroom-tested techniques to the task of teaching astronomy to predominantly non-science students. ISBN 0-13-046630-1

Peer Instruction for Astronomy by Paul Green, *Harvard Smithsonian Center for Astrophysics*

Peer instruction is a simple yet effective method for teaching science. Techniques of peer instruction for introductory physics were developed primarily at Harvard and have aroused interest and excitement in the Physics Education community. This approach involves students in the teaching process, making science more accessible to them. Peer Instruction is a new trend in astronomy that is finding strong interest and is ideally suited to introductory Astronomy classes. This book is an important vehicle for providing a large number of thought-provoking, conceptual short-answer questions aimed at a variety of class levels. While significant numbers of such questions have been published for use in physics, *Peer Instruction for Astronomy* provides the first such compilation for Astronomy. ISBN 0-13-026310-9

For the Student

Observation, Research, and SkyChart III Projects FREE when shrinkwrapped with *Astronomy Today, Fifth Edition*, this student supplement contains observation and research projects authored by Steve McMillan, as well as projects for SkyChart III Student Version planetarium software by Erik Bodegom and Sean P. Goe (Portland State University). ISBN 0-13-144690-8

Lecture Tutorials for Introductory Astronomy by Jeffrey P. Adams, *Montana State University*, Edward E. Prather, *University of Arizona*, Timothy F. Slater, *University of Arizona*, and CAPER, Conceptual Astronomy and Physics Education Research Team

Funded by the National Science Foundation, Lecture-Tutorials for Introductory Astronomy are designed to help make large lecture-format courses more interactive. Each of the 29 Lecture-Tutorials is presented in a classroom-ready format, asks students to work in groups of 2–3, takes between 10–15 minutes, challenges students with a series of carefully designed questions that spark classroom discussion, engages students in critical reasoning, and requires no equipment. ISBN 0-13-147997-0

Norton's Star Atlas and Reference Handbook, 20th edition by Ian Ridpath

Now in a superbly redesigned, two-color landmark 20th edition, the first of a new century, this combination star atlas and reference work has no match in the field. First published in 1910, coinciding with the first of two appearances by Halley's Comet during the book's life, *Norton's* owes much of its legendary success to its unique maps, arranged in slices known as gores, each covering approximately one-fifth of the sky. Every star visible to the naked eye under the clearest skies—down to magnitude 6.5—is charted along with star clusters, nebulae, and galaxies. Extensive tables of data on interesting objects for observation accompany each of the precision drawn maps. Preceding the maps is the unique and authoritative reference handbook covering timekeeping and positional measurements on the celestial sphere; the Sun, Moon, and other bodies of the Solar System; telescopes and other equipment for observing and imaging the sky; and stars, nebulae, and galaxies. Throughout, succinct fundamental principles and practical tips guide the reader into the night sky. The appendices Units and Notation, Astronomical Constants, Symbols and Abbreviations, and Useful Addresses complete what has long been the only essential reference for the stargazer. ISBN 0-13-145164-2

Edmund Scientific Star and Planet Locator

The famous rotating roadmap of the heavens shows the location of the stars, constellations, and planets relative to the horizon for the exact hour and date you determine. This 8″ square star chart was plotted by the late astronomer and cartographer George Lovi. The reverse side of the locator is packed with additional data on the planets, meteor showers, and bright stars. Included with each star chart is a 16-page, fully-illustrated, pocket-size instruction booklet. The Star and Planet Locator is available free through Prentice Hall when packaged with *Astronomy Today*.

Acknowledgments

Throughout the many drafts that have led to this book, we have relied on the critical analysis of many colleagues. Their suggestions ranged from the macroscopic issue of the book's overall organization to the minutiae of the technical accuracy of each and every sentence. We have also benefited from much good advice and feedback from users of the first four editions of the text. To these many helpful colleagues, we offer our sincerest thanks.

Reviewers of the Fifth Edition

Cecilia Barnbaum
Valdosta State University

Peter A. Becker
George Mason University

James R. Dire
U.S. Coast Guard Academy

Michael N. Fanelli
University of North Texas

Harold A. Geller
George Mason University

David Goldberg
Drexel University

Paul Heckert
Western Carolina University

Matthew Malkan
University of California, Los Angeles

Janet McLarty-Schroeder
Cerritos College

Scott Miller
Penn State University

Richard Rand
University of New Mexico

Vicki Sarajedini
University of Florida

George R. Stanley, Jr.
San Antonio College

Craig Tyler
Fort Lewis College

Robert L. Zimmerman
University of Oregon

Reviewers of Previous Editions

Stephen G. Alexander
Miami University of Ohio

Robert H. Allen
University of Wisconsin, La Crosse

Timothy C. Beers
University of Evansville

William J. Boardman
Birmingham Southern College

Donald J. Bord
University of Michigan, Dearborn

Elizabeth P. Bozyan
University of Rhode Island

Anne Cowley
Arizona State University

Bruce Cragin
Richland College

Ed Coppola
Community College of Southern Nevada

David Curott
University of North Alabama

Norman Derby
Bennington College

John Dykla
Loyola University, Chicago

Kimberly Engle
Drexel University

Martin Goodson
Delta College

David G. Griffiths
Oregon State University

Donald Gudehus
Georgia State University

Marilynn Harper
Delaware County Community College

Clint D. Harper
Moorpark College

Susan Hartley
University of Minnesota, Duluth

Joseph Heafner
Catawaba Valley Community College

Fred Hickok
Catonsville Community College

Darren L. Hitt
Loyola College, Maryland

F. Duane Ingram
Rock Valley College

Steven D. Kawaler
Iowa State University

William Keel
University of Alabama

Marvin Kemple
Indiana University-Purdue University, Indianapolis

Mario Klairc
Midlands Technical College

Kristine Larsen
Central Connecticut State University

Andrew R. Lazarewicz
Boston College

Robert J. Leacock
University of Florida

Larry A. Lebofsky
University of Arizona

M. A. Lohdi
Texas Tech University

Michael C. LoPresto
Henry Ford Community College

Phillip Lu
Western Connecticut State University

Fred Marschak
Santa Barbara College

Chris Mihos
Case Western Reserve University

Milan Mijic
California State University, Los Angeles

Richard Nolthenius
Cabrillo College

Edward Oberhofer
University of North Carolina, Charlotte

Andrew P. Odell
Northern Arizona University

Gregory W. Ojakangas
University of Minnesota, Duluth

Ronald Olowin
Saint Mary's College of California

Robert S. Patterson
Southwest Missouri State University

Cynthia W. Peterson
University of Connecticut

Andreas Quirrenback
University of California, San Diego

James A. Roberts
University of North Texas

Gerald Royce
Mary Washington College

Malcolm P. Savedoff
University of Rochester

John C. Schneider
Catonsville Community College

Harry L. Shipman
University of Delaware

C. G. Pete Shugart
Memphis State University

Stephen J. Shulik
Clarion University

Tim Slater
Montana State University

Don Sparks
Los Angeles Pierce College

Maurice Stewart
Williamette University

Jack W. Sulentic
University of Alabama

Andrew Sustich
Arkansas State University

Stephen R. Walton
California State University, Northridge

Peter A. Wehinger
University of Arizona

Louis Winkler
Pennsylvania State University

We would also like to express our gratitude to Ray Villard (Space Telescope Science Institute) for compiling the Comets supplement; to Carl Adler (East Carolina University), Dave McKenzie (Montana State University), David Ziegler (Hannibal-LaGrange College), Philip Langill (University of Calgary), and Chuck Niederriter and Steve Mellema (both of Gustavus Adolphus College) for updating and maintaining the media resources; and to Tim Slater (University of Arizona) for contributing conceptual multiple-choice questions for the end of each chapter.

The publishing team at Prentice Hall has assisted us at every step along the way in creating this text. Much of the credit for getting the project completed on time goes to our Senior Editor, Erik Fahlgren, who has successfully navigated us through the twists, turns, and "absolute final deadlines" of the publishing world, all the while managing the many variables that go into a multifaceted publication such as this. Erin Mulligan, our Development Editor, has read every sentence in the book from the student perspective. Her careful eye and thoughtful suggestions gave us the "student focus" we needed in this edition. Production Editor Shari Toron has done an excellent job of tying to-gether the threads of this very complex project, made all the more complex by the necessity of combining text, art, and electronic media into a coherent whole. Electronic Page Makeup Specialist Joanne Del Ben has done an outstanding job making all the parts of a page fit together and flow easily and sensibly to the next. Special thanks are also in order to John Challice, Editor in Chief, for his insights and feedback; Designer, John Christiana for making the fifth edition look spectacular; Associate Editor, Christian Botting for publishing quality supplements in a timely manner; Media Editor, Michael J. Richards for managing the creation of the eBook and the extensive materials available online; and Editorial Assistant, Andrew Sobel for his attention to detail and deadlines.

Finally, we would like to express our gratitude to renowned space artist Dana Berry for allowing us to use many of his beautiful renditions of astronomical scenes, and to Lola Judith Chaisson for assembling and drawing all the H–R diagrams (including the acetate overlays) for this edition. We are interested in your feedback on this text. Please email us at physics_service@prenhall.com if you find any errors or have comments for the authors.

PART | 1
Astronomy and the Universe

Galileo's sketch of Saturn

It is often said that we live in a golden age of astronomy. Yet the dawn of the 21st century is actually the second such period of rich discovery and rapid exploration. The late Renaissance began the first era of stunning scientific advancement, when modern astronomy was born.

Most notable among those spearheading the rebirth of astronomy at the time was the Italian scientist Galileo Galilei (1564-1642). Although he did not invent the telescope, in 1610 Galileo was the first to record what he saw when he aimed a small (5-centimeter-diameter) lens at the sky. His findings created nothing less than a revolution in astronomy, as well as a breakthrough in human perception of the cosmos.

Viewing for the first time dark blemishes on the Sun, rugged mountains on the Moon, and whole new worlds orbiting Jupiter, he demolished the Aristotelian view of cosmic immutability—the notion that the heavens were perfect and unchanging. To be sure, it was with the philosophers of the day, as much as with the theologians, that Galileo had trouble. In championing the scientific method, he used a tool to test his ideas, and what he found disagreed greatly with the leading thoughts and beliefs of the time.

His advance was simple: He used a telescope to focus, magnify, and study radiation reaching Earth from the heavens—in particular, light from the Sun, the Moon, and the planets. Light is the most familiar kind of radiation to humans on Earth, since it enables us to get around on the surface of our planet. But light also enables telescopes to see objects deep in space, allowing us to probe farther than the eye can alone. With his simple optical telescope, Galileo changed forever the way that the oldest science—astronomy—is pursued.

Among other "wondrous things" he found were crowded star clusters along the Milky Way, moons and rings around the big planets, and colorful nebulae aglow unlike anything anyone had seen before. Some of Galileo's sketches are reproduced here at the left and compared with modern views at the right.

Galileo's sketch of Orion

Galileo Galilei

Galileo's sketch of the Pleiades

Saturn in the ultraviolet (*STScI*)

oday, we are again in the midst of another period of unsurpassed scientific achievement—a revolution in which modern astronomers are revealing the invisible universe as Galileo once spied the visible universe. We have learned how to detect, measure, and analyze invisible radiation streaming to us from dark objects in space. And once again our perceptions are changing.

Astronomy no longer evokes visions of plodding intellectuals peering through long telescope tubes. Nor does the cosmos any longer refer to that seemingly inactive, immutable regime captured visually by gazing at the nighttime sky. Modern astronomers now decipher a more vibrant, changing universe—one in which stars emerge and perish much like living things, galaxies spew forth vast quantities of energy, and life itself is thought to be a natural consequence of the evolution of matter.

New discoveries are rapidly advancing our understanding of the universe, but they also raise new questions. Astronomers will encounter many problems in the decades ahead, but this should neither dismay nor frustrate us, for it is precisely how science operates. Each discovery adds to our storehouse of information, generating a host of questions that lead in turn to more discoveries, and so on, causing an acceleration of basic knowledge.

Most notably, we are beginning to perceive the universe in all its multivaried ways. A single generation—not the generation of our parents and not that of our children, but our generation—has opened up the whole electromagnetic spectrum beyond visible light. And what we, too, have found are "wondrous things."

Emerging largely from studies of the invisible universe, our view of the cosmos in its full splendor is one of many new scientific insights that we have recently been privileged to attain. Historians of the future may well regard our generation as the one that took a great leap forward, providing a whole new glimpse of our richly endowed universe. In all of history, there have been only two periods in which our perception of the universe has been so revolutionized within a single human lifetime. The first occurred four centuries ago at the time of Galileo; the second is now underway.

Orion in the infrared (*Caltech*)

Pleiades in the optical (*AURA*)

Stars are perhaps the most fundamental component of the universe we inhabit. There are roughly as many stars in the observable universe as grains of sand in all the beaches of the world. Here, we see displayed one of the most easily recognizable star fields in the winter sky—the familiar constellation Orion. This field of view spans about 100 light-years, or 10^{15} kilometers. We will return to study Orion many times in this text. (J. Sanford/Astrostock-Sanford) ▶

1

Charting the Heavens

The Foundations of Astronomy

Nature offers no greater splendor than the starry sky on a clear, dark night. Silent and jeweled with the constellations of ancient myth and legend, the night sky has inspired wonder throughout the ages—a wonder that leads our imaginations far from the confines of Earth and the pace of the present day and out into the distant reaches of space and cosmic time itself. Astronomy, born in response to that wonder, is built on two of the most basic traits of human nature: the need to explore and the need to understand. Through the interplay of curiosity, discovery, and analysis—the keys to exploration and understanding—people have sought answers to questions about the universe since the earliest times. Astronomy is the oldest of all the sciences, yet never has it been more exciting than it is today.

LEARNING GOALS

Studying this chapter will enable you to

1 Describe how scientists combine observation, theory, and testing in their study of the universe.

2 Explain the concept of the celestial sphere and how we use angular measurement to locate objects in the sky.

3 Describe how and why the Sun, the Moon, and the stars appear to change their positions from night to night and from month to month.

4 Explain how our clocks and calendars are linked to Earth's rotation and orbit around the Sun.

5 Show how the relative motions of Earth, the Sun, and the Moon lead to eclipses.

6 Explain the simple geometric reasoning that allows astronomers to measure the distances and sizes of otherwise inaccessible objects.

 Visit astro.prenhall.com/chaisson for additional annotated images, animations, and links to related sites for this chapter.

1.1 Our Place in Space

Of all the scientific insights attained to date, one stands out boldly: Earth is neither central nor special. We inhabit no unique place in the universe. Astronomical research, especially within the past few decades, strongly suggests that we live on what seems to be an ordinary rocky *planet* called Earth, one of nine known planets orbiting an average *star* called the Sun, a star near the edge of a huge collection of stars called the Milky Way galaxy, which is one *galaxy* among billions of others spread throughout the observable universe. To begin to get a feel for the relationships among these very different objects, consult Figures 1.1 through 1.4; put them in perspective by studying Figure 1.5.

We are connected to the most distant realms of space and time not only by our imaginations, but also through a common cosmic heritage. Most of the chemical elements that make up our bodies (hydrogen, oxygen, carbon, and many more) were created billions of years ago in the hot centers of long-vanished stars. Their fuel supply spent, these giant stars died in huge explosions, scattering the elements created deep within their cores far and wide. Eventually, this matter collected into clouds of gas that slowly collapsed to give birth to new generations of stars. In this way, the Sun and its family of planets formed nearly five billion years ago. Everything on Earth embodies atoms from other parts of the universe and from a past far more remote than the beginning of human evolution. Elsewhere, other beings—perhaps with intelligence much greater than our own—may at this very moment be gazing in wonder at their own night sky. Our own Sun may be nothing more than an insignificant point of light to them—if it is visible at all. Yet if such beings exist, they must share our cosmic origin.

Simply put, the **universe** is the totality of all space, time, matter, and energy. **Astronomy** is the study of the universe. It is a subject unlike any other, for it requires us to profoundly change our view of the cosmos and to consider matter on scales totally unfamiliar from everyday experience. Look again at the galaxy in Figure 1.3. It is a swarm of about a hundred billion stars—more stars than the number of people who have ever lived on Earth. The entire assemblage is spread across a vast expanse of space 100,000 **light-years** in diameter. Although it sounds like a unit of time, a light-year is in fact the *distance* traveled by light in a year, at a speed of about 300,000 kilometers per second. Multiplying out, it follows that a light-year is equal to 300,000 kilometers/second × 86,400 seconds/day × 365 days, or about 10 trillion kilometers, or roughly 6 trillion miles. Typical galactic systems are truly "astronomi-

15,000 kilometers

R I V U X G

▲ **FIGURE 1.1 Earth** Earth is a planet, a mostly solid object, although it has some liquid in its oceans and core, and gas in its atmosphere. In this view, you can clearly see the North and South American continents. *(NASA)*

1,500,000 kilometers

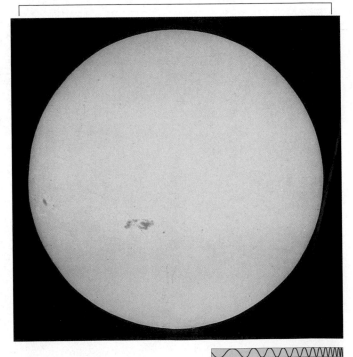

R I V U X G

▲ **FIGURE 1.2 The Sun** The Sun is a star, a very hot ball of gas. Much bigger than Earth—more than 100 times larger in diameter—the Sun is held together by its own gravity. *(AURA)*

cal" in size. For comparison, Earth's roughly 13,000-km diameter is less than one-twentieth of a light-*second*.

The light-year is a unit introduced by astronomers to help them describe immense distances. We will encounter many such "custom" units in our studies. As discussed in more detail in Appendix 1, astronomers frequently augment the standard SI (Système Internationale) metric system with additional units tailored to the particular problem at hand.

A thousand (1000), a million (1,000,000), a billion (1,000,000,000), and even a trillion (1,000,000,000,000)—these words occur regularly in everyday speech. But let's take a moment to understand the magnitude of the numbers and appreciate the differences among them. One thousand is easy enough to understand: At the rate of one number per second, you could count to a thousand in 1000 seconds—about 16 minutes. However, if you wanted to count to a million, you would need more than two weeks of counting at the rate of one number per second, 16 hours per day (allowing 8 hours per day for sleep). To count from one to a billion at the same rate of one number per second and 16 hours per day would take nearly 50 years—the better part of an entire human lifetime.

In this book, we consider spatial domains spanning not just billions of kilometers, but billions of light-years; objects containing not just trillions of atoms, but trillions of stars; and time intervals of not just billions of seconds or hours, but billions of years. You will need to become familiar—and comfortable—with such enormous numbers. A good way to begin is learning to recognize just how much larger than a thousand is a million, and how much larger still is a billion. Appendix 1 explains the convenient method used by scientists for writing and manipulating very large and very small numbers. If you are unfamiliar with this method, please read that appendix carefully—the *scientific notation* described there will be used consistently throughout our text, beginning in Chapter 2.

Lacking any understanding of the astronomical objects they observed, early skywatchers made up stories to explain them: The Sun was pulled across the heavens by a chariot drawn by winged horses, and patterns of stars traced heroes and animals placed in the sky by the gods. Today, of course, we have a radically different conception of the universe. The stars we see are distant, glowing orbs hundreds of times larger than our entire planet, and the patterns they form span hundreds of light years. In this first chapter we present some basic methods used by astronomers to chart the space around us. We describe the slow progress of scientific knowledge, from chariots and gods to today's well-tested theories and physical laws, and explain why we now rely on science rather than on myth to help us explain the universe.

About 1000 quadrillion kilometers, or 100,000 light-years

About 1,000,000 light-years

▲ **FIGURE 1.3 Galaxy** A typical galaxy is a collection of a hundred billion stars, each separated by vast regions of nearly empty space. Our Sun is a rather undistinguished star near the edge of another such galaxy, called the Milky Way. (*NASA*)

▲ **FIGURE 1.4 Galaxy Cluster** This photograph shows a typical cluster of galaxies, roughly a million light-years from Earth. Each galaxy contains hundreds of billions of stars, probably planets, and, possibly, living creatures. (*NASA*)

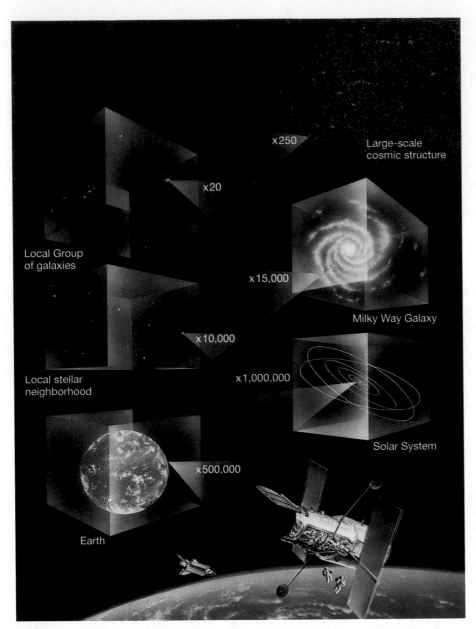

◀ **FIGURE 1.5 Size and Scale** This artist's conception puts each of the previous four figures in perspective. The bottom of this figure shows spacecraft (and astronauts) in Earth orbit, a view that widens progressively in each of the next five cubes drawn from bottom to top—Earth, the planetary system, the local neighborhood of stars, the Milky Way galaxy, and the closest cluster of galaxies. The top image depicts the distribution of galaxies in the universe on extremely large scales—the field of view in this final frame is hundreds of millions of light-years across. The numbers indicate approximately the increase in scale between successive images: Earth is 500,000 times larger than the spacecraft in the foreground, the solar system in turn is some 1,000,000 times larger than Earth, and so on. *(D. Berry)*

1.2 Scientific Theory and the Scientific Method

LEARNING GOAL 1 How have we come to know the universe around us? How do we know the proper perspective sketched in Figure 1.5? The earliest known descriptions of the universe were based largely on imagination and mythology, and made little attempt to explain the workings of the heavens in terms of known earthly experience. However, history shows that some early scientists did come to realize the importance of careful observation and testing to the formulation of their ideas. The success of their approach changed, slowly but surely, the way science was done and opened the door to a fuller understanding of nature.

As knowledge from all sources was sought and embraced for its own sake, the influence of logic and reasoned argument grew and the power of myth diminished. People began to inquire more critically about themselves and the universe. They realized that *thinking* about nature was no longer sufficient—that *looking* at it was also necessary. Experiments and observations became a central part of the process of inquiry. To be effective, a **theory**—the framework of ideas and assumptions used to explain some set of observations and make predictions about the real world—must be continually tested. If experiments and observations favor it, a theory can be further developed and refined. If they do not, the theory must be reformulated or rejected, no matter how appealing it originally seemed. The process is illustrated schematically in Figure 1.6. This new approach to investigation, combining thinking and doing—that is, theory and experiment—is known as the **scientific method**. It lies at the heart of modern science, separating science from pseudoscience, fact from fiction.

Notice that there is no end point to the process depicted in Figure 1.6. A theory can be invalidated by a single wrong prediction, but no amount of observation or experimentation can ever prove it "correct." Theories simply become more and more widely accepted as their predictions are repeatedly confirmed. As a class, modern scientific theories share several important defining characteristics:

- They must be *testable*—that is, they must admit the possibility that their underlying assumptions and their predictions can, in principle, be exposed to experimental verification. This feature separates science from, for example, religion, since, ultimately, divine revelations or scriptures cannot be challenged within a religious framework—we can't design an experiment to "verify the mind of God." Testability also distinguishes science from a pseudoscience such as astrology, whose underlying assumptions and predictions have been repeatedly tested and never verified, with no apparent impact on the views of those who continue to believe in it!

- They must continually be *tested*, and their consequences tested, too. This is the basic circle of scientific progress depicted in Figure 1.6.

- They should be *simple*. Simplicity is less a requirement than a practical outcome of centuries of scientific experience—the most successful theories tend to be the simplest ones that fit the facts. This viewpoint is often encapsulated in a principle known as *Occam's razor*: If two competing theories both explain the facts and make the same predictions, then the simpler one is better. Put another way—"Keep it simple!" A good theory should contain no more complexity than is absolutely necessary.

- Finally, most scientists have the additional bias that a theory should in some sense be *elegant*. When a clearly stated simple principle naturally ties together and explains several phenomena previously thought to be completely distinct, this is widely regarded as a strong point in favor of the new theory.

You may find it instructive to apply these criteria to the many physical theories—some old and well established, others much more recent and still developing—we will encounter throughout the text.

The notion that theories must be tested and may be proven wrong sometimes leads people to dismiss their im-

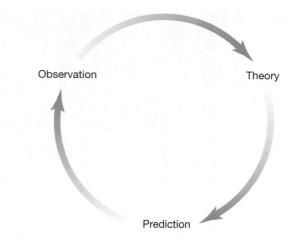

▲ **FIGURE 1.6 Scientific Method** Scientific theories evolve through a combination of observation, theoretical reasoning, and prediction, which in turn suggests new observations. The process can begin at any point in the cycle, and it continues forever.

portance. We have all heard the expression, "Of course, it's only a theory," used to deride or dismiss an idea that someone finds unacceptable. Don't be fooled by this abuse of the concept! Gravity (see ∞ Sec. 2.7) is "only" a theory, but calculations based on it have guided human spacecraft throughout the solar system. Electromagnetism (Chapter 3) and quantum mechanics (Chapter 4) are theories, too, yet they form the foundation for most of 20th- (and 21st) century technology.

The birth of modern science is usually associated with the Renaissance, the historical period from the late 14th to the mid-17th century that saw a rebirth (*renaissance* in French) of artistic, literary, and scientific inquiry in European culture following the chaos of the Dark Ages. However, one of the first documented uses of the scientific method in an astronomical context was made by Aristotle (384–322 B.C.) some 17 centuries earlier. Aristotle is not normally remembered as a strong proponent of this approach—many of his best known ideas were based on pure thought, with no attempt at experimental test or verification. Nevertheless, his brilliance extended into many areas now thought of as modern science. He noted that, during a lunar eclipse (Section 1.6), Earth casts a curved shadow onto the surface of the Moon. Figure 1.7 shows a series of photographs taken during a recent lunar eclipse. Earth's shadow, projected onto the Moon's surface, is indeed slightly curved. This is what Aristotle must have seen and recorded so long ago.

Because the observed shadow seemed always to be an arc of the same circle, Aristotle theorized that Earth, the cause of the shadow, must be round. Don't underestimate the scope of this apparently simple statement. Aristotle also had to reason that the dark region was indeed a shadow and that Earth was its cause—facts we regard as obvious today, but far from clear 25 centuries ago. On the basis of this *hypothesis*—one possible explanation of the observed facts—he then predicted that any and all future lunar

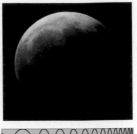

▲ **FIGURE 1.7 A Lunar Eclipse** This series of photographs show Earth's shadow sweeping across the Moon during an eclipse. By observing this behavior, Aristotle reasoned that Earth was the cause of the shadow and concluded that Earth must be round. His theory has yet to be disproved. *(G Schneider)*

R   I   V   U   X   G

eclipses would show Earth's shadow to be curved, regardless of our planet's orientation. That prediction has been tested every time a lunar eclipse has occurred. It has yet to be proved wrong. Aristotle was not the first person to argue that Earth is round, but he was apparently the first to offer observational proof using the lunar-eclipse method.

This basic reasoning forms the basis of all modern scientific inquiry. Armed only with naked-eye observations of the sky (the telescope would not be invented for almost another 2 thousand years), Aristotle first made an observation. Next, he formulated a hypothesis to explain that observation. Then he tested the validity of his hypothesis by making predictions that could be confirmed or refuted by further observations. *Observation, theory, and testing—* these are the cornerstones of the scientific method, a technique whose power will be demonstrated again and again throughout our text.

Today, scientists throughout the world use an approach that relies heavily on testing ideas. They gather data, form a working hypothesis that explains the data, and then proceed to test the implications of the hypothesis using experiment and observation. Eventually, one or more "well-tested" hypotheses may be elevated to the stature of a physical law and come to form the basis of a theory of even broader applicability. The new predictions of the theory will in turn be tested, as scientific knowledge continues to grow. Experiment and observation are integral parts of the process of scientific inquiry. Untestable theories, or theories unsupported by experimental evidence, rarely gain any measure of acceptance in scientific circles. Used properly over a period of time, this rational, methodical approach enables us to arrive at conclusions that are mostly free of the personal bias and human values of any one scientist. The scientific method is designed to yield an objective view of the universe we inhabit.

CONCEPT CHECK

✔ Can a theory ever become a "fact," scientifically speaking?

1.3 The "Obvious" View

To see how astronomers have applied the scientific method to understand the universe around us, let's start with some very basic observations. Our study of the cosmos, the modern science of astronomy, begins simply, with our looking at the night sky.

Constellations in the Sky

Between sunset and sunrise on a clear night, we can see about 3000 points of light. If we include the view from the opposite side of Earth, nearly 6000 stars are visible to the unaided eye. A natural human tendency is to see patterns and relationships among objects even when no true connection exists, and people long ago connected the brightest stars into configurations called **constellations**, which ancient astronomers named after mythological beings, heroes, and animals—whatever was important to them. Figure 1.8 shows a constellation especially prominent in the nighttime sky from October through March: the hunter named Orion. Orion was a mythical Greek hero famed, among other things, for his amorous pursuit of the Pleiades, the seven daughters of the giant Atlas. According to Greek mythology, to protect the Pleiades from Orion, the gods placed them among the stars, where Orion nightly stalks them across the sky. Many constellations have similarly fabulous connections with ancient lore.

Perhaps not surprisingly, the patterns have a strong cultural bias—the astronomers of ancient China saw mythical figures different from those seen by the ancient Greeks, the Babylonians, and the people of other cultures, even though they were all looking at the same stars in the night sky. Interestingly, different cultures often made the same basic *groupings* of stars, despite widely varying interpretations of what they saw. For example, the group of seven stars usually known in North America as "the Dipper" is known as "the Wagon" or "the Plough" in western Europe. The ancient Greeks regarded these same stars as the tail of "the Great Bear," the Egyptians saw them as the leg of an ox, the Siberians as a stag, and some Native Americans as a funeral procession.

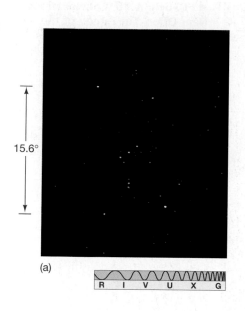

(a)

(b)

◀ **FIGURE 1.8 Constellation Orion** (a) A photograph of the group of bright stars that make up the constellation Orion. (See the preface, p. xv, for an explanation of the icon at the bottom, which simply indicates that this image was made in visible light.) (b) The stars are connected to show the pattern visualized by the Greeks: the outline of a hunter. You can easily find this constellation in the winter sky by identifying the line of three bright stars in the hunter's "belt." *(S. Westphal)*

Early astronomers had very practical reasons for studying the sky. Some constellations served as navigational guides. The star Polaris (part of the Little Dipper) indicates north, and the near constancy of its location in the sky, from hour to hour and night to night, has aided travelers for centuries. Other constellations served as primitive calendars to predict planting and harvesting seasons. For example, many cultures knew that the appearance of certain stars on the horizon just before daybreak signaled the beginning of spring and the end of winter.

In many societies, people came to believe that there were other benefits in being able to trace the regularly changing positions of heavenly bodies. The relative positions of stars and planets at a person's birth were carefully studied by *astrologers*, who used the data to make predictions about that person's destiny. Thus, in a sense, astronomy and astrology arose from the same basic impulse—the desire to "see" into the future—and, indeed, for a long time they were indistinguishable from one another. Today,

most people recognize that astrology is nothing more than an amusing diversion (although millions still study their horoscope in the newspaper every morning!). Nevertheless, the ancient astrological terminology—the names of the constellations and many terms used to describe the locations and motions of the planets—is still used throughout the astronomical world.

Generally speaking, as illustrated in Figure 1.9 for the case of Orion, the stars that make up any particular constellation are not actually close to one another in space, even by astronomical standards. They merely are bright enough to observe with the naked eye and happen to lie in roughly the same direction in the sky as seen from Earth. Still, the constellations provide a convenient means for astronomers to specify large areas of the sky, much as geologists use continents or politicians use voting precincts to identify certain localities on planet Earth. In all, there are 88 constellations, most of them visible from North America at some time during the year.

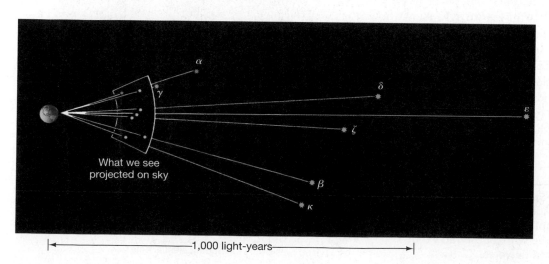

What we see projected on sky

◀1,000 light-years▶

◀ **FIGURE 1.9 Orion in 3-D** The true three-dimensional relationships among the brightest stars. (The Greek letters indicate brightness—see More Precisely 1-2.) The distances were determined by the *Hipparcos* satellite in the early 1990s. (See Chapter 17.)

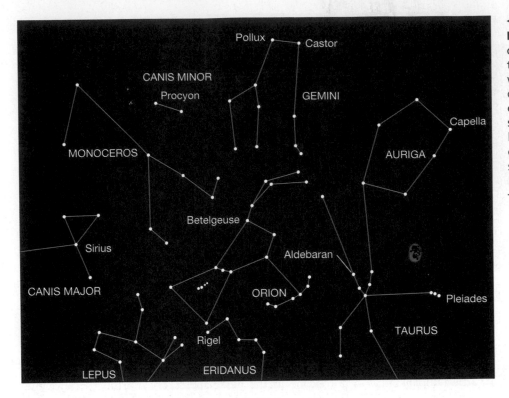

◀ FIGURE 1.10 Constellations **Near Orion** The region of the sky conventionally associated with the constellation Orion, together with some neighboring constellations (labeled in all capital letters). Some prominent stars are also labeled, in lowercase letters. The 88 constellations span the entire sky, so that every astronomical object lies in precisely one of them.

Figure 1.10 shows how the conventionally defined constellations cover a portion of the sky in the vicinity of Orion.

The Celestial Sphere

LEARNING GOAL 2 Over the course of a night, the constellations seem to move smoothly across the sky from east to west, but ancient skywatchers were well aware that the *relative* locations of stars remained unchanged as this nightly march took place. It was natural for those observers to conclude that the stars must be firmly attached to a **celestial sphere** surrounding Earth—a canopy of stars resembling an astronomical painting on a heavenly ceiling. Figure 1.11 shows how early astronomers pictured the stars as moving with this celestial sphere as it turned around a fixed, unmoving Earth. Figure 1.12 shows how all stars appear to move in circles around a point very close to the star Polaris (better known as the Pole Star or North Star). To the ancients, this point represented the axis around which the entire celestial sphere turned.

From our modern standpoint, the apparent motion of the stars is the result of the spin, or **rotation**, not of the celestial sphere, but of Earth. Polaris indicates the direction—due north—in which Earth's rotation axis points.

Even though we now know that the celestial sphere is an incorrect description of the heavens, we still use the idea as a convenient fiction that helps us visualize the positions of stars in the sky. The points where Earth's axis intersects the celestial sphere are called the **celestial poles**. In the Northern Hemisphere, the north celestial pole lies directly above Earth's North Pole. The extension of Earth's axis

▶ **FIGURE 1.11 Celestial Sphere** Planet Earth sits fixed at the hub of the celestial sphere, which contains all the stars. This is one of the simplest possible models of the universe, but it doesn't agree with all the facts that astronomers know about the universe.

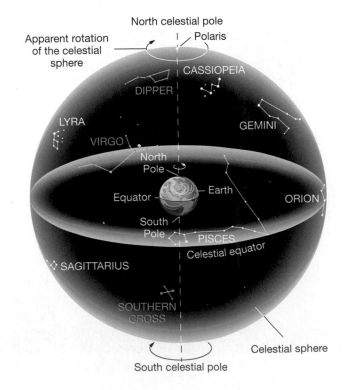

◄**FIGURE 1.12 Northern Sky** A time-lapse photograph of the northern sky. Each trail is the path of a single star across the night sky. The duration of the exposure is about five hours, since each star traces out approximately 20 percent of a circle. The center of the concentric circles is near the North Star, Polaris, whose short, bright arc is prominently visible. *(AURA)*

R I V U X G

in the opposite direction defines the south celestial pole, directly above Earth's South Pole. Midway between the north and south celestial poles lies the **celestial equator**, representing the intersection of Earth's equatorial plane with the celestial sphere. These parts of the celestial sphere are marked on Figure 1.11.

When discussing the locations of stars "on the sky," astronomers naturally talk in terms of *angular* positions and separations. *More Precisely 1-1* presents some basic information on angular measure. *More Precisely 1-2* discusses in more detail some systems of coordinates used to specify stellar positions.

MORE PRECISELY 1-1

Angular Measure

Size and scale are often specified by measuring lengths and angles. The concept of length measurement is fairly intuitive to most of us. The concept of *angular measurement* may be less familiar, but it, too, can become second nature if you remember a few simple facts:

- A full circle contains 360 *degrees* (360°). Thus, the half-circle that stretches from horizon to horizon, passing directly overhead and spanning the portion of the sky visible to one person at any one time, contains 180°.

- Each 1° increment can be further subdivided into fractions of a degree, called *arc minutes*. There are 60 arc minutes (written 60′) in one degree. (The term "arc" is used to distinguish this angular unit from the unit of time.) Both the Sun and the Moon project an angular size of 30 arc minutes (half a degree) on the sky. Your little finger, held at arm's length, has a similar angular size, covering about a 40′ slice of the 180° horizon-to-horizon arc.

- An arc minute can be divided into 60 *arc seconds* (60″). Put another way, an arc minute is 1/60 of a degree, and an arc second is 1/60 × 1/60 = 1/3600 of a degree. An arc second is an extremely small unit of angular measure—the angular size of a centimeter-sized object (a dime, say) at a distance of about 2 kilometers (a little over a mile).

The accompanying figure illustrates this subdivision of the circle into progressively smaller units.

Don't be confused by the units used to measure angles. Arc minutes and arc seconds have nothing to do with the measurement of time, and degrees have nothing to do with tempera-

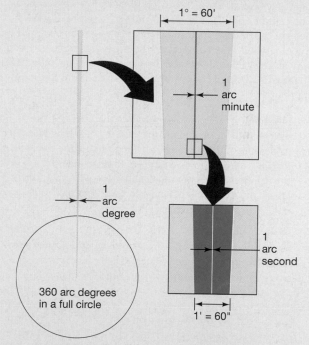

1° = 60′

1 arc minute

1 arc degree

1 arc second

360 arc degrees in a full circle

1′ = 60″

ture. Degrees, arc minutes, and arc seconds are simply ways to measure the size and position of objects in the universe.

The angular size of an object depends both on its actual size and on its distance from us. For example, the Moon, at its present distance from Earth, has an angular diameter of 0.5°, or 30′. If the Moon were twice as far away, it would appear half as big—15′ across—even though its actual size would be the same. Thus, *angular size by itself is not enough to determine the actual diameter of an object—the distance to the object must also be known.* We return to this topic in more detail in *More Precisely 1-3*.

MORE PRECISELY 1-2

Celestial Coordinates

The simplest method of locating stars in the sky is to specify their constellation and then rank the stars in it in order of brightness. The brightest star is denoted by the Greek letter α (alpha), the second brightest by β (beta), and so on. (See Figure 1.8.) Thus, the two brightest stars in the constellation Orion—Betelgeuse and Rigel—are also known as α Orionis and β Orionis, respectively. (More recently, precise observations show that Rigel is actually brighter than Betelgeuse, but the names are now permanent.) Similarly, Sirius, the brightest star in the sky (see Appendix 3), which lies in the constellation Canis Major (the Great Dog), is denoted α Canis Majoris (or α CMa for short); the (present) Pole Star, in Ursa Minor (the Little Bear), is also known as α Ursae Minoris (or α UMi), and so on. Because there are many more stars in any given constellation than there are letters in the Greek alphabet, this method is of limited utility. However, for naked-eye astronomy, where only bright stars are involved, it is quite satisfactory.

For more precise measurements, astronomers find it helpful to lay down a system of *celestial coordinates* on the sky. If we think of the stars as being attached to the celestial sphere centered on Earth, then the familiar system of latitude and longitude on Earth's surface extends naturally to cover the sky. The celestial analogs of latitude and longitude on Earth's surface are called *declination* and *right ascension*, respectively. The accompanying figure illustrates the meanings of right ascension and declination on the celestial sphere and compares them with longitude and latitude on Earth. Note the following points:

- Declination (dec) is measured in degrees (°) north or south of the celestial equator, just as latitude is measured in degrees north or south of Earth's equator. (See *More Precisely 1-1* for a discussion of angular measure.) Thus, the celestial equator is at a declination of 0°, the north celestial pole is at +90°, and the south celestial pole is at −90° (the minus sign here just means "south of the celestial equator").

- Right ascension (RA) is measured in units called *hours*, *minutes*, and *seconds*, and it increases in the eastward direction. The angular units used to measure right ascension are constructed to parallel the units of time, in order to assist astronomical observation. The two sets of units are connected by the rotation of Earth (or of the celestial sphere). In 24 hours, Earth rotates once on its axis, or through 360°. Thus, in a period of 1 hour, Earth rotates through $360°/24 = 15°$, or 1^h. In 1 minute of time, Earth rotates through an angle of $1^m = 15°/60 = 0.25°$, or 15 arc minutes ($15'$). In 1 second of time, Earth rotates through an angle of $1^s = 15'/60 = 15$ arc seconds ($15''$). As with longitude, the choice of zero right ascension (the celestial equivalent of the Greenwich Meridian) is quite arbitrary—it is conventionally taken to be the position of the Sun in the sky *at the instant of the vernal equinox.*

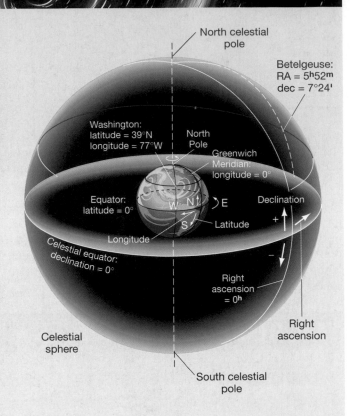

Right ascension and declination specify locations on the sky in much the same way as longitude and latitude allow us to locate a point on Earth's surface. For example, to find Washington on Earth, look 77° west of the Greenwich Meridian (the line on Earth's surface with a longitude of zero) and 39° north of the equator. Similarly, to locate the star Betelgeuse on the celestial sphere, look $5^h52^m0^s$ east of the vernal equinox (the line on the sky with a right ascension of zero) and 7°24′ north of the celestial equator. The star Rigel, also mentioned earlier, lies at $5^h13^m36^s$ (RA), $-8°13'$ (dec). Thus, we have a quantitative alternative to the use of constellations in specifying the positions of stars in the sky. Just as latitude and longitude are tied to Earth, right ascension and declination are fixed on the celestial sphere. Although the stars appear to move across the sky because of Earth's rotation, their celestial coordinates remain *constant* over the course of a night.

Actually, because right ascension is tied to the position of the vernal equinox, the celestial coordinates of any given star slowly drift due to Earth's *precession.* (See Section 1.4.) Since one cycle of precession takes 26,000 years to complete, the shift on a night-to-night basis is a little over 0.1″—a small angle, but one that must be taken into account in high-precision astronomical measurements. Rather than deal with slowly changing coordinates for every object in the sky, astronomers conventionally correct their observations to the location of the vernal equinox at some standard epoch (such as January 1, 1950 or January 1, 2000).

CONCEPT CHECK

✔ Why do astronomers find it useful to retain the fiction of the celestial sphere to describe the sky? What vital piece of information about stars is lost when we talk about their locations "on" the sky?

1.4 Earth's Orbital Motion

Day-To-Day Changes

LEARNING GOAL 3 We measure time by the Sun. Because the rhythm of day and night is central to our lives, it is not surprising that the period from one noon to the next, the 24-hour **solar day**, is our basic social time unit. The daily progress of the Sun and the other stars across the sky is known as *diurnal motion*. As we have just seen, it is a consequence of Earth's rotation. But the stars' positions in the sky do *not* repeat themselves exactly from one night to the next. Each night, the whole celestial sphere appears to be shifted a little relative to the horizon, compared with the night before. The easiest way to confirm this difference is by noticing the stars that are visible just after sunset or just before dawn. You will find that they are in slightly different locations from those of the previous night. Because of this shift, a day measured by the stars—called a **sidereal day** after the Latin word *sidus*, meaning "star"—differs in length from a solar day. Evidently, there is more to the apparent motion of the heavens than simple rotation.

The reason for the difference between a solar day and a sidereal day is sketched in Figure 1.13. It is a result of the fact that Earth moves in two ways simultaneously: It rotates on its central axis while at the same time **revolving** around the Sun. Each time Earth rotates once on its axis, it also moves a small distance along its orbit about the Sun. Earth therefore has to rotate through slightly more than 360° (360 degrees—see *More Precisely 1-1*) for the Sun to return to the same apparent location in the sky. Thus, the interval of time between noon one day and noon the next (a solar day) is slightly greater than one true rotation period (one sidereal day). Our planet takes 365 days to orbit the Sun, so the additional angle is 360°/365 = 0.986°. Because Earth, rotating at a rate of 15° per hour, takes about 3.9 minutes to rotate through this angle, the solar day is 3.9 minutes longer than the sidereal day (i.e., 1 sidereal day is roughly 23$^\text{h}$56$^\text{m}$ long).

Seasonal Changes

Figure 1.14(a) illustrates the major stars visible from most locations in the United States on clear summer evenings. The brightest stars—Vega, Deneb, and Altair—form a conspicuous triangle high above the constellations Sagittarius and Capricornus, which are low on the southern

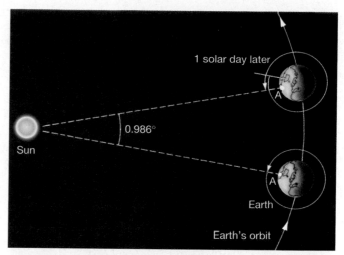

▲ **FIGURE 1.13 Solar and Sidereal Days** The difference between a solar and a sidereal day can easily be explained once we understand that Earth revolves around the Sun at the same time as it rotates on its axis. Because of Earth's orbital motion, between noon at point A on one day and noon at the same point the next day, Earth actually rotates through about 361°. Consequently, the solar day exceeds the sidereal day (360° rotation) by about 4 minutes. Note that the diagram is not drawn to scale, so the true 1° angle is in reality much smaller than shown here.

horizon. In the winter sky, however, these stars are replaced as shown in Figure 1.14(b) by several other, well-known constellations, including Orion, Leo, and Gemini. In the constellation Canis Major lies Sirius (the Dog Star), the brightest star in the sky. Year after year, the same stars and constellations return, each in its proper season. Every winter evening, Orion is high overhead; every summer, it is gone. (For more detailed maps of the sky at different seasons, consult the star charts at the end of the book.)

These regular seasonal changes occur because of Earth's revolution around the Sun: Earth's darkened hemisphere faces in a slightly different direction in space each evening. The change in direction is only about 1° per night (Figure 1.13)—too small to be easily noticed with the naked eye from one evening to the next, but clearly noticeable over the course of weeks and months, as illustrated in Figure 1.15. After 6 months, Earth has reached the opposite side of its orbit, and we face an entirely different group of stars and constellations at night. Because of this motion, the Sun appears (to an observer on Earth) to move relative to the background stars over the course of a year. This apparent motion of the Sun on the sky traces out a path on the celestial sphere known as the **ecliptic**.

The 12 constellations through which the Sun passes as it moves along the ecliptic—that is, the constellations we would see looking in the direction of the Sun if they weren't overwhelmed by the Sun's light—had special significance for astrologers of old. These constellations are collectively known as the **zodiac**.

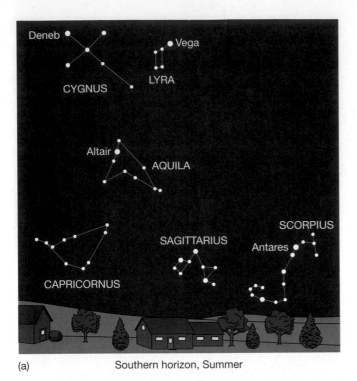

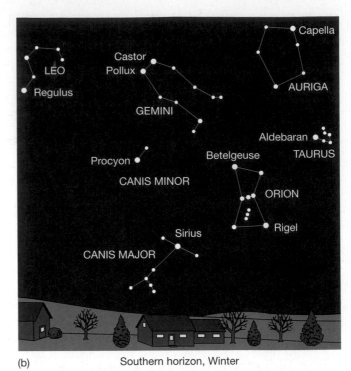

(a) Southern horizon, Summer

(b) Southern horizon, Winter

▲ **FIGURE 1.14 Typical Night Sky** (a) A typical summer sky above the United States. Some prominent stars (labeled in lowercase letters) and constellations (labeled in all capital letters) are shown. (b) A typical winter sky above the United States.

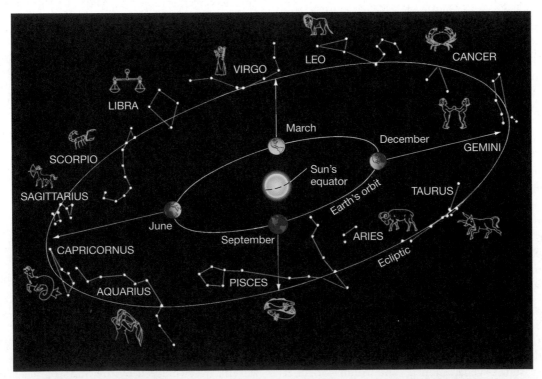

▲ **FIGURE 1.15 The Zodiac** The view of the night sky changes as Earth moves in its orbit about the Sun. As drawn here, the night side of Earth faces a different set of constellations at different times of the year. The 12 constellations named here make up the astrological zodiac.

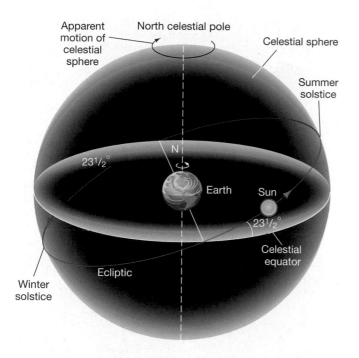

As illustrated in Figure 1.16, the ecliptic forms a great circle on the celestial sphere, inclined at an angle of 23.5° to the celestial equator. In reality, as illustrated in Figure 1.17, the plane of the ecliptic is *the plane of Earth's orbit around the Sun*. Its tilt is a consequence of the *inclination* of our planet's axis rotation to the plane of its orbit.

The point on the ecliptic where the Sun is at its northernmost point above the celestial equator is known as the **summer solstice** (from the Latin words *sol*, meaning "sun," and *stare*, "to stand"). As indicated in Figure 1.17, it represents the location in Earth's orbit where our planet's North Pole points closest to the Sun. This occurs on or near June 21—the exact date varies slightly from year to year because the actual length of a year is not a whole number of days. As Earth rotates, points north of the equator spend the greatest fraction of their time in sunlight on that date, so the summer solstice corresponds to the longest day of the year in the Northern Hemisphere and the shortest day in the Southern Hemisphere.

Six months later, the Sun is at its southernmost point below the celestial equator (Figure 1.16)—or, equivalently, the North Pole is oriented farthest from the Sun (Figure 1.17). We have reached the **winter solstice** (December 21), the shortest day in Earth's Northern Hemisphere and the longest in the Southern Hemisphere.

The Sun's location with respect to the celestial equator and the

◀ **FIGURE 1.16 Ecliptic** The apparent path of the Sun on the celestial sphere over the course of a year is called the ecliptic. As indicated on the diagram, the ecliptic is inclined to the celestial equator at an angle of 23.5°. In this picture of the heavens, the seasons result from the changing height of the Sun above the celestial equator. At the summer solstice, the Sun is at its northernmost point on its path around the ecliptic; it is thus highest in the sky, as seen from Earth's northern hemisphere, and the days are longest. The reverse is true at the winter solstice. At the vernal and autumnal equinoxes, when the Sun crosses the celestial equator, day and night are of equal length.

ANIMATION The Earth's Seasons

▼ **FIGURE 1.17 Seasons** In reality, the Sun's apparent motion along the ecliptic is a consequence of Earth's orbital motion around the Sun. The seasons result from the inclination of our planet's axis of rotation with respect to the plane of the orbit. The summer solstice corresponds to the point on Earth's orbit where our planet's North Pole points most nearly toward the Sun. The opposite is true of the winter solstice. The vernal and autumnal equinoxes correspond to the points in Earth's orbit where our planet's axis is perpendicular to the line joining Earth and the Sun. The insets show how rays of sunlight striking the ground obliquely (e.g., during northern winter) are spread over a much larger area than rays coming nearly straight down (e.g., during northern summer). As a result, the amount of solar heat delivered to a given area of Earth's surface is greatest when the Sun is high in the sky.

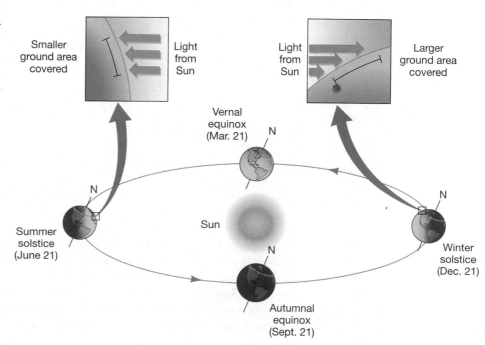

length of the day combine to account for the **seasons** we experience. As illustrated in the insets in Figure 1.17, when the Sun is high in the sky, rays of sunlight striking Earth's surface are more concentrated—falling over a smaller area. As a result, the Sun feels hotter. Thus, in northern summer, when the Sun is highest above the horizon and the days are long, temperatures are generally much higher than in winter, when the Sun is low and the days are short.

The two points where the ecliptic intersects the celestial equator (Figure 1.16)—that is, where Earth's rotation axis is perpendicular to the line joining Earth to the Sun (Figure 1.17)—are known as **equinoxes**. On those dates, day and night are of equal duration. (The word *equinox* derives from the Latin for "equal night.") In the fall (in the Northern Hemisphere), as the Sun crosses from the Northern into the Southern Hemisphere, we have the **autumnal equinox** (on September 21). The **vernal equinox** occurs in northern spring, on or near March 21, as the Sun crosses the celestial equator moving north. Because of its association with the end of winter and the start of a new growing season, the vernal equinox was particularly important to early astronomers and astrologers. It also plays an important role in human timekeeping: The interval of time from one vernal equinox to the next—365.2422 mean solar days (see Section 1.5)—is known as one **tropical year**.

Long-Term Changes

LEARNING GOAL 3 Earth has many motions—it spins on its axis, it travels around the Sun, and it moves with the Sun through our Galaxy. We have just seen how some of these motions can account for the changing nighttime sky and the changing seasons. In fact, the situation is even more complicated. Like a spinning top that rotates rapidly on its own axis while that axis slowly revolves about the vertical, Earth's axis changes its *direction* over the course of time (although the angle between the axis and a line perpendicular to the plane of the ecliptic always remains close to 23.5°). Illustrated in Figure 1.18, this change is called **precession**. It is caused by torques (twisting forces) on Earth due to the gravitational pulls of the Moon and the Sun, which affect our planet in much the same way as the torque due to Earth's own gravity affects a top. During a complete cycle of precession—about 26,000 years—Earth's axis traces out a cone.

The time required for Earth to complete exactly one orbit around the Sun, relative to the stars, is called a **sidereal year**. One sidereal year is 365.256 mean solar days long—about 20 minutes longer than a tropical year. The reason for this slight difference is Earth's precession. Recall that the vernal equinox occurs when Earth's rotation axis is perpendicular to the line joining Earth and the Sun, and the Sun is crossing the celestial equator moving from south to north. In the absence of precession, this combination of events would occur exactly once per sidereal orbit, and the tropical and sidereal years would be

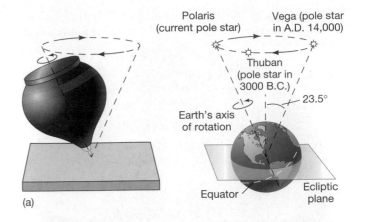

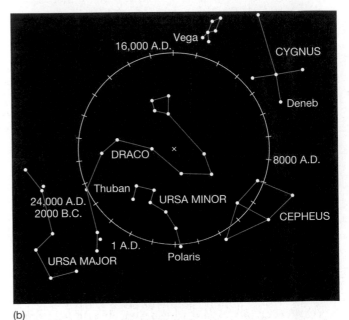

▲ **FIGURE 1.18 Precession** (a) Earth's axis currently points nearly toward the star Polaris. About 12,000 years from now—almost halfway through one cycle of precession—Earth's axis will point toward a star called Vega, which will then be the "North Star." Five thousand years ago, the North Star was a star named Thuban in the constellation Draco. (b) The circle shows the precessional path of the north celestial pole among some prominent northern stars. Tick marks indicate intervals of a thousand years.

identical. However, because of the slow precessional shift in the orientation of Earth's rotation axis, the instant when the axis is next perpendicular to the line from Earth to the Sun occurs slightly *sooner* than we would otherwise expect. Consequently, the vernal equinox drifts slowly around the zodiac over the course of the precession cycle.

The tropical year is the year that our calendars measure. If our timekeeping were tied to the sidereal year, the seasons would slowly march around the calendar as Earth precessed—13,000 years from now, summer in the Northern Hemisphere would be at its height in late February! By using the tropical year, we ensure that

July and August will always be (northern) summer months. However, in 13,000 years' time, Orion will be a summer constellation.

CONCEPT CHECK

✔ In astronomical terms, what are "summer" and "winter," and why do we see different constellations during those seasons?

1.5 Astronomical Timekeeping

4 Earth's rotation and revolution around the Sun define the basic time units—the day and the year—by which we measure our lives. Today the SI unit of time, the second, is defined by ultrahigh-precision atomic clocks maintained in the National Institute of Standards and Technology, in Gaithersburg, Maryland, and in other sites around the world. However, the time those clocks measure is based squarely on astronomical events, and astronomers (or astrologers) have traditionally borne the responsibility for keeping track of the days and the seasons.

As we have just seen, at any location on Earth, a solar day is defined as the time between one noon and the next. Here, "noon" means the instant when the Sun crosses the **meridian**—an imaginary line on the celestial sphere through the north and south celestial poles, passing directly overhead at the given location. This is the time that a sundial would measure. Unfortunately, this most direct measure of time has two serious drawbacks: The length of the solar day varies throughout the year, and the time at which noon occurs varies from place to place.

Recall that the solar day is the result of a "competition" between Earth's rotation and its revolution around the Sun. Earth's revolution means that our planet must rotate through a little more than 360° between one noon and the next. (See Figure 1.13.) However, while Earth's rotation rate is virtually constant, the rate of revolution— or, more specifically, the rate at which the Sun traverses the celestial sphere as it moves along the ecliptic—is not constant—for two reasons, illustrated and exaggerated in Figure 1.19. First, as we will see in Chapter 2, Earth's orbit is not exactly circular, and our orbital speed is not constant—Earth moves more rapidly than average when closer to the Sun, more slowly when farther away—so the speed at which the Sun appears to move along the ecliptic varies with time (Figure 1.19a). Second, because the ecliptic is inclined to the celestial equator, the *eastward* component of the Sun's motion on the celestial sphere depends on the time of year (and note that this would be the case even if Earth's orbit were circular and the first point did not apply). As illustrated in Figure 1.19b, at the equinoxes the Sun's path is inclined to the equator, and only the Sun's eastward progress across the sky contributes to the motion. At the solstices, however, the motion is entirely eastward.

The combination of these effects means that the solar day varies by roughly half a minute over the course

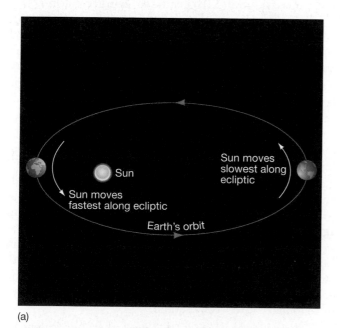

(a)

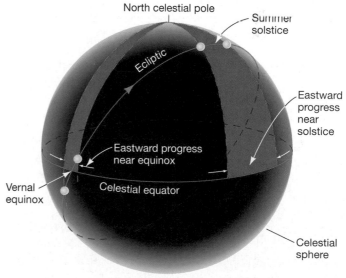

(b)

▲ **FIGURE 1.19 Variations in the Solar Day** (a) Because Earth does not orbit the Sun at a constant speed, the rate at which the Sun appears to move across the sky, and hence the length of the solar day, also varies. (b) In addition, the length of the solar day changes with the season, because the eastward progress of the Sun in, say, 1 hour, (indicated here) is greater at the solstices than at the equinoxes.

of the year—not a large variation, but unacceptable for astronomical and many other purposes. The solution is to define a **mean solar day**, which, in effect, is just the average (or *mean*) of all the solar days over an entire year. This is the day our clocks (atomic or otherwise) measure, and it is constant by definition. One second is $1/24 \times 1/60 \times 1/60 = 1/86,400$ mean solar days.

Now all of our clocks tick at a constant rate, but we are not quite out of the woods in our search for a standard of time. The preceding definition is still *local*—observers at different longitudes see noon at different times, so even though their clocks keep pace, they all tell different times. In 1883, driven by the need for uniform and consistent times in long-distance travel and communications (the railroads and the telegraph), the continental United States was divided into four standard **time zones**. Within each zone, everyone adopted the mean solar time corresponding to a specific meridian inside the zone. Since 1884, **standard time** has been used around the world. The global standard time zones are shown in Figure 1.20. By convention, the reference meridian is taken to be 0° longitude (the Greenwich meridian). In the United States, Eastern, Central, Mountain, and Pacific time zones keep the mean solar time of longitudes 75° W, 90° W, 105° W, and 120° W, respectively. Hawaii and Alaska keep the time of longitude 150° W. In some circumstances, it is even more convenient to adopt a single time zone: **Universal time** (formerly known as *Greenwich mean time*) is simply the mean solar time at the Greenwich meridian.

Having defined the day, let's now turn our attention to how those days fit into the year—in other words, to the construction of a calendar. The year in question is the *tropical year*, tied to the changing seasons. It is conventionally divided into *months*, which were originally defined by the phases of the Moon. (See Section 1.6; the word month derives from *Moon*.) The basic problem with all this is that a lunar month does not contain a whole number of mean solar days and the tropical year does not contain a whole number of either months or days. Many ancient calendars were lunar, and the variable number of days in modern months may be traced back to attempts to approximate the 29.5-day lunar month by alternating 29- and 30-day periods. The Chinese and Islamic calendars in use today are still based squarely on the lunar cycle, each month beginning with the new Moon. The modern Western calendar retains months as convenient subdivisions, but they have no particular lunar significance.

One tropical year is 365.2422 mean solar days long and hence cannot be represented by a whole number of calendar days. The solution is to vary the number of days in the year, in order to ensure the correct *average* length of the year. In 46 B.C., the Roman emperor Julius Caesar decreed that the calendar would include an extra day every fourth year—a **leap year**—ensuring that the average year would be $(3 \times 365 + 366)/4 = 365.25$ days long. (Nowadays, the extra day is inserted at the end of February.) This *Julian calendar* was a great improvement over earlier calendars, which had 354 (=12 × 29.5) days in the year, necessitating the inclusion of an extra month every three years!

Still, the Julian year was not exactly equal to the tropical year, and, over time, the calendar drifted relative to the seasons. By A.D. 1582, the difference amounted to 10 days, and Pope Gregory XIII instituted another reform, first skipping the extra 10 days (resetting the vernal equinox back to March 21) and then changing the rule for leap years to omit the extra day from years that were multiples of 100, but retain it in years that were multiples of 400. The effect of Pope Gregory's reform was that the average year became 365.2425 mean solar days long—good to 1 day in 3300 years. The idea of "losing" 10 days at the behest of the Pope was unacceptable to many countries, with the result that the *Gregorian calendar* was not fully accepted for several centuries. Britain and the American colonies finally adopted it in 1752. Russia abandoned the Julian calendar only in 1917, after the Bolshevik Revolution, at which time the country had to skip 13 days to come into agreement with the rest of the world.

The most recent modern correction to the Gregorian calendar has been to declare that the years 4000, 8000, etc., will *not* be leap years, improving the accuracy of the calendar to 1 day in 20,000 years—good enough for most of us to make it to work on time!

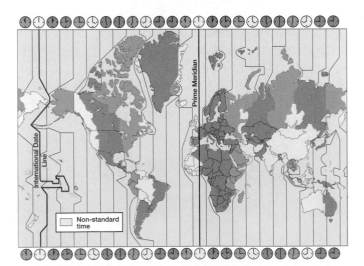

▲ **FIGURE 1.20 Time Zones** The standard time zones were adopted to provide consistent times around the world. Within each time zone, all clocks show the same time, the local solar time of one specific longitude inside the zone.

CONCEPT CHECK

✔ Is the Sun always directly overhead at noon, according to your watch?

1.6 The Motion of the Moon

The Moon is our nearest neighbor in space. Apart from the Sun, it is the brightest object in the sky. Like the Sun, the Moon appears to move relative to the background stars. Unlike the Sun, however, the Moon really does revolve around Earth. It crosses the sky at a rate of about 12° per day, which means that it moves an angular distance equal to its own diameter—30 arc minutes—in about an hour.

Lunar Phases

The Moon's appearance undergoes a regular cycle of changes, or **phases**, taking roughly 29.5 days to complete. Figure 1.21 illustrates the appearance of the Moon at different times in this monthly cycle. Starting from the *new Moon*, which is all but invisible in the sky, the Moon appears to *wax* (or grow) a little each night and is visible as a growing *crescent* (panel 1 of Figure 1.21). One week after new Moon, half of the lunar disk can be seen (panel 2). This phase is known as a *quarter Moon*. During the next

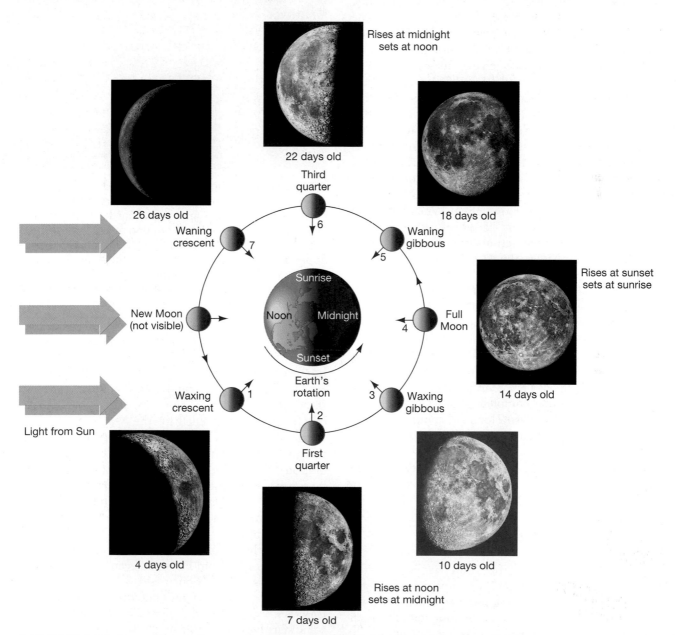

▲ **FIGURE 1.21 Lunar Phases** Because the Moon orbits Earth, the visible fraction of the lunar sunlit face varies from night to night, although the Moon always keeps the same face toward our planet. (Note the location of the arrow, which marks the same point on the lunar surface at each phase shown.) The complete cycle of lunar phases, shown here starting at the waxing crescent phase and following the Moon's orbit counterclockwise, takes 29.5 days to complete. Rising and setting times for some phases are also indicated. *(UC/Lick Observatory)*

week, the Moon continues to wax, passing through the *gibbous* phase (panel 3) until, 2 weeks after new Moon, the *full Moon* (panel 4) is visible. During the next 2 weeks, the Moon *wanes* (or shrinks), passing in turn through the gibbous, quarter, and crescent phases (panels 5–7) and eventually becoming new again.

The location of the Moon in the sky, as seen from Earth, depends on its phase. For example, the full Moon rises in the east as the Sun sets in the west, while the first quarter Moon actually rises at noon, but sometimes becomes visible only late in the day as the Sun's light fades. By this time the Moon is already high in the sky. Some connections between the lunar phase and the rising and setting times of the Moon are indicated in Figure 1.21.

The Moon doesn't actually change its size and shape from night to night, of course. Its full circular disk is present at all times. Why, then, don't we always see a full Moon? The answer to this question lies in the fact that, unlike the Sun and the other stars, the Moon emits no light of its own. Instead, it shines by reflected sunlight. As illustrated in Figure 1.21, half of the Moon's surface is illuminated by the Sun at any instant. However, not all of the Moon's sunlit face can be seen, because of the Moon's position with respect to Earth and the Sun. When the Moon is full, we see the entire "daylit" face because the Sun and the Moon are in opposite directions from Earth in the sky. In the case of a new Moon, the Moon and the Sun are in almost the same part of the sky, and the sunlit side of the Moon is oriented away from us. At new Moon, the Sun must be almost behind the Moon, from our perspective.

As the Moon revolves around Earth, our satellite's position in the sky changes with respect to the stars. In 1 **sidereal month** (27.3 days), the Moon completes one revolution and returns to its starting point on the celestial sphere, having traced out a great circle in the sky. The time required for the Moon to complete a full cycle of phases, one **synodic month**, is a little longer—about 29.5 days. The synodic month is a little longer than the sidereal month for the same reason that a solar day is slightly longer than a sidereal day: Because of Earth's motion around the Sun, the Moon must complete slightly more than one full revolution to return to the same phase in its orbit (Figure 1.22).

Eclipses

From time to time—but only at new or full Moon—the Sun and the Moon line up precisely as seen from Earth, and we observe the spectacular phenomenon known as an **eclipse**. When the Sun and the Moon are in exactly *opposite* directions, as seen from Earth, Earth's shadow sweeps across the Moon, temporarily blocking the Sun's light and darkening the Moon in a **lunar eclipse**, as illustrated in Figure 1.23. From Earth, we see the curved edge of Earth's shadow begin to cut across the face of the full Moon and slowly eat its way into the lunar disk. Usually, the alignment of the Sun, Earth,

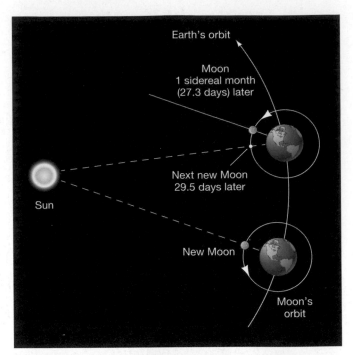

▲ FIGURE 1.22 Sidereal Month The difference between a synodic and a sidereal month stems from the motion of Earth relative to the Sun. Because Earth orbits the Sun in 365 days, in the 29.5 days from one new Moon to the next (1 synodic month), Earth moves through an angle of approximately 29°. Thus, the Moon must revolve more than 360° between new Moons. The sidereal month, which is the time taken for the Moon to revolve through exactly 360°, relative to the stars, is about 2 days shorter.

and Moon is imperfect, so the shadow never completely covers the Moon. Such an occurrence is known as a **partial lunar eclipse**. Occasionally, however, the entire lunar surface is obscured in a **total lunar eclipse**, such as that shown in the inset of Figure 1.23. Total lunar eclipses last only as long as is needed for the Moon to pass through Earth's shadow—no more than about 100 minutes. During that time, the Moon often acquires an eerie, deep red coloration—the result of a small amount of sunlight reddened by Earth's atmosphere (for the same reason that sunsets appear red—see *More Precisely 7-1*) and refracted (bent) onto the lunar surface, preventing the shadow from being completely black.

When the Moon and the Sun are in exactly the *same* direction, as seen from Earth, an even more awe-inspiring event occurs. The Moon passes directly in front of the Sun, briefly turning day into night in a **solar eclipse**. In a *total solar eclipse*, when the alignment is perfect, planets and some stars become visible in the daytime as the Sun's light is reduced to nearly nothing. We can also see the Sun's ghostly outer atmosphere, or *corona* (Figure 1.24).* In a

*Actually, although a total solar eclipse is undeniably a spectacular occurrence, the visibility of the corona is probably the most important astronomical aspect of such an event today. It enables us to study this otherwise hard-to-see part of our Sun (see Chapter 16).

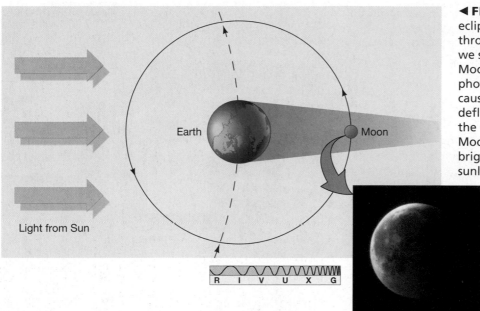

◄ **FIGURE 1.23** Lunar Eclipse A lunar eclipse occurs when the Moon passes through Earth's shadow. At these times, we see a darkened, copper-colored Moon, as shown in the inset photograph. The red coloration is caused by sunlight being reddened and deflected by Earth's atmosphere onto the Moon's surface. An observer on the Moon would see Earth surrounded by a bright, but narrow, ring of orange sunlight. Note that this figure is not drawn to scale, and only Earth's umbra (see text and Figure 1.25) is shown. *(Inset: G. Schneider)*

partial solar eclipse, the Moon's path is slightly "off center," and only a portion of the Sun's face is covered. In either case, the sight of the Sun apparently being swallowed up by the black disk of the Moon is disconcerting even today. It must surely have inspired fear in early observers. Small wonder that the ability to predict such events was a highly prized skill.

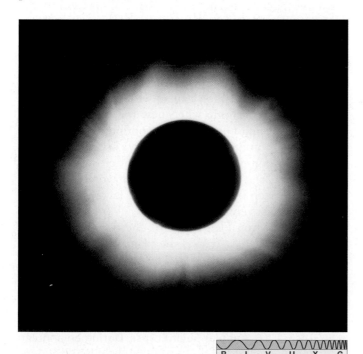

▲ **FIGURE 1.24** Total Solar Eclipse During a total solar eclipse, the Sun's corona becomes visible as an irregularly shaped halo surrounding the blotted-out disk of the Sun. This was the August 1999 eclipse, as seen from the banks of the Danube River near Sofia, Bulgaria. *(Bencho Angelov)*

Unlike a lunar eclipse, which is simultaneously visible from all locations on Earth's night side, a total solar eclipse can be seen from only a small portion of Earth's daytime side. The Moon's shadow on Earth's surface is about 7000 kilometers wide—roughly twice the diameter of the Moon. Outside of that shadow, no eclipse is seen. However, within the central region of the shadow, called the **umbra**, the eclipse is total. Within the shadow, but outside the umbra, in the **penumbra**, the eclipse is partial, with less and less of the Sun obscured the farther one travels from the shadow's center.

The connections among the umbra, the penumbra, and the relative locations of Earth, Sun, and Moon are illustrated in Figure 1.25. The umbra is always very small. Even under the most favorable circumstances, its diameter never exceeds 270 kilometers. Because the shadow sweeps across Earth's surface at over 1700 kilometers per hour, the duration of a total eclipse at any given point on our planet can never exceed 7.5 minutes.

The Moon's orbit around Earth is not exactly circular. Thus, the Moon may be far enough from Earth at the moment of an eclipse that its disk fails to cover the disk of the Sun completely, even though their centers coincide. In that case, there is no region of totality—the umbra never reaches Earth at all, and a thin ring of sunlight can still be seen surrounding the Moon. Such an occurrence, called an **annular eclipse**, is illustrated in Figure 1.25(c) and shown more clearly in Figure 1.26. Roughly half of all solar eclipses are annular.

Eclipse Seasons

Why isn't there a solar eclipse at every new Moon and a lunar eclipse at every full Moon? The answer is that the Moon's orbit is slightly inclined to the ecliptic (at an angle

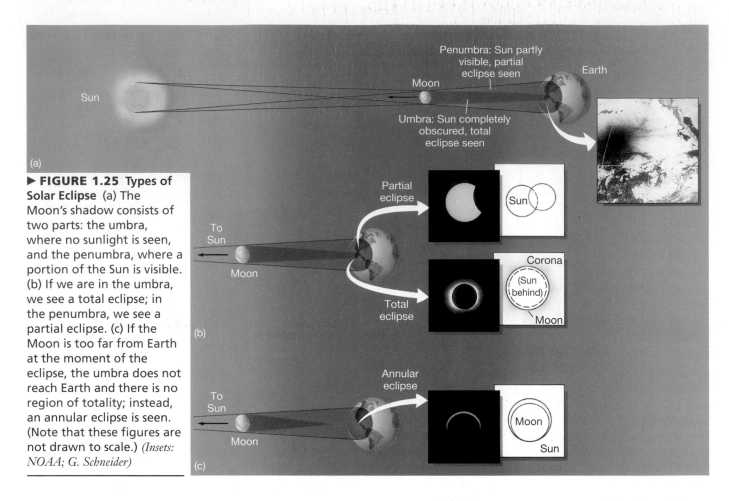

▶ **FIGURE 1.25 Types of Solar Eclipse** (a) The Moon's shadow consists of two parts: the umbra, where no sunlight is seen, and the penumbra, where a portion of the Sun is visible. (b) If we are in the umbra, we see a total eclipse; in the penumbra, we see a partial eclipse. (c) If the Moon is too far from Earth at the moment of the eclipse, the umbra does not reach Earth and there is no region of totality; instead, an annular eclipse is seen. (Note that these figures are not drawn to scale.) *(Insets: NOAA; G. Schneider)*

R I V U X G

of 5.2°), so the chance that a new (or full) Moon will occur just as the Moon happens to cross the plane of the ecliptic (with Earth, Moon, and Sun perfectly aligned) is quite low. Figure 1.27 illustrates some possible configurations of the three bodies. If the Moon happens to lie above or below the plane of the ecliptic when new (or full), a solar (or lunar) eclipse cannot occur. Such a configuration is termed *unfavorable* for producing an eclipse. In a *favorable* configuration, the Moon is new or full just as it crosses the plane of the ecliptic, and eclipses are seen. Unfavorable configurations are much more common than favorable ones, so eclipses are relatively rare events.

As indicated on Figure 1.27(b), the two points on the Moon's orbit where it crosses the plane of the ecliptic are known as the *nodes* of the orbit. The line joining the nodes, which is also the line of intersection of Earth's and the Moon's orbital planes, is known as the *line of nodes*. When the

◀ **FIGURE 1.26 Annular Solar Eclipse** During an annular solar eclipse, the Moon fails to completely hide the Sun, so a thin ring of light remains. No corona is seen in this case because even the small amount of the Sun still visible completely overwhelms the corona's faint glow. This was the December 1973 eclipse, as seen from Algiers. (The gray fuzzy areas at the top left and right are clouds in Earth's atmosphere.) *(G. Schneider)*

line of nodes is not directed toward the Sun, conditions are unfavorable for eclipses. However, when the line of nodes briefly lies along Earth–Sun line, eclipses are possible. These two periods, known as **eclipse seasons**, are the only times at which an eclipse can occur. Notice that there is no guarantee that an eclipse *will* occur. For a solar eclipse, we must have a new Moon during an eclipse season. Similarly, a lunar eclipse can occur only at full Moon during an eclipse season.

Because we know the orbits of Earth and the Moon to great accuracy, we can predict eclipses far into the future.

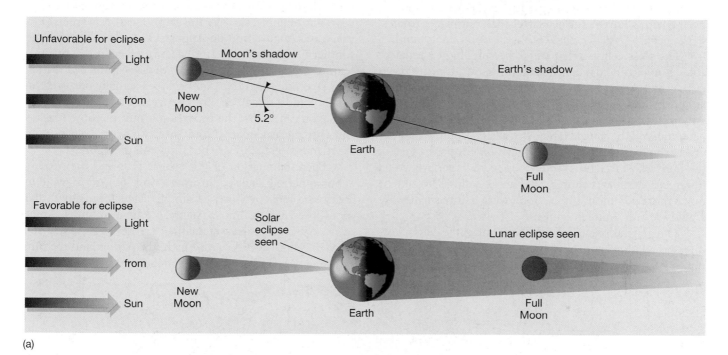

(a)

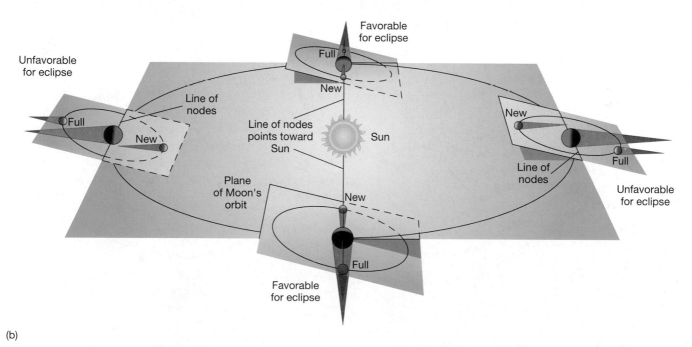

(b)

▲ **FIGURE 1.27 Eclipse Geometry** (a) An eclipse occurs when Earth, Moon, and Sun are precisely aligned. If the Moon's orbital plane lay in exactly the plane of the ecliptic, this alignment would occur once a month. However, the Moon's orbit is inclined at about 5° to the ecliptic, so not all configurations are favorable for producing an eclipse. (b) For an eclipse to occur, the line of intersection of the two planes must lie along the Earth–Sun line. Thus, eclipses can occur just at specific times of the year. Only the umbra of each shadow is shown, for clarity (see Figure 1.25).

Figure 1.28 shows the location and duration of all total and annular eclipses of the Sun between 2000 and 2020. It is interesting to note that the eclipse tracks run from west to east—just the opposite of more familiar phenomena such as sunrise and sunset, which are seen earlier by observers located farther east. The reason is that the Moon's shadow sweeps across Earth's surface faster than our planet rotates, so the eclipse actually *overtakes* observers on the ground.

The solar eclipses that we do see highlight a remarkable cosmic coincidence. Although the Sun is many times farther away from Earth than is the Moon, it is also much larger. In fact, the ratio of distances is almost exactly the same as the ratio of sizes, so the Sun and the Moon both have roughly the *same* angular diameter—about half a degree, seen from Earth. Thus, the Moon covers the face of the Sun almost exactly. If the Moon were larger, we would never see annular eclipses, and total eclipses would be much more common. If the Moon were a little smaller, we would see only annular eclipses.

The gravitational tug of the Sun causes the Moon's orbital orientation, and hence the direction of the line of nodes, to change slowly with time. As a result, the time be-tween one orbital configuration with the line of nodes pointing at the Sun and the next (with the Moon crossing the ecliptic in the same sense in each case) is not exactly 1 year, but instead is 346.6 days—sometimes called one *eclipse year*. Thus, the eclipse seasons gradually progress backward through the calendar, occurring about 19 days earlier each year. For example, in 1999 the eclipse seasons were in February and August, and on August 11 much of Europe and southern Asia was treated to the last total eclipse of the millennium (Figure 1.24). By 2002, those seasons had drifted into December and June, and eclipses actually occurred on June 10 and December 4 of that year. By studying Figure 1.28, you can follow the progression of the eclipse seasons through the calendar. (Note that two partial eclipses in 2004 and two in 2007 are not shown in the figure.)

The combination of the eclipse year and the Moon's synodic period leads to an interesting long-term cycle in solar (and lunar) eclipses. A simple calculation shows that 19 eclipse years is almost exactly 223 lunar months. Thus, every 6585 solar days (actually 18 years, 11.3 days) the "same" eclipse recurs, with Earth, the Moon, and the Sun in the same relative configuration. Several such repetitions

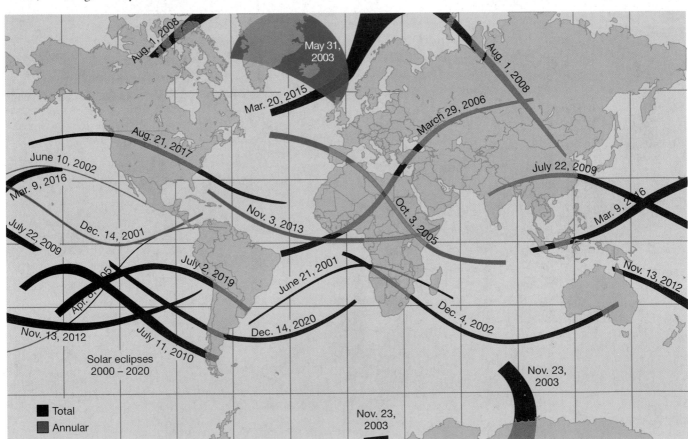

▲ **FIGURE 1.28 Eclipse Tracks** Regions of Earth that saw or will see total or annular solar eclipses between the years 2000 and 2020. Each track represents the path of the Moon's umbra across Earth's surface during an eclipse. The width of the track depends upon the latitude on Earth and the distance from Earth to the Moon during the eclipse. High-latitude tracks are broader because sunlight strikes Earth's surface at an oblique angle near the poles (and also because of the projection of the map). The closer the Moon is to Earth during a total eclipse, the wider is the umbra. (See Figure 1.25.)

are evident in Figure 1.28—see, for example, the similarly shaped December 4, 2002, and December 14, 2020, tracks. (Note that we must take leap years properly into account to get the dates right!) The roughly 120° offset in longitude corresponds to Earth's rotation in 0.3 day. This recurrence is called the *Saros cycle*. Well known to ancient astronomers, it undoubtedly was the key to their "mystical" ability to predict eclipses!

CONCEPT CHECK

✔ What types of solar eclipses would you expect to see if Earth's distance from the Sun were to double? What if the distance became half its present value?

1.7 The Measurement of Distance

6 We have seen a little of how astronomers track and record the positions of the stars in the sky. But knowing the direction to an object is only part of the information needed to locate it in space. Before we can make a systematic study of the heavens, we must find a way of measuring *distances*, too. One distance-measurement method, called **triangulation**, is based on the principles of Euclidean geometry and finds widespread application today in both terrestrial and astronomical settings. Surveyors use these age-old geometric ideas to measure the distance to faraway objects indirectly. Triangulation forms the foundation of the family of distance-measurement techniques making up the **cosmic distance scale**.

Triangulation and Parallax

Imagine trying to measure the distance to a tree on the other side of a river. The most direct method is to lay a tape across the river, but that's not the simplest way (nor, because of the current, may it even be possible). A smart surveyor would make the measurement by visualizing an *imaginary* triangle (hence *triangulation*), sighting the tree on the far side of the river from two positions on the near side, as illustrated in Figure 1.29. The simplest possible triangle is a right triangle, in which one of the angles is exactly 90°, so it is usually convenient to set up one observation position directly opposite the object, as at point *A*. The surveyor then moves to another observation position at point *B*, noting the distance covered between points *A* and *B*. This distance is called the **baseline** of the imaginary triangle. Finally, the surveyor, standing at point *B*, sights toward the tree and notes the angle at point *B* between this line of sight and the baseline. Knowing the value of one side (*AB*) and two angles (the right angle at point *A* and the angle at point *B*) of the right triangle, the surveyor geometrically constructs the remaining sides and angles and establishes the distance from *A* to the tree.

To use triangulation to measure distances, a surveyor must be familiar with *trigonometry*, the mathematics of geometrical angles and distances. However, even if we knew no trigonometry at all, we could still solve the problem by graphical means, as shown in Figure 1.30. Suppose

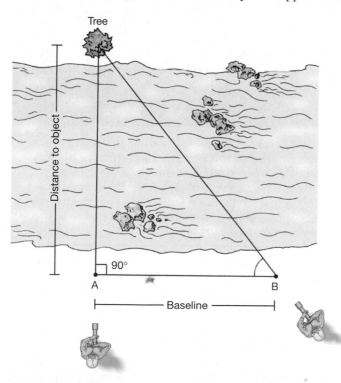

▲ **FIGURE 1.29 Triangulation** Surveyors often use simple geometry and trigonometry to estimate the distance to a faraway object. By measuring the angles at A and B and the length of the baseline, the distance can be calculated without the need for direct measurement.

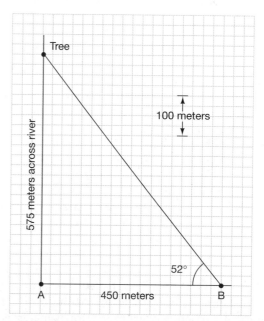

▲ **FIGURE 1.30 Geometric Scaling** We don't even need trigonometry to estimate distances indirectly. Scaled estimates, like this one on a piece of graph paper, often suffice.

that we pace off the baseline *AB*, measuring it to be 450 meters, and measure the angle between the baseline and the line from *B* to the tree to be 52°, as illustrated in the figure. We can transfer the problem to paper by letting one box on our graph represent 25 meters on the ground. Drawing the line *AB* on paper and completing the other two sides of the triangle, at angles of 90° (at A) and 52° (at B), we measure the distance on paper from *A* to the tree to be 23 boxes—that is, 575 meters. We have solved the real problem by *modeling* it on paper. The point to remember here is this: Nothing more complex than basic geometry is needed to infer the distance, the size, and even the shape of an object that is too far away or inaccessible for direct measurement.

Obviously, for a fixed baseline, the triangle becomes longer and narrower as the tree's distance from *A* increases. Narrow triangles cause problems, because it becomes hard to measure the angles at *A* and *B* with sufficient accuracy. The measurements can be made easier by "fattening" the triangle—that is, by lengthening the baseline—but there are limits on how long a baseline we can choose in astronomy. For example, consider an imaginary triangle extending from Earth to a nearby object in space, perhaps a neighboring planet. The triangle is now extremely long and narrow, even for a relatively nearby object (by cosmic standards). Figure 1.31(a) illustrates a case in which the longest baseline possible on Earth—Earth's diameter, measured from point *A* to point *B*—is used. In principle, two observers could sight the planet from opposite sides of Earth, measuring the triangle's angles at *A* and *B*. However, in practice it is easier to measure the third angle of the imaginary triangle. Here's how:

The observers sight toward the planet, taking note of its position *relative to some distant stars* seen on the plane of the sky. The observer at point *A* sees the planet at apparent location *A'* relative to those stars, as indicated in Figure 1.31(a). The observer at *B* sees the planet at point *B'*. If each observer takes a photograph of the appropriate region of the sky, the planet will appear at slightly different places in the two images. The planet's photographic image is slightly displaced, or shifted, relative to the field of distant background stars, as shown in Figure 1.31(b). The background stars themselves appear undisplaced because of their much greater distance from the observer. This apparent displacement of a foreground object relative to the background as the observer's location changes is known as **parallax**. The size of the shift in Figure 1.31(b), measured as an angle on the celestial sphere, is the third, small angle in Figure 1.31(a).

In astronomical contexts, the parallax is usually very small. For example, the parallax of a point on the Moon, viewed using a baseline equal to Earth's diameter, is about 2°; the parallax of the planet Venus at closest approach (45 million km), is just 1'.

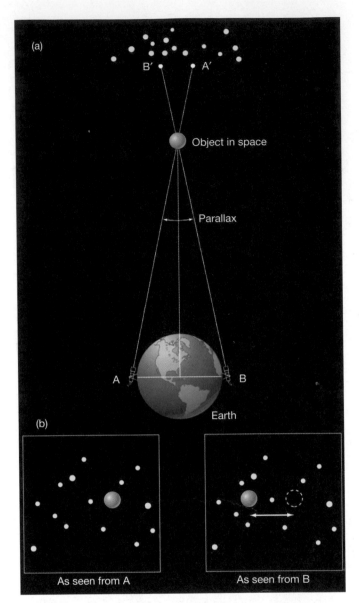

▲ **FIGURE 1.31 Parallax** (a) This imaginary triangle extends from Earth to a nearby object in space (such as a planet). The group of stars at the top represents a background field of very distant stars. (b) Hypothetical photographs of the same star field showing the nearby object's apparent displacement, or shift, relative to the distant, undisplaced stars.

The closer an object is to the observer, the larger is the parallax. Figure 1.32 illustrates how you can see this for yourself. Hold a pencil vertically in front of your nose and concentrate on some far-off object—a distant wall, perhaps. Close one eye, and then open it while closing the other. You should see a large shift in the apparent position of the pencil projected onto the distant wall—a large parallax. In this example, one eye corresponds to point *A*, the other eye to point *B*, the distance between your eyeballs to the baseline, the pencil to the planet, and the distant wall to a remote field of stars. Now hold the pencil at arm's length, corresponding to a more distant object (but still

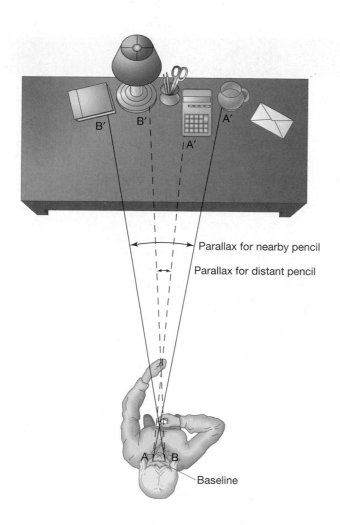

◀ **FIGURE 1.32 Parallax Geometry** Parallax is inversely proportional to an object's distance. An object near your nose has a much larger parallax than an object held at arm's length.

Sizing Up Planet Earth

Now that we have studied some of the tools available to astronomers, let's end the chapter with a classic example of how the scientific method, combined with the basic geometrical techniques just described, enabled an early scientist to perform a calculation of truly "global" proportions.

In about 200 B.C., a Greek philosopher named Eratosthenes (276–194 B.C.) used simple geometric reasoning to calculate the size of our planet. He knew that, at noon on the first day of summer, observers in the city of Syene (now called Aswan), in Egypt, saw the Sun pass directly overhead. This was evident from the fact that vertical objects cast no shadows and sunlight reached to the very bottoms of deep wells, as shown in the insets in Figure 1.33. However, at noon of the same day in Alexandria, a city 5000 *stadia* to the north, the Sun was seen to be displaced slightly from the vertical. (The *stadium* was a Greek unit of length, roughly equal to 0.16 km—the

not as far away as the even more distant stars). The apparent shift of the pencil will be less. You might even be able to verify that the apparent shift is inversely proportional to the distance to the pencil. By moving the pencil farther away, we are narrowing the triangle and decreasing the parallax (and also making accurate measurement more difficult). If you were to paste the pencil to the wall, corresponding to the case where the object of interest is as far away as the background star field, blinking would produce no apparent shift of the pencil at all.

The amount of parallax is thus inversely proportional to an object's distance. Small parallax implies large distance, and large parallax implies small distance. Knowing the amount of parallax (as an angle) and the length of the baseline, we can easily derive the distance through triangulation. *More Precisely 1-3* explores the connection between angular measure and distance in more detail, showing how we can use elementary geometry to determine both the distances and the dimensions of faraway objects.

Surveyors of the land routinely use such simple geometric techniques to map out planet Earth. As surveyors of the sky, astronomers use the same basic principles to chart the universe.

▼ **FIGURE 1.33 Measuring Earth's Radius** The Sun's rays strike different parts of Earth's surface at different angles. The Greek philosopher Eratosthenes realized that the difference was due to Earth's curvature, enabling him to determine Earth's radius by using simple geometry.

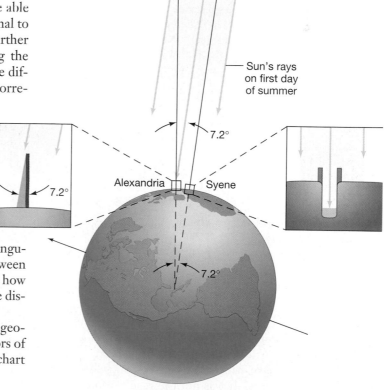

MORE PRECISELY 1-3

Measuring Distances with Geometry

Simple geometrical reasoning forms the basis for almost every statement made in this book about size and scale in the universe. In a very real sense, our modern knowledge of the cosmos depends on the elementary mathematics of ancient Greece. Let's take a moment to look in a little more detail at how astronomers use geometry to measure the distances to, and sizes of, objects near and far.

We can convert baselines and parallaxes into distances by using arguments made by the Greek geometer Euclid. The first figure represents Figure 1.31(a), but we have changed the scale and added the circle centered on the target planet and passing through our baseline on Earth:

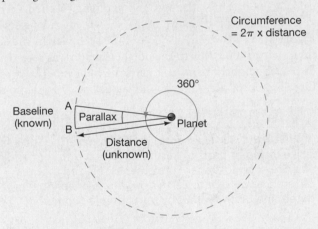

To compute the planet's distance, we note that the ratio of the baseline *AB* to the circumference of the large circle shown in the figure must be equal to the ratio of the parallax to one full revolution, 360°. Recall that the circumference of a circle is always 2π times its radius (where π—the Greek letter "pi"—is approximately equal to 3.142). Applying this relation to the large circle in the figure, we find that

$$\frac{\text{baseline}}{2\pi \times \text{distance}} = \frac{\text{parallax}}{360°},$$

from which it follows that

$$\text{distance} = \text{baseline} \times \frac{(360°/2\pi)}{\text{parallax}}.$$

The angle $360°/2\pi \approx 57.3°$ in the preceding equation is usually called 1 *radian*.

EXAMPLE 1: Two observers 1000 km apart looking at the Moon might measure (via the photographic technique described in the text) a parallax of 9.0 arc minutes—that is, 0.15°. It then follows that the distance to the Moon is 1000 km × (57.3/0.15) ≈ 380,000 km. (More accurate measurements, based on laser ranging using equipment left on the lunar surface by Apollo astronauts, yield a mean distance of 384,000 km.)

Knowing the distance to an object, we can determine many other properties. For example, by measuring the object's *angular diameter*—the angle from one side of the object to the other as we view it in the sky—we can compute its size. The second figure illustrates the geometry involved:

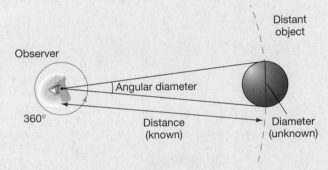

Notice that this is basically the same diagram as the previous one, except that now the angle (the angular diameter) and distance are known, instead of the angle (the parallax) and baseline. Exactly the same reasoning as before then allows us to calculate the diameter. We have

$$\frac{\text{diameter}}{2\pi \times \text{distance}} = \frac{\text{angular diameter}}{360°},$$

so

$$\text{diameter} = \text{distance} \times \frac{\text{angular diameter}}{57.3°}.$$

EXAMPLE 2: The Moon's angular diameter is measured to be about 31 arc minutes—a little over half a degree. From the preceding discussion, it follows that the Moon's actual diameter is 380,000 km × (0.52°/57.3°) ≈ 3450 km. A more precise measurement gives 3476 km.

Study the foregoing reasoning carefully. We will use these simple arguments, in various forms, many times throughout this text.

---

modern town of Aswan lies about 780 km, or 490 miles, south of Alexandria.) By measuring the length of the shadow of a vertical stick and applying elementary trigonometry, Eratosthenes determined the angular displacement of the Sun from the vertical at Alexandria to be 7.2°.

What could have caused this discrepancy between the two measurements? It was not the result of measure-

ment error—the same results were obtained every time the observations were repeated. Instead, as illustrated in Figure 1.33, the explanation is simply that Earth's surface is not flat, but *curved*. Our planet is a sphere. Eratosthenes was not the first person to realize that Earth is spherical—the philosopher Aristotle had done that over 100 years earlier (see Section 1.2), but he was apparently the first to build on this knowledge, combining geometry with direct measurement to infer the size of our planet. Here's how he did it:

Rays of light reaching Earth from a very distant object, such as the Sun, travel almost parallel to one another. Consequently, as shown in the figure, the angle measured at Alexandria between the Sun's rays and the vertical (i.e., the line joining Alexandria to the center of Earth) is equal to the angle between Syene and Alexandria, as seen from Earth's center. (For the sake of clarity, the angle has been exaggerated in the figure.) As discussed in *More Precisely 1-3*, the size of this angle in turn is proportional to the fraction of Earth's circumference that lies between Syene and Alexandria:

$$\frac{7.2° \text{ (angle between Syene and Alexandria)}}{360° \text{ (circumference of a circle)}}$$

$$= \frac{5000 \text{ stadia}}{\text{Earth's circumference}}.$$

Earth's circumference is therefore 50×5000, or 250,000 stadia, or about 40,000 km, so Earth's radius is $250,000/2\pi$ stadia, or 6366 km. The correct values for Earth's circumference and radius, now measured accurately by orbiting spacecraft, are 40,070 km and 6378 km, respectively.

Eratosthenes' reasoning was a remarkable accomplishment. More than 20 centuries ago, he estimated the circumference of Earth to within one percent accuracy, using only simple geometry and basic scientific reasoning. A person making measurements on only a small portion of Earth's surface was able to compute the size of the entire planet on the basis of observation and pure logic—an early triumph of the scientific method.

CONCEPT CHECK

✔ Why is elementary geometry essential for measuring distances in astronomy?

Chapter Review

SUMMARY

The **universe (p. 4)** is the totality of all space, time, matter, and energy. **Astronomy (p. 4)** is the study of the universe. A widely used unit of distance in astronomy is the **light-year (p. 4)**, the distance traveled by a beam of light in 1 year.

The **scientific method (p. 7)** is a methodical approach employed by scientists to explore the universe around us in an objective manner. A **theory (p. 6)** is a framework of ideas and assumptions used to explain some set of observations and make predictions about the real world. These predictions in turn are amenable to further observational testing. In this way, the theory expands and science advances.

Early observers grouped the thousands of stars visible to the naked eye into patterns called **constellations (p. 8)**. These patterns have no physical significance, although they are a useful means of labeling regions of the sky. The nightly motion of the stars across the sky is the result of Earth's **rotation (p. 10)** on its axis. Early astronomers, however, imagined that the stars were attached to a vast **celestial sphere (p. 10)** centered on Earth and that the motions of the heavens were caused by the rotation of the celestial sphere about a fixed Earth. The points where Earth's axis of rotation intersects the celestial sphere are called the north and south **celestial poles (p. 10)**. The line where Earth's equatorial plane cuts the celestial sphere is the **celestial equator (p. 11)**.

The time from one noon to the next is called a **solar day (p. 13)**. The time between successive risings of any given star is one **sidereal day (p. 13)**. Because Earth **revolves (p. 13)** around the Sun, the solar day is a few minutes longer than the sidereal day. Because Earth orbits the Sun, we see different constellations

at different times of the year, and the Sun appears to move relative to the stars. The Sun's yearly path around the celestial sphere, or equivalently, the plane of Earth's orbit around the Sun, is called the **ecliptic (p. 13)**. The constellations lying along the ecliptic are collectively called the **zodiac (p. 13)**. Because Earth's axis is inclined to the plane of the ecliptic, we experience different **seasons (p. 16)**, depending on which hemisphere (Northern or Southern) happens to be "tipped" toward the Sun. At the **summer solstice (p. 15)**, the Sun is highest in the sky, and the length of the day is greatest. At the **winter solstice (p. 15)**, the Sun is lowest, and the day is shortest. At the **vernal (p. 16)** and **autumnal equinoxes (p. 16)**, Earth's axis of rotation is perpendicular to the line joining Earth to the Sun, so day and night are of equal length. The interval of time from one vernal equinox to the next is one **tropical year (p. 16)**.

The speed at which the Sun appears to traverse the ecliptic varies around the year, so the noon-to-noon solar day is not constant. The 24-hour **mean solar day (p. 18)** is the average solar day over the entire year. Since noon occurs when the Sun crosses the local **meridian (p. 17)**, observers at different locations will disagree on the time at which that happens. To standardize timekeeping, Earth is divided into 24 standard **time zones (p. 18)**. Within each zone, all clocks keep the same **standard time (p. 18)**. Standard time in the time zone of 0° longitude is **universal time (p. 18)**. One tropical year is not a whole number of mean solar days. In order to keep the average length of a calendar year equal to 1 tropical year, an extra day is inserted into the calendar on **leap years (p. 18)**.

The Moon emits no light of its own, but instead shines by reflected sunlight. As the Moon orbits Earth, we see **lunar phases (p. 19)** as the amount of the Moon's sunlit face visible to us varies. At full Moon, we can see the entire illuminated side. At quarter Moon, only half the sunlit side can be seen. At new Moon, the sunlit face points away from us, and the Moon is nearly invisible from Earth. The time between successive full Moons is one **synodic month (p. 20)**. The amount of time taken for the Moon to return to the same position in the sky, relative to the stars, is one **sidereal month (p. 20)**. Because of Earth's motion around the Sun, the synodic month is about 2 days longer than the sidereal month.

The time required for Earth to complete one orbit around the Sun, relative to the stars, is one **sidereal year (p. 16)**. In addition to its rotation about its axis and its revolution around the Sun, Earth has many other motions. One of the most important of these is **precession (p. 16)**, the slow "wobble" of Earth's axis due to the influence of the Moon. As a result, the sidereal year is slightly longer than the tropical year, and the particular constellations that happen to be visible during any given season change over the course of thousands of years.

A **lunar eclipse (p. 20)** occurs when the Moon enters Earth's shadow. The eclipse may be **total (p. 20)**, if the entire Moon is (temporarily) darkened, or **partial (p. 20)**, if only a portion of the Moon's surface is affected. A **solar eclipse (p. 20)** occurs when the Moon passes between Earth and the Sun, so that a small part of Earth's surface is plunged into shadow. For observers in the **umbra (p. 21)**, the entire Sun is obscured, and the solar eclipse is total. In the **penumbra (p. 21)**, a partial solar eclipse is seen. If the Moon happens to be too far from Earth for its disk to completely hide the Sun, an **annular eclipse (p. 21)** occurs. Because the Moon's orbit around Earth is slightly inclined with respect to the ecliptic, solar and lunar eclipses do not occur every month, but only during **eclipse seasons (p. 23)** (twice per year).

Surveyors on Earth use **triangulation (p. 25)** to determine the distances to faraway objects. Astronomers use the same technique to measure the distances to planets and stars. The **cosmic distance scale (p. 25)** is the family of distance-measurement techniques by which astronomers chart the universe. **Parallax (p. 26)** is the apparent motion of a foreground object relative to a distant background as the observer's position changes. The larger the **baseline (p. 25)**—the distance between the two observation points—the greater is the parallax. The same basic geometric reasoning is used to determine the sizes of objects whose distances are known. The Greek philosopher Eratosthenes used elementary geometry to determine Earth's radius.

REVIEW AND DISCUSSION

1. Compare the size of Earth with that of the Sun, the Milky Way galaxy, and the entire universe.

2. What does an astronomer mean by "the universe"?

3. How big is a light-year?

4. What is the scientific method, and how does science differ from religion?

5. What is a constellation? Why are constellations useful for mapping the sky?

6. Why does the Sun rise in the east and set in the west each day? Does the Moon also rise in the east and set in the west? Why? Do stars do the same? Why?

7. How and why does a day measured with respect to the Sun differ from a day measured with respect to the stars?

8. How many times in your life have you orbited the Sun?

9. Why do we see different stars at different times of the year?

10. Why are there seasons on Earth?

11. What is precession, and what causes it?

12. Why do we need standard time zones and leap years?

13. If one complete hemisphere of the Moon is always lit by the sun, why do we see different phases of the Moon?

14. What causes a lunar eclipse? A solar eclipse?

15. Why aren't there lunar and solar eclipses every month?

16. Do you think an observer on another planet might see eclipses? Why or why not?

17. What is parallax? Give an everyday example.

18. Why is it necessary to have a long baseline when using triangulation to measure the distances to objects in space?

19. What two pieces of information are needed to determine the diameter of a faraway object?

20. If you traveled to the outermost planet in our solar system, do you think the constellations would appear to change their shapes? What would happen if you traveled to the next-nearest star? If you traveled to the center of our Galaxy, could you still see the familiar constellations found in Earth's night sky?

CONCEPTUAL SELF-TEST: TRUE OR FALSE/MULTIPLE CHOICE

1. The light-year is a measure of distance.

2. The stars in a constellation are physically close to one another.

3. The constellations lying along the ecliptic are collectively referred to as the zodiac.

4. The seasons are caused by the precession of Earth's axis.

5. At the winter solstice, the Sun is at its southernmost point on the celestial sphere.

6. When the Sun, Earth, and Moon are positioned to form a right angle at Earth, the Moon is seen in the new phase.

7. A lunar eclipse can occur only during the full phase.

8. An annular eclipse is a type of eclipse that occurs every year.

9. The parallax of an object is inversely proportional to the object's distance from us.

10. If we know the distance of an object from Earth, we can determine the size of the object by measuring its parallax.

11. If Earth rotated twice as fast as it currently does, but its motion around the Sun stayed the same, then **(a)** the night would be twice as long; **(b)** the night would be half as long;

(c) the year would be half as long; **(d)** the length of the day would be unchanged.

12. A long, thin cloud that stretched from directly overhead to the western horizon would have an angular size of **(a)** 45°; **(b)** 90°; **(c)** 180°; **(d)** 360°.

13. According to Figure 1.15 ("The Zodiac"), in January the Sun is in the constellation **(a)** Cancer; **(b)** Gemini; **(c)** Leo; **(d)** Aquarius.

14. If Earth orbited the Sun in 9 months instead of 12, then, compared with a sidereal day, a solar day would be **(a)** longer; **(b)** shorter; **(c)** unchanged.

15. When a thin crescent of the Moon is visible just before sunrise, the Moon is in its **(a)** waxing phase; **(b)** new phase; **(c)** waning phase; **(d)** quarter phase.

16. If the Moon's orbit were a little larger, solar eclipses would be **(a)** more likely to be annular; **(b)** more likely to be total; **(c)** more frequent; **(d)** unchanged in appearance.

17. If the Moon orbited Earth twice as fast, the frequency of solar eclipses would **(a)** double; **(b)** be cut in half; **(c)** stay the same.

18. In Figure 1.29 ("Triangulation"), using a longer baseline would result in **(a)** a less accurate distance to the tree; **(b)** a more accurate distance to the tree; **(c)** a smaller angle at point B; **(d)** a greater distance across the river.

19. In Figure 1.31 ("Parallax"), a smaller Earth would result in **(a)** a smaller parallax angle; **(b)** a shorter distance measured to the object; **(c)** a larger apparent displacement; **(d)** stars appearing closer together.

20. Today, distances to stars are measured by **(a)** bouncing radar signals; **(b)** reflected laser beams; **(c)** travel time by spacecraft; **(d)** geometry.

PROBLEMS

 Algorithmic versions of these questions are available in the Practice Problems module of the Companion Website at astro.prenhall.com/chaisson.

The number of squares preceding each problem indicates its approximate level of difficulty.

1. ■ In 1 second, light leaving Los Angeles reaches approximately as far as (a) San Francisco, about 500 km; (b) London, roughly 10,000 km; (c) the Moon, 384,000 km; (d) Venus, 45,000,000 km from Earth at closest approach; or (e) the nearest star, about four light-years from Earth. Which is correct?

2. ■ (a) Write the following numbers in scientific notation (see Appendix 1 if you are unfamiliar with this notation): 1000; 0.000001; 1001; 1,000,000,000,000,000; 123,000; 0.000456. (b) Write the following numbers in "normal" numerical form: 3.16×10^7; 2.998×10^5; 6.67×10^{-11}; 2×10^0. (c) Calculate: $(2 \times 10^3) + 10^{-2}$; $(1.99 \times 10^{30})/(5.98 \times 10^{24})$; $(3.16 \times 10^7) \times (2.998 \times 10^5)$.

3. ■■ How, and by roughly how much, would the length of the solar day change if Earth's rotation were suddenly to reverse direction?

4. ■ The vernal equinox is now just entering the constellation Aquarius. In what constellation will it lie in the year A.D. 10,000?

5. ■■ What would be the length of the synodic month if the Moon's sidereal orbital period were (a) 1 week (7 solar days); (b) 1 (sidereal) year?

6. ■ Through how many degrees, arc minutes, or arc seconds does the Moon move in (a) 1 hour of time; (b) 1 minute; (c) 1 second? How long does it take for the Moon to move a distance equal to its own diameter?

7. ■■ Given the data presented in the text, estimate the speed (in km/s) at which the Moon moves in its orbit around Earth.

8. ■ A surveyor wishes to measure the distance between two points on either side of a river, as illustrated in Figure 1.29. She measures the distance *AB* to be 250 m and the angle at *B* to be 30°. What is the distance between the two points?

9. ■ At what distance is an object if its parallax, as measured from either end of a 1000-km baseline, is (a) 1°; (b) 1'; (c) 1"?

10. ■ Given that the angular size of Venus is 55″ when the planet is 45,000,000 km from Earth, calculate Venus's diameter (in kilometers).

11. ■ Calculate the parallax, using Earth's diameter as a baseline, of the Sun's nearest neighbor, Proxima Centauri, which lies 4.3 light-years from Earth.

12. ■ Estimate the angular diameter of your thumb, held at arm's length.

13. ■ The Moon lies roughly 384,000 km from Earth and the Sun lies 150,000,000 km away. If both have the same angular size as seen from Earth, how many times larger than the Moon is the Sun?

14. ■ Given that the distance from Earth to the Sun is 150,000,000 km, through what distance does Earth move in (a) a second, (b) an hour, (c) a day?

15. ■ What angle would Eratosthenes have measured (see Section 1.7) had Earth been flat?

 In addition to the Practice Problems module, the Companion Website at astro.prenhall.com/chaisson provides for each chapter a study guide module with multiple choice questions as well as additional annotated images, animations, and links to related Websites.

2

Astronomy came into its own as a viable and important academic subject during the 20th century. In this collage, clockwise from upper left, are four famous astronomers: Harlow Shapley (1885–1972) discovered our place in the "suburbs" of the Milky Way. Annie Cannon (1863–1941) carefully analyzed photographic plates to classify nearly a million stars over the course of fifty years. Karl Jansky (1905–1950), working at Bell Labs in New Jersey, discovered radiation at radio wavelengths coming from the Milky Way. Edwin Hubble (1889–1953) discovered ▶ the expansion of the universe. (Harvard Observatory; NRAO; Caltech)

The Copernican Revolution

The Birth of Modern Science

Living in the Space Age, we have become accustomed to the modern view of our place in the universe. Images of our planet taken from space leave little doubt that Earth is round, and no one seriously questions the idea that we orbit the Sun. Yet there was a time, not so long ago, when some of our ancestors maintained that Earth was flat and lay at the center of all things. Our view of the universe—and of ourselves—has undergone a radical transformation since those early days. Earth has become a planet like many others, and humankind has been torn from its throne at the center of the cosmos and relegated to a rather unremarkable position on the periphery of the Milky Way galaxy. But we have been amply compensated for our loss of prominence: We have gained a wealth of scientific knowledge in the process. The story of how all this came about is the story of the rise of the scientific method and the genesis of modern astronomy.

LEARNING GOALS

Studying this chapter will enable you to

1 Describe how some ancient civilizations attempted to explain the heavens in terms of Earth-centered models of the universe.

2 Summarize the role of Renaissance science in the history of astronomy.

3 Explain how the observed motions of the planets led to our modern view of a Sun-centered solar system.

4 Describe the major contributions of Galileo and Kepler to the development of our understanding of the solar system.

5 State Kepler's laws of planetary motion.

6 Explain how Kepler's laws allow us to construct a scale model of the solar system, and explain the technique used to determine the actual size of the planetary orbits.

7 State Newton's laws of motion and universal gravitation and explain how they account for Kepler's laws.

8 Explain how the law of gravitation enables us to measure the masses of astronomical bodies.

Visit astro.prenhall.com/chaisson for additional annotated images, animations, and links to related sites for this chapter.

2.1 Ancient Astronomy

LEARNING GOAL 1 Many ancient cultures took a keen interest in the changing nighttime sky. The records and artifacts that have survived until the present make that abundantly clear. But unlike today, the major driving force behind the development of astronomy in those early societies was probably neither scientific nor religious. Instead, it was decidedly practical and down to earth. Seafarers needed to navigate their vessels, and farmers had to know when to plant their crops. In a real sense, then, human survival depended on knowledge of the heavens. The ability to predict the arrival of the seasons, as well as other astronomical events, was undoubtedly a highly prized, perhaps jealously guarded, skill.

In Chapter 1, we saw that the human brain's ability to perceive patterns in the stars led to the "invention" of constellations as a convenient means of labeling regions of the celestial sphere. ∞ (Sec. 1.3) The realization that these patterns returned to the night sky at the same time each year met the need for a practical means of tracking the seasons. Widely separated cultures all over the world built elaborate structures, to serve, at least in part, as primitive calendars. Often the keepers of the secrets of the sky enshrined their knowledge in myth and ritual, and these astronomical sites were also used for religious ceremonies.

Perhaps the best-known such site is *Stonehenge*, located on Salisbury Plain in England, and shown in Figure 2.1. This ancient stone circle, which today is one of the most popular tourist attractions in Britain, dates from the Stone Age. Researchers believe it was an early astronomical ob-servatory of sorts—not in the modern sense of the term (a place for making new observations and discoveries pertaining to the heavens), but rather a kind of three-dimensional calendar or almanac, enabling its builders and their descendants to identify important dates by means of specific celestial events. Its construction apparently spanned a period of about 17 centuries, beginning around 2800 B.C. Additions and modifications continued to about 1100 B.C., indicating its ongoing importance to the Stone Age and, later, Bronze Age people who built, maintained, and used Stonehenge. The largest stones shown in Figure 2.1 weigh up to 50 tons and were transported from quarries many miles away.

Many of the stones are aligned so that they point toward important astronomical events. For example, the line joining the center of the inner circle to the so-called heel stone, set off some distance from the rest of the structure, points in the direction of the rising Sun on the summer solstice. Other alignments are related to the rising and setting of the Sun and the Moon at other times of the year. The accurate alignments (within a degree or so) of the stones of Stonehenge were first noted in the 18th century, but it was only relatively recently—in the second half of the 20th century, in fact—that the scientific community began to credit Stone Age technology with the ability to carry out such a precise feat of engineering. While some of Stonehenge's purposes remain uncertain and controversial, the site's function as an astronomical almanac seems well established. Although Stonehenge is the most impressive and the best preserved, other stone circles, found all over Europe, are believed to have performed similar functions.

◀ **FIGURE 2.1 Stonehenge** This remarkable site in the south of England was probably constructed as a primitive calendar or almanac. The inset shows sunrise at Stonehenge at the summer solstice. As seen from the center of the stone circle, the Sun rose directly over the "heel stone" on the longest day of the year. *(English Heritage)*

Many other cultures are now known to have been capable of similarly precise accomplishments. The Big Horn Medicine Wheel in Wyoming (Figure 2.2a) is similar to Stonehenge in design—and, presumably, intent—although it is somewhat simpler in execution. The Medicine Wheel's alignments with the rising and setting Sun and with some bright stars indicate that its builders—the Plains Indians—had much more than a passing familiarity with the changing nighttime sky. Figure 2.2(b) shows the Caracol temple, built by the Mayans around A.D. 1000 on Mexico's Yucatán peninsula. This temple is much more sophisticated than Stonehenge, but it probably played a similar role as an astronomical observatory. Its many windows are accurately aligned with astronomical events, such as sunrise and sunset at the solstices and equinoxes and the risings and settings of the planet Venus. Astronomy was of more than mere academic interest to the Mayans, however: Caracol was also the site of countless human sacrifices, carried out when Venus appeared in the morning or evening sky.

The ancient Chinese also observed the heavens. Their astrology attached particular importance to "omens" such as comets and "guest stars"—stars that appeared suddenly in the sky and then slowly faded away—and they kept careful and extensive records of such events. Twentieth-century astronomers still turn to the Chinese records to obtain observational data recorded during the Dark Ages (roughly from the 5th to the 10th century A.D.), when turmoil in Europe largely halted the progress of Western science. Perhaps the best-known guest star was one that appeared in A.D. 1054 and was visible in the daytime sky for many months. We now know that the event was actually a *supernova*: the explosion of a giant star, which scattered most of its mass into space (see Chapter 21). It left behind a remnant that is still detectable today, nine centuries later. The Chinese data are a prime source of historical information for supernova research.

A vital link between the astronomy of ancient Greece and that of medieval Europe was provided by astronomers in the Muslim world (see Figure 2.3). For six centuries, from the depths of the Dark Ages to the beginning of the Renaissance, Islamic astronomy flourished and grew, preserving and augmenting the knowledge of the Greeks. Its influence on modern astronomy is subtle, but quite pervasive. Many of the mathematical techniques involved in trigonometry were developed by Islamic astronomers in response to practical problems, such as determining the precise dates of holy days or the direction of Mecca from any given location on Earth. Astronomical terms such as "zenith" and "azimuth" and the names of many stars—for example, Rigel, Betelgeuse, and Vega—all bear witness to this extended period of Muslim scholarship.

Astronomy is not the property of any one culture, civilization, or era. The same ideas, the same tools, and even the same misconceptions have been invented and reinvented by human societies all over the world, in response to the same basic driving forces. Astronomy came into being because people believed that there was a practical benefit in being able to predict the positions of the stars, but its roots go much deeper than that. The need to understand where we came from and how we fit into the cosmos is an integral part of human nature.

(a)

(b)

▲ **FIGURE 2.2 Observatories in the Americas** (a) The Big Horn Medicine Wheel, in Wyoming, was built by the Plains Indians. Its spokes and other features are aligned with risings and settings of the Sun and other stars. (b) Caracol temple in Mexico. The many windows of this Mayan construct are aligned with astronomical events, indicating that at least part of Caracol's function was to keep track of the seasons and the heavens. *(G. Gerster/Comstock)*

2.2 The Geocentric Universe

The Greeks of antiquity, and undoubtedly civilizations before them, built models of the universe. The study of the workings of the universe on the largest scales is called *cosmology*. Today, cosmology entails looking at the universe on scales so large that even entire galaxies can be regarded as mere points of light scattered throughout space. To the Greeks, however, the universe was basically the *solar system*—the Sun, Earth, and Moon, and the planets known at that time. The stars beyond were surely part of the universe, but they were considered to be fixed, unchanging beacons on the celestial sphere. The Greeks did not consider the Sun, the Moon, and the planets to be part of this mammoth celestial dome, however. Those objects had patterns of behavior that set them apart.

Observations of the Planets

Greek astronomers observed that over the course of a night, the stars slid smoothly across the sky. Over the course of a month, the Moon moved smoothly and steadily along its path on the sky relative to the stars, passing through its familiar cycle of phases. Over the course of a year, the Sun progressed along the ecliptic at an almost constant rate, varying little in brightness from day to day. In short, the behavior of both Sun and Moon seemed fairly simple and orderly. But ancient astronomers were also aware of five other bodies in the sky—the planets Mercury, Venus, Mars, Jupiter, and Saturn—whose behavior was not so easy to grasp. Their motions ultimately led to the downfall of an entire theory of the solar system and to a fundamental change in humankind's view of the universe.

To the naked eye (or even through a telescope), planets do not behave in as regular and predictable a fashion as the Sun, Moon, and stars. They vary in brightness, and they don't maintain a fixed position in the sky. Unlike the Sun and Moon, the planets seem to wander around the celestial sphere—indeed, the word *planet* derives from the Greek word *planetes*, meaning "wanderer." Planets never stray far

▲ **FIGURE 2.3 Turkish Astronomers at Work** During the Dark Ages, much scientific information was preserved and new discoveries were made by astronomers in the Islamic world, as depicted in this illustration from a 16th-century manuscript. *(The Granger Collection)*

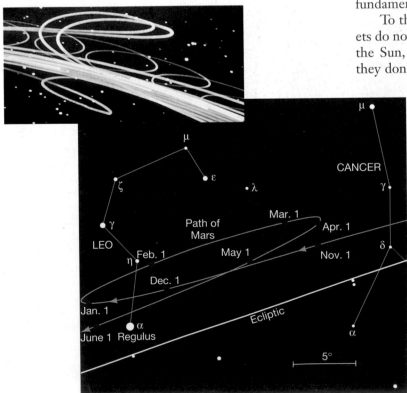

◀ **FIGURE 2.4 Planetary Motion** Most of the time, planets move from west to east relative to the background stars. Occasionally—roughly once per year—however, they change direction and temporarily undergo westward, or retrograde, motion before looping back. The main figure shows an actual retrograde loop in the motion of the planet Mars. The inset depicts the movements of several planets over the course of several years, as reproduced on the inside dome of a planetarium. The motion of the planets relative to the stars (represented as unmoving points) produces continuous streaks on the planetarium "sky." *(Museum of Science, Boston)*

from the ecliptic and generally traverse the celestial sphere from west to east, like the Sun. However, they seem to speed up and slow down during their journeys, and at times they even appear to loop back and forth relative to the stars, as shown in Figure 2.4. In other words, there are periods when a planet's eastward motion (relative to the stars) stops, and the planet appears to move westward in the sky for a month or two before reversing direction again and continuing on its eastward journey. Motion in the eastward sense is usually referred to as *direct*, or *prograde*, motion; the backward (westward) loops are known as **retrograde motion.**

Ancient astronomers knew well that the periods of retrograde motion were closely correlated with other planetary properties, such as apparent brightness and position in the sky. Figure 2.5 (the modern view of the solar system, note!) shows three schematic planetary orbits and defines some time-honored astronomical terminology describing a planet's location relative to Earth and the Sun. Mercury and Venus are referred to as *inferior* ("lower") *planets* because their orbits lie between Earth and the Sun. Mars, Jupiter, and Saturn, whose orbits lie outside Earth's, are known as *superior* ("higher") *planets*. For early astronomers, the key observations of planetary orbits were the following:

- An inferior planet never strays too far from the Sun, as seen from Earth. As illustrated in the inset to Figure 2.5, because its path on the celestial sphere is close to the ecliptic, an inferior planet makes two *conjunctions* (or close approaches) with the Sun during each orbit. (It doesn't actually come close to the Sun, of course.

Conjunction is simply the occasion when the planet and the Sun are in the same direction in the sky.) At *inferior conjunction*, the planet is closest to Earth and moves past the Sun from east to west—that is, in the retrograde sense. At *superior conjunction*, the planet is farthest from Earth and passes the Sun in the opposite (prograde) direction.

- Seen from Earth, the superior planets are not "tied" to the Sun as the inferior planets are. The superior planets make one prograde conjunction with the Sun during each trip around the celestial sphere. However, they exhibit retrograde motion (Figure 2.4) when they are at *opposition*, diametrically opposite the Sun on the celestial sphere.

- The superior planets are brightest at opposition, during retrograde motion. By contrast, the inferior planets are brightest a few weeks before and after inferior conjunction.

The challenge facing astronomers—then as now—was to find a solar-system model that could explain all the existing observations and that could also make testable and reliable predictions of future planetary motions. ∞ (Sec. 1.2)

Ancient astronomers correctly reasoned that the apparent brightness of a planet in the night sky is related to the planet's distance from Earth. Like the Moon, the planets produce no light of their own. Instead, they shine by reflected sunlight and, generally speaking, appear brightest when closest to us. Looking at Figure 2.5, you may already be able to discern

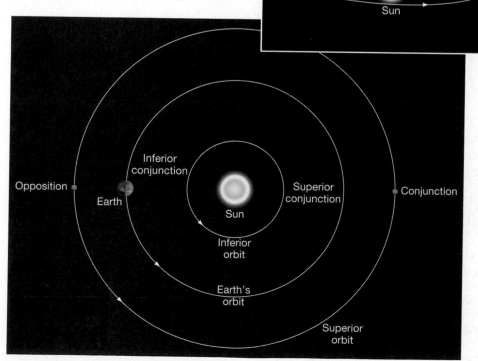

◄ **FIGURE 2.5 Inferior and Superior Orbits** Diagram of Earth's orbit and two other possible planetary orbits. An "inferior" orbit lies between Earth's orbit and the Sun. Mercury and Venus move in such orbits. A "superior" orbit (such as the orbit of Mars, Jupiter, or Saturn) lies outside that of Earth. The points noted on the orbits indicate times when a planet appears to come close to the Sun (conjunction) or is diametrically opposite the Sun on the celestial sphere (opposition); they are discussed further in the text. The inset is a schematic representation of the orbit of an inferior planet relative to the Sun, as seen from Earth.

the basic reasons for some of the planetary properties just listed; we'll return to the "modern" explanation in the next section. However, as we now discuss, the ancients took a very different path in their attempts to explain planetary motion.

A Theoretical Model

The earliest models of the solar system followed the teachings of the Greek philosopher Aristotle (384–322 B.C.) and were **geocentric**, meaning that Earth lay at the center of the universe and all other bodies moved around it. ∞ (Sec. 1.3) (Figures 1.11 and 1.16 illustrate the basic geocentric view.) These models employed what Aristotle, and Plato before him, had taught was the perfect form: the circle. The simplest possible description—uniform motion around a circle with Earth at its center—provided a fairly good approximation to the orbits of the Sun and the Moon, but it could not account for the observed variations in planetary brightness or the retrograde motion of the planets. A more complex model was needed to describe these heavenly "wanderers."

In the first step toward this new model, each planet was taken to move uniformly around a small circle, called an **epicycle**, whose *center* moved uniformly around Earth on a second and larger circle, known as a **deferent** (Figure 2.6). The motion was now composed of two separate circular orbits, creating the possibility that, at some times, the planet's apparent motion could be retrograde. Also, the

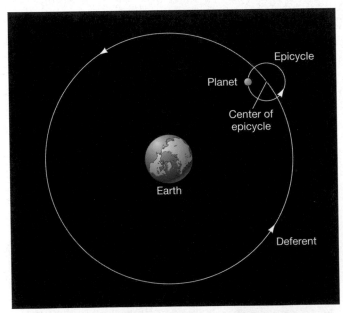

▲ **FIGURE 2.6 Geocentric Model** In the geocentric model of the solar system, the observed motions of the planets made it impossible to assume that they moved on simple circular paths around Earth. Instead, each planet was thought to follow a small circular orbit (the epicycle) about an imaginary point that itself traveled in a large, circular orbit (the deferent) about Earth.

distance from the planet to Earth would vary, accounting for changes in brightness. By tinkering with the relative sizes of the epicycle and deferent, with the planet's speed on the epicycle, and with the epicycle's speed along the deferent, early astronomers were able to bring this "epicyclic" motion into fairly good agreement with the observed paths of the planets in the sky. Moreover, the model had good predictive power, at least to the accuracy of observations at the time.

However, as the number and the quality of observations increased, it became clear that the simple epicyclic model was not perfect. Small corrections had to be introduced to bring it into line with new observations. The center of the deferents had to be shifted slightly from Earth's center, and the motion of the epicycles had to be imagined uniform with respect to yet another point in space, not Earth. Furthermore, in order to explain the motions of the inferior planets, the model simply had to assume that the deferents of Mercury and Venus were, for some (unknown) reason, tied to that of the Sun. Similar assumptions also applied to the superior planets, to ensure that their retrograde motion occurred at opposition.

Around A.D. 140, a Greek astronomer named Ptolemy constructed perhaps the most complete geocentric model of all time. Illustrated in simplified form in Figure 2.7, it explained remarkably well the observed paths of the five planets then known, as well as the paths of the Sun and the Moon. However, to achieve its explanatory and predictive power, the full **Ptolemaic model** required a series of no fewer than 80 distinct circles. To account for the paths of the Sun, Moon, and all nine planets (and their moons) that we know today would require a vastly more complicated set. Nevertheless, Ptolemy's comprehensive text on the topic, *Syntaxis* (better known today by its Arabic name, *Almagest*—"the greatest"), provided the intellectual framework for all discussion of the universe for well over a thousand years.

Today, our scientific training leads us to seek simplicity, because, in the physical sciences, simplicity has so often proved to be an indicator of truth. We would regard the intricacy of a model complicated as the Ptolemaic system as a clear sign of a fundamentally flawed theory. ∞ (Sec. 1.2) With the benefit of hindsight, we now recognize that the major error in the Ptolemaic model lay in its assumption of a geocentric universe. This misconception was compounded by the insistence on uniform circular motion, whose basis was largely philosophical, rather than scientific, in nature.

Actually, history records that some ancient Greek astronomers reasoned differently about the motions of heavenly bodies. Foremost among them was Aristarchus of Samos (310–230 B.C.), who proposed that all the planets, including Earth, revolve around the Sun and, furthermore, that Earth rotates on its axis once each day. This combined revolution and rotation, he argued, would create an *apparent* motion of the sky—a simple

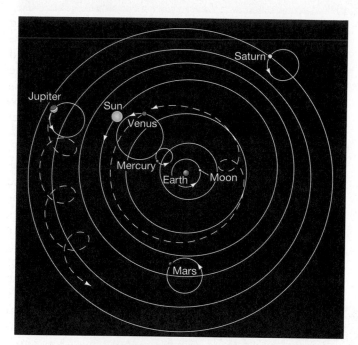

▲ FIGURE 2.7 Ptolemy's Model The basic features, drawn roughly to scale, of the geocentric model of the inner solar system that enjoyed widespread popularity prior to the Renaissance. Only the five planets visible to the naked eye and hence known to the ancients—Mercury, Venus, Mars, Jupiter, and Saturn—are shown. The planets' deferents were considered to move on spheres lying within the celestial sphere that held the stars. The celestial sphere carried all interior spheres around with it, but the planetary (and solar) spheres had additional motions of their own, causing the Sun and planets to move relative to the stars. To avoid confusion, partial paths (dashed) of only two planets—Venus and Jupiter—are drawn here.

idea that is familiar to anyone who has ridden on a merry-go-round and watched the landscape appear to move past in the opposite direction. However, Aristarchus's description of the heavens, though essentially correct, did not gain widespread acceptance during his lifetime. Aristotle's influence was too strong, his followers too numerous, and his writings too comprehensive. The geocentric model went largely unchallenged until the 16th century A.D.

The Aristotelian school did present some simple and (at the time) compelling arguments in favor of their views. First, of course, Earth doesn't *feel* as if it's moving—and if it were moving, wouldn't there be a strong wind as the planet revolves at high speed around the Sun? Also, considering that the vantage point from which we view the stars changes over the course of a year, why don't we see stellar parallax? ∞ (Sec. 1.7)

Nowadays we might dismiss the first points as merely naive, but the last is a valid argument and the reasoning essentially sound. Indeed, we now know that there *is* stellar parallax as Earth orbits the Sun. However, because the

stars are so distant, it amounts to less than 1 arc second, even for the closest stars. Early astronomers simply would not have noticed it. (In fact, stellar parallax was conclusively measured only in the middle of the 19th century.) We will encounter many other instances in astronomy wherein correct reasoning led to the wrong conclusions because it relied on inadequate data. Even when the scientific method is properly applied and theoretical predictions are tested against reality, a theory can be only as good as the observations on which it is based. ∞ (Sec. 1.2)

2.3 The Heliocentric Model of the Solar System

The Ptolemaic picture of the universe survived, more or less intact, for almost 14 centuries, until a 16th-century Polish cleric, Nicolaus Copernicus (Figure 2.8), rediscovered Aristarchus's **heliocentric** (Sun-centered) model and showed how, in its harmony and organization, it provided a more natural explanation of the observed facts than did the tangled geocentric cosmology. Copernicus asserted that Earth spins on its axis and, like the other planets, orbits the Sun. Only the Moon, he said, orbits Earth. As we will see, not only does this model explain the observed daily and seasonal changes in the heavens, but it also naturally accounts for retrograde motion and variations in brightness of the planets. ∞ (Sec. 1.4)

▲ FIGURE 2.8 Nicolaus Copernicus (1473–1543). *(E. Lessing/Art Resource, NY)*

The Foundations of the Copernican Revolution

The following seven points are essentially Copernicus's own words, with the italicized material additional explanation:

1. The celestial spheres do not have just one common center. *Specifically, Earth is not at the center of everything.*

2. The center of Earth is not the center of the universe, but is instead only the center of gravity and of the lunar orbit.

3. All the spheres revolve around the Sun. *By spheres, Copernicus meant the planets.*

4. The ratio of Earth's distance from the Sun to the height of the firmament is so much smaller than the ratio of Earth's radius to the distance to the Sun that the distance to the Sun is imperceptible compared with the height of the firmament. *By firmament, Copernicus meant the distant stars. The point he was making is that the stars are very much farther away than the Sun.*

5. The motions appearing in the firmament are not its motions, but those of Earth. Earth performs a daily rotation around its fixed poles, while the firmament remains immobile as the highest heaven. *Because the stars are so far away, any apparent motion we see in them is the result of Earth's rotation.*

6. The motions of the Sun are not its motions, but the motion of Earth. *Similarly, the Sun's apparent daily and yearly motion are actually due to the various motions of Earth.*

7. What appears to us as retrograde and forward motion of the planets is not their own, but that of Earth. *The heliocentric picture provides a natural explanation for retrograde planetary motion, again as a consequence of Earth's motion.*

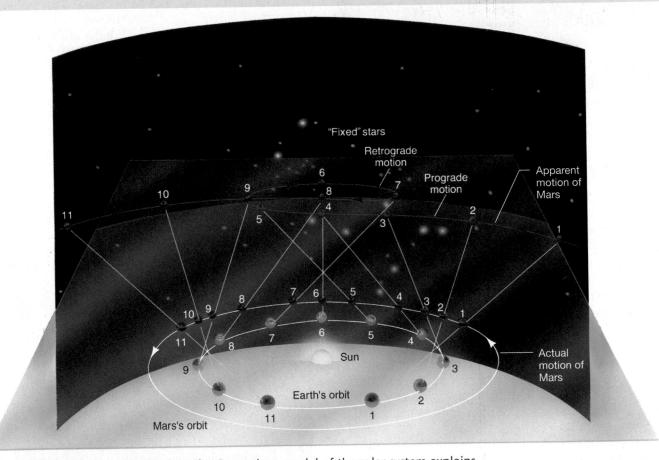

▲ **FIGURE 2.9 Retrograde Motion** The Copernican model of the solar system explains both the varying brightnesses of the planets and the phenomenon of retrograde motion. Here, for example, when Earth and Mars are relatively close to one another in their respective orbits (as at position 6), Mars seems brighter. When they are farther apart (as at position 1), Mars seems dimmer. Also, because the (light blue) line of sight from Earth to Mars changes as the two planets orbit the Sun, Mars appears to loop back and forth in retrograde motion. Follow the lines in numerical order, and note how the line of sight moves backward relative to the stars between locations 5 and 7. The line of sight changes because Earth, on the inside track, moves faster in its orbit than does Mars. The actual planetary orbits are shown as white curves. The apparent motion of Mars, as seen from Earth, is indicated by the red curve.

The critical realization that Earth is not at the center of the universe is now known as the **Copernican revolution.** The seven crucial statements that form its foundation are summarized in *Discovery 2-1.*

Figure 2.9 shows how the Copernican view explains the varying brightness of a planet (in this case, Mars), its observed looping motions, and the fact that the retrograde motion of a superior planet occurs at opposition. If we suppose that Earth moves faster than Mars, then every so often Earth "overtakes" that planet. Mars will then appear to move backward in the sky, in much the same way as a car we overtake on the highway seems to slip backward relative to us. (Replace Mars by Earth and Earth by Venus, and you should also be able to extend the explanation to the inferior planets. To complete the story with a full explanation of their apparent brightnesses, you'll have to wait until Section 9.1!) Notice that, in the Copernican picture, the planet's looping motions are only apparent. In the Ptolemaic view, they are real.

Copernicus's major motivation for introducing the heliocentric model was simplicity. Even so, he was still influenced by Greek thinking and clung to the idea of circles to model the planets' motions. To bring his theory into agreement with observations of the night sky, he was forced to retain the idea of epicyclic motion, although with the deferent centered on the Sun rather than on Earth and with smaller epicycles than in the Ptolemaic picture. Thus, he retained unnecessary complexity and actually gained little in predictive power over the geocentric model. The heliocentric model did rectify some small discrepancies and inconsistencies in the Ptolemaic system, but for Copernicus, the primary attraction of heliocentricity was its simplicity—its being "more pleasing to the mind." His theory was more something he *felt* than he could *prove.* To the present day, scientists still are guided by simplicity, symmetry, and beauty in modeling all aspects of the universe.

Despite the support of some observational data, neither his fellow scholars nor the general public easily accepted Copernicus's model. For the learned, heliocentricity went against the grain of much previous thinking and violated many of the religious teachings of the time, largely because it relegated Earth to a noncentral and undistinguished place within the solar system and the universe. And Copernicus's work had little impact on the general populace of his time, at least in part because it was published in Latin (the standard language of academic discourse at the time), which most people could not read. Only long after Copernicus's death, when others—notably Galileo Galilei—popularized his ideas, did the Roman Catholic Church take them seriously enough to bother banning them. Copernicus's writings on the heliocentric universe were placed on the Church's *Index of Prohibited Books* in 1616, 73 years after they were first published. They remained there until the end of the 18th century.

CONCEPT CHECK

✔ How do the geocentric and heliocentric models of the solar system differ in their explanations of planetary retrograde motion?

2.4 The Birth of Modern Astronomy

In the century following the death of Copernicus and the publication of his theory of the solar system, two scientists—Galileo Galilei and Johannes Kepler—made indelible imprints on the study of astronomy. Contemporaries, they were aware of each other's work and corresponded from time to time about their theories. Each achieved fame for his discoveries and made great strides in popularizing the Copernican viewpoint, yet in their approaches to astronomy they were as different as night and day.

Galileo's Historic Observations

Galileo Galilei (Figure 2.10) was an Italian mathematician and philosopher. By his willingness to perform experiments to test his ideas—a rather radical approach in those days—and by embracing the brand-new technology of the telescope, he revolutionized the way in which science was done, so much so that he is now widely regarded as the father of experimental science.

▲ **FIGURE 2.10** Galileo Galilei (1564–1642). *(Art Resource, NY)*

The telescope was invented in Holland in the early 17th century. Hearing of the invention (but without having seen one), Galileo built a telescope for himself in 1609 and aimed it at the sky. What he saw conflicted greatly with the philosophy of Aristotle and provided much new data to support the ideas of Copernicus.*

Using his telescope, Galileo discovered that the Moon had mountains, valleys, and craters—terrain in many ways reminiscent of that on Earth. Looking at the Sun (something that should *never* be done directly and which may have eventually blinded Galileo), he found imperfections—dark blemishes now known as *sunspots*. These observations ran directly counter to the orthodox wisdom of the day. By noting the changing appearance of sunspots from day to day, Galileo inferred that the Sun *rotates*, approximately once per month, around an axis roughly perpendicular to the ecliptic plane.

Galileo also saw four small points of light, invisible to the naked eye, orbiting the planet Jupiter and realized that they were moons. Figure 2.11 shows some sketches of these moons, taken from Galileo's notes. To Galileo, the fact that another planet had moons provided the strongest support for the Copernican model. Clearly, Earth was not the center of all things. He also found that Venus showed a complete cycle of phases, like those of our Moon (Figure 2.12), a finding that could be explained only by the planet's motion around the Sun. These observations were more strong evidence that Earth is not the center of all things and that at least one planet orbited the Sun.

In 1610, Galileo published a book called *Sidereus Nuncius* (*The Starry Messenger*), detailing his observational findings and his controversial conclusions supporting the Copernican theory. In reporting and interpreting the wondrous observations made with his new telescope, Galileo was directly challenging both the scientific orthodoxy and the religious dogma of his day. He was (literally) playing with fire—he must certainly have been aware that only a few years earlier, in 1600, the astronomer Giordano Bruno had been burned at the stake in Rome, in part for his heretical teaching that Earth orbited the Sun. However, by all accounts, Galileo delighted in publicly ridiculing and irritating his Aristotelian colleagues. In 1616 his ideas were judged heretical, Copernicus's works were banned by the Roman Catholic Church, and Galileo was instructed to abandon his astronomical pursuits.

But Galileo would not desist. In 1632 he raised the stakes by publishing *Dialogue Concerning the Two Chief World Systems*, which compared the Ptolemaic and Copernican models. The book presented a discussion among three people, one of them a dull-witted Aristotelian whose views (which were in fact the stated opinions of the then Pope, Urban VIII) time and again were roundly defeated

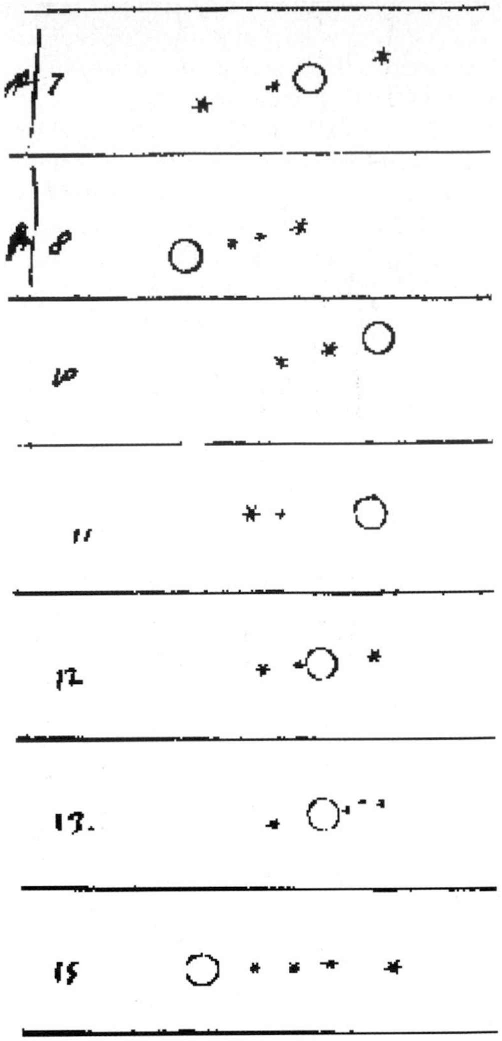

▲ **FIGURE 2.11 Galilean Moons** The four Galilean moons of Jupiter, as sketched by Galileo in his notebook. The sketches show what Galileo saw on seven nights between January 7 and 15, 1610. The orbits of the moons (sketched here as asterisks, and now called Io, Europa, Ganymede, and Callisto) around the planet (open circle) can clearly be seen. Other of Galileo's remarkable sketches of Saturn, star clusters, and the Orion constellation can be seen in the opener to Part I on page 1. (from *Sidereus Nuncius*)

by the arguments of one of his two companions, an articulate proponent of the heliocentric system. To make the book accessible to a wide popular audience, Galileo wrote it in Italian rather than Latin. These actions brought Galileo into direct conflict with the authority of the Church. Eventually, the Inquisition forced him, under threat of torture, to retract his claim that Earth orbits the Sun, and he was placed under house arrest in 1633. He remained imprisoned for the rest of his life. Not until 1992 did the Church publicly forgive Galileo's "crimes." But the damage to the orthodox view of the universe was done, and the Copernican genie was out of the bottle once and for all.

In fact, Galileo had already abandoned Aristotle in favor of Copernicus, although he had not published his beliefs at the time he began his telescopic observations.

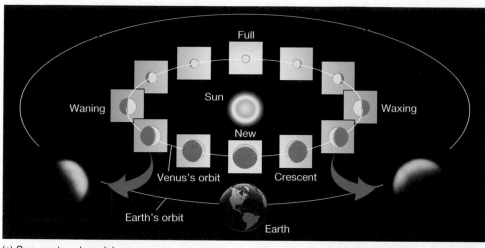

(a) Sun-centered model

R I V U X G

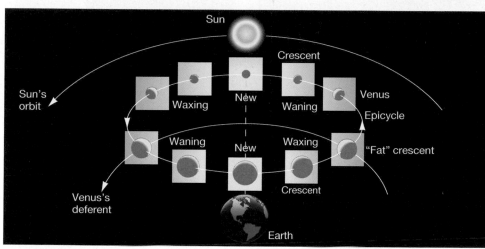

(b) Ptolemy's model

◀ **FIGURE 2.12** Venus Phases (a) The phases of Venus, rendered at different points in the planet's orbit. If Venus orbits the Sun and is closer to the Sun than is Earth, as Copernicus maintained, then Venus should display phases, much as our Moon does. As shown here, when directly between Earth and the Sun, Venus's unlit side faces us, and the planet is invisible to us. As Venus moves in its orbit (at a faster speed than Earth moves in its orbit), progressively more of its illuminated face is visible from Earth. Note also the connection between the orbital phase and the apparent size of the planet. Venus seems much larger in its crescent phase than when it is full because it is much closer to us during its crescent phase. (The insets at bottom left and right are actual photographs of Venus at two of its crescent phases.) (b) The Ptolemaic model (see also Figure 2.7) is unable to account for these observations. In particular, the full phase of the planet cannot be explained. Seen from Earth, Venus reaches only a "fat crescent" phase, then begins to wane as it nears the Sun. (*Images from New Mexico State University Observatory*)

The Ascendancy of the Copernican System

Although Renaissance scholars were correct, they could not *prove* that our planetary system is centered on the Sun or even that Earth moves through space. The observational consequences of Earth's orbital motion were just too small for the technology of the day to detect. Direct evidence of Earth's motion was obtained only in 1728, when English astronomer James Bradley discovered the **aberration of starlight**—a slight (roughly 20″) shift in the observed direction to a star, caused by Earth's motion perpendicular to the line of sight. Figure 2.13 illustrates the phenomenon and compares it with a much more familiar example. Bradley's observation was the first proof that Earth revolves around the Sun; subsequent observations of many stars in many different directions have repeatedly confirmed the effect. Additional proof of Earth's orbital motion came in 1838, with the first unambiguous determination of stellar parallax (see Figure 1.31) by German astronomer Friedrich Bessel. ∞ (Sec. 1.7)

Following those early measurements, support for the heliocentric solar system has grown steadily, as astronomers have subjected the theory to more and more sophisticated observational tests, culminating in the interplanetary expeditions of our unmanned space probes of the 1960s, 1970s, and 1980s. Today, the evidence is overwhelming. The development and eventual acceptance of the heliocentric model were milestones in human thinking. This removal of Earth from any position of great cosmic significance is generally known, even today, as the *Copernican principle*. It has become a cornerstone of modern astrophysics.

The Copernican revolution is a prime example of how the scientific method, though affected at any given time by the subjective whims, human biases, and even sheer luck of researchers, does ultimately lead to a definite degree of objectivity. ∞ (Sec. 1.2) Over time, many groups of scientists checking, confirming, and refining experimental tests can neutralize the subjective attitudes of individuals. Usually, one generation of scientists can

(a)

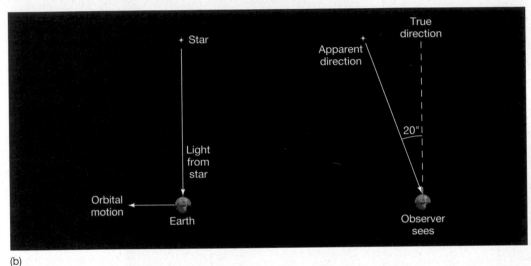

(b)

▲ **FIGURE 2.13 Aberration of Starlight** (a) Rain falling vertically onto the windows of a moving train appears to a passenger on the train to be falling at an angle. The train's motion relative to the raindrops changes the passenger's perception of the direction in which the rain seems to fall. (b) Similarly, because of Earth's motion through space, light from a distant star appears to come from a slightly different direction in the sky than if Earth were at rest. Because the speed of light is large compared with Earth's orbital velocity, the displacement is small—never more than about 20″—but it is easily measurable using a telescope. The maximum displacement occurs when Earth's orbital velocity is exactly perpendicular to the line of sight to the star.

bring sufficient objectivity to bear on a problem, although some especially revolutionary concepts are so swamped by tradition, religion, and politics that more time is necessary. In the case of heliocentricity, objective confirmation was not obtained until about three centuries after Copernicus published his work and more than 2000 years after Aristarchus had proposed the concept. Nonetheless, objectivity *did in fact* eventually prevail, and our knowledge of the universe has expanded immeasurably as a result.

CONCEPT CHECK

✔ In what ways did Galileo's observations of Venus and Jupiter conflict with the prevailing view of the time?

2.5 The Laws of Planetary Motion

At about the same time as Galileo was becoming famous—or notorious—for his pioneering telescopic observations and outspoken promotion of the Copernican system, Johannes Kepler (Figure 2.14), a German mathematician and astronomer, was developing the laws of planetary motion that now bear his name. Galileo was in many ways the first "modern" observer. He used emerging technology, in the form of the telescope, to achieve new insights into the universe. In contrast, Kepler was a pure theorist. His groundbreaking work that so clarified our knowledge of planetary motion was based almost entirely on the observations of others, principally an extensive collection of data compiled by Tycho Brahe (1546–1601), Kepler's employer and arguably one of the greatest observational astronomers that has ever lived.

▲ **FIGURE 2.14 Johannes Kepler** (1571–1630).
(E. Lessing/Art Resource)

Brahe's Complex Data

🔶**2** 🔶**4** Tycho, as he is often called, was both an eccentric aristocrat and a skillful observer. Born in Denmark, he was educated at some of the best universities in Europe, where he studied astrology, alchemy, and medicine. Most of his observations, which predated the invention of the telescope by several decades, were made at his own observatory, named *Uraniborg*, in Denmark (Figure 2.15). There, using instruments of his own design, Tycho maintained meticulous and accurate records of the stars, planets, and other noteworthy celestial events (including a comet and a supernova, the appearance of which helped convince him that the Aristotelian view of the universe could not be correct).

In 1597, having fallen out of favor with the Danish court, Brahe moved to Prague as Imperial Mathematician of the Holy Roman Empire. Prague happens to be fairly close to Graz, in Austria, where Kepler lived and worked. Kepler joined Tycho in Prague in 1600 and was put to work trying to find a theory that could explain Brahe's planetary data. When Tycho died a year later, Kepler inherited not only Brahe's position, but also his most priceless possession: the accumulated observations of the planets, spanning several decades. Tycho's observations, though made with the naked eye, were nevertheless of very high quality. In most cases, his measured positions of stars and planets were accurate to within about 1'. Kepler

set to work seeking a unifying principle to explain in detail the motions of the planets, without the need for epicycles. The effort was to occupy much of the remaining 29 years of his life.

Kepler had already accepted the heliocentric picture of the solar system. His goal was to find a simple and elegant description of planetary motion within the Copernican framework, that fit Tycho's complex mass of detailed observations. In the end, he found it necessary to abandon Copernicus's original simple idea of circular planetary orbits. However, even greater simplicity emerged as a result. After long years of studying Brahe's planetary data, and after many false starts and blind alleys, Kepler developed the laws that now bear his name.

Kepler determined the shape of each planet's orbit by triangulation—not from different points on Earth, but from different points on Earth's orbit, using observations

▲ **FIGURE 2.15 Tycho Brahe** in his observatory Uraniborg, on the island of Hveen in Denmark. Brahe's observations of the positions of stars and planets on the sky were the most accurate and complete set of naked-eye measurements ever made. *(Royal Ontario Museum)*

made at many different times of the year. ∞ (Sec. 1.7) By using a portion of Earth's orbit as a baseline for his triangle, Kepler was able to measure the relative sizes of the other planetary orbits. Noting where the planets were on successive nights, he found the speeds at which the planets move. We do not know how many geometric shapes Kepler tried for the orbits before he hit upon the correct one. His difficult task was made even more complicated because he had to determine Earth's own orbit, too. Nevertheless, he eventually succeeded in summarizing the motions of all the known planets, including Earth, in just three laws: the **laws of planetary motion**.

Kepler's Simple Laws

Kepler's first law has to do with the *shapes* of the planetary orbits:

> I. The orbital paths of the planets are elliptical (*not* circular), with the Sun at one focus.

An **ellipse** is simply a flattened circle. Figure 2.16 illustrates a means of constructing an ellipse with a piece of string and two thumbtacks. Each point at which the string is pinned is called a **focus** (plural: *foci*) of the ellipse. The long axis of the ellipse, containing the two foci, is known as the *major axis*. Half the length of this long axis is referred to as the **semimajor axis**, a measure of the ellipse's size. A circle is a special case in which the two foci happen to coincide; its semimajor axis is simply its radius.

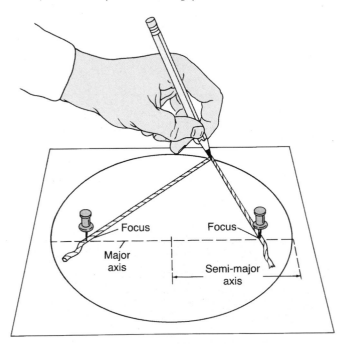

▲ **FIGURE 2.16 Ellipse** An ellipse can be drawn with the aid of a string, a pencil, and two thumbtacks. The wider the separation of the foci, the more elongated, or eccentric, is the ellipse. In the special case where the two foci are at the same place, the curve drawn is a circle.

The **eccentricity** of an ellipse is simply a measure of how flattened it is. Technically, eccentricity is defined as the ratio of the distance between the foci to the length of the major axis, but the most important thing to remember here is that an eccentricity of zero corresponds to no flattening—a perfect circle—while an eccentricity of unity means that the circle has been squashed all the way down to a straight line. Note that, while the Sun resides at one focus of the elliptical orbit, the other focus is empty and has no particular physical significance. (However, we can still figure out where it is, because the two foci are symmetrically placed about the center, along the major axis.)

The length of the semimajor axis and the eccentricity are all we need to describe the size and shape of a planet's orbital path (see *More Precisely 2-1*). In fact, no planet's elliptical orbit is nearly as elongated as the one shown in Figure 2.16. With two exceptions (the paths of Mercury and Pluto), planetary orbits in our solar system have such small eccentricities that our eyes would have trouble distinguishing them from true circles. Only because the orbits are so nearly circular were the Ptolemaic and Copernican models able to come as close as they did to describing reality.

Kepler's substitution of elliptical for circular orbits was no small advance. It amounted to abandoning an aesthetic bias—the Aristotelian belief in the perfection of the circle—that had governed astronomy since Greek antiquity. Even Galileo Galilei, not known for his conservatism in scholarly matters, clung to the idea of circular motion and never accepted the notion that the planets move in elliptical paths.

Kepler's second law, illustrated in Figure 2.17, addresses the speed at which a planet traverses different parts of its orbit:

> II. An imaginary line connecting the Sun to any planet sweeps out equal areas of the ellipse in equal intervals of time.

While orbiting the Sun, a planet traces the arcs labeled *A*, *B*, and *C* in the figure in equal times. Notice, however, that the distance traveled by the planet along arc *C* is greater than the distance traveled along arc *A* or arc *B*. Because the time is the same and the distance is different, the speed must vary. When a planet is close to the Sun, as in sector *C*, it moves much faster than when farther away, as in sector *A*.

By taking into account the relative speeds and positions of the planets in their elliptical orbits about the Sun, Kepler's first two laws explained the variations in planetary brightness and some observed peculiar nonuniform motions that could not be accommodated within the assumption of circular motion, even with the inclusion of epicycles. Gone at last were the circles within circles that rolled across the sky. Kepler's modification of the Copernican theory to allow the possibility of elliptical orbits both greatly simplified the model of the solar system and at the

MORE PRECISELY 2-1

Some Properties of Planetary Orbits

Two numbers—semimajor axis and eccentricity—are all that are needed to describe the size and shape of a planet's orbital path. From them, we can derive many other useful quantities. Two of the most important are the planet's *perihelion* (its point of closest approach to the Sun) and its *aphelion* (point of greatest distance from the Sun). From the definitions presented in the text, it follows that if the planet's orbit has semimajor axis a and eccentricity e, the planet's perihelion is at a distance $a(1-e)$ from the Sun, while its aphelion is at $a(1+e)$. These points and distances are illustrated in the following figure:

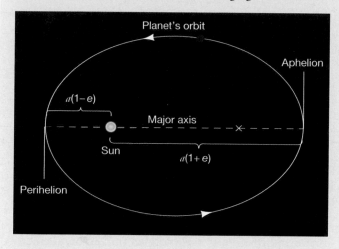

EXAMPLE 1: We can locate the other focus of the ellipse in the diagram, and hence determine the eccentricity from the definition in the text, quite simply. The second focus is placed symmetrically along the major axis, at the point marked with an "X." With a ruler, measure (1) the length of the major axis and (2) the distance between the two foci. Dividing the second distance by the first, you should find an eccentricity of 3.4 cm/6.8 cm = 0.5. Alternatively, we could use the formula given for the perihelion. Measure the perihelion distance to be $a(1-e) = 1.7$ cm. Dividing this by $a = 3.4$ cm, we obtain $1 - e = 0.5$, so once again, $e = 0.5$.

EXAMPLE 2: A (hypothetical) planet with a semimajor axis of 400 million km and an eccentricity of 0.5 (i.e., with an orbit as shown in the diagram) would range between $400 \times (1 - 0.5) = 200$ million km and $400 \times (1 + 0.5) = 600$ million km from the Sun over the course of one complete orbit. With $e = 0.9$, the range in distances would be 40 to 760 million km, and so on.

No planet has an orbital eccentricity as large as 0.5—the planet with the most eccentric orbit is Pluto, with $e = 0.249$. (See Table 2.1.) However, many meteoroids and all comets (see Chapter 14) have eccentricities considerably greater than that. In fact, most comets visible from Earth have eccentricities very close to $e = 1$. Their highly elongated orbits approach within a few astronomical units of the Sun at perihelion, yet these tiny frozen worlds spend most of their time far beyond the orbit of Pluto.

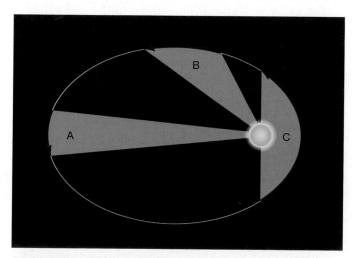

▲ **FIGURE 2.17 Kepler's Second Law** Equal areas are swept out in equal intervals of time. The three shaded areas (A, B, and C) are equal. Note that an object would travel the length of each of the three red arrows in the same amount of time. Therefore, planets move faster when closer to the Sun.

same time provided much greater predictive accuracy than had previously been possible. Note, too, that these laws are not restricted to planets. They apply to *any* orbiting object. Spy satellites, for example, move very rapidly as they swoop close to Earth's surface, not because they are propelled with powerful onboard rockets, but because their highly eccentric orbits are governed by Kepler's laws.

Kepler published his first two laws in 1609, stating that he had proved them only for the orbit of Mars. Ten years later, he extended them to all the then-known planets (Mercury, Venus, Earth, Mars, Jupiter, and Saturn) and added a third law relating the size of a planet's orbit to its sidereal orbital **period**—the time needed for the planet to complete one circuit around the Sun. *Kepler's third law* states,

III. The square of a planet's orbital period is proportional to the cube of its semimajor axis.

This law becomes particularly simple when we choose the (Earth sidereal) year as our unit of time and the *astronomical unit* as our unit of length. One **astronomical unit** (A.U.) is the semimajor axis of Earth's orbit around the Sun—essentially the average distance between Earth and the Sun. Like the light-year, the astronomical unit is custom made for the vast distances encountered in astronomy. Using these units for time and distance, we can write Kepler's third law for any planet as

$$P^2 \text{ (in Earth years)} = a^3 \text{ (in astronomical units)},$$

where P is the planet's sidereal orbital period and a is the length of its semimajor axis. The law implies that a planet's "year" P increases more rapidly than does the size of its orbit, a. For example, Earth, with an orbital semimajor axis of 1 A.U., has an orbital period of 1 Earth year. The planet Venus, orbiting at a distance of roughly 0.7 A.U., takes only 0.6 Earth year—about 225 days—to complete one circuit. By contrast, Saturn, almost 10 A.U. from the Sun, takes considerably more than 10 Earth years—in fact, nearly 30 years—to orbit the Sun just once.

Table 2.1 presents basic data describing the orbits of the nine planets now known. Renaissance astronomers knew these properties for the innermost six planets and used them to construct the currently accepted heliocentric model of the solar system. The second column presents each planet's orbital semimajor axis, measured in astronomical units; the third column gives the orbital period, in Earth years. The fourth column lists the planets' orbital eccentricities. For purposes of verifying Kepler's third law, the fifth column lists the ratio P^2/a^3. As we have just seen, the third law implies that this number should always be unity in the units used in the table.

The main points to be grasped from Table 2.1 are these: (1) With the exception of Mercury and Pluto, the planets' orbits are nearly circular (i.e., their eccentricities are close to zero); and (2) the farther a planet is from the Sun, the greater is its orbital period, in agreement with Kepler's third law to within the accuracy of the numbers in the table. (The small,

but significant, deviations of P^2/a^3 from unity in the cases of Uranus and Neptune are caused by the gravitational attraction between those two planets; see Chapter 13.) Most important, note that Kepler's laws are obeyed by *all* the known planets, *not just by the six on which he based his conclusions*.

The laws developed by Kepler were far more than mere fits to existing data. They also made definite, testable predictions about the future locations of the planets. Those predictions have been borne out to high accuracy every time they have been tested by observation—the hallmark of any scientific theory. ∞ (Sec. 1.2)

CONCEPT CHECK

✔ Why is it significant that Kepler's laws also apply to Uranus, Neptune, and Pluto?

2.6 The Dimensions of the Solar System

Kepler's laws allow us to construct a scale model of the solar system, with the correct shapes and *relative* sizes of all the planetary orbits, but they do not tell us the *actual* size of any orbit. We can express the distance to each planet only in terms of the distance from Earth to the Sun. Why is this? Because Kepler's triangulation measurements all used a portion of Earth's orbit as a baseline, distances could be expressed only relative to the size of that orbit, which was itself not determined by the method. Thus, our model of the solar system would be like a road map of the United States showing the *relative* positions of cities and towns, but lacking the all-important scale marker indicating distances in kilometers or miles. For example, we would know that Kansas City is about three times more distant from New York than it is from Chicago, but we

TABLE 2.1 Some Solar System Dimensions

Planet	Orbital Semimajor Axis, a (astronomical units)	Orbital Period, P (Earth years)	Orbital Eccentricity, e	P^2/a^3
Mercury	0.387	0.241	0.206	1.002
Venus	0.723	0.615	0.007	1.001
Earth	1.000	1.000	0.017	1.000
Mars	1.524	1.881	0.093	1.000
Jupiter	5.203	11.86	0.048	0.999
Saturn	9.537	29.42	0.054	0.998
Uranus	19.19	83.75	0.047	0.993
Neptune	30.07	163.7	0.009	0.986
Pluto	39.48	248.0	0.249	0.999

would not know the actual mileage between any two points on the map.

If we could somehow determine the value of the astronomical unit—in kilometers, say—we would be able to add the vital scale marker to our map of the solar system and compute the precise distances between the Sun and each of the planets. We might propose using triangulation to measure the distance from Earth to the Sun directly. However, we would find it impossible to measure the Sun's parallax using Earth's diameter as a baseline. The Sun is too bright, too big, and too fuzzy for us to distinguish any apparent displacement relative to a field of distant stars. To measure the Sun's distance from Earth, we must resort to some other method.

Before the middle of the 20th century, the most accurate measurements of the astronomical unit were made by using triangulation on the planets Mercury and Venus during their rare *transits* of the Sun—that is, during the brief periods when those planets passed directly between the Sun and Earth (as shown for the case of Mercury in Figure 2.18). Because the time at which a transit occurs can be measured with great precision, astronomers can use this information to make accurate measurements of a planet's position in the sky. They can then employ simple geometry to compute the distance to the planet by combining observations made from different locations on Earth, as discussed earlier in Chapter

1. ∞ (Sec. 1.7) For example, the parallax of Venus at closest approach to Earth, as seen from two diametrically opposite points on Earth (separated by about 13,000 km), is about 1 arc minute (1/60°)—at the limit of naked-eye capabilities, but easily measurable telescopically. Using the second formula presented in *More Precisely 1-3*, we find that this parallax represents a distance of 13,000 km × 57.3°/(1/60°), or approximately 45,000,000 km.

Knowing the distance to Venus, we can compute the magnitude of the astronomical unit. Figure 2.19 is an idealized diagram of the Sun–Earth–Venus orbital geometry. The planetary orbits are drawn as circles here, but in reality they are slight ellipses. This is a subtle difference, and we can correct for it using detailed knowledge of orbital motions. Assuming for the sake of simplicity that the orbits are perfect circles, we see from the figure that the distance from Earth to Venus at closest approach is approximately 0.3 A.U. Knowing that 0.3 A.U. is 45,000,000 km makes determining 1 A.U. straightforward—the answer is 45,000,000/0.3, or 150,000,000 km.

The modern method for deriving the absolute scale (that is, the scale expressed in kilometers, rather than just relative to Earth's orbit) of the solar system uses radar rather than triangulation. The word **radar** is an acronym for **ra**dio **d**etection **a**nd **r**anging. In this technique, radio

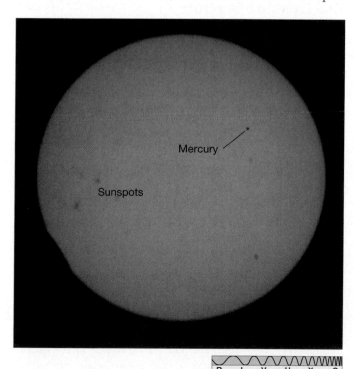

▲ **FIGURE 2.18 Solar Transit** The transit of Mercury across the face of the Sun. Such transits happen only about once per decade, because Mercury's orbit does not quite coincide with the plane of the ecliptic. Transits of Venus are even rarer, occurring only about twice per century. The most recent took place in June 2004. (*AURA*)

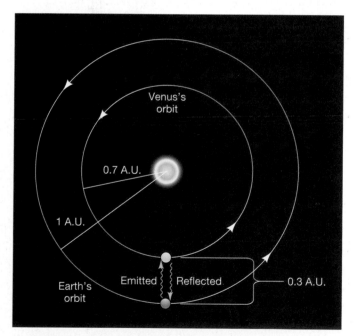

▲ **FIGURE 2.19 Astronomical Unit** Simplified geometry of the orbits of Earth and Venus as they move around the Sun. The wavy lines represent the paths along which radar signals might be transmitted toward Venus and received back at Earth at the particular moment (chosen for simplicity) when Venus is at its minimum distance from Earth. Because the radius of Earth's orbit is 1 A.U. and that of Venus is about 0.7 A.U., we know that this distance is 0.3 A.U. Thus, radar measurements allow us to determine the astronomical unit in kilometers.

waves are transmitted toward an astronomical body, such as a planet. (We cannot use radar ranging to measure the distance to the Sun directly, because radio signals are absorbed at the solar surface and are not reflected to Earth.) The returning echo indicates the body's direction and range, or distance, in absolute terms—that is, in kilometers rather than in astronomical units. Multiplying the 300-second round-trip travel time of the radar signal (the time elapsed between transmission of the signal and reception of the echo) by the speed of light (300,000 km/s, which is also the speed of radio waves), we obtain twice the distance to the target planet.

Venus, whose orbit periodically brings it closest to Earth, is the most common target for radar ranging. The round-trip travel time (for example, at closest approach, as indicated by the wavy lines in Figure 2.19) can be measured with high precision—in fact, well enough to determine the planet's distance to an accuracy of about 1 km. In this way, the astronomical unit is now known to be 149,597,870 km. We will use the rounded-off value of 1.5×10^8 km in this text.

Having determined the value of the astronomical unit, we can reexpress the sizes of the other planetary orbits in terms of more familiar units, such as miles or kilometers. The entire scale of the solar system can then be calibrated to high precision.

CONCEPT CHECK

✔ Why don't Kepler's laws tell us the value of the astronomical unit?

2.7 Newton's Laws

Kepler's three laws, which so simplified the solar system, were discovered *empirically*. In other words, they resulted solely from the analysis of observational data and were not derived from any theory or mathematical model. Indeed, Kepler did not have any appreciation of the physics underlying his laws. Nor did Copernicus understand *why* his heliocentric model of the solar system worked. Even Galileo, often called the father of modern physics, failed to understand why the planets orbit the Sun (although Galileo's work laid vital groundwork for Newton's theories).

What prevents the planets from flying off into space or from falling into the Sun? What causes them to revolve about the Sun, apparently endlessly? To be sure, the motions of the planets obey Kepler's three laws, but only by considering something more fundamental than those laws can we really understand planetary motion. The heliocentric system was secured when, in the 17th century, the British mathematician Isaac Newton (Figure 2.20) developed a deeper understanding of the way *all* objects move and interact with one another.

▲ **FIGURE 2.20** Isaac Newton (1642–1727). *(The Granger Collection)*

The Laws of Motion

Isaac Newton was born in Lincolnshire, England, on Christmas Day in 1642, the year Galileo died. Newton studied at Trinity College of Cambridge University, but when the bubonic plague reached Cambridge in 1665, he returned to the relative safety of his home for two years. During that time he made probably the most famous of his discoveries, the law of gravity (although it is but one of the many major scientific advances for which Newton was responsible). However, either because he regarded the theory as incomplete or possibly because he was afraid that he would be attacked or plagiarized by his colleagues, he did not tell anyone of his monumental achievement for almost 20 years. It was not until 1684, when Newton was discussing the leading astronomical problem of the day—*Why do the planets move according to Kepler's laws?*—with Edmund Halley (of Halley's comet fame) that he astounded his companion by revealing that he had solved the problem in its entirety nearly two decades before!

Prompted by Halley, Newton published his theories in perhaps the most influential physics book ever written: *Philosophiae Naturalis Principia Mathematica* (*The Mathematical Principles of Natural Philosophy*—what we would today call "science"), usually known simply as Newton's *Principia*. The ideas expressed in that work form the basis for what is now known as **Newtonian mechanics**. Three basic laws of motion, the law of gravity, and a little calculus (which Newton also developed) are sufficient to explain and quantify virtually all the complex dynamic behavior we see on Earth and throughout the universe.

Figure 2.21 illustrates *Newton's first law of motion*:

I. Every body continues in a state of rest or in a state of uniform motion in a straight line, unless it is compelled to change that state by a force acting on it.

The first law simply states that a moving object will move forever in a straight line, unless some external **force**—a push or a pull—changes its direction of motion. For example, the object might glance off a brick wall or be hit with a baseball bat; in either case, a force changes the original motion of the object. Another example of a force, well known to most of us, is **weight**—the force (commonly measured in pounds in the United States) with which gravity pulls you toward Earth's center.

The tendency of an object to keep moving at the same speed and in the same direction unless acted upon by a force is known as **inertia**. A familiar measure of an object's inertia is its **mass**—loosely speaking, the total amount of matter the object contains. The greater an object's mass, the more inertia it has, and the greater is the force needed to change its state of motion.

Newton's first law implies that it requires no force to maintain motion in a straight line with constant speed—that is, motion with constant **velocity**. An object's velocity includes both its speed (in miles per hour or meters per second, say) *and* its direction in space (up, down, northwest, and so on). In everyday speech, we tend to use the terms "speed" and "velocity" more or less interchangeably, but we must realize that they are actually different quantities and that Newton's laws are always stated in terms of the latter. As a specific illustration of the difference, consider a rock tied to a string, moving at a constant rate in a circle as you whirl it around your head. The rock's *speed* is constant, but its *direction* of motion, and hence its velocity, is continually changing. Thus, according to Newton's first law, a force must be acting. That force is the tension you feel in the string. In a moment, we'll see reasoning similar to this applied to the problem of planetary motion.

Newton's first law contrasts sharply with the view of Aristotle, who maintained (incorrectly) that the natural state of an object was to be *at rest*—most probably an opinion based on Aristotle's observations of the effect of friction. In our discussion, we will neglect friction—the force that slows balls rolling along the ground, blocks sliding across tabletops, and baseballs moving through the air. In any case, it is not an issue for the planets because there is no appreciable friction in outer space—there is no air or any other matter to impede a planet's motion. The fallacy in Aristotle's argument was first realized and exposed by Galileo, who conceived of the notion of inertia long before Newton formalized it into a law.

The rate of change of the velocity of an object—speeding up, slowing down, or simply changing direction—is called the object's **acceleration** and is the subject of *Newton's second law*, which states that the acceleration of an object is directly proportional to the applied force and inversely proportional to its mass:

II. When a force F acts on a body of mass m, it produces in it an acceleration a equal to the force divided by the mass. Thus, $a = F/m$, or $F = ma$.

Hence, the greater the force acting on the object or the smaller the mass of the object, the greater is the acceleration of the object. If two objects are pulled with the same force, the more massive one will accelerate less; if two identical objects are pulled with different forces, the one acted on by the greater force will accelerate more.

Acceleration is the rate of change of velocity, so its units are velocity units per unit time, such as meters per second *per second* (usually written as m/s^2). In honor of Newton, the SI unit of force is named after him. By definition, 1 newton (N) is the force required to cause a mass of 1 kilogram to accelerate at a rate of 1 meter per second every second (1 m/s^2). One newton is approximately 0.22 pound.

At Earth's surface, the force of gravity produces a downward acceleration of approximately 9.8 m/s^2 on *all* bodies, regardless of mass. According to Newton's second law, this means that your weight (in newtons) is directly proportional to your mass (in kilograms). We will return to this very important point later.

Finally, *Newton's third law* simply tells us that forces cannot occur in isolation:

III. To every action, there is an equal and opposite reaction.

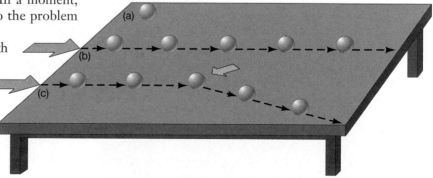

▲ **FIGURE 2.21 Newton's First Law** An object at rest will remain at rest (a) until some force (pink arrow) acts on it (b). It will then remain in that state of uniform motion until another force acts on it. The green arrow in (c) shows a second force acting at a direction different from that in which the first force acts, causing the object to change its direction of motion.

In other words, if body *A* exerts a force on body *B*, then body *B* necessarily exerts a force on body *A* that is equal in magnitude, but oppositely directed.

Only in extreme circumstances—when speeds approach the speed of light—do Newton's laws break down, and this fact was not realized until the 20th century, when Albert Einstein's theories of relativity once again revolutionized our view of the universe (see Chapter 22). Most of the time, however, Newtonian mechanics provides an excellent description of the motions of planets, stars, and galaxies through the cosmos.

Gravity

Forces may act *instantaneously* or *continuously*. The force from a baseball bat that hits a home run can reasonably be thought of as being instantaneous. A good example of a continuous force is the one that prevents the baseball from zooming off into space—**gravity**, the phenomenon that started Newton on the path to the discovery of his laws. Newton hypothesized that any object having mass always exerts an attractive **gravitational force** on all other massive objects. The more massive an object, the stronger is its gravitational pull.

Consider a baseball thrown upward from Earth's surface, as illustrated in Figure 2.22. In accordance with Newton's first law, the downward force of Earth's gravity continuously modifies the baseball's velocity, slowing the

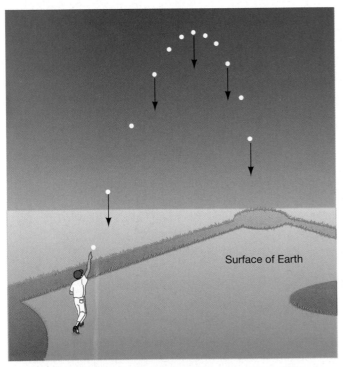

Surface of Earth

▲ **FIGURE 2.22 Gravity** A ball thrown up from the surface of a massive object, such as a planet, is pulled continuously downward (arrows) by the gravity of that planet (and, conversely, the gravity of the ball continuously pulls the planet).

initial upward motion and eventually causing the ball to fall back to the ground. Of course, the baseball, having some mass of its own, also exerts a gravitational pull on Earth. By Newton's third law, this force is equal and opposite to the weight of the ball (the force with which Earth attracts it). But, by Newton's second law, Earth has a much greater effect on the light baseball than the baseball has on the much more massive Earth. The ball and Earth act upon each other with the same gravitational force, but Earth's *acceleration* is much smaller.

Now consider the trajectory of the same baseball batted from the surface of the Moon. The pull of gravity is about one-sixth as great on the Moon as on Earth, so the baseball's velocity changes more slowly—a typical home run in a ballpark on Earth would travel nearly half a mile on the Moon. Less massive than Earth, the Moon has less gravitational influence on the baseball. The magnitude of the gravitational force, then, depends on the *masses* of the attracting bodies. In fact, the force is *directly proportional* to the product of the two masses.

Studying the motions of the planets uncovers a second aspect of the gravitational force. At locations equidistant from the Sun's center, the gravitational force has the same strength and is always directed toward the Sun. Furthermore, detailed calculation of the planets' accelerations as they orbit the Sun reveals that the strength of the Sun's gravitational pull decreases in proportion to the *square* of the distance from the Sun. The force of gravity is said to obey an **inverse-square law**. (See *More Precisely 2-2* for more on the line of reasoning that originally led Newton to this conclusion.) As shown in Figure 2.23, inverse-square forces decrease rapidly with distance from their source. For example, tripling the distance makes the force $3^2 = 9$ times weaker, while multiplying the distance by five results in a force that is $5^2 = 25$ times weaker. Despite this rapid decrease, the force never quite reaches zero. The gravitational pull of an object having some mass can never be completely extinguished.

We can combine the preceding statements about mass and distance to form a law of gravity that dictates the way in which *all* massive objects (i.e., objects having some mass) attract one another:

> Every particle of matter in the universe attracts every other particle with a force that is directly proportional to the product of the masses of the particles and inversely proportional to the square of the distance between them.

As a proportionality, Newton's law of gravity is

$$\text{gravitational force} \propto \frac{\text{mass of object 1} \times \text{mass of object 2}}{\text{distance}^2}.$$

(The symbol $\propto$ here means "is proportional to.") The rule for computing the force *F* between two bodies of

MORE PRECISELY 2-2

The Moon Is Falling!

The story of Isaac Newton seeing an apple fall to the ground and "discovering" gravity is well known, in one form or another, to most high school students. However, the real importance of Newton's observation was his realization that, by observing falling bodies on Earth and elsewhere, he could *quantify* the properties of the gravitational force and deduce the mathematical form of his law of gravitation.

Galileo Galilei had demonstrated some years earlier, by the simple experiment of dropping different objects from a great height (the top of the Tower of Pisa, according to lore) and noting that they hit the ground at the same time (at least, to the extent that air resistance was unimportant), that gravity causes the *same* acceleration in all bodies, regardless of mass. Since acceleration is proportional to force divided by mass (Newton's second law), this meant that the gravitational force on one body due to another had to be directly proportional to the first body's mass. Applying the same reasoning to the other body and using Newton's third law, we find that the force must also be proportional to the mass of the second body. This is the origin of the two "mass" terms in the law of gravity on p. 52. (The experimental finding that the gravitational force is precisely proportional to mass is now known as the *equivalence principle*. This principle forms an essential part of the modern theory of gravity; see Section 22.6).

What about the inverse-square part of Newton's law? At Earth's surface, the acceleration due to gravity, denoted by the letter g, is approximately 9.80 m/s^2. Where else other than on Earth could a falling body be seen? As illustrated in the accompanying figures, Newton realized that he could tell how gravity varies with distance by studying another object influenced by our planet's gravity: the Moon. Here's how he did it:

Let's assume for the sake of simplicity that the Moon's orbit around Earth is circular. As shown in the second figure, even though the Moon's orbital *speed* is constant, its *velocity* (red arrows) is not—the direction of the Moon's motion is steadily changing. In other words, the Moon is *accelerating*, constantly falling toward Earth. In fact, the acceleration of any body moving with speed v in a circular orbit of radius r may be shown to be

$$a = \frac{v^2}{r},$$

always directed toward the center of the circle (toward Earth in the case of the Moon). This acceleration is sometimes called *centripetal* ("center-seeking") acceleration. You probably already have an intuitive experience of this kind of acceleration—just think of the acceleration you feel as you take a tight corner (small r) at high speed (large v) in your car.

EXAMPLE: Knowing the Moon's distance $r = 384{,}000$ km (measured by triangulation) and the Moon's sidereal orbit period $P = 27.3$ days, Newton computed the Moon's orbital speed to be (distance $2\pi r$ in time P) $v = 2\pi r/P = 1.02$ km/s and hence determined its acceleration to be $a = 0.00272$ m/s^2, or $0.000278\ g$. ∞ *(More Precisely 1-3)*

Thus, Newton found, the Moon, lying 60 times farther from Earth's center than the apple falling from the tree in his garden (taking Earth's radius to be 6400 km), experiences an acceleration 3600, or 60^2, times smaller. In other words, *the acceleration due to gravity is inversely proportional to the square of the distance*. Isaac Newton's application of simple geometric reasoning and some very basic laws of motion resulted in a breakthrough that would revolutionize astronomers' view of the solar system and ultimately, through its influence on spaceflight and interplanetary navigation, pave the way for humanity's exploration of the universe.

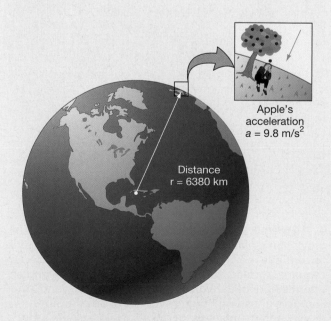

Apple's
acceleration
$a = 9.8$ m/s^2

Distance
$r = 6380$ km

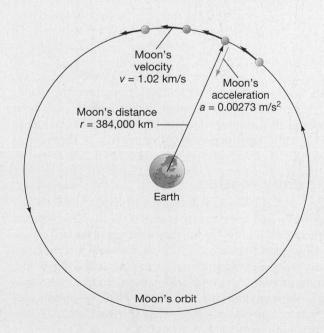

Moon's
velocity
$v = 1.02$ km/s

Moon's
acceleration
$a = 0.00273$ m/s^2

Moon's distance
$r = 384{,}000$ km

Earth

Moon's orbit

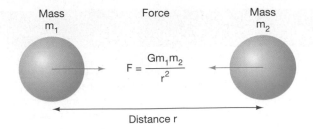

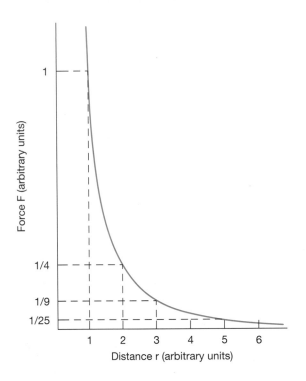

The Sun–planet interaction sketched here is analogous to our earlier example of the rock whirling on a string. The Sun's gravitational pull is your hand and the string, and the planet is the rock at the end of that string. The tension in the string provides the force necessary for the rock to move in a circular path. If you were suddenly to release the string—which would be like eliminating the Sun's gravity—the rock would fly away along a tangent to the circle, in accordance with Newton's first law.

In the solar system, at this very moment, Earth is moving under the combined influence of gravity and inertia. The net result is a stable orbit, despite our continuous rapid motion through space. (In fact, Earth orbits the Sun at a speed of about 30 km/s, or approximately 70,000 mph. You can verify this for yourself by calculating how fast Earth must move to complete a circle of radius 1 A.U.—and hence of circumference 2π A.U., or 940 million km—in 1 year, or 3.2×10^7 seconds. The answer is 9.4×10^8 km/3.2×10^7 s, or 29.4 km/s.) *More Precisely 2-3* describes how astronomers can use Newtonian mechanics and the law of gravity to quantify planetary mo-

masses m_1 and m_2, separated by a distance r, is usually written more compactly as

$$F = \frac{Gm_1m_2}{r^2}.$$

The quantity G is known as the *gravitational constant*, or, often, simply as Newton's constant. It is one of the fundamental constants of the universe. The value of G has been measured in extremely delicate laboratory experiments as 6.67×10^{-11} newton meter^2/kilogram^2 (N m^2/kg^2).

Planetary Motion

The mutual gravitational attraction between the Sun and the planets, as expressed by Newton's law of gravity, is responsible for the observed planetary orbits. As depicted in Figure 2.24, this gravitational force continuously pulls each planet toward the Sun, deflecting its forward motion into a curved orbital path. Because the Sun is much more massive than any of the planets, it dominates the interaction. We might say that the Sun "controls" the planets, not the other way around.

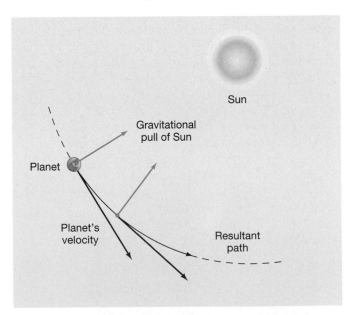

▲ **FIGURE 2.24 Solar Gravity** The Sun's inward pull of gravity on a planet competes with the planet's tendency to continue moving in a straight line. These two effects combine, causing the planet to move smoothly along an intermediate path, which continually "falls around" the Sun. This unending "tug-of-war" between the Sun's gravity and the planet's inertia results in a stable orbit.

tion and measure the masses of Earth, the Sun, and many other astronomical objects by studying the orbits of objects near them.

Kepler's Laws Reconsidered

Newton's laws of motion and law of universal gravitation provided a theoretical explanation for Kepler's empirical laws of planetary motion. Kepler's three laws follow directly from Newtonian mechanics, as solutions of the equations describing the motion of a body moving in response to an inverse-square force. However, just as Kepler modified the Copernican model by introducing ellipses rather than circles, so, too, did Newton make corrections to Kepler's first and third laws. It turns out that a planet does not orbit the exact center of the Sun. Instead, both the planet and the Sun orbit their common **center of mass**—the "average" position of all the matter making up the two bodies. Because the Sun and the planet are acted upon by equal and opposite gravitational forces (by Newton's third law), the Sun must also move (by Newton's first law), driven by the gravitational influence of the planet. The Sun, however, is so much more massive than any planet, that the center of mass of the planet–Sun system is very close to the center of the Sun, which is why Kepler's laws are so accurate. Thus, Kepler's first law becomes

    I. The orbit of a planet around the Sun is an ellipse, with the *center of mass of the planet–Sun system* at one focus.

As shown in Figure 2.25, the center of mass of two objects of comparable mass does not lie within either object. For identical masses (Figure 2.25a), the orbits are identical ellipses, with a common focus located midway between the two objects. For unequal masses (as in Figure 2.25b), the elliptical orbits still share a focus, and both have the same eccentricity, but the more massive object moves more slowly and on a tighter orbit. (Note that Kepler's second law, as stated earlier, continues to apply without modification to each orbit separately, but the *rates* at which the two orbits sweep out areas are different.) In the extreme case of a planet orbiting the much more massive Sun (see Figure 2.25c), the path traced out by the Sun's center lies entirely within the Sun itself.

The change to Kepler's third law is also small in the case of a planet orbiting the Sun, but very important in other circumstances, such as the orbital motion of two stars that are gravitationally bound to each other. Following through the mathematics of Newton's theory, we find that the true relationship between the semimajor axis *a* (measured in astronomical units) of the planet's orbit relative to the Sun and its orbital period *P* (in Earth years) is

$$P^2 \text{ (in Earth years)} = \frac{a^3 \text{ (in astronomical units)}}{M_{\text{total}} \text{ (in solar units)}},$$

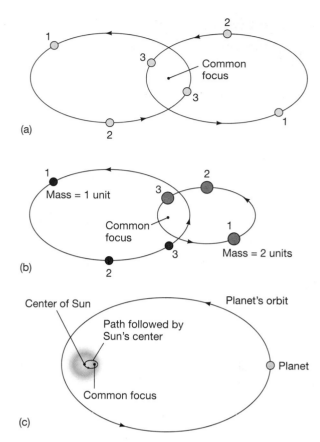

▲ **FIGURE 2.25 Orbits** (a) The orbits of two bodies (stars, for example) with equal masses, under the influence of their mutual gravity, are identical ellipses with a common focus. That focus is not at the center of either star, but instead is located at the center of mass of the pair, midway between them. The positions of the two bodies at three different times are indicated by the pairs of numbers. (Notice that a line joining the bodies always passes through the common focus.) (b) The orbits of two bodies, one of which is twice as massive as the other. Again, the elliptical orbits have a common focus, and the two ellipses have the same eccentricity. However, in accordance with Newton's laws of motion, the more massive body moves more slowly and in a smaller orbit, staying closer to the center of mass (at the common focus). In this particular case, the larger ellipse is twice the size of the smaller one. (c) In this extreme case of a hypothetical planet orbiting the Sun, the common focus of the two orbits lies inside the Sun.

where M_{total} is the *combined* mass of the two objects. Notice that Newton's restatement of Kepler's third law preserves the proportionality between P^2 and a^3, but now the proportionality includes M_{total}, so it is *not* quite the same for all the planets. The Sun's mass is so great, however, that the differences in M_{total} among the various combinations of the Sun and the other planets are almost unnoticeable, so Kepler's third law, as originally stated, is a very good approximation. This modified form of Kepler's third law is true in all circumstances, inside or outside the solar system.

Escaping Forever

The law of gravity that describes the orbits of planets around the Sun applies equally well to natural moons and artificial satellites orbiting any planet. All our Earth-orbiting, human-made satellites move along paths governed by a combination of the inward pull of Earth's gravity and the forward motion gained during the rocket launch. If the rocket initially imparts enough speed to the satellite, it can go into orbit. Satellites not given enough speed at launch, by accident or design (e.g., intercontinental ballistic missiles) fail to achieve orbit and fall back to Earth (see Figure 2.26).

Some space vehicles, such as the robot probes that visit other planets, attain enough speed to escape our planet's gravity and move away from Earth forever. This speed, known as the **escape speed**, is about 41% greater (actually, $\sqrt{2} = 1.414\ldots$ times greater) than the speed of an object traveling in a circular orbit at any given radius.* At less than the escape speed, the old adage "What goes up must come down" (or at least stay in orbit) still applies. At more than the escape speed, however, a spacecraft will leave Earth for good. Planets, stars, galaxies—all gravitating bodies—have escape speeds. No matter how massive the body, gravity decreases with distance. As a result, the escape speed diminishes with increasing separation. The far-

*In terms of the formula presented in More Precisely 2-3, the escape speed is given by $v_{escape} = \sqrt{2GM/r}$.

ther we go from Earth (or any gravitating body), the easier it becomes to escape.

The speed of a satellite in a circular orbit just above Earth's atmosphere is 7.9 km/s (roughly 18,000 mph). The satellite would have to travel at 11.2 km/s (about 25,000 mph) to escape from Earth altogether. If an object exceeds the escape speed, its motion is said to be **unbound**, and the orbit is no longer an ellipse. In fact, the path of the spacecraft relative to Earth is a related geometric figure called a *hyperbola*. If we simply change the word *ellipse* to *hyperbola*, the modified version of Kepler's first law still applies, as does Kepler's second law. (Kepler's third law does not extend to unbound orbits, because it doesn't make sense to talk about a period in those cases.)

Newton's laws explain the paths of objects moving at any point in space near any gravitating body. These laws provide a firm physical and mathematical foundation for Copernicus's heliocentric model of the solar system and for Kepler's laws of planetary motions. But they do much more than that: Newtonian gravitation governs not only the planets, moons, and satellites in their elliptical orbits, but also the stars and galaxies in their motion throughout our universe—as well as apples falling to the ground.

CONCEPT CHECK

✔ Explain, in terms of Newton's laws of motion and gravity, why planets orbit the Sun.

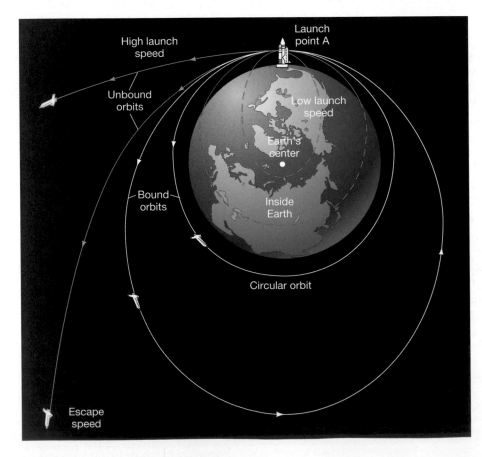

◀ **FIGURE 2.26 Escape Speed**
The effect of launch speed on the trajectory of a satellite. With too low a speed at point *A*, the satellite will simply fall back to Earth. Given enough speed, however, the satellite will go into orbit—it "falls around Earth." As the initial speed at point *A* is increased, the orbit will become more and more elongated. When the initial speed exceeds the escape speed, the satellite will become unbound from Earth and will escape along a hyperbolic trajectory.

MORE PRECISELY 2-3

Weighing the Sun

We can use Newtonian mechanics to calculate some useful formulae relating the properties of planetary orbits to the mass of the Sun. Again for simplicity, let's assume that the orbits are circular (not a bad approximation in most cases, and Newton's laws easily extend to cover the more general case of eccentric orbits). Consider a planet of mass m moving at speed v in an orbit of radius r around the Sun, of mass M. The planet's acceleration (see *More Precisely 2-2*) is

$$a = \frac{v^2}{r},$$

so, by Newton's second law, the force required to keep the planet in orbit is

$$F = ma = \frac{mv^2}{r}.$$

Setting this equation equal to the gravitational force due to the Sun, we obtain

$$\frac{mv^2}{r} = \frac{GmM}{r^2},$$

so the speed of the planet in the circular orbit is

$$v = \sqrt{\frac{GM}{r}}.$$

Now let's turn the problem around. Because we have measured G in the laboratory on Earth, and because we know the length of a year and the size of the astronomical unit, we can use Newtonian mechanics to *weigh* the Sun—that is, find its mass by measuring its gravitational influence on another body (in this case, Earth). Rearranging the last equation to read

$$M = \frac{rv^2}{G}$$

and substituting the known values of $v = 30$ km/s, $r = 1$ A.U. $= 1.5 \times 10^{11}$ m, and $G = 6.7 \times 10^{-11}$ N m^2/kg^2, we calculate the mass of the Sun to be 2.0×10^{30} kg—an enormous mass by terrestrial standards.

EXAMPLE: Similarly, knowing the distance from Earth to the Moon ($r = 384{,}000$ km) and the length of the (sidereal) month ($P = 27.3$ days), we can calculate the Moon's orbital speed to be $v = 2\pi r/P = 1.02$ km/s, and hence, using the preceding formula, measure Earth's mass to be 6.0×10^{24} kg.

In fact, this is how basically *all* masses are measured in astronomy. Because we can't just go out and attach a scale to an astronomical object when we need to know its mass, we must look for its gravitational influence on something else. This principle applies to planets, stars, galaxies, and even clusters of galaxies—very different objects, but all subject to the same physical laws.

Chapter Review

SUMMARY

Geocentric (p. 38) models of the universe were based on the assumption that the Sun, the Moon, and the planets all orbit Earth. The most successful and long lived of these was the **Ptolemaic model (p. 38)**. Unlike the Sun and the Moon, planets sometimes appear to temporarily reverse their direction of motion (from night to night) relative to the stars and then resume their normal "forward" course. This phenomenon is called **retrograde motion (p. 37)**. To account for retrograde motion within the geocentric picture, it was necessary to suppose that planets moved on small circles called **epicycles (p. 38)**, whose centers orbited Earth on larger circles called **deferents (p. 38)**.

The **heliocentric (p. 39)** view of the solar system holds that Earth, like all the planets, orbits the Sun. This model accounts for retrograde motion and the observed sizes and variations in brightness of the planets in a much more straightforward way than the geocentric model. The widespread realization during the Renaissance that the solar system is Sun centered, and not Earth centered, is known as the **Copernican revolution (p. 41)**, in honor of Nicolaus Copernicus, who laid the foundations of the modern heliocentric model.

Direct observational evidence for Earth's motion around the Sun was obtained in the 18th century, with the first measurement of the **aberration of starlight (p. 43)**.

Galileo Galilei is often regarded as the father of experimental science. His telescopic observations of the Moon, the Sun, Venus, and Jupiter played a crucial role in supporting and strengthening the Copernican picture of the solar system. Johannes Kepler improved on Copernicus's model with his three **laws of planetary motion (p. 46)**: (1) Planetary orbits are **ellipses (p. 46)**, with the Sun at one **focus (p. 46)**. (2) A planet moves faster as its orbit takes it closer to the Sun. (3) The **semimajor axis (p. 46)** of the orbit is related in a simple way to the planet's orbital **period (p. 47)**. Most planets move on orbits whose **eccentricities (p. 46)** are quite small, so their paths differ only slightly from perfect circles. The distance from Earth to the Sun is called the **astronomical unit (p. 48)**. Nowadays, the astronomical unit is determined by bouncing **radar (p. 49)** signals off the planet Venus and measuring the time the signal takes to return.

Isaac Newton succeeded in explaining Kepler's laws in terms of a few general physical principles, now known as **Newtonian** mechanics (p. 50). The tendency of a body to keep moving at constant velocity is called **inertia (p. 51)**. The greater the body's **mass (p. 51)**, the greater is its inertia. To change the velocity, a **force (p. 51)** must be applied. **Weight (p. 51)** is another name for the force of gravity acting on an object. The rate of change of velocity, called **acceleration (p. 51)**, is equal to the applied force divided by the body's mass.

To explain planetary orbits, Newton postulated that **gravity (p. 52)** attracts the planets to the Sun. Every object with any mass exerts a **gravitational force (p. 52)** on all other objects in the universe. The strength of this force decreases with distance according to an **inverse-square law (p. 52)**. Newton's laws imply that a planet does not orbit the precise center of the Sun, but instead that both the planet and the Sun orbit their common **center of mass (p. 55)**. For one object to escape from the gravitational pull of another, its speed must exceed the **escape speed (p. 56)** of that other object. In this case, the motion is said to be **unbound (p. 56)**, and the orbital path is no longer an ellipse, although it is still described by Newton's laws.

REVIEW AND DISCUSSION

1. What contributions to modern astronomy were made by Chinese and Islamic astronomers during the Dark Ages of medieval Europe?

2. Briefly describe the geocentric model of the universe.

3. The benefit of our current knowledge lets us see flaws in the Ptolemaic model of the universe. What is its basic flaw?

4. What was the great contribution of Copernicus to our knowledge of the solar system? What was still a flaw in the Copernican model?

5. What is a theory? Can a theory ever be proved to be true?

6. When were Copernicus's ideas finally accepted?

7. What is the Copernican principle?

8. What discoveries of Galileo helped confirm the views of Copernicus, and how did they do so?

9. Briefly describe Kepler's three laws of planetary motion.

10. How did Tycho Brahe contribute to Kepler's laws?

11. If radio waves cannot be reflected from the Sun, how can radar be used to find the distance from Earth to the Sun?

12. How did astronomers determine the scale of the solar system prior to the invention of radar?

13. What does it mean to say that Kepler's laws are empirical?

14. What are Newton's laws of motion and gravity?

15. List the two modifications made by Newton to Kepler's laws.

16. Why do we say that a baseball falls toward Earth, and not Earth toward the baseball?

17. Why would a baseball go higher if it were thrown up from the surface of the Moon than if it were thrown with the same velocity from the surface of Earth?

18. In what sense is the Moon falling toward Earth?

19. What is the meaning of the term *escape speed*?

20. What would happen to Earth if the Sun's gravity were suddenly "turned off"?

CONCEPTUAL SELF-TEST: TRUE OR FALSE/MULTIPLE CHOICE

1. Aristotle was the first to propose that all planets revolve around the Sun.

2. Retrograde motion is the apparent "backward" (westward) motion of a planet relative to the stars.

3. The heliocentric model of the universe holds that Earth is at the center and everything else moves around it.

4. Galileo's observations of the sky were made with the naked eye.

5. Kepler discovered that the shape of an orbit is a circle, not an ellipse, as had previously been believed.

6. Kepler's discoveries regarding the orbital motion of the planets were based mainly on observations made by Copernicus.

7. The Sun lies at the center of a planet's orbit.

8. Kepler never knew the true distances between the planets and the Sun, only their relative distances.

9. According to Newton's laws, a moving object will come to rest unless acted upon by a force.

10. You throw a baseball to someone; before the ball is caught, it is temporarily in orbit around Earth's center.

11. Planets near opposition **(a)** rise in the east; **(b)** rise in the west; **(c)** do not rise or set; **(d)** have larger deferents.

12. A major flaw in Copernicus's model was that it still had **(a)** the Sun at the center; **(b)** Earth at the center **(c)** retrograde loops; **(d)** circular orbits.

13. As shown in Figure 2.12 ("Venus Phases"), Galileo's observations of Venus demonstrated that Venus must be **(a)** orbiting Earth; **(b)** orbiting the Sun; **(c)** about the same diameter as Earth; **(d)** similar to the Moon.

14. An accurate sketch of Mars's orbit around the Sun would show **(a)** the Sun far off center; **(b)** an oval twice as long as it is wide; **(c)** a nearly perfect circle; **(d)** phases.

15. A calculation of how long it takes a planet to orbit the Sun would be most closely related to Kepler's **(a)** first law of orbital shapes; **(b)** second law of orbital speeds; **(c)** third law of planetary distances; **(d)** first law of inertia.

16. An asteroid with an orbit lying entirely inside Earth's **(a)** has an orbital semimajor axis of less than 1 A.U.; **(b)** has a longer orbital period than Earth's; **(c)** moves more slowly than Earth; **(d)** has a highly eccentric orbit.

17. If Earth's orbit around the Sun were twice as large as it is now, the orbit would take **(a)** less than twice as long; **(b)** two times longer; **(c)** more than two times longer to traverse.

18. Figure 2.22 ("Gravity"), showing the motion of a ball near Earth's surface, depicts how gravity **(a)** increases with altitude; **(b)** causes the ball to accelerate downward; **(c)** causes the ball to accelerate upward; **(d)** has no effect on the ball.

19. If the Sun *and its mass* were suddenly to disappear, Earth would **(a)** continue in its current orbit; **(b)** suddenly change its orbital speed; **(c)** fly off into space; **(d)** stop spinning.

20. Figure 2.25(b) ("Orbits") shows the orbits of two stars of unequal masses. If one star has twice the mass of the other, then the more massive star **(a)** moves more slowly than; **(b)** moves more rapidly than; **(c)** has half the gravity of; **(d)** has twice the eccentricity of the less massive star.

PROBLEMS

 Algorithmic versions of these questions are available in the Practice Problems module of the Companion Website at astro.prenhall.com/chaisson.

The number of squares preceding each problem indicates its approximate level of difficulty.

1. ■ Tycho Brahe's observations of the stars and planets were accurate to about 1 arc minute (1′). To what distance does this angle correspond at the distance of (a) the Moon; (b) the Sun; and (c) Saturn (at closest approach)?

2. ■■ To an observer on Earth, through what angle will Mars appear to move relative to the stars over the course of 24 hours when the two planets are at closest approach? Assume for simplicity that Earth and Mars move on circular orbits of radii 1.0 A.U. and 1.5 A.U., respectively, in exactly the same plane. Will the apparent motion be prograde or retrograde?

3. ■ Using the data in Table 2.1, show that Pluto is closer to the Sun at perihelion (the point of closest approach to the Sun in its orbit) than Neptune is at any point in its orbit.

4. ■■ An asteroid has a perihelion distance of 2.0 A.U. and an aphelion distance of 4.0 A.U. Calculate its orbital semimajor axis, eccentricity, and period.

5. ■■ A spacecraft has an orbit that just grazes Earth's orbit at aphelion and just grazes Venus's orbit at perihelion. Assuming that Earth and Venus are in the right places at the right times, how long will the spacecraft take to travel from Earth to Venus?

6. ■■ Halley's comet has a perihelion distance of 0.6 A.U. and an orbital period of 76 years. What is the aphelion distance of Halley's comet from the Sun?

7. ■■ What is the maximum possible parallax of Mercury during a solar transit, as seen from either end of a 3000-km baseline on Earth?

8. ■ How long would a radar signal take to complete a round-trip between Earth and Mars when the two planets are 0.7 A.U. apart?

9. ■ Jupiter's moon Callisto orbits the planet at a distance of 1.88 million km. Callisto's orbital period about Jupiter is 16.7 days. What is the mass of Jupiter? [Assume that Callisto's mass is negligible compared with that of Jupiter, and use the modified version of Kepler's third law (Section 2.7).]

10. ■■ The Sun moves in a roughly circular orbit around the center of the Milky Way galaxy at a distance of 26,000 light-years. The orbit speed is approximately 220 km/s. Calculate the Sun's orbital period and centripetal acceleration, and use these numbers to estimate the mass of our galaxy.

11. ■■ At what distance from the Sun would a planet's orbital period be 1 million years? What would be the orbital period at a distance of one light-year?

12. ■ The acceleration due to gravity at Earth's surface is 9.80 m/s². What is the gravitational acceleration at altitudes of (a) 100 km; (b) 1000 km; (c) 10,000 km? Take Earth's radius to be 6400 km.

13. ■■ What would be the speed of a spacecraft moving in a circular orbit at each of the three altitudes listed in the previous problem? In each case, how does the centripetal acceleration (*More Precisely 2-2*) compare with the gravitational acceleration?

14. ■ Use Newton's law of gravity to calculate the force of gravity between you and Earth. Convert your answer, which will be in newtons, to pounds, using the conversion 4.45 N equals 1 pound. What do you normally call this force?

15. ■ The Moon's mass is 7.4×10^{22} kg and its radius is 1700 km. What is the speed of a spacecraft moving in a circular orbit just above the lunar surface? What is the escape speed from the Moon?

 In addition to the Practice Problems module, the Companion Website at astro.prenhall.com/chaisson provides for each chapter a study guide module with multiple choice questions as well as additional annotated images, animations, and links to related Websites.

Stars change from birth to maturity to death, much like living things, but on vastly longer timescales. Our own star, the Sun, is about mid-way through its evolutionary cycle. About 5 billion years ago, it emerged from a stellar nursery much like the one shown here. This infrared image from the new Spitzer Space Telescope captures radiation longer in wavelength than light, enabling us to peer inside young star clusters—this one called the Tarantula Nebula about 160,000 light-years away. (NASA) ▶

Radiation

Information from the Cosmos

Astronomical objects are more than just things of beauty in the night sky. Planets, stars, and galaxies are of vital significance if we are to fully understand our place in the big picture—the "grand design" of the universe. Each object is a source of information about the material aspects of our universe—its state of motion, its temperature, its chemical composition, and even its past history. When we look at the stars, the light we see actually began its journey to Earth decades, centuries—even millennia—ago. The faint rays from the most distant galaxies have taken billions of years to reach us. The stars and galaxies in the night sky show us the far away and the long ago. In this chapter, we begin our study of how astronomers extract information from the light emitted by astronomical objects. These basic concepts of radiation are central to modern astronomy.

LEARNING GOALS

Studying this chapter will enable you to

1 Discuss the nature of electromagnetic radiation and tell how that radiation transfers energy and information through interstellar space.

2 Describe the major regions of the electromagnetic spectrum and explain how Earth's atmosphere affects our ability to make astronomical observations at different wavelengths.

3 Explain what is meant by the term "blackbody radiation" and describe the basic properties of such radiation.

4 Tell how we can determine the temperature of an object by observing the radiation that it emits.

5 Show how the relative motion between a source of radiation and its observer can change the perceived wavelength of the radiation, and explain the importance of this phenomenon to astronomy.

Visit astro.prenhall.com/chaisson for additional annotated images, animations, and links to related sites for this chapter.

3.1 Information from the Skies

Figure 3.1 shows a galaxy in the constellation Andromeda. On a dark, clear night, far from cities or other sources of light, the Andromeda Galaxy, as it is generally called, can be seen with the naked eye as a faint, fuzzy patch on the sky, comparable in diameter to the full Moon. Yet the fact that it is visible from Earth belies this galaxy's enormous distance from us: It lies roughly 2.5 million *light-years* away.

An object at such a distance is truly inaccessible in any realistic human sense. Even if a space probe could miraculously travel at the speed of light, it would need 2.5 million years to reach this galaxy and 2.5 million more to return with its findings. Considering that civilization has existed on Earth for less than 10,000 years, and its prospects for the next 10,000 are far from certain, even this unattainable technological feat would not provide us with a practical means of exploring other galaxies. Even the farthest reaches of our own galaxy, "only" a few tens of thousands of light-years distant, are effectively off limits to visitors from Earth, at least for the foreseeable future.

Light and Radiation

Given the practical impossibility of traveling to such remote parts of the universe, how do astronomers know anything about objects far from Earth? How do we obtain detailed information about any planet, star, or galaxy too distant for a personal visit or any kind of controlled experiment? The answer is that we use the laws of physics, as we know them here on Earth, to interpret the **electromagnetic radiation** emitted by those objects.

Radiation is any way in which energy is transmitted through space from one point to another without the need for any physical connection between the two locations. The term *electromagnetic* just means that the energy is carried in the form of rapidly fluctuating *electric* and *magnetic* fields (to be discussed in more detail later in Section 3.2).

Virtually all we know about the universe beyond Earth's atmosphere has been gleaned from painstaking analysis of electromagnetic radiation received from afar. Our understanding depends completely on our ability to decipher this steady stream of data reaching us from space.

How bright are the stars (or galaxies, or planets), and how hot? What are their masses? How rapidly do they spin, and what is their motion through space? What are they made of, and in what proportion? The list of questions is long, but one fact is clear: Electromagnetic theory is vital to providing the answers—without it, we would have no way of testing our models of the cosmos, and the modern science of astronomy simply would not exist. ∞ (Sec. 1.2)

Visible light is the particular type of electromagnetic radiation to which our human eyes happen to be sensitive. As light enters our eye, small chemical reactions triggered by the incoming energy send electrical impulses to the brain, producing the sensation of sight. But there is also *invisible* electromagnetic radiation, which goes completely undetected by our eyes. **Radio, infrared**, and **ultraviolet** waves, as well as **X rays** and **gamma rays**, all fall into this category. Note that, despite the different names, the words *light*, *rays*, *radiation*, and *waves* all really refer to the same thing. The names are just historical accidents, reflecting the fact that it took many years for scientists to realize that these apparently very different types of radiation are in re-

◄ **FIGURE 3.1**
Andromeda The pancake-shaped Andromeda Galaxy lies about 2.5 million light-years away, according to the most recent distance measurements. It contains a few hundred billion stars. *(T. Hallas)*

ality one and the same physical phenomenon. Throughout this text, we will use the general terms "light" and "electromagnetic radiation" more or less interchangeably.

Wave Motion

Despite the early confusion still reflected in current terminology, scientists now know that all types of electromagnetic radiation travel through space in the form of **waves**. To understand the behavior of light, then, we must know a little about wave motion.

Simply stated, a wave is a way in which energy is transferred from place to place without the physical movement of material from one location to another. In wave motion, the energy is carried by a *disturbance* of some sort. This disturbance, whatever its nature, occurs in a distinctive repeating pattern. Ripples on the surface of a pond, sound waves in air, and electromagnetic waves in space, despite their many obvious differences, all share this basic defining property.

Imagine a twig floating in a pond (Figure 3.2). A pebble thrown into the pond at some distance from the twig disturbs the surface of the water, setting it into up-and-down motion. This disturbance will move outward from the point of impact in the form of waves. When the waves reach the twig, some of the pebble's energy will be imparted to it, causing the twig to bob up and down. In this way, both energy and *information*—the fact that the pebble entered the water—are transferred from the place where the pebble landed to the location of the twig. We could tell that a pebble (or, at least, some object) had entered the water just by observing the twig. With a little additional physics, we could even estimate the pebble's energy.

A wave is not a physical object. No water traveled from the point of impact of the pebble to the twig—at any location on the surface, the water surface simply moved up and down as the wave passed. What, then, *did* move across the surface of the pond? As illustrated in the figure, the answer is that the wave was the *pattern* of up-and-down mo-

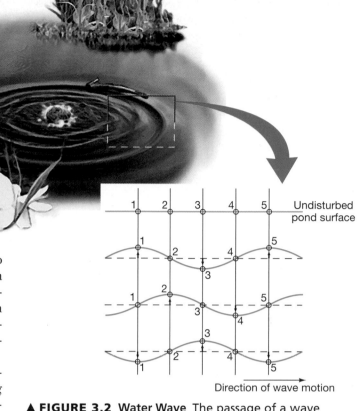

▲ **FIGURE 3.2 Water Wave** The passage of a wave across a pond causes the surface of the water to bob up and down, but there is no movement of water from one part of the pond to another. Here waves ripple out from the point where a pebble has hit the water to the point where a twig is floating. The inset shows a simplified series of "snapshots" of part of the pond surface as the wave passes by. The points numbered 1 through 5 represent surface locations that move up and down with passage of the wave.

tion. This pattern was transmitted from one point to the next as the disturbance moved across the water.

Figure 3.3 shows how wave properties are quantified and illustrates some standard terminology. The wave's **period** is the number of seconds needed for the wave to repeat itself at any given point in space. The **wavelength** is

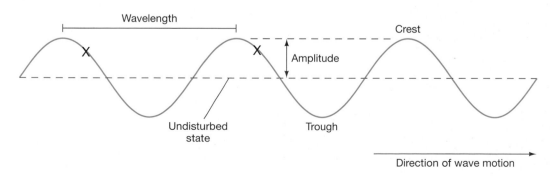

◀ **FIGURE 3.3 Wave Properties** Representation of a typical wave, showing its direction of motion, wavelength, and amplitude. In one period, the entire pattern shown here moves one wavelength to the right.

the number of meters needed for the wave to repeat itself at a given moment in time. It can be measured as the distance between two adjacent wave *crests*, two adjacent wave *troughs*, or any other two similar points on adjacent wave cycles (e.g., the points marked X in the figure). A wave moves a distance equal to one wavelength in one period. The maximum departure of the wave from the undisturbed state—still air, say, or a flat pond surface—is called the wave's **amplitude**.

The number of wave crests passing any given point per unit time is called the wave's **frequency**. If a wave of a given wavelength moves at high speed, then many crests pass per second and the frequency is high. Conversely, if the same wave moves slowly, then its frequency will be low. The frequency of a wave is just the reciprocal of the wave's period; that is,

$$\text{frequency} = \frac{1}{\text{period}}.$$

Frequency is expressed in units of inverse time (that is, 1/second, or cycles per second), called hertz (Hz) in honor of the 19th-century German scientist Heinrich Hertz, who studied the properties of radio waves. Thus, a wave with a period of 5 seconds (5 s) has a frequency of $(1/5)$ cycles/s = 0.2 Hz, meaning that one wave crest passes a given point in space every five seconds.

Because a wave travels one wavelength in one period, it follows that the *wave velocity* is simply equal to the wavelength divided by the period:

$$\text{velocity} = \frac{\text{wavelength}}{\text{period}}.$$

Since the period is the reciprocal of the frequency, we can equivalently (and more commonly) write this relationship as

$$\text{velocity} = \text{wavelength} \times \text{frequency}.$$

Thus, if the wave in our earlier example had a wavelength of 0.5 m, its velocity would be (0.5 m) / (5 s), or (0.5 m) × (0.2 Hz) = 0.1 m/s. Notice that wavelength and wave frequency are *inversely* related—doubling one halves the other.

The Components of Visible Light

White light is a mixture of colors, which we conventionally divide into six major hues: red, orange, yellow, green, blue, and violet. As shown in Figure 3.4, we can separate a beam of white light into a rainbow of these basic colors—called a *spectrum* (plural, *spectra*)—by passing it through a prism. This experiment was first reported by Isaac Newton over 300 years ago. In principle, the original beam of white light could be recovered by passing the spectrum through a second prism to recombine the colored beams.

What determines the color of a beam of light? The answer is its frequency (or alternatively, its wavelength). We see different colors because our eyes react differently to electromagnetic waves of different frequencies. A prism splits a beam of light up into separate colors because light rays of different frequencies are bent, or *refracted*, slightly differently as they pass through the prism—red light the least, violet light the most. Red light has a frequency of roughly 4.3×10^{14} Hz, corresponding to a wavelength of

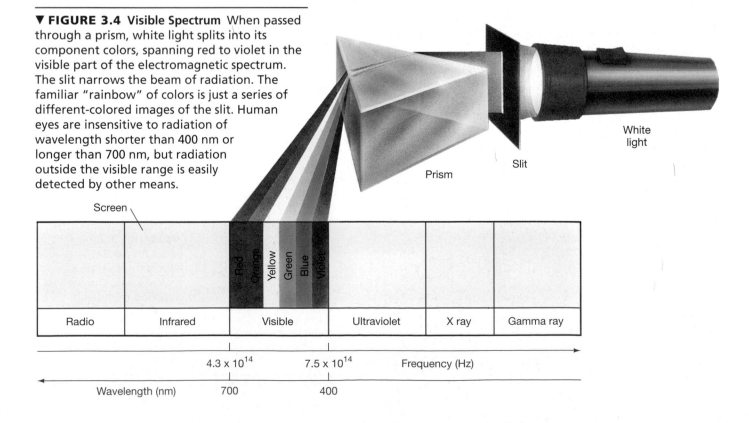

▼ **FIGURE 3.4 Visible Spectrum** When passed through a prism, white light splits into its component colors, spanning red to violet in the visible part of the electromagnetic spectrum. The slit narrows the beam of radiation. The familiar "rainbow" of colors is just a series of different-colored images of the slit. Human eyes are insensitive to radiation of wavelength shorter than 400 nm or longer than 700 nm, but radiation outside the visible range is easily detected by other means.

about 7.0×10^{-7} m. Violet light, at the other end of the visible range, has nearly double the frequency—7.5×10^{14} Hz—and (since the speed of light is always the same) just over half the wavelength—4.0×10^{-7} m. The other colors we see have frequencies and wavelengths intermediate between these two extremes, spanning the entire *visible spectrum* shown in Figure 3.4. Radiation outside this range is invisible to human eyes.

Scientists often use a unit called the *nanometer* (nm) in describing the wavelength of light. (See Appendix 2.) There are 10^9 nanometers in 1 meter. An older unit called the *angstrom* ($1 \text{ Å} = 10^{-10}$ m $= 0.1$ nm) is also widely used. (The unit is named after the 19th-century Swedish physicist Anders Ångstrom—pronounced "ong·strem.") However, in SI units, the nanometer is preferred. Thus, the visible spectrum covers the range of wavelengths from 400 nm to 700 nm (4000 Å to 7000 Å). The radiation to which our eyes are most sensitive has a wavelength near the middle of this range, at about 550 nm (5500 Å), in the yellow-green region of the spectrum. It is no coincidence that this wavelength falls within the range of wavelengths at which the Sun emits most of its electromagnetic energy—our eyes have evolved to take greatest advantage of the available light.

3.2 Waves in What?

LEARNING GOAL 1 Waves of radiation differ fundamentally from water waves, sound waves, or any other waves that travel through a material medium. Radiation needs *no* such medium. When light travels from a distant galaxy, or from any other cosmic object, it moves through the virtual vacuum of space. Sound waves, by contrast, cannot do this, despite what you have probably heard in almost every sci-fi movie ever made! If we were to remove all the air from a room, conversation would be impossible (even with suitable breathing apparatus to keep our test subjects alive!) because sound waves cannot exist without air or some other physical medium to support them. Communication by flashlight or radio, however, would be entirely feasible.

The ability of light to travel through empty space was once a great mystery. The idea that light, or any other kind of radiation, could move as a wave through nothing at all seemed to violate common sense, yet it is now a cornerstone of modern physics.

Interactions Between Charged Particles

To understand more about the nature of light, consider for a moment an *electrically charged* particle, such as an **electron** or a **proton**. Like mass, electrical charge is a fundamental property of matter. Electrons and protons are elementary particles—"building blocks" of atoms and all matter—that carry the basic unit of charge. Electrons are said to carry a *negative* charge, whereas protons carry an equal and opposite *positive* charge.

Just as a massive object exerts a gravitational force on every other massive body, an electrically charged particle exerts an *electrical* force on every other charged particle in the universe. ∞ (Sec. 2.7) The buildup of electrical charge (a net excess of positive over negative, or vice versa) is what causes "static cling" on your clothes when you take them out of a hot clothes dryer; it also causes the shock you sometimes feel when you touch a metal door frame on a particularly dry day.

Unlike the gravitational force, which is always attractive, electrical forces can be either attractive or repulsive. As illustrated in Figure 3.5(a), particles with *like* charges (i.e., both negative or both positive—for example, two electrons or two protons) repel one another. Particles with *unlike* charges (i.e., having opposite signs—an electron and a proton, say) attract.

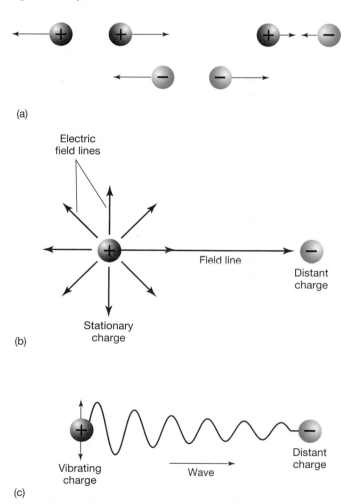

▲ **FIGURE 3.5 Charged Particles** (a) Particles carrying like electrical charges repel one another, whereas particles carrying unlike charges attract. (b) A charged particle is surrounded by an electric field, which determines the particle's influence on other charged particles. We represent the field by a series of field lines. (c) If a charged particle begins to vibrate back and forth, its electric field changes. The resulting disturbance travels through space as a wave.

How is the electrical force transmitted through space? Extending outward in all directions from any charged particle is an **electric field**, which determines the electrical force exerted by the particle on all other charged particles in the universe (Figure 3.5b). The strength of the electric field, like the strength of the gravitational field, decreases with increasing distance from the charge according to an inverse-square law. By means of the electric field, the particle's presence is "felt" by all other charged particles, near and far.

Now, suppose our particle begins to vibrate, perhaps because it becomes heated or collides with some other particle. Its changing position causes its associated electric field to change, and this changing field in turn causes the electrical force exerted on other charges to vary (Figure 3.5c). If we measure the change in the force on these other charges, we learn about our original particle. Thus, *information about the particle's state of motion is transmitted through space via a changing electric field*. This *disturbance* in the particle's electric field travels through space as a wave.

Electromagnetic Waves

The laws of physics tell us that a **magnetic field** must accompany every changing electric field. Magnetic fields govern the influence of *magnetized* objects on one another, much as electric fields govern interactions among charged particles. The fact that a compass needle always points to magnetic north is the result of the interaction between the magnetized needle and Earth's magnetic field (Figure 3.6). Magnetic fields also exert forces on *moving* electric charges (i.e., electric currents)—electric meters and motors rely on this basic fact. Conversely, moving charges *create* magnetic fields (electromagnets are a familiar example). In short, electric and magnetic fields are inextricably linked to one another: A change in either one necessarily creates the other.

Thus, as illustrated in Figure 3.7, the disturbance produced by the moving charge in Figure 3.5(c) actually consists of vibrating electric *and* magnetic fields, moving together through space. Furthermore, as shown in the diagram, these fields are always oriented *perpendicular* to one another and to the direction in which the wave is traveling. The fields do not exist as independent entities; rather, they are different aspects of a single physical phenomenon: **electromagnetism**. Together, they constitute an *electromagnetic wave* that carries energy and information from one part of the universe to another.

Now consider a real cosmic object—a star, say. When some of its charged contents move around, their electric fields change, and we can detect that change. The resulting electromagnetic ripples propagate (travel) outward as

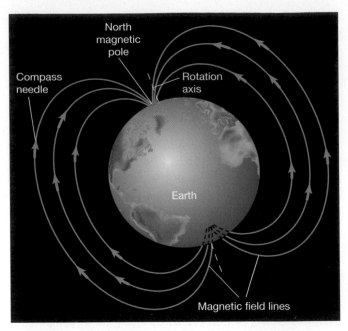

▲ **FIGURE 3.6 Magnetism** Earth's magnetic field interacts with a magnetic compass needle, causing the needle to become aligned with the field—that is, to point toward Earth's north (magnetic) pole. The north magnetic pole lies at latitude 80° N, longitude 107° W, some 1140 km from the geographic North Pole.

waves through space, requiring no material medium in which to move. Small charged particles, either in our eyes or in our experimental equipment, eventually respond to the electromagnetic field changes by vibrating in tune with the radiation that is received. This response is how we de-

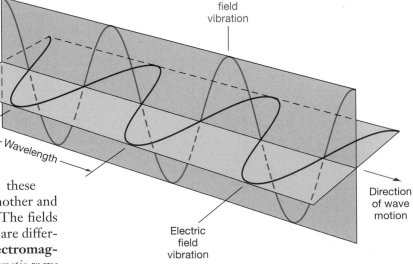

▲ **FIGURE 3.7 Electromagnetic Wave** Electric and magnetic fields vibrate perpendicularly to each other. Together, they form an electromagnetic wave that moves through space at the speed of light in the direction perpendicular to both the electric and the magnetic fields comprising it.

tect the radiation—and how we see. Figure 3.8 shows a more familiar example of information being transferred by electromagnetic radiation. A cellular transmitter causes electric charges to oscillate up and down in a metal rod mounted at the top of a tower, thereby generating electromagnetic radiation. This radiation is detected by the antenna in your cell phone. Within the metal core of the receiving antenna, electric charges respond to the incoming radiation by vibrating in time with the transmitted wave frequency. The information carried by the pattern of vibrations is then reconverted into sound and images by your phone.

How *quickly* is one charge influenced by the change in the electromagnetic field when another charge begins to move? This is an important question, because it is equivalent to asking how fast an electromagnetic wave travels. Does it propagate at some measurable speed, or is it instantaneous? Both theory and experiment tell us that all electromagnetic waves move at a very specific speed—the **speed of light** (always denoted by the letter c). Its exact value is 299,792.458 km/s in a vacuum (and somewhat less in material substances such as air or water). We will round this value off to $c = 3.00 \times 10^5$ km/s, an extremely high speed. In the time needed to snap your fingers (about a tenth of a second), light can travel three-quarters of the way around our planet! If the currently known laws of physics are correct, then the speed of light is the fastest speed possible. (See *More Precisely 22-1*.)

The speed of light is very large, but it is still *finite*. That is, light does not travel instantaneously from place to place. This fact has some interesting consequences for our study of distant objects. It takes time—often lots of time—for light to travel through space. The light we see from the nearest large galaxy—the Andromeda Galaxy, shown in Figure 3.1—left that object about 2.5 million years ago, around the time our first human ancestors appeared on planet Earth. We can know nothing about this galaxy as it exists today. For all we know, it may no longer even exist! Only our descendants, 2.5 million years into the future, will know whether it exists now. So as we study objects in the cosmos, remember that the light we see left those objects long ago. We can never observe the universe as it is—only as it was.

The Wave Theory of Radiation

The description presented in this chapter of light and other forms of radiation as electromagnetic waves traveling through space is known as the **wave theory of radiation**. It is a spectacularly successful scientific theory, full of explanatory and predictive power and deep insight into the complex interplay between light and matter—a cornerstone of modern physics.

Two centuries ago, however, the wave theory stood on much less solid scientific ground. Before about 1800, scientists were divided in their opinions about the nature of light. Some thought that light was a wave phenomenon

◄ FIGURE 3.8 Cellular Signal Charged particles in a cell-phone antenna vibrate in response to electromagnetic radiation broadcast by a distant transmitter. The radiation is produced when electric charges are made to oscillate in the transmitter's emitting antenna. The vibrations in the receiving antenna "echo" the oscillations in the transmitter, allowing the original information to be retrieved.

(although at the time, electromagnetism was unknown), while others maintained that light was in reality a stream of particles that moved in straight lines. Given the experimental apparatus available at the time, neither camp could find conclusive evidence to disprove the other's theory. *Discovery 3-1* discusses some wave properties that are of particular importance to modern astronomers and describes how their detection in experiments using visible light early in the 19th century tilted the balance of scientific opinion in favor of the wave theory.

But that's not the end of the story. The wave theory, like all good scientific theories, can and must continually be tested by experiment and observation. ∞ (Sec. 1.2) Around the turn of the 20th century, physicists made a se-

ries of discoveries about the behavior of radiation and matter on very small (atomic) scales that could not be explained by the "classical" wave theory just described. Changes had to be made. As we will see in Chapter 4, the modern theory of radiation is actually a "hybrid" of the once-rival wave and particle views, combining key elements of each in a unified and—for now—undisputed whole.

CONCEPT CHECK

✔ What is light? List some similarities and differences between light waves and waves on water or in air.

DISCOVERY 3-1

The Wave Nature of Radiation

Until the early 19th century, debate raged in scientific circles regarding the true nature of light. On the one hand, the particle, or *corpuscular*, theory, first expounded in detail by Isaac Newton, held that light consisted of tiny particles moving in straight lines at the speed of light. Different colors were presumed to correspond to different particles. On the other hand, the *wave* theory, championed by the 17th-century Dutch astronomer Christian Huygens, viewed light as a wave phenomenon, in which color was determined by frequency, or wavelength. During the first few decades of the 19th century, growing experimental evidence that light displayed three key wave properties—*diffraction*, *interference*, and *polarization*—argued strongly in favor of the wave theory.

Diffraction is the deflection, or "bending," of a wave as it passes a corner or moves through a narrow gap. As depicted in the first figure, a sharp-edged hole in a barrier seems at first glance to produce a sharp shadow, as we might expect if radiation were composed of rays or particles moving in perfectly straight lines. Closer inspection, however, reveals that the shadow actually has a "fuzzy" edge, as shown in the photograph at the right of the diffraction pattern produced by a small circular opening. We are not normally aware of such effects in everyday life, because diffraction is generally very small for visible light. For any wave, the amount of diffraction is proportional to the ratio of the wavelength to the width of the gap. The longer the wavelength or the small-

er the gap, the greater is the angle through which the wave is diffracted. Thus, visible light, with its extremely short wavelengths, shows perceptible diffraction only when passing through very narrow openings. (The effect is much more noticeable for sound waves: No one thinks twice about our abili-

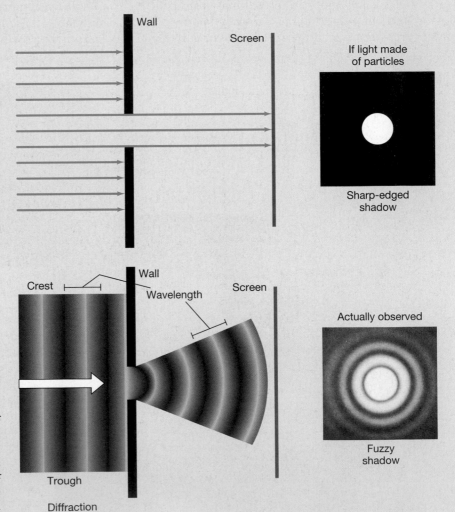

3.3 The Electromagnetic Spectrum

Figure 3.9 plots the entire range of electromagnetic radiation, illustrating the relationships among the different "types" of electromagnetic radiation listed earlier. Notice that the only characteristic distinguishing one from another is wavelength, or frequency. To the low-frequency, long-wavelength side of visible light lie *radio* and *infrared* radiation. Radio frequencies include radar, microwave radiation, and the familiar AM, FM, and TV bands. We perceive infrared radiation as heat. At higher frequencies (shorter wavelengths) are the domains of *ultraviolet, X-ray*, and *gamma-ray* radiation. Ultraviolet radiation, lying just beyond the violet end of the visible spectrum, is responsible for suntans and sunburns. The shorter-wavelength X rays are perhaps best known for their ability to penetrate human tissue and reveal the state of our insides without resorting to surgery. Gamma rays are the shortest-wavelength radiation. They are often associated with radioactivity and are invariably damaging to living cells they encounter.

All these spectral regions, including the visible spectrum, collectively make up the **electromagnetic spectrum**. Remember that, despite their greatly differing wavelengths and the different roles they play in everyday life on Earth, all are basically the same phenomenon, and all move at the same speed—the speed of light, *c*.

ty to hear people even when they are around a corner and out of our line of sight.)

Interference is the ability of two or more waves to reinforce or cancel each other. The second figure shows two identical waves moving through the same region of space. In the first part, the waves are positioned so that their crests and troughs exactly coincide. The net effect is that the two wave motions reinforce each other, resulting in a wave of greater amplitude. This phenomenon is known as *constructive interference*. In the second part of the figure, the two waves exactly cancel, so no net motion remains. This effect is known as *destructive interference*. As with diffraction, interference between waves of visible light is not noticeable in everyday experience; however, today it is easily measured in the laboratory.

Finally, the phenomenon known as *polarization* of light is also readily understood in terms of the description of electromagnetic waves presented in the text. Normally, light waves are randomly oriented—the electric field in Figure 3.7 may vibrate in any direction perpendicular to the direction of wave motion—and we say that the radiation is unpolarized. Most natural objects emit unpolarized radiation. Under some circumstances, however, the electric fields can become aligned, all vibrating in the same plane as the radiation moves through space, and the radiation is said to be polarized. On Earth, we can produce polarized light by passing unpolarized light through a Polaroid™ filter, which has specially aligned elongated molecules that allow the passage of only those waves having electric fields oriented in some specific direction. Reflected light is often polarized, which is why sunglasses constructed with suitably oriented Polaroid™ filters can be effective in blocking reflected glare.

Diffraction and interference play critical roles in many areas of observational astronomy, including telescope design (Chapter 5). The polarization of starlight provides astronomers with an important technique for probing the properties of interstellar gas (Chapter 18). All three phenomena are predicted by the wave theory of light. The particle theory did not predict them; in fact, it predicted that they should *not* occur. Until the early 1800s, the technology was inadequate to resolve the issue. However, by 1830, experimenters had reported the unequivocal measurement of each, convincing most scientists that the wave theory was the proper description of electromagnetic radiation. It would be almost a century before the particle description of radiation would resurface, but in a radically different form, as we will see in Chapter 4.

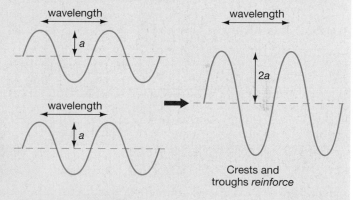

Constructive interference

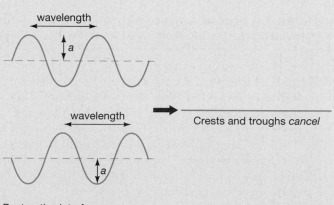

Destructive interference

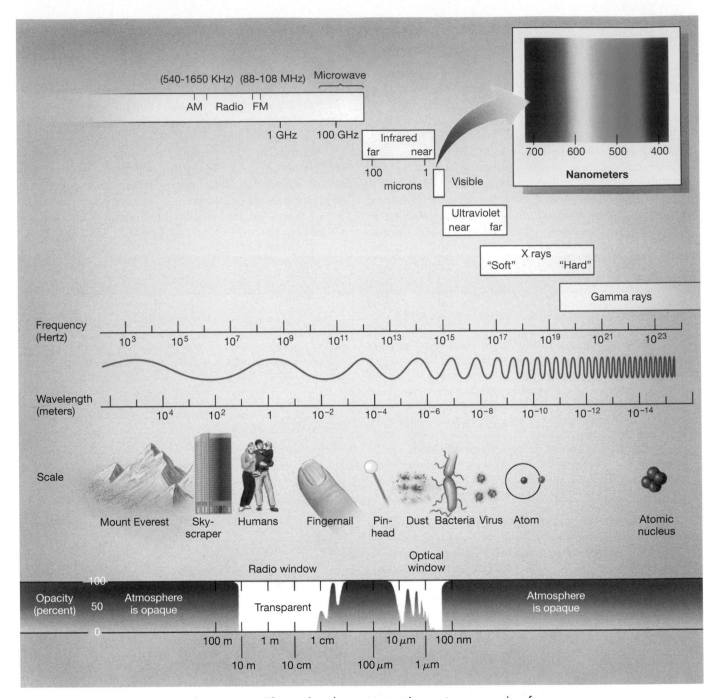

▲ **FIGURE 3.9 Electromagnetic Spectrum** The entire electromagnetic spectrum, running from long-wavelength, low-frequency radio waves to short-wavelength, high-frequency gamma rays.

Figure 3.9 is worth studying carefully, as it contains a great deal of information. Note that wave frequency (in hertz) increases from left to right, and wavelength (in meters) increases from right to left. Scientists often disagree on the "correct" way to display wavelengths and frequencies in diagrams of this type. In picturing wavelengths and frequencies, this book consistently adheres to the convention that frequency increases toward the *right*.

Notice also that the wavelength and frequency scales in Figure 3.9 do not increase by equal increments of 10.

Instead, successive values marked on the horizontal axis differ by *factors of 10*—each is 10 times greater than its neighbor. This type of scale, called a *logarithmic* scale, is often used in science to condense a large range of some quantity into a manageable size. Had we used a linear scale for the wavelength range shown in the figure, it would have been many light-years long! Throughout the text, we will often find it convenient to use a logarithmic scale to compress a wide range of some quantity onto a single, easy-to-view plot.

Figure 3.9 shows that wavelengths extend from the size of mountains (radio radiation) to the size of an atomic nucleus (gamma-ray radiation). The box at the upper right emphasizes how small the visible portion of the electromagnetic spectrum is. Most objects in the universe emit large amounts of invisible radiation. Indeed, many of them emit only a tiny fraction of their total energy in the visible range. A wealth of extra knowledge can be gained by studying the invisible regions of the electromagnetic spectrum.

To remind you of this important fact and to identify the region of the electromagnetic spectrum in which a particular observation was made, we have attached a spectrum icon—an idealized version of the wavelength scale in Figure 3.9—to every astronomical image presented in this text. Hence, we can tell at a glance from the highlighted "V" that, for example, Figure 3.1 (p. 62) is an image made with the use of visible light, while the first image in Figure 3.12 was captured in the radio ("R") part of the spectrum. Chapter 5 discusses in more detail how astronomers actually make such observations, using telescopes and sensitive detectors tailored to different electromagnetic waves.

Only a small fraction of the radiation produced by astronomical objects actually reaches Earth's surface, because of the *opacity* of our planet's atmosphere. Opacity is the extent to which radiation is blocked by the material through which it is passing—in this case, air. The more opaque an object is, the less radiation gets through it: Opacity is just the opposite of transparency. Earth's atmospheric opacity is plotted along the wavelength and frequency scales at the bottom of Figure 3.9. The extent of shading is proportional to the opacity. Where the shading is greatest (such as at the X-ray or "far" infrared regions of the spectrum), no radiation can get in or out. Where there is no shading at all (in the optical and part of the radio domain), the atmosphere is almost completely transparent. In some parts of the spectrum (e.g., the microwave band and much of the infrared portion), Earth's atmosphere is partly transparent, meaning that some, but not all, incoming radiation makes it to the surface.

What causes opacity to vary along the spectrum? Certain atmospheric gases absorb radiation very efficiently at some wavelengths. For example, water vapor (H_2O) and oxygen (O_2) absorb radio waves having wavelengths less than about a centimeter, while water vapor and carbon dioxide (CO_2) are strong absorbers of infrared radiation. Ultraviolet, X-ray, and gamma-ray radiation are completely blocked by the *ozone* (O_3) *layer* high in Earth's atmosphere (see Section 7.3). A passing, but unpredictable, source of atmospheric opacity in the visible part of the spectrum is the blockage of light by atmospheric clouds.

In addition, the interaction between the Sun's ultraviolet radiation and the upper atmosphere produces a thin, electrically conducting layer at an altitude of about 100 km. The *ionosphere*, as this layer is known, reflects long-wavelength radio waves (wavelengths greater than about 10 m) as well as a mirror reflects visible light. In this way, extraterrestrial waves are kept out, and terrestrial waves—such as those produced by AM radio stations—are kept in. (That's why it is possible to transmit some radio frequencies beyond the horizon—the broadcast waves bounce off the ionosphere.)

The effect of atmospheric opacity is that there are only a few *spectral windows*, at well-defined locations in the electromagnetic spectrum, where Earth's atmosphere is transparent. In much of the radio domain and in the visible portions of the spectrum, the opacity is low and we can study the universe at those wavelengths from ground level. In parts of the infrared range, the atmosphere is partially transparent, so we can make certain infrared observations from the ground. Moving to the tops of mountains, above as much of the atmosphere as possible, improves observations. In the rest of the spectrum, however, the atmosphere is opaque: Ultraviolet, X-ray, and gamma-ray observations can be made only from above the atmosphere, from orbiting satellites.

CONCEPT CHECK

✔ In what sense are radio waves, visible light, and X rays one and the same phenomenon?

3.4 Thermal Radiation

All macroscopic objects—fires, ice cubes, people, stars—emit radiation at all times, regardless of their size, shape, or chemical composition. They radiate mainly because the microscopic charged particles they are made up of are in constantly varying random motion, and whenever charges interact ("collide") and change their state of motion, electromagnetic radiation is emitted. The **temperature** of an object is a direct measure of the amount of microscopic motion within it. (See *More Precisely 3-1*.) The hotter the object—that is, the higher its temperature—the faster its component particles move, the more violent are their collisions, and the more energy they radiate.

The Blackbody Spectrum

Intensity is a term often used to specify the amount or strength of radiation at any point in space. Like frequency and wavelength, intensity is a basic property of radiation. No natural object emits all its radiation at just one frequency. Instead, because particles collide at many different speeds—some gently, others more violently—the energy is generally spread out over a range of frequencies. By studying how the intensity of this radiation is distributed across the electromagnetic spectrum, we can learn much about the object's properties.

Figure 3.10 sketches the distribution of radiation emitted by an object. The curve peaks at a single, well-defined frequency and falls off to lesser values above and below that frequency. Note that the curve is not shaped like a symmetrical bell that declines evenly on either side of the peak. Instead, the intensity falls off more slowly from the peak to lower frequencies than it does on the high-frequency side. This overall shape is characteristic of the thermal radiation emitted by *any* object, regardless of its size, shape, composition, or temperature.

The curve drawn in Figure 3.10(a) is the radiation-distribution curve for a mathematical idealization known as a *blackbody*—an object that absorbs all radiation falling on it. In a steady state, a blackbody must reemit the same amount of energy it absorbs. The **blackbody curve** shown in the figure describes the distribution of that reemitted radiation. (The curve is also known as the *Planck curve*, after Max Planck, the German physicist whose mathematical analysis of such thermal emission in 1900 played a key role in the development of modern physics.) No real object absorbs and radiates as a perfect blackbody. For example, the Sun's actual curve of emission is shown in Figure 3.10(b). However, in many cases, the blackbody curve is a good approximation to reality, and the properties of blackbodies provide important insights into the behavior of real objects.

The Radiation Laws

LEARNING GOAL 4 The blackbody curve shifts toward higher frequencies (shorter wavelengths) and greater intensities as an object's temperature increases. Even so, the *shape* of the curve remains the same. This shifting of radiation's peak frequency with temperature is familiar to us all: Very hot glowing objects, such as toaster filaments or stars, emit visible light. Cooler objects, such as warm rocks, household radiators, or people, produce invisible radiation—warm to the touch, but not glowing hot to the eye. These latter objects emit most of their radiation in the lower frequency infrared part of the electromagnetic spectrum (Figure 3.9).

MORE PRECISELY 3-1

The Kelvin Temperature Scale

The atoms and molecules that make up any piece of matter are in constant random motion. This motion represents a form of energy known as *thermal energy*, or, more commonly, *heat*. The quantity we call *temperature* is a direct measure of an object's internal motion: The higher the object's temperature, the faster, on average, is the random motion of its constituent particles. The temperature of a piece of matter specifies the average thermal energy of the particles it contains.

Our familiar Fahrenheit temperature scale, like the archaic English system in which length is measured in feet and weight in pounds, is of somewhat dubious value. In fact, the "degree Fahrenheit" is now a peculiarity of American society. Most of the world uses the Celsius scale of temperature measurement (also called the centigrade scale). In the Celsius system, water freezes at 0 degrees (0°C) and boils at 100 degrees (100°C), as illustrated in the accompanying figure.

There are, of course, temperatures below the freezing point of water. In principle, temperatures can reach as low as −273.15°C (although we know of no matter anywhere in the universe that is actually that cold). Known as *absolute zero*, this is the temperature at which, theoretically, all thermal atomic and molecular motion ceases. Since no object can have a temperature below that value, scientists find it convenient to use a temperature scale that takes absolute zero as its starting point. This scale is called the *Kelvin scale*, in honor of the 19th-century British physicist Lord Kelvin. Since it starts at absolute zero, the Kelvin scale differs from the Celsius scale by 273.15°. In this book, we round off the decimal places and simply use

$$\text{kelvins} = \text{degrees Celsius} + 273.$$

Thus,

- All thermal motion ceases at 0 kelvins (0 K).

- Water freezes at 273 kelvins (273 K).

- Water boils at 373 kelvins (373 K).

Note that the unit is "kelvins," or "K," *not* "degrees kelvin" or "°K." (Occasionally, the term "degrees absolute" is used instead.)

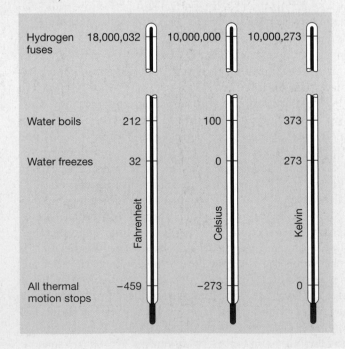

	Fahrenheit	Celsius	Kelvin
Hydrogen fuses	18,000,032	10,000,000	10,000,273
Water boils	212	100	373
Water freezes	32	0	273
All thermal motion stops	−459	−273	0

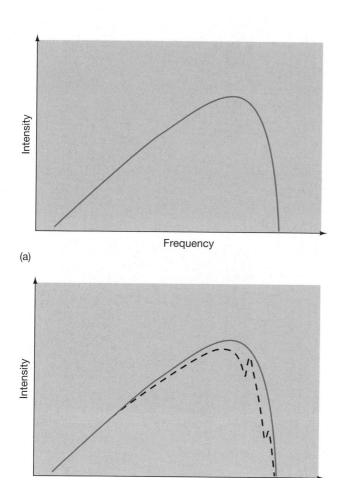

(a)

(b)

▲ FIGURE 3.10 Blackbody Curves, Ideal vs. Reality The blackbody, or Planck, curve represents the spread of the intensity of radiation emitted by any object over all possible frequencies. The clean, "textbook" case (a) contrasts with a real graph (dashed) of the Sun's emission (b). Absorption in the atmospheres of the Sun and Earth causes the difference.

Imagine a piece of metal placed in a hot furnace. At first, the metal becomes warm, although its visual appearance doesn't change. As it heats up, the metal begins to glow dull red, then orange, brilliant yellow, and finally white. How do we explain this phenomenon? As illustrated in Figure 3.11, when the metal is at room temperature (300 K—see *More Precisely 3-1* for a discussion of the Kelvin temperature scale), it emits only invisible infrared radiation. As the metal becomes hotter, the peak of its blackbody curve shifts toward higher frequencies. At 1000 K, for instance, most of the emitted radiation is still infrared, but now there is also a small amount of visible (dull red) radiation being emitted. (Note that the high-frequency portion of the 1000 K curve just overlaps the visible region of the graph.)

As the temperature continues to rise, the peak of the metal's blackbody curve moves through the visible spectrum, from red (the 4000 K curve) through yellow. Even-

tually, the metal becomes white hot because, when its blackbody curve peaks in the blue or violet part of the spectrum (the 7000 K curve), the low-frequency tail of the curve extends through the entire visible spectrum (to the left in the figure), meaning that substantial amounts of green, yellow, orange, and red light are also emitted. Together, all these colors combine to produce white.

From studies of the precise form of the blackbody curve, we obtain a very simple connection between the wavelength at which most radiation is emitted and the absolute temperature (i.e., the temperature measured in kelvins) of the emitting object:

$$\text{wavelength of peak emission} \propto \frac{1}{\text{temperature}}.$$

(Recall that the symbol "$\propto$" here just means "is proportional to.") This relationship is called **Wien's law**, after Wilhelm Wien, the German scientist who formulated it in 1897.

Simply put, Wien's law tells us that the hotter the object, the bluer is its radiation. For example, an object with a temperature of 6000 K emits most of its energy in the visible part of the spectrum, with a peak wavelength of 480 nm. At 600 K, the object's emission would peak at a wavelength of 4800 nm, well into the infrared portion of the spectrum. At a temperature of 60,000 K, the peak would move all the way through the visible spectrum to a

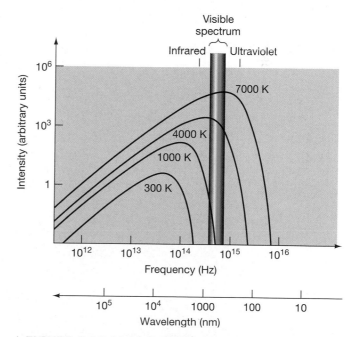

▲ FIGURE 3.11 Multiple Blackbody Curves As an object is heated, the radiation it emits peaks at higher and higher frequencies. Shown here are curves corresponding to temperatures of 300 K (room temperature), 1000 K (beginning to glow dull red), 4000 K (red hot), and 7000 K (white hot).

TUTORIAL The Planck Spectrum

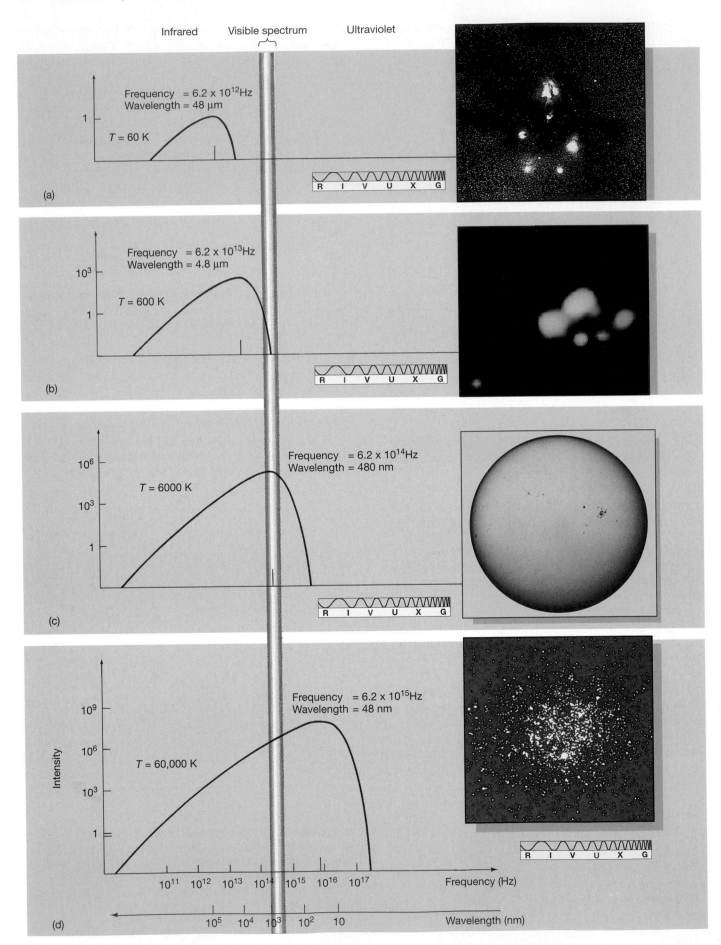

Infrared   Visible spectrum   Ultraviolet

(a)
Frequency = 6.2 x 10^{12}Hz
Wavelength = 48 μm

T = 60 K

1

R I V U X G

(b)
Frequency = 6.2 x 10^{13}Hz
Wavelength = 4.8 μm

T = 600 K

10^3
1

R I V U X G

(c)
Frequency = 6.2 x 10^{14}Hz
Wavelength = 480 nm

T = 6000 K

10^6
10^3
1

R I V U X G

(d)
Frequency = 6.2 x 10^{15}Hz
Wavelength = 48 nm

T = 60,000 K

Intensity

10^9
10^6
10^3
1

R I V U X G

10^{11} 10^{12} 10^{13} 10^{14} 10^{15} 10^{16} 10^{17} Frequency (Hz)

10^5 10^4 10^3 10^2 10   Wavelength (nm)

◀ **FIGURE 3.12 Astronomical Thermometer** Comparison of blackbody curves of four cosmic objects. The frequencies and wavelengths corresponding to peak emission are marked. (a) A cool, invisible galactic gas cloud called Rho Ophiuchi. At a temperature of 60 K, it emits mostly low-frequency radio radiation. (b) A dim, young star (shown red in the inset photograph) near the center of the Orion Nebula. The star's atmosphere, at 600 K, radiates primarily in the infrared, here falsely colored to represent differences in temperature. (c) The Sun's surface, at approximately 6000 K, is brightest in the visible region of the electromagnetic spectrum. (d) Some very hot, bright stars in a cluster called Omega Centauri, as observed by a telescope aboard a space shuttle. At a temperature of 60,000 K, these stars radiate strongly in the ultraviolet. *(Harvard College Observatory; J. Moran; AURA; NASA)*

wavelength of 48 nm, in the ultraviolet range. (See Figure 3.12.)

It is also a matter of everyday experience that, as the temperature of an object increases, the *total* amount of energy it radiates (summed over all frequencies) increases rapidly. For example, the heat given off by an electric heater increases very sharply as it warms up and begins to emit visible light. Careful experimentation leads to the conclusion that the total amount of energy radiated per unit time is actually proportional to the fourth power of the object's temperature:

$$\text{total energy emission} \propto \text{temperature}^4.$$

This relation is called **Stefan's law**, after the 19th-century Austrian physicist Josef Stefan. From the form of Stefan's law, we can see that the energy emitted by a body rises dramatically as its temperature increases. Doubling the temperature causes the total energy radiated to increase by a factor of $2^4 = 16$; tripling the temperature increases the emission by $3^4 = 81$, and so on.

The radiation laws are presented in more detail in *More Precisely 3-2*.

Astronomical Applications

No known natural terrestrial objects reach temperatures high enough to emit very high frequency radiation. Only human-made thermonuclear explosions are hot enough for their spectra to peak in the X-ray or gamma-ray range. (Most human inventions that produce short-wavelength, high-frequency radiation, such as X-ray machines, are designed to emit only a specific range of wavelengths and do not operate at high temperatures. They are said to produce a *nonthermal* spectrum of radiation.) Many extraterrestrial objects, however, do emit copious quantities of ultraviolet, X-ray, and even gamma-ray radiation. Figure 3.13 shows a familiar object—our Sun—as it appears when viewed with the use of radiation from different parts of the electromagnetic spectrum.

Astronomers often use blackbody curves as thermometers to determine the temperatures of distant objects. For example, an examination of the solar spectrum indicates the temperature of the Sun's surface. Observations of the radiation from the Sun at many frequencies yield a curve shaped somewhat like that shown in Figure

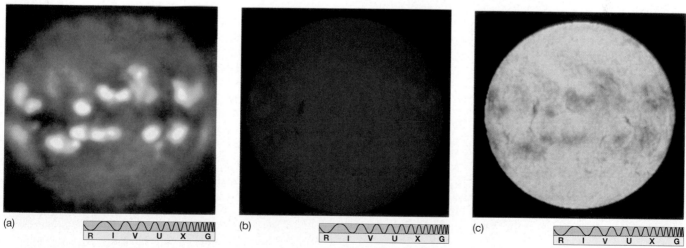

▲ **FIGURE 3.13 The Sun at Many Wavelengths** Three images of the Sun, obtained using telescopes sensitive to (a) radio waves (note the highlighted "R" in the spectrum icon), (b) infrared radiation ("I"), and (c) visible light ("V"). These images are shown here in "false color," a technique commonly used for displaying intensity, especially with nonvisible radiation. By studying the similarities and differences among various views of the same object acquired on the same day, astronomers can find important clues to the object's structure, composition, and surface activity. Although most sunlight is emitted in the form of infrared and visible radiation, a wealth of information about our parent star can be obtained by studying it in other regions of the electromagnetic spectrum. *(NRAO; AURA)*

3.10. The Sun's curve peaks in the visible part of the electromagnetic spectrum; the Sun also emits a lot of infrared and a little ultraviolet radiation. Using Wien's law, we find that the temperature of the Sun's surface is approximately 6000 K. (A more precise measurement, applying Wien's law to the blackbody curve that best fits the solar spectrum, yields a temperature of 5800 K.)

Other cosmic objects have surfaces very much cooler or hotter than the Sun's, emitting most of their radiation in invisible parts of the spectrum. For example, the relatively cool surface of a very young star may measure 600 K and emit mostly infrared radiation. Cooler still is the interstellar gas cloud from which the star formed; at a temperature of 60 K, such a cloud emits mainly long-

---

MORE PRECISELY 3-2

More about the Radiation Laws

As mentioned in Section 3.4, Wien's law relates the temperature T of an object to the wavelength λ_{max} at which the object emits the most radiation. (The Greek letter λ—lambda—is conventionally used to denote wavelength.) Mathematically, if we measure T in kelvins and λ_{max} in centimeters, we can determine the constant of proportionality in the relation presented in the text, to find that

$$\lambda_{max} = \frac{0.29 \text{ cm}}{T}.$$

We could also convert Wien's law into an equivalent statement about frequency f, using the relation $f = c/\lambda$ (see Section 3.1), where c is the speed of light, but the law is most commonly stated in terms of wavelength and is probably easier to remember that way.

EXAMPLE 1: For a blackbody with the same temperature T as the surface of the Sun ($\approx$6000 K), the wavelength of maximum intensity is $\lambda_{max} = (0.29/6000)$ cm, or 480 nm, corresponding to the yellow-green part of the visible spectrum. A cooler star with a temperature of $T = 3000$ K has a peak wavelength of $\lambda_{max} = (0.29/3000)$ cm ≈ 970 nm, just beyond the red end of the visible spectrum, in the near infrared. The blackbody curve of a hotter star with a temperature of 12,000 K peaks at 242 nm, in the near ultraviolet, and so on.

In fact, this application—simply looking at the spectrum and determining where it peaks—is an important way of estimating the temperature of planets, stars, and other objects throughout the universe and will be used extensively throughout the text.

We can also give Stefan's law a more precise mathematical formulation. With T measured in kelvins, the total amount of energy emitted per square meter of the body's surface per second (a quantity known as the *energy flux F*) is given by

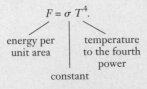

$$F = \sigma T^4.$$

energy per unit area | temperature to the fourth power

constant

This equation is usually referred to as the *Stefan–Boltzmann* equation. Stefan's student, Ludwig Boltzmann, was an Austri-

an physicist who played a central role in the development of the laws of thermodynamics during the late 19th and early 20th centuries. The constant σ (the Greek letter sigma) is known as the Stefan–Boltzmann constant.

The SI unit of energy is the *joule* (J). Probably more familiar is the closely related unit called the *watt* (W), which measures power—the *rate* at which energy is emitted or expended by an object. One watt is the emission of 1 joule per second. For example, a 100-W lightbulb emits energy (mostly in the form of infrared and visible light) at a rate of 100 J/s. In SI units, the Stefan–Boltzmann constant has the value $\sigma = 5.67 \times 10^{-8}$ W/m$^2 \cdot$K^4.

EXAMPLE 2: Notice just how rapidly the energy flux increases with increasing temperature. A piece of metal in a furnace, when at a temperature of $T = 3000$ K, radiates energy at a rate of $\sigma T^4 \times (1 \text{ cm})^2 = 5.67 \times 10^{-8}$ W/m$^2 \cdot$K$^4 \times (3000 \text{ K})^4 \times (0.01 \text{ m})^2 = 460$ W for every square centimeter of its surface area. Doubling this temperature to 6000 K, the surface temperature of the Sun (so that the metal becomes yellow hot, by Wien's law), increases the energy emitted by a factor of 16 (four "doublings"), to 7.3 *kilo*watts (7300 W) per square centimeter.

Finally, note that Stefan's law relates to energy emitted *per unit area*. The flame of a blowtorch is considerably hotter than a bonfire, but the bonfire emits far more energy *in total* because it is much larger. Thus, in computing the total energy emitted from a hot object, both the object's temperature *and* its surface area must be taken into account. This fact is of great importance in determining the "energy budget" of planets and stars, as we will see in later chapters. The next example illustrates the point in the case of the Sun.

EXAMPLE 3: The Sun's temperature is approximately $T = 5800$ K. (The earlier example used a rounded-off version of this number.) Thus, by Stefan's law, each square meter of the Sun's surface radiates energy at a rate of $\sigma T^4 = 6.4 \times 10^7$ W (64 megawatts). By measuring the Sun's angular size and knowing its distance, we can employ simple geometry to determine the solar radius. ∞ (Secs. 1.7, 2.6) The answer is $R = 700,000$ km, or 7×10^8 m, allowing us to calculate the Sun's total surface area as $4\pi R^2 = 6.2 \times 10^{18}$ m^2. Multiplying by the energy emitted per unit area, we find that the Sun's total energy emission (or *luminosity*) is 4×10^{26} W—a remarkable number that we obtained without ever leaving Earth!

wavelength radiation in the radio and infrared parts of the spectrum. The brightest stars, by contrast, have surface temperatures as high as 60,000 K and hence emit mostly ultraviolet radiation. (See Figure 3.12.)

CONCEPT CHECK

✔ Describe, in terms of the radiation laws, how and why the appearance of an incandescent lightbulb changes as you turn a dimmer switch to increase its brightness from "off" to "maximum."

3.5 The Doppler Effect

Imagine a rocket ship launched from Earth with enough fuel to allow it to accelerate to speeds approaching that of light. As the ship's speed increased, a remarkable thing would happen (Figure 3.14). Passengers would notice that the light from the star system toward which they were traveling seemed to be getting *bluer*. In fact, *all* stars in front of the ship would appear bluer than normal, and the greater the ship's speed, the greater the color change would be. Furthermore, stars behind the vessel would seem *redder* than normal, while stars to either side would be unchanged in appearance. As the spacecraft slowed down and came to rest relative to Earth, all stars would resume their usual appearance. The travelers would have to conclude that the stars had changed their colors not because of any real change in their physical properties, but because of the spacecraft's own *motion*.

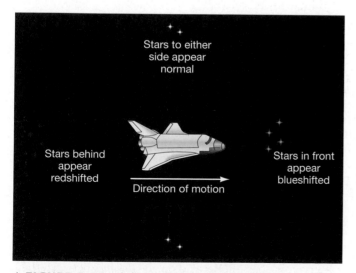

▲ **FIGURE 3.14 High-Speed Observers** Observers in a fast-moving spacecraft will see the stars ahead of them seem bluer than normal, while those behind are reddened. The stars have not changed their properties— the color changes are the result of the observers' motion relative to the stars.

This phenomenon is not restricted to electromagnetic radiation and fast-moving spacecraft. Waiting at a railroad crossing for an express train to pass, most of us have had the experience of hearing the pitch of a train whistle change from high shrill (high frequency, short wavelength) to low blare (low frequency, long wavelength) as the train approaches and then recedes. This motion-induced change in the observed frequency of a wave is known as the **Doppler effect**, in honor of Christian Doppler, the 19th-century Austrian physicist who first explained it in 1842. Applied to cosmic sources of electromagnetic radiation, it has become one of the most important measurement techniques in all of modern astronomy. Here's how it works:

Imagine a wave moving from the place where it is created toward an observer who is not moving with respect to the source of the wave, as shown in Figure 3.15(a). By noting the distances between successive crests, the observer can determine the wavelength of the emitted wave. Now suppose that not just the wave, but the *source* of the wave, also is moving. As illustrated in Figure 3.15(b), because the source moves between the times of emission of one crest and the next, successive crests in the direction of motion of the source will be seen to be *closer together* than normal, whereas crests behind the source will be more widely spaced. An observer in front of the source will therefore measure a *shorter* wavelength than normal, while one behind will see a *longer* wavelength. (The numbers indicate (a) successive crests emitted by the source and (b) the location of the source at the instant each crest was emitted.)

The greater the relative speed between source and observer, the greater is the observed shift. If the other velocities involved are not too large compared with the wave speed—less than a few percent, say—we can write down a particularly simple formula for what the observer sees. In terms of the net velocity of *recession* between source and observer, the apparent wavelength and frequency (measured by the observer) are related to the true quantities (emitted by the source) as follows:

$$\frac{\text{apparent wavelength}}{\text{true wavelength}} = \frac{\text{true frequency}}{\text{apparent frequency}}$$

$$= 1 + \frac{\text{recession velocity}}{\text{wave speed}}.$$

The recession velocity measures the rate at which the distance between the source and the observer is changing. A positive recession velocity means that the two are moving apart; a negative velocity means that they are approaching.

The wave speed is the speed of light, c, in the case of electromagnetic radiation. For most of this text, the assumption that the recession velocity is small compared to the speed of light will be a good one. Only when we discuss the properties of black holes (Chapter 22) and the structure

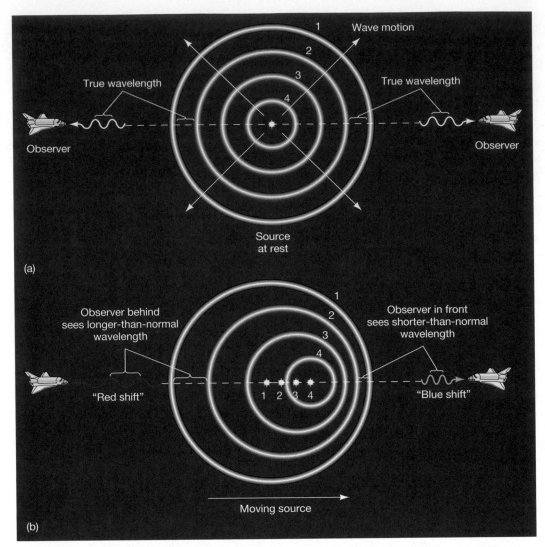

▲ FIGURE 3.15 Doppler Effect (a) Wave motion from a source toward an observer at rest with respect to the source. The four numbered circles represent successive wave crests emitted by the source. At the instant shown, the fifth crest is just about to be emitted. As seen by the observer, the source is not moving, so the wave crests are just concentric spheres (shown here as circles). (b) Waves from a moving source tend to "pile up" in the direction of motion and be "stretched out" on the other side. (The numbered points indicate the location of the source at the instant each wave crest was emitted.) As a result, an observer situated in front of the source measures a shorter-than-normal wavelength—a "blueshift"—while an observer behind the source sees a "redshift." In this diagram, the source is shown in motion. However, the same general statements hold whenever there is any relative motion between source and observer.

of the universe on the largest scales (Chapters 25 and 26) will we have to reconsider this formula.

Note that in Figure 3.15 the *source* is shown in motion (as in our train analogy), whereas in our earlier spaceship example (Figure 3.14) the *observers* were in motion. For electromagnetic radiation, the result is the same in either case—only the *relative* motion between source and observer matters. Note also that only motion along the line joining source and observer—known as *radial* motion—appears in the foregoing equation. Motion that is

transverse (perpendicular) to the line of sight has no significant effect.* Notice, incidentally, that the Doppler effect depends only on the *relative motion* between source and observer; it does not depend on the *distance* between them in any way.

---

*In fact, Einstein's theory of relativity (see Chapter 22) implies that when the transverse velocity is comparable to the speed of light, a change in wavelength, called the transverse Doppler shift, does occur. For most terrestrial and astronomical applications, however, this shift is negligibly small, and we will ignore it here.

A wave measured by an observer situated in front of a moving source is said to be *blueshifted*, because blue light has a shorter wavelength than red light. Similarly, an observer situated behind the source will measure a longer-than-normal wavelength—the radiation is said to be *redshifted*. This terminology is used even for invisible radiation, for which "red" and "blue" have no meaning. Any shift toward shorter wavelengths is called a blueshift, and any shift toward longer wavelengths is called a redshift. For example, ultraviolet radiation might be blueshifted into the X-ray part of the spectrum or redshifted into the visible; infrared radiation could be redshifted into the microwave range, and so on.

Because *c* is so large—300,000 km/s—the Doppler effect is extremely small for everyday terrestrial velocities. For example, consider a source receding from the observer at Earth's orbital speed of 30 km/s, a velocity much greater than any encountered in day-to-day life. A beam of blue light would be shifted by only 30 km/s/300,000 km/s = 0.01 percent, from 400 nm to 400.04 nm—a very small change indeed, and one that the human eye cannot distinguish. (It is easily detectable with modern instruments, though.)

The importance of the Doppler effect to astronomers is that it allows them to find the speed of any cosmic object along the line of sight simply by measuring the extent to which its light is redshifted or blueshifted. Suppose that the beam of blue light just mentioned is observed to have a wavelength of 401 nm, instead of the 400 nm with which it was emitted. (Let's defer until the next chapter the question of *how* an observer might know the wavelength of the emitted light.) Using the earlier equation, the observer could calculate the source's recession velocity to be 401/400 − 1 = 0.0025 times the speed of light. In other words, the source is receding from the observer at a speed of 0.0025*c*, or 750 km/s. The basic reasoning is simple, but very powerful. The motions of nearby stars and distant galaxies—even the expansion of the universe itself—have all been measured in this way.

Motorists stopped for speeding on the highway have experienced another, much more down-to-earth, application: Police radar measures speed by means of the Doppler effect, as do the radar guns used to clock the velocity of a pitcher's fastball or a tennis player's serve.

In practice, it is hard to measure the Doppler shift of an entire blackbody curve, simply because it is spread over many wavelengths, making small shifts hard to determine with any accuracy. However, if the radiation were more narrowly defined and took up just a narrow "sliver" of the spectrum, then precise measurements of Doppler effect *could* be made. We will see in the next chapter that in many circumstances this is precisely what does happen, making the Doppler effect one of the observational astronomer's most powerful tools.

CONCEPT CHECK

✔ Astronomers observe two stars orbiting one another. How might the Doppler effect be useful in determining the masses of the stars?

Chapter Review

SUMMARY

Visible light (p. 62) is a particular type of **electromagnetic radiation (p. 62)** and travels through space in the form of a **wave (p. 63)**. A wave is characterized by its **period (p. 63)**, the length of time taken for one complete cycle; its **wavelength (p. 63)**, the distance between successive wave crests; and its **amplitude (p. 64)**, which measures the size of the disturbance associated with the wave. A wave's **frequency (p. 64)** is the reciprocal of the period—it counts the number of wave crests that pass a given point in one second.

Electrons (p. 65) and **protons (p. 65)** are elementary particles that carry equal and opposite electrical charges. Any electrically charged object is surrounded by an **electric field (p. 66)** that determines the force the object exerts on other charged objects. Like the gravitational force, electric fields decrease as the square of the distance from their source. When a charged particle moves, information about its motion is transmitted throughout the universe by the particle's changing electric and magnetic fields. According to the **wave theory of radiation (p. 67)**, the information travels through space in the form of a wave at the **speed of light (p. 67)**. Because both electric and **magnetic fields (p. 66)** are involved, the phenomenon is known as **electromagnetism (p. 66)**. Diffraction, interference, and polarization are properties of radiation that mark it as a wave phenomenon.

A beam of white light is bent, or refracted, as it passes through a prism. Different frequencies of light within the beam are refracted by different amounts, so the beam is split up into its component colors—the visible spectrum. The color of visible light is simply a measure of its wavelength—red light has a longer wavelength than blue light. The entire **electromagnetic spectrum (p. 69)** consists of (in order of increasing frequency) **radio waves**, **infrared radiation**, visible light, **ultraviolet radiation**, **X rays**, and **gamma rays (p. 62)**. The opacity of Earth's atmosphere—the extent to which it absorbs radiation—varies greatly with wavelength. Only radio waves, some infrared wavelengths, and visible light can penetrate the atmosphere and reach the ground from space.

The **temperature (p. 71)** of an object is a measure of the speed with which its constituent particles move. The intensity of radiation of different frequencies emitted by a hot object has a characteristic distribution, called a **blackbody curve (p. 72)**, that depends only on the temperature of the object. **Wien's law (p. 73)** tells us that the wavelength at which the object radiates most of its energy is inversely proportional to the temperature of the object. **Stefan's law (p. 75)** states that the total amount of energy radiated is proportional to the fourth power of the temperature.

Our perception of the wavelength of a beam of light can be altered by our velocity relative to the source. This motion-induced change in the observed frequency of a wave is called the **Doppler effect (p. 77)**. Any net motion away from the source causes a redshift—a shift to lower frequencies—in the received beam. Motion toward the source causes a blueshift. The extent of the shift is directly proportional to the observer's recession velocity relative to the source.

REVIEW AND DISCUSSION

1. What is a wave?

2. Define the following wave properties: period, wavelength, amplitude, and frequency.

3. What is the relationship between wavelength, wave frequency, and wave velocity?

4. What is diffraction, and how does it relate to the behavior of light as a wave?

5. What's so special about c?

6. Name the colors that combine to make white light. What is it about the various colors that causes us to perceive them differently?

7. What effect does a positive charge have on a nearby negatively charged particle?

8. Compare and contrast the gravitational force with the electric force.

9. Describe the way in which light leaves a star, travels through the vacuum of space, and finally is seen by someone on Earth.

10. Why is light referred to as an electromagnetic wave?

11. What do radio waves, infrared radiation, visible light, ultraviolet radiation, X rays, and gamma rays have in common? How do they differ?

12. In what regions of the electromagnetic spectrum is the atmosphere transparent enough to allow observations from the ground?

13. What is a blackbody? What are the main characteristics of the radiation it emits?

14. What does Wien's law reveal about stars in the sky?

15. What does Stefan's law tell us about the radiation emitted by a blackbody?

16. In terms of its blackbody curve, describe what happens as a red-hot glowing coal cools.

17. What is the Doppler effect, and how does it alter the way in which we perceive radiation?

18. How do astronomers use the Doppler effect to determine the velocities of astronomical objects?

19. A source of radiation and an observer are traveling through space at precisely the same velocity, as seen by a second observer. Would you expect the first observer to measure a Doppler shift in the light received from the source?

20. If Earth were completely blanketed with clouds and we couldn't see the sky, could we learn about the realm beyond the clouds? What forms of radiation might be received?

CONCEPTUAL SELF-TEST: TRUE OR FALSE/MULTIPLE CHOICE

1. Electromagnetic waves can travel through a perfect vacuum.

2. Sound is a familiar type of electromagnetic wave.

3. Interference occurs when one wave is brighter than another and the fainter wave cannot be observed.

4. A blackbody emits all its radiation at a single frequency.

5. Earth's atmosphere is transparent to all forms of electromagnetic radiation.

6. The peak of an object's emitted radiation occurs at a frequency determined by the object's temperature.

7. The lowest possible temperature is 0 K.

8. Two otherwise identical objects have temperatures of 1000 K and 1200 K, respectively. The object at 1200 K emits roughly twice as much radiation as the object at 1000 K.

9. As you drive away from a radio transmitter, the radio signal you receive from the station is shifted to longer wavelengths.

10. The Doppler effect occurs for all types of wave motion.

11. Compared with ultraviolet radiation, infrared radiation has a greater **(a)** wavelength; **(b)** amplitude; **(c)** frequency; **(d)** energy.

12. Compared with red light, blue wavelengths of visible light travel **(a)** faster; **(b)** slower; **(c)** at the same speed.

13. An electron that collides with an atom will **(a)** cease to have an electric field; **(b)** produce an electromagnetic wave; **(c)** change its electric charge; **(d)** become magnetized.

14. A wavelength of green light is about the size of **(a)** an atom; **(b)** a bacterium; **(c)** a fingernail; **(d)** a skyscraper.

15. An X-ray telescope located in Antarctica would not work well because of **(a)** the extreme cold; **(b)** the ozone hole; **(c)** continuous daylight; **(d)** Earth's atmosphere.

16. In Figure 3.11 ("Blackbody Curves"), an object at 1000 K emits mostly **(a)** infrared light; **(b)** red light; **(c)** green light; **(d)** blue light.

17. According to Wien's law, the hottest stars also have **(a)** the longest peak wavelength; **(b)** the shortest peak wavelength; **(c)** maximum emission in the infrared region of the spectrum; **(d)** the largest diameters.

18. Stefan's law says that if the Sun's temperature were to double, its energy emission would **(a)** become half its present value; **(b)** double; **(c)** increase four times; **(d)** increase 16 times.

19. A star much cooler than the Sun would appear **(a)** red; **(b)** blue; **(c)** smaller; **(d)** larger.

20. The blackbody curve of a star moving toward Earth would have its peak shifted **(a)** to a higher intensity; **(b)** toward higher energies; **(c)** toward longer wavelengths; **(d)** to a lower intensity.

PROBLEMS

 Algorithmic versions of these questions are available in the Practice Problems module of the Companion Website at astro.prenhall.com/chaisson.

The number of squares preceding each problem indicates its approximate level of difficulty.

1. ■ A sound wave moving through water has a frequency of 256 Hz and a wavelength of 5.77 m. What is the speed of sound in water?

2. ■ What is the wavelength of a 100-MHz ("FM 100") radio signal?

3. ■ What would be the frequency of an electromagnetic wave having a wavelength equal to Earth's diameter? In what part of the electromagnetic spectrum would such a wave lie?

4. ■ Estimate the frequency of an electromagnetic wave having a wavelength equal to the size of the period at the end of this sentence. In what part of the electromagnetic spectrum would such a wave lie?

5. ■ What would be the wavelength of an electromagnetic wave having a frequency equal to the clock speed of a 3.2 GHz personal computer? In what part of the electromagnetic spectrum would such a wave lie?

6. ■ The blackbody emission spectrum of object *A* peaks in the ultraviolet region of the electromagnetic spectrum, at a wavelength of 200 nm. That of object *B* peaks in the red region, at 650 nm. Which object is hotter and, according to Wien's law, how many times hotter is it? According to Stefan's law, how many times more energy per unit area does the hotter body radiate per second?

7. ■ Normal human body temperature is about 37°C. What is this temperature in kelvins? What is the peak wavelength emitted by a person with this temperature? In what part of the spectrum does this lie?

8. ■■ Estimate the total amount of energy you radiate to your surroundings.

9. ■ The Sun has a temperature of 5800 K, and its blackbody emission peaks at a wavelength of approximately 500 nm. At what wavelength does a protostar with a temperature of 1000 K radiate most strongly?

10. ■ Two otherwise identical bodies have temperatures of 300 K and 1500 K, respectively. Which one radiates more energy, and by what factor does its emission exceed the emission of the other body?

11. ■■ According to the Stefan–Boltzmann law, how much energy is radiated into space per unit time by each square meter of the Sun's surface? (See *More Precisely 3-2*.) If the Sun's radius is 696,000 km, what is the total power output of the Sun?

12. ■ Radiation from the nearby star Alpha Centauri is observed to be reduced in wavelength (after correction for Earth's orbital motion) by a factor of 0.999933. What is the recession velocity of Alpha Centauri relative to the Sun?

13. ■ At what velocity and in what direction would a spacecraft have to be moving for a radio station on Earth transmitting at 100 MHz to be picked up by a radio tuned to 99.9 MHz?

14. ■■ A space traveler is approaching the Sun at a speed of 100 km/s and is observing a 700-nm red laser beam coming from Earth. If the traveler's trajectory lies in the same plane as Earth's orbit, what will be the minimum and maximum wavelengths he observes as Earth orbits the Sun?

15. ■■■ Imagine that you are observing a spacecraft moving in a circular orbit of radius 100,000 km around a distant planet. You happen to be located in the plane of the spacecraft's orbit. You find that the spacecraft's radio signal varies periodically in wavelength between 2.99964 m and 3.00036 m. Assuming that the radio is broadcasting normally, at a constant wavelength, what is the mass of the planet?

 In addition to the Practice Problems module, the Companion Website at astro.prenhall.com/chaisson provides for each chapter a study guide module with multiple choice questions as well as additional annotated images, animations, and links to related Websites.

Spectroscopy

The Inner Workings of Atoms

The wave description of radiation allowed 19th-century astronomers to begin to decipher the information reaching Earth from the cosmos in the form of visible and invisible light. However, early in the 20th century, it became clear that the wave theory of electromagnetic phenomena was incomplete—some aspects of light simply could not be explained in purely wave terms. When radiation interacts with matter on atomic scales, it does so not as a continuous wave, but in a jerky, discontinuous way—in fact, as a particle. With this discovery, scientists quickly realized that atoms, too, must behave in a discontinuous way, and the stage was set for a scientific revolution that has affected virtually every area of modern life. In astronomy, the observational and theoretical techniques that enable researchers to determine the nature of distant atoms by the way they emit and absorb radiation are now the indispensable foundation of modern astrophysics.

LEARNING GOALS

Studying this chapter will enable you to

1 Describe the characteristics of continuous, emission, and absorption spectra and the conditions under which each is produced.

2 Explain the relation between emission and absorption lines and what we can learn from those lines.

3 Specify the basic components of the atom and describe our modern conception of its structure.

4 Discuss the observations that led scientists to conclude that light has particle as well as wave properties.

5 Explain how electron transitions within atoms produce unique emission and absorption features in the spectra of those atoms.

6 Describe the general features of spectra produced by molecules.

7 List and explain the kinds of information that can be obtained by analyzing the spectra of astronomical objects.

Visit astro.prenhall.com/chaisson for additional annotated images, animations, and links to related sites for this chapter.

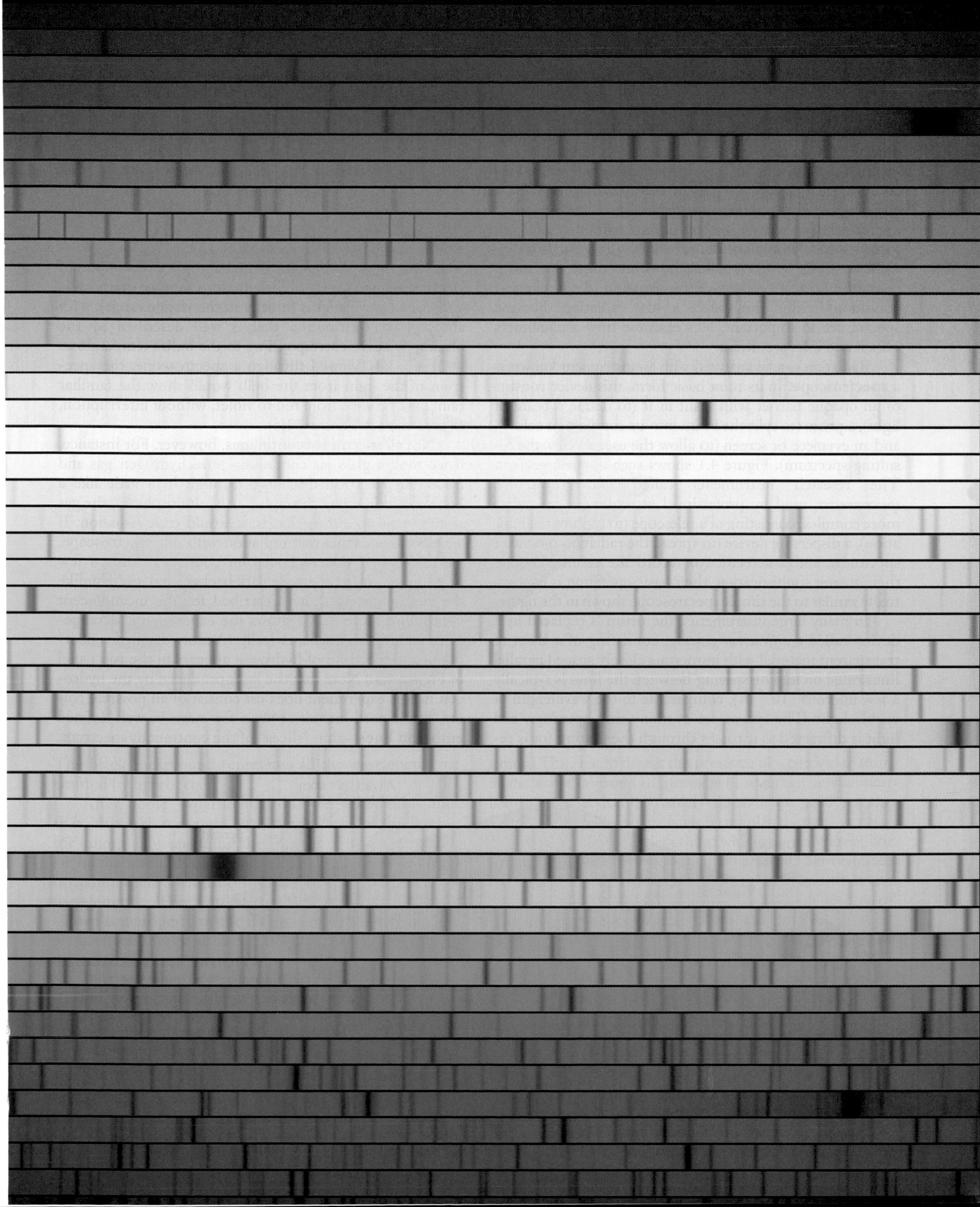

Telescopes are time machines and astronomers, in a sense, are historians. Telescopes aid our eyes and senses, enabling us not only to see much greater distances than are accessible to the naked eye alone, but also to perceive cosmic objects that emit radiation at wavelengths to which our human senses are completely blind. Without telescopes, we would know only a tiny fraction of the information presented in this textbook. Here, a 2-hour time exposure captures one of the world's premier optical telescopes available to astronomers today—the Gemini North Telescope atop Mauna Kea in Hawaii. (P. Michaud, Gemini, AURA) ▶

Telescopes

The Tools of Astronomy

At its heart, astronomy is an observational science. More often than not, observations of cosmic phenomena precede any clear theoretical understanding of their nature. As a result, our detecting instruments—our telescopes—have evolved to observe as broad a range of wavelengths as possible. Until the middle of the 20th century, telescopes were limited to collecting visible light. Since then, technological advances have expanded our view of the universe to all regions of the electromagnetic spectrum. Some telescopes are sited on Earth, whereas others must be placed in space, and design considerations vary widely from one part of the spectrum to another. Whatever the details of their construction, however, telescopes are devices whose basic purpose is to collect electromagnetic radiation and deliver it to a detector for detailed study.

LEARNING GOALS

Studying this chapter will enable you to

1 Sketch and describe the basic designs of the major types of optical telescopes used by astronomers.

2 Explain the purpose of some of the detectors used in astronomical telescopes.

3 Explain the particular advantages of reflecting telescopes for astronomical use, and specify why very large telescopes are needed for most astronomical studies.

4 Describe how Earth's atmosphere affects astronomical observations, and discuss some of the current efforts to improve ground-based astronomy.

5 Discuss the advantages and disadvantages of radio astronomy compared with optical observations.

6 Explain how interferometry can enhance the usefulness of astronomical observations.

7 Explain why some astronomical observations are best done from space, and discuss the advantages and limitations of space-based astronomy.

8 Tell why it is important to make astronomical observations in different regions of the electromagnetic spectrum.

 Visit astro.prenhall.com/chaisson for additional annotated images, animations, and links to related sites for this chapter.

5.1 Telescope Design

LEARNING GOAL 1 In essence, a **telescope** is a "light bucket" whose primary function is to capture as many photons as possible from a given region of the sky and concentrate them into a focused beam for analysis.

Optical telescopes are designed specifically to collect the wavelengths that are visible to the human eye. These telescopes have a long history, stretching back to the days of Galileo in the early 17th century, and for most of the past 4 centuries astronomers have built their instruments primarily for use in the narrow, visible, portion of the electromagnetic spectrum. ∞ (Sec. 2.4) Optical telescopes are probably also the best-known type of astronomical hardware, so it is perhaps fitting that we begin our study with them.

Although the various telescope designs presented in this section all come to us from optical astronomy, the discussion applies equally well to many instruments designed to capture *invisible* radiation, particularly in the infrared and ultraviolet regimes. Many large ground-based optical facilities are also used extensively for infrared work.* Indeed, many ground-based observatories have recently been constructed with infrared observing as their principal function.

Reflecting and Refracting Telescopes

Optical telescopes fall into two basic categories: *reflectors* and *refractors*. Figure 5.1 shows how a **reflecting telescope** uses a curved mirror to gather and concentrate a beam of light. The mirror, usually called the *primary mirror* because telescopes often contain more than one mirror, is constructed so that all light rays arriving parallel to its axis (the imaginary line through the center of, and perpendicular to, the mirror), regardless of their distance from that axis, are reflected to pass through a single point, called the *focus*. The distance between the primary mirror and the focus is the *focal length*. In astronomical contexts, the focus of the primary mirror is referred to as the **prime focus**.

A **refracting telescope** uses a lens instead of a mirror to focus the incoming light, relying on refraction rather than reflection to achieve its purpose. **Refraction** is the bending of a beam of light as it passes from one transparent medium (e.g., air) into another (e.g., glass). For instance, consider how a pencil that is half immersed in a glass of water looks bent. The pencil is straight, of course, but the light by which we see it is bent—refracted—as that light leaves the water and enters the air. When the light then enters our eyes, we perceive the pencil as being bent. Figure 5.2(a) illustrates the process and shows how a prism can be used to change the direction of a beam of light. As illustrated in Figure 5.2(b), we can think of a lens as a series of prisms combined in such a way that all light rays

Recall from Chapter 3 that, while Earth's atmosphere effectively blocks all ultraviolet, and most infrared, radiation, there remain several fairly broad spectral windows through which ground-based infrared observations can be made. ∞ (Sec. 3.3)

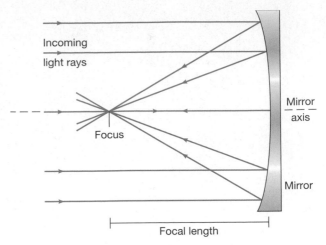

▲ **FIGURE 5.1 Reflecting Mirror** A curved mirror can be used to focus all rays of light arriving parallel to the mirror axis to a single point. Light rays traveling along the axis are reflected back along it, as indicated by the arrowheads pointing in both directions. Off-axis rays are reflected through greater and greater angles the farther they are from the axis, so that they all pass through the same point—the focus.

striking the lens parallel to the axis are refracted to pass through the focus.

Astronomical telescopes are often used to make **images** of their field of view (simply, the portion of the sky that the telescope "sees"). Figure 5.3 illustrates how that is accomplished, in this case by the mirror in a reflecting telescope. Light from a distant object (here, a comet) reaches us as parallel, or very nearly parallel, rays. Any ray of light entering the instrument parallel to the telescope's axis strikes the mirror and is reflected through the prime focus. Light coming from a slightly different direction—inclined slightly to the axis—is focused to a slightly different point. In this way, an image is formed near the prime focus. Each point on the image corresponds to a different point in the field of view.

The prime-focus images produced by large telescopes are actually quite small—the image of the entire field of view may be as little as 1 cm across. Often, the image is magnified with a lens known as an *eyepiece* before being observed by eye or, more likely, recorded as a photograph or digital image. The angular diameter of the magnified image is much greater than the telescope's field of view, allowing much more detail to be discerned. Figure 5.4(a) shows the basic design of a simple reflecting telescope, illustrating how a small secondary mirror and eyepiece are used to view the image. Figure 5.4(b) shows how a refracting telescope accomplishes the same function.

The two telescope designs shown in Figure 5.4 achieve the same result: Light from a distant object is captured and focused to form an image. On the face of it, then, it might appear that there is little to choose between the two in deciding which type to buy or build. However, as the sizes of telescopes have increased steadily over the

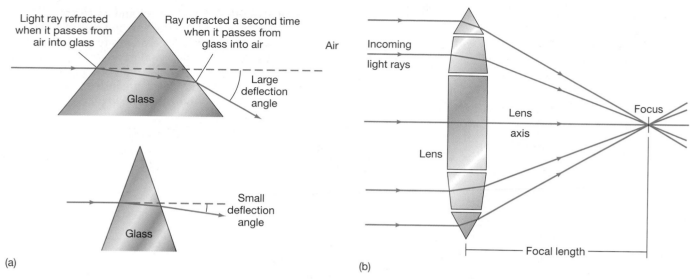

▲ **FIGURE 5.2 Refracting Lens** (a) Refraction by a prism changes the direction of a light ray by an amount that depends on the angle between the faces of the prism. When the angle between the faces is large, the deflection is large; when the angle is small, so is the deflection. (b) A lens can be thought of as a series of prisms. A light ray traveling along the axis of a lens is unrefracted as it passes through the lens. Parallel rays arriving at progressively greater distances from the axis are refracted by increasing amounts, in such a way that all are focused to a single point.

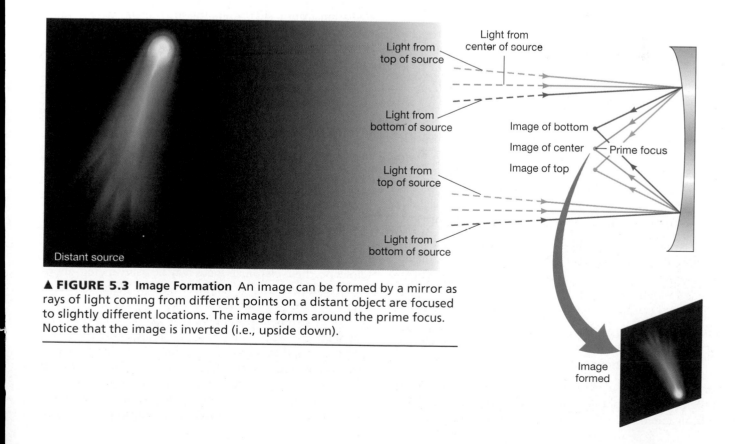

▲ **FIGURE 5.3 Image Formation** An image can be formed by a mirror as rays of light coming from different points on a distant object are focused to slightly different locations. The image forms around the prime focus. Notice that the image is inverted (i.e., upside down).

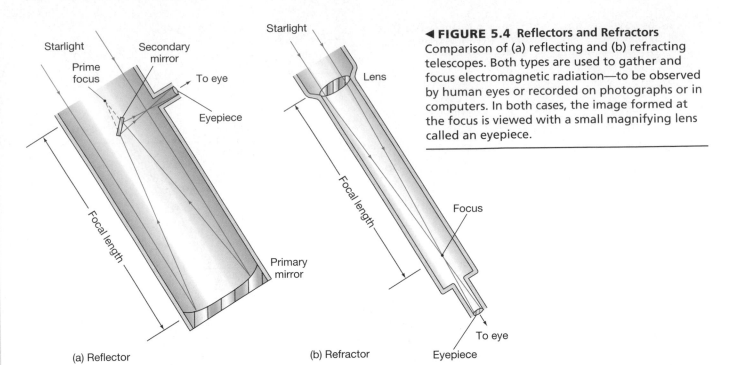

(a) Reflector

(b) Refractor

◄ **FIGURE 5.4 Reflectors and Refractors** Comparison of (a) reflecting and (b) refracting telescopes. Both types are used to gather and focus electromagnetic radiation—to be observed by human eyes or recorded on photographs or in computers. In both cases, the image formed at the focus is viewed with a small magnifying lens called an eyepiece.

years (for reasons to be discussed in Section 5.3), a number of important factors have tended to favor reflecting instruments over refractors:

1. The fact that light must pass through the lens is a major disadvantage of refracting telescopes. Just as a prism disperses white light into its component colors, the lens in a refracting telescope tends to focus red and blue light differently. This deficiency is known as *chromatic aberration*. Figure 5.5 shows how chromatic aberration occurs and indicates how it affects the image of a star. Careful design and choice of materials can largely correct this deficiency, but it is very difficult to eliminate entirely. Obviously, the problem does not occur with mirrors.

2. As light passes through the lens, some of it is absorbed by the glass. This absorption is a relatively minor problem for visible radiation, but it can be severe for infrared and ultraviolet observations, because glass blocks most of the radiation in those regions of the electromagnetic spectrum. Again, the problem does not affect mirrors.

3. A large lens can be quite heavy. Because it can be supported only around its edge (so as not to block the incoming radiation), the lens tends to deform under its own weight. A mirror does not have this drawback, because it can be supported over its entire back surface.

4. A lens has two surfaces that must be accurately machined and polished—a task that can be very difficult indeed—but a mirror has only one.

For these reasons, *all* large modern telescopes use mirrors as their primary light gatherers. The largest refractor ever built, installed in 1897 at the Yerkes Observatory in Wis-

consin and still in use today, has a lens diameter of just over 1 m (40 inches). By contrast, many recently constructed reflecting telescopes have mirror diameters in the 10-m range, and still larger instruments are on the way.

Types of Reflecting Telescope

Figure 5.6 shows some basic reflecting-telescope designs. Radiation from a star enters the instrument, passes down the main tube, strikes the primary mirror, and is reflected back toward the prime focus, near the top of the tube. Sometimes astronomers place their recording instruments at the prime focus; however, it can be inconvenient, or even impossible, to suspend bulky pieces of equipment there. More often, the light is intercepted on its path to the focus by a *secondary mirror* and redirected to a more convenient location, as in Figure 5.6(b) through (d).

In a **Newtonian telescope** (named after Sir Isaac Newton, who invented this particular design), the light is intercepted before it reaches the prime focus, and then is deflected by 90°, usually to an eyepiece at the side of the instrument. This is a popular design for smaller reflecting telescopes, such as those used by amateur astronomers, but it is relatively uncommon in large instruments. On a large telescope, the Newtonian focus may be many meters above the ground, making it an inconvenient place to attach equipment (or place an observer).

Alternatively, astronomers may choose to work on a rear platform where they can use equipment, such as a spectroscope, that is too heavy to hoist to the prime focus. In this case, light reflected by the primary mirror toward the prime focus is intercepted by a smaller secondary mirror, which reflects it back down through a small hole at the center of the primary mirror. This arrangement is known

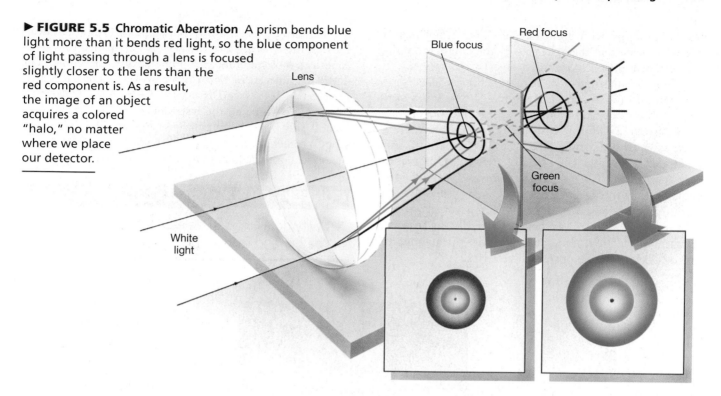

► **FIGURE 5.5 Chromatic Aberration** A prism bends blue light more than it bends red light, so the blue component of light passing through a lens is focused slightly closer to the lens than the red component is. As a result, the image of an object acquires a colored "halo," no matter where we place our detector.

as a **Cassegrain telescope** (after Guillaume Cassegrain, a French lensmaker). The point behind the primary mirror where the light from the star finally converges is called the *Cassegrain focus*.

A more complex observational configuration requires starlight to be reflected by several mirrors. As in the Cassegrain design, light is first reflected by the primary mirror toward the prime focus and is then reflected back down the tube by a secondary mirror. Next, a third, much smaller,

mirror reflects the light out of the telescope, where (depending on the details of the telescope's construction) the beam may be analyzed by a detector mounted alongside, at the *Nasmyth focus*, or it may be directed via a further series of mirrors into an environmentally controlled laboratory known as the *coudé* room (from the French word for "bent"). This laboratory is separate from the telescope itself, enabling astronomers to use very heavy and finely tuned equipment that cannot be placed at any of the other foci (all of which

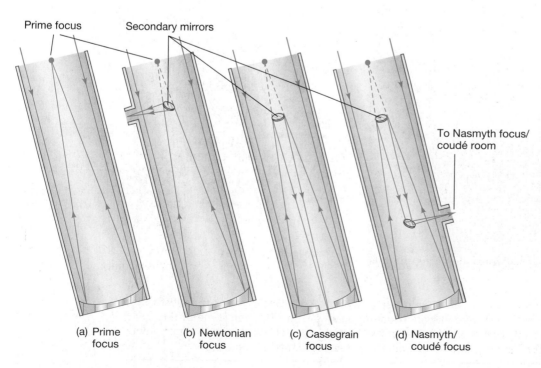

(a) Prime focus

(b) Newtonian focus

(c) Cassegrain focus

(d) Nasmyth/ coudé focus

◄ **FIGURE 5.6 Reflecting Telescopes** Four reflecting telescope designs: (a) prime focus, (b) Newtonian focus, (c) Cassegrain focus, and (d) Nasmyth/coudé focus. Each uses a primary mirror at the bottom of the telescope to capture radiation, which is then directed along different paths for analysis. Notice that the secondary mirrors shown in (c) and (d) are actually slightly diverging, so that they move the focus outside the telescope.

necessarily move with the telescope). The arrangement of mirrors is such that the light path to the coudé room does not change as the telescope tracks objects across the sky.

To illustrate some of these points, Figure 5.7(a) shows the twin 10-m-diameter optical/infrared telescopes of the Keck Observatory on Mauna Kea in Hawaii, operated jointly by the California Institute of Technology and the University of California. The diagram in part (b) illustrates the light paths and some of the foci. Observations may be made at the Cassegrain, Nasmyth, or coudé focus, depending on the needs of the user. As the size of the person in part (c) indicates, this is indeed a very large telescope—in fact, the two mirrors are currently the largest on Earth. We will see numerous examples throughout this text of Keck's many important discoveries.

Perhaps the best-known telescope on (or, rather, near) Earth is the *Hubble Space Telescope (HST)*, named for one of America's most notable astronomers, Edwin Hubble. The device was placed in Earth orbit by NASA's space shuttle *Discovery* in 1990 (see *Discovery 5-1*). *HST* is a Cassegrain telescope in which all the instruments are located directly behind the primary mirror (see Figure 5.8). The telescope's detectors are capable of making measurements in the optical, infrared, and ultraviolet parts of the spectrum.

CONCEPT CHECK

✔ Why do all modern telescopes use mirrors to gather and focus light?

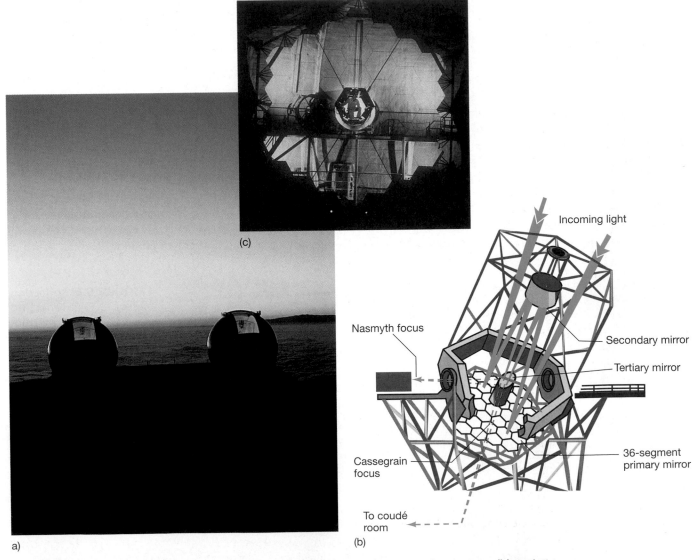

▲ **FIGURE 5.7 Keck Telescope** (a) The two 10-m telescopes of the Keck Observatory. (b) Artist's illustration of the telescope, the path taken by an incoming beam of starlight, and some of the locations where instruments may be placed. (c) One of the 10-m mirrors. (The odd shape is explained in Section 5.3.) Note the technician in orange coveralls at center. *(W. M. Keck Observatory)*

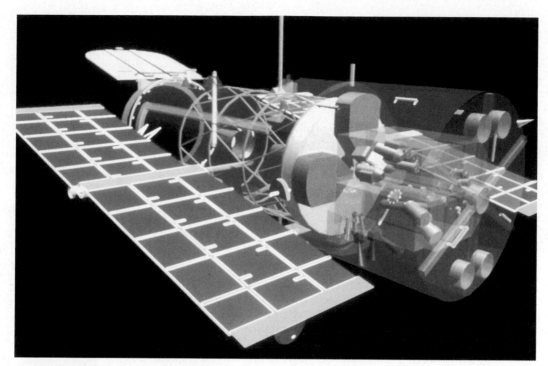

▲ **FIGURE 5.8 HST Detectors** The principal features inside the *Hubble Space Telescope* are displayed in this "see-through" illustration. The telescope's Cassegrain design focuses light through a small opening in the center of the primary mirror (large bluish disk at center), from where it can be directed to any of several instruments arrayed behind the mirror (shown here in various colors at right). The large red objects are sensors that guide the pointing of the telescope, and the huge blue panels collect light from the Sun to power everything onboard. *(D. Berry)*

5.2 Images and Detectors

In the previous section, we saw how telescopes gather and focus light to form an image of their field of view. In fact, most large observatories use many different instruments to analyze the radiation received from space. These devices may be placed at various points along the light path outside the telescope—see, for example, the multiple foci and light paths in Figure 5.7(b). Figure 5.8 shows the more compact arrangement of detectors within the *Hubble Space Telescope*. In this section, we look in a little more detail at how telescopic images are actually produced and at some other types of detectors that are in widespread use.

Image Acquisition

Computers play a vital role in observational astronomy. Most large telescopes today are controlled either by computers or by operators who rely heavily on computer assistance, and images and data are recorded in a form that can be easily read and manipulated by computer programs.

It is becoming rare for photographic equipment to be used as the primary means of data acquisition at large observatories. Instead, electronic detectors known as **charge-coupled devices**, or **CCDs**, are in widespread use. Their output goes directly to a computer. A CCD (Figure 5.9(a) and (b)) consists of a wafer of silicon divided into a two-dimensional array of many tiny picture elements, known as **pixels**. When light strikes a pixel, an electric charge builds up on it. The amount of charge is directly proportional to the number of photons striking each pixel—in other words, to the intensity of the light at that point. The buildup of charge is monitored electronically, and a two-dimensional image is obtained (Figure 5.9(c) and (d)).

A CCD is typically a few square centimeters in area and may contain several million pixels, generally arranged on a square grid. As the technology improves, both the areas of CCDs and the number of pixels they contain continue to increase. Incidentally, the technology is not limited to astronomy: Many home video cameras contain CCD chips similar in basic design to those in use at the great astronomical observatories of the world.

CCDs have two important advantages over photographic plates, which were the staple of astronomers for over a century. First, CCDs are much more *efficient* than photographic plates, recording as many as 90 percent of the photons striking them, compared with less than 5 percent for photographic methods. This difference means

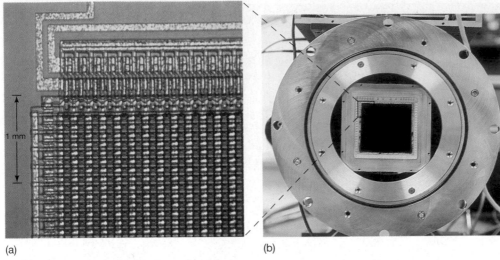

(a)

(b)

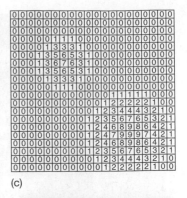

(c)

(d)

▲ **FIGURE 5.9 CCD Chip** A charge-coupled device (CCD) consists of hundreds of thousands, or even millions, of tiny light-sensitive cells called pixels, usually arranged in a square array. Light striking a pixel causes an electrical charge to build up on it. By electronically reading out the charge on each pixel, a computer can reconstruct the pattern of light—the image—falling on the chip. (a) Detail of a CCD array. (b) A CCD chip mounted for use at the focus of a telescope. (c) Typical data from the chip consist of an array of numbers, running from 0 to 9 in this simplified example. Each number represents the intensity of the radiation striking that particular pixel. (d) When interpreted as intensity levels on a computer screen, an image of the field of view results. *(AURA; R. Wainscoat/Peter Arnold)*

that a CCD image can show objects 10 to 20 times fainter than a photograph made with the same telescope and the same exposure time. Alternatively, a CCD can record the same level of detail in less than a tenth of the time required by photographic techniques, or it can record that detail with a much smaller telescope. Second, CCDs produce a faithful representation of an image in a digital format that can be placed directly on magnetic tape or disk or, more commonly, sent across a computer network to an observer's home institution.

Image Processing

Computers are also widely used to reduce *background noise* in astronomical images. Noise is anything that corrupts the integrity of a message, such as static on an AM radio or "snow" on a television screen. The noise corrupting telescopic images has many causes. In part, it results from faint, unresolved sources in the telescope's field of view and from light scattered into the line of sight by Earth's atmosphere. It can also be caused by imperfections within the detector itself, which may result in an electronic "hiss" similar to the faint background hiss you might hear when you listen to a particularly quiet piece of music on your stereo.

Even though astronomers often cannot determine the origin of the noise in their observations, they can at least measure its characteristics. For example, if we observe a part of the sky where there are no known sources of radia-

tion, then whatever signal we do receive is (almost by definition) noise. Once the properties of the signal have been measured, the effects of noise can be partially removed with the aid of high-speed computers, allowing astronomers to see features in their data that would otherwise remain hidden.

Using computer processing, astronomers can also compensate for known instrumental defects. In addition, the computer can often carry out many of the relatively simple, but tedious and time-consuming, chores that must be performed before an image (or spectrum) reaches its final "clean" form. Figure 5.10 illustrates how computerized image-processing techniques were used to correct for known instrumental problems in the *Hubble Space Telescope*, allowing much of the planned resolution of the telescope to be recovered even before its repair in 1993. ∞ (see *Discovery 3-1*)

Wide-Angle Views

Large reflectors are very good at forming images of narrow fields of view, wherein all the light that strikes the mirror surface moves almost parallel to the axis of the instrument. However, as the angle at which the light enters increases, the accuracy of the focus decreases, degrading the overall quality of the image. The effect (called *coma*) worsens as we move farther from the center of the field of view. Eventually, the quality of the image is reduced to the point where the image is no longer usable. The distance from the center to where the image becomes unacceptable

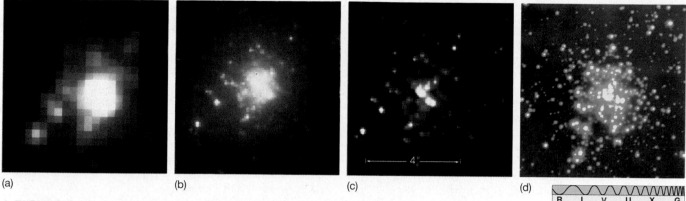

(a)   (b)   (c)   (d)

▲ **FIGURE 5.10** **Image Processing** (a) Ground-based view of the star cluster R136, a group of stars in the Large Magellanic Cloud (a nearby galaxy). (b) The "raw" image of this same region as seen by the *Hubble Space* Telescope in 1990, before the repair mission. (c) The same image after computer processing that partly compensated for imperfections in the mirror. (d) The same region as seen by the repaired HST in 1994, here observed at a somewhat bluer wavelength. (*AURA/NASA*)

defines the useful field of view of the telescope—typically, only a few arc minutes for large instruments.

A design that overcomes this problem is the *Schmidt telescope*, named after its inventor, Bernhard Schmidt, who built the first such instrument in the 1930s. The telescope uses a correcting lens that sharpens the final image of the entire field of view. Consequently, a Schmidt telescope is well suited to producing wide-angle photographs, covering several degrees of the sky. The design of the Schmidt telescope results in a curved image that is not suitable for viewing with an eyepiece, so the image is recorded on a specially shaped piece of photographic film. For this reason, the instrument is often called a *Schmidt camera* (Figure 5.11).

Photometry

When a CCD is placed at the focus of a telescope to record an image of the instrument's field of view, the telescope is acting, in effect, as a high-powered camera. However, as-

tronomers often want to carry out more specific measurements of the radiation received from space.

One very fundamental property of a star (or any other astronomical object) is its *brightness*—the amount of light energy from the star striking our detector every second. The measurement of brightness is called **photometry** (literally, "light measurement"). In principle, determining a star's brightness is just a matter of adding up the values in all the CCD pixels corresponding to the star (see Figure 5.9c). However, in practice, the process is more complicated, as stellar images may overlap, and computer assistance is generally need to disentangle them.

Astronomers often combine photometric measurements with the use of colored *filters* in order to limit the wavelengths they measure. (A filter simply blocks out all incoming radiation, except in some specific range of wavelengths; ∞ see Sec. 17.4 for a more detailed discussion.) Many standard filters exist, covering various "slices" of the

◄ **FIGURE 5.11** **Schmidt Telescope** The Schmidt camera of the European Southern Observatory in Chile. The astronomer in the photograph is not looking through the optics of the main telescope, but instead is using a smaller "finding telescope" to verify the direction in which the instrument is pointed. (*ESO*)

DISCOVERY 5-1

The Hubble Space Telescope

The *Hubble Space Telescope* (*HST*) is the largest, most complex, most sensitive observatory ever deployed in space. At over $7.8 billion (including the cost of several missions to service and refurbish the system), it is also the most expensive scientific instrument ever built and operated. A joint project of NASA and the European Space Agency, *HST* was designed to allow astronomers to probe the universe with at least 10 times finer resolution and some 30 times greater sensitivity to light than existing Earth-based devices. *HST* is operated remotely from the ground; there are no astronauts aboard the telescope, which orbits Earth about once every 95 minutes at an altitude of some 600 km (380 miles).

The telescope's overall dimensions approximate those of a city bus or railroad tank car: 13 m (43 feet) long, 12 m (39 feet) across with solar arrays extended, and 11,000 kg (12.5 tons when weighed on Earth). At the heart of *HST* is a 2.4-m-diameter mirror designed to capture optical, ultraviolet, and infrared radiation before it reaches Earth's murky atmosphere. The accompanying figure shows the telescope being lifted out of the cargo bay of the space shuttle *Discovery* in the spring of 1990.

The telescope reflects light from its large mirror back to a smaller, 0.3-m secondary mirror, which sends the light through a small hole in the main mirror and into the rear portion of the spacecraft. There, any of five major scientific instruments wait to analyze the incoming radiation. Most of these instruments (see Figure 5.8) are about the size of a telephone booth. They are designed to be maintained by NASA

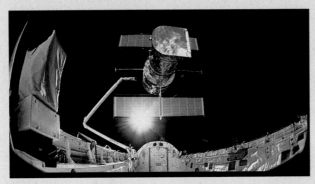

(NASA)

astronauts, and indeed, most of the telescope's instruments have been upgraded or replaced since *HST* was launched. The current detectors on the telescope span the visible, near-infrared, and near-ultraviolet regions of the electromagnetic spectrum, from about 100 nm (UV) to 2200 nm (IR).

Soon after the launch of *HST*, astronomers discovered that the telescope's primary mirror had been polished to the wrong shape. The mirror was too flat by 2 μm, about 1/50 the width of a human hair, making it impossible to focus light as well as expected. The optical flaw (known as *spherical aberration*) produced by this design error meant that *HST* was not as sensitive as designed, although it could still see many objects in the universe with unprecedented resolution. In 1993, astronauts aboard the space shuttle *Endeavour* visited *HST* and succeeded in repairing some of its ailing equipment. They replaced *Hubble's* gyroscopes to help the telescope point more accurately, installed sturdier versions of the solar panels that power the telescope's electronics, and—most importantly—inserted an intricate set of small mirrors (each about the size of a coin) to compensate for the faulty primary mirror. *Hubble's* resolution is now close to the original design specifications, and the telescope has regained much of its lost sensitivity. Additional service missions were performed in 1997, 1999, and 2002 to replace instruments and repair faulty systems.

A good example of *Hubble's* scientific capabilities today can be seen by comparing the two images of the spiral galaxy M100, shown in the accompanying figure. On the left is perhaps the best ground-based photograph of this beautiful galaxy, showing rich detail and color in its spiral arms. On the right, to the same scale and orientation, is an *HST* image showing improvement in both resolution and sensitivity. (The chevron-shaped field of view is caused by the corrective optics inserted into the telescope; an additional trade-off is that *Hubble's* field of view is smaller than those of ground-based telescopes.) The inset shows *Hubble's* exquisite resolution of small fields of view.

During its first decade of operation, *Hubble* has revolutionized our view of the sky, helping to rewrite some theories of the universe along the way. The space-based instrument has studied newborn galaxies almost at the limit of the observable universe with unprecedented clarity, allowing astronomers to see the interactions and collisions that may have shaped the evolution of our Milky Way. Turning its gaze to the hearts of

spectrum, from near-infrared through visible to near-ultraviolet wavelengths. By confining their attention to these relatively narrow ranges, astronomers can often estimate the shape of an object's blackbody curve and hence determine, at least approximately, the object's temperature. ∞ (Sec. 3.4) Filters are also used with CCD images in order to simulate natural color. For example, most of the visible-light *HST* images in this text are actually composites of *three* raw images, taken through red, green, and blue filters, respectively, and combined afterwards to reconstruct a single color frame.

Astronomical objects are generally faint, and most astronomical images entail long exposures—minutes to hours—in order to see fine detail (Section 5.3). Thus, the brightness we measure from an image is really an average over the entire exposure. Short-term fluctuations (if any) cannot be seen. When highly accurate and rapid measurements of light intensity are required, a specialized device known as a *photometer* is used. The photometer measures the total amount of light received in all or part of the field of view. When only a portion of the field is of interest, that region is selected simply by masking (blocking) out the

(D. Malin/Anglo-Australian Telescope)

(NASA)

R I V U X G

galaxies closer to home, *Hubble* has provided strong evidence for supermassive black holes in their cores. Within our own Galaxy, it has given astronomers stunning new insights into the physics of star formation and the evolution of stellar systems and stars of all sizes, from superluminous giants to objects barely more massive than planets. Finally, in our solar system, *Hubble* has afforded scientists new views of the planets, their moons, and the tiny fragments from which they formed long ago. Many spectacular examples of the telescope's remarkable capabilities appear throughout this book.

With *Hubble* now well into its second decade of a highly successful career, NASA scientists are planning the telescope's successor. The "Next Generation" Space Telescope (now known as the James Webb Space Telescope, or JWST, after the administrator who led NASA's *Apollo* program during the 1960s and 1970s) will dwarf *Hubble* in both scale and capability. Sporting a 6-m segmented mirror and containing a formidable array of detectors optimized for use at visible and infrared wavelengths, JWST will orbit at a distance of 1.5 million km from Earth, far beyond the Moon. NASA hopes to launch the instrument by 2010. The telescope's primary mission will involve examining the formation of the first stars and galaxies, measuring the large-scale structure of the universe, and investigating the evolution of planets, stars, and galaxies.

In early 2004 NASA announced that, for budgetary and safety reasons, they had canceled the planned 2006 servicing mission which would have extended the telescope's lifetime. Not surprisingly, the decision is actively opposed by many astronomers. Without servicing, *Hubble* is expected to become inoperative by 2007.

rest of the field of view. Using a photometer often means "throwing away" spatial detail—usually no image is produced—but in return, more information is obtained about the intensity and time variability of a source, such as a pulsating star or a supernova explosion.

Spectroscopy

Often, astronomers want to study the *spectrum* of the incoming light. Large **spectrometers** work in tandem with optical telescopes. Light collected by the primary mirror may be redirected to the coudé room, defined by a narrow slit, dispersed (split into its component colors) by means of a prism or a diffraction grating, and then sent on to a detector—a process not so different in concept from the operation of the simple spectroscope described in Chapter 4. ∞ (Sec. 4.1) The spectrum can be studied in real time (i.e., as it is being received at the telescope) or recorded using a CCD (or, less commonly nowadays, a photographic plate) for later analysis. Astronomers can then apply the analysis techniques discussed in Chapter 4 to extract detailed information from the spectral lines they record. ∞ (Sec. 4.4)

CONCEPT CHECK

✔ Why aren't astronomers satisfied with just taking photographs of the sky?

5.3 Telescope Size

③ Modern astronomical telescopes have come a long way from Galileo's simple apparatus. Their development over the years has seen a steady increase in *size*, for two main reasons. The first has to do with the amount of light a telescope can collect—its *light-gathering power*. The second is related to the amount of detail that can be seen—the telescope's *resolving power*. Simply put, large telescopes can gather and focus more radiation than can their smaller counterparts, allowing astronomers to study fainter objects and to obtain more detailed information about bright ones. This fact has played a central role in determining the design of contemporary instruments.

Light-Gathering Power

One important reason for using a larger telescope is simply that it has a greater **collecting area**, which is the total area capable of gathering radiation. The larger the telescope's reflecting mirror (or refracting lens), the more light it collects, and the easier it is to measure and study an object's radiative properties. Astronomers spend much of their time observing very distant—and hence very *faint*—cosmic sources. In order to make detailed observations of such objects, very large telescopes are essential. Figure 5.12 illustrates the effect of increasing the size of a telescope by comparing images of the Andromeda Galaxy taken with two different instruments. A large collecting area is particularly important for spectroscopic work, as the radiation received in that case must be split into its component wavelengths for further analysis.

The observed brightness of an astronomical object is directly proportional to the area of our telescope's mirror and therefore to the *square* of the mirror diameter. Thus, a 5-m telescope will produce an image 25 times as bright as a 1-m instrument, because a 5-m mirror has $5^2 = 25$ times the collecting area of a 1-m mirror. We can also think of this relationship in terms of the length of *time* required for a telescope to collect enough energy to create a recognizable image on a photographic plate. Our 5-m telescope will produce an image 25 times faster than the 1-m device because it gathers energy at a rate 25 times greater. Put another way, a 1-hour exposure with a 1-m telescope is roughly equivalent to a 2.4-minute exposure with a 5-m instrument.

Until the 1980s, the conventional wisdom was that telescopes with mirrors larger than 5 or 6 meters in diameter were simply too expensive and impractical to build. The

(a)

(b)

R I V U X G

▲ **FIGURE 5.12 Sensitivity** Effect of increasing the size of a telescope on an image of the Andromeda Galaxy. Both photographs had the same exposure time, but image (b) was taken with a telescope twice the size of that used to make (a). Fainter detail can be seen as the diameter of the telescope mirror increases because larger telescopes are able to collect more photons per unit time. (*Adapted from AURA*)

problems involved in casting, cooling, and polishing a huge block of quartz or glass to very high precision (typically less than the width of a human hair) were just too great. However, new high-tech manufacturing techniques, coupled with radically new mirror designs, make the construction of telescopes in the 8- to 12-m range almost a routine matter. Experts can now make large mirrors much lighter for their size than had previously been believed feasible and can combine many smaller mirrors into the equivalent of a much larger single-mirror telescope. Several large-diameter instruments now exist, and many more are planned.

The Keck telescopes, shown in detail in Figure 5.7 and in a larger view in Figure 5.13, are a case in point. Each telescope combines 36 hexagonal 1.8-m mirrors into the equivalent collecting area of a single 10-m reflector. The first Keck telescope became fully operational in 1992; the second was completed in 1996. The large size of these devices and the high altitude at which they operate make them particularly well suited to detailed spectroscopic studies of very faint objects, in both the optical and infrared parts of the spectrum. Mauna Kea's 4.2-km (13,800 feet) altitude minimizes atmospheric absorption of infrared radiation, making this site one of the finest locations on Earth for infrared astronomy.

Numerous other large telescopes can be seen in the figure. Some are designed exclusively for infrared work; others, like Keck, operate in both the optical and the infrared. To the right of the Keck domes is the 8.3-m Subaru (the Japanese name for the Pleiades) telescope, part of the National Astronomical Observatory of Japan. Its mir-

ror, shown in Figure 5.13(b), is the largest single mirror (as opposed to the segmented design used in Keck) yet built. Subaru saw "first light" in 1999. In the distance is another large single-mirror instrument: the 8.1-m Gemini North telescope, completed in 1999 by a consortium of seven nations, including the United States. Its twin, Gemini South, in the Chilean Andes, went into service in 2002.

In terms of total available collecting area, the largest telescope currently available is the European Southern Observatory's optical-infrared Very Large Telescope (VLT), located at Cerro Paranal, in Chile (Figure 5.14). The VLT consists of four separate 8.2-m mirrors that can function as a single instrument. The last of its four mirrors was completed in 2001.

Resolving Power

A second advantage of large telescopes is their finer **angular resolution**. In general, *resolution* refers to the ability of any device, such as a camera or telescope, to form distinct, separate images of objects lying close together in

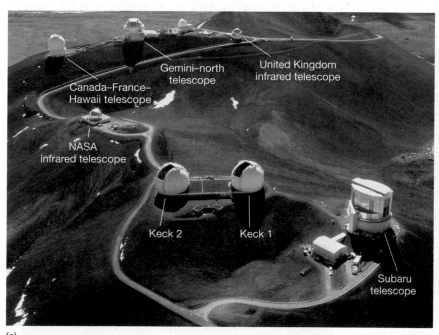

(a)

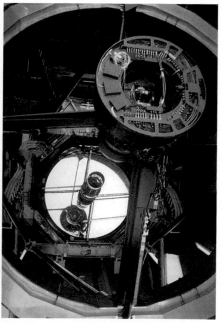

(b)

▲ **FIGURE 5.13 Mauna Kea Observatory** (a) The world's highest ground-based observatory, at Mauna Kea, Hawaii, is perched atop an extinct volcano more than 4 km (nearly 14,000 feet) above sea level. Among the domes visible in the picture are those housing the Canada–France–Hawaii 3.6-m telescope, the 8.1-m Gemini North instrument, the 2.2-m telescope of the University of Hawaii, Britain's 3.8-m infrared facility, and the twin 10-m Keck telescopes. To the right of the twin Kecks is the Japanese 8.3-m Subaru telescope. The thin air at this high-altitude site guarantees less atmospheric absorption of incoming radiation and hence a clearer view than at sea level, but the air is so thin that astronomers must occasionally wear oxygen masks while working. (b) The mirror in the Subaru telescope, currently the largest one-piece mirror in any astronomical instrument. (*R. Wainscoat; R. Underwood/Keck Observatory; NAOJ*)

where 1 μm (1 micron) = 10^{-6} m (see Appendix 2).

For a given size of telescope, the amount of diffraction increases in proportion to the wavelength used. Observations in the infrared or radio range are often limited by the effects of diffraction. For example, according to the preceding formula, in an otherwise perfect observing environment the best possible angular resolution of blue light (with a wavelength of 400 nm) that can be obtained using a 1-m telescope is about $0.25'' \times (0.4/1) = 0.1''$. This quantity is known as the *diffraction-limited* resolution of the telescope. But if we were to use our 1-m telescope to make observations in the near infrared, at a wavelength of 10 μm (10,000 nm), the best resolution we could obtain would be only 2.5". A 1-m radio telescope operating at a wavelength of 1 cm would have an angular resolution of just under 1°.

For light of any given wavelength, large telescopes produce less diffraction than small ones. A 5-m telescope observing in blue light would have a diffraction-limited resolution five times finer than the 1-m telescope just discussed—about 0.02". A 0.1-m (10-cm) telescope would

the field of view. The finer the resolution, the better we can distinguish the objects and the more detail we can see. In astronomy, where we are always concerned with angular measurement, "close together" means "separated by a small angle in the sky," so angular resolution is the factor that determines our ability to see fine structure. Figure 5.15 illustrates how the appearance of two objects—stars, say—might change as the angular resolution of our telescope varies. Figure 5.16 shows the result of increasing resolving power with views of the Andromeda Galaxy at several different resolutions.

What limits a telescope's resolution? One important factor is *diffraction*, the tendency of light—and all other waves, for that matter—to bend around corners. ∞ *(Discovery 3-1)* Because of diffraction, when a parallel beam of light enters a telescope, the rays spread out slightly, making it impossible to focus the beam to a sharp point, even with a perfectly constructed mirror. Diffraction introduces a certain "fuzziness," or loss of resolution, into any optical system. The degree of fuzziness—the minimum angular separation that can be distinguished—determines the angular resolution of the telescope. The amount of diffraction is proportional to the wavelength of the radiation divided by the diameter of the telescope mirror. As a result, we can write, in convenient units,

$$\text{angular resolution (arc sec)} = 0.25 \frac{\text{wavelength } (\mu\text{m})}{\text{mirror diameter (m)}},$$

▶ **FIGURE 5.15** Resolving Power Two comparably bright light sources become progressively clearer when viewed at finer and finer angular resolution. When the angular resolution is much poorer than the separation of the objects, as in (a), the objects appear as a single fuzzy "blob." As the resolution improves, through (b) and (c), the two sources become discernible as separate objects.

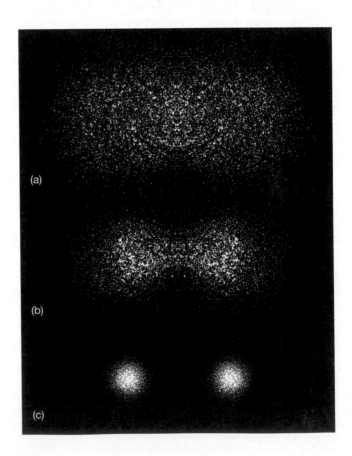

(a)

(b)

(c)

(a)

(b)

(c)

(d)

R I V U X G

◀ **FIGURE 5.16 Resolution** Detail becomes clearer in the Andromeda Galaxy as the angular resolution is improved some 600 times, from (a) 10′, to (b) 1′, (c) 5″, and (d) 1″. *(Adapted from AURA)*

have a diffraction limit of 1″, and so on. For comparison, the angular resolution of the human eye in the middle of the visual range is about 0.5′.

CONCEPT CHECK

✔ Give two reasons why astronomers need to build very large telescopes.

5.4 High-Resolution Astronomy

Even large telescopes have limitations. For example, according to the discussion in the preceding section, the 10-m Keck telescope should have an angular resolution of around 0.01″ in blue light. In practice, however, without the technological advances discussed in this section, it could not do better than about 1″. In fact, apart from instruments using special techniques developed to examine some particularly bright stars, *no* ground-based optical telescope built before 1990 can resolve astronomical objects to much better than 1″. The reason is *turbulence* in Earth's atmosphere—small-scale eddies of swirling air all along the line of sight, which blur the image of a star even before the light reaches our instruments.

Atmospheric Blurring

As we observe a star, atmospheric turbulence produces continual small changes in the optical properties of the air between the star and our telescope (or eye). As a result, the light from the star is refracted slightly as it travels toward us, so the stellar image dances around on our detector (or on our retina). This continual deflection is the cause of the well-known "twinkling" of stars. It occurs for the same reason that objects appear to shimmer when viewed across a hot roadway on a summer day: The constantly shifting rays of light reaching our eyes produce the illusion of motion.

On a good night at the best observing sites, the maximum amount of deflection produced by the atmosphere is slightly less than 1″. Consider taking a photograph of a star. After a few minutes of exposure (long enough for the intervening atmosphere to have undergone many small random changes), the image of the star has been smeared out over a roughly circular region an arc second or so in diameter. Astronomers use the term **seeing** to describe the effects of atmospheric turbulence. The circle over which a star's light

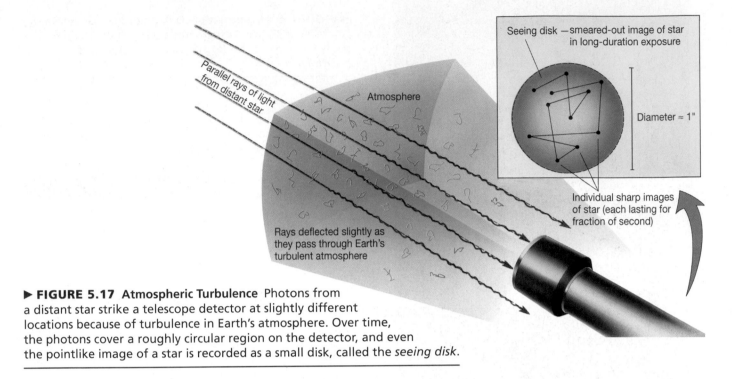

▶ **FIGURE 5.17 Atmospheric Turbulence** Photons from
a distant star strike a telescope detector at slightly different
locations because of turbulence in Earth's atmosphere. Over time,
the photons cover a roughly circular region on the detector, and even
the pointlike image of a star is recorded as a small disk, called the *seeing disk*.

(or the light from any other astronomical source) is spread
is called the **seeing disk**. Figure 5.17 illustrates the formation of the seeing disk for a small telescope.*

Atmospheric turbulence has less effect on light of
longer wavelengths—ground-based astronomers generally
"see" better in the infrared. However, offsetting this improvement in image quality is the fact that the atmosphere
is wholly or partially opaque over much of the infrared
range. ∞ (Sec. 3.3) For these reasons, to achieve the best
possible observing conditions, telescopes are sited on
mountaintops (to get above as much of the atmosphere as
possible) in regions of the world where the atmosphere is
known to be fairly stable and relatively free of dust, moisture, and light pollution from cities.

In the continental United States, these sites tend to
be in the desert Southwest. The U.S. National Observatory for optical astronomy in the Northern Hemisphere,
completed in 1973, is located high on Kitt Peak near
Tucson, Arizona. The site was chosen because of its
many dry, clear nights. Seeing of less than 1″ from such a
location is regarded as good, and seeing of a few arc seconds is tolerable for many purposes. Even better conditions are found on Mauna Kea, Hawaii (Figure 5.13), and
at numerous sites in the Andes Mountains of Chile
(Figures 5.14 and 5.18), which is why many large telescopes have recently been constructed at these exceptionally clear locations.

*In fact, for a large instrument—more than about 1m in diameter—the situa-
tion is more complicated, because rays striking different parts of the mirror have
actually passed through different turbulent regions of the atmosphere. The end
result is still a seeing disk, however.*

An optical telescope placed in orbit about Earth or on
the Moon could obviously overcome the limitations imposed by the atmosphere on ground-based instruments.
Without atmospheric blurring, extremely fine resolution—
close to the diffraction limit—can be achieved, subject only
to the engineering restrictions associated with building or
placing large structures in space. The 2.4-m mirror in the
Hubble Space Telescope has a (blue-light) diffraction limit of
only 0.05″, giving astronomers a view of the universe as
much as 20 times sharper than that normally available from
even much larger ground-based instruments.

New Telescope Design

The latest techniques for producing ultrasharp images take
the ideas of computer control and image processing (see
Section 5.2) several stages further. By analyzing the image
formed by a telescope *while the light is still being collected*, it
is now possible to adjust the telescope from moment to
moment to avoid or compensate for the effects of mirror
distortion, temperature changes in the dome, and even atmospheric turbulence. By these means, some recently constructed telescopes have achieved resolutions very close to
their theoretical (diffraction) limits.

Even under conditions of perfect seeing, most telescopes would not achieve diffraction-limited resolution.
The temperature of the mirror or in the dome may fluctu-
ate slightly during the many minutes or even hours required for the image to be exposed, and the precise shape
of the mirror may change slightly as the telescope tracks a
source across the sky. The effect of these changes is that
the mirror's focus may shift from minute to minute, blur-

▲ **FIGURE 5.18 European Southern Observatory** Located in the Andes Mountains of Chile, the European Southern Observatory at La Silla is run by a consortium of European nations. Numerous domes house optical telescopes of different sizes, each with varied support equipment, making this one of the most versatile observatories south of the equator. The largest telescope at La Silla—the square building to the right of center—is the New Technology Telescope, a 3.5-m state-of-the-art active optics device. *(ESO)*

ring the eventual image in much the same way as atmospheric turbulence creates a seeing disk (Figure 5.17). At the best observing sites, the seeing is often so good that these tiny effects may be the main cause of image blurring. The collection of techniques aimed at controlling such environmental and mechanical fluctuations is known as **active optics**.

The first telescope designed to incorporate active optics was the New Technology Telescope (NTT), constructed in 1989 at the European Southern Observatory in Chile and upgraded in 1997. (NTT is the most prominent instrument visible in Figure 5.18.) This 3.5-m instrument, employing the latest in real-time telescope controls, can achieve a resolution as sharp as 0.2″ by making minute

modifications to the tilt of its mirror as its temperature and orientation change, thus maintaining the best possible focus at all times. Figure 5.19 shows how active optics can dramatically improve the resolution of an image. Active-optics techniques now include improved dome design to control airflow, precise control of the mirror temperature, and the use of pistons behind the mirror to maintain its precise shape. All of the large telescopes described earlier include active-optics systems, improving their resolution to a few tenths of an arc second.

Real-Time Control

With active-optics systems in place, Earth's atmosphere once again becomes the main agent limiting a telescope's

(a)

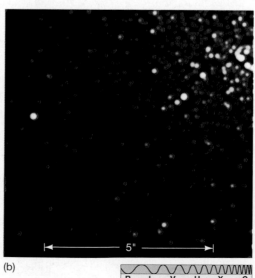

(b)

|←——— 5″ ———→|

R I V U X G

◄ **FIGURE 5.19 Active Optics** These false-color infrared photographs of part of the star cluster R136—the same object shown in Figure 5.10—contrast the resolution obtained (a) without and (b) with an active-optics system. Both images were taken with the New Technology Telescope shown in Figure 5.18. *(ESO)*

resolution. Remarkably, even this problem can now be addressed, using an approach known as **adaptive optics**, a technique that actually deforms the shape of a mirror's surface, under computer control, while the image is being exposed, in order to undo the effects of atmospheric turbulence. The mirror in question is generally not the large primary mirror of the telescope. Rather, for both economic and technical reasons, a much smaller (typically 20–50-cm-diameter) mirror is inserted into the light path and manipulated to achieve the desired effect. Adaptive optics presents formidable theoretical and practical problems, but the rewards are so great that it has been the subject of intense research since the 1970s.

The effort received an enormous boost in the 1990s from declassified military technology from the Strategic Defensive Initiative, a Reagan-era missile defense program (dubbed "Star Wars" by its detractors) intended to target and shoot down incoming ballistic missiles. In the experimental military system shown in Figure 5.20, lasers probe the atmosphere above the telescope, returning information about the air's swirling motion to a computer that modifies the mirror thousands of times per second to compensate for poor seeing.

Astronomical adaptive-optics systems often do not use lasers to gauge atmospheric conditions. Instead, they monitor standard "natural guide stars" in the field of view, constantly adjusting the mirror's shape to preserve those stars' appearance. In some systems, if no bright standard stars are available, then artificial "laser guide stars" are employed instead.

Adaptive corrections are somewhat easier to apply in the infrared than in the optical, because atmospheric distortions are smaller (stars "twinkle" less in the infrared) and because the longer infrared wavelengths impose less stringent requirements on the precise shape of the mirror. Infrared adaptive-optics systems already exist in many large telescopes. For example, Gemini and Subaru

▲ **FIGURE 5.20 Adaptive Optics** Until the 1990s, the Starfire Optical Range at Kirtland Air Force Base in Albuquerque, New Mexico, was one of the U.S. Air Force's most closely guarded secrets. Here, laser beams probe the atmosphere above the system's 1.5-m telescope, allowing the effects of air turbulence to be greatly reduced by means of minute computer-controlled changes in the shape of the mirror surface thousands of times each second. *(R. Ressmeyer)*

have reported adaptive-optics resolutions of around 0.06″ in the near infrared—not quite at the diffraction limit (0.03″ for an 8-m telescope at 1 μm, according to the equation on p. 118), but already better than the resolution of *HST* at the same wavelengths (see Figure 5.21). Both Keck and the VLT incorporate adaptive-optics instrumentation that will ultimately be capable of producing diffraction-limited images at near-infrared wavelengths.

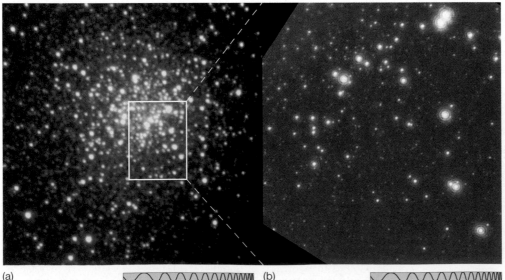

(a)   R I V U X G   (b)   R I V U X G

◀ **FIGURE 5.21 Infrared Adaptive Optics** Many large infrared telescopes now employ adaptive-optics systems. Here, the rich star cluster NGC 6934, some 50,000 light-years away, is seen with and without the use of adaptive optics at the Gemini 8-m telescope atop Mauna Kea. (a) A visible-light image whose resolution is a little less than 1″. (b) An infrared image of the central regions of the cluster, showing more stars more clearly, with resolution nearly 10 times better. *(NOAO)*

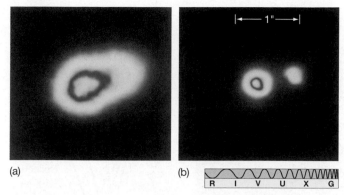

(a)   (b)

R I V U X G

▲ **FIGURE 5.22 Visible-Light Adaptive Optics** The improvement in seeing produced by adaptive-optics systems can be dramatic, as shown in these visible-light images acquired at a military observatory atop Mount Haleakala in Maui, Hawaii. The uncorrected image (a) of the double star Castor is a blur spread over several arc seconds, giving little hint of its binary nature. With adaptive compensation applied (b), the resolution is improved to a mere 0.1″, and the two components are clearly resolved. *(MIT Lincoln Laboratory)*

Visible-light adaptive optics has been demonstrated experimentally, and some astronomical telescopes may incorporate the technology within the next few years. Figure 5.22 compares a pair of visible-light observations of a nearby double star called Castor. The observations were made with a relatively modest 1.5-m telescope. The adaptive-optics system clearly distinguishes the two stars. Remarkably, it may soon be possible to have the "best of both worlds," achieving with large ground-based optical telescopes the kind of resolution presently attainable only from space.

CONCEPT CHECK

✔ What steps do optical astronomers take to overcome the obscuring and blurring effects of Earth's atmosphere?

5.5 Radio Astronomy

In addition to the visible radiation that penetrates Earth's atmosphere on a clear day, radio radiation also reaches the ground. Indeed, as indicated in Figure 3.9, the radio window in the electromagnetic spectrum is much wider than the optical window. ∞ (Sec. 3.3) Because the atmosphere is no hindrance to long-wavelength radiation, radio astronomers have built many ground-based **radio telescopes** capable of detecting radio waves reaching us from space. These devices have all been constructed since the 1950s—radio astronomy is a much younger subject than optical astronomy.

Early Observations

The field of radio astronomy originated with the work of Karl Jansky at Bell Labs in 1931. Jansky was engaged in a study of shortwave-radio interference when he discovered a faint static "hiss" that had no apparent terrestrial (Earthly) source. He noticed that the strength of the hiss varied in time and that its peak occurred about four minutes earlier each day. He soon realized that the peaks were coming exactly one *sidereal day* apart, and he concluded that the hiss was indeed not of terrestrial origin, but came from a definite direction in space. ∞ (Sec. 1.4) That direction is now known to correspond to the center of our Galaxy.

Some astronomers were intrigued by Jansky's discovery, but with the limited technology of the day—and even more limited Depression-era budgets—progress was slow. Jansky himself was moved to another project at Bell Labs and never returned to astronomical studies. However, by 1940, the first systematic surveys of the radio sky were being carried out. After a series of technological breakthroughs made during World War II, these studies rapidly grew into a distinct branch of astronomy.

During the 1930s, astronomers became aware that the space between the stars in our Galaxy is not empty, but instead is filled with extremely diffuse (low-density) gas (see Chapter 18). The growing realization in the 1940s that this otherwise completely invisible part of the Galaxy could be observed and mapped in detail at radio wavelengths established the true importance of Jansky's pioneering work. Today he is regarded as the father of radio astronomy.

Essentials of Radio Telescopes

Figure 5.23(a) shows the world's largest steerable radio telescope: the large 105-m- (340-foot-) diameter telescope located at the National Radio Astronomy Observatory in West Virginia. Although much larger than reflecting optical telescopes, most radio telescopes are built in basically the same way. They have a large, horseshoe-shaped mount supporting a huge curved metal dish that serves as the collecting area. As illustrated in Figure 5.23(b), the dish captures incoming radio waves and reflects them to the focus, where a receiver detects the signals and channels them to a computer.

Conceptually, the operation of a radio telescope is similar to the operation of an optical reflector with the detecting instruments placed at the prime focus (Figure 5.6a). However, unlike optical instruments, which can detect all visible wavelengths simultaneously, radio detectors normally register only a narrow band of wavelengths at any one time. To observe radiation at another radio frequency, we must retune the equipment, much as we tune a television set to a different channel.

Radio telescopes must be built large partly because cosmic radio sources are extremely faint. In fact, the total amount of radio energy received by Earth's entire surface is less than a trillionth of a watt. Compare this with the

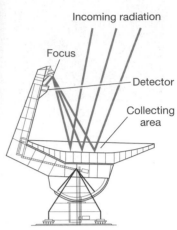

▲ **FIGURE 5.23** Radio Telescope (a) The 105-m-diameter device at the National Radio Astronomy Observatory in Green Bank, West Virginia, is 150 meters high—taller than the Statue of Liberty and nearly as tall as the Washington Monument. (b) Schematic diagram of the telescope, showing the path taken by an incoming beam of radio radiation. *(NRAO)*

roughly 10 *million* watts our planet's surface receives in the form of infrared and visible light from any of the bright stars seen in the night sky. To capture enough radio energy to allow detailed measurements to be made, a large collecting area is essential.

Because of diffraction, the angular resolution of radio telescopes is generally quite poor compared with that of their optical counterparts. Typical wavelengths of radio waves are about a million times longer than those of visible light, and these longer wavelengths impose a corresponding crudeness in angular resolution. (Recall from Section 5.3 that the longer the wavelength, the greater the amount of diffraction.) Even the enormous sizes of radio dishes only partly offset this effect. The radio telescope shown in Figure 5.23 can achieve a resolution of about 1′ when receiving radio waves having wavelengths of around 3 cm. However, it was designed to operate most efficiently (i.e., it is most sensitive to radio signals) at wavelengths closer to 1 cm, where the resolution is approximately 20″. The best angular resolution obtainable with a single radio telescope is about 10″ (for the largest instruments operating at millimeter wavelengths)—at least 100 times coarser than the capabilities of some large optical systems.

Radio telescopes can be built so much larger than their optical counterparts because their reflecting surfaces need not be as smooth as is necessary for light waves of shorter wavelength. Provided that surface irregularities (dents, bumps, and the like) are much smaller than the wavelength of the waves to be detected, the surface will reflect them without distortion. Because the wavelength of visible radiation is short (less than 10^{-6} m), extremely smooth mirrors are needed to reflect the waves properly, and it is difficult to construct very large mirrors to such exacting tolerances. However, even rough metal surfaces can focus 1-cm waves accurately, and radio waves of wavelength a meter or more can be reflected and focused perfectly well by surfaces having irregularities even as large as your fist.

Figure 5.24 shows the world's largest and most sensitive radio telescope, strung among the hills of Arecibo, Puerto Rico. Approximately 300 m (1000 feet) in diameter, the surface of the Arecibo telescope spans nearly 20 acres. Constructed in 1963 in a natural depression in the hillside, the dish was originally surfaced with chicken wire, which was lightweight and cheap. Although fairly rough, the chicken wire was adequate for proper reflection because the openings between adjacent strands of wire were much smaller than the long-wavelength radio waves that were to be detected.

The entire Arecibo dish was resurfaced with thin metal panels in 1974 and was upgraded in 1997, so that it can now be used to study radio radiation of shorter wavelength. Since the 1997 upgrade, the panels can be adjusted to maintain a precise spherical shape to an accuracy of about 3 mm over the entire surface. At a frequency of 5 GHz (corresponding to a wavelength of 6 cm—the short-

▲ **FIGURE 5.24 Arecibo Observatory** An aerial photograph of the 300-m-diameter dish at the National Astronomy and Ionospheric Center near Arecibo, Puerto Rico. The receivers that detect the focused radiation are suspended nearly 150 m (about 45 stories) above the center of the dish. The left inset shows a close-up of the radio receivers hanging high above the dish. The right inset shows technicians adjusting the dish surface to make it smoother. *(D. Parker/T. Acevedo/NAIC; Cornell)*

est wavelength that can be studied, given the properties of the dish surface), the telescope's angular resolution is about 1′. The huge size of the dish creates one distinct disadvantage, however: The Arecibo telescope cannot be pointed very well to follow cosmic objects across the sky. The detectors can move roughly 10° on either side of the focus, restricting the telescope's observations to those objects which happen to pass within about 20° of overhead as Earth rotates.

Arecibo is an example of a rough-surfaced telescope capable of detecting long-wavelength radio radiation. At the other extreme, Figure 5.25 shows the 36-m-diameter Haystack dish in northeastern Massachusetts. Constructed of polished aluminum, this telescope maintains a parabolic curve to an accuracy of about a millimeter all the way across its solid surface. It can reflect and accurately focus radio radiation with wavelengths as short as a few millimeters. The telescope is contained within a protective shell, or radome, that protects the surface from the harsh New

England weather. The radome acts much like the protective dome of an optical telescope, except that there is no slit through which the telescope "sees." Incoming cosmic radio signals pass virtually unimpeded through the radome's fiberglass construction.

The Value of Radio Astronomy

Despite the inherent disadvantage of relatively poor angular resolution, radio astronomy enjoys many advantages. Radio telescopes can observe 24 hours a day. Darkness is not needed for receiving radio signals, because the Sun is a relatively weak source of radio energy, so its emission does not swamp radio signals arriving at Earth from elsewhere in the sky. In addition, radio observations can often be made through cloudy skies, and radio telescopes can detect the longest-wavelength radio waves even during rain or snowstorms. Poor weather causes few problems, because the wavelength of most radio waves is much larger than

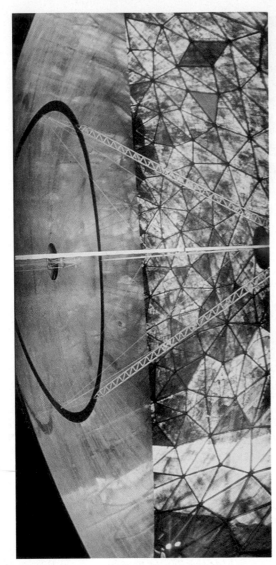

▲ **FIGURE 5.25 Haystack Observatory** Photograph of the Haystack dish, inside its protective radome. For scale, note the engineer standing at the bottom. Note also the dull shine on the telescope surface, indicating its smooth construction. Haystack is a poor optical mirror, but a superb radio telescope. Accordingly, it can be used to reflect and accurately focus radiation having short radio wavelengths, even as small as a fraction of a centimeter. (*MIT*)

not necessarily strong radio emitters, many of the strongest radio sources in the universe emit little or no visible light. Second, visible light may be strongly absorbed by interstellar dust along the line of sight to a source. Radio waves, by contrast, are generally unaffected by intervening matter. Third, as mentioned earlier, many parts of the universe cannot be seen at all by optical means, but are easily detectable at longer wavelengths. The center of the Milky Way galaxy is a prime example of such a totally invisible region—our knowledge of the Galactic center is based almost entirely on radio and infrared observations. Thus, these observations not only afford us the opportunity to study the same objects at different wavelengths, but also allow us to see whole new classes of objects that would otherwise be completely unknown.

Figure 5.26 shows an optical photograph of the Orion Nebula (a huge cloud of interstellar gas) taken with the 4-m telescope on Kitt Peak. Superimposed on the optical image is a radio map of the same region, obtained by scanning the Haystack radio telescope (Figure 5.25) back and forth across the nebula and taking many measurements of radio intensity. The map is drawn as a series of contour

R I V U X G

▲ **FIGURE 5.26 Orion Nebula in Radio and Visible** The Orion Nebula is a star-forming region about 1500 light-years from Earth. (The nebula is located in the constellation Orion and can be seen in Figure 1.8.) The bright regions in this photograph are stars and clouds of glowing gas. The dark regions are not empty, but their visible emission is obscured by interstellar matter. Superimposed on the optical image is a radio contour map (white lines) of the same region. Each curve of the contour map represents a different intensity of radio emission. The resolution of the optical image is about 1″; that of the radio map is 1′. (*background photo: AURA*)

the typical size of atmospheric raindrops or snowflakes. Optical astronomy cannot be performed under these conditions, because the wavelength of visible light is smaller than a raindrop, a snowflake, or even a minute water droplet in a cloud.

However, perhaps the greatest value of radio astronomy (and, in fact, of all astronomies concerned with nonvisible regions of the electromagnetic spectrum) is that it opens up a whole new window on the universe. There are three main reasons for this. First, just as objects that are bright in the visible part of the spectrum (e.g., the Sun) are

lines connecting locations of equal radio brightness, similar to pressure contours drawn by meteorologists on weather maps or height contours drawn by cartographers on topographic maps. The inner contours represent stronger radio signals, the outside contours weaker signals. Note, however, how the radio map is much less detailed than its optical counterpart; that's because the acquired radio radiation has such a long wavelength compared to light.

The radio map in Figure 5.26 has many similarities to the visible-light image of the nebula. For instance, the radio emission is strongest near the center of the optical image and declines toward the nebular edge. But there are also subtle differences between the radio and optical images. The two differ chiefly toward the upper left of the main cloud, where visible light seems to be absent, despite the existence of radio waves. How can radio waves be detected from locations not showing any emission of light? The answer is that this particular nebular region is known to be especially dusty in its top left quadrant. The dust obscures the short-wavelength visible light, but not the long-wavelength radio radiation. Thus, we have a trade-off typical of many in astronomy: Although the long-wave radio signals provide a less resolved map of the region, those same radio signals can pass relatively unhindered through dusty regions, in this case allowing us to see the true extent of the Orion Nebula.

CONCEPT CHECK

✔ In what ways does radio astronomy complement optical observations?

5.6 Interferometry

The main disadvantage of radio astronomy compared with optical work is its relatively poor angular resolution. However, in some circumstances, radio astronomers can overcome this limitation with a technique known as **interferometry**. This technique makes it possible to produce radio images of angular resolution higher than can be achieved with even the best optical telescopes, on Earth or in space.

In interferometry, two or more radio telescopes are used in tandem to observe the *same* object at the *same* wavelength and at the *same* time. The combined instruments together make up an **interferometer**. Figure 5.27 shows a large interferometer—many separate radio telescopes working together as a team. By means of electronic cables or radio links, the signals received by each antenna in the array making up the interferometer are sent to a central computer that combines and stores the data.

Interferometry works by analyzing how the signals interfere with each other when added together. ∞ *(Discovery 3-1)* Consider an incoming wave striking two detectors (Figure 5.28). Because the detectors lie at different distances from the source, the signals they record will, in general, be out of step with one another. In that case, when the signals are combined, they will interfere destructively, partly canceling each other out. Only if the detected radio waves happen to be exactly in step will the signals combine constructively to produce a strong signal. Notice that the amount of interference depends on the *direction* in which the wave is traveling relative to the line joining the detectors. Thus—in principle, at least—careful analysis of the

(a)

(b)

▲ **FIGURE 5.27 VLA Interferometer** (a) This large interferometer is made up of 27 separate dishes spread along a Y-shaped pattern about 30 km across on the Plain of San Augustin near Socorro, New Mexico. The most sensitive radio device in the world, it is called the Very Large Array, or VLA for short. (b) A close-up view from ground level shows how some of the VLA dishes are mounted on railroad tracks so that they can be moved easily. *(NRAO)*

▶ **FIGURE 5.28 Interferometry** Two detectors, *A* and *B*, record different signals from the same incoming wave because of the time it takes the radiation to traverse the distance between them. When the signals are combined, interference occurs. The amount of interference depends on the wave's direction of motion, providing a means of measuring the position of the source in the sky. Here, the brown-colored waves come from a source high in the sky and interfere destructively when captured by antennas *A* and *B*. But when the same source has moved because of Earth's rotation (orange-colored waves), the interference can be constructive.

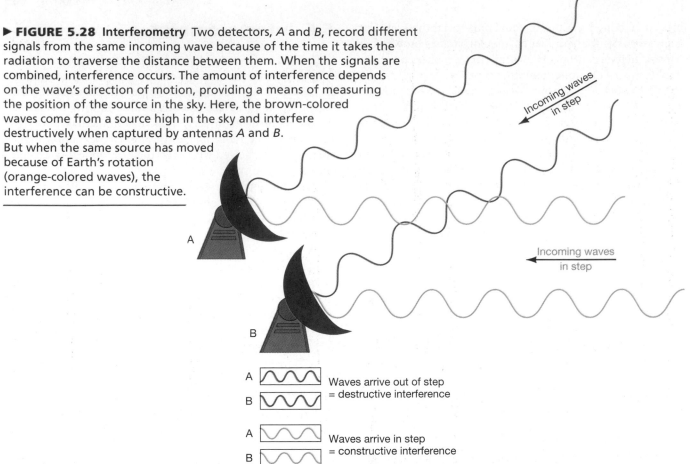

A [waveform] B [waveform] Waves arrive out of step = destructive interference

A [waveform] B [waveform] Waves arrive in step = constructive interference

strength of the combined signal can provide an accurate measurement of the source's position in the sky.

As Earth rotates and the antennae track their target, the interferometer's orientation relative to the source changes, and a pattern of peaks and troughs emerges. In practice, extracting positional information from the data is a complex task, as multiple antennae and several sources are usually involved. Suffice it to say that, after extensive computer processing, the interference pattern translates into a high-resolution image of the target object.

An interferometer is, in essence, a substitute for a single huge antenna. As far as resolving power is concerned, the effective diameter of an interferometer is the distance between its outermost dishes. In other words, two small dishes can act as opposite ends of an imaginary, but huge, single radio telescope, dramatically improving angular resolution. For example, a resolution of a few arc seconds can be achieved at typical radio wavelengths (such as 10 cm), either by using a single radio telescope 5 km in diameter (which is impossible to build) or by using two or more much smaller dishes separated by the same 5 km, but connected electronically. The larger the distance separating the telescopes—that is, the longer the *baseline* of the interferometer—the better is the resolution attainable.

Large interferometers like the instrument shown in Figure 5.27 now routinely attain radio resolution comparable to that of optical images. Figure 5.29 shows an inter-

ferometric radio map of a nearby galaxy and a photograph of that same galaxy, made with a large optical telescope. The radio clarity is much better than in the contour map of Figure 5.26—in fact, the radio resolution in part (a) is comparable to that of the optical image in part (b).

Astronomers have created radio interferometers spanning great distances, first across North America and later between continents. A typical very-long-baseline interferometry experiment (usually known by its abbreviation, VLBI) might use radio telescopes in North America, Europe, Australia, and Russia to achieve an angular resolution on the order of 0.001″. It seems that even Earth's diameter is no limit: Radio astronomers have successfully used an antenna in orbit, together with several antennae on the ground, to construct an even longer baseline and achieve still better resolution. Proposals exist to place interferometers entirely in Earth orbit and even on the Moon.

Although the technique was originally developed by radio astronomers, interferometry is no longer restricted to the radio domain. Radio interferometry became feasible when electronic equipment and computers achieved speeds great enough to combine and analyze radio signals from separate radio detectors without loss of data. As the technology has improved, it has become possible to apply the same methods to radiation of higher frequency. Millimeter-wavelength interferometry has already become an established and important observational technique, and both

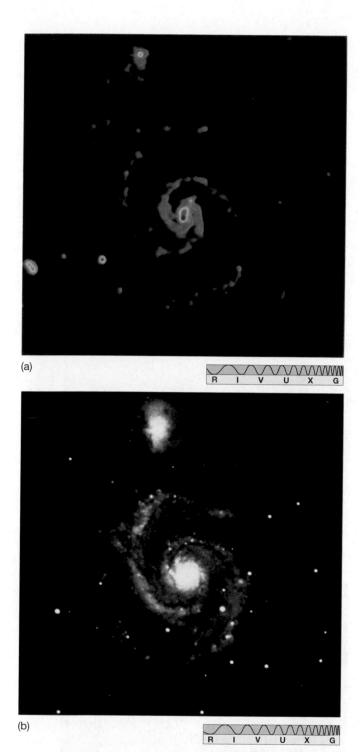

(a)

(b)

◄ **FIGURE 5.29** **Radio–Optical Comparison** (a) VLA radio "image" (or radiograph) of the spiral galaxy M51, observed at radio frequencies with an angular resolution of a few arc seconds. (b) Visible-light photograph of that same galaxy, made with the 4-m Kitt Peak optical telescope and displayed on the same scale as (a). *(NRAO/AURA)*

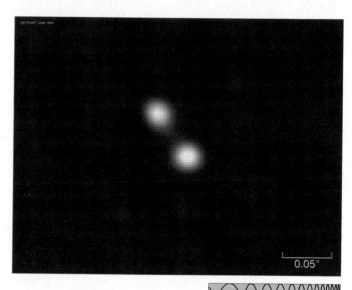

0.05"

▲ **FIGURE 5.30** **Optical Interferometry** This extremely high resolution image of the double-star Capella was made by the COAST interferometer in Cambridge, England, at the infrared wavelength of 0.83 μm. Capella's two member stars are normally seen from the ground only as a slight oblong blur, but with the COAST array the two stars are clearly resolved. *(MRAO)*

the Keck telescopes and the VLT are expected to be used for near-infrared interferometry within the next few years.

Optical interferometry is currently the subject of intensive research. In 1997, a group of astronomers in Cambridge, England, succeeded in combining the light from three small optical telescopes to produce a single, remarkably clear, image. Each telescope of the Cambridge Optical Aperture Synthesis Telescope (COAST) was only 0.4 m in diameter, but with the mirrors positioned 6 m apart, the resulting resolution was a stunning 0.01″—better than the resolution of the best adaptive-optics systems on the ground or of *HST* operating above Earth's atmosphere. The three telescopes working together enabled astronomers to "split" the binary star Capella, whose two member stars are separated by only 0.05″. Normally Capella is seen from the ground only as a slightly oblong blur. With the COAST array, the two stars that make up the system are cleanly and individually separated (Figure 5.30). NASA has ambitious plans to place an optical interferometer—the *Space Interferometry Mission*, or *SIM*—in Earth orbit by 2010.

CONCEPT CHECK

✔ What is the main reason for the poor angular resolution of radio telescopes? How do radio astronomers overcome this problem?

5.7 Space-Based Astronomy

Optical and radio astronomy are the oldest branches of astronomy, but since the 1970s there has been a virtual explosion of observational techniques spanning the rest of the electromagnetic spectrum. Today, all portions of the spectrum are studied, from radio waves to gamma rays, to maximize the amount of information available about astronomical objects. As noted earlier, the types of astronomical objects that can be observed differ quite markedly from one wavelength range to another. Full-spectrum coverage is essential not only to see things more clearly, but even to see some things at all. Because of the transmission characteristics of Earth's atmosphere, astronomers must study practically all regions of the electromagnetic spectrum, from gamma rays through X rays to visible light, and on down to infrared and radio waves, from space. The rise of these "other astronomies" has therefore been closely tied to the development of the space program.

Infrared Astronomy

Infrared studies are a vital component of modern observational astronomy. Infrared astronomy spans a broad range of phenomena in the universe, from planets and their parent stars to the vast regions of interstellar space where new stars are forming, to explosive events occurring in faraway galaxies. Generally, **infrared telescopes** resemble optical telescopes, but their detectors are designed to be sensitive to radiation of longer wavelengths. Indeed, as we have seen, many ground-based "optical" telescopes are also used for infrared work, and some of the most useful infrared observing is done from the ground (e.g., from

(a)

(b)

(c)

(d)

▲ **FIGURE 5.31 Smog Revealed** (a) An optical photograph taken near San Jose, California, and (b) an infrared photo of the same area taken at the same time. Infrared radiation of longer wavelength can penetrate smog much better than short-wavelength visible light can. The same advantage pertains to astronomical observations: (c) An optical view of an especially dusty part of our Galaxy, the central region of the Orion Nebula, is more clearly revealed (d) in this infrared image showing a cluster of stars behind the obscuring dust. *(Lick Observatory; NASA)*

Mauna Kea—see Figure 5.13), even though the radiation is somewhat diminished in intensity by our atmosphere.

As with radio observations, the longer wavelength of infrared radiation often enables us to perceive objects that are partially hidden from optical view. As a terrestrial example of the penetrating properties of infrared radiation, Figure 5.31(a) shows a dusty and hazy region in California, hardly viewable optically, but easily seen in the infrared (Figure 5.31b). Figures 5.31(c) and (d) show a similar comparison for an astronomical object—the dusty regions of the Orion Nebula, where much visible light is hidden behind interstellar clouds, but which is clearly distinguishable in the infrared.

Astronomers can make better infrared observations if they can place their instruments above most or all of Earth's atmosphere, using balloon-, aircraft-, rocket-, and satellite-based telescopes (see Figure 5.32). However, as might be expected, the infrared telescopes that can be carried above the atmosphere are considerably smaller than the massive instruments found in ground-based observatories.

Figure 5.33 shows an infrared image of the Orion region as seen by the 0.6-m British–Dutch–U.S. *Infrared Astronomy Satellite (IRAS)* in 1983 and compares it with the same region made in visible light. Figure 5.33(a) is a composite of IRAS images taken at three different infrared wavelengths (12 μm, 60 μm, and 100 μm). It is represented in *false color*, a technique commonly used for displaying images taken in nonvisible light. The colors do not represent the actual wavelength of the radiation emitted, but instead some other property of the source, in this case temperature, descending from white to orange to black. The whiter regions thus denote higher temperatures and hence greater strength of infrared radiation. At about 1' angular resolution, the fine details of the Orion nebula evident in the visible portion of Figure 5.26 cannot be perceived. Nonetheless, the infrared image shows clouds of warm dust and gas, believed to play a critical role in the formation of stars, and also shows groups of bright young stars that are completely obscured at visible wavelengths.

Much of the material between the stars has a temperature between a few tens and a few hundreds of kelvins. Accordingly, Wien's law tells us that, the infrared domain is the natural portion of the electromagnetic spectrum in which to study it. ∞ (Sec. 3.4) Unfortunately, also by Wien's law, telescopes themselves also radiate strongly in the infrared, unless they are cooled to nearly absolute zero. The end of *IRAS's* mission came not because of any equipment malfunction or unexpected mishap, but simply because its supply of liquid helium coolant ran out.

(a)

◄ **FIGURE 5.32 Infrared Telescopes** (a) A gondola containing a 1-m infrared telescope (lower left) is readied for its balloonborne ascent to an altitude of about 30 km, where it will capture infrared radiation that cannot penetrate Earth's atmosphere. (b) An artist's conception of the *Spitzer Space Telescope*, placed in orbit in 2003. This 0.8-m telescope surveys the infrared sky at wavelengths ranging from 3 to 200 μm. During its five-year mission, the instrument is expected to greatly increase astronomers' understanding of many aspects of the universe, from the formation of stars and planets to the evolution of galaxies. *(SAO; NASA)*

(b)

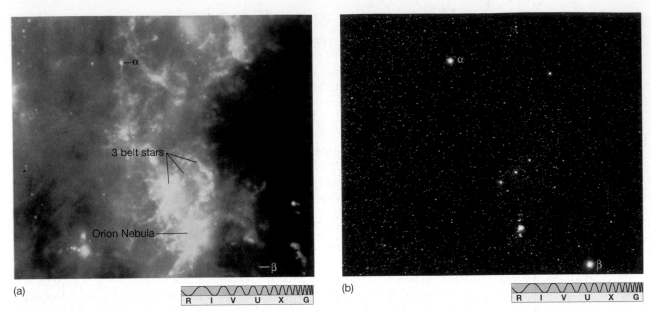

▲ FIGURE 5.33 Infrared Image (a) This infrared image of the Orion region was made by the *Infrared Astronomy Satellite*. In the false-color image, colors denote different temperatures, descending from white to orange to black. The resolution is about 1′. (b) The same region photographed in visible light with 1″ resolution. The labels α and β refer, respectively, to Betelgeuse and Rigel, the two brightest stars in the constellation. Note how the red star Betelgeuse is easily seen in the infrared (part a), while the blue star Rigel is very faint. *(NASA; J. Sanford)*

IRAS's own thermal emission then overwhelmed the radiation it was built to detect.

In August 2003, NASA launched the 0.85-m *Space Infrared Telescope Facility* (*SIRTF*, shown in Figure 5.32b). In December of that year, when the first images from the instrument were made public, NASA renamed the spacecraft the *Spitzer Space Telescope* (*SST*), in honor of Lyman Spitzer, Jr., renowned astrophysicist and the first person to propose (in 1946) that a large telescope be located in space. The facility's detectors are designed to operate at wavelengths in the 3-μm to 200-μm range. Unlike previous space-based observatories, the *SST* does not orbit Earth, but instead follows our planet in its orbit around the Sun, trailing millions of kilometers behind in order to minimize Earth's heating effect on the detectors. The spacecraft is currently drifting away from Earth at the rate of 0.1 AU per year. The detectors are cooled to near absolute zero in order to observe infrared signals from space without interference from the telescope's own heat. Figure 5.34 shows some early imagery from NASA's latest eye on the universe.

The present instrument package aboard the *Hubble Space Telescope (see Discovery 5-1)* also includes a high-resolution (0.1″) near-infrared camera and spectroscope.

Ultraviolet Astronomy

On the short-wavelength side of the visible spectrum lies the ultraviolet domain. This region of the spectrum, extending in wavelength from 400 nm (blue light) down to a few nanometers ("soft" X rays), has only recently begun to be explored. Because Earth's atmosphere is partially opaque to radiation below 400 nm and is totally opaque below about 300 nm (due in part to the ozone layer), astronomers cannot conduct any useful ultraviolet observations from the ground, not even from the highest mountaintop. Rockets, balloons, or satellites are therefore essential to any **ultraviolet telescope**—a device designed to capture and analyze that high-frequency radiation.

One of the most successful ultraviolet space missions was the *International Ultraviolet Explorer* (*IUE*), placed in Earth orbit in 1978 and shut down for budgetary reasons in late 1996 (see *Discovery 18-1*). Like all ultraviolet telescopes, its basic appearance and construction were quite similar to those of optical and infrared devices. Several hundred astronomers from all over the world used *IUE's* near-ultraviolet spectroscopes to explore a variety of phenomena in planets, stars, and galaxies. In subsequent chapters, we will learn what this relatively new window on the universe has shown us about the activity and even the violence that seems to pervade the cosmos. Today, the *Far Ultraviolet Spectrographic Explorer* (*FUSE*) satellite, launched in 1999, is extending *IUE's* work into the "far" ultraviolet (around 100 nm) regime. In addition, the *Hubble Space Telescope*, best known as an optical telescope, is also a superb imaging and spectroscopic ultraviolet instrument.

Figure 5.35 shows an image of a supernova remnant—the remains of a violent stellar explosion that occurred

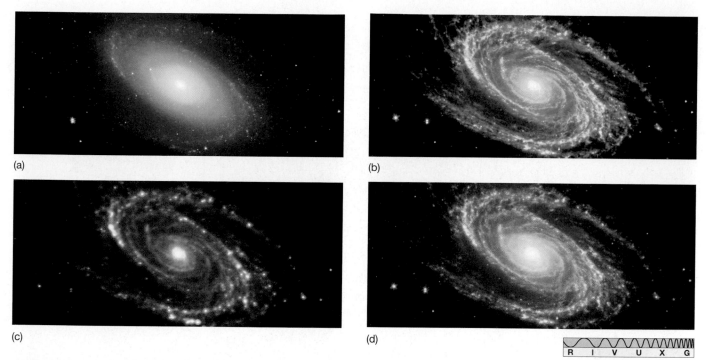

(a)

(b)

(c)

(d)

R I V U X G

▲ **FIGURE 5.34** *Spitzer* Images Early images from the new *Spitzer Space Telescope*, now in orbit around the Sun, clearly show its camera's capabilities while focusing on the magnificent spiral galaxy, M81, about 12 million light-years away. Frames (a)–(c) show the galaxy at increasing infrared wavelengths, namely 4, 8 and 24 microns, while (d) shows the multiwavelength composite made from combining the three frames. *(JPL)*

millennia ago—obtained by Europe's *X-ray Multi-Mirror* (*XMM-Newton*) satellite. Launched in 1999 from French Guiana, *XMM* operates in the ultraviolet and X-ray parts of the spectrum, making it sensitive to phenomena involving high temperatures (hundreds of thousands to millions of kelvins) or other energetic events. ∞ (Sec. 3.4) Now

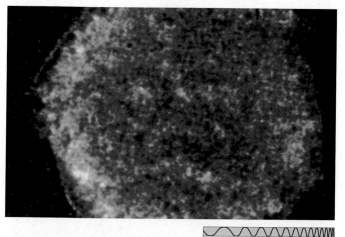

R I V U X G

▲ **FIGURE 5.35** Ultraviolet Image A camera on board the European *XMM-Newton* captured this image of the Tycho supernova remnant. This was a massive star that blew itself virtually to smithereens, as seen on Earth in the year 1572. The release of energy was prodigious and the afterglow has lingered for centuries.

mapping our local cosmic neighborhood as it presents itself in the far ultraviolet, *XMM* is changing astronomers' conception of interstellar space in the vicinity of the Sun.

High-Energy Astronomy

High-energy astronomy studies the universe as it presents itself to us in X rays and gamma rays—the types of radiation whose photons have the highest frequencies and hence the greatest energies. How do we detect radiation of such short wavelengths? First, it must be captured high above Earth's atmosphere, because none of it reaches the ground. Second, its detection requires the use of equipment fundamentally different in design from that used to capture the relatively low energy radiation discussed up to this point.

The difference in the design of **high-energy telescopes** comes about because X rays and gamma rays cannot be reflected easily by any kind of surface. Rather, these rays tend to pass straight through, or be absorbed by, any material they strike. When X rays barely graze a surface, however, they can be reflected from it in a way that yields an image, although the mirror design is fairly complex. As illustrated in Figure 5.36, to ensure that all incoming rays are reflected at grazing angles, the telescope is constructed as a series of nested cylindrical mirrors, carefully shaped to bring the X rays to a sharp focus. For gamma rays, no such method of producing an image has yet been devised; present-day gamma-ray telescopes simply point in a specified direction and count the photons they collect.

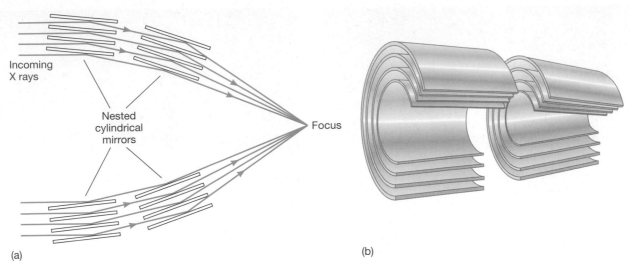

(a)

(b)

▲ **FIGURE 5.36 X-Ray Telescope** (a) The arrangement of nested mirrors in an X-ray telescope allows X rays to be reflected at grazing angles and focused to form an image. (b) A cutaway 3-D rendition of the mirrors, showing their shape more clearly.

In addition, detection methods using photographic plates or CCD devices do not work well for hard (high-frequency) X rays and gamma rays. Instead, individual photons are counted by electronic detectors on board an orbiting device, and the results are then transmitted to the ground for further processing and analysis. Furthermore, the number of photons in the universe seems to be inversely related to their frequency. Trillions of visible (starlight) photons reach the detector of an optical telescope on Earth each second, but hours or even days are sometimes needed for a single gamma-ray photon to be recorded. Not only are these photons hard to focus and measure, but they are also very scarce.

The *Einstein Observatory*, launched by NASA in 1978, was the first X-ray telescope capable of forming an image of its field of view. During its two-year lifetime, this spacecraft drove major advances in our understanding of high-energy phenomena throughout the universe. Its observational database is still heavily used. More recently, the German *ROSAT* (short for "Röntgen Satellite," after

Wilhelm Röntgen, the discoverer of X rays) was launched in 1991. During its seven-year lifetime, this instrument generated a wealth of high-quality observational data. It was turned off in 1999, a few months after its electronics were irreversibly damaged when the telescope was accidentally pointed too close to the Sun.

In July 1999, NASA launched the *Chandra X-Ray Observatory* (named in honor of the Indian astrophysicist

▶ **FIGURE 5.37** *Chandra* **Observatory** The *Chandra* X-ray telescope is shown here during the final stages of its construction in 1998. The left end of the mirror arrangement, depicted in Figure 5.36, is at the bottom of the satellite as oriented here. *Chandra*'s effective angular resolution is 1″, allowing this spacecraft to produce images of quality comparable to that of optical photographs. *Chandra* now occupies an elliptical orbit high above Earth; its farthest point from our planet, 140,000 km out, reaches almost one-third of the way to the Moon. Such an orbit takes *Chandra* above most of the interfering radiation belts girdling Earth (see Section 7.5) and allows more efficient observations of the cosmos. *(NASA)*

Subramanyan Chandrasekhar and shown in Figure 5.37). With greater sensitivity, a wider field of view, and better resolution than either *Einstein* or *ROSAT*, *Chandra* is providing high-energy astronomers with new levels of observational detail. Figure 5.38 shows the first image returned by *Chandra*: a supernova remnant in the constellation Cassiopeia. Known as Cas A, the ejected gas is all that now remains of a star that was observed to explode about 320 years ago. The false-color image shows 50-million-K gas in wisps of ejected stellar material; the bright white point at the very center of the debris may be a black hole. The *XMM-Newton* satellite is more sensitive to X rays than is *Chandra* (that is, it can detect fainter X-ray sources), but it has significantly poorer angular resolution (5″, compared to 0.5″ for *Chandra*), making the two missions complementary to one another.

Gamma-ray astronomy is the youngest entrant into the observational arena. As mentioned a moment ago, imaging gamma-ray telescopes do not exist, so only fairly coarse (1° resolution) observations can be made. Nevertheless, even at this resolution, there is much to be learned. Cosmic gamma rays were originally detected in the 1960s by the U.S. *Vela* series of satellites, whose primary mission was to monitor illegal nuclear detonations on Earth. Since then, several X-ray telescopes have also been equipped with gamma-ray detectors.

By far the most advanced instrument was the *Compton Gamma-Ray Observatory* (*CGRO*), launched by the space shuttle in 1991 (Figure 5.39(a)). At 17 tons, it was at the time the largest astronomical instrument ever launched by NASA. *CGRO* scanned the sky and studied individual

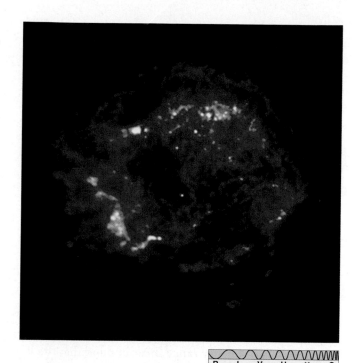

R I V U X G

▲ **FIGURE 5.38** **X-Ray Image** A false-color *Chandra* X-ray image of the supernova remnant Cassiopeia A, a debris field of scattered, glowing gases that were once part of a massive star. Here, color represents the intensity of the X rays observed, from white (brightest) through red (faintest). Cas A itself is roughly 10,000 light-years from Earth. The region, now barely visible in the optical part of the spectrum, is awash in brilliantly glowing X rays spread across some 10 light-years. (*NASA*)

ANIMATION Light and Data Paths

(a)

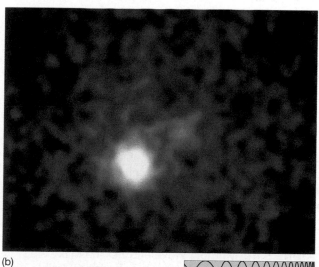

(b)

R I V U X G

▲ **FIGURE 5.39** **Gamma-Ray Astronomy** (a) This photograph of the 17-ton *Compton Gamma-Ray Observatory* (*CGRO*; named after an American pioneer in gamma-ray telescopy) was taken by an astronaut during the satellite's deployment from the space shuttle *Atlantis* over the Pacific Coast of the United States. (b) A typical false-color gamma-ray image—this one showing a violent event in the distant galaxy 3C279, also known as a "gamma-ray blazar." (*NASA*)

objects in much greater detail than had previously been attempted. Figure 5.39(b) shows a false-color gamma-ray image of a highly energetic outburst in the center of a distant galaxy. The mission ended on June 4, 2000, when, following a failure of one of the satellite's three gyroscopes, NASA opted for a controlled reentry and dropped *CGRO* into the Pacific Ocean.

CONCEPT CHECK

✔ List some scientific benefits of placing telescopes in space. What are the drawbacks of space-based astronomy?

5.8 Full-Spectrum Coverage

Table 5.1 lists the basic regions of the electromagnetic spectrum and describes objects typically studied in each frequency range. Bear in mind that the list is far from exhaustive and that many astronomical objects are now routinely observed at many different electromagnetic wavelengths. As we proceed through the text, we will discuss more fully the wealth of information that high-precision astronomical instruments can provide us.

It is reasonable to suppose that the future holds many further improvements in both the quality and the availability of astronomical data and that many new discoveries will be made. The current and proposed pace of technological progress presents us with the following exciting prospect: Within the next decade, if all goes according to plan, it will be possible, for the first time ever, to make *simultaneous* high-quality measurements of any astronomical object at *all* wavelengths, from radio to gamma ray. The consequences of this development for our understanding of the workings of the universe may be little short of revolutionary.

As a preview of the sort of comparison that full-spectrum coverage allows, Figure 5.40 shows a series of images of our own Milky Way galaxy. The images were made by several different instruments, at wavelengths ranging from radio to gamma ray, over a period of about five years. By comparing the features visible in each, we immediately see how multiwavelength observations can complement one another, greatly extending our perception of the dynamic universe around us.

▶ **FIGURE 5.40 Multiple Wavelengths** The Milky Way Galaxy as it appears at (a) radio, (b) infrared, (c) visible, (d) X-ray, and (e) gamma-ray wavelengths. Each frame is a panoramic view covering the entire sky. The center of our Galaxy, which lies in the direction of the constellation Sagittarius, is at the center of each map. *(NRAO; NASA; Lund Observatory; K. Dennuerl and W. Voges, MPI; NASA)*

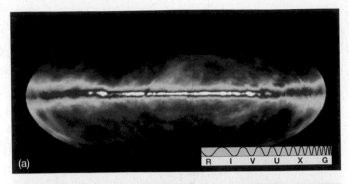

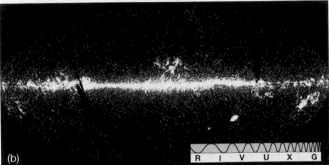

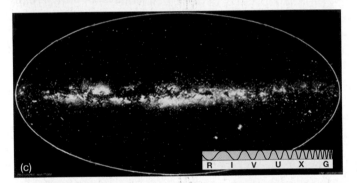

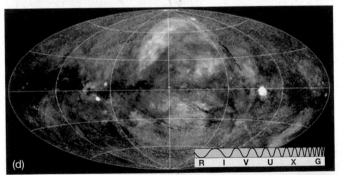

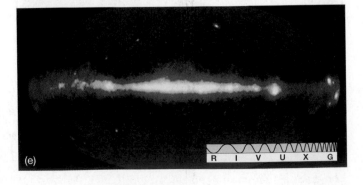

TABLE 5.1 Astronomy at Many Wavelengths

	General Considerations	Common Applications (Chapter Reference)
Radio	Can penetrate dusty regions of interstellar space Earth's atmosphere largely transparent to these wavelengths Can be detected in the daytime as well as at night High resolution at long wavelengths requires very large telescopes or interferometers	Radar studies of planets (2,9) Planetary magnetic fields (11) Interstellar gas clouds and molecules (18) Galactic structure (23, 24) Galactic nuclei and active galaxies (23, 24) Cosmic background radiation (27)
Infrared	Can penetrate dusty regions of interstellar space Earth's atmosphere only partially transparent to infrared radiation, so some observations must be made from space	Star formation (19,20); ec stars (20) Center of the Milky Way galaxy (23) Active galaxies (24) Large-scale structure of the universe (25, 27)
Visible	Earth's atmosphere transparent to visible light	Planets (7–14) Stars and stellar evolution (17, 20, 21) Normal and Active Galaxies (23, 24) Large-scale structure of the universe (25)
Ultraviolet	Earth's atmosphere is opaque to ultraviolet radiation, so observations must be made from space	Interstellar medium (19) Hot stars (21)
X ray	Earth's atmosphere is opaque to X rays, so observations must be made from space. Special mirror configurations are needed to form images	Stellar atmospheres (16) Neutron stars and black holes (22) Active galactic nuclei (24) Hot gas in galaxy clusters (25)
Gamma ray	Earth's atmosphere is opaque to gamma rays, so observations must be made from space. Cannot form images	Neutron stars (22) Active galactic nuclei (24)

Chapter Review

SUMMARY

A **telescope (p. 106)** is a device designed to collect as much light as possible from some distant source and deliver the light to a detector for detailed study. **Reflecting telescopes (p. 106)** use a mirror to concentrate and focus the light. **Refracting telescopes (p. 106)** use a lens; **refraction (p. 106)** is the bending of light as it passes from one medium to another. All large astronomical telescopes use mirrors in their design. The **prime focus (p. 106)** of a telescope is the point on which the incoming beam is focused and at which instruments for analysis may be placed. The **Newtonian (p. 108)** and **Cassegrain telescope (p. 109)** designs employ secondary mirrors to avoid placing detectors at the prime focus. More complex light paths are also used, to allow the use of large or heavy equipment that cannot be placed near the telescope.

Most modern telescopes now use **charge-coupled devices**, or **CCDs (p. 111)**, instead of photographic plates, to collect their data. The field of view is divided into an array of millions of **pixels (p. 111)** that accumulate an electric charge when light strikes them. CCDs are many times more sensitive than photographic plates, and the resultant data are easily saved directly on disk or tape for later image processing. The light collected by a telescope may be processed in a number of ways. It can be made

to form an **image (p. 106)**. **Photometry (p. 113)** may be performed either on a stored image or during the observation itself, using a specialized detector, or a **spectrometer (p. 115)** may be used to analyze the spectrum of the radiation received.

The light-gathering power of a telescope depends on its **collecting area (p. 116)**, which is proportional to the square of the mirror diameter. To study the faintest sources of radiation, astronomers must use large telescopes. Another important aspect of a telescope is its **angular resolution (p. 117)**—the ability to distinguish between light sources lying close together in the sky. Diffraction limits a telescope's resolution by making it impossible to focus a beam perfectly. The amount of diffraction is proportional to the wavelength of the radiation under study and inversely proportional to the size of the mirror. Thus, at a given wavelength, larger telescopes suffer least from the effects of diffraction. The resolution of most ground-based optical telescopes is actually limited by **seeing (p. 119)**—Earth's turbulent atmosphere smears the pointlike images of stars out into **seeing disks (p. 120)** a few arc seconds in diameter. Radio and space-based telescopes do not suffer from this effect, so their resolution is determined by the effects of diffraction.

Using **active optics (p. 121)**, in which a telescope's environment and focus are carefully monitored and controlled, and **adaptive optics (p. 122)**, in which the blurring effects of atmospheric turbulence are corrected for in real time, astronomers can now come close to diffraction-limited resolution in some ground-based instruments.

Radio telescopes (p. 123) are conceptually similar in construction to optical reflectors. However, they are generally much larger than optical instruments, for two reasons: first, the amount of radio radiation reaching Earth from space is tiny compared with optical wavelengths, so a large collecting area is essential; second, the long wavelengths of radio waves mean that diffraction severely limits the resolution, unless large instruments are used.

In order to increase the effective area of a telescope, and hence improve its resolution, several separate instruments may be combined into a device called an **interferometer (p. 127)**. Using **interferometry (p. 127)**, radio telescopes can produce images sharper than those from the best optical telescopes. Infrared interferometers are under construction, and optical interferometric systems are under active development.

Infrared telescopes (p. 130) and **ultraviolet telescopes (p. 132)** are generally similar in design to optical systems. Studies undertaken in some parts of the infrared range can be carried out using large ground-based systems. Ultraviolet astronomy, by contrast, *must* be carried out from space. **High-energy telescopes (p. 133)** study the X-ray and gamma-ray regions of the electromagnetic spectrum. X-ray telescopes can form images of their field of view, although the mirror design is more complex than that of optical instruments. Gamma-ray telescopes simply point in a certain direction and count the photons they collect. Because the atmosphere is opaque at these short wavelengths, both types of telescopes must be placed in space.

Radio and other nonoptical telescopes are essential to studies of the universe because they allow astronomers to probe regions of space that are completely opaque to visible light and to study objects that emit little or no optical radiation at all.

REVIEW AND DISCUSSION

1. Cite two reasons that astronomers are continually building larger and larger telescopes.

2. List three advantages of reflecting telescopes over refractors.

3. What and where are the largest optical telescopes in use today?

4. How does Earth's atmosphere affect what is seen through an optical telescope?

5. What advantages does the *Hubble Space Telescope* have over ground-based telescopes? List some disadvantages.

6. What are the advantages of a CCD over a photograph?

7. What is image processing?

8. What determines the resolution of a ground-based telescope?

9. How do astronomers use active optics to improve the resolution of telescopes?

10. How do astronomers use adaptive optics to improve the resolution of telescopes?

11. Why do radio telescopes have to be very large?

12. Which astronomical objects are best studied with radio techniques?

13. What is interferometry, and what problem in radio astronomy does it address?

14. Is interferometry limited to radio astronomy?

15. Compare the highest resolution attainable with optical telescopes with the highest resolution attainable with radio telescopes (including interferometers).

16. Why do infrared satellites have to be cooled?

17. Are there any ground-based ultraviolet observatories?

18. In what ways do the mirrors in X-ray telescopes differ from those found in optical instruments?

19. What are the main advantages of studying objects at many different wavelengths of radiation?

20. Our eyes can see light with an angular resolution of $1'$. Suppose our eyes detected only infrared radiation, with $1°$ angular resolution. Would we be able to make our way around on Earth's surface? To read? To sculpt? To create technology?

CONCEPTUAL SELF-TEST: TRUE OR FALSE/MULTIPLE CHOICE

1. The primary purpose of any telescope is to produce an enormously magnified image of the field of view.

2. The main advantage to using the *Hubble Space Telescope* is the increased amount of "night time" viewing it affords.

3. The Keck telescopes contain the largest single mirrors ever constructed.

4. The term "seeing" is used to describe how faint an object can be detected by a telescope.

5. Radio telescopes are large in part to improve their angular resolution, which is poor because of the long wavelengths at which they are used to observe the skies.

6. As a rule, larger telescopes can detect fainter objects.

7. The angular resolution of a 2-m ground-based optical telescope is limited more by seeing than by diffraction.

8. An object having a temperature of 300 K would be best observed with an infrared telescope.

9. X-ray telescopes cannot form images of their fields of view.

10. Gamma-ray telescopes employ the same basic design that optical instruments use.

11. The thickest lenses deflect and bend light **(a)** the fastest; **(b)** the slowest; **(c)** the most; **(d)** the least.

12. The main reason that most professional research telescopes are reflectors is that **(a)** mirrors produce sharper images than lenses do; **(b)** their images are inverted; **(c)** they do not suffer from the effects of seeing; **(d)** large mirrors are easier to build than large lenses.

13. If telescope mirrors could be made of odd sizes, the one with the *most* light-gathering power would be **(a)** a triangle with

1-m sides; **(b)** a square with 1-m sides; **(c)** a circle 1 m in diameter; **(d)** a rectangle with two 1-m sides and two 2-m sides.

14. The image shown in Figure 5.16 ("Resolution") is sharpest when the ratio of wavelength to telescope size is **(a)** large; **(b)** small; **(c)** close to unity; **(d)** none of these.

15. The primary reason professional observatories are built on the highest mountaintops is to **(a)** get away from city lights; **(b)** be above the rain clouds; **(c)** reduce atmospheric blurring; **(d)** improve chromatic aberration.

16. Compared with radio telescopes, optical telescopes can **(a)** see through clouds; **(b)** be used during the daytime; **(c)** resolve finer detail; **(d)** penetrate interstellar dust.

17. When multiple radio telescopes are used for interferometry, resolving power is most improved by increasing **(a)** the distance between telescopes; **(b)** the number of telescopes in a given area; **(c)** the diameter of each telescope; **(d)** the electrical power supplied to each telescope.

18. The *Spitzer Space Telescope* is stationed far from Earth because **(a)** this increases the telescope's field of view; **(b)** the telescope is sensitive to electromagnetic interference from terrestrial radio stations; **(c)** doing so avoids the obscuring effects of Earth's atmosphere; **(d)** Earth is a heat source and the telescope must be kept very cool.

19. The best way to study young stars hidden behind interstellar dust clouds would be to use **(a)** X rays; **(b)** infrared light; **(c)** ultraviolet light; **(d)** blue light.

20. Table 5.1 ("Astronomy at Many Wavelengths") suggests that the best frequency range in which to study the hot (million-kelvin) gas found among the galaxies in the Virgo cluster would be **(a)** the radio frequencies; **(b)** the infrared region of the electromagnetic spectrum; **(c)** the X-ray region of the spectrum; **(d)** the gamma-ray region of the spectrum.

PROBLEMS

Algorithmic versions of these questions are available in the Practice Problems module of the Companion Website at astro.prenhall.com/chaisson.

The number of squares preceding each problem indicates its approximate level of difficulty.

1. ■ A certain telescope has a 10′ × 10′ field of view that is recorded using a CCD chip having 2048 × 2048 pixels. What angle on the sky corresponds to 1 pixel? What would be the diameter, in pixels, of a typical seeing disk (1″ radius)?

2. ■ The *SST*'s planned operating temperature is 5.5 K. At what wavelength (in micrometers) does the telescope's own blackbody emission peak? How does this wavelength compare with the wavelength range in which the telescope is designed to operate? ∞ *(More Precisely 3-2)*

3. ■ A 2-m telescope can collect a given amount of light in 1 hour. Under the same observing conditions, how much time would be required for a 6-m telescope to perform the same task? A 12-m telescope?

4. ■ A space-based telescope can achieve a diffraction-limited angular resolution of 0.05″ for red light (wavelength 700 nm). What would the resolution of the instrument be **(a)** in the infrared, at 3.5 μm, and **(b)** in the ultraviolet, at 140 nm?

5. ■ Based on the numbers given in the text, estimate the angular resolution of the Gemini North telescope **(a)** in red light (700 nm) and **(b)** in the near infrared (2 μm).

6. ■■ Two identical stars are moving in a circular orbit around one another, with an orbital separation of 2 A.U. ∞ *(Sec. 2.6)* The system lies 200 light-years from Earth. If we happen to view the orbit head-on, how large a telescope would we need to resolve the stars, assuming diffraction-limited optics at a wavelength of 2 μm?

7. ■■ What is the greatest distance at which *HST*, in blue light (400 nm), could resolve the stars in the previous question?

8. ■ The photographic equipment on a telescope is replaced by a CCD. If the photographic plate records 5 percent of the light reaching it, while the CCD records 90 percent, how much time will the new system take to collect as much information as the old detector recorded in a 1-hr exposure?

9. ■ The Moon lies about 380,000 km away. To what distances do the angular resolutions of *SST* (3″), HST (0.05″), and a radio interferometer (0.001″) correspond at that distance?

10. ■ The Andromeda Galaxy lies about 2.5 million light-years away. To what distances do the angular resolutions of SST (3″), HST (0.05″), and a radio interferometer (0.001″) correspond at that distance?

11. ■ Based on collecting areas, how much more sensitive would you expect the Arecibo telescope (Figure 5.24) to be, compared with the large telescope in Figure 5.23?

12. ■ What is the equivalent single-mirror diameter of a telescope constructed from two separate 10-m mirrors? Four separate 8-m mirrors?

13. ■ Estimate the angular resolutions of **(a)** a radio interferometer with a 5000-km baseline, operating at a frequency of 5 GHz, and **(b)** an infrared interferometer with a baseline of 50 m, operating at a wavelength of 1 μm.

14. ■■ During a particular observation, the *Chandra* telescope detects photons with an average energy of 1.5 keV ∞ *(More Precisely 4-1)*, at a rate of 0.001 per second. What is the total power received, in watts?

15. ■■ The resolution of the *Chandra* telescope is about 1″ for photons of energy 1 keV. The effective collecting diameter of the mirror assembly is 1.2 m. Based on these numbers, is *Chandra* diffraction limited?

In addition to the Practice Problems module, the Companion Website at astro.prenhall.com/chaisson provides for each chapter a study guide module with multiple choice questions as well as additional annotated images, animations, and links to related Websites.

PART | 2
Our Planetary System

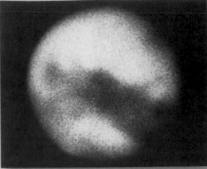
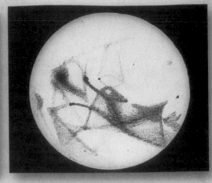

Early photo and sketch of Mars (*Lowell Observatory*)

The year 1877 was an important one in the study of the planet Mars. The Red Planet came unusually close to Earth, affording astronomers an especially good view. Of particular note was the discovery, by U.S. Naval Observatory astronomer Asaph Hall, of the two moons circling Mars. But most exciting was the report of the Italian astronomer Giovanni Schiaparelli on his observations of a network of linear markings that he termed *canali*. In Italian, this word usually means "grooves" or "channels," but it can also mean "canals." Schiaparelli probably did not intend to imply that the *canali* were anything other than natural, but the word was translated into English as "canals," suggesting that the grooves had been constructed by intelligent beings. As is often the case with the media—then as now—Schiaparelli's observations became sensationalized in the world's press (especially in the United States), and some astronomers began drawing elaborate maps of Mars, showing oases and lakes where canals met in desert areas.

Percival Lowell (1855-1916), a successful Boston businessman (and brother of the poet Amy Lowell and Harvard president Abbott Lawrence Lowell), became so fascinated by these reports that he abandoned his business and purchased a clear-sky site at Flagstaff, Arizona, where he built a private observatory. He devoted his fortune and energies until the day he died to achieving a better understanding of the Martian "canals." In doing so, he championed the idea that intelligent inhabitants of a drying (and dying) Mars had constructed a planetwide system of canals to transport water from the polar ice caps to the arid equatorial deserts.

Lowell was no slouch. He had earlier served as a distinguished diplomat and wrote extensively about the Far East. Later in life, he also made an elaborate mathematical study of the orbit of Uranus, thereby predicting the presence of an unseen planet beyond Neptune—which was eventually found and named Pluto by Lowell Observatory's Clyde Tombaugh in 1930. And Lowell offered support to other astronomers, including Vesto Slipher, a young observer who undertook pioneering research on the recession of the galaxies and helped found modern cosmology. Even so, Lowell is best remembered for his passionate belief in advanced Martian civilizations—all of which fueled the widespread idea a hundred years ago that intelligent life exists on Mars.

Percival Lowell (*Lowell Observatory*)

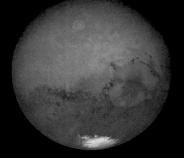

Mars today, via *Hubble* (*STScI*)

oday, we know that Mars is cool, dry, and probably life-less; it certainly houses no intelligent beings and almost certainly never has. The Martian valleys and channels photographed by robot spacecraft in the 1970s are far too small to be the *canali* that Schiaparelli, Lowell, and others thought they saw on Mars. And the most recent armada of spacecraft—including the European *Mars Express* orbiter and the U.S. *Spirit* and *Opportunity* landers—that arrived at Mars in the early years of the 21st century reconfirm there is no liquid water there now.

The entire episode of Lowell's "canals" represents a classic case in which well-intentioned observers, perhaps obsessed with the notion of life on other worlds, let their personal opinions and prejudices seriously affect their analysis of reasonable data. The pair of globes of Mars opposite show how surface features, which were probably genuinely observed by astronomers a century ago, might have been imagined to be connected. The figure on the left of the pair is a photograph of how Mars actually looked in the best telescopes at the end of 19th century. The sketch at its right is an interpretation, done at the height of the canal hoopla, of that same view. The human eye, under physio-logical stress, tends to connect dimly observed, yet distinctly separated, features, causing old maps of Mars to have been as much a work of art as of science. Humans saw patterns and canals where none in fact existed—as noted by today's higher quality image of Mars above.

The chronicle of the Martian canals illustrates how the sci-entific method demands that we acquire new data to sort out sense from nonsense, fact from fiction. Rather than simply believing the claims about the Martian canals, scientists sought further observations to test Lowell's hypothesis. Eventually, improved observations, climaxing in several robotic missions to the Red Planet more than a century after all the fuss began, totally disproved the existence of artificial canals. It often takes time, but the scientific method does lead to progress in under-standing reality.

Equatorial Mars, via *Opportunity Rover* (*JPL*)

Panoramic View of Mars, via *Opportunity Rover* (*JPL*)

6

The engineering feats of the modern space age allow us to probe many of the diverse worlds of the solar system in great detail. Here, we see a view of Mars's landscape from the Opportunity rover, a golf-cart-sized robot delivered to the (obviously) red planet by the Mars Exploration spacecraft that landed in a small, unnamed crater near the Martian equator in early 2004. The advantage of ending up in a crater is that geologists on Earth can explore what's below the surface ▶ without having to dig. (JPL)

The Solar System

An Introduction to Comparative Planetology

In less than a single generation, we have learned more about the solar system—the Sun and everything that orbits it—than in all the centuries that went before. By studying the planets, their moons, and the countless fragments of material that orbit in interplanetary space, astronomers have gained a richer outlook on our own home in space. Instruments aboard unmanned robots have taken close-up photographs of the planets and their moons and in some cases have made on-site measurements. The discoveries of the past few decades have revolutionized our understanding not only of our present cosmic neighborhood, but also of its history, for our solar system is filled with clues to its own origin and evolution.

LEARNING GOALS

Studying this chapter will enable you to

1 Discuss the importance of comparative planetology to solar system studies.

2 Describe the overall scale and structure of the solar system.

3 Summarize the basic differences between the terrestrial and the jovian planets.

4 Identify and describe the major nonplanetary components of the solar system.

5 Describe some of the spacecraft missions that have contributed significantly to our knowledge of the solar system.

6 Outline the theory of solar-system formation that accounts for the overall properties of our planetary system.

7 Account for the differences between the terrestrial and the jovian planets.

 Visit astro.prenhall.com/chaisson for additional annotated images, animations, and links to related sites for this chapter.

6.1 An Inventory of the Solar System

The Greeks and other astronomers of old were aware of the Moon, the stars, and five planets—Mercury, Venus, Mars, Jupiter, and Saturn—in the night sky. ∞ (Sec. 2.2) They also knew of two other types of heavenly objects that were clearly neither stars nor planets. *Comets* appear as long, wispy strands of light in the night sky that remain visible for periods of up to several weeks and then slowly fade from view. *Meteors*, or "shooting stars," are sudden bright streaks of light that flash across the sky, usually vanishing less than a second after they first appear. These transient phenomena must have been familiar to ancient astronomers, but their role in the "big picture" of the solar system was not understood until much later.

Human knowledge of the basic content of the solar system remained largely unchanged from ancient times until the early 17th century, when the invention of the telescope made more detailed observations possible. Galileo Galilei was the first to capitalize on this new technology. (His simple telescope is shown in Figure 6.1.) Galileo's discovery of the phases of Venus and of four moons orbiting Jupiter early in the 17th century helped change forever humankind's vision of the universe. ∞ (Sec. 2.4)

As technological advances continued, knowledge of the solar system improved rapidly. Astronomers began discovering objects invisible to the unaided human eye. By the end of the 19th century, astronomers had found Sat-

urn's rings (1659), the planets Uranus (1758) and Neptune (1846), many planetary moons, and the first *asteroids*—"minor planets" orbiting the Sun, mostly in a broad band (called the *asteroid belt*) lying between Mars and Jupiter. Ceres, the largest asteroid and the first to be sighted, was discovered in 1801. A large telescope of mid-19th-century vintage is shown in Figure 6.2.

The 20th century brought continued improvements in optical telescopes. One more planet (Pluto) was discovered, along with three more planetary ring systems, dozens of moons, and thousands of asteroids. The century also saw the rise of both nonoptical—especially radio and infrared—astronomy and spacecraft exploration, each of which has made vitally important contributions to the field of planetary science. Astronauts have carried out experiments on the Moon (see Figure 6.3), and numerous unmanned probes have left Earth and traveled to all but one (Pluto) of the other planets. Figure 6.4 shows a view from the *Spirit* robot having just landed on the Martian surface in 2004, its parachute and airbags still surrounding the vehicle.

As currently explored, our **solar system** is known to contain one star (the Sun), nine planets, 135 moons (at last count) orbiting those planets, six asteroids larger than 300 km in diameter, tens of thousands of smaller (but well-studied) asteroids, myriad comets a few kilometers in diameter, and countless *meteoroids* less than 100 m across. The list will undoubtedly grow as we continue to explore our cosmic neighborhood.

As we proceed through the solar system in the next few chapters, we will seek to understand how each planet compares with our own and what each contributes to our knowledge of the solar system as a

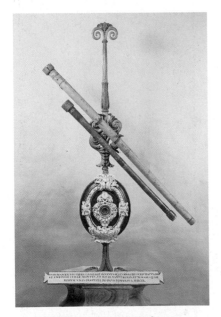

▲ **FIGURE 6.1 Early Telescope** The refracting telescope with which Galileo made his first observations was simple, but its influence on astronomy was immeasurable. (*Museo della Scienza; Scala/Art Resource, NY*)

▲ **FIGURE 6.2 Nineteenth-Century Telescope** By the mid-19th century, telescopes had improved enormously in both size and quality. Shown here is the Newtonian reflector built and used by Irish nobleman and amateur astronomer the Earl of Rosse. For 75 years, this 72-inch-diameter instrument was the largest telescope on Earth. (*Birr Scientific & Heritage Foundation*)

whole. We will use the powerful perspective of **comparative planetology**—comparing and contrasting the properties of the diverse worlds we encounter—to understand better the conditions under which planets form and evolve.

Our goal will be to develop (starting here, and concluding in Chapter 15) a comprehensive theory of the origin and evolution of our planetary system—a theory that explains all, or at least most, of the solar system's observed properties. We will seek to answer basic questions such as Why did planet X evolve in one way, while planet Y turned out completely different and why are the planets' orbits so orderly when their individual properties are not? In addressing these issues, we will find many similarities and common features among planets. However, each planet will also present new questions and afford unique insights into the ways planets work.

As we unravel the origin of our solar system, we may hope to learn something about planetary systems beyond our own. Since the mid-1990s, astronomers have detected more than 100 *extrasolar planets*—planets orbiting stars other than our own Sun. Many new planets are discovered each year (see Chapter 15), and our observations of them provide critical tests of modern theories of planet formation. Before the discovery of extrasolar planets, those theories had necessarily been based on observations only of our own solar system. Now astronomers have a whole new set of "proving grounds" in which to compare theory with reality.

Curiously, current data suggest that many of the newly discovered systems have properties rather different from those of our own, adding fuel to the long-standing debate among astronomers on the prevalence of planets like Earth and the possible existence of life as we know it elsewhere in

▲ **FIGURE 6.3 Lunar Exploration** An *Apollo* astronaut does some lunar geology—prospecting near a huge boulder in the Mare Serenitatis during the final manned mission to the Moon in 1972. *(NASA)*

the universe. It will be some time before astronomers can make definitive statements about the existence (or nonexistence) of planetary systems like our own.

CONCEPT CHECK

✔ In what ways might observations of extrasolar planets help us understand our own solar system?

▲ **FIGURE 6.4** *Spirit* **on Mars** The Mars rover *Spirit* took this group of images shortly after reaching Mars in 2004. The most prominent surface feature on the left, a circular topographic depression dubbed Sleepy Hollow, may be a small impact crater or perhaps a dried up pond. *(JPL)*

6.2 Planetary Properties

Table 6.1 lists some basic orbital and physical properties of the nine planets, with a few other solar-system objects (the Sun, the Moon, an asteroid, and a comet) included for comparison. Most of the quantities listed can be determined using methods described in Chapters 1 and 2. We present here some of the simpler techniques for making these measurements. Note that the Sun, with more than a thousand times the mass of the next most massive object (the planet Jupiter), is clearly the dominant member of the solar system. In fact, the Sun contains about 99.9 percent of all solar-system material. The planets—including our own—are insignificant in comparison. *More Precisely 6-1* makes the methods presented a little more concrete by applying them to observations of the planet Jupiter. Here is a brief summary of the properties listed in Table 6.1 and the techniques used to measure them:

- The *distance* of each planet from the Sun is known from Kepler's laws once the scale of the solar system is set by radar ranging on Venus. ∞ (Sec. 2.6)

- A planet's (sidereal) *orbital period* can be measured from repeated observations of its location on the sky, so long as Earth's own motion around the Sun is properly taken into account.

- A planet's *radius* is found by measuring the angular size of planet—the angle from one side to the other as we view it on the sky—and then applying elementary geometry. ∞ (*More Precisely 1-3*)

- The *masses* of planets with moons may be calculated by applying Newton's laws of motion and gravity, just by observing the moons' orbits around the planets. ∞ (*More Precisely 2-3*) The *sizes* of those orbits, like the sizes of the planets themselves, are determined by geometry.

- The masses of Mercury and Venus (as well as those of our own Moon and the asteroid Ceres) are a little harder to determine accurately, because these bodies have no natural satellites of their own. Nevertheless, it is possible to measure their masses by careful observations of their gravitational influence on other planets or nearby bodies. Mercury and Venus produce small, but measurable, effects on each other's orbits, as well as on that of Earth. The Moon also causes small "wobbles" in Earth's motion as the two bodies orbit their common center of mass.

- These techniques for determining mass were available to astronomers well over a century ago. Today, the masses of most of the objects listed in Table 6.1 have been accurately measured through their gravitational interaction with artificial satellites and space probes launched from Earth. Only in the case of Ceres is the mass still poorly known, mainly because that asteroid's gravity is so weak.

TABLE 6.1	Properties of Some Solar-System Objects							
Object	Orbital Semimajor Axis (A.U.)	Orbital Period (Earth Years)	Mass (Earth Masses)	Radius (Earth Radii)	Number of Known Satellites	Rotation Period[*] (days)	Average Density (kg/m³)	(g/cm³)
Mercury	0.39	0.24	0.055	0.38	0	59	5400	5.4
Venus	0.72	0.62	0.82	0.95	0	−243	5200	5.2
Earth	1.0	1.0	1.0	1.0	1	1.0	5500	5.5
Moon	—	—	0.012	0.27	—	27.3	3300	3.3
Mars	1.52	1.9	0.11	0.53	2	1.0	3900	3.9
Ceres (asteroid)	2.8	4.7	0.00015	0.073	0	0.38	2700	2.7
Jupiter	5.2	11.9	318	11.2	61	0.41	1300	1.3
Saturn	9.5	29.4	95	9.5	31	0.44	700	0.7
Uranus	19.2	84	15	4.0	27	−0.72	1300	1.3
Neptune	30.1	164	17	3.9	12	0.67	1600	1.6
Pluto	39.5	248	0.002	0.2	1	−6.4	2100	2.1
Comet Hale–Bopp	180	2400	1.0×10^{-9}	0.004	—	0.47	100	0.1
Sun	—	—	332,000	109	—	25.8	1400	1.4

[*]*A negative rotation period indicates retrograde (backward) rotation relative to the sense in which all planets orbit the Sun.*

MORE PRECISELY 6-1

Computing Planetary Properties

Let's look a little more closely at some of the methods used to determine the physical properties of a planet. The accompanying figure shows the planet Jupiter and Europa, one of its inner moons. The orbits of both Jupiter and Earth have been measured very precisely so, at any instant, the distance between the two planets is accurately known. However, since Jupiter's orbital semimajor axis is 5.2 A.U. and that of Earth is 1 A.U., for definiteness in what follows we will simply assume a distance of 4.2 A.U., or about 630,000,000 km. ∞ (Sec. 2.6) This distance corresponds roughly to the point of closest approach between the two planets.

As indicated in the figure, Jupiter's angular diameter at a distance of 4.2 A. U. from Earth is 46.8 arc seconds. (Recall that 1 arc second is 1/3600 of a degree.) From ∞ More Precisely 1-3, it follows that the planet's actual diameter is

$$\text{diameter } (2R) = \text{distance} \times \frac{\text{angular diameter}}{57.3°}$$

$$= (4.2 \times 150{,}000{,}000 \text{ km}) \times \frac{(46.8/3600)°}{57.3°}$$

$$= 143{,}000 \text{ km},$$

so Jupiter's radius is $R = 71{,}500$ km.

Jupiter's moon Europa is observed to orbit the planet with an orbital period $P = 3.55$ days. The orbit is circular, with an angular radius of 3.66 arc minutes, as seen from Earth. Converting this quantity to a distance in kilometers, as before, we find that the radius of Europa's orbit around Jupiter is $r = 671{,}000$ km. The moon thus travels a distance equal to the circumference of its orbit, $2\pi r = 4.22$ million km, in a time $P = 3.55$ days $= 307{,}000$ s. The orbital speed of Europa is $V = 2\pi r/P = 13.7$ km/s. Applying Newton's laws, we then determine Jupiter's mass to be ∞ (More Precisely 2-3)

$$M = \frac{rV^2}{G} = 1.9 \times 10^{27} \text{ kg}.$$

Finally, dividing Jupiter's mass by the volume of a sphere of radius R, namely, $^4/_3 \pi R^3$, we obtain a density of 1240 kg/m^3. This number differs from the figure listed in Table 6.1 because Jupiter is, in reality, not perfectly spherical. It is *flattened* somewhat at the poles, so our expression for the volume is not quite correct. When we take the flattening into account, the actual volume is a little lower than the number given here, and the density is correspondingly higher.

We have determined several important physical properties of Jupiter by combining observations made from Earth with our knowledge of simple geometry and Newton's laws of motion. In fact, Jupiter is perhaps the easiest planet to study in this way, as it is big, is relatively close, and has several easy-to-see satellites. Nevertheless, these fundamental techniques are applicable throughout the solar system—indeed, throughout the entire universe. Prior to the Space Age, they formed the basis for virtually all astronomical measurements of planetary properties.

Europa

50"

(NASA)

R I V U X G

• A planet's *rotation period* may, in principle, be determined simply by watching surface features alternately appear and disappear as the planet rotates. However, with most planets, this is difficult to do, as their surfaces are hard to see or may even be nonexistent. Mercury's surface features are hard to distinguish, the surface of Venus is completely obscured by clouds, and Jupiter, Saturn, Uranus, and Neptune have no solid surfaces at all—their atmospheres simply thicken and eventually become liquid as we descend deeper and deeper below the visible clouds. We will describe the methods used to measure these planets' rotation periods in later chapters.

• The final two columns in Table 6.1 list the *average density* of each object. **Density** is a measure of the "compactness" of matter. Average density is computed by dividing an object's mass (in kilograms, say) by its volume (in cubic meters, for instance). For example, we can easily compute Earth's average density. Earth's mass, as determined from observations of the Moon's orbit, is approximately 6.0×10^{24} kg. ∞ (More Precisely 2-3) Earth's radius R is roughly 6400 km, so its volume is $\frac{4}{3}\pi R^3 \approx 1.1 \times 10^{12}$ km^3, or 1.1×10^{21} m^3. ∞ (Sec. 1.7) Dividing Earth's mass by its volume, we obtain an average density of approximately 5500 kg/m^3. *On average*, then, there are about 5500 kilograms of Earth matter in every cubic meter of Earth volume. For comparison, the density of ordinary water is 1000 kg/m^3, rocks on Earth's surface have densities in the range 2000–3000 kg/m^3, and iron has a density of some 8000 kg/m³. Earth's atmosphere (at sea level) has a density of only a few kilograms per cubic meter. Because many working astronomers are more familiar with the CGS (centimeter–gram–second) unit of density (grams per cubic centimeter, abbreviated g/cm³, where 1 kg/m³ = 1000 g/cm³), Table 6.1 lists density in both SI and CGS units.

CONCEPT CHECK

✔ How do astronomers go about determining the bulk properties (i.e., masses, radii, and densities) of distant planets?

6.3 The Overall Layout of the Solar System

By earthly standards, the solar system is immense. The distance from the Sun to Pluto is about 40 A.U., almost a million times Earth's radius and roughly 15,000 times the distance from Earth to the Moon. Yet, despite the solar system's vast extent, the planets all lie very close to the Sun, astronomically speaking. Even the diameter of Pluto's orbit is less than 1/1000 of a light-year, whereas the next nearest star is several light-years distant.

The planet closest to the Sun is Mercury. Moving outward, we encounter, in turn, Venus, Earth, Mars, Jupiter, Saturn, Uranus, Neptune, and Pluto. In Chapter 2, we saw the basic properties of the planets' orbits. Their paths are all ellipses, with the Sun at (or very near) one focus. ∞ (Sec. 2.5) Most planetary orbits have low eccentricities. The exceptions are the innermost and the outermost worlds, Mercury and Pluto. Accordingly, we can reasonably think of most planets' orbits as circles centered on the Sun. The orbits of the major bodies in the solar system are illustrated in Figure 6.5. Note that the planetary orbits are not evenly spaced, becoming farther and farther apart as we move outward from the Sun.

All the planets orbit the Sun counterclockwise as seen from above Earth's North Pole, and in nearly the same plane as Earth (i.e., the plane of the ecliptic). Mercury and Pluto deviate somewhat from the latter condition: Their orbital planes lie at 7° and 17° to the ecliptic, respectively. Still, we can think of the solar system as being quite flat. Its "thickness" perpendicular to the plane of the ecliptic is less than 1/50 the diameter of Pluto's orbit. If we were to view the planets' orbits from a vantage point in the plane of the ecliptic about 50 A.U. from the Sun, only Pluto's orbit would be noticeably tilted. Figure 6.6 is a photograph of the planets Mercury, Venus, Mars, Jupiter, and Saturn taken during a chance planetary alignment in April 2002. These five planets can (occasionally) be found in the same region of the sky, in large part because their orbits lie nearly in the same plane in space.

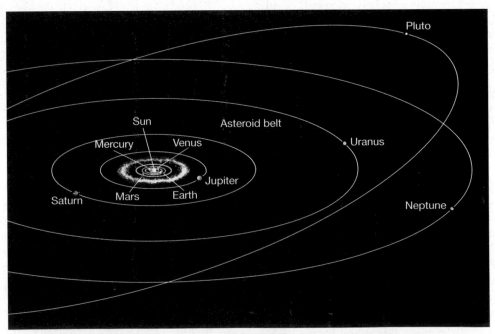

▲ FIGURE 6.5 Solar System Might future space voyagers travel far enough from Earth to gain this perspective on our solar system? Except for Mercury and Pluto, the orbits of the planets lie nearly in the same plane. As we move out from the Sun, the distance between the orbits of the planets increases. The entire solar system spans nearly 80 A.U.

6.4 Terrestrial and Jovian Planets

LEARNING GOAL 3 On large scales, the solar system presents us with a sense of orderly motion. The planets move nearly in a plane, on almost concentric (and nearly circular) elliptical paths, in the same direction around the Sun, at steadily increasing orbital intervals. However, the individual properties of the planets are much less regular.

Figure 6.7 compares the planets with one another and with the Sun. A clear distinction can be drawn between the inner and the outer members of our planetary system based on densities and other physical properties. The inner planets—Mercury, Venus, Earth, and Mars—are small, dense, and *rocky* in composition. The outer worlds—Jupiter, Saturn, Uranus, and Neptune (but not Pluto)—are large, of low density, and *gaseous*.

Because the physical and chemical properties of Mercury, Venus, and Mars are somewhat similar to Earth's, the four innermost planets are called the **terrestrial planets**. (The word *terrestrial* derives from the Latin word *terra*, meaning "land" or "earth.") The larger outer planets—Jupiter, Saturn, Uranus, and Neptune—are all similar to one another chemically and physically (and very different from the terrestrial worlds). They are labeled the **jovian planets**, after Jupiter, the largest member of the group. (The word *jovian* comes from *Jove*, another name for the Roman god Jupiter.) The jovian worlds are all much larger

▲ **FIGURE 6.6 Planetary Alignment** This image shows six planets—Mercury, Venus, Mars, Jupiter, Saturn, and Earth—during a planetary alignment in April 2002. Because the planets all orbit in nearly the same plane, it is possible for them all to appear (by chance) in the same region of the sky, as seen from Earth. The Sun and the new Moon are just below the horizon. As usual, the popular press contained many sensationalized predictions of catastrophes that would occur during this rare astronomical event. Also as usual, none came true. *(J. Lodriguss)*

◀ **FIGURE 6.7 Sun and Planets** Diagram, drawn to scale, of the relative sizes of the planets and our Sun. Notice how much larger the Jovian planets are than Earth and the other terrestrials and how much larger still is the Sun.

ANIMATION *Size and Scale of the Terrestrial Planets I/The Gas Giants*

149

than the terrestrial planets and quite different from them in both composition and structure.

The four terrestrial planets all lie within about 1.5 A.U. of the Sun. All are small and of relatively low mass, and all have a generally rocky composition and solid surfaces. Beyond that, however, the similarities end:

- All four terrestrial planets have atmospheres, but the atmospheres are about as dissimilar as we could imagine, ranging from a near vacuum on Mercury to a hot, dense inferno on Venus.

- Earth alone has oxygen in its atmosphere and liquid water on its surface.

- Surface conditions on the four planets are quite distinct from one another, ranging from barren, heavily cratered terrain on Mercury to widespread volcanic activity on Venus.

- Earth and Mars spin at roughly the same rate—one rotation every 24 (Earth) hours—but Mercury and Venus both take months to rotate just once, and Venus rotates in the opposite sense from the others.

- Earth and Mars have moons, but Mercury and Venus do not.

- Earth and Mercury have measurable magnetic fields, of very different strengths, whereas Venus and Mars have none.

Finding the common threads in the evolution of these four diverse worlds is no simple task! Comparative planetology will be our indispensable guide as we proceed through the coming chapters.

Comparing the average densities of the terrestrial planets allows us to say something about their overall *compositions*. However, before making the comparison, we must take into account how the weight of overlying layers compresses the interiors of the planets to different extents. When we do this, we find that the *uncompressed densities* of the terrestrial worlds—the densities they would have in the absence of any compression due to their own gravity—decrease as we move outward from the Sun: 5300, 4400, 4400, and 3800 kg/m^3 for Mercury, Venus, Earth, and Mars, respectively. The amount of compression is greatest for the most massive planets, Earth and Venus, and much less for Mercury and Mars. Partly on the basis of these figures, planetary scientists conclude that Earth and Venus are quite similar in overall composition. Mercury's higher density implies that it contains a higher proportion of some dense material—most likely nickel or iron. The lower density of Mars probably means that it is deficient in that same material.

Yet, for all their differences, the terrestrial worlds still seem similar compared with the jovian planets. Perhaps the simplest way to express the major differences between

the terrestrial and jovian worlds is to say that the jovian planets are everything the terrestrial planets are not. Table 6.2 compares and contrasts some key properties of these two planetary classes.

The terrestrial worlds lie close together, near the Sun; the jovian worlds are widely spaced through the outer solar system. The terrestrial worlds are small, dense, and rocky; the jovian worlds are large and gaseous, made predominantly of hydrogen and helium (the lightest elements), which are rare on the inner planets. The terrestrial worlds have solid surfaces; the jovian worlds have none (their dense atmospheres thicken with depth, eventually merging with their liquid interiors). The terrestrial worlds have weak magnetic fields, if any; the jovian worlds all have strong magnetic fields. The terrestrial worlds have only three moons among them; the jovian worlds have many moons each, no two of them alike and none of them like our own. Furthermore, all the jovian planets have *rings*, a feature unknown on the terrestrial planets. Finally, all four jovian worlds are thought to contain large, dense "terrestrial" cores some 10 to 15 times the mass of Earth. These cores account for an increasing fraction of each planet's total mass as we move outward from the Sun.

Beyond the outermost jovian planet, Neptune, lies one more small world, frozen and mysterious. Pluto doesn't fit well into either planetary category. Indeed, there is ongoing debate among planetary scientists as to whether it should be classified as a planet at all. In both mass and composition, Pluto has much more in common with the icy jovian moons than with any terrestrial or jovian planet. Many astronomers suspect that it may in fact be the largest member of a recently recognized class of solar system objects that reside beyond the jovian worlds (see Chapter 14).

TABLE 6.2 Comparison of the Terrestrial and Jovian Planets

Terrestrial Planets	Jovian Planets
close to the Sun	far from the Sun
closely spaced orbits	widely spaced orbits
small masses	large masses
small radii	large radii
predominantly rocky	predominantly gaseous
solid surface	no solid surface
high density	low density
slower rotation	faster rotation
weak magnetic fields	strong magnetic fields
few moons	many moons
no rings	many rings

► **FIGURE 6.8 Asteroid and Comet** (a) Asteroids, like meteoroids, are generally composed of rocky material. This asteroid, Eros, is about 34 km long. It was photographed by the *NEAR* (*Near Earth Asteroid Rendezvous*) spacecraft that actually landed on the asteroid in 2001. (b) Comet Hale–Bopp, seen as it approached the Sun in 1997. Most comets are composed largely of ice and so tend to be relatively fragile. The comet's vaporized gas and dust form the tail, here extending away from the Sun for nearly a quarter of the way across the sky. (*JHU/APL; J. Lodriguss*)

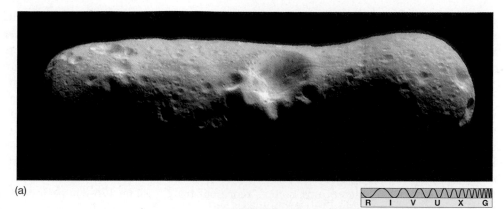

(a)

(b)

In 1999, the International Astronomical Union, which oversees the rules for classifications in astronomy, decided that Pluto should, for now at least, still be called a planet. However, that status may well change as the makeup of the outer solar system becomes better understood.

CONCEPT CHECK

✔ Why do astronomers draw such a clear distinction between the inner and the outer planets?

6.5 Interplanetary Debris

In the vast space among the nine known major planets move countless chunks of rock and ice, all orbiting the Sun, many on highly eccentric paths. This final component of the solar system is the collection of *interplanetary matter*—cosmic "debris" ranging in size from the relatively large asteroids, through the smaller comets and even smaller meteoroids, down to the smallest grains of interplanetary dust that litter our cosmic environment.

The dust arises when larger bodies collide and break apart into smaller pieces that, in turn, collide again and are slowly ground into microscopic fragments, which eventually settle into the Sun or are swept away by the *solar wind*, a stream of energetic charged particles that continually flows outward from the Sun and pervades the entire solar system. The dust is quite difficult to detect in visible light, but infrared studies reveal that interplanetary space contains surprisingly large amounts of it. Our solar system is an extremely good vacuum by terrestrial standards, but positively dirty by the standards of interstellar or intergalactic space.

Asteroids (Figure 6.8a) and meteoroids are generally rocky in composition, somewhat like the outer layers of the terrestrial planets. The distinction between the two is simply a matter of size: Anything larger than 100 m in diameter (corresponding to a mass of about 10,000 tons) is conventionally termed an asteroid; anything smaller is a meteoroid. Their total mass is much less than that of Earth's Moon, so these objects play no important role in the present-day workings of the planets or their moons. Yet they are of crucial importance to our studies, for they provide the keys to answering some very fundamental questions about our planetary environment and what the solar system was like soon after its birth. Many of these bodies are made of material that has hardly evolved since the early days of the solar system. In addition, they often conveniently deliver themselves right to our doorstep, in the form of meteorites (the name we give them if they happen to survive the plunge through Earth's atmosphere and find their way to the ground—see Section 14.3), allowing us to study them in detail without having to fetch them from space.

Comets are quite distinct from the other small bodies in the solar system. They are generally icy rather than rocky in composition (although they do contain some rocky material) and typically have diameters in the 1–10-km range. They are quite similar in chemical makeup to some of the icy moons of the outer planets. Even more so than the asteroids and meteoroids, comets represent truly ancient material—the vast majority probably have not changed in any significant way since their formation long ago along with the rest of the solar system (see Chapter 15). Comets striking Earth's atmosphere do not reach the surface intact, so we do not have actual samples of cometary material. However, they do vaporize and emit radiation as their highly elongated orbits take them near the Sun. (See Figure 6.8b.) Astronomers can determine a comet's makeup by spectroscopic study of the radiation it gives off as it is destroyed. ∞ (Sec. 4.2)

CONCEPT CHECK

✔ Why are astronomers so interested in interplanetary matter?

6.6 Spacecraft Exploration of the Solar System

Since the 1960s, dozens of unmanned space missions have traveled throughout the solar system. All of the planets but Pluto have been visited and probed at close range, and spacecraft have visited numerous comets and asteroids. The first landing on an asteroid (by the *NEAR* spacecraft) occurred in February 2001. The impact of these missions on our understanding of our planetary system has been nothing short of revolutionary.

In the next eight chapters, we will see many examples of the marvelous images radioed back to Earth, and we will discover how they fit into our modern picture of the solar system. Here, we focus on just a few of these remarkable technological achievements; Table 6.3 summarizes the various successful missions highlighted in this section.

The *Mariner 10* Flybys of Mercury

In 1974, the U.S. spacecraft *Mariner 10* came within 10,000 km of the surface of Mercury, sending back high-resolution images of the planet. (A "flyby," in NASA parlance, is any space mission in which a probe passes relatively close to a planet—within a few planetary radii, say—but does not go into orbit around it.) The photo-

graphs, which showed surface features as small as 150 m across, dramatically increased our knowledge of the planet. For the first time, we saw Mercury as a heavily cratered world, in many ways reminiscent of our own Moon.

Mariner 10 was launched from Earth in November 1973 and was placed in an eccentric 176-day orbit about the Sun, aided by a gravitational assist (see *Discovery 6-1*) from the planet Venus (Figure 6.9). In that orbit, *Mariner 10*'s nearest point to the Sun (perihelion) is close to Mercury's path, and its farthest point away (aphelion) lies between the orbits of Venus and Earth (*More Precisely 2-1*). The 176-day period is exactly two Mercury years, so the spacecraft revisits Mercury roughly every six months. However, only on the first three encounters—in March 1974, September 1974, and March 1975—did the spacecraft return data. After that, the craft's supply of maneuvering fuel was exhausted; the craft still orbits the Sun today, but out of control and silent. In total, over 4000 photographs, covering about 45 percent of the planet's surface, were radioed back to Earth during the mission's active lifetime. The remaining 55 percent of Mercury is still unexplored.

No new missions have been sent to Mercury since *Mariner 10*. NASA plans a return in 2009, when the *Messenger* probe will be placed in orbit around the planet to map its entire surface at much higher resolution than was possible with *Mariner*. Both the European and the Japanese Space Agencies also have plans to place spacecraft in orbit around Mercury at roughly the same time as the *Messenger* mission.

Exploration of Venus

In all, some 20 spacecraft have visited Venus since the 1970s, far more than have spied on any other planet in the solar system. The Soviet space program took the lead in exploring Venus's atmosphere and surface, while American spacecraft have performed extensive radar mapping of the planet from orbit.

The Soviet *Venera* (derived from the Russian word for Venus) program began in the mid-1960s, and the Soviet *Venera 4* through *Venera 12* probes parachuted into the planet's atmosphere between 1967 and 1978. The early spacecraft were destroyed by enormous atmospheric pressures before reaching the surface, but in 1970, *Venera 7* (Figure 6.10) became the first spacecraft to soft-land on the planet. During the 23 minutes it survived on the surface, it radioed back information on the planet's atmospheric pressure and temperature. Since then, a number of *Venera* landers have transmitted photographs of the surface back to Earth and have analyzed the atmosphere and the soil. None of them survived for more than an hour in the planet's hot, dense atmosphere. The data they sent back make up the entirety of our direct knowledge of Venus's surface. In 1983, the *Venera 15* and *Venera 16* orbiters sent back detailed radar maps (at about 2-km resolution) of large portions of Venus's northern hemisphere.

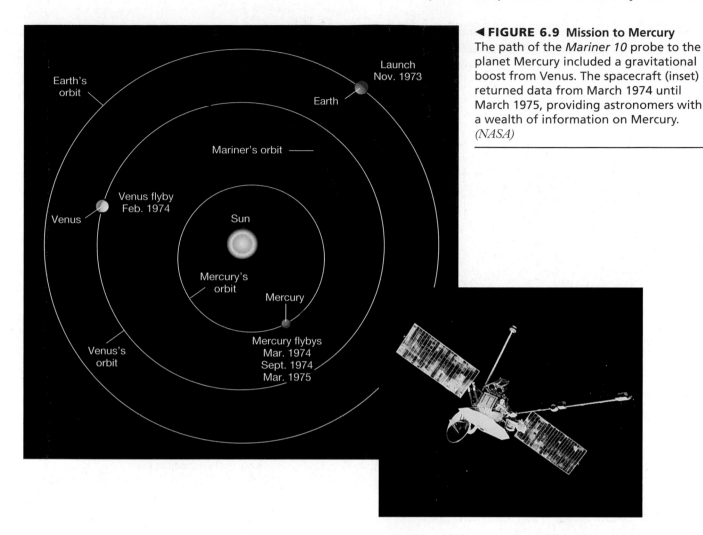

◀ **FIGURE 6.9 Mission to Mercury**
The path of the *Mariner 10* probe to the planet Mercury included a gravitational boost from Venus. The spacecraft (inset) returned data from March 1974 until March 1975, providing astronomers with a wealth of information on Mercury. *(NASA)*

In 1978, the U.S. *Pioneer Venus* mission placed an orbiter at an altitude of some 150 km above Venus's surface and dispatched a "multiprobe" consisting of five separate instrument packages into the planet's atmosphere. During its hour-long descent to the surface, the probe returned information on the variation in density, temperature, and chemical composition with altitude in the atmosphere. The orbiter's radar produced images of most of the planet's surface.

The most recent U.S. mission was the *Magellan* probe (shown in Figure 6.11), which entered orbit around Venus in August 1990. Its radar imaging system could distinguish objects as small as 120 m across. Between 1991 and 1994, the probe mapped 98 percent of the surface of Venus with unprecedented clarity and made detailed measurements of the planet's gravity, rendering all previous data virtually obsolete. The mission ended in October 1994 with a (planned) plunge into the planet's dense atmosphere, sending back one final stream of high-quality data. Many theories of the processes shaping the planet's surface have had to be radically altered or abandoned completely because of *Magellan*'s data.

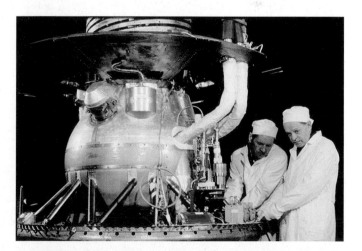

▲ **FIGURE 6.10 Venus Lander** One of the Soviet *Venera* landers that reached the surface of Venus. The design was essentially similar for all the surface missions. Note the heavily armored construction, necessary to withstand the harsh conditions of high temperature and crushing atmospheric pressure on the planet's surface. *(Sovfoto/Eastfoto)*

TABLE 6.3   Some Missions to the Other Planets

Target Planet	Year	Project	Launched by	Type of Mission	Scientific Achievements or Goals, and other Comments
Mercury	1974–1975	Mariner 10	U.S.	flyby	Photographed 45 percent of the planet's surface; still orbiting the Sun
Venus	1967	Venera 4	U.S.S.R.	atmospheric probe	First probe to enter the planet's atmosphere
	1970	Venera 7	U.S.S.R.	lander	First landing on the planet's surface
	1978–1992	Pioneer Venus	U.S.	orbiter	Radar mapping of the entire planet
	1983	Venera 15, 16	U.S.S.R.	orbiter	Radar mapping of northern hemisphere
	1990–1994	Magellan	U.S.	atmospheric probe, orbiter	High-resolution radar mapping of the entire planet
Mars	1965	Mariner 4	U.S.	flyby	Photographs showed cratered surface
	1969	Mariner 6, 7	U.S.	flyby	More photographs of the planet's surface
	1971	Mariner 9	U.S.	orbiter	First complete survey of the surface; revealed complex terrain and evidence of past geological activity
	1976–1982	Viking 1, 2	U.S.	orbiter, lander	Detailed surface maps, implying geological and climatic change; first surface landings, first atmospheric and soil measurements; search for life
	1997	Mars Pathfinder	U.S.	lander	First Mars rover, local geological survey
	1997–	Mars Global Surveyor	U.S.	orbiter	High-resolution surface mapping, remote analysis of surface composition
	2001–	Mars Odyssey	U.S.	orbiter	Remote surface chemical analysis, search for subsurface water, measurement of radiation levels
	2003–2004	Mars Express	European Space Agency	atmospheric probe, orbiter	Study of Martian atmosphere and geology; search for water and evidence of life
	2004	Mars Exploration Rover	U.S.	two landers	Assessment of likelihood that life arose on Mars, measurement of Martian climate and detailed geological surveys near the landing sites
Eros[1]	2001–2002	NEAR	U.S.	orbiter	First mission to an asteroid; detailed geological study
Jupiter	1973	Pioneer 10	U.S.	flyby	First mission to the outer planets
	1974	Pioneer 11	U.S.	flyby	Detailed close-up images; gravity assist from Jupiter to reach Saturn
	1979	Voyager 1	U.S.	flyby	Detailed observations of planet and moons
	1979	Voyager 2	U.S.	flyby	Continued reconnaissance of the Jovian system
	1995–2003	Galileo	U.S.	atmospheric probe, orbiter	Atmospheric studies; long-term precision measurements of the planet's moon system
Saturn	1979	Pioneer 11	U.S.	flyby	First close-up observations of Saturn; now leaving the solar system
	1981	Voyager 1	U.S.	flyby	Observations of planet, rings and moons; close-up measurements of the moon Titan
	1982	Voyager 2	U.S.	flyby	Observations of planet and moons
	2004–	Cassini	U.S., European Space Agency	atmospheric probe, orbiter	Atmospheric studies; repeated and detailed measurements of the planet's moons
Uranus	1986	Voyager 2	U.S.	flyby	Observations of planet and moons; "Grand Tour" of the outer planets
Neptune	1989	Voyager 2	U.S.	flyby	Observations of planet and moons; now leaving the solar system

[1]An asteroid.

Exploration of Mars

Both NASA and the Soviet (now Russian) Space Agency have Mars exploration programs that began in the 1960s. However, the Soviet effort was plagued by a string of technical problems, along with a liberal measure of plain bad luck. As a result, almost all of the detailed planetary data we have on Mars has come from unmanned U.S. probes.

The first spacecraft to reach the Red Planet was *Mariner 4*, which flew by Mars in July 1965. The images sent back by the craft showed large numbers of craters caused by impacts of meteoroids with the planet's surface, instead of the Earthlike terrain some scientists had expected to find. Flybys in 1969 by *Mariner 6* and *Mariner 7* confirmed these findings, leading to the conclusion that Mars was a geologically dead planet having a heavily cratered, old surface. This conclusion was soon reversed by the arrival in 1971 of the *Mariner 9* orbiter. The craft mapped the entire Martian surface at a resolution of about 1 km, and it rapidly became clear that here was a world far more complex than the dead planet imagined only a year or two previously. *Mariner 9*'s maps revealed vast plains, volcanoes, drainage channels, and canyons. All these features were completely unexpected, given the data provided by the earlier missions. These new findings paved the way for the next step: actual landings on the planet's surface.

The two spacecraft of the U.S. *Viking* mission arrived at Mars in mid-1976. *Viking 1* and *Viking 2* each consisted of two parts. An orbiter mapped the surface at a resolution of about 100 m (about the same as the resolution achieved by *Magellan* for Venus), and a lander (see Figure 6.12) descended to the surface and performed a wide array of geological and biological experiments. *Viking 1* touched down on Mars on July 20, 1976. *Viking 2* arrived in September of the same year. By any standards, the *Viking* mission was a complete success: The orbiters and landers returned a wealth of long-term data on the Martian surface and atmosphere. *Viking 2* stopped transmitting data in April 1980. *Viking 1* continued to operate until November 1982.

The relative positions of Mars and Earth in their respective orbits are such that it is most favorable to launch a spacecraft from Earth roughly every 26 months. It then takes roughly eight to nine months for the craft to arrive at Mars (Figure 6.13). In August 1993, the first U.S. probe since *Viking*—*Mars Observer*—exploded just before entering Mars orbit, most probably due to a fuel leak as its engine fired to slow the craft. *Mars Observer* had been designed to radio back detailed images of the planet's surface and provide data on the Martian atmosphere, gravity, and magnetic field. Its replacement, called *Mars Global Surveyor*, was launched from Earth in 1996 and arrived at Mars in late 1997. This spacecraft is currently orbiting the planet, scanning it with cameras and other sensors and charting the Martian landscape. Originally scheduled to end in 2002, the mission (as of early 2004) continues to return high-quality images and data.

Mars Global Surveyor was followed (and in fact overtaken) by *Mars Pathfinder*, which arrived at Mars in late June 1997. On July 4, *Pathfinder* parachuted an instrument package to the Martian surface. Near the ground, the parachute

▲ **FIGURE 6.11** *Magellan* **Orbiter** The U.S. *Magellan* spacecraft was launched from the space shuttle *Atlantis* (at bottom) in May 1989 on a mission to explore the planet Venus. The large radio antenna at the top was used both for mapping the surface of Venus and for communicating with Earth. Contrast the relatively delicate structure of this craft, designed to operate in space, with the much more bulky design of the Venus lander shown in Figure 6.10. (*NASA*)

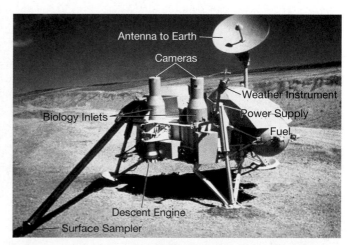

▲ **FIGURE 6.12** *Viking* **Lander** A *Viking* lander, here being tested in the Mojave Desert prior to launch. For an idea of the scale, the reach of its extended arm at left is about 1 m. (*NASA*)

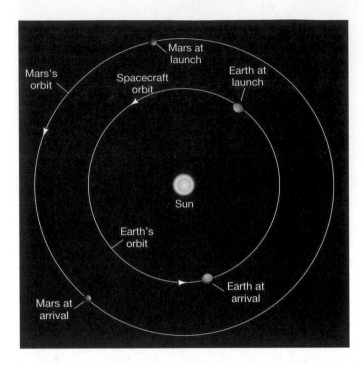

◀ **FIGURE 6.13 Mission to Mars** A typical orbital path from Earth to Mars. A spacecraft orbiting in the path shown can take anywhere from eight to nine months to make the trip, depending on precisely where Mars happens to be in its orbit, which is illustrated here as circular, but which, in reality, is somewhat eccentric.

its proper orbit insertion point and instead burned up in the Martian atmosphere, apparently as a result of navigational commands being sent to the spacecraft's onboard computer in English, rather than in metric, units. In December of the same year, the sister mission, *Mars Polar Lander*, failed to reestablish contact with Earth after entering the Martian atmosphere en route to a planned soft landing on the planet's south polar ice cap. The reason is still unknown.

Despite these setbacks, the U.S. Mars program has now moved into high gear, with an ambitious program of missions spanning the first decade of the 21st century. A new orbiter, called *Mars Odyssey*, reached Mars in October 2001. Its sensors are designed to probe the chemical make-up of the Martian surface and look for possible water ice below. In the summer of 2003, no fewer than three Mars-bound missions left Earth. The European Space Agency launched *Mars Express*, consisting of an orbiter and a lander, while NASA launched *Mars Exploration Rover*, comprising two *Sojourner*-like landers, named *Spirit* and *Opportunity*. *Mars Express* reached the planet in December 2003; its orbiter worked fine and began transmitting data back to Earth, but its surface probe apparently crash land-ed as no signals were heard from it. A month later, the two NASA probes successfully soft-landed by bouncing across the surface until they rolled to a stop. One ended up, as

fell away and huge air bags deployed, enabling the robot to bounce softly to a safe landing. Side panels opened, and out came a small six-wheeled minirover, called *Sojourner* (Figure 6.14). During its three-month lifetime, the lander took measurements of the Martian atmosphere and atmospheric dust, while *Sojourner* roamed the Martian countryside at a rate of a few meters per day, carrying out chemical analyses of the soil and rocks within about 50 m of the parent craft. In addition, over 16,000 images of the region were returned to Earth.

NASA's next two Mars exploration missions met with failure. In September 1999, *Mars Climate Orbiter* missed

◀ **FIGURE 6.14 Martian Lander** The Mars rover *Sojourner* was a small self-propelled vehicle used to explore the Martian surface. It is seen at left center using a device called an alpha proton X-ray spectrometer to determine the chemical composition of a large Martian rock, nicknamed Yogi by mission scientists. For scale, the rock is about 1 m tall and lies some 10 m from *Sojourner*'s mother ship, *Pathfinder* (foreground). *(NASA)*

planned, in a huge crater that might be a dried-up lake; the other mistakenly rolled into an equatorial depression that surprisingly displayed evidence for subsurface layering. Both craft began a two-month odyssey searching for water on Mars.

In 2007, if all goes according to plan, a NASA landing craft will return samples of Martian surface material to Earth (or at least Earth orbit—the issues of possible contamination of the sample by Earth bacteria, or vice versa, are still being debated). NASA also has plans for manned missions to Mars. Indeed, parts of the *Mars Odyssey* and *Mars Exploration Rover* missions are devoted specifically to determining the feasibility of human exploration of the planet. However, the enormous expense (and danger) of such an undertaking, coupled with the belief of many astronomers that unmanned missions are economically and scientifically preferable to manned missions, makes the future of these projects uncertain at best.

Missions to the Outer Planets

Two pairs of U.S. spacecraft launched in the 1970s—*Pioneer* and *Voyager*—revolutionized our knowledge of Jupiter and the Jovian planets. *Pioneer 10* and *Pioneer 11* were launched in March 1972 and April 1973, respectively, and arrived at Jupiter in December 1973 and December 1974.

The *Pioneer* spacecraft took many photographs and made numerous scientific discoveries. Their orbital trajectories also allowed them to observe the polar regions of Jupiter in much greater detail than later missions would achieve. In addition to their many scientific accomplishments, the *Pioneer* craft also played an important role as "scouts" for the later *Voyager* missions. The *Pioneer* series demonstrated that spacecraft could travel the long route from Earth to Jupiter without colliding with debris in the solar system. They also discovered—and survived—the perils of Jupiter's extensive radiation belts (somewhat like the Van Allen radiation belts that surround Earth, but on a much larger scale). In addition, *Pioneer 11* used Jupiter's gravity to propel it along the same trajectory to Saturn that the *Voyager* controllers planned for *Voyager 2*'s visit to Saturn's rings.

The two *Voyager* spacecraft (see Figure 6.15) left Earth in 1977 and reached Jupiter in March (*Voyager 1*) and July (*Voyager 2*) of 1979 to study the planet and its major satellites in detail. Each craft carried sophisticated equipment to investigate the planet's magnetic field, as well as radio, visible-light, and infrared sensors to analyze its reflected and emitted radiation. Both *Voyager 1* and *Voyager 2* used Jupiter's gravity to send them on to Saturn. *Voyager 1* was programmed to visit Titan, Saturn's largest moon, and so did not come close enough to the planet to receive a gravity-assisted boost to Uranus. However, *Voyager 2* went on to visit both Uranus and Neptune in a spectacularly successful "Grand Tour" of the outer planets. The data returned by the two craft are still being analyzed today. Like *Pioneer 11*, the two *Voyager* craft are now

headed out of the solar system, still sending data as they race toward interstellar space. The figure in *Discovery 6-1* shows the past and present trajectories of the *Voyager* spacecraft.

The most recent mission to Jupiter is the U.S. *Galileo* probe, launched by NASA in 1989. The craft arrived at its target in 1995 after a rather roundabout route involving a gravity assist from Venus and two from Earth itself. The mission consisted of an orbiter and an atmospheric probe. The probe descended into Jupiter's atmosphere in December 1995, slowed by a heat shield and a parachute, taking measurements and performing chemical analyses as it went. The orbiter executed a complex series of gravity-assisted maneuvers through Jupiter's system of moons, returning to some already studied by *Voyager* and visiting others for the first time. Some of *Galileo*'s main findings are described in Chapter 11.

The *Galileo* program was originally scheduled to last until December 1997, but NASA extended its lifetime for six more years to obtain even more detailed data on Jupiter's inner moons. The mission finally ended in September 2003. With the fuel supply dwindling, *Galileo*'s controllers elected to steer the craft directly into Jupiter, rather than run any risk that it might collide with, and possibly contaminate, Jupiter's moon Europa, which (thanks largely to data returned by *Galileo*) is now a leading candidate in the search for extraterrestrial life. NASA has ambitious (but as yet unfunded) plans for a return to the Jupiter system, possibly as soon as 2010.

In October 1997, NASA launched the *Cassini* mission to Saturn. The launch (from Cape Canaveral) sparked controversy because of fears that the craft's plutonium power source might contaminate parts of our planet following an accident either during launch or during a subsequent gravity assist from Earth in 1999—one of four assists

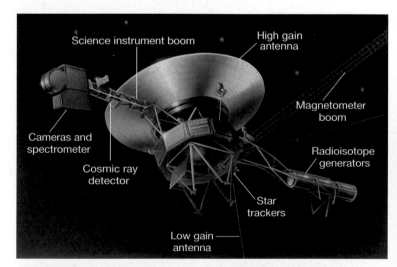

▲ **FIGURE 6.15** *Voyager* The two *Voyager* spacecraft that swung by several of the outer planets were identically constructed. Their main features are shown here. (*NASA*)

needed for the craft to reach Saturn's distant orbit (Figure 6.16). When the craft reaches its destination in 2004, it will dispatch a probe built by the European Space Agency into the atmosphere of Titan, Saturn's largest moon, and orbit among the planet's moons for four years, much as *Galileo* did at Jupiter. If the experience with *Galileo* is any guide, *Cassini* will likely resolve many outstanding questions about the Saturn system, but it is sure to pose many new ones, too.

▶ **FIGURE 6.16** *Cassini* **Mission to Saturn** To reach Saturn by July 2004, the *Cassini* spacecraft, shown under construction in part (a), was launched in 1997 and increased its velocity progressively by flying by Venus in 1998 and 1999 and also by Earth in 1999 (part b). Finally, it flew by Jupiter in 2000, picking up one final gravity boost. *(NASA)*

(a)

DISCOVERY 6-1

Gravitational "Slingshots"

Celestial mechanics is the study of the motions of gravitationally interacting objects, such as planets and stars, applying Newton's laws of motion to understand the intricate movements of astronomical bodies. ∞ (Sec. 2.7) Computerized celestial mechanics lets astronomers calculate planetary orbits to high precision, taking the planets' small gravitational influences on one another into account. Even before the computer age, the discovery of one of the outermost planets, Neptune, came about almost entirely through studies of the distortions of Uranus's orbit that were caused by Neptune's gravity.

Celestial mechanics is also an essential tool for scientists and engineers who wish to navigate manned and unmanned spacecraft throughout the solar system. Robot probes can now be sent on stunningly accurate trajectories, expressed in the trade with such slang phrases as "sinking a corner shot on a billion-kilometer pool table." Near-flawless rocket launches, aided by occasional midcourse changes in flight paths, now enable interplanetary navigators to steer remotely controlled spacecraft through an imaginary "window" of space just a few kilometers wide and a billion kilometers away.

However, sending a spacecraft to another planet requires a lot of energy—often more than can be conveniently provided by a rocket launched from Earth or safely transported in a shuttle for launch from orbit. Faced with these limitations, mission scientists often use their knowledge of celestial mechanics to carry out "slingshot" maneuvers, which can boost an interplanetary probe into a more energetic orbit and also aid navigation toward the target, all at no additional cost!

The accompanying figure illustrates a gravitational slingshot, or *gravity assist*, in action. A spacecraft approaches a planet, passes close by, and then escapes along a new trajectory. Obviously, the spacecraft's *direction* of motion is changed

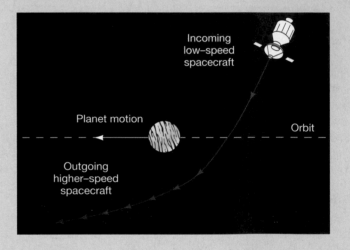

by the encounter. Less obviously, the spacecraft's *speed* is also altered as the planet's gravity propels the spacecraft in the direction of the planet's motion. By a careful choice of incoming trajectory, the craft can speed up (by passing "behind" the planet, as shown) *or* slow down (by passing in front), by as much as twice the planet's orbital speed. Of course, there is no "free lunch"—the spacecraft gains energy from, or loses it to, the planet's motion, causing the planet's own orbit to change ever so slightly. However, since planets are so much more massive than spacecraft, the effect on the planet is insignificant.

Such a slingshot maneuver has been used many times in missions to both the inner and the outer planets, as illustrated in the second figure, which shows the trajectories of the *Voyager* spacecraft through the outer solar system. The gravitational pulls of these giant worlds whipped the craft around at each visitation, enabling flight controllers to get consider-

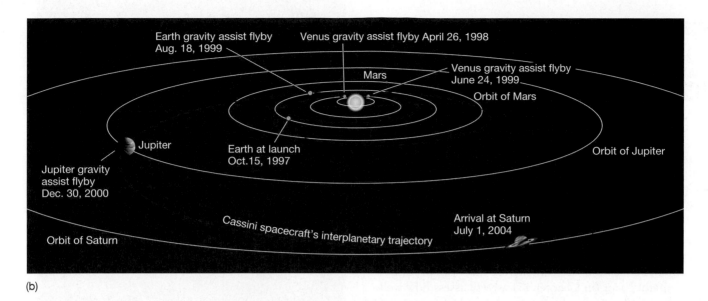

(b)

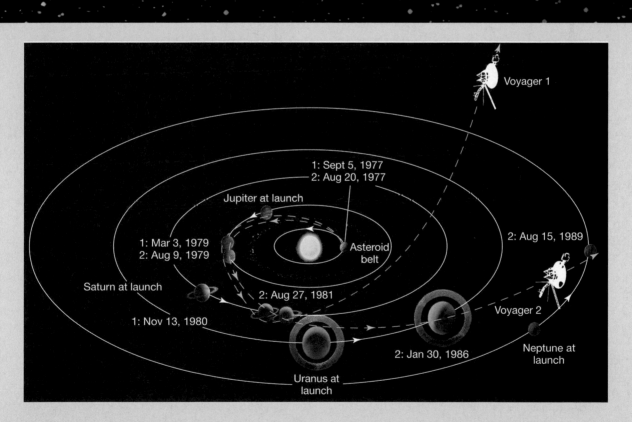

able extra "mileage" out of the probes. *Voyager 1* is now high above the plane of the solar system, having been deflected up and out following its encounter with Saturn. *Voyager 2* continued on for a "Grand Tour" of the four Jovian planets. It is now outside the orbit of Pluto.

More recently, the *Galileo* mission to Jupiter, which was launched in 1989 and arrived at its target in 1995, received *three* gravitational assists en route—one from Venus and two

from Earth. Once in the Jupiter system, *Galileo* used the gravity of Jupiter and its moons to propel it through a complex series of maneuvers designed to bring it close to all the major moons, as well as to the planet itself. Every encounter with a moon had a slingshot effect—sometimes accelerating and sometimes slowing the probe, but each time moving it into a different orbit—and every one of these effects was carefully calculated long before *Galileo* ever left Earth.

6.7 How Did the Solar System Form?

6 During the past four decades, interplanetary probes have vastly increased our knowledge of the solar system, and their data form the foundation for much of the discussion in the next eight chapters. However, we can understand the overall organization of our planetary system—the basic properties presented in Sections 6.3 and 6.4—without dwelling on these details. Indeed, some key elements of the modern theory of planetary formation predate the Space Age by many years.

We present here (in simplified form) the "standard" view of how the solar system came into being. This picture will underlie much of our upcoming discussion of the planets, their moons, and the contents of the vast spaces between them. Later (in Chapter 15) we will assess how well our theory holds up in the face of detailed observational data—including evidence for the existence of well over 100 extrasolar planets orbiting neighboring stars—and examine what it means for the prospects of finding Earth-like planets, and maybe even life, elsewhere in the Galaxy.

Nebular Contraction

One of the earliest heliocentric models of solar-system formation may be traced back to the 17th-century French philosopher René Descartes. Imagine a large cloud of interstellar dust and gas (called a *nebula*) a light-year or so across. Now suppose that, due to some external influence, such as a collision with another interstellar cloud or perhaps the explosion of a nearby star, the nebula starts to contract under the influence of its own gravity. As it contracts, it becomes denser and hotter, eventually forming a star—the Sun—at its center (see Chapter 19). Descartes suggested that, while the Sun was forming in the cloud's hot core, the planets and their moons formed in the cloud's cooler outer regions. In other words, planets are by-products of the process of star formation.

In 1796, the French mathematician–astronomer Pierre Simon de Laplace developed Descartes's ideas in a more quantitative way. He showed mathematically that conservation of angular momentum (see *More Precisely 6-2*) demands that our hypothetical nebula spin faster as it contracts. A decrease in the size of a rotating mass must be balanced by an increase in its rotational speed. The latter, in turn, causes the nebula's *shape* to change as it collapses. Centrifugal forces (due to rotation) tend to oppose the contraction in directions perpendicular to the rotation axis, with the result that the nebula collapses most rapidly along that axis. As shown in Figure 6.17, the fragment eventually flattens into a pancake-shaped primitive solar system. This swirling mass destined to become our solar system is usually referred to as the **solar nebula**.

If we now simply suppose that the planets formed out of this spinning material, then we can already understand

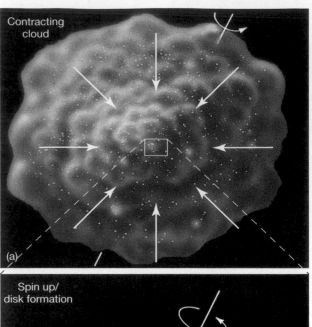

▲ **FIGURE 6.17 Nebular Contraction** (a) Conservation of angular momentum demands that a contracting, rotating cloud spin faster as its size decreases. (b) Eventually, the primitive solar system comes to resemble a giant pancake. In the case of our solar system, the large blob at the center ultimately became the Sun.

the origin of much of the large-scale architecture observed in our planetary system today, such as the circularity of the planets' orbits and the fact that they move in the same sense in nearly the same plane. The planets inherited all these properties from the rotating disk in which they were born. The idea that the planets formed from such a disk is called the **nebular theory**.

Astronomers are fairly confident that the solar nebula formed a disk, because similar disks have been observed (or inferred) around other stars. Figure 6.18(a) shows visible-light images of the region around a star called Beta Pictoris, lying about 60 light-years from the Sun. When the light from Beta Pictoris itself is suppressed and the resulting image enhanced by a computer, a faint disk of warm matter (viewed almost edge-on here) can be seen. This particular disk is roughly 1000 A.U. across—about 10 times the diameter of Pluto's orbit. Astronomers believe that Beta Pictoris is a very young star, perhaps only 20 million years old, and that we are witnessing it pass through

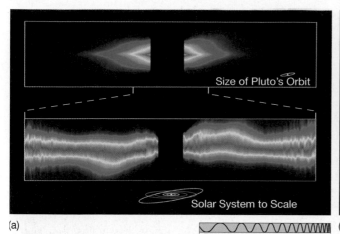

(a)   R I V U X G   (b)

▲ **FIGURE 6.18 Beta Pictoris** (a) A computer-enhanced view of a disk of warm matter surrounding the star Beta Pictoris. Both images in (a) display data taken at visible wavelengths, but are presented here in false color to accentuate the details; the bottom image is a close-up of the inner part of the disk, indicating the presence of a warp, possibly caused by the gravitational pull of unseen companions. In both images, the overwhelmingly bright central star has been covered to let us see the much fainter disk surrounding it. The disk is nearly edge-on to our line of sight. For scale, the dimension of Pluto's orbit (78 A.U.) has been drawn adjacent to the images. (b) An artist's conception of the disk of clumped matter, showing the warm disk with a young star at the center and several comet-sized or larger bodies already forming. The colors are thought to be accurate: At the outer edges of the disk, the temperature is low and the color is dull red. Progressing inward, the colors brighten and shift to a more yellowish tint as the temperature increases. Mottled dust is seen throughout—such protoplanetary regions are probably very "dirty." (*NASA; D. Berry*)

an evolutionary stage similar to the one our own Sun underwent long ago. Figure 6.18(b) shows an artist's conception of the disk.

The Condensation Theory

Scientific theories must continually be tested and refined as new data become available. ∞ (Sec. 1.2) Unfortunately for the original nebular theory, while Laplace's description of the collapse and flattening of the solar nebula was basically correct, we now know that a disk of warm gas would *not* form clumps of matter that would subsequently evolve into planets. In fact, modern computer calculations predict just the opposite: Clumps in the gas would tend to disperse, not contract further. However, the model currently favored by most astronomers, known as the **condensation theory**, rests squarely on the old nebular theory, combining its basic physical reasoning with new information about interstellar chemistry to avoid most of the original theory's problems.

The key new ingredient is the presence of *interstellar dust* in the solar nebula. Astronomers now recognize that the space between the stars is strewn with microscopic dust grains, an accumulation of the ejected matter of many long-dead stars (see Chapter 22). These dust particles probably formed in the cool atmospheres of old stars and then grew by accumulating more atoms and molecules from the interstellar gas within the Milky Way galaxy. The end result is that our entire Galaxy is littered with miniature chunks of icy and rocky matter having typical sizes of

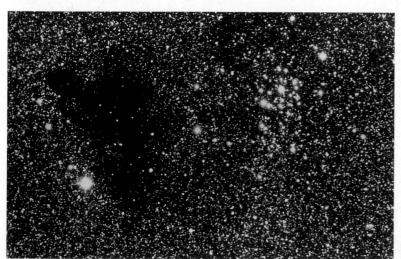

R I V U X G

◄ **FIGURE 6.19 Dark Cloud** Interstellar gas and dark dust lanes mark this region of star formation. The dark cloud known as Barnard 86 (left) flanks a cluster of young blue stars called NGC 6520 (right). Barnard 86 may be part of a larger interstellar cloud that gave rise to these stars. (*D. Malin/Anglo-Australian Telescope*)

The Concept of Angular Momentum

Most celestial objects rotate. Planets, moons, stars, and galaxies all have some *angular momentum*, which we can define as the tendency of a body to keep spinning or moving in a circle. Angular momentum is as important a property of an object as its mass or its energy.

Consider first a simpler motion—*linear momentum*, which is defined as the product of an object's mass and its velocity:

$$\text{linear momentum} = \text{mass} \times \text{velocity}.$$

Linear momentum is the tendency of an object to keep moving in a straight line in the absence of external forces. Picture a truck and a bicycle rolling equally fast down a street. Each has some linear momentum, but you would obviously find it easier to stop the less massive bicycle. Although the two vehicles have the same speed, the truck has more momentum. We see, then, that the linear momentum of an object depends on the mass of the object. It also depends on the speed of the object: If two bicycles were rolling down the street at different speeds, the slower one could be stopped more easily.

Angular momentum is an analogous property of objects that are rotating or revolving. It is a measure of the object's tendency to keep spinning, or, equivalently, of how much effort must be expended to stop the object from spinning. However, in addition to mass and (angular) speed, angular momentum also depends on the way in which an object's mass is distributed. Intuitively, we know that the more massive an object, or the larger it is, or the faster it spins, the harder it is to stop. In fact, angular momentum depends on the object's *mass*, *rotation rate* (measured in, say, revolutions per second), and *radius*, in a very specific way:

$$\text{angular momentum} \propto \text{mass} \times \text{rotation rate} \times \text{radius}^2.$$

(Recall that the symbol "$\propto$" means "is proportional to"; the constant of proportionality depends on the details of how the object's mass is distributed.)

According to Newton's laws of motion, both types of momentum—linear and angular—must be conserved at all times. ∞ (Sec. 2.7) In other words, both linear and angular momentum must remain constant before, during, and after a physical change in any object (so long as no external forces act on the object). For example, as illustrated in the first figure, if a spherical object having some spin begins to contract, the previous relationship demands that it spin faster, so that the product mass × angular speed × radius² remains constant. The sphere's mass does not change during the contraction, yet the size of the object clearly decreases. Its rotation speed must therefore increase in order to keep the total angular momentum unchanged.

Figure skaters use the principle of conservation of angular momentum, too. They spin faster by drawing in their arms (as shown in the second pair of figures) and slow down by extending them. Here, the mass of the human body remains the same, but its lateral size changes, causing the body's rotation speed to increase or decrease, as the case may be, to keep its angular momentum unchanged.

EXAMPLE 1: Suppose the sphere has radius 1 m and starts off rotating at 1 revolution per minute. It then contracts to one-tenth its initial size. Conservation of angular momentum entails that the sphere's final angular speed A must satisfy the relationship

$$\text{mass} \times A \times (0.1 \text{ m})^2 = \text{mass} \times (1 \text{ rev/min}) \times (1 \text{ m})^2.$$

The mass is the same on either side of the equation and therefore cancels, so we find that $A = (1 \text{ rev/min}) \times (1 \text{ m})^2/(0.1 \text{ m})^2 = 100$ rev/min, or about 1.7 rev/s.

EXAMPLE 2: Now suppose that the "sphere" is a large interstellar gas cloud that is about to collapse and form the solar nebula. Initially, let's imagine that it has a diameter of 1 light-year and that it rotates very slowly—once every 10 million years. Assuming that the cloud's mass stays constant, its rotation rate must *increase* to conserve angular momentum as the radius decreases. By the time it has collapsed to a diameter of 100 A.U., the cloud has shrunk by a factor of (1 light-year/100 A.U.) ≈ 630. Conservation of angular momentum then implies that the cloud's (average) spin rate has increased by a factor of $630^2 \approx 400{,}000$, to roughly 1 revolution every 25 years—about the orbital period of Saturn.

Incidentally, the law of conservation of angular momentum also applies to planetary orbits (where the radius is now the distance from the planet to the Sun). In fact, Kepler's second law *is* just conservation of angular momentum, expressed another way. ∞ (Sec. 2.5)

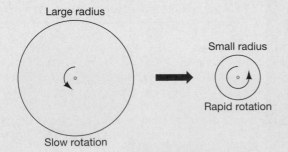

Large radius

Small radius

Rapid rotation

Slow rotation

(Orban/Corbis/Sygma)

about 10^{-5} m. Figure 6.19 shows one of many such dusty regions found in the vicinity of the Sun.

Dust grains play two important roles in the evolution of a gas cloud. First, dust helps to cool warm matter by efficiently radiating its heat away in the form of infrared radiation. ∞ (Sec. 3.4) As the cloud cools, its molecules move around more slowly, reducing the internal pressure and allowing the nebula to collapse more easily under the influence of gravity. ∞ (*More Precisely 3-1*) Second, acting as **condensation nuclei**—microscopic platforms to which other atoms can attach, forming larger and larger balls of matter—the dust grains greatly speed up the process of collecting enough atoms to form a planet. This mechanism is similar to the way raindrops form in Earth's atmosphere: Dust and soot in the air act as condensation nuclei around which water molecules cluster.

Thus, according to the modern condensation theory, once the solar nebula had formed and begun to cool (Figure 6.20a), dust grains formed condensation nuclei around which matter began to accumulate (Figure 6.20b). This vital step greatly hastened the critical process of forming the first small clumps of matter. Once these clumps formed, they grew rapidly by sticking to other clumps. (Imagine a snowball thrown through a fierce snowstorm, growing bigger as it encounters more snowflakes.) As the clumps grew larger, their surface areas increased and consequently, the rate at which they swept up new material accelerated. Gradually, the accreted matter grew into objects of pebble size, baseball size, basketball size, and larger.

Eventually, this process of **accretion**—the gradual growth of small objects by collision and sticking—created objects a few hundred kilometers across (Figure 6.20c). By that time, their gravity was strong enough to sweep up material that would otherwise not have collided with them, and their rate of growth accelerated, allowing them to form still larger objects. Because larger bodies have stronger gravity, eventually (Figure 6.20d) almost all the original material was swept up into a few large **protoplanets**—the accumulations of matter that would in time evolve into the planets we know today (Figure 6.20e).

The Role of Heat

The condensation theory just described can account—in broad terms, at least—for the formation of the planets and the large-scale architecture of the solar system. What does it say about the differences between the terrestrial and the jovian planets? Specifically, why are smaller, rocky planets found close to the Sun, while the larger gas giants orbit at much greater distances? To understand why a planet's composition depends on its location in the solar system, we must consider the *temperature* profile of the solar nebula. Indeed, it is in this context that the term *condensation* derives its true meaning.

As the primitive solar system contracted under the influence of gravity, it heated up as it flattened into a disk. The density and temperature were greatest near the center

(a) initially

(b)

(c) few million years

(d)

(e) 100 million years

◄ **FIGURE 6.20 Solar-System Formation**
The condensation theory of planetary formation (not drawn to scale, nor is Pluto shown in part e). (a) The solar nebula after it has contracted and flattened to form a spinning disk (Figure 6.17b). The large blob in the center will become the Sun. (b) Dust grains act as condensation nuclei, forming clumps of matter that collide, stick together, and grow into moon-sized bodies. The composition of the grains—and hence the composition of the bodies that form—depends on location within the nebula. (c) After a few million years, strong winds from the still-forming Sun begin to expel the nebular gas. By this time, some particularly massive bodies in the outer solar system have already captured large amounts of gas from the nebula. (d) With the gas ejected, objects in the inner solar system continue to collide and grow. The outer gas giant planets are already formed. (e) Over the next hundred million years or so, most of the remaining small bodies are accreted or ejected, leaving a few large planets that travel in roughly circular orbits.

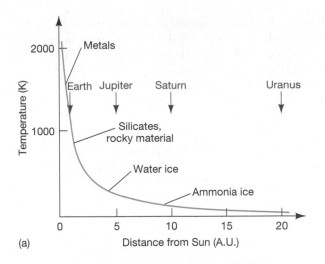

(a)

(b)

◀ **FIGURE 6.21 Temperature in the Early Solar Nebula**
(a) Theoretically computed variation of temperature across the primitive solar nebula, illustrated in (b), which shows half of the disk in Figure 6.20(b). In the hot central regions, only metals could condense out of the gaseous state to form grains. At greater distances from the central proto-Sun, the temperature was lower, so rocky and icy grains could also form. The labels indicate the minimum radii at which grains of various types could condense out of the nebula.

and much lower in the outlying regions. Detailed calculations indicate that the gas temperature near the core of the contracting system was several thousand kelvins. At a distance of 10 A.U., out where Saturn now resides, the temperature was only about 100 K.

The high temperatures in the warmer regions of the cloud caused dust grains to break apart into molecules, which in turn split into excited atoms. Because the extent to which the dust was destroyed depended on temperature, it therefore also depended on location in the solar nebula. Most of the original dust in the inner solar system disappeared at this stage, while the grains in the outermost parts probably remained largely intact.

As the dusty nebula radiated away its heat, its temperature decreased everywhere except in the very core, where the Sun was forming. As the gas cooled, new dust grains began to condense (or crystallize) out, much as raindrops, snowflakes, and hailstones condense from moist, cooling air here on Earth. It may seem strange that although there was plenty of interstellar dust early on, it was partly destroyed, only to form again later. However, a critical change had occurred. Initially, the nebular gas was uniformly peppered with dust grains of all compositions; when the dust re-formed later, the distribution of grains was very different.

Figure 6.21 plots the temperature in various parts of the primitive solar system just before accretion began. At any given location, the only materials to condense out were those able to survive the temperature there. As marked on the figure, in the innermost regions, around Mercury's present orbit, only metallic grains could form; it was simply too hot for anything else to exist. A little farther out, at about 1 A.U., it was possible for rocky, silicate grains to form, too. Beyond about 3 or 4 A.U., water ice could exist, and so on, with the condensation of more and more material possible at greater and greater distances from the Sun. The composition of the material that could condense out at any given radius would determine the type of planet that formed there.

Terrestrial and Jovian Planets

In the inner regions of the primitive solar system, condensation from gas to solid began when the average temperature was about 1000 K. The environment there was too hot for ice to survive. Many of the more abundant heavier elements, such as silicon, iron, magnesium, and aluminum, combined with oxygen to produce a variety of rocky materials. The dust grains in the inner solar system were therefore predominantly rocky or metallic in nature, as were the protoplanets and planets they eventually became.

In the middle and outer regions of the primitive planetary system, beyond about 5 A.U. from the center, the temperature was low enough for the condensation of several abundant gases—water (H_2O), ammonia (NH_3), and methane (CH_4)—into solid form. After hydrogen (H) and helium (He), the elements carbon (C), nitrogen (N), and oxygen (O) are the most common materials in the universe. As a result, wherever icy grains could form, they greatly outnumbered rocky and metallic particles. Consequently, the objects that formed at these distances were formed under cold conditions out of predominantly low-density, icy material. These ancestral fragments were destined to form the cores of the jovian planets.

Because more material could condense out of the solar nebula at large radii than in the inner regions near the Sun, accretion began sooner, with more resources to draw on. The outer solar system thus had a "head start" in the accretion process, and the outer planets grew rapidly, eventually becoming massive enough to enter a new phase of planet formation, in which their strong gravitational fields swept up large amounts of gas directly from the solar nebula. The smaller, inner protoplanets never reached this stage, and, as a result, their masses remained relatively low. This is the chain of events depicted in Figures 6.20(c) and (d).

Formation Stops

The events just described did not take long, astronomically speaking. The giant planets formed within a few million years of the appearance of the flattened solar nebula—the blink of an eye compared with the 4.6-*billion*-year age of the solar system. At that point, intense radiation and strong winds from violent activity on the surface of the newborn Sun ejected the nebular gas, effectively halting

further growth of the outer worlds. Accretion in the inner solar system proceeded more slowly, taking perhaps 100 million years to form the planets we know today (Figure 6.20e). The rocky asteroids and icy comets are all that remain of the matter that originally condensed out of the solar nebula—the last surviving witnesses to the birth of our planetary system.

CONCEPT CHECK

✔ Why was interstellar dust so important to the formation of our solar system?

Chapter Review

SUMMARY

The **solar system (p. 144)** consists of the Sun and everything that orbits it, including the nine major planets, the moons that orbit them, and the many small bodies found in interplanetary space. The science of **comparative planetology (p. 145)** compares and contrasts the properties of the diverse bodies found in the solar system and elsewhere in order to understand better the conditions under which planets form and develop. The asteroids, or "minor planets," are small bodies, none of them larger than Earth's Moon and most of which orbit in a broad band called the asteroid belt between the orbits of Mars and Jupiter. Comets are chunks of ice found chiefly in the outer solar system. Their importance to planetary astronomy lies in the fact that they are thought to be "leftover" material from the formation of the solar system and therefore contain clues to the very earliest stages of its development.

The major planets orbit the Sun in the same sense—counterclockwise as viewed from above Earth's North Pole—on roughly circular orbits that lie close to the plane of the ecliptic. The orbits of the innermost planet, Mercury, and the outermost, Pluto, are the most eccentric and have the greatest orbital inclination. The spacing between planetary orbits increases as we move outward from the Sun.

Density (p. 147) is a convenient measure of the compactness of any object. The average density of a planet is obtained by dividing the planet's total mass by its volume. The innermost four planets in the solar system have average densities comparable to Earth's and are generally rocky in composition. The outermost planets have much lower densities than the terrestrial worlds and, with the exception of Pluto, are made up mostly of gaseous or liquid hydrogen and helium.

Planetary scientists divide the eight large planets (excluding Pluto) in the solar system, on the basis of their densities and composition, into the rocky **terrestrial planets (p. 150)**—Mercury, Venus, Earth, and Mars—which lie closest to the Sun, and the gaseous **Jovian planets (p. 150)**—Jupiter, Saturn, Uranus, and Neptune—which lie at greater distances. Compared with the ter-

restrial worlds, the Jovian planets are larger and more massive, rotate more rapidly, and have stronger magnetic fields. In addition, the Jovian planets all have ring systems and many moons orbiting them. All the major planets, with the exception of Pluto, have been visited by unmanned space probes. Spacecraft have landed on Venus and Mars. In many cases, to reach their destinations, the spacecraft were set on trajectories that included "gravitational assists" from one or more planets.

According to the **nebular theory (p. 160)** of the formation of the solar system, a large cloud of dust and gas—the **solar nebula (p. 160)**—began to collapse under its own gravity. As it did so, it began to spin faster, to conserve angular momentum, eventually forming a disk out of which the planets arose. The **condensation theory (p. 161)** builds on the nebular theory by incorporating the effects of particles of interstellar dust, which helped cool the nebula and acted as **condensation nuclei (p. 163)**, allowing the planet-building process to begin. Small clumps of matter grew by **accretion (p. 163)**, gradually sticking together and growing into moon-sized bodies whose gravitational fields were strong enough to accelerate the accretion process, subsequently causing the bodies to grow into **protoplanets (p. 163)**. Eventually, only a few planet-sized objects remained. In the outer solar system, protoplanet cores became so large that they could capture hydrogen and helium gas directly from the solar nebula.

At any given location, the temperature in the solar nebula determined which materials could condense out and hence also determined the composition of any planets forming there. The terrestrial planets are rocky because they formed in the hot inner regions of the solar nebula, near the Sun, where only rocky and metallic materials condensed out. Farther out, the nebula was cooler, and ices of water and ammonia could also form, ultimately leading to the observed differences in composition between the inner and outer solar system.

When the Sun became a star, its strong winds blew away any remaining gas in the solar nebula. Leftover small bodies that never became part of a planet are seen today as the asteroids and comets.

REVIEW AND DISCUSSION

1. Name and describe all the different types of objects found in the solar system. Give one distinguishing characteristic of each. Include a mention of interplanetary space.

2. What is comparative planetology? Why is it useful? What is its ultimate goal?

3. Why is it necessary to know the distance to a planet in order to determine the planet's mass?

4. List some ways in which the solar system is an "orderly" place.

5. What are some "disorderly" characteristics of the solar system?

6. Which are the terrestrial planets? Why are they given that name?

7. Which are the Jovian planets? Why are they given that name?

8. Name three important differences between the terrestrial planets and the Jovian planets.

9. Compare the properties of Pluto given in Table 6.1 with the properties of the terrestrial and Jovian planets presented in Table 6.2. What do you conclude regarding the classification of Pluto as either a terrestrial or Jovian planet?

10. Why are asteroids and meteoroids important to planetary scientists?

11. Comets generally vaporize upon striking Earth's atmosphere. How, then, do we know their composition?

12. Why has our knowledge of the solar system increased greatly in recent years?

13. How and why do scientists use gravity assists to propel spacecraft through the solar system?

14. Which planets have been visited by spacecraft from Earth? On which ones have spacecraft actually landed?

15. Why do you think *Galileo* and *Cassini* took such circuitous routes to Jupiter and Saturn, while *Pioneer* and *Voyager* did not?

16. How do you think NASA's new policy of building less complex, smaller, and cheaper spacecraft—with shorter times between design and launch—will affect the future exploration of the outer planets? Will missions like *Galileo* and *Cassini* be possible in the future?

17. What is the key ingredient in the modern condensation theory of the solar system's origin that was missing or unknown in the nebular theory?

18. Give three examples of how the condensation theory explains the observed features of the present-day solar system.

19. Why are the Jovian planets so much more massive than the terrestrial planets?

20. How did the temperature structure of the solar nebula determine planetary composition?

CONCEPTUAL SELF-TEST: TRUE OR FALSE/MULTIPLE CHOICE

1. Most planets orbit the Sun in nearly the same plane as Earth does.

2. The largest planets also have the largest densities.

3. The total mass of all the planets is much less than the mass of the Sun.

4. The jovian planets all rotate more rapidly than Earth.

5. Saturn is the largest and most massive planet in the solar system.

6. Asteroids are similar in overall composition to the terrestrial planets.

7. Comets have compositions similar to the icy moons of the jovian planets.

8. All planets have moons.

9. Both *Voyager* missions used gravity assists to visit all four jovian planets.

10. Interstellar dust plays a key role in the formation of a planetary system.

11. A planet's mass can most easily be determined by measuring the planet's **(a)** moon's orbits; **(b)** angular diameter; **(c)** position in the sky; **(d)** orbital speed around the Sun.

12. If we were to construct an accurate scale model of the solar system on a football field with the Sun at one end and Pluto at the other, the planet closest to the center of the field would be **(a)** Earth; **(b)** Jupiter; **(c)** Saturn; **(d)** Uranus.

13. The inner planets tend to have **(a)** fewer moons; **(b)** faster rotation rates; **(c)** stronger magnetic fields; **(d)** higher gravity than the outer planets have.

14. The planets that have rings also tend to have **(a)** solid surfaces; **(b)** many moons; **(c)** slow rotation rates; **(d)** weak gravitational fields.

15. A solar system object of rocky composition and comparable in size to a small city is most likely **(a)** a meteoroid; **(b)** a comet; **(c)** an asteroid; **(d)** a planet.

16. The asteroids are mostly **(a)** found between Mars and Jupiter; **(b)** just like other planets, only younger; **(c)** just like other planets, only smaller; **(d)** found at the very edge of our solar system.

17. The *Sojourner* Mars rover, part of the Mars *Pathfinder* mission in 1997, was able to travel over an area about the size of **(a)** a soccer field; **(b)** a small U.S. city; **(c)** a very large U.S. city; **(d)** a small U.S. state.

18. To travel from Earth to the planet Neptune at more than 30,000 mph, the *Voyager 2* spacecraft took nearly **(a)** a year; **(b)** a decade; **(c)** three decades; **(d)** a century.

19. In the leading theory of solar-system formation, the planets **(a)** were ejected from the Sun following a close encounter with another star; **(b)** formed from the same flattened, swirling gas cloud that formed the Sun; **(c)** are much younger than the Sun; **(d)** are much older than the Sun.

20. The solar system is differentiated because **(a)** all the heavy elements in the outer solar system have sunk to the center; **(b)** all the light elements in the inner solar system became part of the Sun; **(c)** all the light elements in the inner solar system were carried off in the form of comets; **(d)** only rocky and metallic particles could form close to the Sun.

PROBLEMS

 Algorithmic versions of these questions are available in the Practice Problems module of the Companion Website at astro.prenhall.com/chaisson.

The number of squares preceding each problem indicates its approximate level of difficulty.

1. ■■■ Use the data given in Table 6.1 to calculate the angular diameter of Saturn when it lies 9 A.U. from Earth. Saturn's moon Titan is observed to orbit 3.1′ from the planet. What is Titan's orbital period?

2. ■ Use Newton's law of gravity to compute your weight (a) on Earth, (b) on Mars, (c) on the asteroid Ceres, and (d) on Jupiter (neglecting temporarily the absence of a solid surface on this planet!). ∞ (Sec. 2.7)

3. ■ Only Mercury, Mars, and Pluto have orbits that deviate significantly from circles. Calculate the perihelion and aphelion distances of these planets from the Sun. ∞ (More Precisely 2-1)

4. ■■■ At closest approach, the planet Neptune lies roughly 29.1 A.U. from Earth. At that distance, Neptune's angular diameter is 2.3″. Its moon Triton moves in a circular orbit with an angular diameter of 33.6″ and a period of 5.9 days. Use these data to compute the radius, mass, and density of Neptune, and compare your results with the figures given in Table 6.1.

5. ■ Suppose the average mass of each of the 7000 asteroids in the solar system is about 10^{17} kg. Compare the total mass of all asteroids with the mass of Earth.

6. ■ Assuming a roughly spherical shape and a density of 3000 kg/m^3, estimate the diameter of an asteroid having the average mass given in the previous question.

7. ■■ A *short-period* comet is conventionally defined as a comet having an orbital period of less than 200 years. What is the maximum possible aphelion distance for a short-period comet with a perihelion of 0.5 A.U.? Where does this place the comet relative to the outer planets?

8. ■ How many times has *Mariner 10* now orbited the Sun?

9. ■■■ A spacecraft has an orbit that just grazes Earth's orbit at perihelion and the orbit of Mars at aphelion. What are the orbital eccentricity and semimajor axis of the orbit? How long does it take to go from Earth to Mars? (The orbit given is the so-called *minimum-energy orbit* for a craft leaving Earth and reaching Mars. Assume circular planetary orbits for simplicity.)

10. ■■ The asteroid Icarus has a perihelion distance of 0.2 A.U. and an orbital eccentricity of 0.7. What is Icarus's aphelion distance from the Sun?

11. ■■■ Earth and Mars were at closest approach in September 2003. The first *Mars Exploration Rover* was launched in June 2003 and arrived at Mars in January 2004. Sketch the orbits of the two planets and the trajectory of the spacecraft. Be sure to indicate the location of Mars at launch and of Earth when the spacecraft reached Mars. (For simplicity, neglect the eccentricity of Mars's orbit in your sketch.)

12. ■■ How long would it take for a radio signal to complete the round-trip between Earth and Saturn? Assume that Saturn is at its closest point to Earth. How far would a spacecraft orbiting the planet in a circular orbit of radius 100,000 km travel in that time? Do you think that mission control could maneuver the spacecraft in real time—that is, control all its functions directly from Earth?

13. ■■ An interstellar cloud fragment 0.2 light-year in diameter is rotating at a rate of 1 revolution per million years. It now begins to collapse. Assuming that the mass remains constant, estimate the cloud's rotation period when it has shrunk to (a) the size of the solar nebula, 100 A.U. across, and (b) the size of Earth's orbit, 2 A.U. across.

14. ■■ By what factor would Earth's rotational angular momentum change if the planet's spin rate were to double? By what factor would Earth's orbital angular momentum change if the planet's distance from the Sun were to double (assuming that the orbit remained circular)? The orbital angular momentum of a planet in a circular orbit is simply the product of the planet's mass, orbital speed, and distance from the Sun.

15. ■■ Consider a planet growing by the accretion of material from the solar nebula. As the planet grows, its density remains roughly constant. Does the force of gravity at the surface of the planet increase, decrease, or stay the same? Specifically, what would happen to the surface gravity and escape speed as the radius of the planet doubled? Give reasons for your answer.

 In addition to the Practice Problems module, the Companion Website at astro.prenhall.com/chaisson provides for each chapter a study guide module with multiple choice questions as well as additional annotated images, animations, and links to related Websites.

7

Photographs like this one, showing Earth hovering in space like a "small blue marble," help us appreciate our place in space. Akin to the phenomenon of sunrise from our vantage point on Earth, this image neatly captures lunar "Earthrise"— quite a different perspective for residents of Earth. Only during the space age— less than 50 years old—has humanity begun to leave its parent planet and venture into the nearby cosmos. This photo was taken by Apollo 11 astronauts while orbiting the Moon just prior to humankind's first lunar landing in 1969. Don't just ▶ *look at this photo; mentally place yourself on that fragile planet. (NASA)*

Earth

Our Home in Space

Earth is the best-studied terrestrial planet. From the matter of our world sprang life, intelligence, culture, and all the technology we now use to explore the cosmos. We ourselves are "Earthstuff" as much as are rocks, trees, and air. Now, as humanity begins to explore the solar system, we can draw on our knowledge of Earth to aid our understanding of the other planets. By cataloging Earth's properties and attempting to explain them, we set the stage for our comparative study of the solar system. Every piece of information we glean about the structure and history of our own world may play a vital role in helping us understand the planetary system in which we live. If we are to appreciate the universe, we must first come to know our own planet. Our study of astronomy begins at home.

LEARNING GOALS

Studying this chapter will enable you to

1 Summarize the physical properties of planet Earth.

2 Explain how Earth's atmosphere helps to heat us, as well as protect us.

3 Outline our current model of Earth's interior and describe some of the experimental techniques used to establish the model.

4 Summarize the evidence for the phenomenon of "continental drift" and discuss the physical processes that drive it.

5 Discuss the nature and origin of Earth's magnetosphere.

6 Describe how both the Moon and the Sun influence Earth's oceans.

Visit astro.prenhall.com/chaisson for additional annotated images, animations, and links to related sites for this chapter.

7.1 Overall Structure of Planet Earth

The Earth Data box on p. 172 lists some of Earth's physical and orbital properties in detail. The data shown are determined with techniques that are conceptually similar to those presented in Chapter 6: using simple geometry to determine Earth's radius, the orbit of the Moon to measure our planet's mass, and so on. ∞ (Sec. 6.2) Throughout the body of this text, we will use rounded-off numbers whenever possible, taking our planet's mass and radius to be 6.0×10^{24} kg and 6400 km, respectively.

Dividing mass by volume, we find that Earth's average density is around 5500 kg/m^3. This simple calculation allows us to make a very important deduction about the interior of our planet. From direct measurements, we know that the water which makes up much of Earth's surface has a density of 1000 kg/m^3, and the rock beneath us on the continents, as well as on the seafloor, has a density in the range from 2000 to 4000 kg/m^3. We can immediately conclude that, because the surface layers have densities much less than the average, much denser material must lie deeper, under the surface. Hence, we should expect that much of Earth's interior is made up of very dense matter, far more compact than the densest continental rocks on the surface.

On the basis of measurements made in many different ways—using aircraft in the atmosphere, satellites in orbit, gauges on the land, submarines in the ocean, and drilling gear below the rocky crust—scientists have built up the following overall picture of our planet: As indicated in Figure 7.1, Earth may be divided into six main regions. In Earth's interior, a thick **mantle** surrounds a smaller, two-part **core.** At the surface, we have (1) a relatively thin **crust,** comprising the solid continents and the seafloor, and (2) the **hydrosphere,** which contains the liquid oceans and accounts for some 70 percent of our planet's total surface area. An **atmosphere** of air lies just above the surface. At much greater altitudes, a zone of charged particles trapped by Earth's magnetic field forms the **magnetosphere.** Virtually all our planet's mass is contained within the surface and the interior. The gaseous atmosphere and the magnetosphere contribute hardly anything—less than 0.1 percent—to the total.

7.2 Earth's Atmosphere

From a human perspective, probably the most important aspect of Earth's atmosphere is that we can breathe it. Air is a mixture of gases, the most common of which are nitrogen (78 percent by volume), oxygen (21 percent), argon (0.9 percent), and carbon dioxide (0.03 percent). The amount of water vapor varies from 0.1 to 3 percent, depending on location and climate. The presence of a large amount of oxygen makes our atmosphere unique in the solar system, and the presence of even trace amounts of water and carbon dioxide play vital roles in the workings of our planet.

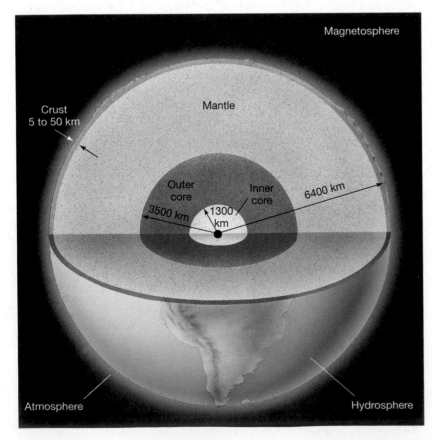

◀ **FIGURE 7.1 Earth** The main regions of planet Earth. At the center lies our planet's inner core, about 2600 km in diameter. Surrounding the inner core is an outer core, some 7000 km across. Most of the rest of Earth's 13,000-km diameter is taken up by the mantle, which is topped by a thin crust only a few tens of kilometers thick. The liquid portions of Earth's surface make up the hydrosphere. Above the hydrosphere and solid crust lies the atmosphere, most of it within 50 km of the surface. Earth's outermost region is the magnetosphere, extending thousands of kilometers into space.

Magnetosphere

Mantle

Crust
5 to 50 km

Outer core

Inner core

6400 km

3500 km   1300 km

Atmosphere

Hydrosphere

Atmospheric Structure

Figure 7.2 shows a cross section of our planet's atmosphere. Compared with Earth's overall dimensions, the extent of the atmosphere is not great. Half of it lies within 5 km of the surface, and all but 1 percent is found below 30 km. The portion of the atmosphere below about 12 km is called the *troposphere*. Above it, extending up to an altitude of 40 to 50 km, lies the *stratosphere*. Between 50 and 80 km from the surface is the *mesosphere*. Above about 80 km, in the *ionosphere*, the atmosphere is kept partly ionized by solar ultraviolet radiation. Note how the temperature gradient (decreasing or increasing with altitude) changes from one atmospheric region to the next.

Atmospheric density decreases steadily with increasing altitude, and as the right-hand vertical axis in Figure 7.2 shows, so does pressure. Climbing even a modest mountain—4 or 5 km high, say—clearly demonstrates the thinning of the air in the troposphere. Climbers must wear oxygen masks when scaling the tallest peaks on Earth.

The troposphere is the region of Earth's (or any other planet's) atmosphere where *convection* occurs, driven by the heat of Earth's warm surface. **Convection** is the constant upwelling of warm air and the concurrent downward flow of cooler air to take its place, a process that physically transfers heat from a lower (hotter) to a higher (cooler) level. In Figure 7.3, part of Earth's surface is heated by the Sun. The air immediately above the warmed surface is heated, expands a little, and becomes less dense. As a result, the hot air becomes buoyant and starts to rise. At

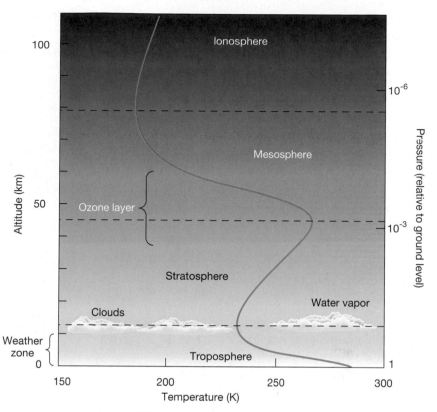

▲ **FIGURE 7.2 Earth's Atmosphere** Diagram of Earth's atmosphere, showing the changes in temperature (blue line, bottom axis) and pressure (blue line, right-hand axis) from the planet's surface to the bottom of the ionosphere. Pressure decreases steadily with increasing altitude, but temperature may fall or rise, depending on height above the ground.

◄ **FIGURE 7.3 Convection** Convection occurs whenever cool matter overlies warm matter. The resulting circulation currents are familiar to us as the winds in Earth's atmosphere, caused by the solar-heated ground. Hot air rises, cools, and falls repeatedly. Eventually, steady circulation patterns are established and maintained, provided that the source of heat (the Sun in the case of Earth) remains intact.

higher altitudes, the opposite effect occurs: The air gradually cools, grows denser, and sinks back to the ground. Cool air at the surface rushes in to replace the hot buoyant air. In this way, a circulation pattern is established. These *convection cells* of rising and falling air not only contribute to atmospheric heating, but also are responsible for surface winds. The constant churning motion in convection cells is responsible for all the weather we experience.

Atmospheric convection can also create clear-air turbulence—the bumpiness we sometimes experience on aircraft flights. Ascending and descending parcels of air, especially below fluffy clouds (themselves the result of convective processes, when water vapor condenses out at the cool tops of convection cells), can cause a choppy ride. For this reason, passenger aircraft tend to fly above most of the turbulence, at the top of the troposphere or in the lower stratosphere, where the atmosphere is stable and the air is calm.

Above about 100 km, in the ionosphere, the atmosphere is significantly ionized by the high-energy portion of the Sun's radiation spectrum, which breaks down molecules into atoms and atoms into ions. The degree of ionization increases with altitude. The presence of many free electrons makes this region of the upper atmosphere a good conductor of electricity, and the conductivity renders the ionosphere highly reflective to certain radio wavelengths. ∞ (Sec. 3.3) The reason that AM radio stations can be heard well beyond the horizon is that their signals bounce off the ionosphere before reaching a receiver. FM signals cannot be received from stations over the horizon, however, because the ionosphere is transparent to the somewhat shorter wavelengths of radio waves in the FM band.

Atmospheric Ozone

Straddling the boundary between the stratosphere and the mesosphere is the **ozone layer,** where, at an altitude of around 50 km, incoming solar ultraviolet radiation is absorbed by atmospheric ozone and nitrogen. (Ozone [O_3] consists of three oxygen atoms combined into a single molecule. Ultraviolet radiation breaks ozone down, forming molecular oxygen [O_2] again.) The ozone layer is one of the insulating spheres that serve to shield life on Earth from the harsh realities of outer space. Not so long ago, scientists judged space to be hostile to advanced life-forms because of what is missing out there: breathable air and a warm environment. Now most scientists regard outer space as harsh because of what is *present* out there: fierce radiation and energetic particles, both of which are injurious to human health. Without the protection of the ozone layer, advanced life (at least on Earth's surface) would be at best unlikely and at worst impossible.

Human technology has reached the point where it has begun to produce measurable—and possibly permanent—changes to our planet. One particularly undesirable by-product of our ingenuity is a group of chemicals known as

PLANETARY DATA:

EARTH

(NASA)

Orbital semimajor axis[1]	1.00 A.U. 149.6 million km
Orbital eccentricity[1]	0.017
Perihelion[2]	0.98 A.U. 147.1 million km
Aphelion[3]	1.02 A.U. 152.1 million km
Mean orbital speed	29.79 km/s
Sidereal[4] orbital period	1.000038 tropical years
Orbital inclination to the ecliptic	0.01°
Mass	5.976×10^{24} kg
Equatorial radius	6378 km
Mean density	5520 kg/m^3
Surface gravity[5]	9.80 m/s^2
Escape speed	11.2 km/s
Sidereal rotation period	0.9973 solar day
Axial tilt[6]	23.45°
Magnetic axis tilt relative to rotation axis	11.5°
Mean surface temperature	290 K
Number of moons	1

[1] See Section 2.5.
[2] *Closest approach to the Sun.* ∞ *(More Precisely 2-1)*
[3] *Greatest distance from the Sun.*
[4] *Relative to the distant stars.*
[5] *Acceleration due to gravity at Earth's surface.*
[6] *Angle between Earth's axis of rotation and perpendicular to Earth's orbital plane (the plane of the ecliptic).*

chlorofluorocarbons (CFCs), relatively simple compounds once widely used for a variety of purposes—propellant in aerosol cans, solvents in dry-cleaning products, and coolant in air conditioners and refrigerators. In the 1970s, it was discovered that, instead of quickly breaking down after use, as had previously been thought, CFCs accumulate in the atmosphere and are carried high into the stratosphere by convection. There, they are broken down by sunlight, releasing chlorine, which quickly reacts with ozone, turning it into oxygen. In chemical terms, the chlorine is said to act as a *catalyst*—it is not consumed in the reaction, so it survives to react with many more ozone molecules. A single chlorine atom can destroy up to 100,000 ozone molecules before being removed by other, less frequent chemical reactions.

Thus, even a small amount of CFCs is extraordinarily efficient at destroying atmospheric ozone, and the net result of CFC emission is a substantial increase in ultraviolet radiation levels at Earth's surface, with detrimental effects to most living organisms. Figure 7.4 shows a vast ozone "hole" over the Antarctic. The hole is a region where atmospheric circulation and low temperatures conspire each Antarctic spring to create a vast circumpolar cloud of ice crystals that

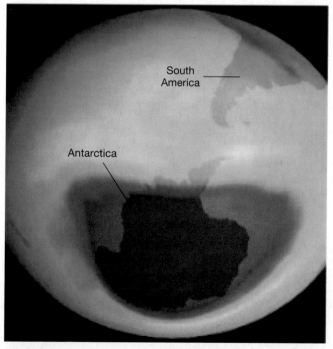

South America

Antarctica

▲ FIGURE 7.4 Antarctic Ozone Hole This composite image constructed from satellite observations shows (in pink) a huge "hole" in the ozone layer over the Antarctic continent. The hole is a region where climatic conditions and human-made chemicals combine to rob our atmosphere of its protective ozone blanket. The depth and area of the hole have grown significantly since the hole was discovered in the 1980s. Its maximum size is now larger than North America.

act to promote the ozone-destroying reactions, resulting in ozone levels about 50 percent below normal for the region. Ozone depletion is not confined to the Antarctic, although the effect is greatest there. Smaller holes have been observed in the Arctic, and occasional ozone depletions of up to 20 percent have been reported at lower northern latitudes.

In the late 1980s, when the effects of CFCs on the atmosphere were realized, the world moved rapidly to curtail their production and use, with the goal of phasing them out entirely by 2030. Substantial cuts have already been made, and the agreement to do so has become a model of international cooperation. Still, scientists think that, even if all remaining CFC emissions were to stop today, it would nonetheless take several decades for CFCs to leave the atmosphere completely.

Surface Heating

Much of the Sun's radiation manages to penetrate Earth's atmosphere, eventually reaching the ground. (See *More Precisely 7-1* for more on how the atmosphere affects incoming sunlight before it reaches the surface.) Most of this energy takes the form of visible and near-infrared radiation—ordinary sunlight. ∞ (Sec. 3.3) Essentially all of the solar radiation that is not absorbed by or reflected from clouds in the upper atmosphere is absorbed by Earth's surface. The result is that our planet's surface and most objects on it heat up considerably during the day. Earth cannot absorb this solar energy indefinitely, however. If it did, the surface would soon become hot enough to melt, and life on our planet would not exist.

As it heats up, Earth's surface reradiates much of the absorbed energy. This reemitted radiation follows the usual blackbody curve discussed in Chapter 3. ∞ (Sec. 3.4) As the surface temperature rises, the amount of energy radiated increases rapidly, in accordance with Stefan's law. Eventually, Earth radiates as much energy back into space as it receives from the Sun, and a stable balance is struck. In the absence of any complicating effects, this balance would be achieved at an average surface temperature of about 250 K (−23°C). Wien's law tells us that, at that temperature, most of the reemitted energy is in the form of infrared (heat) radiation.

But there *are* complications. Infrared radiation is partially blocked by Earth's atmosphere, primarily because of the presence of molecules of water vapor and carbon dioxide, which absorb very efficiently in the infrared portion of the spectrum. Even though these two gases are only trace constituents of our atmosphere, they manage to absorb a large fraction of all the infrared radiation emitted from the surface. Consequently, only some of that radiation escapes back into space. The remainder is trapped within our atmosphere, causing the temperature to increase.

This partial trapping of solar radiation is known as the **greenhouse effect.** The name comes from the fact that a similar process operates in a greenhouse, where sunlight

Why Is the Sky Blue?

Is the sky blue because it reflects the color of the ocean, or is the ocean blue because it reflects the color of the surrounding sky? The answer is the latter, and the reason has to do with the way that light is *scattered* by air molecules and minute dust particles. By *scattering*, we mean the process by which radiation is absorbed and then reradiated by the material through which it passes.

As sunlight passes through our atmosphere, it is scattered by gas molecules in the air. The British physicist Lord Rayleigh first investigated this phenomenon about a century ago, and today it bears his name—it is known as *Rayleigh scattering*. The process turns out to be highly sensitive to the wavelength of the light involved.

Rayleigh found that blue light is much more easily scattered than red light, basically because the wavelength of blue light (400 nm) is closer to the size of air molecules than the wavelength of red light (700 nm). He went on to prove mathematically, on the basis of the laws of electromagnetism, that the amount of scattering is inversely proportional to the *fourth* power of the wavelength:

$$\text{scattering by molecules} \propto \frac{1}{\text{wavelength}^4}.$$

Rayleigh's formula applies to scattering by particles (such as molecules) that are smaller than the wavelength of the light involved. Larger particles, such as dust, also preferentially scatter blue light, but by an amount that depends only inversely on the wavelength:

$$\text{scattering by dust} \propto \frac{1}{\text{wavelength}}.$$

EXAMPLE: Let's compare the relative scattering of blue (400 nm) and red (700 nm) light by atmospheric molecules and dust. For Rayleigh scattering, blue light is scattered $(700/400)^4 \approx 9.4$ times more efficiently than red light. That is, blue photons are almost 10 times more likely to be scattered out of a beam of sunlight (taken out of the forward beam and redirected to the side) than are red photons. For scattering by dust, the corresponding factor is $(700/400) = 1.75$—not as big a differential, but still enough to have a large effect when the air happens to be particularly dirty.

When the Sun is high in the sky, the blue component of incoming sunlight is scattered much more than any other-color component. Thus, some blue light is removed from the line of sight between us and the Sun and may scatter many times in the atmosphere before eventually entering our eyes, as shown in the first figure. Red or yellow light is scattered relatively little and arrives at our eyes predominantly along the line of sight to the Sun. The net effect is that the Sun is "reddened" slightly, because of the removal of blue light, while the sky away from the Sun appears blue. In outer space, where there is no atmosphere, there is no Rayleigh scattering of sunlight, and the sky is black (although, as we will see in Chapter 18, light from distant stars is reddened in precisely the same way as it passes through clouds of interstellar gas and dust).

At dawn or dusk, with the Sun near the horizon, sunlight must pass through much more atmosphere before reaching our eyes—so much so, in fact, that the blue component of the Sun's light is almost entirely scattered out of the line of sight, and even the red component is diminished in intensity. Accordingly, the Sun itself appears orange—a combination of its normal yellow color and a reddishness caused by the subtraction of virtually all of the blue end of the spectrum—and dimmer than at noon.

At the end of a particularly dusty day (second figure), when weather conditions or human activities during the daytime hours have raised excess particles into the air, short-wavelength Rayleigh scattering can be so heavy that the Sun appears brilliantly red. Reddening is often especially evident when we look at the westerly "sinking" summer Sun over the ocean, where seawater molecules have evaporated into the air, or during the weeks and months after an active volcano has released huge quantities of gas and dust particles into the air—as was the case in North America when the Philippine volcano Mount Pinatubo erupted in 1991.

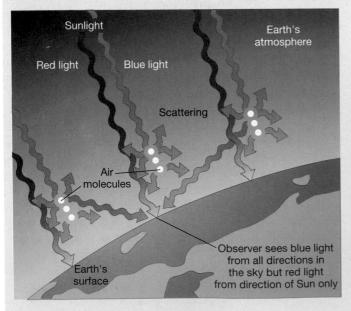

(NCAR)

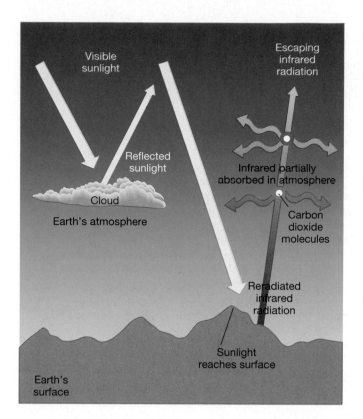

◄ **FIGURE 7.5 Greenhouse Effect** Sunlight that is not reflected by clouds reaches Earth's surface, warming it up. Infrared radiation reradiated from the surface is partially absorbed by the carbon dioxide (and also water vapor, not shown here) in the atmosphere, causing the overall surface temperature to rise.

passes relatively unhindered through glass panes, but much of the infrared radiation reemitted by the plants is blocked by the glass and cannot get out. Consequently, the interior of the greenhouse heats up, and flowers, fruits, and vegetables can grow even on cold wintry days.* The radiative processes that determine the temperature of Earth's atmosphere are illustrated in Figure 7.5. Earth's greenhouse effect makes our planet almost 40 K (40°C) hotter than would otherwise be the case.

The magnitude of the greenhouse effect is highly sensitive to the concentration of so-called *greenhouse gases* (i.e., gases that absorb infrared radiation efficiently) in the atmosphere. Carbon dioxide and water vapor are the most important of these, although other atmospheric gases (such as methane) also contribute. The amount of carbon dioxide in Earth's atmosphere is increasing, largely as a result of the burning of fossil fuels (principally oil and coal) in the industrialized world. Carbon dioxide levels have increased by over 20 percent in the last century, and they are continuing to rise at a present rate of 4 percent per decade. *Discovery 7-1* discusses the causes and some possible consequences of rising carbon dioxide levels in Earth's atmosphere.

In Chapter 9, we will see how a runaway increase in carbon dioxide levels in the atmosphere of the planet Venus radically altered conditions on its surface, causing its temperature to rise to over 700 K. Although no one is predicting that Earth's temperature will ever reach that of Venus, many scientists now think that our planet's increase in carbon dioxide levels, if left unchecked, may result in

*Note that, although this process does contribute to warming the interior of a greenhouse, it is not the most important effect. A greenhouse works mainly because its glass panes prevent convection from carrying heat up and away from the interior. Nevertheless, the name "greenhouse effect" to describe the heating effect due to Earth's atmosphere has stuck.

global temperature increases of several kelvins over the next half century—enough to cause dramatic, and possibly catastrophic, changes in Earth's climate.

Origin of Earth's Atmosphere

Why is our atmosphere made up of its present constituents? Why is it not composed entirely of nitrogen, say, or of carbon dioxide, like the atmospheres of Venus and Mars? The origin and development of Earth's atmosphere was a fairly complex and lengthy process.

When Earth first formed, any *primary atmosphere* it might have had would have consisted of the gases most common in the early solar system: hydrogen, helium, methane, ammonia, and water vapor—a far cry from the atmosphere we enjoy today. Almost all this low-density material, and especially any hydrogen or helium, escaped into space during the first half-billion or so years after Earth was formed. (For more information on how planets retain or lose their atmospheres, see *More Precisely 8-1*.)

Subsequently, Earth developed a *secondary atmosphere*, which was *outgassed* (expelled) from the planet's interior as a result of volcanic activity. Volcanic gases are rich in water vapor, methane, carbon dioxide, sulfur dioxide, and compounds containing nitrogen (such as nitrogen gas, ammonia, and nitric oxide). Solar ultraviolet radiation split the lighter, hydrogen-rich gases into their component atoms, allowing the hydrogen to escape and liberating much of the nitrogen from its bonds with other elements. As Earth's surface temperature fell, the water vapor condensed and oceans formed. Much of the carbon dioxide and sulfur dioxide became dissolved in the oceans or combined with surface rocks. Oxygen is such a reactive gas that any free oxygen that appeared at early times was removed as quickly as it formed. An atmosphere consisting largely of nitrogen slowly appeared.

The final major development in the story of our planet's atmosphere is known so far to have occurred only on Earth. *Life* appeared in the oceans more than 3.5 billion years ago, and organisms eventually began to produce atmospheric oxygen. The ozone layer formed, shielding the surface from the Sun's harmful radiation. Eventually, life spread to the land and flourished. The fact that oxygen is a major constituent of the present-day atmosphere is a direct consequence of the evolution of life on Earth.

CONCEPT CHECK

✔ Why is the greenhouse effect important for life on Earth?

DISCOVERY 7-1

The Greenhouse Effect and Global Warming

We saw in the text how greenhouse gases in Earth's atmosphere—notably, water vapor and carbon dioxide (CO_2)—tend to trap heat leaving the surface, raising our planet's temperature by several tens of degrees Celsius. In and of itself, this *greenhouse effect* is not a bad thing—in fact, it is the reason that water exists in the liquid state on Earth's surface, and thus it is crucial to the existence and survival of life on our planet (see Chapter 28). However, if atmospheric greenhouse gas levels rise unchecked, the consequences could be catastrophic.

Since the Industrial Revolution in the 18th century, and particularly over the past few decades, human activities on Earth have steadily raised the level of carbon dioxide in our atmosphere. Fossil fuels (coal, oil, and gas), still the dominant energy source of modern industry, all release CO_2 when burned. At the same time, the extensive forests that once covered much of our planet are being systematically destroyed to make room for human expansion. Forests play an important role in this situation because vegetation absorbs carbon dioxide, thus providing a natural control mechanism for atmospheric CO_2. Deforestation therefore also tends to increase the amount of greenhouse gases in Earth's atmosphere. The first figure shows atmospheric CO_2 levels over the past thousand years. Note the dramatic increase during the past two centuries.

Global warming is the slow rise in Earth's surface temperature caused by the increased greenhouse effect resulting from higher levels of atmospheric carbon dioxide. As shown in the second figure, average global temperatures have risen by about 0.5° C during the past century. This may not seem like much, but climate models predict that, if CO_2 levels continue to rise, a further increase of as much as 5° C is possible by the end of the 21st century. Such a rise would be enough to cause serious climatic change on a global scale. Among the possible (some would say likely) consequences of such a temperature increase are the following phenomena.

- Melting of glaciers and the polar ice caps, leading to a rise in sea level of up to a meter by the year 2100, with the potential for widespread coastal flooding.

- Longer and more extreme periods of severe weather—heat waves, droughts and wildfires—yet with more precipitation (rain and snow) between them.

- Crop failures as Earth's temperate zones move toward the poles.

- Expansion of deserts in heavily populated equatorial regions.

- Increased numbers of mosquitoes and other pests spreading tropical diseases into unprotected populations.

The fossil record shows that major climatic changes have occurred many times before in Earth's history, but never at the rate predicted by these dire warnings. The unprecedented speed of the forecasted events may well be too rapid for many species (and some human societies) to survive.

In Section 7.2, we described the danger to Earth's ozone layer posed by CFCs—another product of modern technology with unexpected global consequences—and saw how, once their environmental impact was identified, rapid steps were taken to curb their use. A concerted international response to global warming has been much slower in coming. Most scientists see the human-enhanced greenhouse effect as a real threat to Earth's climate, and they urge prompt and deep reductions in CO_2 emissions, along with steps to slow and ultimately reverse deforestation. Some, however—particularly those connected with the industries most responsible for the production of greenhouse gases—argue that Earth's long-term response to increased greenhouse emissions is too complex for simple conclusions to be drawn and that immediate action is unnecessary. They suggest that the current temperature trend may be part of some much longer cycle or that natural environmental factors may in time stabilize, or even reduce, the level of CO_2 in the atmosphere without human intervention.

Given the stakes, it is perhaps not surprising that these debates have become far more political than scientific in tone—not at all like the deliberative scientific method presented elsewhere in this text! The basic observations and much of the basic science are generally not seriously questioned, but the interpretation, long-term consequences, and proper response are all hotly debated. Separating the two sometimes is not easy, but the outcome may be of vital importance to life on Earth.

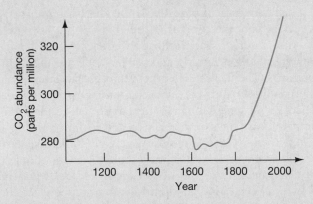

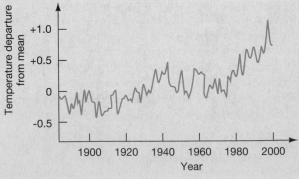

7.3 Earth's Interior

3 Although we reside on Earth, we cannot easily probe our planet's interior. Drilling gear can penetrate rock only so far before breaking. No substance used for drilling—even diamond, the hardest known material—can withstand the pressure below a depth of about 10 km. That's rather shallow compared with Earth's 6400-km radius. Fortunately, geologists have developed other techniques that indirectly probe the deep recesses of our planet.

Seismic Waves

A sudden dislocation of rocky material near Earth's surface—an *earthquake*—causes the entire planet to vibrate a little. Earth rings like a giant bell. These vibrations are not random, however. They are systematic waves, called **seismic waves** (after the Greek word for "earthquake"), that move outward from the site of the quake. Like all waves, they carry information. This information can be detected and recorded with sensitive equipment—a *seismograph*—designed to monitor Earth tremors.

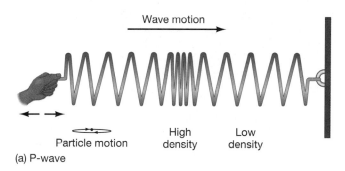

Wave motion

High density   Low density

Particle motion

(a) P-wave

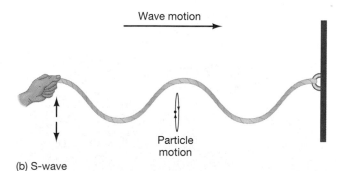

Wave motion

Particle motion

(b) S-wave

▲ **FIGURE 7.6 P- and S-waves** (a) A pressure (P-) wave traveling through Earth's interior causes material to vibrate in a direction parallel to the direction of motion of the wave. Material is alternately compressed and expanded. (b) A shear (S-) wave produces motion perpendicular to the direction in which the wave travels, pushing material from side to side. Also shown is the motion of one typical particle. In case (a), the particle oscillates forward and backward about its initial position. In (b), the particle moves up and down.

Decades of earthquake research have demonstrated the existence of many kinds of seismic waves. Two are of particular importance to the study of Earth's internal structure. First to arrive at a monitoring site after a distant earthquake are the primary waves, or *P-waves*. These are *pressure* waves, a little like ordinary sound waves in air, that alternately expand and compress the medium (the core or mantle) through which they move. Seismic P-waves usually travel at speeds ranging from 5 to 6 km/s and can travel through both liquids and solids. Some time later (the actual delay depends on the distance from the earthquake site), secondary waves, or *S-waves*, arrive. These are *shear* waves. Unlike P-waves, which vibrate the material through which they pass back and forth along the direction of travel of the wave, S-waves cause side-to-side motion, more like waves in a guitar string. The two types of waves are illustrated in Figure 7.6. S-waves normally travel through Earth's interior at 3 to 4 km/s; however, they cannot travel through liquid, which absorbs them.

The speeds of both P- and S-waves depend on the density of the matter through which the waves are traveling. Consequently, if we can measure the time taken for the waves to move from the site of an earthquake to one or more monitoring stations on Earth's surface, we can determine the density of matter in the interior. Figure 7.7 illustrates some P- and S-wave paths away from the site of an

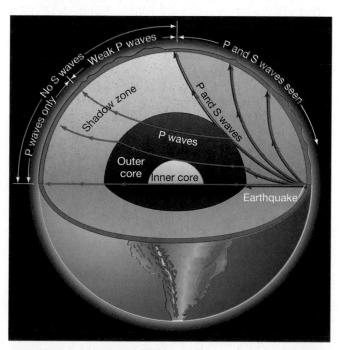

▲ **FIGURE 7.7 Seismic Waves** Earthquakes generate pressure (P, or primary) and shear (S, or secondary) waves that can be detected at seismographic stations around the world. S-waves are not detected by stations "shadowed" by the liquid core of Earth. P-waves do reach the side of Earth opposite the earthquake, but their interaction with Earth's core produces another shadow zone, where almost no P-waves are seen.

earthquake. Seismographs located around the world measure the times of arrival, as well as the strengths, of the seismic waves. Both observations contain much useful information—about the earthquake itself and about Earth's interior through which the waves pass. Notice that the waves do not travel in straight lines through the planet. Because the wave velocity increases with depth, waves that travel deeper tend to overtake those on slightly shallower paths, and the waves bend as they move through the interior.

A particularly important result emerged after numerous quakes were monitored several decades ago: Seismic stations on the side of Earth opposite a quake never detect S-waves—these waves are blocked by material within Earth's interior. Further, while P-waves always arrive at stations diametrically opposite the quake, parts of Earth's surface receive almost none (see Figure 7.7). Most geologists think that S-waves are absorbed by a liquid core at Earth's center and that P-waves are refracted at the core boundary, much as light is refracted by a lens. The result is the S- and P-wave "shadow zones" we observe. The fact that every earthquake exhibits these shadow zones is the best evidence that the core of our planet is hot enough to be in the liquid state. *Discovery* 7-2 presents some more seismic data on the structure of Earth's core.

DISCOVERY 7-2

Earth's "Rapidly" Spinning Core

Seismic waves racing out from the site of earthquakes allow geologists to map the interior makeup of planet Earth (Section 7.3). Recent analyses of old earthquake data suggest that Earth's inner core is spinning slightly *faster* than the rest of our planet. The whole planet rotates in the same direction (west to east), but the inner core takes about two-thirds of a second less to complete its daily rotation than the bulk of the matter surrounding it—which means that it gains a quarter turn each century.

Geologists arrived at this rather surprising result by carefully timing seismic P-waves moving through Earth's interior. Actually, the scientists used historical data acquired over the past few decades, mainly from 38 earthquakes that occurred in the south Atlantic Ocean and whose seismic vibrations had to pierce the inner core on their way to monitoring stations in Alaska. As discussed in the text, only P-waves can make it through the liquid outer core, to be detected on the opposite side of Earth.

For many years, geologists have known that P-waves travel through the inner core about 3 to 4 percent faster along north–south pathways than along those close to the (east–west) plane of the equator. This difference probably results from the orientation of the iron atoms in the solid inner core; the atoms there must be aligned, much like the regular arrangement of atoms found in a crystal. Where the atomic alignment parallels the motion of the P-waves, those P-waves travel just a little faster—a bit like going "with the grain" rather than against it.

The new finding is that the route of the fastest seismic waves through Earth's deep interior is gradually shifting eastward, implying that the inner core has some motion slightly different from the rest of the planet. In the accompanying figure, the red region is the solid inner core, the orange region the liquid outer core. The dashed lines passing through the core depict the paths of the fastest seismic waves from 1900 to 1996. The data indicate that the axis of the "fast track" for the P-waves is shifting by about 1° per year relative to the crust above. That axis now intersects Earth's surface in the Arctic Ocean northeast of Siberia; some 30 years ago, it was 33° farther west. Gradual as it may be, this shift amounts to about 20 km at the edge of the inner core—around a million times faster than the typical rates (centimeters per year) at which the continents creep across Earth's surface.

These discoveries may help us understand both how the inner core is continuing to grow by condensing from the molten outer core as Earth cools and how these two regions interact to produce Earth's magnetic field. As noted in Section 7.5, the dynamo theory of magnetism requires rotating, electrically conducting matter within Earth's interior. Detailed modeling of Earth's interior had also predicted that a rapidly moving jet stream of partly molten matter—a "mushy zone"—is probably established at the base of the outer core, causing the inner core to "superrotate"—rotate faster than the overlying layers. The shift in the axis of the fastest seismic waves indicates that this differential rotation pattern does indeed exist.

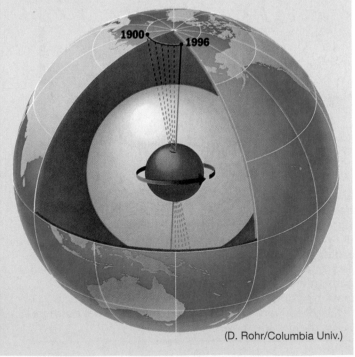

(D. Rohr/Columbia Univ.)

The sizes of the shadow zones depend on the radius of the core, and careful analysis of the seismic data yield a core radius of about 3500 km. In fact, very faint P-waves *are* observed in the P-wave shadow zone indicated in Figure 7.7. These are thought to be reflected off the surface of a solid **inner core**, of radius 1300 km, lying at the center of the liquid **outer core**.

Modeling Earth's Interior

Because earthquakes occur often and at widespread places across the globe, geologists have accumulated a large amount of data about shadow zones and seismic-wave properties. They have used these data, along with direct knowledge of surface rocks, to build mathematical models of Earth's interior. Our knowledge of the deepest recesses of our planet is based almost entirely on modeling and indirect observation. We will find many more examples of this powerful combination throughout the text.

Figure 7.8 presents a model that most scientists accept. According to this model, Earth's outer core is surrounded by a thick mantle and topped with a thin crust. The mantle is about 3000 km thick and accounts for the bulk (80 percent) of our planet's volume. The crust has an average thickness of only 15 km—a little less (around 8 km) under the oceans and somewhat more (20–50 km) under the continents. The average density of crust material is around 3000 kg/m^3. Density and temperature both increase with depth. Specifically, from Earth's surface to its very center, the density increases from roughly 3000 kg/m^3 to a little more than 12,000 kg/m^3, while the temperature rises from just under 300 K to well over 5000 K. Much of the mantle has a density midway between the densities of the core and crust: about 5000 kg/m^3.

The high central density suggests to geologists that the inner parts of Earth must be rich in nickel and iron. Under the heavy pressure of the overlying layers, these metals (whose densities under surface conditions are around 8000 kg/m^3) can be compressed to the high densities predicted by the model. The sharp increase in density at the mantle–core boundary results from the difference in composition between the two regions. The mantle is composed of dense, but *rocky*, material—compounds of silicon and oxygen. The core consists primarily of even denser *metallic* elements. There is no similar jump in density or temperature at the inner core boundary—the material there simply changes from the liquid to the solid state.

The model suggests that the core must be a mixture of nickel, iron, and some other lighter element, possibly sulfur. Without direct observations, it is difficult to be absolutely certain of the light component's identity. All geologists agree that much of the core must be liquid. The existence of the shadow zone demands this (and, as we will see, our current explanation of Earth's magnetic field relies on it). However, despite the high temperature, the pressure near the center—about 4 million times the atmospheric pressure at Earth's surface—is high enough to force the material there into the solid state.

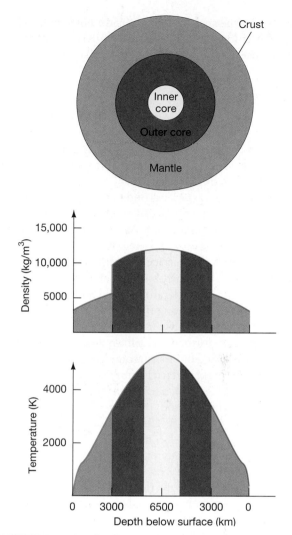

▲ **FIGURE 7.8 Earth's Interior** Computer models of Earth's interior imply that the density and temperature vary considerably through the mantle and the core. Note the sharp discontinuity in density between Earth's core and mantle.

Because geologists have been unable to drill deeper than about 10 km, no experiment has yet recovered a sample of Earth's mantle. However, we are not entirely ignorant of the mantle's properties. In a *volcano*, hot lava upwells from below the crust, bringing a little of the mantle to us and providing some inkling of Earth's interior. Observations of the chemical and physical properties of newly emerged lava are generally consistent with the model sketched in Figure 7.8.

The composition of the upper mantle is probably quite similar to the iron–magnesium–silicate mixtures known as *basalt*. You may have seen some dark gray basaltic rocks scattered across Earth's surface, especially near volcanoes. Basalt is formed as material from the mantle upwells from Earth's interior as lava, cools, and then solidifies. With a density between 3000 kg/m^3 and 3300 kg/m^3, basalt contrasts with the lighter *granite* (density 2700–3000 kg/m^3) that constitutes much of the rest of Earth's crust. Granite is richer than basalt in the light elements silicon and aluminum, which explains why the surface continents do not sink into the interior. Their low-density composition lets the crust "float" atop the denser matter of the mantle and core below.

Differentiation

Earth, then, is not a homogeneous ball of rock. Instead, it has a layered structure, with a low-density crust at the surface, an intermediate-density material in the mantle, and a high-density core. Such variation in density and composition is known as **differentiation.**

Why isn't our planet just one big, rocky ball of uniform density? The answer appears to be that much of Earth was *molten* at some time in the past. As a result, the higher density matter sank to the core and the lower density material was displaced toward the surface. A remnant of this ancient heating exists today: Earth's central temperature is nearly equal to the surface temperature of the Sun. What processes were responsible for heating the entire planet to that extent? To answer this question, we must try to visualize the past.

When Earth formed 4.6 billion years ago, it did so by capturing material from its surroundings, growing in mass as it swept up "preplanetary" chunks of matter in its vicinity. ∞ (Sec. 6.7) As the young planet grew, its gravitational field strengthened and the speed with which newly captured matter struck its surface increased. This process generated a lot of heat—so much, in fact, that Earth may already have been partially or wholly molten by the time it reached its present size. As Earth began to differentiate and heavy material sank to the center, even more gravitational energy was released, and the interior temperature must have increased still further.

Later, Earth continued to be bombarded with debris left over from the formation process. At its peak some 4 billion years ago, this secondary bombardment was probably intense enough to keep the surface molten, but only down to a depth of a few tens of kilometers. Erosion by wind and water has long since removed all trace of this early period from the surface of Earth, but the Moon still bears visible scars of the onslaught.

A second important process for heating Earth soon after its formation was **radioactivity**—the release of energy by certain rare heavy elements, such as uranium, thorium, and plutonium (see *More Precisely 7-2*). These elements release energy and heat their surroundings as their complex heavy nuclei decay (break up) into simpler lighter ones. While the energy produced by the decay of a single radioactive atom is tiny, Earth contained a lot of radioactive atoms, and a lot of time was available. Rock is such a poor conductor of heat that the energy would have taken a very long time to reach the surface and leak away into space, so the heat built up in the interior, adding to the energy left there by Earth's formation.

Provided that enough radioactive elements were originally spread throughout the primitive Earth, rather like raisins in a cake, the entire planet—from crust to core—could have melted and remained molten for about a billion years. That's a long time by human standards, but not so long in the cosmic scheme of things. Measurements of the ages of some surface rocks indicate that Earth's crust finally began to solidify roughly 700 million years after it originally formed. Radioactive heating did not stop at that point, of course; it continued even after Earth's surface cooled and solidified. But radioactive decay works in only one direction, always producing lighter elements from heavier ones. Once gone, the heavy and rare radioactive elements cannot be replenished.

So the early source of heat diminished with time, allowing the planet to cool over the past 4 billion years. In this process, Earth has cooled from the outside in, much like a hot potato, since regions closest to the surface can most easily unload their excess heat into space. In that way, the surface developed a solid crust, and the differentiated interior attained the layered structure now implied by seismic studies.

CONCEPT CHECK

✔ How might our knowledge of Earth's interior be changed if our planet were geologically inactive, with no volcanos or earthquakes?

7.4 Surface Activity

Earth is geologically alive today. Its interior seethes and its surface constantly changes. Figure 7.9 shows two indicators of surface geological activity: a volcano, whereby molten rock and hot ash upwell through fissures or cracks in the surface, and (the aftermath of) an earthquake, which occurs when the crust suddenly dislodges under great pressure. Catastrophic volcanoes and earthquakes are relatively rare events these days, but geological studies of rocks, lava, and other surface features imply that surface activity must have been more frequent, and probably more violent, long ago.

Continental Drift

Many traces of past geological events are scattered across our globe. Erosion by wind and water has wiped away much of the evidence for ancient activity, but modern exploration has documented the sites of most of the recent activity, such as earthquakes and volcanic eruptions. Figure 7.10 is a map of the currently active areas of our planet. The red dots represent sites of volcanism or earthquakes. Nearly all these sites have experienced surface activity within this century, some of them suffering much damage and the loss of many lives.

The intriguing aspect of Figure 7.10 is that the active sites are not spread evenly across our planet. Instead, they trace well-defined lines of activity, where crustal rocks dislodge (as in earthquakes) or mantle material upwells (as in volcanoes). In the mid-1960s, scientists realized that these

(a)

(b)

▲ **FIGURE 7.9 Geological Activity** (a) An active volcano on Kilauea in Hawaii. Kilauea seems to be a virtually ongoing eruption. Other eruptions, such as that of Mount St. Helens in Washington State on May 18, 1980, are rare catastrophic events that can release more energy than the detonation of a thousand nuclear bombs. (b) The aftermath of an earthquake that claimed more than 5000 lives and caused billions of dollars' worth of damage in Kobe, Japan, in January 1995. *(P. Chesley/Getty Images, Inc.; H. Yamaguchi/Sygma)*

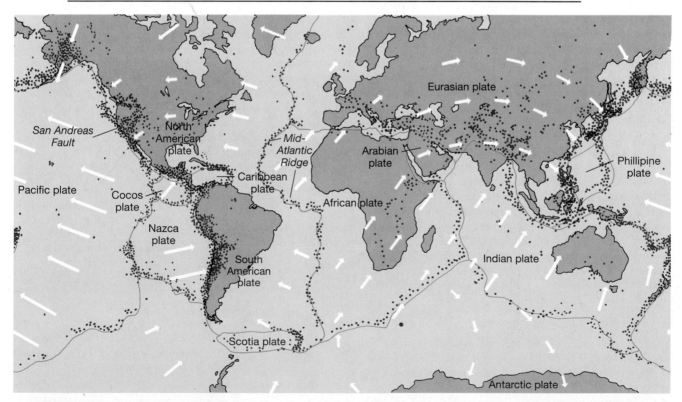

▲ **FIGURE 7.10 Global Plates** Red dots represent active sites where major volcanoes or earthquakes have occurred in the 20th century. Taken together, the sites reveal vast "plates" (outlined by dark blue lines) that drift around on the surface of our planet. The white arrows show the general directions and speeds of the plate motions.

lines are really the outlines of gigantic "plates," or slabs of Earth's surface.* Most startling of all, the plates are slowly moving—literally drifting around the surface of our planet. These plate motions have created the surface moun-

tains, oceanic trenches, and other large-scale features across the face of planet Earth. In fact, plate motions have shaped the continents themselves. The process is popularly known as "continental drift." The technical term for the study of plate movement and its causes is **plate tectonics.** The major plates of the world are marked on Figure 7.10.

The idea of continental drift was first suggested in 1912 by a German meteorologist named Alfred Wegener, who pointed out the remarkable geographic fit between the

*Not all volcanoes are associated with plate boundaries. The Hawaiian Islands, located near the center of the Pacific plate, are a case in point. They are associated with a "hot spot" in Earth's upper mantle that melts the crust above it. Over millions of years, the motion of the Pacific plate across the underlying hot spot has resulted in a chain of volcanic islands.

MORE PRECISELY 7-2

Radioactive Dating

In Chapter 4, we saw that atoms are made up of electrons and nuclei and that nuclei are composed of protons and neutrons. ∞ (Sec. 4.2) The number of protons in a nucleus determines which element it represents. However, the number of neutrons can vary. In fact, most elements can exist in several *isotopic* forms, all containing the same number of protons, but different numbers of neutrons in their nuclei. The particular nuclei we have encountered so far—the most common forms of hydrogen, helium, carbon, and iron—are all *stable*. For example, left alone, a carbon-12 nucleus, consisting of six protons and six neutrons, will remain unchanged forever. It will not break up into smaller pieces, nor will it turn into anything else.

Not all nuclei are stable, however. Many nuclei—such as carbon-14 (containing 6 protons and 8 neutrons), thorium-232 (90 protons, 142 neutrons), uranium-235 (92 protons, 143 neutrons), uranium-238 (92 protons, 146 neutrons), and plutonium-241 (94 protons, 147 neutrons)—are inherently *unstable*. Left alone, they will eventually break up into lighter "daughter" nuclei, emitting some elementary particles and releasing some energy in the process. The change happens spontaneously, without any external influence. This instability is known as *radioactivity*. The energy released by the disintegration of the radioactive elements just listed is the basis for nuclear fission reactors (and atomic bombs).

Unstable heavy nuclei achieve greater stability by disintegrating into lighter nuclei, but they do not do so immediately. Each type of "parent" nucleus takes a characteristic amount of time to decay. The *half-life* is the name given to the time required for half of a sample of parent nuclei to disintegrate. Notice that this is really a statement of probability. We cannot say *which* nuclei of a given element will decay in any given half-life interval; we can say only that half of them are expected to do so. If a given sample of material has half-life T, then we can write down a simple expression for the amount of material remaining after time t:

$$\text{fraction of material remaining} = (1/2)^{t/T}.$$

Thus, if we start with a billion radioactive nuclei embedded in a sample of rock, a half-billion nuclei will remain after one half-life, a quarter-billion after two half-lives, and so on. The first figure illustrates the decline in the number of parent nuclei as a function of time.

Every radioactive isotope has its own half-life, and most of their half-lives are now well known from studies conducted

since the 1950s. For example, the half-life of uranium-235 is 713 million years, and that of uranium-238 is 4.5 billion years. Some radioactive elements decay much more rapidly, others much more slowly, but these two types of uranium are particularly important to geologists because their half-lives are comparable to the age of the solar system. The second figure illustrates the half-lives and decay reactions for four unstable heavy nuclei.

The decay of unstable radioactive nuclei into more stable *daughter* nuclei provides us with a useful tool for measuring the ages of any rocks we can get our hands on. The first step is to measure the amount of stable nuclei of a given kind (e.g., lead-206, which results from the decay of uranium-238). This amount is then compared with the amount of remaining unstable parent nuclei (in this case, uranium-238) from which the daughter nuclei descended. Knowing the rate (or half-life) at which the disintegration occurs, the age of the rock then follows directly. If half of the parent nuclei of some element have decayed, so that the number of daughter nuclei equals the number of parents, the age of the rock must be equal to the half-life of the radioactive nucleus studied. Similarly, if only a quarter of the parent nuclei remain (three times as many daughters as parents), the rock's age is twice the half-life of that element, and so on.

In practice, ages can be determined by these means to within an accuracy of a few percent. The most ancient rocks on Earth are dated at 3.9 billion years old. These rare specimens have been found in Greenland and Labrador.

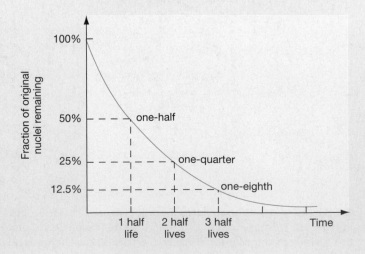

continents on either side of the Atlantic ocean. Note in Figure 7.10 how the Brazilian coast meshes nicely with the Ivory Coast of Africa. In fact, most of the continental landmasses in the Southern Hemisphere fit together remarkably well. Following the arrows in the figure backwards, we can see that the fits are in fact roughly consistent with the present motions of the plates involved. The fit appears to be not so good in the Northern Hemisphere, but it improves markedly if we consider the entire continental shelves (the

continental borders, which are under water) instead of just the portions that happen to stick up above sea level.

Few took Wegener's ideas seriously at the time, in part because there was no known mechanism that could drive the plates' motions. Nearly all scientists thought it preposterous that large segments of rocky crust could be drifting across the surface of our planet. These skeptical views persisted for more than half a century, when the accumulation of data in support of continental drift became overwhelming.

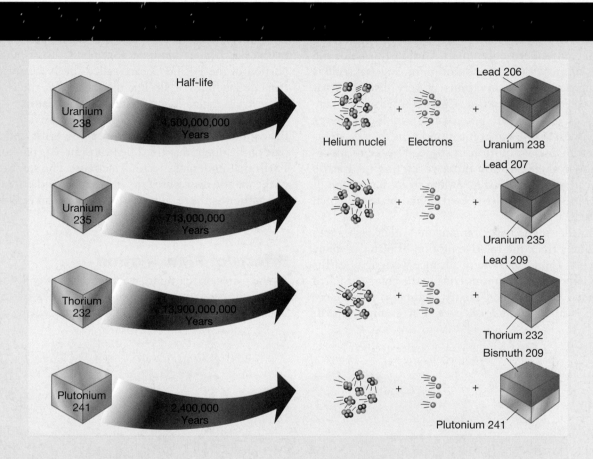

EXAMPLE: Suppose that careful chemical analysis of a sample of rock reveals that, for every nucleus of uranium-238 remaining in the sample, there is 0.41 of a lead-206 nucleus. If we assume that there was no lead-206 initially present, and hence that every lead-206 nucleus is the decay product of a nucleus of uranium-238, we can easily calculate the fraction of uranium-238 nuclei remaining. The answer is

$$\text{fraction of uranium-238} = \frac{1}{1 + 0.41} =$$

$$0.71 \approx \sqrt{\frac{1}{2}} = \left(\frac{1}{2}\right)^{0.5}.$$

From this equation, it follows that the elapsed time must be 0.5 times the half-life of uranium-238, or 2.25 billion years. If we were to repeat the analysis with uranium-235 and lead-

207, we would expect to find consistent results, within the measurement errors: 2.25 billion years is 2250/713 = 3.2 uranium-235 half-lives, so only $(1/2)^{3.2} \approx 11$ percent of any uranium-235 should remain—daughter lead-207 nuclei should outnumber parent uranium-235 nuclei by more than eight to one.

The radioactive-dating technique rests on the assumption that the rock has remained *solid* while the radioactive decays have been going on. If the rock melts, there is no particular reason to expect the daughter nuclei to remain in the same locations their parents had occupied, and the whole method fails. Thus, radioactive dating indicates the time that has elapsed since the last time the rock in question solidified. Hence, the 3.9-billion-year value represents only a portion—a lower limit—of the true age of our planet. It does not measure the duration of Earth's molten existence.

Similar-looking fossils were found on opposite sides of the Atlantic Ocean, at just the locations where the continents "fit together," and studies of the sea floor near the center of the Atlantic (to be discussed in more detail in a moment) strongly suggested the formation of new crust as plates separated. Today Wegener's "crazy" theory forms the foundation for all geological studies of our planet's outer layers.

The plates are not simply slowing to a stop after some ancient initial movements. Rather, they are still drifting today, although at an extremely slow rate. Typically, the speeds of the plates amount to only a few centimeters per year—about the same rate as your fingernails grow. Still, this is well within the measuring capabilities of modern equipment. Curiously, one of the best ways of monitoring plate motion on a global scale is by making accurate observations of very distant astronomical objects. Quasars (see Chapter 25), lying many hundreds of millions of light-years from Earth, never show any measurable apparent motion on the sky stemming from their own motion in space. Thus, any apparent change in their position (after correction for Earth's motion) can be interpreted as arising from the motion of the telescope—that is, of the continental plate on which it is located!

On smaller scales, laser-ranging and other techniques now routinely track the relative motion of plates in many areas, such as California, where advance warning of earthquake activity is at a premium. During the course of Earth history, each plate has had plenty of time to move large distances, even at its sluggish pace. For example, a drift rate of only 2 cm per year can cause two continents (e.g., Europe and North America) to separate by some 4000 km over the course of 200 million years. That may be a long time by human standards, but it represents only about 5 percent of the age of Earth.

A common misconception is that the plates are the continents themselves. Some plates are indeed made mostly of continental landmasses, but other plates are made of a continent plus a large part of an ocean. For example, the Indian plate includes all of India, much of the Indian Ocean, and all of Australia and its surrounding south seas. (See Figure 7.10.) Still other plates are mostly ocean. The seafloor itself is a slowly drifting plate, and the oceanic water merely fills in the depressions between continents. The southeastern portion of the Pacific Ocean, called the Nazca plate, contains no landmass at all. For the most part, the continents are just passengers riding on much larger plates.

Taken together, the plates make up Earth's **lithosphere,** which contains both the crust and a small part of the upper mantle. The lithosphere is the portion of Earth that undergoes tectonic activity. The semisolid part of the mantle over which the lithosphere slides is known as the **asthenosphere.** The relationships between these regions of Earth are shown in Figure 7.11.

Effects of Plate Motion

As the plates drift around, we might expect collisions to be routine. Indeed, plates do collide, but unlike two automobiles that collide and then stop, the surface plates are driv-

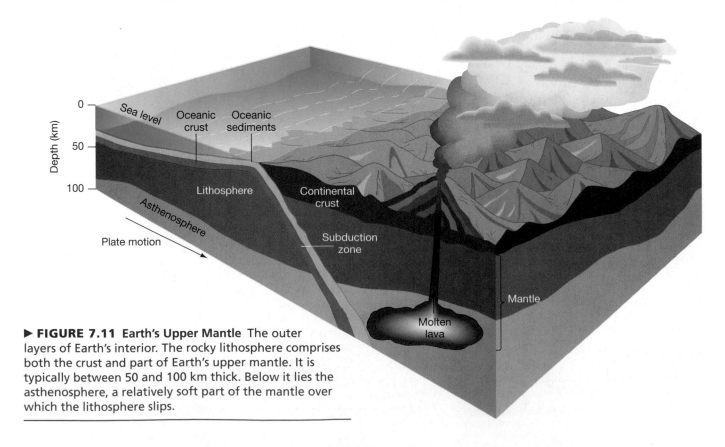

► **FIGURE 7.11 Earth's Upper Mantle** The outer layers of Earth's interior. The rocky lithosphere comprises both the crust and part of Earth's upper mantle. It is typically between 50 and 100 km thick. Below it lies the asthenosphere, a relatively soft part of the mantle over which the lithosphere slips.

en by enormous forces. They do not stop easily. Instead, they just keep crunching into one another. Figure 7.12(a) shows a collision currently occurring between two continental landmasses: The subcontinent of India, on the prow of the northward-moving Indian plate, is crashing into the landmass of Asia, located on the Eurasian plate (see Figure 7.10). The resulting folds of rocky crust create mountains—in this case, the snow-covered Himalayan mountain range at the upper right. A peak like Mount Everest (Figure 7.12b) represents a portion of Earth's crust that has been lifted over 8800 m by the slow, but inexorable, force produced when one plate plows into another.

Not all colliding plates produce mountain ranges. At other locations, called *subduction zones*, one plate slides under the other, ultimately to be destroyed as it sinks into the mantle. Subduction zones are responsible for most of the deep trenches in the world's oceans.

Nor do all plates experience head-on collisions. As noted by the arrows of Figure 7.10, many plates slide or shear past one another. A good example is the most famous active region in North America: the San Andreas Fault in California (Figure 7.13). The site of much earthquake activity, this fault marks the boundary where the Pacific and North American plates are rubbing past each other. The

motion of these two plates, like that of moving parts in a poorly oiled machine, is neither steady nor smooth. The sudden jerks that occur when they do move against each other are often strong enough to cause major earthquakes.

At still other locations, the plates are moving apart. As they recede, new material from the mantle wells up between them, forming *midocean ridges*. Notice in Figure 7.10 the major boundary separating the North and South American plates from the Eurasian and African plates, marked by the thin strip down the middle of the Atlantic Ocean. Discovered after World War II by oceanographic ships studying the geography of the seafloor, this giant fault is called the *Mid-Atlantic Ridge*. It extends, like a seam on a giant baseball, all the way from Scandinavia in the North Atlantic to the latitude of Cape Horn at the southern tip of South America. The entire ridge is a region of seismic and volcanic activity, but the only major part of it that rises above sea level is the island of Iceland.

Robot submarines have retrieved samples of the ocean floor at a variety of locations on either side of the Mid-Atlantic Ridge, and the ages of the samples have been measured by means of radioactive dating techniques. As depicted in Figure 7.14, the ocean floor closest to the ridge is relatively young, whereas material farther away, on

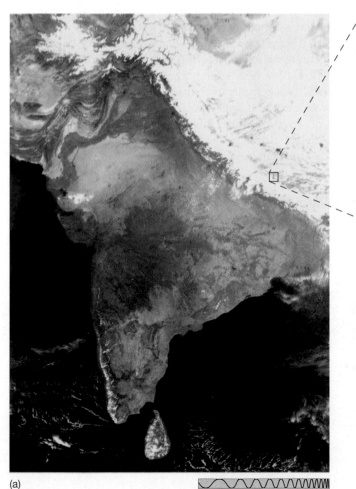

(a)

(b)

◄ **FIGURE 7.12 Himalayas** Mountain building results largely from collisions between plates. (a) The subcontinent of India, imaged here in infrared light from orbit, lies at the northernmost tip of the Indian plate. As this plate drifts northward, the Indian landmass collides with Asia, on the Eurasian plate. The impact causes Earth's crust to buckle and fold, thrusting up the Himalayan mountain range (covered with snow at the upper right). (b) The results of the ongoing process depicted in (a) can be seen in this view of Mount Everest (the dark peak in the background). *(Earth Satellite Corporation/Photo Researchers, Inc.; L. Day/Black Star)*

◀ **FIGURE 7.13 California Fault** A small portion of the San Andreas fault in California. The fault is the result of the North American and Pacific plates sliding past one another. The Pacific plate, which includes a large slice of the California coast, is drifting to the northwest with respect to the North American plate. The directions of motion of the two plates are indicated by arrows whose lengths are proportional to the speeds of the plates. *(U.S. Department of the Interior)*

either side, is older—exactly as we would expect if hot molten matter is upwelling and solidifying as the plates on either side drift apart. The Atlantic Ocean has apparently been growing in this way for the past 200 million years, the age of the oldest rocks found on any part of the Atlantic seafloor.

Other studies of the Mid-Atlantic Ridge have yielded important information about Earth's magnetic field. As hot material (carrying traces of iron) from the mantle emerges from cracks in the oceanic ridges and solidifies, it becomes slightly magnetized, retaining an imprint of Earth's magnetic field *at the time it cooled*. Thus, the ocean floor has preserved within it a record of Earth's magnetism during past times, rather like a tape recording. Figure 7.15 is a diagram of a small portion of the Atlantic Ocean floor near Iceland.

Earth's current magnetism is oriented in the familiar north–south fashion, and when samples of ocean floor close to the ridge are examined, the iron deposits are oriented just as expected: north–south. This material is the "young" basalt that upwelled and cooled fairly recently. However, samples retrieved farther from the ridge, corresponding to older material that upwelled long ago, are often magnetized with the *opposite* orientation. As we move away from the ridge, the imprinted magnetic field flips back and forth, more or less regularly and symmetrically on either side of the ridge.

The leading explanation of these different magnetic orientations is that they were caused by reversals in Earth's magnetic field that occurred as the plates drifted away from the central ridge. Working backward, we can use the fossil magnetic field to infer the past positions of the plates, as well as the orientation of Earth's magnetism. In addition to providing strong support for the idea of seafloor spreading, these measurements, when taken in conjunction with the data on the age of the seafloor, allow us to time our planet's magnetic reversals. On average, Earth's magnetic field reverses itself roughly every half-million years. Current theory suggests that such reversals are part of the way in which all planetary magnetic fields are generated. As we will see in Chapter 16, a similar phenomenon (with a reversal time of approximately 11 years) is also observed on the Sun.

What Drives the Plates?

What process is responsible for the enormous forces that drag plates apart in some locations and ram them together in others? The answer is probably convection—the same physical process we encountered earlier in our study of the

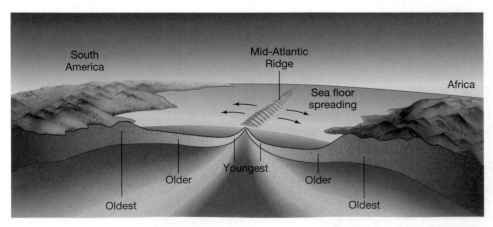

◀ **FIGURE 7.14 Seafloor Spreading** Samples of ocean floor retrieved by oceanographic vessels are youngest close to the Mid-Atlantic Ridge and progressively older farther away.

▶ **FIGURE 7.15 Magnetic Reversals** Samples of basalt retrieved from the ocean floor often reveal Earth's magnetism to have been oriented oppositely from the current north–south magnetic field. This simplified diagram shows the ages of some of the regions in the vicinity of the Mid-Atlantic Ridge (see Figure 7.10), together with the direction of the fossil magnetic field. The colored areas have the current orientation; they are separated by regions of reversed magnetic polarity.

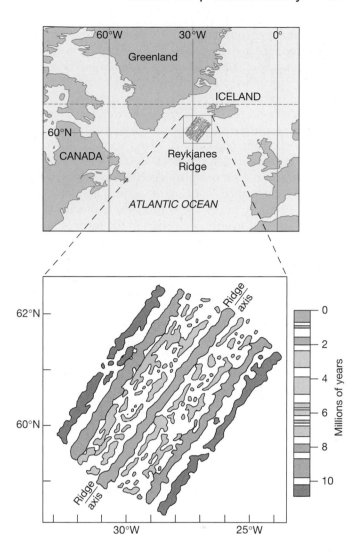

atmosphere. Figure 7.16 is a cross-sectional diagram of the top few hundred kilometers of our planet's interior. It depicts roughly the region in and around a midocean ridge. There, the ocean floor is covered with a layer of sediment—dirt, sand, and dead sea organisms that have fallen through the seawater for millions of years. Below the sediment lies about 10 km of granite, the low-density rock that makes up the crust. Deeper still lies the upper mantle, whose temperature increases with depth. Below the base of the lithosphere, at a depth of perhaps 50 km, the temperature is sufficiently high that the mantle is soft enough to flow very slowly, although it is not molten. This region is the asthenosphere.

The setting is a perfect one for convection—warm matter underlying cool matter. The warm mantle rock rises, just as hot air rises in our atmosphere. Sometimes, the rock squeezes up through cracks in the granite crust. Every so often, such a fissure may open in the midst of a continental landmass, producing a volcano such as Mount St. Helens or possibly a geyser like those at Yellowstone National Park. However, most such cracks are on the ocean floor. The Mid-Atlantic Ridge is a prime example.

Not all the rising warm rock in the upper mantle can squeeze through cracks and fissures. Some warm rock cools and falls back down to lower levels. In this way, large circulation patterns become established within the upper

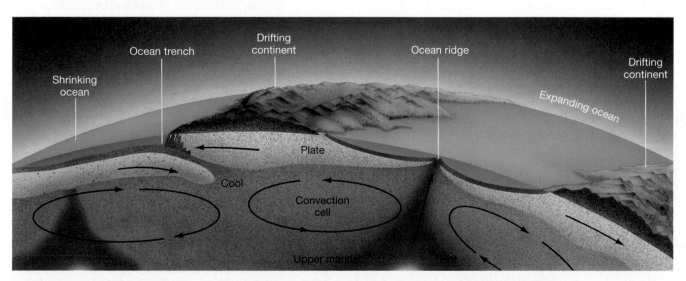

▲ **FIGURE 7.16 Plate Drift** The motion of Earth's tectonic plates is probably caused by convection—in this case, giant circulation patterns in the upper mantle that drag the plates across the surface.

mantle, as depicted in Figure 7.16. Riding atop these convection patterns are the plates. The circulation is extraordinarily sluggish. Semisolid rock takes millions of years to complete one convection cycle. Although the details are far from certain and remain controversial, many researchers suspect that the large-scale circulation patterns near plate boundaries drive the motions of the plates.

This constant recycling of plate material provides a natural explanation for the **rock cycle**—the process by which surface rock on our planet is continuously redistributed and transformed from one type into another. Deep below the surface, in the asthenosphere, temperatures are high enough that mantle rock exists in the form of molten *magma*. When this material cools and hardens, it forms *igneous rocks*. (Granite and basalt are familiar examples.) Igneous rocks are associated with volcanic activity (in which the magma is called *lava*) and spreading regions such as the Mid-Atlantic ridge, where magma emerges as two plates separate. The weathering and erosion of surface rocks produce sandy grains that are deposited as sediments and may eventually become compacted into *sedimentary rocks* such as sandstone and shale. Subsequently, at high temperatures or pressures, igneous or sedimentary rocks may be physically or chemically transformed into *metamorphic rocks* (e.g., marble and slate). Such conditions occur as plates collide and form mountain ranges or as a plate dives deep into a subduction zone.

Past Continental Drift

Figure 7.17 illustrates how all the continents nearly fit together like pieces of a puzzle. Geologists think that sometime in the past a single gargantuan landmass dominated our planet. This ancestral supercontinent, known as *Pangaea* (meaning "all lands"), is shown in Figure 7.17(a). The rest of the planet was presumably covered with water. The present locations of the continents, along with measurements of their current drift rates, suggest that Pangaea was the major land feature on Earth approximately 200 million years ago. Dinosaurs, which were then the dominant form of life, could have sauntered from Russia to Texas via Boston without getting their feet wet. Pangaea explains the geographical and fossil evidence (cited earlier) that first led scientists to the idea of continental drift. The other frames in Figure 7.17 show how Pangaea split apart, its separate pieces drifting across Earth's surface, eventually becoming the familiar continents we know today.

There is nothing particularly special about a time 200 million years in the past. We do not suppose that Pangaea remained intact for some 4 billion years after the crust first formed, only to break up so suddenly and so recently. It is much more likely that Pangaea itself came into existence after an earlier period during which other plates, carrying widely separated continental masses, were driven together by tectonic forces, merging their landmasses to form a single supercontinent. It is quite possible that there has been a long series of "Pangaeas" stretching back in time over

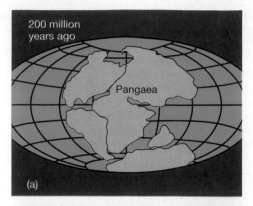

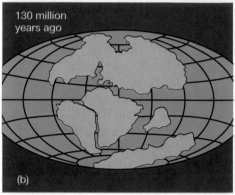

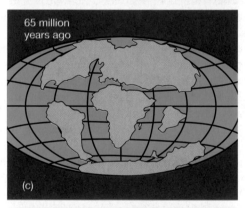

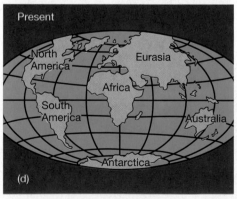

▲ **FIGURE 7.17 Pangaea** Given the current estimated drift rates and directions of the plates, we can trace their movements back into the past. About 200 million years ago, they would have been at the approximate positions shown in (a). The continents' current positions are shown in (d).

much of Earth's history, as tectonic forces have continually formed, destroyed, and re-formed our planet's landmasses. There probably will be many more.

CONCEPT CHECK

✔ Describe the causes, and some consequences, of plate tectonics on Earth.

7.5 Earth's Magnetosphere

Discovered by artificial satellites launched in the late 1950s and sketched in Figure 7.18, Earth's magnetosphere extends far above the atmosphere, completely surrounding our planet. Simply put, the magnetosphere is the region around a planet that is influenced by that planet's magnetic field. Close to Earth, the magnetic field is similar in overall structure to the field of a gigantic bar magnet (Figure 7.19). The *magnetic field lines*, which indicate the strength and direction of the field at any point in space, run from south to north, as indicated by the blue arrowheads in these figures.

The north and south *magnetic poles*, where the magnetic field lines intersect Earth's surface vertically, are roughly aligned with Earth's spin axis. Neither pole is fixed relative to our planet, however—both drift at a rate of some 10 km per year—nor are the poles symmetrically placed. At present, Earth's magnetic north pole lies in northern Canada, at a latitude of about 80°N, almost due north of the center of North America; the magnetic south pole lies at a latitude of about 60°S, just off the coast of Antarctica south of Adelaide, Australia.

Earth's magnetosphere contains two doughnut-shaped zones of high-energy charged particles, one located about 3000 km, and the other 20,000 km, above Earth's surface. These zones are named the **Van Allen belts,** after the American physicist whose instruments on board one of the first artificial satellites initially detected them. We call them "belts" because they are most pronounced near Earth's equator and because they completely surround the planet. Figure 7.19 shows how these invisible regions envelop Earth except near the North and South Poles.

The particles that make up the Van Allen belts originate in the solar wind—the steady stream of charged particles flowing from the Sun. ∞ (Sec. 6.5) Traveling through space, neutral particles and electromagnetic radiation are unaffected by Earth's magnetism, but electrically charged particles are strongly influenced. As illustrated in the inset to Figure 7.19, a magnetic field exerts a force on a moving charged particle, causing the particle to spiral around the magnetic field lines. In this way, charged particles—mainly electrons and protons—from the solar wind can become trapped by Earth's magnetism. Earth's magnetic field exerts electromagnetic control over these particles, herding them into the Van Allen belts. The outer belt contains

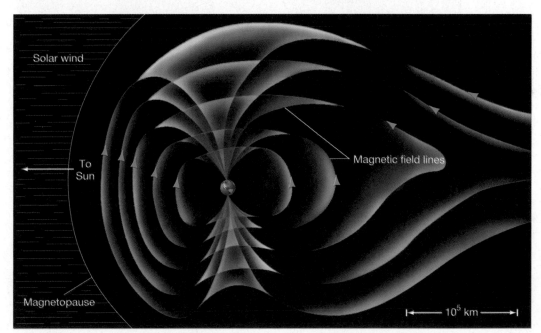

▲ **FIGURE 7.18 Earth's Magnetosphere** The magnetosphere is the region surrounding a planet wherein particles from the solar wind are trapped by the planet's magnetic field. Shown here is Earth's magnetosphere, with blue arrowheads on the field lines indicating the direction in which a compass needle would point. Far from Earth, the magnetosphere is greatly distorted by the solar wind, with a long tail extending from the nighttime side of Earth far into space. The magnetopause is the boundary of the magnetosphere in the sunward direction.

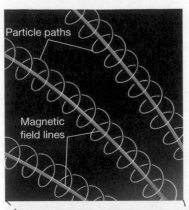

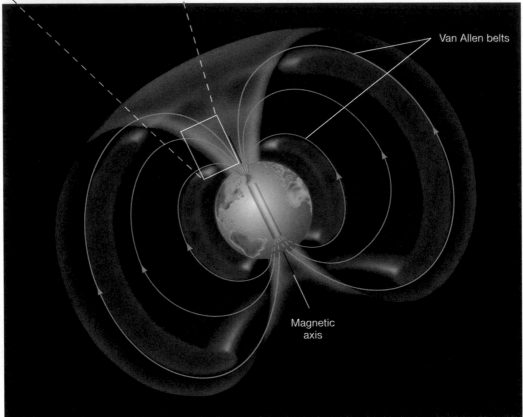

▲ **FIGURE 7.19 Van Allen Belts** Closer to home, Earth's magnetic field resembles somewhat the field of an enormous bar magnet buried deep inside our planet, offset slightly from Earth's axis of rotation. High above the atmosphere, the magnetosphere (light blue-green area) contains two doughnut-shaped regions (purple areas) of magnetically trapped charged particles. These are the Van Allen belts. The inset illustrates how charged particles in a magnetic field spiral around the field lines. The convergence of the field lines near Earth's magnetic poles causes the particles to be reflected back toward the other pole. Thus, charged particles tend to become "trapped" by Earth's magnetic field.

mostly electrons; the much heavier protons accumulate in the inner belt.

We could never survive unprotected in the Van Allen belts. Unlike the lower atmosphere, on which humans and other life-forms rely for warmth and protection, much of the magnetosphere is subject to intense bombardment by large numbers of high-velocity, and potentially very harmful, charged particles. Colliding violently with an unpro-

tected human body, these particles would deposit large amounts of energy wherever they made contact, causing severe damage to living organisms. Without sufficient shielding on the *Apollo* spacecraft, for example, the astronauts might not have survived the passage through the magnetosphere on their journey to the Moon.

Particles from the Van Allen belts often escape from the magnetosphere near Earth's north and south magnetic

(a)

(b)

R I V U X G

▲ **FIGURE 7.20 Aurorae** (a) A colorful aurora flashes rapidly across the sky like huge windblown curtains glowing in the dark. Aurorae result from the emission of light radiation after magnetospheric particles collide with atmospheric molecules. The colors are produced as excited ions, atoms, and molecules recombine and cascade back to their ground states. (b) An aurora high above Earth, as photographed from a space shuttle (visible at left). *(NCAR; NASA)*

poles, where the field lines intersect the atmosphere. Their collisions with air molecules create a spectacular light show called an **aurora** (pl. *aurorae*; Figure 7.20). This colorful display results when atmospheric molecules, excited upon collision with the charged particles, fall back to their ground states and emit visible light. Many different colors are produced because each type of atom or molecule can take one of several possible paths as it returns to its ground state. ∞ (Sec. 4.2) Aurorae are most brilliant at high latitudes, especially inside the Arctic and Antarctic circles. In the north, the spectacle is called the *aurora borealis*, or *northern lights*. In the south, it is called the *aurora australis*, or *southern lights*.

Occasionally, particularly after a storm on the Sun (see Chapter 16), the Van Allen belts can become distorted by the solar wind and overloaded with many more particles than normal, allowing some particles to escape prematurely and at lower latitudes. For example, in North America, the aurora borealis is normally seen with any regularity only in northern Canada and Alaska. However, at times of greatest solar activity, the display has occasionally been seen as far south as the southern United States.

As is evident from Figure 7.18, Earth's magnetosphere is not symmetrical. Satellite mapping reveals that it is quite distorted, forming a teardrop-shaped cavity. On the sunlit (daytime) side of Earth, the magnetosphere is compressed by the flow of high-energy particles in the solar wind. The boundary between the magnetosphere and this flow, known as the *magnetopause*, is found at about 10 Earth radii from our planet. On the side opposite the Sun, the field lines are extended away from Earth, with a long tail often reaching beyond the orbit of the Moon.

What is the origin of the magnetosphere and the Van Allen belts within it? Earth's magnetism is not really the result of a huge bar magnet lying within our planet. In fact, geophysicists think that Earth's magnetic field is not a "permanent" part of our planet at all. Instead, it is thought to be continuously generated within the outer core and to exist only because Earth is rotating. As in the dynamos that run industrial machines, Earth's magnetism is produced by the spinning, electrically conducting, liquid metal core deep within our planet. The theory that explains planetary (and other) magnetic fields in terms of rotating, conducting material flowing in the planet's interior is known as *dynamo theory*. Both rapid rotation *and* a conducting liquid core are needed for such a mechanism to work. This connection between internal structure and magnetism is very important for studies of the other planets in the solar system: We can tell a lot about a planet's interior simply by measuring its magnetic field.

Earth's magnetic field plays an important role in controlling many of the potentially destructive charged particles that venture near our planet. Without the magnetosphere, Earth's atmosphere—and perhaps the surface, too—would be bombarded by harmful particles, possibly damaging many forms of life on our planet. Some researchers have even suggested that, had the magnetosphere not existed in the first place, life might never have arisen at all on our planet.

CONCEPT CHECK

✔ What does the existence of a planetary magnetic field tell us about a planet's interior?

7.6 The Tides

Earth is unique among the planets in that it has large quantities of liquid water on its surface. Approximately three-quarters of Earth's surface is covered by water, to an average depth of about 3.6 km. Only 2 percent of the water is contained within lakes, rivers, clouds, and glaciers. The remaining 98 percent is in the oceans.

Most people are familiar with the daily fluctuation in ocean level known as the **tides.** At most coastal locations on Earth, there are two low tides and two high tides each day. The "height" of the tides—the magnitude of the variation in sea level—can range from a few centimeters to many meters, depending on the location on Earth and the time of year. The height of a typical tide on the open ocean is about a meter, but if this tide is funneled into a narrow opening such as the mouth of a river, it can become much higher. For example, at the Bay of Fundy, on the U.S.–Canada border between Maine and New Brunswick, the high tide can reach nearly 20 m (approximately 60 feet, or the height of a six-story building) above the low-tide level. An enormous amount of energy is contained in the daily motion of the oceans. This energy is constantly eroding and reshaping our planet's coastlines. In some locations, it has been harnessed as a source of electrical power for human activities.

Gravitational Deformation

LEARNING GOAL 6 What causes the tides? A clue comes from the observation that they exhibit daily, monthly, and yearly cycles. In fact, the tides are a direct result of the gravitational influence of the Moon and the Sun on Earth. We have already seen how gravity keeps Earth and the Moon in orbit about each other, and both in orbit around the Sun. ∞ (Sec. 2.7) For simplicity, let's first consider just the interaction between Earth and the Moon.

Recall that the strength of the gravitational force depends on the distance separating any two objects. Thus, the Moon's gravitational attraction is greater on the side of Earth that faces the Moon than on the opposite side, some 12,800 km (Earth's diameter) farther away. This difference in the gravitational force is small—only about 3 percent—but it produces a noticeable effect—a **tidal bulge.** As illustrated in Figure 7.21, Earth becomes slightly elongated, with the long axis of the distortion pointing toward the Moon.

Earth's oceans undergo the greatest deformation, because liquid can most easily move around on our planet's surface. (A bulge *is* actually raised in the solid material of Earth, but it is about a hundred times smaller than the oceanic bulge.) Thus, the ocean becomes a little deeper in some places (along the line joining Earth to the Moon) and shallower in others (perpendicular to this line). The daily tides we experience result as Earth rotates beneath this deformation.

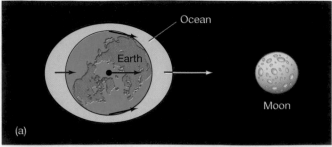
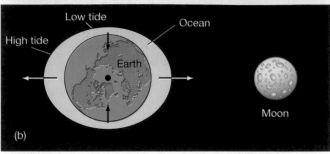

▲ **FIGURE 7.21** Lunar Tides This exaggerated illustration shows how the Moon induces tides on the near and far sides of Earth. The lengths of the arrows indicate the relative strengths of the Moon's gravitational pull on various parts of Earth. (a) The lunar gravitational forces acting on several locations on and inside Earth. The force is greatest on the side nearest the Moon and smallest on the opposite side. (b) The *differences* between the lunar forces acting at the locations shown in part (a) and the force exerted by the Moon on Earth's center. Closest to the Moon, the relative force is toward the Moon because the Moon's gravitational pull is stronger at the surface than it is at the center. However, on the opposite side of Earth, the force at the surface is weaker than at the center, so the relative force is away from the Moon. The arrows thus represent the force with which the Moon tends to either pull matter away from or squeeze it toward Earth's center. Closest to the Moon, the oceans tend to be "pulled away from Earth"; on the far side, Earth tends to be "pulled away from the oceans." The result is a tidal bulge.

The variation in the Moon's gravity across Earth is an example of a *differential force*, or **tidal force.** The average gravitational force between two bodies determines their orbit around one another. However, the *tidal* force, superimposed on that average, tends to deform the bodies. The tidal influence of one body on another diminishes very rapidly with increasing distance—in fact, as the inverse *cube* of the separation. For example, if the distance from Earth to the Moon were to double, the tides resulting from the Moon's gravity would decrease by a factor of eight. This rapid decline with increasing distance means that one object has to be very close or very massive in order to have a significant tidal effect on another.

We will see many situations in this book where tidal forces are critically important in understanding astronomical phenomena. We still use the word *tidal* in these other contexts, even though we are not discussing oceanic tides and, possibly, not even planets at all. In general astronomical use, the term refers to the deforming effect that the gravity of one body has on another.

Notice in Figure 7.21 that the side of Earth opposite the Moon also exhibits a tidal bulge. The different gravitational pulls—greatest on that part of Earth closest to the Moon, weaker at Earth's center, and weakest of all on Earth's opposite side—cause average tides on opposite sides of our planet to be approximately equal in height. On the side nearer the Moon, the ocean water is pulled slightly toward the Moon. On the opposite side, the ocean water is left behind as Earth is pulled closer to the Moon. Thus, high tide occurs *twice*, not once, each day at any given location.

Both the Moon and the Sun exert tidal forces on our planet. Thus, instead of one tidal bulge, there are actually two—one pointing toward the Moon, the other toward the Sun. Even though the Sun is 375 times farther away from Earth than is the Moon, the Sun's mass is so much greater (by a factor of 27 million) that its tidal influence is still significant—about half that of the Moon. The interaction between them accounts for the changes in the height of the tides over the course of a month or a year. When Earth, the Moon, and the Sun are roughly lined up (Figure 7.22a), the gravitational effects reinforce one another, so the highest tides are generally found at times of new and full moons. These tides are known as *spring tides*. When the Earth–Moon line is perpendicular to the Earth–Sun line (at the first and third quarters; Figure 7.22b), the daily tides are smallest. These are termed *neap tides*.

Effect of Tides on Earth's Rotation

Earth rotates once on its axis (relative to the stars) in $23^h 56^m$—one sidereal day. However, we know from fossil measurements that Earth's rotation is gradually slowing down, causing the length of the day to increase by about 1.5 milliseconds (ms) every century—not much on the scale of a human lifetime, but over millions of years, this steady slowing of Earth's spin adds up. At this rate, half a billion years ago, the day was just over 22 hours long and the year contained 397 days.

A number of natural biological clocks lead us to the conclusion that Earth's spin rate is decreasing. For example, each day, a growth mark is deposited on a certain type of coral in the reefs off the Bahamas. These growth marks are similar to the annual rings found in tree trunks, except that in the case of coral, the marks are made daily, in response to the day–night cycle of solar illumination. However, they

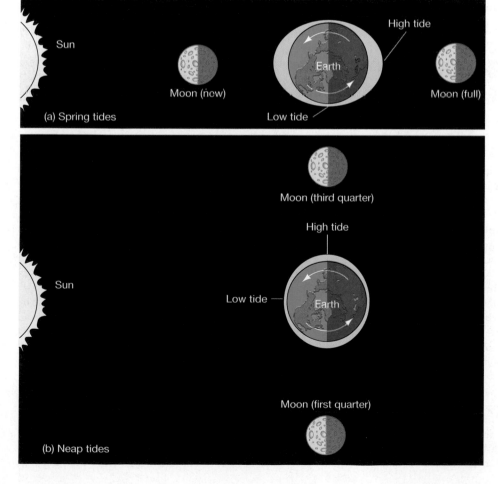

◀ **FIGURE 7.22 Solar and Lunar Tides** The combined effects of the Sun and the Moon produce variations in the high and low tides. (a) When the Moon is either full or new, Earth, Moon, and Sun are approximately aligned, and the tidal bulges raised in Earth's oceans by the Moon and the Sun reinforce one another. (b) At first- or third-quarter Moon, the tidal effects of the Moon and the Sun partially cancel each other, and the tides are smallest. Because the Moon's tidal effect is greater than that of the Sun (since the Moon's distance is much closer to us), the net bulge points toward the Moon.

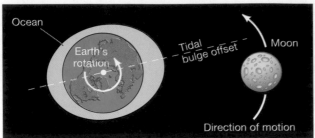

◄ **FIGURE 7.23 Tidal Bulge** The tidal bulge raised in Earth by the Moon does not point directly at the Moon. Instead, because of the effects of friction, the bulge points slightly "ahead" of the Moon, in the direction of Earth's rotation. (The magnitude of the effect is greatly exaggerated in this diagram.) Because the Moon's gravitational pull on the near-side part of the bulge is greater than the pull on the far side, the overall effect is to decrease Earth's rotation rate.

also show yearly variations as the coral's growth responds to Earth's seasonal changes, allowing us to perceive annual cycles. Coral growing today shows 365 marks per year, but ancient coral shows many more growth deposits per year. Fossilized reefs that are five hundred million years old contain coral with nearly 400 deposits per year of growth.

Why is Earth's spin slowing? The main reason is the tidal effect of the Moon. In reality, the tidal bulge raised in Earth by the Moon does *not* point directly at the Moon, as was shown in Figure 7.21. Instead, because of the effects of friction, both between the crust and the oceans and within Earth itself, Earth's rotation tends to drag the tidal bulge around with it, causing the bulge to be displaced by a small angle from the Earth–Moon line, in the same direction as Earth's spin (Figure 7.23). The net effect of the Moon's gravitational pull on this slightly offset bulge is to *reduce* our planet's rotation rate. At the same

time, the Moon is spiraling slowly away from Earth, increasing its average distance from our planet by about 4 cm per year.

This process will continue until Earth rotates on its axis at exactly the same rate as the Moon orbits Earth. At that time, the Moon will always be above the same point on Earth and will no longer lag behind the bulge it raises. Earth's rotation period will be 47 of our present days, and the distance to the Moon will be 550,000 km (about 43 percent greater than at present). However, this will take a very long time—many billions of years—to occur.

CONCEPT CHECK

✔ In what ways do tidal forces differ from the familiar inverse-square force of gravity?

Chapter Review

SUMMARY

The six main regions of Earth are (from inside to outside) a central metallic **core (p. 170)**, which is surrounded by a thick rocky **mantle (p. 170)**, topped with a thin **crust (p. 170)**. The liquid oceans on our planet's surface make up the **hydrosphere (p. 170)**. Above the surface is the **atmosphere (p. 170)**, which is composed primarily of nitrogen and oxygen. Higher still lies the **magnetosphere (p. 170)**, where charged particles from the Sun are trapped by Earth's magnetic field.

Earth's atmosphere, composed primarily of nitrogen (78 percent), oxygen (21 percent), argon (0.9 percent), and carbon dioxide (0.03 percent), thins rapidly with altitude. **Convection (p. 171)** is the process by which heat is moved from one place to another by the upwelling or downflow of a fluid, such as air or water. Convection occurs in the troposphere, the lowest region of Earth's atmosphere and is the cause of surface winds and weather. Incoming solar ultraviolet radiation is absorbed in the **ozone layer (p. 172)**. The ionosphere is kept ionized by high-energy radiation and particles from the Sun.

The **greenhouse effect (p. 173)** is the absorption and trapping by atmospheric gases (primarily carbon dioxide and water vapor) of infrared radiation emitted by Earth's surface. Incoming

visible light from the Sun is not significantly absorbed by these gases. By making it more difficult for Earth to radiate its energy back into space, the greenhouse effect makes our planet's surface some 40 K warmer than would otherwise be the case. The air we breathe is not Earth's original atmosphere. That atmosphere was outgassed from our planet's interior by volcanoes and was then altered by solar radiation and, finally, by the emergence of life.

We study Earth's interior by observing how **seismic waves (p. 177)**, produced by earthquakes just below Earth's surface, travel through the mantle. We can also study the upper mantle by analyzing the material brought to the surface when a volcano erupts. Seismic studies and mathematical modeling indicate that Earth's iron core consists of a solid **inner core (p. 179)** surrounded by a liquid **outer core (p. 179)**. Earth's center is extremely hot—about the same temperature as the surface of the Sun. The density at Earth's center is much greater than the density of surface rocks.

The process by which heavy material sinks to the center of a planet while lighter material rises to the surface is called **differentiation (p. 180)**. The differentiation of Earth implies that our planet must have been at least partially molten in the past. One way in which this could have occurred is by the heat released during Earth's formation and subsequent bombardment by material from interplanetary space. Another possibility is the energy released by the decay of **radioactive (p. 180)** elements present in the material from which Earth formed.

Earth's surface is made up of about a dozen enormous slabs, or plates. The slow movement of these plates across the surface is called continental drift or **plate tectonics (p. 182)**. Earthquakes, volcanism, and mountain building are associated with plate boundaries, where plates may collide, move apart, or rub against one another. The motion of the plates is thought to be driven by convection in Earth's mantle. The rocky upper layer of Earth that makes up the plates is the **lithosphere (p. 184)**. The semisolid region in the upper mantle over which the plates slide is called the **asthenosphere (p. 184)**. The constant recycling and transformation of crust material as plates separate, collide, and sink into the mantle is called the **rock cycle (p. 188)**. Evidence for past plate motion can be found in the geographical fit of continents, in the fossil record, and in the ages and magnetism of surface rocks.

Earth's magnetic field extends far beyond the surface of our planet. Charged particles from the solar wind are trapped by Earth's magnetic field lines to form the **Van Allen belts (p. 189)** that surround our planet. When particles from the Van Allen belts hit Earth's atmosphere, they heat and ionize the atoms there, causing them to glow in an **aurora (p. 191)**. According to dynamo theory, planetary magnetic fields are produced by the motion of rapidly rotating, electrically conducting fluid (such as molten iron) in the planet's core.

The daily **tides (p. 192)** in Earth's oceans are caused by the gravitational effect of the Moon and the Sun, which raise **tidal bulges (p. 192)** in the hydrosphere. The tidal effect of the Moon is almost twice that of the Sun. The size of the tides depends on the orientations of the Sun and the Moon relative to Earth. A differential gravitational force is always called a **tidal force (p. 192)**, even when no oceans or planets are involved. The tidal interaction between Earth and the Moon is causing Earth's spin to slow.

REVIEW AND DISCUSSION

1. By comparison with Earth's average density, what do the densities of the water and rocks in Earth's crust tell us about Earth's interior?

2. What is Rayleigh scattering? What is its most noticeable effect for us on Earth?

3. How do geologists use earthquakes to obtain information about Earth's interior?

4. Compare and contrast P-waves and S-waves, and explain how they are useful to geologists.

5. What is the greenhouse effect, and what effect does it have on Earth's surface temperature?

6. Give two reasons geologists think that part of Earth's core is liquid.

7. What clue to our planet's history does Earth's differentiation provide?

8. What is convection? What effect does it have on (a) Earth's atmosphere? (b) Earth's interior?

9. How does radioactivity allow us to estimate Earth's age?

10. How did radioactive decay heat Earth early in its history? When did this heating end?

11. What process is responsible for the surface mountains, oceanic trenches, and other large-scale features on Earth's surface?

12. Discuss how distant quasars, lying hundreds of millions of light-years from Earth, are used to monitor the motion of Earth's tectonic plates.

13. What conditions are needed to create a dynamo in Earth's interior? What effect does this dynamo have?

14. Give a brief description of Earth's magnetosphere, and tell how it was discovered.

15. How does Earth's magnetosphere protect us from the harsh realities of interplanetary space?

16. How do we know that Earth's magnetic field has undergone reversals in the past? How do you think Earth's magnetic field reversals might have affected the evolution of life on our planet?

17. Explain how the Moon produces tides in Earth's oceans.

18. If the Moon had oceans like Earth's, what would the tidal effect be like there? How many high and low tides would there be during a "day"? How would the variations in height compare with those on Earth?

19. If Earth had no moon, do you think we would know anything about tidal forces?

20. Is the greenhouse effect operating in Earth's atmosphere helpful or harmful? Give examples. What are the consequences of an enhanced greenhouse effect?

CONCEPTUAL SELF-TEST: TRUE OR FALSE/MULTIPLE CHOICE

1. The troposphere is the part of the atmosphere in which convection occurs.

2. The oxygen in Earth's atmosphere is the result of the appearance of life on our planet.

3. Sunlight is absorbed by Earth's surface and then reemitted in the form of ultraviolet radiation.

4. Geologists obtain most of their information about Earth's mantle by drilling into our planet's interior.

5. Earth's core temperature is comparable to the surface temperature of the Sun.

6. When plates collide, they simply come to rest and fuse together.

7. Earth's magnetic field is the result of our planet's large, permanently magnetized iron core.

8. An aurora occurs when trapped electrons and protons in the magnetosphere collide with the upper atmosphere.

9. There is one high tide and one low tide per day at any given coastal location on Earth.

10. Tides are caused by the differences in the gravitational pulls of the planets from one side of Earth to the other.

11. If you were making a scale model of Earth, representing our planet by a 12-inch basketball, the inner core would be about the size of **(a)** a $\frac{1}{2}$-inch ball bearing; **(b)** a 2-inch golf ball; **(c)** a 4-inch tangerine; **(d)** a 7-inch grapefruit.

12. Earth's average density is about the same as that of **(a)** a glass of water; **(b)** a heavy iron meteorite; **(c)** an ice cube; **(d)** a chunk of black volcanic rock.

13. According to Figure 7.2 ("Earth's Atmosphere"), commercial jet airplanes flying at 10 km are in **(a)** the troposphere; **(b)** the stratosphere; **(c)** the ozone; **(d)** the mesosphere.

14. If there were significantly more greenhouse gases, such as CO_2, in Earth's atmosphere, then **(a)** the ozone hole would close; **(b)** the ozone hole would get larger; **(c)** Earth's average temperature would change; **(d)** plants would grow faster than animals could eat them.

15. If seismometers registered P- and S-waves everywhere on the Moon, they would suggest that the Moon had **(a)** the same layered structure as Earth; **(b)** no molten core; **(c)** no moonquakes; **(d)** the same density throughout.

16. The deepest that geologists have drilled into Earth is about the same as **(a)** the height of the Statue of Liberty; **(b)** the altitude most commercial jet airplanes fly; **(c)** the distance between New York and Los Angeles; **(d)** the distance between the United States and China.

17. Due to plate tectonics, the width of the Atlantic Ocean is separating at a rate about the same as the growth of **(a)** grass; **(b)** human hair; **(c)** human fingernails; **(d)** dust in a typical home.

18. At Earth's geographic North Pole, a magnetic compass needle would point (approximately) **(a)** toward Alaska; **(b)** toward Kansas City; **(c)** toward Paris; **(d)** straight down.

19. If Earth had no Moon, then tides would **(a)** not occur; **(b)** occur more often and with more intensity; **(c)** still occur, but not really be measurable; **(d)** occur with the same frequency, but would not be as strong.

20. Which of the following statements is true? Because of the tides, **(a)** Earth's rotation rate is increasing; **(b)** the Moon is spiraling away from Earth; **(c)** Earth will eventually drift away from the Sun; **(d)** earthquake activity is increasing.

PROBLEMS

 Algorithmic versions of these questions are available in the Practice Problems module of the Companion Website at astro.prenhall.com/chaisson.

The number of squares preceding each problem indicates its approximate level of difficulty.

1. ■ Verify that Earth's orbital perihelion and aphelion, mean orbital speed, surface gravity, and escape speed are correct as listed in the Earth Data box on p. 172.

2. ■ What would Earth's surface gravity and escape speed be if the entire planet had a density equal to that of the crust, say, 3000 kg/m³?

3. ■ Approximating Earth's atmosphere as a layer of gas 7.5 km thick, with uniform density 1.3 kg/m³, calculate the total mass of the atmosphere. Compare your result with Earth's mass.

4. ■■■ As discussed in the text, without the greenhouse effect, Earth's average surface temperature would be about 250 K. With the greenhouse effect, it is some 40 K higher. Use this information and Stefan's law to calculate the fraction of infrared radiation leaving Earth's surface that is absorbed by greenhouse gases in the atmosphere. ∞ (Sec. 3.3)

5. ■■ Most of Earth's ice is found in Antarctica, where permanent ice caps cover approximately 0.5 percent of Earth's total surface area and are 3 km thick, on average. Earth's oceans cover roughly 71 percent of our planet, to an average depth of 3.6 km. Assuming that water and ice have roughly the

same density, estimate by how much sea level would rise if global warming were to cause the Antarctic ice caps to melt.

6. ■ Following an earthquake, how long would it take a P-wave, moving in a straight line with a speed of 5 km/s, to reach Earth's opposite side?

7. ■ On the basis of the data presented in the text, estimate the fractions of Earth's volume represented by (a) the inner core, (b) the outer core, (c) the mantle, and (d) the crust.

8. ■ At 3 cm/yr, how long would it take a typical plate to traverse the present width of the Atlantic Ocean, about 6000 km?

9. ■ In a certain sample of rock, it is found that 25 percent of uranium-238 nuclei have decayed into lead-206. On the basis of the data given in *More Precisely* 7-2, estimate the age of the rock sample.

10. ■■ A second sample of rock is found to contain three times as many lead-207 nuclei as uranium-235 nuclei. On the basis of the data given in *More Precisely* 7-2, what ratio of uranium-238 to lead-206 nuclei would you expect?

11. ■■■ Astronauts in orbit are weightless because they are falling freely in Earth's gravitational field, but they are still subject to tidal forces. Calculate the relative acceleration due to tidal forces of two masses in low Earth orbit, placed 1 m apart along a line extending radially outward from Earth's center. Compare this acceleration with the acceleration due to gravity at Earth's surface.

12. ■■ You are standing on Earth's surface during a total eclipse, and both the Moon and the Sun are directly overhead. By what fraction is your weight changed due to their combined tidal gravitational force?

13. ■■ You are standing on Earth's surface, and the full Moon is directly overhead. By what fraction is your weight decreased due to the combination of the Sun's and the Moon's tidal gravitational forces?

14. ■■■ The planet Jupiter exerts a strong tidal force on its innermost moon, Io. Compare Jupiter's tidal force on an object on Io's surface with the gravitational force on the body due to Io's own gravity. (For definiteness, perform the calculation for a 1-kg mass resting on Io's surface, on the line joining the center of the planet to the center of the moon.) Io orbits at a distance of 422,000 km. Its mass and radius are 9.0×10^{22} kg and 3600 km, respectively. Jupiter's mass is 1.9×10^{27} kg.

15. ■■ Compare the magnitude of the tidal gravitational force on Earth due to Jupiter with that due to the Moon. Assume an Earth–Jupiter distance of 4.2 A.U. On the basis of your answer, do you think that the tidal stresses caused by a "cosmic convergence"—a chance alignment of the four Jovian planets so that they all appear from Earth to be in exactly the same direction in the sky—would have any noticeable effect on our planet?

 In addition to the Practice Problems module, the Companion Website at astro.prenhall.com/chaisson provides for each chapter a study guide module with multiple choice questions as well as additional annotated images, animations, and links to related Websites.

8

America's manned exploration of the Moon was arguably the greatest engineering feat of the twentieth century, indeed one of the greatest of all time. Nine missions were launched to the Moon, a dozen astronauts were landed, and all returned safely to Earth. The Apollo program ended in 1972, as quickly as it had begun a decade before—largely because of political posturing at the height of the Cold War. Here, an Apollo 15 astronaut near Mount Hadley (see also Figure 8.22) is adjusting some instruments for testing the soil. (NASA) ▶

The Moon and Mercury

Scorched and Battered Worlds

The Moon is Earth's only natural satellite. Mercury, the smallest terrestrial world, is the planet closest to the Sun. Despite their different environments, these two bodies have many similarities—indeed, at first glance, you might even mistake one for the other. Both have heavily cratered, ancient surfaces, littered with boulders and pulverized dust. Both lack atmospheres to moderate day-to-night variations in solar heating and experience wild temperature swings as a result. Both are geologically dead. In short, the Moon and Mercury differ greatly from Earth, but it is precisely those differences which make these desolate worlds so interesting to planetary scientists. Why is the Moon so unlike our own planet, despite its nearness to us, and why does planet Mercury apparently have so much more in common with Earth's Moon than with Earth itself? In this chapter, we explore the properties of these two worlds as we begin our comparative study of the planets and moons that make up our solar system.

LEARNING GOALS

Studying this chapter will enable you to

1 Specify the general characteristics of the Moon and Mercury, and compare them with those of Earth.

2 Describe the surface features of the Moon and Mercury, and recount how those two bodies were formed by dynamic events early in their history.

3 Explain how the Moon's rotation is influenced by its orbit around Earth and Mercury's by its orbit around the Sun.

4 Explain how observations of cratering can be used to estimate the age of a body's surface.

5 Compare the Moon's interior structure with that of Mercury.

6 Summarize the various theories of the formation of the Moon, and indicate which is currently considered most likely.

7 Discuss how astronomers have pieced together the story of the Moon's evolution, and compare its evolutionary history with that of Mercury.

 Visit astro.prenhall.com/chaisson for additional annotated images, animations, and links to related sites for this chapter.

8.1 Orbital Properties

We begin our study of the Moon and Mercury by examining their orbits, the knowledge of which will, in turn, aid us in determining and explaining the other properties of these worlds. Detailed orbital and physical data are presented in the Moon Data box (p. 202) and the Mercury Data box (p. 220).

The Moon

Parallax methods, described in Chapter 1, can provide us with quite accurate measurements of the distance to the Moon, using Earth's diameter as a baseline. ∞ (Sec. 1.7) Radar ranging yields more accurate distances. The Moon is much closer than any of the planets, and the radar echo bounced off the Moon's surface is strong. A radio telescope receives the echo after about a 2.56-second wait. Dividing this time by 2 (to account for the round-trip taken by the signal) and multiplying it by the speed of light (300,000 km/s) gives us a mean distance of 384,000 km. (The actual distance at any specific time depends on the Moon's location in its slightly elliptical orbit around Earth.)

Current laser-ranging technology, using reflectors placed on the lunar surface by *Apollo* astronauts (see *Discovery 8-1*, pp. 208–209) to reflect laser beams fired from Earth, allows astronomers to measure the round-trip time with submicrosecond accuracy. Repeated measurements have allowed astronomers to determine the Moon's orbit to within a few centimeters. This precision is necessary for programming unmanned spacecraft to land successfully on the lunar surface.

Mercury

Viewed from Earth, Mercury never strays far from the Sun. As illustrated in Figure 8.1(a), the planet's 0.4-A.U. orbital semimajor axis means that its angular distance from the Sun never exceeds 28°. Consequently, the planet is visible to the naked eye only when the Sun's light is blotted out—just before dawn or just after sunset (or, much less frequently, during a total solar eclipse)—and it is not possible to follow Mercury through a full cycle of phases. In fact, although Mercury was well known to ancient astronomers, they originally believed that this companion to the Sun was two different objects, and the connection between the planet's morning and evening appearances took some time to establish. However, later Greek astronomers were certainly aware that the "two planets" were really different alignments of a single body. Figure 8.1(b), a photograph taken just after sunset, shows Mercury above the western horizon, along with three other planets and the Moon.

Because Earth rotates at a rate of 15° per hour, Mercury is visible for at most 2 hours on any given night, even under the most favorable circumstances. For most observers at most times of the year, Mercury is considerably

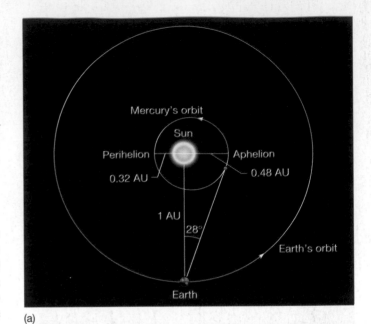

(a)

(b)

R I V U X G

▲ **FIGURE 8.1 Evening Sky** (a) Mercury's orbit has a semimajor axis of just 0.4 A.U., so the planet can never be farther than 28° from the Sun, as seen from Earth. Mercury's eccentric orbit means that this maximum separation is achieved only for the special configuration shown here, in which the Earth–Sun line is perpendicular to the long axis of Mercury's orbit and Mercury is near aphelion (its greatest distance from the Sun). (b) Four planets, together with the Moon, are visible in this photograph taken shortly after sunset. To the right of the Moon (top left) is the brightest planet, Venus. A little farther to the right is Mars, with the star Regulus just below and to its left. At the lower right, at the edge of the Sun's glare, are Jupiter and Mercury. (The Moon appears round rather than crescent shaped because the "dark" portion of its disk is indirectly illuminated by sunlight reflected from Earth. This "earthshine," relatively faint to the naked eye, is exaggerated in the overexposed photographic image.) *(J. Sanford/Photo Researchers, Inc.)*

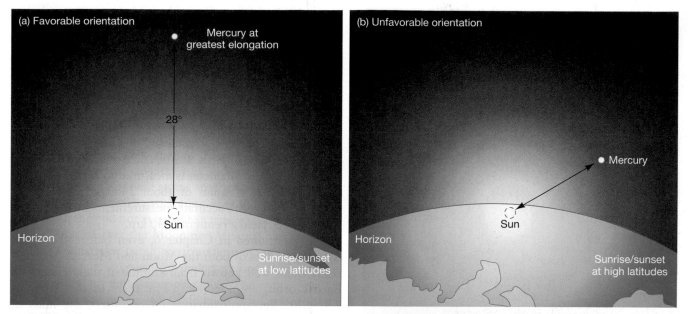

▲ **FIGURE 8.2 Visibility of Mercury** Favorable and unfavorable orientations of Mercury's orbit result from different orientations of Earth and different locations of the observer, both of which affect the angle between the horizon and Mercury's orbital plane. (a) At greatest elongation, Mercury lies some 28° from the Sun. The planet is most easily visible when it also lies high above the horizon. (b) At the most unfavorable orientations, Mercury is close to both the Sun and the horizon.

less than 28° above the horizon, so it is generally visible for a much shorter period (see Figure 8.2). Nowadays, large telescopes can filter out the Sun's glare and observe Mercury even during the daytime, when the planet is higher in the sky and atmospheric effects are reduced. (The amount of air that the light from the planet has to traverse before reaching our telescope decreases as the height of the planet above the horizon increases.) In fact, some of the best views of Mercury have been obtained in this way. The

naked-eye or amateur astronomer is generally limited to nighttime observations, however.

In all cases, it becomes progressively more difficult to view Mercury the closer (in the sky) its orbit takes it to the Sun. The best images of the planet therefore show a "half Mercury," close to its maximum angular separation from the Sun, or *maximum elongation*, as illustrated in Figure 8.3. (A planet's elongation is just its angular distance from the Sun, as seen from Earth.)

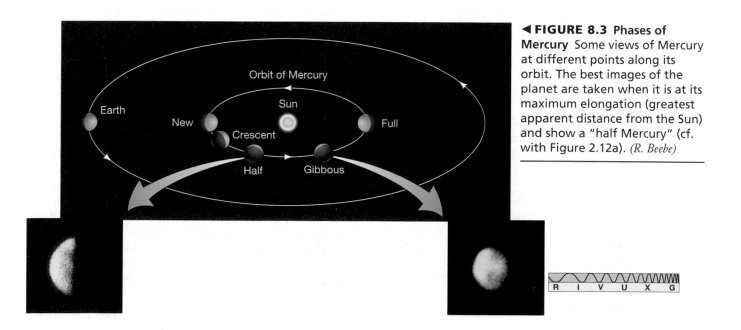

◄ **FIGURE 8.3 Phases of Mercury** Some views of Mercury at different points along its orbit. The best images of the planet are taken when it is at its maximum elongation (greatest apparent distance from the Sun) and show a "half Mercury" (cf. with Figure 2.12a). *(R. Beebe)*

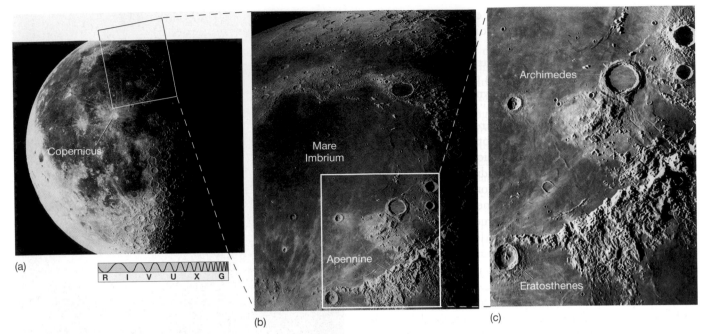

(a)

R I V U X G

Mare Imbrium

Apennine

(b)

Archimedes

Eratosthenes

(c)

▲ **FIGURE 8.5 Moon, Close Up** (a) The Moon near third quarter. Surface features are much more visible near the *terminator*, the line separating light from dark, where sunlight strikes at a sharp angle and shadows highlight the landscape. The light-colored crater just below the lower left corner of the breakout box is Copernicus. (b) Magnified view of a region near the terminator, as seen from Earth through a large telescope. The central dark area is Mare Imbrium, ringed at the bottom right by the Apennine mountains. (c) An enlargement of a portion of (b). The crater at the bottom left is Eratosthenes; Archimedes is at top center. The smallest craters visible here have diameters of about 2 km. *(UC/Lick Observatory; Caltech)*

Based on studies of lunar rock brought back to Earth by *Apollo* astronauts and unmanned Soviet landers, geologists have identified important differences in both *composition* and *age* between the highlands and the maria. The highlands are made largely of rocks rich in aluminum, making them lighter in color and lower in density (2900 kg/m^3) than the material in the maria, which contains more iron, giving it a darker color and greater density (3300 kg/m^3). Loosely speaking, the highlands represent the Moon's crust, while the maria are made of mantle material. Maria rock is quite similar to terrestrial basalt, and geologists think that it arose on the Moon

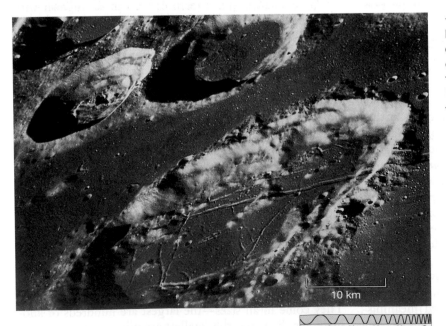

10 km

R I V U X G

◄ **FIGURE 8.6 Moon from *Apollo*** The Moon, as seen from the *Apollo 8* orbiter during the first human circumnavigation of our satellite in 1968. Craters ranging in size from 50 km to 500 m (also the width of the long fault lines) can be seen. *(NASA)*

much as basalt did on Earth, from the upwelling of molten material through the crust. ∞ (Sec. 7.3) Radioactive dating indicates ages of 4 to 4.4 billion years for highland rocks and from 3.2 to 3.9 billion years for those from the maria. ∞ (*More Precisely 7-2*)

All of the Moon's significant surface features have names. The 14 maria bear fanciful Latin names—Mare Imbrium ("Sea of Showers"), Mare Nubium ("Sea of Clouds"), Mare Nectaris ("Sea of Nectar"), and so on. Most mountain ranges in the highlands bear the names of terrestrial mountain ranges—the Alps, the Carpathians, the Apennines, the Pyrenees, and so on. Most of the craters are named after great scientists or philosophers, such as Plato, Aristotle, Eratosthenes, and Copernicus.

Because the Moon rotates once on its axis in exactly the same time it takes to complete one orbit around Earth, the Moon has a "near" side, which is always visible from Earth, and a "far" side, which never is (see Section 8.4). To the surprise of most astronomers, when the far side of the Moon was mapped, first by Soviet and later by U.S. spacecraft (see *Discovery 8-1*), no major maria were found there. The lunar far side (Figure 8.7) is composed almost entirely of highlands. This fact has great bearing on our theory of how the Moon's surface terrain came into being, for it implies that the processes which were involved could *not* have been entirely internal in nature. Earth's presence must somehow have played a role.

CONCEPT CHECK

✔ Describe three important ways in which the lunar maria differ from the highlands.

The Surface of Mercury

LEARNING GOAL 2 Mercury is difficult to observe from Earth because of Mercury's closeness to the Sun. Even with a fairly large telescope, we see it only as a slightly pinkish disk. Figure 8.8 is one of the few photographs of Mercury taken from Earth that shows any evidence of surface markings. Astronomers could only speculate about the faint, dark markings in the days before *Mariner 10*'s arrival. We now know that these markings are much like those seen by an observer gazing casually at Earth's Moon. The largest ground-based telescopes can resolve surface features on Mercury about as well as we can perceive features on the Moon with our unaided eyes.

In 1974, *Mariner 10* approached within 10,000 km of the surface of Mercury, sending back high-resolution images of the planet. ∞ (Sec. 6.6) These photographs, which showed surface features as small as 150 m across, revolutionized our knowledge of the planet. Figures 8.9(a) and (b) show views of Mercury taken by *Mariner 10* from a distance of about 200,000 km.

As discussed in Chapter 6, *Mariner 10* did not go into orbit around Mercury, but instead was placed in a somewhat

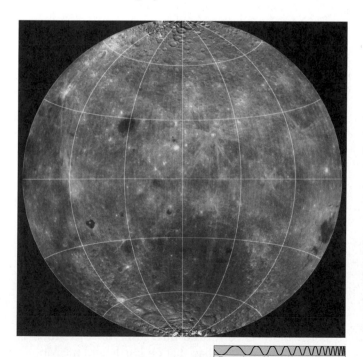

▲ **FIGURE 8.7 Full Moon, Far Side** The far side of the Moon, as seen by the *Clementine* military spacecraft. The large, dark region at the center bottom outlines the south pole–Aitken Basin, the largest and deepest impact feature known in the solar system. This image shows only one or two small maria on the far side. *(Dept. of Defense)*

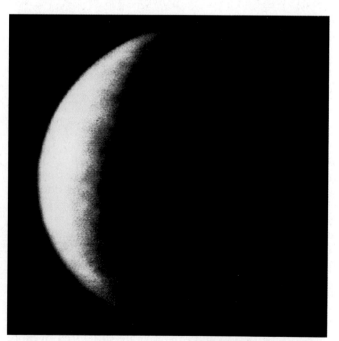

▲ **FIGURE 8.8 Mercury** Photograph of Mercury taken from Earth with one of the largest ground-based optical telescopes. Only a few surface features are discernible. *(Palomar Observatory/Caltech)*

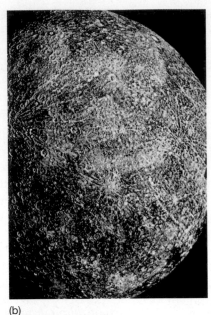

(a)

(b)

◀ **FIGURE 8.9** Mercury, Up Close
(a) Mercury is imaged here through a mosaic of photographs taken by the *Mariner 10* spacecraft in the mid-1970s during its approach to the planet. At the time, the spacecraft was some 200,000 km away. (b) *Mariner 10*'s view of Mercury as it sped away from the planet after each encounter. Again, the spacecraft was about 200,000 km away when the photographs making up this mosaic were taken. *(NASA)*

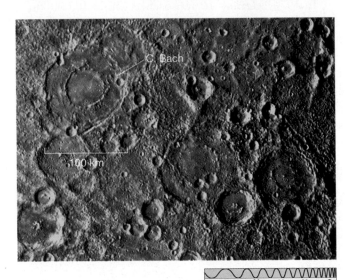

▲ **FIGURE 8.10** Mercury, Very Close Another photograph of Mercury by *Mariner 10*, this time from about 20,000 km above the planet's surface. The double-ringed crater at the upper left, named C. Bach, is about 100 km across; it exemplifies many of the large craters on Mercury, which tend to have double, rather than single, rings. The reason is not yet understood. *(NASA)*

eccentric orbit around the Sun that brought it close to the planet every 176 days—exactly two Mercury years. ∞ (Sec. 6.6) However, the peculiar combination of Mercury's orbital period and rotation rate (discussed in more detail in Sec. 8.4) meant that *Mariner* saw the *same* face of the planet at each approach. As a result, less than half of the planet's surface has been mapped. Together, the two mosaics in Figure 8.9 cover the known surface of Mercury. No similar photographs exist of the hemisphere that happened to be in shadow during each encounter.

Figure 8.10 shows a higher resolution photograph of the planet from a distance of 20,000 km. The similarities to the Moon are striking. We see no sign of clouds, rivers, dust storms, or other aspects of weather. Much of the cratered surface bears a strong resemblance to the Moon's highlands. Mercury, however, shows few extensive lava flow regions akin to the lunar maria.

8.4 Rotation Rates

The Rotation of the Moon

As mentioned earlier, the Moon's rotation period is precisely equal to its period of revolution about Earth—27.3 days—so the Moon keeps the same side facing Earth at all times (see Figure 8.11). To an astronaut standing on the Moon's near-side surface, Earth would appear almost stationary in the sky (although our planet's daily rotation would be clearly evident). This condition, in which the spin of one body is precisely equal to (or *synchronized* with) its revolution around another body, is known as a **synchronous orbit**. The fact that the Moon is in a synchronous orbit around Earth is no accident. It is an inevitable consequence of the gravitational interaction between those two bodies.

Just as the Moon raises tides on Earth, Earth also produces a tidal bulge in the Moon. Indeed, because Earth is so much more massive, the tidal force on the Moon is about 20 times greater than that on Earth, and the Moon's tidal bulge is correspondingly larger. In Chapter 7, we saw how lunar tidal forces are causing Earth's spin to slow and how, as a result, Earth will eventually rotate on its axis at the same rate as the Moon revolves around Earth. ∞ (Sec. 7.6) Earth's rotation will not become synchronous with the Earth–Moon orbital period for hundreds of billions of years. In the case of the Moon, however, the process has al-

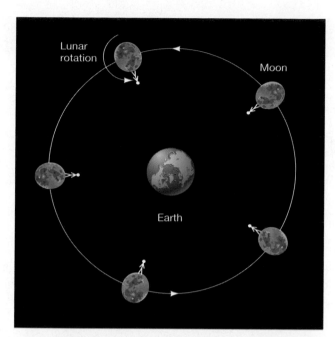

▲ **FIGURE 8.11 The Moon's Synchronous Rotation** As the Moon orbits Earth, it keeps one face permanently pointed toward our planet. To the astronaut shown here, Earth is always directly overhead. In fact, the Moon is slightly elongated in shape, with its long axis perpetually pointing toward Earth. (The elongation is highly exaggerated in this diagram.)

ready gone to completion. The Moon's much larger tidal deformation caused it to evolve into a synchronous orbit long ago, and the Moon is said to have become *tidally locked* to Earth. Most of the moons in the solar system are similarly locked by the tidal fields of their parent planets.

Actually, the size of the lunar bulge is too great to be produced by Earth's present-day tidal influence. The explanation seems to be that, long ago, the distance from Earth to the Moon may have been as little as two-thirds of its current value, or about 250,000 km. Earth's tidal force on the Moon would then have been more than three times greater than it is today and could have accounted for the Moon's elongated shape. The resulting distortion could have "set" when the Moon solidified, thus surviving to the present day, while at the same time accelerating the synchronization of the Moon's orbit.

Measurement of Mercury's Spin

In principle, the ability to discern surface features on Mercury should allow us to measure its rotation rate simply by watching the motion of a particular region around the planet. In the mid-19th century, an Italian astronomer named Giovanni Schiaparelli did just that. He concluded that Mercury always keeps one side facing the Sun, much as our Moon perpetually presents only one face to Earth. The explanation suggested for this synchronous rotation was the same as that for the Moon: The tidal bulge raised in Mercury by the Sun had modified the plan-

et's rotation rate until the bulge always pointed directly at the Sun. Although the surface features could not be seen clearly, the combination of Schiaparelli's observations and a plausible physical explanation was enough to convince most astronomers, and the belief that Mercury rotates synchronously with its revolution about the Sun (i.e., once every 88 Earth days) persisted for almost half a century.

In 1965, astronomers making radar observations of Mercury from the Arecibo radio telescope in Puerto Rico (see Figure 5.24) discovered that this long-held view was in error. The technique they used is illustrated in Figure 8.12, which shows a radar signal reflecting from the surface of a hypothetical planet. Let's imagine, for the purpose of this discussion, that the pulse of outgoing radiation is of a single frequency. The returning pulse bounced off the planet is very much weaker than the outgoing signal. Beyond this change, the reflected signal can be modified in two important ways. First, the signal as a whole may be redshifted or blueshifted as a consequence of the Doppler effect, depending on the overall radial velocity of the planet with respect to Earth. ∞ (Sec. 3.5) We will assume for simplicity that this velocity is zero, so that, on average, the frequency of the reflected signal is the same as that of the outgoing beam.

Second, if the planet is rotating, the radiation reflected from the side of the planet moving toward us returns at a slightly higher frequency than the radiation reflected from the receding side. (Think of the two hemispheres as being separate sources of radiation and moving at slightly different velocities, one toward us and one away.) The effect is very similar to the rotational line broadening discussed in Chapter 4 (see Figure 4.18), except that in this case the radiation we are measuring was not emitted by the planet, but only reflected from its surface. ∞ (Sec. 4.4) What we

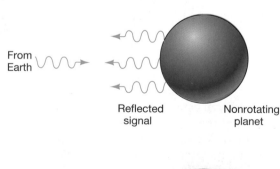

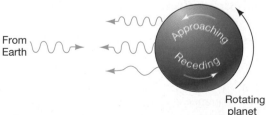

▲ **FIGURE 8.12 Planetary Radar** A radar beam reflected from a rotating planet yields information about both the planet's line-of-sight motion and its rotation rate. (Cf. Figure 4.18.)

DISCOVERY 8-1

Lunar Exploration

The Space Age began in earnest on October 4, 1957, with the launch of the Soviet satellite *Sputnik 1*. Thirteen months later, on January 4, 1959, the Soviet *Luna 1*, the first human-made craft to escape Earth's gravity, passed the Moon. *Luna 2* crash-landed on the surface in September of that year, and *Luna 3* returned the first pictures of the far side a month later. The long-running *Luna* series established a clear Soviet lead in the early "space race" and returned volumes of detailed information about the Moon's surface. Several of the *Luna* missions landed and returned surface material to Earth.

The U.S. lunar exploration program got off to a rocky start. The first six attempts in the *Ranger* series, between 1961 and 1964, failed to accomplish their objective of just hitting the Moon. The last three were successful, however. *Ranger 7* collided with the lunar surface (as intended) on June 28, 1964. Five U.S. *Lunar Orbiter* spacecraft, launched in 1966 and 1967, were successfully placed in orbit around the Moon, and they relayed high-resolution images of much of the lunar surface back to Earth. Between 1966 and 1968, seven *Surveyor* missions soft-landed on the Moon and performed detailed analyses of the surface.

Many of these unmanned U.S. missions were performed in support of the manned *Apollo* program. On May 25, 1961, at a time when the U.S. space program was in great disarray, President John F. Kennedy declared that the United States would "send a man to the Moon and return him safely to Earth" before the end of the decade, and the *Apollo* program was born. On July 20, 1969, less than 12 years after *Sputnik* and only 8 years after the statement of the program's goal, *Apollo 11* commander Neil Armstrong became the first human to set foot on the Moon, in Mare Tranquilitatis (the Sea of Tranquility). Three-and-a-half years later, on December 14, 1972, scientist–astronaut Harrison Schmitt, of *Apollo 17*, was the last.

The astronauts who traveled in pairs to the lunar surface in each lunar lander (shown in the first photograph) performed numerous geological and other scientific studies on the surface. The later landers brought with them a "lunar rover"—a small golf cart–sized vehicle that greatly expanded the area the astronauts could cover. Probably the most important single aspect of the *Apollo* program was the collection of samples of surface rock from various locations on the Moon. In all, some 382 kg of material was returned to Earth. Chemical analysis and radioactive dating of these samples revolutionized our understanding of the Moon's surface history. No amount of Earth-based observations could have achieved the same results.

Each *Apollo* lander left behind a nuclear-powered package of scientific instruments called the *Apollo Lunar Surface Experiments Package* (*ALSEP*, second photograph) to monitor the solar wind, measure heat flow in the Moon's interior, and, per-

see in the reflected signal is a spread of frequencies on either side of the original frequency. By measuring the extent of that spread, we can determine the planet's rotational speed.

In this way, the Arecibo researchers found that the rotation period of Mercury is not 88 days, as had previously been thought, but 59 days, exactly two-thirds of the planet's orbital period. Because there are exactly three rotations for every two revolutions, we say that there is a 3:2 *spin–orbit resonance* in Mercury's motion. In this context, the term **resonance** just means that two characteristic times—here, Mercury's day and year—are related to each other in a simple way. An even simpler example of a spin–orbit resonance is the Moon's orbit around Earth. In that case, the rotation is synchronous with the revolution, and the resonance is said to be 1:1.

Figure 8.13 illustrates some implications of Mercury's curious rotation for a hypothetical inhabitant of the planet. Mercury's solar day—the time from noon to noon, say—is

two Mercury years long! The Sun stays "up" in the black Mercury sky for almost three Earth months at a time, after which follow nearly three Earth months of darkness. At any given point in its orbit, Mercury presents the same face to the Sun, not every time it revolves, but *every other* time.

▶ **FIGURE 8.13 Mercury's Rotation** Mercury's orbital and rotational motions combine to produce a day that is two Mercury years long. The red arrow represents an observer standing on the surface of the planet. At day 0 (center right in Year 1 drawing), it is noon for our observer and the Sun is directly overhead. By the time Mercury has completed one full orbit around the Sun and moved from day 0 to day 88, it has rotated on its axis exactly 1.5 times, so that it is now midnight at the observer's location. After another complete orbit, it is noon once again on day 176 (center right in Year 3 drawing). The eccentricity of Mercury's orbit is not shown in this simplified diagram.

haps most important, record lunar seismic activity. With several *ALSEP*s on the surface, scientists could determine the location of "moonquakes" by triangulation and map the Moon's inner structure, obtaining information critical to our understanding of the Moon's evolution.

By any standards, the *Apollo* program was a spectacular success. It represents a towering achievement of the human race. The project's goals were met on schedule and within budget, and our knowledge of the Moon, Earth, and the solar system increased enormously. But the "Age of *Apollo*" was short lived. Public interest quickly waned. Over half a billion people breathlessly watched on television as Neil Armstrong set foot on the Moon, yet barely three years later, when the program was abruptly canceled for largely political (rather than scientific, technological, or economic) reasons, the landings had become so routine that they no longer excited the interest of the American public. Unmanned space science moved away from the Moon and toward the other planets, and the manned space program foundered. Perhaps one of the most amazing—and saddest—aspects of the *Apollo* program is that only now, some three decades later, is the U.S. (and perhaps China) gearing up for new crewed missions to the Moon in the coming decade. Yet it is not clear if either country has the desire or money to repeat the feat.

In 1994, the small U.S. military satellite *Clementine* was placed in lunar orbit, to perform a detailed survey of the lunar surface. In January 1998, NASA returned to the Moon for the first time in a quarter century with the launch of *Lunar Prospector*, another small satellite on a one-year mission to study the Moon's structure and origins. As discussed in more detail in *Discovery 8-2*, both missions were successful and have amply demonstrated the wealth of information that can be obtained by low-budget spacecraft. Following the spectacular end of the *Lunar Prospector* mission in 1999 (see Section 8.5), there are currently no active probes in lunar orbit.

Plans do exist to establish permanent human colonies on the Moon, either for commercial ventures, such as mining, or for scientific research. Proposals have also been made to site large optical, radio, and other telescopes on the lunar surface. Such instruments, which could be constructed larger than Earth-based devices, would benefit from perfect seeing and no light pollution. None of these projects is scheduled to become reality in the near future, although the possible discovery of water on the lunar surface (Section 8.5) alleviates at least one major logistical problem associated with such an undertaking. After a brief encounter with humankind, the Moon is once again a lifeless, unchanging world.

(NASA)

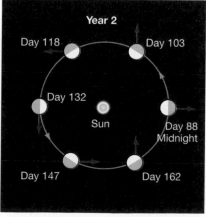
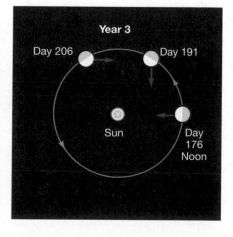

MORE PRECISELY 8-1

Why Air Sticks Around

Why do some planets and moons have atmospheres, while others do not, and what determines the composition of the atmosphere if one exists? Why does a layer of air, made up mostly of nitrogen and oxygen, lie just above Earth's surface? After all, experience shows that most gas naturally expands to fill all the volume available. Perfume in a room, fumes from a poorly running engine, and steam from a teakettle all disperse rapidly until we can hardly sense them. Why doesn't our planet's atmosphere similarly disperse by floating away into space?

The answer is that *gravity* holds it down. Earth's gravitational field exerts a pull on all the atoms and molecules in our atmosphere, preventing them from escaping. However, gravity is not the only influence acting, for if it were, all of Earth's air would have fallen to the surface long ago. *Heat* competes with gravity to keep the atmosphere buoyant. Let's explore this competition between gravity and heat in a little more detail.

All gas molecules are in constant random motion. The temperature of any gas is a direct measure of this motion: The hotter the gas, the faster the molecules are moving. ∞ *(More Precisely 3-1)* The Sun continuously supplies heat to our planet's atmosphere, and the resulting rapid movement of heated molecules produces *pressure*, which tends to oppose the force of gravity, preventing our atmosphere from collapsing under its own weight.

An important measure of the strength of a body's gravity is the body's *escape speed*—the speed needed for any object to escape forever from its surface. ∞ (Sec. 2.7) This speed increases with increased mass or decreased radius of the parent body (often a moon or a planet). In convenient (Earth) units, it can be expressed as

escape speed (in km/s)

$$= 11.2 \sqrt{\frac{\text{mass of body (in Earth masses)}}{\text{radius of body (in Earth radii)}}}.$$

(See the Earth Data box on p. 172.) Thus, Earth's escape speed is $11.2\sqrt{1/1} = 11.2$ km/s. If the *mass* of the parent body is quadrupled, the escape speed doubles. If the parent body's *radius* quadruples, then the escape speed is halved. In other words, you need high speed to escape the gravitational attraction of a very massive or very small body, but you can escape from a less massive or larger body at lower speeds.

To determine whether a planet will retain an atmosphere, we must compare the planet's escape speed with the *molecular speed*, which is the average speed of the gas particles making up the planet's atmosphere. This speed actually depends not only on the temperature of the gas, but also on the mass of the individual molecules—the hotter the gas or the smaller the molecular mass, the higher is the average speed of the molecules:

average molecular speed (in km/s)

$$= 0.157 \sqrt{\frac{\text{gas temperature (K)}}{\text{molecular mass (hydrogen atom masses)}}}.$$

Thus, increasing the absolute temperature of a sample of gas by a factor of four—for example, from 100 K to 400 K—doubles the average speed of its constituent molecules, and, at a given temperature, molecules of hydrogen (H_2: molecular mass = 2) in air move, on average, four times faster than molecules of oxygen (O_2: molecular mass = 32), which are 16 times heavier.

EXAMPLE 1: For nitrogen (N_2: molecular mass = 28) and oxygen (O_2: molecular mass = 32) in Earth's atmosphere, where the temperature near the surface is nearly 300 K, the preceding formula yields the following average molecular speeds:

nitrogen: speed $= 0.157$ km/s $\times \sqrt{\dfrac{300}{28}} = 0.51$ km/s;

oxygen: speed $= 0.157$ km/s $\times \sqrt{\dfrac{300}{32}} = 0.48$ km/s.

These speeds are far smaller than the 11.2 km/s needed for a molecule to escape into space. As a result, Earth is able to

Explanation of Mercury's Rotation

Mercury's 3:2 spin–orbit resonance did not occur by chance. What mechanism establishes and maintains it? In the case of the Moon orbiting Earth, the 1:1 resonance is explained as the result of tidal forces. In essence, the lunar rotation period, which probably started off much shorter than its present value, has lengthened so that the tidal bulge created by Earth is fixed relative to the body of the Moon. Tidal forces (this time due to the Sun) are also responsible for Mercury's 3:2 resonance, but in a much more subtle way.

Mercury cannot settle into a 1:1 resonance because its orbit around the Sun is quite eccentric. By Kepler's second law, Mercury's orbital speed is greatest at perihelion (clos-est approach to the Sun) and least at aphelion (greatest distance from the Sun). ∞ *(More Precisely 2-1)* A moment's thought shows that, because of these variations in the planet's orbital speed, there is no way that the planet (rotating at a constant rate) can remain in a synchronous orbit. If its rotation were synchronous near perihelion, it would be too rapid at aphelion, while synchronism at aphelion would produce too slow a rotation at perihelion.

Tidal forces always act so as to synchronize the rotation rate with the instantaneous orbital speed, but such synchronization cannot be maintained over Mercury's entire orbit. What happens? The answer is found when we realize that tidal effects diminish very rapidly with increasing distance. The tidal forces acting on Mercury at perihe-

retain its nitrogen–oxygen atmosphere. On the whole, our planet's gravity simply has more influence than the heat of our atmosphere.

In reality, the situation is a little more complicated than a simple comparison of speeds. Atmospheric molecules can gain or lose speed by bumping into one another or by colliding with objects near the ground. Thus, although we can characterize a gas by its average molecular speed, the molecules do not *all* move at the same speed, as illustrated in the accompanying figure. A tiny fraction of the molecules in any gas have speeds much greater than average—one molecule in two million has a speed more than three times the average, and one in 10^{16} exceeds the average by more than a factor of five. This means that at any instant, *some* molecules are moving fast enough to escape, even when the average molecular speed is much less than the escape speed. The result is that all planetary atmospheres slowly leak away into space.

Don't be alarmed—the leakage is usually very gradual! As a rule of thumb, if the escape speed from a planet exceeds the

average speed of a given type of molecule by a factor of six or more, then molecules of that type will not have escaped from the planet's atmosphere in significant quantities in the 4.6 billion years since the solar system formed. Conversely, if the escape speed is less than six times the average speed of molecules of a given type, then most of them will have escaped by now, and we should not expect to find them in the atmosphere.

For air on Earth, the mean molecular speeds of oxygen and nitrogen that we just computed are comfortably below one-sixth of the escape speed. However, if the Moon originally had an Earthlike atmosphere, that lunar atmosphere would have been heated by the Sun to much the same temperature as Earth's air today, so the average molecular speed would have been about 0.5 km/s. Because the Moon's escape speed is only $11.2\sqrt{0.012/0.27} = 2.4$ km/s —less than six times the average molecular speed—any original lunar atmosphere long ago dispersed into interplanetary space. Mercury's escape speed is $11.2\sqrt{0.055/0.38} = 4.2$ km/s. However, its peak surface temperature is around 700 K, corresponding to an average molecular speed for nitrogen or oxygen of about 0.8 km/s, more than one-sixth of the escape speed, so there has been ample time for those gases to escape.

Average molecular speed

99% of all molecules

$\frac{1}{2}$%

$\frac{1}{2}$%

Number (arbitrary units)

Molecular speed

EXAMPLE 2: We can use the foregoing arguments to understand some aspects of atmospheric *composition*. Hydrogen molecules (H_2: molecular mass = 2) move, on average, at about 1.9 km/s in Earth's atmosphere at sea level, so they have had time to escape since our planet formed (6×1.9 km/s = 11.4 km/s, which is greater than Earth's 11.2-km/s escape speed). Consequently, we find very little hydrogen in Earth's atmosphere today. However, on the planet Jupiter, with a lower temperature (about 100 K), the speed of hydrogen molecules is correspondingly slower—about 1.1 km/s. At the same time, Jupiter's escape speed is 60 km/s, over five times higher than Earth's. For those reasons, Jupiter has retained its hydrogen—in fact, hydrogen is the dominant ingredient of Jupiter's atmosphere.

lion are much greater than those at aphelion, so perihelion "won" the struggle to determine the rotation rate. In the 3:2 resonance, Mercury's orbital and rotational motion are almost exactly synchronous *at perihelion*, so that particular rotation rate was naturally "picked out" by the Sun's tidal influence on the planet. Notice that even though Mercury rotates through only 180° between one perihelion and the next (see Figure 8.13), the appearance of the tidal bulge is the *same* each time around.

The motion of Mercury is one of the simplest nonsynchronous resonances known in the solar system. Astronomers now know that these intricate dynamic interactions are responsible for much of the fine detail observed in the motion of the solar system. Examples of res-

onances can be found in the orbits of many of the planets, their moons, and their rings, as well as in the asteroid belt.

The Sun's tidal influence also causes Mercury's rotation axis to be exactly perpendicular to its orbital plane. As a result, and because of Mercury's eccentric orbit and the spin–orbit resonance, some points on the surface get much hotter than others. In particular, the two (diametrically opposite) points on the equator where the Sun is directly overhead at perihelion get hottest of all. They are called the *hot longitudes*. The peak temperature of 700 K mentioned earlier occurs at noon at those two locations. At the *warm longitudes*, where the Sun is directly overhead at aphelion, the peak temperature is about 150 K cooler—a mere 550 K.

By contrast, the Sun is always on the horizon as seen from the planet's poles, so temperatures there never reach the sizzling levels of the equatorial regions. Earth-based radar studies carried out during the 1990s suggest that Mercury's polar temperatures may be as low as 125 K and that, despite the planet's scorched equator, the poles may be covered with extensive sheets of water ice. (See Section 8.5 for similar findings regarding the Moon.)

CONCEPT CHECK

✔ How has gravity influenced the rotation rates of the Moon and Mercury?

8.5 Lunar Cratering and Surface Composition

On Earth, the combined actions of wind and water erode our planet's surface and reshape its appearance almost daily. Coupled with the never-ending motion of Earth's surface plates, the result is that most of the ancient history of our planet's surface is lost to us. The Moon, in contrast, has no air, no water, no plate tectonics, and no ongoing volcanic or seismic activity. Consequently, features dating back almost to its formation are still visible today.

Meteoritic Impacts

The primary agent of change on the lunar surface is interplanetary debris, in the form of *meteoroids*. This material, much of it rocky or metallic in composition, is strewn throughout the solar system, orbiting the Sun in interplanetary space, perhaps for billions of years, until it happens to collide with some planet or moon. ∞ (Sec. 6.5) On Earth, most meteoroids burn up in the atmosphere, producing the streaks of light known as *meteors*, or "shooting stars." But the Moon, without an atmosphere, has no protection against this onslaught. Large and small meteoroids zoom in and collide with the surface, sometimes producing huge craters. Over billions of years, these collisions have scarred, cratered, and sculpted the lunar landscape. Craters are still being formed today—even as you read this—all across the surface of the Moon.

▶ **FIGURE 8.14 Meteoroid Impact** Several stages in the formation of a crater by meteoritic impact. (a) A meteoroid strikes the surface, releasing a large amount of energy. (b, c) The resulting explosion ejects material from the site of the impact and sends shock waves through the underlying surface. (d) Eventually, a characteristic crater surrounded by a blanket of ejected material results.

PHYSLET® ILLUSTRATION Meteor Energy

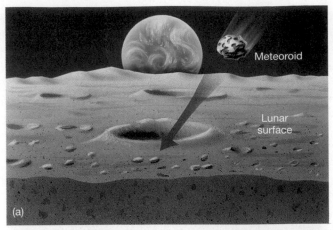

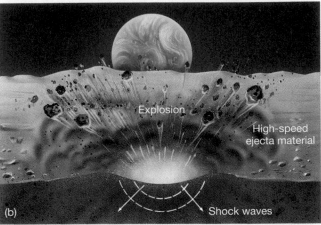

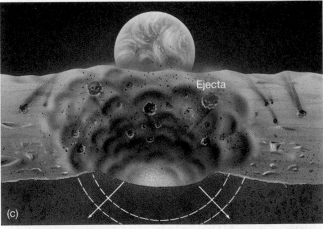

Meteoroids generally strike the Moon at speeds of several kilometers per second. At these speeds, even a small piece of matter carries an enormous amount of energy. For example, a 1-kg object hitting the Moon's surface at 10 km/s releases as much energy as the detonation of 10 kg of TNT! As illustrated in Figure 8.14, the impact of a meteoroid with the surface causes sudden and tremendous pressures to build up, heating the normally brittle rock and deforming the ground like heated plastic. The ensuing explosion pushes previously flat layers of rock up and out, forming a crater.

The diameter of the eventual crater is typically 10 times that of the incoming meteoroid; the depth of the crater is about twice the meteoroid's diameter. Thus, our 1-kg meteoroid, measuring perhaps 10 cm across, would produce a crater about 1 m in diameter and 20 cm deep. Shock waves from the impact pulverize the lunar surface to a depth many times that of the crater itself. Numerous rock samples brought back by the *Apollo* astronauts show patterns of repeated shattering and melting—direct evi-

dence of the violent shock waves and high temperatures produced in meteoritic impacts. The material thrown out by the explosion surrounds the crater in a layer called an *ejecta blanket*. The ejected debris ranges in size from fine dust to large boulders. Figure 8.15(a) shows the result of one particularly large meteoritic impact on the Moon. As shown in Figure 8.15(b), the larger pieces of ejecta may themselves form secondary craters.

In addition to the bombardment by meteoroids with masses of a gram or more, a steady "rain" of *micrometeoroids* (debris with masses ranging from a few micrograms up to about 1 gram) also eats away at the structure of the lunar surface. Some examples can be seen in Figure 8.16, a photomicrograph (a photograph taken through a microscope) of some glassy "beads" brought back to Earth by *Apollo* astronauts. The beads themselves were formed during the explosion following the impact of a meteoroid, when surface rock was melted, ejected, and rapidly cooled. Note how several of the beads also display fresh miniature

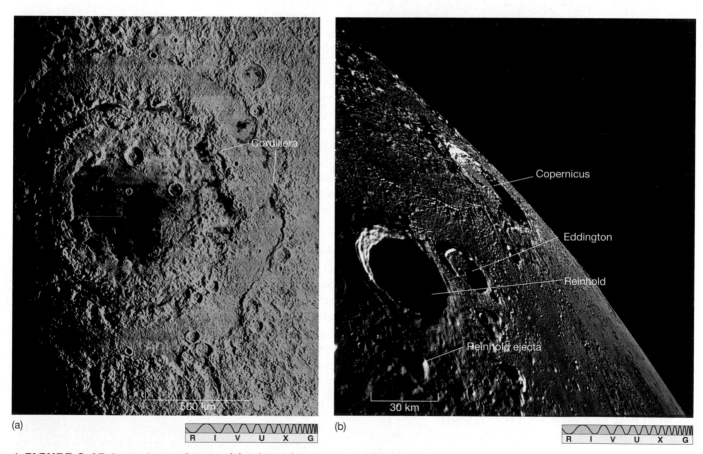

(a)   R I V U X G   (b)   R I V U X G

▲ **FIGURE 8.15 Large Lunar Craters** (a) A large lunar crater, called the Orientale Basin. The impact that produced this crater thrust up much surrounding matter, which can be seen as concentric rings of cliffs called the Cordillera Mountains. The outermost ring is nearly 1000 km in diameter. (b) Two smaller craters called Reinhold and Eddington sit amid the secondary cratering resulting from the impact that created the 90-km-wide Copernicus crater (near the horizon) about a billion years ago. The ejecta blanket from crater Reinhold, 40 km across and in the foreground, can be seen clearly. The view is obtained by looking northeast from the lunar module during the *Apollo 12* mission. *(NASA)*

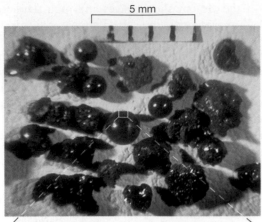

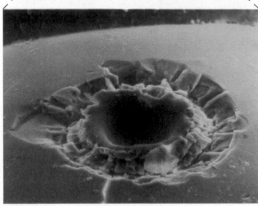

▲ **FIGURE 8.16 Microcraters** Craters of all sizes litter the lunar landscape. Some shown here, embedded in glassy beads retrieved by *Apollo* astronauts, measure only 0.01 mm across. (The scale at the top is in millimeters.) *(NASA)*

craters, caused by micrometeoroids that struck the beads after they had cooled and solidified.

In fact, the *rate* of cratering decreases rapidly with the size of the crater—fresh large craters are scarce, but small craters are common. The reason for this is simple: There just aren't very many large chunks of debris in interplanetary space, so their collisions with the Moon are rare. At present average rates, one new 10-km (diameter) lunar crater is formed roughly every 10 million years, a new meter-sized crater is created about once a month, and centimeter-sized craters are formed every few minutes.

Cratering History of the Moon

Astronomers can use the known ages (from radioactive dating) of Moon rocks to estimate the rate of cratering in the past. One very important result of this work is the discovery that the Moon was subjected to an extended period of intense meteoritic bombardment roughly 4 billion years ago. Indeed, this is a key piece of evidence supporting the condensation theory of solar-system formation. ∞ (Sec. 6.7)

As we have seen, the heavily cratered highlands are older than the less cratered maria, but the difference in cratering is not simply a matter of exposure time. Astronomers

now think that the Moon, and presumably the entire inner solar system, experienced a sudden drop in meteoritic bombardment about 3.9 billion years ago. The highlands solidified and received most of their craters before that time, whereas the maria solidified afterward. The rate of cratering has been declining slowly ever since.*

The great basins that comprise the maria are thought to have been created during the final stages of the heavy bombardment, between about 4.1 and 3.9 billion years ago. Subsequent volcanic activity filled the craters with lava, ultimately creating the formations we see today as the lava turned into solid rock. In a sense, then, the maria *are* oceans—ancient seas of molten lava, now solidified.

Not all these great craters became flooded with lava, however. One of the youngest craters is the Orientale Basin (Figure 8.15a), which formed about 3.9 billion years ago. This crater did not undergo much subsequent volcanism, and we can recognize its structure as an impact crater rather than as another mare. Similar "unflooded" basins are seen on the lunar far side (Figure 8.7).

Apart from meteorites found on Earth, the Moon is the only solar-system object for which we have accurate age measurements. However, studies of lunar cratering provide astronomers with an important alternative means of estimating ages in the solar system. By counting craters on a planet, moon, or asteroid and using the Moon to calibrate the numbers, an approximate age for the surface can be obtained. In fact, this is how most of the ages presented in the next few chapters are determined. Note that, as with radioactive dating, the technique measures only the time since the surface in question last solidified—all cratering is erased and the clock is reset if the rock melts. ∞ (*More Precisely 7-2*)

Lunar Dust

Meteoroid collisions with the Moon are the main cause of the layer of pulverized ejecta—also called lunar dust, or *regolith* (meaning "fine rocky layer")—that covers the lunar landscape to an average depth of about 20 m. This microscopic dust has a typical particle size of about 0.01 mm. In consistency, it is rather like talcum powder or ready-mix dry mortar. Figure 8.17 shows an *Apollo* astronaut's boot print in the regolith, which is thinnest on the maria (10 m) and thickest on the highlands (over 100 m deep in places).

The constant barrage from space results in a slow, but steady, erosion of the lunar surface. The soft edges of the craters visible in the foreground of Figure 8.18 are the result of this process. In the absence of erosion, those features would still be as jagged and angular today as they

*Recent detailed studies of lunar rock samples returned by the Apollo missions now suggest that the slow decline may have ended about 400–500 million years ago, when the cratering rate increased by about a factor of 4, back to the levels of about 3 billion years earlier. The cause of the increase—if it is real—is unknown, but the fact that it can be inferred at all illustrates the importance of the Apollo samples.

▲ **FIGURE 8.17 Regolith** Photograph of an *Apollo* astronaut's boot print in the lunar dust. The astronaut's weight has compacted the regolith to a depth of a few centimeters. *(NASA)*

R I V U X G

▲ **FIGURE 8.18 Lunar Surface** The lunar surface is not entirely changeless. Despite the complete lack of wind and water on the airless Moon, the surface has still eroded a little under the constant "rain" of meteoroids, especially micrometeoroids. Note the soft edges of the craters visible in the foreground of this image. In the absence of erosion, these features would be as jagged and angular today as they were when they formed. (The twin tracks were made by the *Apollo* lunar rover.) *(NASA)*

were just after they formed. Instead, the steady buildup of dust due to innumerable impacts has smoothed their outlines and will probably erase them completely in about 100 million years.

From the known dependence of the cratering rate on the size of a crater, planetary scientists can calculate how many small craters they would expect to find, given the numbers of large craters actually observed. When they make this calculation, they find a shortage of craters less than about 20 m deep. These "missing" craters have been filled in by erosion over the lifetime of the Moon. This gives us a very rough estimate of the average erosion rate: about 5 m per billion years, or roughly 1/10,000 the rate on Earth.

The current lunar erosion rate is very low because meteoritic bombardment on the Moon is a much less effective erosive agent than are wind and water on Earth. For comparison, the Barringer Meteor Crater (Figure 8.19) in the Arizona desert, one of the largest meteoroid craters on Earth, is only 50,000 years old, but has already undergone noticeable erosion. It will probably disappear completely in a mere million years, quite a short time geologically. If a crater that size had formed on the Moon even 4 billion years ago, it would still be plainly visible today. Even the shallow boot print shown in Figure 8.17 is likely to remain intact for several million years.

Lunar Ice?

In contrast to Earth's soil, the lunar regolith contains no organic matter like that produced by biological organisms. No life whatsoever exists on the Moon. Nor were any fossils found in *Apollo* samples. Lunar rocks are barren of life and apparently always have been. NASA was so confident of this fact that the astronauts were not even quarantined on their return from the last few *Apollo* landings. Furthermore, all the lunar samples returned by the U.S. and Soviet Moon programs were bone dry—they didn't even contain minerals having water molecules locked within

their crystal structure. Terrestrial rocks, by contrast, are almost always one or two percent water. The main reason for this lack of water is the high (up to 400 K) daytime temperatures found over most of the lunar surface.

Some regions of the Moon *are* thought to contain water, however—in the form of ice. As early as the 1960s, some scientists had considered the theoretical possibility that ice might be found near the lunar poles. Since the Sun never rises more than a few degrees above the horizon, as seen from the Moon's polar regions, temperatures on the permanently shaded floors of craters near the poles never exceed about 100 K. Consequently, those scientists theorized, any water ice there could have remained permanently frozen since the very early days of the solar system, never melting or vaporizing and hence never escaping into space.

In November 1996, mission controllers of the *Clementine* spacecraft (see *Discovery 8-2*) reported that radar echoes captured by *Clementine* from an old, deep crater near the lunar south pole suggested deposits of low-density material, probably water ice, at a depth of a

◄ **FIGURE 8.19 Barringer Crater** The Barringer Meteor Crater, near Winslow, Arizona, is 1.2 km in diameter and 0.2 km deep. Note the access road leading to the crater at the right. Geologists think that a large meteoroid formed this crater about 50,000 years ago. The meteoroid probably was about 50 m across and weighed around 200,000 tons. (*U.S. Geological Survey*)

few meters. In early March 1998, NASA announced that sensitive equipment on board the *Lunar Prospector* mission had confirmed *Clementine*'s findings and in fact had detected large amounts of ice—possibly totaling trillions of tons—at both lunar poles. At first, it appeared that the ice was mainly in the form of tiny crystals mixed with the lunar regolith, spread over many tens of thousands of square kilometers of deeply shadowed crater floors. However, subsequent analysis of the data suggests that much of the ice may exist in the form of smaller, but more concentrated, "lakes" of nearly pure material lying perhaps half a meter below the surface.

The *Lunar Prospector* discovery of lunar ice was indirect; the instruments on board the spacecraft actually detected the presence of hydrogen (H), whose existence was taken as evidence of water (H_2O). In an attempt to gain more direct information about lunar ice, NASA scientists decided to end the *Lunar Prospector* mission in a very spectacular way. As the spacecraft neared the end of its lifetime, it was directed to crash into one of the deep craters in which the ice was suspected to hide. Figure 8.20 shows the intended site of the impact and the trajectory to be taken by the satellite as it approached the surface. The hope was that the *Hubble Space Telescope* and ground-based telescopes on Earth might detect spectroscopic signatures of water vapor released by the impact. No water vapor was seen, although mission planners had stressed in advance that there were so many uncertainties involved in the effort that the probability of success was low—less than 10 percent. Thus, no conclusion can be drawn from the fact that water was not directly observed. Lunar ice remains a strong possibility, but its existence has not yet been definitively proven.

Assuming that it does exist, where did all this ice come from? Most likely, it was brought to the lunar surface by meteoroids and comets. (We will see in Chapter 15 that this is the likely origin of Earth's water, too.) Any ice that survived the impact would have been scattered across the surface. Over most of the Moon, that ice would have rapidly vaporized and escaped, but in the deep basins near the poles, it survived and built up over time. Whatever its origin, the polar ice may be a crucial component of any serious attempt at human colonization of the Moon: The anticipated cost of transporting a kilogram of water from Earth to the Moon is between $2,000 and $20,000, prompting one *Clementine* scientist to describe the lunar ice deposits as "possibly the most valuable piece of real estate in the solar system."

Lunar Volcanism

Only a few decades ago, debate raged in scientific circles about the origin of lunar craters, with most scientists of the opinion that the craters were the result of volcanic activity. We now know that almost all lunar craters are actually meteoritic in origin. However, a few apparently are not. Figure 8.21 shows an intriguing alignment of several craters in a *crater-chain* pattern so straight that it is highly unlikely to have been produced by the random collision of meteoroids with the surface. Instead, the chain probably marks the location of a subsurface fault—a place where cracking or shearing of the surface once allowed molten matter to well up from below. As the lava cooled, it formed a solid "dome" above each fissure. Subsequently, the underlying lava receded and the centers of the domes collapsed, forming the craters we see today. Similar features have been observed on Venus by the orbiting *Magellan* probe (see Chapter 9).

Many other examples of lunar volcanism are known, both in telescopic observations from Earth and in the close-up photographs taken during the *Apollo* missions. Figure 8.22 shows a volcanic **rille,** a ditch where molten lava once flowed. There is good evidence for surface volcanism early in the Moon's history, and volcanism explains the presence of the lava that formed the maria. However, whatever volcanic activity once existed on the Moon ended long ago. The measured ages for rock samples returned from the Moon are all greater than 3 billion years. (Recall from *More Precisely 7-2* that the radioactivity clock starts "ticking" when the rock solidifies.) Apparently, the maria solidified over 3 billion years ago, and the Moon has been dormant ever since.

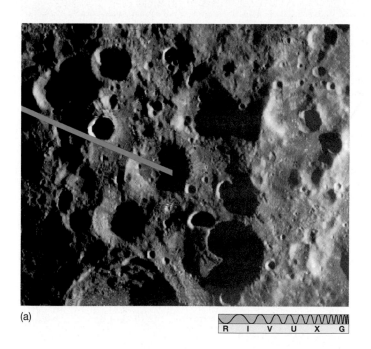

(a)

R I V U X G

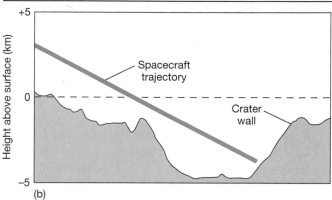

◀ **FIGURE 8.20** *Prospector* **Impact** (a) The intended impact site for the *Lunar Prospector* spacecraft was a deep crater close to the Moon's south pole. The purpose of the impact was to release water vapor for spectroscopic study by telescopes on or near Earth. (b) The trajectory of the spacecraft was designed to have the craft hit near the crater floor. (*Dept. of Defense*)

(b)

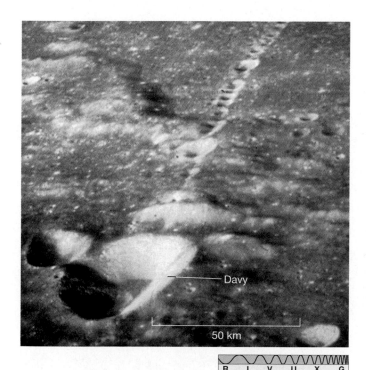

Davy

50 km

R I V U X G

▲ **FIGURE 8.21 Crater Chain** This "chain" of well-ordered craters was photographed by an *Apollo 14* astronaut. The largest crater, called Davy, is located on the western edge of Mare Nubium. The entire field of view measures about 100 km across. (*NASA*)

Autolycus

Apennine

Hadley rille

Mare
Imbrium

Mount
Hadley

50 km

R I V U X G

▲ **FIGURE 8.22 Lunar Volcanism** A volcanic rille, photographed from the *Apollo 15* spacecraft orbiting the Moon, can be seen clearly here (bottom and center) winding its way through one of the maria. Called Hadley Rille, this system of valleys runs along the base of the Apennine Mountains (lower right) at the edge of the Mare Imbrium (to the left). Autolycus, the large crater closest to the center, spans 40 km. The shadow-sided, most prominent peak at the lower right, Mount Hadley, rises almost 5 km high. (*NASA*)

The Moon on a Shoestring

Since the early 1990s, the watchword for unmanned exploration of space has been "smaller, faster, cheaper." Unlike previous generations of space missions, the emphasis now is on creating small-scale, lightweight systems that can be designed and built rapidly and cheaply, affording mission planners much greater flexibility, both in designing follow-up missions and in quickly changing mission parameters as circumstances warrant. Two lunar exploration satellites—*Clementine* and *Lunar Prospector*—have demonstrated the power of this approach. Both have returned high-quality science for total costs of about $70 million each, a small fraction of the cost of most other planetary spacecraft, which normally carry price tags of hundreds of millions, and sometimes even billions, of dollars.

The *Clementine* satellite was sent to the Moon in 1994 by the U.S. Defense Department, largely to test some new sensing devices developed for the ballistic missile defense program. This was the first lunar mission by any nation since the crew of *Apollo 17* left the Moon in 1972. *Clementine* originally was a code word for a military-classified space project known as the Deep-Space Program Science Experiment. The vehicle and its onboard suite of instruments were designed to test the feasibility of miniaturizing a complex spacecraft, its engineering subsystems, and its sophisticated sensors for use in deep space. The spacecraft's total mass was less than 150 kg. *Clementine*'s technical design was a product of the Strategic Defense Initiative—the Star Wars program—now known as the Ballistic Missile Defense Organization. Its target—the Moon—was of no interest to the military, other than being a convenient, known object in the cosmic neighborhood.

The lunar portion of the mission was a spectacular success, although a follow-up mission to map an asteroid had to be canceled after a computer malfunction caused an onboard thruster to fire until it had used up all its fuel, leaving the spacecraft spinning out of control. *Clementine* made the first digital global map of the Moon, at very high resolution. In two months of operation, its sensors took over 2.5 million images with a clarity at least 10 times better than NASA's most sophisticated planetary camera in the 1990s—the one aboard the *Galileo* mission to Jupiter. *Clementine* was able to obtain global coverage of the Moon at visible, ultraviolet, and infrared wavelengths. In addition, the small craft carried lidar devices (the visible-light equivalent of radar), able to pulse the lunar surface and listen for echoes. In all, more than 50 advanced lightweight technologies were demonstrated on this powerful dwarf spacecraft.

The accompanying figure is a mosaic of about 1500 images centered on the Moon's south pole. The bottom half is part of the near side of the Moon, as seen from Earth; the top half is the far side that we never see from home. For scale, the double-ringed crater at the upper left, called Schroedinger, has an outer diameter of 320 km. (Note that the circular image is rather misleading—only a portion of the Moon is shown here.) The dark region at the pole is the center of an old, permanently shadowed depression called Aitken Basin. Its rim (outlined with a red dashed line) spans some 2000 km; the basin itself averages 10 km deep. This huge depression is the largest impact basin known in the solar system.

NASA's *Lunar Prospector* was launched in January 1998. Designed and built in just 22 months and weighing 295 kg fully fueled, the spacecraft orbited the Moon for 18 months, probing the lunar surface and interior with an array of onboard instruments: A *gamma-ray spectrometer* mapped the abundances of certain elements on the Moon's surface; two sensitive *magnetometers* probed the Moon's extremely weak magnetic field; a *neutron spectrometer* searched for water ice by detecting the element hydrogen on the Moon's surface; an *alpha-particle spectrometer* searched for particles emitted by radioactive gases leaking out of the lunar interior; and a *Doppler gravity experiment* made detailed measurements of the Moon's gravitational field by carefully monitoring small fluctuations in the craft's orbital velocity. *Lunar Prospector* did not carry a camera. The spacecraft's instrument package was designed in part to complement the imaging and radar capabilities of *Clementine*.

By far the most publicized aspect of the *Lunar Prospector* mission was its search for water ice at the lunar poles. Radar observations made by *Clementine* of the deep depression at the Moon's south pole (at the center of the image) had suggested deposits of water ice at a depth of a few meters. *Lunar Prospector*'s neutron spectrometer confirmed the presence of large amounts of hydrogen (and, presumably, therefore, water) there, although, as discussed in more detail in the text, a follow-up attempt to detect water directly was unsuccessful.

The data returned by *Clementine* and *Lunar Prospector* have allowed scientists to construct much more accurate models of the Moon's interior and to probe its past history with greater precision. But beyond the scientific results, important as they are, these spacecraft may have done something even more far reaching: By demonstrating that major scientific findings can come from low-cost, fast-turnaround missions, they may have changed forever the way planetary scientists explore the solar system.

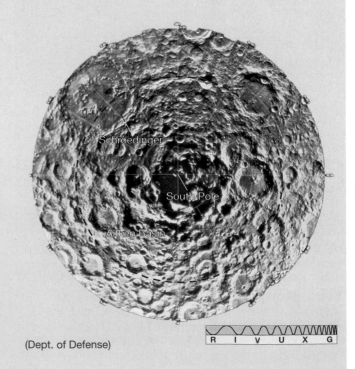

(Dept. of Defense)

R I V U X G

▲ **FIGURE 8.23 Mercury's Surface** Discovery Scarp on Mercury's surface (arrow), as photographed by *Mariner 10*. This scarp appears to be a compressional feature that formed when the planet's crust cooled and contracted early in its history, causing a crease in the surface. Running diagonally across the center of the frame, the scarp is several hundred kilometers long and up to 3 km high in places. *(NASA)*

off

▲ **FIGURE 8.24 Mercury's Basin** Mercury's most prominent geological feature—the Caloris Basin—measures about 1400 km across and is ringed by concentric mountain ranges (marked by dashed lines) that reach more than 3 km high in places. This huge circular basin, only half of which can be seen (at left) in this *Mariner 10* photo, is similar in size to the Moon's Mare Imbrium and spans more than half of Mercury's radius. *(NASA)*

CONCEPT CHECK

✔ How has meteoritic bombardment affected the surface of the Moon?

8.6 The Surface of Mercury

Like craters on the Moon, almost all craters on Mercury are the result of meteoritic bombardment. However, Mercury's craters are less densely packed than their lunar counterparts, and there are extensive **intercrater plains.** The crater walls are generally not as high as those on the Moon, and the ejected material appears to have landed closer to the impact site exactly as we would expect on the basis of Mercury's stronger surface gravity.

One likely explanation for Mercury's relative lack of craters is that the older craters were filled in by volcanic activity, in much the same way as the Moon's maria filled in older craters as they formed. However, the intercrater plains do not look much like maria—they are much lighter in color and not as flat. Still, most geologists think that volcanism did occur in Mercury's past, obscuring the old craters. The details of how Mercury's landscape came to look the way it does remain unexplained. The apparent absence of rilles or other obvious features associated with very large scale lava flows, along with the light color of the lava-flooded regions, suggests that Mercury's volcanic past was different from the Moon's.

Mercury has at least one type of surface feature not found on the Moon. Figure 8.23 shows a **scarp,** or cliff, on the surface that does not appear to be the result of volcanic or any other familiar geological activity. The scarp cuts across several craters, which indicates that whatever produced it occurred *after* most of the meteoritic bombardment was over. Mercury shows no evidence of crustal motions like plate tectonics on Earth—no fault lines, spreading sites, or indications of plate collisions are seen. ∞ (Sec. 7.4) The scarps, of which several are known from the *Mariner* images, probably formed when the planet's interior cooled and shrank long ago, much as wrinkles form on the skin of an old, shrunken apple. On the basis of the amount of cratering observed in the surrounding terrain (as discussed in the previous section), astronomers estimate that the scarps probably formed about 4 billion years ago.

Figure 8.24 shows what may have been a result of the last great geological event in the history of Mercury: an immense bull's-eye crater called the Caloris Basin, formed eons ago by the impact of a large asteroid. (The basin is so called because it lies in Mercury's "hot longitudes"—see Section 8.3—close to the planet's equator; *calor* is the Latin word for "heat.") Because of the orientation of the planet during *Mariner 10*'s flybys, only half of the basin was visible. The center of the crater is off the left-hand side of the

▲ **FIGURE 8.25 Weird Terrain** The refocusing of seismic waves after the Caloris Basin impact may have created the weird terrain on the opposite side of the planet.

photograph. Compare this basin with the Orientale Basin on the Moon (Figure 8.15a). The impact-crater structures are quite similar, but even here there is a mystery: The patterns visible on the Caloris floor are unlike any seen on the Moon. Their origin, like the composition of the floor itself, is unknown.

So large was the impact that created the Caloris Basin that it apparently sent strong seismic waves reverberating throughout the entire planet. On the opposite side of Mercury from Caloris, there is a region of oddly rippled and wavy surface features, often referred to as *weird* (or *jumbled*) terrain. Scientists theorize that this terrain was produced when seismic waves from the Caloris impact traveled around the planet and converged on the diametrically opposite point, causing large-scale disruption of the surface there, as illustrated in Figure 8.25.

CONCEPT CHECK

✔ How do scarps on Mercury differ from geological faults on Earth?

8.7 Interiors

The Moon

The Moon's average density, about 3300 kg/m³, is similar to the measured density of lunar surface rock, virtually eliminating any chance that the Moon has a large, massive, and very dense nickel–iron core like that of Earth. In fact, the low density implies that the entire Moon is actually deficient in iron and other heavy metals compared with their abundance on our planet.

There is no evidence for any large-scale lunar magnetic field. *Lunar Prospector* detected some very weak surface magnetic fields—less than a thousandth of Earth's field—apparently associated with some large impact basins, but these are not thought to be related to condi-

PLANETARY DATA:
MERCURY

(NASA)

Orbital semimajor axis	0.39 A.U. 57.9 million km
Orbital eccentricity	0.206
Perihelion	0.31 A.U. 46 million km
Aphelion	0.47 A.U. 69.8 million km
Mean orbital speed	47.9 km/s
Sidereal orbital period	88.0 solar days 0.241 tropical years
Synodic orbital period*	115.9 solar days
Orbital inclination to the ecliptic	7.00°
Greatest angular diameter, as seen from Earth	13″
Mass	3.30×10^{23} kg 0.055 (Earth = 1)
Equatorial radius	2440 km 0.38 (Earth = 1)
Mean density	5430 kg/m³ 0.98 (Earth = 1)
Surface gravity	3.70 m/s² 0.38 (Earth = 1)
Escape speed	4.2 km/s
Sidereal rotation period	58.6 solar days
Axial tilt	0.0°
Surface magnetic field	0.011 (Earth = 1)
Magnetic axis tilt relative to rotation axis	<10°
Mean surface temperature	100–700 K
Number of moons	0

* The planet's apparent orbital period, taking Earth's own motion into account; specifically, the mean time between one closest approach to Earth and the next. (See ∞ More Precisely 9-1.)

tions in the lunar core. As we saw in Chapter 7, researchers think that planetary magnetism requires a rapidly rotating liquid metal core, like Earth's. ∞ (Sec. 7.5) Thus, the absence of a lunar magnetic field could be a consequence of the Moon's slow rotation, the absence of a liquid core, or both.

Data from the gravity experiment aboard *Lunar Prospector*, combined with measurements made by the probe's magnetometers as the Moon passed through Earth's magnetic "tail" (see Figure 7.21), imply that the Moon may have a small iron core perhaps 300 km in radius. Near the center, the temperature may be as low as 1500 K, too cool to melt rock. However, seismic data collected by sensitive equipment left on the surface by *Apollo* astronauts (see *Discovery 8-1*) suggest that the inner parts of the core may be at least partially molten, implying a somewhat higher temperature. Our knowledge of the Moon's deep interior is still quite limited.

Based on a combination of seismic data, gravitational and magnetic measurements, and a good deal of mathematical modeling resting on assumptions about the Moon's interior composition, Figure 8.26 presents a schematic diagram of the Moon's interior structure. The central core is surrounded by a roughly 400-km-thick inner mantle of semisolid rock having properties similar to Earth's asthenosphere. ∞ (Sec. 7.4) Above these regions lies an outer mantle of solid rock, some 900–950 km thick, topped by a 60–150-km crust (considerably thicker than that of Earth). Together, these layers constitute the Moon's lithosphere. Outside the core, the mantle seems to be of almost uniform density, although it is chemically differentiated (i.e., its chemical properties change from the deep interior to near the surface). The crust material, which forms the lunar highlands, is lighter than the mantle, which is similar in composition to the lunar maria.

The crust on the lunar far side is *thicker* than that on the side facing Earth. If we assume that lava takes the line of least resistance in getting to the surface, then we can readily understand why the far side of the Moon has no large maria: Volcanic activity did not occur on the far side simply because the crust was too thick to allow it to occur there. But *why* is the far-side crust thicker? The answer is probably related to Earth's gravitational pull. Just as heavier material tends to sink to the center of Earth, the denser lunar mantle tended to sink below the lighter crust in Earth's gravitational field. The effect of this tendency was that the crust and the mantle became slightly off center with respect to each other. The mantle was pulled a little closer to Earth, while the crust moved slightly away. Thus, the crust became thinner on the near side and thicker on the far side.

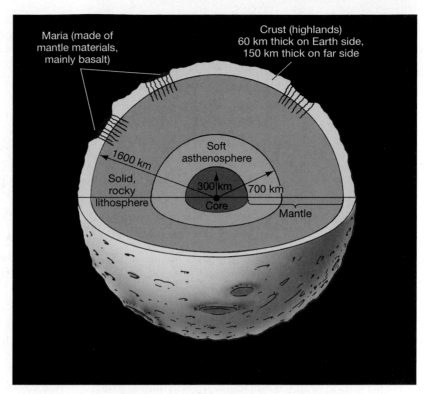

▲ **FIGURE 8.26 Lunar Interior**
Cross-sectional diagram of the Moon. Unlike Earth's rocky lithosphere, the Moon's is very thick—about 1000 km. Below the lithosphere is the inner mantle, or lunar asthenosphere, similar in properties to that of Earth. At the center lies the core, which may be partly molten.

Mercury

Mercury's magnetic field, discovered by *Mariner 10*, is about a hundredth that of Earth. Actually, the discovery that Mercury has any magnetic field at all came as a surprise to planetary scientists. Having detected no magnetic field in the Moon (and, in fact, none in Venus or Mars, either), they had expected Mercury to have no measurable magnetism. Certainly, Mercury does not rotate rapidly, and it may lack a liquid metal core, yet a magnetic field undeniably surrounds it. Although weak, the field is strong enough to deflect the solar wind and create a small magnetosphere around the planet.

Scientists have no clear understanding of the origin of Mercury's magnetic field. If it is produced by ongoing dynamo action, as in Earth, then Mercury's core must be at least partially molten. ∞ (Sec. 7.5) Yet the absence of any recent surface geological activity suggests that the outer layers are solid to a considerable depth, as on the Moon. It is difficult to reconcile these two considerations in a single theoretical model of Mercury's interior. If the field is being generated dynamically, Mercury's slow rotation may at least account for the field's weakness. Alternatively, Mercury's current weak magnetism may simply be the remnant of an extinct dynamo—the planet's iron core may have solidified long ago, but still bears a permanent magnetic

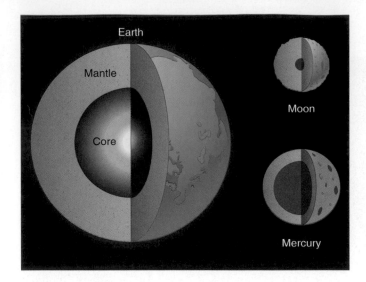

◄ **FIGURE 8.27** Terrestrial Interiors The internal structures of Earth, the Moon, and Mercury, drawn to the same scale. Note how large a fraction of Mercury's interior is the planet's core.

imprint of the past. The models are inconclusive on this issue, and no spacecraft is scheduled to revisit Mercury until at least 2009.

Mercury's magnetic field and large average density together imply that the planet is differentiated. Even without the luxury of seismographs on the surface, we can infer that most of its interior must be dominated by a large, heavy, iron-rich core with a radius of perhaps 1800 km. Whether that core is solid or liquid remains to be determined. Probably a less dense lunarlike mantle lies above this core, to a depth of about 500 to 600 km. Thus, about 40 percent of the volume of Mercury, or 60 percent of its mass, is contained in its iron core. The ratio of core volume to total planet volume is greater for Mercury than for any other object in the solar system. Figure 8.27 illustrates the relative sizes and internal structures of Earth, the Moon, and Mercury.

CONCEPT CHECK

✔ Why would we not expect strong magnetic fields on the Moon or Mercury?

8.8 The Origin of the Moon

6 Over the years, many theories have been advanced to account for the origin of the Moon. However, both the similarities *and* the differences between the Moon and Earth conspire to confound many promising attempts to explain the Moon's existence.

One theory (the *sister*, or *coformation*, theory) suggests that the Moon formed as a separate object near Earth in much the same way as our own planet formed—the "blob" of material that eventually coalesced into Earth gave rise to the Moon at about the same time. The two objects thus formed as a double-planet system, each revolving about a common center of mass. Although once favored by many astronomers, this idea suffers from a major flaw: The Moon differs in both density and composition from Earth, making it hard to understand how both could have originated from the same preplanetary material.

A second theory (the *capture* theory) maintains that the Moon formed far from Earth and was later captured by it. In this way, the density and composition of the two objects need not be similar, for the Moon presumably materialized in a quite different region of the early solar system. The objection to this theory is that the Moon's capture would be an extraordinarily difficult event; it might even be an impossible one. Why? Because the mass of our Moon is so large relative to that of Earth. It is not that our Moon is the largest natural satellite in the solar system, but it is unusually large compared with its parent planet. Mathematical modeling suggests that it is quite implausible that Earth and the Moon could have interacted in just the right way for the Moon to have been captured during a close encounter sometime in the past. Furthermore, although there are indeed significant differences in composition between our world and its companion, there are also many similarities—particularly between the mantles of the two bodies—that make it unlikely that they formed entirely independently of one another.

A third, older, theory (the *daughter*, or fission, theory) speculates that the Moon originated out of Earth itself. The Pacific Ocean basin has often been mentioned as the place from which protolunar matter may have been torn—the result, perhaps, of the rapid spin of a young, molten Earth. Indeed, there are some chemical similarities between the matter in the Moon's outer mantle and that in Earth's Pacific basin. However, this theory offers no solution to the fundamental mystery of how Earth could have been spinning so fast that it ejected an object as large as our Moon. Also, computer simulations indicate that the ejection of the Moon into a stable orbit simply would not have occurred. As a result, the daughter theory, in this form at least, is no longer taken seriously.

Today, many astronomers favor a hybrid of the capture and daughter themes. This idea—often called the *impact* theory—postulates a collision by a large, Mars-sized object with a youthful and molten Earth. Such collisions may have been quite frequent in the early solar system. ∞ (Sec. 6.7) The collision presumed by the impact theory would have been more a glancing blow than a direct impact. The matter dislodged from our planet then reassembled to form the Moon.

Computer simulations of such a catastrophic event show that most of the bits and pieces of splattered Earth could have coalesced into a stable orbit. Figure 8.28 shows some of the stages of one such calculation. If Earth had already formed an iron core by the time the collision occurred, then the Moon would indeed have ended up with a composition similar to that of Earth's mantle. During the collision, any iron core in the colliding object itself would have been left behind in Earth, eventually to become part of Earth's core. Thus, both the Moon's overall similarity to that of Earth's mantle and its lack of a dense central core are naturally explained.

Over the past two decades, planetary scientists have come to realize that collisions like this probably played important roles in the formation of all the terrestrial planets (see Chapter 15). Because of the randomness inherent in such events, as well as the Moon's unique status as the only large satellite in the inner solar system, it seems that the Moon may not provide a particularly useful model for studies of the other moons in the solar system. Instead, as we will see, a moon's properties depend greatly on the characteristics of its parent planet.

Nevertheless, the quest to understand the origin of the Moon highlights the interplay between theory and observation that characterizes modern science. ∞ (Sec. 1.2) Detailed data from generations of unmanned and manned lunar missions have allowed astronomers to discriminate between competing theories of the formation of the Moon, discarding some and modifying others. At the same time, the condensation theory of solar-system formation provides a natural context in which the currently favored impact theory can occur. ∞ (Sec. 6.7) Indeed, without the idea that planets formed by collisions of smaller bodies, such an impact might well have been viewed as so improbable that the theory would never have gained ground.

Finally, do not think that every last detail of the Moon's formation is understood or agreed upon by experts. That is far from the case. Some important aspects of the Moon's physical and chemical makeup are still inadequately explained—for example, the degree to which the Moon melted during its formation and whether current models are actually consistent with the observed lunar composition. The impact theory may well not be the last word on the subject. Still, past experience of the scientific method gives us confidence that the many twists and turns still to come will in the end lead us to a more complete understanding of our nearest neighbor in space.

CONCEPT CHECK

✔ How does the currently favored theory of the Moon's origin account for the Moon's lack of heavy materials compared to Earth's relative abundance, and for the similarity in composition between the lunar crust and that of Earth?

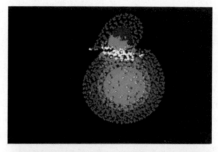
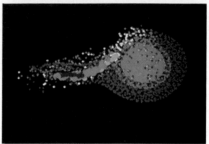

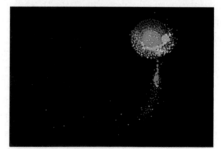
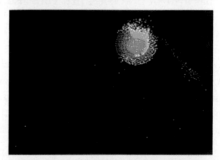

▶ **FIGURE 8.28 Lunar Formation** This sequence shows a simulated collision between Earth and an object the size of Mars. The sequence proceeds from top to bottom and zooms out dramatically. The arrow in the final frame shows the newly formed Moon. Red and blue colors represent rocky and metallic regions, respectively, and the direction of motion of the blue matter in the bottom few frames is toward Earth. Note how most of the impactor's metallic core becomes part of Earth, leaving the Moon composed mainly of rocky material. *(W. Benz)*

8.9 Evolutionary History of the Moon and Mercury

The Moon

![LEARNING GOAL 7] Given all the data, can we construct a reasonably consistent history of the Moon? The answer seems to be yes. Many specifics are still debated, but a consensus exists. (Examine Figure 8.29 while studying the details that follow.)

The Moon formed about 4.6 billion years ago (see Chapter 15). The approximate age of the oldest rocks discovered in the lunar highlands is 4.4 billion years, so we know that at least part of the crust must already have solidified by that time and survived to the present. At its formation, the Moon was already depleted in heavy metals compared with Earth.

During the earliest phases of the Moon's existence—roughly the first half billion years or so—meteoritic bombardment must have been frequent enough to heat and remelt most of the *surface* layers of the Moon, perhaps to a depth of 400 km in places. The early solar system was surely populated with lots of interplanetary matter, much of it in the form of boulder-sized fragments that were capable of generating large amounts of energy upon colliding with planets and their moons. But the intense heat derived from such collisions could not have penetrated very far into the lunar interior: Rock simply does not conduct heat well.

This situation resembles the surface melting we suspect occurred on Earth from meteoritic impacts during the first billion years or so. But the Moon is much less massive than Earth and did not contain enough radioactive elements to heat it much further. Radioactivity probably heated the Moon a little, but not sufficiently to transform it from a warm, semisolid object to a completely liquid one. The chemical differentiation now inferred in the Moon's interior must have occurred during this period. If the Moon has a small iron core, that core also formed at this time.

About 3.9 billion years ago, around the time that Earth's crust solidified, the heaviest phase of the meteoritic bombardment ceased. The Moon was left with a solid crust, which would ultimately become the highlands, dented with numerous large basins, soon to flood with lava and become the maria (Figure 8.29a). Between 3.9 and 3.2 billion years ago, lunar volcanism filled the maria with the

▶ **FIGURE 8.29 Lunar Evolution** Paintings of how the Moon might have looked (a) about 4 billion years ago, after much of the meteoritic bombardment had subsided and the surface had solidified somewhat; (b) about 3 billion years ago, after molten lava had made its way up through surface fissures to fill the low-lying impact basins and create the smooth maria; and (c) today, with much of the originally smooth maria now heavily pitted with craters formed at various times within the past 3 billion years. (*U.S. Geological Survey*)

(a) 4 billion years ago

(b) 3 billion years ago

(c) Today

basaltic material we see today. The age of the youngest maria—3.2 billion years—indicates the time when the volcanic activity subsided. The maria are the sites of the last extensive lava flows on the Moon, over 3 billion years ago. Their smoothness, compared with the older, more rugged highlands, disguises their great age.

Small objects cool more rapidly than large ones because their interior is closer to the surface, on average. Being so small, the Moon rapidly lost its internal heat to space. As a consequence, it cooled much faster than Earth. As the Moon cooled, the volcanic activity ended and the thickness of the solid surface layer increased. With the exception of a few meters of surface erosion from eons of meteoritic bombardment (Figure 8.29c), the lunar landscape has remained more or less structurally frozen for the past 3 billion years. The Moon is dead now, and it has been dead for a long time.

Mercury

Like the Moon, Mercury seems to have been a geologically dead world for much of the past 4 billion years. On both the Moon and Mercury, the absence of ongoing geological activity is a consequence of a thick, solid mantle that prevents volcanism or tectonic motion. Because of the *Apollo* program, the Moon's early history is much better understood than Mercury's, which remains somewhat speculative. Indeed, what we do know about Mercury's history is gleaned mostly through comparison with the Moon.

When Mercury formed some 4.6 billion years ago, it was already depleted of lighter, rocky material. We will see later that this was largely a consequence of its location in the hot inner regions of the early solar system, although it is possible that a collision stripped away some of the planet's light mantle. During the next half-billion years, Mercury melted and differentiated, like the other terrestrial worlds. It suffered the same intense meteoritic bombardment as the Moon. Being more massive than the Moon, Mercury cooled more slowly, so its crust was thinner and volcanic activity more common at early times. More craters were erased, resulting in the intercrater plains found by *Mariner 10*.

As Mercury's large iron core formed and then cooled, the planet began to shrink, compressing the crust. This compression produced the scarps seen on Mercury's surface and may have prematurely terminated volcanic activity by squeezing shut the cracks and fissures on the surface. Thus, the extensive volcanic outflows that formed the lunar maria did not take place on Mercury. Despite its larger mass and greater internal temperature, Mercury has probably been geologically inactive even longer than the Moon.

Chapter Review

SUMMARY

Both the Moon and Mercury are airless, virtually unchanging worlds that exhibit extremes in temperature. Mercury has no permanent atmosphere, although it does have a thin envelope of gas temporarily trapped from the solar wind. The main surface features on the Moon are the dark **maria (p. 203)** and the lighter colored **highlands (p. 203)**. Highland rocks are less dense than rocks from the maria and are thought to represent the Moon's crust. Maria rocks are thought to have originated in the lunar mantle. The surfaces of both the Moon and Mercury are covered with **craters (p. 203)** of all sizes, caused by meteoroids striking from space. Meteoritic impacts are the main source of erosion on the surfaces of both worlds. The lunar highlands are older than the maria and are much more heavily cratered. The rate at which craters are formed decreases rapidly with increasing crater size.

The hot dayside temperatures and cold nightside temperatures on the Moon and Mercury result from the absence of significant heat conduction or atmospheric blanketing. Sunlight strikes the polar regions of both the Moon and Mercury at such an oblique angle that temperatures there are very low, with the result that both bodies may have significant amounts of water ice near the poles.

The tidal interaction between Earth and the Moon is responsible for the Moon's **synchronous orbit (p. 206)**, in which the same side of the Moon always faces our planet. The large lunar equatorial bulge probably indicates that the Moon once rotated more rapidly and orbited closer to Earth. Mercury's rotation rate is strongly influenced by the tidal effect of the Sun. Because of Mercury's eccentric orbit, the planet rotates not synchronously, but exactly three times for every two orbits around the Sun. The condition in which a body's rotation rate is simply related to its orbital period around some other body is known as spin–orbit **resonance (p. 208)**.

The Moon's surface consists of both rocky and dusty material. Lunar dust, called regolith, is made mostly of pulverized lunar rock, mixed with a small amount of material from meteorites that have struck the surface. Evidence for past volcanic activity on the Moon is found in the form of solidified lava channels called **rilles (p. 216)**. Mercury's surface features bear a striking similarity to those of the Moon. The planet is heavily cratered, much like the lunar highlands. Among the differences between Mercury and the Moon are Mercury's lack of lunarlike maria, its extensive **intercrater plains (p. 219)**, and the great cracks, or **scarps (p. 219)**, in its crust. The plains were caused by extensive lava flows early in Mercury's history. The scarps were apparently formed when the planet's core cooled and shrank, causing the surface to crack. Mercury's evolutionary path was similar to that of the Moon for half a billion years after they both formed. Mercury's volcanic period probably ended before that of the Moon.

The absence of a lunar atmosphere and any present-day lunar volcanic activity are both consequences of the Moon's small size. Lunar gravity is too weak to retain any gases, and lunar volcanism was stifled by the Moon's cooling mantle shortly after extensive lava flows formed the maria more than 3 billion years ago. The crust on the far side of the Moon is substantially thicker than the crust on the near side. As a result, there are almost no maria on the lunar far side. Mercury has a large impact crater called the Caloris Basin, whose diameter is comparable to the radius of the planet. The impact that formed the crater apparently sent violent shock waves around the entire planet, buckling the crust on the opposite side.

The Moon's average density is not much greater than that of its surface rocks, probably because the Moon cooled more rapidly than the larger Earth and solidified sooner, so there was less time for differentiation to occur, although the Moon likely has a small iron-rich core. The lunar crust is too thick and the mantle too cool for plate tectonics to occur. Mercury's average density is considerably greater—similar to that of Earth—implying that Mercury contains a large high-density core, probably composed primarily of iron. The Moon has no measurable large-scale magnetic field, a consequence of its slow rotation and lack of a molten metallic core. Mercury's weak magnetic field seems to have been "frozen in" long ago, when the planet's iron core solidified.

The most likely explanation for the formation of the Moon is that the newly formed Earth was struck by a large (Mars-sized) object. Part of the colliding body remained behind as part of our planet. The rest ended up in orbit as the Moon.

REVIEW AND DISCUSSION

1. How is the distance to the Moon most accurately measured?

2. Why is Mercury seldom seen with the naked eye?

3. Why did early astronomers think that Mercury was two separate planets?

4. Employ the concept of escape speed to explain why the Moon and Mercury have no significant atmospheres.

5. In what sense are the lunar maria "seas"?

6. Why is the surface of Mercury often compared with that of the Moon? List two similarities and two differences between the surfaces of Mercury and the Moon.

7. What does it mean to say that the Moon is in a synchronous orbit around Earth? How did the Moon come to be in such an orbit?

8. What does it mean to say that Mercury has a 3:2 spin–orbit resonance? Why didn't Mercury settle into a 1:1 spin–orbit resonance with the Sun, as the Moon did with Earth?

9. What is a scarp? How are scarps thought to have formed? Why do scientists think that the scarps on Mercury formed after most meteoritic bombardment ended?

10. What is the primary source of erosion on the Moon? Why is the average rate of lunar erosion so much less than on Earth?

11. What evidence do we have for ice on the Moon?

12. Name two pieces of evidence indicating that the lunar highlands are older than the maria.

13. In contrast with Earth, the Moon and Mercury undergo extremes in temperature. Why?

14. How is Mercury's evolutionary history like that of the Moon? How is it different?

15. Describe the theory of the Moon's origin favored by many astronomers.

16. Because the Moon always keeps one face toward Earth, an observer on the moon's near side would see Earth appear almost stationary in the lunar sky. Still, Earth would change its appearance as the Moon orbited Earth. How would Earth's appearance change?

17. The best place to aim a telescope or binoculars on the Moon is along the terminator line—the line between the Moon's light and dark hemispheres. Why? If you were standing on the lunar terminator, where would the Sun be in your sky? What time of day would it be if you were standing on Earth's terminator line?

18. Where on the Moon would be the best place from which to make astronomical observations? What would be this location's advantage over locations on Earth?

19. Explain why Mercury is never seen overhead at midnight in Earth's sky.

20. How is the varying thickness of the lunar crust related to the presence or absence of maria on the Moon?

CONCEPTUAL SELF-TEST: TRUE OR FALSE/MULTIPLE CHOICE

1. Laser ranging can determine the distance to the Moon to an accuracy of a few centimeters.

2. Mercury can sometimes be seen at midnight.

3. Mercury's solar day is longer than its solar year.

4. The most accurate method for determining the distance to the Moon is by parallax.

5. Mercury's daytime temperature is higher than the Moon's because Mercury is more massive than the Moon.

6. Craters on the Moon and Mercury are primarily the result of volcanic activity.

7. There is no volcanic activity today on the surface of the Moon.

8. Although daytime temperatures on the Moon and Mercury are very high, it may still be possible for those two bodies to have large amounts of water ice at their poles.

9. The lunar maria's dark, dense rock originally was part of the lunar mantle.

10. The most likely scenario for the formation of the Moon is a collision between Earth and another planet-sized body.

11. Compared with the diameter of Earth's Moon, the diameter of Mercury is **(a)** larger; **(b)** smaller; **(c)** nearly the same.

12. In relation to the density of Earth's Moon, Mercury's density suggests that the planet **(a)** has an interior structure similar to that of the Moon; **(b)** has a dense metal core; **(c)** has a stronger magnetic field than the Moon; **(d)** is younger than the Moon.

13. Compared with the phases of Earth's Moon, Mercury goes from new phase to full phase **(a)** faster; **(b)** more slowly; **(c)** in about the same time.

14. Compared with the surface of Mercury, the surface of Earth's Moon has significantly **(a)** bigger craters; **(b)** more atmosphere; **(c)** more maria; **(d)** deeper craters.

15. Every two times Earth's Moon rotates on its axis, it orbits Earth **(a)** less than twice; **(b)** exactly two times; **(c)** more than twice; **(d)** three times.

16. Planets and moons showing the most craters have **(a)** the oldest surfaces; **(b)** been hit by meteors the most times; **(c)** the strongest gravity; **(d)** molten cores.

17. Compared with the Moon, Mercury has **(a)** a much smaller core; **(b)** a much larger core; **(c)** a similar-sized core.

18. The most likely theory of the formation of Earth's Moon is that it **(a)** was formed by the gravitational capture of a large asteroid; **(b)** formed simultaneously with Earth's formation; **(c)** was created from a collision scooping out the Pacific Ocean; **(d)** formed from a collision of Earth with a Mars'-sized object.

19. Mercury, being smaller than Mars, probably cooled and solidified **(a)** faster, because it is smaller; **(b)** slower, because it is closer to the Sun; **(c)** in about the same time, because space is generally cold.

20. On the scale of the 5-billion-year age of the solar system, the Moon is **(a)** about the same age as Earth; **(b)** much younger than Earth; **(c)** much older than Earth.

PROBLEMS

 Algorithmic versions of these questions are available in the Practice Problems module of the Companion Website at astro.prenhall.com/chaisson.

The number of squares preceding each problem indicates its approximate level of difficulty.

1. ■ How long does a radar signal take to travel from Earth to Mercury and back when Mercury is at its closest point to Earth?

2. ■ The Moon's mass is one-eightieth that of Earth, and the lunar radius is one-fourth Earth's radius. On the basis of these figures, calculate the total weight on the Moon of a 100-kg astronaut with a 50-kg space suit and backpack, relative to his weight on Earth.

3. ■ What would be the same astronaut's weight on Mercury?

4. ■ Based on the data presented in the Moon Data box (p. 202), verify the values given for the Moon's perigee (minimum distance from Earth) and apogee (maximum distance from Earth), and estimate the Moon's minimum and maximum angular diameter, as seen from Earth. Compare these values with the angular diameter of the Sun (of actual diameter 1.4 million km), as seen from a distance of 1 A.U.

5. ■ What is the angular diameter of the Sun, as seen from Mercury, at perihelion? At aphelion?

6. ■ The *Hubble Space Telescope* has a resolution of about 0.05″. What is the size of the smallest feature it can distinguish on the surface of the Moon (distance = 380,000 km)? On Mercury, at closest approach to Earth?

7. ■ What was the orbital period of the *Apollo 11* command module, orbiting 10 km above the lunar surface?

8. ■■ Compare the gravitational tidal acceleration of the Sun on Mercury (at perihelion; solar mass = 2×10^{30} kg) with the tidal effect of Earth on the Moon (at perigee). ∞ (Sec. 7.6)

9. ■■■ Mercury's average orbital speed around the Sun is 47.9 km/s. Use Kepler's second law to calculate Mercury's speed **(a)** at perihelion and **(b)** at aphelion. ∞ (Sec. 2.5, *More Precisely 2-1*) Convert these speeds to angular speeds (in degrees per day), and compare them with Mercury's 6.1°-per-day rotation rate.

10. ■■ What would the lengths of a sidereal and a solar day on Mercury be if the planet were in a 4:3 spin–orbit resonance instead of the 3:2 resonance actually observed?

11. ■■ Assume that a planet will have lost its initial atmosphere by the present time if the average molecular speed exceeds one-sixth of the escape speed (see *More Precisely 8-1*). What would Mercury's mass have to be in order for it to still have a nitrogen atmosphere? The molecular weight of nitrogen is 28.

12. ■■ With the same assumptions as in the previous question, estimate the minimum molecular mass that might still be found in Mercury's atmosphere.

13. ■■ Using the rate given in the text for the formation of 10-km craters on the Moon, estimate how long would be needed for the entire Moon to be covered with new craters of that size. How much higher must the cratering rate have been in the past to cover the entire lunar surface with such craters in the 4.6 billion years since the Moon formed?

14. ■■ Repeat the previous question, for meter-sized craters.

15. ■ Using the data given in the text, calculate how long erosion would take to obliterate **(a)** the boot print in Figure 8.17, **(b)** the Barringer Meteor Crater in Figure 8.19, **(c)** lunar crater Reinhold in Figure 8.15(b).

 In addition to the Practice Problems module, the Companion Website at astro.prenhall.com/chaisson provides for each chapter a study guide module with multiple choice questions as well as additional annotated images, animations, and links to related Websites.

9

Often called Earth's sister planet, Venus is nothing like Earth when it comes to surface temperature; it's hot enough there (730 K) to melt lead. We now know that Venus's climate, like Earth's, has varied over time—largely the result of geological activity and atmospheric change. What we do not know well is why Venus became so very much hotter than Earth—or if Earth could someday heat up similarly. Here, in a series of radar scans of the Venusian surface in the 1990s, the Magellan robot spacecraft captured this image of the area near Maat Mons (to left rear), an ▶ *8-kilometer-high volcano that is now apparently dormant. (JPL)*

Venus

Earth's Sister Planet

Venus seems almost a carbon copy of our own world. The two planets are similar in size, density, and chemical composition. They orbit at comparable distances from the Sun. At formation, they must have been almost indistinguishable from one another. Yet they are now about as different as two terrestrial planets can be. Whereas Earth is a vibrant world, teeming with life, Venus is an uninhabitable inferno, with a dense, hot atmosphere of carbon dioxide, lacking any trace of oxygen or water. Somewhere along their respective evolutionary paths, Venus and Earth diverged, and diverged radically. How did this occur? What were the factors leading to Venus's present condition? Why are Venus's surface, atmosphere, and interior so different from Earth's? In answering these questions, we will discover that a planet's environment, as well as its composition, can play a critical role in determining its future.

LEARNING GOALS

Studying this chapter will enable you to

1 Summarize Venus's general orbital and physical properties.

2 Explain why Venus is hard to observe from Earth and how we have obtained more detailed knowledge of the planet.

3 Compare the surface features and geology of Venus with those of Earth and the Moon.

4 Describe the characteristics of Venus's atmosphere and contrast it with that of Earth.

5 Explain why the greenhouse effect has produced conditions on Venus very different from those on Earth.

6 Describe Venus's magnetic field and internal structure.

 Visit astro.prenhall.com/chaisson for additional images, animations, and links to related sites for this chapter.

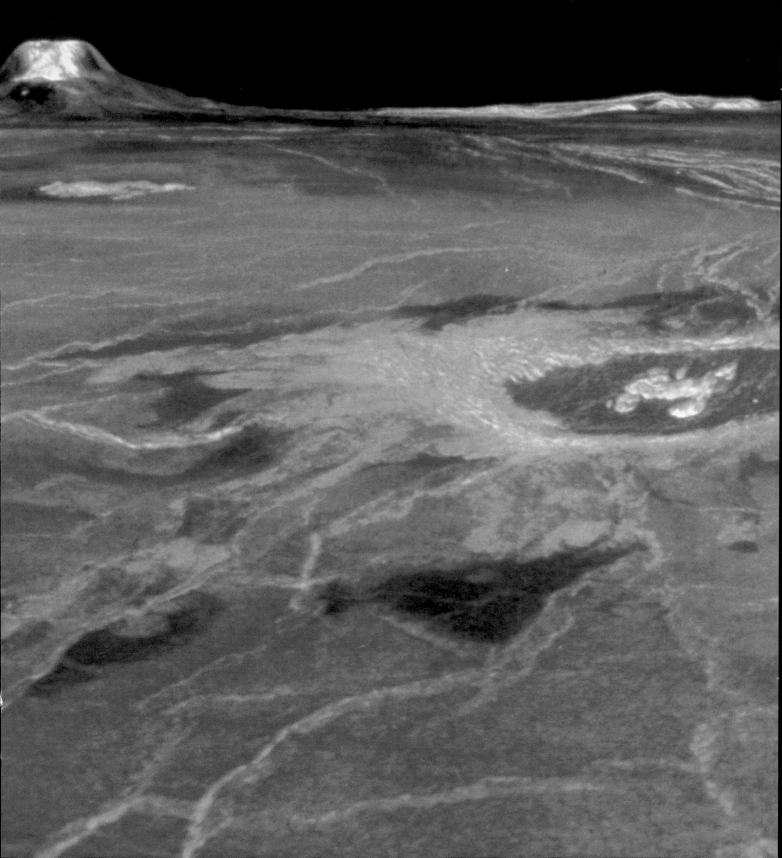

9.1 Orbital Properties

Venus is the second planet from the Sun. Its orbit lies within Earth's, so Venus, like Mercury, is always found fairly close to the Sun in the sky—our sister planet is never seen more than 47° from the Sun. Given Earth's rotation rate of 15° per hour, this means that Venus is visible above the horizon for at most three hours before the Sun rises or after it sets. Because we can see Venus from Earth only just before sunrise or just after sunset, the planet is often called the "morning star" or the "evening star," depending on where it happens to be in its orbit. Figure 9.1 shows Venus in the western sky just after sunset. The Venus Data box on p. 235 lists some of the planet's orbital and physical properties.

Venus is the third-brightest object in the entire sky (after the Sun and the Moon). It appears more than 10 times brighter than the brightest star, Sirius. You can see Venus even in the daytime if you know just where to look. On a moonless night away from city lights, Venus casts a faint shadow. The planet's brightness stems from the fact that Venus is highly reflective. Nearly 70 percent of the sunlight reaching Venus is reflected back into space. (Compare this percentage with roughly 10 percent in the case of Mercury and the Moon.) Most of the sunlight is reflected from clouds high in the planet's atmosphere.

We might expect Venus to appear brightest when it is "full"—that is, when we can see the entire sunlit side. However, because Venus orbits between Earth and the Sun, Venus is full when it is at its greatest distance from us—1.7 A.U. away on the other side of the Sun, as illustrated in Figure 9.2. Recall from Chapter 2 that this alignment is known as *superior conjunction*, where the term

"conjunction" simply indicates that two objects are close together in the sky. ∞ (Sec. 2.2)

When Venus is closest to us, the planet is in the new phase, lying between Earth and the Sun (at *inferior conjunction*), and we again can't see it, because now the sunlit side faces away from us; only a thin ring of sunlight, caused by refraction in Venus's atmosphere, surrounds the planet. As Venus moves away from inferior conjunction, more and more of it becomes visible, but its distance from us also continues to increase. Venus's maximum brightness, as seen from Earth, actually occurs about 36 days before or after its closest approach to our planet. At that time, Venus is about 39° from the Sun and 0.47 A.U. from Earth, and we see it as a rather fat crescent.

9.2 Physical Properties

Radius, Mass, and Density

We can determine Venus's radius from simple geometry, just as we did for Mercury and the Moon. ∞ (Sec. 8.2) At closest approach, when Venus is only 0.28 A.U. from us, its angular diameter is 64″. From this observation, we can determine the planet's radius to be about 6000 km. More accurate measurements from spacecraft give a value of 6052 km, or 0.95 Earth radii.

Like Mercury, Venus has no moon. Before the Space Age, astronomers calculated its mass by indirect means—through studies of its small gravitational effect on the orbits of the other planets, especially Earth. Now that spacecraft have orbited the planet, we know Venus's mass very accurately from measurements of its gravitational pull: Venus has a mass of 4.9×10^{24} kg, or 0.82 the mass of Earth.

From its mass and radius, we find that Venus's average density is 5200 kg/m^3. As far as these bulk properties are concerned, then, Venus seems similar to Earth. If the planet's overall composition were similar to Earth's as well, we could then reasonably conclude that Venus's internal structure and evolution were basically Earthlike. We will review what evidence there is on this subject later in the chapter.

Rotation Rate

The same clouds whose reflectivity makes Venus so easy to see in the night sky also make it impossible for us to discern any surface features on the planet, at least in visible light. As a result, until the advent of suitable radar techniques in the 1960s, astronomers did not know the rotation period of Venus. Even when viewed through a large optical telescope, the planet's cloud cover shows few features, and attempts to determine Venus's period of rotation by observing the cloud layer were frustrated by the rapidly changing nature of the clouds themselves. Some astronomers argued for a 25-day period, while others favored a 24-hour cycle. Controversy raged until, to the sur-

▲ **FIGURE 9.1 Venus at Sunset** The Moon and Venus in the western sky just after sunset. Venus clearly outshines even the brightest stars in the sky. (*J. Schad/Photo Researchers, Inc.*)

▶ **FIGURE 9.2** **Venus's Brightness** Venus appears full when it is at its greatest distance from Earth, on the opposite side of the Sun from us (superior conjunction). As its distance decreases, less and less of its sunlit side becomes visible. When it is closest to Earth, it lies between us and the Sun (inferior conjunction), so we cannot see the sunlit side of the planet at all. Venus appears brightest when it is about 39° from the Sun. (Cf. Figure 2.12.) *(UC/Lick Observatory)*

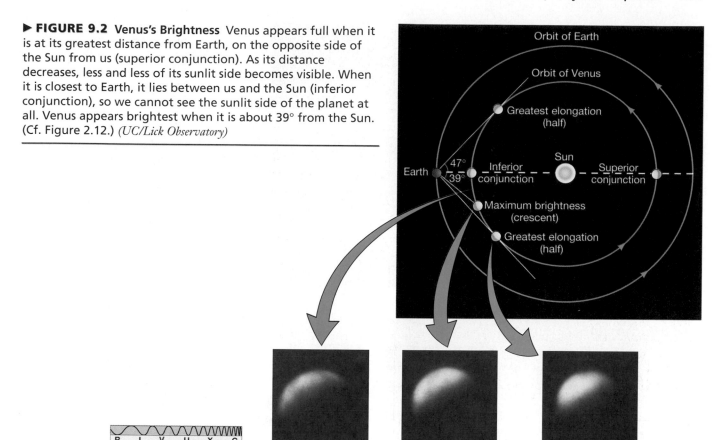

ANIMATION The Phases of Venus

prise of all, radar observers announced that the Doppler broadening of their returned echoes implied a sluggish 243-day rotation period! ∞ (Sec. 8.3) Furthermore, Venus's spin was found to be *retrograde*—that is, in a sense opposite that of Earth and most other solar system objects and opposite that of Venus's orbital motion.

Planetary astronomers define "north" and "south" for each planet in the solar system by the convention that planets *always* rotate from west to east. With this definition, Venus's retrograde spin means that the planet's north pole lies *below* the plane of the ecliptic, unlike any of the other terrestrial worlds. Venus's axial tilt—the angle between its equatorial and orbital planes—is 177.4° (compared with 23.5° in the case of Earth). However, astronomical images of solar-system objects conventionally place objects lying above the ecliptic at the top of the frame. Thus, with the preceding definition of north and south, all the images of Venus shown in this chapter have the *south* pole at the top.

Figure 9.3 illustrates Venus's retrograde rotation and compares it with the rotation of its neighbors Mercury, Earth, and Mars. Because of the planet's slow retrograde rotation, its solar day (from noon to noon) is quite different from its sidereal rotation period of 243 Earth days (the time for one "true" rotation relative to the stars). ∞ (Sec. 1.4) In fact, as illustrated in Figure 9.4, one Venus day is a little more than half a Venus year (225 Earth days).

Why is Venus rotating "backward" and why so slowly? At present, the best explanation planetary scientists can offer is that early in Venus's evolution, the planet was struck by a large body, much like the one that may have hit Earth and formed the Moon, and that impact was sufficient to reduce the planet's spin almost to zero. ∞ (Sec. 8.8) Whatever its cause, the planet's rotation poses practical problems for Earth-bound observers. It turns out that Venus rotates almost exactly five times between one closest approach to Earth and the next. As a result, *Venus always presents nearly the same face to Earth at closest approach.* This means that observations of the planet's surface cover one side—the one facing us at closest approach—much more thoroughly than the other side, which we can see only when the planet is close to its maximum distance from Earth.

This nearly perfect 5:1 resonance between Venus's rotation and orbital motion is reminiscent of the Moon's synchronous orbit around Earth and Mercury's 3:2 spin–orbit resonance with the Sun. ∞ (Sec. 8.3) However, no known interaction between Earth and Venus can account for such an odd state of affairs. Earth's tidal effect on Venus is tiny and is much less than the Sun's tidal effect in any case.

Furthermore, the key word in the preceding paragraph is *nearly*. A resonance, if it existed, would require that the number of rotations per relative orbit be *exactly* five. The discrepancy amounts to less than 3 hours in 584

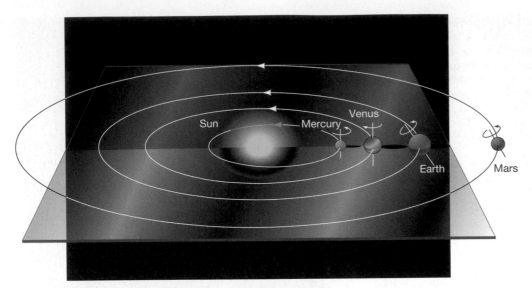

◄ **FIGURE 9.3 Terrestrial Planets' Spins** The inner planets of the solar system—Mercury, Venus, Earth, and Mars—display widely differing rotational properties. Although all orbit the Sun in the same direction and in nearly the same plane, Mercury's rotation is slow and prograde (in the same sense as the orbital motion), the rotations of Earth and Mars are fast and prograde, and Venus's rotation is slow and retrograde. Venus rotates clockwise as seen from above the plane of the ecliptic, but Mercury, Earth, and Mars all spin counterclockwise.

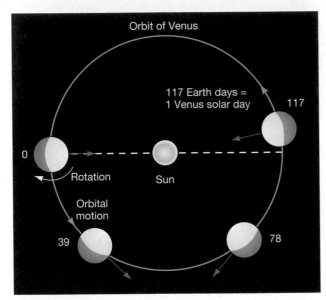

▲ **FIGURE 9.4 Venus's Solar Day** Venus's orbit and retrograde rotation combine to produce a solar day on Venus equal to 117 Earth days, or slightly more than half a Venus year. The red arrows represent a fixed location on the planet's surface. The numbers in the figure mark time in Earth days.

days (Venus's *synodic period* relative to Earth—see *More Precisely 9-1*), but it appears to be real, and if that is so, then no true resonance exists. Astronomers hate to appeal to coincidence to explain their observations, but the case of Venus's rotation appears to be just that. For now, we are simply compelled to accept this strange coincidence without explanation.

CONCEPT CHECK

✔ What is peculiar about Venus's rotation, and why does Venus rotate that way?

9.3 Long-Distance Observations of Venus

Because Venus, of all the other planets, most nearly matches Earth in size, mass, and density, and because its orbit is closest to us, it is often called Earth's sister planet. But unlike Earth, Venus has a dense atmosphere and thick clouds that are opaque to visible radiation, making its surface completely invisible from the outside at optical wavelengths. Figure 9.5 shows one of the best photographs of Venus taken with a large telescope on Earth. The planet presents an almost featureless white-yellow disk, although it shows occasional hints of cloud circulation.

Atmospheric patterns on Venus are much more evident when the planet is examined with equipment capable of detecting ultraviolet radiation. Some of Venus's atmospheric constituents absorb this high-frequency radiation, greatly increasing the cloud contrast. Figure 9.6 is an ultraviolet image taken in 1979 by the U.S. *Pioneer Venus* spacecraft at a distance of 200,000 km from the planet's surface. The large, fast-moving cloud patterns resemble Earth's high-altitude jet stream more than the great whirls characteristic of Earth's low-altitude clouds. The upper deck of clouds on Venus move at almost 400 km/h, encircling the planet in just 4 days—much faster than the planet itself rotates!

Early spectroscopic studies of sunlight reflected from Venus's clouds revealed the presence of large amounts of carbon dioxide, but provided little evidence for any other atmospheric gases. Until the 1950s, astronomers generally believed that observational difficulties alone prevented them from seeing other atmospheric components. The

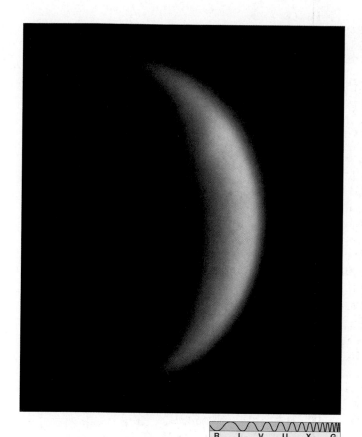

R I V U X G

▲ **FIGURE 9.5 Venus** This photograph, taken from Earth, shows Venus with its creamy yellow mask of clouds. No surface detail can be seen, because the clouds completely obscure our view of whatever lies beneath them. *(AURA)*

R I V U X G

▲ **FIGURE 9.6 Venus, Up Close** Venus as photographed by the *Pioneer* spacecraft's cameras some 200,000 km away from the planet's surface. This image was made by capturing solar ultraviolet radiation reflected from the planet's clouds, which are probably composed mostly of sulfuric acid droplets (much like the highly corrosive acid in a car battery). *(Edmund Scientific Co.)*

hope lingered that Venus's clouds were actually predominantly water vapor, like those on Earth, and that below the cloud cover Venus might be a habitable planet similar to our own. Indeed, in the 1930s, scientists had measured the temperature of the atmosphere spectroscopically at about 240 K, not much different from the temperature of our own upper atmosphere. ∞ (Sec. 4.4) Calculations of the planet's surface temperature—taking into account the cloud cover and Venus's proximity to the Sun, and assuming an atmosphere much like our own—suggested that Venus should have a surface temperature only 10 or 20 degrees higher than Earth's.

These hopes for an Earth-like Venus were dashed in 1956, when radio observations of the planet were used to measure its thermal energy emission. Unlike visible light, radio waves easily penetrate the cloud layer—and they gave the first indication of conditions on or near the surface: The radiation emitted by the planet has a blackbody spectrum characteristic of a temperature near 730 K! ∞ (Sec. 3.4) Almost overnight, the popular conception of Venus changed from that of a lush tropical jungle to an arid, uninhabitable desert.

Radar observations of the surface of Venus are routinely carried out from Earth with the Arecibo radio tele-

scope. ∞ (Sec. 5.4) With careful signal processing, this instrument can achieve a resolution of a few kilometers, but it can adequately cover only a fraction (roughly 25 percent) of the planet. The telescope's view of Venus is limited by the planet's peculiar near resonance described in the previous section (which means that only one side of the planet can be studied) and also because radar reflections from regions near the "edge" of the planet are hard to obtain. However, the Arecibo data can usefully be combined with information received from probes orbiting Venus to build up a detailed picture of the planet's surface. Only with the arrival of the *Magellan* probe were more accurate data obtained.

CONCEPT CHECK

✔ Why did early studies of Venus lead astronomers to such an inaccurate picture of the planet's surface conditions?

MORE PRECISELY 9-1

Synodic Periods and Solar Days

Recall from Chapter 1 that a body's *sidereal* orbital (or rotational) period is the time taken for the body to complete one orbit (or rotation) relative to the "fixed" stars. ∞ (Sec. 1.3) In many cases, however, we are more interested in how things look from our vantage point on Earth. The *synodic* orbital period is defined to be the time taken for the body to return to the same configuration relative to the Sun, *taking Earth's own motion into account*. Examples are the time from one full Moon to the next or between successive inferior conjunctions (times of closest approach) of Venus. ∞ (Sec. 1.4)

Similarly, a planet's *solar day*—the time from noon to noon—in general differs from its sidereal day because of the planet's motion around the Sun. In the case of Earth, the difference between a sidereal and a solar day is relatively small, as is the difference between a sidereal and a synodic month. ∞ (Sec. 1.3) However, as we have seen, Mercury's sidereal and solar days differ significantly, as do the sidereal and synodic periods of all the planets. ∞ (Sec. 8.4) Let's take a moment to look at this topic in a little more detail.

Calculating the synodic period is easy. With Earth's sidereal orbital period (365.26 solar days) denoted by P_E and that of the body in question by P, it follows that the *angular speeds* of Earth and the other body (in degrees per day) are, respectively, $360°/P_E$ and $360°/P$, since there are 360 degrees in one complete revolution. Thus, as illustrated in the accompanying figure, the rate at which the body "outruns" Earth is $360°/P - 360°/P_E$. The body's synodic period S is, by definition, the time taken for the body to "lap" Earth (i.e., outstrip it by 360°), so it follows that

$$\frac{1}{S} = \frac{1}{P} - \frac{1}{P_E},$$

where S is positive for the interior planets Mercury and Venus (having P less than P_E, as in the figure) and negative for the exterior planets Mars, Jupiter, and so on (meaning that Earth overtakes them).

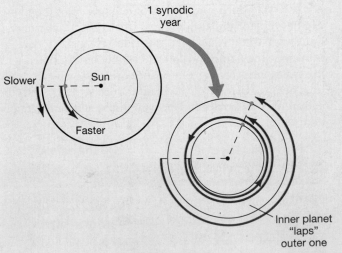

EXAMPLE 1: The Moon has $P = 27.3$ days, so its synodic period S is given by

$$\frac{1}{S} = \frac{1}{27.3} - \frac{1}{365.26} = \frac{1}{29.5}.$$

Hence, $S = 29.5$ days. Similarly, for Venus, $P = 224.7$ days, yielding the synodic period $S = 583.9$ days mentioned in the text. During this time, Earth orbits the Sun 1.6 times, and Venus 2.6, so the actual planetary orbits are a little more complicated than those sketched in the figure. For Mars, we have $P = 687$ days, so $S = 780$ days, accounting for the roughly two-year interval between launch windows for space probes to that planet, as mentioned in Chapter 6. ∞ (Sec. 6.6)

Now let's consider how the competition between the planet's sidereal rotation period R and its sidereal orbital period P determines the length of its solar day. ∞ (Fig. 1.13) Reasoning similar to the preceding leads to the following expression for the planetary solar day D:

$$\frac{1}{D} = \frac{1}{R} - \frac{1}{P}.$$

For example, Earth has $R = 0.9973$ solar day and $P = P_E = 365.26$ solar days, which gives $D = 1$ solar day, as expected. The difference between Earth's solar and sidereal days is small because Earth rotates on its axis much more rapidly than it revolves around the Sun. For Mercury and Venus, however, the rotation and revolution periods are comparable, and consequently the difference between solar and sidereal days is much greater.

EXAMPLE 2: Using the data from the Mercury Data box, p. 220, we find that, for Mercury, $R = 58.6$ (Earth) solar days and $P = 88.0$ solar days, so $D = 176.0$ solar days, or two sidereal years. ∞ (Sec. 8.3) Venus has $R = -243.0$ solar days and $P = 224.7$ solar days, so

$$\frac{1}{D} = \frac{1}{58.6} - \frac{1}{88.0} = -\frac{1}{116.7}.$$

Thus, $D = -116.7$ solar days. (The negative signs, as usual, indicate retrograde rotation: The calculation implies that the Sun would move "backwards" through Venus's sky if someone on the surface could see it.) The extreme example would be a planet in a synchronous orbit around the Sun (or the Moon orbiting Earth), with $R = P$. In that case, D would be infinite—the day would last forever, and the Sun would never set!

Sidereal periods are "physical," in the sense that they are the quantities to which the laws of Kepler and Newton refer. However, from the point of view of timekeeping, skywatching, or scheduling a mission to another planet, synodic periods, which tell us when a satellite returns to the same phase, or when a planet "returns to the same position" in the night sky, are just as important!

9.4 The Surface of Venus

LEARNING GOAL 3 Although the planet's clouds are thick and the terrain below them totally shrouded, we are by no means ignorant of Venus's surface. Detailed radar observations have been made both from Earth and from the *Venera*, *Pioneer Venus*, and *Magellan* spacecraft. ∞ (Sec. 6.6) Analysis of the radar echoes yields a map of the planet's surface. Except for the last two figures, all the views of Venus in this section are "radargraphs" (as opposed to photographs) created in this way. As Figure 9.7(a) illustrates, the early maps of Venus suffered from poor resolution; however, more recent probes—especially *Magellan*—have provided much sharper views. As in all the *Magellan* images, the light areas in Figure 9.7(b) represent regions where the surface is rough and efficiently scatters *Magellan*'s sideways-looking radar beam back to the detector. Smooth areas tend to reflect the beam off into space instead and so appear dark.

Large-Scale Topography

Figure 9.8(a) shows basically the same *Pioneer Venus* data of Venus as Figure 9.7(a), except that the figure has been flattened out into a more conventional map. The altitude of the surface relative to the average radius of the planet is indicated by the use of color, with white representing the highest elevations and blue the lowest. (Note that the blue has nothing to do with oceans, nor does white indicate snow-capped mountains!) Figure 9.8(b) shows a map of Earth to the same scale and at the same spatial resolution. Some of Venus's main features are labeled in Figure 9.8(c).

The surface of Venus appears to be relatively smooth, resembling rolling plains with modest highlands and lowlands. Two continent-sized features, called Ishtar Terra and Aphrodite Terra (named after the Babylonian and Greek counterparts, respectively, of Venus, the Roman goddess of love), adorn the landscape and contain mountains comparable in height to those on Earth. The elevated "continents" occupy only 8 percent of Venus's total surface area. For comparison, continents on Earth make up about 25 percent of the surface. The remainder of Venus's surface is classified as lowlands (27 percent) or rolling plains (65 percent), although there is probably little geological difference between the two terrains.

Note that, while Earth's tectonic plate boundaries are evident in Figure 9.8(b), no similar features can be seen in Figure 9.8(a). ∞ (Sec. 7.4) There simply appears to be no large-scale plate tectonics on Venus. Ishtar Terra ("Land of Ishtar") lies in the southern high latitudes (at the *tops* of Figures 9.7a and 9.8a—recall our earlier discussion of Venus's retrograde rotation). The projection used in Figure 9.8 makes Ishtar Terra appear larger than it really is—it is actually about the same size as Australia. This landmass is dominated by a great plateau known as Lakshmi Planum (Figure 9.9), some 1500 km across at its widest

PLANETARY DATA:
VENUS

(NASA)

Orbital semimajor axis	0.72 A.U. 108.2 million km
Orbital eccentricity	0.007
Perihelion	0.72 A.U. 107.5 million km
Aphelion	0.73 A.U. 108.9 million km
Mean orbital speed	35.0 km/s
Sidereal orbital period	224.7 solar days 0.615 tropical year
Synodic orbital period	583.9 solar days
Orbital inclination to the ecliptic	3.39°
Greatest angular diameter, as seen from Earth	64″
Mass	4.87×10^{24} kg 0.82 (Earth = 1)
Equatorial radius	6052 km 0.95 (Earth = 1)
Mean density	5240 kg/m^3 0.95 (Earth = 1)
Surface gravity	8.87 m/s^2 0.91 (Earth = 1)
Escape speed	10.4 km/s
Sidereal rotation period	−243.0 solar days[*]
Axial tilt	177.4°
Surface magnetic field	<0.001 (Earth = 1)
Magnetic axis tilt relative to rotation axis	____
Mean surface temperature	730 K
Number of moons	0

A negative sign denotes retrograde rotation.

(a)

(b)

R I V U X G

▲ **FIGURE 9.7 Venus Mosaics** (a) This image of the surface of Venus was made by a radar transmitter and receiver on board the *Pioneer* spacecraft, which is still in orbit about the planet, but is now inoperative. The two continent-sized landmasses are named Ishtar Terra (upper left) and Aphrodite (lower right). Colors represent altitude: Blue is lowest, red highest. The spatial resolution is about 25 km. (b) A planetwide mosaic of *Magellan* images, colored in roughly the same way as part (a). Aphrodite Terra is at the center of this image. *(NASA)*

point and ringed by mountain ranges, including the Maxwell Montes range, which contains the highest peak on the planet, rising some 14 km above the level of Venus's deepest surface depressions. Again for comparison, the highest point on Earth (the summit of Mount Everest) lies about 20 km above the deepest section of Earth's ocean floor (Challenger Deep, at the bottom of the Marianas Trench on the eastern edge of the Philippines plate).

Figure 9.9(a) shows a large-scale *Venera* image of Lakshmi Planum, at a resolution of about 2 km. The "wrinkles" are actually chains of mountains, hundreds of kilometers long and tens of kilometers apart. The red area immediately to the right of the plain is Maxwell Montes. On the western (right-hand) slope of the Maxwell range lies a great crater, called Cleopatra, about 100 km across. Figure 9.9(b) shows a *Magellan* image of Cleopatra, which was originally thought to be volcanic in origin. Close-up views of the crater's structure, however, have led planetary scientists to conclude that the crater is meteoritic in origin, although some volcanic activity was apparently associated with its formation when the colliding body temporarily breached the planet's crust. Notice the dark (smooth) lava flow emerging from within the inner ring and cutting across the outer rim at the upper right.

It is now conventional to name features on Venus after famous women—Aphrodite, Ishtar, Cleopatra, and so on.

However, the early nonfemale names (e.g., Maxwell Montes, named after the Scottish physicist James Clerk Maxwell) predating this convention have stuck, and they are unlikely to change. Venus's other continent-sized formation, Aphrodite Terra, is located on the planet's equator and is comparable in size to Africa. Before *Magellan*'s arrival, some researchers had speculated that Aphrodite Terra might have been the site of something akin to seafloor spreading at the Mid-Atlantic ridge on Earth—a region where two lithospheric plates moved apart and molten rock rose to the surface in the gap between them, forming an extended ridge. ∞ (Sec. 7.4) With the low-resolution data then available, the issue could not be settled at the time.

The *Magellan* images now seem to rule out even this small-scale tectonic activity, and the Aphrodite region gives no indication of spreading. Figure 9.10(a) shows a portion of Aphrodite Terra called Ovda Regio. The crust appears buckled and fractured, with ridges running in two distinct directions across the image, suggesting that large compressive forces are distorting the crust. There seem to have been repeated periods of extensive lava flows. The dark regions are probably solidified lava flows. Some narrow lava channels, akin to rilles on the Moon, also appear. ∞ (Sec. 8.5) Such lava channels appear to be quite common on Venus. Unlike lunar rilles, however, they can be extremely long—hundreds or even thousands of kilometers (Figure 9.10b). These lava "rivers" often have lava "deltas" at their mouths, where they deposited their contents into the surrounding plains.

Figure 9.11 shows a series of angular cracks in the crust, thought to have formed when lava welled up from a deep fissure, flooded the surrounding area, and then retreated below the planet's surface. As the molten lava withdrew, the thin, new crust of solidified material collapsed under its own weight, forming the cracks we now see. Even taking into account the differences in temperature and composition between Venus's crust and Earth's, this terrain is not at all what we would expect at a spreading site similar to the Mid-Atlantic Ridge. ∞ (Sec. 7.4) Although there is no evidence for plate tectonics on Venus, it is likely that the stresses in the crust which led to the large mountain ranges were caused by convective motion within Venus's mantle—the same basic process that drives Earth's plates. Lakshmi Planum, for example, is probably the result of a "plume" of upwelling mantle material that raised and buckled the planet's surface.

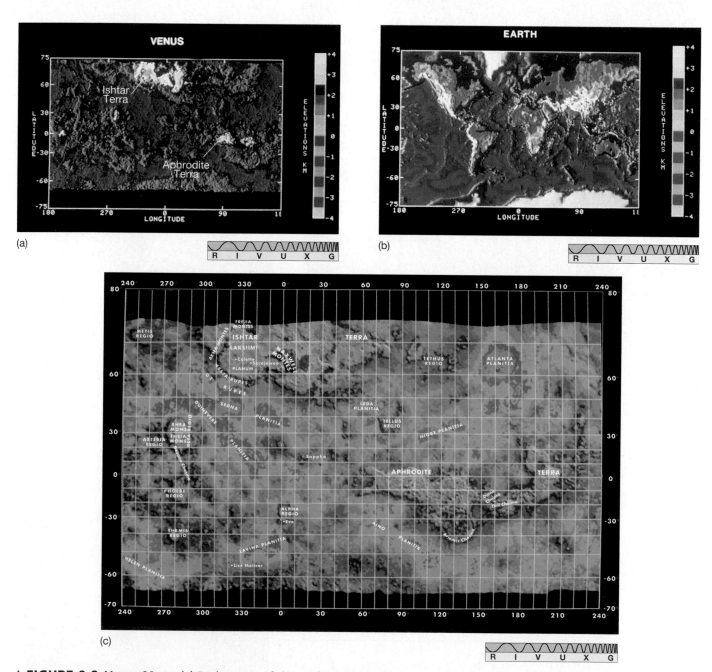

▲ **FIGURE 9.8 Venus Maps** (a) Radar map of the surface of Venus, based on *Pioneer Venus* data. Color represents elevation according to the scale at the right. (b) A similar map of Earth, at the same spatial resolution. (c) Another version of (a), with major surface features labeled. Compare with Figure 9.7, and notice how the projection exaggerates the size of surface features near the poles. *(NASA)*

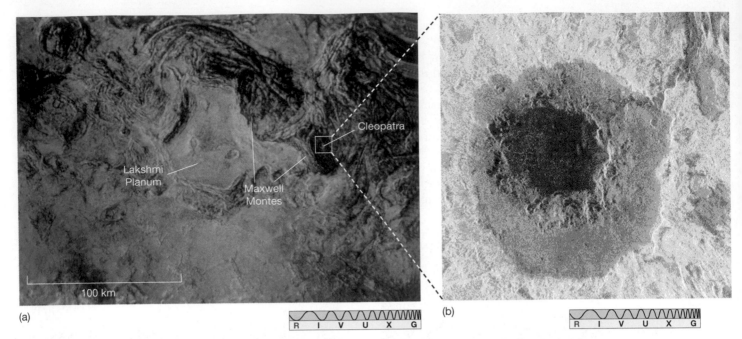

(a)

R I V U X G

(b)

R I V U X G

▲ **FIGURE 9.9 Ishtar Terra** (a) A *Venera* orbiter image of a plateau known as Lakshmi Planum in Ishtar Terra. The Maxwell Montes mountain range (red) lies on the western margin of the plain, near the right-hand edge of the image. A meteor crater named Cleopatra is visible on the western slope of the Maxwell range. Note the two larger craters in the center of the plain itself. (b) A *Magellan* image of Cleopatra showing a double-ringed structure that identifies the feature to geologists as an impact crater. *(NASA)*

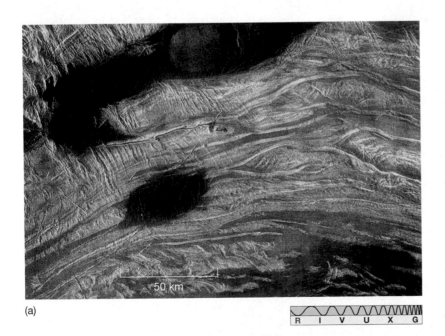

(a)

R I V U X G

▲ **FIGURE 9.10 Aphrodite Terra** *(a)* A *Magellan* image of Ovda Regio, part of Aphrodite Terra. The intersecting ridges indicate repeated compression and buckling of the surface. The dark areas represent regions that have been flooded by lava upwelling from cracks like those shown in Figure 9.11. (b) This lava channel in Venus's south polar region, known as Lada Terra, extends for nearly 200 km. *(NASA)*

(b)

R I V U X G

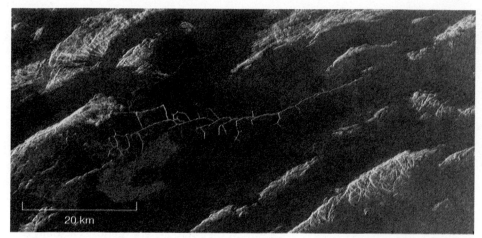

◄ **FIGURE 9.11** **Lava Flows** These cracks in Venus's surface, detected by *Magellan* in another part of Aphrodite Terra, have allowed lava to reach the surface and flood the surrounding terrain. The dark regions are smooth lava flows. The network of fissures visible here is about 50 km long. (*NASA*)

Volcanism and Cratering

Many areas of Venus have extensive volcanic features. Figure 9.12 shows a series of seven pancake-shaped **lava domes**, each about 25 km across. They probably formed when lava oozed out of the surface and then withdrew, leaving the crust to crack and subside. Lava domes such as these are found in numerous locations on Venus.

Most volcanoes on the planet are of the type known as **shield volcanoes**. Two large shield volcanoes, called Sif Mons and Gula Mons, are shown in Figure 9.13. Shield volcanoes, such as the Hawaiian Islands on Earth, are not associated with plate boundaries. Instead, they form when lava wells up through a "hot spot" in the crust and are built up over long periods by successive eruptions and lava flows. A characteristic of shield volcanoes is the formation

of a *caldera*, or crater, at the summit when the underlying lava withdraws and the surface collapses. The distribution of volcanoes over the surface of Venus appears random—quite different from the distribution on Earth, where volcanic activity clearly traces out plate boundaries (see Figure 7.9)—strongly supporting the idea that plate tectonics is absent on Venus.

The largest volcanic structures on Venus are huge, roughly circular regions known as **coronae** (singular: *corona*). A large corona, called Aine, can be seen in Figure 9.14, another large-scale mosaic of *Magellan* images. Unique to Venus, coronae appear to have been caused by upwelling mantle material, perhaps similar to the uplift that resulted in Lakshmi Planum, but on a somewhat smaller scale. They generally have volcanoes both in and

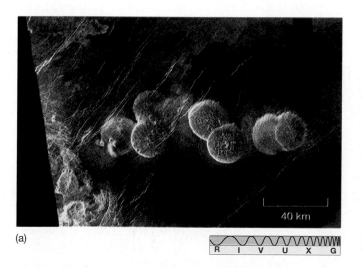

(a)

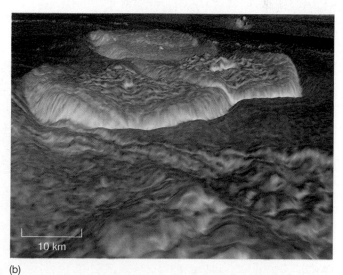

(b)

▲ **FIGURE 9.12** **Lava Dome** (a) These dome-shaped structures resulted when viscous molten rock bulged out of the ground and then retreated, leaving behind a thin, solid crust that subsequently cracked and subsided. *Magellan* has found features like this in several locations on Venus. (b) A three-dimensional representation of four of the domes. The computer view is looking toward the right from near the center of the image in part (a). Colors in (b) are based on data returned by Soviet *Venera* landers. (*NASA*)

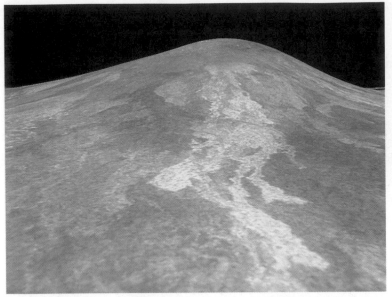

(b)

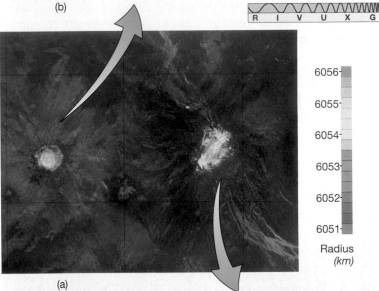

(a)

6056
6055
6054
6053
6052
6051

Radius
(km)

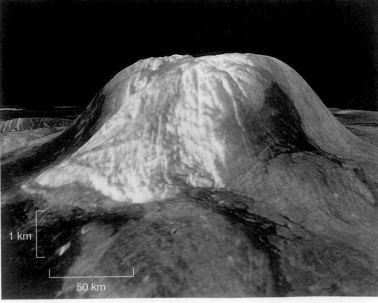

1 km

50 km

(c)

◄ **FIGURE 9.13 Volcanism on Venus** (a) Two larger volcanoes, known as Sif Mons (left) and Gula Mons, appear in this *Magellan* image. Color indicates height above a nominal planetary radius of 6052 km and ranges from purple (−1 km, the level of the surrounding plain) to orange (corresponding to an altitude of about 4 km). The two volcanic calderas at the summits are about 100 km across. (b) A computer-generated view of Sif Mons, as seen from ground level. (c) Gula Mons, as seen from ground level. In (b) and (c), the colors are based on data returned from Soviet landers, and the vertical scales have been greatly exaggerated (by about a factor of 40), so these mountains look much taller relative to their widths than they actually are; Venus is actually a remarkably flat place. *(NASA)*

around them, and closer inspection of the rims usually shows evidence for extensive lava flows into the plains below.

There is overwhelming evidence for past surface activity on Venus. Has this activity now stopped, or is it still going on? Two pieces of indirect evidence suggest that volcanism continues today. First, the level of sulfur dioxide above Venus's clouds shows large and fairly frequent fluctuations. It is quite possible that these variations result from volcanic eruptions on the surface. If so, volcanism may be the primary cause of Venus's thick cloud cover. Second, both the *Pioneer Venus* and the *Venera* orbiters observed bursts of radio energy from Aphrodite and other regions of the planet's surface. The bursts are similar to those produced by lightning discharges that often occur in the plumes of erupting volcanoes on Earth, again suggesting ongoing activity. However, while these pieces of evidence are quite persuasive, they are still only circumstantial. No "smoking gun" (or to be more precise, erupting volcano) has yet been seen, so the case for active volcanism is not yet complete.

Not all the craters on Venus are volcanic in origin: Some, like Cleopatra (Figure 9.9b), were formed by meteoritic impact. Large impact craters on Venus are generally circular, but those less than about 15 km in diameter can be quite asymmetric in appearance. Figure 9.15(a) shows a *Magellan* image of a relatively small meteoritic impact crater, about 10 km across, in Venus's southern hemisphere. Geologists think that the light-colored region is the ejecta blanket—material ejected from the crater following the impact. The odd shape may be the result of a large meteoroid's breaking up just before impact into pieces that hit the surface near one another. Making craters such as these seems to be a fairly common fate for

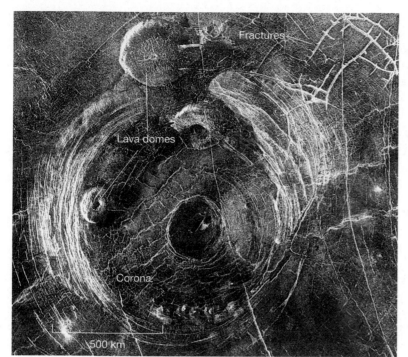

Fractures

Lava domes

Corona

500 km

◄ **FIGURE 9.14 Venus Corona** This corona, called Aine, lies in the plains south of Aphrodite Terra and is about 300 km across. It is probably the result of mantle material that upwelled, causing the surface to bulge outward. Note the pancake-shaped lava domes at the top, the many fractures in the crust around the corona, and the large impact craters with their surrounding white (rough) ejecta blankets that stud the region. *(NASA)*

R I V U X G

medium-sized bodies (1 km or so in diameter) that plow through Venus's dense atmosphere. Figure 9.15(b) shows the largest known impact feature on Venus: the 280-km-diameter crater called Mead. Its double-ringed structure is in many ways similar to the Moon's Mare Orientale (Figure 8.15). Numerous impact craters (identifiable by their ejecta blankets) can also be discerned in Figure 9.14.

Venus's atmosphere is sufficiently thick that small meteoroids do not reach the ground, so there are no impact craters smaller than about 3 km across. Atmospheric effects probably also account for the observed scarcity of impact craters less than 25 km in diameter. Overall, the rate of formation of large-diameter craters on Venus's surface seems to be only about one-tenth that in the lunar maria.

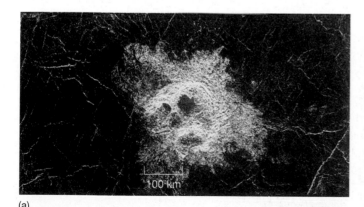

(a)

▲ **FIGURE 9.15 Impact Cratering on Venus** (a) A *Magellan* image of an apparent multiple-impact crater in Venus's southern hemisphere. The irregular shape of the light-colored ejecta seems to be the result of a meteoroid that fragmented just prior to impact. The dark regions in the crater may be pools of solidified lava. (b) Venus's largest crater, named Mead after anthropologist Margaret Mead, is about 280 km across. Bright (rough) regions clearly show its double-ringed structure. *(NASA)*

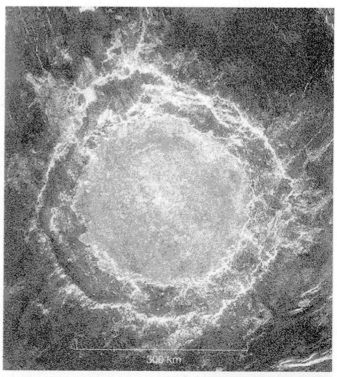

(b)

R I V U X G

Applying the same crater-age estimates to Venus as we do to Earth and the Moon suggests that much of the surface of Venus is quite young—less than a billion years old. Some planetary scientists have suggested that some areas, such as the region shown in Figure 9.13, are even younger—perhaps as little as 200 or 300 million years. Although erosion by the planet's atmosphere may play some part in obliterating surface features, the main agent is volcanism, which appears to have "resurfaced" much of the planet about 500 hundred million years ago.

Data from the Soviet Landers

The 1975 soft landings of the Soviet *Venera 9* and *Venera 10* spacecraft directly established that Venus's surface is dry and dusty. Figure 9.16(a) shows one of the first photographs of the surface of Venus radioed back to Earth. Each craft lasted only about an hour before overheating, their electronic circuitry literally melting in this planetary oven. Typical rocks in the photo measure about 50 cm by 20 cm across—a little like flagstones on Earth. Sharp-edged and slablike, these rocks show little evidence of erosion. Apparently they are quite young rocks, again supporting the idea of ongoing surface activity of some kind on Venus.

Later *Venera* missions took more detailed photographs, as shown in Figure 9.16(b). The presence of small rocks and finer material indicates the effects of erosive processes. These later missions also performed simple chemical analyses of the surface of Venus. The samples studied by *Venera 13* and *Venera 14* were predominantly basaltic in nature, again implying a volcanic past. However, not all the rocks were found to be basaltic: The *Venera 17* and *Venera 18* landers also found surface material resembling terrestrial granite, probably (as on Earth) part of the planet's ancient crust.

CONCEPT CHECK

✔ Are volcanoes on Venus associated mainly with the movement of tectonic plates, as on Earth?

(a)

R I V U X G

(b)

▲ **FIGURE 9.16** **Venus In Situ** (a) The first direct view of the surface of Venus, radioed back to Earth from the Soviet *Venera 9* spacecraft, which made a soft landing on the planet in 1975. The amount of sunlight penetrating Venus's cloud cover apparently resembles that on a heavily overcast day on Earth. (b) Another view of Venus, in true color, from *Venera 14.* Flat rocks like those visible in part (a) are seen among many smaller rocks and even fine soil on the surface. This landing site is not far from the *Venera 9* site shown in (a). The peculiar filtering effects of whatever light does penetrate the clouds make Venus's air and ground appear peach colored—in reality, they are most likely gray, like rocks on Earth. *(Russian Space Agency)*

9.5 The Atmosphere of Venus

Atmospheric Structure

Measurements made by the *Venera* and *Pioneer Venus* spacecraft have allowed us to paint a fairly detailed picture of Venus's atmosphere. ∞ (Sec. 6.6) Figure 9.17 shows the variation of temperature and pressure with height. Compare this figure with Figure 7.4, which gives similar information for Earth. The atmosphere of Venus is about 90 times more massive than Earth's, and it extends to a much greater height above the surface. On Earth, 90 percent of the atmosphere lies within about 10 km of sea level. On Venus, the 90 percent level is found at an altitude of 50 km instead. The surface temperature and pressure of Venus's atmosphere are much greater than Earth's. However, the temperature drops more rapidly with altitude, and the upper atmosphere of Venus is actually colder than our own.

Venus's troposphere extends up to an altitude of nearly 100 km. The reflective clouds that block our view of the surface lie between 50 and 70 km above the surface. Data from the *Pioneer Venus* multiprobe indicate that the clouds may actually be separated into three distinct layers within that altitude range. Below the clouds, extending down to an altitude of some 30 km, is a layer of haze. Below 30 km, the air is clear. Above the clouds, a high-speed "jet stream" blows from west to east at about 300–400 km/h, fastest at the equator and slowest at the poles. This high-altitude flow is responsible for the rapidly moving cloud patterns seen in ultraviolet light. Figure 9.18 shows a sequence of three ultraviolet images of Venus in which the variations in the cloud patterns can be seen. Note the characteristic V-shaped appearance of the clouds—a consequence of the fact that, despite their slightly lower speeds, the winds near the poles have a shorter distance to travel in circling the planet and so are always forging ahead of winds at the equator. Near the surface, the dense atmosphere moves more sluggishly—indeed, the fluid flow bears more resemblance to that in Earth's oceans than to the flow in Earth's air. Surface wind speeds on Venus are typically less than 2 m/s (roughly 4 mph).

Atmospheric Composition

Carbon dioxide is the dominant component of Venus's atmosphere, accounting for 96.5 percent of it by volume. Almost all of the remaining 3.5 percent is nitrogen. Trace amounts of other gases, such as water vapor, carbon monoxide, sulfur dioxide, and argon, are also present. This composition is clearly radically different from that of Earth's atmosphere. The absence of oxygen is perhaps not surprising, given the absence of life. (Recall our discussion of Earth's atmosphere in Chapter 7.) ∞ (Sec. 7.2) However, there is no sign of the large amount of water vapor we would expect to find if a volume of water equivalent to Earth's oceans had evaporated and remained in the planet's atmosphere. If Venus started off with an Earth-like composition, then something has happened to its water—Venus is now a very dry planet.

For a long time, the chemical makeup of the reflective cloud layer surrounding Venus was unknown. At first, scientists assumed that the clouds were water vapor or ice, as on Earth, but the reflectivity of the clouds at different wavelengths didn't match that of water ice. Later infrared observations carried out in the 1970s showed that the clouds (or at least the top layer of clouds) are actually composed of sulfuric acid, created by reactions between water and sulfur dioxide. Sulfur dioxide is an excellent absorber of ultraviolet radiation and could be responsible for many of the cloud patterns seen in ultraviolet light. Spacecraft observations confirmed the presence of all three compounds in the atmosphere and also indicated that there may be particles of sulfur suspended in and near the cloud layers, which may account for Venus's characteristic yellowish hue.

The Greenhouse Effect on Venus

Given the distance of Venus from the Sun, the planet was not expected to be such a pressure cooker. As mentioned earlier, calculations based on Venus's orbit and reflectivity indicated a temperature not much different from Earth's, and early measurements of the cloud temperatures seemed to concur. Certainly, scientists

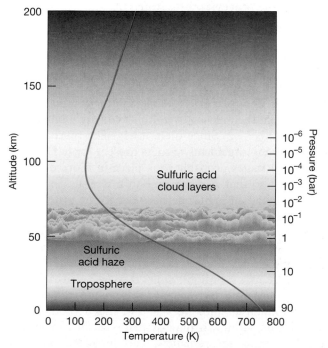

▲ **FIGURE 9.17 Venus's Atmosphere** The structure of the atmosphere of Venus, as determined by U.S. and Soviet probes. (One bar is the atmospheric pressure at sea level on Earth.)

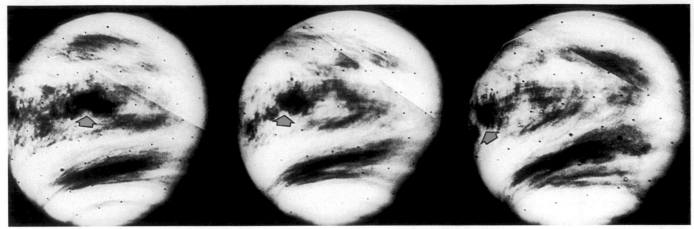

▲ **FIGURE 9.18 Atmospheric Circulation** Three ultraviolet views of Venus, taken by the *Pioneer Venus* orbiter, showing the changing cloud patterns in the planet's upper atmosphere. The wind flow is from right to left, in the direction opposite the sideways "V" in the clouds. Notice the motion of the dark region marked by the blue arrow. Venus's retrograde rotation means that north is at the bottom of these images and west is to the right. The time difference between the left and right photographs is about 20 hours. *(NASA)*

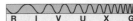

reasoned, Venus could be no hotter than the sunward side of Mercury, and it should probably be much cooler. This reasoning was obviously seriously in error.

Why is Venus's atmosphere so hot? And if, as we think, Venus started off like Earth, why is it now so different? The answer to the first question is fairly easy: Given the present composition of its atmosphere, Venus is hot because of the greenhouse effect. Recall from our discussion in Chapter 7 that "greenhouse gases" in Earth's atmosphere—particularly water vapor and carbon dioxide—serve to trap heat from the Sun. ∞ (Sec. 7.2) By inhibiting the escape of infrared radiation reradiated from Earth's surface, these gases increase the planet's equilibrium temperature, in much the same way as an extra blanket keeps you warm on a cold night. Continuing the analogy a little further, the more blankets you place on the bed, the warmer you will become. Similarly, the more greenhouse gases there are in the atmosphere, the hotter the surface will be.

The same effect occurs naturally on Venus, whose dense atmosphere is made up almost entirely of a primary greenhouse gas, carbon dioxide. As illustrated schematically in Figure 9.19, the thick carbon dioxide blanket absorbs nearly 99 percent of all the infrared radiation released from the surface of Venus and is the immediate cause of the planet's sweltering 730 K surface temperature. Furthermore, the temperature is nearly as high at the poles as at the equator, and there is not much difference between the temperatures on the day and night sides. The circulation of the atmosphere spreads energy efficiently around the planet, making it impossible to escape the blazing heat, even during the planet's two-month-long night.

The Runaway Greenhouse Effect

But *why* is Venus's atmosphere so different from Earth's? Let's assume that the two planets started off with basically similar compositions. Why, then, is there so much carbon dioxide in the atmosphere of Venus, and why is the planet's atmosphere so dense? To address these questions, we must consider the processes that created the atmospheres of the terrestrial planets and then determined their evolution. In fact, we can turn the question around and ask instead, "Why is there so *little* carbon dioxide in Earth's atmosphere compared with that of Venus?"

Earth's atmosphere has evolved greatly since it first appeared. Our planet's secondary atmosphere was outgassed from the interior by volcanic activity 4 billion years ago. ∞ (Sec. 7.2) Since then, it has been reprocessed, in part by living organisms, into its present form. On Venus, the initial stages probably took place in more or less the same way, so that at some time in the past, Venus might well have had an atmosphere similar to the primitive secondary atmosphere on Earth, containing water, carbon dioxide, sulfur dioxide, and nitrogen-rich compounds. What happened on Venus to cause such a major divergence from subsequent events on our own planet?

On Earth, sunlight split the nitrogen-rich compounds, releasing nitrogen into the air. Meanwhile, the water condensed into oceans, and much of the carbon dioxide and sulfur dioxide eventually became dissolved in them. Most of the remaining carbon dioxide combined with surface rocks. Thus, much of the secondary outgassed atmosphere quickly became part of the surface of the planet. If all the dissolved or chemically combined carbon dioxide were released back into Earth's present-day

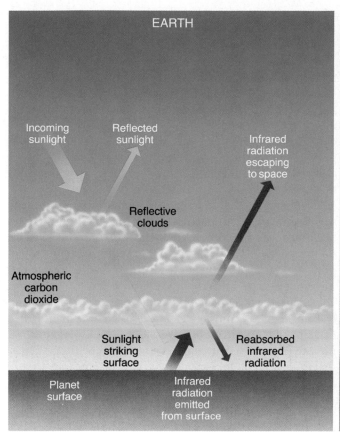

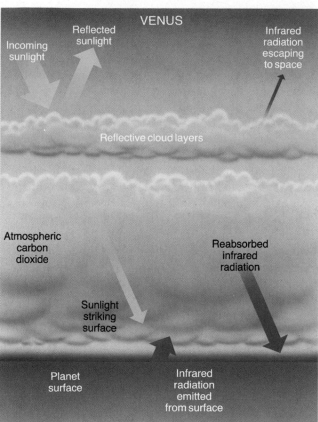

▲ **FIGURE 9.19 Greenhouse Effect on Earth and Venus** Because Venus's atmosphere is much thicker and denser than Earth's, a much smaller fraction of the infrared radiation leaving the planet's surface actually escapes into space. The result is a much stronger greenhouse effect than on Earth and a correspondingly hotter planet. The outgoing infrared radiation is not absorbed at a single point in the atmosphere; instead, absorption occurs at all atmospheric levels. (The arrows are meant to indicate only that absorption occurs, not that it occurs at one specific level.)

atmosphere, its new composition would be 98 percent carbon dioxide and 2 percent nitrogen, and it would have a pressure about 70 times its current value. In other words, apart from the presence of oxygen (which appeared on Earth only after the development of life) and water (the absence of which on Venus will be explained shortly), Earth's atmosphere would be a lot like that of Venus! The real difference between Earth and Venus, then, is that Venus's greenhouse gases never left the atmosphere, the way they did on Earth.

When Venus's secondary atmosphere appeared, the temperature was higher than on Earth, simply because Venus is closer to the Sun. However, the Sun was probably somewhat dimmer then (see Chapter 22)—perhaps only half its present brightness—so there is some uncertainty as to exactly how much hotter than Earth Venus actually was. If the temperature was already so high that no oceans condensed, the outgassed water vapor and carbon dioxide would have remained in the atmosphere, and the full greenhouse effect would have gone into operation immediately. If oceans did form and most of the greenhouse

gases left the atmosphere, as they did on Earth,[*] the temperature must still have been sufficiently high that a process known as the **runaway greenhouse effect** came into play.

To understand the runaway greenhouse effect, imagine that we took Earth from its present orbit and placed it in Venus's orbit, some 30 percent closer to the Sun. At that distance from the Sun, the amount of sunlight striking Earth's surface would be about twice its present level, so the planet would warm up. More water would evaporate from the oceans, leading to an increase in atmospheric water vapor. At the same time, the ability of both the oceans and surface rocks to hold carbon dioxide would diminish, allowing more carbon dioxide to enter the atmosphere. As a result, greenhouse heating would increase, and the planet would warm still further, leading to a further increase in atmospheric greenhouse gases, and so on. Once

*In fact, careful study of the Magellan images reveals no sign of ancient seashores or ocean basins, nor evidence of erosion by rivers on Venus. However, it is unclear whether such features would have survived the heavy volcanism known to have occurred in the planet's more recent past.

started, the process would "run away," eventually leading to the complete evaporation of the oceans, restoring all the original greenhouse gases to the atmosphere. Although the details are quite complex, basically the same thing would have happened on Venus long ago, ultimately resulting in the planetary inferno we see today.

The presence of atmospheric water vapor meant that the greenhouse effect on Venus was even more extreme in the past. By intensifying the blanketing effect of the carbon dioxide, the water vapor helped the surface of Venus reach temperatures perhaps twice as hot as current temperatures. At the high temperatures of the past, the water vapor was able to rise high into the planet's upper atmosphere—so high that it was broken up by solar ultraviolet radiation into its components, hydrogen and oxygen. The light hydrogen rapidly escaped, the reactive oxygen quickly combined with other atmospheric gases, and all water on Venus was lost forever. This is the reason that Venus lacks water today.

Although it is highly unlikely that global warming will ever send Earth down the path taken by Venus, this episode highlights the relative fragility of the planetary environment. No one knows how close to the Sun Earth could have formed before a runaway greenhouse effect would have occurred. But in comparing our planet with Venus, we have come to understand that there is an orbital limit, presumably between 0.7 and 1.0 A.U., inside of which Earth would have suffered a similar catastrophic runaway. We must consider this "greenhouse limit" when we assess the likelihood that planets harboring life formed elsewhere in our Galaxy.

CONCEPT CHECK

✔ If Venus had formed at Earth's distance from the Sun, what might its climate be like today?

9.6 Venus's Magnetic Field and Internal Structure

In 1962, *Mariner 2* flew by Venus, carrying, among other instruments, magnetometers to measure the strength of the planet's magnetic field. None was detected, and subsequent Soviet and U.S. missions, carrying more sensitive detectors, have confirmed this finding. Venus,

with an average density similar to Earth's, probably has a similar overall composition and a partially molten iron-rich core. The lack of any detectable magnetic field on Venus, then, is almost surely the result of the planet's extremely slow rotation and consequent lack of dynamo action. ∞ (Sec. 7.5) Having no magnetosphere, Venus has no protection from the solar wind. The planet's upper atmosphere is continually bombarded by high-energy particles from the Sun, keeping the topmost layers permanently ionized. However, the great thickness of the atmosphere prevents any of these particles from reaching the surface.

None of the *Venera* landers carried seismic equipment, so no direct measurements of the planet's interior have been made, and theoretical models of the interior have little hard data to constrain them. However, to many geologists, the surface of Venus resembles that of the young Earth, at an age of perhaps a billion years. At that time, volcanic activity had begun, but the crust was still relatively thin and the convective processes in the mantle that drive plate tectonic motion were not yet fully established. Measurements of the planet's gravitational field suggest that Venus lacks an asthenosphere, the semisolid part of the upper mantle over which Earth's lithosphere slides. ∞ (Sec. 7.4)

Why has Venus remained in that immature state and not developed plate tectonics as Earth did? That question remains to be answered. Some planetary geologists have speculated that the high surface temperature has inhibited Venus's evolution by slowing the planet's cooling. Possibly, the high surface temperature has made the crust too soft for Earth-style plates to develop—or perhaps the high temperature and soft crust have led to more volcanism, tapping the energy that might otherwise go into convective motion. It may also be that the presence of water plays an important role in lubricating convection in the mantle and plate motion, so that arid Venus could never have evolved along the same path as Earth.

CONCEPT CHECK

✔ If the interior of Venus is quite Earth-like, and Venus has a molten iron core, why doesn't the planet have a magnetic field as Earth does?

Chapter Review

SUMMARY

The interior orbit of Venus with respect to Earth's means that Venus never strays far from the Sun in the sky. Because of its highly reflective cloud cover, Venus is brighter than any star in the sky, as seen from Earth. Indeed, the planet is so bright that it can be seen even in the daytime.

The extremely thick atmosphere of Venus is nearly opaque to visible radiation, making the planet's surface invisible from the outside. Spectroscopic examination of sunlight reflected from the planet's cloud tops shows the presence of large amounts of carbon dioxide. Venus's atmosphere is nearly 100 times denser than Earth's. The temperature of the upper atmosphere is much like that of Earth's upper atmosphere, but the surface temperature of Venus is 730 K. The planet's rotation is slow and retrograde, most likely because of a collision between Venus and some other solar system body during the late stages of the planet's formation.

Venus's surface has been thoroughly mapped by radar from Earth-based radio telescopes and orbiting satellites. The most recent and most thorough survey has been carried out by the U.S. *Magellan* satellite. The planet's surface is mostly smooth, resembling rolling plains with modest highlands and lowlands. Two elevated continent-sized regions are called Ishtar Terra and Aphrodite Terra. There is no evidence for plate tectonic activity as on Earth.

Many **lava domes (p. 239)** and **shield volcanoes (p. 239)** have been found by *Magellan* on Venus's surface, but none of the volcanoes has yet been proven to be currently active. The planet's surface shows no sign of plate tectonics. Features called **coronae (p. 239)** are thought to have been caused by an upwelling of mantle material. For unknown reasons, the upwelling never developed into full convective motion. The surface of the planet appears to be relatively young, resurfaced by volcanism every few hundred million years. Some craters on Venus are due to mete-

oritic impact, but the majority appear to be volcanic in origin. The evidence for currently active volcanoes on Venus includes surface features resembling those produced in Earthly volcanism, fluctuating levels of sulfur dioxide in Venus's atmosphere, and bursts of radio energy similar to those produced by lightning discharges that often occur in the plumes of erupting volcanoes on Earth. However, no actual eruptions have been seen.

Soviet spacecraft that landed on Venus photographed surface rocks with sharp edges and a slablike character. Some rocks on Venus appear predominantly basaltic in nature, implying a volcanic past. Other rocks resemble terrestrial granite and are probably part of the planet's ancient crust.

Venus is comparable in both mass and radius to Earth, suggesting that the two planets started off with fairly similar surface conditions. However, the atmospheres of Earth and Venus are now very different. The total mass of Venus's atmosphere is about 90 times greater than Earth's. The greenhouse effect stemming from the large amount of carbon dioxide in Venus's atmosphere is the basic cause of the planet's current high temperatures. Almost all the water vapor and carbon dioxide initially present in Earth's early atmosphere quickly became part of the oceans or surface rocks. Because Venus orbits closer to the Sun than does Earth, surface temperatures on Venus were initially higher, and the planet's greenhouse gases never left the atmosphere. The **runaway greenhouse effect (p. 245)** caused all the planet's greenhouse gases—carbon dioxide and water vapor—to end up in the atmosphere, leading to the extreme conditions we observe today.

Venus has no detectable magnetic field, almost certainly because the planet's rotation is too slow for any appreciable dynamo effect to have developed. To some planetary geologists, Venus's interior structure suggests that of the young Earth, before convection became established in the mantle.

REVIEW AND DISCUSSION

1. Why does Venus appear so bright to the eye? Upon what factors does the planet's brightness depend?

2. Explain why Venus is always found in the same general part of the sky as the Sun.

3. Why do astronomers think that the near resonance between Venus's rotation and revolution, as seen from Earth, is not a true resonance?

4. Describe one observational problem associated with Venus's near resonance of rotation and revolution.

5. What is our current best explanation of Venus's slow, retrograde spin?

6. If you were standing on Venus, how would Earth look?

7. How did radio observations of Venus made in the 1950s change our conception of the planet?

8. What did ultraviolet images returned by *Pioneer Venus* show about the planet's high-level clouds?

9. Name three ways in which the atmosphere of Venus differs from that of Earth.

10. What are the main constituents of Venus's atmosphere? What are clouds in the upper atmosphere made of?

11. What component of Venus's atmosphere causes the planet to be so hot? Explain why there is so much of this gas in the atmosphere of Venus, compared with its presence in Earth's atmosphere. What happened to all the water that Venus must have had when the planet formed?

12. Earth and Venus are nearly alike in size and density. What primary fact caused one planet to evolve as an oasis for life, while the other became a dry and inhospitable inferno?

13. If Venus had formed at Earth's distance from the Sun, what do you imagine its climate would be like today? Why do you think so?

14. How do the "continents" of Venus differ from Earth's continents?

15. How are the impact craters of Venus different from those found on other bodies in the solar system?

16. What evidence exists that volcanism of various types has changed the surface of Venus?

17. What is the evidence for active volcanoes on Venus?

18. Given that Venus, like Earth, probably has a partially molten iron-rich core, why doesn't Venus also have a magnetic field?

19. Do you think there might be life on Venus? Explain your answer.

20. Do you think that Earth is in any danger of being subject to a runaway greenhouse effect like that on Venus?

CONCEPTUAL SELF-TEST: TRUE OR FALSE/MULTIPLE CHOICE

1. Venus has a retrograde orbit around the Sun.

2. Venus is brightest when it is in its full phase.

3. The entire surface of Venus has been mapped by means of radar observations from Earth.

4. The atmosphere of Venus is quite similar to that of Earth.

5. Venus has roughly the same temperature at its equator as at its poles.

6. Images from Magellan confirm that Venus has tectonic activity like that on Earth.

7. Lava flows are common on the surface of Venus.

8. Venus's mass has been accurately measured by orbiting spacecraft.

9. Most craters on the surface of Venus are the result of volcanism.

10. The greenhouse effect in the early atmosphere of Venus was most likely intensified by the presence of water vapor.

11. Venus is never seen at midnight because (a) it is closer to the Sun than is Earth; (b) it will be in its new phase then; (c) it is visible only at sunset; (d) it will be at superior conjunction.

12. Venus's permanent retrograde rotation about its axis results in the planet's (a) always rising in the western sky; (b) orbiting the Sun in the opposite direction from Earth; (c) having its north pole below the plane of the ecliptic; (d) being brighter than any other planet.

13. Compared with Earth, Venus is (a) much smaller; (b) much larger; (c) about the same size.

14. Venus's surface is permanently obscured by clouds. As a result, the surface has been studied primarily by (a) robotic landers; (b) orbiting satellites using radar; (c) spectroscopy; (d) radar signals from Earth.

15. Compared with Earth, Venus has a level of plate tectonic activity that is (a) much more rapid; (b) virtually nonexistent; (c) about the same.

16. Venus's atmosphere (a) has almost the same chemical composition as Earth's; (b) shows very high levels of humidity; (c) is composed mostly of carbon dioxide; (d) is predominantly made of acid droplets.

17. Compared with Earth's atmosphere, most of Venus's atmosphere is (a) compressed much closer to the surface; (b) spread out much farther from the surface; (c) similar in extent and structure.

18. Venus's atmospheric temperature is (a) about the same as Earth's; (b) cooler than temperatures on the planet Mercury; (c) hotter than temperatures on Mercury; (d) high due to the presence of sulfuric acid.

19. Carbon dioxide on Venus (a) is all in the atmosphere; (b) was absorbed in surface water and has evaporated into space; (c) has dissolved in the atmospheric acid; (d) is integrated into the surface rocks.

20. Venus lacks a planetary magnetic field because (a) it rotates very slowly; (b) it does not have a molten core; (c) there are no plate tectonics on the planet; (d) the core contains little or no iron.

PROBLEMS

 Algorithmic versions of these questions are available in the Practice Problems module of the Companion Website at astro.prenhall.com/chaisson.

The number of squares preceding each problem indicates its approximate level of difficulty.

1. ■■ Using the data given in the text, calculate Venus's angular diameter, as seen by an observer on Earth, when the planet is (a) at its brightest, (b) at greatest elongation, and (c) at the most distant point in its orbit.

2. ■■ Seen from Earth, through how many degrees per night (relative to the stars) does Venus move around the time of inferior conjunction (closest approach to Earth)?

3. ■ How long does a radar signal take to travel from Earth to Venus and back when Venus is brightest? Compare this time with the round-trip time when Venus is at its closest point to Earth.

4. ■■ What would be the length of a solar day on Venus if the planet's rotation were prograde rather than retrograde?

5. ■■■ Calculate the mean angular orbital speeds of Venus and Mercury, in degrees per day. (For simplicity, imagine that the planets have circular orbits.) On the basis of these speeds,

how long does it take for Mercury to "lap" Venus—that is, to complete exactly one extra revolution around the Sun? This length of time is the synodic period of Mercury, as seen from Venus.

6. ■ Draw a diagram showing the positions of Earth and Venus in their orbits over the course of one Venus synodic year (584 Earth days), starting at the point of closest approach of the planets. Mark the locations of the planets at intervals of 73 days, and indicate with an arrow (as in Figure 9.4) the orientation of some point on Venus's surface at each instant.

7. ■■ Compare the magnitude of the tidal gravitational acceleration on Venus (the difference between the accelerations at center and surface) due to Earth at closest approach with Venus's surface gravitational acceleration. Assume circular orbits for both planets. Repeat the question for the tidal acceleration on Venus due to the Sun. What do you conclude about the possibility that Venus's near resonance with Earth is the result of Earth's tidal influence?

8. ■ What is the size of the smallest feature that could be distinguished on the surface of Venus (at closest approach) by the Arecibo radio telescope at an angular resolution of 1′?

9. ■ Could an infrared telescope with an angular resolution of 0.1″ distinguish impact craters on the surface of Venus?

10. ■ When used as a radar instrument, the Arecibo installation can distinguish echoes received as little as 10^{-5} s apart. To what distance does this correspond? That distance is the effective resolution of Arecibo in making radar observations of Venus.

11. ■ *Pioneer Venus* observed high-level clouds moving around Venus's equator in four days. What was their speed in km/h? In mph?

12. ■ Approximating Venus's atmosphere as a layer of gas 50 km thick, with uniform density 21 kg/m^3, calculate the total mass of the atmosphere. Compare your answer with the mass of Earth's atmosphere (Chapter 7, Problem 3) and with the mass of Venus.

13. ■ According to Stefan's law (see Section 3.4), how much more radiation—per square meter, say—is emitted by Venus's surface at 730 K than is emitted by Earth's surface at 300 K?

14. ■■■ In the absence of any greenhouse effect, Venus's average surface temperature, like Earth's, would be about 250 K. In fact, it is about 730 K. Use this information and Stefan's law to estimate the fraction of infrared radiation leaving Venus's surface that is absorbed by carbon dioxide in the planet's atmosphere.

15. ■■ Calculate the orbital period of the *Magellan* spacecraft, moving around Venus on an elliptical orbit with a minimum altitude of 294 km and a maximum altitude of 8543 km above the planet's surface. In 1993, the spacecraft's orbit was changed to have minimum and maximum altitudes of 180 km and 541 km, respectively. What was the new period?

 In addition to the Practice Problems module, the Companion Website at astro.prenhall.com/chaisson provides for each chapter a study guide module with multiple choice questions as well as additional annotated images, animations, and links to related Websites.

10

The search for life on Mars continues unabated. One key ingredient for life any-where in the universe is probably water, therefore much of the current controver-sy centers around whether Mars—now drier than any of Earth's deserts—was ever much wetter than it is today. Here, the Mars Global Surveyor spacecraft took this remarkable image in 2001 of what seems to be extensive layers and gullies of sand and rock near the Mariner valley. This view measures only 1.5 kilometers across and its smallest features are resolved to about 10 meters. The question is: Were those features really caused by flowing water long ago or perhaps by wind action ▶ that carved them over eons of time? (NASA)

Mars

A Near Miss for Life?

Named by the ancient Romans for their bloody god of war, Mars is for many people the most intriguing of all celestial objects. Over the years, it has inspired speculation that life—perhaps intelligent and possibly hostile—may exist there. With the dawn of the Space Age, those notions had to be abandoned. Visits by robot spacecraft have revealed no signs of life of any sort, even at the microbial level, on Mars. Still, the planet's properties are close enough to those of Earth that Mars is even now widely regarded as the second most hospitable environment for the appearance of life in the solar system, after Earth itself. At about the same time as Earth's "twin," Venus, was evolving into a searing inferno, the Mars of long ago may have had running water and blue skies. If life ever arose there, however, it must be long extinct. The Mars of today appears to be a dry, dead world.

LEARNING GOALS

Studying this chapter will enable you to

1 Summarize the general orbital and physical properties of Mars.

2 Describe the observational evidence for seasonal changes on Mars.

3 Compare the surface features and geology of Mars with those of the Moon and Earth, and account for these characteristics in terms of Martian history.

4 Discuss the evidence that Mars once had a much denser atmosphere and running water on its surface.

5 Compare the atmosphere of Mars with those of Earth and Venus, and explain why the evolutionary histories of these three worlds diverged so sharply.

6 Describe the characteristics of Mars's moons, and explain their probable origin.

 Visit astro.prenhall.com/chaisson for additional images, animations, and links to related sites for this chapter.

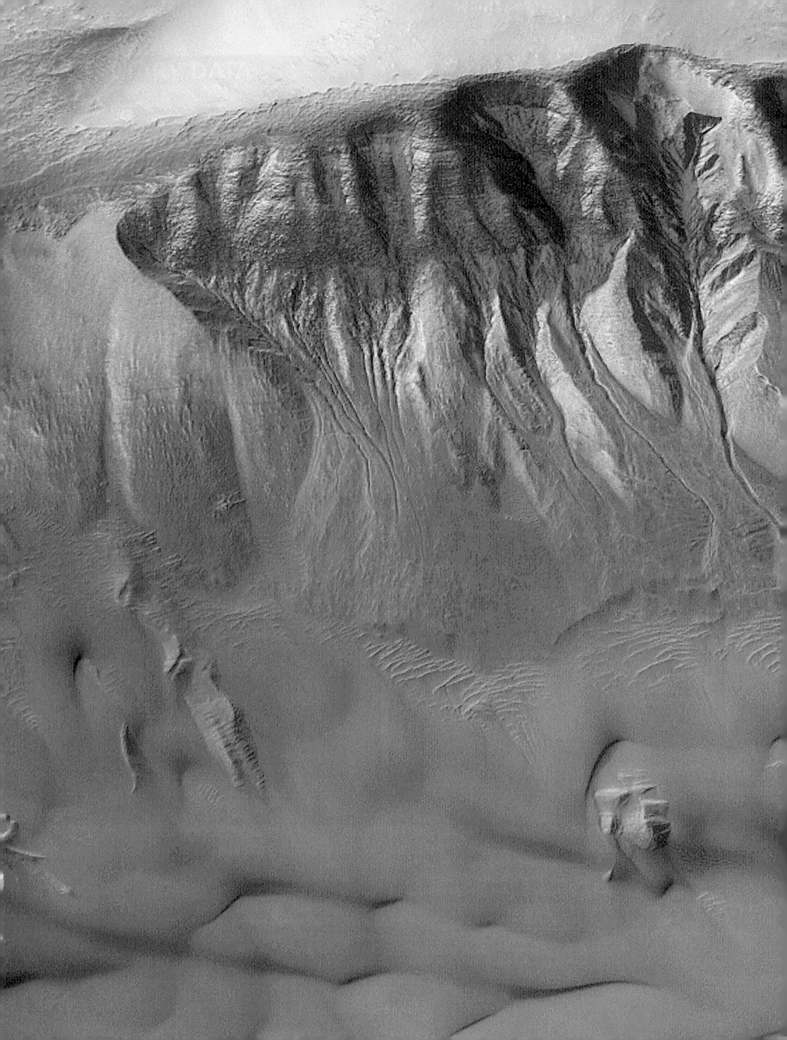

their individual properties at the end of the chapter. The larger of the two, Phobos, orbits at a distance of just 9378 km from the center of the planet once every 459 minutes. Applying the modified version of Kepler's third law (which states that the square of a moon's orbital period is proportional to the cube of its orbital semimajor axis divided by the mass of the planet it orbits), we find that the mass of Mars is 6.4×10^{23} kg, or 0.11 times that of Earth. ∞ (Sec. 2.7) Naturally, the orbit of Deimos yields the same result.

From the planet's mass and radius, it follows that the average density of Mars is 3900 kg/m^3, only slightly greater than that of the Moon. If we assume that Martian surface rocks are similar to those on the other terrestrial planets, this average density suggests the existence of a substantial higher density core within the planet. Planetary scientists suspect that this core is composed largely of iron sulfide (a compound about twice as dense as Martian surface rock) and has a diameter of about 2500 km.

Surface markings easily seen on Mars allow astronomers to track the planet's rotation. Mars rotates once on its axis every 24.6 hours. One Martian day is thus similar in length to one Earth day. The planet's equator is inclined to the orbital plane at an angle of 24.0°, again similar to Earth's inclination of 23.5°. Thus, as Mars orbits the Sun, we find both daily and seasonal cycles, just as on Earth. In the case of Mars, however, the seasons are complicated somewhat by variations in solar heating due to the planet's eccentric orbit—southern summer occurs around the time of Martian perihelion and so is significantly warmer than summer in the north.

10.3 Long-Distance Observations of Mars

At opposition, when Mars is closest to us and most easily observed, we see it as full, so the Sun's light strikes the surface almost vertically, casting few shadows and preventing us from seeing any topographic detail, such as craters or mountains. Even through a large telescope, Mars appears only as a reddish disk, with some light and dark patches and prominent polar caps. These surface features undergo slow seasonal changes over the course of a Martian year. We saw in Chapter 1 how the inclination of Earth's axis produces similar seasonal changes. ∞ (Sec. 1.3) Figure 10.2 shows some of the best images of Mars ever made from Earth or Earth orbit, along with a photograph taken by one of the U.S. *Viking* spacecraft en route to the planet.

When the planet is viewed from Earth, the most obvious Martian surface features are the bright polar caps (see Figure 10.2a), growing and diminishing according to the seasons and almost disappearing at the time of the Martian summer. The dark surface features on Mars also change from season to season, although their variability probably has little to do with the melting of the polar ice caps. To the more fanciful observers around the start of the 20th century, these changes suggested the seasonal growth of vegetation on the planet. It was but a small step from seeing polar ice caps and speculating about teeming vegetation to imagining a planet harboring intelligent life, perhaps not unlike us.

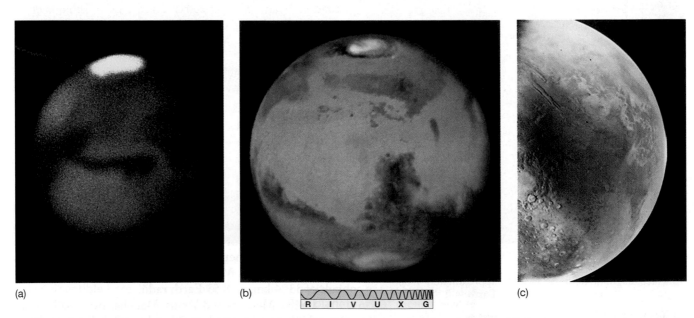

(a)   (b)   R I V U X G   (c)

▲ **FIGURE 10.2 Mars** (a) A deep-red (800-nm) image of Mars, taken in 1991 at Pic-du-Midi, an exceptionally clear site in the Pyrenees. One of the planet's polar caps appears at the top. (b) A visible-light *Hubble Space Telescope* image of Mars, taken while the planet was near opposition in 2003. (c) A view of Mars taken from a *Viking* spacecraft during its approach in 1976. The planet's surface features can be seen clearly at a level of detail completely invisible from Earth. (*Pic-du-Midi Observatory; NASA*)

But those speculations and imaginings were not to be confirmed. Spectroscopic observations from Earth and from Earth orbit revealed that the changing caps are mostly frozen carbon dioxide (i.e., dry ice), not water ice, as at Earth's North and South poles. ∞ (Sec. 4.4) The polar caps do contain water, but it remains permanently frozen, and the dark markings seen in Figures 10.2, once thought (by some) to be part of a network of canals dug by Martians for irrigation purposes, are actually highly cratered and eroded areas around which surface dust occasionally blows. From a distance, the repeated covering and uncovering of these landmarks gives the impression of surface variability, but it's only the thin dust cover that changes.

The powdery Martian surface dust is borne aloft by strong winds that often reach hurricane proportions (hundreds of kilometers per hour). In fact, when the U.S. *Mariner 9* spacecraft went into orbit around Mars in 1971, a planetwide dust storm obscured the entire landscape. Had the craft been on a flyby mission (for a quick look) instead of an orbiting mission (for a longer view), its visit would have been a failure. Fortunately, the storm subsided, enabling the craft to radio home detailed information about the planet's surface.

CONCEPT CHECK

✔ Does Mars have seasons like those on Earth?

10.4 The Surface of Mars

③ Maps of the surface of Mars returned by orbiting spacecraft show a wide range of geological features. Mars has huge volcanoes, deep canyons, vast dune fields, and many other geological wonders. Orbiters have performed large-scale surveys of much of the planet's surface, and lander data have complemented these planetwide studies with detailed information on (so far) three specific sites. ∞ (Sec. 6.6)

Large-Scale Topography

Figure 10.3 shows a planetwide mosaic of thousands of images taken in the 1970s by the *Viking* orbiters. The images show some of the planet's topographic features in true color. More recently, *Mars Global Surveyor* has mapped out the Martian surface to an accuracy of a few meters, using

Chryse
Planitia

Volcanoes

Tharsis
region

Valles Marineris

R I V U X G

◀ **FIGURE 10.3 Tharsis** A mosaic of Mars based on images from a *Viking* spacecraft in orbit around the planet. Some 5000 km across, the Tharsis region bulges out from Mars's equatorial zone, rising to a height of about 10 km. The large volcanoes on the left mark the approximate peak of the bulge. One of the plains flanking the Tharsis bulge, Chryse Planitia, is toward the right. Dominating the center of the field of view is a vast "canyon" known as Valles Marineris—the Mariner Valley. (*NASA*)

VIDEO Rotation of Mars

an instrument called a *laser altimeter*, which analyzes pulses of laser light to measure the distance between the space-craft and the planet's surface. Figure 10.4 shows some results of those measurements, projected onto a Martian globe. Figure 10.5 flattens things out into a more conventional map and marks some prominent surface features.

A striking feature of the terrain of Mars is the marked difference between the northern and southern hemispheres. The northern hemisphere is made up largely of rolling volcanic plains not unlike the lunar maria—indeed, this similarity was key to their identification as lava-flow features. Much larger than their counterparts on Earth or the Moon, these extensive lava plains were formed by eruptions involving enormous volumes of material. They are strewn with blocks of volcanic rock, as well as with boulders blasted out of impact areas by infalling meteoroids. (The Martian atmosphere is too thin to offer much resistance to incoming debris.) The southern hemisphere consists of heavily cratered highlands lying some 5 kilometers above the level of the lowland north. Most of the dark regions visible from Earth are mountainous regions in the south. Figure 10.6 contrasts typical terrains in the two hemispheres.

The northern plains are much less cratered than the southern highlands. On the basis of the arguments presented in Chapter 8, this smoother surface suggests that the northern surface is younger. ∞ (Sec. 8.5) Its age is perhaps 3 billion years, compared with 4 billion in the south. In places, the boundary between the southern highlands and the northern plains is quite sharp—the surface level can drop by as much as 4 km in a horizontal distance of 100 km or so. Most scientists assume that the southern terrain is the original crust of the planet. How most of the northern hemisphere could have been lowered in elevation and subsequently flooded with lava remains a mystery.

The major geological feature on the planet is the Tharsis bulge (marked in Figure 10.3). Roughly the size of North America, Tharsis lies on the Martian equator and rises some 10 km higher than the rest of the Martian surface. To its east lies Chryse Planitia (the "Plains of Gold"), to the west a region called Isidis Planitia (the "Plains of Isis," an Egyptian goddess). These features are wide depressions, hundreds of kilometers across and up to 3 km deep. If we wished to extend the idea of "continents" from Earth and Venus to Mars, we would conclude that Tharsis

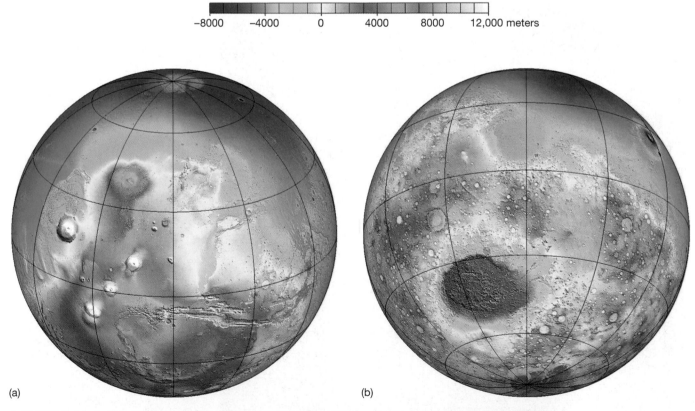

▲ **FIGURE 10.4 Mars Globes** Two computer-generated globes of planet Mars, based on detailed measurements made by *Mars Global Surveyor*. Color represents height above (or below) the mean planetary radius, ranging from dark blue (−8 km), through green, yellow and red (+4 km), to white (over 8 km in altitude), as given on the scale above the images. Frame (a) shows roughly the same hemisphere as that in Figure 10.3, containing the Tharsis region of Mars. Frame (b) shows the planet's other hemisphere, dominated by the giant Hellas impact basin. *(NASA)*

is the only continent on the Martian surface. However, as on Venus, there is no sign of plate tectonics on Mars—the absence of fault lines or other evidence of plate motion tells geologists that the "continent" of Tharsis is not drifting as its Earthly counterparts are. ∞ (Sec. 7.4) Tharsis appears to be even less heavily cratered than the northern plains, making it the youngest region on the planet, an estimated 2 to 3 billion years old.

Almost diametrically opposite Tharsis, in the southern highlands, lies the Hellas Basin, which, paradoxically, contains the *lowest* point on Mars. (Hellas is clearly visible in Figure 10.4 and is labeled in Figure 10.5.) Some 3000 km across, the floor of the basin lies nearly 9 km below the basin's rim and over 6 km below the average level of the planet's surface. Its shape and structure identify the Hellas Basin as an impact feature, similar in many ways to the

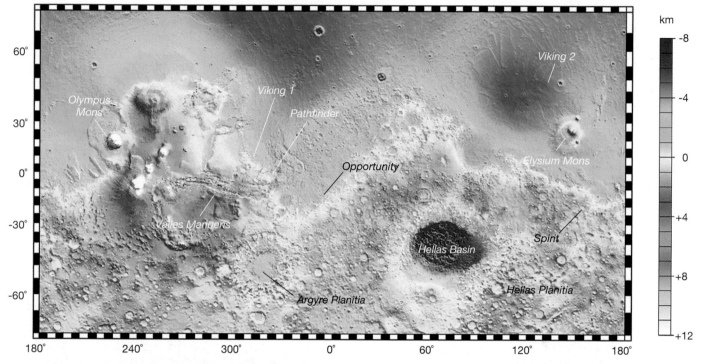

▲ **FIGURE 10.5 Mars Map** The *Mars Global Surveyor* data of Figure 10.4, now displayed as a flat map, with some surface features labeled. Landing sites of several US robot craft are also marked. *(NASA)*

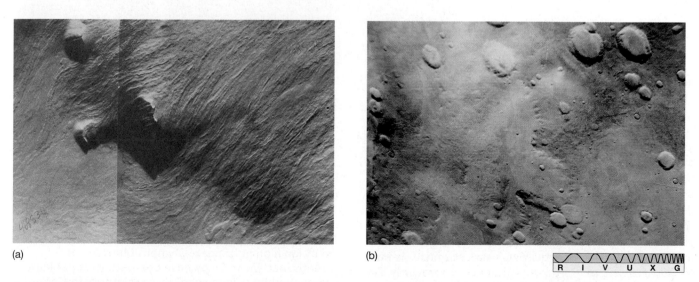

▲ **FIGURE 10.6 Mars Up Close** (a) The northern hemisphere of Mars consists of rolling, volcanic plains (false-color image). (b) The southern Martian highlands are heavily cratered (true color). Both photographs show roughly the same scale, nearly 1000 km across. *(NASA)*

south pole–Aitken Basin on the far side of Earth's Moon. ∞ *(Discovery 8-2)* The formation of the Hellas Basin must have caused a major redistribution of the young Martian crust—perhaps even enough to account for a substantial portion of the highlands around it, according to some researchers. The basin's heavily cratered floor indicates that the impact occurred very early on in Martian history—some 4 billion years ago—during the heavy bombardment that accompanied the formation of the terrestrial planets. ∞ (Secs. 6.7, 8.5)

Volcanism

Mars contains the largest known volcanoes in the solar system. Three very large volcanoes are found on the Tharsis bulge, two of them visible on the left-hand side of Figure 10.3. The largest volcano of all is Olympus Mons (Figure 10.7), northwest of Tharsis, lying just over the left (western) horizon of Figure 10.3. This volcano measures some 700 km in diameter at its base—only slightly smaller than the state of Texas—and rises to a height of 25 km above the surrounding plains. The caldera, or crater, at its summit, measures 80 km across. The other three large volcanoes are a little smaller— a "mere" 18 km high—and lie near the top of the bulge.

Like Maxwell Mons on Venus, none of these volcanoes is associated with plate motion on Mars—as just mentioned, there is none. Instead, they are shield volcanoes, sitting atop a hot spot in the underlying Martian mantle. ∞ (Sec. 9.4) All four show distinctive lava channels and other flow features similar to those found on shield volcanoes on Earth. *Viking* and *Mars Global Surveyor* images of the Martian surface reveal many hundreds of volcanoes. Most of the largest are associated with the Tharsis bulge, but many smaller volcanoes are found in the northern plains.

The great height of Martian volcanoes is a direct consequence of the planet's low surface gravity. (See the Mars Data box on p. 253.) As lava flows and spreads to form a shield volcano, its eventual height depends on the new mountain's ability to support its own weight. The lower the gravity, the less is the weight and the higher is the mountain. It is no accident that Maxwell Mons on Venus and the Hawaiian shield volcanoes on Earth rise to roughly the same height (about 10 km) above their respective bases—Earth and Venus have similar surface gravity. Mars's surface gravity is only 40 percent that of Earth, so volcanoes rise roughly 2.5 times as high.

Are the Martian shield volcanoes still active? Scientists have found no direct evidence for recent or ongoing eruptions. However, if these volcanoes have been around since the Tharsis uplift (as the formation of the Tharsis bulge is known) and were active as recently as 100 million years ago (an age estimate based on the extent of impact cratering on their slopes), some of them may still be at least intermittently active. Millions of years, though, may pass between eruptions.

Impact Cratering

The *Mariner* spacecraft found that the surfaces of Mars and its two moons are pitted with impact craters formed by meteoroids falling in from space. As on our Moon, the smaller craters are often filled with surface matter—mostly dust—confirming that Mars is a dry desert world. However, Martian craters are filled in considerably faster than their lunar counterparts. On the Moon, ancient craters less than 100 m across (corresponding to depths of about 20 m) have been obliterated, primarily by meteoritic erosion. ∞ (Sec. 8.5) On Mars, there are relatively few small craters less than about 5 km in diameter. The Martian atmosphere is an efficient erosive agent, transporting dust from place to place and erasing surface features much faster than meteoritic impacts alone can obliterate them.

As on the Moon, the extent of large impact cratering (i.e., producing craters too big to have been filled in by erosion since they formed) serves as an age indicator for

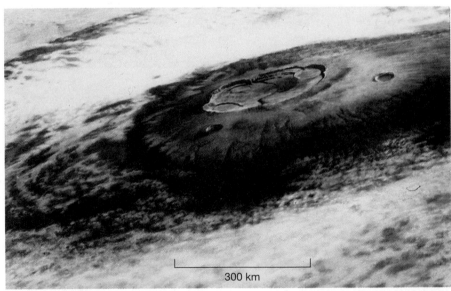

300 km

R I V U X G

▲ **FIGURE 10.7 Olympus Mons** The largest volcano known on Mars or anywhere else in the solar system and nearly three times taller than Mount Everest on Earth, this Martian mountain measures about 700 km across its base and 25 km high at its peak. Olympus Mons seems currently inactive and may have been extinct for at least several hundred million years. By comparison, the largest volcano on Earth, Hawaii's Mauna Loa, measures a mere 120 km across and peaks just 9 km above the Pacific Ocean floor. *(NASA)*

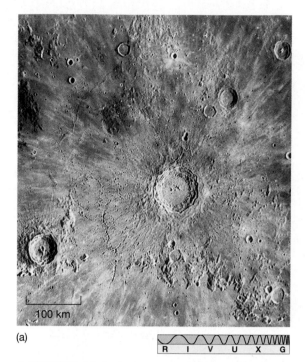

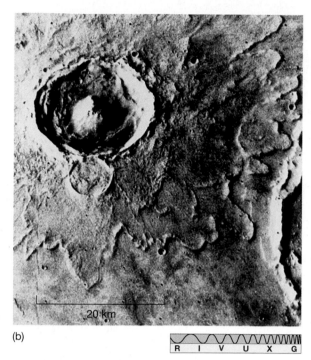

(a)

(b)

▲ **FIGURE 10.8 Crater Comparison** (a) The large lunar impact crater Copernicus is quite typical of those found on Earth's Moon. Its ejecta blanket appears to be composed of dry, powdery material. (b) The ejecta from Mars's crater Yuty (18 km in diameter) evidently was liquid in nature. This type of crater is sometimes called a "splosh" crater. *(NASA)*

the Martian surface. The ages quoted earlier, ranging from 4 billion years for the southern highlands to a few hundred million years in the youngest volcanic areas, were obtained in this way.

The detailed appearance of Martian impact craters provides an important piece of information about conditions just below the planet's surface. The ejecta blankets surrounding many Martian craters look quite different from their lunar counterparts. Figure 10.8 compares the Copernicus crater on the Moon with the (fairly typical) crater Yuty on Mars. The material surrounding the lunar crater is just what one would expect from an explosion ejecting a large volume of dust, soil, and boulders. However, the ejecta blanket on Mars gives the distinct impression of a liquid that has splashed or flowed out of the crater. Geologists think that this *fluidized ejecta* crater indicates that a layer of **permafrost**, or water ice, lies just a few meters under the surface. The explosive impact heated and liquefied the ice, resulting in the fluid appearance of the ejecta.

Prior to the arrival of *Mars Global Surveyor*, astronomers thought that all the water below the Martian surface existed in the form of ice. However, since 2000, *Surveyor* mission scientists have reported the discovery of numerous small-scale "gullies" in Martian cliffs and crater walls that apparently were carved by running water in the relatively recent past. These features are too small to have been resolved by *Viking* cameras. Figure 10.9 shows a few such gullies, found on the inner rim of a Martian impact crater in the southern highlands. These gullies have many similarities to the channels carved by flash floods on Earth.

The ages of these intriguing features are uncertain and might be as great as a million years in some cases, but the *Surveyor* team speculates that some of them may still be active today, implying that liquid water could exist in a number of regions of Mars at depths of less than 500 meters. However, some scientists dispute this interpretation, arguing that the "fluid" responsible for the gullies might well have been solid (granular) or liquid carbon dioxide, expelled under great pressure from the Martian crust. It seems that a lot more study—perhaps even a human visit—will be needed before this issue is settled.

CONCEPT CHECK

✔ How do we know that the northern Martian lowlands are younger than the southern highlands?

The Martian "Grand Canyon"

Yet another feature associated with the Tharsis bulge is a great "canyon" known as Valles Marineris (the Mariner Valley). Shown in its entirety in Figure 10.3 and in more detail in Figure 10.10, this feature is not really a canyon in the terrestrial sense, because running water played no part in its formation. Planetary astronomers theorize that it was formed by the same crustal forces that caused the entire Tharsis region to bulge outward, making the surface split and crack. The resulting cracks, called *tectonic fractures*, are found all around the Tharsis bulge. The Valles

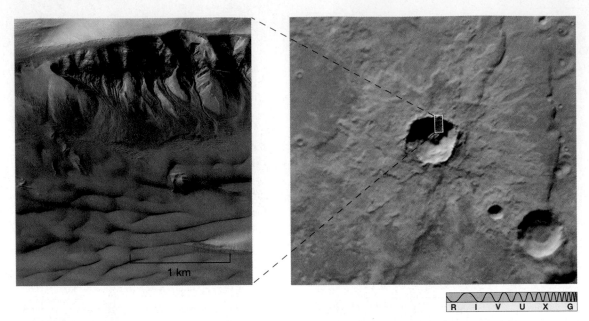

R I V U X G

▲ **FIGURE 10.9 Running Water on Mars?** This high-resolution *Mars Global Surveyor* view (left) of a crater wall (right) near the Mariner Valley shows evidence of "gullies" apparently formed by running water in the relatively recent past. If this interpretation is correct, it suggests that liquid water may still reside below the Martian surface. *(NASA)*

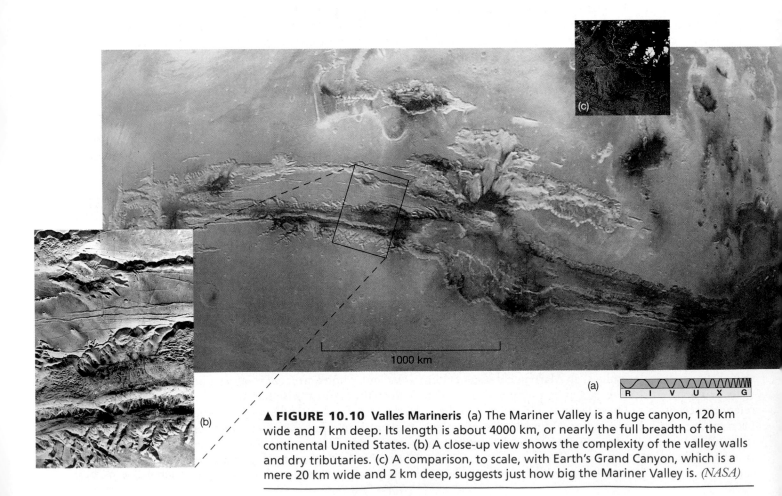

R I V U X G

▲ **FIGURE 10.10 Valles Marineris** (a) The Mariner Valley is a huge canyon, 120 km wide and 7 km deep. Its length is about 4000 km, or nearly the full breadth of the continental United States. (b) A close-up view shows the complexity of the valley walls and dry tributaries. (c) A comparison, to scale, with Earth's Grand Canyon, which is a mere 20 km wide and 2 km deep, suggests just how big the Mariner Valley is. *(NASA)*

Marineris is the largest of them. Cratering studies suggest that the cracks are at least 2 billion years old. Similar (but much smaller) cracks, originating from similar causes, have been found in the Aphrodite Terra region of Venus. ∞ (Sec. 9.4)

Valles Marineris runs for almost 4000 km along the Martian equator, about one-fifth of the way around the planet. At its widest, it is some 120 km across, and it is as deep as 7 km in places. Like many Martian surface features, it simply dwarfs Earthly competition. The Grand Canyon in Arizona would easily fit into one of its side "tributary" cracks. Valles Marineris is so large that it can even be seen from Earth—in fact, it was one of the few "canals" observed by 19th-century astronomers that actually corresponded to a real feature on the planet's surface. (It was known as the Coprates canal.) We must reemphasize, however, that this Martian feature was not constructed by intelligent beings, nor was it carved by a river, nor is it a result of Martian plate tectonics. For some reason, the crustal forces that formed it never developed into full-fledged plate motion as exists on Earth.

Evidence for Running Water

Although the great surface cracks in the Tharsis region are not really canyons and were not formed by running water, photographic evidence reveals that liquid water once existed in great quantity on the surface of Mars. Two types of flow feature are seen: *runoff channels* and *outflow channels*.

Runoff channels (one of which is shown in Figure 10.11a) are found in the southern highlands. These flow features are extensive systems—sometimes hundreds of kilometers in total length—of interconnecting, twisting channels that seem to merge into larger, wider channels. They bear a strong resemblance to river systems on Earth, and geologists think that that is just what they are—the dried-up beds of long-gone rivers that once carried rainfall on Mars from the mountains down into the valleys. Runoff channels on Mars speak of a time 4 billion years ago (the age of the Martian highlands), when the atmosphere was thicker, the surface warmer, and liquid water widespread.

Outflow channels (Figure 10.12a) are probably relics of catastrophic flooding on Mars long ago. They appear only in equatorial regions and generally do not form the extensive interconnected networks that characterize runoff channels. Instead, they are probably the paths taken by huge volumes of water draining

from the southern highlands into the northern plains. The onrushing water arising from these flash floods likely also formed the odd teardrop-shaped "islands" (resembling the miniature versions seen in the wet sand of our beaches at low tide) that have been found on the plains close to the ends of the outflow channels (Figure 10.12b). Judging from the width and depth of the channels, the flow rates must have been truly enormous—perhaps as much as a hundred times greater than the 10^5 tons per second carried by the Amazon river, the largest river system on Earth. Flooding shaped the outflow channels approximately 3 billion years ago, about the same time as the northern volcanic plains formed.

Some scientists speculate that Mars may have enjoyed an extended early period during which rivers, lakes, and perhaps even oceans adorned its surface. Figure 10.13 is a 2003 *Mars Global Surveyor* image showing what mission specialists think may be a *delta*—a fan-shaped network of channels and sediments where a river once flowed into a larger body of water, in this case a lake filling a crater in the southern highlands. Other researchers go even further, suggesting that the data provide evidence for large open expanses of water on the early Martian surface. Figure 10.14(a) is a computer-generated view of the Martian north polar region, showing the extent of what may have

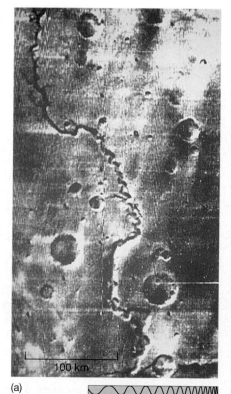

▲ **FIGURE 10.11 Martian Channel** (a) This runoff channel on Mars measures about 400 km long and 5 km wide. (b) The Red River running from Shreveport, Louisiana, to the Mississippi River. The two differ mainly in that there is currently no liquid water in this, or any other, Martian valley. *(NASA)*

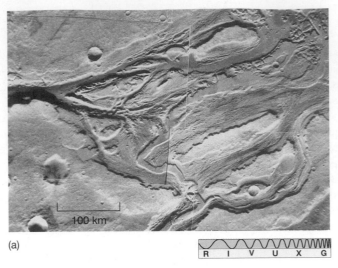

(a)

▲ **FIGURE 10.12 Martian Outflow** (a) An outflow channel near the Martian equator bears witness to a catastrophic flood that occurred about 3 billion years ago. (b) The onrushing water that carved out the outflow channels was responsible for forming these oddly shaped "islands" as the flow encountered obstacles—impact craters—in its path. Each "island" is about 40 km long. *(NASA)*

(b)

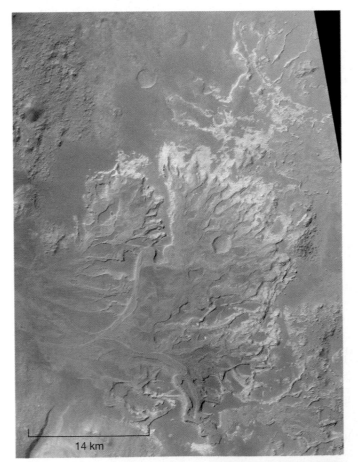

been an ancient ocean covering much of the northern low-lands. The Hellas basin (Figure 10.5) is another candidate for an ancient Martian sea.

These ideas remain controversial. Proponents point to features such as the terraced "beaches" shown in Figure 10.14b, which could conceivably have been left behind as a lake or ocean evaporated and the shoreline receded. But detractors maintain that the terraces could also have been created by geological activity, perhaps related to the tectonic forces that depressed the northern hemisphere far below the level of the south, in which case they have nothing whatever to do with Martian water. Furthermore, *Mars Global Surveyor* data released in 2003 seem to indicate that the Martian surface contains too few *carbonate* rock layers—layers containing compounds of carbon and oxygen that should have been formed in abundance in an ancient ocean. Their absence supports the picture of a cold, dry Mars which never experienced the extended mild period required to form lakes and oceans. However, as discussed below, data from the most recent *NASA* landers imply that at least some parts of the planet did in fact experience long

◀ **FIGURE 10.13 Martian River Delta** Did this fan-shaped region of twisted streams form as a river flowed into a larger sea? If it did, the *Mars Global Surveyor* image supports the idea that Mars once had large bodies of liquid water on its surface. Not all scientists agree with this interpretation, however. *(NASA)*

periods in the past during which liquid water existed on the surface. Obviously, the debate is far from over.

Aside from the gullies mentioned earlier, which are highly suggestive, but by no means conclusive, astronomers have no direct evidence for liquid water anywhere on the surface of Mars today, and the amount of water vapor in the Martian atmosphere is tiny. Yet even setting aside the unproven hints of ancient oceans, the extent of the outflow channels suggests that a huge total volume of water existed on Mars in the past. Where did all that water go? The answer may be that virtually all the

water on Mars is now locked in the permafrost layer under the surface, with more contained in the planet's polar caps.

Scientists have speculated about this possibility for years, and in 2002 a gamma-ray spectrometer aboard the *Mars Odyssey* orbiter detected extensive deposits of water ice crystals (actually, the hydrogen they contain) mixed with the Martian surface layers. The instrument was similar in design to the one carried by *Lunar Prospector* that searched for (and found) ice crystals in the regolith near the lunar poles. ∞ (Sec. 8.5, *Discovery 8-2*) Ironically, one site where ice was found had in fact been the target of the *Mars Polar Lander* probe, which disappeared just as it reached the planet in 1999. ∞ (Sec. 6.5) Unfortunately, none of the other five successful Martian landers to date has put down in a region where ice has been detected.

Four billion years ago, as climatic conditions changed, the running water that formed the runoff channels began to freeze, creating the permafrost and drying out the river beds. Mars remained frozen for about a billion years, until volcanic (or some other) activity heated large regions of the surface, melting the subsurface ice and causing the

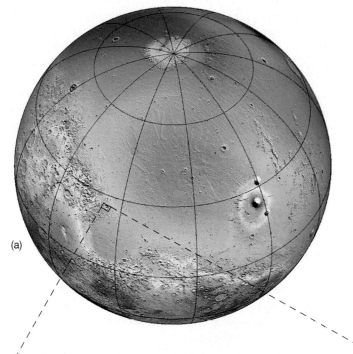

(a)

◄ **FIGURE 10.14 An Ancient Ocean?** (a) An ancient Martian ocean once may have spanned the polar regions. The blue color in this computer-generated map actually indicates depth below the average radius of the planet, but it also outlines quite accurately the extent of the ocean suggested by *Mars Global Surveyor* data. (The color elevation scale is nearly the same as in Figure 10.4.) (b) This high-resolution image shows tentative evidence for erosion by standing water in the floor of Holden Crater, about 140 km across. *(NASA)*

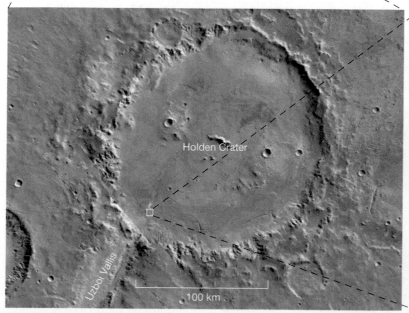

(b)

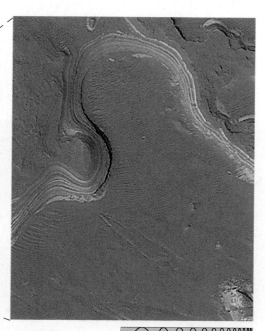

R   I   U   X   G

flash floods that created the outflow channels. Subsequently, volcanic activity subsided, the water refroze, and Mars once again became a dry world.

Polar Caps

We have already noted that the Martian polar caps are composed predominantly of carbon dioxide frost—dry ice—and show seasonal variations. Each cap in fact consists of two distinct parts: the **seasonal cap**, which grows and shrinks each year, and the **residual cap**, which remains permanently frozen. At maximum size, in southern midwinter, the southern seasonal cap is some 4000 km across. Half a Martian year later, the northern cap is at its largest, reaching a diameter of roughly 3000 km. The two seasonal polar caps do not have the same maximum size because of the eccentricity of Mars's orbit around the Sun. During southern winter, Mars is considerably farther from the Sun than half a year later, in northern winter. The southern winter season is longer and colder than that of the north, and the polar cap grows correspondingly larger.

The seasonal caps are composed entirely of carbon dioxide. Their temperatures are never greater than about 150 K($-120°$C), the point at which dry ice can form. During the Martian summer, when sunlight striking a cap is most intense, carbon dioxide evaporates into the atmosphere, and the cap shrinks. In the winter, atmospheric carbon dioxide refreezes, and the cap re-forms. As the caps grow and shrink, they cause substantial variations (up to 30 percent) in the Martian atmospheric pressure—a large fraction of the planet's atmosphere freezes out and evaporates again each year. From studies of these atmospheric fluctuations, scientists can estimate the amount of carbon dioxide in the seasonal polar caps. The maximum thickness of the seasonal caps is thought to be about 1 m.

The residual caps (Figure 10.15) are smaller and brighter than the seasonal caps and show an even more marked north–south asymmetry. The southern residual cap is about 350 km across and, like the seasonal caps, is probably made mostly of carbon dioxide, although it may contain some water ice. Its temperature remains below 150 K at all times. The northern residual cap is much larger— about 1000 km across—and warmer, with a temperature that can exceed 200 K in northern summertime. Planetary scientists think that the northern residual cap is made mostly of water ice, an opinion strengthened by spectroscopic observations, which show an increase in the concentration of water vapor above the north pole in northern summer as some small fraction of the residual cap's water ice evaporates in the Sun's heat. It is quite possible that the northern residual polar cap is a major storehouse for water on Mars.

Why is there such a temperature difference (at least 50 K) between the two residual polar caps, and why is the northern cap warmer, despite the fact that the planet's northern hemisphere is generally cooler than the south? (See Section 10.2.) The reason is not fully understood, but it seems to be related to the giant dust storms that envelop the planet during southern summer. These storms, which last for a quarter of a Martian year (about six Earth months), tend to blow the dust from the warmer south into the cooler northern hemisphere. As a result, the northern ice cap becomes dusty and less reflective, absorbing more sunlight and warming up.

The View from the Martian Landers

Viking 1 landed in Chryse Planitia, a broad depression to the east of Tharsis. The view that greeted its cameras (Figure 10.16) was a windswept, gently rolling, rather des-

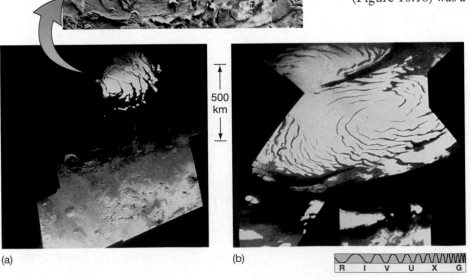

(a)　　　(b)　　　R I V U X G

◀ **FIGURE 10.15 Martian Polar Caps** The southern (a) and northern (b) polar caps of Mars are shown to scale in these mosaics of *Mariner 9* images. These are the residual caps, seen here during their respective summers half a Martian year apart. The southern cap is some 350 km across and is made up mostly of frozen carbon dioxide. The northern cap is about 1000 km across and is composed mostly of water ice. The inset shows greater detail in the southern cap. (*NASA*)

500 km

R I V U X G

◀ **FIGURE 10.16** *Viking 1* This is the view from the *Viking 1* spacecraft now parked on the surface of Mars. The fine-grained soil and the rock-strewn terrain stretching toward the horizon are reddish. Containing substantial amounts of iron ore, the surface of Mars is literally rusting away. The sky is a pale pink color, the result of airborne dust. *(NASA)*

Sojourner lander found that the soil at its landing site was similar to that found by the *Viking* landers. However analyses of nearby rocks revealed a chemical makeup different from that of the Martian meteorites found on Earth. (See *Discovery 10-1.*)

The most recent visitors to the Martian surface were the twin landers *Spirit* and *Opportunity* of the *Mars Exploration Rover* mission. ∞ (Sec. 6.6) During their operating lifetimes on the planet, these two robots made extensive chemical and geological studies of rocks within a kilometer or so of their landing sites on opposite sides of the planet, with the goal of finding evidence for liquid water on the surface at some time in the past.

Figure 10.18 shows the view from *Spirit* just before it embarked on its mission across the Martian surface. *Spirit*'s landing site was rocky and similar in many ways to the terrains encountered by earlier landers. In contrast,

olate plain, littered with rocks of all sizes, not unlike a high desert on Earth. The surface rocks visible in the figure are probably part of the ejecta blanket of a nearby impact crater. *Viking 2* landed somewhat farther north, in a region of Mars called Utopia, chosen in part because mission planners anticipated greater seasonal climatic variations there. The plain on which *Viking 2* landed was flat and featureless. From space, the landing site appeared smooth and dusty. In fact, the surface turned out to be very rocky, even rockier than the Chryse site, and without the dust layer the mission directors had expected (Figure 10.17). The views that the two landers recorded may turn out to be quite typical of the low-latitude northern plains.

The *Viking* landers performed numerous chemical analyses of the Martian regolith. One important finding of these studies was the high iron content of the planet's surface. Chemical reactions between the iron-rich surface soil and free oxygen in the atmosphere is responsible for the iron oxide ("rust") that gives Mars its characteristic color. Although the surface layers are rich in iron relative to Earth's surface, the overall abundance is similar to Earth's average iron content. On Earth, much of the iron has differentiated to the center. Chemical differentiation does not appear to have been nearly so complete on Mars. The later

R I V U X G

▲ **FIGURE 10.17** *Viking 2* Another view of the Martian surface, this one rock strewn and flat, as seen through the camera aboard the *Viking 2* robot that soft-landed on the northern Utopian plains. The discarded canister is about 20 cm long. The 0.5-m scars in the dirt were made by the robot's shovel. *(NASA)*

R I V U X G

▲ **FIGURE 10.18** *Spirit* This 360° panorama was taken by the *Mars Exploration Rover* lander in 2004. Its golf-cart size vehicle, *Spirit*, had not yet rolled off its mothership (see also Fig. 6.4). This view is from inside Gusev crater, a relatively smooth depression about 150 km wide and 15° south of the Martian equator; the largest rocks seen are about 20 cm across. *(JPL)*

Opportunity hit the jackpot in its quest, finding itself surrounded by rocks showing every indication of having been very wet—possibly immersed in salt water—at one time. These findings are the best evidence yet for standing water on the ancient Martian surface.

CONCEPT CHECK

✔ Where has all the Martian water gone?

10.5 The Martian Atmosphere

Composition

4 Long before the arrival of the *Mariner* and *Viking* spacecraft, astronomers knew from Earth-based spectroscopy that the Martian atmosphere was quite thin and composed primarily of carbon dioxide. In 1964, *Mariner 4* confirmed these results, finding that the atmospheric pressure was only about 1/150 the pressure of Earth's atmosphere at sea level and that carbon dioxide made up at least 95 percent of the total atmosphere. With the arrival of *Viking*, more detailed measurements of the Martian atmosphere could be made. Its composition is now known to be 95.3 percent carbon dioxide, 2.7 percent nitrogen, 1.6 percent argon, 0.13 percent oxygen, 0.07 percent carbon monoxide, and about 0.03 percent water vapor. The level of water vapor is quite variable. Weather conditions encountered by *Mars Pathfinder* were quite similar to those found by *Viking 1*.

As the *Viking* landers descended to the surface, they made measurements of the temperature and pressure at various heights. The results are shown in Figure 10.19. The Martian atmosphere contains a troposphere (the lowest-lying atmospheric zone, where convection and "weather" occur), which varies both from place to place and from season to season. ∞ (Sec. 7.2) The variability of the tro-

posphere arises from the variability of the Martian surface temperature. At noon in the summertime, surface temperatures may reach 300 K. Atmospheric convection is strong, and the top of the troposphere can reach an altitude of 30 km. At night, the atmosphere retains little heat, and the temperature can drop by as much as 100 K. Convection then ceases and the troposphere vanishes.

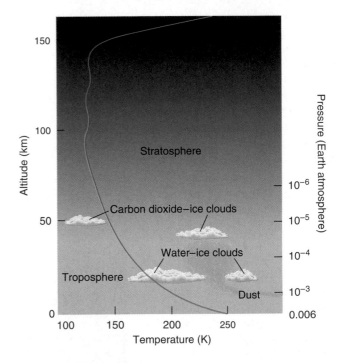

▲ **FIGURE 10.19** **Martian Atmosphere** Structure of the Martian atmosphere, as determined by *Viking* and *Mars Global Surveyor*. The troposphere, which rises to an altitude of about 30 km in the daytime, occasionally contains clouds of water ice or, more frequently, dust during the planetwide dust storms that occur each year. Above the troposphere lies the stratosphere. Note the absence of a higher temperature zone in the stratosphere, indicating the absence of an ozone layer.

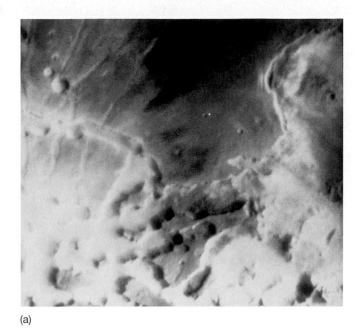

(a)

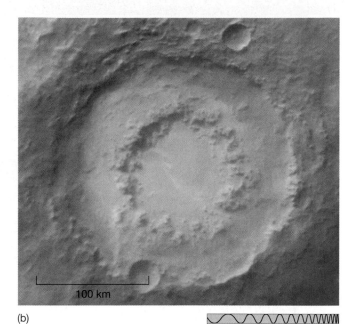

(b)

R   I   V   U   X   G

▲ **FIGURE 10.20 Fog in the Canyons** (a) As the Sun's light reaches and heats the canyon floor, it drives water vapor from the surface. When this vapor comes in contact with the colder air above the surface, it condenses again, and a temporary water-ice "fog" results, as seen here, near Mars's northern polar cap. (b) Fog also shrouds the floor of the 200-km-wide Lowell Crater, imaged here by *Mars Global Surveyor* in the autumn of 2000. *(NASA)*

On average, surface temperatures on Mars are about 50 K cooler than on Earth. The low early-morning temperatures often produce water-ice "fog" in the Martian canyons (Figure 10.20). Higher in the atmosphere, in the stratosphere, temperatures are low enough for carbon dioxide to solidify, giving rise to a high-level layer of carbon dioxide clouds and haze.

For most of the year, there is little day-to-day variation in the Martian weather: The Sun rises, the surface warms up, and light winds blow until sunset, when the temperature drops again. Only in the southern summer does the daily routine change. Strong surface winds (without rain or snow) sweep up the dry dust, carry it high into the stratosphere, and eventually deposit it elsewhere on the planet. At its greatest fury, a Martian storm floods the atmosphere with dust, making the worst storm we could imagine on Earth's Sahara Desert seem inconsequential by comparison. The dust can remain airborne for months at a time. The blown dust forms systems of sand dunes similar in appearance to those found on Earth.

Evolution

Although there is some superficial similarity in composition between the atmospheres of Mars and Venus, the two planets obviously have quite different atmospheric histories—Mars's "air" is over 10,000 times thinner than that on Venus. As with the other planets we have studied, we can ask *why* the Martian atmosphere is as it is.

Presumably, Mars acquired a secondary outgassed atmosphere quite early in its history, just as the other terrestrial worlds did. Around 4 billion years ago, as indicated by the runoff channels in the highlands, Mars may have had a fairly dense atmosphere, complete with blue skies and rain. Despite Mars's distance from the Sun, the greenhouse effect would have kept conditions fairly comfortable, and a surface temperature of over 0°C seems quite possible. Sometime during the next billion years, most of the Martian atmosphere disappeared. Possibly, some of it was lost because of violent impacts with other large (Moon-sized) bodies in the early solar system. More likely, the Martian atmosphere became unstable, in a kind of reverse runaway greenhouse effect.

On Venus, as we have seen, the familiar greenhouse effect ran away to high temperatures and pressures. ∞ (Sec. 9.4) Planetary scientists theorize that the early Martian atmosphere also ran away, but in the opposite direction. The presence of liquid water would have caused much of the atmospheric carbon dioxide to dissolve in Martian rivers and lakes (and oceans, if any), ultimately to combine with surface rocks. Recall that on Earth most carbon dioxide is found in surface rocks. Calculations show that much of the Martian atmospheric carbon dioxide could have been depleted in this way in a relatively short period—perhaps as little as a few hundred million years—although some of it might have been replenished by volcanic activity, possibly extending the "comfortable" lifetime of the planet to a half-billion years or so.

As the level of carbon dioxide declined and the greenhouse-heating effect diminished, the planet cooled. The water froze out of the atmosphere, lowering the level of

atmospheric greenhouse gases still further and accelerating the cooling. (Recall from Section 9.4 that water vapor also contributes to the greenhouse effect.)

Since those early times, the overall density of the Martian atmosphere has continued to decline through the steady loss of carbon dioxide, nitrogen, and water vapor as solar ultraviolet radiation in the upper atmosphere splits the molecules of these gases into their component atoms, providing some with enough energy to escape. The current level of water vapor in the Martian atmosphere is the maximum possible, given the atmosphere's present density and temperature. Estimates of the total amount of water stored as permafrost or in the polar caps are quite uncertain, but it is likely that if all the water on Mars were to become liquid, it would cover the surface to a depth of several meters.

CONCEPT CHECK

✔ Where did the Martian atmosphere go?

10.6 Martian Internal Structure

The *Viking* landers carried seismometers to probe the internal structure of Mars. However, one failed to work, and the other was unable to clearly distinguish seismic activity from the buffeting of the Martian wind. As a result, no seismic studies of the Martian interior have yet been carried out. On the basis of studies of the stresses that occurred during the Tharsis uplift, astronomers estimate the thickness of the crust to be about 100 km.

During its visit to Mars in 1965, *Mariner 4* detected no planetary magnetic field, and for many years the most that could be said about the Martian magnetic field was that its strength was no more than a few thousandths the strength of Earth's field (to the level of sensitivity of *Mariner*'s instruments). The *Viking* spacecraft were not designed to make magnetic measurements. In September 1997, *Mars Global Surveyor* detected a very weak Martian field, about 1/800 times that of Earth. However, this is probably a local anomaly, akin to the magnetic fluctuations detected by *Lunar Prospector* at certain locations on the surface of Earth's Moon, and not a global field. ∞ (Sec. 8.7)

Because Mars rotates rapidly, the absence of a global magnetic field is taken to mean that the planet's core is nonmetallic, nonliquid, or both. ∞ (Sec. 7.5) The small size of Mars indicates that any radioactive (or other internal) heating of its interior would have been less effective at heating and melting the planet than similar heating on Earth. The heat was able to reach the surface and escape more easily than on a larger planet such as Earth or Venus. The evidence we noted earlier for ancient surface activity, especially volcanism, suggests that at least parts of the planet's interior must have melted and possibly differentiated at some time in the past. But the lack of current activity, the absence of any significant magnetic field, the relatively low density (3900 kg/m^3), and an abnormally high abundance of iron at the surface all suggest that Mars never melted as extensively as did Earth.

The history of Mars appears to be that of a planet on which large-scale tectonic activity almost started, but was stifled by the planet's rapidly cooling outer layers. The large upwelling of material that formed the Tharsis bulge might have developed into full-fledged plate tectonic motion on a larger, warmer planet, but the Martian mantle became too rigid and the crust too thick for that to occur. Instead, the upwelling continued to fire volcanic activity, almost up to the present day, but, geologically, much of the planet apparently died 2 billion years ago.

CONCEPT CHECK

✔ What is the principal reason for the lack of geological activity on Mars today?

10.7 The Moons of Mars

Unlike Earth's moon, Mars's moons are tiny compared with their parent planet and orbit very close to it, relative to the planet's radius. Discovered by American astronomer Asaph Hall in 1877, the two Martian moons—Phobos ("fear") and Deimos ("panic")—are only a few tens of kilometers across. Their composition is quite unlike that of the planet. Astronomers regard it as unlikely that Phobos and Deimos formed along with Mars. Instead, it is more likely that they are asteroids which were slowed and captured by the outer fringes of the early Martian atmosphere (which, as we have just seen, was probably much denser than the atmosphere today). It is even possible that they are remnants of a single object that broke up during capture. They are quite difficult to study from Earth because their proximity to Mars makes it hard to distinguish them from their much brighter parent. The *Mariner* and *Viking* orbiters, however, studied both in great detail.

Mars's moons, shown in Figure 10.21, are quite irregularly shaped and heavily cratered. The larger of the two is Phobos (Figure 10.21a), which is about 28 km long and 20 km wide and is dominated by an enormous 10-km-wide crater named Stickney (after Angelina Stickney, Asaph Hall's wife, who encouraged him to persevere in his observations). The smaller Deimos (Figure 10.21b) is only 16 km long by 10 km wide. Its largest crater is 2.3 km in diameter. The fact that both moons have quite dark surfaces, reflecting no more than 6 percent of the light falling on them, contributes to the difficulty in observing them from Earth.

Phobos and Deimos move in circular, equatorial orbits, and they rotate synchronously (i.e., they each keep the same face permanently turned toward the planet). These characteristics are direct consequences of the tidal influence of Mars. Both moons orbit Mars in the prograde

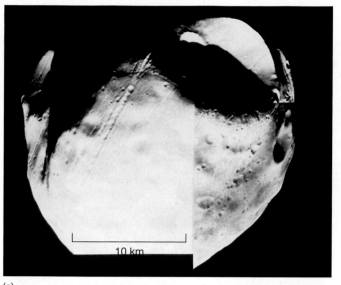

(a)

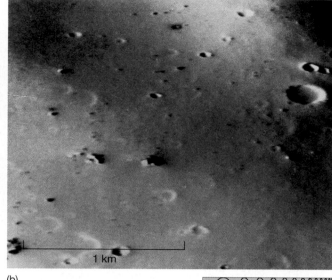

(b)

| R | I | V | U | X | G |

▲ **FIGURE 10.21 Martian Moons** (a) A *Mariner 9* photograph of the potato-shaped Phobos, not much larger than Manhattan Island. The prominent crater (called Stickney) at the top right is about 10 km across. (b) Like Phobos, the smaller moon, Deimos, has a composition unlike that of Mars. Both moons are probably captured asteroids. This close-up photograph of Deimos was taken by a *Viking* orbiter. The field of view is only 2 km across, and most of the boulders shown are about the size of a house. *(NASA)*

sense—that is, in the same sense (counterclockwise, as seen from above the north celestial pole) as the planet orbits the Sun and rotates on its axis.

Phobos lies only 9378 km (less than three planetary radii) from the center of Mars and, as we saw earlier, has an orbital period of 7 hours and 39 minutes. This period is much less than a Martian day, so an observer standing on the Martian surface would see Phobos move "backward" across the Martian sky—that is, in a direction opposite that of the apparent daily motion of the Sun. Because the moon moves faster than the observer, it overtakes the planet's rotation, rising in the west and setting in the east, crossing the sky from horizon to horizon in about 5.5 hours. Deimos lies somewhat farther out, at 23,459 km, or slightly less than seven planetary radii, and orbits in 30 hours and 18 minutes. Because it completes its orbit in more than a Martian day, it moves "normally," as seen from the ground (i.e., from east to west), taking almost 3 days to traverse the sky.

Astronomers have estimated the masses of the two moons on the basis of measurements of their gravitational effect on the *Viking* orbiters. The density of the Martian moons is around 2000 kg/m³, far less than that of any world we have yet encountered in our outward journey through the solar system. This uniqueness is an important reason that astronomers do not think that these moons formed along with Mars. If they are indeed captured asteroids, Phobos and Deimos represent material left over from the earliest stages of the solar system. Astronomers study them not to gain insight into Martian evolution, but rather because the moons contain information about the very early solar system, before the major planets had formed.

CONCEPT CHECK

✔ In what ways do Phobos and Deimos differ from Earth's Moon?

Chapter Review

SUMMARY

Mars lies outside Earth's orbit, so it traverses the entire plane of the ecliptic, as seen from Earth. Its orbit is more elliptical than Earth's, so its distance from the Sun varies more. Mars rotates at almost the same rate as Earth, and its axis of rotation is inclined to the ecliptic at almost the same angle as Earth's axis. As a result of its axial tilt, Mars has daily and seasonal cycles much like those on our own planet, but they are more complex than those on

Earth because of Mars's eccentric orbit. From Earth, the most obvious Martian surface features are the polar caps, which grow and diminish as the seasons change on Mars. The appearance of the planet also changes because of seasonal dust storms that obscure its surface.

Like the atmosphere of Venus, Mars's atmosphere is composed primarily of carbon dioxide. However, unlike Venus's atmosphere, the cool Martian atmosphere has a density less than 1 percent that of Earth's. Mars may once have had a dense atmosphere, but it was lost, partly to space and partly to surface rocks and subsurface **permafrost (p. 259)** and polar caps. Even today, the thin atmosphere is slowly leaking away. Surface temperatures on Mars average about 50 K cooler than those on Earth. Otherwise, Martian weather is reminiscent of that on Earth, with dust storms, clouds, and fog. The two polar caps on Mars each consist of a **seasonal cap (p. 264)**, composed of carbon dioxide, which grows and shrinks, and a **residual cap (p. 264)**, of water ice, which remains permanently frozen.

In 1971, *Mariner 9* mapped the entire Martian surface, revealing plains, volcanoes, channels, and canyons. *Viking 1* and *Viking 2* reached Mars in 1976 and returned a wealth of data on the planet's surface and atmosphere. Experiments onboard the *Viking* landers detected no evidence for Martian life. There is a marked difference between the two Martian hemispheres. The northern hemisphere consists of rolling volcanic plains and lies several kilometers below the level of the heavily cratered southern hemisphere. The lack of craters in the north suggests that this region is younger. The cause of the north–south asymmetry is not known.

Mars's major surface feature is the Tharsis bulge, located on the planet's equator. This feature may have been caused by a "plume" of upwelling material in the youthful Martian mantle. Associated with the bulge are Olympus Mons, the largest known volcano in the solar system, and a huge crack, called the Valles Marineris, in the planet's surface. The height of the Martian volcanoes is a direct consequence of Mars's low surface gravity. No evidence for recent or ongoing eruptions has been found. Convection in the Martian interior seems to have been stifled 2 billion years ago by the planet's rapidly cooling and solidifying mantle. On the other side of Mars from Tharsis lies the Hellas basin, the site of a violent meteoritic impact early in the planet's history.

Martian craters differ from those on the Moon by the presence of fluidized ejecta, which provide direct evidence for a permafrost layer beneath the surface, and *Viking* and *Mars Global Surveyor* images indicate that liquid water once existed in great quantity on Mars. The planet may have enjoyed a relatively brief "Earth-like" phase early on in its evolution, but that possibility remains controversial. **Runoff channels (p. 261)** are the remains of ancient Martian rivers, whereas **outflow channels (p. 261)** are the paths taken by flash floods that cascaded from the southern highlands into the northern plains. Today, a large amount of that water may be locked up in the polar caps and in the layer of permafrost lying under the Martian surface. *Mars Odyssey* has detected ice crystals mixed with the surface layers. *Surveyor* images also suggest that there may still be liquid water below the surface in some places.

Mars has an extremely weak magnetic field, which, together with the planet's rapid rotation, implies that its core is nonmetallic, nonliquid, or both. The lack of current volcanism, the absence of any significant magnetic field, the planet's relatively low density, and a high abundance of surface iron all suggest that Mars never melted and differentiated as extensively as did Earth.

The Martian moons Phobos and Deimos are probably asteroids captured by Mars early in its history. Their densities are far less than that of any planet in the inner solar system. These moons may be representative of conditions in the early solar system.

REVIEW AND DISCUSSION

1. Why is opposition the best time to see Mars from Earth?

2. Why are some Martian oppositions better than others for viewing Mars?

3. For a century, there was speculation that intelligent life had constructed irrigation canals on Mars. What did the "canals" turn out to be?

4. Imagine that you will be visiting the southern hemisphere of Mars during its summer. Describe the atmospheric conditions you might face.

5. Describe the two Martian polar caps, their seasonal and permanent composition, and the differences between them.

6. Why is Mars red?

7. Describe the major large-scale surface features of Mars.

8. Why were Martian volcanoes able to grow so large?

9. Why couldn't you breathe on Mars?

10. What is the evidence that water once flowed on Mars?

11. Is there liquid water on Mars today?

12. Is there water on Mars today, in any form?

13. Why do some scientists think Mars once had an extensive ocean? Where was it located?

14. How were the masses of Mars's moons measured, and what did these measurements tell us about their origin?

15. What do measurements of Martian magnetism tell us about the planet's interior?

16. What is the evidence that Mars never melted as extensively as did Earth?

17. How would Earth look from Mars?

18. If humans were sent to Mars to live, what environmental factors would have to be considered? What resources might Mars provide, and which would have to come from Earth?

19. Since Mars has an atmosphere, and it is composed mostly of a greenhouse gas, why isn't there a significant greenhouse effect to warm its surface?

20. Compare and contrast the evolution of the atmospheres of Mars, Venus, and Earth.

CONCEPTUAL SELF-TEST: TRUE OR FALSE/MULTIPLE CHOICE

1. Seen from Mars, Earth would go through phases, just as Venus and Mercury do.

2. Because Mars has such a thin atmosphere, the planet has no significant surface winds.

3. Seasonal changes in the appearance of Mars are caused by vegetation on the surface.

4. Olympus Mons is the largest impact crater in the solar system.

5. There are many indications of past plate tectonics on Mars.

6. Valles Marineris is comparable in size to Earth's Grand Canyon.

7. The polar caps of Mars are composed entirely of frozen carbon dioxide.

8. The great height of Martian volcanoes is a direct result of the planet's low gravity.

9. NASA's *Opportunity* rover detected liquid water just under the Martian surface.

10. Water once flowed on the surface of Mars.

11. Compared with the Earth's orbit, the orbit of Mars (a) has the same eccentricity; (b) is more eccentric; (c) is less eccentric; (d) is smaller.

12. As seen from Earth, Mars exhibits a retrograde loop about once every (a) week; (b) six months; (c) two years; (d) decade.

13. Compared with Earth's diameter, the diameter of Mars is (a) significantly larger; (b) significantly smaller; (c) nearly the same size; (d) unknown.

14. The lengths of the seasons on Mars can be determined by observing the planet's (a) tilt; (b) eccentricity; (c) polar caps; (d) moons.

15. In terms of area, the extinct Martian volcano Olympus Mons is about the size of (a) Mt. Everest; (b) Colorado; (c) North America; (d) Earth's Moon.

16. Figure 10.5 ("Mars Map") clearly shows (a) surface water and ice at northern latitudes; (b) a giant canyon stretching all the way across the planet; (c) iron deposits in the mid-latitudes; (d) cratered terrain in the south.

17. The best evidence for the existence of liquid water on an ancient Mars is Figure (a) 10.12; (b) 10.14; (c) 10.15; (d) 10.17.

18. Compared with the atmosphere of Venus, the Martian atmosphere has (a) a significantly higher temperature; (b) significantly more carbon dioxide; (c) a significantly lower atmospheric pressure; (d) significantly more acidic compounds.

19. In comparison to the atmosphere of Venus, the vastly different atmospheric character of Mars is likely due to a/an (a) ineffective greenhouse effect; (b) reverse greenhouse effect; (c) absence of greenhouse gases that would hold in heat; (d) greater distance from the Sun.

20. The moons of Mars (a) are probably captured asteroids; (b) formed following a collision with Earth; (c) are the remnants of a larger moon; (d) formed simultaneouly with Mars.

PROBLEMS

 Algorithmic versions of these questions are available in the Practice Problems module of the Companion Website at astro.prenhall.com/chaisson.

The number of squares preceding each problem indicates its approximate level of difficulty.

1. ■■ By calculating the rate at which Earth overtakes Mars in its orbit (see *More Precisely 9-1*), verify the value of Mars's synodic period given in the Mars Data box on p. 253.

2. ■ Calculate the minimum and maximum angular diameters of the Sun, as seen from Mars.

3. ■■ What is the maximum elongation of Earth, as seen from Mars? (For simplicity, assume circular orbits for both planets.)

4. ■ What will be the minimum size of a Martian surface feature resolvable during the 2003 opposition by an Earth-based telescope with an angular resolution of 0.05"?

5. ■■ Use the reasoning presented in Chapter 1 to calculate the difference in length between the mean Martian solar day and the Martian sidereal day. ∞ (Sec. 1.3)

6. ■ Verify that the surface gravity on Mars is 40 percent that of Earth.

7. ■ What would you weigh on Mars?

8. ■ How long would it take the wind in a Martian dust storm, moving at a speed of 150 km/h, to encircle the planet's equator?

9. ■ The mass of the Martian atmosphere is about 1/150 the mass of Earth's atmosphere and is composed mainly (95 percent) of carbon dioxide. Using the result of Problem 3 in Chapter 7 to determine the mass of Earth's atmosphere, estimate the total mass of carbon dioxide in the atmosphere of Mars. Compare your answer with the mass of a seasonal polar cap, approximated as a circular sheet of frozen carbon dioxide ("dry ice," having a density of 1600 kg/m^3) of diameter 3000 km and thickness 1 m.

10. ■ Compare the mass of a seasonal polar cap (see previous question) with that of a residual cap of diameter 1000 km, thickness 1 km, and density 1000 kg/m^3.

11. ■ The Hellas impact basin is roughly circular, 3000 km across, and 6 km deep. Taking the Martian crust to have a density of 3000 kg/m^3, estimate how much mass was blasted off the Martian surface when the basin formed. Compare your answer with the present total mass of the Martian atmosphere. (See Problem 9.)

12. ■■ The outflow channel shown in Figure 10.12 is about 10 km across and 100 m deep. If it carried 10^7 metric tons (10^{10} kg) of water per second, as stated in the text, estimate the speed at which the water must have flowed.

13. ■■ Calculate the total mass of a uniform layer of water covering the entire Martian surface to a depth of 2 m. (See Section 10.5.) Compare your answer with the mass of Mars.

14. ■■ Using the data from Section 10.7, compute the time interval between each rising of the moons Phobos and Deimos and the next, as seen from the Martian surface.

15. ■ Using the data given in the text, calculate the maximum angular sizes of Phobos and Deimos, as seen by an observer standing on the Martian surface directly under their orbits. Would a Martian observer ever see a total solar eclipse? (See Problem 2.)

 In addition to the Practice Problems module, the Companion Website at astro.prenhall.com/chaisson provides for each chapter a study guide module with multiple choice questions as well as additional annotated images, animations, and links to related Websites.

11

Astronomers study weather phenomena on other planets partly to gain a better understanding of our own planet, but we also study other planets specifically to learn more about them. Each one seems to display different conditions than those on Earth, and until all of them are well probed and understood, our inventory of the local part of the universe will be incomplete. Here, in this image taken in 2000 as the Cassini spacecraft glided past on its way to Saturn, Jupiter's north polar region is seen up close and in true color. The intricate structures represent clouds of ▶ different heights, thicknesses, and chemical compositions. (JPL)

Jupiter

Giant of the Solar System

Beyond the orbit of Mars, the solar system is very different from our own backyard. The outer solar system presents us with a totally unfamiliar environment: huge gas balls, peculiar moons, complex ring systems, and a wide variety of physical and chemical phenomena, many of which are still only poorly understood. Although the jovian planets—Jupiter, Saturn, Uranus, and Neptune—differ from one another in many ways, we will find that they have much in common, too. As with the terrestrial planets, we will learn from their differences as well as from their similarities. Our study of these alien places begins with the jovian planet closest to Earth: Jupiter, the largest planet in the solar system and a model for the other jovian worlds.

LEARNING GOALS

Studying this chapter will enable you to

1 Specify the ways in which Jupiter differs from the terrestrial planets in its physical and orbital properties.

2 Discuss the processes responsible for the appearance of Jupiter's atmosphere.

3 Describe Jupiter's internal structure and composition, and explain how their properties are inferred from external measurements.

4 Summarize the characteristics of Jupiter's magnetosphere.

5 Discuss the orbital properties of the Galilean moons of Jupiter, and describe the appearance and physical properties of each moon.

6 Explain how tidal forces can produce enormous internal stresses in a jovian moon, and discuss some effects of those stresses.

 Visit astro.prenhall.com/chaisson for additional annotated images, animations, and links to related sites for this chapter.

11.1 Orbital and Physical Properties

🎯 **1** Named after the most powerful god of the Roman pantheon, Jupiter is by far the largest planet in the solar system. Ancient astronomers could not have known the planet's true size, but their choice of names was apt. The Jupiter Data box presents some orbital and physical data on the planet.

The View from Earth

Jupiter is the third-brightest object in the night sky (after the Moon and Venus), making it easy to locate and study. As in the case of Mars, Jupiter is brightest when it is near opposition. When this happens to occur close to perihelion, the planet can be up to 50″ across, and a lot of detail can be discerned through even a small telescope.

Figure 11.1(a) is a photograph of Jupiter, taken through a telescope on Earth. In contrast to the terrestrial worlds, Jupiter has many moons that vary greatly in size and other properties. The four largest, visible in this telescopic view (and, to a few people, with the naked eye), are known as the **Galilean moons** after Galileo Galilei, who discovered them in 1610. ∞ (Sec. 2.4) Figure 11.1(b) is a *Hubble Space Telescope* image of Jupiter taken during the opposition of December 1990. Notice both the alternating light and dark bands that cross the planet parallel to its equator and also the large oval at the lower right. These atmospheric features are quite unlike anything found on the inner planets. Figure 11.1(c) is an up-close, true-color image of Jupiter's north polar region, taken by the *Cassini* spacecraft while it was gliding past in 2001.

Mass and Radius

Since astronomers have been able to study the motion of the Galilean moons for quite some time, Jupiter's mass has long been known to high accuracy. It is 1.9×10^{27} kg, or 318 Earth masses. ∞ (*More Precisely 6-1*) Indeed, Jupiter

(a)

(b)

(c)

R I V U X G

▲ **FIGURE 11.1 Jupiter** (a) Photograph of Jupiter made through a ground-based telescope, showing the planet and several of its Galilean moons. (b) A *Hubble Space Telescope* image of Jupiter, in true color. Features as small as a few hundred kilometers across are resolved. (c) A *Cassini* spacecraft image of Jupiter, taken while the vehicle was on its way to Saturn, shows intricate clouds of different heights, thicknesses, and chemical composition. (*NASA; AURA*)

PLANETARY DATA:

JUPITER

Orbital semimajor axis	5.20 A.U. 778.4 million km
Orbital eccentricity	0.048
Perihelion	4.95 A.U. 740.7 million km
Aphelion	5.46 A.U. 816.1 million km
Mean orbital speed	13.1 km/s
Sidereal orbital period	11.86 tropical years
Synodic orbital period	398.88 solar days
Orbital inclination to the ecliptic	1.31°
Greatest angular diameter, as seen from Earth	50″
Mass	1.90×10^{27} kg 317.8 (Earth = 1)
Equatorial radius	71,492 km 11.21 (Earth = 1)
Mean density	1330 kg/m³ 0.241 (Earth = 1)
Surface gravity (at cloud tops)	24.8 m/s² 2.53 (Earth = 1)
Escape speed	59.5 km/s
Sidereal rotation period	0.41 solar day
Axial tilt	3.08°
Surface magnetic field	13.89 (Earth = 1)
Magnetic axis tilt relative to rotation axis	9.6°
Surface temperature	124 K (at cloud tops)
Number of moons	16 (more than 10 km in diameter) 61 (total)

has more than twice the mass of all the other planets combined, and it is such a large planet that many celestial mechanicians—those researchers concerned with the motions of interacting cosmic objects—regard our solar system as containing only two important objects: the Sun and Jupiter. To be sure, in this age of sophisticated and precise spacecraft navigation, the gravitational influence of all the planets must be considered, but in the broadest sense, our solar system is a two-object system with a lot of additional debris. Nonetheless, as massive as Jupiter is, it is only a thousandth the mass of the Sun—still enough to make studies of Jupiter all the more important, for here we have an object intermediate in size between the Sun and the terrestrial planets.

Knowing Jupiter's distance and angular size, we can easily determine the planet's radius, which turns out to be 71,500 km, or 11.2 Earth radii. More dramatically stated, more than 1400 Earths would be needed to equal the volume of Jupiter. From the planet's size and mass, we derive an average density of 1300 kg/m³ for Jupiter. Here (as if we needed it) is yet another indicator that Jupiter is radically different from the terrestrial worlds: It is clear that, whatever Jupiter's composition, it cannot possibly be made up of the same material as the inner planets. (Recall from Chapter 7 that Earth's average density is 5500 kg/m³). ∞ (Sec. 7.1) In fact, theoretical studies of the planet's internal structure indicate that Jupiter must be composed primarily of hydrogen and helium. The enormous pressure in the planet's interior due to Jupiter's strong gravity greatly compresses these light gases, whose densities on Earth (at room temperature and sea level) are 0.08 and 0.16 kg/m³, respectively, producing the relatively high average density we observe.

Rotation Rate

As with other planets, we can attempt to determine Jupiter's rotation rate simply by timing a surface feature as it moves around the planet. However, in the case of Jupiter (and, indeed, all the gaseous outer planets), there is a catch: Jupiter has no solid surface. All we see are the features of clouds in the planet's upper atmosphere. With no solid surface to "tie them down," different parts of Jupiter's atmosphere move independently of one another. Visual observations and Doppler-shifted spectral lines indicate that the equatorial zones rotate a little faster (with a period of 9^h50^m) than the higher latitudes (with a period of 9^h55^m). Jupiter thus exhibits **differential rotation**—the rotation rate is not constant from one location to another. Differential rotation is not possible in solid objects like the terrestrial planets, but it is normal for fluid bodies such as Jupiter.

Observations of Jupiter's magnetosphere provide a more meaningful measurement of the rotation period. The planet's magnetic field is strong and emits radiation at radio wavelengths as charged particles accelerate in

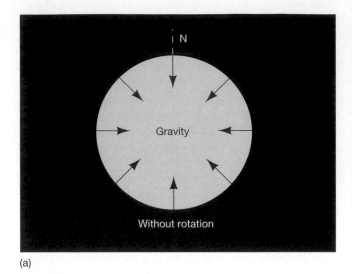

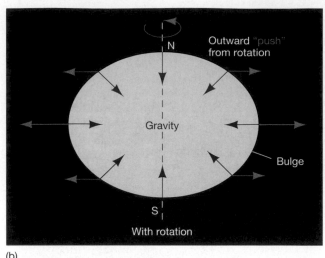

(a)   (b)

▲ **FIGURE 11.2 Rotational Flattening** All spinning objects tend to develop an equatorial bulge because rotation causes matter to push outward against the inward-pulling gravity. The size of the bulge depends on the mechanical strength of the matter and the rate of rotation. The inward-pointing arrows denote gravity, the outward arrows the "push" due to rotation.

response to Jupiter's magnetic field. Careful studies show a periodicity of 9^h55^m at these radio wavelengths. We assume that this measurement matches the rotation of the planet's interior, where the magnetic field arises. ∞ (Sec. 7.5) Thus, Jupiter's interior rotates at the same rate as the clouds at the planet's poles. The equatorial zones rotate more rapidly.

A rotation period of 9^h55^m is fast for such a large object. In fact, Jupiter has the fastest rotation rate of any planet in the solar system, and this rapid spin has altered Jupiter's shape. As illustrated in Figure 11.2, a spinning object tends to develop a bulge around its midsection. The more loosely the object's matter is bound together, or the faster it spins, the larger the bulge becomes. In objects like Jupiter, which are made up of gas or loosely packed matter, high spin rates can produce a quite pronounced bulge. Jupiter's equatorial radius (71,500 km) exceeds its polar radius (66,900 km) by about 6.5 percent.*

But there is more to the story of Jupiter's shape. Jupiter's observed equatorial bulge also tells us something important about the planet's deep interior. Careful calculations indicate that Jupiter would be *more* flattened than it actually is if its core were composed of hydrogen and helium alone. To account for the planet's observed shape, we must assume that Jupiter has a dense, compact core, probably of rocky composition, about 5–10 times the mass of Earth. This is one of the few pieces of data we have on Jupiter's internal structure.

Earth also bulges slightly at the equator because of rotation. However, our planet is much more rigid than Jupiter, and the effect is much smaller—the equatorial diameter is only about 40 km larger than the distance from pole to pole, a tiny difference compared with Earth's full diameter of nearly 13,000 km. Relative to its overall dimensions, Earth is smoother and more spherical than a billiard ball.

CONCEPT CHECK

✔ How do observations of a planet's magnetosphere allow astronomers to measure the rotation rate of the interior?

11.2 The Atmosphere of Jupiter

Jupiter is visually dominated by two features: a series of ever-changing atmospheric bands arranged parallel to the equator; and an oval atmospheric blob called the **Great Red Spot**, or, often, just the "Red Spot." The bands of clouds, clearly visible in Figure 11.1, display many colors—pale yellows, light blues, deep browns, drab tans, and vivid reds, among others. Shown in more detail in Figure 11.3, a close-up photograph taken as *Voyager 1* sped past in 1979, the Red Spot is the largest of many features associated with Jupiter's weather. It seems to be a hurricane twice the size of planet Earth that has persisted for hundreds of years.

Atmospheric Composition

Spectroscopic studies of sunlight reflected from Jupiter gave astronomers their first look at the planet's atmospheric composition. Radio, infrared, and ultraviolet observations provided more details later. The most abundant gas is molecular hydrogen (86.1 percent by number of molecules), followed by helium (13.8 percent). Together, these two gases make up over 99 percent of Jupiter's atmosphere. Small amounts of atmospheric methane, ammonia, and water vapor are also found. Researchers think that hydro-

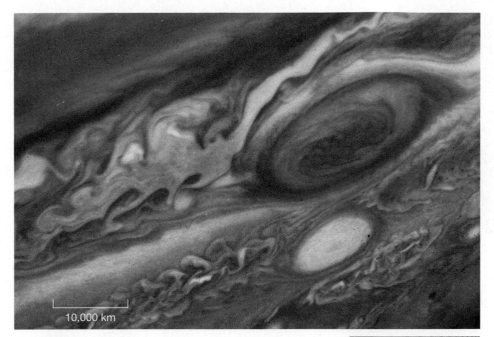

◄ **FIGURE 11.3** Jupiter's Red Spot *Voyager 1* took this photograph of Jupiter's Red Spot (upper right) from a distance of about 100,000 km. The resolution is about 100 km. Note the complex turbulence to the left of both the Red Spot and the smaller white oval vortex below it. (For scale, planet Earth is about the size of the white oval.) *(NASA)*

10,000 km

R I U V X G

gen and helium in those same proportions make up the bulk of the planet's interior as well.

The abundance of hydrogen and helium on Jupiter is a direct consequence of the planet's strong gravity. Unlike the gravitational pull of the terrestrial planets, the gravity of the much more massive jovian planets is powerful enough to have retained even hydrogen. ∞ *(More Precisely 8-1)* Little, if any, of Jupiter's original atmosphere has escaped since the planet formed 4.6 billion years ago.

Atmospheric Bands

Astronomers generally describe Jupiter's banded appearance—and, to a lesser extent, the appearance of the other jovian worlds as well—as a series of bright *zones* and dark *belts* crossing the planet. The light-colored zones lie above upward-moving convective currents in Jupiter's atmosphere. The dark belts are regions representing the other part of the convection cycle, during which material is generally sinking downward, as illustrated schematically in Figure 11.4. Because of the upwelling material below them, the zones are regions of high pressure; the belts, conversely, are low-pressure regions.

Thus, belts and zones are Jupiter's equivalents of the familiar high- and low-pressure systems that cause our weather on Earth. However, a major difference between Jupiter and Earth is that Jupiter's rapid rotation has caused these systems to wrap all the way around the planet, instead of forming localized circulating storms, as on our own world. Because of the pressure difference between the two, the zones lie slightly higher in the atmosphere than the belts do. The associated temperature differences (we will see in a moment that the temperature increases the

deeper we descend into the atmosphere) and the resulting differences in chemical reactions are the basic reasons for the different colors of these jovian features.

Underlying the bands is an apparently very stable pattern of eastward and westward wind flow, known as Jupiter's **zonal flow**. The pattern is evident in Figure 11.5, which shows the wind speed at different planetary latitudes, measured relative to the rotation of the planet's interior (determined from studies of Jupiter's magnetic field). As mentioned earlier, the equatorial regions of the atmosphere rotate faster than the planet; their average flow speed is some 85 m/s, or about 300 km/h, in the easterly direction. The speed of this equatorial flow is quite similar to that of the jet stream on Earth. At higher latitudes, there are alternating regions of westward and eastward flow, roughly symmetric about the equator, with the flow speed generally diminishing toward the poles.

As Figure 11.5 shows, the belts and zones are closely related to Jupiter's zonal flow pattern. However, closer inspection reveals that the simplified picture presented in Figure 11.4, with wind direction alternating between adjacent bands as Jupiter's rotation deflects surface winds into eastward or westward streams, is really too crude to describe the actual flow. Scientists think that the interaction between convective motion in Jupiter's atmosphere and the planet's rapid rotation channels the largest eddies into the observed zonal pattern, but that smaller eddies tend to cause irregularities in the flow. Near the poles, where the zonal flow disappears, the band structure vanishes also.

The zones and belts vary in both latitude and intensity during the year, although the general pattern remains. The variations are not seasonal in nature: Having a low-eccentricity orbit and a rotation axis almost exactly perpendicular

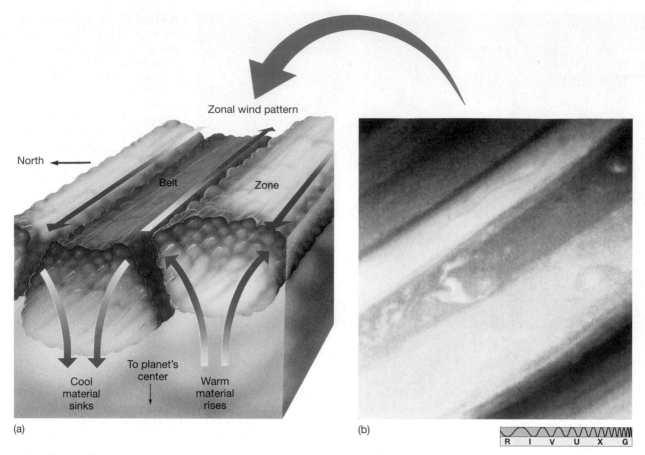

▲ **FIGURE 11.4 Jupiter's Convection** (a) The colored bands in Jupiter's atmosphere are associated with vertical convective motion. Upwelling warm gas results in zones of lighter color; the darker bands overlie regions of lower pressure where cooler gas sinks back down into the atmosphere. As on Earth, surface winds tend to blow from high- to low-pressure regions. Jupiter's rapid rotation channels those winds into an east–west flow pattern, as indicated by the three yellow-red arrows drawn atop the belts and zones. (b) A *Voyager* photo of part of Jupiter's cloud layer, as seen from above, showing the banded structure. *(NASA)*

to its orbital plane, Jupiter has no seasons. ∞ (Sec. 1.4) Instead, the annual changes appear to be the result of dynamic motion in the planet's atmosphere.

Atmospheric Structure and Color

None of the atmospheric gases listed earlier can, by itself, account for Jupiter's observed coloration. For example, frozen ammonia and water vapor would simply produce white clouds, not the many colors actually seen. Scientists suspect that the colors of the clouds are the result of complex chemical processes occurring in the planet's turbulent upper atmosphere, although the details are still not fully understood. When we observe Jupiter's colors, we are actually looking down to many different depths in the planet's atmosphere.

Based on the best available data and mathematical models, Figure 11.6 is a cross-sectional diagram of Jupiter's atmosphere. Since the planet lacks a solid surface to use as a reference level for measuring altitude, the top of

the troposphere is conventionally taken to lie at 0 km. As on all planets, weather on Jupiter is the result of convection in the troposphere, so the clouds, which are associated with planetary weather systems, all lie at negative altitudes in the diagram. Just above the troposphere lies a thin, faint layer of haze created by photochemical reactions (i.e., reactions involving sunlight) similar to those which cause smog on Earth. The temperature at this level is about 110 K; it increases with altitude as the atmosphere absorbs solar ultraviolet radiation.

Jupiter's clouds are arranged in three main layers. Below the haze, at a depth of about 40 km (shown as −40 km in Figure 11.6), lies a layer of white, wispy clouds made up of ammonia ice. The temperature here is approximately 125–150 K; it increases quite rapidly with increasing depth. A few tens of kilometers below the ammonia clouds, the temperature is a little warmer—over 200 K—and the clouds are probably made up mostly of droplets or crystals of ammonium hydrosulfide, produced by reactions between ammonia and hydrogen sulfide in the planet's at-

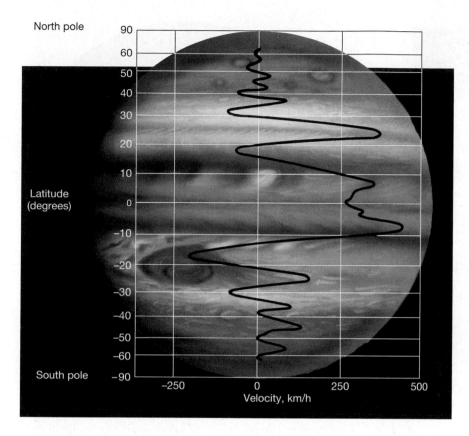

North pole

South pole

Latitude (degrees)

Velocity, km/h

◄ **FIGURE 11.5 Zonal Flow** The wind speed in Jupiter's atmosphere, measured relative to the planet's internal rotation rate. Alternations in wind direction are associated with the atmospheric band structure.

mosphere. At deeper levels in the atmosphere, the ammonium hydrosulfide clouds give way to clouds of water ice or water vapor. This lowest cloud layer, which is not seen in visible-light images of Jupiter, lies some 80 km below the top of the troposphere.

Instead of being white (the color of ammonium hydrosulfide on Earth), Jupiter's middle cloud layer is tawny in color. This is the level at which atmospheric chemistry begins to play a role in determining Jupiter's appearance. Many planetary scientists think that molecules containing the element sulfur, and perhaps even sulfur itself, are important in influencing the cloud colors—particularly the reds, browns, and yellows, all colors associated with sulfur or its compounds. It is also possible that compounds containing the element phosphorus contribute to the coloration.

Deciphering the detailed causes of Jupiter's distinctive colors is a difficult task, however. The cloud chemistry is complex and highly sensitive to small changes in atmospheric conditions, such as pressure and temperature, as well as to chemical composition. The atmosphere is in incessant, churning motion, causing conditions to change from place to place and from hour to hour. In addition, the energy that powers the reactions comes in many different forms: the planet's own interior heat, solar ultraviolet radiation, aurorae in the planet's magnetosphere, and lightning discharges within the clouds themselves. All of these factors combine to keep a complete explanation of Jupiter's appearance beyond our present grasp.

The preceding description of Jupiter's atmosphere, based largely on *Voyager* data, was put to the test in December 1995, when the *Galileo* atmospheric probe arrived at the planet. ∞ (Sec. 6.6) The probe survived for about an hour before being crushed by atmospheric pressure at an altitude of −150 km (i.e., right at the bottom of Figure 11.6). Overall, *Galileo*'s findings on wind speed, temperature, and composition were in good agreement with the picture just presented. However, the probe's entry location was in Jupiter's equatorial zone and, as luck would have it, coincided with an atypical "hole" almost devoid of upper-level clouds (see Figure 11.7). The probe measured a temperature of 425 K at 150 km depth—a little higher than indicated in Figure 11.6, but consistent with the craft's having entered a clearing in Jupiter's cloud decks, where convective heat can more readily rise (and thus be detected). The probe also measured a slightly lower than expected water content, but that, too, may be normal for the hot, windy regions near Jupiter's equator.

The experts were somewhat surprised by the depth to which Jupiter's winds continued. *Galileo*'s probe measured high wind speeds throughout its descent into the clouds, not just at the cloud tops, implying that heat deep within the planet, rather than sunlight, drives Jupiter's weather patterns. Finally, complex organic molecules were sought, but not found. Some simple carbon-based molecules, such as ethane (C_2H_6), were detected by one of the onboard spectrometers, but nothing suggesting prebiotic compounds

ANIMATION Galileo Mission to Jupiter

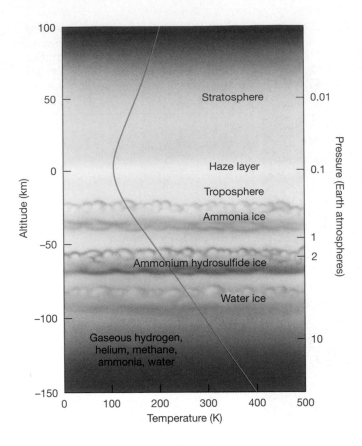

▲ **FIGURE 11.6 Jupiter's Atmosphere** The vertical structure of Jupiter's atmosphere contains clouds arranged in three main layers, each with quite different colors and chemistry. The colors we see in photographs of the planet depend on the cloud cover. The white regions are the tops of the upper ammonia clouds. The yellows, reds, and browns are associated with the second cloud layer, which is made of ammonium hydrosulfide ice. The lowest (bluish) cloud layer is water ice, but the overlying layers are sufficiently thick that this level is not seen in visible light.

▲ **FIGURE 11.7** *Galileo's* **Entry Site** The arrow on this image shows where the *Galileo* atmospheric probe plunged into Jupiter's cloud deck on December 7, 1995. The entry location was in Jupiter's equatorial zone and apparently almost devoid of upper-level clouds. Until its demise, the probe took numerous meteorological measurements, transmitting those signals to the orbiting mother ship, which then relayed them to Earth. *(NASA)*

(molecules that could combine to form the building blocks of life—see Section 28.1) or bacteria floating in the atmosphere was found. That same instrument also detected traces of phosphine (PH_3), which may be a key coloring agent for Jupiter's clouds.

Weather on Jupiter

In addition to the zonal flow pattern, Jupiter has many "small-scale" weather patterns. The Great Red Spot (Figure 11.3) is a prime example. The spot was first reported by British scientist Robert Hooke in the mid-17th century, and we can be reasonably sure that it has existed continuously, in one form or another, for over 300 years. It may well be much older. *Voyager* observations showed the spot to be a region of swirling, circulating winds, rather like a whirlpool or a terrestrial hurricane—a persistent and

vast atmospheric storm. The size of the spot varies, although it averages about twice the diameter of Earth. Its present dimensions are roughly 25,000 km by 15,000 km. The spot rotates around Jupiter at a rate similar to that of the planet's interior, perhaps suggesting that the roots of the Great Red Spot lie far below the atmosphere.

The origin of the spot's red color is unknown, as is its source of energy, although it is generally supposed that the spot is somehow sustained by Jupiter's large-scale atmospheric motion. Repeated observations show that the gas flow around the spot is counterclockwise, with a period of about six days. Turbulent eddies form and drift away from its edge. The spot's center, however, remains quite tranquil in appearance, like the eye of a hurricane on Earth. The zonal motion north of the Great Red Spot is westward, whereas that to the south is eastward (see Figure 11.8), supporting the idea that the spot is confined and powered by the zonal flow. However, the details of how it is so confined are still a matter of conjecture. Computer simulations of the complex fluid dynamics of Jupiter's atmosphere are only now beginning to hint at answers.

Storms, which as a rule are much smaller than the Great Red Spot, may be quite common on Jupiter. Spacecraft photographs of the dark side of the planet reveal

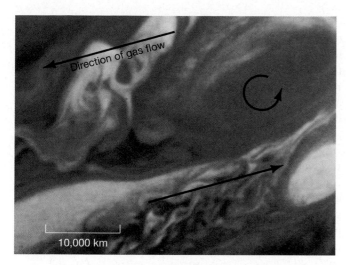

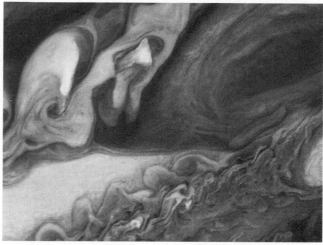

▲ **FIGURE 11.8 Red Spot Details** These *Voyager 2* close-up views of the Great Red Spot, taken four hours apart, show clearly the turbulent flow around its edges. The general direction of motion of the gas north of (above) the spot is westward (to the left), whereas gas south of the spot flows east. The spot itself rotates counterclockwise, suggesting that it is being "rolled" between the two oppositely directed flows. The colors have been exaggerated somewhat to enhance the contrast. *(NASA)*

bright flashes resembling lightning. The *Voyager* mission discovered many smaller light- and dark-colored spots that are also apparently circulating storm systems. Note the **white ovals** in Figures 11.3 and 11.8, south of the spot. Like the spot itself, they rotate counterclockwise. Their high cloud tops give them their color. These particular white ovals are known to be at least 40 years old. Figure 11.9 shows a **brown oval**, a "hole" in the clouds that allows us to look down into Jupiter's lower atmosphere. For unknown reasons, brown ovals appear only at latitudes around 20°N. Although not as long lived as the Great Red Spot, these systems can persist for many years or even decades.

We cannot explain the formation of brown ovals, but we can offer at least a partial explanation for the longevity of storm systems on Jupiter. On Earth, a large storm, such as a hurricane, forms over the ocean and may survive for many days, but it dies quickly once it encounters land. Earth's continental landmasses disrupt the flow patterns that sustain the storm. Jupiter has no continents, so once a storm becomes established and reaches a size at which other storm systems cannot destroy it, apparently little affects it. The larger the system, the longer is its lifetime.

CONCEPT CHECK

✔ List some similarities and differences between Jupiter's belts, zones, and spots, on the one hand, and weather systems on Earth, on the other.

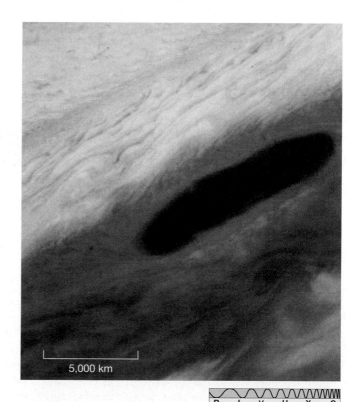

▲ **FIGURE 11.9 Brown Oval** This close-up view of an oval in Jupiter's northern hemisphere actually shows a break in the upper cloud layer, allowing us to see deeper into the atmosphere, where the clouds are brown. The oval's length is approximately equal to the diameter of Earth. *(NASA)*

11.3 Internal Structure

On the basis of Jupiter's distance from the Sun, astronomers had expected to find the temperature of the cloud tops to be around 105 K. At that temperature, they reasoned, Jupiter would radiate back into space exactly the same amount of energy as it received from the Sun. When radio and infrared observations were first made of the planet, however, astronomers found that its blackbody spectrum corresponded to a temperature of 125 K instead. Subsequent measurements, including those made by *Voyager* and *Galileo*, have verified that finding. Although a difference of 20 K may seem small, recall from Chapter 5 that the energy emitted by a planet grows as the *fourth* power of the surface temperature (in Jupiter's case, the temperature of the cloud tops). ∞(Sec. 3.4) A planet at 125 K therefore radiates $(125/105)^4$, or about twice as much energy as a planet at 105 K radiates. Put another way, Jupiter actually emits about twice as much energy as it receives from the Sun. Thus, unlike any of the terrestrial planets, Jupiter must have its own internal heat source.

What is responsible for Jupiter's extra energy? It is not the decay of radioactive elements within the planet. That process must be occurring, as in Earth, but estimates of the total amount of energy released into Jupiter's interior are far below the levels needed to account for the temperature we measure. ∞ (Sec. 7.3) Nor is it nuclear fusion, the process that generates energy in the Sun. The temperature in Jupiter's interior, high as it is, is still far too low for that (see *Discovery 11-1*). Instead, astronomers theorize that the source of Jupiter's excess energy is the slow escape of gravitational energy released during the planet's formation. As the planet took shape, some of its gravitational energy was converted into heat in the interior. That heat is still slowly leaking out through the planet's heavy atmospheric blanket, resulting in the excess emission we observe.

Despite the huge amounts of energy involved—Jupiter emits about 4×10^{17} watts more energy than it receives from the Sun—the loss is slight compared with the planet's total energy. On the basis of the planet's mass and temperature, as well as the rate at which thermal energy is leaving the planet, astronomers calculate that the average temperature of the interior of Jupiter decreases by only about a millionth of a kelvin per year. ∞ (More Precisely 3-1)

Jupiter's clouds, with their complex chemistry, are probably less than 200 km thick. Below them, the temperature and pressure steadily increase as the atmosphere becomes the "interior" of the planet. Much of our knowledge of Jupiter's interior comes from theoretical modeling. Indeed, apart from data gained following the collision of a comet with Jupiter in 1994, we have very little direct evidence of the planet's internal properties. Planetary scientists use all available bulk data on the planet—its mass, radius, composition, rotation, temperature, and so on—to construct a model of the interior that agrees with observations. Modeling is an integral part of the scientific method, and our statements about Jupiter's structure are really statements about the model that best

DISCOVERY 11-1

Almost a Star?

Jupiter has a starlike composition—predominantly hydrogen and helium, with a trace of heavier elements. Did Jupiter ever come close to becoming a star itself? Might the solar system have formed as a double-star system? Probably not. Unlike a star, Jupiter is cold. Its central temperature is far too low to ignite the nuclear fires that power our Sun. Jupiter's mass would have to increase eightyfold before its central temperature would rise to the point where nuclear reactions could begin, converting Jupiter into a small, dim star. Even so, it is interesting to note that, although Jupiter's present-day energy output is very small (by solar standards, at least), it must have been much greater in the distant past, while the planet was still contracting rapidly toward its present size. For a brief period—a few hundred million years—Jupiter might actually have been as bright as a faint star, although its brightness never came within a factor of 100 of the Sun's. Still, seen from Earth at that time, Jupiter would have been about 100 times brighter than the Moon!

What might have happened had our solar system formed as a double-star system? Conceivably, had Jupiter been mas-sive enough, its radiation might have produced severe temperature fluctuations on all the planets, perhaps to the point of making life on Earth impossible. Even if Jupiter's brightness were too low to cause us any problems, its gravitational pull (which would be one-twelfth that of the Sun if its mass were eighty times its present value) might have made the establishment of stable, roughly circular planetary orbits in the inner solar system an improbable event, again to the detriment of life on Earth.

Curiously, in recent years astronomers have come to realize that, had Jupiter been too *small*, that also could have adversely affected the chances for life on our planet! As we will see in Chapter 15, Jupiter played a crucial role in clearing cometary debris from the outer solar system during and after the period when the planets formed. ∞ (Sec. 6.7) Had that not occurred, the meteoritic bombardment of our planet might have been too severe and too extended for complex life ever to have evolved. ∞ (Sec. 8.5) Many stars near the Sun are now known to have Jupiter-sized planets orbiting them. It seems that the size of the "Jupiter," or second-largest body, in a newborn planetary system may be a critical factor in determining the likelihood of the appearance of life there.

fits the facts. ∞ (Sec. 1.2) However, because the planet consists largely of hydrogen and helium—two simple gases whose physics we think we understand well—we can be fairly confident that Jupiter's internal structure is now understood.

Both the temperature and the density of Jupiter's atmosphere increase with depth below the cloud cover. However, no "surface" of any kind exists anywhere inside. Instead, Jupiter's atmosphere just becomes denser and denser because of the pressure of the overlying layers. At a depth of a few thousand kilometers, the gas makes a gradual transition into the liquid state (see Figure 11.10). By a depth of about 20,000 km, the pressure is about 3 million times greater than atmospheric pressure on Earth. Under those conditions, the hot liquid hydrogen is compressed so much that it undergoes another transition, this time to a "metallic" state with properties in many ways similar to those of a liquid metal. Of particular importance for Jupiter's magnetic field (see Section 11.4) is that this metallic hydrogen is an excellent conductor of electricity.

As mentioned earlier, Jupiter's observed flattening requires that there be a relatively small (i.e., relatively small compared with the size of Jupiter), dense core at its center. On the basis of *Voyager* data, scientists once thought that the core might contain as much as 20 Earth masses of ma-

terial. However, following *Galileo's* arrival, it now appears that the core's mass could be as low as 5 Earth masses and perhaps even less. The precise composition of the core is unknown, but planetary scientists think that it contains much denser materials than the rest of the planet. Current best estimates indicate that the core consists of "rocky" materials, similar to those found on the terrestrial worlds. (Note that the term *rocky* here refers to the *chemical composition* of the core, not to its physical state. At the high temperatures and pressures found deep in the jovian interiors, the core material bears little resemblance to rocks found on Earth's surface.) In fact, it now appears that all four jovian planets contain similarly large rocky cores and that the formation of such a large "terrestrial" planetary core may be a necessary stage in the process of building up a gas giant (see Section 15.2).

Because of the enormous pressure at the center of Jupiter—approximately 50 million times that on Earth's surface, or 10 times that at Earth's center—the core must be compressed to a very high density (perhaps twice the density of Earth's core). The jovian core is probably not much more than 20,000 km in diameter (still big enough for Earth to fit inside, with plenty of room left over), and the central temperature may be as high as 40,000 K.

CONCEPT CHECK

✔ How have astronomers determined the properties of Jupiter's core?

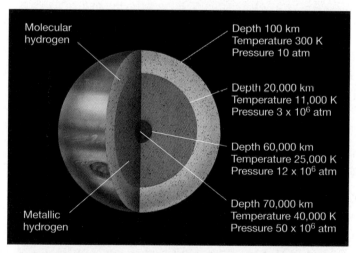

▲ FIGURE 11.10 Jupiter's Interior Jupiter's internal structure, as deduced from *Voyager* measurements and theoretical modeling. The outer radius represents the top of the cloud layers, some 70,000 km from the planet's center. The density and temperature increase with depth, and the atmosphere gradually liquefies at a depth of a few thousand kilometers. Below a depth of 20,000 km, the hydrogen behaves like a liquid metal. At the center of the planet lies a large rocky core, somewhat terrestrial in composition, but much larger than any of the inner planets. Although the values are uncertain, the temperature and pressure at the center are probably about 40,000 K and 50 million (Earth) atmospheres, respectively.

Labels in figure:
Molecular hydrogen
Depth 100 km / Temperature 300 K / Pressure 10 atm
Depth 20,000 km / Temperature 11,000 K / Pressure 3×10^6 atm
Depth 60,000 km / Temperature 25,000 K / Pressure 12×10^6 atm
Depth 70,000 km / Temperature 40,000 K / Pressure 50×10^6 atm
Metallic hydrogen

11.4 Jupiter's Magnetosphere

LEARNING GOAL 4 For decades, ground-based radio telescopes monitored radiation leaking from Jupiter's magnetosphere, but only when the *Pioneer* and *Voyager* spacecraft reconnoitered the planet in the mid-1970s did astronomers realize the full extent of its magnetic field. The *Galileo* probe spent many years orbiting within Jupiter's magnetosphere, returning a wealth of detailed information about its structure.

Jupiter, it turns out, is surrounded by a vast sea of energetic charged particles, mostly electrons and protons, somewhat similar to Earth's Van Allen belts, but much, much larger. The radio radiation detected on Earth is emitted when these particles are accelerated to very high speeds—close to the speed of light—by Jupiter's powerful magnetic field. This radiation is several thousand times more intense than that produced by Earth's magnetic field. The particles present a serious hazard to manned and unmanned space vehicles alike. Sensitive electronic equipment (not to mention even more sensitive human bodies) requires special protective shielding to operate for long in this hostile environment. *Galileo* was not expected to survive as long as it did.

Direct measurements from spacecraft show Jupiter's magnetosphere to be almost 30 million km across, roughly a million times more voluminous than Earth's magnetosphere and far larger than the entire Sun. As with Earth's, the size and shape of Jupiter's magnetosphere are determined by the interaction between the planet's magnetic field and the solar wind. Jupiter's magnetosphere has a long tail extending away from the Sun at least as far as Saturn's orbit (over 4 A.U. farther out from the Sun), as sketched in Figure 11.11. However, on the sunward side, the *magnetopause*—the boundary of Jupiter's magnetic influence on the solar wind—lies only 3 million km from the planet. Near Jupiter's surface, the magnetic field channels particles from the magnetosphere into the upper atmosphere, forming aurorae vastly larger and more energetic than those observed on Earth (Figure 11.12). ∞ (Sec. 7.5)

The outer magnetosphere of Jupiter appears to be quite unstable, sometimes deflating in response to "gusts" in the solar wind and then reexpanding as the wind subsides. In the inner magnetosphere, Jupiter's rapid rotation has forced most of the charged particles into a flat *current sheet*, lying on the planet's magnetic equator, quite unlike the Van Allen belts surrounding Earth. ∞ (Sec. 7.5) The

portion of the magnetosphere close to Jupiter is sketched in Figure 11.13. Notice that the planet's magnetic axis is not exactly aligned with its rotation axis, but is inclined to it at an angle of approximately 10°. Jupiter's magnetic field happens to be oriented opposite Earth's, with field lines running from north to south, rather than south to north as in the case of our own planet (see Figure 7.21).

Both ground- and space-based observations of the radiation emitted from Jupiter's magnetosphere imply that the *intrinsic* strength of the planet's magnetic field is nearly 20,000 times greater than Earth's. The existence of such a strong field further supports our theoretical model of Jupiter's internal structure. The conducting liquid interior that is thought to make up most of the planet should combine with Jupiter's rapid rotation to produce a large dynamo effect and a strong magnetic field, just as are observed. ∞ (Sec. 7.4)

CONCEPT CHECK

✔ Why is Jupiter's magnetosphere so much larger than Earth's?

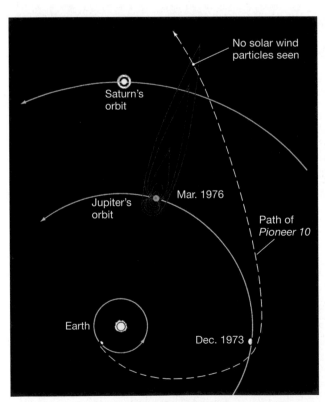

▲ **FIGURE 11.11** *Pioneer 10* **Mission** The *Pioneer 10* spacecraft (a forerunner of the *Voyager* missions) did not detect any solar particles while moving far behind Jupiter in 1976. Accordingly, as sketched here, Jupiter's magnetosphere apparently extends beyond the orbit of Saturn.

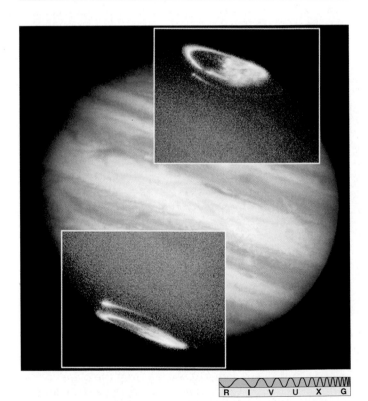

▲ **FIGURE 11.12 Aurorae on Jupiter** Aurorae on Jupiter, as seen by the *Hubble Space Telescope*. The main image was taken in visible (true-color) light, but the two insets at the poles were taken in the ultraviolet part of the spectrum. The oval-shaped aurorae, extending hundreds of kilometers above Jupiter's surface, result from charged particles escaping the jovian magnetosphere and colliding with the atmosphere, causing the gas to glow. *(NASA)*

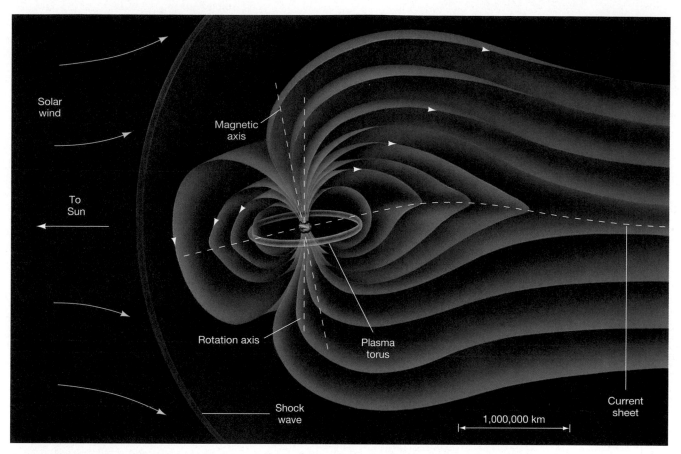

▲ **FIGURE 11.13 Jupiter's Magnetosphere** Jupiter's inner magnetosphere is characterized by a flat current sheet consisting of charged particles squeezed into the magnetic equatorial plane by the planet's rapid rotation. The plasma torus, a ring of charged particles associated with the moon Io, is discussed in Section 11.5.

11.5 The Moons of Jupiter

As of early 2004, Jupiter's official satellite count stands at 61. Table 11.1 presents some properties of the 16 largest members of Jupiter's moon system—those with diameters of 10 km or more. The 45 small bodies not in the table are also the most recently detected of Jupiter's moons; all have been found since 1999 by systematic surveys (made from Earth) of the space around the giant planet. As discussed in *Discovery 11-2*, their sizes, orbits, and sheer growing numbers are causing some astronomers to reconsider just what the definition of *moon* should be.

The four largest satellites—the Galilean moons—are each comparable in size to Earth's Moon. ∞ (Sec. 2.5) Moving outward from Jupiter, the four are named Io, Europa, Ganymede, and Callisto, after the mythical attendants of the Roman god Jupiter. They move in nearly circular orbits about their parent planet. When the *Voyager 1* spacecraft passed close to the Galilean moons in 1979, it sent some remarkably detailed photographs back to Earth, allowing planetary scientists to discern fine surface features on each moon. ∞ (Sec. 6.6) More recently, in the late 1990s, the *Galileo* mission expanded our knowledge of

these small, but complex, worlds still further. We will consider the Galilean satellites in more detail momentarily.

Within the orbit of Io lie four small satellites, all but one discovered by *Voyager* cameras. The largest of the four, Amalthea, is less than 300 km across and is irregularly shaped. E. E. Barnard discovered it in 1892. Amalthea orbits at a distance of 181,000 km from Jupiter's center—only 110,000 km above the cloud tops. Its rotation, like that of most of Jupiter's satellites, is synchronous with its orbit because of Jupiter's strong tidal field. Amalthea rotates once per orbital period—every 11.7 hours.

Beyond the Galilean moons lie eight more small satellites, all discovered in the 20th century, but before the *Voyager* missions. They fall into two groups of four moons each. The moons in the inner group move in eccentric, inclined orbits, about 11 million km from the planet. The outer four moons lie about 22 million km from Jupiter. Their orbits, too, are fairly eccentric, but *retrograde*, moving in a sense opposite that of all the other moons' orbits (and Jupiter's rotation). It is very likely that each group represents a single body that was captured by Jupiter's strong gravitational field long after the planet and its larger moons originally formed. Both bodies subsequently

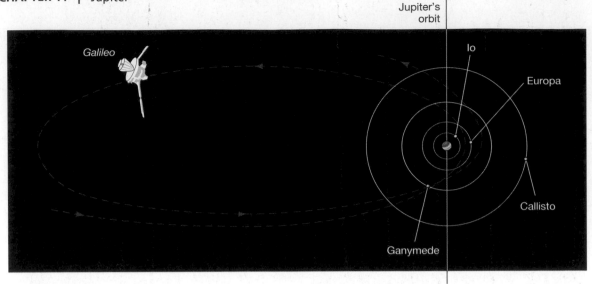

▲ **FIGURE 11.14** *Galileo* **at Jupiter** The orbits of Jupiter's Galilean moons, to scale, as seen from above the planet's north pole. The orbits are prograde and circular and lie in the planet's equatorial plane. Also drawn are some orbits of the *Galileo* probe (dashed red), showing it swinging around the interior moons of Jupiter. Each time it reconnoitered a moon, the probe would enter a slightly different orbit, allowing it to see more detail during hundreds of close encounters throughout the mission. Here, overtaking Europa on the "inside track," *Galileo* is slowed by that moon's gravity, placing the spacecraft in a lower orbit. Subsequent encounters with this and other moons would further modify *Galileo*'s trajectory, allowing the vehicle to visit all of the inner moons in the Jupiter system repeatedly.

broke up, either during or after the capture, resulting in the two families of similar orbits we see today. The masses, and hence the densities, of these small worlds are unknown. However, their appearance and sizes suggest compositions more like asteroids or comets than their larger Galilean companions.

The Galilean Moons as a Model of the Inner Solar System

Jupiter's Galilean moons have several interesting parallels with the terrestrial planets. Their orbits are direct (i.e., in the same sense as Jupiter's rotation), are roughly circular, and lie close to Jupiter's equatorial plane. Figure 11.14 shows the moons' orbits, with Jupiter to scale. Also shown in the figure are some representative orbits of the *Galileo* spacecraft, illustrating schematically how the probe used gravity assists from Galilean moons to maneuver through Jupiter's system of satellites. ∞ (Sec. 6.6)

The four Galilean moons range in size from slightly smaller than Earth's Moon (Europa) to slightly larger than Mercury (Ganymede). Figure 11.15 is a *Voyager 1* image of Io and Europa, with Jupiter providing a spectacular backdrop. Figure 11.16 shows the four Galilean moons to scale.

The similarity to the inner solar system continues with the fact that the moons' densities decrease with increasing distance from Jupiter. ∞ (Sec. 6.4) Largely on the basis of detailed measurements made by *Galileo* of the moons'

gravitational fields, together with mathematical models of the interiors, researchers have built up fairly detailed pictures of each moon's composition and internal structure (Figure 11.17). The innermost two Galilean moons, Io and Europa, have thick rocky mantles, possibly similar to the

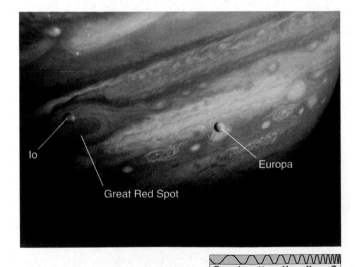

▲ **FIGURE 11.15** **Jupiter, Up Close** *Voyager 1* took this photo of Jupiter with ruddy Io on the left and pearllike Europa toward the right. Note the scale of objects here: Both Io and Europa are comparable in size to our Moon, and the Red Spot is roughly twice as big as Earth. (*NASA*)

Io   Europa   Ganymede   Callisto

R I V U X G

▲ **FIGURE 11.16 Galilean Moons** The *Galileo* spacecraft photographed each of the four Galilean moons of Jupiter. Shown here to scale, as they would appear from a distance of about 1 million km, they are, from left to right, Io, Europa, Ganymede, and Callisto. For scale, Earth's Moon is slightly smaller than Io. *(NASA)*

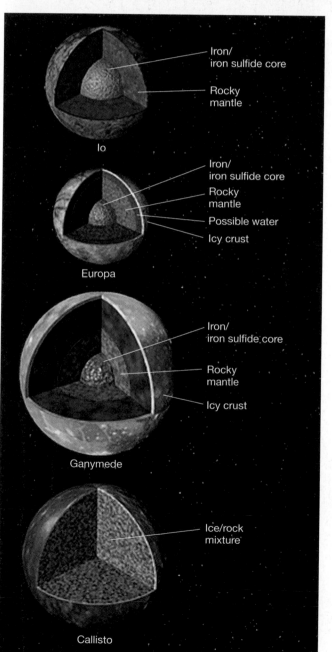

Iron/ iron sulfide core

Rocky mantle

Io

Iron/ iron sulfide core

Rocky mantle

Possible water

Icy crust

Europa

Iron/ iron sulfide core

Rocky mantle

Icy crust

Ganymede

Ice/rock mixture

Callisto

crusts of the terrestrial planets, surrounding iron–iron sulfide cores. Io's core accounts for about half that moon's total radius. Europa has a water–ice outer shell between 100 and 200 km thick. The two outer moons, Ganymede and Callisto, are clearly deficient in rocky materials. Lighter materials, such as water and ice, may account for as much as half of their total mass. Ganymede appears to have a relatively small metallic core topped by a rocky mantle and a thick icy outer shell. Callisto seems to be a largely undifferentiated mixture of rock and ice.

Many astronomers think that the formation of Jupiter and the Galilean satellites may in fact have mimicked, on a small scale, the formation of the Sun and the inner planets. For that reason, studies of the Galilean moon system could provide us with valuable insight into the processes that created our own world. We will return to this parallel in Chapter 15. So interested were mission planners in learning more about the Galilean moon system that the already highly successful *Galileo* mission was extended for six more years to allow for even more detailed study, particularly of Europa. ∞ (Sec. 6.6) The Galilean moons were scrutinized at resolutions as fine as a few meters during numerous extremely close passages by the spacecraft.

◀ **FIGURE 11.17 Galilean Moon Interiors** Cutaway diagrams showing the interior structure of the four Galilean satellites. Moving outward from Io to Callisto, we see that the moons' densities steadily decrease as the composition shifts from rocky mantles and metallic cores in Io and Europa, to a thick icy crust and smaller core in Ganymede, to an almost uniform rock-and-ice mix in Callisto. Both Ganymede and Europa are thought to have layers of liquid water beneath their icy surfaces.

DISCOVERY 11-2

Jupiter's Many Moons

The moons of Jupiter listed in Table 11.1 were all discovered long before *Galileo* reached the planet in 1995. The spacecraft focused on studying Jupiter and its inner moons and discovered no new moons itself, so the number of known satellites of Jupiter stayed fixed at 16. Since 1999, however, the number has exploded, reaching 61 at the time of writing and very likely to increase further. How did that happen?

Remarkably, this host of new moons was discovered not via close-up spacecraft exploration, but rather by painstaking long-distance observations made from Earth, using large ground-based telescopes and specially designed instruments and software to scan large areas of the sky for very faint objects. Steadily improving technology means that small bodies once far too faint for Earth-based telescopes to see are now being detected and cataloged almost routinely. The first pair of figures shows a typical discovery image sequence, wherein a faint and eminently undistinguished-looking moon (marked by the arrow) is revealed by its motion relative to the background stars. The two images were taken 40 minutes apart in 2003 by the Canada–France–Hawaii telescope on Mauna Kea.

These latest additions to Jupiter's family of satellites have characteristics comparable to those of the outermost eight

moons listed in Table 11.1. All are very small—less than 10 km, and in some cases as little as 1 km, in diameter—and their masses are unknown. Their orbits are all moderately eccentric, with semimajor axes between 10 and 25 million km and inclined at 15–40° to Jupiter's equatorial plane. Most of the orbits are retrograde. Very likely, these newcomers have the same origin as do the eight outlying satellites just mentioned—space "junk" captured by Jupiter, probably long ago. ∞ (Sec. 6.7) The second figure shows the orbits of all the known moons of Jupiter, to scale. The orbits of the Galilean satellites can be seen at the center. All the other orbits are those of irregular, small moons that may very well not be original members of the Jupiter system.

While the origin of Jupiter's moons is an interesting puzzle and may have much to teach us about the early solar system, the rapidly escalating number of tiny bodies orbiting Jupiter and the other giant planets is beginning to pose a classification problem for astronomers. At the moment, any object—no matter how small—orbiting a planet is called a moon. This simple criterion works well when the list of known moons consists of a relatively small number of relatively large objects, hundreds or thousands of kilometers across.

However, as observations improve and more and more kilometer-sized jovian moons are discovered, some researchers have begun suggesting that perhaps a size limit should be imposed to separate "real" moons (like the Galilean satellites) from captured cometary material. In this way, we might distinguish objects that formed along with Jupiter from material that formed elsewhere and that now—by chance—happens to orbit Jupiter. For the present, though, no such limit exists, so we can expect to hear reports of a steadily increasing number of "moons" in the coming years.

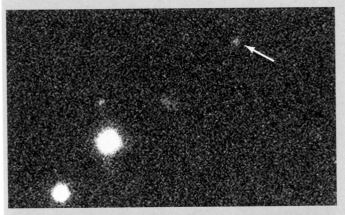

(U. Hawaii)

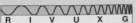

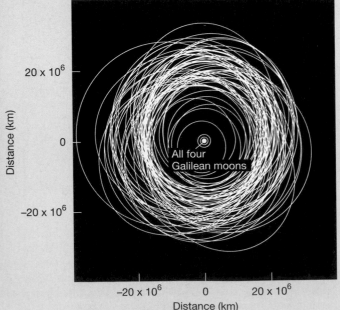

TABLE 11.1   The Major Moons of Jupiter[*]

Name	Distance from Jupiter (km)	(planetary radii)	Orbital Period (days)	Size (longest diameter, km)	Mass[**] (Earth Moon masses)	Density (kg/m³)	(g/cm³)
Metis	128,000	1.79	0.29	40			
Adrastea	129,000	1.80	0.30	20			
Amalthea	181,000	2.54	0.50	260			
Thebe	222,000	3.10	0.67	100			
Io	422,000	5.90	1.77	3640	1.22	3500	3.5
Europa	671,000	9.38	3.55	3130	0.65	3000	3.0
Ganymede	1,070,000	15.0	7.15	5270	2.02	1900	1.9
Callisto	1,880,000	26.3	16.7	4800	1.46	1900	1.9
Leda	11,100,000	155	239	10			
Himalia	11,500,000	161	251	170			
Lysithea	11,700,000	164	259	24			
Elara	11,700,000	164	260	80			
Ananke	21,200,000	297	−631[†]	20			
Carme	22,600,000	316	−692[†]	30			
Pasiphae	23,500,000	329	−735[†]	36			
Sinope	23,700,000	332	−758[†]	28			

[*] *Moons larger than 10 km in diameter. This table does not include the 45 recently discovered small moons described in the text. All of these small moons move on inclined, eccentric, mainly retrograde orbits some 10–25 million km from the planet.*

[**] *Mass of Earth's Moon = 7.4×10^{22} kg = 3.9×10^{-5} Jupiter masses.*

[†] *Retrograde orbit.*

Not all the properties of the Galilean moons find analogs in the inner solar system, however. For example, all four Galilean satellites are locked into states of synchronous rotation by Jupiter's strong tidal field, so they all keep one face permanently pointing toward their parent planet. By contrast, of the terrestrial planets, only Mercury is strongly influenced by the Sun's tidal force, and even its orbit is not synchronous. ∞ (Sec. 8.4) Finally, inspection of Table 11.1 shows a remarkable coincidence in the orbital periods of the three inner Galilean moons: Their periods are almost exactly in the ratio 1:2:4 (and the fourth moon Callisto is not too far from being the "8" in the sequence). This configuration may be the result of a complex, but poorly understood, three-body (or perhaps even four-body) resonance in the Galilean moon system, something not found among the terrestrial worlds.

Io: The Most Active Moon

Io, the densest of the Galilean moons, is the most geologically active object in the entire solar system. Its mass and radius are fairly similar to those of Earth's Moon, but there the resemblance ends. Shown in Figure 11.18, Io's surface is a collage of reds, yellows, and blackish browns—resembling a giant pizza in the minds of some startled *Voyager* scientists. As the spacecraft sped past Io, it made an outstanding discovery: Io has active volcanoes! *Voyager 1* photographed eight erupting volcanoes. Six were still erupting when *Voyager 2* passed by four months later.

By the time *Galileo* arrived in 1995, several of the eruptions observed by *Voyager* had subsided. However, many new ones were seen—in fact, *Galileo* found that Io's surface features can change significantly in as little as a few weeks. In all, more than 80 active volcanoes have been identified on Io. The largest, called Loki (on the far side of Figure 11.19), is larger than the state of Maryland and emits more energy than all of Earth's volcanoes combined.

The top right inset in Figure 11.19 shows a volcano called Prometheus ejecting matter at speeds of up to 2 km/s to an altitude of about 150 km. These high-speed gases are quite unlike the (relatively) sluggish ooze that emanates from Earth's volcanoes. According to *Galileo*'s instruments, lava temperatures on Io generally range from 650 to 900 K, with the higher end of the range implying that at least some of the volcanism is similar to that found on Earth. However, temperatures as high as 2000 K—far hotter than any earthly volcano—have been measured at

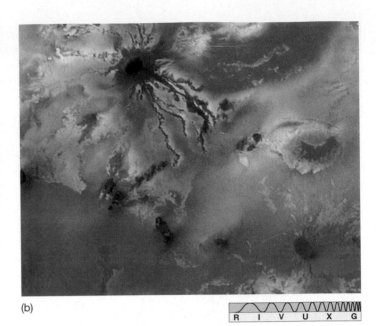

(a)

(b)

R I V U X G

▲ **FIGURE 11.18 Io** Jupiter's innermost moon, Io, is quite different in character from the other three Galilean satellites. Its surface is kept smooth and brightly colored by the moon's constant volcanism. The resolution of the *Galileo* photograph in (a) is about 7 km. In the more detailed *Voyager* image (b), features as small as 2 km across can be seen. *(NASA)*

some locations. Mission scientists speculate that these "superhot" volcanoes may be similar to those which occurred on Earth more than 3 billion years ago.

The orange color immediately surrounding the volcanoes most likely results from sulfur compounds in the ejected material. In stark contrast to the surfaces of the other Galilean moons, Io's surface is neither cratered nor streaked (the circular features visible in Figures 11.17 and 11.18 are volcanoes), but is instead exceptionally smooth, mostly varying in altitude by less than about 1 km, although some volcanoes are several kilometers high. The smoothness is apparently the result of molten matter that constantly fills in any "dents and cracks." This remarkable moon has the youngest surface of any known object in the solar system. Io also has a thin, temporary atmosphere made up primarily of sulfur dioxide, presumably the result of gases ejected by volcanic activity.

Io's volcanism has a major effect on Jupiter's magnetosphere. All the Galilean moons orbit within the magnetosphere and play some part in modifying its properties, but Io's influence is particularly marked. Although many of the charged particles in Jupiter's magnetosphere come from the solar wind, there is strong evidence that Io's volcanism is the primary source of heavy ions in the inner regions. Jupiter's magnetic field continually sweeps past Io, gathering up the particles its volcanoes spew into space and accelerating them to high speed. The result is the *Io plasma torus* (Figure 11.20; see also Figure 11.13), a dough-

nut-shaped region of energetic heavy ions that follows Io's orbital track, completely encircling Jupiter. (A plasma is a gas that has been heated to such high temperatures that all its atoms are ionized. A few neutral atoms have also been observed in the Io plasma torus.)

The plasma torus is quite easily detectable from Earth, but before *Voyager* its origin was unclear. *Galileo* made detailed studies of the plasma's dynamic and rapidly varying magnetic field. Spectroscopic analysis shows that sulfur is indeed one of the torus's major constituents, strongly implicating Io's volcanoes as its source. As a hazard to spacecraft—manned or unmanned—the plasma torus is formidable, with lethal radiation levels.

What causes such astounding volcanic activity on Io? The moon is far too small to have geological activity like that on Earth. Io should be long dead, like our own Moon. At one time, some scientists suggested that Jupiter's magnetosphere might be the culprit: Perhaps the (then-unknown) processes creating the plasma torus were somehow also stressing the moon. We now know that this is not the case. The real source of Io's energy is *gravity*—Jupiter's gravity. Io orbits very close to Jupiter—only 422,000 km, or 5.9 Jupiter radii, from the center of the planet. As a result, Jupiter's huge gravitational field exerts strong tidal forces on the moon. If Io were the only satellite in the Jupiter system, it would long ago have come into a state of synchronous rotation with the planet, just as our own Moon has with Earth, for the reasons discussed in Chapter

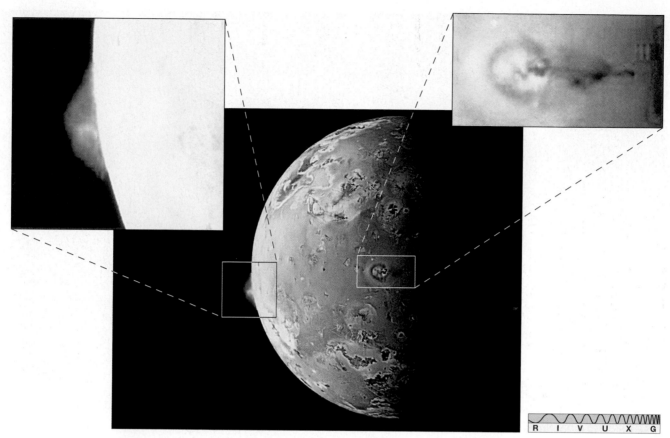

▲ **FIGURE 11.19 Volcanoes on Io** The main image shows a *Galileo* view of Io with a resolution of about 6 km. The dark, circular features are volcanoes. The left inset shows an umbrellalike eruption of one of Io's volcanoes, captured as *Galileo* flew past this fascinating moon in 1997. The right inset shows another volcano, this time face-on. Surface features here are resolved to a few kilometers. The plume at Io's western edge measures about 150 km high and 300 km across. *(NASA)*

8. ∞ (Sec. 8.4) In that case, Io would move in a perfectly circular orbit, with one face permanently turned toward Jupiter, and the tidal bulge would be stationary with respect to the moon.

But Io is not alone. As it orbits, it is constantly tugged by the gravity of its nearest large neighbor, Europa. The tugs are small and not enough to cause any great tidal effect, but they are sufficient to make Io's orbit slightly noncircular, preventing the moon from settling into a precisely synchronous state. The reason for this effect is exactly the same as in the case of Mercury, also as discussed in Chapter 8. ∞ (Sec. 8.4) In a noncircular orbit, the moon's speed varies from place to place as it revolves around its planet, but its rate of rotation on its axis remains constant. Thus, it cannot keep one face always turned toward Jupiter. Instead, as seen from Jupiter, Io rocks or "wobbles" slightly from side to side as it moves. The large (100-m) tidal bulge, however, always points directly toward Jupiter, so it moves back and forth across Io's surface as the moon wobbles. These conflicting forces result in enormous tidal stresses that continually flex and squeeze Io's interior.

Just as the repeated back-and-forth bending of a piece of wire can produce heat through friction, the ever-changing distortion of Io's interior constantly energizes the moon. This generation of large amounts of heat within Io ultimately causes huge jets of gas and molten rock to squirt out of the moon's surface. *Galileo's* sensors indicated extremely high temperatures in the outflowing material. It is likely that much of Io's interior is soft or molten, with only a relatively thin solid crust overlying it. Researchers estimate that the total amount of heat generated within Io as a result of tidal flexing is about 100 million megawatts. This phenomenon makes Io one of the most fascinating objects in our solar system.

Europa: Liquid Water Locked in Ice

Europa (Figure 11.21) is a world very different from Io. Lying outside Io's orbit, 671,000 km (9.4 Jupiter radii) from Jupiter, Europa showed relatively few craters on its surface in images taken by *Voyager*, suggesting geologic youth—perhaps just a few million years. Recent activity

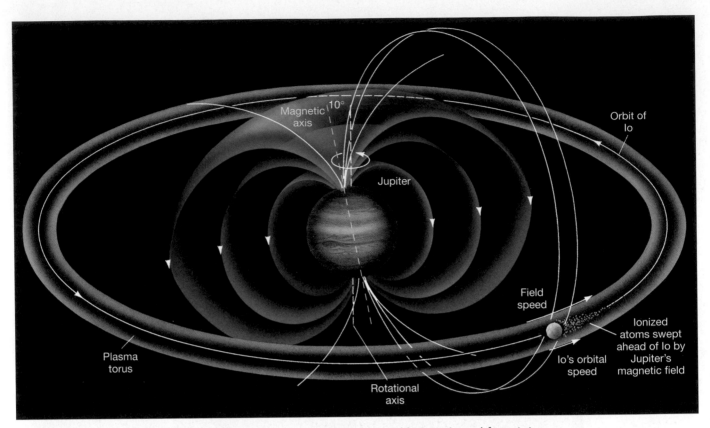

▲ **FIGURE 11.20 Io Plasma Torus** The torus is the result of material being ejected from Io's volcanoes and swept up by Jupiter's rapidly rotating magnetic field. Spectroscopic analysis indicates that the torus is made mainly of sodium and sulfur atoms and ions.

has erased any scars of ancient meteoritic impacts. The dark areas are rocky deposits that may have come from the moon's interior or that may have been swept up by Europa as it moved in its orbit. Europa's surface also displays a vast network of lines crisscrossing bright, clear fields of water ice. Some of these linear "bands," or fractures, extend halfway around the satellite and resemble, in some ways, the pressure ridges that develop in ice floes on Earth's polar oceans.

On the basis of *Voyager* images, planetary scientists had theorized that Europa might be completely covered by an ocean of liquid water with its top frozen at the low temperatures prevailing so far from the Sun. In this view, the cracks in the surface are attributed to the tidal influence of Jupiter and the gravitational pulls of the other Galilean satellites, although these forces are considerably weaker than those powering Io's violent volcanic activity. However, other researchers had contended that Europa's fractured surface was instead related to some form of tectonic activity—involving ice rather than rock. High-resolution *Galileo* observations now appear to strongly support the former idea. Figure 11.21(d) is a *Galileo* image of this weird moon, showing what look like icebergs—flat chunks of ice that have been broken apart, moved several kilometers, and reassembled, perhaps by the action of water currents below. Mission scientists estimate that Europa's surface ice

may be several kilometers thick and that there may be a 100-km-deep liquid ocean below it.

Other detailed images of Europa's surface lend further support to this hypothesis. Figure 11.22(a) shows a region where Europa's icy crust appears to have been pulled apart and new material has filled in the gaps between the separating ice sheets. Elsewhere on the surface, *Galileo* found what appeared to be the icy equivalent of lava flows on Earth—regions where water apparently "erupted" through the surface and flowed for many kilometers before solidifying. The smooth "trenches" shown in Figure 11.22(b) strongly suggest local flooding of the terrain. The scarcity of impact craters on Europa implies that the processes responsible for these features did not stop long ago. Rather, they must be ongoing.

Further evidence comes from studies of Europa's magnetic field. Measurements made by *Galileo* on repeated flybys of the moon revealed that Europa has a weak magnetic field that constantly changes strength and direction. This finding is entirely consistent with the idea that the field is generated by the action of Jupiter's magnetism on a shell of electrically conducting fluid about 100 km below Europa's surface—in other words, the salty layer of liquid water suggested by the surface observations. *Galileo's* observations convinced quite a few skeptical scientists of the reality of Europa's ocean.

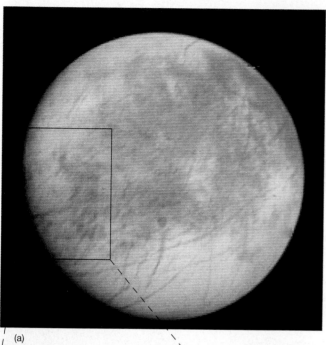

(a)

◄ **FIGURE 11.21 Europa** The second Galilean moon is Europa. Its icy surface is only lightly cratered, indicating that some ongoing process must be obliterating impact craters soon after they form. The cracks crisscrossing the surface are most likely caused by the tidal effect of Jupiter. The resolution of the *Voyager 2* mosaic in (a) is about 5 km. The two images below it (b and c) display even finer detail. (d) At 50-m resolution, this image from the *Galileo* spacecraft shows a smooth, yet tangled, surface resembling the huge ice floes that cover Earth's polar regions. The region of Io that is shown is called Conamara Chaos. (*NASA*)

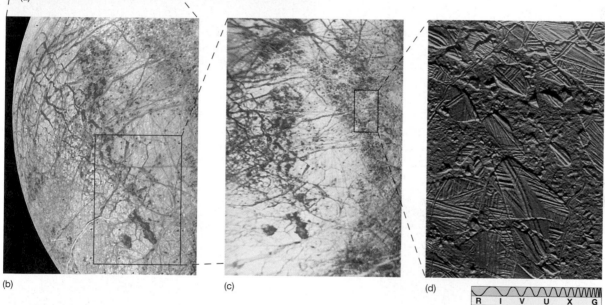

(b)   (c)   (d)

R I V U X G

The likelihood that Europa has an extensive layer of liquid water below its surface ice opens up many interesting avenues of speculation about the possible development of life there. In the rest of the solar system, only Earth has liquid water on or near its surface, and most scientists agree that water played a key role in the appearance of life on Earth (see Chapter 28). Europa may well contain more liquid water than exists on our entire planet! Of course, the existence of water does not *necessarily* imply the emergence of life: Europa, even in its liquid ocean, is still a hostile environment compared with Earth. Nevertheless, the possibility—even a remote one—of life on Europa was an important motivating factor in the decision to extend the *Galileo* mission for six more years.

Ganymede and Callisto: Fraternal Twins

The two outermost Galilean moons are Ganymede (at 1.1 million km, or 15 planetary radii, from the center of Jupiter) and Callisto (at 1.9 million km, or 26 Jupiter radii). The density of each is only about 2000 kg/m³, suggesting that they harbor substantial amounts of ice throughout and are not just covered by thin icy or snowy surfaces. Ganymede, shown in Figure 11.23, is the largest moon in the solar system, exceeding not only Earth's Moon, but also the planets Mercury and Pluto, in size. It has many impact craters on its surface and patterns of dark and light markings that are reminiscent of the highlands and maria on Earth's own Moon. In fact, Ganymede's his-

(a)

(b)

▲ **FIGURE 11.22 Europa Surface Detail** Detailed *Galileo* images of Europa, showing (a) "pulled-apart" terrain that suggests upwelling material filling in the gaps between separating surface ice sheets and (b) the Conamara Chaos region, where liquid water appears to have flooded a portion of the surface. (*NASA*)

tory has many parallels with that of the Moon (with water ice replacing lunar rock). The large, dark region clearly visible in Figure 11.23 is called Galileo Regio.

As on the inner planets, we can estimate ages on Ganymede by counting craters. We learn that the darker regions, such as Galileo Regio, are the oldest parts of Ganymede's surface. These regions are the original icy surface of the moon, just as the ancient highlands on our own Moon are its original crust. The surface darkens with

age as micrometeorite dust slowly covers it. The light-colored parts of Ganymede are much less heavily cratered, so they must be younger. They are Ganymede's "maria" and probably formed in a manner similar to the way that maria on the Moon were created. ∞ (Sec. 8.5) Intense meteoritic bombardment caused liquid water—Ganymede's counterpart to our own Moon's molten lava—to upwell from the interior and flood the affected regions before solidifying.

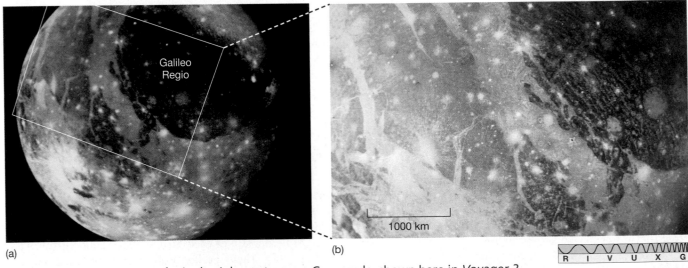

(a)

Galileo Regio

1000 km

(b)

▲ **FIGURE 11.23 Ganymede** Jupiter's largest moon, Ganymede, shown here in *Voyager 2* images, is also the largest satellite in the solar system. The dark regions are the oldest parts of the moon's surface and probably represent its original icy crust. The largest dark region visible in (a), spanning some 3200 km, is called Galileo Regio. The lighter, younger regions are the result of flooding and freezing that occurred within a billion years or so of Ganymede's formation. The light-colored spots are recent impact craters. The resolution of the detailed image in (b) is about 3 km. (*NASA*)

Not all of Ganymede's surface features follow the lunar analogy. Ganymede has a system of grooves and ridges (shown in Figure 11.24) that may have resulted from crustal tectonic motion, much as Earth's surface undergoes mountain building and faulting at plate boundaries. ∞ (Sec. 7.4) Ganymede's large size indicates that its original radioactivity probably helped heat and differentiate its interior, after which the moon cooled and the crust cracked. Ganymede seems to have had some early plate tectonic activity, but the process stopped about 3 billion years ago, when the cooling crust became too thick for such activity to continue. The *Galileo* data suggest that the surface of Ganymede may be older than was previously thought. With the improved resolution of that spacecraft's images (Figure 11.24c), some regions thought to have been smooth, and hence young, are now seen to be heavily splintered by fractures and thus probably very old.

In 1996, *Galileo* detected a weak magnetosphere surrounding Ganymede, making it the first moon in the solar system on which a magnetic field had been observed and implying that Ganymede has a modest iron-rich core. Ganymede's magnetic field is about 1 percent that of Earth. In December 2000, the magnetometer team reported fluctuations in the field strength similar to those near Europa, suggesting that Ganymede, too, may have liquid or perhaps "slushy" water under its surface. Recent observations of surface formations similar to those attributed to flowing water "lava" on Europa appear to support this view.

Callisto, shown in Figure 11.25, is in many ways similar in appearance to Ganymede, although it has more craters and fewer fault lines. Its most obvious feature is a huge series of concentric ridges surrounding each of two large basins. The larger of the two, on Callisto's Jupiter-facing side, is named Valhalla and measures some 3000 km across. It is clearly visible in the figure. The ridges resemble the ripples made as a stone hits water, but on Callisto they probably resulted from a cataclysmic impact with an asteroid or comet. The upthrust ice was partially melted, but it resolidified quickly, before the ripples had a chance to subside.

Today, both the ridges and the rest of the crust are frigid ice and show no obvious signs of geological activity (such as the grooved terrain on Ganymede). Apparently, Callisto froze before plate tectonic or other activity could start. The density of impact craters on the Valhalla basin indicates that it formed long ago, perhaps 4 billion years in the past. Yet, even on this frozen world, there are hints from *Galileo*'s magnetometers that there might be a thin layer of water, or more likely slush, deep below the surface.

Ganymede's internal differentiation indicates that the moon was largely molten at some time in the past; Callisto is undifferentiated and hence apparently never melted. Researchers are uncertain why two such similar bodies should have evolved so differently. Complicating matters further is Ganymede's magnetic field and possible subsurface liquid water, which suggest that the moon's interior may still be relatively warm. If that is so, then Ganymede's heating and differentiation must have happened quite recently—less than a billion years ago, based on recent estimates of how rapidly the moon's heat escapes into space.

Scientists have no clear explanation for how Ganymede could have evolved in this manner. Heating by meteoritic bombardment ended too early, and radioactivity probably could not have provided enough energy at

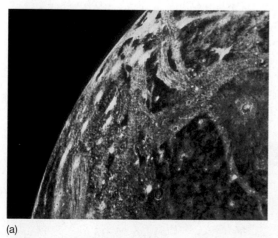

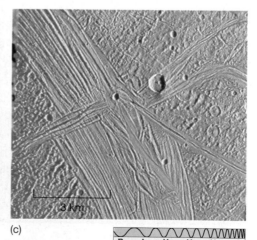

(a)   (b)   (c)

▲ **FIGURE 11.24 Ganymede Surface Detail** (a and b) Grooved terrain on Ganymede may have been caused by a process similar to plate tectonics on Earth. These images were captured by the *Galileo* spacecraft in 1997. The area shown in (b) is about 50 km across and reveals a multitude of ever-smaller ridges, valleys, and craters, right down to the resolution limit of *Galileo*'s camera (about 300 m, or about three times the length of a football field). Part (c) shows the grooves at even higher resolution, suggesting erosion of some sort, possibly even caused by water. (*NASA*)

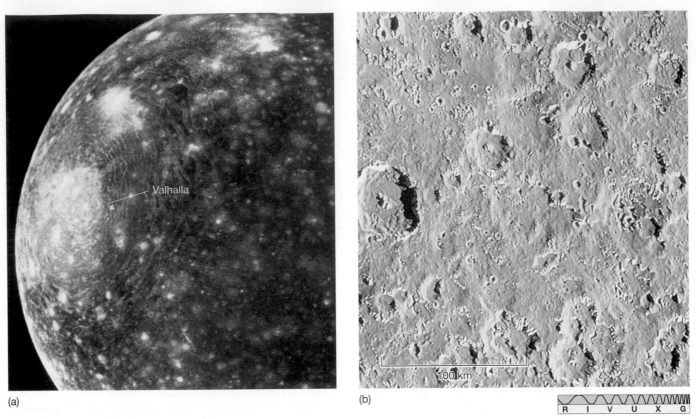

(a)

(b)

▲ FIGURE 11.25 Callisto Callisto, the outermost Galilean moon of Jupiter, is similar to Ganymede in composition, but is more heavily cratered. (a) The large series of concentric ridges visible on the left of the image is known as Valhalla. Extending nearly 1500 km from center of the basin, the ridges formed when "ripples" from a large meteoritic impact froze before they could disperse completely. The resolution in this *Voyager 2* image is around 10 km. (b) The higher resolution *Galileo* image of Callisto's equatorial region displays its heavy cratering even more clearly. *(NASA)*

such a late time. ∞ (Sec. 7.3) Some astronomers speculate that interactions among the inner moons, possibly related to the 1:2:4 near resonance mentioned earlier, may have been responsible. These interactions might have caused Ganymede's orbit to change significantly about a billion years ago, and prior tidal heating by Jupiter could have helped melt the moon's interior.

CONCEPT CHECKS

✔ What is the ultimate source of all the activity observed on Jupiter's Galilean satellites?

✔ Why are scientists so interested in the existence of liquid water on Europa and Ganymede?

▶ FIGURE 11.26 Jupiter's Ring Jupiter's faint ring, as photographed (nearly edge-on) by *Voyager 2*. Made of dark fragments of rock and dust possibly chipped off the innermost moons by meteorites, the ring was unknown before the two *Voyager* spacecraft arrived at the planet. It lies in Jupiter's equatorial plane, only 50,000 km above the cloud tops. *(NASA)*

11.6 Jupiter's Ring

Yet another remarkable finding of the 1979 *Voyager* missions was the discovery of a faint ring of matter encircling Jupiter in the plane of the planet's equator (see Figure 11.26). The ring lies roughly 50,000 km above the top cloud layer of the planet, inside the orbit of the innermost moon. A thin sheet of material may extend all the way down to Jupiter's cloud tops, but most of the ring is con-

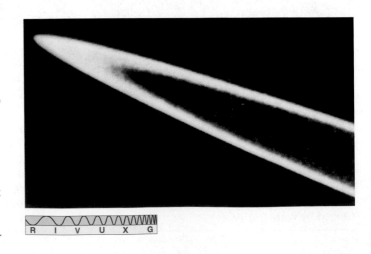

fined within a region only a few thousand kilometers across. The outer edge of the ring is quite sharply defined. In the direction perpendicular to the equatorial plane, the ring is only a few tens of kilometers thick. The small, dark particles that make up the ring may be fragments chipped off by meteoritic impact from two small moons—Metis and Adrastea, discovered by *Voyager*—that lie very close to the ring itself.

Despite differences in appearance and structure, Jupiter's ring can perhaps be best understood by studying the most famous planetary ring system—that of Saturn—so we will postpone further discussion of ring properties until the next chapter.

Chapter Review

SUMMARY

Jupiter is the largest planet in the solar system. Its mass is more than twice the mass of all the other planets combined, although it is still only about a thousandth the mass of the Sun. Composed primarily of hydrogen and helium, Jupiter rotates rapidly, producing a pronounced equatorial bulge. The planet's flattened shape allows astronomers to infer the presence of a large rocky core in its interior. Jupiter displays **differential rotation (p. 277)**: Because the planet has no solid surface, the rotation rate varies from place to place in the atmosphere. Measurements of radio emission from Jupiter's magnetosphere provide a measure of the planet's interior rotation rate.

Jupiter's atmosphere consists of three main cloud layers. The colors we see are the result of chemical reactions, fueled by the planet's interior heat, solar ultraviolet radiation, auroral phenomena, and lightning, at varying depths below the cloud tops, seen through "holes" in the overlying clouds. The cloud layers on Jupiter, as on all the jovian worlds, are arranged into bands of bright zones and darker belts crossing the planets parallel to the equator. The bands are the result of both convection in Jupiter's interior and the planet's rapid rotation. The lighter zones are the tops of upwelling, warm currents, and the darker belts are cooler regions where gas is sinking. Underlying them is a stable pattern of eastward or westward wind flow called a **zonal flow (p. 279)**. The wind direction alternates as we move north or south away from the equator. The main weather pattern on Jupiter is the **Great Red Spot (p. 278)**, an Earth-sized hurricane that has been raging for at least three centuries. Other, smaller, weather systems—**white ovals (p. 283)** and **brown ovals (p. 283)**—are also observed and can persist for decades.

Jupiter's atmosphere becomes hotter and denser with depth, eventually becoming liquid. Interior pressures are so high that the hydrogen is "metallic" in nature near the center. The planet has a large "terrestrial" core about 10 times the mass of Earth. Jupiter radiates about twice as much energy into space as it receives from the Sun. The source of this energy is most likely heat released into the planet's interior when Jupiter formed 4.6 billion years ago, now slowly leaking from the surface.

The magnetosphere of Jupiter is about a million times more voluminous than Earth's magnetosphere, and the planet has a long magnetic "tail" extending away from the Sun to at least the distance of Saturn's orbit. Energetic particles spiral around magnetic field lines, accelerated by Jupiter's rotating magnetic field, producing intense radio radiation.

Jupiter and its system of moons resemble a small solar system. Sixty-one moons have been discovered so far. The outermost eight resemble asteroids and have retrograde orbits, suggesting that they may have been captured by Jupiter's gravity long after the planets and largest moons formed. Jupiter's four major moons are called the **Galilean moons (p. 276)**, after their discoverer, Galileo Galilei. Their densities decrease with increasing distance from the planet.

The innermost Galilean moon, Io, has active volcanoes powered by the constant flexing of the moon by Jupiter's tidal forces. As Io orbits Jupiter, the moon "wobbles" because of the gravitational pull of Europa. The ever-changing distortion of its interior energizes Io, and geyserlike volcanoes keep its surface smooth with constant eruptions. The material ejected by these volcanoes forms the Io plasma torus in Jupiter's inner magnetosphere. Europa has a cracked, icy surface that probably conceals an ocean of liquid water, making the moon an interesting candidate for life in the solar system. Its fields of ice are nearly devoid of craters, but have extensive large-scale fractures, due most likely to the tidal influence of Jupiter, the gravitational effects of the other Galilean satellites, and small-scale "chaos" caused by the action of the underlying ocean.

Ganymede and Callisto have ancient, heavily cratered surfaces. Ganymede, the largest moon in the solar system, shows evidence of past geological activity, but now appears to be unmoving rock and ice, although recent evidence suggests that it, too, may have subsurface liquid water. Callisto apparently froze before tectonic activity could start there.

Jupiter has a faint, dark ring extending down to the planet's cloud tops. The ring was discovered in 1979 by *Voyager 1.*

REVIEW AND DISCUSSION

1. In what sense does our solar system consist of only two important objects?

2. What is differential rotation, and how is it observed on Jupiter?

3. What does Jupiter's degree of flattening tell us about the planet's interior?

4. Describe some of the ways in which the *Voyager* mission changed our perception of Jupiter.

5. Describe some of the ways in which the *Galileo* mission changed our perception of Jupiter.

6. What is the Great Red Spot? What is known about the source of its energy?

7. What is the cause of the colors in Jupiter's atmosphere?

8. Why has Jupiter retained most of its original atmosphere?

9. Explain the theory that accounts for Jupiter's internal heat source.

10. What is Jupiter thought to be like beneath its clouds? Why do we think this?

11. What is responsible for Jupiter's enormous magnetic field?

12. In what sense are Jupiter and its moons like a miniature solar system?

13. How does the density of the Galilean moons vary with increasing distance from Jupiter? Is there a trend to this variation? If so, why?

14. What is the cause of Io's volcanic activity?

15. What evidence do we have for liquid water below Europa's surface?

16. Why do scientists think that Ganymede's interior may have been heated as recently as a billion years ago?

17. How does the amount of cratering vary among the Galilean moons? Does it depend on their location? If so, why?

18. Why is there speculation that the Galilean moon Europa might be an abode for life?

19. What might be the consequences of the discovery of life on Europa?

20. Water is relatively uncommon among the terrestrial planets. Is it common among the moons of Jupiter?

CONCEPTUAL SELF-TEST: TRUE OR FALSE/MULTIPLE CHOICE

1. The solid surface of Jupiter lies just below the cloud layers that are visible from Earth.

2. Storms in Jupiter's atmosphere are generally much longer lived than storms on Earth.

3. The element helium plays an important role in producing the colors in Jupiter's atmosphere.

4. Jupiter is noticeably flattened due to its rapid rotation.

5. Jupiter emits more energy than it receives from the Sun.

6. Although often referred to as a gaseous planet, Jupiter is mostly liquid in its interior.

7. Because of Jupiter's strong magnetic field, the Galilean satellites rotate synchronously with their orbits around the planet.

8. Io is the only moon in the solar system with active volcanoes.

9. Scientists speculate that Europa may have liquid water below its frozen surface.

10. Ganymede shows evidence of ancient plate tectonics.

11. Compared with Earth's orbit, the orbit of Jupiter is approximately (a) half as large; (b) twice as large; (c) 5 times larger; (d) 10 times larger.

12. Compared with Earth's density, the density of Jupiter is (a) much greater; (b) much less; (c) about the same.

13. The main constituent of Jupiter's atmosphere is (a) hydrogen; (b) helium; (c) ammonia; (d) carbon dioxide.

14. Figure 11.5 ("Zonal Flow") shows that the most rapid westerly wind flows on Jupiter occur at (a) northern mid-latitudes; (b) equatorial latitudes; (c) southern mid-latitudes; (d) polar latitudes.

15. According to Figure 11.6 ("Jupiter's Atmosphere"), if ammonia and ammonium hydrosulfide ice were transparent to visible light, Jupiter would appear (a) bluish; (b) red; (c) tawny brown; (d) exactly as it does now.

16. Jupiter's rocky core is (a) smaller than Earth's Moon; (b) comparable in size to Mars; (c) almost the same size as Venus; (d) larger than Earth.

17. Jupiter's magnetosphere extends far into space, stretching about (a) 1 AU; (b) 5 AU; (c) 10 AU; (d) 20 AU beyond the planet.

18. The moon of Jupiter most similar in size to Earth's Moon is (a) Io; (b) Europa; (c) Ganymede; (d) Callisto.

19. Io's surface appears very smooth because it (a) is continually resurfaced by volcanic activity; (b) is covered with ice; (c) has been shielded by Jupiter from meteorite impacts; (d) is liquid.

20. The Galilean moons of Jupiter are sometimes described as a miniature inner solar system because (a) there are the same number of Galilean moons as there are terrestrial planets; (b) the moons have generally "terrestrial" composition; (c) the moons' densities decrease with increasing distance from Jupiter; (d) the moons all move on circular, synchronous orbits.

PROBLEMS

 Algorithmic versions of these questions are available in the Practice Problems module of the Companion Website at astro.prenhall.com/chaisson.

The number of squares preceding each problem indicates its approximate level of difficulty.

1. ■ How does the force of gravity at Jupiter's cloud tops compare with the force of gravity at Earth's surface?

2. ■ What are the angular diameters of the orbits of Jupiter's four Galilean satellites, as seen from Earth at closest approach (assuming, for definiteness, that opposition occurs near perihelion)?

3. ■ Using the figures given in the text, calculate how long it takes Jupiter's equatorial winds to circle the planet, relative to the interior.

4. ■■ Calculate the rotational speed (in km/s) of a point on Jupiter's equator, at the level of the cloud tops. Compare it with the orbital speed just above the cloud tops.

5. ■ Given Jupiter's age and current atmospheric temperature, what is the smallest possible mass the planet could have and still have retained its hydrogen atmosphere? ∞ (*More Precisely 8-1*)

6. ■ If Jupiter had been just massive enough to fuse hydrogen (see *Discovery 11-1*), calculate what the planet's gravitational force on Earth would have been at closest approach, relative to the gravitational pull of the Sun. Assume circular orbits. Also, estimate what the magnitude of the planet's tidal effect on our planet would have been, again relative to that of the Sun.

7. ■ Calculate the ratio of Jupiter's mass to the total mass of the Galilean moons. Compare your answer with the ratio of Earth's mass to that of the Moon.

8. ■■■ At what distance would a satellite orbit Jupiter in the time taken for Jupiter to rotate exactly once, so that the satellite would appear stationary above the planet? ∞ (*More Precisely 2-3*)

9. ■ Illustrate the 1:2:4 "resonance" mentioned in the text by drawing a diagram showing the locations of Io, Europa, and Ganymede at various times over the course of one Ganymede orbit. Show the moons' locations at intervals of Io's orbital period.

10. ■■ Estimate the strength of Jupiter's gravitational tidal acceleration on Io, and compare it with the moon's own surface gravity. ∞ (*More Precisely 7-3*) Compare your answer with the strength of Earth's gravitational tidal force on the Moon, relative to the Moon's own surface gravity.

11. ■■ Estimate the strength of Jupiter's gravitational tidal force on Europa, relative to the moon's own surface gravity.

12. ■■ Estimate the strength of Jupiter's gravitational tidal force on Ganymede, relative to the moon's own surface gravity.

13. ■■ Calculate the strength of Europa's gravitational pull on Io at closest approach, relative to Jupiter's gravitational attraction on Io.

14. ■■ What are the surface gravity and escape speed of Europa?

15. ■■ Compare the apparent sizes of the Galilean moons, as seen from Jupiter's cloud tops, with the angular diameter of the Sun as seen from Jupiter. Would you expect ever to see a total solar eclipse from Jupiter's cloud tops?

 In addition to the Practice Problems module, the Companion Website at astro.prenhall.com/chaisson provides for each chapter a study guide module with multiple choice questions as well as additional annotated images, animations, and links to related Websites.

12

Hardly more than a few decades ago, Saturn was the only planet known to have rings about it. Now, astronomers realize that all the big jovian planets have ring systems, though not as spectacular as Saturn's, shown here as imaged by the Hubble telescope. Saturn is often portrayed in garish colors to highlight certain features, as in the ultraviolet view at the top or the infrared view at the bottom, but this planet actually has very little color; the pastels shown in the visible image at center are the closest approximation of Saturn's natural colors. These images were taken in 2003 when the planet was at its maximum tilt (27°) toward Earth in ▶ *its 30-year orbit about the Sun. (STScI)*

Saturn

Spectacular Rings and Mysterious Moons

aturn is one of the most beautiful and enchanting of all astronomical objects. Its rings are a breathtaking sight when viewed through even a small telescope, and they are probably the planet's best-known feature. Aside from its famous rings, however, Saturn presents us with another good example of a giant gaseous planet. Saturn is in many ways similar to its larger neighbor, Jupiter, in terms of composition, size, and structure. Yet when we study the two planets in detail, we find that there are important differences as well. A comparison between Saturn and Jupiter provides us with valuable insight into the structure and evolution of all the jovian worlds.

LEARNING GOALS

Studying this chapter will enable you to

1 Summarize the orbital and physical properties of Saturn, and compare them with those of Jupiter.

2 Describe the composition and structure of Saturn's atmosphere and interior.

3 Explain why Saturn's internal heat source and magnetosphere differ from those of Jupiter.

4 Describe the structure and composition of Saturn's rings.

5 Define the Roche limit, and explain its relevance to the origin of Saturn's rings.

6 Summarize the general characteristics of Titan, and discuss the chemical processes in its atmosphere.

7 Discuss some of the orbital and geological properties of Saturn's smaller moons.

 Visit astro.prenhall.com/chaisson for additional annotated images, animations, and links to related sites for this chapter.

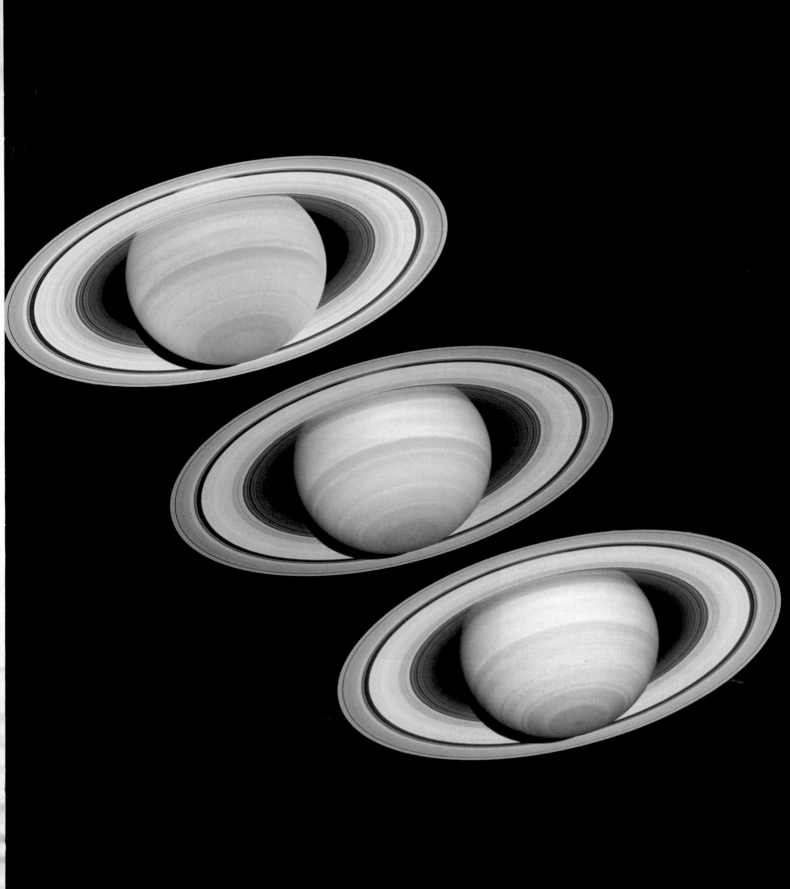

◄ **FIGURE 12.2 Saturn** Saturn as seen by the *Hubble Space Telescope* in 1994, when a huge and rare storm was visible near the planet's equator—the bright part at the center. The bland colors are approximately true—that is, as the human eye sees things. The inset shows the northern polar region at an earlier time. (Note the ring segment in the background.) *(NASA)*

remained behind. Instead, astronomers think that, at some time in Saturn's past, the heavier helium began to sink toward the center of the planet, reducing its abundance in the outer layers and leaving them relatively rich in hydrogen. We will return to the reasons for this differentiation and its consequences in a moment.

Figure 12.4 illustrates Saturn's atmospheric structure. (Cf. the corresponding diagram for Jupiter, Figure 11.6.) In many respects, Saturn's atmosphere is quite similar to Jupiter's, except that the temperature is a little lower because of Saturn's greater distance from the Sun and because its clouds are somewhat thicker. Since Saturn, like Jupiter, lacks a solid surface, we take the top of the troposphere as our ref-

erence level and set it to 0 km. The top of the visible clouds lies about 50 km below this level. As on Jupiter, the clouds are arranged in three distinct layers, composed (in order of increasing depth) of ammonia, ammonium hydrosulfide, and water ice. Above the clouds lies a layer of haze formed by the action of sunlight on Saturn's upper atmosphere.

The total thickness of the three cloud layers in Saturn's atmosphere is roughly 200 km, compared with about 80 km on Jupiter, and each layer is itself somewhat thicker than its counterpart on Jupiter. The reason for this difference is Saturn's weaker gravity (due to its lower mass). At the haze level, Jupiter's gravitational field is nearly two-and-a-half times stronger than Saturn's, so Jupiter's atmosphere is

▲ **FIGURE 12.3 Saturn, False Color** This image was obtained with an infrared camera on board the *Hubble Space Telescope* in 1998. The false colors were created for greater contrast by combining observations made at three different wavelengths: 1.0 μm (blue), 1.8 μm (green), and 2.1 μm (red). Blue coloration indicates regions where the atmosphere is relatively free of haze (see Figure 12.4); green and yellow indicate increasing levels of haze; red and orange indicate high-level clouds. Two small storm systems near the equator appear white. Two of Saturn's moons are also visible in the image, as noted. *(NASA)*

VIDEO Saturn Storm

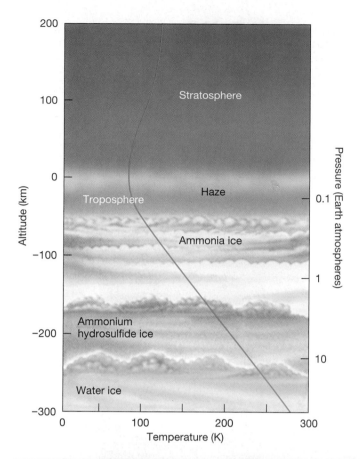

pulled much more powerfully toward the center of the planet. Thus, Jupiter's atmosphere is compressed more than Saturn's, and the clouds are squeezed more closely together. The colors of Saturn's cloud layers, as well as the planet's overall butterscotch hue, are due to the same basic cloud chemistry as on Jupiter. However, because Saturn's clouds are thicker, there are few holes and gaps in the top layer, so we rarely glimpse the more colorful levels below. Instead, we see different levels only in the topmost layer, which accounts for Saturn's rather uniform appearance.

Weather

Saturn has atmospheric wind patterns that are in many ways reminiscent of those on Jupiter. There is an overall east–west zonal flow, which is apparently quite stable. Computer-enhanced images of the planet that bring out more cloud contrast (see Figure 12.5) clearly show the existence of bands, oval storm systems, and turbulent flow

▲ **FIGURE 12.5 Saturn's Cloud Structure** More structure is seen in Saturn's cloud cover when computer processing and artificial color are used to enhance the contrast of the image, as in these *Voyager* images of the entire gas ball and a smaller, magnified piece of it. *(NASA)*

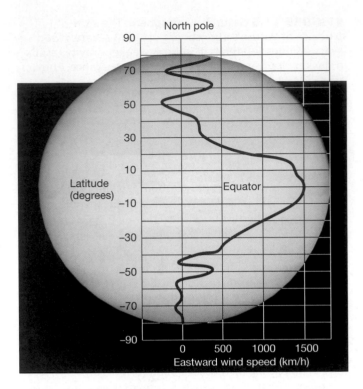

▲ **FIGURE 12.6 Saturn's Zonal Flow** Winds on Saturn reach speeds even greater than those on Jupiter. As on Jupiter, the visible bands appear to be associated with variations in wind speed.

patterns looking very much like those seen on Jupiter. Scientists think that Saturn's bands and storms have essentially the same cause as does Jupiter's weather. Ultimately, the large-scale flows and small-scale storm systems are powered by convective motion in Saturn's interior and by the planet's rapid rotation.

The zonal flow on Saturn is considerably faster than on Jupiter and shows fewer east–west alternations, as can be seen from Figure 12.6 (Cf. Figure 11.5). The equatorial eastward jet stream, which reaches a speed of about 400 km/h on Jupiter, moves at a brisk 1500 km/h on Saturn, and extends to much higher latitudes. Not until latitudes 40° north and south of the equator are the first westward flows found. Latitude 40° north also marks the strongest bands on Saturn and the most obvious ovals and turbulent eddies. Astronomers still do not fully understand the reasons for the differences between Jupiter's and Saturn's flow patterns.

In September 1990, amateur astronomers detected a large white spot in Saturn's southern hemisphere, just below the equator. In November of that year, when the *Hubble Space Telescope* imaged the phenomenon in more detail, the spot had developed into a band of clouds completely encircling the planet's equator. Some of the *Hubble* images are shown in Figure 12.7. Astronomers suspect that the white coloration arose from crystals of ammonia ice formed when an upwelling plume of warm gas penetrated the cool upper cloud layers. Because the

crystals were freshly formed, they had not yet been affected by the chemical reactions that still color the planet's other clouds.

Such spots are relatively rare on Saturn. The previous one visible from Earth appeared in 1933, but it was much smaller than the 1990 system and much shorter lived, last-

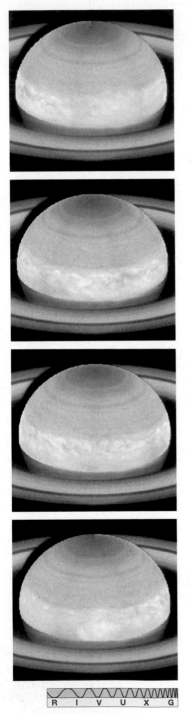

R I V U X G

▲ **FIGURE 12.7 Saturn's Clouds** Circulating and evolving cloud systems on Saturn, imaged by the *Hubble Space Telescope* at approximately two-hour intervals and shown in true color. *(NASA)*

ing for only a few weeks. Another large storm system is visible in Figure 12.2, taken in 1994. Figure 12.3 (from 1998) shows two somewhat smaller systems. The turbulent flow patterns seen around the 1990 white spot have many similarities to the flow around Jupiter's Great Red Spot. Scientists speculate that these white spots represent long-lived weather systems on Saturn and hope that routine observations of such temporary atmospheric phenomena on the outer worlds will enable them to gain greater insight into the dynamics of planetary atmospheres.

CONCEPT CHECK

✔ Why are atmospheric features on Saturn generally less vivid than those on Jupiter?

12.3 Saturn's Interior and Magnetosphere

Interior Structure and Internal Heating

LEARNING GOAL 3 Figure 12.8 depicts Saturn's internal structure. (Cf. Figure 11.10 of Jupiter.) The picture has been pieced together by planetary scientists using the same tools—*Voyager* observations and theoretical modeling—that they employed to infer Jupiter's inner workings. Saturn has the same basic internal parts as Jupiter, but their relative proportions are somewhat different: Saturn's metallic hydrogen layer is thinner and its core is larger. Because of its lower mass, Saturn has a less extreme core temperature, density, and pressure than does Jupiter. The central pressure is around a tenth of Jupiter's—not too different from the pressure at the center of Earth.

Infrared measurements indicate that Saturn's surface (i.e., cloud-top) temperature is 97 K, substantially higher than the temperature at which Saturn would reradiate all the energy it receives from the Sun. In fact, Saturn radiates almost three times more energy than it absorbs. Thus, Saturn, like Jupiter, has an internal energy source. ∞ (Sec. 11.3) But the explanation behind Jupiter's excess energy—that the planet has a large reservoir of heat left over from its formation—doesn't work for Saturn. Smaller than Jupiter, Saturn must have cooled more rapidly—rapidly enough that its original supply of energy was used up long ago. What, then, is happening inside Saturn to produce this extra heat?

The explanation for the origin of Saturn's extra heat also explains the mystery of the planet's apparent helium deficit. At the temperatures and high pressures found in Jupiter's interior, liquid helium *dissolves* in liquid hydrogen. Inside Saturn, where the internal temperature is lower, the helium doesn't dissolve so easily and tends to form droplets instead. The phenomenon is familiar to cooks, who know that it is generally much easier to dissolve ingredients in hot liquids than in cold ones. Saturn probably started out with a fairly uniform solution of helium dissolved in hydrogen, but the helium tended to condense out of the surrounding hydrogen, much as water vapor condenses out of Earth's atmosphere to form a mist. The amount of helium condensation was greatest in the planet's cool outer layers, where the mist turned to rain about 2 billion years ago. A light shower of liquid helium has been falling through Saturn's interior ever since. This **helium precipitation** is responsible for depleting the outer layers of their helium content.

So we can account for the unusually low abundance of helium in Saturn's atmosphere: Much of it has rained down to lower levels. But what about the excess heating? The answer is simple: As the helium sinks toward the center, the planet's gravitational field compresses it and heats it up. The gravitational energy thus released is the source of Saturn's internal heat. In the distant future the helium rain will stop, and Saturn will cool until its outermost layers radiate only as much energy as they receive from the Sun. When that happens, the temperature at Saturn's cloud tops will be 74 K. As Jupiter cools, it, too, may someday experience precipitate helium in its interior, causing its surface temperature to rise once again.

Magnetospheric Activity

Saturn's electrically conducting interior and rapid rotation produce a strong magnetic field and an extensive magnetosphere. Probably because of the considerably smaller mass of Saturn's metallic hydrogen zone, the planet's basic magnetic field strength is only about one-twentieth that of Jupiter, or about a thousand times greater than that of Earth. The magnetic field at Saturn's cloud tops (roughly 10 Earth radii from the planet's center) is approximately the same as that at Earth's surface. *Voyager* measurements indicate that, unlike Jupiter's and Earth's magnetic axes,

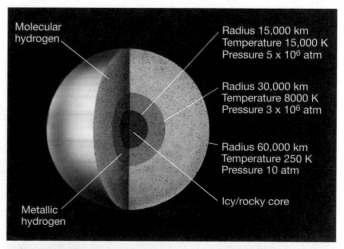

Molecular hydrogen

Radius 15,000 km
Temperature 15,000 K
Pressure 5 × 10⁶ atm

Radius 30,000 km
Temperature 8000 K
Pressure 3 × 10⁶ atm

Radius 60,000 km
Temperature 250 K
Pressure 10 atm

Icy/rocky core

Metallic hydrogen

▲ **FIGURE 12.8 Saturn's Interior** Saturn's internal structure, as deduced from *Voyager* observations and computer modeling.

R I V U X G

▲ **FIGURE 12.9 Aurora on Saturn** An ultraviolet camera aboard the *Hubble Space Telescope* recorded this image of a remarkably symmetrical (orange) aurora on Saturn during a solar storm in 1998. *(NASA)*

which are slightly tilted, Saturn's magnetic field is not inclined with respect to its axis of rotation. Saturn's magnetic field, like Jupiter's, is oriented opposite that of Earth: that is, an Earth compass needle would point toward Saturn's south pole rather than its north. Figure 12.9 shows an aurora on Saturn, imaged in 1998 by the *Hubble Space Telescope*.

Saturn's magnetosphere extends about 1 million km toward the Sun and is large enough to contain the planet's ring system and the innermost 16 small moons. Saturn's largest moon, Titan, orbits about 1.2 million km from the planet, so it is found sometimes just inside the outer magnetosphere and sometimes just outside, depending on the intensity of the solar wind (which tends to push the sunward side of the magnetosphere closer to the planet). Because no major moons lie deep within Saturn's magnetosphere, the details of its structure are different from those of Jupiter's magnetosphere. For example, there is no equivalent of the Io plasma torus. ∞ (Sec. 11.5) Like Jupiter, Saturn emits radio waves, but as luck would have it, they are reflected from Earth's ionosphere (they lie in the AM band) and were not detected until the *Voyager* craft approached the planet.

CONCEPT CHECK

✔ Where did Saturn's atmospheric helium go?

12.4 Saturn's Spectacular Ring System

The View from Earth

LEARNING GOAL 4 The most obvious aspect of Saturn's appearance is, of course, its *planetary ring system*. Astronomers now know that all the jovian planets have rings, but Saturn's are by far the brightest, the most extensive, and the most beautiful. Galileo saw them first in 1610, but he did not recognize what he saw as a planet with a ring. At the resolution of his small telescope, the rings looked like bumps on the planet, or perhaps (he speculated) parts of a triple planet of some sort. Figure 12.10(a) and (b) shows two of Galileo's early sketches of Saturn. By 1616, Galileo had already realized that the "bumps" were not round, but rather elliptical in shape. In 1655, the Dutch astronomer Christian Huygens realized what the bumps were: a thin, flat ring, completely encircling the planet (Figure 12.10c).

In 1675, the French–Italian astronomer Giovanni Domenico Cassini discovered the first ring feature: a dark band about two-thirds of the way out from the inner edge. From Earth, the band looks like a gap in the ring (an observation that is not too far from the truth, although we

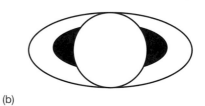

(a)

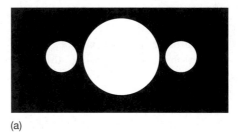

(b)

(c)

▲ **FIGURE 12.10 Sketches of Saturn's Rings** Three sketches of Saturn's rings, made (a) by Galileo in 1610, (b) by Galileo in 1616, and (c) by Huygens in 1655.

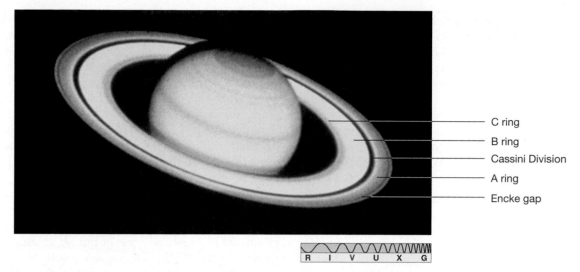

▲ FIGURE 12.11 Saturn, Up Close Much fine structure, especially in the rings, appears in this image of Saturn taken with the *Hubble* telescope. The main ring features long known from Earth-based observations—the A, B, and C rings, the Cassini Division, and the Encke gap—are marked. The other rings listed in Table 12.1 are not visible here. *(NASA)*

now know that there is actually some ring material within it). This "gap" is named the **Cassini division**, in honor of its discoverer. Careful observations from Earth show that the inner "ring" is in reality also composed of two rings. From the outside in, the three rings are known somewhat prosaically as the **A, B,** and **C rings**. The Cassini division lies between the A and B rings. The much narrower **Encke gap**, some 300 km wide, is found in the outer part of the A ring. These ring features are marked on Figure 12.11. No finer ring details are visible from our Earthly vantage point. Of the three main rings, the B ring is brightest, followed by the somewhat fainter A ring, and then by the almost translucent C ring. A more complete list of ring properties appears in Table 12.1. (The D, E, F, and G rings listed in the table are discussed later in this section.)

What Are Saturn's Rings?

A fairly obvious question—and one that perplexed the best scientists and mathematicians on Earth for almost two centuries—is "What are Saturn's rings made of?" By the middle of the 19th century, various dynamical and thermodynamic arguments had conclusively proved that the rings could be neither solid, liquid, nor gas! What is left? In 1857, after showing that a solid ring would become unstable and break up, Scottish physicist James Clerk Maxwell suggested that the rings are composed of a great number of small particles, all independently orbiting Saturn, like so many tiny moons. That inspired speculation was verified in 1895, when Lick Observatory astronomers measured the Doppler shift of sunlight reflected from the rings and showed that the velocities thus determined were exactly what would be expected from separate particles moving in circular orbits in accordance with Newton's law of gravity.

What sort of particles make up the rings? The fact that they reflect most (over 80 percent) of the sunlight striking them had long suggested to astronomers that they were made of ice, and infrared observations in the 1970s confirmed that water ice is indeed a prime constituent of the rings. Radar observations and later *Voyager* studies of scattered sunlight showed that the diameters of the particles range from fractions of a millimeter to tens of meters, with most particles being about the size (and composition) of a large snowball on Earth.

TABLE 12.1	The Rings of Saturn				
Ring	**Inner Radius**		**Outer Radius**		**Width**
	(km)	**(planetary radii)**	**(km)**	**(planetary radii)**	**(km)**
D	67,000	1.11	74,700	1.24	7700
C	74,700	1.24	92,000	1.53	17,300
B	92,000	1.53	117,500	1.95	25,500
Cassini Division	117,500	1.95	122,300	2.03	4800
A	122,300	2.03	136,800	2.27	14,500
Encke gap*	133,400	2.22	133,700	2.22	300
F	140,300	2.33	140,400	2.33	100
G	165,800	2.75	173,800	2.89	8000
E	180,000	3.00	480,000	8.00	300,000

The Encke gap lies within the A ring.

We now know that the rings are truly thin—perhaps only a few tens of meters thick in places. Stars can occasionally be seen through them, like automobile headlights penetrating a snowstorm. Why are the rings so thin? The answer seems to be that collisions between ring particles tend to keep them all moving in circular orbits in a single plane. Any particle that tries to stray from this orderly motion finds itself in an orbit that soon runs into other ring particles. Over long periods, the ensuing jostling serves to keep all of the particles moving in circular, planar orbits. The asymmetric gravitational field of Saturn (a result of its flattened shape) sees to it that the rings lie in the planet's equatorial plane.

The Roche Limit

5 But why a ring of particles at all? What process produced the rings in the first place? To answer these questions, consider the fate of a small moon orbiting close to a massive planet such as Saturn. The moon is held together by internal forces—its own gravity, for example. As we bring our hypothetical moon closer to the planet, the tidal force on it increases. Recall from Chapter 7 that the effect of such a tidal force is to stretch the moon along the direction toward the planet—that is, to create a tidal bulge. Recall also that the tidal force increases rapidly with decreasing distance from the planet. ∞ (Sec. 7.6) As the moon is brought closer to the planet, it reaches a point where the tidal force tending to stretch it out becomes *greater* than the internal forces holding it together. At that point, the moon is torn apart by the planet's gravity, as shown in Figure 12.12. The pieces of the satellite then pursue their own individual orbits around that planet, eventually spreading all the way around it in the form of a ring.

For any given planet and any given moon, the critical distance inside of which the moon is destroyed is known as the *tidal stability limit*, or the **Roche limit**, after the 19th-century French mathematician Edouard Roche, who first calculated it. As a handy rule of thumb, if our hypothetical moon is held together by its own gravity and its average density is comparable to that of the parent planet (both reasonably good approximations for Saturn's larger moons), then the Roche limit is roughly 2.4 times the radius of the planet. Thus, for Saturn, no moon can survive within a distance of 144,000 km of the planet's center, about 7000 km beyond the outer edge of the A ring. The main (A, B, C, D, and F) rings of Saturn occupy the region inside Saturn's Roche limit.

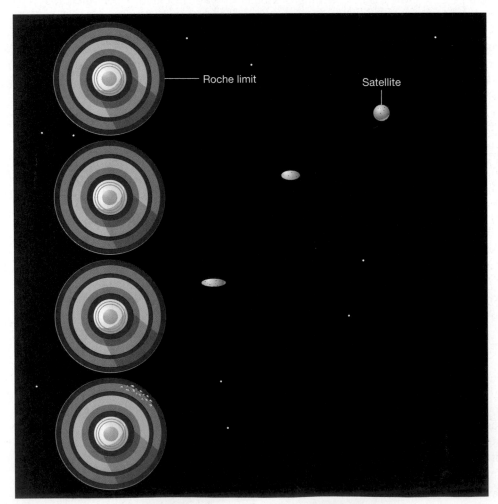

◀ **FIGURE 12.12 Roche Limit** From top to bottom, these four frames illustrate how the increasing tidal field of a planet first distorts, and then destroys, a moon that strays too close. (The distortion is exaggerated in the second and third panels.)

Roche limit

Satellite

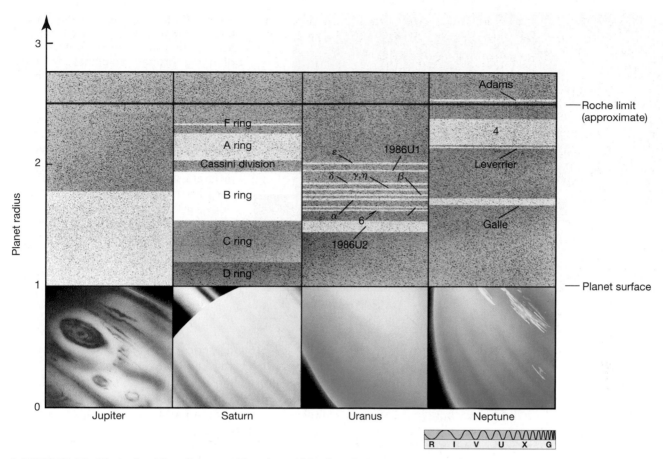

▲ **FIGURE 12.13 Jovian Ring Systems** The rings of Jupiter, Saturn, Uranus, and Neptune. All distances are expressed in planetary radii. The red line represents the Roche limit. All the rings lie within (or very close to) the Roche limit of their parent planets. Note that the red line is just an approximation—the details depend on the internal structure of the planet, and Neptune's Adams ring is actually consistent with Roche's calculations.

These considerations apply equally well to the other jovian worlds. Figure 12.13 shows the location of the ring system of each jovian planet relative to the planet's Roche limit. Given the approximations in our assumptions, we can conclude that all the major planetary rings are found within the Roche limit of their parent planet. Notice that, strictly speaking, the calculation of this limit applies only to low-density moons massive enough for their own gravity to be the dominant force binding them together. Sufficiently small moons (less than 10 km or so in diameter) can survive even within the Roche limit because they are held together mostly by interatomic (electromagnetic) forces, not by gravity.

The View From *Voyager*

Thus it was, as *Voyagers 1* and *2* approached Saturn, that scientists on Earth were fairly confident that they understood the nature of the rings. However, there were many surprises in store. The *Voyager* missions changed forever our view of this spectacular region in our cosmic backyard, revealing the rings to be vastly more complex than

astronomers had imagined. The four-year *Cassini* mission, which arrived at Saturn in mid-2004, will undoubtedly provide many new insights into this fascinating system.

As the *Voyager* probes approached Saturn, it became obvious that the main rings were actually composed of tens of thousands of narrow **ringlets** (shown in Figure 12.14). Although *Voyager* cameras did find several new gaps in the rings, the ringlets are generally not separated from one another by empty space. Instead, the rings contain concentric regions of alternating high and low concentrations of ring particles—the ringlets are just the high-density peaks. Although the process is not fully understood, it seems that the mutual gravitational attraction of the ring particles (as well as the effects of Saturn's inner moons) enables waves of matter to form and move in the plane of the rings, rather like ripples on the surface of a pond. The wave crests typically wrap around the rings, forming tightly wound spiral patterns called *spiral density waves* that resemble grooves in a huge celestial phonograph record.

Although the ringlets are probably the result of spiral waves in the rings, the true gaps are not. The narrower

◄ **FIGURE 12.14 Saturn's Rings, Up Close** *Voyager 2* took this close-up of Saturn's ring structure just before plunging through the planet's tenuous outer rings. The ringlets in the B ring, spread over several thousand kilometers, are resolved here to about 10 km. As *Voyager* approached Saturn, more and more of these tiny ringlets became noticeable in the main rings. The enhanced color variations indicate different sizes and compositions of the particles making up the thousands of rings. Earth is superposed, to proper scale, for a size comparison. *(NASA)*

R I V U X G

gaps—about 20 of them—are most likely swept clean by the action of small moonlets embedded in them. These moonlets are larger (perhaps 10 or 20 km in diameter) than the largest true ring particles, and they simply "sweep up" ring material through collisions as they go. Despite many careful searches of *Voyager* images, only one of these moonlets has so far been found—in 1991, after five years of exhaustive study, NASA scientists confirmed the discovery of the 18th moon of Saturn (now named Pan) in the Encke gap. Astronomers have found indirect evidence for embedded moonlets, in the form of "wakes" that they leave behind them in the rings, but no other direct sightings have occurred. Despite their elusiveness, moonlets are still regarded as the best explanation for the small gaps.

Voyager 2 found a series of faint rings, now known collectively as the **D ring**, inside the inner edge of the C ring, stretching down almost to Saturn's cloud tops. The D ring contains relatively few particles and is so dark that it is completely invisible from Earth. Another faint ring, also a *Voyager 2* discovery, lies well outside the main ring structure. Known as the **E ring**, it appears to be associated with volcanism on the moon Enceladus (see Section 12.5).

The *Voyager 2* cameras revealed one other completely unexpected feature. A series of dark radial "spokes" formed on the B ring, moved around the planet for about one ring orbital period, and then disappeared (Figure 12.15). Careful scrutiny of these peculiar drifters showed that they were composed of very fine (micron-sized) dust hovering a few tens of meters *above* the plane of the rings. Scientists think that this dust was held in place by electrostatic forces generated in the ring plane, perhaps resulting from collisions among particles there. The electrical fields slowly dispersed, and the spokes faded as the ring revolved. We expect that the creation and dissolution of such spokes are regular occurrences in the Saturn ring system.

Orbital Resonances and Shepherd Satellites

Voyager images show that the largest gap in the rings, the Cassini division, is not completely empty of matter. In fact, as shown in Figure 12.16, the division contains a series of faint ringlets and gaps (and, presumably, embedded moon-

R I V U X G

◄ **FIGURE 12.15 Spokes in the Rings** Saturn's B ring showed a series of dark temporary "spokes" as *Voyager 2* flew by at a distance of about 4 million km. The spokes were caused by small particles suspended just above the ring plane. *(NASA)*

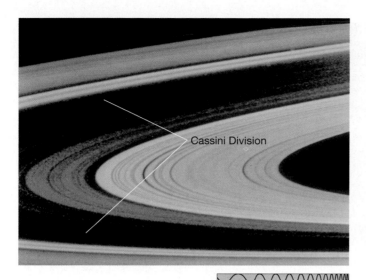

R   I   V   U   X   G

▲ FIGURE 12.16 Cassini Division Close inspection by *Voyager 2* revealed that the Cassini division (shown here as the darker color) is not completely empty. Rather, it contains a series of faint ringlets and gaps, assumed to be caused by unseen embedded satellites. The density of material in the division is very low, accounting for its dark appearance from Earth. *(NASA)*

lets, too). The overall concentration of ring particles in the division as a whole is, however, much lower than in the A and B rings. Although its small internal gaps probably result from embedded satellites, the division itself does not. Instead, it owes its existence to another solar-system *resonance*, this time involving particles orbiting in the division, on the one hand, and Saturn's innermost major moon, Mimas, on the other. ∞ (Sec. 8.4) A ring particle moving in an orbit within the Cassini division has an orbital period exactly half that of Mimas. Particles in the division thus complete exactly two orbits around Saturn in the time taken for Mimas to orbit once—a configuration known as a 2:1 resonance. Applying Kepler's third law (recast to refer not to the planets, but to Saturn's moons), we can show that this 2:1 resonance with Mimas corresponds to a radius of 117,000 km, the inner edge of the division. ∞ *(More Precisely 2-3)* The effect of this resonance is that particles in the division receive a gravitational tug from Mimas at exactly the same location in their orbit every other time around. Successive tugs reinforce one another, and the initially circular trajectories of the ring particles soon get stretched out into ellipses. In their new orbits, these particles collide with other particles and eventually find their way into new circular orbits at other radii. The net effect is that the number of ring particles in the Cassini division is greatly reduced.

Particles in "nonresonant" orbits (i.e., at radii whose orbital period is not simply related to the period of Mimas) also are acted upon by Mimas's gravitational pull. But the times when the force is greatest are spread uniformly around the orbit, and the tugs cancel out. It's a little like pushing a child on a swing: Pushing at the same point in the swing's motion each time produces much better results than do random shoves. Thus, Mimas (or any other moon) has a large effect on the ring at those radii at which a resonance exists and little or no effect elsewhere.

We now know that resonances between ring particles and moons play an important role in shaping the fine structure of Saturn's rings. For example, the sharp outer edge of the A ring is thought to be governed by a 3:2 resonance with Mimas (three ring orbits in two Mimas orbital periods). Most theories of planetary rings predict that the ring system should spread out with time, basically because of collisions among ring particles. Instead, the A ring's outer edge is "patrolled" by a small satellite named Atlas, held in place by the gravity of Mimas, that prevents ring particles from diffusing away. Compare Tables 12.1 and 12.2, and see if you can identify other resonant connections between Saturn's moons or between the moons and the rings. (You should be able to find quite a few—Saturn is quite a complex place!)

Outside the A ring lies the strangest ring of all. The faint, narrow **F ring** (shown in Figure 12.17) was discovered by *Pioneer 11* in 1979, but its full complexity became evident only when *Voyager 1* took a closer look. Unlike the inner major rings, the F ring is narrow—less than a hundred kilometers wide. It lies just inside Saturn's Roche limit, separated from the A ring by about 3500 km. Its narrowness by itself is unusual, as is its slightly eccentric shape, but the F ring's oddest feature is that it looks as though it is made up of several separate strands braided together! This remarkable discovery sent dynamicists

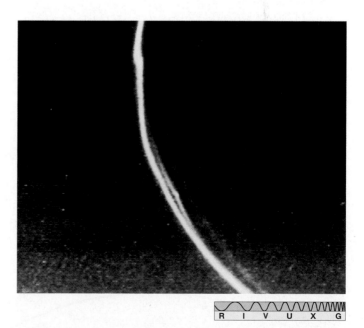

R   I   V   U   X   G

▲ FIGURE 12.17 F Ring Saturn's narrow F ring appears to contain kinks and braids, making it unlike any of Saturn's other rings. *(NASA)*

scrambling in search of an explanation. It now seems as though the ring's intricate structure, as well as its thinness, arise from the influence of two small moons, known as **shepherd satellites**, that orbit on either side of it (Figure 12.18).

These two small, dark satellites, each little more than 100 km in diameter, are called Prometheus and Pandora. They orbit about 1000 km on either side of the F ring, and their gravitational influence on the F-ring particles keeps the ring tightly confined in its narrow orbit. As shown in Figure 12.19, any particle straying too far out of the F ring is gently guided back into the fold by one or the other of the moons. (The moon Atlas confines the A ring in a somewhat similar way.) However, the details of how Prometheus and Pandora produce the braids in the F ring and why the two moons are there at all, in such similar orbits, remain unclear. There is some evidence that other eccentric rings found in the gaps in the A, B, and C rings may also result from the effects of shepherding moonlets.

Outside the F ring lies another faint, narrow ring: the **G ring**. Discovered by *Pioneer 11* and imaged in more detail by *Voyager 2*, the G ring apparently lacks ringlets as well as the peculiar internal structure found in the F ring. The narrowness and sharp edges of the G ring suggest the presence of shepherd satellites, but so far none has been found.

The Origin of the Rings

Two possible origins have been suggested for Saturn's rings. Astronomers estimate that the total mass of ring material is no more than 10^{15} tons—enough to make a satellite about 250 km in diameter. If such a satellite

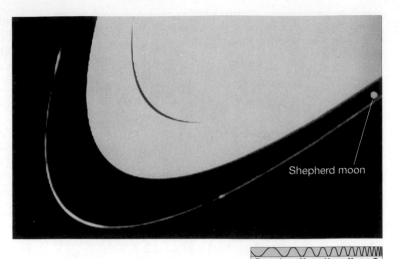

R   I   V   U   X   G

▲ **FIGURE 12.18 Shepherd Moon** The F ring's thinness, and possibly its other peculiarities, can be explained by the effects of two shepherd satellites that orbit a few hundred kilometers inside and outside the ring. This photo shows one of the shepherding satellites, roughly 100 km in length, at the right.

strayed inside Saturn's Roche limit or was destroyed (perhaps by a collision) near that radius, a ring could have resulted. An alternative view is that the rings represent material left over from Saturn's formation stage 4.6 billion years ago. In this scenario, Saturn's tidal field prevented any moon from forming inside the Roche limit, so the material has remained a ring ever since. Which view is correct?

All the dynamic activity observed in Saturn's rings suggests to many researchers that the rings must be quite

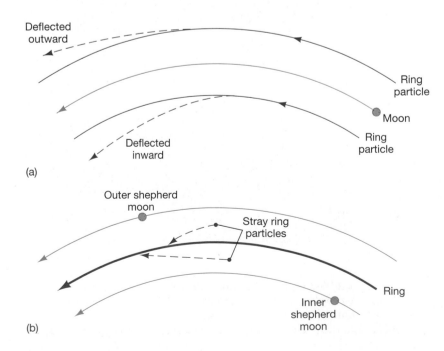

◀ **FIGURE 12.19 Moon–Ring Interaction** (a) Strange as it may seem, the net effect of the interactions between a moon and ring particles is that the moon tends to push those particles away from it. (b) The F-ring shepherd satellites operate by forcing errant F-ring particles back into the main ring. Each moon operates as in part (a), so that the ring is confined between the two moons. As a consequence of Newton's third law of motion, the satellites themselves slowly drift away from the ring (unless they, in turn, are held in place by interactions with other, larger moons).

young—perhaps no more than 50 million years old, or 100 times younger than the solar system. There is just too much going on for the ring to have remained stable for billions of years, so they probably aren't left over from the planet's formative stages. If this is so, then either the rings are continuously replenished, perhaps by fragments of Saturn's moons chipped off by meteorites, or they are the result of a relatively recent, possibly catastrophic, event in the planet's system—a small moon that may have been hit by a large comet or even by another moon.

Astronomers prefer not to invoke catastrophic events to explain specific phenomena, but the more we learn of the universe, the more we realize that catastrophe probably plays an important role. For now, the details of the formation of Saturn's ring system simply aren't known.

CONCEPT CHECK

✔ What do the Roche limit and orbital resonances have to do with planetary rings?

12.5 The Moons of Saturn

General Features

Saturn has the most extensive, and in many ways the most complex, system of natural satellites of all the planets. The planet's 18 largest moons are listed in Table 12.2. Observations of sunlight reflected from them suggests that most are covered with snow and ice. Many of them are probably made almost entirely of water ice. Even so, they are a curious and varied lot, and many aspects of their structure and history are simply not well understood. Much of our detailed knowledge of these moons comes from the *Pioneer* and *Voyager* flybys in the late 1970s and early 1980s. ∞ (Sec. 6.6) Scientists eagerly await new data from the extended *Cassini* orbital mission, scheduled to run from 2004 until 2008.

The moons fall into three fairly natural groups. First, there are the many "small" moons—irregularly shaped chunks of ice, all less than 400 km across—that exhibit a bewildering variety of complex and fascinating motion. Second, there are six "medium-sized" moons—spherical

TABLE 12.2   The Major Moons of Saturn*

Name	Distance from Saturn (km)	(planetary radii)	Orbit Period (days)	Size (longest diameter, km)	Mass** (Earth moon masses)	Density (kg/m³)	(g/cm³)
Pan	133,600	2.22	0.57	20	$4. \times 10^{-8}$		
Atlas	138,000	2.28	0.60	37			
Prometheus	139,000	2.31	0.61	148	1.9×10^{-6}		
Pandora	142,000	2.35	0.63	110	1.8×10^{-6}		
Epimetheus	151,000	2.51	0.69	138	7.5×10^{-6}		
Janus	151,000	2.51	0.69	199	2.7×10^{-5}		
Mimas	186,000	3.08	0.94	398	0.00051	1100	1.1
Enceladus	238,000	3.95	1.37	498	0.00099	1100	1.1
Calypso	295,000	4.89	1.89	30			
Telesto	295,000	4.89	1.89	30			
Tethys	295,000	4.89	1.89	1060	0.0085	1000	1.0
Dione	377,000	6.26	2.74	1120	0.014	1400	1.4
Helene	377,000	6.26	2.74	32			
Rhea	527,000	8.74	4.52	1530	0.032	1200	1.2
Titan	1,220,000	20.3	16.0	5150	1.83	1900	1.9
Hyperion	1,480,000	24.6	21.3	370			
Iapetus	3,560,000	59.1	79.3	1440	0.022	1000	1.0
Phoebe	13,000,000	215	−550[†]	230			

*Moons larger than 10 km in diameter only; does not include the 13 recently discovered small moons described in the text.

** Mass of Earth's Moon = 7.4×10^{22} kg = 1.3×10^{-4} Saturn masses.

[†]Retrograde orbit.

bodies with diameters ranging from about 400 to 1500 km—that offer clues to the past and present state of the environment of Saturn while presenting many puzzles regarding their own appearance and history. Finally, there is Saturn's single "large" moon—Titan—which, at 5150 km in diameter, is the second-largest satellite in the solar system. (Jupiter's Ganymede is a little bigger.) Titan has an atmosphere denser than Earth's and (some scientists think) surface conditions that may be conducive to life. Notice, incidentally, that Jupiter has no "medium-sized" moons, as just defined: The Galilean satellites are large, like Titan, and all of Jupiter's other satellites are small—no more than 200 km in diameter.

In 1995, researchers using the *Hubble Space Telescope* reported the sighting of two more moons. However, these observations now seem to have been in error—easy to understand when dealing with faint objects at the limits of detectability (see *Discovery 12-1*). More recently, since 2000, astronomers working at observatories on Mauna Kea and at the VLT have reported the discovery of 13

DISCOVERY 12-1

New Moons or Ring Debris at Saturn?

The last 10 years have seen an unprecedented number of reports of new moons orbiting the outer planets. At least 60 previously unknown satellites have been observed. These observations involve searching images of regions of space near the planet for very faint specks of light, usually very close to the telescope's limit of sensitivity. The recent slew of discoveries has come about in large part because of better detectors and improved search techniques. ∞ (*Discovery 11-2*)

Such undertakings are not without their problems, however. In 1995, researchers using the *Hubble Space Telescope* announced the discovery of several new moons orbiting Saturn. The findings were reported with great fanfare in the press at the time, for these would have been the first such moons spotted anywhere by *Hubble*. Now, however, it seems unlikely that these sightings were actually new moons. They might not even have been real objects at all!

The original images were processed by removing stray light from near the edge of Saturn's ring, which was minimal since the observations were made during the ring crossing in 1995, when the rings virtually disappeared from view (see Figure 12.1). After the positions of all of Saturn's known moons were accounted for, the field of view still displayed four blobs of light, in addition to the body of Saturn itself. The accompanying four-picture sequence spans 30 minutes of observing time and shows how astronomers might have been fooled into thinking they had made a discovery. Saturn itself appears as a bright white disk at the far right, heavily overexposed in the search for faint new moons; its edge-on rings extend diagonally to the upper left and are barely discernible. What remains is one known moon—Epimetheus, the brightest small blob at the left—and several other fainter blobs. The blobs in these images formed the basis for the astronomers' claims of new moons in the Saturn system.

Subsequent observations by another team of astronomers failed to confirm some of the faint blobs of light in these images, but reported some new blobs. What is going on here? It seems that if the blobs were indeed real, then they are probably just clouds of ice and dust orbiting near the planet's rings. Conceivably, the strong tidal pull of Saturn on some of its small moons broke off a piece large enough to be spotted for a time as a transient piece of debris before ultimately dispersing.

Perhaps this is even how Saturn's rings are maintained. As for the blobs that cannot be confirmed, they may be only instrumental artifacts, caused by small-scale imperfections in *Hubble*'s mirror, by "hot pixels" within the onboard cameras, or by the computer methods used by astronomers to massage their data.

In any case, most astronomers are now of the opinion that these faint blobs, even if real, do not represent new moons around Saturn. Note that this episode highlights again the concern of some astronomers as to what the definition of a "real" moon should be. ∞ (*Discovery 11-2*) The faintest moon candidates will always be those most prone to observational errors. They are also most likely to be randomly captured pieces of interplanetary "junk."

Are the many recent observations more reliable? Probably. They were based on a search of a portion of the sky far from Saturn, much less affected by the planet's light or possible ring debris and—very important—in at least some cases have already been reproduced. Still, we can be sure that new candidate moons will be carefully scrutinized before being declared "official."

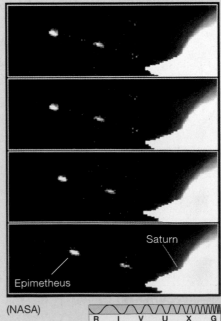

(NASA)

R I V U X G

more moons orbiting Saturn, bringing the planet's total (for now) to 31. As with the newly discovered small satellites of Jupiter, these moons are all very faint (and hence small) and revolve around Saturn quite far from the planet on rather inclined, often retrograde, orbits, much as Saturn's other "small" moons do. ∞ (*Discovery 11-2*) Most likely, they are chunks of debris captured from interplanetary space after close encounters with Saturn.

Titan

Perhaps the most intriguing of all Saturn's moons, Titan was discovered by Christian Huygens in 1655. Even through a large Earth-based telescope, this moon is visible only as a barely resolved reddish disk. However, long before the *Voyager* missions, astronomers already knew (from spectroscopic observations) that the moon's reddish coloration was caused by something quite special—an atmosphere. So eager were mission planners to obtain a closer look that they programmed *Voyager 1* to pass very close to Titan, even though it meant that the spacecraft could not then use Saturn's gravity to continue on to Uranus and Neptune. (Instead, *Voyager 1* left the Saturn system on a path taking the craft out of the solar system well above the plane of the ecliptic.)

Scientists expect that Titan's internal composition and structure must be similar to those of Ganymede and Callisto, because all three moons have quite similar masses, radii, and, hence, average densities. (Titan's density is 1900 kg/m³.) ∞ (*Sec.11.5*) Titan probably contains a rocky core surrounded by a thick mantle of water ice. However, in view of *Galileo*'s discovery of significant differences between the internal structures of Ganymede and Callisto, the degree of differentiation within Titan is currently unknown. In addition to releasing the *Huygens* atmospheric probe and lander in 2005 to study Titan's atmosphere and surface, another important objective of the *Cassini* mission is to make detailed gravitational and other measurements of the moon's interior. ∞ (*Sec.6.6*)

A *Voyager 1* image of Titan is shown in Figure 12.20. Unfortunately, despite the spacecraft's close pass, the moon's surface remains a mystery. A thick, uniform layer of haze, similar to the photochemical smog (created by chemical reactions powered by light) found over many

cities on Earth, that envelops the moon completely obscured the spacecraft's view. Still, *Voyager 1* was able to provide mission specialists with detailed atmospheric data.

Titan's atmosphere is thicker and denser even than Earth's, and it is certainly far more substantial than that of any other moon. Prior to *Voyager 1*'s arrival in 1980, only methane and a few other simple hydrocarbons had been conclusively detected on Titan. (Hydrocarbons are molecules consisting solely of hydrogen and carbon atoms; methane, CH_4, is the simplest.) Radio and infrared observations from *Voyager 1* showed that the atmosphere is

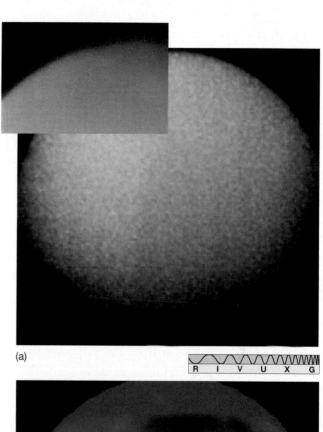

(a)

R I V U X G

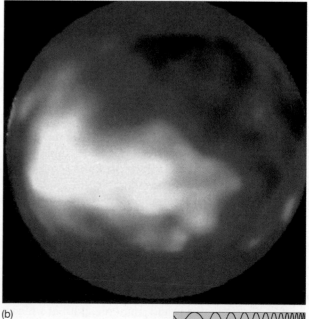

(b)

R I V U X G

▶ **FIGURE 12.20 Titan** (a) Larger than the planet Mercury and roughly half the size of Earth, Titan, Saturn's largest moon, was photographed in visible light from only 4000 km away as *Voyager 1* passed by in 1980. All we can see here is Titan's upper cloud deck. The inset shows a contrast-enhanced image of the haze layer (falsely colored blue). (b) In infrared, as captured with the adaptive-optics system on the Canada–France–Hawaii telescope on Mauna Kea, Titan displays large-scale surface features. The bright regions are thought to be highlands, possibly covered with frozen methane. The brightest area is nearly 4000 km across—about the size of Australia. (*NASA; CFHT*)

actually made up mostly of nitrogen (roughly 90 percent) and argon (at most 10 percent), with a small percentage of methane. In addition, complex chemistry in Titan's atmosphere maintains steady (but trace) levels of hydrogen gas, the hydrocarbons ethane and propane, and carbon monoxide.

Titan's atmosphere seems to act like a gigantic chemical factory. Powered by the energy of sunlight, the atmosphere is undergoing a complex series of chemical reactions that ultimately result in the observed smog and trace chemical composition. The upper atmosphere is thick with aerosol haze (droplets so small that they remain suspended in the atmosphere), and the unseen surface may be covered with organic sediment that has settled down from the clouds. Speculation runs the gamut from oceans of liquid ethane to icy valleys laden with hydrocarbon sludge. In 2003, radio astronomers using the Arecibo telescope reported the first evidence (as yet unconfirmed) for liquid hydrocarbon lakes on Titan's surface. ∞ (Sec. 5.5) Exploration of Titan by *Cassini–Huygens* (see Figure 12.21) and future spacecraft present scientists with an opportunity to study the kind of chemistry thought to have occurred billions of years ago on Earth—the prebiotic chemical reactions that eventually led to life on our own planet (see Chapter 28).

Based largely on *Voyager* measurements, Figure 12.22 shows the probable structure of Titan's atmosphere. Despite the moon's low mass (a little less than twice that of Earth's Moon), and hence its low surface gravity (one-seventh of Earth's), the atmospheric pressure at ground level is 60 percent greater than on Earth. Titan's atmosphere contains about 10 times more gas than Earth's atmosphere. Because of Titan's weaker gravitational pull, the atmosphere extends some 10 times farther into space than does our own. The top of the main haze layer lies about 200 km above the surface, although there are additional layers, seen primarily through their absorption of ultraviolet radiation, at 300 km and 400 km (Figure 12.23). Below the haze the atmosphere is reasonably clear, although rather gloomy, because so little sunlight gets through.

Titan's surface temperature is a frigid 94 K, roughly what we would expect on the basis of that moon's distance from the Sun. At the temperatures typical of the lower atmosphere, methane and ethane may behave rather like water on Earth, raising the possibility of methane rain, snow, and fog and even ethane oceans! At higher levels in the atmosphere, the temperature rises, the result of photochemical absorption of solar radiation.

Why does Titan have such a thick atmosphere, when similar moons of Jupiter, such as Ganymede and Callisto, have none? The answer seems to be a direct result of Titan's greater distance from the Sun. The moons of Saturn formed at considerably lower temperature than did those of Jupiter. Such conditions would have enhanced the ability of the water ice that makes up the bulk of Titan's interior to absorb methane and ammonia, both of which were present in abundance at those early times. As a result, Titan

▲ **FIGURE 12.21 Cassini–Huygens Probe** In January 2005, the *Huygens* probe, part of the *Cassini* mission, will descend into Titan's atmosphere and land on the moon's surface, as illustrated in this artist's conception. During the descent and on the surface, *Huygens* will make detailed measurements of Titan's atmospheric structure, composition, and chemistry; its wind speed and cloud physics; and the composition and physical state (solid or liquid) of the moon's surface and interior structure. *(D. Ducros/ESA)*

was initially laden with much more methane and ammonia gas than was either Ganymede or Callisto. As Titan's internal radioactivity warmed the moon, the ice released the trapped gases, forming a thick methane–ammonia atmosphere. Sunlight split the ammonia into hydrogen, which escaped into space, and nitrogen, which remained in the atmosphere. The methane, which is more tightly bound and so was less easily broken apart, survived intact. Together with argon outgassed from Titan's interior, these gases form the basis of the atmosphere we see on Titan today.

CONCEPT CHECK

✔ What is unusual about Titan?

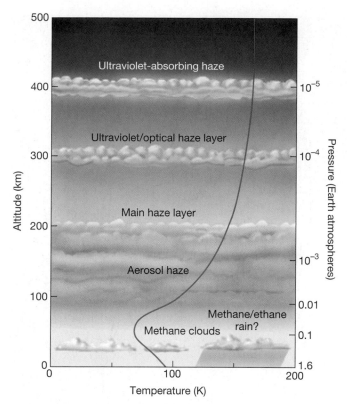

▲ FIGURE 12.22 Titan's Atmosphere The structure of Titan's atmosphere, as deduced from *Voyager 1* observations. The solid line represents temperature at different altitudes.

show any correlation with distance from Saturn. The densities of Saturn's midsized moons are all between 1000 and 1400 kg/m^3, implying that nearness to the central planetary heat source was a less important influence during their formation than it was in the Jupiter system. Scientists think that the midsized moons are composed largely of rock and water ice, as is Titan. Their densities are lower than Titan's primarily because their lesser masses produce less compression of their interiors.

The largest of the six, Rhea, has a mass only one-thirtieth that of Earth's Moon, and its icy surface is highly reflective and heavily cratered (Figure 12.25). At the low temperatures found on the surface of this moon, water ice is very hard and behaves rather like rock on the inner planets. For that reason, Rhea's surface craters look very much like craters on the Moon or Mercury. The density of craters is similar to that in the lunar highlands, indicating that the surface is old, and there is no evidence of extensive geological activity. Rhea's only real riddle is the presence of so-called *wispy terrain*—prominent light-colored streaks—on its trailing side (see Figure 12.24). The leading face, by contrast, shows no such markings, but only craters. Astronomers theorize that the wisps were caused by some event in the distant past during which water was released from the interior and condensed on the surface. Any similar markings on the leading side have presumably been obliterated by cratering, which should be more frequent on the satellite's forward-facing surface.

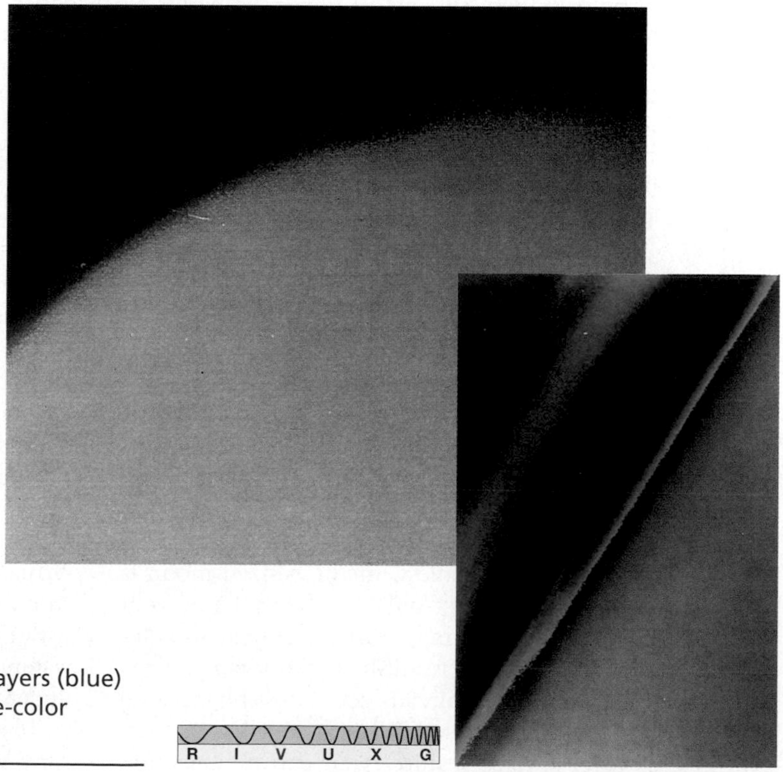

Saturn's Medium-Sized Moons

LEARNING GOAL 7 Saturn's complement of midsized moons consists (in order of increasing distance from the planet) of Mimas (at 3.1 planetary radii), Enceladus (4.0), Tethys (4.9), Dione (6.3), Rhea (8.7), and Iapetus (59.1). These moons are shown, to proper scale, in Figure 12.24. All six were known from Earth-based observations long before the Space Age. The inner five move on circular trajectories, and all are tidally locked into synchronous rotation (so that one side always faces the planet) by Saturn's gravity. They therefore all have permanently "leading" and "trailing" faces as they move in their orbits, a fact that is important in understanding their often asymmetrical surface markings.

Unlike the densities of the Galilean satellites of Jupiter, the densities of these six moons do not

▶ FIGURE 12.23 Haze Layers on Titan The haze layers (blue) of Titan's upper atmosphere are visible in these false-color *Voyager 1* images. *(NASA)*

R I V U X G

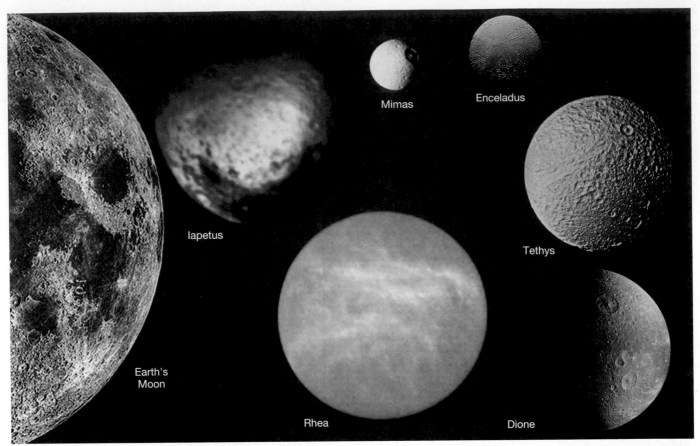

▲ **FIGURE 12.24 Moons of Saturn** Saturn's six medium-sized satellites, to scale. All are heavily cratered. Iapetus shows a clear contrast between its light-colored (icy) trailing surface (top and center in this image) and its dark leading hemisphere (at bottom). The light-colored "wisps" on much of Rhea are thought to be water and ice released from the moon's interior during some long-ago period of activity. Dione and Tethys show possible evidence of ancient geological activity. Enceladus appears to still be volcanically active, but the cause of the volcanism is unexplained. Mimas's main surface feature is the large crater Herschel, plainly visible at the upper right in this *Voyager 2* image; the impact that caused this crater must have come very close to shattering the moon. For scale, part of Earth's Moon is shown at left. *(NASA)*

Inside Rhea's orbit lie the orbits of Tethys and Dione. These two moons are comparable to each other in size and have masses somewhat less than half the mass of Rhea. Like Rhea, they have reflective surfaces that are heavily cratered, but each shows signs of surface activity, too. Dione's trailing face has prominent bright streaks that are probably similar to Rhea's wispy terrain. Dione also has "maria" of sorts, where flooding appears to have obliterated the older craters. The cracks on Tethys may have been caused by cooling and shrinking of the surface layers or, more probably, by meteoritic bombardment.

The innermost, and smallest, medium-sized moon is Mimas. Despite its low mass—only 1 percent the mass of Rhea—its closeness to the rings causes resonant interactions with the ring particles, resulting most notably in the Cassini division, as we have already seen. Possibly because of its proximity to the rings, Mimas is heavily cratered. The moon's chief surface feature is an enormous crater,

known as Herschel, on the leading face. The diameter of this crater is almost one-third that of the moon itself. The impact that formed Herschel must have come very close to destroying Mimas completely. It is quite possible that the debris produced by such impacts is responsible for creating or maintaining the spectacular rings we see.

Enceladus orbits just outside Mimas. Its size, mass, composition, and orbit are so similar to those of Mimas that one might guess that the two moons would also be similar to each other in appearance and history. However, this is not so. Enceladus is so bright and shiny—it reflects virtually 100 percent of the sunlight falling on it—that astronomers think that its surface must be completely coated with fine crystals of pure ice, which may be the icy "ash" of water "volcanoes," formed when liquid water emerged under pressure from the moon's interior. The moon bears visible evidence of large-scale volcanic activity of some sort. Much of its surface is devoid of impact craters, which

◄ **FIGURE 12.25** Rhea's Polar Cap The north polar region of Rhea, seen here at a superb resolution of only 1 km. The heavily cratered surface resembles that of the Moon, except that we see here craters in bright ice, rather than dark lunar rock. *(NASA)*

100 km

R I V U X G

seem to have been erased by what look like lava flows, except that the "lava" is water, temporarily liquefied during recent internal upheavals and now frozen again. Arguing that the processes involved may actually be more similar to the geothermal activity found in many volcanic regions on Earth, some astronomers prefer to describe these features as *geysers*, rather than volcanoes. Similar activity has been found on Neptune's moon Triton. ∞ (Sec. 13.6)

Although no volcanoes (or geysers) have actually been observed on Enceladus, there seems to be strong circumstantial evidence of volcanism on the satellite. In addition, the nearby thin cloud of small, reflective particles that makes up Saturn's E ring is known to be densest near Enceladus. Calculations indicate that the E ring is unstable because of the disruptive effects of the solar wind, supporting the view that volcanism on Enceladus continually supplies new particles to maintain the ring.

Why is there so much activity on a moon so small? No one knows. Attempts have been made to explain Enceladus's water volcanism in terms of tidal stresses. (Recall the role that Jupiter's tidal stresses play in creating volcanism on Io.) ∞ (Sec. 11.5) However, Saturn's tidal force on Enceladus is only one-quarter of the force exerted by Jupiter on Io, and there are no nearby large satellites to force Enceladus away from a circular trajectory. Thus, the ingredients that power Io's volcanoes may not be present on Enceladus. For now, the mystery of Enceladus's internal activity remains unresolved.

The outermost midsized moon is Iapetus (Figure 12.24). This moon orbits Saturn on a somewhat eccentric, inclined orbit with a semimajor axis of 3.6 million km. Its mass is about three-quarters that of Rhea. Iapetus is a two-faced moon. The dark, leading face reflects only about 3 percent of the sunlight reaching it, whereas the icy trailing side reflects 50 percent. Spectroscopic studies of the dark regions seem to indicate that the material originates on Ia-

petus, in which case the moon is not simply sweeping up dark material as it orbits. Similar dark deposits seen elsewhere in the solar system are thought to be organic (containing carbon) in nature; they can be produced by the action of solar radiation on hydrocarbon (e.g., methane) ice. But how the dark markings can adorn only one side of Iapetus in that case is still unknown.

The Small Satellites

Finally we come to Saturn's dozen or so small moons. Their masses are poorly known (they are inferred mainly from their gravitational effects on the rings), but they are thought to be similar in composition to the small moons of Jupiter. The outermost small moons, Hyperion and Phoebe, were discovered in the 19th century, in 1848 and 1898, respectively. The others were first detected in the second half of the 20th century. Only the moons in or near the rings themselves were actually discovered by the *Voyager* spacecraft.

Just 10,000 km beyond the F ring lie the so-called *co-orbital satellites* Janus and Epimetheus. As the name implies, these two satellites "share" an orbit, but in a very strange way. At any given instant, both moons are in circular orbits about Saturn, but one of them has a slightly smaller orbital radius than the other. Each satellite obeys Kepler's laws, so the inner satellite orbits slightly faster than the outer one and slowly catches up to it. The inner moon takes about four Earth years to "lap" the outer one. As the inner satellite gains ground on the outer one, a strange thing happens: As illustrated in Figure 12.26, when the two get close enough to begin to feel each other's weak gravity, they switch orbits—the new inner moon (which used to be the outer one) begins to pull away from its companion, and the whole process begins again! No one knows how the co-orbital satellites came to be engaged in this curious dance. Possibly they are portions of a single moon

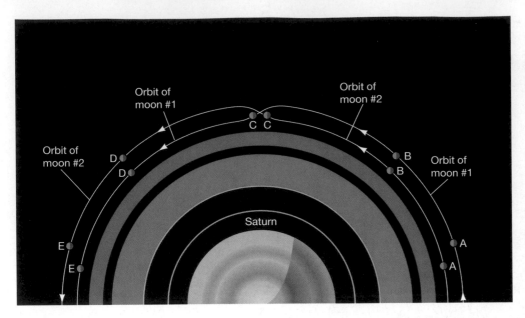

that broke up, perhaps after a meteoritic impact, leaving the two pieces in almost the same orbit.

In fact, several of the other small moons also share orbits, this time with larger moons. Telesto and Calypso have orbits that are synchronized with the orbit of Tethys, so that the two smaller moons always remain fixed relative to the larger moon, lying precisely 60° ahead of and 60° behind it as it travels around Saturn (see Figure 12.27). The moon Helene is similarly tied to Dione. These 60° points are known as **Lagrangian points**, after the French mathematician Joseph Louis Lagrange, who first studied them. Later we will see further examples of this special 1:1 orbital resonance in the motion of some asteroids about the Sun, trapped in the Lagrangian points of Jupiter's orbit.

The strangest motion of all is that of the moon Hyperion, which orbits between Titan and Iapetus, at a distance of 1.5 million km from the planet. Unlike most of Saturn's moons, Hyperion has a rotation that is not synchronous with its orbital motion. Because of the gravitational effect of Titan, Hyperion's orbit is not circular, so synchronous rotation cannot occur. In response to the competing gravitational influences of Titan and Saturn, this irregularly shaped satellite constantly changes both its rotational speed and its axis of rotation, in a condition known as *chaotic rotation*. As Hyperion orbits Saturn, it tumbles apparently at random, never stopping and never repeating itself, in a completely unpredictable way. Since the 1970s, the study of chaos on Earth has revealed new classes of unexpected behavior in even very simple systems. Hyperion is one of the few other places in the universe where this behavior has been unambiguously observed.

CONCEPT CHECK

✔ Why do Saturn's midsized moons show asymmetric surface markings?

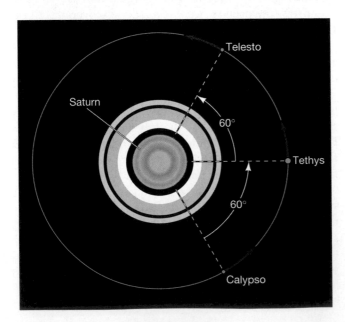

◄ **FIGURE 12.27 Synchronous Orbits** The orbits of the moons Telesto and Calypso are tied to the motion of the moon Tethys. The combined gravitational pulls of Saturn and Tethys keep the small moons exactly 60° ahead and behind the larger moon at all times, so all three moons share an orbit and never change their relative positions.

Chapter Review

SUMMARY

Saturn was the outermost planet known to ancient astronomers. Its rings and moons were not discovered until after the invention of the telescope. Saturn is smaller than Jupiter, but still much larger than any of the terrestrial worlds. Like Jupiter, Saturn rotates rapidly, producing a pronounced flattening, and displays differential rotation. Strong radio emission from the planet's magnetosphere allows the rotation rate of the interior to be determined.

As on Jupiter, weather systems are seen on Saturn, although they are less distinct. Short-lived storms are occasionally seen. Saturn has weaker gravity and a more extended atmosphere than Jupiter. The planet's overall butterscotch hue is due to cloud chemistry similar to that occurring in Jupiter's atmosphere. Saturn, like Jupiter, has bands, ovals, and turbulent flow patterns powered by convective motion in the interior.

Also like Jupiter, Saturn emits far more radiation into space than it receives from the Sun. Unlike Jupiter's, Saturn's excess energy emission is the result of **helium precipitation (p. 309)** in the planet's interior, where helium liquefies and forms droplets that then fall toward the center of the planet. This process is also responsible for Saturn's observed helium deficit. Saturn's interior is theoretically similar to that of Jupiter, but with a thinner layer of metallic hydrogen and a larger core. Its lower mass gives Saturn a less extreme core temperature, density, and pressure than Jupiter's core has. Saturn's conducting interior and rapid rotation produce a strong magnetic field and an extensive magnetosphere that contains the planet's ring system and the innermost 16 moons.

Saturn's rings lie in the planet's equatorial plane, so their appearance from Earth changes as Saturn orbits the Sun. From Earth, the main visible features of the rings are the **A, B,** and **C rings (p. 311)**, the **Cassini division (p. 311)**, and the **Encke gap (p. 311)**. The Cassini division is a dark region between the A and B rings. The Encke gap lies near the outer edge of the A ring. The rings are made up of trillions of icy particles ranging in size from dust grains to boulders, all orbiting Saturn like so many tiny moons. Their total mass is comparable to that of a small moon. Both divisions are dark because they are almost empty of ring particles. When the *Pioneer* and *Voyager* probes reached Saturn, they found that the rings were actually made up of tens of thousands of narrow **ringlets (p. 313)**. Interactions between the ring particles and the planet's inner moons are responsible for much of the fine structure observed in the main rings.

The **Roche limit (p. 312)** of a planet is the distance within which the planet's tidal field would overwhelm the internal gravity of a moon, tearing it apart and forming a ring. All known planetary ring systems lie inside their parent planets' Roche limits. Saturn's narrow **F ring (p. 315)**, discovered by *Pioneer 11*, lies just outside the A ring and has a kinked, braided structure, apparently caused by two small **shepherd satellites (p. 316)** that orbit close to the ring and prevent it from breaking up. Beyond the F ring is the faint, narrow **G ring (p. 316)**, also discovered by *Pioneer 11*. *Voyager 2* discovered the faint **D ring (p. 314)**, lying between the C ring and Saturn's cloud layer, and the **E ring (p. 314)**, associated with the moon Enceladus. Planetary rings may have lifetimes of only a few tens of millions of years. If so, the fact that we see rings around all four jovian planets means that they must constantly be re-formed or replenished, perhaps by material chipped off moons by meteoritic impact or by the tidal destruction of entire moons.

Saturn's single large moon Titan is the second-largest moon in the solar system. Its thick atmosphere obscures the moon's surface and may be the site of complex cloud and surface chemistry. The existence of Titan's atmosphere is a direct consequence of the cold conditions that prevailed at the time of the moon's formation.

The medium-sized moons of Saturn are made up predominantly of rock and water ice. They show a wide variety of surface terrains, are heavily cratered, and are tidally locked into synchronous orbits by the planet's gravity. The innermost mid-sized moon Mimas exerts an influence over the structure of the rings. The Cassini division, now known to contain faint ringlets and gaps, is the result of resonance between its particles and Mimas. The moon Iapetus has a marked contrast between its leading and trailing faces, while Enceladus has a highly reflective appearance, possibly the result of water "volcanoes" on its surface.

Saturn's small moons exhibit a wide variety of complex motion. Several moons "share" orbits, in some cases lying at the **Lagrangian points (p. 324)** 60° ahead of and 60° behind the orbit of a larger moon. The moon Hyperion tumbles in an unpredictable way as it orbits the planet.

REVIEW AND DISCUSSION

1. Seen from Earth, Saturn's rings sometimes appear broad and brilliant, but at other times seem to disappear. Why?

2. What is a ring crossing? When will the next one occur?

3. Why does Saturn have a less varied appearance than Jupiter?

4. What does Saturn's shape tell us about its deep interior?

5. Compare and contrast the atmospheres and weather systems of Saturn and Jupiter, and tell how the differences affect each planet's appearance.

6. Compare the thicknesses of Saturn's various layers (clouds, molecular hydrogen, metallic hydrogen, and core) with the equivalent layers in Jupiter. Why do the thicknesses differ?

7. What mechanism is responsible for the relative absence of helium in Saturn's atmosphere, compared with Jupiter's atmosphere?

8. Is Saturn as a whole deficient in helium relative to Jupiter?

9. When were Saturn's rings discovered? When did astronomers realize what they were?

10. What would happen to a satellite if it came too close to Saturn?

11. What evidence supports the idea that a relatively recent catastrophic event was responsible for Saturn's rings?

12. What effect does Mimas have on Saturn's rings?

13. What are shepherd satellites?

14. When *Voyager 1* passed Saturn in 1980, why didn't it see the surface of Titan, Saturn's largest moon?

15. Compare and contrast Titan with Jupiter's Galilean moons.

16. Why does Titan have a dense atmosphere, whereas other large moons in the solar system don't?

17. What is the evidence for geological activity on Enceladus?

18. What mystery is associated with Iapetus?

19. Describe the behavior of Saturn's co-orbital satellites.

20. Imagine what the sky would look like from Saturn's moon Hyperion. Would the Sun rise and set in the same way it does on Earth? How do you imagine Saturn might look?

CONCEPTUAL SELF-TEST: TRUE OR FALSE/MULTIPLE CHOICE

1. Saturn probably does not have a rocky core.

2. Unlike the weather on Jupiter, no storm systems have ever been seen on Saturn.

3. Relative to Jupiter's atmosphere, Saturn's atmosphere is deficient in helium.

4. Saturn is the only planet with a ring system.

5. The composition of Saturn's ring particles is predominantly water ice.

6. Although Saturn's ring system is tens of thousands of kilometers wide, it is only a few tens of meters thick.

7. Saturn's rings exist because they lie within the planet's Roche limit.

8. Two small shepherd satellites are responsible for the unusually complex form of Saturn's F ring.

9. Titan's surface is obscured by thick clouds of ammonia ice.

10. Astronomers think that Titan's surface may be covered with water.

11. From Figure 12.1 ("Ring Orientation"), the next time Saturn's rings will appear roughly edge-on as seen from Earth will be around (a) 2004; (b) 2010; (c) 2020; (d) 2035.

12. Compared with the time it takes Jupiter to orbit the Sun once, the time it takes Saturn, which is twice as far away, to orbit the Sun is (a) significantly less than twice as long; (b) about twice as long; (c) significantly more than twice as long.

13. Saturn's cloud layers are much thicker than those of Jupiter because Saturn has (a) more moons; (b) lower density; (c) a weaker magnetic field; (d) weaker surface gravity.

14. According to Figure 12.6 ("Saturn's Zonal Flow"), the winds on Saturn are fastest at (a) the north pole; (b) 50° N latitude; (c) the equator; (d) 50° S latitude.

15. Saturn's icy–rocky core is roughly (a) half the mass of; (b) the same mass as; (c) twice as massive as; (d) 10 times more massive than planet Earth.

16. Of the following, which are most like the particles found in Saturn's rings? (a) house-sized rocky boulders; (b) grains of silicate sand; (c) asteroids from the asteroid belt; (d) fist-sized snowballs.

17. A moon placed at a planet's Roche limit will (a) change color; (b) break into smaller pieces; (c) develop a magnetic field; (d) flatten into a disk.

18. The atmospheric pressure at the surface of Titan is (a) less than; (b) about the same as; (c) about one-and-a-half times greater than; (d) about 16 times greater than the atmospheric pressure at Earth's surface.

19. A tidally locked moon of Saturn (a) always presents the same face to the planet; (b) does not rotate; (c) always stays above the same point on the planet's surface; (d) maintains a constant distance from all the other moons.

20. The moons Telesto and Calypso, orbiting at the Lagrangian points of Saturn and the moon Tethys (a) orbit twice as far from Saturn as does Tethys; (b) orbit closer to Saturn than does Tethys; (c) always stay the same distance apart; (d) always stay between Saturn and the Sun.

PROBLEMS

Algorithmic versions of these questions are available in the Practice Problems module of the Companion Website at astro.prenhall.com/chaisson.

The number of squares preceding each problem indicates its approximate level of difficulty.

1. ■ What is the angular diameter of Saturn's A ring, as seen from Earth at closest approach?

2. ■ What is the size of the smallest feature visible in Saturn's rings, as seen from Earth at closest approach with a resolution of 0.05"?

3. ■■ What would be the mass of Saturn if it were composed entirely of hydrogen at a density of 0.08 kg/m³, the density of hydrogen at sea level on Earth? Assume for simplicity that Saturn is spherical. Compare your answer with Saturn's actual mass and with the mass of Earth.

4. ■ How long does it take for Saturn's equatorial flow, moving at 1500 km/h, to encircle the planet? Compare your answer with the wind-circulation time on Jupiter.

5. ■■ Use Stefan's law to calculate what the surface temperature on Saturn would be in the absence of any internal heat source if Saturn's actual surface temperature is 97 K and the planet radiates three times more energy than it receives from the Sun.

6. ■ On the basis of the data given in Sections 12.1 and 12.3 (Figure 12.8), estimate the average density of Saturn's core.

7. ■■ The text states that the total mass of material in Saturn's rings is about 10^{15} tons (10^{18} kg). Suppose the average ring particle is 6 cm in radius (the size of a large snowball) and has a density of 1000 kg/m³. How many ring particles are there?

8. ■ What is the orbital speed of ring particles at the inner edge of the B ring, in km/s? Compare your answer with the speed of a satellite in low Earth orbit (500 km altitude, say). Why are these speeds so different?

9. ■ Show that Titan's surface gravity is about one-seventh of Earth's, as stated in the text. What is Titan's escape speed?

10. ■■ Assuming a spherical shape and a uniform density of 2000 kg/m³, calculate how small an icy moon would have to be before a fastball pitched at 40 m/s (about 90 mph) could escape.

11. ■■■ Calculate the orbital radii of particles having the following properties: (a) a 3:1 orbital resonance with Tethys—that is, orbiting Saturn three times for every orbit of Tethys; (b) a 2:1 resonance with Mimas (two orbits for every orbit of Mimas); (c) a 3:2 resonance with Mimas (three orbits for every two of Mimas); (d) a 2:1 resonance with Dione.

12. ■■ Compare Saturn's tidal gravitational effect on Mimas with Mimas's own surface gravity.

13. ■■ Compare Saturn's tidal gravitational effect on Titan with Titan's own surface gravity. On the basis of these numbers and the corresponding numbers for Jupiter's Galilean moons, would you expect significant internal heating in Titan?

14. ■■■ Sunlight reflected back to Earth from a particle in Saturn's rings is Doppler shifted twice—first because of the relative motion between the source of the radiation (the Sun) and the ring particle and then again by the relative motion between the particle and the observer on Earth (see Section 3.5). As a result, if Earth, Saturn, and the Sun are roughly aligned (i.e., Saturn is near opposition), the observed Doppler shift corresponds to *twice* the particle's orbital speed. A certain solar spectral line, of wavelength 656.112 nm, is reflected from the rings and observed on Earth. If the rings happen to be seen almost edge-on, what is the line's observed wavelength in light reflected from (a) the approaching inner edge of the B ring? (b) the receding inner edge of the B ring? (c) the approaching outer edge of the A ring? and (d) the receding outer edge of the A ring?

15. ■■■ On the basis of the data given in the text, estimate the difference in orbital radii between Saturn's two co-orbital satellites.

In addition to the Practice Problems module, the Companion Website at astro.prenhall.com/chaisson provides for each chapter a study guide module with multiple choice questions as well as additional annotated images, animations, and links to related Websites.

13

Uranus, Neptune, and Pluto

The Outer Worlds of the Solar System

The three outermost planets were unknown to the ancients. All were discovered by telescopic observations: Uranus in 1781, Neptune in 1846, and Pluto in 1930. Uranus and Neptune have similar bulk properties, so it is natural to consider them together; they are part of the jovian family of planets. Pluto, by contrast, is not a jovian world. It is much smaller than even the terrestrial planets and generally seems much more moonlike than planetlike. Indeed, at one time, astronomers even speculated that Pluto was an erstwhile moon that had somehow escaped from one of the outer planets, most likely Neptune. However, it now seems more probable that Pluto is really the best-known representative of a newly recognized class of objects residing in the outer solar system. Whatever its origin, because of Pluto's location and its similarity to the jovian moons, we study it here along with its larger jovian neighbors.

LEARNING GOALS

Studying this chapter will enable you to

① Describe how both calculation and chance played major roles in the discoveries of the outer planets.

② Summarize the similarities and differences between Uranus and Neptune, and compare these planets with the other two jovian worlds.

③ Explain what the moons of the outer planets tell us about their past.

④ Contrast the rings of Uranus and Neptune with those of Jupiter and Saturn.

⑤ Summarize the orbital and physical properties of Pluto, and explain how the Pluto–Charon system differs fundamentally from all the other planets.

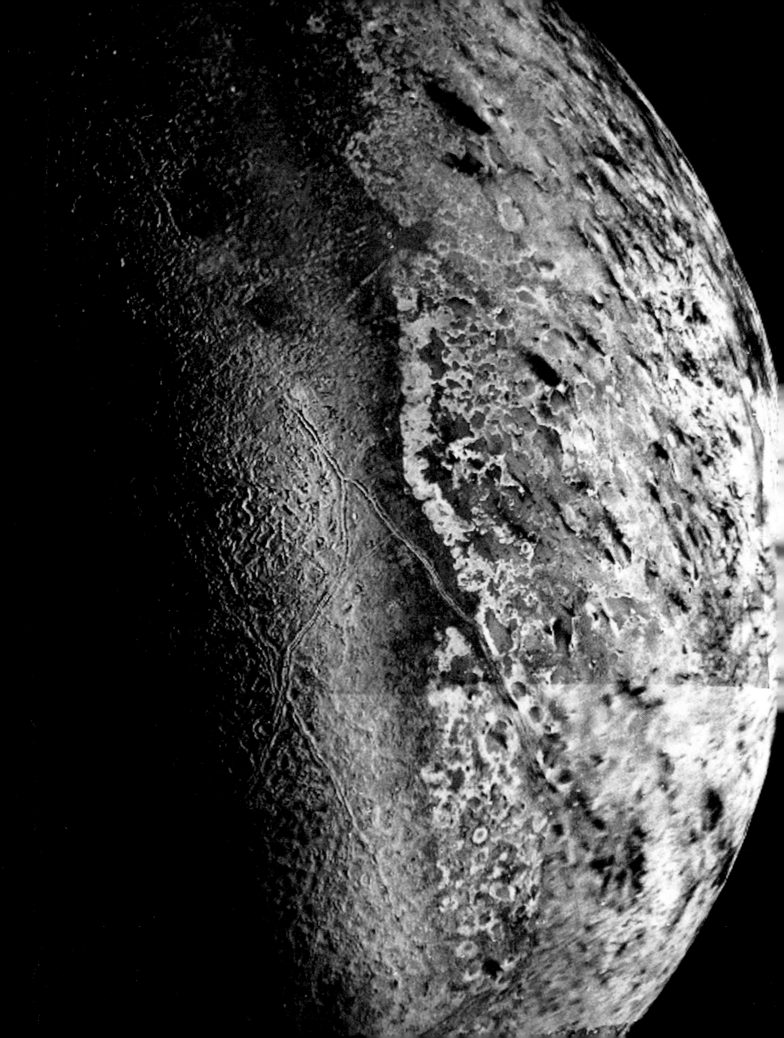

13.1 The Discovery of Uranus

The planet Uranus was discovered by British astronomer William Herschel in 1781. Herschel was engaged in charting the faint stars in the sky when he came across an odd-looking object that he described as "a curious either nebulous star or perhaps a comet." Repeated observations showed that it was neither. The object appeared as a disk in Herschel's 6-inch telescope and moved relative to the stars, but it traveled too slowly to be a comet. Herschel soon realized that he had found the seventh planet in the solar system.

Since this was the first new planet discovered in well over 2000 years, the event caused quite a stir at the time. The story goes that Herschel's first instinct was to name the new planet "Sidus Georgium" (Latin for "George's star"), after his king, George III of England. The world was saved from a planet named George by the wise advice of another astronomer, Johann Bode, who suggested instead that the tradition of using names from Greco–Roman mythology be continued and that the planet be named Uranus, after the father of Saturn.

Uranus is in fact just barely visible to the naked eye if you know exactly where to look. At opposition, it has a maximum angular diameter of 4.1″ and shines just above the unaided eye's threshold of visibility. It looks like a faint, undistinguished star. No wonder it went unnoticed by the ancients. Even today, few astronomers have seen it without a telescope.

Through a large Earth-based optical telescope (Figure 13.1), Uranus appears hardly more than a tiny pale greenish disk. With the flyby of *Voyager 2* in 1986, our knowledge of Uranus increased dramatically, although close-up

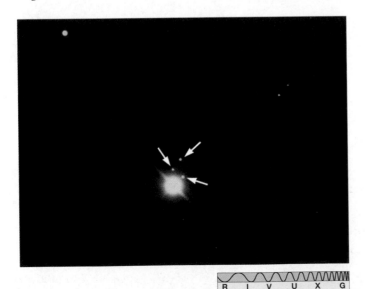

▲ **FIGURE 13.1 Uranus from Earth** Details are virtually invisible on photographs of Uranus made with large Earth-based telescopes. (Arrows point to three of the planet's moons.) (*UC/Lick Observatory*)

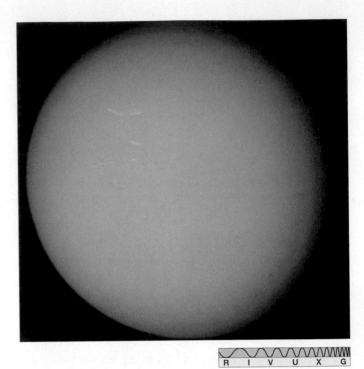

▲ **FIGURE 13.2 Uranus, Close Up** This image of Uranus, taken from a distance of about 1 million km, was sent back to Earth by the *Voyager 2* spacecraft as it whizzed past the giant planet at 10 times the speed of a rifle bullet. The image approximates the planet's true color, but shows virtually no detail in the largely featureless upper atmosphere, except for a few wispy clouds in the northern hemisphere. (*NASA*)

images of the planet still showed virtually no surface detail (Figure 13.2). The apparently featureless atmosphere of Uranus contrasts sharply with the bands and spots visible on all the other jovian worlds. Some orbital and physical properties of Uranus are presented in the Uranus Data box on p. 334.

13.2 The Discovery of Neptune

Once Uranus was discovered, astronomers set about charting its orbit and quickly discovered a small discrepancy between the planet's predicted position and where they actually observed it. Try as they might, astronomers could not find an elliptical orbit that fit the planet's trajectory to within the accuracy of their measurements. Half a century after Uranus's discovery, the discrepancy had grown to a quarter of an arc minute, far too big to be explained away as observational error.

The logical conclusion was that an unknown body must be exerting a gravitational force on Uranus—much weaker than that of the Sun, but still measurable. But what body could that be? Astronomers realized that there had to be *another* planet in the solar system perturbing Uranus's motion.

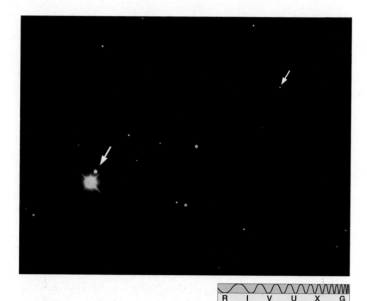

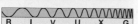

▲ **FIGURE 13.3 Neptune from Earth** Neptune and two of its moons, Triton (left arrow) and Nereid (right), imaged with a large Earth-based telescope. *(UC/Lick Observatory)*

In the 1840s, two mathematicians independently solved the difficult problem of determining the new planet's mass and orbit. A British astronomer, John Adams, reached the solution in September 1845; in June of the following year, the French mathematician Urbain Leverrier came up with essentially the same answer. British astronomers seeking the new planet found nothing during the summer of 1846. In September, a German astronomer named Johann Galle began his own search from the Berlin Observatory, using a newly completed set of more accurate sky charts. He found the new planet within one or two degrees of the predicted position—on his first attempt. After some wrangling over names and credits, the new planet was named Neptune, and Adams and Leverrier (but not Galle!) are now jointly credited with its discovery.

Neptune's orbital and physical properties are listed in the Neptune Data box on p. 343. With an orbital semimajor axis of 30.1 A.U. and an orbital period of 163.7 Earth years, Neptune has not yet completed one revolution since its discovery. Unlike Uranus, distant Neptune cannot be seen with the naked eye, although it can be seen with a small telescope—in fact, according to his notes, Galileo might actually have seen Neptune, although he had no idea what it really was at the time. Through a large telescope, Neptune appears as a bluish disk, with a maximum angular diameter of 2.4″ at opposition.

Figure 13.3 shows a long Earth-based exposure of Neptune and its largest moon, Triton. Neptune is so distant that surface features on the planet are virtually impossible to discern. Even under the best observational conditions, only a few markings can be seen. These features are suggestive of multicolored cloud bands, with light bluish hues seeming to dominate. With *Voyager 2*'s arrival, much more detail emerged, as shown in Figure 13.4. Superficially, at least, Neptune resembles a blue-tinted Jupiter, with atmospheric bands and spots clearly evident.

CONCEPT CHECK

✔ How did observations of the orbit of Uranus lead to the discovery of Neptune?

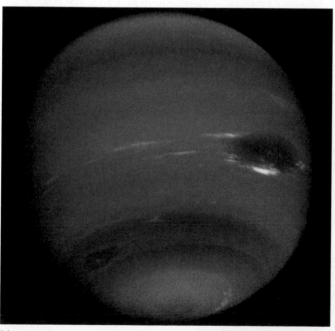

◀ **FIGURE 13.4 Neptune, Close Up** (a) Neptune as seen by *Voyager 2*, from a distance of roughly 1 million km. A closer view (b), resolved to about 10 km, shows cloud streaks ranging in width from 50 km to 200 km. *(NASA)*

(a)

(b)

13.3 Physical Properties of Uranus and Neptune

Figure 13.5 shows Uranus and Neptune to scale, along with Earth for comparison. The two giant planets are quite similar in their bulk properties. The radius of Uranus is 4.0 times that of Earth, that of Neptune 3.9 Earth radii. Their masses (first determined from terrestrial observations of their larger moons and later refined by *Voyager 2*) are 14.5 Earth masses for Uranus and 17.1 Earth masses for Neptune. Uranus's average density is 1300 kg/m³, and Neptune's is 1600 kg/m³. These densities imply that large rocky cores constitute a greater fraction of the planets' masses than do the cores of either Jupiter or Saturn. The cores themselves are probably comparable in size, mass, and composition to those of the two larger giants.

Like the other jovian planets, Uranus has a short rotation period. Earth-based observations of the Doppler shifts in spectral lines first indicated that Uranus's "day" was between 10 and 20 hours long. The precise value of the planet's rotation period—accurately determined when *Voyager 2* timed radio signals associated with Uranus's magnetosphere—is now known to be 17.2 hours. Again, as with Jupiter and Saturn, the planet's atmosphere rotates differentially. However, Uranus's atmosphere actually rotates *faster* at the poles (where the period is 14.2 hours) than near the equator (where the period is 16.5 hours).

Each planet in our solar system seems to have some outstanding peculiarity, and Uranus is no exception. Un-like all the other planets, whose spin axes are roughly perpendicular to the plane of the ecliptic, Uranus's axis of rotation lies almost within that plane—98° from the perpendicular, to be precise. (Because the north pole lies below the ecliptic plane, the rotation of Uranus, like that of Venus, is classified as retrograde.) We might say that, relative to the other planets, Uranus lies tipped over on its side. As a result, the "north" (spin) pole of Uranus, at some time in its orbit, points almost directly toward the Sun.* Half a "year" later, its "south" pole faces the Sun, as illustrated in Figure 13.6. When *Voyager 2* encountered the planet in 1986, the north pole happened to be pointing nearly at the Sun, so it was midsummer in the northern hemisphere.

The strange orientation of Uranus's rotation axis produces some extreme seasonal effects. Starting at the height of northern summer, when the north pole points closest to the Sun, an observer near that pole would see the Sun move in gradually increasing circles in the sky, completing one circuit (counterclockwise) every 17 hours and dipping slightly lower in the sky each day. Eventually, the Sun would begin to set and rise again in a daily cycle, and the nights would grow progressively longer with each passing day. Twenty-one Earth years after the summer solstice, the autumnal equinox would occur, with day and night each 8.5 hours long.

As in Chapter 9, we adopt the convention that a planet's rotation is always counterclockwise as seen from above the north pole (i.e., planets always rotate from west to east). ∞ (Sec. 9.2)

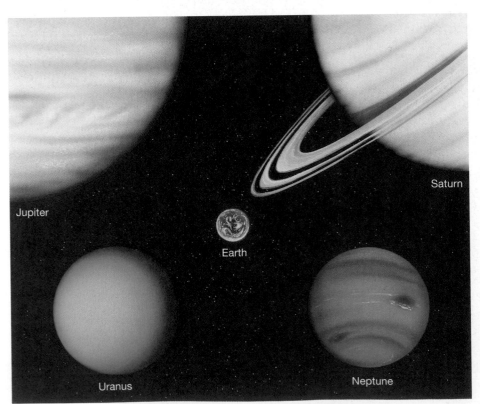

◀ **FIGURE 13.5 Jovian Planets** Jupiter, Saturn, Uranus, and Neptune, drawn to scale and compared with Earth. Uranus and Neptune are quite similar in their bulk properties. Each probably contains a core about 10 times more massive than Earth. Jupiter and Saturn are each substantially larger, but their rocky cores are probably comparable in mass to those of Uranus and Neptune. *(NASA)*

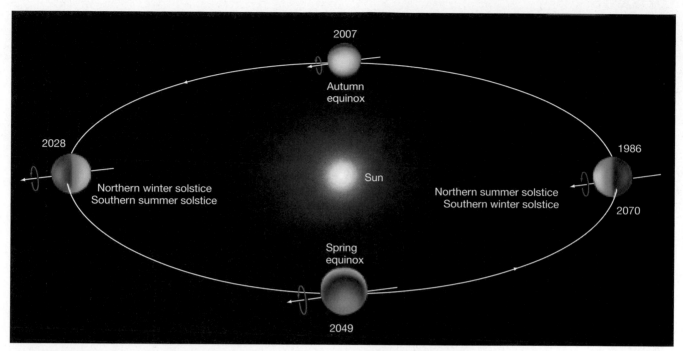

▲ **FIGURE 13.6 Seasons on Uranus** The 98° axial tilt of Uranus places its equator almost perpendicular to the ecliptic. As a result, the planet is characterized by the most extreme seasons known in the solar system. The equatorial regions have two "summers" (warm seasons around the times of the two equinoxes) and two "winters" (cold seasons at the solstices) each year; the poles are alternately plunged into darkness for 42 Earth years at a time.

The days would continue to shorten, until one day the Sun would fail to rise at all. The ensuing period of total darkness would be equal in length to the earlier period of constant daylight, plunging the northern hemisphere into the depths of winter. Eventually, the Sun would rise again; the days would lengthen through the vernal equinox and beyond, and in time the observer would again experience a long summer of uninterrupted (though dim) sunshine. From the point of view of an observer on the equator, by contrast, summer and winter would be almost equally cold seasons, with the Sun never rising far above the horizon. Spring and fall would be the warmest times of year, with the Sun passing almost overhead each day.

No one knows why Uranus is tilted in this way. Some scientists have speculated that a catastrophic event, such as a glancing collision between the planet and another planet-sized body, might have altered Uranus's spin axis. There is no direct evidence for such an occurrence, however, and no theory to tell us how we should seek to confirm it.

Neptune's clouds show more variety and contrast than do those of Uranus, and Earth-based astronomers studying them determined a rotation rate for Neptune even before *Voyager 2*'s flyby in 1989. The average rotation period of Neptune's atmosphere is 17.3 hours (quite similar to that of Uranus). Measurements of Neptune's radio emissions by *Voyager 2* showed that the magnetic field of the planet, and presumably also its interior, rotates once every

16.1 hours. Thus, Neptune is unique among the jovian worlds in that its atmosphere rotates *more slowly* than its interior. Neptune's axis of rotation is inclined 29.6° to a line perpendicular to the planet's orbital plane, quite similar to the 27° tilt of Saturn.

CONCEPT CHECK

✔ What is unusual about the rotation of Uranus?

13.4 The Atmospheres of Uranus and Neptune

Composition

Spectroscopic studies of sunlight reflected from Uranus's and Neptune's dense clouds indicate that the two planets' outer atmospheres (the parts we actually measure spectroscopically) are quite similar to the atmospheres of Jupiter and Saturn. The most abundant element is molecular hydrogen (84 percent), followed by helium (about 14 percent) and methane, which is more abundant on Neptune (about 3 percent) than on Uranus (2 percent). Ammonia, which plays such an important role in the Jupiter and Saturn systems, is not present in any significant quantity in the outermost jovian worlds.

PLANETARY DATA:

URANUS

Orbital semimajor axis	19.19 A.U. 2871 million km
Orbital eccentricity	0.047
Perihelion	18.29 A.U. 2736 million km
Aphelion	20.10 A.U. 3006 million km
Mean orbital speed	6.80 km/s
Sidereal orbital period	83.75 tropical years
Synodic orbital period	369.66 solar days
Orbital inclination to the ecliptic	0.77°
Greatest angular diameter, as seen from Earth	4.1″
Mass	8.68×10^{25} kg 14.54 (Earth = 1)
Equatorial radius	25,559 km 4.01 (Earth = 1)
Mean density	1271 kg/m³ 0.230 (Earth = 1)
Surface gravity (at cloud tops)	8.87 m/s² 0.91 (Earth = 1)
Escape speed	21.3 km/s
Sidereal rotation period	−0.72 solar day (retrograde)
Axial tilt	97.92°
Surface magnetic field	0.74 (Earth = 1)
Magnetic axis tilt relative to rotation axis	58.6°
Mean surface temperature	58 K
Number of moons	27 (more than 10 km in diameter) 27 (total)

The abundances of gaseous ammonia and methane vary systematically among the jovian planets. Jupiter has much more gaseous ammonia than methane, but moving outward from the Sun, we find that the more distant planets have steadily decreasing amounts of ammonia and relatively greater amounts of methane. The reason for this variation is temperature. Ammonia gas freezes into ammonia ice crystals at about 70 K. This temperature is cooler than the cloud-top temperatures of Jupiter and Saturn, but warmer than those of Uranus (58 K) and Neptune (59 K). Thus, the outermost jovian planets have little or no *gaseous* ammonia in their atmospheres, so their spectra (which record atmospheric gases only) show only traces of ammonia.

The increasing amounts of methane are largely responsible for the outer jovian planets' blue coloration. Methane absorbs long-wavelength red light quite efficiently, so sunlight reflected from the planets' atmospheres is deficient in red and yellow photons and appears bluish-green or blue. As the concentration of methane increases, the reflected light should appear bluer—just the trend that is observed: Uranus, with less methane, looks bluish-green, while Neptune, with more, looks distinctly blue.

Weather

Voyager 2 detected just a few cloud features in Uranus's atmosphere (Figure 13.2), and even those became visible only after extensive computer enhancement. Figure 13.7 shows some more recent *Hubble Space Telescope* views of the planet. Parts (a) through (c) are heavily processed optical images that show the progress of a pair of bright clouds around the planet. Part (d) shows a false-color, near-infrared rendition of Uranus. The colors in this image generally indicate the depth to which we can see into the atmosphere. Blue-green regions are clear atmospheric regions where astronomers can study conditions down to the lower cloud levels. Yellow-gray colors show sunlight reflecting from higher cloud layers or from atmospheric haze. Orange-red colors, such as the prominent "spots" on the south (right) edge of this image, indicate very high clouds, much like the wispy, white cirrus clouds often seen at high altitudes in Earth's atmosphere. Like cirrus clouds on Earth, these Uranian clouds are made up predominantly of ice crystals, formed in the planet's cold upper atmosphere.

Uranus apparently lacks any significant internal heat source, and because the planet has a low surface temperature, its clouds are found only at low-lying, warmer levels in the atmosphere. The absence of high-level clouds means that we must look deep into the planet's atmosphere to see any structure, so the bands and spots that characterize flow patterns on the other jovian worlds are largely "washed out" on Uranus by intervening stratospheric haze.

From computer-processed images such as those shown in Figure 13.7, astronomers have learned that Uranus's atmospheric clouds and flow patterns move around the planet in the same sense as the planet's rota-

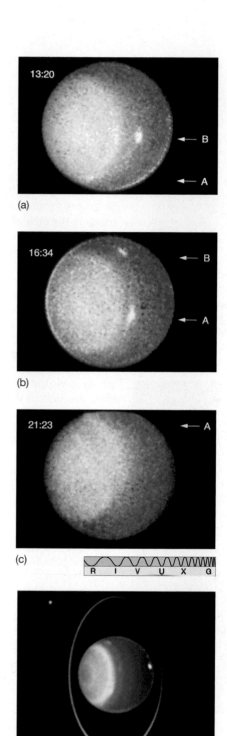

(a)

(b)

(c)

R   I   V   U   X   G

(d)

R   I   V   U   X   G

▲ **FIGURE 13.7 Uranus's Rotation** (a), (b), and (c) These computer-enhanced *Hubble Space Telescope* images, taken at roughly four-hour intervals, show the motion of a pair of bright clouds (labeled A and B) in the planet's southern hemisphere. (The numbers at the top give the time of each photo.) (d) A near-infrared image of Uranus, also from the *Hubble* instrument, shows the planet's ring system, as well as a number of clouds (pink and red regions) in the upper atmosphere. The clouds circle the planet at speeds of up to 500 km/h. *(NASA)*

tion, with wind speeds ranging from 200 to 500 km/h. In fact, tracking these clouds allowed Uranus's differential rotation, mentioned earlier, to be measured. Despite the odd angle at which sunlight is currently striking the surface (recall that it is now late summer in the northern hemisphere), the planet's rapid rotation still channels the wind flow into bands reminiscent of those found on Jupiter and Saturn. Wind speeds are greater near the north pole, possibly because that part of the planet currently receives the greatest amount of solar heating. Even though the predominant wind flow is in the east–west direction, the atmosphere seems to be quite efficient at transporting energy from the heated north to the unheated southern hemisphere. Although much of the south is currently in total darkness, the temperature there is only a few kelvins less than in the north.

Neptune's clouds and band structure are much more easily seen than Uranus's. Although Neptune lies at a greater distance from the Sun, the planet's upper atmosphere is actually slightly warmer than that of Uranus. Like Jupiter and Saturn, but unlike Uranus, Neptune has an internal energy source—in fact, Neptune radiates 2.7 times more heat than it receives from the Sun. The cause of this heating is still uncertain. Some scientists have suggested that Neptune's excess methane has helped "insulate" the planet, tending to maintain its initially high internal temperature. If that is so, then the source of Neptune's internal heat is the same as Jupiter's: energy left over from the planet's formation. ∞ (Sec. 11.3) The combination of extra heat and less haze may be responsible for the greater visibility of Neptune's atmospheric features (see Figure 13.8), as its cloud layers lie at higher levels in the atmosphere than do those of Uranus.

Neptune sports several storm systems similar in appearance to those seen on Jupiter (and assumed to be produced and sustained by the same basic processes). The largest such storm, known simply as the **Great Dark Spot**, is shown in Figure 13.8(a). ∞ (Sec. 11.2) Discovered by *Voyager 2* in 1989, the spot was about the size of Earth, was located near the planet's equator, and exhibited many of the same general characteristics as the Great Red Spot on Jupiter. The flow around it was counterclockwise, as with the Red Spot, and there appeared to be turbulence where the winds associated with the Great Dark Spot interacted with the zonal flow to its north and south. The flow around this and other dark spots may drive updrafts to high altitudes, where methane crystallizes out of the atmosphere to form high-lying cirrus clouds—those visible in Figure 13.8(a) lie some 50 km above the main cloud tops. Astronomers did not have long to study the Dark Spot's properties, however: As shown in Figure 13.8(b), when the *Hubble Space Telescope* viewed Neptune after the mid-1990s, the spot had vanished, although several new storms (bright spots) had appeared.

Infrared views such as those shown in Figure 13.8(b) reveal Neptune's dynamic weather patterns. The planet's weather can change in as little as a few rotation periods,

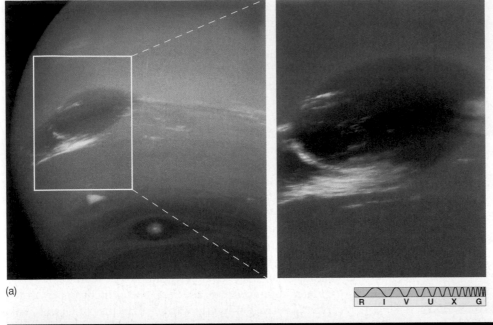

(a)

R I V U X G

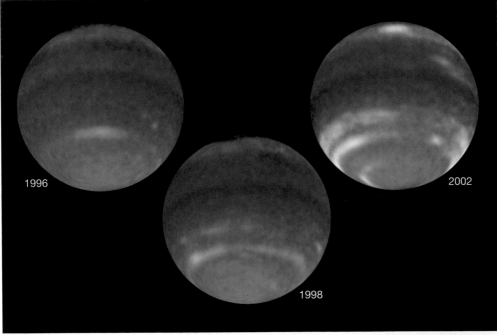

1996

1998

2002

(b)

R I V U X G

◄ **FIGURE 13.8** Neptune's **Dark Spot** (a) Close-up views, taken by *Voyager 2* in 1989, of the Great Dark Spot of Neptune, a large storm system in the planet's atmosphere, possibly similar in structure to Jupiter's Great Red Spot. Resolution in the photo on the right is about 50 km; the entire dark spot is roughly the size of planet Earth. (b) These *Hubble Space Telescope* views of Neptune were taken in 1996, 1998, and 2002, with the planet some 4.5 billion km from Earth. The cloud features (mostly methane ice crystals) are tinted pink here because they were imaged in the infrared, but are really white in visible light. By the time the first of these images was taken, the Great Dark Spot had disappeared. Note that Neptune appears to be getting brighter over the period shown here—apparently, the planet's southern hemisphere is responding to the tiny, but increasing, amount of sunshine during southern spring. *(NASA)*

and winds blow at speeds in excess of 1500 km/h—almost half the speed of sound in Neptune's upper atmosphere—with storms the size of Earth more the rule than the exception. The planet's stormy disposition is well established, but very difficult to understand. On Earth, weather systems are driven by the heat of the Sun. However, Neptune lies far from the Sun, in the outer solar system, and the Sun's heating effect is minuscule—nearly a thousand times less than at Earth. How can Neptune be so cold, yet so active?

Intriguingly, the three images of Neptune shown in Figure 13.8(b) reveal that the planet's southern hemi-sphere has brightened significantly over the period shown. Apparently, despite the Sun's faint heating, the planet is responding to the increase in solar energy as its southern half slowly moves from winter into spring.

CONCEPT CHECK

✔ Why are planetary scientists puzzled by the strong winds and rapidly changing storm systems on Neptune?

13.5 Magnetospheres and Internal Structure

2 *Voyager 2* found that both Uranus and Neptune have fairly strong internal magnetic fields—about a hundred times stronger than Earth's field and one-tenth as strong as Saturn's. However, because Uranus and Neptune are so much larger than Earth, the magnetic fields at the cloud tops—spread out over far larger volumes than is the field on Earth—are actually comparable in strength to Earth's field. Uranus and Neptune each have substantial magnetospheres, populated largely by electrons and protons either captured from the solar wind or created from ionized hydrogen gas escaping from the planets themselves.

When *Voyager 2* arrived at Uranus, it discovered that the planet's magnetic field was tilted at about 60° to the axis of rotation. On Earth, such a tilt would put the north magnetic pole somewhere in the Caribbean. Furthermore, on Uranus, the magnetic field lines are *not* centered on the planet. It is as though Uranus's field were due to a bar magnet that is tilted with respect to the planet's rotation axis and displaced from the center by about one-third the radius of the planet. Figure 13.9 shows the magnetic field structures of the four jovian planets, with Earth's also shown for comparison. The locations and orientations of the bar magnets represent the observed planetary fields, and the sizes of the bars indicate magnetic field strength.

Because dynamo theories generally predict that a planet's magnetic axis should be roughly aligned with its

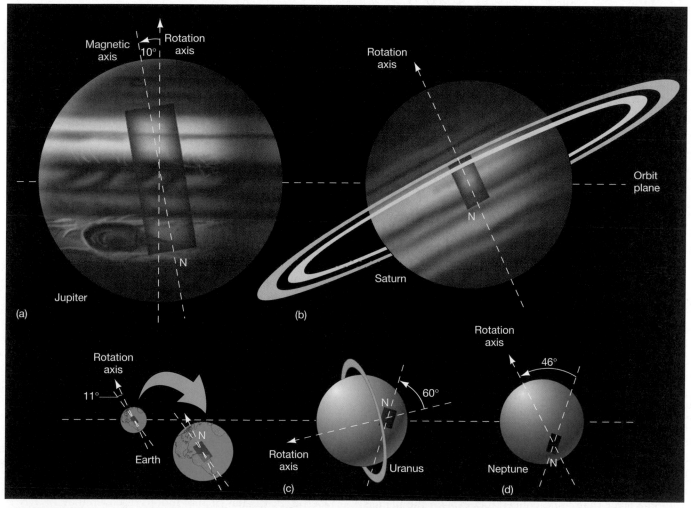

▲ **FIGURE 13.9 Jovian Magnetic Fields** A comparison of the magnetic field strengths, orientations, and offsets in the four jovian planets: (a) Jupiter, (b) Saturn, (c) Uranus, and (d) Neptune. The planets are drawn to scale, and in each case the magnetic field is represented as though it came from a bar magnet (a simplification, for purposes of illustration only). The size and location of each magnet represent the strength and orientation of the planetary field. Notice that the fields of Uranus and Neptune are significantly offset from the center of the planet and are inclined considerably to the planet's axis of rotation. Earth's magnetic field is shown for comparison. One end of each magnet is marked N to indicate the polarity of Earth's field.

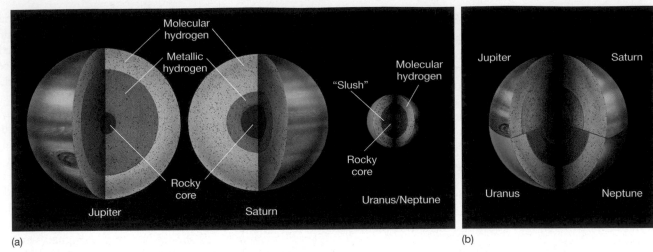

▲ FIGURE 13.10 Jovian Interiors A comparison of the interior structures of the four jovian planets. (a) The planets drawn to scale. (b) The relative proportions of the various internal zones.

rotation axis—as on Earth, Jupiter, Saturn, and the Sun—the misalignment on Uranus suggested to some researchers that perhaps the planet's field had been caught in the act of reversing. ∞ (Sec. 7.5) Another possibility was that the oddly tilted field was in some way related to the planet's axial tilt—perhaps one catastrophic collision skewed both axes at the same time. Those ideas evaporated in 1989 when *Voyager 2* found that Neptune's field is also inclined to the planet's axis of rotation, at an angle of 46° (see Figure 13.9d), and also substantially offset from the center of the planet. It now appears that the internal structures of Uranus and Neptune are different from those of Jupiter and Saturn, and this difference changes how the former planets' magnetic fields are generated.

Theoretical models indicate that Uranus and Neptune have rocky cores similar to those found in Jupiter and Saturn—about the size of Earth and perhaps 10 times more massive. However, the pressure outside the cores of Uranus and Neptune (unlike the pressure within Jupiter and Saturn) is too low to force hydrogen into the metallic state, so hydrogen stays in its molecular form all the way into the planets' cores. Astronomers theorize that deep below the cloud layers, Uranus and Neptune may have high-density, "slushy" interiors containing thick layers of water clouds. It is also possible that much of the planets' ammonia is dissolved in the hypothetical water, accounting for the absence of ammonia at higher levels. Such an ammonia solution would provide a thick, electrically conducting ionic layer that could conceivably explain the planets' misaligned magnetic fields if the circulating electrical currents that generate the fields occur mainly in regions far from the planets' centers and rotation axes.

At present, we simply don't know enough about the interiors of Uranus and Neptune to assess the correctness of this picture. Our current state of knowledge is summarized in Figure 13.10, which compares the internal structures of the four jovian worlds.

CONCEPT CHECK

✔ What is odd about the magnetic fields of Uranus and Neptune?

13.6 The Moon Systems of Uranus and Neptune

LEARNING GOAL 3 Like Jupiter and Saturn, both Uranus and Neptune have extensive moon systems, each consisting of a few large moons, long known from ground-based observations, and many smaller moonlets, discovered by *Voyager 2* or recently detected from Earth.

Uranus's Moons

As of early 2004, some 27 moons are known to orbit Uranus. The properties of those more than 25 km in diameter are listed in Table 13.1.

William Herschel discovered and named Titania and Oberon, the two largest of Uranus's five major moons, in 1789. British astronomer William Lassell found Ariel and Umbriel, the next-largest moons, in 1851. Gerard Kuiper found Miranda, the smallest, in 1948. In order of increasing distance from the planet, they are Miranda (at 5.1 planetary radii), Ariel (7.5), Umbriel (10.4), Titania (17.1), and Oberon (22.8). Ten smaller moons discovered by *Voyager 2* all lie inside the orbit of Miranda. Many of them are intimately related to the Uranian ring system. All of these moons revolve in the planet's skewed equatorial plane, almost perpendicular to the ecliptic, in circular, tidally locked orbits, sharing their parent's extreme seasons.

Of the remaining 12 moons, one, orbiting close to the planet, was found after careful reanalysis of *Voyager 2* images. All the rest were discovered via systematic ground-based searches made since 1997, with techniques similar to those which have been so successful in identifying new moons of Jupiter and Saturn. ∞ (*More Precisely 11-3, Sec. 12.5*) These small bodies orbit far from Uranus, mostly on retrograde, highly inclined orbits. Like the outer moons of Jupiter and Saturn, and like Phobos and Deimos of Mars, each is thought to be interplanetary debris captured following a glancing encounter with the planet's atmosphere.

The five largest Uranian moons are similar in many respects to the six midsized moons of Saturn. ∞ (*Sec. 12.5*) Their densities lie in the range from 1100 to 1700 kg/m³, suggesting a composition of ice and rock, like Saturn's moons, and their diameters range from 1600 km for Titania and Oberon, to 1200 km for Umbriel and Ariel, to 480 km for Miranda. Uranus has no moons comparable to the Galilean satellites of Jupiter or to Saturn's single large moon, Titan. Figure 13.11 shows Uranus's five large moons to scale, along with Earth's Moon and Neptune's only midsized moon (named Proteus) for comparison.

The outermost of the five moons, Titania and Oberon, are heavily cratered and show little indication of geological activity. Their overall appearance (and quite possibly their history) is comparable to that of Saturn's moon Rhea, except that they lack Rhea's wispy streaks. Also, like all Uranian moons, they are considerably less reflective than Saturn's satellites, suggesting that their icy surfaces are quite dirty.

One possible reason for the lesser reflectivity may simply be that the planetary environment in the vicinity of Uranus and Neptune contains more small "sooty" particles than do the parts of the solar system that are closer to the Sun. An alternative explanation, now considered more likely by many planetary scientists, cites the effects of radiation and high-energy particles that strike the surfaces of these moons. The impacts tend to break up the molecules on the moons' surfaces, eventually leading to chemical reactions that slowly build up a layer of dark, organic material. This **radiation darkening** is thought to contribute to the generally darker coloration of many of the moons and rings in the outer solar system. In either case, the

▲ **FIGURE 13.11 Moons of Uranus and Neptune** The five largest moons of Uranus, and Proteus, the sole midsized moon of Neptune, are shown to scale, with part of Earth's Moon (also to scale) for comparison. In order of increasing distance from the planet, the Uranian moons are Miranda, Ariel, Umbriel, Titania, and Oberon. The appearance, structure, and history of Titania and Oberon may be quite similar to those of Saturn's moon Rhea. The smallest details visible on both moons are about 15 km across. Umbriel is one of the darkest bodies in the solar system, although it has a bright white spot on its sunward side. Ariel is similar in size, but has a brighter surface, with signs of past geological activity. The resolution is approximately 10 km. (*NASA; Lick Observatory*)

longer a moon has been inactive and untouched by meteoritic impact, the darker its surface should be.

The darkest of the moons of Uranus is Umbriel. This moon displays little evidence of any past surface activity; its only mark of distinction is a bright spot about 30 km across, of unknown origin, in its northern hemisphere. By contrast, Ariel, similar in size to Umbriel, but closer to Uranus, does appear to have undergone some activity in the past. Ariel shows signs of resurfacing in places and exhibits surface cracks a little like those seen on another of Saturn's moons, Tethys. However, unlike Tethys, whose cracks are probably due to meteoritic impact, Ariel's activity likely occurred when internal forces and external tidal stresses (due to the gravitational pull of Uranus) distorted the moon and cracked its surface.

Strangest of all of Uranus's icy moons is Miranda, shown in Figure 13.12. Before the *Voyager 2* encounter, astronomers expected that Miranda would resemble Mimas, the moon of Saturn whose size and location it most closely approximates. However, instead of being a relatively uninteresting, cratered, geologically inactive world, Miranda displays a wide range of surface terrains, including ridges, valleys, large oval faults, and many other tortuous geological features.

To explain why Miranda seems to combine so many different types of surface features, some researchers have hypothesized that this baffling object has been catastrophically disrupted several times (from within or without), with the pieces falling back together in a chaotic, jumbled way. Certainly, the frequency of large craters on the outer moons suggests that destructive impacts may once have been quite common in the Uranian system. It will be a long time, though, before we can obtain more detailed information to test this theory.

Neptune's Moons

From Earth, we can see only two moons orbiting Neptune. William Lassell discovered Triton, the inner moon, in 1846. The moon Nereid was located by Gerard Kuiper in

TABLE 13.1 The Major Moons of Uranus*

Name	Distance from Uranus (km)	Distance from Uranus (planetary radii)	Orbital Period (days)	Size (longest diameter, km)	Mass** (Earth Moon masses)	Density (kg/m³)	Density (g/cm³)
Cordelia	49,800	1.95	0.34	26			
Ophelia	53,800	2.10	0.38	32			
Bianca	59,200	2.31	0.43	44			
Cressida	61,800	2.42	0.46	66			
Desdemona	62,700	2.45	0.47	58			
Juliet	64,400	2.52	0.49	84			
Portia	66,100	2.59	0.51	110			
Rosalind	69,900	2.74	0.56	58			
Belinda	75,300	2.94	0.62	68			
Puck	86,000	3.36	0.76	150			
Miranda	130,000	5.08	1.41	480	0.00090	1100	1.1
Ariel	191,000	7.48	2.52	1160	0.018	1600	1.6
Umbriel	266,000	10.4	4.14	1170	0.016	1400	1.4
Titania	436,000	17.1	8.71	1580	0.048	1700	1.7
Oberon	583,000	22.8	13.5	1520	0.041	1600	1.6
Caliban (S/1997U1)	7,231,000	283	−580‡	100			
Sycorax (S/1997 U2)	12,179,000	477	−1290‡	190			
Prospero (S/1999 U3)	16,256,000	636	−1980‡	30			
Setebos (S/1999 U1)	17,418,000	681	−2230‡	30			

*For reasons of space, only moons more than 25 km in diameter are listed. We have omitted 8 moons with diameters between 10 and 25 km.

**Mass of Earth's Moon = 7.4×10^{22} kg = 8.5×10^{-4} Uranus mass.

†Orbital paramers not well determined.

‡Retrograde orbit.

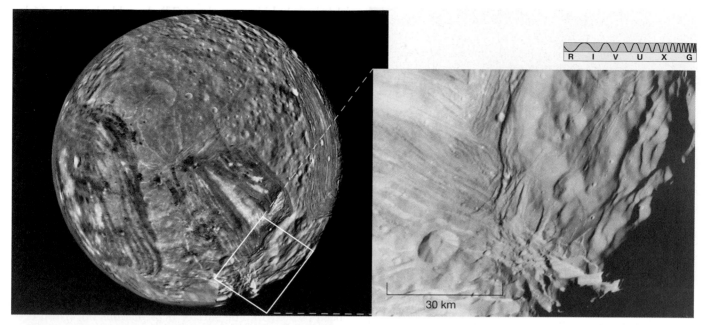

R I V U X G

▲ FIGURE 13.12 Miranda The asteroid-sized innermost moon of Uranus, photographed by *Voyager 2*. Miranda has a strange, fractured surface suggestive of a violent past, but the cause of the grooves and cracks is currently unknown. The resolution in the inset is about 2 km. The long "canyon" near the bottom of the inset is about 20 km deep. *(NASA)*

1949. *Voyager 2* discovered six additional moons, all less than a few hundred kilometers across and all lying within Nereid's orbit. Five more small moons, on wide, eccentric orbits, have been discovered by ground-based surveys since 2002. The planet's 13 known moons are listed in Table 13.2. Proteus, Neptune's only midsized moon (by our previous definition) is shown in Figure 13.11. ⚬ (Sec. 12.5)

In its moons, we find Neptune's contribution to our list of solar system peculiarities. Unlike the other jovian worlds, Neptune has no regular moon system—that is, no moons on roughly circular, equatorial, prograde orbits. The largest moon, Triton, is 2700 km in diameter and occupies a circular retrograde orbit 355,000 km (14.3 planetary radii) from the planet, inclined at about 20° to Neptune's equatorial plane. Triton is the only large moon in our solar system to have a retrograde orbit. The other moon visible from Earth, Nereid, is only 340 km across. This moon orbits Neptune in the prograde sense, but on an elongated trajectory that brings it as close as 1.4 million km to the planet and as far away as 9.7 million km. Nereid is probably similar in both size and composition to Neptune's small inner moons.

Voyager 2 approached to within 24,000 km of Triton's surface, providing us with virtually all that we now know about that distant, icy world. Astronomers redetermined the moon's radius (correcting it downward by about 20 percent) and measured its mass for the first time. Along with Saturn's Titan and the four Galilean moons of Jupiter, Triton is one of the six large moons in the outer solar system. Triton is the smallest of them, with about half the mass of the next smallest, Jupiter's Europa.

Lying 4.5 billion km from the Sun, and with a fairly reflective surface, Triton has a surface temperature of just 37 K. It has a tenuous nitrogen atmosphere, perhaps a hundred thousand times thinner than Earth's, and a surface that most likely consists primarily of water ice. A *Voyager 2* mosaic of Triton's south polar region is shown in Figure 13.13. The moon's low temperatures produce a layer of nitrogen frost that forms and evaporates over the polar caps, a little like the carbon dioxide frost responsible for the seasonal caps on Mars. The frost is visible as the pinkish region at the right of the figure.

Overall, Triton exhibits a marked lack of cratering, presumably indicating that surface activity has obliterated the evidence of most impacts. There are many other signs of an active past. For example, Triton's face is scarred by large fissures similar to those seen on Ganymede, and Triton's odd cantaloupelike terrain may indicate repeated faulting and deformation over the moon's lifetime. In addition, Triton has numerous frozen "lakes" of water ice (Figure 13.14), which may be volcanic in origin. The basic process may be similar to the water volcanism inferred (but not directly observed) on Saturn's moon Enceladus. ⚬ (Sec. 12.5)

Triton's surface activity is not just a thing of the past. As *Voyager 2* passed the moon, its cameras detected two great jets of nitrogen gas erupting from below the surface and rising several kilometers above it. It is thought that these "geysers" form when liquid nitrogen below Triton's surface is heated and vaporized by some internal energy source or perhaps even by the Sun's feeble light. Vaporization produces high pressure, which forces the gas through fissures in the crust, creating the displays *Voyager 2* saw. Scientists conjecture that nitrogen geysers may be common on Triton and are perhaps responsible for much of

13.7 The Rings of the Outermost Jovian Planets

The Rings of Uranus

All the jovian planets have rings. The ring system surrounding Uranus was discovered in 1977, when astronomers observed it passing in front of a bright star, momentarily dimming the star's light. Such a **stellar occultation** (Figure 13.15) happens a few times per decade and allows astronomers to measure planetary structures that are too small and faint to be detected directly. The 1977 observation was actually aimed at studying the planet's atmosphere by watching how it absorbed starlight. However, 40 minutes before and after Uranus itself occulted (passed in front of) the star, the flickering starlight revealed the presence of a set of rings. The discovery was particularly exciting because, at the time, only Saturn was known to have rings. Jupiter's rings went unseen until *Voyager 1* arrived there in 1979, and those of Neptune were unambiguously detected only in 1989, by *Voyager 2*.

The ground-based observations revealed the presence of a total of nine thin rings around Uranus. The main rings, in order of increasing radius, are named Alpha, Beta, Gamma, Delta, and Epsilon, and they range from 44,000 to 51,000 km from the planet's center. All lie within the Roche limit of Uranus, which is about 62,000 km from the planet's center. A fainter ring, known as the Eta ring, lies between the Beta and Gamma rings, and three other faint rings, known as 4, 5, and 6, lie between the Alpha ring and the planet itself. In 1986, *Voyager 2* discovered two more even fainter rings, one between Delta and Epsilon and one between ring 6 and Uranus. The main rings are shown in Figure 13.16. More details on the rings are given in Table 13.3.

The rings of Uranus are quite different from those of Saturn. Whereas Saturn's rings are bright and wide, with relatively narrow gaps between them, the rings of Uranus are dark, narrow, and widely spaced. With the exception of the Epsilon ring and the diffuse innermost ring, the rings of Uranus are all less than about 10 km wide, and the spacing between them ranges from a few hundred to about a thou-

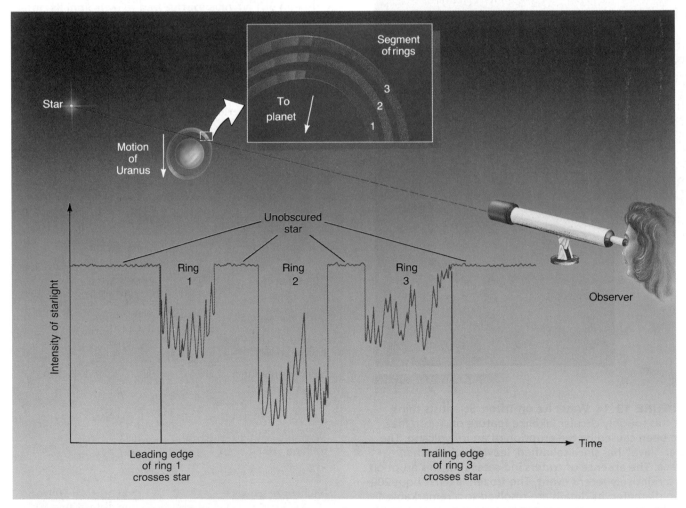

▲ **FIGURE 13.15 Occultation of Starlight** By carefully watching the dimming of distant starlight as a planet crosses the line of sight, astronomers can infer fine details about that planet. The rings of Uranus were discovered with this technique.

sand kilometers. However, like Saturn's rings, all Uranus's rings are less than a few tens of meters thick (that is, measured in the direction perpendicular to the ring plane).

The density of particles within Uranus's rings is comparable to that found in Saturn's A and B rings. The particles that make up Saturn's rings range in size from dust grains to boulders, but in the case of Uranus, the particles show a much smaller spread—few, if any, are smaller than a centimeter or so in diameter. The ring particles are also considerably less reflective than Saturn's ring particles, possibly because they are covered with the same dark material as Uranus's moons. The Epsilon ring (shown in detail in Figure 13.17) exhibits properties a little like those of Saturn's F ring. It has a slight eccentricity of 0.008 and is of variable width, although no braids were found in it. It also appears to be composed of ringlets.

Like the F ring of Saturn, Uranus's narrow rings require shepherding satellites to keep them from diffusing. In fact, the theory of shepherd satellites was first worked out to explain the rings of Uranus, which had been detected by stellar occultation even before *Voyager 2*'s encounter with Saturn. Thus, the existence of the F ring did not come as quite such a surprise as it might have otherwise! Presumably, many of the small inner satellites of Uranus play some role in governing the appearance of the rings. *Voyager 2* detected Cordelia and Ophelia, the shepherds of the Epsilon ring (see Figure 13.18). Many other, undetected, shepherd satellites must also exist.

The Rings of Neptune

As shown in Figure 13.19 and presented in more detail in Table 13.4, Neptune is surrounded by five dark rings.

TABLE 13.3	The Rings of Uranus				
Ring	**Inner Radius**		**Outer Radius***		**Width**
	(km)	(planetary radii)	(km)	(planetary radii)	(km)
1986U2R	37,000	1.45	39,500	1.55	2500
6	41,800	1.64			2
5	42,200	1.65			2
4	42,600	1.67			3
Alpha	44,700	1.75			4–10
Beta	45,700	1.79			5–11
Eta	47,200	1.83			2
Gamma	47,600	1.86			1–4
Delta	48,300	1.90			3–7
1986U1R	50,000	1.96			2
Epsilon	51,200	2.00			20–100

**Most of Uranus's rings are so thin that there is little difference between their inner and outer radii.*

Three are quite narrow, like the rings of Uranus; the other two are broad and diffuse, more like Jupiter's ring. The dark coloration probably results from radiation darkening, as discussed earlier in the context of the moons of Uranus. All the rings lie within Neptune's Roche limit. The outermost (Adams) ring is noticeably clumped in places. From Earth, we see not a complete ring, but only partial arcs—the unseen parts of the ring arc simply too thin (unclumped) to be detected. The connection between the rings and the planet's small inner satellites has not yet been firmly established, but many astronomers think that the clumping is caused by shepherd satellites.

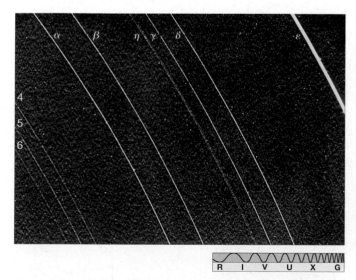

▲ FIGURE 13.16 Uranus's Rings The main rings of Uranus, as imaged by *Voyager 2*—all nine of them known before the spacecraft's arrival—can be seen in this photo. From the inside out, they are labeled from 6 to Epsilon. The resolution is about 10 km, which is just about the width of most of these rings. The two rings discovered by *Voyager 2* are too faint to be seen here. *(NASA)*

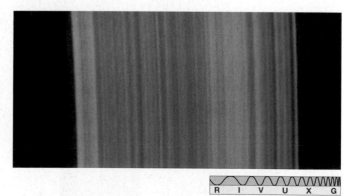

▲ FIGURE 13.17 Epsilon Ring A close-up of Uranus's Epsilon ring reveals some of its internal structure. The width of the ring averages 30 km; special image processing has magnified the resolution to about 100 m. *(NASA)*

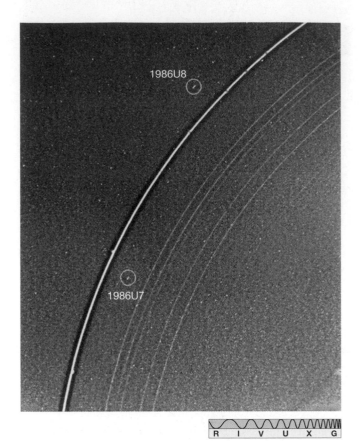

▲ **FIGURE 13.18 Uranian Shepherd Moons** These two small moons, discovered by *Voyager 2* in 1986 and now named Cordelia (U7) and Ophelia (U8), tend to "shepherd" the Epsilon ring, keeping it from diffusing. *(NASA)*

Although all the jovian worlds have ring systems, the rings themselves differ widely from planet to planet. Is there some "standard" way in which rings form around a planet? Also, is there a standard manner in which ring systems evolve? Or do the processes of ring formation and evolution depend entirely on the particular planet in question? If, as now appears to be the case, ring systems are relatively short lived, their formation must be a fairly common event. Otherwise, we would not expect to find rings around all four jovian planets at once. There are many indications that the individual planetary environment plays an important role in determining a ring system's appearance and longevity. Although many aspects of ring formation and evolution are now understood, it must be admitted that no comprehensive theory yet exists.

TABLE 13.4	The Rings of Neptune				
Ring	**Inner Radius** (planetary		**Outer Radius*** (planetary		**Width**
	(km)	radii)	(km)	radii)	(km)
Galle (1989N3R)	40,900	1.65	42,900	1.73	2000
Leverrier (1989N2R)	53,200	2.15			100
Lassell (1989N4R)[†]	53,200	2.15	57,200	2.31	4000
Arago (1989N4R)[†]	57,200	2.31			100
Adams (1989N1R)	62,900	2.54			50

* *Three of Neptune's rings are so thin that there is little difference between their inner and outer radii.*
[†] *Lassell and Arago were originally identified as a single ring.*

◀ **FIGURE 13.19 Neptune's Faint Rings** In this long-exposure image, Neptune (center) is heavily overexposed and has been artificially blocked to make the rings easier to see. One of the two fainter rings lies between the inner bright ring (Leverrier) and the planet. The others lie between the Leverrier ring and the outer bright ring (known as the Adams ring). *(NASA)*

CONCEPT CHECK

✔ What does the Epsilon ring of Uranus have in common with the F ring of Saturn?

13.8 The Discovery of Pluto

LEARNING GOAL 1 By the end of the 19th century, observations of the orbits of Uranus and Neptune suggested that Neptune's influence was not sufficient to account for all the irregularities in Uranus's motion. Furthermore, it seemed that Neptune itself might be affected by some other unknown body. Following their success in the discovery of Neptune, astronomers hoped to pinpoint the location of this new planet by using similar techniques. One of the most ardent searchers was Percival Lowell, a capable, persistent observer and one of the best-known astronomers of his day. (Recall that he was also the leading proponent of the theory that the "canals" on Mars were constructed by an intelligent race of Martians—see *Discovery 10-1*.)

Basing his investigation primarily on the motion of Uranus (Neptune's orbit was still relatively poorly determined at the time), and using techniques similar to those developed earlier in the search for Neptune by Adams and Leverrier, Lowell set about calculating where the supposed ninth planet should be. He sought it, without success, during the decade preceding his death in 1916. Not until 14 years later did American astronomer Clyde Tombaugh, working with improved equipment and better photographic techniques at the Lowell Observatory, finally succeed in finding Lowell's ninth planet, only 6° away from Lowell's predicted position. The new planet was named Pluto, for the Roman god of the dead who presided over eternal darkness (and also because its first two letters and its astrological symbol, ♇, are Lowell's initials). The discovery of Pluto was announced on March 13, 1930, Percival Lowell's birthday (and also the anniversary of Herschel's discovery of Uranus).

On the face of it, the discovery of Pluto looked like another spectacular success for celestial mechanics. Unfortunately, it now appears that the supposed irregularities in the motions of Uranus and Neptune did not exist and that the mass of Pluto, not measured accurately until the 1980s, is far too small to have caused them anyway. The discovery of Pluto thus owed much more to simple luck than to elegant mathematics!

Some orbital and physical data on Pluto are presented in the Pluto Data box. Unlike the paths of the other outer planets, Pluto's orbit is quite elongated, with an eccentricity of 0.25. It is also inclined at 17.2° to the plane of the ecliptic. Here, already, we have some indication that Pluto is unlike its jovian neighbors. Because of the planet's substantial orbital eccentricity, Pluto's distance from the Sun varies considerably. At perihelion, Pluto lies 29.7 A.U. (4.4 billion km) from the Sun, inside the orbit of Neptune. At aphelion, the distance is 49.3 A.U. (7.4 billion km), well

PLANETARY DATA:
PLUTO

Orbital semimajor axis	39.48 A.U. 5906 million km
Orbital eccentricity	0.249
Perihelion	29.66 A.U. 4437 million km
Aphelion	49.31 A.U. 7376 million km
Mean orbital speed	4.74 km/s
Sidereal orbital period	248.0 tropical years
Synodic orbital period	366.72 solar days
Orbital inclination to the ecliptic	17.15°
Greatest angular diameter, as seen from Earth	0.11″
Mass	1.27×10^{22} kg 0.0021 (Earth = 1)
Equatorial radius	1137 km 0.18 (Earth = 1)
Mean density	2060 kg/m^3 0.374 (Earth = 1)
Surface gravity	0.66 m/s^2 0.067 (Earth = 1)
Escape speed	1.2 km/s
Sidereal rotation period	−6.387 solar days (retrograde)
Axial tilt	118°
Surface magnetic field	unknown
Magnetic axis tilt relative to rotation axis	—
Surface temperature	40 to 60 K
Number of moons	1

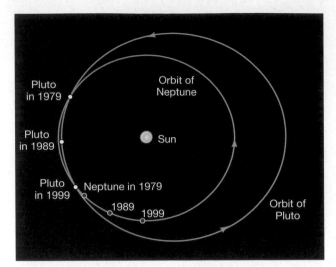

▲ **FIGURE 13.20 Neptune and Pluto** The orbits of Neptune and Pluto cross, although Pluto's orbital inclination and a 3:2 resonance prevent the planets from actually coming close to each other. Between 1979 and 1999, Pluto was inside Neptune's orbit, making Neptune the most distant planet from the Sun during those years.

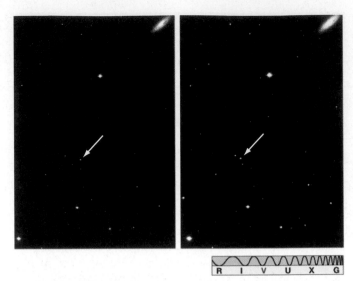

▲ **FIGURE 13.21 Motion of Pluto** These two photographs, taken one night apart, show the motion of the planet Pluto (arrow) against a field of much more distant stars. (Note the galaxy in the upper right corner.) Most of Pluto's apparent motion in these two frames is actually due to the orbital motion of Earth rather than that of Pluto. *(UC/Lick Observatory)*

outside Neptune's orbit. Pluto last passed perihelion in 1989 and remained inside Neptune's orbit until February 1999. The sidereal period of Pluto is 248.0 years, so the next perihelion passage will not occur until the middle of the 23rd century.

Pluto's orbital period is apparently exactly 1.5 times that of Neptune; in other words, the two planets are locked into a 3:2 resonance (two orbits of Pluto for every three of Neptune) as they orbit the Sun. As a result, even though their orbits appear to cross, Pluto and Neptune are in no danger of colliding with each other. Because of the orbital resonance and Pluto's tilted orbital plane, the distance between the two planets at closest approach is actually about 17 A.U. (compare with Pluto's closest approach to Uranus of just 11 A.U.). The orbits of Neptune and Pluto are sketched in Figure 13.20.

At nearly 40 A.U. from the Sun, Pluto is often hard to distinguish from the background stars. As the two photographs in Figure 13.21 indicate, the planet is actually considerably fainter than many stars in the sky. Like Neptune, Pluto is never visible to the naked eye.

13.9 Physical Properties of Pluto

Pluto is so far away that little is known of its physical nature. Until the late 1970s, studies of sunlight reflected from its surface suggested a rotation period of nearly a week, but measurements of its mass and radius were uncertain. All this changed in July 1978, when astronomers at the U.S. Naval Observatory discovered that Pluto has a satellite. It is now named Charon, after the mythical boatman who ferried the dead across the river Styx into Hades, Pluto's domain. The discovery photo-

graph of Charon is shown in Figure 13.22(a). Charon is the small bump near the top of the image. Knowing the moon's orbital period of 6.4 days, astronomers could determine the mass of Pluto to much greater accuracy than had previously been possible. Pluto's mass is 0.0021 Earth mass (1.3×10^{22} kg), far smaller than any earlier estimate and more like the mass of a moon than of a planet. In 1990, the *Hubble Space Telescope* imaged the Pluto–Charon system (Figure 13.22b). The improved resolution of that instrument clearly resolved the two bodies and allowed even more accurate measurements of their properties.

The discovery of Charon also permitted astronomers to measure Pluto's radius more precisely. Pluto's angular size is much less than 1″, so its true diameter is blurred by the effects of Earth's turbulent atmosphere. ∞ (Sec. 5.3) However, Charon's orbit has given astronomers new insight into the system. By pure chance, over the 6-year period from 1985 to 1991 (less than 10 years after the moon was discovered), Charon's orbit happened to be oriented in such a way that viewers on Earth saw a series of eclipses. Pluto and Charon repeatedly passed in front of each other, as seen from our vantage point. Figure 13.23 sketches this orbital configuration. With more good fortune, these eclipses took place while Pluto was closest to the Sun, making for the best possible Earth-based observations.

Basing their calculations on the variations in reflected light as Pluto and Charon periodically hid each other, astronomers computed the masses and radii of both bodies and determined their orbital plane. Additional studies of Pluto's surface brightness indicate that the two are tidally locked as they orbit each other. Pluto's diameter is 2270 km, about one-fifth the size of Earth. Charon is about

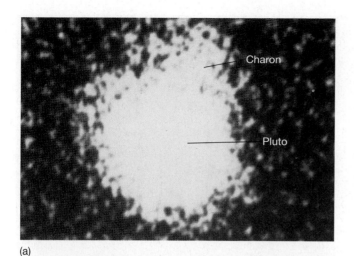

(a)

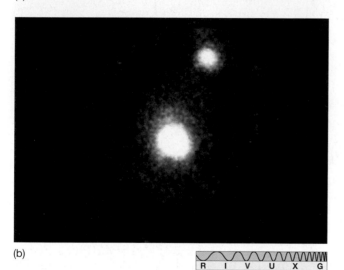

(b)

R I V U X G

▲ **FIGURE 13.22 Pluto and Charon** (a) The photograph from which Pluto's moon, Charon, was discovered. The moon is the small blotch of light at the top right portion of the image. The larger blob of reflected sunlight is Pluto itself. (b) The Pluto–Charon system, shown to the same scale as in part (a), as seen by the *Hubble Space Telescope*. The angular separation of the planet and its moon is about 0.9″. (*US Naval Observatory; NASA*)

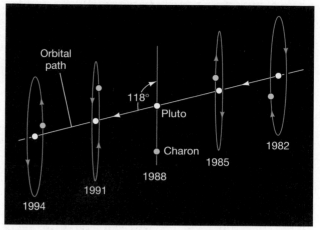

▲ **FIGURE 13.23 Pluto–Charon Eclipses** The orbital orientation of Charon produced a series of eclipses between 1985 and 1991. Observations of eclipses of Charon by Pluto and of Pluto by Charon have provided detailed information about the sizes and orbits of both bodies.

a terrestrial planet, but far too high for a mixture of hydrogen and helium of that mass. Instead, the mass, radius, and density of Pluto are just what we would expect for one of the icy moons of a jovian planet. In fact, Pluto is quite similar in both mass and radius to Neptune's large moon, Triton. The planet is almost certainly made up mostly of water ice.

Spectroscopy reveals the presence of *frozen* methane as a major surface constituent of the planet. Pluto is thus the only planet in the solar system on which methane exists in the solid state, implying that the surface temperature on Pluto is no more than 50 K. Pluto may also have a thin methane atmosphere, associated with the methane ice on its surface. Recent computer-generated maps have begun to hint at surface features on Pluto (see *Discovery 13-1*). Similar studies indicate that Charon may have bright polar caps, but their composition and nature are as yet unknown.

13.10 The Origin of Pluto

Because Pluto is neither terrestrial nor jovian in its make-up, and because it is similar to the ice moons of the outer planets, some researchers suspect that Pluto is not a "true" planet at all (see *Discovery 14-2*). This view is bolstered by Pluto's eccentric, inclined orbit, which is quite unlike the orbits of the other known planets.

Soon after Pluto was discovered, its orbital association with Neptune suggested to some theorists that a catastrophic encounter of some sort might have ejected Pluto from its original orbit around Neptune and perhaps even simultaneously knocked Triton onto its present retrograde path. Although this was an attractive theory at the time, a plausible sequence of events that could account in detail for the present orbits of those bodies proved elusive. Furthermore, in 1978, the picture was greatly complicated by the discovery of Charon. It was much easier to suppose that Pluto was an escaped moon before we learned that it

1300 km across and orbits at a distance of 19,700 km from Pluto. If planet and moon have the same composition (probably a reasonable assumption), Charon's mass must be about one-sixth that of Pluto, giving the Pluto–Charon system by far the largest satellite-to-planet mass ratio in the solar system.

As shown in Figure 13.24, Charon's orbit is inclined at an angle of 118° to the plane of Pluto's orbit around the Sun. Since the spins of both planet and moon are perpendicular to the plane of Charon's orbit around Pluto, the geographic "north" poles of both bodies lie below the plane of Pluto's orbit. Thus, Pluto is the third planet in the solar system (along with Venus and Uranus) found to have retrograde rotation.

The known mass and radius of Pluto allow us to determine its average density, which is 2100 kg/m³—too low for

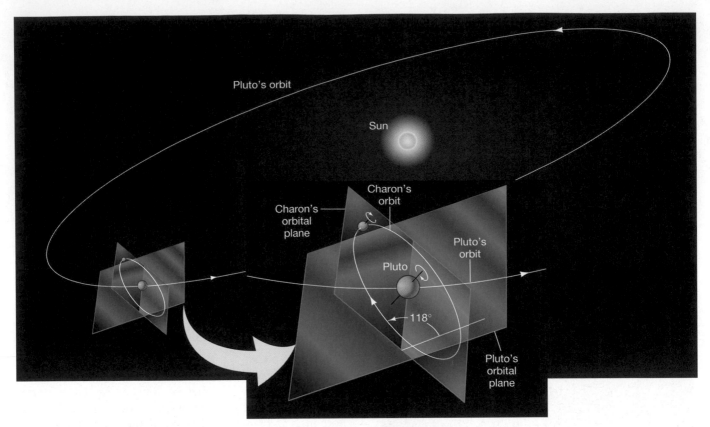

▲ **FIGURE 13.24 Pluto–Charon Orbit** Charon's path around Pluto is circular, synchronous, and inclined at 118° to the orbital plane of the Pluto–Charon system about the Sun, which is itself inclined at 17° to the plane of the ecliptic.

had a moon of its own. As a result, few astronomers still regard this theory as a likely explanation of Pluto's origin.

Conceivably, Pluto may be just what it seems: a planet that formed in its current orbit, possibly even with its own moon right from the outset. However, there is mounting evidence that this is not the case. Astronomers now know of numerous large chunks of ice circulating in interplanetary space beyond the orbits of the jovian planets in a region called the *Kuiper belt* (see Section 14.2), and many researchers think that there may have been thousands of Pluto-sized objects initially present in the outer solar system. Indeed, as we saw in Chapter 6, such objects are the natural building blocks of planets, according to current theories of planetary formation. ∞ (Sec. 6.7)

Almost all of these Pluto-sized objects were "kicked" out to large distances from the Sun following gravitational interactions with Uranus and Neptune (see Chapter 15). The capture of a few of them by the giant planets could explain some of the strange moons of the outer worlds—especially Triton—and if there were enough moon-sized chunks originally orbiting beyond Neptune, it is quite plausible that Pluto could have captured Charon following a collision (or near miss) between the two. Pluto remained behind after most other bodies were captured or ejected, perhaps because its orbital resonance with Neptune prevented it from ever having a close encounter with that (or any other) planet.

At present, our scant knowledge of the Pluto–Charon system prevents us from testing this theory directly. NASA has ambitious plans for a mission to Pluto and the outer solar system, with a possible launch date as early as 2006 and arrival at Pluto by 2015, although the project has neither been given final approval nor actually been funded. Nevertheless, the study of Pluto, Charon, and their environment is a top priority of the space science community, and hopes remain high that the next two decades will see a dramatic improvement in our understanding of these distant worlds.

Paradoxically, if all goes according to plan, and in 2020 we have a comprehensive theory of Pluto's origin, this newfound understanding might have one unfortunate side effect: If current theory holds up in the face of new observations, then Pluto will be "demoted" to just one among many large preplanetary objects present in the early solar system. It is quite conceivable that Pluto's status as a major planet will have been "revoked" long before we get to celebrate the first centennial of Tombaugh and Lowell's remarkable discovery!

CONCEPT CHECK

✔ What do Pluto and Triton have in common?

DISCOVERY 13-1

Surface Detail on Planet Pluto

Pluto is the only planet on whose surface we have little detailed information, because Pluto is the only planet that has not yet been visited by a robot spacecraft. Now new models of Pluto's surface have been constructed, based mainly on images taken by the *Hubble Space Telescope*. The first figure shows one sample of a dozen snapshots of Pluto, taken at visible and ultraviolet wavelengths throughout the planet's 6.4-day rotation period. This is not an image of Pluto, but a modeled view, that is, a computer-generated sketch of the planet's surface based on a compilation of all available data.

The second figure is a complete surface map of Pluto, created by carefully collating and analyzing all the *Hubble* data. The map is good confirmation of rougher maps made over the previous decade by using ground-based observations of eclipses of Pluto by its moon, Charon. Clearly, we have now progressed beyond the stage of seeing Pluto merely as a fuzzy, distant dot of light. With maps like these, astronomers can monitor the planet's surface detail about as well as you can study nearby Mars with a small backyard telescope.

These models reveal quite a lot of large-scale contrast; surprisingly, for a body composed almost entirely of ice, there is more contrast than on any other planet except Earth. About a dozen "provinces" can be identified, including icy-bright polar cap regions, another bright spot seen rotating with the planet itself, a cluster of dark spots, and a few peculiar linear markings. Some of these spots could turn out to be craters or impact basins, as exist on Earth's Moon, but much of Pluto's contrast is probably caused by frosts and snows that migrate across the planet as it goes through its seasonal cycles. The reflectivities of the lighter areas resemble that of fresh snow, whereas the darker areas are more akin to the subdued brightness of dirty snow. Let's just hope that people don't start talking about the odd linear markings as being reminiscent of canals!

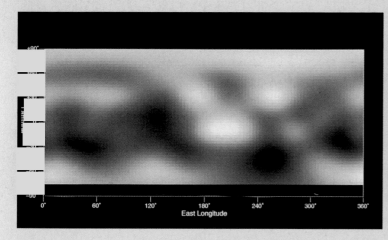

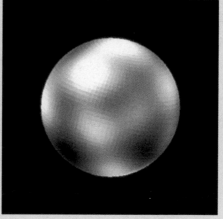

(NASA)

Chapter Review

SUMMARY

The outer planets Uranus, Neptune, and Pluto were unknown to ancient astronomers. Uranus was discovered in the 18th century, by chance. Neptune was discovered after mathematical calculations of Uranus's slightly non-Keplerian orbit revealed the presence of an eighth planet. At opposition, Uranus is barely visible to the unaided eye. Through a telescope, the planet appears as a pale green disk. Neptune cannot be seen with the naked eye, but a telescope shows it as a tiny bluish disk. Today, we know the giant planets Uranus and Neptune mainly through data taken by *Voyager 2*. Small, remote Pluto has not been visited by a spacecraft, and our knowledge of it stems from painstaking observations from Earth and the *Hubble Space Telescope*.

The masses of the outer planets are determined from measurements of their orbiting moons. The radii of Uranus and Neptune were relatively poorly known until the *Voyager 2* flybys in the 1980s. Uranus and Neptune have similar bulk properties; their densities imply large, rocky cores making up a greater fraction of the planets' masses than in either Jupiter or Saturn. For unknown reasons, Uranus's spin axis lies nearly in the plane of the ecliptic, leading to extreme seasonal variations in solar heating on the planet as it orbits the Sun. Surface features are barely discernible on Uranus, but computer-enhanced images from *Voyager 2* revealed atmospheric clouds and flow patterns moving beneath the planet's haze. Neptune, although farther away from us, has atmospheric features that are clearer because of warmer temperatures and less haze. The **Great Dark Spot (p. 335)** on Neptune had many similarities to Jupiter's Red Spot, but disappeared in 1994.

Unlike the other jovian planets, Uranus has no excess heat emission. The source of Neptune's excess energy, like that of Jupiter's, is most likely heat left over from the planet's formation. Both Uranus and Neptune have substantial magnetospheres. *Voyager 2* discovered that the magnetic fields of the two planets are tilted at large angles to the planets' rotation axes. The reason for large tilts is not known.

All but two of Uranus's moons revolve in the planet's equatorial plane, almost perpendicular to the ecliptic, in circular synchronous orbits. Like the moons of Saturn, the medium-sized moons of Uranus are made up predominantly of rock and water ice. Many of them are heavily cratered and in some cases must have come close to being destroyed by the meteoritic impacts whose craters we now see. The strange moon Miranda has geological features that suggest repeated violent impacts in the past. Neptune's moon Triton has a fractured surface of water ice and a thin atmosphere of nitrogen, probably produced by nitrogen "geysers" on its surface. Triton is the only large moon in the solar system to have a retrograde orbit around its parent planet. This orbit is unstable and will eventually cause Triton to be torn apart by Neptune's gravity.

Uranus has a series of dark, narrow rings, first detected from Earth by **stellar occultation (p. 344)**—their obscuration of the light received from background stars. Shepherd satellites are responsible for the rings' thinness. Neptune has three narrow rings like Uranus's and one broad ring like Jupiter's. The four were discovered by *Voyager 2*. The dark coloration of both the rings and the moons of the outer giant planets may be due to **radiation darkening (p. 339)**, whereby exposure to solar high-energy radiation slowly causes a dark hydrocarbon layer to build up on a body's icy surface.

Pluto was discovered in the 20th century after a laborious search for a planet that was supposedly affecting Uranus's orbital motion. We now know that Pluto is far too small to have any detectable influence on Uranus's path. Pluto has a moon, Charon, whose mass is about one-sixth that of the planet itself. Studies of Charon's orbit around Pluto have allowed the masses and radii of both bodies to be accurately determined. Pluto is too small to be a terrestrial planet; its properties are far more moonlike than planetlike. Most astronomers view Pluto as the largest (or nearest) member of a class of icy asteroids found in the outer solar system.

REVIEW AND DISCUSSION

1. How was Uranus discovered?
2. Why did astronomers suspect an eighth planet beyond Uranus?
3. How did Uranus come to be spinning "on its side"?
4. What is responsible for the colors of Uranus and Neptune?
5. How are the interiors of Uranus and Neptune thought to differ from those of Jupiter and Saturn?
6. How do the magnetic fields of Uranus and Neptune compare with that of Earth?
7. Describe a day on Titania.
8. What is unique about Miranda? Give a possible explanation.
9. Why are the icy moons of the outer planets so dark?
10. How does Neptune's moon system differ from those of the other jovian worlds? What do these differences suggest about the origin of Neptune's moon system?
11. What causes Triton's geysers?

12. What is the predicted fate of Triton?
13. The rings of Uranus are dark, narrow, and widely spaced. Which of these properties makes them different from the rings of Saturn?
14. Why are the Uranian rings so narrow and sharply defined?
15. How do the rings of Neptune differ from those of Uranus and Saturn?
16. Will Pluto and Neptune ever collide?
17. How were the mass and radius of Pluto determined?
18. In what respect is Pluto more like a moon than a jovian or terrestrial planet?
19. Why was the discovery of Uranus in 1781 so surprising? Might there be similar surprises in store for today's astronomers?
20. In what sense were astronomers fortunate to discover Pluto?

CONCEPTUAL SELF-TEST: TRUE OR FALSE/MULTIPLE CHOICE

1. Since its discovery, Uranus has completed just two-and-a-half orbits of the Sun.
2. During the northern summer of Uranus, an observer near the north pole would observe the Sun high and almost stationary in the sky.
3. Uranus's rotation axis is almost parallel to the plane of the ecliptic.
4. *Voyager 2* observed nitrogen geysers on the surface of Neptune.
5. The magnetic fields of both Uranus and Neptune are highly tilted relative to their rotation axes and significantly offset from the planets' centers.
6. Triton's orbit is unusual because it is retrograde.
7. Pluto was discovered via its gravitational effect on Neptune.

8. The radii of Pluto and Charon were determined accurately by observing the two pass in front of a background star during the late 1980s.
9. Pluto is smaller than Earth's Moon.
10. Astronomers have firm evidence that Pluto is an escaped moon of Neptune.
11. The discovery of new planets mostly requires **(a)** complex calculations and supercomputers; **(b)** the patient use of improving technology; **(c)** an astronomy degree from a large university; **(d)** luck.
12. Uranus was discovered about the same time as **(a)** Columbus reached North America; **(b)** the US Declaration of Indepen-

dence; **(c)** the American Civil War; **(d)** the Great Depression in the United States.

13. Compared with Uranus, the planet Neptune is **(a)** much smaller; **(b)** much larger; **(c)** roughly the same size; **(d)** tilted on its side.

14. The jovian planets with the largest diameters also tend to **(a)** have the slowest rotation rates; **(b)** move most slowly in their orbit around the Sun; **(c)** have the fewest moons; **(d)** have magnetic field axes most closely aligned with their axes of rotation.

15. The five largest moons of Uranus **(a)** all orbit in the ecliptic plane; **(b)** can never come between Uranus and the Sun; **(c)** all orbit directly above the planet's equator; **(d)** all have significantly eccentric orbits.

16. Moons that show few craters probably **(a)** are captured asteroids; **(b)** have been shielded from impacts by their host planet; **(c)** have had their smaller craters obliterated by larger impacts; **(d)** have warm interiors.

17. A gas-giant planet orbiting a distant star would be expected to have **(a)** a ring system like that of Saturn; **(b)** a density less than water; **(c)** many large moons orbiting in different directions; **(d)** evidence for hydrogen in its spectrum.

18. Uranus's rings were discovered by the occultation of starlight, as shown in Figure 13.15 ("Occultation of Starlight"). If Uranus were moving more rapidly relative to Earth, the graph in that figure would appear **(a)** more compressed horizontally; **(b)** the same; **(c)** more stretched out horizontally.

19. The discovery of a moon orbiting a planet allows astronomers to measure **(a)** the planet's mass; **(b)** the moon's mass and density; **(c)** the planet's ring structure; **(d)** the planet's cratering history.

20. The solar system object most similar to Pluto is **(a)** Mercury; **(b)** the Moon; **(c)** Titan; **(d)** Triton.

PROBLEMS

 Algorithmic versions of these questions are available in the Practice Problems module of the Companion Website at astro.prenhall.com/chaisson.

The number of squares preceding each problem indicates its approximate level of difficulty.

1. ■■ Calculate the time between successive closest approaches of (a) Neptune and Uranus and (b) Pluto and Neptune. For simplicity, assume circular orbits and calculate the time taken for the inner plate to "lap" the other one, as in Problem 5 of Chapter 9. Compare your answers with Neptune's sidereal orbital period.

2. ■ What is the gravitational force exerted on Uranus by Neptune, at closest approach? Compare your answer with the Sun's gravitational force on Uranus.

3. ■■ What is the angular diameter of the Sun, as seen from Uranus? Compare your answer with the angular diameter of Titania, as seen from the planet's cloud tops. Would you expect solar eclipses to occur on Uranus?

4. ■■ If the core of Uranus has a radius twice that of planet Earth and an average density of 8000 kg/m³, what is the mass of Uranus outside the core? What fraction of the planet's total mass is core?

5. ■ Estimate the speed of cloud A in Figure 13.7, assuming that it lies near the equator. Is your estimate consistent with the rotation speed of the planet?

6. ■ Estimate the strength of Neptune's gravitational tidal acceleration on Triton, relative to the moon's own surface gravity. Compare your answer with the corresponding ratio for Jupiter and Io. (See Chapter 11, Problem 10.)

7. ■ Add up the masses of all the moons of Uranus, Neptune, and Pluto. (Neglect the masses of the small moons—they contribute little to the result.) How does this sum compare with the mass of Earth's Moon and with the mass of Pluto?

8. ■■■ Astronomers on Earth are observing the occultation of a star by Uranus and its rings (see Figure 13.18). It so happens that the event is occurring when Uranus is at opposition, and the center of the planet appears to pass directly across the star. Assuming circular planetary orbits and, for simplicity, taking the rings to be face-on, calculate (a) how long the 90-km-wide Epsilon ring takes to cross the line of sight and (b) the time interval between the passage of the Alpha ring and that of the Epsilon ring. (*Hint*: The apparent motion of Uranus is due to a combination of both the planet's own motion and Earth's motion in their respective orbits around the Sun.)

9. ■■ On the basis of the earlier discussion of planetary atmospheres, would you expect Triton to have retained a nitrogen atmosphere? ∞ (*More Precisely 8-1*)

10. ■■ How long does it take for the inner shepherd moon in Figure 13.18 to "lap" the outer one?

11. ■■■ If Pluto's apparent motion relative to the stars is indeed due mainly to Earth's orbital motion, estimate the angle through which Pluto has moved between the two frames in Figure 13.21.

12. ■ What would be your weight on Pluto? On Charon?

13. ■ How close is Charon to Pluto's Roche limit?

14. ■ From Wien's law, at what wavelength does Pluto's thermal emission peak? In what part of the electromagnetic spectrum does this wavelength lie? ∞ (*More Precisely 3-2*)

15. ■ What is the round-trip travel time of light from Earth to Pluto (at a distance of 40 A.U.)? How far would a spacecraft orbiting Pluto at a speed of 0.5 km/s travel during that time?

 In addition to the Practice Problems module, the Companion Website at astro.prenhall.com/chaisson provides for each chapter a study guide module with multiple choice questions as well as additional annotated images, animations, and links to related Websites.

14

Only within the past couple of decades have scientists taken seriously the idea that life on Earth has been disrupted over the course of billions of years by asteroid and comet impacts. We now have a much better appreciation for Earth's presence in an occasionally hostile environment, and for Earth's fragility in the face of external cosmic dangers. Here, the 5-km-diameter nucleus of comet Wild2 was imaged in 2004 as part of the Stardust mission designed to fly through a comet's tail—and ultimately to return samples of the tail to Earth. (JPL) ▶

Solar System Debris

Keys to Our Origin

According to standard definitions, there are only nine planets in the solar system. But several thousand other celestial bodies are also known to revolve around the Sun. These minor bodies—the asteroids and comets—are small and of negligible mass compared with the planets and their major moons. Yet each is a separate world, with its own story to tell about the early solar system. On the basis of statistical deductions, astronomers estimate that there are more than a billion such objects still to be discovered. They may seem to be only rocky and icy "debris," but more than the planets themselves, they hold a record of the formative stages of our planetary system. Many are nearly pristine, unevolved bodies with much to teach us about our local origins.

LEARNING GOALS

Studying this chapter will enable you to

1 Describe the orbital properties of the major groups of asteroids.

2 Summarize the composition and physical properties of a typical asteroid.

3 Explain the effect of orbital resonances on the structure of the asteroid belt.

4 Detail the composition and structure of a typical comet, and explain the formation and appearance of its tail.

5 Discuss the characteristics of cometary orbits and what they tell us about the probable origin of comets.

6 Distinguish among the terms *meteor, meteoroid,* and *meteorite.*

7 Summarize the orbital and physical properties of meteoroids, and explain what these properties suggest about the probable origin of meteoroids.

 Visit astro.prenhall.com/chaisson for additional annotated images, animations, and links to related sites for this chapter.

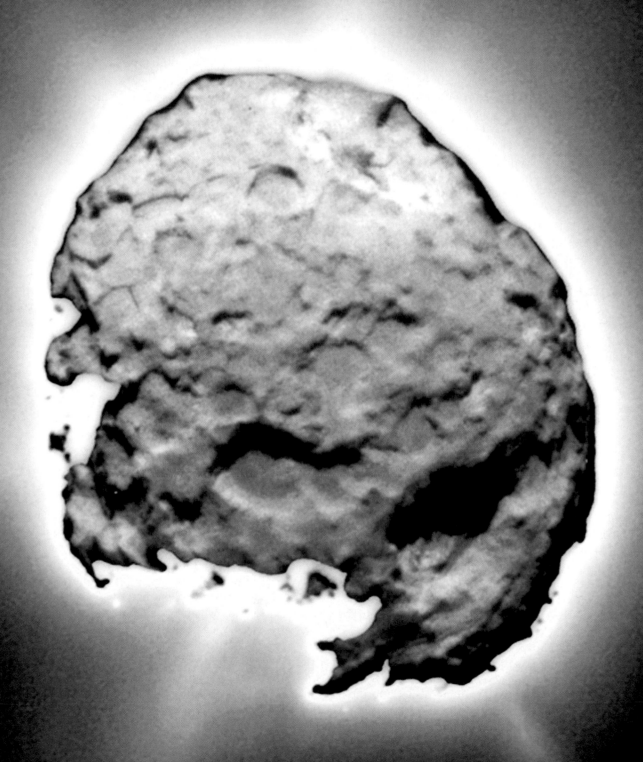

14.1 Asteroids

Asteroids are relatively small, predominantly rocky objects that revolve around the Sun. Their name literally means "starlike bodies," but asteroids are definitely not stars. They are too small even to be classified as planets. Astronomers often refer to them as "minor planets" or, sometimes, "planetoids."

Asteroids differ from planets in both their orbits and their size. They generally move on quite eccentric trajectories between Mars and Jupiter, unlike the almost circular paths of the major planets. Few asteroids are larger than 300 km in diameter, and most are far smaller—as small as a tenth of a kilometer across. The largest known asteroid, Ceres, is just 1/10,000 the mass of Earth and measures only 940 km across. Taken together, the known asteroids amount to less than the mass of the Moon, so they do not contribute significantly to the total mass of the solar system.

Orbital Properties

European astronomers discovered the first asteroids early in the 19th century as they searched the sky for an additional planet orbiting between Mars and Jupiter. Italian astronomer Giuseppe Piazzi was the first to discover an asteroid. He detected Ceres in 1801 and measured its orbital semimajor axis to be 2.8 A.U. Within a few years, three more asteroids—Pallas (2.8 A.U.), Juno (2.7 A.U.), and Vesta (3.4 A.U.)—were discovered.

By the start of the 20th century, astronomers had cataloged several hundred asteroids with well-determined orbits. Now, at the beginning of the 21st century, the list has grown to over 75,000. The total number of known asteroids (including those whose orbits are not yet known with sufficient accuracy to make them "official") now exceeds 200,000. The vast majority of these bodies are found in a region of the solar system known as the **asteroid belt**, located between 2.1 and 3.3 A.U. from the Sun—roughly midway between the orbits of Mars (1.5 A.U.) and Jupiter (5.2 A.U.). All but one of the known asteroids revolve about the Sun in prograde orbits, in the same sense as the planets. The overall layout of the asteroid belt is sketched in Figure 14.1.

Such a compact concentration of asteroids in a well-defined belt suggests that they are either the fragments of a planet broken up long ago or primal rocks that never managed to accumulate into a genuine planet. On the basis of the best evidence currently available, researchers favor the latter view: There is far too little mass in the belt to constitute a planet, and the marked chemical differences among individual asteroids strongly suggest that they could not all have originated in a single body. Instead, astronomers think that the strong gravitational field of Jupiter continuously disturbs the motions of these chunks of primitive matter, nudging and pulling at them and preventing them from aggregating into a planet.

Physical Properties

With few exceptions, asteroids are too small to be resolved by Earth-based telescopes, so astronomers must rely on indirect methods to find their sizes, shapes, and composition. Consequently, only a few of their physical and chemical properties are accurately known. To the extent that astronomers can determine their compositions, asteroids have been found to differ not only from the nine known planets and their many moons, but also among themselves.

Asteroids are classified by their spectroscopic properties. The darkest, or least reflective, asteroids contain a large fraction of carbon in their makeup. These asteroids are known as *C-type* (or *carbonaceous*) asteroids. The more reflective *S-type* asteroids contain silicate, or rocky, material. Generally speaking, S-type asteroids predominate in the inner portions of the asteroid belt, and the fraction of C-type bodies steadily increases as we move outward. Overall, about 15 percent of all asteroids are S-type, 75 percent are C-type, and 10 percent are other types (mainly the *M-type* asteroids, containing large fractions of nickel and iron). Many planetary scientists think that the carbonaceous asteroids consist of very primitive material representative of the earliest stages of the solar system. Carbonaceous asteroids have not been subject to significant heating or undergone chemical evolution since they first formed 4.6 billion years ago.

In most cases, astronomers estimate the sizes of asteroids from the amount of sunlight they reflect and the amount of heat they radiate. These observations are difficult, but size measurements have been obtained in this way for more than 1000 asteroids. On rare occasions, astronomers witness an asteroid occulting a star, allowing them to determine the asteroid's size and shape with great accuracy. The largest asteroids are roughly spherical, but the smaller ones can be highly irregular.

The three largest asteroids—Ceres, Pallas, and Vesta—have diameters of 940 km, 580 km, and 540 km, respectively. Only two dozen or so asteroids are more than 200 km across, and most are much smaller. Almost assuredly, many hundreds of thousands more await discovery. However, observers estimate that they are mostly very small. Probably 99 percent of all asteroids larger than 100 km are known and cataloged, and at least 50 percent of asteroids larger than 10 km are accounted for. Although the vast majority of asteroids are probably less than a few kilometers across, most of the *mass* in the asteroid belt resides in objects greater than a few tens of kilometers in diameter.

Vesta is unique among asteroids in that, despite its small size, it appears to have undergone *volcanism* in its distant past. On the basis of their orbits and their overall spectral similarities to Vesta, numerous meteorites (Section 14.3) found on Earth are thought to have been chipped off that asteroid following collisions with other members of the asteroid belt. Remarkably, these meteorites have compositions similar to that of terrestrial basalt, indicating that they were subject to ancient volcanic activity. Why Vesta should exhibit such past activity, while the larger Ceres and

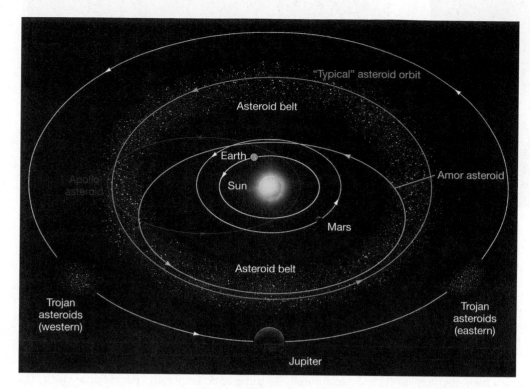

◀ **FIGURE 14.1** Inner Solar System The asteroid belt, along with the orbits of Earth, Mars, and Jupiter (not drawn to scale). The main belt, the Trojan asteroids, and some Apollo (Earth-crossing) and Amor (Mars-crossing) orbits are shown. (We will learn more about these classes of asteroids later in the chapter.)

Pallas do not, is not known. Possibly, Vesta was once part of a larger body that subsequently broke up. No other fragments of that body have yet been conclusively identified, although the spectrum and surface properties of a tiny 2-km-long asteroid discovered in 1999 suggests that it may once have been part of Vesta.

Asteroid Observations from Space

The first close-up views of asteroids were provided by the Jupiter probe *Galileo*, which, on its rather roundabout path to the giant planet, passed twice through the asteroid belt, making close encounters with asteroid Gaspra in October 1991 and asteroid Ida in August 1993 (Figure 14.2). ∞ (Sec. 6.6) Both Gaspra and Ida are S-type asteroids. Technical problems limited the amount of data that could be sent back from the spacecraft during the flybys. Nevertheless, the images produced by *Galileo* showed far more detail than any photographs made from Earth.

Gaspra and Ida are irregularly shaped bodies with maximum diameters of about 20 km and 60 km, respectively. They are pitted with craters ranging in size from a few hundred meters to 2 km across and are covered with a layer of dust of variable thickness. Ida is much more heavily cratered than Gaspra, in part because it resides in a denser part of the asteroid belt. Also, scientists think that Ida has suffered more from the ravages of time. Ida is about a billion years old, far older than Gaspra, which is estimated to have an age of just 200 million years, based on the extent of cratering on its surface. Both asteroids are thought to be fragments of much larger objects that broke up into many smaller pieces following violent collisions long ago.

To the surprise of most mission scientists, closer inspection of the Ida image (Figure 14.2b) revealed the presence of a tiny moon, now named Dactyl, just 1.5 km across, orbiting the asteroid at a distance of about 90 km. A few such *binary asteroids* (two asteroids orbiting one another as they circle the Sun) had previously been observed from Earth. However, the rare binary systems known before the Ida flyby were all much larger than Ida—a moon the size of Dactyl cannot be detected from the ground. Scientists think that, given the relative congestion of the asteroid belt, collisions between asteroids may be quite common, providing a source of both interplanetary dust and smaller asteroids and possibly deflecting one or both of the bodies involved onto eccentric, Earth-crossing orbits. The less violent collisions may be responsible for the binary systems we see.

By studying the *Galileo* images, astronomers were able to obtain limited information on Dactyl's orbit around Ida and hence (using Newton's law of gravity— see *More Precisely 2-3*) to estimate Ida's mass at about $5–10 \times 10^{16}$ kg. This information in turn allowed them to measure Ida's density as 2200–2900 kg/m³, a range consistent with the asteroid's rocky, S-type classification.

In June 1997, the *Near Earth Asteroid Rendezvous* (*NEAR*) spacecraft visited the C-type asteroid Mathilde on its way to the mission's main target: the S-type asteroid Eros. Shown in Figure 14.3, Mathilde is some 60 km across. By sensing its gravitational pull, *NEAR* measured Mathilde's mass to be about 10¹⁷ kg, implying a density of just 1400 kg/m³. To account for this low density, scientists speculate that the asteroid's interior must be quite porous. Indeed, many smaller asteroids seem to be more like loosely bound "rubble piles" than pieces of solid rock. The

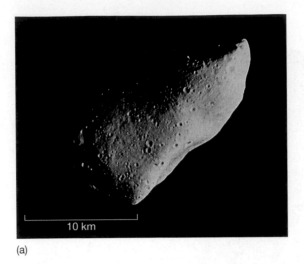

(a)

(b)

R I V U X G

▲ **FIGURE 14.2 Gaspra and Ida** (a) The S-type asteroid Gaspra, as seen from a distance of 1600 km by the space probe *Galileo* on its way to Jupiter. (b) The S-type asteroid Ida, photographed by *Galileo* from a distance of 3400 km. (Ida's moon, Dactyl, is visible at the right of the photo.) The resolution in these photographs is about 100 m. True-color images showed the surfaces of both bodies to be a fairly uniform shade of gray. Sensors on board the spacecraft indicated that the amount of infrared radiation absorbed by these surfaces varies from place to place, probably because of variations in the thickness of the dust layer blanketing them. *(NASA)*

interior's relatively soft consistency may also explain the unexpectedly large size of many of the craters observed on Mathilde's surface. A solid object would probably have shattered after an impact violent enough to cause such large craters. However, like crumple zones in a car, Mathilde's porous interior could have absorbed and dissipated the impactor's energy, allowing the asteroid to survive the event.

Upon its arrival at Eros on February 14, 2000, *NEAR* (now renamed *NEAR–Shoemaker*) went into orbit around the asteroid, changing the trajectory of the vehicle several times and coming as close as 5.5 km to the surface (Figure 14.4a). For one year, the spacecraft sent back high-resolution images of Eros (Figure 14.4b) and made detailed measurements of its size, shape, gravitational and magnetic fields, composition, and structure. The craft's various sensors revealed Eros to be a heavily cratered body of mass 7×10^{15} kg and roughly uniform density around 2700 kg/m³. The asteroid's interior seems to be solid rock—not rubble, as in the case of Mathilde—although it is extensively fractured due to innumerable impacts in the past. All in all, the measurements are consistent with Eros being a primitive, unevolved sample of material from the early solar system. In February 2001, *NEAR–Shoemaker* landed on Eros, sending back a series of close-up images as the craft descended to the surface. Remarkably, despite its having no landing gear, the spacecraft survived the low-velocity impact. While no further images were obtained, the probe maintained radio contact with Earth for 16 more days before communication finally ceased.

Apart from Ida, Mathilde, and Eros, the masses of most asteroids are unknown. However, a few of the largest asteroids do have strong enough gravitational fields for their effects on their neighbors to be measured and their masses thereby determined to reasonable accuracy. Their computed densities are generally compatible with the rocky or carbonaceous compositions just described.

CONCEPT CHECK

✔ Describe some basic similarities and differences between asteroids and the inner planets.

R I V U X G

▲ **FIGURE 14.3 Asteroid Mathilde** The C-type asteroid Mathilde, imaged by the *NEAR* spacecraft en route to the near-Earth asteroid Eros. Mathilde measures some 60×50 km and rotates every 17.5 days. The largest craters visible in this image are about 20 km across—much larger than the craters seen on either Gaspra or Ida. The cause may be the asteroid's low density (approximately 1400 kg/m³) and rather soft composition. *(NASA)*

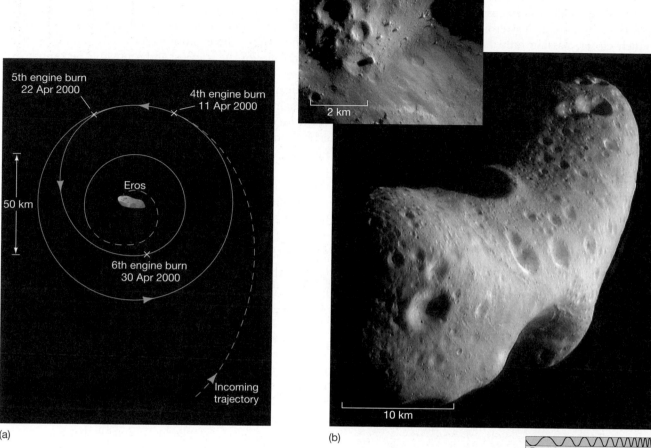

▲ **FIGURE 14.4** *NEAR* **at Eros** (a) The *NEAR–Shoemaker* spacecraft entered orbit around asteroid Eros in February 2000, making a series of orbital corrections (engine burns) during April to bring it closer and closer to the surface before landing on the asteroid in February 2001. (b) A mosaic of detailed images showing the entire asteroid, which has a very odd shape, 34 × 11 × 11 km. Craters of all sizes, ranging from 50 m (the resolution of the image) to 5 km, pit Eros's surface. The inset shows a close-up image of a "young" section of the surface, where loose material from recent impacts has apparently filled in and erased all trace of older craters. *(JHU/NASA)*

Earth-Crossing Asteroids

The orbits of most asteroids have eccentricities lying in the range from 0.05 to 0.3, ensuring that they always remain between the orbits of Mars and Jupiter. Very few asteroids have eccentricities greater than 0.4. Those which do are of particular interest to us, however, as their paths may intersect Earth's orbit, leading to the possibility of a collision with our planet. These bodies are collectively known as **Earth-crossing asteroids**. Those stray asteroids having highly elliptical orbits or orbits that do not lie in the main asteroid belt, have probably been influenced by nearby Mars and especially Jupiter. The gravitational fields of those two planets can disturb normal asteroid orbits, deflecting them into the inner solar system. Earth-crossing asteroids are termed *Apollo asteroids* (after the first known Earth-crossing asteroid, Apollo) if their orbital semimajor axes exceed 1 A.U. and *Aten asteroids* otherwise. Asteroids whose orbits cross only the orbit of Mars are known as *Amor asteroids*. (See Figure 14.1.)

As of early 2004, more than 2600 Earth-crossing asteroids are known. Most were discovered beginning in the late 1990s, when, spurred in large part by concerns over a possible collision between an asteroid or a comet and Earth, systematic searches for such objects began. Several large, dedicated telescopes now scan the skies for faint objects in our neighborhood. Almost 600 Earth crossers are officially designated "potentially hazardous," meaning that they are more than about 150 m in diameter (three times the size of the impactor responsible for the Barringer crater shown in Figure 8.19) and move in orbits that could bring them within 0.05 A.U. (7.5 million km) of our planet.

The potential for collision with Earth is real. For example, the perihelion of the Apollo asteroid Icarus (shown in Figure 14.5) is within 0.2 A.U. of the Sun. On its way past Earth in 1968, Icarus missed our planet by "only" 6 million km—a very close call by cosmic standards. More recently, in 1991, an unnamed asteroid (designated 1991 BA) came much closer, passing only 170,000 km from Earth, less than half the distance to the Moon. In December 1994, the Apollo asteroid 1994 XM1 missed us by a mere 105,000 km. All told, between 1994 and 2004, more than 850 asteroids (that we know of!) passed within 15 million km (0.1 A.U.) of our planet. At least 200 more are predicted to pass within that same distance between 2004 and 2014.

Calculations imply that most Earth-crossing asteroids will eventually collide with Earth. On average, during any given million-year period, our planet is struck by about three asteroids. Because Earth is largely covered with water, two of those impacts are likely to occur in the ocean and only one on land. Several dozen large land basins and eroded craters on our planet are suspected to be sites of ancient asteroid collisions (see, for example, Figure 14.23, later in the chapter). The many large impact craters on the Moon, Venus, and Mars are direct evidence of similar events on other worlds.

Most known Earth-crossing asteroids are relatively small—about 1 km in diameter (although one 10 km in diameter has been identified). Even so, a visit of even a kilometer-sized asteroid to Earth could be catastrophic by human standards. Such an object packs enough energy to

RIVUXG

▲ FIGURE 14.5 Asteroid Icarus The asteroid Icarus has an orbit that passes within 0.2 A.U. of the Sun, well within Earth's orbit. Icarus occasionally comes close to Earth, making it one of the best-studied asteroids in the solar system. Its motion relative to the stars makes it appear as a streak (marked) in this long-exposure photograph.
(Palomar/Caltech)

devastate an area some 100 km in diameter. The explosive power would be equivalent to about a million 1-megaton nuclear bombs—a hundred times more than all the nuclear weapons currently in existence on Earth. A fatal blast wave (the shock from the explosion, spreading rapidly outward from the site of the impact) and a possible accompanying tsunami (tidal wave) from an ocean impact would doubtless affect a much larger area still. Should an asteroid hit our planet hard enough, it might even cause the extinction of entire species—indeed, many scientists think that the extinction of the dinosaurs was the result of just such an impact (see *Discovery 14-1*). Some astronomers take the prospect of an asteroid impact sufficiently seriously that they maintain an "asteroid watch"—an effort to catalog and monitor all Earth-crossing asteroids in order to maximize our warning time of any impending collision.

Orbital Resonances

Although most asteroids orbit in the main belt, between about 2 and 3 A.U. from the Sun, an additional class of asteroids, called the **Trojan asteroids**, orbits at the distance of Jupiter. Several hundred such asteroids are now known. They are locked into a 1:1 orbital resonance with Jupiter by that planet's strong gravity, just as some of the small moons of Saturn share orbits with the medium-sized moons Tethys and Dione, as described in Chapter 12. ∞ (Sec. 12.5)

Calculations first performed by the French mathematician Joseph Louis Lagrange in 1772 show that there are exactly five places in the solar system where a small body can orbit the Sun in synchrony with Jupiter, subject to the combined gravitational influence of both large bodies. (Lagrange in fact demonstrated that five such points exist for any planet.) These places are now known as the *Lagrangian points* of the planet's orbit. As illustrated in Figure 14.6, three of the points (referred to as L_1, L_2, and L_3) lie on the line joining Jupiter and the Sun (or its extension in either direction). The other two—L_4 and L_5—are located on Jupiter's orbit, exactly 60° ahead of and 60° behind the planet. All five Lagrangian points revolve around the Sun at the same rate as Jupiter.

In principle, an asteroid placed at any of the Lagrangian points will circle the Sun in lockstep with Jupiter, always maintaining the same position relative to the planet. However, the three Lagrangian points that are in line with Jupiter and the Sun are known to be *unstable*—a body displaced, however slightly, from any of those points will tend to drift slowly away from it, not back toward it. Since matter in the solar system is constantly subjected to small perturbations—by the planets, the asteroids, and even the solar wind—matter does not accumulate in these regions. Thus, no asteroids orbit near the L_1, L_2, or L_3 point of Jupiter's orbit.

This is not the case for the other two Lagrangian points, L_4 and L_5. They are both *stable*—matter placed near them tends to remain in their vicinity. Consequently, asteroids tend to accumulate near these points (see Figure

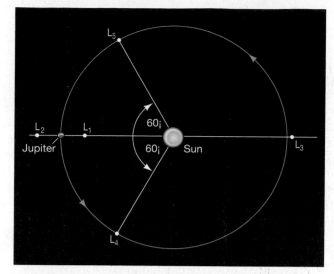

▲ **FIGURE 14.6 Lagrangian Points** The Lagrangian points of the Jupiter–Sun system, where a third body could orbit in synchrony with Jupiter on a circular trajectory. Only the L_4 and L_5 points are stable. They are the locations of the Trojan asteroids.

14.1). For unknown reasons, Trojan asteroids tend to be found near Jupiter's leading (L_4, or *eastern*, as it lies to the east of Jupiter in the sky) Lagrangian point, rather than the trailing (L_5) point. Recently, a few small asteroids have been found similarly trapped in the Lagrangian points of Venus, Earth, and Mars.

The main asteroid belt also has structure—not so obvious as the Trojan orbits or the prominent gaps and ringlets in Saturn's ring system, but nevertheless of great dynamic significance. A graph of the number of asteroids having various orbital semimajor axes (Figure 14.7a) shows that there are several prominently underpopulated regions in the distribution. These "holes" are known as the **Kirkwood gaps**, after their discoverer, the 19th-century American astronomer Daniel Kirkwood.

The Trojan asteroids share an orbit with Jupiter, orbiting in 1:1 resonance with the planet. The Kirkwood gaps result from other, more complex, orbital resonances with

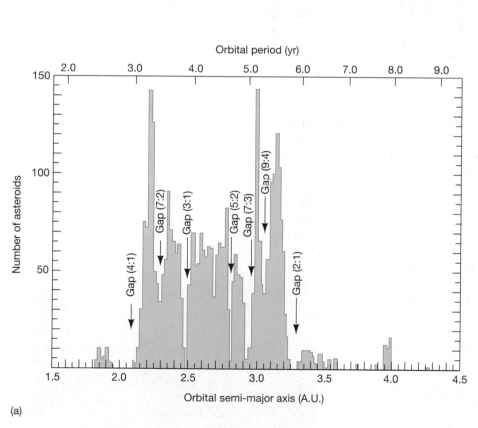

(a)

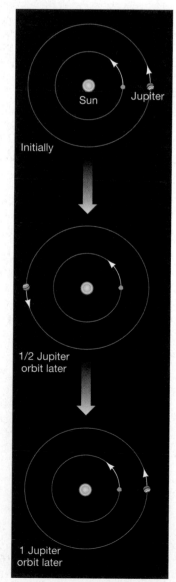

(b)

▲ **FIGURE 14.7 Kirkwood Gaps** (a) The distribution of asteroid semimajor axes shows some prominent gaps caused by resonances with Jupiter's orbital motion. Note, for example, the prominent gap at 3.3 A.U., which corresponds to the 2:1 resonance—the orbital period is 5.9 years, exactly half that of Jupiter. (b) An asteroid in a 2:1 resonance with Jupiter receives a strong gravitational tug from the planet each time they are closest together (as in panels 1 and 3). Because the asteroid's period is precisely half that of Jupiter, the tugs come at exactly the same point in every other orbit and their effects reinforce each other.

Jupiter. For example, an asteroid with a semimajor axis of 3.3 A.U. would (by Kepler's third law) orbit the Sun in exactly half the time taken by Jupiter. ∞ (Sec. 2.5) The gap at 3.3 A.U., then, corresponds to a 2:1 resonance. An asteroid at that particular resonance receives a regular, periodic tug from Jupiter at the same point in every other orbit (Figure 14.7b). The cumulative effect of all the tugs is to deflect the asteroid into an elongated orbit—one that crosses the orbit of Mars or Earth. Eventually, the asteroid collides with one of those two planets or comes close enough that it is pushed onto an entirely different trajectory. In this way, Jupiter's gravity creates the Kirkwood gaps, and some of the cleared-out asteroids become Apollo or Amor asteroids.

Notice that, although there are many similarities between this mechanism and the resonances that produce the gaps in Saturn's rings (see Chapter 12), there are differences, too. ∞ (Sec. 12.4) Unlike Saturn's rings, where eccentric orbits are rapidly circularized by collisions among ring particles, there are no *physical* gaps in the asteroid belt. The in-and-out motion of the belt asteroids as they travel in their eccentric orbits around the Sun means that no part of the belt is actually empty. Only when we look at semimajor axes (or, equivalently, at orbital *energies*) do the gaps become apparent.

CONCEPT CHECK

✔ Why are astronomers so interested in Earth-crossing asteroids?

DISCOVERY 14-1

What Killed the Dinosaurs?

The name *dinosaur* derives from the Greek words *deinos* ("terrible") and *sauros* ("lizard"). Dinosaurs were no ordinary reptiles: In their prime, roughly 100 million years ago, the dinosaurs were the all-powerful rulers of Earth. Their fossilized remains have been uncovered on all the world's continents. Despite their dominance, according to the fossil record, these creatures vanished from Earth quite suddenly about 65 million years ago. What happened to them?

Until fairly recently, the prevailing view among paleontologists—scientists who study prehistoric life—was that dinosaurs were rather small-brained, cold-blooded creatures. In chilly climates, or even at night, the metabolisms of these huge reptiles would have become sluggish, making it difficult for them to move around and secure food. The suggestion was that they were poorly equipped to adapt to sudden changes in Earth's climate, so they eventually died out. However, a competing, and still controversial, view of dinosaurs has emerged: Recent fossil evidence suggests that many of these monsters may in fact have been warm-blooded and relatively fast-moving creatures—not at all the dull-witted, slow-moving giants of earlier conception. In any case, no species able to dominate Earth for more than 100 million years could have been too poorly equipped for survival. For comparison, humans have thus far dominated for a little over 2 million years.

If the dinosaurs didn't die out simply because of stupidity and inflexibility, then what happened to cause their sudden and complete disappearance? Many explanations have been offered for the extinction of the dinosaurs. Devastating plagues, magnetic field reversals, increased tectonic activity, severe climate changes, and supernova explosions have all been proposed. ∞ (Secs. 7.4, 7.5) In the 1980s, it was suggested that a huge extraterrestrial object collided with Earth 65 million years ago, and this is now (arguably) the leading explanation for the demise of the dinosaurs.

According to this idea, a 10- to 15-km-wide asteroid or comet struck Earth, releasing as much energy as 10 million or more of the largest hydrogen bombs humans have ever constructed and kicking huge quantities of dust (including the pulverized remnants of the impactor itself) high into the atmosphere. (See the first figure.) The dust may have shrouded our planet for many years, virtually extinguishing the Sun's rays during that time. On the darkened surface, plants could not survive. The entire food chain was disrupted, and the dinosaurs, at the top of that chain, eventually became extinct.

Although we have no direct astronomical evidence to confirm or refute this idea, we can estimate the chances that a large asteroid or comet will strike Earth today, on the basis of observations of the number of objects that are presently on Earth-crossing orbits. The second figure shows the likelihood of an impact as a function of the size of the colliding body. The horizontal scale indicates the energy released by the collision, measured in *megatons* of TNT. The megaton— 4.2×10^{16} joules, the explosive yield of a large nuclear

(C. Butler/Astrostock-Sanford)

14.2 Comets

Comets are usually discovered as faint, fuzzy patches of light on the sky while they are still several astronomical units away from the Sun. Traveling in a highly elliptical orbit with the Sun at one focus (Figure 14.8), a comet brightens and develops an extended **tail** as it nears the Sun. (The name "comet" derives from the Greek word *kome*, meaning "hair.") As the comet departs the Sun's vicinity, its brightness and tail diminish until it once again becomes a faint point of light receding into the distance. Like the planets, comets emit no visible light of their own—they shine by reflected (or reemitted) sunlight. Each year, a few dozen are detected as they pass through the inner solar system. Many more must pass by unseen.

Comet Appearance And Structure

LEARNING GOAL 4 The various parts of a typical comet are shown in Figure 14.9. Even through a large telescope, the **nucleus**, or main solid body, of a comet is no more than a minute point of light. A typical cometary nucleus is extremely small—only a few kilometers in diameter. During most of the comet's orbit, far from the Sun, only this frozen nucleus exists. When a comet comes within a few astronomical units of the Sun, however, its icy surface becomes too warm to remain stable. Part of the comet becomes gaseous and expands into space, forming a diffuse **coma** ("halo") of dust and evaporated gas around the nucleus. The coma becomes larger and brighter as the comet nears the Sun. At maximum size, the coma can measure 100,000 km in diameter—almost as large as Saturn or Jupiter.

---

warhead—is the only common terrestrial measure of energy adequate to describe the violence of these occurrences.

We see that 100-million-megaton events, like the planetwide catastrophe that supposedly wiped out the dinosaurs, are very rare, occurring only once every 10 million years or so. However, smaller impacts, equivalent to "only" a few tens of kilotons of TNT (roughly equivalent to the bomb that destroyed Hiroshima in 1945), could happen every few years—and we may be long overdue for one. The most recent large impact was the Tunguska explosion in Siberia, in 1908, which packed a roughly 1-megaton punch. (See Figure 14.24.)

The main geological evidence supporting the theory that the dinosaurs' extinction was the result of an asteroid impact is a layer of clay enriched with the element iridium. The layer is found in 65-million-year-old rocky sediments all around our planet. Iridium on Earth's surface is rare, because most of it sank into our planet's interior long ago. The abundance of iridium in this one layer of clay is about 10 times greater than

in other terrestrial rocks, but it matches closely the abundance of iridium found in meteorites (and, we assume, in asteroids and comets, too). The site of the catastrophic impact has also been tentatively identified as being near Chicxulub, in the Yucatán Peninsula in Mexico, where evidence of a heavily eroded, but not completely obliterated, crater of just the right size and age has been found.

The theory is not without its detractors, however. Perhaps predictably, the idea of catastrophic change on Earth being precipitated by events in interplanetary space was rapidly accepted by most astronomers, but it remains controversial among some paleontologists and geologists. Opponents argue that the amount of iridium in the clay layer varies greatly from place to place across the globe, and there is no complete explanation of why that should be so. Perhaps, they suggest, the iridium was produced by volcanoes, and has nothing to do with an extraterrestrial impact at all.

Still, in the 20 years since the idea was first suggested, the focus of the debate seems to have shifted. The reality of a major impact 65 million years ago has become widely accepted, and much of the argument now revolves around the question of whether that event actually caused the extinction of the dinosaurs or merely accelerated a process that was already underway. Either way, the realization that such catastrophic events can and do occur marks an important milestone in our understanding of evolution on our planet. This realization was bolstered by the Shoemaker–Levy 9 impact on Jupiter in 1994. ∞ (*Discovery 11-2*) In addition, there is growing evidence for even larger impacts in the more distant past, with yet more sweeping evolutionary consequences. As is often the case in science, the debate has evolved, sometimes erratically, as new data have been obtained, but a real measure of consensus has already been achieved, and important new insights into our planetary environment have been gained. ∞ (Sec. 1.2)

As a general rule, we can expect that global catastrophes are bad for the dominant species on a planet. As the dominant species on Earth, we are the ones who now stand to lose the most.

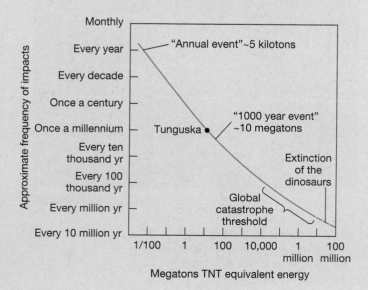

Monthly / Every year — "Annual event"~5 kilotons / Every decade / Once a century / Once a millennium — Tunguska, "1000 year event" ~10 megatons / Every ten thousand yr / Every 100 thousand yr — Extinction of the dinosaurs / Every million yr — Global catastrophe threshold / Every 10 million yr

Approximate frequency of impacts

1/100   1   100   10,000   1 million   100 million

Megatons TNT equivalent energy

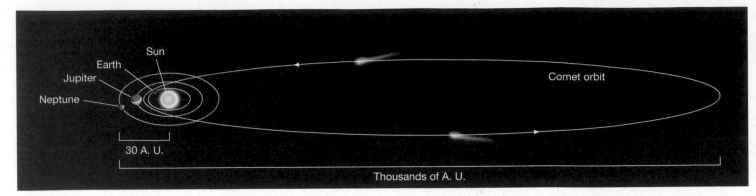

▲ **FIGURE 14.8** **Distant Orbit** Comets move in highly eccentric paths that carry them far beyond the known planets.

Engulfing the coma, an invisible **hydrogen envelope**, usually distorted by the solar wind, stretches across millions of kilometers of space. The comet's tail, which is most pronounced when the comet is closest to the Sun and the rate of sublimation* of material from the nucleus is greatest, is much larger still, sometimes spanning as much as 1 A.U. From Earth, only the coma and tail of a comet are visible to the naked eye. Despite the size of the tail, most of the light comes from the coma; most of the comet's mass resides in the nucleus.

*Sublimation *is the process by which a solid changes directly into a gas without passing through the liquid phase. Frozen carbon dioxide (dry ice) is an example of a solid that undergoes sublimation rather than melting and subsequent evaporation. In space, sublimation is the rule, rather than the exception, for the behavior of ice when it is exposed to heat.*

Two types of comet tails may be distinguished. **Ion tails** are approximately straight and are often made up of glowing, linear streamers like those seen in Figure 14.10(a). Their spectra show emission lines of numerous ionized molecules—molecules that have lost some of their normal complement of electrons—including carbon monoxide, nitrogen, and water, among many others. ∞ (Sec. 4.2) **Dust tails** are usually broad, diffuse, and gently curved (Figure 14.10b). They are rich in microscopic dust particles that reflect sunlight, making the tail visible from afar. (Ion and dust tails are also referred to as *Type I* and *Type II* tails, respectively.)

Comets' tails are in all cases directed *away* from the Sun by the solar wind (the invisible stream of matter and radiation escaping the Sun). Consequently, as depicted in

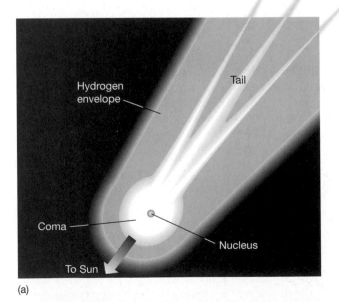

(a)

(b)

R I V U X G

ANIMATION Comet Hale–Bopp Nucleus Animation

▲ **FIGURE 14.9** **Comet Structure** (a) Diagram of a typical comet, showing the nucleus, coma, hydrogen envelope, and tail. The tail is not a sudden short-lived streak across the sky, as in the case of meteors or fireworks. Instead, it travels through space along with the rest of the comet (as long as the comet is sufficiently close to the Sun for the tail to exist). (b) Halley's comet in 1986, about one month before it rounded the Sun at perihelion. *(NOAO)*

(a)

◀ **FIGURE 14.10** Comet Tails (a) A comet with a primarily ion tail. Called comet Giacobini–Zinner and seen here in 1959, its coma measured 70,000 km across; its tail was well over 500,000 km long. (b) Photograph of a comet having both an ion tail (dark blue) and a dust tail (white blue), showing the gentle curvature and inherent fuzziness of the dust. (See also *Discovery 14-3*.) This is comet Hale–Bopp in 1997. At the comet's closest approach to the Sun, its tail stretched nearly 40° across the sky. (*US Naval Observatory; Aaron Horowitz/Corbis*)

Ion tail

Dust tail

(b)

R I V U X G

Figure 14.11, the tail always lies outside the comet's orbit and actually leads the comet during the portion of the orbit that is outbound from the Sun.

Ion tails and dust tails differ in shape because of the different responses of gas and dust to the forces acting in interplanetary space. Every tiny particle in space in our solar system—including those in comets' tails—follows an orbit determined by gravity and the solar wind. If gravity alone were acting, the particle would follow the same curved path as its parent comet, in accordance with Newton's laws of motion. ∞ (Sec. 2.7) If the solar wind were the only influence, the tail would be swept up by it and would trail radially outward from the Sun. Ion tails are much more strongly influenced by the solar wind than by the Sun's gravity, so those tails always point directly away

from the Sun. The heavier dust particles have more of a tendency to follow the comet's orbit, giving rise to the slightly curved dust tails.

Comet Orbits

Comets that survive their close encounter with the Sun—some break up entirely—continue their outward journey to the edge of the solar system. Their highly elliptical orbits take many comets far beyond Pluto, perhaps even as far as 50,000 A.U., where, in accord with Kepler's second law, they move more slowly and so spend most of their time. ∞ (Sec. 2.5) Most comets take hundreds of thousands, and some even take millions, of years to complete a single orbit around the Sun. These comets are known as *long-period comets*. A few *short-period comets*,

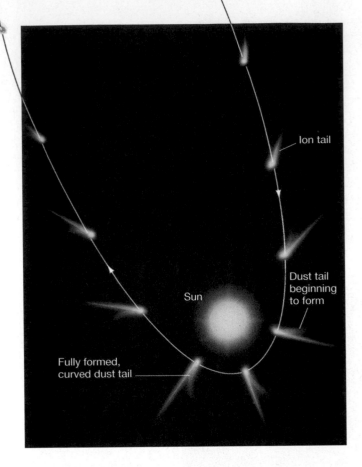

◀ **FIGURE 14.11 Comet Trajectory** As it approaches the Sun, a comet develops an ion tail, which is always directed away from the Sun. Closer in, a curved dust tail, also directed generally away from the Sun, may appear. Notice that although the ion tail always points directly away from the Sun on both the inbound and the outgoing portions of the orbit, the dust tail has a marked curvature, always tending to "lag behind" the ion tail. (Cf. with photo of a real comet, Figure 4.10.)

tiny portion of a typical long-period cometary orbit lies within the inner solar system, so it follows that, for every comet we see, there must be many more similar objects at great distances from the Sun. On these general grounds, many astronomers reason that there must be a huge "cloud" of comets far beyond the orbit of Pluto, completely surrounding the Sun. This region, which may contain trillions of comets, with a total mass comparable to the mass of the inner planets, is named the **Oort cloud**, after the Dutch astronomer Jan Oort, who first wrote (in the 1950s) of the possibility of such a vast reservoir of inactive, frozen comets orbiting far from the Sun. The Kuiper belt and the orbits of some typical Oort-cloud comets are sketched in Figure 14.12.

The observed orbital properties of long-period comets have led researchers to conclude that the Oort cloud may be up to 100,000 A.U. in diameter. Like those of the Kuiper belt, however, most of the comets of the Oort cloud never come anywhere near the Sun. Indeed, Oort-cloud comets rarely approach even the orbit of Pluto, let alone that of Earth. Only when the gravitational field of a passing star happens to deflect a comet into an extremely eccentric orbit that passes through the inner solar system do we actually get to see the comet. Because the Oort cloud surrounds the Sun in all directions, instead of being confined near the plane of the ecliptic like the Kuiper belt, the long-period comets we see can come from any direction in the sky. Despite their great distances and long orbital periods, Oort-cloud comets are still gravitationally bound to the Sun. Their orbits are governed by precisely the same laws of motion that control the planets' orbits.

CONCEPT CHECK

✔ In what sense are the comets we see *un*representative of comets in general?

conventionally defined as those having orbital periods of less than 200 years, return for another encounter within a relatively short time. According to Kepler's third law, short-period comets do not venture far beyond the distance of Pluto at aphelion.

Unlike the orbits of the other solar-system objects we have studied so far, the orbits of comets are not necessarily confined to within a few degrees of the ecliptic plane. Short-period comets do tend to have prograde orbits lying close to the ecliptic, but long-period comets exhibit all inclinations and all orientations, both prograde and retrograde, roughly uniformly distributed in all directions from the Sun.

The short-period comets originate beyond the orbit of Neptune, in a region of the solar system called the **Kuiper belt** (named after Gerard Kuiper, a pioneer in infrared and planetary astronomy). A little like the asteroids in the inner solar system, most Kuiper-belt comets move in roughly circular orbits between about 30 and 100 A.U. from the Sun, never venturing inside the orbits of the jovian planets. Occasionally, however, a close encounter between two comets, or (more likely) the cumulative gravitational influence of one of the outer planets, "kicks" a Kuiper-belt comet into an eccentric orbit that brings it into the inner solar system and into our view. The observed orbits of these comets reflect the flattened structure of the Kuiper belt.

What of the long-period comets? How do we account for their apparently random orbital orientations? Only a

A Visit to Halley's Comet

Probably the most famous comet of all is Halley's comet. (Two more recent and widely publicized contenders for the title "most famous" are described in *Discovery 14-2*.) In 1705, the British astronomer Edmund Halley realized that the appearance of a certain comet in 1682 was not a one-time event. Basing his work on previous sightings of the

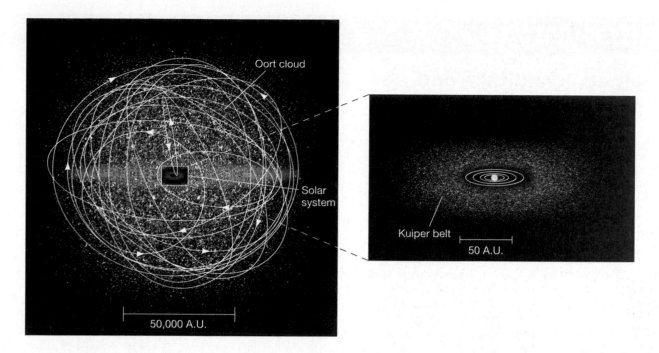

▲ **FIGURE 14.12 Comet Reservoirs** (a) Diagram of the Oort cloud, showing a few cometary orbits. Most Oort-cloud comets never come close to the Sun. Of the orbits shown, only the most elongated ellipse represents a comet that will actually enter the solar system (which, on the scale of this drawing, is much smaller than the dot at the center of the figure) and possibly become visible from Earth. (See also Figure 14.8.) (b) The Kuiper belt, the source of short-period comets, whose orbits tend to hug the plane of the ecliptic.

comet, Halley calculated its path and found that the comet orbited the Sun with a period of 76 years. He predicted its reappearance in 1758. Halley's successful determination of the comet's trajectory and his prediction of its return was an early triumph of Newton's laws of motion and gravity. ∞ (Sec. 2.7) Although Halley did not live to see his calculations proved correct, the comet was named in his honor.

Once astronomers knew the comet's period, they traced the appearances of the comet backwards in time. Historical records from many ancient cultures show that Halley's comet has been observed at every one of its passages since 240 B.C. A spectacular show, the tail of Halley's comet can reach almost a full astronomical unit in length, stretching many tens of degrees across the sky. Figure 14.13(a) shows Halley's comet as seen from Earth in 1910. Its most recent appearance, in 1986 (Figure 14.13b and also Figure 14.9b), was not ideal for terrestrial viewing, as the perihelion happened to occur on roughly the opposite side of the Sun

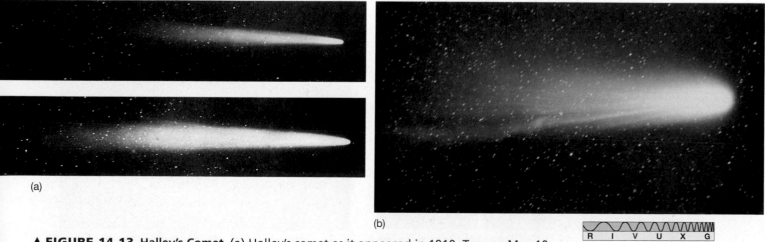

▲ **FIGURE 14.13 Halley's Comet** (a) Halley's comet as it appeared in 1910. Top, on May 10, with a 30° tail, bottom, on May 12, with a 40° tail. (b) Halley on its return and photographed with higher resolution on March 14, 1986. *(Caltech; Mt. Stromlo and Siding Springs Observatories)*

DISCOVERY 14-2

Comets Hyakutake and Hale–Bopp

One of the most spectacular comets in recent years was comet Hyakutake 1996. Named after a Japanese amateur astronomer who noticed it as "something odd and out of place" while scanning the skies with a pair of binoculars, Hyakutake grew from a small smudge while still far from the Sun into a splendid display comprising a huge coma nearly the apparent size of the Moon and a tail that eventually stretched a third of the way across the sky. The first figure shows *Hubble Space Telescope* images of Hyakutake taken in March 1996, when the comet passed closest to Earth—only 15 million km (0.1 A.U.) away. The comet's icy nucleus, the brightest point in the left image, is unresolved here (the field of view is about 1000 km across), but radar pulses sent toward Hyakutake did return an echo indicating that the diameter of the nucleus was 1–3 km. In these images, the Sun is out of the frame at the bottom right, and the innermost part of the comet's tail is at the upper left (on the side opposite the Sun, as explained in the text).

The other two images show sporadic jets pointing mostly sunward. These jets were gases gushing from the side of the comet closest to the Sun before wrapping around to become part of the tail. The comet was examined at every conceivable wavelength, but perhaps the most surprising result was that it emitted intense X rays from its head. Even the sunward side of the comet was far too cool to emit X rays, which are usually associated with very-high-temperature phenomena. Astronomers speculate that the X rays were produced by shock waves created as the solar wind hit the leading edge of the comet's coma.

Hot on the heels of Hyakutake came another interplanetary vagabond: comet Hale–Bopp 1995. Discovered in 1995 by two American amateur astronomers, this comet reached its maximum brightness in the spring of 1997. Outshining everything in the night sky except the Moon and the brightest planets and stars, Hale–Bopp was probably the most widely viewed and studied comet in history.

The comet's unusual brightness and long (20°) tail were probably caused by a huge nucleus, about 30–40 km in diameter. That's a very large ball of dirty ice, compared with the average comet core, which measures some 3–5 km across. The comet or asteroid that struck the Earth 65 million years ago (see *Discovery 14-1*), perhaps causing the extinction of the dinosaurs, is thought to have been about 10–15 km in size. At perihelion, Hale–Bopp's hydrogen envelope was enormous—1 A.U. across. The second figure shows a spectacular ground-based view of the comet around the time of perihelion.

Apart from its unusual size, Hale–Bopp seems to be quite representative of many long-period comets. Just about all the molecules previously known from other comets were seen in Hale–Bopp's spectrum. New infrared observing techniques allowed many of the minerals making up the comet's dust to be identified for the first time. That dust was found to be similar to the interplanetary dust populating near-Earth space and collected in Earth's upper atmosphere. What was unique about Hale–Bopp was that, being so large, it could be observed for a long period of time, even when it was far from the Sun, allowing observers to keep a continuous and detailed record of its evolution.

Hale–Bopp is now moving out of the inner solar system. Astronomers plan to monitor the comet for several more years as it moves far from the Sun's warmth and its activity winds down. The solar-heated gases rushing away from the head of a comet act in much the same way as a rocket engine on a spacecraft, and change the comet's orbit each time it rounds the Sun. Before its last perihelion passage, Hale–Bopp's orbital period was approximately 4200 years. Now, however, its orbit has been significantly modified by the forces from its jets. After a trip well outside the orbit of Pluto, extending some 350 A.U. from the Sun, the comet will return to the inner solar system in about 2400 years.

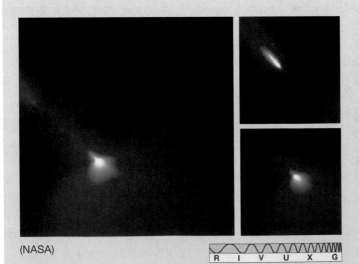

(NASA)

R I V U X G

(H. Mikuz, B. Kambi)

R I V U X G

from Earth, but the comet was closely scrutinized by spacecraft. The comet's orbit is shown in Figure 14.14; its next scheduled visit to the inner solar system is in 2061.

When Halley's comet rounded the Sun in 1986, a small armada of spacecraft launched by the (former) USSR, Japan, and a group of western European countries went to meet it. One of the Soviet craft, *Vega 2*, traveled through the comet's coma and came within 8000 km of the nucleus. Using positional knowledge of the comet gained from the Soviet craft encounter, the European *Giotto* spacecraft (named after the Italian artist who painted an image of Halley's comet not long after its appearance in the year 1301) was navigated to within 600 km of the nucleus. This was a daring trajectory, since at 70 km/s—the speed of the craft relative to the comet—a colliding dust particle becomes a devastating bullet. Debris did in fact damage *Giotto*'s camera, but not before it sent home a wealth of data. Figure 14.15 shows *Giotto*'s view of the comet's nucleus, along with a sketch of its structure.

The results of the Halley encounters were somewhat surprising. Halley's nucleus is an irregular, potato-shaped object, somewhat larger than astronomers had estimated. Spacecraft measurements showed it to be 15 km long by as much as 10 km wide. Also, the nucleus appeared almost jet black—as dark as finely ground charcoal or soot. The solid nucleus was enveloped by a cloud of dust, which scattered light throughout the coma. Partly because of this scatter-

ing and partly because of dimming by the dust, none of the visiting spacecraft were able to discern much surface detail on the nucleus.

The spacecraft found direct evidence for several jets of matter streaming from the nucleus. Instead of evaporating uniformly from the whole surface to form the comet's coma and tail, gas and dust apparently vent from small areas on the sunlit side of Halley's nucleus. The force of these jets may be largely responsible for the comet's observed 53-hour rotation period. Like maneuvering rockets on a spacecraft, such jets can cause a comet to change its rotation rate and even to veer away from a perfectly elliptical orbit. Astronomers had hypothesized the existence of these nongravitational forces on the basis of slight deviations from Kepler's laws observed in some cometary trajectories. However, only during the Halley encounter did astronomers actually see the jets at work.

Physical Properties of Comets

The mass of a comet can sometimes be estimated by watching how the comet interacts with other solar-system objects or by determining the size of the nucleus and assuming a density characteristic of icy composition. These methods yield typical cometary masses ranging from 10^{12} to 10^{16} kg, comparable to the masses of small asteroids. A comet's mass decreases with time, because some material is lost each time the comet rounds the

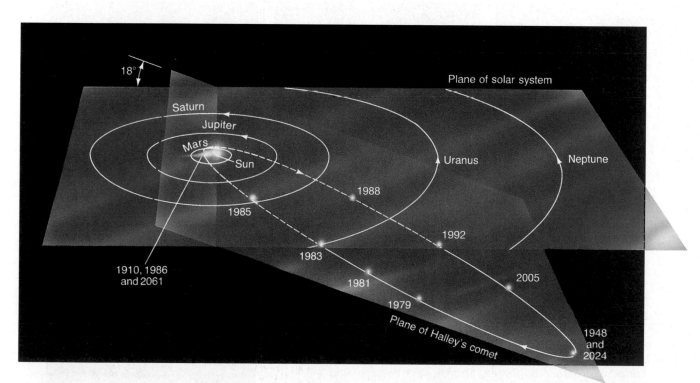

▲ **FIGURE 14.14 Halley's Orbit** Halley's comet has a smaller orbital path and a shorter period than most comets, but its orbital orientation is not typical of a short-period comet. Sometime in the past, the comet must have encountered a jovian planet (probably Jupiter itself), which threw it into a tighter orbit that extends not to the Oort cloud, but merely a little beyond Neptune. Edmund Halley applied Newton's law of gravity to predict this comet's return.

ANIMATION Anatomy of a Comet Part II

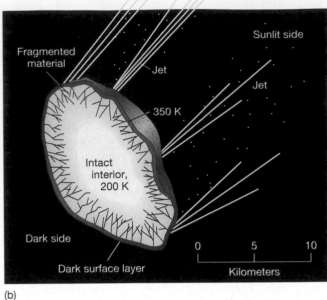

▲ **FIGURE 14.15 Halley, Up Close** (a) The *Giotto* spacecraft resolved the nucleus of Halley's comet, showing it to be very dark, although heavy dust in the area obscured any surface features. The resolution here is about 50 m—half the length of a football field. At the time this image was made, in March 1986, the comet was within days of perihelion, and the Sun was toward the right. The brightest areas are jets of evaporated gas and dust spewing from the comet's nucleus. (b) A diagram of Halley's nucleus, showing its size, shape, jets, and other physical and chemical properties. *(ESA/Max Planck Institute)*

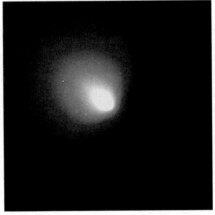

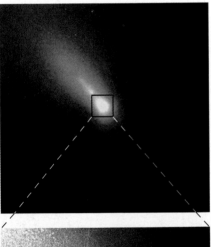

▲ **FIGURE 14.16 Comet Flare-up** As a comet approaches the heat of the Sun, its icy material vaporizes, causing the spectacular tail we observe. Sometimes, as in this sequence of *Hubble Space Telescope* images captured on three consecutive days in July 2000, the vaporization can be explosive, causing a sudden flare-up in brightness. Shortly after these images were taken, the comet's brightness diminished rapidly as its nucleus disintegrated. This was comet C/1999 S4 LINEAR, named (in part) after its "discoverer," the automated Lincoln Lab telescope in New Mexico, conducting a search for Earth-crossing asteroids. The inset shows a dramatic shower of "minicomets" as the nucleus of the comet fragmented. *(NASA)*

Sun as material evaporates from its surface. For comets that travel within an astronomical unit of the Sun, the evaporation rate can reach as high as 10^{30} molecules per second—about 30 tons of cometary material lost for every second the comet spends near the Sun (within Earth's orbit, say). Astronomers have estimated that this loss of material will destroy Halley's comet in about 5000 orbits, or 40,000 years. Occasionally, as shown in Figure 14.16, the process can be much more violent, causing a comet suddenly to flare up in brightness and then rapidly fade as its nucleus disintegrates.

In seeking the physical makeup of a cometary body itself, astronomers are guided by the observation that comets have dust that reflects light and also certain gas that emits spectral lines of hydrogen, nitrogen, carbon, and oxygen. Even as the atoms, molecules, and dust particles boil off, creating the coma and tail, the nucleus itself remains a cold mixture of gas and dust, hardly more than a ball of loosely packed ice with a density of about $100 \ kg/m^3$ and a temperature of only a few tens of kelvins. Experts now consider cometary nuclei to be made up largely of dust particles trapped within a mixture of methane, ammonia, carbon dioxide, and ordinary water ice. (These constituents should be fairly familiar to you as the main components of most of the small moons in the outer solar system, discussed in Chapters 12 and 13.) Because of this composition, comets are often described as "dirty snowballs." *Discovery 14-3* describes a spectacular event that provided astronomers with a unique alternative means of probing the makeup of one particular comet.

In February 1999, NASA launched the *Stardust* mission, with the objective of collecting the first ever samples of cometary material and returning them to Earth. In January 2001, the spacecraft used a gravity assist from Earth to boost it onto a path designed so that the craft would approach within 150 km of the nucleus of comet P/Wild 2 ("Wild" is German, pronounced "Vilt") in January 2004 (see Figure 14.17). The comet was chosen because it is a relative newcomer to the inner solar system, having been deflected onto its present orbit by an encounter with Jupiter in 1974. It therefore has not been subject to much solar heating or loss of mass by evaporation. If all goes as planned, *Stardust* will return to Earth in January 2006 with debris from a comet! Mission scientists eagerly await this unprecedented opportunity to study in detail the physical, chemical, and biological properties of a body that most probably has not changed significantly since our solar system formed.

The Kuiper Belt and Pluto

No one has ever observed any comets in the faraway Oort cloud—they are just too small and dim for us to see from Earth. But in the 1990s such faint objects began to be inventoried in the relatively nearby Kuiper belt, just beyond Neptune's orbit, some 40 to 50 A.U. from the Sun. They

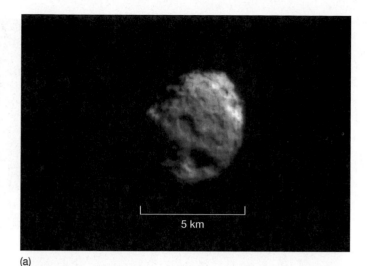

(a)

(b)

R I V U X G

▲ **FIGURE 14.17 Stardust at Wild-2** NASA's *Stardust* spacecraft captured this image (a) of comet Wild-2 in January 2004, just before the craft passed through the comet's coma. Onboard is a detector made of a foamlike gel (b) that is 99.8% air, yet is strong enough to stop and store cometary dust particles for study upon return of the craft to Earth in 2006. *(NASA)*

are collectively referred to as **Kuiper-belt objects** (or KBOs).* Ground-based telescopes have led the way in the painstaking effort to capture the meager amounts of sunlight reflected from these dark inhabitants of the outer solar system. Some of the best available images of a Kuiper-belt object are shown in Figure 14.18.

As of early 2004, the current count of Kuiper-belt objects was just 710. Astronomers estimate that the total number of KBOs larger than 100 km may exceed 100,000, so the combined mass of all the debris in the Kuiper belt could well be hundreds of times larger than the mass of the inner asteroid belt (although still less than the mass of

*The term trans-Neptunian object (TNO) is also widely used.

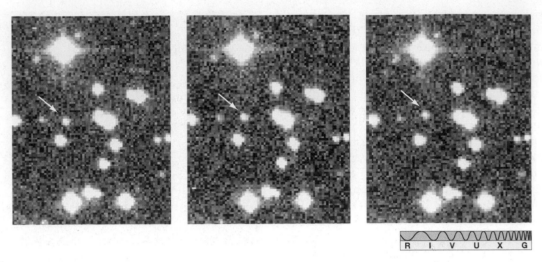

▲ **FIGURE 14.18 Kuiper-Belt Object** Some of the best available images of a Kuiper-belt object. Known as Pholus, the object itself is the fuzzy blob (marked with an arrow) that changes position between one frame and the next. It may be almost 1000 km across and lies more than 40 A.U. from Earth. *(LPL)*

DISCOVERY 14-3

A Cometary Impact

In July 1994, skywatchers were treated to an exceedingly rare event that greatly increased our knowledge of comet composition and structure: the collision of a comet (called Shoemaker–Levy 9, after its discoverers) with the planet! When it was discovered in March 1993, comet Shoemaker–Levy 9 appeared to have a curious, "squashed" appearance. Higher resolution images (see the first accompanying figure) revealed that the comet was really made up of several pieces, the largest no more than 1 km across. All the pieces were following the same orbit, but they were spread out along the comet's path, like a string of pearls a million kilometers long.

"normal" comet was captured by Jupiter and torn apart by its strong gravitational field. The data revealed an even more remarkable fact: On its next approach, roughly a year later, the comet would collide with Jupiter!

Between July 16 and July 22, 1994, fragments from Shoemaker–Levy 9 struck Jupiter's upper atmosphere, plowing into it at a speed of more than 60 km/s and causing a series of enormous explosions. Every major telescope on Earth, the *Hubble Space Telescope, Galileo* (which was only 1.5 A.U. from the planet at the time), and even *Voyager 2* were watching. Each impact created, for a period of a few minutes, a brilliant fireball hundreds of kilometers across and having a temperature of many thousands of kelvins. The largest of the fireballs

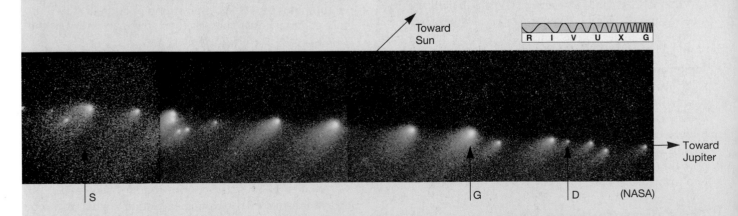

How could such an unusual object have originated? Tracing the orbit backward in time, researchers calculated that early in July 1992 the comet had approached within about 100,000 km of Jupiter. They realized that the objects shown in the figure were the fragments produced when a previously

were bigger than planet Earth. The energy released in each explosion was comparable to a billion terrestrial nuclear detonations, rivaling in violence the prehistoric impact suspected of causing the extinction of the dinosaurs on Earth 65 million years ago (see *Discovery 14-1*). One of the largest pieces of the

Earth). As the number of observed KBOs has grown, researchers have built up a better understanding of their properties. These bodies range from about 50 km to more than 1000 km in diameter—still considerably smaller than Pluto (2300 km), but now extending almost up to Pluto's moon Charon (1100 km) in size. ∞ (Sec. 13.9) The largest known, called Quaoar, was discovered in 2002.

Figure 14.19 shows the diameters of a few of the largest known members of the Kuiper belt and also includes Pluto, Charon, and Neptune's moon Triton for scale. As the details are filled in, it is becoming clear that there may not be a significant gap in size between the largest KBOs and Pluto. Indeed, many astronomers have already concluded that Pluto is not a different type of object at all, but simply the largest member of the KBO class—the "King of the Kuiper belt." In this view, Charon and Triton are Kuiper-belt objects, too, now moons of larger bodies following encounters in the early solar system. ∞ (Sec. 13.10) Officially, Pluto is still the ninth planet of the solar system, but most professional astronomers view it as a *minor planet*, having much the same status in the Kuiper belt as Ceres among the asteroids.

Are more Pluto-sized objects still out there, waiting to be discovered? The possibility of other "Plutos" orbiting at larger radii has not been conclusively ruled out. Systematic faint surveys of a broad swath of the sky that includes the entire plane of the ecliptic (as are planned within the next decade) will be needed before a definitive statement can be made.

Is Pluto a Planet?

Some people are disturbed at Pluto's apparent demotion from the solar-system "A list." After all, Pluto orbits the Sun, has enough mass to be spherical due to its own gravity, and even has its own moon. So what is the problem with calling it a planet? The answer is that Pluto *is* still a planet, just not a major one. Simply put, it is no longer sufficiently different from the other known KBOs to warrant inclusion

comet, fragment G, produced the spectacular fireball shown in the second image.

The effects on the planet's atmosphere and the vibrations produced throughout Jupiter's interior were observable for days after the impact. The fallen material from the impacts spread slowly around Jupiter's bands and reached completely around the planet after five months. It took years for all the cometary matter to settle into Jupiter's interior.

As best we can determine, none of the cometary fragments breached the jovian clouds. Only *Galileo* had a direct view of the impacts on the back side of Jupiter, and in every case the explosions seemed to occur high in the atmosphere, above the uppermost cloud layer. Most of the dark material seen in the images is probably pieces of the comet rather than parts of Jupiter. Spectral lines from silicon, magnesium, and iron were detected in the aftermath of the collisions, and the presence of these metals might explain the dark material observed near some of the impact sites (third image). Water vapor was also detected spectroscopically, again apparently from the melted and vaporized comet—which really did resemble a loosely packed snowball.

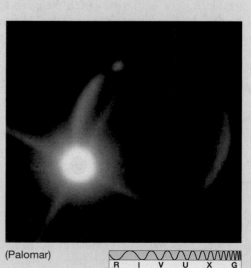

(Palomar)

R I V U X G

R I V U X G

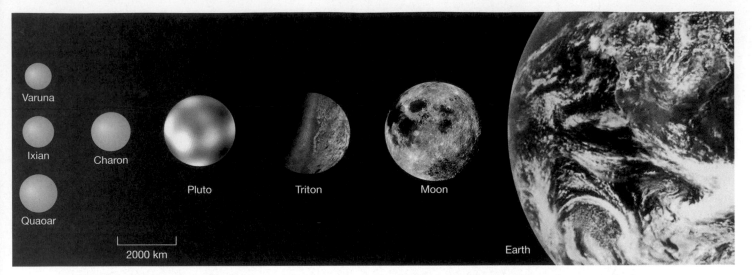

▲ **FIGURE 14.19 KBOs Compared** Some large Kuiper-belt objects, including the largest known, called Quaoar, with Pluto, Charon, Neptune's moon Triton, and the Earth–Moon system added for scale. All KBO diameters are approximate, as they must be estimated from the object's observed brightness. *(NASA; Caltech)*

in the "major planet" category. Few astronomers doubt that, if Pluto were discovered today, it would be immediately classified as a member (the largest yet, the headlines would say!) of the Kuiper belt.

Recall from Chapter 13 that Pluto was discovered in the 1930s after a search for a hypothetical object perturbing Neptune's orbit. ∞ (Sec. 13.8) We now know that Pluto is much smaller and lighter than was originally supposed and could not possibly perturb Neptune's orbit in the way imagined. Nevertheless, despite the misgivings of many astronomers of the day, Pluto became established as "the ninth planet" of our solar system. This situation parallels the discovery of the first asteroids in the early 19th century. They, too, were initially classified as planets—in fact, in the 1840s, leading astronomy texts listed no fewer than 11 planets in our solar system, including as numbers 5 through 8 the asteroids Vesta, Juno, Ceres, and Pallas! Within a couple of decades, however, the discovery of several dozen more asteroids had made it clear that these small bodies represented a whole new class of solar-system objects, separate from the major planets, and the number of major planets fell to eight (including newly discovered Neptune).

Pluto's potential (some would say inevitable) reclassification illustrates the way in which science evolves. Our conception of the cosmos has undergone many changes—some radical, others more gradual—since the time of Copernicus. ∞ (Sec. 1.2) We have seen several examples already, and we will see many more later in this book. As our understanding grows, our terminology and classifications change. Much of the observational work on the Kuiper belt began as a search for a tenth planet. It is ironic that the end result of these efforts may well be a *reduction* in the number of "true" (that is, major) solar-system planets back to eight!

14.3 Meteoroids

On a clear night, it is possible to see a few *meteors*—"shooting stars"—every hour. A **meteor** is a sudden streak of light in the night sky caused by friction between air molecules in Earth's atmosphere and an incoming piece of interplanetary matter—an asteroid, a comet, or a **meteoroid**. The friction heats and excites the air molecules, which then emit light as they return to their ground states, producing the characteristic bright streak shown in Figure 14.20. Note that the brief flash that is a meteor is in no way similar to the broad, steady swath of light associated with a comet's tail. A meteor is a fleeting event in Earth's atmosphere, whereas a comet tail exists in deep space and can be visible in the sky for weeks or even months. (Recall from Section 6.5 that the distinction between an asteroid and a meteoroid is simply a matter of size. Both are chunks of rocky interplanetary debris; meteoroids are conventionally taken to be less than 100 m in diameter.)

Before encountering the atmosphere, the piece of debris causing a meteor was almost certainly a meteoroid, simply because these small interplanetary fragments are far more common than either asteroids or comets. Any piece of interplanetary debris that survives its fiery passage through our atmosphere and finds its way to the ground is called a **meteorite**.

R   I   V   U   X

▲ **FIGURE 14.20 Meteor Trails** A bright streak called a meteor is produced when a fragment of interplanetary debris plunges into the atmosphere, heating the air to incandescence. (a) Distant stars and the northern lights provide a stunning background for a bright meteor trail. (b) These meteors (and a red smoke trail) streak across the sky during the height of the Leonid meteor storm of November 2001. *(P. Parviainen; J. Lodriguss)*

Cometary Fragments

Smaller meteoroids are mainly the rocky remains of broken-up comets. Each time a comet passes near the Sun, some cometary fragments are dislodged from the main body. The fragments initially travel in a tightly knit group of dust or pebble-sized objects, called a **meteoroid swarm**, moving in nearly the same orbit as the parent comet. Over the course of time, the swarm gradually disperses along the orbit, and eventually the **micrometeoroids**, as these small meteoroids are known, become more or less smoothly spread all the way around the parent comet's orbit. If Earth's orbit happens to intersect the orbit of such a young cluster of meteoroids, a spectacular **meteor shower** can result. Earth's motion takes our planet across a given comet's orbit at most twice a year (depending on the precise orbit of each body). Intersection occurs at the same time each year (see Figure 14.21), so the appearance of certain meteor showers is a regular and (fairly) predictable event.

▶ **FIGURE 14.21 Meteor Showers** A meteoroid swarm associated with a given comet intersects Earth's orbit at specific locations, giving rise to meteor showers at certain fixed times of the year. A portion of the comet breaks up near perihelion, at the point marked 1. The fragments continue along the comet's orbit, gradually spreading (points 2 and 3). The rate at which the debris disperses around the orbit is actually much slower than depicted here—it takes many orbits for the material to spread out as shown, but eventually the fragments extend all around the orbit, more or less uniformly. If the orbit happens to intersect Earth's, a meteor shower is seen each time Earth passes through the intersection (point 4).

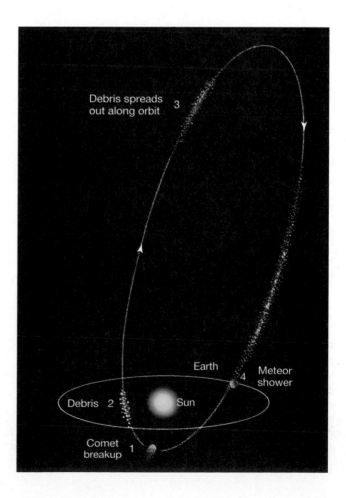

TABLE 14.1   Some Prominent Meteor Showers

Morning of Maximum Activity	Name of Shower	Rough Hourly Count	Parent Comet
Jan. 3	Quadrantid	40	—
Apr. 21	Lyrid	10	1861I (Thatcher)
May 4	Eta Aquarid	20	Halley
June 30	Beta Taurid	25[‡]	Encke
July 30	Delta Aquarid	20	—
Aug. 11	Perseid	50	1862III (Swift–Tuttle)
Oct. 9	Draconid	up to 500	Giacobini–Zinner
Oct. 20	Orionid	30	Halley
Nov. 7	Taurid	10	Encke
Nov. 16	Leonid	12[*]	1866I (Tuttle)
Dec. 13	Geminid	50	3200 (Phaeton)[†]

[*]*Every 33 years, as Earth passes through the densest region of this meteoroid swarm, we see intense showers that can exceed 1000 meteors per minute for brief periods. This intense activity is next expected to occur in 2032.*
[†]*Phaeton is actually an asteroid and shows no signs of cometary activity, but its orbit matches the meteoroid paths very well.*
[‡]*Meteor count peaks after sunrise.*

Table 14.1 lists some prominent meteor showers, the dates they are visible from Earth, and the comets from which they are thought to originate. Meteor showers are usually named for their *radiant*, the constellation from whose direction they appear to come (Figure 14.22). For example, the Perseid shower is seen to emanate from the constellation Perseus. It can last for several days, but reaches maximum every year on the morning of August 12, when upward of 50 meteors per hour can be observed. Astronomers can use the speed and direction of a meteor's flight to compute the meteor's interplanetary trajectory. This is how certain meteoroid swarms have come to be identified with well-known comet orbits. For example, the Perseid shower shares the same orbit as comet 1862III (also known as comet Swift–Tuttle), the third comet discovered in the year 1862.

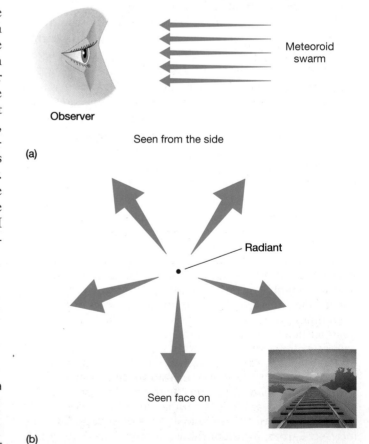

▶ **FIGURE 14.22 Radiant** (a) A group of meteoroids approaches an observer. All are moving in the same direction at the same speed. (b) From the observer's viewpoint, the trajectories of the meteoroids (and the meteor shower they produce) appear to spread out from a central point, called the radiant, in much the same way as parallel railroad tracks seem to converge in the distance (inset).

Stray Asteroids

Larger meteoroids—more than a few centimeters in diameter—are usually *not* associated with swarms of cometary debris. Generally regarded as small bodies that have strayed from the asteroid belt, possibly as the result of collisions with or between asteroids, these objects have produced most of the cratering on the surfaces of the Moon, Mercury, Venus, Mars, and some of the moons of the jovian planets. When these large meteoroids enter Earth's atmosphere with a typical velocity of nearly 20 km/s, they produce energetic shock waves, or "sonic booms," as well as bright streaks in the sky and dusty trails of discarded debris. Such large meteors are sometimes known as *fireballs*. The greater the speed of the incoming object, the hotter its surface becomes and the faster it burns up. A few large meteoroids enter the atmosphere at such high speed (about 75 km/s) that they either fragment or disperse entirely at high altitudes.

The more massive meteoroids (at least a ton in mass and a meter across) do make it to Earth's surface, producing a crater such as the kilometer-wide Barringer Crater shown in Figure 8.19. From the size of this crater, we can estimate that the meteoroid responsible for its formation must have had a mass of about 200,000 tons. Since only 25 tons of iron meteorite fragments have been found at the crash site, the remaining mass must have been scattered by the explosion at impact, broken down by subsequent erosion, or buried in the ground.

Currently, Earth is scarred with nearly 100 craters larger than 0.1 km in diameter. Most of these craters are so heavily eroded by weather and distorted by crustal activity that they can be identified only in satellite photography, as shown in Figure 14.23. Fortunately, such major collisions between Earth and large meteoroids are thought to be rare events now. Researchers estimate that, on average, they occur only once every few hundred thousand years (see *Discovery 14-1*).

The orbits of large meteorites that survive their plunge through Earth's atmosphere can be reconstructed in a manner similar to that used to determine the orbits of meteor showers. In most cases, their computed orbits do indeed intersect the asteroid belt, providing the strongest evidence we have that they were once part of the belt before being redirected, probably by a collision with another asteroid, into the Earth-crossing orbit that led to the impact with our planet. Not all meteoroids come from the asteroid belt, however: As we have already seen, some are known to have originated on the surface of Mars. ∞ *(Discovery 10-2)* In addition, detailed composition studies reveal that others most likely came from the Moon, blasted off the lunar surface by violent impacts long ago.

Not all encounters of meteoroids with Earth result in an impact. One of the most recent meteoritic events occurred in central Siberia on June 30, 1908 (Figure 14.24). The presence of only a shallow depression, as well as a complete absence of fragments, implies that this Siberian intruder exploded several kilometers above the ground, leaving a blasted depression at ground level, but no well-formed crater. Recent calculations suggest that the object in question was a rocky meteoroid about 30 m across. The explosion, estimated to have been equal in energy to a 10-megaton nuclear detonation, was heard hundreds of kilometers away and produced measurable increases in atmospheric dust levels all across the Northern Hemisphere.

▲ FIGURE 14.23 Manicouagan Reservoir This photograph, taken from orbit by the U.S. *Skylab* space station, clearly shows the ancient impact basin that forms Quebec's Manicouagan Reservoir. A large meteorite landed there about 200 million years ago. The central floor of the crater rebounded after the impact, forming an elevated central peak. The lake, 70 km in diameter, now fills the resulting ring-shaped depression. *(NASA)*

Meteorite Properties

One feature that distinguishes small micrometeoroids, which burn up in Earth's atmosphere, from larger meteoroids, which reach the ground, is their composition. The average density of meteoritic fireballs that are too small to reach the ground (but that can be captured by high-flying aircraft) is about 500–1000 kg/m³. Such a low density is typical of comets, which are made of loosely packed ice and dust. In contrast, the meteorites that reach Earth's surface are often much denser—up to 5000 kg/m³—suggesting a composition more like that of the asteroids. Meteorites like those shown in Figure 14.25 have received close scrutiny from planetary scientists—prior to the Space Age, they were the only type of extraterrestrial matter we could touch and examine in terrestrial laboratories.

◀ **FIGURE 14.24** Tunguska Debris The Tunguska event of 1908 leveled trees over a vast area. Although the impact of the blast was tremendous and its sound audible for hundreds of kilometers, the Siberian site is so remote that little was known about the event until scientific expeditions arrived to study it many years later. *(Sovfoto/Eastfoto)*

Most meteorites are rocky in composition (Figure 14.26a), although a few percent are composed mainly of iron and nickel (Figure 14.26b). The basic composition of the rocky meteorites is much like that of the inner planets and the Moon, except that some of their lighter elements—such as hydrogen and oxygen—appear to have boiled away long ago when the bodies from which the meteorites originated were molten. Some meteorites show clear evidence of strong heating at some time in their past, most likely indicating that they originated on a larger body that either underwent some geological activity or was partially melted during the collision which liberated the fragments that eventually became the meteorites. Others show no such evidence and probably date from the formation of the solar system.

Most primitive of all are the *carbonaceous* meteorites, so called because of their relatively high carbon content. These meteorites are black or dark gray and may well be related to the carbon-rich C-type asteroids that populate the outer asteroid belt. (Similarly, the silicate-rich stony meteorites are probably associated with the inner S-type asteroids.) Many carbonaceous meteorites contain significant amounts of ice and other volatile substances, and they are usually rich in organic molecules.

Finally, almost all meteorites are *old*. Direct radioactive dating shows most of them to be between 4.4 and 4.6 billion years old—roughly the age of the oldest lunar rocks. Meteorites, along with some lunar rocks, comets, and perhaps the planet Pluto, provide essential clues to the original state of matter in the solar neighborhood.

CONCEPT CHECK

✔ What are meteoroids, and why are they important to planetary scientists?

(a)

(b)

▲ **FIGURE 14.25** Large Meteorites (a) The world's second largest meteorite, the Ahnighito, on display at the American Museum of Natural History in New York, serves as a jungle gym for curious children. This 34-ton rock is so heavy that the Museum floor had to be specially reinforced to support its weight. (b) The Wabar meteorite, discovered in the Arabian desert. Although small fragments of the original meteor had been collected more than a century before, the 2000-kg main body was not found until 1965. *(Corbis-Blair; Jon Mandaville/Aramco World)*

(a)   (b)

▲ **FIGURE 14.26 Meteorite Samples** (a) A stony (silicate) meteorite often has a dark fusion crust, created when the surface of the incoming meteoroid is melted by the tremendous heat generated during its passage through the atmosphere. (b) Iron meteorites, much rarer than stony ones, usually contain some nickel as well. Most show characteristic crystalline patterns when their surfaces are cut, polished, and etched with weak acid. *(Science Graphics)*

Chapter Review

SUMMARY

More than 100,000 **asteroids (p. 356)** have been cataloged. Most orbit in a broad band called the **asteroid belt (p. 356)** between the orbits of Mars and Jupiter. These asteroids are probably primal rocks that never clumped together to form a planet. The largest ones are a few hundred kilometers across. Most are much smaller. The total mass of all asteroids combined is less than the mass of Earth's Moon. Asteroids are classified according to the properties of their reflected light: Brighter, S-type (silicate) asteroids dominate the inner asteroid belt, whereas darker, C-type (carbonaceous) asteroids are more plentiful in the outer regions. The C-type asteroids are thought to have changed little since the solar system formed. Smaller asteroids tend to be irregular in shape and may have undergone violent collisions in the past.

A few **Earth-crossing asteroids (p. 359)** have orbits that intersect Earth's orbit and will probably collide with our planet one day. The **Trojan asteroids (p. 360)** share Jupiter's orbit, remaining 60° ahead of or behind that planet as it moves around the Sun. The **Kirkwood gaps (p. 361)** in the main asteroid belt have been cleared by Jupiter's gravity.

Comets (p. 363) are fragments of icy material that normally orbit far from the Sun. Unlike the orbits of most other bodies in the solar system, comets' orbits are often highly elongated and not confined to the ecliptic plane. Most comets are thought to reside in the **Oort cloud (p. 366)**, a vast "reservoir" of cometary material, tens of thousands of astronomical units across, completely surrounding the Sun. A very small fraction of comets happen to have highly elliptical orbits that bring them into the inner solar system. Comets with orbital periods less than about 200 years are thought to originate not in the Oort cloud, but in the

Kuiper belt (p. 366), a broad band lying roughly in the plane of the ecliptic, beyond the orbit of Neptune. More than 700 **Kuiper-belt objects (p. 371)** are now known. Pluto may well be the largest member of the class.

As a comet approaches the Sun, the comet's surface ice begins to vaporize. We see the comet by the sunlight reflected from the dust and vapor that are released. The **nucleus (p. 363)**, or core, of a comet may be only a few kilometers in diameter. It is surrounded by a **coma (p. 363)** of dust and gas. Surrounding this in turn is an extensive invisible **hydrogen envelope (p. 364)**. Stretching behind the comet is a long **tail (p. 363)**, formed by the interaction between the cometary material and the solar wind. The comet's **ion tail (p. 364)** consists of ionized gas particles and always points directly away from the Sun. The comet's **dust tail (p. 364)** is less affected by the solar wind and has a somewhat curved shape. Various spacecraft visited Halley's comet in 1986 and studied its nucleus. All other studies of comets have been indirect, usually based on spectroscopic measurements. Comets are icy, dusty bodies, sometimes called "dirty snowballs," that are thought to be leftover material unchanged since the formation of the solar system. Their masses are comparable to the masses of small asteroids.

Meteors (p. 374), or "shooting stars," are bright streaks of light that flash across the sky as **meteoroids (p. 374)**—pieces of interplanetary debris—enter Earth's atmosphere. If a meteoroid reaches the ground, it is called a **meteorite (p. 374)**. The major difference between meteoroids and asteroids is their size: The dividing line between them is conventionally taken to be 100 m. Comets and stray asteroids are responsible for most of the cratering on the various worlds in the solar system. Earth is still subject to these sorts of collisions. The most recent large impact occurred in 1908, when an asteroid apparently exploded several miles above Siberia. Comet Shoemaker–Levy 9 struck Jupiter in 1994, causing violent explosions in that planet's atmosphere.

Each time a comet rounds the Sun, some cometary material becomes dislodged, forming a **meteoroid swarm (p. 375)**—a group of small **micrometeoroids (p. 375)** following the comet's

original orbit. If Earth happens to pass through the comet's orbit, a **meteor shower (p. 375)** occurs. Larger meteoroids are probably pieces of material chipped off asteroids following collisions in the asteroid belt. Meteorites are thought to be composed of the same material that makes up the asteroids, and the few orbits that have been determined are consistent with an origin in the asteroid belt. Some meteorites show evidence of heating, but the oldest ones do not. Most meteorites are between 4.4 and 4.6 billion years old.

REVIEW AND DISCUSSION

1. What are the Trojan, Apollo, and Amor asteroids?

2. How are asteroid masses measured?

3. How have the best photographs of asteroids been obtained?

4. What are the Kirkwood gaps? How did they form?

5. How do the C-type and S-type asteroids differ?

6. Are all asteroids found in the asteroid belt?

7. What are comets like when they are far from the Sun? What happens when they enter the inner solar system?

8. Where in the solar system do most comets reside?

9. Describe the various parts of a comet while it is near the Sun.

10. What are the typical ingredients of a comet nucleus?

11. How do we know what comets are made of?

12. What are some possible fates of comets?

13. Describe two ways in which a comet's orbit may change.

14. In what ways is the Kuiper belt similar to the asteroid belt? In what ways do they differ?

15. Explain the difference between a meteor, a meteoroid, and a meteorite.

16. What causes a meteor shower?

17. What do meteorites reveal about the age of the solar system?

18. Why can comets approach the Sun from any direction, but asteroids generally orbit close to the plane of the ecliptic?

19. Why do meteorites contain information about the early solar system, yet Earth does not?

20. What might be the consequences if a 10-km-diameter meteorite struck Earth today?

CONCEPTUAL SELF-TEST: TRUE OR FALSE/MULTIPLE CHOICE

1. Asteroids, meteoroids, and comets are remnants of the early solar system.

2. Most asteroids move on roughly circular orbits between the orbits of Earth and Mars.

3. The Kirkwood gaps are two broad zones within the asteroid belt in which no asteroids are found.

4. The passage of a comet near the Sun may leave a meteoroid swarm moving in the comet's orbit.

5. Some comets travel in orbits that take them up to 50,000 A.U. from the Sun.

6. Cometary orbits always lie close to the plane of the ecliptic.

7. The Oort cloud is the large cloud of gas surrounding a comet while it is near the Sun.

8. Some meteorites found on Earth originally came from the Moon or Mars.

9. Comets are the sources of many meteor showers.

10. Astronomers have succeeded in tracing the orbits of some meteorites back into the asteroid belt.

11. According to Figure 14.1 ("Inner Solar System"), the asteroid groups with the smallest perihelion distances also tend to have orbits that **(a)** are slowest; **(b)** are nearly circular; **(c)** are most eccentric; **(d)** extend nearly to Jupiter.

12. Most main-belt asteroids are about the size of **(a)** the Moon; **(b)** North America; **(c)** a US state; **(d)** a small US city.

13. Spectroscopic studies indicate that the majority of asteroids contain large fractions of **(a)** carbon; **(b)** silicate rocks; **(c)** iron and nickel; **(d)** ice.

14. Trojan asteroids orbiting at Jupiter's Lagrangian points are located **(a)** far outside Jupiter's orbit; **(b)** close to Jupiter; **(c)** behind and in front of Jupiter, sharing its orbit; **(d)** between Mars and Jupiter.

15. The tails of a comet **(a)** point away from the Sun; **(b)** point opposite the direction of motion of the comet; **(c)** curve from right to left; **(d)** curve clockwise with the interplanetary magnetic field.

16. Compared with the orbits of the short-period comets, the orbits of long-period comets **(a)** tend to lie in the plane of the ecliptic; **(b)** look like short-period orbits, but are simply much larger; **(c)** are much less eccentric; **(d)** can come from all directions.

17. A typical comet has a composition most similar to that of **(a)** the Moon; **(b)** Pluto; **(c)** a meteorite; **(d)** an asteroid.

18. According to the chart in *Discovery 14-1*, an impact resulting in global catastrophe is expected to occur roughly once per **(a)** year; **(b)** century, **(c)** millennium, **(d)** million years.

19. A meteorite is a piece of interplanetary debris that **(a)** burns up in Earth's atmosphere; **(b)** misses Earth's surface; **(c)** glances off Earth's atmosphere; **(d)** survives the trip to the surface.

20. According to Table 14.1, the meteor shower that occurs closest to the autumnal equinox is the **(a)** Lyrids; **(b)** Beta Taurids; **(c)** Perseids; **(d)** Orionids.

PROBLEMS

Algorithmic versions of these questions are available in the Practice Problems module of the Companion Website at astro.prenhall.com/chaisson.

The number of squares preceding each problem indicates its approximate level of difficulty.

1. ■ (a) The asteroid Pallas has an average diameter of 520 km and a mass of 3.2×10^{20} kg. How much would a 100-kg astronaut weigh there? (b) What is the asteroid's escape speed?

2. ■ You are standing on the surface of a spherical asteroid 10 km in diameter, of density 3000 kg/m³. Could you throw a small rock fast enough that it escapes? Give the speed required in km/s and mph.

3. ■■ How large would the asteroid in the previous problem have to be for your weight to be 1 percent of your weight on Earth?

4. ■ Can you find a simple orbital resonance with Jupiter that can account for the small Kirkwood gap evident in Figure 14.7(a) at a semimajor axis around 2.7 A.U.?

5. ■■ The asteroid Icarus (Figure 14.5) has a perihelion of 0.19 A.U. and an orbital eccentricity of 0.83. Calculate the asteroid's orbital semimajor axis and aphelion distance from the Sun. Do these figures, by themselves, *necessarily* imply that Icarus will one day collide with Earth?

6. ■■ Calculate the minimum and maximum angular diameter of the Sun, as seen from Icarus (Problem 5). How does your answer compare with the Sun's diameter as seen from Earth?

7. ■■ Using the data given in the text, estimate Dactyl's orbital period as it orbits Ida.

8. ■■■ (a) *NEAR*'s initial orbit around Icarus had a periapsis (distance of closest approach) of about 100 km from the asteroid's center. If the mass of Eros is 6.7×10^{15} kg and the orbit has an eccentricity of 0.3, calculate the spacecraft's orbital period. (b) Subsequently, *NEAR* moved into a closer orbit, with periapsis 14 km and apoapsis (greatest distance) 60 km. What was the new orbital period?

9. ■■ (a) Calculate the orbital period of a comet with a perihelion distance of 0.5 A.U. and aphelion in the Oort cloud, at a distance of 50,000 A.U. from the Sun. (b) A short-period comet has a perihelion distance of 1 A.U. and an orbital period of 125 years. What is its maximum distance from the Sun?

10. ■■■ As comet Hale–Bopp rounded the Sun, nongravitational forces changed its orbital period from 4200 years to 2400 years. By what factor did the comet's semimajor axis change? Given that the perihelion remained unchanged at 0.914 A.U., calculate the old and new orbital eccentricities.

11. ■ Astronomers estimate that comet Hale–Bopp lost mass at an average rate of about 350,000 kg/s during the time it spent close to the Sun—a total of about 100 days. Estimate the total amount of mass lost and compare it with the comet's estimated mass of 5×10^{15} kg.

12. ■■ It has been hypothesized that Earth is under continuous bombardment by house-sized "minicomets" with typical diameters of 10 m, at the rate of some 30,000 per day. Assuming spherical shapes and average densities of 100 kg/m³, calculate the total mass of material reaching Earth each year. Compare the total mass received in the past one billion years (assuming that all rates were the same in the past) with the mass of Earth's oceans (see Chapter 7, Problem 5).

13. ■ A particular comet has a total mass of 10^{13} kg, 95 percent of which is ice and dust. The remaining 5 percent is in the form of rocky fragments with an average mass of 100 g. How many meteoroids would you expect to find in the swarm formed by the breakup of this comet?

14. ■■ It is observed that the number of asteroids or meteoroids of a given diameter is roughly inversely proportional to the square of the diameter. Approximating the actual distribution of asteroids first as a single 1000-km body (e.g., Ceres), then as one hundred 100-km bodies, then as ten thousand 10-km asteroids, and so on, and assuming constant densities of 3000 kg/m³, calculate the total mass (in units of Ceres's mass) in the form of 1000-km bodies, 100-km bodies, 10-km bodies, 1-km bodies, and 100-m bodies.

15. ■■ A meteoroid was created by a collision in the asteroid belt 2 billion years ago. Its greatest distance from the Sun is 3 A.U., and its eccentricity is 0.8. How many times has the meteoroid crossed Earth's orbital radius?

In addition to the Practice Problems module, the Companion Website at astro.prenhall.com/chaisson provides for each chapter a study guide module with multiple choice questions as well as additional annotated images, animations, and links to related Websites.

15

The formation of our solar system was a long-ago event, with much of the matter of our primordial galactic cloud eventually either comprising the Sun and planets or ejected back into deep space. Now, some 4.5 billion years later, it is not easy to reconstruct what exactly did happen here. Astronomers therefore observe other young star systems, hoping to gain some insight about the origins of our own solar system. Here, the new Spitzer Space Telescope has taken this infrared image of Herbig-Haro 46/47, where an embryonic star about 1200 light-years away has ▶ formed. (NASA)

The Formation of Planetary Systems

The Solar System and Beyond

Having completed the chapters on the planets, you may be struck by the vast range of physical and chemical properties found in the solar system. The planets present a long list of interesting features and bizarre peculiarities, and the list grows even longer when we also consider their moons. Every object has its idiosyncrasies, some of them due to particular circumstances, others the result of planetary evolution. Each time a new discovery is made, we learn a little more about the properties and history of our planetary system. Still, our astronomical neighborhood might seem more like a great junkyard than a smoothly running planetary system. Can we really make any sense of the collection of solar-system matter? Is there some underlying principle that unifies the knowledge we have gained? And if there is, Does it extend to planetary systems beyond our own? The answer, as we will see, is "Maybe...."

LEARNING GOALS

Studying this chapter will enable you to

1 Summarize the major planetary features that a theory of solar-system origins must explain.

2 Outline the process by which planets form as natural by-products of star formation, and explain how that process accounts for the overall properties of our solar system.

3 Describe how comets and asteroids formed, and explain their role in determining planetary properties.

4 Discuss the role of collisions in determining specific characteristics of the solar system.

5 Outline the properties of known extrasolar planets, and explain how they differ from planets in the solar system.

6 Discuss how observations of extrasolar planets challenge current theories of solar-system formation.

 Visit astro.prenhall.com/chaisson for additional annotated images, animations, and links to related sites for this chapter.

15.1 Modeling Planet Formation

The origin of the planets and their moons is a complex and as yet incompletely solved puzzle, although the basic outlines of the processes involved are becoming understood. ∞ (Sec. 6.7) Most of our knowledge of the solar system's formative stages has emerged from studies of interstellar gas clouds, fallen meteorites, and Earth's Moon, as well as of the various planets observed with ground-based telescopes and planetary space probes. Ironically, studies of Earth itself do not help much, because information about our planet's early stages eroded away long ago. Meteorites and comets provide perhaps the most useful information, for nearly all have preserved within them traces of solid and gaseous matter from the earliest times.

Until the mid-1990s, theories of the formation of planetary systems concentrated almost exclusively on our own solar system, for the very good reason that astronomers had no other examples of planetary systems against which to test their ideas. However, all that has now changed. As of early 2004, we know of more than a hundred **extrasolar planets**—planets orbiting stars other than the Sun—to challenge our theories. And challenge them they do! As we will see, the other planetary systems discovered to date seem to have properties quite different from our own and may well require us to radically rethink our conception of how stars and planets form.

Still, although we now know of many such extrasolar systems, we currently have only limited information on each—little more than estimates of orbits and masses for the largest planets. Accordingly, we begin our study by outlining the comprehensive theory that accounts, in detail, for most of the observed properties of our own planetary system: the solar system. Later we will return to the observations of extrasolar planets and assess how our theory holds up in the face of these new data.

Model Requirements

Any theory of the origin and architecture of our planetary system must adhere to the known facts. We know of 10 outstanding properties of our solar system as a whole:

1. *Each planet is relatively isolated in space.* The planets exist as independent bodies at progressively larger distances from the central Sun; they are not bunched together. In rough terms, each planet tends to be twice as far from the Sun as its next inward neighbor.

2. *The orbits of the planets are nearly circular.* In fact, with the exceptions of Mercury and Pluto, which we will argue are special cases, each planetary orbit closely describes a perfect circle.

3. *The orbits of the planets all lie in nearly the same plane.* The planes swept out by the planets' orbits are accu-

rately aligned to within a few degrees. Again, Mercury and Pluto are slight exceptions.

4. *The direction in which the planets orbit the Sun (counterclockwise as viewed from above Earth's North Pole) is the same as the direction in which the Sun rotates on its axis.* Virtually all the large-scale motions in the solar system (other than comets' orbits) are in the same plane and in the same sense. The plane is that of the Sun's equator, and the sense is that of the Sun's rotation.

5. *The direction in which most planets rotate on their axis is roughly the same as the direction in which the Sun rotates on its axis.* This property is less general than the one just described for revolution, as three planets—Venus, Uranus, and Pluto—do not share it.

6. *Most of the known moons revolve about their parent planets in the same direction that the planets rotate on their axes.*

7. *Our planetary system is highly differentiated.* The inner, terrestrial planets are characterized by high densities, moderate atmospheres, slow rotation rates, and few or no moons. By contrast, the jovian planets, farther from the Sun, have low densities, thick atmospheres, rapid rotation rates, and many moons.

8. *The asteroids are very old and exhibit a range of properties not characteristic of either the inner or the outer planets or their moons.* The asteroid belt shares, in rough terms, the bulk orbital properties of the planets. However, it appears to be made of primitive, unevolved material, and the meteorites that strike Earth are the oldest rocks known.

9. *The Kuiper belt is a collection of asteroid-sized icy bodies orbiting beyond Neptune.* Pluto may well be a member of the class of Kuiper-belt objects.

10. *The Oort-cloud comets are primitive, icy fragments that do not orbit in the plane of the ecliptic and reside primarily at large distances from the Sun.* While similar to the Kuiper belt in composition, the Oort cloud is a completely distinct part of the outer solar system.

All these observed facts, taken together, strongly suggest a high degree of order within our solar system, at least on large scales. The whole system is not a random assortment of objects spinning or orbiting this way or that. Rather, the overall organization points toward a single origin—an ancient, but one-time, event that occurred 4.6 billion years ago. ∞ (Sec. 14.3) A convincing theory that explains all the features just listed has been a goal of astronomers for centuries.

In Chapter 6, we saw how the modern condensation theory can account for much of the basic architecture of our planetary system. ∞ (Sec. 6.7) In this chapter, we consider some of the ramifications of that theory and try to connect it with the growing list of planetary systems beyond our own.

Planetary Irregularities

While our theory of the solar system must explain the facts just listed, it is equally important to recognize what it does *not* have to explain. There is plenty of scope for planets to evolve after their formation, so things that may have happened after the initial state of the solar system was established need not be included in our list. Examples are Mercury's 3:2 spin–orbit coupling, Venus's runaway greenhouse effect, the Moon's synchronous rotation, the emergence of life on Earth and its apparent absence on Mars, the Kirkwood gaps in the asteroid belt, and the rings and atmospheric appearance of the jovian planets. There are many more. Indeed, all the properties of the planets for which we have already provided an *evolutionary* explanation need not be included as items that our theory must account for at the outset.

In addition to its many regularities, our solar system has many notable *irregularities*, some of which we have already mentioned. Far from threatening our theory, however, these irregularities are important facts for us to consider in shaping our explanations. For example, it is necessary that the explanation for the solar system not insist that *all* planets rotate in the same sense or have *only* prograde moons, because that is not what we observe. Instead, the theory of the solar system should provide strong reasons for the observed planetary characteristics, yet be flexible enough to allow for and explain the deviations, too. And, of course, the existence of the asteroids and comets that tell us so much about our past must be an integral part of the picture. That's quite a tall order, yet many researchers now believe that we are close to that level of understanding.

CONCEPT CHECK

✔ Why is it important that a theory of solar-system formation make clear statements about how planets arose, yet not be too rigid in its predictions?

15.2 Planets in the Solar System

2 Armed with our new knowledge of the planets and their moons, let's take another look at how the solar system formed. We begin with a brief review of the condensation theory. Figure 15.1 is an adaptation of Figure 6.20 and illustrates the various processes at work.

Condensation and Accretion

Modern models trace the formative stages of our solar system along the following broad lines. The story starts with a dusty interstellar cloud fragment perhaps 100,000 A.U.—roughly a light-year—across (Figure 15.1a). The cloud is composed mainly of hydrogen and helium, the dominant constituents of the universe, but intermingled with these gases are microscopic dust grains made up of heavier elements—carbon, nitrogen, silicon, and iron. The dust will play a crucial role in the formation of planets.

Some external influence, such as the passage of another interstellar cloud or perhaps the explosion of a nearby star, starts the fragment contracting. As the cloud collapses, it rotates faster (because of the law of conservation of angular momentum) and begins to flatten. ∞ (*More Precisely 6-2*) By the time it has shrunk to a diameter of 100 A.U., the *solar nebula*, as it is known at this stage, has formed an extended, rotating disk (Figure 15.1b).

As the disk cools, dust grains form *condensation nuclei*, providing the means by which the first clumps of solid matter form within the solar nebula. These clumps inherit the spinning motion of the nebula and move in roughly circular orbits around the Sun. Subsequently, they grow rapidly by *accretion*, colliding with and sticking to other clumps, ultimately to form the planets we know today (Figure 15.1c–d).

Simulations indicate that, in perhaps as little as 100,000 years, the clumps grow to the size of small moons (a few hundred kilometers across). At that time, their gravitational pulls become just strong enough to affect their neighbors. Astronomers call these objects **planetesimals**—the building blocks of the solar system (Figure 15.1d–e). Figure 15.2 shows two infrared views of relatively nearby stars thought to be surrounded by disks in which planetesimals are growing.

Gravitational forces between planetesimals cause them to collide and merge, forming larger and larger objects. Because larger objects have stronger gravity, the rich become richer in the early solar system, and eventually, almost all the planetesimal material is swept up into a few large protoplanets—the accumulations of matter that will ultimately evolve into the planets we know today (Figure 15.1f).

The condensation theory is an example of an *evolutionary* theory—a theory that describes the development of the solar system as a series of gradual and natural steps, understandable in terms of well-established physical principles. Evolutionary theories may be contrasted with *catastrophic* theories, which invoke accidental or unlikely celestial events to interpret observations.* Scientists generally try not to invoke catastrophes to explain the universe. However, as we will see, there are instances where pure chance has played a critical role in determining the present state of the solar system.

---

*A good example of such a theory is the collision hypothesis, which imagines that the planets were torn from the Sun by a close encounter with a passing star. This hypothesis enjoyed some measure of popularity during the 19th century, due in large part to the inability of the nebular theory to account for the observed properties of the solar system, but no scientist takes it seriously today (see Discovery 15-1). Aside from its extreme improbability, the collision hypothesis is completely unable to explain the orbits, the rotations, or the composition of the planets and their moons.

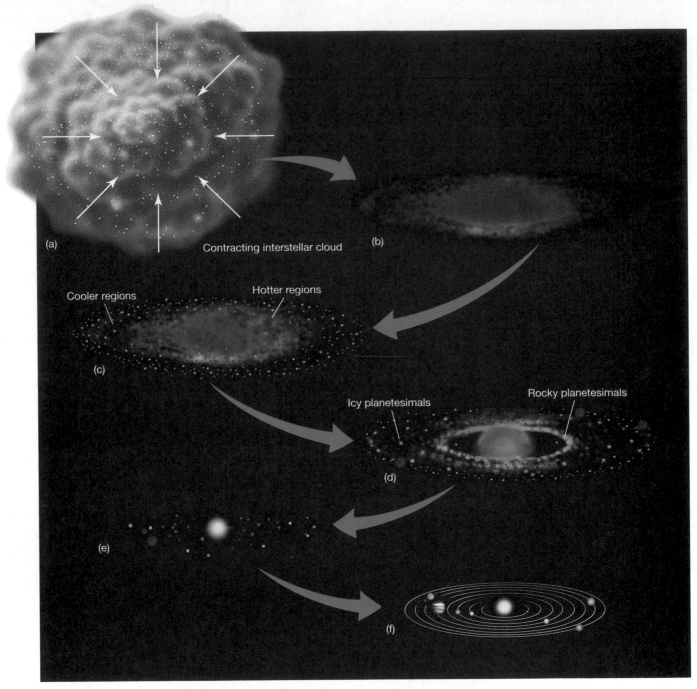

▲ **FIGURE 15.1 Solar-System Formation** The "standard" condensation theory of planet formation (adapted from Figure 6.20). (a) An infalling interstellar cloud, actually very much larger than shown here. (b) The solar nebula after it has contracted and flattened to form a spinning disk. The temperature is greatest in the center, near the still-forming Sun, and coolest at the edges. (c) Dust grains act as condensation nuclei, forming clumps of matter that collide, stick together, and grow into moon-sized (and larger) planetesimals. The composition of any given planetesimal (and hence of the grains of which it is composed) depends on the location of the planetesimal within the nebula. (d) As strong winds from the Sun start to expel the nebular gas, some large planetesimals in the outer solar system have already begun to accrete gas from the nebula. (e) With the gas ejected, planetesimals continue to collide and grow. The gas giant planets are already formed. (f) Over the course of a few hundred million years, planetesimals are accreted or ejected, leaving a few large planets that travel in roughly circular orbits.

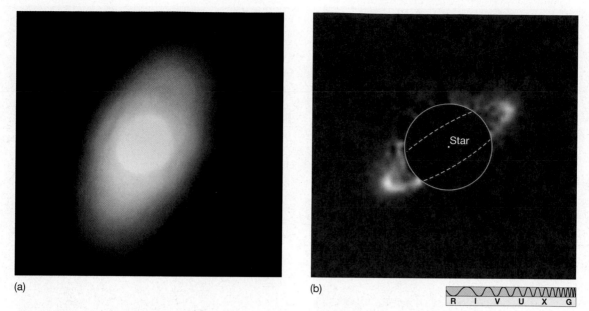

(a)

(b)

R I V U X G

▲ **FIGURE 15.2 Newborn Solar Systems?** (a) This infrared image, taken by the *Spitzer Space Telescope*, of the bright star Fomalhaut, some 25 light-years from Earth, shows a circumstellar disk in which the process of accretion is underway. The star itself is well inside the yellowish blob at center. The outer disk, which is falsely colored orange to match the dust emission, is about three times the diameter of our solar system. (b) This higher-resolution *Hubble Space Telescope* image blocks out the central (circled) parts of another such disk around a more distant star HR4796A, but shows the edges more clearly. *(NASA)*

Making Terrestrial Planets

The high temperatures near the Sun meant that only rocky and metallic grains could survive in the inner part of the solar nebula, leading to the formation of predominantly rocky planetesimals and, in time, of the terrestrial planets. These heavier materials condensed into grains in the outer solar system, too. However, they were vastly outnumbered by the far more abundant light elements there. Thus, the outer solar system is not deficient in heavy elements; rather, the inner solar system is *underrepresented in light material.*

Figure 15.3 shows a computer simulation of accretion in the inner solar system over the course of about 100 million years, spanning parts (d–f) of Figure 15.1. Notice how, as the number of bodies decreases, the orbits of the remainder become more widely spaced and more nearly circular. The fact that this particular simulation produced exactly four terrestrial planets is strictly a matter of chance. The nature of the accretion process means that the eventual outcome is controlled by random events—which objects happen to collide and when. However, regardless of the precise number of planets formed, the computer models do generally reproduce both the planets' approximately circular orbits and their increasing orbital spacing as one moves outward from the Sun.

As the protoplanets grew, another process became important. The ever-stronger gravitational fields of the growing protoplanets produced many high-speed collisions between planetesimals and protoplanets. These collisions led to **fragmentation**, as small objects broke into still smaller chunks, which were then swept up by the proto-planets. Not only did the rich get richer, but the poor were mostly driven to destruction! Some of the fragments produced the intense meteoritic bombardment we know occurred during the early evolution of the planets and moons, as we have seen repeatedly in the last few chapters. ∞ (Secs. 8.9, 11.5, 13.6)

The relative scarcity of the rocky and metallic elements that formed grains in the hot inner solar nebula is one important reason that the terrestrial planets never became as massive as the jovian worlds. In addition, the temperature of the inner parts of the nebula had to drop to the point where grains appeared and accretion could begin, whereas the temperature of the outer regions was initially low enough that accretion began almost with the formation of the nebula itself.

Making Jovian Planets

The accretion picture just described has become the accepted model for the formation of the terrestrial planets. However, while similar processes may also have occurred in the outer solar system, the origin of the giant jovian worlds is decidedly less clear. Two somewhat different views, with potentially important consequences for our understanding of extrasolar planets, have emerged.

The first, more conventional, scenario is the chain of events described in Section 6.7 and illustrated in Figure 15.1. With raw material readily available in the form of abundant icy grains, protoplanets in the outer solar system grew quickly and soon became massive enough for their strong gravitational fields to capture large amounts of gas directly from the solar nebula. In this view, called the

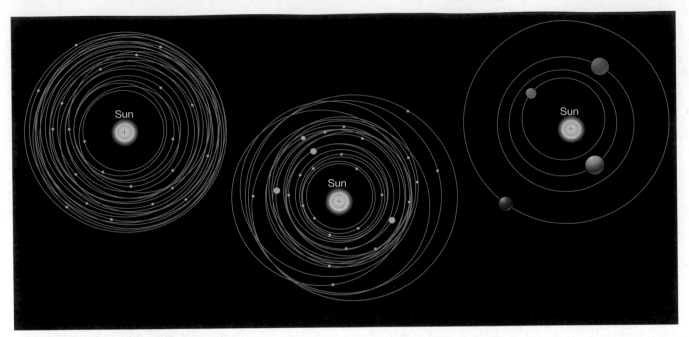

▲ FIGURE 15.3 Making the Inner Planets Accretion in the inner solar system: Initially, many moon-sized planetesimals orbited the Sun. Over the course of about 100 million years, they gradually collided and coalesced, forming a few large planets in roughly circular orbits.

core-accretion theory, four large protoplanets became the cores of the jovian worlds. ∞ (Secs. 11.1, 12.1, 13.5)

Recently, some astronomers have highlighted a potentially serious snag in this picture: There may not have been enough *time* for these events to have taken place. Most young stars apparently go through a highly active evolutionary stage known as the *T Tauri* phase (see Chapter 19), in which their radiation and stellar winds become very intense. This was the period mentioned in Chapter 6 (and indicated in Figure 15.1d) during which much of the nebular gas between the planets was blown away into interstellar space (Figure 15.4).

The problem is that the nebular disk was probably at most a few million years old when all this occurred, leaving very little time for the large jovian cores to grow and capture gas from the nebula before it was destroyed. Furthermore, some researchers argue that, in the relatively dense stellar environments in which most stars are born (see Chapter 19), close encounters between still-forming stars may destroy many disks even sooner than that, giving giant planets perhaps as little as a few hundred thousand years in which to form.

The second formation scenario suggests that the giant planets formed through *instabilities* in the cool outer regions of the solar nebula, where portions of the cloud began to collapse under their own gravity—a picture not so far removed from Laplace's basic idea—mimicking, on small scales, the collapse of the initial interstellar cloud. ∞ (Sec. 6.7) In this alternative **gravitational instability theory**, illustrated in Figure 15.5, the jovian protoplanets formed directly from the nebular gas, skipping the initial condensation-and-accretion stage and perhaps taking no more than a thousand years to acquire much of their mass. Right from the start, these first protoplanets had gravitational fields strong enough to scoop up more gas and dust from the solar nebula, allowing them to grow into the giants we see today before the gas supply dispersed.

If both these theories eventually lead to gas-rich jovian planets, how can we distinguish between them? One possible way involves the composition of the rocky jovian cores. ∞ (Secs. 11.1, 12.2, 13.3) Because the planets formed so quickly in the instability theory, computer models suggest that their cores should contain no more than about six Earth masses of rocky material. The core-accretion theory, by contrast, predicts much larger core masses—up to 20 times that of Earth. Detailed measurements of the jovian interiors by future space missions could settle the argument. Another decisive observation would be the detection of a Jupiter-sized extrasolar planet orbiting far from its parent star—out where Neptune is in our solar system. The accretion theory suggests that such a planet would take too long (100 million years or more) to form, so finding one would argue strongly in favor of the instability model.

Many of the moons of the jovian planets presumably also formed through accretion, but on a smaller scale, in the gravitational fields of their parent planets. Once the nebular gas began to accrete onto the large jovian protoplanets, conditions probably resembled a miniature solar nebula, with condensation and accretion continuing to occur. The large moons of the outer planets (with the possible exception of Triton) almost certainly formed in this way. ∞ (Secs. 11.5, 13.6) The smaller moons are more likely captured planetesimals.

Giant-Planet Migration

Many aspects of the formation of the giant planets remain unresolved. Interactions among the growing planets, and between the planets and their environment, probably played a critical role in determining just how and where the planets formed. One particularly intriguing scenario, accepted by some—but not all—planetary scientists, is the possibility that Jupiter—and maybe all four giant planets—formed considerably farther from the Sun than its present orbit and subsequently "migrated" inward.

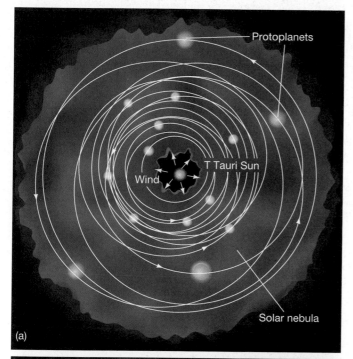

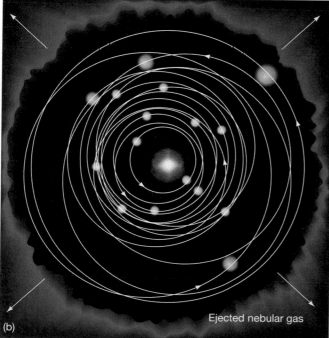

▲ **FIGURE 15.4 T Tauri Star** (a) Strong stellar winds from the newborn Sun sweep away the gas disk of the solar nebula, (b) leaving only giant planets and planetesimals behind. This stage of stellar evolution occurs only a few million years after the formation of the nebula.

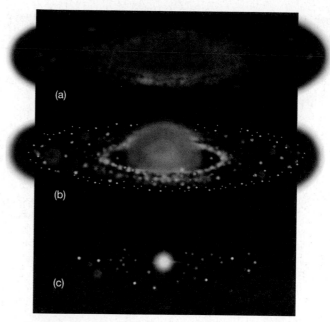

▲ **FIGURE 15.5 Jovian Condensation** As an alternative to the growth of massive protoplanetary cores followed by the accretion of nebular gas, it is possible that some or all of the giant planets formed directly via instabilities in the cool gas of the outer solar nebula. Part (a) shows the same instant as Figure 15.1(b). (b) Only a few thousand years later, four gas giants have already formed, preceding and circumventing the accretion process sketched in Figure 15.1. With the nebula gone (c), the giant planets have taken their place in the outer solar system. (See Figure 15.1e.)

This supposed migration is indicated schematically in Figures 15.1 and 15.5 by the changing locations of the jovian protoplanets.

The idea of planetary migration has been around since the mid-1980s, when theorists realized that friction between massive planets and the nebula in which they moved would have caused just such an inward drift. Observational support came in 1999, when *Galileo* scientists announced much higher than expected concentrations of the gases nitrogen, argon, krypton, and xenon in Jupiter's atmosphere. ∞ (Sec. 11.2) These gases, which are thought to have been carried to the planet by captured planetesimals, could not have been retained in the planetesimal ice at temperatures typical of Jupiter's current orbit. Instead, they imply that the planetesimals—and, presumably, Jupiter too—formed at much lower temperatures. Either the nebula was cooler than previously thought, or Jupiter formed out in what is now the Kuiper belt!

CONCEPT CHECK

✔ Why is the rate at which the Sun formed important in a theory of the formation of the jovian planets?

DISCOVERY 15-1

The Angular Momentum Problem

According to Laplace's nebular hypothesis, the Sun and solar system formed from a contracting, spinning cloud of interstellar gas. ∞ (Sec. 6.7) Conservation of angular momentum caused the cloud to spin faster as it collapsed, and this in turn caused it to flatten into a disk, from which the planets eventually formed. ∞ (*More Precisely 6-2*) This simple, yet elegant, idea underlies the modern condensation theory of planetary formation. However, not long after Laplace proposed his explanation for the basic architecture of the solar system, astronomers uncovered what appeared to be a serious flaw in the theory. It has to do with the distribution of angular momentum in the solar system.

Although our Sun contains about a thousand times more mass than all the planets combined, it possesses a mere 0.3 percent of the total angular momentum of the solar system. Jupiter, for example, has a lot more angular momentum than does our Sun. In fact, because of its large mass and great distance from the Sun, Jupiter holds about 60 percent of the solar system's total angular momentum. All told, the four jovian planets account for well over 99 percent of the angular momentum of the entire present-day solar system. (The lighter and closer terrestrial planets contribute negligibly to the total.)

The problem here is that the law of conservation of angular momentum predicts that the Sun should have been spinning very rapidly during the earliest epochs of the solar system. The Sun should command most of the solar system's angular momentum basically because it contains most of the mass. However, as we have just seen, the reverse is true. Indeed, if all the planets, with their large amounts of orbital angular momentum, were placed inside the Sun, it would spin on its axis about a hundred times faster than it does at present. Somehow, the Sun must have lost (or perhaps never gained) most of this angular momentum. This discrepancy between observations and theoretical expectations is known as the *angular momentum problem*.

Nineteenth-century scientists had no ready explanation of how the Sun could have shed its angular momentum, leading some to abandon the nebular theory even though it provides a natural explanation of many of the planets' orbital properties. For a time, the *collision hypothesis*, in which the planets (or preplanetary blobs) were ripped from the Sun by the gravitational pull of a passing star, gained some degree of prominence. The fact that this latter theory persisted despite the extreme improbability of such a stellar encounter, as well as the theory's own difficulties in explaining planetary orbits, is a measure of how seriously the angular momentum problem was taken by scientists of the time.

Although the details remain uncertain, astronomers today surmise that the Sun transferred much of its spin angular momentum to the orbital angular momentum of the planets via the solar nebular disk. Friction within the disk would have caused the rapidly spinning inner regions (i.e., the embryonic Sun) to slow, while causing the slowly spinning outer regions (where the planets would someday form) to speed up. The net effect was to move angular momentum outward from the Sun to the planets.

In addition, many researchers speculate that the solar wind, moving away from the Sun into interplanetary space, may have carried away much of the Sun's remaining angular momentum. ∞ (Sec. 6.5) The early Sun probably produced more of a dense solar gale than the relatively gentle "breezes" now measured by our spacecraft. High-velocity particles leaving the Sun followed the solar magnetic field lines. As the rotating magnetic field of the Sun tended to drag those particles around with it, they acted as a brake on the Sun's spin. The accompanying figure illustrates the process. This interaction between the solar wind and the Sun's magnetic field was completely unknown to 19th-century astronomers.

Although each particle that is boiled off the Sun carries only a tiny amount of the Sun's angular momentum with it, over the course of nearly 5 billion years the vast numbers of escaping particles have probably robbed the Sun of most of its initial spin momentum. Even today, our Sun's spin continues to slow. A similar mechanism operating while the Sun was expelling the relatively dense nebular disk could also have transported a lot of solar angular momentum into interstellar space.

Today, the angular momentum problem represents a minor source of uncertainty, rather than a fundamental challenge, to the nebular–condensation theory. With new physical insights, the problem has been greatly reduced in severity—if not completely explained away—and the nebular hypothesis is once again central to theories of solar-system formation.

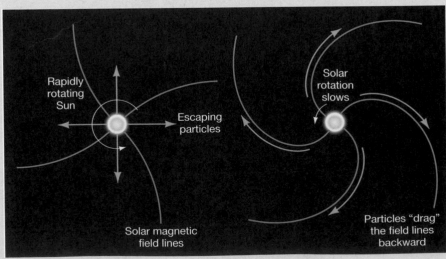

Rapidly rotating Sun

Escaping particles

Solar magnetic field lines

Solar rotation slows

Particles "drag" the field lines backward

15.3 Interplanetary Debris

LEARNING GOAL 3 Regardless of the precise chain of events that led to the formation of the gas-rich jovian planets, we know that, after the solar nebula was ejected into interstellar space, all that remained in orbit around the Sun were protoplanets and planetesimal fragments, ready to continue their long evolution into the solar system we know today. To place all these formative processes in perspective, Figure 15.6 presents a simplified time line of the first billion years after the formation of the solar nebula.

The Asteroid Belt

In the inner solar system, planetesimal fragments that escaped capture by one of the terrestrial planets received repeated "gravity assists" from those bodies and were eventually boosted beyond the orbit of Mars. ∞ (*Discovery 6-1*) Roughly a billion years were required to sweep the inner solar system clear of interplanetary "trash." This was the period that saw the heaviest meteoritic bombardment, most evident on Earth's Moon and tapering off as the number of planetesimals decreased. ∞ (Sec. 8.5)

The myriad rocks of the asteroid belt between Mars and Jupiter failed to accumulate into a planet. Probably, nearby Jupiter's huge gravitational field caused them to collide too destructively to coalesce. Strong tidal forces from Jupiter on the planetesimals in the belt would also have hindered the development of a protoplanet. The result is a band of rocky planetesimals, still colliding and occasionally fragmenting, but never coalescing into a larger body.

Comets and the Kuiper Belt

In the outer solar system, with the formation of the four giant jovian planets, the remaining planetesimals were subject to those planets' strong gravitational fields. Over a period of hundreds of millions of years, interactions with the giant planets, especially Uranus and Neptune, flung many of the outer region's interplanetary fragments into orbits taking them far from the Sun (Figure 15.7). Astronomers think that those icy bodies now make up the Oort cloud, whose members occasionally visit the inner solar system as comets. ∞ (Sec. 14.2)

A key prediction of this model is that some of the original planetesimals remained behind, forming the broad band known as the Kuiper belt, lying beyond the orbit of Neptune, some 30 to 40 A.U. from the Sun. ∞ (Sec. 14.2) Some 900 Kuiper-belt objects, having diameters ranging from 50 km to 1000 km, are now known. Their existence

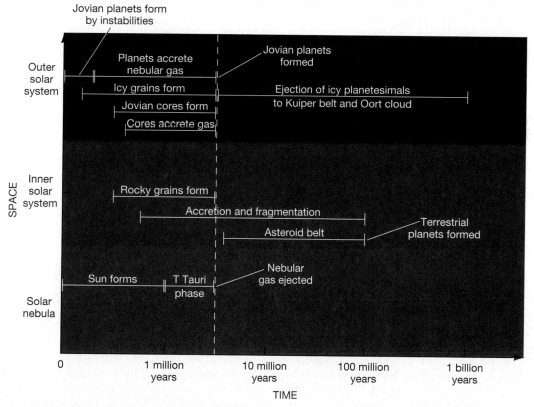

▲ **FIGURE 15.6 Solar-System Formation** Schematic time line of some key events occurring during the first billion years of our solar system. The various tracks show the evolution of the Sun and the solar nebula, as well as that of the inner and outer solar system. Note that the tracks are intended to illustrate approximate relationships between events, not the precise times at which they occurred.

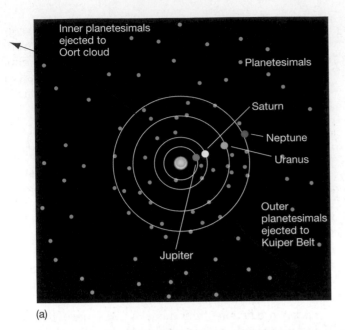

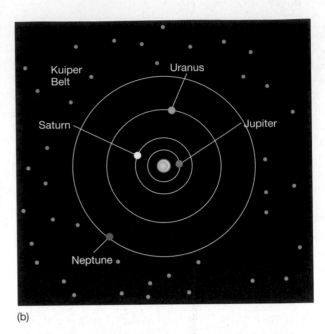

▲ **FIGURE 15.7 Planetesimal Ejection** The ejection of icy planetesimals to form the Oort cloud and Kuiper belt. (a) After the giant planets had formed, leftover planetesimals were found throughout the solar system. It is thought that interactions with Jupiter and Saturn "kicked" planetesimals out to very large radii (the Oort cloud). Interactions with Uranus and especially Neptune tended to keep the Kuiper belt populated, but also deflected some planetesimals inward to interact with Jupiter and Saturn. (b) As a result of the inward and outward "traffic," the orbits of all four giant planets were significantly modified by the time the planetesimals interior to Neptune's orbit had been ejected. Neptune was affected most and may have moved outward by as much as 10 A.U.

lends strong support to the condensation theory of planetary formation.

Computer simulations reveal that the ejection of the planetesimals involved a remarkably complex interplay among the jovian planets, whereby Uranus and Neptune kicked some bodies out into the Kuiper belt, but deflected others *inward* toward Jupiter and Saturn, whose strong gravitational fields then propelled the planetesimals out into the distant Oort cloud. As shown in Figure 15.7, the orbits of all four giant planets were significantly modified by these interactions. By the time the outer solar system had been cleared of comets, Jupiter had moved slightly closer to the Sun, its orbital semimajor axis decreasing by a few tenths of an A.U. The other giant planets moved outward—Saturn by about 1 A.U., Uranus by 3 or 4 A.U., and Neptune by some 7–10 A.U. Note that these orbital changes occurred long *after* the supposed inward migrations mentioned earlier. Life as a jovian planet is far from simple!

Strong support for the preceding ideas comes from one curious feature of the Kuiper belt: A large fraction of its members—perhaps 15 percent of the entire belt—orbit in a *3:2 resonance* with Neptune. That is, they orbit the Sun twice for every three orbits of Neptune. Pluto shares this resonance, and the Kuiper-belt objects orbiting in this manner have accordingly been dubbed **plutinos**. ∞ (Sec. 13.8) The presence of one object (Pluto) in such an exceptional orbit might conceivably be attributed to chance: It is possible for a planet such as Neptune to "capture" a Pluto-sized

object into a resonant orbit, if the object's orbit started out close to resonance. However, to account for more than 100 similar orbits, a more comprehensive theory is needed.

The leading explanation of the plutinos is that, as Neptune's orbit moved slowly outward, the radius corresponding to the 3:2 resonance also swept outward through the surviving planetesimals. Apparently, this process was slow enough that many, if not most, of the planetesimals on near-resonant orbits (Pluto included) were captured and subsequently carried outward, locked forever in synchrony with Neptune as that planet's orbit drifted outward to its present location. These are the plutinos we see today.

During this period, many icy planetesimals were also deflected into the inner solar system, where they played an important role in the evolution of the inner planets. A long-standing puzzle in the condensation theory's account of the formation of the inner planets has been where the water and other volatile gases on Earth and elsewhere originated. At the inner planets' formation, their surface temperatures were far too high, and their gravity too low, to capture or retain those gases. The most likely explanation seems to be that the water and other light gases found on Earth and elsewhere in the inner solar system arrived there in the form of comets from the outer solar system. Kicked into eccentric orbits as the gravitational fields of the jovian planets cleared the outer solar system of leftover planetesimals, these icy fragments bombarded the newborn terrestrial worlds, supplying them with water *after* their formation.

✔ Might you expect to find comets and asteroids orbiting other stars?

15.4 The Role of Catastrophes

The condensation theory accounts for the 10 "characteristic" points listed at the start of this chapter. Specifically, the planets' orbits are nearly circular (2), in the same plane (3), and in the same direction as the Sun's rotation on its axis (4) as a direct consequence of the nebula's shape and rotation. The rotation of the planets (5) and the orbits of the moon systems (6) are due to the tendency of the smaller scale condensations to inherit the nebula's overall sense of rotation. The growth of planetesimals throughout the nebula, with each protoplanet ultimately sweeping up the material near it, accounts for (1) the fact that the planets are widely spaced (even if the theory does not quite explain the regularity of the spacing). The heating of the nebula and the Sun's eventual ignition resulted in the observed differentiation (7), and the debris from the accretion–fragmentation stage naturally accounts for the asteroids (8), the Kuiper belt (9), and the Oort-cloud comets (10).

Earlier, we stressed the fact that an important aspect of any solar-system theory is its capability to allow for the possibility of imperfections—deviations from the otherwise well-ordered scheme of things. In the condensation theory, that capability is provided by the randomness inherent in the encounters that ultimately combined the planetesimals into protoplanets. As the numbers of large bodies decreased and their masses increased, individual collisions acquired greater and greater importance. The effects of these collisions can still be seen today in many parts of the solar system—for example, the large craters on many of the moons we have studied thus far.

Having started with 10 regular points to explain, we end our discussion of solar-system formation with the following eight irregular features that still fall within the theory's scope:

1. Mercury's exceptionally large nickel–iron core may be the result of a collision between two partially differentiated protoplanets. The cores may have merged, and much of the mantle material may have been lost. ∞ (Sec. 8.7)

2. Two large bodies could have merged to form Venus, giving it its abnormally low rotation rate. ∞ (Sec. 9.2)

3. The Earth–Moon system may have formed from a collision between the proto-Earth and a Mars-sized object. ∞ (Sec. 8.8)

4. A late collision with a large planetesimal may have caused Mars's curious north–south asymmetry and ejected much of the planet's atmosphere. ∞ (Sec. 10.4)

5. The tilted rotation axis of Uranus may have been caused by a grazing collision with a sufficiently large planetesimal, or by a merger of two smaller planets. ∞ (Sec. 13.3)

6. Uranus's moon Miranda may have been almost destroyed by a planetesimal collision, accounting for its bizarre surface terrain. ∞ (Sec. 13.6)

7. Interactions between the jovian protoplanets and one or more planetesimals may account for the irregular moons of those planets and, in particular, Triton's retrograde motion. ∞ (Sec. 13.6)

8. Pluto may simply be a large representative of the Kuiper belt, and the Pluto–Charon system may be the result of a collision or near-miss between two icy planetesimals before most were ejected by interactions with the jovian planets. ∞ (Sec. 13.9)

Note that it is impossible to test any of these assertions directly, but it is reasonable to suppose that some (or even all) of the preceding "odd" aspects of the solar system can be explained in terms of collisions late in the formative stages of the protoplanetary system. Not all astronomers agree with all of the explanations. However, most would accept at least some.

✔ What is the key "random" element in the condensation theory?

15.5 Planets beyond the Solar System

The test of any scientific theory is how well it holds up in situations different from those in which it was originally conceived. ∞ (Sec. 1.2) With the discovery in recent years of numerous planets orbiting other stars, astronomers now have the opportunity—indeed, the scientific obligation—to test their theories of solar-system formation.

The Discovery of Extrasolar Planets

The detection of planets orbiting other stars has been a goal of astronomers for decades if not centuries. Many claims of extrasolar planets have been made since the middle of the 20th century, but before 1995 none had been confirmed, and most have been discredited. Only since the mid-1990s have we seen genuine advances in this fascinating area of astronomy. The advances have come, not because of dramatic scientific or technical breakthroughs, but rather through steady improvements in telescope and detector technology and in computerized data analysis.

It is still not possible to image extrasolar planets: They are just too faint and too close to their parent stars for us to resolve them with current equipment. Instead, the techniques used to find them are *indirect*, based on analyses of the light from the parent star, not from the planet itself. As a planet orbits a star, gravitationally pulling first one way and then the other, the star "wobbles" slightly. The more massive the planet, or the less massive the star, the greater is the star's movement. If the wobble happens to occur along our line of sight to the star (Figure 15.8), then we see small fluctuations in the star's radial velocity, which can be measured using the Doppler effect. ∞ (Sec. 3.5) Those fluctuations allow us to estimate the planet's mass.

Figure 15.9 shows two sets of radial-velocity data that betray the presence of planets orbiting other stars. Part (a) shows the line-of-sight velocity of the star 51 Pegasi, a near twin to our Sun lying some 40 light-years away. The data, acquired in 1994 by Swiss astronomers using the 1.9-m telescope at Haute-Provence Observatory in France, were the first substantiated evidence for an extrasolar planet orbiting a Sun-like star.*

The regular 50-m/s fluctuations in the star's velocity have since been confirmed by several groups of astronomers and imply that a planet of at least half the mass of Jupiter orbits 51 Pegasi in a circular orbit with a period of just 4.2 days. (For comparison, the corresponding fluctuation in the Sun's velocity due to Jupiter is roughly 12 m/s.) Note that we say "at least half" here because Doppler observations suffer from a fundamental limitation: They cannot distinguish between low-speed orbits seen edge-on

As we will see in Chapter 22, two other planets having masses comparable to Earth, and one planet with a mass comparable to that of Earth's Moon, had previously been detected orbiting a particular kind of collapsed star called a pulsar. However, their formation was the result of a chain of events very different from those which formed Earth and the solar system.

and high-speed orbits seen almost face-on (so only a small component of the orbital motion contributes to the line-of-sight Doppler effect). As a result, only lower limits to planetary masses can be obtained.

Figure 15.9(b) shows another set of Doppler data, this time revealing the most complex system of planets discovered to date: a triple-planet system orbiting another nearby Sun-like star named Upsilon Andromedae. The three planets have minimum masses of 0.7, 2.1, and 4.3 times the mass of Jupiter and orbital semimajor axes of 0.06, 0.83, and 2.6 A.U., respectively. Figure 15.9(c) sketches their orbits, with the orbits of the solar terrestrial planets shown for scale. Well over 100 planets have been detected by means of radial-velocity searches.

If the wobble produced in a star's motion is predominantly *perpendicular* to our line of sight, then little or no Doppler effect will be observed, so the radial-velocity technique cannot be used to detect a planet. However, in this case, the star's *position* in the sky changes slightly from night to night, and, in principle, measuring this transverse motion provides an alternative means of detecting extrasolar planets. Unfortunately, these side-to-side wobbles have proven difficult to measure accurately, as the angles involved are very small and the star in question has to be quite close to the Sun for useful observation to be possible. On the basis of observations of this type, several candidate planetary systems have been proposed, but none has yet been placed on the "official" list of confirmed observations.

As just noted, the Doppler technique suffers from the limitation that the angle between the line of sight and the planet's orbital plane cannot be determined. However, in one system originally discovered through Doppler measurements, that is not the case. Observations of a distant solar-type star (known only by its catalog name of HD

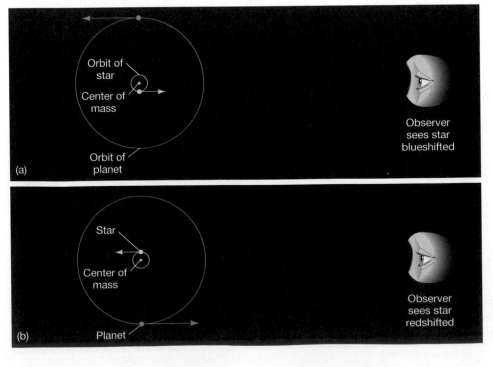

(a) Orbit of star / Center of mass / Orbit of planet

Observer sees star blueshifted

(b) Star / Center of mass / Planet

Observer sees star redshifted

◀ **FIGURE 15.8 Detecting Extrasolar Planets** As a planet orbits its parent star, it causes the star to "wobble" back and forth. The greater the mass of the planet, the larger is the wobble. The center of mass of the planet–star system stays fixed. If the wobble happens to occur along our line of sight to the star, as shown by the yellow arrow, we can detect it by the Doppler effect. (In principle, side-to-side motion perpendicular to the line of sight is also measurable, although there are as yet no confirmed cases of planets being detected this way.)

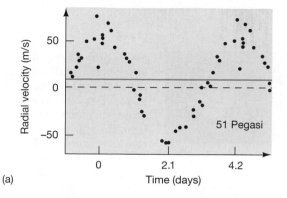

(a)

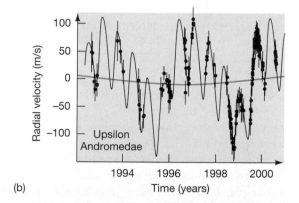

(b)

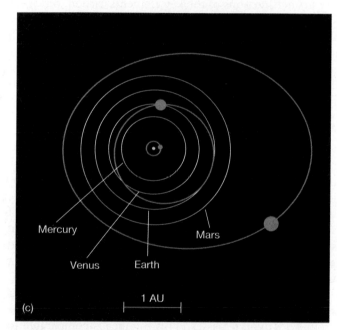

(c)

▲ **FIGURE 15.9 Planets Revealed** (a) Measurements of the Doppler shift of the star 51 Pegasi reveal a clear periodic signal indicating the presence of a planetary companion with mass at least half the mass of Jupiter. (b) Radial-velocity data for Upsilon Andromedae are much more complex, but are well fit (solid line) by a three-planet system orbiting the star. For reference in parts (a) and (b), the maximum possible signal produced by Jupiter orbiting the Sun (i.e., the wobble our Sun would have, as seen by a distant observer looking edge-on at our solar system) is also shown in blue. (c) A sketch of the inferred orbits of three planets from the Upsilon Andromedae system (in orange), with the orbits of the terrestrial planets superimposed for comparison (in white).

209458 and lying some 150 light-years from Earth) reveal a clear drop in brightness each time its 0.6-Jupiter-mass companion, orbiting at a distance of just 7 million km (0.05 A.U.), passes between the star and Earth (Figure 15.10). The drop in brightness is just 1.7 percent, but it occurs precisely on schedule every 3.5 days, the orbital period inferred from radial-velocity measurements.

Such planetary *transits*, similar to the transit of Mercury shown in Figure 2.15, are rare, as they require us to see the orbit almost exactly edge-on. When they do occur, however, they allow an unambiguous determination of the planet's mass and radius. In this case, the planet's density is just 200 kg/m^3, consistent with a high-temperature gas-giant planet orbiting very close to its parent star. Attempts to detect similar fluctuations in brightness in other nearby stars due to planetary transits have so far proven unsuccessful.

Planetary Properties

As of mid-2004, some 120 extrasolar planets have been detected orbiting more than 100 stars within a few hundred light years of the Sun. Overall, only a relatively small fraction (about 5 percent) of the nearby stars surveyed to date have shown evidence for planetary companions. Most observed planetary systems consist of a single massive planet orbiting its parent star. About a dozen two-planet systems and two three-planet systems (one of them shown in Figure 15.9c) have been confirmed.

The basic properties of the "official" list of extrasolar planets may be summarized as follows:

1. *All of the planets observed so far have masses comparable to that of Jupiter*, ranging from about one-third to 10 Jupiter masses. However, planets at the low end of this range are far more common—most of the measured masses are less than twice that of Jupiter, and a few have masses comparable to Saturn.

2. *The observed orbits are generally much smaller than those of Jupiter or Saturn*—less than a few A.U. across. A substantial fraction of these planets (about 15 percent) orbit very close to their parent star, with semimajor axes of 0.1 A.U. or less.

3. *The observed orbits are generally much more eccentric than those of Jupiter and Saturn*. Only 30 percent of the planets detected so far have eccentricities less than 0.1. (Recall that no jovian planet in our solar system has an eccentricity greater than 0.06.)

Since, in most cases, we see only a single giant planet, we cannot say much about the overall properties of these planetary "systems" in comparison with our own. So far, no terrestrial planets nor any evidence of interplanetary matter has been found orbiting any of these stars.

Figure 15.11 shows the orbital semimajor axes and eccentricities of the known extrasolar planets. Each dot represents a planet, and we have added points corresponding to

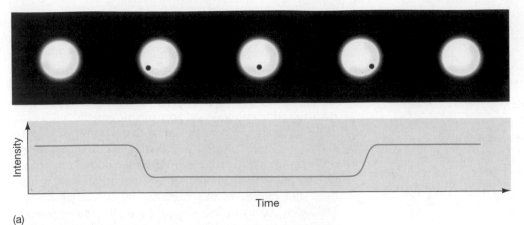

(a)

◀ **FIGURE 15.10 An Extrasolar Transit** (a) If an extrasolar planet happens to pass between us and its parent star, the light from the star dims in a characteristic way. (b) Artist's conception of the planet orbiting a star known as HD 209458. The planet is 200,000 km across and transits every 3.5 days, blocking about 2 percent of the star's light each time it does so.

(b)

Earth and Jupiter in our own solar system. The tight orbits and broad range of eccentricities of the extrasolar planets contrast markedly with those in our own planetary system. The numerous jovian-mass planets orbiting close (within 0.1 A.U.) to their parent stars (at the left of the diagram) have come to be known as **hot Jupiters**. Figure 15.12 plots the actual orbits of some of the known extrasolar planets (excluding the hot Jupiters), with Earth's orbit superimposed for comparison.

Notice in Figure 15.11 that the hot Jupiters tend to have more circular (low-eccentricity) orbits. This is not a matter of chance. These bodies reside so close to their stars that their orbits are thought to have been circularized by tidal effects, similar to those controlling the orbits of Earth's Moon or the Galilean moons of Jupiter. ∞ (Secs. 8.4, 11.5) Thus, the shape of a hot-Jupiter orbit is a direct evolutionary consequence of the planet's proximity to the star. As we will see in the next section, the manner in which the planet got into such a tight orbit in the first place may

provide a much-needed connection between extrasolar planetary systems and our own.

Finally, spectroscopic observations of the parent stars reveal what may be a crucial piece in the puzzle of extrasolar-planet formation. ∞ (Sec. 4.4) Stars having compositions similar to that of the Sun are statistically *much more likely* to have planets orbiting them than are stars containing smaller fractions of the key elements carbon, nitrogen, silicon, and iron. Because the elements found in a star reflect the composition of the nebula from which it formed, and the elements just listed are the main ingredients of interstellar dust, this finding provides strong support for the condensation theory. Dusty disks really are more likely to form planets.

Are They Really Planets?

Given that these systems seem so alien from our own, for a time some astronomers questioned whether the mass measurements, and hence the identification of these objects as "planets," could be trusted. Eccentric orbits are known to be common among double-star systems (pairs of stars in orbit around one another—see Chapter 17), and some researchers suggested that many of the newly found planets were actually *brown dwarfs*—"failed stars" having insufficient mass to become true stars (see *Discovery 19-1*).

The dividing line between genuine Jupiter-like planets and starlike brown dwarfs is unclear, but it is thought to be around 15 Jupiter masses. This number is comparable to the largest extrasolar planet mass yet measured. Planet proponents would say that this is not a coincidence, but rather indicates that planets up to the maximum possible mass have in fact been observed. Detractors would argue that we just happen to see the orbits almost face-on, greatly reducing the parent star's radial velocity and fooling us into thinking that we are observing low-mass planets instead of higher mass brown dwarfs.

However, the latter view has a serious problem, which worsens with every new low-mass extrasolar planet reported: Since the orientations of the actual orbits are presumably random, it is extremely unlikely that we would just happen to see *all* of them face-on—and even if we did,

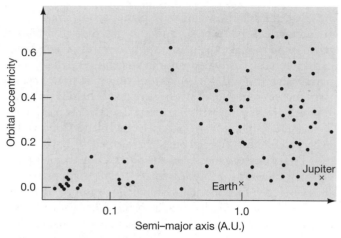

▲ FIGURE 15.11 Extrasolar Orbital Parameters Orbital semimajor axes and eccentricities of 75 of the 120 known extrasolar planets. Each point represents one planetary orbit. The corresponding points for Earth and Jupiter in our solar system are also shown. The known extrasolar planets generally move on smaller, much more eccentric orbits than do the planets circling the Sun.

there is no observational evidence for the many edge-on (and much easier to detect) systems we would then also expect to see on statistical grounds. Consequently, most astronomers agree that, while there may well be a few brown dwarfs lurking among the list of extrasolar planets, they probably do not constitute a significant fraction of the total.

CONCEPT CHECK

✔ Describe three ways in which observed extrasolar planetary systems differ from the solar system.

15.6 Is Our Solar System Unusual?

6 Not so long ago, many astronomers argued that the condensation scenario described earlier in this chapter was in no way unique to our own system. The same basic processes could have occurred, and perhaps *did* occur, during the formative stages of many of the stars in our Galaxy, so planetary systems like our own should be common. Today we know that planetary systems *are* apparently quite common, but, by and large, the ones we see don't look at all like ours! We can thus legitimately ask whether our solar system really is as unusual as recent observations seem to imply and whether those observations undermine our current theory of solar-system formation.

Observational Limitations

Let's start by asking whether the planets we observe really are representative of extrasolar planets in general. The fact that we don't see low-mass planets, or more massive planets on wide orbits, is not surprising. It is what astronomers call a **selection effect**: Lightweight or distant planets simply don't produce large enough velocity fluctuations for them to be detectable. The methods employed so far are heavily biased toward finding massive objects orbiting close to their parent stars. Those systems would be expected to give the strongest signal, and they are precisely what have been observed.

Almost all of the planets detected so far produce stellar radial velocities substantially greater than the 12 m/s that would be produced (under the best circumstances) by Jupiter's orbit around the Sun. Furthermore, while the Sun's wobble could be detected with current technology, it is close enough to the instrumental limits that several orbits—that

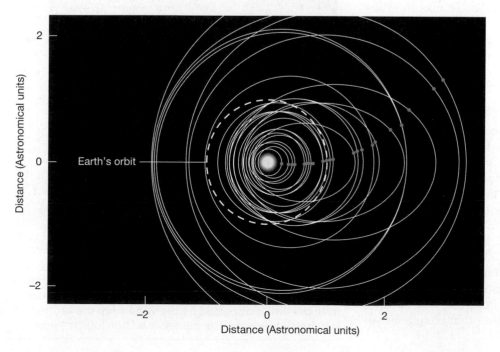

◄ FIGURE 15.12 Extrasolar Orbits The orbits of 60 extrasolar planets residing beyond 0.15 A.U. from their parent star, superimposed on a single plot, with Earth's orbit shown for comparison. All are comparable in mass to Jupiter. A plot of all known extrasolar planets would be very cluttered, but the message would be much the same: These planetary systems don't look much like ours!

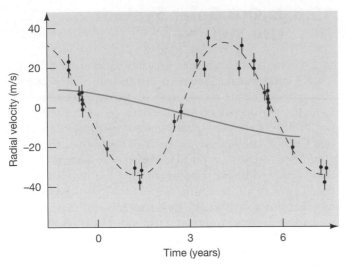

▲ FIGURE 15.13 Jupiter-like Planet? Velocity "wobbles" in the star HD 70642 reveal the presence of the extrasolar planet with the most "Jupiter-like" orbit yet discovered. The parent star is almost identical to the Sun, and the 2-Jupiter-mass planet orbits at a distance of 3.3 A.U. with an orbital eccentricity of 0.1. Again, the blue line marks the corresponding plot for Jupiter itself.

is, several decades' worth of observations—would be needed before a definitive detection could be claimed.

In fact, as search techniques improve, astronomers are finding more and more Jupiter- (and lower) mass planets on wider and less eccentric orbits. Figure 15.13 shows recent (2003) evidence for one of the most "Jupiter-like" planets yet detected: a 2-Jupiter-mass planet moving on a roughly circular orbit around a near twin of our own Sun. The planet's period is 6 years. Planet hunters are quietly confident that advances during the next decade will either bring numerous detections of extrasolar planets in orbits comparable to those in the solar system or allow astronomers to conclude that systems like our own are indeed in the minority. Either way, the consequences are profound.

Making Eccentric Jupiters

Are the extrasolar orbits we do see *inconsistent* with the condensation theory? Probably not. Current theory in fact provides many ways in which massive planets can end up in short-period or eccentric orbits. Indeed, an important aspect of solar-system formation not mentioned in our earlier discussion is the fact that many theorists worry about how Jupiter could have remained in a stable orbit after it formed in the protosolar disk! Jupiter-sized planets may be knocked into eccentric orbits by interactions with other Jupiter-sized planets or by the tidal effects of nearby stars. If these planets formed by gravitational instability, they could have eccentric orbits right from the start (and we then would have to explain how those orbits circularized in the case of the solar system).

Regardless of how a massive planet forms, gravitational interactions between it and the gas disk in which it moves tend to make the planet spiral inward, as mentioned earlier, and can easily deposit it in an orbit very close to the parent star (Figure 15.14). Interestingly, it now appears that the presence of Saturn may have helped stabilize Jupiter's orbit against this last effect. Isolated or particularly massive "Jupiters" are precisely the planets one would expect to find on "hot" orbits.

The orbits and masses of the observed extrasolar planets spell disaster for any low-mass "terrestrial" planets in these systems. In our solar system, the presence of a massive Jupiter on a nearly circular orbit is known to have a stabilizing influence on the other planetary orbits, tending to preserve the relative tranquility of our planetary environment. In the known extrasolar systems, not only is this stabilization absent, but having a Jupiter-sized planet repeatedly plow through the inner parts of the system means that any terrestrial planets or planetesimals have almost certainly been ejected from the system.

Searching for Earth-like Planets

To summarize the current state of extrasolar-planet searches, only a small percentage (perhaps 5 percent) of the stars surveyed to date show evidence for extrasolar planets. The planets that are observed are precisely those that could have been detected, given today's technology,

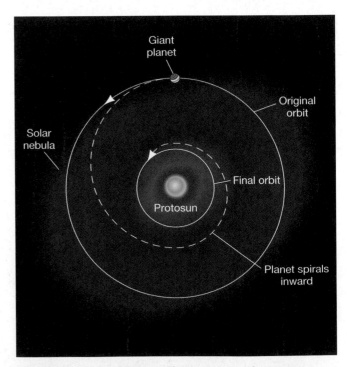

▲ FIGURE 15.14 Sinking Planet Friction between a giant planet and the nebular disk in which it formed tends to make the planet spiral inward. The process continues until the disk is dispersed by the wind from the central star, possibly leaving the planet in a "hot-Jupiter" orbit.

and their orbits are allowed, if not actually predicted, by current theories of planet formation. The properties of the systems we can see reveal little about those we can't, so for now, the planetary properties of the vast majority of stars remain a mystery. Still, there seems to be no pressing reason to conclude just yet that our solar system is unique, or even unusual, in its structure.

Assuming, then, that planetary systems like our own do exist, with Jupiter-like planets in circular, or at least nondisruptive, orbits and some terrestrial planets on stable orbits close to their parent star, how might we go about searching for planets like our own? Current estimates suggest that looking for "wobbles" is not likely to be successful. Variations in radial velocities will be so small as to be lost within the natural internal motions of the parent star itself, and the side-to-side movement of a star due to Earth-sized objects will be very hard to measure, even from space. Direct imaging of terrestrial planets remains a long-term goal of NASA, but tangible results are probably decades away.

Perhaps surprisingly, the approach given the best chance of success entails searching for planets *transiting* their parent stars. The effect of a transit on a star's brightness is tiny—less than one part in 10^4 for an Earth-like planet crossing the face of the Sun—and also very unlikely to be detected, as only nearly edge-on systems will show it. Nevertheless, it is measurable with existing technology. NASA's upcoming *Kepler* mission, approved in 2001, will monitor some 100,000 Sun-like stars over a four-year period for brightness fluctuations caused by Earth-like planets.

Under optimistic assumptions, some 100 terrestrial transits are expected. From the change in brightness as the planet passes in front of the star, the planet's size can be calculated. The size of the orbit can be calculated from the orbital period and an application of Newton's laws. ∞ (Sec. 2.7) Once the distance from the planet to the star is known, the planet's surface temperature may be estimated (as in our own solar system) by determining the temperature at which the planet radiates as much energy back into space as it receives in the form of starlight. ∞ (Secs. 3.4, 7.2)

The detection by *Kepler* of even one planet with properties similar to our own could revolutionize our view of the universe, as it would suggest that Earth-like planets are common in our Galaxy. *Kepler* is currently scheduled for launch in 2007.

CONCEPT CHECK

✔ Why is it not too surprising that the extrasolar planetary systems detected thus far have properties quite different from those of planets in our solar system?

Chapter Review

SUMMARY

Our solar system is an orderly place, making it unlikely that the planets were simply and "accidentally" captured by the Sun. The overall organization points toward the solar system's formation as the product of an ancient, one-time event 4.6 billion years ago. An ideal theory of the solar system should provide strong reasons for the observed characteristics of the planets, yet be flexible enough to allow for deviations.

According to the condensation theory, as the solar nebula collapsed under its own gravity, it began to spin faster, eventually forming a disk. Small clumps of matter appeared around condensation nuclei and grew by accretion, sticking together and expanding into moon-sized **planetesimals (p. 385)**, whose gravitational fields were strong enough to accelerate the accretion process. Competing with accretion in the solar nebula was **fragmentation (p. 387)**—the breaking up of small bodies following collisions with larger ones. Eventually, only a few planet-sized objects remained.

At any given location, the temperature would determine which materials could condense out of the nebula; thus, temperature controlled the future composition of the planets. The condensation period ended when the Sun's strong winds expelled the nebular gas. In the hot inner solar system, rocky and metallic planetesimals eventually formed the terrestrial planets. The asteroid belt is a collection of planetesimals that never managed to form a planet, because of Jupiter's gravitational influence.

In the outer solar system, the nebula was cooler, and ices of water and ammonia could also form. According to the **gravitational instability theory (p. 388)**, the jovian planets formed directly and very rapidly via instabilities in the nebular disk. In the more standard **core-accretion theory (p. 388)**, icy protoplanet cores became so large that they could capture hydrogen and helium gas from the nebula. Before the nebula was ejected, interactions between the giant planets and the gas probably caused the former to migrate inward from their initial orbits.

Many leftover planetesimals in the outer solar system were ejected into the Oort cloud and the Kuiper belt by the gravitational fields of the outer planets. Some now occasionally revisit our part of the solar system as comets. In the inner solar system, light elements such as hydrogen and helium would have escaped into space. Much, if not all, of Earth's water was carried to our world by comets deflected from the outer solar system. The expulsion of the icy planetesimals may have significantly changed the giant planets' orbits. In particular, Uranus and Neptune probably moved outward during this period. As Neptune's orbit expanded, it captured Pluto and the Kuiper-belt **plutinos (p. 392)**, sending them into resonant orbits.

Many "odd" aspects of the solar system may conceivably be explained in terms of collisions late in the formation stages of the protoplanetary system.

Some 120 **extrasolar planets (p. 384)** are now known. All have been discovered by observing their parent star wobble back and forth as the planet orbits, although one has since been observed passing in front of the star, reducing the star's brightness slightly. Most systems found so far contain a single, massive planet comparable in mass to Jupiter in an orbit taking the planet close to the central star. None look much like our solar system. Many contain **hot Jupiters (p. 396)**, jovian-sized planets orbiting within a fraction of an astronomical unit of the star. Several theories have been advanced to explain how hot Jupiters might form, including migration in the nebular disk.

It is currently not known whether the observed extrasolar systems or our own solar system represent the norm in planetary systems. The fact that we see the extrasolar planets we do may well be a **selection effect (p. 397)**—these are the only planets we *can* see with current techniques. The orbits of the observed Jupiter-mass planets would probably have ejected any terrestrial planets long ago. The observations indicate that stars containing larger fractions of "dusty" elements, such as carbon and silicon, are more likely to have planets, lending support to the condensation theory. Searches for terrestrial planetary systems are in the planning stage.

REVIEW AND DISCUSSION

1. List at least six properties of the solar system that any model of its formation must be able to explain.

2. Give three examples of present-day properties that our solar-system model does not have to explain, and say why no explanation is necessary.

3. Explain the difference between evolutionary theories and catastrophic theories of the solar system's origin.

4. Describe the basic features of the condensation theory of solar-system formation.

5. Why are the jovian planets so much larger than the terrestrial planets?

6. Describe two possible ways in which the jovian planets may have formed.

7. What role did the Sun play in governing the formation of the giant planets?

8. Why do giant planets "migrate"?

9. What solar system objects, still observable today, resulted from the process of fragmentation?

10. What influence did Earth's location in the solar nebula have on our planet's final composition?

11. Why could Earth not have formed out of material containing water? How might Earth's water have gotten here?

12. What happened to the outer planets as the solar system was cleared of icy planetesimals?

13. What are plutinos, and how did they come to be in their present orbits?

14. How did the Kuiper belt and the Oort cloud form?

15. Describe some ways in which random processes played a role in the determination of planetary properties.

16. Describe a possible history of a single comet now visible from Earth, starting with its birth in the solar nebula somewhere near the planet Jupiter.

17. How do astronomers set about looking for extrasolar planets?

18. In what ways do extrasolar planetary systems differ from our own solar system?

19. What is a "hot Jupiter"?

20. Do the observed extrasolar planets imply that Earth-like planets are rare?

CONCEPTUAL SELF-TEST: TRUE OR FALSE/MULTIPLE CHOICE

1. Theory predicts that planets must always rotate in the same sense as the Sun's rotation.

2. Moons usually revolve in the same direction as their parent planet.

3. The asteroids were recently formed from the collision and breakup of an object orbiting within the asteroid belt.

4. The large number of leftover planetesimals beyond about 5 A.U. were destined to become asteroids.

5. The condensation theory offers an evolutionary explanation of the slow retrograde rotation of Venus.

6. Random collisions can explain many of the odd properties found in the solar system.

7. The ejection of the comets from the outer solar system caused all four giant planets to move inward toward the Sun.

8. Astronomers can detect extrasolar planets by observing the spectra of their parent stars.

9. Many of the extrasolar planets observed so far have masses and orbits similar to those of Jupiter.

10. Astronomers have no theoretical explanation for the "hot Jupiters" observed orbiting some other stars.

11. A successful scientific model of the origin of planetary systems must account for all of the following solar-system features, except for **(a)** intelligent life. **(b)** the roughly circular planetary orbits. **(c)** the roughly coplanar planetary orbits. **(d)** the extremely distant orbits of the comets.

12. The initial gas cloud that formed the Sun and solar system must have been about as large as **(a)** the present-day Sun; **(b)** Jupiter's orbit; **(c)** Pluto's orbit; **(d)** one light-year across.

13. If interstellar dust did not exist, then **(a)** the terrestrial planets would not have formed; **(b)** the solar nebulae would not have collapsed; **(c)** there would be many more jovian planets; **(d)** new stars would serve as condensation nuclei.

14. After the central star has formed, the gas making up the original nebular cloud is **(a)** blown away into interstellar space; **(b)** absorbed into the star; **(c)** ignited by the star's heat; **(d)** incorporated into jovian planet atmospheres.

15. The inner planets formed **(a)** when the Sun's heat destroyed all the smaller bodies in the inner solar system; **(b)** in the outer solar system and then were deflected inward by inter-

actions with Jupiter and Saturn; **(c)** by collisions and mergers of planetesimals; **(d)** when a larger planet broke into pieces.

16. Water on Earth **(a)** was transported there by comets; **(b)** was accreted from the solar nebula; **(c)** was outgassed from volcanoes in the form of steam; **(d)** was created by chemical reactions involving hydrogen and oxygen shortly after Earth formed.

17. Using the standard model of planetary system formation, scientists invoke catastrophic events to explain why **(a)** Mercury has no moon; **(b)** Pluto is not a gas giant; **(c)** Uranus has an extremely tilted rotation axis; **(d)** there is no planet between Mars and Jupiter.

18. Astronomers have confirmed the existence of at least **(a)** one; **(b)** ten; **(c)** one hundred; **(d)** one thousand planets beyond our own solar system.

19. So far, most of the planets discovered orbiting other stars have orbits that **(a)** are larger than that of Pluto; **(b)** are larger than that of Saturn; **(c)** are very similar to that of Jupiter; **(d)** approach within 1 A.U. of the parent star.

20. Astronomers have not yet detected any Earth-like planets orbiting other stars because **(a)** there are none; **(b)** they are not detectable with current technology; **(c)** no nearby stars are of the type expected to have Earth-like planets; **(d)** the government is preventing them from reporting their discovery.

PROBLEMS

 Algorithmic versions of these questions are available in the Practice Problems module of the Companion Website at astro.prenhall.com/chaisson.

The number of squares preceding each problem indicates its approximate level of difficulty.

1. ■■ The orbital angular momentum of a planet in a circular orbit is simply the product of the planet's mass, its orbital speed, and its distance from the Sun. ∞ *(More Precisely 6-2)* **(a)** Compare the orbital angular momenta of Jupiter, Saturn, and Earth. **(b)** Calculate the orbital angular momentum of an Oort-cloud comet with mass 10^{13} kg, moving in a circular orbit 50,000 A.U. from the Sun.

2. ■■■ We can make a rough model of accretion in the inner solar nebula by imagining a 1-km-diameter body moving at a relative speed of 500 m/s through a collection of similar bodies having a roughly uniform spatial density of 10^{-10} body per cubic kilometer. Neglecting any gravitational forces (and hence assuming that the body moves in a straight line until it collides with something), estimate how much time, on average, it will take before the body collides.

3. ■ If Neptune formed by gravitational instability in just a thousand years at a distance of 25 A.U. from the Sun, how many orbits did Neptune complete while it was forming?

4. ■ How many 100-km-diameter rocky (3000 kg/m^3) planetesimals would have been needed to form Earth?

5. ■■ Two asteroids, each of mass 10^{18} kg, orbit near the center of the asteroid belt in the plane of the ecliptic on circular paths of radii 2.80 and 2.81 A.U., respectively. Calculate the gravitational force between them at closest approach, and compare it with the tidal force exerted by Jupiter if that planet were also at closest approach at the time.

6. ■■ The temperature in the early solar nebula at a distance of 1 A.U. from the Sun was about 1100 K. On the basis of the discussion of surface temperature in Chapter 7, estimate the factor by which the Sun's current energy output would have to increase in order for Earth's present temperature to have that value. ∞ *(Sec. 7.2)*

7. ■ A typical comet contains some 10^{13} kg of water ice. How many comets would have to strike Earth in order to account

for the roughly 2×10^{21} kg of water presently found on our planet? If this amount of water accumulated over a period of 0.5 billion years, how frequently must Earth have been hit by comets during that time?

8. ■ Use the data given in the text to calculate Neptune's orbital period before interactions with planetesimals expanded the orbit to its present size.

9. ■■ What was the period of a plutino's orbit when Neptune's orbital semimajor axis was 25 A.U.?

10. ■ How many comet-sized planetesimals (10 km in diameter, density of 100 kg/m^3) would have been needed to form Pluto?

11. ■■ The two planets orbiting the nearby star Gliese 876 are observed to be in a 2:1 resonance (i.e., the period of one is twice that of the other). The inner planet has an orbital period of 30 days. If the star's mass is the mass of the Sun, calculate the semimajor axis of the outer planet's orbit.

12. ■■ The planet orbiting star HD187123 has a semimajor axis of 0.042 A.U. If the star's mass is 1.06 times the mass of the Sun, calculate how many times the planet has orbited its star since the paper announcing its discovery was published on December 1, 1998.

13. ■■■ Given that Mercury's noontime surface temperature at perihelion is 700K, estimate the temperature of a "hot Jupiter" moving on a 3-day circular orbit around a Sun-like star.

14. ■■ How close to the Sun would an identical star have to come in order to exert a 0.1 percent tidal perturbation on Jupiter (i.e., a tidal force equal to 0.1 percent of the Sun's gravitational attraction on the planet)?

15. ■■■ Not surprisingly, as the mass of a planet decreases, the planet becomes harder to detect. Current Doppler techniques can reliably detect a planet having 0.25 times the mass of Jupiter orbiting a Sun-like star only if the planet's *maximum* orbital speed exceeds 40 km/s. Under these conditions, what is the minimum detectable eccentricity for a planet with a 5-year orbital period?

 In addition to the Practice Problems module, the Companion Website at astro.prenhall.com/chaisson provides for each chapter a study guide module with multiple choice questions as well as additional annotated images, animations, and links to related Websites.

PART | 3
Stars and Stellar Evolution

Portrait of Cecilia Payne-Gaposchkin (*Harvard*)

Graduate students in astronomy lead a risky, uncertain life. They take some tough courses, usually in physics, and they assist in teaching undergraduate courses, but mostly they strive to do original research. Ideally, on the (typically five- or six-year) road to their Ph.D. degree, they make a discovery or gain some unique insight that they then write up as part of their doctoral dissertation. The process is exhilarating for some, and they often go on to productive careers in astronomy; for others, though, the grad-school grind is so traumatic that they leave the field and never again publish a paper in a scientific journal.

It is widely regarded that the one of the most brilliant doctoral theses in astronomy was written in 1925 by a student at Harvard— and she did it in two years. Cecilia Payne (1900-1979) was an English student who had crossed the ocean to pursue graduate studies at Radcliffe College, and she quickly gravitated to the nearby Harvard Observatory, then perhaps the leading center for research on stars. It was also a place where women, though they often did not get the credit at the time, were making some of the most fundamental advances in stellar astronomy. It was Nirvana for her, and she never left.

Cecilia knew far more physics than most astronomers of the time. She was one of the first to apply the then revolutionary quantum theory of atoms to the spectra of stars, thereby ascertaining stellar temperatures and chemical abundances. Of fundamental importance, her work proved that hydrogen and helium—not the heavy elements, as was then supposed—are the most abundant elements in stars and, therefore, in the universe. Her findings were so revolutionary that the leading theorist of the time, Henry Norris Russell of Princeton, declared her work to be "clearly impossible." It took years to convince the astronomical community that hydrogen is about a million times more abundant in stars than are most of the common elements found on Earth—a fact we take for granted today.

In collaboration with her husband, the exiled Russian astronomer Sergei Gaposchkin, Cecilia spent decades making literally several million observations of thousands of star clusters, variable stars, and galactic novae. Her analysis provided a firm theoretical basis for many properties of stars and their use as distance indicators in the universe. Much of her work has stood the test of time and, despite a flood of new data and new theoretical ideas, is still correct today.

Women "computers" at work: Cecelia at inclined desk (*Harvard*)

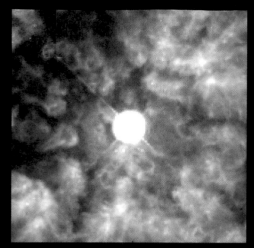

Variable Star Wr 124 (*STScI*)

Pistol Star (*STScI*)

In 1976, well past her retirement, Cecilia was chosen (perhaps ironically, given Russell's initial reaction to her work half a century earlier) as the Henry Norris Russell Lecturer, the highest accolade of the American Astronomical Society. Her talk was an enthusiastic summary of a lifetime of astronomical research—an encyclopedic talk with no notes and no prompts, given in perfectly punctuated English. It was very clear that she knew some individual giant stars as well as she knew her best friends. Her talk ended with the following advice to astronomers young and old:

> "The reward of the young scientist is the emotional thrill of being the first person in the history of the world to see something or to understand something. Nothing can compare with that experience ...The reward of the old scientist is the sense of having seen a vague sketch grow into a masterly landscape. Not a finished picture, of course; a picture that is still growing in scope and detail with the application of new techniques and new skills. The old scientist cannot claim that the masterpiece is his own work. He may have roughed out part of the design, laid on a few strokes, but he has learned to accept the discoveries of others with the same delight that he experienced on his own when he was young."

Rosebud Nebula (*JPL*)

oday, the landscape of stellar research is even richer than Cecilia Payne-Gaposchkin ever knew. Yet the picture is very much unfinished, as 21st-century astronomy continues to uncover new and exciting features of stars and stellar systems. We now see stars much more clearly, their spectra in much finer detail, and all of it at much greater distances.

Illustrated on this page are some recent findings in stellar research—work that undoubtedly would have caused Cecilia to express more of her trademark enthusiasm. Today's research also would have made her justly proud, for so much of it relies on the insights gained by her and her colleagues in the first half of the 20th century.

Henize Nebula (*JPL*)

16

The Sun is our star. It is not merely the center of the solar system or the source of light that distinguishes day from night; it is also the main source of energy that powers weather, climate, and life on Earth. Humans simply would not exist without the Sun. Although we take it for granted each and every day, the Sun is of great importance to us in the cosmic scheme of things. Here, in this composite ultraviolet image made from several observations with the SOHO spacecraft, many active features are revealed on and around the Sun's surface; the various colors are artificial, indicating different wavelengths. (ESA/NASA) ▶

The Sun

Our Parent Star

Living in the solar system, we have the chance to study, at close range, perhaps the most common type of cosmic object—a star. Our Sun is a star, and a fairly average one at that, but with a unique feature: It is very close to us—some 300,000 times closer than our next nearest neighbor, Alpha Centauri. Whereas Alpha Centauri is 4.3 light-years distant, the Sun is only 8 light-minutes away from us. Consequently, astronomers know far more about the properties of the Sun than about any of the other distant points of light in the universe. A good fraction of all our astronomical knowledge is based on modern studies of the Sun. Just as we studied our parent planet, Earth, to set the stage for our exploration of the solar system, we now examine our parent star, the Sun, as the next step in our exploration of the universe.

LEARNING GOALS

Studying this chapter will enable you to

1 Summarize the overall properties of the Sun.

2 Outline the process by which energy is produced in the Sun's interior.

3 Explain how energy travels from the solar core, through the interior, and out into space.

4 Name the Sun's outer layers and describe what those layers tell us about the Sun's surface composition and temperature.

5 Discuss the nature of the Sun's magnetic field and its relationship to the various types of solar activity.

6 Explain how observations of the Sun's core challenge our present understanding of fundamental physics.

WWW Visit astro.prenhall.com/chaisson for additional annotated images, animations, and links to related sites for this chapter.

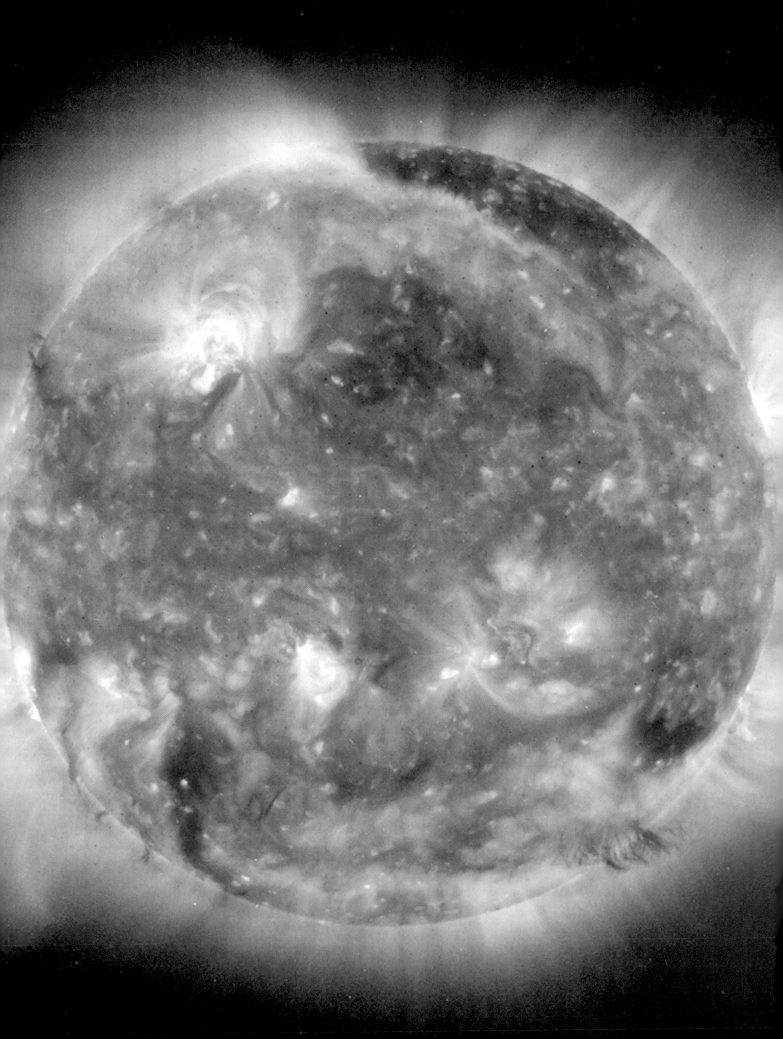

16.1 Physical Properties of the Sun

The Sun is the sole source of light and heat for the maintenance of life on Earth. The Sun is a **star**—a glowing ball of gas held together by its own gravity and powered by nuclear fusion at its center. In its physical and chemical properties, the Sun is similar to most other stars, regardless of when and where they formed. Indeed, our Sun appears to be a rather typical star, lying right in the middle of the observed ranges of stellar mass, radius, brightness, and composition. Far from detracting from our interest in the Sun, this very mediocrity is one of the main reasons that astronomers study it—they can apply knowledge of solar phenomena to many other stars in the universe.

Overall Properties

The Sun Data box at the right lists some basic orbital and physical solar data. The Sun's radius, roughly 700,000 km, is determined most directly by measuring the angular size (0.5°) of the Sun and then employing elementary geometry. ∞ (Sec. 1.7) The Sun's mass, 2.0×10^{30} kg, follows from Newton's laws of motion and gravity, applied to the observed orbits of the planets. ∞ (*More Precisely 2-3*) The average solar density derived from its mass and volume, approximately 1400 kg/m^3, is quite similar to that of the jovian planets and about one-quarter the average density of Earth.

Solar rotation can be measured by timing sunspots and other surface features as they traverse the solar disk. ∞ (Sec. 2.4) These observations indicate that the Sun rotates in about a month, but it does not do so as a solid body. Instead, it spins differentially—faster at the equator and slower at the poles, like Jupiter and Saturn. The equatorial rotation period at the equator is about 25 days. Sunspots are never seen above latitude 60° (north or south), but at that latitude they indicate a 31-day period. Other measurement techniques, such as those discussed in Section 16.3, reveal that the Sun's rotation period continues to increase as we approach the poles. The polar rotation period is not known with certainty, but it may be as long as 36 days.

The Sun's surface temperature is measured by applying the radiation laws to the observed solar spectrum. ∞ (Sec. 3.4) The distribution of solar radiation has the approximate shape of a blackbody curve for an object at about 5800 K. The average solar temperature obtained in this way is known as the Sun's *effective temperature*.

Having a radius of more than 100 Earth radii, a mass of more than 300,000 Earth masses, and a surface temperature well above the melting point of any known material, the Sun is clearly a body that is very different from any other we have encountered so far.

Solar Structure

The Sun has a surface of sorts—not a solid surface (the Sun contains no solid material), but rather that part of the brilliant gas ball we perceive with our eyes or view through a heavily filtered telescope. This "surface"—the part of the Sun that emits the radiation we see—is called the **photosphere**. Its radius (listed in the Data box as the equatorial radius of the Sun) is about 700,000 km. However, the thickness of the photosphere is probably no more than 500 km, less than 0.1 percent of the radius, which is why we perceive the Sun as having a well-defined, sharp edge (Figure 16.1).

The main regions of the Sun are illustrated in Figure 16.2 and summarized in Table 16.1. Just above the photo-

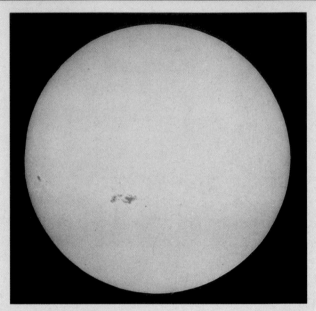

SUN DATA

Greatest angular diameter, as seen from Earth	32.5′
Mass	1.99×10^{30} kg 332,000 (Earth = 1)
Equatorial radius	696,000 km 109 (Earth = 1)
Mean density	1410 kg/m^3 0.255 (Earth = 1)
Surface gravity	274 m/s^2 28.0 (Earth = 1)
Escape speed	618 km/s
Sidereal rotation period	25.1 solar days (equator) 30.8 solar days (60° latitude) 36 solar days (poles) 26.9 solar days (interior)
Axial tilt	7.25° (relative to ecliptic)
Surface temperature	5780 K (effective temperature)
Luminosity	3.85×10^{26} W

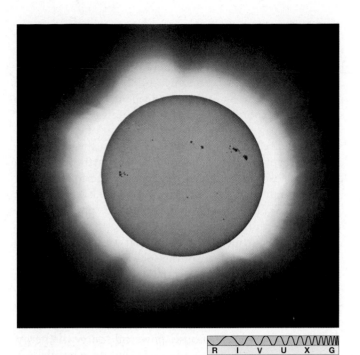

▲ FIGURE 16.1 The Sun The inner part of this composite, filtered image of the Sun shows a sharp solar limb, although our star, like all stars, is made of a gradually thinning gas. The edge appears sharp because the solar photosphere is so thin. The outer portion of the image is the solar corona, normally too faint to be seen, but visible during an eclipse, when the light from the solar disk is blotted out. Note the blemishes—they are sunspots. ∞ (Sec. 2.4) *(NOAO)*

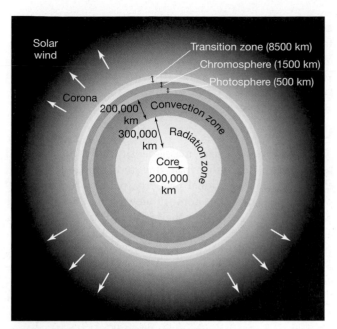

▲ FIGURE 16.2 Solar Structure The main regions of the Sun, not drawn to scale, with some physical dimensions labeled. The photosphere is the visible "surface" of the Sun. Below it lie the convection zone, the radiation zone, and the core. Above the photosphere, the solar atmosphere consists of the chromosphere, the transition zone, and the corona.

sphere is the Sun's lower atmosphere, called the **chromosphere**, about 1500 km thick. From 1500 km to 10,000 km above the top of the photosphere lies a region called the **transition zone**, in which the temperature rises dramati-cally. Above 10,000 km, and stretching far beyond, is a ten-uous (thin), hot upper atmosphere: the solar **corona**. At still greater distances, the corona turns into the *solar wind*, which flows away from the Sun and permeates the entire solar system. ∞ (Sec. 6.5) Extending down some 200,000 km below the photosphere is the **convection zone**, a re-gion where the material of the Sun is in constant convec-tive motion. Below the convection zone lies the **radiation**

Region	Inner Radius (km)	Temperature (K)	Density (kg/m^3)	Defining Properties
Core	0	15,000,000	150,000	Energy generated by nuclear fusion
Radiation zone	200,000	7,000,000	15,000	Energy transported by electromagnetic radiation
Convection zone	500,000	2,000,000	150	Energy carried by convection
Photosphere	696,000*	5800	2×10^{-4}	Electromagnetic radiation can escape—the part of the Sun we see
Chromosphere	696,500*	4500	5×10^{-6}	Cool lower atmosphere
Transition zone	698,000*	8000	2×10^{-10}	Rapid increase in temperature
Corona	706,000	1,000,000	10^{-12}	Hot, low-density upper atmosphere
Solar wind	10,000,000	2,000,000	10^{-23}	Solar material escapes into space and flows outward through the solar system

TABLE 16.1 The Standard Solar Model

*These radii are based on the accurately determined radius of the photosphere. The other radii quoted are approximate, round numbers.

zone, in which solar energy is transported toward the surface by radiation rather than by convection. The term *solar interior* is often used to mean both the radiation and convection zones. The central **core**, roughly 200,000 km in radius, is the site of powerful nuclear reactions that generate the Sun's enormous energy output.

Luminosity

The properties of size, mass, density, rotation rate, and temperature are familiar from our study of the planets. But the Sun has an additional property, perhaps the most important of all from the point of view of life on Earth: The Sun *radiates* a great deal of energy into space, uniformly (we assume) in all directions. By holding a light-sensitive device—a photoelectric cell, perhaps—perpendicular to the Sun's rays, we can measure how much solar energy is received per square meter of surface area every second. Imagine our detector as having a surface area of 1 square meter (1 m^2) and as being placed at the top of Earth's atmosphere. The amount of solar energy reaching this surface each second is a quantity known as the **solar constant**, whose value is approximately 1400 watts per square meter (W/m^2).

About 50 to 70 percent of the incoming energy from the Sun reaches Earth's surface; the rest is intercepted by

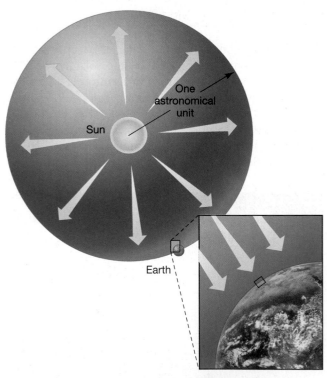

▲ **FIGURE 16.3 Solar Luminosity** We can draw an imaginary sphere around the Sun so that the sphere's edge passes through Earth's center. The radius of this imaginary sphere equals 1 A.U. The "solar constant" is the amount of power striking a 1-m^2 detector at Earth's distance, as suggested by the inset. By multiplying the sphere's surface area by the solar constant, we can measure the Sun's luminosity—the amount of energy it emits each second.

the atmosphere (30 percent) or reflected away by clouds (0 to 20 percent). Thus, on a clear day, a sunbather's body having a total surface area of about 0.5 m^2 receives solar energy at a rate of roughly 1400 W/m^2 × 0.70 (70 percent) × 0.5 m^2 ≈ 500 W, equivalent to the output of a small electric room heater or five 100-watt lightbulbs.

Let us now ask about the *total* amount of energy radiated in all directions from the Sun, not just the small fraction intercepted by our detector or by Earth. Imagine a three-dimensional sphere is centered on the Sun and just large enough that its surface intersects Earth's center (Figure 16.3). The sphere's radius is 1 A.U., and its surface area is therefore 4π × (1 A.U.)^2, or approximately 2.8 × 10^{23} m^2. Multiplying the rate at which solar energy falls on each square meter of the sphere (i.e., the solar constant) by the total surface area of our imaginary sphere, we can determine the total rate at which energy leaves the Sun's surface. This quantity is known as the **luminosity** of the Sun. It turns out to be just under 4 × 10^{26} W.

The Sun is an enormously powerful source of energy. *Every second*, it produces an amount of energy equivalent to the detonation of about 10 billion 1-megaton nuclear bombs. Six seconds worth of solar energy output, suitably focused, would evaporate all of Earth's oceans. Three minutes would melt our planet's crust. The scale on which the Sun operates simply defies Earthly comparison. Let's begin our more detailed study with a look at where all this energy comes from.

CONCEPT CHECK

✔ Why must we assume that the Sun radiates equally in all directions when we compute the solar luminosity from the solar constant?

16.2 The Heart of the Sun

What powers the Sun? What forces are at work in the Sun's core to produce such energy? By what process does the Sun shine, day after day, year after year, eon after eon? Answers to these questions are central to all astronomy. Without them, we can understand neither the physical existence of stars and galaxies in the universe nor the biological existence of life on Earth.

Solar Energy Production

In round numbers, the Sun's luminosity is 4 × 10^{26} W and its mass is 2 × 10^{30} kg. We can quantify how efficiently the Sun generates energy by dividing the solar luminosity by the solar mass:

$$\frac{\text{solar luminosity}}{\text{solar mass}} = 2 \times 10^{-4} \text{ W/kg}.$$

This simply means that, on average, every kilogram of solar material yields about 0.2 milliwatt of energy—0.0002 joule (J) of energy every second. This is not much energy—a piece of burning wood generates about a million times more energy per unit mass per unit time than does our Sun. But there is an important difference: The wood will not burn for billions of years.

To appreciate the magnitude of the energy generated by our Sun, we must consider not the ratio of the solar luminosity to the solar mass, but instead the total amount of energy generated by each gram of solar matter *over the entire lifetime of the Sun as a star*. This is easy to do. We simply multiply the rate at which the Sun generates energy by the age of the Sun, about 5 billion years. We obtain a value of 3×10^{13} J/kg. This is the average amount of energy radiated by every kilogram of solar material since the Sun formed. It represents a *minimum* value for the total energy radiated by the Sun, for more energy will be needed for every additional day the Sun shines. Should the Sun endure for another 5 billion years (as is predicted by theory), we would have to double this value.

Either way, this energy-to-mass ratio is very large. At least 60 trillion joules (on average) of energy must arise from every kilogram of solar matter to power the Sun throughout its lifetime. But the Sun's generation of energy is not explosive, releasing large amounts of energy in a short period. Instead, it is slow and steady, providing a uniform and long-lived rate of energy production. Only one known energy-generation mechanism can conceivably power the Sun in this way: **nuclear fusion**—the combining of light nuclei into heavier ones.

Nuclear Fusion

We can represent a typical fusion reaction symbolically as

nucleus 1 + nucleus 2 → nucleus 3 + energy.

In terms of powering the Sun, the most important piece of this equation is the energy produced. Where does it come from?

The key point is that, during a fusion reaction, the total mass *decreases*—the mass of nucleus 3 is less than the combined masses of nuclei 1 and 2. To understand the consequences of this decrease in mass, we use a very important law of modern physics: the **law of conservation of mass and energy**. At the beginning of the 20th century, Albert Einstein showed that matter and energy are interchangeable. One can be converted into the other, in accordance with Einstein's famous equation

$$E = mc^2,$$

or

energy = mass × (speed of light)^2.

This equation says that, to determine the amount of energy corresponding to a given mass, simply multiply the mass by the square of the speed of light (c in the equation). For example, the energy equivalent of 1 kg of matter is $1 \times (3 \times 10^8)^2$, or 9×10^{16} J. The speed of light is so large that even small amounts of mass translate into enormous amounts of energy.

The law of conservation of mass and energy states that the *sum* of mass and energy (properly converted to the same units) must always remain constant in any physical process. *There are no known exceptions.* According to this law, an object can literally disappear, provided that some energy appears in its place. If magicians really made rabbits disappear, the result would be a flash of energy equaling the product of the rabbit's mass and the square of the speed of light—enough to destroy the magician, everyone in the audience, and probably all of the surrounding state as well! In the case of fusion reactions in the solar core, the energy is produced primarily in the form of electromagnetic radiation. The light we see coming from the Sun means that the Sun's mass must be slowly, but steadily, decreasing with time.

The Proton–Proton Chain

All atomic nuclei are positively charged, so they repel one another. Furthermore, by the inverse-square law, the closer two nuclei come to one another, the greater is the repulsive force between them (Figure 16.4a). ∞ (Sec. 3.2) How, then, do nuclei—two protons, say—ever manage to fuse into anything heavier? The answer is that if they collide at high enough speeds, one proton can momentarily plow deep into the other, eventually coming within the exceedingly short range of the **strong nuclear force**, which binds nuclei together (see *More Precisely 16-1*). At distances less than about 10^{-15} m, the attraction of the nuclear force overwhelms the electromagnetic repulsion, and fusion occurs. Speeds in excess of a few hundred kilometers per second, corresponding to a gas temperature of 10^7 K or more, are needed to slam protons together fast enough to initiate fusion. Such conditions are found in the core of the Sun and at the centers of all stars.

The fusion of two protons is illustrated schematically in Figure 16.4(b). In effect, one of the protons turns into a neutron, creating new particles in the process, and combines with the other proton to form a **deuteron**, the nucleus of a special form of hydrogen called *deuterium*. Deuterium (also referred to as "heavy hydrogen") differs from ordinary hydrogen by virtue of an extra neutron in its nucleus. We can represent the reaction as follows:

proton + proton → deuteron + positron + neutrino.

The **positron** in this reaction is a positively charged electron. Its properties are identical to those of a normal, negatively charged electron, except for the positive charge. Scientists call the electron and the positron a "matter–antimatter pair"—the positron is said to be the *antiparticle* of the electron. The newly created positrons find themselves in the midst of a sea of electrons, with which they interact immediately and violently. The particles and antiparticles annihilate (destroy) one another, producing pure energy in the form of gamma-ray photons.

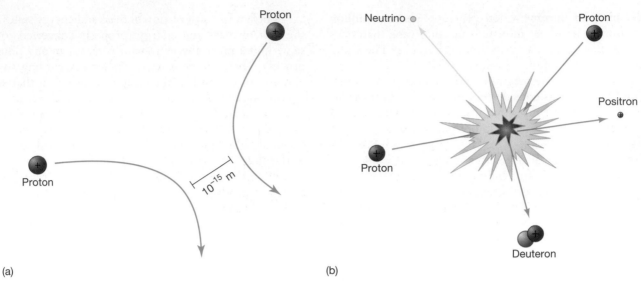

(a)   (b)

▲ **FIGURE 16.4 Proton Interactions** (a) Since like charges repel, two low-speed protons veer away from one another, never coming close enough for fusion to occur. (b) Sufficiently high-speed protons may succeed in overcoming their mutual repulsion, approaching close enough for the strong force to bind them together—in which case they collide violently, initiating the chain of nuclear fusion that powers the Sun.

The final product of the reaction is a particle known as a **neutrino**, a word derived from the Italian for "little neutral one." Neutrinos carry no electrical charge and are of very low mass—at most 1/100,000 the mass of an electron, which itself has only 1/2000 the mass of a proton. (The exact mass of the neutrino is uncertain, although recent evidence suggests that it is not zero.) Neutrinos move at (or nearly at) the speed of light and interact with hardly anything. They can penetrate, without stopping, several light-years of lead (a very dense material, widely used in terrestrial laboratories as an effective shield against radiation). Their interactions with matter are governed by the **weak nuclear force**, described in *More Precisely 16-1*. Despite their elusiveness, neutrinos can be detected with carefully constructed instruments. (See Section 16.6.)

Nuclei such as normal hydrogen and deuterium, containing the same number of protons, but different numbers of neutrons, represent different forms of the same element—they are known as **isotopes** of that element. Usually, there are about as many neutrons in a nucleus as protons, but the exact number of neutrons can vary, and most elements exist in a number of isotopic forms. To avoid confusion when talking about isotopes of the same element, nuclear physicists attach a number to the symbol representing the element. This number indicates the total number of particles (protons plus neutrons) in the nucleus of an atom of the element. Thus, ordinary hydrogen is denoted by ^{1}H, deuterium by ^{2}H. Normal helium (two protons plus two neutrons) is ^{4}He (also referred to as helium-4), and so on. We will adopt this convention for the rest of the book. We can now rewrite the preceding proton–proton reaction as

$$^1\text{H} + {}^1\text{H} \rightarrow {}^2\text{H} + \text{positron} + \text{neutrino}. \qquad \text{(I)}$$

This equation is labeled (I) because the production of a deuteron by the fusion of two protons is the first step in the fusion process powering most stars. It is the start of the **proton–proton chain**. Gargantuan quantities of protons are fused in this way within the core of the Sun each second. The positrons produced in the reaction quickly encounter electrons and are annihilated, releasing energy in the form of gamma rays.

The next step in solar fusion is the formation of an isotope of helium. In this step, a proton interacts with the deuteron produced in step (I):

$$^2\text{H} + {}^1\text{H} \rightarrow {}^3\text{He} + \text{energy}. \qquad \text{(II)}$$

Step II begins as soon as deuterons appear. The reaction product is an isotope of helium—helium-3 (^{3}He)—lacking one of the neutrons contained in the normal helium-4 nucleus. Energy is also emitted, again in the form of gamma-ray photons.

The third and final step in the proton–proton chain involves the production of a nucleus of helium-4 (^{4}He), which comes about most often through the fusion of two of the helium-3 nuclei created in step (II):

$$^3\text{He} + {}^3\text{He} \rightarrow {}^4\text{He} + {}^1\text{H} + {}^1\text{H} + \text{energy}. \qquad \text{(III)}$$

The result is a helium-4 nucleus plus two more protons, as well as another gamma ray.

The *net* effect of steps (I) through (III) is this: Four hydrogen nuclei (six protons consumed, two returned) combine to create one helium-4 nucleus, plus some gamma-ray radiation and two neutrinos. The whole process is illustrated in Figure 16.5. Symbolically, we have

$$4\,(^1\text{H}) \rightarrow {}^4\text{He} + \text{energy} + 2\ \text{neutrinos}.$$

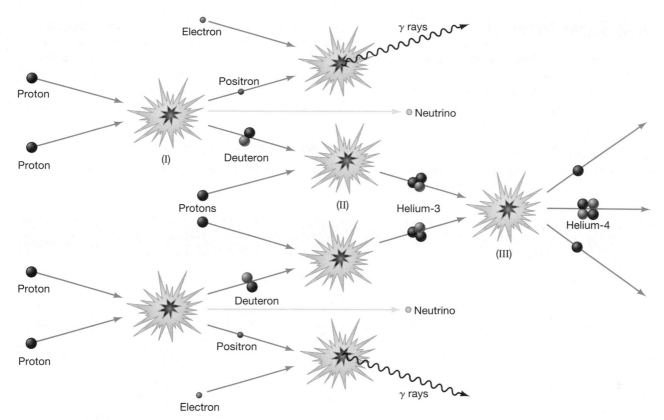

▲ **FIGURE 16.5 Solar Fusion** A total of six protons are converted into two protons, one helium-4 nucleus, and two neutrinos. The two leftover protons are available as fuel for new proton–proton reactions, so the net effect is that four protons are fused to form one helium-4 nucleus. Energy, in the form of gamma rays, is produced in each reaction. (Most of the photons are omitted for clarity.) The fusion events shown at the left, center, and right correspond to reactions (I), (II), and (III), respectively, in the text.

The gamma rays are slowly degraded in energy as they pass through the solar interior. Photons are repeatedly absorbed by electrons and ions and are then reemitted at a wavelength that reflects the temperature of the surrounding gas, in accordance with Wien's law. ∞ (Sec. 3.4) Thus, as the radiation makes its way to the surface through ever-cooler layers, its wavelength increases. Eventually, it leaves the surface in the form of visible light. The neutrinos escape unhindered into space. The helium stays put in the core.

Energy Generated by the Proton–Proton Chain

Now let's calculate the energy produced in the fusion process and compare it with the 6×10^{13} J/kg needed to account for the Sun's luminosity. Careful laboratory experiments have determined the masses of all the particles involved in the conversion of four protons into a helium-4 nucleus: The total mass of the protons is 6.6943×10^{-27} kg, the mass of the helium-4 nucleus is 6.6466×10^{-27} kg, and the neutrinos are (virtually) massless. The difference between the total mass of the protons and the mass of the helium nucleus is 0.048×10^{-27} kg—not much mass, just 0.7 percent of the total, but easily measurable.

Multiplying the vanished mass by the square of the speed of light yields 4.3×10^{-12} J—the energy produced by the fusion of 6.7×10^{-27} kg (the rounded-off mass of the four protons) of hydrogen into helium. It follows that the fusion of 1 kg of hydrogen would generate 6.4×10^{14} J. Thus, to fuel the Sun's present energy output, hydrogen must be fused into helium in the core at a rate of 600 million ton/s, converting 4.3 million tons of matter into energy every second (where 1 ton = 1000 kg). As we will see in Chapter 20, the supply of hydrogen in the solar core will be enough to power the Sun for about another 5 billion years.

This basic process—the slow, but steady, transformation of hydrogen into helium—is responsible for the light emitted by almost all the stars we see. Alternative reactions leading to the same final result exist (see *More Precisely 20-1*), but they are relatively rare in stars like the Sun.

CONCEPT CHECK

✔ Why does the fact that we see sunlight imply that the Sun's mass is slowly decreasing?

16.3 The Solar Interior

How do astronomers know about conditions in the interior of the Sun? As we have just seen, the fact that the Sun shines tells us that its center must be very hot, but our direct knowledge of the solar interior is actually quite limited. (See Section 16.6 for a discussion of one important, if problematic, "window" we do have into the solar core.) Lacking direct measurements, researchers must use other means to probe the inner workings of our parent star. To this end, they construct *mathematical models* of the Sun,

combining all available data with theoretical insight into solar physics to find the model that agrees most closely with observations. ∞ (Sec. 1.2) Recall from Chapter 11 how similar techniques are used to infer the structures of the jovian planets. ∞ (Sec. 11.3) The result in the case of the Sun is the **standard solar model**, which has gained widespread acceptance among astronomers.

Modeling the Structure of the Sun

The Sun's bulk properties—its mass, radius, temperature, and luminosity—do not vary much from day to day or

---

MORE PRECISELY 16-1

Fundamental Forces

Our studies of nuclear reactions have uncovered new ways in which matter can interact with matter at the subatomic level. Let's pause to consider in a slightly more systematic fashion the relationships among the various forces of nature.

As best we can tell, the behavior of all matter in the universe—from elementary particles to clusters of galaxies—is ruled by just four (or fewer) basic forces, which are *fundamental* to everything in the universe. In a sense, the search to understand the nature of the universe is the quest to understand the nature of these forces.

The *gravitational force* is probably the best known of the four. Gravity binds galaxies, stars, and planets together and holds humans on the surface of Earth. As we saw in Chapter 2, its magnitude decreases with distance according to an inverse-square law. ∞ (Sec. 2.7) Its strength is also proportional to the masses of each of the two objects involved. Thus, the gravitational field of an atom is extremely weak, but that of a galaxy, consisting of huge numbers of atoms, is very powerful. Gravity is by far the weakest of the forces of nature, but its effect accumulates as we move to larger and larger volumes of space, and nothing can cancel its attractive pull. As a result, gravity is the dominant force in the universe on all scales larger than that of Earth.

The *electromagnetic force* is another of nature's basic agents. Any particle having a net electric charge, such as an electron or a proton in an atom, exerts an electromagnetic force on any other charged particle. The everyday things we see around us are held together by this force. As with gravity, the strength of the electromagnetic force decreases with distance according to an inverse-square law. ∞ (Sec. 3.2) However, for subatomic particles, electromagnetism is much stronger than gravity. For example, the electromagnetic force between two protons exceeds their gravitational attraction by a factor of about 10^{36}. Unlike gravity, electromagnetic forces can repel (like charges) as well as attract (opposite charges). Positive and negative charges tend to neutralize each other, greatly diminishing their net electromagnetic influence. Above the microscopic level, most objects are in fact very close to being electrically neutral. Thus, except in unusual circumstances, the electromagnetic force is relatively unimportant on macroscopic scales.

A third fundamental force of nature is simply termed the *weak nuclear force*. This force is much weaker than electromagnetism, and its influence is somewhat more subtle. The weak nuclear force governs the emission of radiation from some radioactive atoms; the emission of a neutrino during the first stage of the proton–proton reaction is also the result of a weak interaction. The weak nuclear force does not obey the inverse-square law. Its effective range is less than the size of an atomic nucleus, about 10^{-15} m.

It is now known that electromagnetism and the weak force are not really separate forces at all, but rather two different aspects of a more basic *electroweak* force. At "low" temperatures, such as those found on Earth or even in stars, the electromagnetic and weak forces have quite distinct properties. However, as we will see in Chapter 27, at very high temperatures, such as those that prevailed in the universe when it was much less than a second old, the two are indistinguishable. Under those conditions, electromagnetism and the weak force are said to be "unified" into the electroweak force, and the universe had only three fundamental forces, rather than four.

Strongest of all the forces is the *strong nuclear force*. This force binds atomic nuclei together, binds the intranuclear particles together and governs the generation of energy in the Sun and all other stars. Like the weak force, and unlike the forces of gravity and electromagnetism, the strong force operates only at very close range. It is unimportant outside a distance of a hundredth of a millionth of a millionth (10^{-14}) of a meter. However, within that range (e.g., in atomic nuclei), it binds particles with enormous strength. In fact, it is the range of the strong force that determines the typical sizes of atomic nuclei. Only when two protons are brought within about 10^{-15} m of one another can the attractive strong force overcome their electromagnetic repulsion.

Not all particles are subject to all types of force. All particles interact through gravity because all have mass. However, only charged particles interact electromagnetically. Protons and neutrons are affected by the strong nuclear force, but electrons are not. Finally, under the right circumstances, the weak force can affect any type of subatomic particle, regardless of its charge.

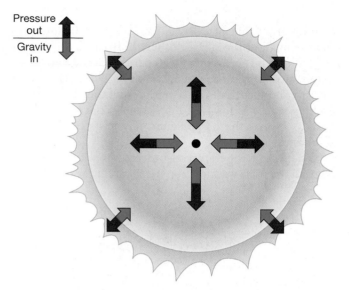

Pressure
out
Gravity
in

◄ **FIGURE 16.6 Hydrostatic Equilibrium** In the interior of a star such as the Sun, the outward pressure of hot gas exactly balances the inward pull of gravity. This is true at every point within the star, guaranteeing its stability.

from year to year. Although we will see in Chapter 20 that stars like the Sun do change significantly over periods of *billions* of years, for our purposes here this slow evolution may be ignored. On "human" time scales, the Sun may reasonably be thought of as unchanging.

Accordingly, as illustrated in Figure 16.6, theoretical models generally begin by assuming that the Sun is in a state of **hydrostatic equilibrium**, in which pressure's outward push exactly counteracts gravity's inward pull. This stable balance between opposing forces is the basic reason that the Sun neither collapses under its own weight nor explodes into interstellar space. ∞ (*More Precisely 8-1*) It also has an im-

portant consequence for the solar interior. Because the Sun is very massive, its gravitational pull is very strong, so very high internal pressure is needed to maintain hydrostatic equilibrium. This high pressure in turn requires a very high central temperature—hot enough, in fact, for hydrogen fusion to occur. In other words, the observation that the Sun is in hydrostatic equilibrium implies the extreme core conditions described in the previous section. Indeed, calculations of this sort carried out by British astrophysicist Sir Arthur Eddington around 1920 provided astronomers with the first inkling that fusion might be the process that powers the Sun.

To test and refine the standard solar model, astronomers are eager to obtain information about the solar interior. However, with so little direct information about conditions below the photosphere, we must rely on more indirect techniques. In the 1960s, measurements of the Doppler shifts of solar spectral lines revealed that the surface of the Sun oscillates, or vibrates, like a complex set of bells. ∞ (Secs. 3.5, 4.4) These vibrations, illustrated in Figure 16.7(a), are the result of internal pressure waves (somewhat like sound waves in air) that reflect off the photosphere and repeatedly cross the solar interior (Figure 16.7b). Because the waves can penetrate deep inside the

(a)

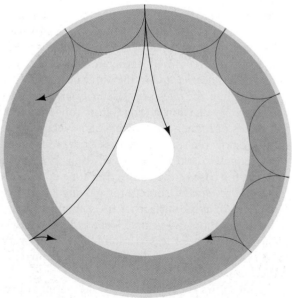

(b)

▲ **FIGURE 16.7 Solar Oscillations** (a) The Sun has been found to vibrate in a complex way. By observing the motion of the solar surface, scientists can determine the wavelengths and the frequencies of the individual waves and deduce information about the Sun that is not obtainable by other means. The alternating patches represent gas moving down (red) and up (blue). (See also *Discovery 16-1*.) (b) Depending on their initial directions, the waves contributing to the observed oscillations may travel deep inside the Sun, providing vital information about the solar interior. The wave shown closest to the surface here corresponds approximately to the vibration pattern depicted in part (a). (*National Solar Observatory*)

Sun, analysis of their surface patterns allows scientists to study conditions far below the Sun's surface. The process is similar to the way in which seismologists learn about the interior of Earth by observing the P- and S-waves produced by earthquakes. ∞ (Sec. 7.3) For this reason, the study of solar surface patterns is usually called **helioseismology**, even though solar pressure waves have nothing whatever to do with solar seismic activity—there is no such thing.

The most extensive study of solar oscillations is the ongoing Global Oscillations Network Group (GONG) project. By making continuous observations of the Sun from many clear sites around Earth, solar astronomers can obtain uninterrupted high-quality solar data spanning many days and even weeks—almost as though Earth were not rotating and the Sun never set. The *Solar and Heliospheric Observatory* (*SOHO*), launched by the European Space Agency in 1995 and now permanently stationed between Earth and the Sun some 1.5 million km from our planet (see *Discovery 16-1*), also provides continuous monitoring of the Sun's surface and atmosphere. Analysis of ground- and space-based data provides important additional information about the temperature, density, rotation, and convective state of the solar interior, allowing detailed comparisons between theory and reality to be made. Direct comparison is possible throughout a large portion of the Sun, and the agreement between model and observations is spectacular: The frequencies and wavelengths of observed solar oscillations are within 0.1 percent of the predictions of the standard solar model.

Some mysteries remain, however. For example, helioseismology indicates that the Sun's rotation speed varies with depth—perhaps not too surprising, given the surface differential rotation mentioned earlier and the fact that similar behavior has been noted in the outer planets. What is puzzling, though, is the *complexity* of the differential motion. The surface layers show a "zonal flow" of sorts, with alternating bands of higher- and lower-than-average rotation rates. Just below the surface are wide "rivers" of lower speed (at the equator) and higher speed (polar) rotation. The material at the base of the convection zone appears to oscillate in rotation speed, sometimes moving faster (by about 10 percent) than the surface layers, sometimes slower, with a period of about 1.3 years. Deeper still, the radiative interior rotates more or less as a solid body, once every 26.9 days. A full explanation of the Sun's rotation currently eludes theorists.

Figure 16.8 shows the solar density and temperature, plotted as functions of distance from the Sun's center, according to the standard solar model. Notice how the density drops rather sharply at first and then decreases more slowly near the solar photosphere, some 700,000 km from the center. The variation in density is large, ranging from a core value of about 150,000 kg/m³, 20 times the density of iron, to an intermediate value (at 350,000 km) of about 1000 kg/m³, the density of water, to an extremely small

photospheric value of 2×10^{-4} kg/m³, 10,000 times less dense than air at the surface of Earth. Because the density is so high in the core, roughly 90 percent of the Sun's mass is contained within the inner half of its radius. The solar density continues to decrease out beyond the photosphere, reaching values as low as 10^{-23} kg/m³ in the far corona—about as thin as the best vacuum physicists can create in laboratories on Earth.

The solar temperature also decreases with increasing radius in the solar interior, but not as rapidly as the density.

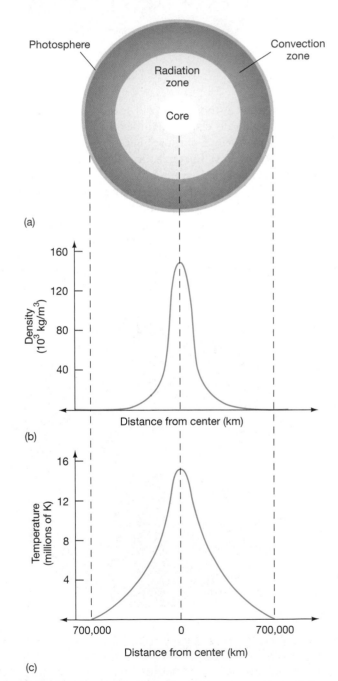

▲ **FIGURE 16.8 Solar Interior** (a) A cross section of the Sun, with corresponding graphs of (b) density and (c) temperature, according to the standard solar model.

SOHO: Eavesdropping on the Sun

Throughout the few decades of the Space Age, various nations, led by the United States, have sent spacecraft to all but two of the major bodies in the solar system. One of these as-yet-unexplored bodies is Pluto, the most distant planet from the Sun, which has never been visited by a robot orbiter or even a flyby craft—and such a mission is not likely anytime soon. The other unexplored body is the Sun. Currently, the next best thing to a dedicated reconnoitering spacecraft is the *Solar and Heliospheric Observatory* (*SOHO*), which has been radioing back to Earth volumes of new data—and a few new puzzles—about our parent star since the vehicle's launch in late 1995.

SOHO is a billion-dollar mission operated primarily by the European Space Agency. The 2-ton robot is now on station about 1.5 million km sunward of Earth—about 1 percent of the distance from Earth to the Sun. This is the so-called L_1 Lagrangian point, where the gravitational pull of the Sun and Earth are precisely equal—a good place to park a monitoring platform. ∞ (Sec. 14.1) There, *SOHO* looks unblinkingly at our star 24 hours a day, eavesdropping on the Sun's surface, atmosphere, and interior. The automated vehicle carries a dozen instruments, capable of measuring almost everything from the Sun's corona and magnetic field to its solar wind and internal vibrations. The first figure shows a false-color image of the Sun's lower corona, obtained by combining *SOHO* data captured at three different extreme ultraviolet wavelengths.

SOHO is positioned just beyond Earth's magnetosphere, so its instruments can cleanly study the charged particles of the solar wind—high-speed matter escaping from the Sun, flowing outward from the corona. Coordinating these on-site measurements with *SOHO* images of the Sun itself, astronomers now think they can follow solar magnetic field loops expanding and breaking as the Sun prepares itself for

mass ejections several days before they actually occur. (These topics will be discussed in Section 16.5.) Given that such coronal storms can wreak havoc on communications, power grids, satellite electronics, and other human activities, the prospect of having accurate forecasts of disruptive solar events is a welcome development.

Section 16.2 discussed how astronomers can "take the pulse" of the Sun by measuring its complex rhythmic motions. *SOHO* also has the ability to study the weak sound waves that echo and resonate inside the Sun and can map these vibrations with much higher resolution than was previously possible. It does so not by sensing sound itself, but by watching the Sun's surface move up and down ever so slightly. The Sun's "loudest" vibrations are extremely low pitched (0.003 Hz, or one oscillation every 5 minutes), more like rolling rumbles, just as we might expect from such a huge and massive object.

The second figure is not an image, but a radial-velocity map of the Sun's surface (at much higher resolution than the diagram in Figure 16.7), obtained by measuring the Doppler shift of different portions of the solar photosphere. Bright areas are moving toward us, dark areas away. Observations of this sort enable researchers to refine models of the solar interior. For example, they have pinpointed the exact boundary between the convection zone and the radiation zone at 71.3 percent of the radius of the photosphere. The hope is to extend this work to discover how sharply defined the core is and just where its outer boundary lies.

As of mid-2004, *SOHO* is still operating, six years beyond its planned lifetime, having survived assaults from both the Sun and Earth. Twice so far, skillful engineers have brought the spacecraft back from apparent death after mission controllers mistakenly sent incorrect commands from the ground. It is a good thing that they did, for this remarkable spacecraft has radioed back to Earth a wealth of new scientific insight into our parent star.

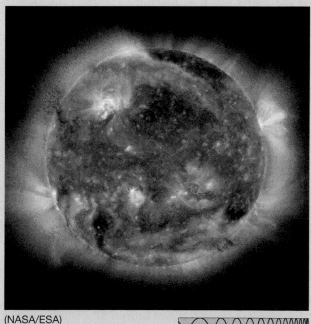

(NASA/ESA)

R I V U X G

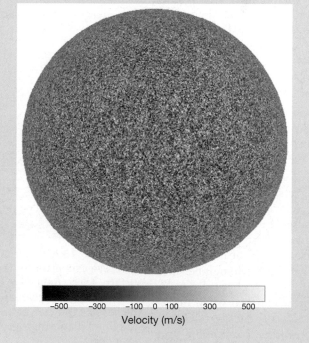

Velocity (m/s)

-500 -300 -100 0 100 300 500

Computer models indicate a temperature of about 15 million K at the core, consistent with the minimum 10 million K needed to initiate the nuclear reactions known to power most stars, decreasing to the observed value of about 5800 K at the photosphere.

Energy Transport

LEARNING GOAL 3 The very hot solar interior ensures violent and frequent collisions among gas particles. Particles move in all directions at high speeds, bumping into one another unceasingly. In and near the core, the extremely high temperatures guarantee that the gas is completely ionized. Recall from Chapter 4 that, under less extreme conditions, atoms absorb photons that can boost their electrons to more excited states. ∞ (Sec. 4.2) With no electrons left on atoms to capture the photons, however, the deep solar interior is relatively transparent to radiation. Only occasionally does a photon encounter and scatter off of a free electron or proton. As a result, the energy produced by nuclear reactions in the core travels outward toward the surface in the form of radiation with relative ease.

As we move outward from the core, the temperature falls, atoms collide less frequently and less violently, and more and more electrons manage to remain bound to their parent nuclei. With more and more atoms retaining electrons that can absorb the outgoing radiation, the gas in the interior changes from being relatively transparent to being almost totally opaque. By the outer edge of the radiation zone, 200,000 km below the photosphere, *all* the photons produced in the Sun's core have been absorbed. Not one of them reaches the surface. But what happens to the energy they carry?

The photons' energy must travel beyond the Sun's interior: That we see sunlight—visible energy—proves that energy escapes. The escaping energy reaches the surface by *convection*—the same basic physical process we saw in our study of Earth's atmosphere, although it operates in a very different environment in the Sun. ∞ (Sec. 7.2) Hot solar gas moves outward while cooler gas above it sinks, creating a characteristic pattern of convection cells. All through the convection zone, energy is transported to the surface by physical motion of the solar gas. (Remember that there is no physical movement of material when radiation is the energy-transport mechanism; convection and radiation are *fundamentally*

different ways in which energy can be transported from one place to another.)

In reality, the zone of convection is much more complex than we have just described. As illustrated in Figure 16.9, there is a hierarchy of convection cells, organized in tiers of many different sizes at different depths. The deepest tier, lying approximately 200,000 km below the photosphere, is thought to contain large cells some tens of thousands of kilometers in diameter. Heat is then successively carried upward through a series of progressively smaller cells, stacked one on another, until, at a depth of about 1000 km, the individual cells are about 1000 km across. The top of this uppermost tier of convection is the visible surface of the Sun, where astronomers can directly observe the cell sizes. Information about convection below that level is inferred mostly from computer models of the solar interior.

At some distance from the core, the solar gas becomes too thin to sustain further upwelling by convection. Theory suggests that this distance roughly coincides with the photospheric surface we see. Convection does not proceed into the solar atmosphere; there is simply not enough gas there—the density is so low that there are too few atoms or ions to intercept much sunlight, so the gas becomes transparent again and radiation once more becomes the mechanism of energy transport. Photons reaching the photosphere escape more or less freely into space, and the photosphere emits thermal radiation, like any other hot object. The photosphere is narrow, and the "edge" of the Sun sharp, because this transition from opacity to complete transparency is very rapid. Just below the bottom of the photosphere the gas is still convective, and radiation does not reach us directly. A few hundred kilometers higher, the gas is too thin to emit or absorb any significant amount of radiation.

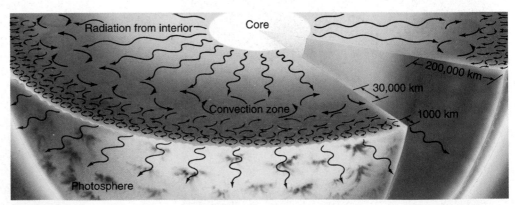

▲ **FIGURE 16.9 Solar Convection** Physical transport of energy in the Sun's convection zone. We can visualize the upper interior as a boiling, seething sea of gas. Near the surface, each convective cell is about 1000 km across. The sizes of the convective cells become progressively larger at greater depths, reaching some 30,000 km in diameter at the base of the convection zone, 200,000 km below the photosphere. (This is a highly simplified diagram; there are many different cell sizes, and they are not so neatly arranged.)

PHYSLET® ILLUSTRATION Random Walk

Granulation

Figure 16.10 is a high-resolution photograph of the solar surface. The visible surface is highly mottled, or **granulated**, with regions of bright and dark gas known as *granules*. Each bright granule measures about 1000 km across—comparable in size to a continent on Earth—and has a lifetime of between 5 and 10 minutes. Together, several million granules constitute the top layer of the convection zone, immediately below the photosphere.

Each granule forms the topmost part of a solar convection cell. Spectroscopic observation within and around the bright regions shows direct evidence for the upward motion of gas as it "boils" up from within—evidence that convection really does occur just below the photosphere. Spectral lines detected from the bright granules appear slightly bluer than normal, indicating Doppler-shifted matter approaching us at about 1 km/s. ∞ (Sec. 3.5) Spectroscopes focused on the darker portions of the granulated photosphere show the same spectral lines to be redshifted, indicating matter moving away from us.

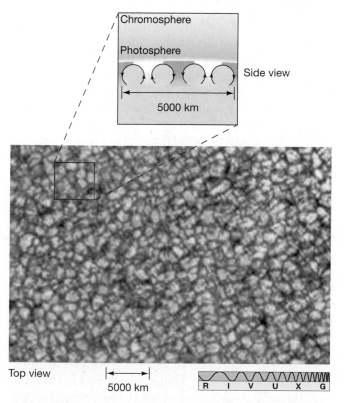

▲ **FIGURE 16.10 Solar Granulation** Photograph of the granulated solar photosphere, taken from the *Skylab* space station looking directly down on the Sun's surface. Typical solar granules are comparable in size to a large U.S. state. The bright portions of the image are regions where hot material is upwelling from below. The dark regions correspond to cooler gas that is sinking back down into the interior. The inset shows a perpendicular cut through the solar surface. *(Big Bear Solar Observatory)*

The variations in brightness of the granules result strictly from differences in temperature. The upwelling gas is hotter and therefore emits more radiation than the cooler, downward-moving gas. The adjacent bright and dark gases appear to contrast considerably, but in reality their temperature difference is less than about 500 K. Careful measurements also reveal a much larger-scale flow on the solar surface. *Supergranulation* is a flow pattern quite similar to granulation, except that supergranulation cells measure some 30,000 km across. As with granulation, material upwells at the center of the cells, flows across the surface, then sinks down again at the edges. Scientists suspect that supergranules are the imprint on the photosphere of a deeper tier of large convective cells, like those depicted in Figure 16.9.

CONCEPT CHECK

✔ What are the two distinct ways in which energy moves outward from the solar core to the photosphere?

16.4 The Solar Atmosphere

Composition

LEARNING GOAL 4 Astronomers can glean an enormous amount of information about the Sun from an analysis of the absorption lines that arise in the photosphere and lower atmosphere. ∞ (Sec. 4.4) Figure 16.11 (see also Figure 4.4) is a detailed spectrum of the Sun spanning a range of wavelengths from 360 to 690 nm. Notice the intricate dark Fraunhofer absorption lines superposed on the background continuous spectrum.

As discussed in Chapter 4, spectral lines arise when electrons in atoms or ions make transitions between states of well-defined energies, emitting or absorbing photons of specific energies (i.e., wavelengths or colors) in the process. ∞ (Sec. 4.2) However, to explain the spectrum of the Sun (and, indeed, the spectra of all stars), we must slightly modify our earlier description of the formation of absorption lines. We explained these lines in terms of cool foreground gas intercepting light from a hot background source. In actuality, both the bright background and the dark absorption lines in Figure 16.11 form at roughly the same locations in the Sun—the solar photosphere and lower chromosphere. To understand how these lines are formed, consider again the solar energy–emission process in a little more detail.

Below the photosphere, the solar gas is sufficiently dense, and interactions among photons, electrons, and ions sufficiently common, that radiation cannot escape directly into space. In the solar atmosphere, however, the probability that a photon will escape without further interaction

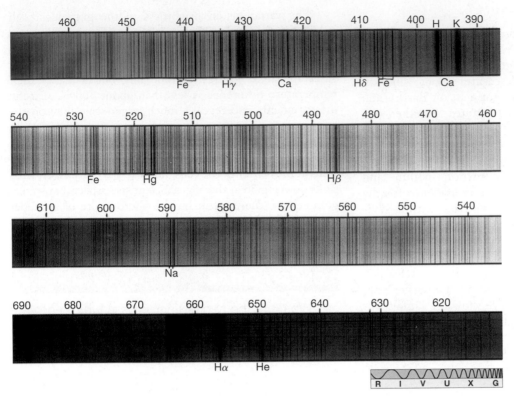

◄**FIGURE 16.11** Solar
Spectrum A detailed visible
spectrum of our Sun shows
thousands of dark Fraunhofer
spectral lines indicating the
presence of some 67 different
elements in various stages of
excitation and ionization in
the lower solar atmosphere.
The numbers give
wavelengths, in nanometers.
(Palomar Observatory/Caltech)

with matter depends on the photon's energy. If that energy happens to correspond to some electronic transition in one of the atoms or ions that are present in the gas, then the photon may be absorbed again before it can travel very far—the more elements present of the type suitable for absorption, the lower is the escape probability. Conversely, if the photon's energy does not coincide with any such transition, then the photon cannot interact further with the gas, and it leaves the Sun headed for interstellar space or perhaps the detector of an astronomer on Earth.

Thus, as illustrated in Figure 16.12, when we look at the Sun, we are actually peering down into the solar atmosphere to a depth that depends on the wavelength of the light under consideration. Photons with wavelengths far from any absorption feature (i.e., having energies far from any atomic transition) are less likely to interact with matter as they travel through the solar gas and so tend to come from deep in the photosphere. However, photons with wavelengths near the centers of absorption lines are much more likely to be captured by an atom or ion and therefore escape mainly from higher (and cooler) levels. The lines are darker than their surroundings because the temperature at the level of the atmosphere where they form is lower than the 5800-K temperature at the base of the photosphere, where most of the continuous emission originates. (Recall that, by Stefan's law, the brightness of a radiating object depends on its temperature—the cooler the gas, the less energy it radiates.) ∞ (Sec. 3.4) Thus, the existence of Fraunhofer lines is direct evidence that the temperature in the Sun's atmosphere decreases with height above the photosphere.

Tens of thousands of spectral lines have been observed and cataloged in the solar spectrum. In all, some 67 elements have been identified in the Sun, in various states of ionization and excitation. ∞ (Sec. 4.2) More elements probably exist there, but they are present in such small quantities that our instruments are simply not sensitive enough to detect them. Table 16.2 lists the 10 most common elements in the Sun. Notice that hydrogen is by far the most abundant element, followed by helium. This distribution is just what we saw on the jovian planets, and it is what we will find for the universe as a whole.

Strictly speaking, spectral analysis allows us to draw conclusions only about the part of the Sun where the lines form—the photosphere and chromosphere. However, most astronomers think that, with the exception of the solar core (where nuclear reactions are steadily changing the composition—see Section 16.2), the data in Table 16.2 are representative of the entire Sun. That assumption is strongly supported by the excellent agreement between the standard solar model, which makes the same assumption, and helioseismological observations of the solar interior.

The Chromosphere

Above the photosphere lies the cooler chromosphere, the inner part of the solar atmosphere. This region emits very little light of its own and cannot be observed visually under normal conditions. The photosphere is just too bright, dominating the chromosphere's radiation. The relative dimness of the chromosphere results from its low

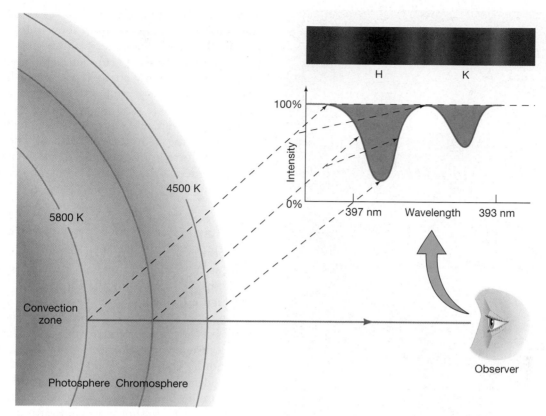

▲ **FIGURE 16.12 Spectral Line Formation** Photons with energies well away from any atomic transition can escape from relatively deep in the photosphere, but those with energies close to a transition are more likely to be reabsorbed before escaping, so those we see on Earth tend to come from higher, cooler levels in the solar atmosphere. The dashed lines indicate schematically the levels in the atmosphere where photons corresponding to different parts of the absorption line originate. The inset shows a close-up tracing of two of the thousands of solar absorption lines: the "H" and "K" lines of calcium at about 395 nm.

density—large numbers of photons simply cannot be emitted by a tenuous gas containing very few atoms per unit volume. Still, although it is not normally seen, astronomers have long been aware of the chromosphere's existence. Figure 16.13 shows the Sun during an eclipse in which the photosphere—but not the chromosphere—is obscured by the Moon. The chromosphere's characteristic reddish hue is plainly visible. This coloration is due to the red Hα (hydrogen alpha) emission line of hydrogen, which dominates the chromospheric spectrum. ∞ (*More Precisely 4-1*)

The chromosphere is far from tranquil. Every few minutes, small solar storms erupt, expelling jets of hot matter known as *spicules* into the Sun's upper atmosphere

▶ **FIGURE 16.13 Solar Chromosphere** This photograph of a total solar eclipse shows the solar chromosphere, a few thousand kilometers above the Sun's surface. (*G. Schneider*)

R   I   V   U   X   G

R I V U X G

3000 km

▲ **FIGURE 16.14 Solar Spicules** Short-lived narrow jets of gas that typically last mere minutes, spicules can be seen sprouting up from the solar chromosphere in this Hα image of the Sun. The spicules are the thin, dark, spikelike regions. They appear dark against the face of the Sun because they are cooler than the solar photosphere. *(NOAO)*

TABLE 16.2	The Composition of the Sun	
Element	Abundance (percentage of total number of atoms)	Abundance (percentage of total mass)
Hydrogen	91.2	71.0
Helium	8.7	27.1
Oxygen	0.078	0.97
Carbon	0.043	0.40
Nitrogen	0.0088	0.096
Silicon	0.0045	0.099
Magnesium	0.0038	0.076
Neon	0.0035	0.058
Iron	0.0030	0.14
Sulfur	0.0015	0.040

(Figure 16.14). These long, thin spikes of matter leave the Sun's surface at typical speeds of about 100 km/s and reach several thousand kilometers above the photosphere. Spicules are not spread evenly across the solar surface. Instead, they cover only about 1 percent of the total area, tending to accumulate around the edges of supergranules. The Sun's magnetic field is also known to be somewhat stronger than average in those regions. Scientists speculate that the downward-moving material there tends to strengthen the solar magnetic field, and spicules are the result of magnetic disturbances in the Sun's churning outer layers.

The Transition Zone and the Corona

During the brief moments of an eclipse, if the Moon's angular size is large enough that both the photosphere and the chromosphere are blocked, the ghostly solar corona can be seen (Figure 16.15). With the photospheric light removed, the pattern of spectral lines changes dramatically. The intensities of the usual lines alter (suggesting changes in composition or temperature, or both), the spectrum shifts from absorption to emission, and an entirely new set of spectral lines suddenly appears. These new coronal (and in some cases chromospheric) lines were first observed during eclipses in the 1920s. For years afterward, some researchers (for want of any better explanation) attributed them to a new nonterrestrial element, which they dubbed "coronium."

R I V U X G

▲ **FIGURE 16.15 Solar Corona** When both the photosphere and the chromosphere are obscured by the Moon during a solar eclipse, the faint corona becomes visible. This photograph clearly shows the emission of radiation from the relatively inactive solar corona. *(Bencho Angelov)*

We now recognize that these new spectral lines do not indicate any new kind of atom. Coronium does not exist. Rather, the new lines arise because atoms in the corona have lost several more electrons than atoms in the photosphere—that is, the coronal atoms are much more highly ionized. Therefore, their internal electronic structures, and hence their spectra, are quite different from the structure and spectra of atoms and ions in the photosphere. For example, astronomers have identified coronal lines corresponding to iron ions with as many as 13 of their normal 26 electrons missing. In the photosphere, most iron atoms have lost only 1 or 2 of their electrons. The cause of this extensive electron stripping is the high coronal temperature. The degree of ionization inferred from spectra observed during solar eclipses tells us that the temperature of the upper chromosphere exceeds that of the photosphere. Furthermore, the temperature of the solar corona, where even more ionization is seen, is higher still.

Figure 16.16 shows how the temperature of the Sun's atmosphere varies with altitude. The temperature decreases to a minimum of about 4500 K some 500 km above the photosphere, after which it rises steadily. About 1500 km above the photosphere, in the transition zone, the temperature begins to rise rapidly, reaching more than 1 million K at an altitude of 10,000 km. Thereafter, in the corona, the temperature remains roughly constant at around 3 million K, although *SOHO* and other orbiting instruments have detected coronal "hot spots" having temperatures many times higher than this average value.

The cause of the rapid temperature rise is not fully understood. The temperature profile runs contrary to intuition: Moving away from a heat source, we would normally expect the heat to diminish, but this is not the case in the lower atmosphere of the Sun. The corona must have another energy source. Astronomers now think that magnetic disturbances in the solar photosphere are ultimately responsible for heating the corona. (See Section 16.5.)

The Solar Wind

Electromagnetic radiation and fast-moving particles—mostly protons and electrons—escape from the Sun all the time. The radiation moves away from the photosphere at the speed of light, taking 8 minutes to reach Earth. The particles travel more slowly, although at the still considerable speed of about 500 km/s, reaching Earth in a few days. This constant stream of escaping solar particles is the **solar wind**.

The solar wind results from the high temperature of the corona. About 10 million km above the photosphere, the coronal gas is hot enough to escape the Sun's gravity, and it begins to flow outward into space. At the same time, the solar atmosphere is continuously replenished from below. If that were not the case, the corona would disappear in about a day. The Sun is, in effect, "evaporating"—constantly shedding mass through the solar wind. The wind is an extremely thin medium, however, so, although it carries away over a million tons of solar matter each second, less than 0.1 percent of the Sun's mass has been lost since the solar system formed 4.6 billion years ago.

CONCEPT CHECK

✔ Describe two ways in which the spectrum of the solar corona differs from that of the photosphere.

16.5 The Active Sun

Most of the Sun's luminosity results from continuous emission from the photosphere. However, superimposed on this steady, predictable aspect of our star's energy output is a much more irregular component, characterized by explosive and unpredictable surface activity. Solar activity contributes little to the Sun's total luminosity and probably has no significant bearing on the evolution of the Sun, but it does affect us here on Earth. The size and duration of coronal holes are strongly influenced by the level of solar activity. Hence, so is the strength of the solar wind, and that in turn directly affects Earth's magnetosphere.

Sunspots

Figure 16.17 is an optical photograph of the entire Sun, showing numerous dark blemishes on its surface. First studied in detail by Galileo, these "spots" provided one of the first clues that the Sun was not a perfect, unvarying creation,

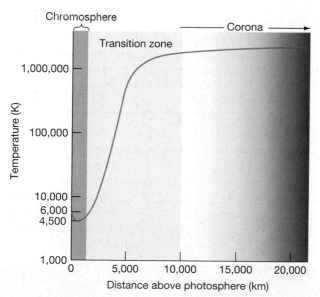

▲ **FIGURE 16.16 Solar Atmospheric Temperature** The change of gas temperature in the lower solar atmosphere is dramatic. The temperature reaches a minimum of 4500 K in the chromosphere and then rises sharply in the transition zone, finally leveling off at around 3 million K in the corona.

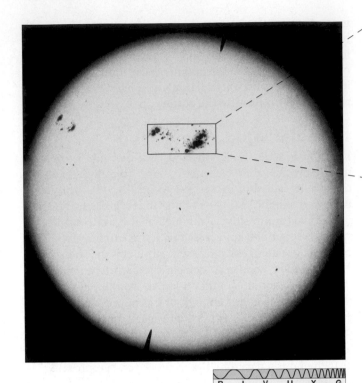

R I V U X G

▲ **FIGURE 16.17 Sunspots** This photograph of the entire Sun, taken during a period of maximum solar activity, shows several groups of sunspots. The largest spots in the image are more than 20,000 km across—nearly twice the diameter of Earth. Typical sunspots are only about half that size. *(Palomar Observatory/Caltech)*

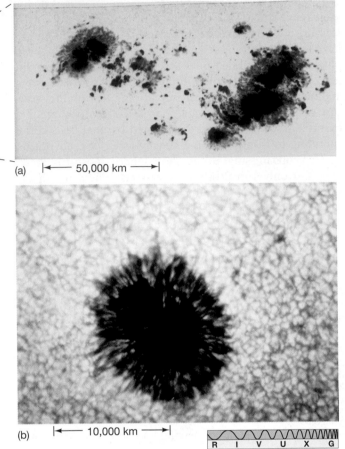

(a) |◄— 50,000 km —►|

(b) |◄— 10,000 km —►|

R I V U X G

▲ **FIGURE 16.18 Sunspots, Up Close** (a) An enlarged photograph of the largest pair of sunspots in Figure 16.17. Each spot consists of a cool, dark inner region called the umbra, surrounded by a warmer, brighter region called the penumbra. The spots appear dark because they are slightly cooler than the surrounding photosphere. (b) A high-resolution image of a single sunspot shows details of its structure as well as the surface granules surrounding it. This spot is about the size of Earth. *(Palomar Observatory/ Caltech; National Solar Observatory)*

but a place of constant change. ∞ (Sec. 2.4) The dark areas are called **sunspots** and typically measure about 10,000 km across, approximately the size of Earth. As shown in the figure, they often occur in groups. At any given time, the Sun may have hundreds of sunspots, or it may have none at all.

Studies of sunspots show an *umbra*, or dark center, surrounded by a grayish *penumbra*. The close-up views in Figure 16.18 show each of these dark areas and the brighter undisturbed photosphere nearby. This gradation in darkness is really a gradual change in photospheric temperature—sunspots are simply *cooler* regions of the photospheric gas. The temperature of the umbra is about 4500 K, compared with the penumbra's 5500 K. The spots, then, are certainly composed of hot gases. They seem dark only because they appear against an even brighter background (the 5800 K photosphere). If we could magically remove a sunspot from the Sun (or just block out the rest of the

Sun's emission), the spot would glow brightly, just like any other hot object having a temperature of roughly 5000 K.

Sunspots are not steady. Most change their size and shape, and all come and go. Figure 16.19 shows a time sequence in which a number of spots varied—sometimes growing, sometimes dissipating—over a period of several days. Individual spots may last anywhere from 1 to 100 days. A large group of spots typically lasts 50 days.

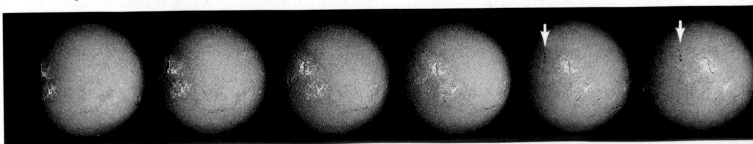

Solar Magnetism

What causes a sunspot? Why is it cooler than the surrounding photosphere? The answers to these questions involve the Sun's magnetism. As we saw in Chapter 4, an analysis of spectral lines can yield information about the magnetic field in which the lines originate. ∞ (Sec. 4.4) The magnetic field in a typical sunspot is about a thousand times greater than the field in neighboring, undisturbed photospheric regions (which is itself several times stronger than Earth's magnetic field). Scientists theorize that sunspots are cooler than their surroundings because these abnormally strong fields tend to block (or redirect) the convective flow of hot gas, which is normally toward the surface of the Sun.

Another indicator of the magnetic nature of sunspots is their grouping. The *polarity* of a sunspot simply indicates which way its magnetic field is directed. We conventionally label spots where field lines emerge from the interior as "S" and those where the lines dive below the photosphere as "N" (so field lines above the surface always run from S to N, as on Earth). Sunspots almost always come in pairs whose members lie at roughly the same latitude and have opposite magnetic polarities. Figure 16.20 illustrates how magnetic field lines emerge from the solar interior through one member (S) of a sunspot pair, loop through the solar atmosphere, and then reenter the photosphere through the other member (N).

Despite the irregular appearance of the sunspots themselves, there is a great deal of order in the underlying solar field. *All* the sunspot pairs in the same solar hemisphere (north or south) at any instant have the *same* magnetic configuration. That is, if the leading spot (measured in the direction of the Sun's rotation) of one pair has N polarity, as shown in the figure, then all leading spots in that hemisphere have the same polarity. What's more, in the other hemisphere at the same time, all sunspot pairs have the *opposite* magnetic configuration (S polarity leading). To understand these regularities in sunspot polarities, we must look at the Sun's magnetic field in more detail.

As illustrated in Figure 16.21, the Sun's differential rotation greatly distorts the solar magnetic field, "wrapping"

it around the solar equator and eventually causing any originally north–south magnetic field to reorient itself in an east–west direction. At the same time, convection causes the magnetized gas to well up toward the surface, twisting and tangling the magnetic field pattern. In some places, the field lines become kinked like a twisted garden hose, causing the field strength to increase. Occasionally, the field becomes so strong that it overwhelms the Sun's gravity, and a "tube" of field lines bursts out of the surface and loops through the lower atmosphere, forming a sunspot pair. The general east–west orientation of the underlying solar field accounts for the observed polarities of the resulting sunspot pairs in each hemisphere.

Don't be fooled by the simplicity of the schematic diagram in Figure 16.20(a). Figure 16.20(b) shows a far-ultraviolet image taken by NASA's *Transition Region and Coronal Explorer* (*TRACE*) satellite in 1999. The image reveals high-temperature gas flowing along a complex network of magnetic field lines connecting two sunspot groups. The gas is thought to be heated by explosive events occurring in or near the sunspots themselves, as their ever-changing magnetic field lines continually merge and re-form, releasing large amounts of energy in the process.

The Solar Cycle

Not only do sunspots come and go with time, but their numbers and distribution across the face of the Sun also change fairly regularly. Centuries of observations have established a clear **sunspot cycle**. Figure 16.22(a) shows the number of sunspots observed each year during the 20th century. The average number of spots reaches a maximum every 11 or so years and then falls off almost to zero before the cycle begins afresh.

The latitudes at which sunspots appear vary as the sunspot cycle progresses. Individual sunspots do not move up or down in latitude, but new spots appear closer to the equator as older ones at higher latitudes fade away. Figure 16.22(b) is a plot of observed sunspot latitude as a function of time. At the start of each cycle, at *solar minimum*, only a few spots are seen, and these are generally confined to two

▼ **FIGURE 16.19 Sunspot Rotation** The evolution of some sunspots and lower chromospheric activity over a period of 12 days. The sequence runs from left to right. An Hα filter was used to make these photographs, taken from the *Skylab* space station in 1975. An arrow follows one set of sunspots over the course of a week as they are carried around the Sun by its rotation. (*NASA*)

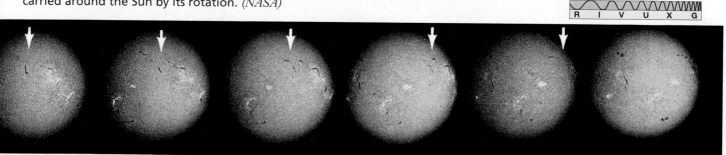
R I V U X G

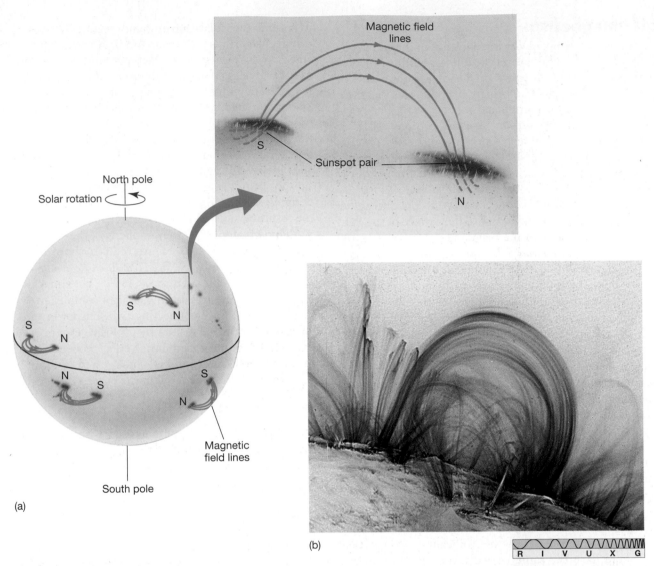

▲ **FIGURE 16.20 Solar Magnetism** (a) Sunspot pairs are linked by magnetic field lines. The Sun's magnetic field lines emerge from the surface through one member of a pair and reenter the Sun through the other member. The leading members of all sunspot pairs in the solar northern hemisphere have the same polarity (labeled N or S, as described in the text). If the magnetic field lines are directed into the Sun in one leading spot, they are inwardly directed in all other leading spots in that hemisphere. The same is true in the southern hemisphere, except that the polarities are always opposite those in the north. The overall direction of the magnetic field reverses itself roughly every 11 years. (b) A far-ultraviolet image taken by NASA's *Transition Region and Coronal Explorer* (*TRACE*) satellite in 1999, showing magnetic field lines arching between two sunspot groups. Note the complex structure of the lines, which are seen here via the radiation emitted by superheated gas flowing along them. The resolution is about 700 km. In this negative image (which shows the lines more clearly), the darkest regions have temperatures of about 2 million K. (*NASA*)

narrow zones about 25° to 30° north and south of the solar equator. Approximately four years into the cycle, around *solar maximum*, the number of spots has increased markedly, and they are found within about 15° to 20° of the equator. Finally, by the end of the cycle, at solar minimum, the number has fallen again, and most sunspots lie within about 10° of the solar equator. The beginning of each new cycle appears to overlap the end of the last.

Complicating this picture further, the 11-year sunspot cycle is actually only half of a longer 22-year **solar cycle**. During the first 11 years of the cycle, the leading spots of all the pairs in the northern hemisphere have the same polarity, while spots in the southern hemisphere have the opposite polarity (Figure 16.20). These polarities then reverse their signs for the next 11 years, so the full solar cycle takes 22 years.

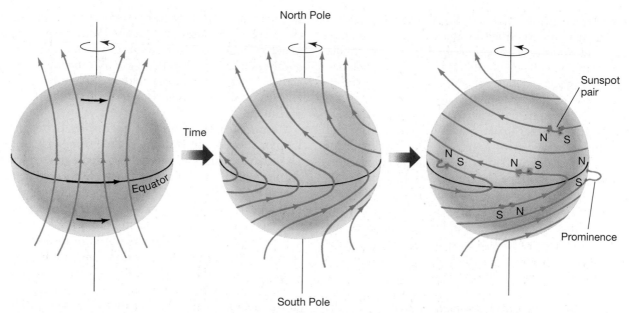

▲ **FIGURE 16.21 Solar Rotation** The Sun's differential rotation wraps and distorts the solar magnetic field. Occasionally, the field lines burst out of the surface and loop through the lower atmosphere, thereby creating a sunspot pair. The underlying pattern of the solar field lines explains the observed pattern of sunspot polarities. If the loop happens to occur on the limb of the Sun and is seen against the blackness of space, we see a phenomenon called a prominence, described in Section 16.5.

▼ **FIGURE 16.22 Sunspot Cycle** (a) Annual number of sunspots throughout the 20th century, showing yearly averages of data to make long-term trends more evident. The (roughly) 11-year solar cycle is clearly visible. At the time of solar minimum, hardly any sunspots are seen. About 4 years later, at solar maximum, as many as 100 to 200 spots are observed per year. The most recent solar maximum occurred in 2001. (b) Sunspots cluster at high latitudes when solar activity is at a minimum. They appear at lower latitudes as the number of sunspots peaks. They are again prominent near the Sun's equator as solar minimum is approached once more. The red lines at the top indicate how the "average" sunspot latitude varies over the course of the cycle.

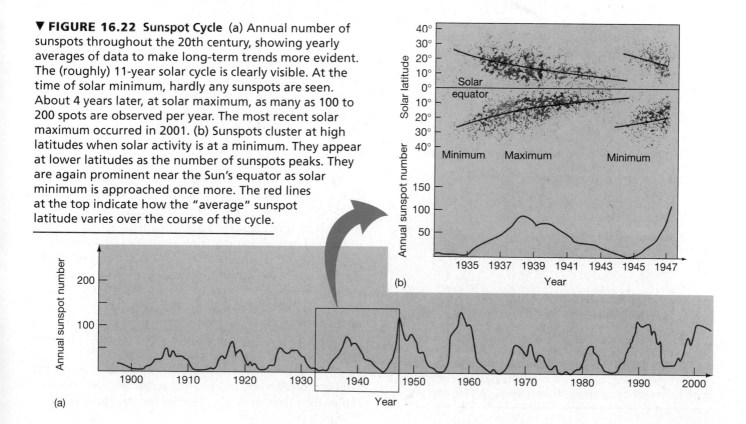

Astronomers think that the Sun's magnetic field is both generated and amplified by the constant stretching, twisting, and folding of magnetic field lines that results from the combined effects of differential rotation and convection, although the details are still not well understood. The theory is similar to the "dynamo" theory that accounts for the magnetic fields of Earth and the jovian planets, except that the solar dynamo operates much faster and on a much larger scale. ∞ (Sec. 7.5) One prediction of this theory is that the Sun's magnetic field should rise to a maximum, then fall to zero, and reverse itself, more or less periodically, just as is observed. Solar surface activity, such as the sunspot cycle, simply follows the variations in the magnetic field. The changing numbers of sunspots and their migration to lower latitudes are both consequences of the strengthening and eventual decay of the field lines as they become more and more tightly wrapped around the solar equator.

Figure 16.23 plots sunspot data extending back to the invention of the telescope. As can be seen, the 11-year "periodicity" of the solar sunspot cycle is far from regular. Not only does the period range from 7 to 15 years, but the sunspot cycle disappeared entirely over a number of years in the relatively recent past. The lengthy period of solar inactivity that extended from 1645 to 1715 is called the *Maunder minimum*, after the British astronomer who drew attention to these historical records. The corona was apparently also less prominent during total solar eclipses around that time, and Earth aurorae were sparse throughout the late 17th century. Lacking a complete understanding of the solar cycle, we cannot easily explain how it could shut down entirely. Most astronomers suspect changes in the Sun's convection zone or rotation pattern, but the specific causes of the Sun's century-long variations, as well as the details of the connection between solar activity and Earth's climate, remain a mystery (see *Discovery 16-2*).

Active Regions

The photosphere surrounding a pair or group of sunspots can be a violent place, sometimes erupting explosively,

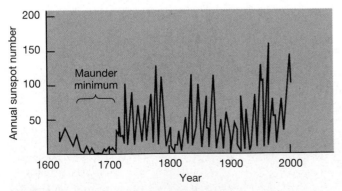

▲ **FIGURE 16.23 Maunder Minimum** Number of sunspots occurring each year over the past 4 centuries. Note the absence of spots during the late 17th century.

spewing forth large quantities of energetic particles into the corona. The sites of these energetic events are known as **active regions**. Most groups of sunspots have active regions associated with them. Like all other aspects of solar activity, these phenomena tend to follow the solar cycle and are most frequent and violent around the time of solar maximum.

Figure 16.24(a) shows a large solar **prominence**—a loop or sheet of glowing gas ejected from an active region on the solar surface, moving through the inner parts of the corona under the influence of the Sun's magnetic field. Magnetic instabilities in the strong fields found in and near sunspot groups may cause the prominences, although the details are not fully understood. Figure 16.24(b) shows a filament of hot material soaring high into the solar atmosphere above an active region. The arching magnetic field lines in and around the active region are also easily seen. (See Figure 16.20b.) The rapidly changing structure of the field lines and the fact that they can quickly transport mass and energy from one part of the solar surface to another, possibly tens of thousands of kilometers away, make the theoretical study of active regions an extraordinarily difficult task.

Quiescent prominences persist for days or even weeks, hovering high above the photosphere, suspended by the Sun's magnetic field. *Active prominences* come and go much more erratically, changing their appearance in a matter of hours or surging up from the solar photosphere and then immediately falling back on themselves. A typical solar prominence measures some 100,000 km in extent, nearly 10 times the diameter of planet Earth. Prominences as large as the one shown in Figure 16.24(a) (which traversed almost half a million kilometers of the solar surface) are less common and usually appear only at times of greatest solar activity. The largest prominences can release up to 10^{25} joules of energy, counting both particles and radiation—not much compared with the total solar luminosity of 4×10^{26} W, but still enormous by terrestrial standards. (All the power plants on Earth would take a billion years to produce that much energy.)

Flares are another type of solar activity observed low in the Sun's atmosphere near active regions. Also the result of magnetic instabilities, flares, like that shown in Figure 16.25(a), are even more violent (and even less well understood) than prominences. They often flash across a region of the Sun in minutes, releasing enormous amounts of energy as they go. Space-based observations indicate that X-ray and ultraviolet emissions are especially intense in the extremely compact hearts of flares, where temperatures can reach 100 million K. So energetic are these cataclysmic explosions that some researchers have likened flares to bombs exploding in the lower regions of the Sun's atmosphere.

A major flare can release as much energy as the largest prominences, but in a matter of minutes or hours rather than days or weeks. Unlike the gas that makes up the characteristic loop of a prominence, the particles produced by a flare are so energetic that the Sun's magnetic field is unable to hold them and shepherd them back to the surface. Instead, the particles are simply blasted into space by the violence of the explosion. Figure 16.25(b) shows the vicinity of a solar active region immediately before a major flare. The bright green area is high-temperature gas barely confined by the solar magnetic field. The flare occurred

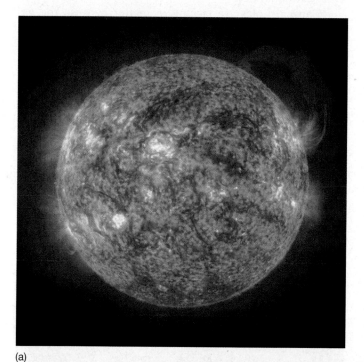

(a)

just a few minutes later, ejecting billions of tons of material into interplanetary space.

Figure 16.26 shows a **coronal mass ejection** from the Sun. Sometimes (but not always) associated with flares and prominences, these phenomena are giant magnetic "bubbles" of ionized gas that separate from the rest of the solar atmosphere and escape into interplanetary space. Carrying an enormous amount of energy, they can—if their fields are properly oriented—connect with Earth's magnetic field, dump some of their energy into the magnetosphere, and cause communications and power disruptions on our planet. (See *Discovery 16-2.*) Such ejections occur about once per week at times of sunspot minimum, but up to two or three times per day at solar maximum.

The Sun in X Rays

Unlike the 5800-K photosphere, which emits most strongly in the visible part of the electromagnetic spectrum, the hot coronal gas radiates at much higher frequencies—primarily in the X-ray range. ∞ (Sec. 3.4) For this reason, X-ray telescopes have become important tools in the study of the solar corona. Figure 16.27(a) shows several X-ray images of the Sun. The full corona extends well beyond the regions shown, but the density of coronal particles emitting the radiation diminishes rapidly with distance from the Sun. The intensity of X-ray radiation farther out is too dim to be seen here.

In the mid-1970s, instruments aboard NASA's *Skylab* space station revealed that the solar wind escapes mostly through solar "windows" called **coronal holes**. The dark

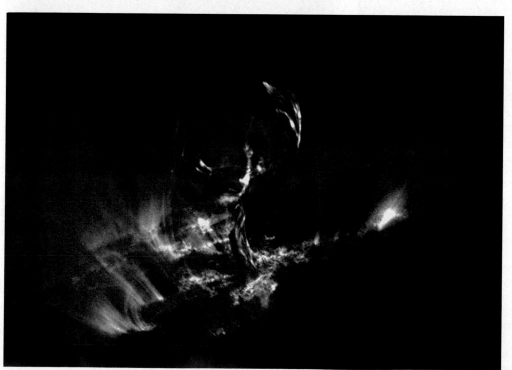

(b)

◀ **FIGURE 16.24 Solar Prominences** (a) This particularly large solar prominence was observed by ultraviolet detectors aboard the *SOHO* spacecraft in June 2002. (b) Like a phoenix rising from the solar surface, this filament of hot gas measures more than 100,000 km in length. Earth could easily fit between its outstretched "arms." Dark regions in this *TRACE* image have temperatures less than 20,000 K; the brightest regions are about 1 million K. The ionized gas follows the solar magnetic field lines away from the Sun. Most of the gas will subsequently cool and fall back into the photosphere. (*NASA*)

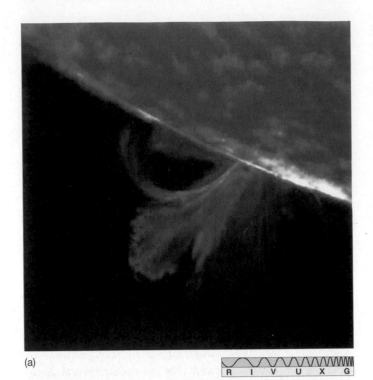

(a)

R I V U X G

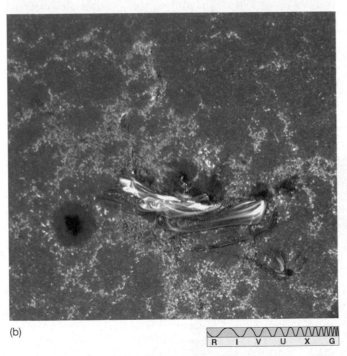

(b)

R I V U X G

▲ **FIGURE 16.25** **Solar Flares** (a) Much more violent than a prominence, a solar flare is an explosion on the Sun's surface that sweeps across an active region in a matter of minutes, accelerating solar material to high speeds and blasting it into space. (b) A flare occurs when hot gas breaks free of the magnetic field confining it and bursts into space. This composite image shows a dark sunspot group (captured in visible light), the surrounding solar photosphere (in the ultraviolet, shown in red here) and a collection of magnetic loops (extreme ultraviolet, colored green here) confining million-degree gas a few minutes before a major flare in June 2000. *(USAF; NASA)*

area moving from left to right in Figure 16.27(a), which shows more recent data from the Japanese *Yohkoh* X-ray solar observatory, represents a coronal hole. Not really holes, such structures are simply deficient in matter—vast regions of the Sun's atmosphere where the density is about 10 times lower than the already tenuous, normal corona. Note that the underlying solar photosphere looks black in these images because it is far too cool to emit X rays in any significant quantity.

Coronal holes are lacking in matter because the gas there is able to stream freely into space at high speeds, driven by disturbances in the Sun's atmosphere and magnetic field. Figure 16.27(b) illustrates how, in coronal holes, the solar magnetic field lines extend from the surface far out into interplanetary space. Charged particles tend to follow the field lines, so they can escape, particularly from the Sun's polar regions, according to recent *SOHO* findings. In other regions of the corona, the solar magnetic field lines stay close to the Sun, keeping charged particles near the surface and inhibiting the outward flow of the solar wind (just as Earth's magnetic field tends to prevent the incoming solar wind from striking Earth), so the density remains relatively high. Because of the "open" field structure in coronal holes, flares and other magnetic activ-

R I V U X G

▲ **FIGURE 16.26** **Coronal Mass Ejection** A few times per week, on average, a giant magnetized "bubble" of solar material detaches itself from the Sun and escapes rapidly into space, as shown in this *SOHO* image taken in 2002. (The Sun itself was intentionally blocked by the disk at the center.) Should such a coronal mass ejection encounter Earth, it could severely disrupt our planet's magnetosphere. *(NASA/ESA)*

VIDEO May 12, 1997 Solar Flare Event

ANIMATION Coronal Mass Ejection

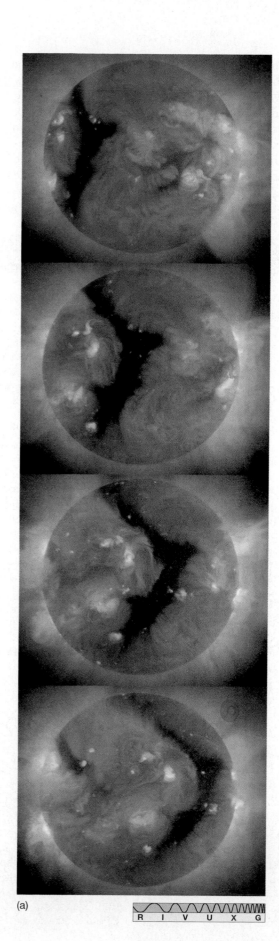

(b)

◀ **FIGURE 16.27 Coronal Holes** (a) Images of X-ray emission from the Sun observed by the *Yohkoh* satellite. These frames were taken at roughly two-day intervals, starting at the top. Note the dark, V-shaped coronal hole traveling from left to right, where the X-ray observations outline in dramatic detail the abnormally thin regions through which the high-speed solar wind streams forth. (b) Charged particles follow magnetic field lines that compete with gravity. When the field is trapped and loops back toward the photosphere, the particles are also trapped; otherwise, they can escape as part of the solar wind. (*Lockheed Martin*)

ity (which, as we have seen, are related to magnetic loops near the solar photosphere) tend to be suppressed there.

The largest coronal holes, like that shown in Figure 16.27a, can be hundreds of thousands of kilometers across and may survive for many months. Structures of this size are seen only a few times per decade. Smaller holes—perhaps only a few tens of thousand kilometers in size—are much more common, appearing every few hours.

Coronal holes appear to be an integral part of the process by which the Sun's large-scale field reverses and replenishes itself over the course of the solar cycle. Long-lived holes persist at the Sun's polar regions over much of the magnetic cycle, and the numbers and locations of other holes appear to change in step with solar activity. However, like many aspects of the solar magnetic field, the structure and evolution of coronal holes are not fully understood; they are currently the subject of intense research.

The Changing Solar Corona

The solar corona varies with the sunspot cycle. The photograph of the corona in Figure 16.15 shows the quiet Sun, at sunspot minimum. At such times, the corona is fairly regular in appearance and seems to surround the Sun more or less uniformly. Compare that image with Figure 16.28, which was taken in 1991 near a peak in the sunspot cycle.

(a)

R I V U X G

The active corona is much more irregular in appearance and extends farther from the solar surface. The "streamers" of coronal material pointing away from the Sun are characteristic of this phase.

Astronomers think that the corona is heated primarily by solar surface activity, which can inject large amounts of energy into the upper solar atmosphere. *SOHO* observations have implicated a "magnetic carpet" as the source of much of the heating. Using the craft's Doppler imager, astronomers have observed small magnetic loops perpetu-ally sprouting up and then disappearing all over the solar photosphere—resembling a kind of celestial shag carpet. The whole carpet replenishes itself rapidly, roughly every 40 hours or so, but the loops don't just sink back beneath the surface. Rather, they tend to break open, dumping vast amounts of energy into the lower solar atmosphere. Along with the myriad spicules, these small-scale magnetic loops probably provide most of the energy needed to heat the corona. In addition, more extensive disturbances often move through the corona above an active site in the photo-

DISCOVERY 16-2

Solar–Terrestrial Relations

Our Sun has often been worshipped as a god with power over human destinies. Obviously, the steady stream of solar energy arriving at our planet every day is essential to our lives, but over the past century there have also been repeated claims of a correlation between the Sun's activity and Earth's weather. Only recently, however, has the subject become scientifically respectable—that is, more natural than supernatural.

In fact, there do seem to be some correlations between the 22-year solar cycle (two sunspot cycles with oppositely di-rected magnetic fields) and periods of climatic dryness here on Earth. For example, near the start of the past eight cycles, there have been droughts in North America—at least within the middle and western plains from South Dakota to New Mexico. Another possible Sun–Earth connection is a link be-tween solar activity and increased atmospheric circulation on our planet. As circulation increases, terrestrial storm systems deepen, extend over wider ranges of latitude, and carry more moisture. The relationship is complex and the subject contro-versial, because no one has yet shown any physical mechanism (other than the Sun's heat, which does not vary much during the solar cycle) that would allow solar activity to stir our ter-restrial atmosphere. Without a better understanding of the physical mechanism involved, none of these effects can be in-corporated into our weather-forecasting models.

Solar activity may also influence long-term climate on Earth. For example, the Maunder minimum (see Section 16.5) seems to correspond fairly well with the coldest years of the so-called Little Ice Age that chilled northern Europe and North America during the late 1600s. The accompanying "win-ter" scene actually captured one summer season in 17th-century Holland. How the active Sun and its abundance of sunspots may affect Earth's climate is a frontier problem in terrestrial climatology.

Measurements of the solar constant made over the past two decades indicate that the Sun's energy output varies with the solar cycle. Paradoxically, the Sun's luminosity is greatest when many dark sunspots cover its surface! Thus, the Maunder minimum does correspond to an extended period of lower-than-average solar emission. How-ever, recent observed changes in the Sun's luminosity have been small—no more than 0.2 or 0.3 percent. It is not known by how much, if at all, the Sun's output declined during the Maunder minimum, nor how large a change would be needed to account for the alterations in climate that occurred.

One correlation that is definitely established, and also better understood, is that between solar activity and geomag-netic disturbances at Earth. The extra radiation and particles thrown off by flares or coronal mass ejections impinge on Earth's environment, overloading the Van Allen belts, thereby causing brilliant auroras in our atmosphere and degrading our communication networks. We are only beginning to under-stand how the radiation and particles emitted by solar phe-nomena also interfere with terrestrial radars, power networks, and other technological equipment. Some power outages on Earth are actually caused, not by increased customer demand or malfunctioning equipment, but by weather on the Sun!

We cannot yet predict just when and where solar flares or coronal mass ejections will occur. However, it would certainly be to our advantage to be able to do so, as that aspect of the ac-tive Sun affects our lives. This is a highly fertile area of astro-nomical research and one with clear terrestrial applications.

(Toledo Museum of Art)

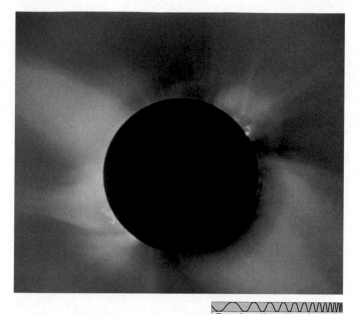

▲ FIGURE 16.28 Active Corona Photograph of the solar corona during the July 1991 eclipse, at the peak of the sunspot cycle. At these times, the corona is much less regular and much more extended than at sunspot minimum. (cf. Figure 16.15.) Astronomers think that coronal heating is caused by surface activity on the Sun. The changing shape and size of the corona is the direct result of variations in prominence and flare activity over the course of the solar cycle. *(National Solar Observatory)*

sphere, distributing the energy throughout the coronal gas. Given this connection, it is hardly surprising that both the appearance of the corona and the strength of the solar wind are closely correlated with the solar cycle.

CONCEPT CHECK

✔ What do observations of sunspot polarities tell us about the solar magnetic field?

16.6 Observations of Solar Neutrinos

LEARNING GOAL 6 Theorists are quite sure that the proton–proton chain described in Section 16.2 operates in the Sun. However, because the gamma-ray energy created in the solar core chain has been transformed into visible and infrared radiation by the time it emerges from the photosphere, astronomers have no direct electromagnetic evidence of the nuclear reactions in the solar core. Fortunately, there are other ways of probing the Sun's deep interior.

The Solar Neutrino Problem

Electromagnetic radiation has been the source of virtually all the information presented thus far in the text. However, the *neutrinos* created in the proton–proton chain are our

best bet for learning about conditions in the heart of the Sun. Core neutrinos travel cleanly out of the Sun, interacting with virtually nothing, and escape into space a few seconds after being created. They arrive at Earth's orbit about eight minutes later. Of course, the fact that they can pass through the entire Sun without interacting also makes neutrinos rather difficult to detect here on Earth! Nevertheless, with some knowledge of neutrino physics, it is possible to construct neutrino detectors.

Over the past four decades, several major experiments have sought to detect solar neutrinos reaching Earth's surface. The first, built in the late 1960s, was sited 1.5 km below ground near the bottom of the Homestake gold mine in South Dakota (to shield it from interference due to cosmic rays and other sources, most of which are unable to penetrate Earth's crust to such a depth). The Homestake detector was designed to measure the changes occurring when a neutrino interacts with a nucleus of chlorine-37, converting it into argon-37. The researchers left a tank containing some 400,000 liters (about 100,000 gallons) of a chlorine-containing chemical—the common cleaning fluid used by dry cleaners—in the mine for months at a time, periodically checking to see if any of the chlorine had been converted into argon, which would signal the absorption of a neutrino.

Given the size of the detector and the physical conditions in the Sun's core implied by the standard solar model, about one solar neutrino of the roughly 10^{16} that streamed through the tank each day should have been detected. The experiment did succeed in detecting solar neutrinos—in itself a remarkable achievement—but the numbers were not as great as predicted. Over the course of the entire experiment, neutrinos were detected about twice per week, on average, not once per day. The neutrino deficit persisted over two decades of almost continuous monitoring, until the experiment was terminated in 1993. This clear disagreement between theory and observation has come to be known as the **solar neutrino problem**.

The other experiments are more recent and have quite different detector designs. The 1990s saw the construction of the Soviet–American Gallium Experiment (SAGE, for short) and the U.S.–European GALLEX collaborations. Each experiment used the element gallium instead of chlorine to capture and detect solar neutrinos. Gallium was chosen despite its considerable expense, because it happens to be much more likely than chlorine to interact with neutrinos of the sort expected from the Sun. Both experiments found a roughly 50 percent shortfall in the numbers of neutrinos detected, relative to theoretical expectations.

Other experiments use a quite different approach to neutrino detection. Figure 16.29(a) shows a large detector located in Kamioka, Japan, constructed in the 1980s (and upgraded in the 1990s to greatly increase its sensitivity) to measure the telltale light emitted as neutrinos stream through it. The light does not come directly from the neutrinos. Rather, it is produced when a high-energy neutrino occasionally collides with an electron in a water molecule, accelerating the neutrino to almost the speed of light. The large photomultiplier tubes (light-amplification devices) detect the resultant faint glow that betrays the neutrino's passage. The instrument's operation was (temporarily) halted in 2001 by a freak accident in which more than half of its detectors were destroyed. However, during its four years of upgraded operation before the accident, the "Super Kamiokande" device, like the Gallium Experiments, also detected just half the number of neutrinos predicted by theory.

Interestingly, by careful measurement of the light pattern, the Kamioka detector and similar instruments can determine the direction of motion of the electron producing the radiation and hence estimate the direction from which the original neutrino came. In other words, the instrument can function as a *neutrino telescope*, albeit one of very low resolution. Figure 16.29(b) shows an "image" of the Sun produced by this instrument. As technology improves, such directional detectors will likely come to play a central role in neutrino astronomy.

Neutrino Physics and the Standard Model

Thus, while the four neutrino-detection experiments just described disagree somewhat on the extent of the deficit, each sees significantly fewer than the predicted number of solar neutrinos—only 30 to 50 percent. The Homestake and Kamioka experiments were actually sensitive not to neutrinos produced in reaction (I) on page 410 and in Figure 16.5—the initial step in the proton–proton chain—but to those created by a much less probable sequence of events, occurring only about 0.25 percent of the time, in principle leaving a little room for theorists to maneuver in their attempts to explain the observations. (It should be noted, however, that, to many workers, the disagreement between the Homestake and Kamioka results constitutes a second, as yet unresolved, solar neutrino problem.) By contrast, SAGE and GALLEX could detect neutrinos produced by reaction (I), providing a more direct probe of energy generation in the solar core.

The inescapable conclusion is that, although solar neutrinos are observed (and in fact their measured energies do lie in the range predicted by the standard solar model), there is a real discrepancy between the Sun's theoretical neutrino output and the number of neutrinos we detect on Earth. How can we explain this discrepancy? If,

(a)

(b)

▲ **FIGURE 16.29 Neutrino Telescope** (a) This swimming-pool-sized detector is a "neutrino telescope" of sorts, buried beneath a mountain near Tokyo, Japan. Called Super Kamiokande, it is filled (in operation) with 50,000 tons of purified water and contains 13,000 light detectors (some shown here being inspected by technicians) to sense the telltale signature of a neutrino passing through the apparatus. The instrument was badly damaged by a freak accident in November 2001, when one detector apparently imploded, creating a shock wave and chain reaction that destroyed several thousand others. It is still under repair. (b) An "image" of the Sun made by the Kamiokande instrument. Current neutrino detectors have very low resolution, and the field of view here is about 90° across. The small black dot indicates the actual size of the Sun. *(Institute for Cosmic Ray Research)*

as we think, the detectors are working correctly—and the lead scientists of the Homestake and Kamioka experiments have received the strongest possible scientific endorsement of their work, in the form of the 2002 Nobel prize!—there are really only two possibilities: Either solar neutrinos are not produced as frequently as we think, or not all of them make it to Earth.

Most theorists think it very unlikely that the resolution of the solar neutrino problem will be found in the physics of the Sun's interior. In principle, we could reduce the theoretical number of neutrinos by postulating a lower temperature in the solar core. However, the nuclear reactions described earlier are just too well known, and the agreement between the standard solar model and helioseismological observations is far too close to permit conditions in the core to deviate much from the predictions of the model.

Instead, the most likely explanation seems to lie in the properties of the neutrinos themselves. Theory predicts that, if neutrinos have even a minute amount of mass, it should be possible for them to change their properties—even to transform into other particles—during their eight-minute flight from the solar core to Earth, through a process known as **neutrino oscillations**. In this picture, neutrinos are produced in the Sun at the rate required by the standard solar model, but some turn into something else—actually, other types of neutrinos—on their way to Earth and hence go undetected in the experiments just described. (In the jargon of the field, the neutrinos are said to "oscillate" into other particles.)

In 1998, the Super Kamiokande group reported the first experimental evidence of neutrino oscillations (and hence of nonzero neutrino masses), although the observed oscillations did not involve neutrinos of the type produced in the Sun. More recently, in 2001, measurements made at the Sudbury Neutrino Observatory (SNO) in Ontario, Canada, revealed strong evidence for the "other" neutrinos into which the Sun's neutrinos have been transformed. The SNO detector is similar in design to the Kamioka device, but, by using "heavy" water (with hydrogen replaced by deuterium) instead of ordinary water (H_2O), it also becomes sensitive to other neutrino species. The total numbers of neutrinos observed are completely consistent with the standard solar model. Apparently, the solar neutrino problem has at last been solved—and neutrino astronomy can claim its first major success!

CONCEPT CHECK

✔ Why do neutrinos give us direct information about conditions in the solar core, while electromagnetic radiation does not?

Chapter Review

SUMMARY

A **star (p. 406)** is a glowing ball of gas held together by its own gravity and powered by nuclear fusion at its center. The main interior regions of the Sun are the **core (p. 408)**, where nuclear reactions generate energy, the **radiation zone (p. 407)**, in which the energy travels outward in the form of electromagnetic radiation, and the **convection zone (p. 407)**, where the Sun's matter is in constant convective motion. The sharp solar disk visible from Earth marks the solar **photosphere (p. 406)**—the thin surface layer from which the Sun's light is emitted. Above the photosphere lies the **chromosphere (p. 407)**, which is separated from the solar **corona (p. 407)** by a thin **transition zone (p. 407)** in which the temperature increases from a few thousand to over a million kelvins. At about 10 to 15 solar radii, the gas in the corona is hot enough to escape the Sun's gravity, and the corona begins to flow outward as the **solar wind (p. 421)**.

The Sun's **luminosity (p. 408)** is the total amount of energy radiated from the solar surface per second. It is determined by measuring the Earth's **solar constant (p. 408)**—the amount of solar radiation reaching each square meter at Earth's distance from the Sun—and multiplying that amount by the area of an imaginary sphere of radius 1 A.U.

The Sun generates energy by "burning" hydrogen into helium in its core through **nuclear fusion (p. 409)**. Nuclei are held together by the **strong nuclear force (p. 409)**. When four protons overcome their electromagnetic repulsion and fuse into a helium nucleus in the **proton–proton chain (p. 410)**, some mass is lost. The **law of conservation of mass and energy (p. 409)** requires that this mass appear as energy, resulting in the light we see. Very high temperatures are needed for fusion to occur.

Some particles produced during the solar fusion process are (1) the **positron (p. 409)**, or antielectron, which is quickly annihilated by electrons it encounters in the Sun's core to generate gamma rays, (2) the **deuteron (p. 409)**, an **isotope (p. 410)** of hydrogen consisting of a proton and a neutron, and (3) the **neutrino (p. 410)**, a near-massless particle that escapes from the Sun without any further interactions once it is created in the core. Neutrinos interact via the **weak nuclear force (p. 410)**.

The mathematical model that best fits the observed properties of the Sun is the **standard solar model (p. 412)**, based on the observation that the Sun is in **hydrostatic eq[uilibrium]** meaning that gravity and pressure are balanc[ed] the sun. Ground- and space-based **heliosei[smology]** the study of oscillations of the solar surface ca[n] in the interior—provides considerable ins[ight] structure. The effect of the solar convection the surface in the form of **granulation (p.** phere: As hotter (and therefore brighter) (dimmer) gas sinks, a characteristic "mottle[d]

The total number of stars, even in our local neighborhood, is virtually beyond our ability to count. Relatively few of them have been studied in detail. Yet astronomers have learned a great deal about stars in general, and their range of properties—their masses, their temperatures, their luminosities, even their ages and (as we shall see in later chapters) their destinies. Here, the Hubble Space Telescope has imaged the magnificent globular star cluster M80, a beehive-like collection of nearly a million stars some 28,000 light-years away. (STScI) ▶

Red Giants and White Dwarfs

A Field Guide to the Stars

e have now studied Earth, the Moon, the solar system, and the Sun. To continue our inventory of the contents of the universe, we must move away from our local environment into the depths of space. In this chapter, we take a great leap in distance and consider stars in general. Our primary goal is to comprehend the nature of the stars that make up the constellations, as well as the myriad more distant stars we cannot perceive with our unaided eyes. Rather than studying their individual peculiarities, however, we will concentrate on determining the physical and chemical properties they share. There is order in the legions of stars scattered across the sky. Like comparative planetology in the solar system, the study of the stars plays a vital role in furthering our understanding of the Galaxy and the universe we inhabit.

LEARNING GOALS

Studying this chapter will enable you to

❶ Explain how stellar distances are determined.

❷ Discuss the motions of the stars through space and how those motions are measured from Earth.

❸ Distinguish between luminosity and apparent brightness, and explain how stellar luminosity is determined.

❹ Explain the usefulness of classifying stars according to their colors, surface temperatures, and spectral characteristics.

❺ Explain how physical laws are used to estimate stellar sizes.

❻ Describe how an H–R diagram is constructed and used to identify stellar properties.

❼ Explain how the masses of stars are measured and how mass is related to other stellar properties.

Visit astro.prenhall.com/chaisson for additional annotated images, animations, and links to related sites for this chapter.

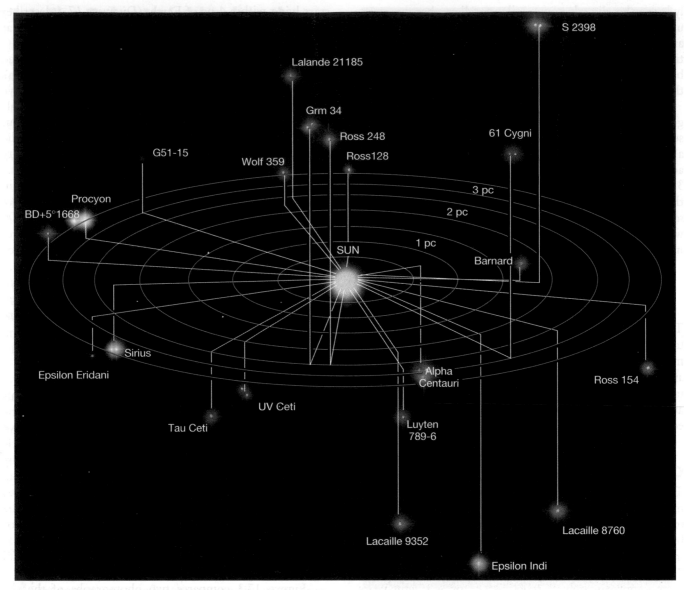

▲ **FIGURE 17.2 The Solar Neighborhood** A plot of the 30 closest stars to the Sun, projected so as to reveal their three-dimensional relationships. The circles represent distances from the Sun in an imaginary plane that extends through Earth's equator; the vertical lines denote distances perpendicular to that plane. Notice that many stars are members of double—or even multiple—star systems. All lie within 4 pc (about 13 light-years) of Earth.

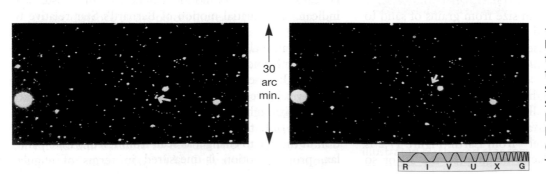

◀ **FIGURE 17.3 Proper Motion** A comparison of two photographic plates taken 22 years apart shows evidence of real spatial motion for Barnard's Star (denoted by an arrow). (*Harvard College Observatory*)

DISCOVERY 17-1

Naming the Stars

Looking at Figure 17.2, you might well wonder how the stars shown there got their names. In fact, there is no coherent convention for naming stars—or rather, there are many, generally leading to many names for any given object! The following is a brief (and roughly chronological) description of some of the more common naming schemes:

1. The brightest stars in the sky were well known to ancient astronomers, and their names often reflect that fact. Many of the names are of Arabic origin, dating back to around the 10th century A.D., when Muslim astronomy flourished, preserving the scientific knowledge of Greek and other early cultures while European science floundered in the Dark Ages. ∞ (Sec. 2.1) Examples are the well-known stars Betelgeuse (from Ibt al Jauzah, the "Armpit of the Central One," corrupted to Bed Elgueze), Rigel (from Rijl Jauzah al Yusra, "Left Foot of the Giant"), Aldebaran (from Al Dabaran, "the Follower"), and Deneb (from Al Dhanab al Dulfin, "the Dolphin's Tail"). The names Procyon and Sirius (see Figure 17.2) are Greek.

2. A more systematic scheme was introduced in 1603 by German lawyer Johann Bayer. In Bayer's scheme, stars in a given constellation (Orion, say) are ranked by brightness and are labeled with Greek letters: Alpha Orionis (Betelgeuse), Beta Orionis (Rigel), or Epsilon Eridani (Figure 17.2). After the 24 Greek letters were exhausted, Bayer used the lowercase roman letters "a" through "z," followed by uppercase letters.

3. Eventually, the Bayer scheme broke down. In the early 18th century, British Astronomer Royal John Flamsteed suggested simply numbering the stars from west to east within a constellation. In the Flamsteed scheme, Betelgeuse is 58 Ori, Rigel is 19 Ori, and so on. (See also 61 Cygni in Figure 17.2.) Sometimes the Bayer and Flamsteed designations are combined, as in "58 α Ori."

4. As telescopes improved and more and more stars were discovered, astronomers started compiling their own catalogs, each with their own numbering system, and the confusion really started! Some catalogs listed stars by celestial coordinates—for example, star BD + 5°1668 (see Figure 17.2), which appears in the 19th-century German *Bonner Durchmusterung* (BD) catalog and lies just over 5 degrees above the celestial equator. ∞ (*More Precisely 1-2*) Other catalogs simply named stars by the order in which they were discovered (or added to the growing list)—for example, the star Lalande 21185 in Figure 17.2 is the 21,185th entry in a catalog compiled by 18th-century French astronomer Joseph de Lalande. Unfortunately, the catalogs often overlap, so any given star probably appears in many different catalogs with very different names. For example, Lalande 21185 is also known as HD 95735 (in the Henry Draper catalog), while Betelgeuse is also known as HD 39801 and SAO 113271 (in the Smithsonian Astrophysical Observatory catalog), among many other aliases.

5. Today, newly discovered stars are defined simply by their coordinates in the sky and are not given any special name. If you like, you can (for a fee, of course) register your own star name, but don't expect any astronomer ever to use that name, or for that matter to take the registry seriously!

Many other naming schemes are in use. For example, the "UV" in UV Ceti signifies that it is a *variable star*, one whose physical properties are observed to change with time. (In fact, it is the 45th such star discovered in the constellation Cetus—the convention in this case was developed by German astronomer Friedrich Wilhelm Argelander and runs from R through Z, then RR through RZ, SR, through SZ, etc.) Still other schemes exist for galaxies, but we won't go into them here. By now you are probably completely perplexed—our job as astronomers is done!

---

displacement. Since the angles involved are typically very small, proper motion is usually expressed in arc seconds per year. Barnard's Star moved 228″ in 22 years, so its proper motion is 228″/22 years, or 10.4″/yr.

A star's transverse velocity is easily calculated once its proper motion and its distance are known. At the distance of Barnard's Star (1.8 pc), an angle of 10.4″ corresponds to a physical displacement of 0.000091 pc, or about 2.8 billion km. Barnard's Star takes a year (3.2×10^7 s) to travel this distance, so its transverse velocity is 2.8 billion km/3.2×10^7 s, or 89 km/s. ∞ (*More Precisely 1-3*) Even though stars' transverse velocities are often quite large—tens or even hundreds of kilometers per second—their great distances from the Sun mean that their proper motion is small, and it usually takes many years for

us to discern their movement across the sky. In all probability, every star in Figure 17.3 has some transverse motion relative to the Sun. However, only Barnard's Star has proper motion large enough to be visible in these frames. In fact, Barnard's Star has the largest known proper motion of any star. Only a few hundred stars have proper motions greater than 1″/yr.

Now consider the three-dimensional motion of our nearest neighbor, the Alpha Centauri system, sketched in Figure 17.4 in relation to our own solar system. Alpha Centauri's proper motion has been measured to be 3.7″/yr. At Alpha Centauri's distance of 1.35 pc, that measurement implies a transverse velocity of 24 km/s. We can determine the other component of motion—the radial velocity—by means of the Doppler effect. Spectral

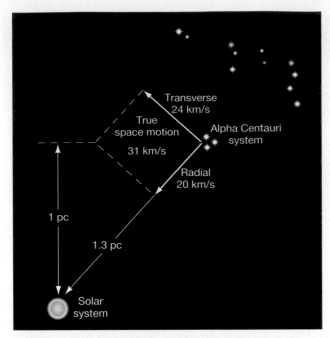

▲ **FIGURE 17.4 Real Spatial Motion** The motion of the Alpha Centauri star system drawn relative to our solar system. The transverse component of the velocity has been determined by observing the system's proper motion. The radial component is measured by using the Doppler shift of lines in Alpha Centauri's spectrum. The true spatial velocity, indicated by the red arrow, results from the combination of the two.

lines from Alpha Centauri are blueshifted by a tiny amount—about 0.0067 percent—allowing astronomers to measure the star system's radial velocity (relative to the Sun) as $300{,}000 \text{ km/s} \times 6.7 \times 10^{-5} = 20 \text{ km/s}$ toward us. ∞ (Sec. 3.5)

What is the true spatial motion of Alpha Centauri? Will this alien system collide with our own sometime in the future? The answer is no: Alpha Centauri's transverse velocity will steer it well clear of the Sun. We can combine the transverse (24 km/s) and radial (20 km/s) velocities according to the Pythagorean theorem, as indicated in Figure 17.4. The total velocity is $\sqrt{24^2 + 20^2}$, or about 31 km/s, in the direction shown by the horizontal red arrow. As the figure indicates, Alpha Centauri will get no closer to us than about 1 pc, and that won't happen until 280 centuries from now.

CONCEPT CHECK

✔ Why can't astronomers use simultaneous observations from different parts of Earth's surface to determine stellar distances?

✔ Why are the spatial velocities of distant stars generally poorly known?

17.2 Luminosity and Apparent Brightness

LEARNING GOAL 3 Luminosity is an *intrinsic* property of a star—it does not depend in any way on the location or motion of the observer. Luminosity is sometimes referred to as the star's *absolute brightness*. However, when we look at a star, we see, not its luminosity, but rather its **apparent brightness**—the amount of energy striking a unit area of some light-sensitive surface or device (such as a CCD chip or a human eye) per unit time. Apparent brightness is a measure, not of a star's luminosity, but of the *energy flux* (energy per unit area per unit time) produced by the star, as seen from Earth. A star's apparent brightness depends on our *distance* from the star. In this section, we discuss in more detail how these important quantities are related to one another.

Another Inverse-Square Law

Figure 17.5 shows light leaving a star and traveling through space. Moving outward, the radiation passes through imaginary spheres of increasing radius surrounding the source. The amount of radiation leaving the star per unit time—the star's luminosity—is constant, so the farther the light travels from the source, the less energy passes through each unit of area. Think of the energy as being spread out over an ever-larger area and therefore spread more thinly, or "diluted," as it expands into space. Because the area of a sphere grows as the square of the radius, the energy per unit area—the star's apparent brightness, as seen by our eye or our telescope—is inversely proportional to the square of the distance from the star.

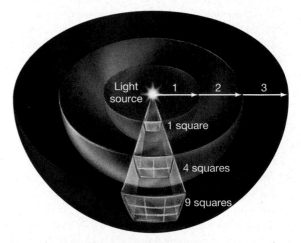

▲ **FIGURE 17.5 Inverse-square Law** As radiation moves away from a source such as a star, it is steadily diluted while spreading over progressively larger surface areas (depicted here as sections of spherical shells). Thus, the amount of radiation received by a detector (the source's apparent brightness) varies inversely as the square of its distance from the source.

Doubling the distance from a star makes it appear 2^2, or 4, times dimmer. Tripling the distance reduces the apparent brightness by a factor of 3^2, or 9, and so on.

Of course, the star's luminosity also affects its apparent brightness. Doubling the luminosity doubles the energy crossing any spherical shell surrounding the star and hence doubles the apparent brightness. We can therefore say that the apparent brightness of a star is directly proportional to the star's luminosity and inversely proportional to the square of its distance:

$$\text{apparent brightness (energy flux)} \propto \frac{\text{luminosity}}{\text{distance}^2}.$$

Thus, two identical stars can have the same apparent brightness if (and only if) they lie at the same distance from Earth. However, as illustrated in Figure 17.6, two nonidentical stars can also have the same apparent brightness if the more luminous one lies farther away. A bright star (i.e., a star with large apparent brightness) is a powerful emitter of radiation (high luminosity), is near Earth, or both. Without additional information, we cannot distinguish between the effects of increasing luminosity and decreasing distance. Similarly, a faint star (a star with small apparent brightness) is a weak emitter (low luminosity), is far from Earth, or both.

Determining a star's luminosity is a twofold task. First, the astronomer must determine the star's apparent brightness by measuring the amount of energy detected through a telescope in a given amount of time. Second, the star's distance must be measured—by parallax for nearby stars

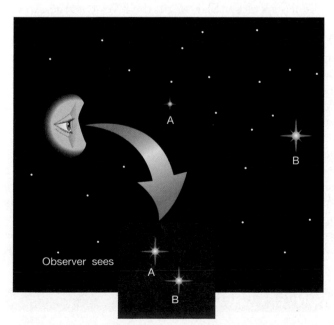

▲ **FIGURE 17.6 Luminosity** Two stars A and B of different luminosities can appear equally bright to an observer on Earth if the brighter star B is more distant than the fainter star A.

and by other means (to be discussed later) for more distant stars. The luminosity can then be found from the inverse-square law. This is basically the same reasoning we used earlier in Chapter 16, in our discussion of how astronomers measure solar luminosity. (In our new terminology, the solar constant is just the apparent brightness of the Sun.) ∞ (Sec. 16.1)

The Magnitude Scale

Instead of measuring apparent brightness in SI units (e.g., watts per square meter, the unit used for the solar constant in Chapter 16), astronomers often find it more convenient to work in terms of a construct called the **magnitude scale**. ∞ (Sec. 16.1) The scale dates from the second century B.C., when the Greek astronomer Hipparchus classified the naked-eye stars into six groups. The brightest stars were categorized as first magnitude. The next brightest stars were labeled second magnitude, and so on, down to the faintest stars visible to the naked eye, which were classified as sixth magnitude. The range 1 (brightest) through 6 (faintest) spanned all the stars known to the ancients. Notice that magnitudes are really *rankings* in terms of apparent brightness (energy flux)—a *large* magnitude means a *faint* star. Just as "first rate" means "good" in everyday speech, "first magnitude" in astronomy means "bright."

When astronomers began using telescopes with sophisticated detectors to measure the light received from stars, they quickly discovered two important facts about the magnitude scale. First, the 1–6 magnitude range defined by Hipparchus spans about a factor of 100 in apparent brightness—a first-magnitude star is approximately 100 times brighter than a sixth-magnitude star. Second, the physiological characteristics of the human eye are such that each change in magnitude of 1 corresponds to a factor of about 2.5 in apparent brightness. In other words, to the human eye, a first-magnitude star is roughly 2.5 times brighter than a second-magnitude star, which is roughly 2.5 times brighter than a third-magnitude star, and so on. (By combining factors of 2.5, you can confirm that a first-magnitude star is indeed $(2.5)^5 \approx 100$ times brighter than a sixth-magnitude star.)

Modern astronomers have modified and extended the magnitude scale in a number of ways. First, we now *define* a change of 5 in the magnitude of an object (going from magnitude 1 to magnitude 6, say, or from magnitude 7 to magnitude 2) to correspond to *exactly* a factor of 100 in apparent brightness. Second, because we are really talking about apparent (rather than absolute) brightnesses, the numbers in Hipparchus's ranking system are called **apparent magnitudes**. Third, the scale is no longer limited to whole numbers: A star of apparent magnitude 4.5 is intermediate in apparent brightness between a star of apparent magnitude 4 and one of apparent magnitude 5. Finally, magnitudes outside the range from 1 to 6 are

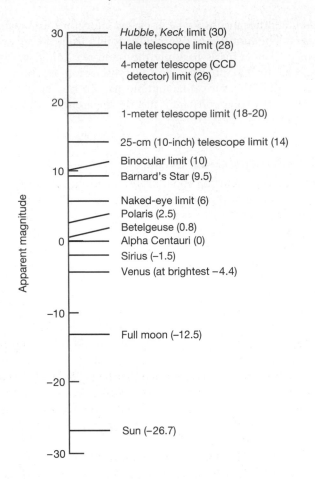

◀ **FIGURE 17.7 Apparent Magnitude** This graph illustrates the apparent magnitudes of some astronomical objects and the limiting magnitudes of some telescopes used to observe them. The original magnitude scale was defined so that the brightest stars in the night sky have magnitude 1, whereas the faintest stars visible to the naked eye have magnitude 6. The scale has since been extended to cover much brighter and much fainter objects. An increase of 1 in apparent magnitude corresponds to a decrease by a factor of approximately 2.5 in apparent brightness.

when it is placed at a distance of 10 pc from the observer. Because the distance to the star is fixed in this definition, *absolute magnitude is a measure of a star's absolute brightness, or luminosity.*

We can use the earlier discussion of the inverse-square law to relate absolute and apparent magnitudes if the distance to the star is known. When a star farther than 10 pc away from us is moved to a point 10 pc away, its apparent brightness increases and hence its apparent magnitude decreases. Stars more than 10 pc from Earth therefore have apparent magnitudes that are greater than their absolute magnitudes. For example, if a star at a distance of 100 pc were moved to the standard 10-pc distance, its distance would decrease by a factor of 10, so (by the inverse-square law) its apparent brightness would *increase* by a factor of $10^2 = 100$. Its apparent magnitude (by definition) would therefore decrease by 5. In other words, at 100 pc distance, the star's apparent magnitude exceeds its absolute magnitude by 5.

For stars closer than 10 pc, the reverse is true. An extreme example is our Sun. Because of its proximity to Earth, it appears very bright and thus has a large negative apparent magnitude (Figure 17.7). However, the Sun's absolute magnitude is 4.83. If the Sun were moved to a distance of 10 pc from Earth, it would be only slightly brighter than the faintest stars visible to the naked eye in the night sky.

Knowledge of a star's apparent magnitude and distance allows us to compute its absolute magnitude (luminosity). Conversely, the numerical difference between a star's absolute and apparent magnitudes is a direct measure of the distance to the star. *More Precisely 17-1* presents more detail and some examples on the connection between absolute magnitude and luminosity and on the use of the magnitude scale in computing stellar luminosities and distances.

allowed: Very bright objects can have apparent magnitudes much less than 1, and very faint objects can have apparent magnitudes far greater than 6.

Figure 17.7 illustrates the apparent magnitudes of some astronomical objects, ranging from the Sun, at −26.7, to the faintest object detectable by the *Hubble* or *Keck* telescopes, an object having an apparent magnitude of 30—about as faint as a firefly seen from a distance equal to Earth's diameter. Note that this range in magnitudes corresponds to a very large factor (actually, of $10^{56.7/2.5} = 10^{22.7} \approx 5 \times 10^{22}$) in apparent brightness. Indeed, one of the main reasons that astronomers use this (admittedly rather intimidating) scale is that it allows them to compress a large spread in observed stellar properties into more "manageable" form.*

Apparent magnitude measures a star's apparent brightness when the star is seen at its actual distance from the Sun. To compare intrinsic, or absolute, properties of stars, however, astronomers imagine looking at all stars from a standard distance of 10 pc. There is no particular reason to use 10 pc—it is simply convenient. A star's **absolute magnitude** is its apparent magnitude

---

Putting in the numbers, we can calculate that magnitude 1 corresponds to a flux of 1.1×10^{-8} W/m², magnitude 20 to 2.9×10^{-16} W/m², and so on, but astronomers find the "magnitude" versions more intuitive and much easier to remember.

CONCEPT CHECK

✔ Two stars are observed to have the same apparent magnitude. On the basis of this information, what, if anything, can be said about their luminosities?

More on the Magnitude Scale

The entire discussion of luminosity, brightness, and energy flux in Section 17.2 can equally well be written in terms of magnitudes. Here we present a little more detail on the "magnitude versions" of two important topics: stellar luminosity and the inverse-square law.

Recall that a star's *absolute magnitude* is the apparent magnitude we would measure if the star were located at a standard distance of 10 pc from us. Absolute magnitude is equivalent to luminosity, which is an intrinsic property of a star. Given that the Sun's absolute magnitude is 4.83, we can construct a "conversion chart" (shown below) relating the two quantities. Since an *increase* in brightness by a factor of 100 corresponds, by definition, to a *reduction* in a star's magnitude by five units, it follows that a star with luminosity 100 times that of the Sun has absolute magnitude $4.83 - 5 = -0.17$, and a star with 0.01 solar luminosity has absolute magnitude $4.83 + 5 = 9.83$. We can fill in the gaps by realizing that one magnitude corresponds to a factor of $100^{1/5} \approx 2.512$, two magnitudes to $100^{2/5} \approx 6.310$, and so on. A factor of 10 in brightness corresponds to 2.5 magnitudes. You can use this chart to convert between solar luminosities and absolute magnitudes in many of the figures in this and later chapters.

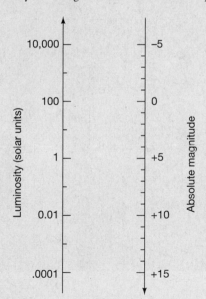

EXAMPLE 1: Let's calculate the luminosity (in solar units) of a star having absolute magnitude M (the conventional symbol for absolute magnitude, not to be confused with mass!). The star's absolute magnitude differs from that of the Sun by $(M - 4.83)$ magnitudes, so, in accordance with the reasoning just presented, the luminosity L differs from the solar luminosity by a factor of $100^{-(M-4.83)/5}$, or $10^{-(M-4.83)/2.5}$. We can therefore write

$$L \text{ (solar units)} = 10^{-(M-4.83)/2.5}.$$

Plugging in some numbers (taken from Appendix 3), we find that the Sun, with $M = 4.83$, of course has $L = 10^0 = 1$. Sirius A, with $M = 1.45$, has luminosity

$10^{1.35} = 22$ solar units, Barnard's Star, with $M = 13.24$, has luminosity $10^{-3.35} = 4.3 \times 10^{-4}$ solar units, Betelgeuse has $M = -5.14$ and luminosity 9,700 Suns, and so on.

EXAMPLE 2: Often, we need to convert from luminosity to absolute magnitude. To do so, we invert the preceding relationship and write

$$M = 4.83 - 2.5 \log_{10} L \text{ (solar units)}$$

(where the *logarithm* function—the "LOG" key on your calculator—is defined by the property that if $a = \log_{10}(b)$, then $b = 10^a$). Thus, a star (Vega) of luminosity 50 times that of the Sun has $M = 4.83 - 2.5 \log_{10} 50 = 4.83 - 2.5(1.70) = 0.58$, while a star of 0.3 solar luminosity (ε Eridani) has $M = 4.83 - 2.5(-0.52) = 6.2$, and so on.

Now let's turn to the relationship between apparent luminosity (energy flux), absolute magnitude (luminosity), and distance. The *inverse-square law* (Section 17.2) can also be recast in terms of magnitudes. Increasing the distance to a star by a factor of 10 decreases its energy flux by a factor of $10^2 = 100$ and hence increases the star's apparent magnitude by five units. Increasing the distance by a factor of 100 increases the apparent magnitude by 10. Since absolute magnitude is a measure of the energy flux received from a star at a distance of 10 pc, denoting the apparent magnitude by m and the distance by D, we can write

$$m = M + 5 \log_{10}\left(\frac{D}{10 \text{ pc}}\right).$$

This doesn't look much like the inverse-square law presented in the text! Nevertheless, it contains exactly the same information, and it is in widespread use in this form throughout astronomy. Notice once again that, if D is greater than 10 pc, then m is greater than M, and vice versa.

Finally, if we want D on the left-hand side, we can rearrange the equation to read

$$D = 10 \text{ pc} \times 10^{(m-M)/5}.$$

We can use this version whenever we need to turn apparent and absolute magnitudes into a distance. Notice that knowing the difference $m - M$ between the apparent and absolute magnitudes of an object is completely equivalent to knowing the object's distance from us. Indeed, the quantity $m - M$ is often called the object's *distance modulus*.

EXAMPLE 3: Again using stars listed in Appendix 3, the star Rigel is observed to have apparent magnitude $m = 0.18$ and has a measured distance (using parallax) $D = 240$ pc. From the preceding equation, the absolute magnitude of Rigel is then $M = 0.18 - 5 \log_{10} 24 = -6.7$, corresponding to a luminosity $10^{4.61} \approx 41,000$ times that of the Sun. Conversely, the star Rigel Kentaurus (Alpha Centauri) has absolute magnitude $M = 4.34$ and is observed to have apparent magnitude $m = -0.01$. Its distance must therefore be $D = 10 \text{ pc} \times 10^{(-0.01-4.34)/5} = 1.35$ pc, in agreement with the result (obtained by parallax!) given in the text.

17.3 Stellar Temperatures

Looking at the night sky, you can tell at a glance which stars are hot and which are cool. In Figure 17.8, which shows the constellation Orion as it appears through a small telescope, the colors of the cool red star Betelgeuse (α) and the hot blue star Rigel (β) are clearly evident. Note that these colors are intrinsic properties of the stars and have *nothing* to do with Doppler redshifts or blueshifts. However, to obtain these stars' temperatures (3200 K for Betelgeuse and 11,000 K for Rigel), more detailed observations are required.

Color and the Blackbody Curve

Astronomers can determine a star's surface temperature by measuring the star's apparent brightness (energy flux) at several frequencies and then matching the observations to the appropriate blackbody curve. ∞ (Sec. 3.4) In the case of the Sun, the theoretical curve that best fits the emission describes a 5800-K emitter. ∞ (Sec. 16.1) The same technique works for any other star, regardless of its distance from Earth.

Because the basic shape of the blackbody curve is so well understood, astronomers can estimate a star's temperature using as few as *two* measurements at selected wavelengths (which is fortunate, as detailed spectra of faint stars are often difficult and time consuming to obtain). This is accomplished through the use of telescope filters that block out all radiation except that within specific wavelength ranges. For example, a B (blue) filter rejects all radiation except for a certain range from violet to blue light. Defined by international agreement to extend from 380 to 480 nm, this range corresponds to wavelengths to which photographic film happens to be most sensitive. Similarly, a V (visual) filter passes only radiation within the 490-to-590-nm range (green to yellow), corresponding to that part of the spectrum to which human eyes are particularly sensitive. Many other filters are also in routine use—a U (ultraviolet) filter covers the near ultraviolet, and infrared filters span longer wavelength parts of the spectrum.

Figure 17.9 shows how the B and V filters admit different amounts of light for objects radiating at different temperatures. In curve (a), corresponding to a very hot 30,000-K emitter, considerably more radiation (about twice as much) is received through the B filter than through the V filter, so this object looks brighter in B than in V. In curve (b), the temperature is 10,000 K, and the B and V fluxes are about the same. In the cool 3000-K curve (c), about five times more energy is received in the V range than in the B range, so the B image now is much fainter than the V image. In each case, it is possible to reconstruct the entire blackbody curve on the basis of only those two measurements, because no other blackbody curve can be drawn through both measured points. To the extent that a star's spectrum is well approximated as a

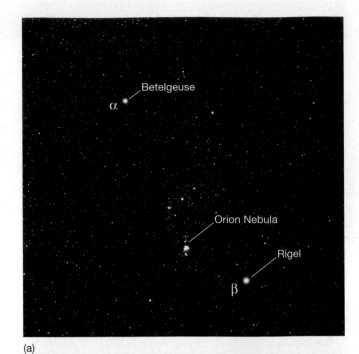

(a)

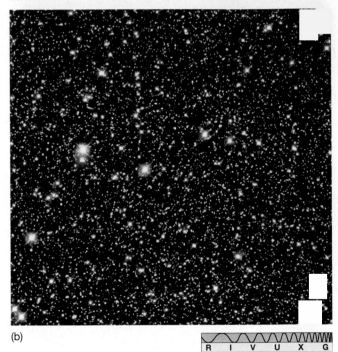

(b)

R   I   V   U   X   G

▲ **FIGURE 17.8 Star Colors** (a) Different colors of the stars of the constellation Orion are easily distinguished in this photograph taken through a wide-field camera attached to a small telescope. The bright red star at the upper left is Betelgeuse (α); the bright blue-white star at the lower right is Rigel (β). (Compare with Figure 1.8 and the large opening photo for Chapter 1.) The scale of the photograph is about 20° across. (b) An incredibly rich field of colorful stars in the direction of the center of the Milky Way. Here, the field of view is just 2 arc minutes across—much smaller than in (a). The image is heavily populated with stars of many different temperatures. (*J Sanford/ Astrostock-Sanford; NASA*)

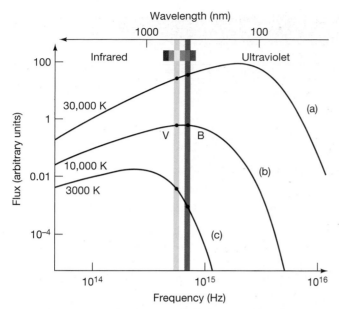

▲ **FIGURE 17.9 Blackbody Curves** The locations of the B (blue) and V (visual) filters, with blackbody curves for three different temperatures. Star (a) is very hot—30,000 K—so its B flux is greater than that its V flux (as is actually the case for Rigel in Figure 17.8a). Star (b) has roughly equal B and V readings and so appears white. Its temperature is about 10,000 K. Star (c) is red; its V flux greatly exceeds the B value, and its temperature is 3000 K (much as for Betelgeuse in Figure 17.8a).

blackbody, measurements of the B and V fluxes are enough to specify the star's blackbody curve and thus yield its surface temperature.

Thus, astronomers can estimate a star's temperature simply by measuring and comparing the amount of light received through different colored filters. As discussed in Chapter 5, this type of non-spectral-line analysis using a standard set of filters is known as photometry. ∞ (Sec. 5.2) Table 17.1 lists, for several prominent stars, the surface temperatures derived by photometric means, along with the dominant color that would be perceived in the absence of filters. Astronomers commonly refer to the ratio of the B flux to the V flux as the *color index*, or often just the *color*, of a star.

Stellar Spectra

Color is a useful way to describe a star, but astronomers often use a more detailed scheme to classify stellar properties, incorporating additional knowledge of stellar physics obtained through spectroscopy. Figure 17.10 compares the spectra of several different stars, arranged in order of decreasing surface temperature (as determined from measurements of their colors). All the spectra extend from 400 to 650 nm, and each shows a series of dark absorption lines superimposed on a background of continuous color, like the spectrum of the Sun. ∞ (Sec. 16.4) However, the precise patterns of lines reveal many differences. Some stars

display strong lines in the long-wavelength part of the spectrum (to the left in the figure). Other stars have their strongest lines at short wavelengths (to the right). Still others show strong absorption lines spread across the whole visible spectrum. What do these differences tell us?

Although spectral lines of many elements are present with widely varying strengths, the differences among the spectra in Figure 17.10 are not due to differences in composition. Detailed spectral analysis indicates that the seven stars shown have similar elemental abundances—all are more or less solar in makeup. ∞ (Sec. 16.4) Rather, as discussed in Chapter 4, the differences are due almost entirely to the stars' *temperatures*. ∞ (Sec. 4.4) The spectrum at the top of the figure is exactly what we would expect from a star with solar composition and a surface temperature of about 30,000 K, the second spectrum is what we would anticipate from a 20,000-K star, and so on, down to the 3000-K star at the bottom.

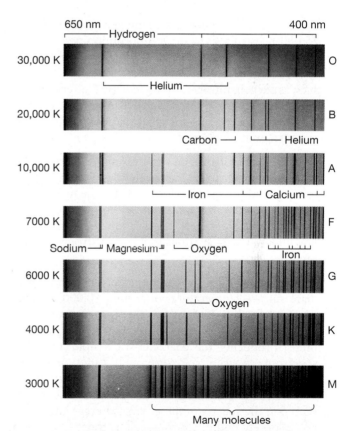

▲ **FIGURE 17.10 Stellar Spectra** Comparison of spectra of seven different stars having a range of surface temperatures. These are not actual spectra, which are messy and complex, but simplified artist's renderings highlighting some special features. The spectra of the hottest stars, at the top, show lines of helium and multiply ionized heavy elements. In the coolest stars, at the bottom, helium lines are not seen, but lines of neutral atoms and molecules are plentiful. At intermediate temperatures, hydrogen lines are strongest. All seven stars have about the same chemical composition.

TABLE 17.1	Stellar Colors and Temperatures		
B flux / V flux	**Approximate Surface Temperature (K)**	**Color**	**Familiar Examples**
1.3	30,000	blue-violet	Mintaka (δ Orionis)
1.2	20,000	blue	Rigel
1.00	10,000	white	Vega, Sirius
0.72	7000	yellow-white	Canopus
0.55	6000	yellow	Sun, Alpha Centauri
0.33	4000	orange	Arcturus, Aldebaran
0.21	3000	red	Betelgeuse, Barnard's Star

The main differences among the spectra in Figure 17.10 are as follows:

- Spectra of stars having surface temperatures exceeding 25,000 K usually show *strong* absorption lines of singly ionized helium (i.e., helium atoms that have lost one orbiting electron) and multiply ionized heavier elements, such as oxygen, nitrogen, and silicon (the latter lines are not shown in the figure). These strong lines are not seen in the spectra of cooler stars because only very hot stars can excite and ionize such tightly bound atoms.

- In contrast, the hydrogen absorption lines in the spectra of very hot stars are relatively *weak*. The reason is not a lack of hydrogen, which is by far the most abundant element in all stars. At these high temperatures, however, much of the hydrogen is ionized, so there are few intact hydrogen atoms to produce strong spectral lines.

- Hydrogen lines are strongest in stars having intermediate surface temperatures of around 10,000 K. This temperature is just right for electrons to move frequently between hydrogen's second and higher orbitals, producing the characteristic visible hydrogen spectrum. ∞ *(More Precisely 4-1)* Lines of tightly bound atoms—for example, of helium and nitrogen—which need lots of energy for excitation, are rarely observed in the spectra of these stars, whereas lines from more loosely bound atoms—such as those of calcium and titanium—are relatively common.

- Hydrogen lines are again weak in stars with surface temperatures below about 4000 K, but now because the temperature is too low to boost many electrons out of the ground state. ∞ (Sec. 4.2) The most intense spectral lines in these stars are due to weakly excited heavy atoms; no lines from ionized elements are seen. Temperatures in the coolest stars are low enough for molecules to survive, and many of the observed absorption lines are produced by molecules rather than by atoms. ∞ (Sec. 4.3)

Stellar spectra are the source of all the detailed information we have on stellar composition, and they do in fact reveal significant differences in composition among stars, particularly in the abundances of carbon, nitrogen, oxygen, and heavier elements. However, as we have just seen, these differences are *not* the primary reason for the different spectra that are observed. Instead, the main determinant of a star's spectral appearance is its temperature, and stellar spectroscopy is a powerful and precise tool for measuring this important stellar property.

Spectral Classification

Stellar spectra like those shown in Figure 17.10 were obtained for many stars well before the start of the 20th century as observatories around the world amassed spectra from stars in both hemispheres of the sky. Between 1880 and 1920, researchers correctly identified some of the observed spectral lines on the basis of comparisons between those lines and lines obtained in the laboratory. The researchers, though, had no firm understanding of how the lines were produced. Modern atomic theory had not yet been developed, so the correct interpretation of the line strengths, as just described, was impossible at the time.

Lacking a full understanding of how atoms produce spectra, early workers classified stars primarily according to their hydrogen-line intensities. They adopted an alphabetic A, B, C, D, E . . . scheme in which A stars, with the strongest hydrogen lines, were thought to have more hydrogen than did B stars, and so on. The classification extended as far as the letter P.

In the 1920s, scientists began to understand the intricacies of atomic structure and the causes of spectral lines. Astronomers quickly realized that stars could be more meaningfully classified according to their surface temperature. Instead of adopting an entirely new scheme, however, they chose to shuffle the existing alphabetical categories—those based on the strengths of the hydrogen lines—into a new sequence based on temperature. In the modern scheme, the hottest stars are designated O, because they have very weak absorption lines of hydrogen and were

TABLE 17.2   Stellar Spectral Classes

Spectral Class	Approximate Surface Temperature (K)	Noteworthy Absorption Lines	Familiar Examples
O	30,000	Ionized helium strong; multiply ionized heavy elements; hydrogen faint	Mintaka (O9)
B	20,000	Neutral helium moderate; singly ionized heavy elements; hydrogen moderate	Rigel (B8)
A	10,000	Neutral helium very faint; singly ionized heavy elements; hydrogen strong	Vega (A0), Sirius (A1)
F	7000	Singly ionized heavy elements; neutral metals; hydrogen moderate	Canopus (F0)
G	6000	Singly ionized heavy elements; neutral metals; hydrogen relatively faint	Sun (G2), Alpha Centauri (G2)
K	4000	Singly ionized heavy elements; neutral metals strong; hydrogen faint	Arcturus (K2), Aldebaran (K5)
M	3000	Neutral atoms strong; molecules moderate; hydrogen very faint	Betelgeuse (M2), Barnard's Star (M5)

classified toward the end of the original scheme. In order of decreasing temperature, the surviving letters now run O, B, A, F, G, K, M. (The other letter classes have been dropped.) These stellar designations are called **spectral classes** (or *spectral types*). Use the time-honored (and politically incorrect) mnemonic "**O**h, **B**e **A** **F**ine **G**irl, **K**iss **M**e" to remember them in the correct order.*

Astronomers further subdivide each lettered spectral classification into 10 subdivisions, denoted by the numbers 0–9. By convention, the lower the number, the hotter is the star. For example, our Sun is classified as a G2 star (a little cooler than G1 and a little hotter than G3), Vega is a type A0, Barnard's Star is M5, Betelgeuse is M2, and so on. Table 17.2 lists the main properties of each stellar spectral class for the stars presented in Table 17.1.

We should not underestimate the importance of the early work in classifying stellar spectra. Even though the original classification was based on erroneous assumptions, the painstaking accumulation of large quantities of accurate data paved the way for rapid improvements in understanding once a theory came along that explained the observations.

CONCEPT CHECK

✔ Why does a star's spectral classification depend on its temperature?

Some astronomers have proposed the addition of two new spectral classes—L and T—for low-mass, low-temperature stars whose odd spectra distinguish them from the M-class stars in the current scheme. For now, at least, the new classification has not been widely adopted. Astronomers are still uncertain whether these new objects are "true" stars, fusing hydrogen into helium in their cores, or whether they are "brown dwarfs" (see Chapter 20) that never achieved high enough central temperatures for fusion to begin.

17.4 Stellar Sizes

Direct and Indirect Measurements

Most stars are unresolved points of light in the sky, even when viewed through the largest telescopes. However, a few are big enough, bright enough, and close enough to allow us to measure their sizes *directly*.

One well-known example is the bright red star Betelgeuse, a prominent member of the constellation Orion (Figure 17.8). In a technique known as *speckle interferometry*, many short-exposure images of a star, each too brief for Earth's turbulent atmosphere to smear it out into a disk, are combined to make a high-resolution map of the star's surface. ∞ (Sec. 5.4) In some cases, the results are detailed enough to allow a few surface features to be distinguished (Figure 17.11a). As shown in Figure 17.11(b), Betelgeuse is also (barely) large enough to be resolvable by the *Hubble Space Telescope* at short wavelengths. Steadily improving speckle and adaptive-optics techniques have allowed astronomers to construct very-high-resolution stellar images in a small number of cases. ∞ (Sec. 5.4)

Once a star's angular size has been measured, if its distance is also known, we can determine its radius by simple geometry. ∞ (Sec. 1.7) For example, with a distance of 130 pc and an angular diameter of up to 0.045″, Betelgeuse's maximum radius is 630 times that of the Sun. (We say "maximum radius" here because, as it happens, Betelgeuse is a *variable star*—its radius and luminosity vary somewhat irregularly, with a period of roughly 6 years.) All told, the sizes of perhaps a few dozen stars have been measured in this way.

Most stars are too distant or too small for such direct measurements to be made. Instead, their sizes must be inferred by indirect means, using the radiation laws. ∞ (Sec. 3.4) The radiation emitted by a star is governed by the

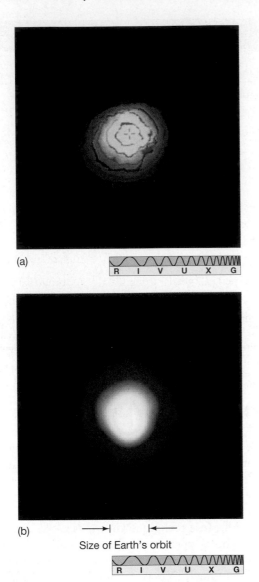

(a)

(b)

Size of Earth's orbit

▲ **FIGURE 17.11 Betelgeuse** (a) The swollen star Betelgeuse (shown here in false color) is close enough for us to resolve its size directly, along with some surface features thought to be storms similar to those which occur on the Sun. Betelgeuse is such a huge star (some 600 times the size of the Sun) that its photosphere exceeds the size of Mars's orbit. Most of the surface features discernible here are larger than the entire Sun. (The dark lines are drawn contours, not part of the star itself.) (b) An ultraviolet view of Betelgeuse, in false color, as seen by a European camera onboard the *Hubble Space Telescope*, shows more clearly just how large this huge star is. *(NOAO; NASA/ESA)*

Stefan–Boltzmann law, which states that the energy emitted *per unit area* per unit time increases as the fourth power of the star's surface temperature. ∞ *(More Precisely 3-2)* To determine the star's luminosity, we must multiply by its surface area—large bodies radiate more energy than do small bodies at the same temperature. Because the surface area is proportional to the square of the radius, we have

$$\text{luminosity} \propto \text{radius}^2 \times \text{temperature}^4.$$

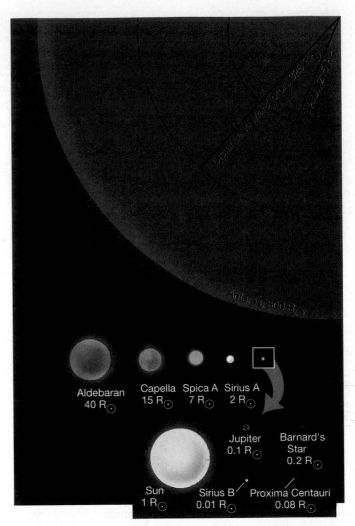

▲ **FIGURE 17.12 Stellar Sizes** Star sizes vary greatly. Shown here are the estimated sizes of several well-known stars, including a few of those discussed in this chapter. Only part of the red-giant star Antares can be shown on this scale, and the supergiant Betelgeuse would fill the entire page. (The symbol "$R_\odot$" period with a circle around it means "solar radius.")

This **radius–luminosity–temperature relationship** is important because it demonstrates that knowledge of a star's luminosity and temperature can yield an estimate of the star's radius—an *indirect* determination of stellar size.

Giants and Dwarfs

Let's consider some examples to clarify these ideas. The star known as Aldebaran (the orange-red "eye of the bull" in the constellation Taurus) has a surface temperature of about 4000 K and a luminosity of 1.3×10^{29} W. Thus, its surface temperature is 0.7 times, and its luminosity about 330 times, the corresponding quantities for our Sun. The radius–luminosity–temperature relationship (see *More Precisely 17-2*) then implies that the star's radius is almost 40 times the solar value. If our Sun were that large, its photosphere would extend halfway to the orbit of Mercury and, seen from Earth, would cover more than 20 degrees on the sky. A star as large as Aldebaran is known as a *giant*. More precisely, **giants** are stars having radii between 10 and 100 times that of the Sun. Since any 4000-K object is reddish in color, Aldebaran is known as a **red giant**. Even larger stars, ranging up to 1000 solar radii in size, are known as **supergiants**. Betelgeuse is a prime example of a **red supergiant**.

Now consider Sirius B, a faint companion to Sirius A, the brightest star in the night sky. Sirius B's surface temperature is roughly 27,000 K, some four-and-a-half times that of the Sun. The star's total luminosity is 10^{25} W, about 0.025 times the solar value. Again using the radius–luminosity–temperature relationship, we obtain a radius of 0.007 solar radii—slightly smaller than that of Earth. Sirius B is much hotter, but smaller and far less lu-minous, than our Sun. Such a star is known as a *dwarf*. In astronomical parlance, the term **dwarf** refers to any star of radius comparable to or smaller than the Sun (including the Sun itself). Because any 27,000 K object glows blue-white, Sirius B is an example of a **white dwarf**.

The radii of the vast majority of stars (measured mostly with the radius–luminosity–temperature relationship) range from less than 0.01 to over 100 times the radius of the Sun. Figure 17.12 illustrates the estimated sizes of a few well-known stars.

CONCEPT CHECK

✔ Can we measure the radius of a star without knowing the star's distance from Earth?

MORE PRECISELY 17-2

Estimating Stellar Radii

We can combine the Stefan–Boltzmann law $F = \sigma T^4$ with the formula for the area of a sphere, $A = 4\pi R^2$, to obtain the relationship between a star's radius (R), luminosity (L), and temperature (T) described in the text: ∞ (*More Precisely 3-2*)

$$L = 4\pi\sigma R^2 T^4,$$

or

$$\text{luminosity} \propto \text{radius}^2 \times \text{temperature}^4.$$

If we adopt convenient "solar" units, in which L is measured in solar luminosities (3.9×10^{26} W), R in solar radii (696,000 km), and T in units of the solar temperature (5800 K), we can eliminate the constant $4\pi\sigma$ and write this equation as

L (in solar luminosities) =

R^2 (in solar radii) × T^4 (in units of 5800 K).

To compute the radius of a star from its luminosity and temperature, we rearrange terms so that the equation reads (in the same units)

$$R = \sqrt{L}/T^2.$$

This simple application of the radiation laws is the basis for almost every estimate of stellar size made in this text.

EXAMPLE: Let's compute the radii of the two stars discussed in the text. The star Aldebaran has luminosity $L = 330$ units and temperature $T = 4000/5800 = 0.69$ unit. According to the foregoing equation for R, its radius is therefore 18/0.48 = 39 solar radii—that of a giant star. At the opposite extreme, Sirius B has $L = 0.025$ and $T = 4.7$, so its radius is 0.007 times the radius of the Sun—that of a dwarf. The accompanying table lists luminosities, temperatures, and calculated radii (all in solar units) for some of the other stars mentioned in this chapter.

Star	Luminosity, L	Temperature, T	Radius, R
Barnard's Star	0.0045	0.56	0.2
Sun	1	1	1
Sirius A	23	2.1	1.9
Vega	55	1.6	2.8
Arcturus	160	0.78	21
Rigel	63,000	1.9	70
Betelgeuse	36,000	0.55	630

Note that some of the luminosities shown in the table differ from those in Appendix 3 and elsewhere in the text. This is because the values used elsewhere refer to visible light only, whereas the radiation laws (and the values in the table) refer to *total* luminosities (i.e., luminosities at all wavelengths).

Finally, let's follow the chain of reasoning leading from observations of a star to a fairly complete determination of its properties. We use the giant star Canopus, the second brightest star in the (southern) sky, as an example. Canopus is observed to have an apparent magnitude of −0.62. *Hipparcos* (*Discovery 17-2*) measured its parallax as 0.0104 arc second, so Canopus's distance from Earth is 96 pc. Its absolute magnitude is then −5.5, making its luminosity 14,000 times that of the Sun (*More Precisely 17-1*). The spectral type of Canopus is F0; that information and *Hipparcos* color measurements imply a surface temperature of approximately 7400 K, or 1.3 times the surface temperature of the Sun. The radius–luminosity–temperature relationship then gives a radius of 70 solar radii. Thanks to *Hipparcos*, the same series of calculations has been performed for each of the more than 100,000 stars whose properties are now known.

17.5 The Hertzsprung–Russell Diagram

🏅 LEARNING GOAL 6 Astronomers use luminosity and surface temperature to classify stars in much the same way that height and weight serve to classify the bulk properties of human beings. We know that people's height and weight are well correlated: Tall people tend to weigh more than short ones. We might naturally wonder if the two basic stellar properties are also related in some way.

Figure 17.13 plots luminosity versus temperature for a few well-known stars. Figures of this sort are called *Hertzsprung–Russell diagrams*, or **H–R diagrams**, after Danish astronomer Ejnar Hertzsprung and U.S. astronomer Henry Norris Russell, who independently pioneered the use of such plots in the second decade of the 20th century. The vertical scale, expressed in units of the solar luminosity (3.9×10^{26} W), extends over a large range, from 10^{-4} to 10^{4}; the Sun appears right in the middle of the luminosity range, at a luminosity of 1. Surface temperature is plotted on the horizontal axis, although in the unconventional sense of temperature increasing to the *left* (so that the spectral sequence O, B, A ... reads from left to right). To change the horizontal scale so that temperature would increase conventionally to the right would play havoc with historical precedent.

As we have just seen, astronomers often use a star's *color* to measure its temperature. Indeed, the spectral classes plotted along the horizontal axis of Figure 17.13 are equivalent to the B/V color index. Also, because astronomers commonly express a star's luminosity as an absolute magnitude, stellar *magnitude* instead of stellar luminosity could be plotted on the vertical axis. (See *More Precisely 17-1*.) For these reasons, many astronomers refer to diagrams such as Figure 17.13 as **color–magnitude diagrams**. In this book, however, we will cast our discussion mainly in terms of temperature and luminosity measurements.

The Main Sequence

The few stars plotted in Figure 17.13 give little indication of any particular connection between stellar properties. However, as Hertzsprung and Russell plotted more and more stellar temperatures and luminosities, they found that a relationship does in fact exist: Stars are *not* uniformly scattered across the H–R diagram; instead, most are confined to a fairly well-defined band stretching diagonally from the top left (high temperature, high luminosity) to the bottom right (low temperature, low luminosity). In other words, cool stars tend to be faint (less luminous) and hot stars tend to be bright (more luminous). This band of stars spanning the H–R diagram is known as the **main sequence**.

Figure 17.14 shows a more systematic study of stellar properties, covering the 80 or so stars that lie within 5 pc of the Sun. As more points are included in the diagram, the main sequence "fills up," and the pattern becomes more evident. The vast majority of stars in the immediate vicinity of the Sun lie on the main sequence.

The surface temperatures of main-sequence stars range from about 3000 K (spectral class M) to over 30,000 K (spectral class O). This relatively small temperature range—a difference of only a factor of 10—is determined mainly by the rates at which nuclear reactions occur in stellar cores. ∞ (Sec. 16.3) In contrast, the observed range in luminosities is very large, covering some eight orders of magnitude (i.e., a factor of 100 million), ranging from 10^{-4} to 10^{4} times the luminosity of the Sun.

Using the radius–luminosity–temperature relationship (Section 17.4), astronomers find that stellar radii also vary along the main sequence. The faint, red M-type stars at the bottom right of the H–R diagram are only about one-tenth the size of the Sun, whereas the bright, blue

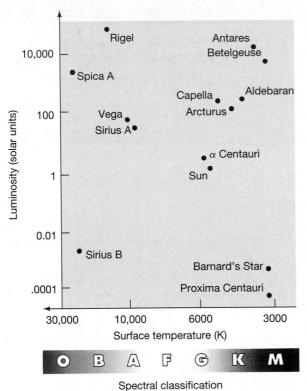

▲ **FIGURE 17.13 H–R Diagram of Prominent Stars**
Known as an H–R diagram, a plot of luminosity against surface temperature (or spectral classification) is a useful way to compare stars. Plotted here are the data for some stars mentioned earlier in the text. The Sun, of course, has a luminosity of 1 solar unit. Its temperature, read off the bottom scale, is 5800 K—that of a G-type star. Similarly, the B-type star Rigel, at the top left, has a temperature of about 11,000 K and a luminosity more than 10,000 times that of the Sun. The M-type star Proxima Centauri, at the bottom right, has a temperature of about 3000 K and a luminosity less than 1/10,000 that of the Sun.

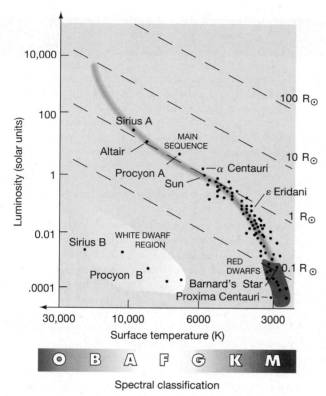

▲ **FIGURE 17.14 H–R Diagram of Nearby Stars** Most stars have properties within the shaded region known as the main sequence. The points plotted here are for stars lying within about 5 pc of the Sun. Each dashed diagonal line corresponds to a constant stellar radius, so that stellar size can be indicated on the same diagram as luminosity and temperature. (Here and elsewhere, the symbol ☉ stands for the Sun.)

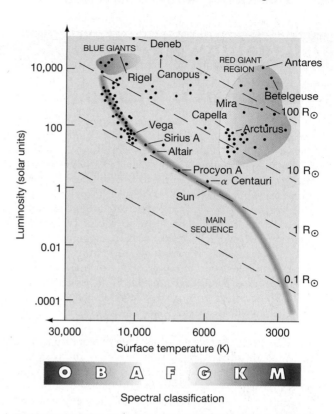

▲ **FIGURE 17.15 H–R Diagram of Bright Stars** An H–R diagram for the 100 brightest stars in the sky is biased in favor of the most luminous stars—which appear toward the upper left—because we can see them more easily than we can the faintest stars. (Compare with Figure 17.14, which shows only the closest stars.)

O-type stars in the upper left are about 10 times larger than the Sun. The diagonal dashed lines in Figure 17.14 represent constant stellar radii, meaning that any star lying on a given line has the same radius, regardless of its luminosity or temperature. Along a constant-radius line, the radius–luminosity–temperature relationship implies that

$$\text{luminosity} \propto \text{temperature}^4.$$

By including such lines on our H–R diagrams, we can indicate stellar temperatures, luminosities, and radii on a single plot.

We see a very clear trend as we traverse the main sequence from top to bottom. At one end, the stars are large, hot, and bright. Because of their size and color, they are referred to as **blue giants**. The very largest are called **blue supergiants**. At the other end, stars are small, cool, and faint. They are known as **red dwarfs**. Our Sun lies right in the middle.

Figure 17.15 shows an H–R diagram for a different group of stars—the 100 stars of known distance having the greatest apparent brightness, as seen from Earth. Notice

the much larger number of very luminous stars at the upper end of the main sequence than at the lower end. The reason for this excess of blue giants is simple: We can see very luminous stars a long way off. The stars shown in this figure are scattered through a much greater volume of space than those depicted in Figure 17.14, but the sample is heavily biased toward the brightest objects. In fact, of the 20 brightest stars in the sky, only 6 lie within 10 pc of us; the rest are visible, despite their great distances, because of their high luminosities.

If very luminous blue giants are overrepresented in Figure 17.15, low-luminosity red dwarfs are surely underrepresented. In fact, no dwarfs appear on the diagram. This absence is not surprising, because low-luminosity stars are difficult to observe from Earth. In the 1970s, astronomers began to realize that they had greatly underestimated the number of red dwarfs in our galaxy. As hinted at by the H–R diagram in Figure 17.14, which shows an unbiased sample of stars in the solar neighborhood, red dwarfs are actually the most common type of star in the sky. In fact, they probably account for upward of 80 percent of all stars in the universe. In contrast, O- and B-type supergiants are extremely rare, with only about one star in 10,000 falling into these categories.

DISCOVERY 17-2

The *Hipparcos* Mission

In 1989, shortly before the large, complex, and expensive *Hubble Space Telescope* was launched by NASA into Earth orbit, a small, simple, and cheap satellite was also deployed by the European Space Agency (ESA). Called *Hipparcos*—a tortured attempt to honor the ancient Greek astronomer Hipparchus of Nicaea (who made the first known star map) by creating an acronym out of *High Precision Parallax Collecting Satellite*—this spacecraft was initially thought to be lost in space. *Hipparcos* went into the wrong orbit, its onboard engine failed to fire, and its solar panels were heavily pelted with particles in Earth's Van Allen belts. However, before the mission ended in 1993, the engineers managed to control the errant spacecraft, and ESA scientists revamped its mission. Nearly a decade after its launch, its massive database was cataloged and released.

The heart of *Hipparcos* (shown here before launch) was a small 29-cm mirror designed to determine the positions of stars to a high degree of accuracy. By measuring those positions from two vantage points in orbit (above Earth's turbulent atmosphere), the satellite was able to determine stellar parallaxes with about 10 times greater accuracy than is possible by using telescopes on the ground. Since volume scales as the cube of the size, *Hipparcos*'s prime mission was to obtain accurate distances for roughly a million stars out to some 200 pc. The instrument also measured the colors (i.e., temperatures) and apparent brightnesses (and hence luminosities, since the distances are known) of the stars it surveyed.

The impact of the *Hipparcos* mission extends far beyond the solar neighborhood. Parallax calibrates other "yardsticks" used by astronomers to measure distances in increasingly larger realms and, eventually, to estimate the size and scale of the entire observable universe. As a result, *Hipparcos* has affected distances far beyond those it was able to measure directly. In particular, the recalculation of distances from Earth to a certain class of variable stars (called Cepheids—see Chapter 23) has put astronomy's entire distance scale on a much firmer foundation.

Because of these recalibrations, some distances have changed a little, while others needed substantial revision. For example, the Large Magellanic Cloud (a small satellite galaxy orbiting our own—see Chapter 24) is now thought to be about 10 percent farther away than had previously been estimated. Likewise, the distance to the Andromeda Galaxy has been increased by about a third and is now pegged at 0.8 Mpc (or 2.5 million light-years). ∞ (Sec. 3.1) All distances throughout this text reflect the new values determined by *Hipparcos*.

In addition, a major revision of the Hertzsprung–Russell diagram (see Figures 17.13–17.16) based on *Hipparcos* data implies that the oldest stars are a little younger than was previously thought. This new analysis of stellar evolutionary tracks seems to reconcile the age of the universe (obtained using methods to be discussed in Chapter 26) with that of the oldest stars (Chapter 20). Prior to the *Hipparcos* mission, astronomers had faced the embarrassing problem that some stars seemed to be older than the universe itself. But no more—this major breakthrough in a previously logjammed subject may be *Hipparcos*'s greatest contribution to astronomy.

The *Hipparcos* mission was not glamorous: The little satellite sent back no pictures for European citizens (who paid for it) and others to marvel at, and it grabbed few headlines in the world's press. Rather, its catalogs, comprising some 1000 gigabytes of data (or about 1600 CD-ROMs), resemble huge phone books—unexciting to look at, but of enormous value for accessing the vital statistics of stars and the entire universe beyond.

Following *Hipparcos*, both NASA and ESA have ambitious plans to expand the scope of stellar measurements enormously. NASA's *Space Interferometry Mission* (*SIM*) and ESA's *GAIA* project, both planned for launch around 2010, will each have the astonishing range of 25,000 pc—covering our entire Galaxy and encompassing roughly 1 billion stars! The science goals are wide-ranging. In addition to mapping out the structure of the Milky Way Galaxy to unprecedented precision, these missions will allow astronomers to study in detail the properties of nearby stars of all masses—including some of the numerous but extremely hard-to-detect red and brown dwarfs—and will also greatly expand our knowledge of extrasolar planetary systems. ∞ (Sec. 15.4) In a time span of just three decades, the fundamental stellar database upon which almost all of astronomy depends will have increased in size by a factor of a million. The results may be nothing short of revolutionary.

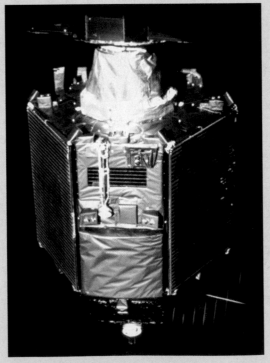

(ESA)

The White-Dwarf and Red-Giant Regions

Most stars lie on the main sequence. However, some of the points plotted in Figures 17.13 through 17.15 clearly do not. One such point in Figure 17.13 represents Sirius B, the white dwarf discussed earlier (Section 17.4), with surface temperature 27,000 K and luminosity about 0.025 times the solar value. A few more such faint, hot stars can be seen in Figure 17.14 in the bottom left-hand corner of the H–R diagram. This region, known as the **white-dwarf region**, is marked on Figure 17.14.

Also shown in Figure 17.13 is Aldebaran (discussed in Section 17.4), whose surface temperature is 4000 K and whose luminosity is some 300 times greater than the Sun's. Another point represents Betelgeuse (Alpha Orionis), the ninth-brightest star in the sky, a little cooler than Aldebaran, but more than 100 times brighter. The upper right-hand corner of the H–R diagram, where these stars lie (marked on Figure 17.15), is called the **red-giant region**. No red giants are found within 5 pc of the Sun (Figure 17.14), but many of the brightest stars seen in the sky are in fact red giants (Figure 17.15). Though relatively rare, red giants are so bright that they are visible to great distances. They form a third distinct class of stars on the H–R diagram, very different in their properties from both main-sequence stars and white dwarfs.

The *Hipparcos* mission (*Discovery 17-2*), in addition to determining hundreds of thousands of stellar parallaxes with unprecedented accuracy, also measured the colors and luminosities of more than 2 million stars. Figure 17.16 shows an H–R diagram based on a tiny portion of the enormous *Hipparcos* data set. The main-sequence and red-giant regions are clearly evident. Few white dwarfs appear, however, simply because the telescope was limited to observations of relatively bright objects—brighter than apparent magnitude 12. Almost no white dwarfs lie close enough to Earth that their magnitudes fall below this limit.

About 90 percent of all stars in our solar neighborhood, and probably a similar percentage elsewhere in the universe, are main-sequence stars. About 9 percent of stars are white dwarfs, and 1 percent are red giants.

CONCEPT CHECK

✔ Only a tiny fraction of all stars are giants. Why, then, do giants account for so many of the brightest stars in the night sky?

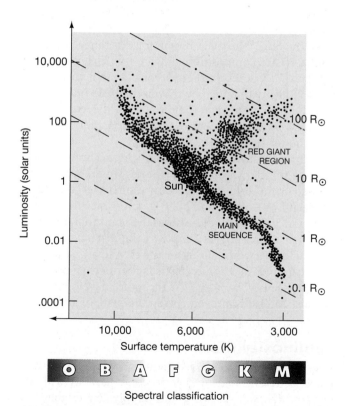

▲ **FIGURE 17.16** *Hipparcos* **H–R Diagram** This simplified version of the most complete H–R diagram ever compiled represents more than 20,000 data points, as measured by the European *Hipparcos* spacecraft for stars within a few hundred parsecs of the Sun.

17.6 Extending the Cosmic Distance Scale

In Chapter 2, we introduced the first "rung" on a ladder of distance-measurement techniques that will ultimately carry us to the edge of the observable universe. That rung is radar ranging on the inner planets. ∞ (Sec. 2.6) It establishes the scale of the solar system to great accuracy and, in doing so, defines the astronomical unit. In Section 17.1, we discussed a second rung in the cosmic distance ladder—stellar parallax—which is based on the first. Now, having used the first two rungs to determine the distances and other physical properties of many nearby stars, we can employ that knowledge in turn to construct a third rung in the ladder: **spectroscopic parallax**.* As illustrated schematically in Figure 17.17, this new rung expands our cosmic field of view still deeper into space.

Spectroscopic Parallax

How can we use the knowledge we have just gained to measure the distance to a star? Consider an analogy. Most of us have a rough idea of the approximate brightness and size of a red traffic signal. Suppose we are driving down an

This unfortunate name is rather misleading, as the method has nothing in common with stellar (geometric) parallax other than its use as a means of determining stellar distances.

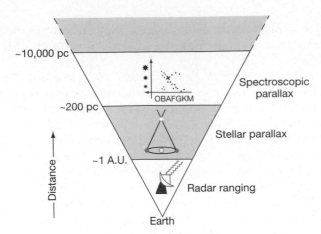

▲ FIGURE 17.17 Stellar Distances Knowledge of a star's luminosity and apparent brightness can yield an estimate of its distance. Astronomers use this third "rung" in our distance ladder, called spectroscopic parallax, to measure distances as far out as individual stars can be clearly discerned—several thousand parsecs.

unfamiliar street and see a red traffic light in the distance. Our knowledge of the intrinsic luminosity of the light often enables us immediately to make a mental estimate of its distance. A normal traffic light that is relatively dim must be quite distant (assuming that it's not just dirty); a bright one must be relatively close. Thus, *a measurement of the apparent brightness of a light source, combined with some knowledge of its intrinsic properties, can yield an estimate of the source's distance.*

For a star, the trick is to find an independent measure of the luminosity without knowing the distance. The H–R diagram can provide just that. For example, suppose we observe a star and determine its apparent magnitude to be 10. By itself, that doesn't tell us much—the star could equally well be faint and close, or bright and distant (Figure 17.6). But suppose we have some additional information: The star lies on the main sequence and has spectral type A0. Then we can read the star's luminosity off Figure 17.15. A main-sequence A0 star has a luminosity of approximately 100 solar units. According to *More Precisely 17-1*, this corresponds to an absolute magnitude of 0 and hence to a distance of 1000 pc.

In Section 17.2, we described how astronomers use the inverse-square law to obtain the luminosities of stars whose distances are known by stellar parallax. Now we have turned the calculation around: In spectroscopic parallax,

1. we measure the star's apparent brightness and spectral type *without* knowing how far away it is;

2. then we use the spectral type to estimate the star's luminosity; and

3. finally, we apply the inverse-square law to determine the distance to the star.

The main sequence represents a fairly close correlation between temperature and luminosity for most stars (with the exception of a few giants and dwarfs, to be discussed in a moment). Thus, its existence allows us to make a connection between an easily measured quantity (spectral type) and the star's luminosity, which would otherwise be unknown. The term *spectroscopic parallax* refers to the specific process of using stellar spectra to infer luminosities and hence distances. However, as we will see in upcoming chapters, this basic logic (with a variety of different techniques replacing step 2) is used again and again as a means of distance measurement in astronomy.

Spectroscopic parallax can be used to determine stellar distances out to several thousand parsecs. Beyond that, spectra and colors of individual stars are difficult to obtain. The "standard" main sequence is obtained from H–R diagrams of stars whose distances can be measured by (geometric) parallax, so the method of spectroscopic parallax is calibrated by using nearby stars. Note that, in employing this method, we are assuming (without proof) that distant stars are basically similar to nearby stars and that *they fall on the same main sequence as nearby stars.* Only by making this assumption can we expand the boundaries of our distance-measurement techniques.

Of course, the main sequence is not really a line in the H–R diagram: It has some thickness. For example, the luminosity of a main-sequence A0-type star (such as Vega) can actually range from about 30 to 100 times the luminosity of the Sun. The main reason for this range is the variation in stellar composition and age from place to place in the Galaxy. As a result, there is considerable uncertainty in the luminosity obtained by this method and hence a corresponding uncertainty in the distance of the star. Distances obtained by spectroscopic parallax are generally accurate to no better than about 25 percent.

Although this may not seem very accurate—a cross-country traveler in the United States would hardly be impressed to be told that the best estimate of the distance between Los Angeles and New York is somewhere between 3000 and 5000 km—it illustrates the point that, in astronomy, even something as simple as the distance to another star can be very difficult to measure. Still, an estimate with an uncertainty of ±25 percent is far better than no estimate at all. (See also *Discovery 17-2* for future developments that promise radical improvements in this state of affairs.)

Luminosity Class

If a star happens to be a red giant or a white dwarf, its distance as determined by spectroscopic parallax will be incorrect. Of course, we could simply argue that, since roughly 90 percent of all stars are on the main sequence, the assumption that a star is a main-sequence star is valid 9 out of 10 times; but in fact, astronomers can do much better. Recall from Chapter 4 that the *width* of a spectral line can provide information on the *density* of the gas where the

TABLE 17.3 Stellar Luminosity Classes	
Class	**Description**
Ia	Bright supergiants
Ib	Supergiants
II	Bright giants
III	Giants
IV	Subgiants
V	Main-sequence stars and dwarfs

line formed. ∞ (Sec. 4.4) Models of stellar structure clearly show that the atmosphere of a red giant is much less dense than that of a main-sequence star, which in turn is much less dense than the atmosphere of a white dwarf. ∞ (Sec. 16.3) Thus, by studying the width of a star's spectral lines, astronomers can usually tell, with a high degree of confidence, whether the star is on the main sequence.

Astronomers have developed a system for classifying stars according to the width of their spectral lines. Because the line width is particularly sensitive to density in the stellar photosphere, and the atmospheric density in turn is well correlated with luminosity, the class into which a star is categorized has come to be known as the star's **luminosity class.** This classification provides a means for astronomers to distinguish supergiants from giants, giants from main-sequence stars, and main-sequence stars from white dwarfs by studying a single spectral property—line broadening—of the radiation received.

The standard stellar luminosity classes are given in Table 17.3. Their locations on the H–R diagram are indicated in Figure 17.18. Now we have a way of specifying a star's location in the diagram in terms of properties that are measurable by purely spectroscopic means. *Spectral type and luminosity class define a star just as surely as do temperature and luminosity.* The full specification of a star's spectral properties includes its luminosity class. For example, the Sun, on the main sequence, is of class G2V, Vega is A0V, the red dwarf Barnard's Star is M5V, the red supergiant Betelgeuse is M2Ia, and so on.

Consider, for example, a K2-type star (Table 17.4) with a surface temperature of approximately 4500 K. If the widths of the star's spectral lines tell us that it lies on

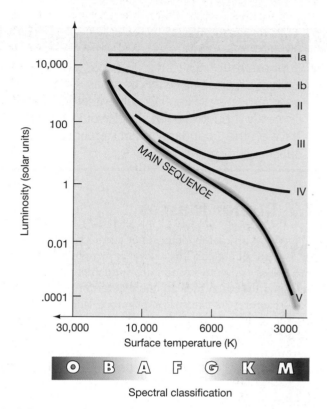

▲ FIGURE 17.18 Luminosity Classes Stellar luminosity classes in the H–R diagram. A star's location on the diagram can be specified by the star's spectral type and luminosity class instead of its temperature and luminosity.

the main sequence (i.e., it is a K2V star), then its luminosity is about 0.3 times the solar value. If the star's spectral lines are observed to be narrower than lines normally found in main-sequence stars, the star may be recognized as a K2III giant, with a luminosity 100 times that of the Sun (Figure 17.18). If the lines are very narrow, the star may instead be classified as a K2Ib supergiant, brighter by a further factor of 40, at 4000 solar luminosities. In this way, the observed widths of the star's spectral lines translate directly into a measure of the star's physical state. Thus, our knowledge of luminosity classes allows us to use spectroscopic parallax with some degree of confidence that we are not accidentally counting a red giant or a white dwarf as a main-sequence star and making a huge error in our distance estimate as a result.

TABLE 17.4 Variation in Stellar Properties within a Spectral Class				
Approximate Surface Temperature (K)	**Luminosity (solar luminosities)**	**Radius (solar radii)**	**Object**	**Example**
4900	0.3	0.8	K2V main-sequence star	ε Eridani
4500	110	21	K2III red giant	Arcturus
4300	4000	140	K2Ib red supergiant	ε Pegasi

CONCEPT CHECK

✔ Suppose astronomers discover that, due to a calibration error, *all* distances measured by geometric parallax are 10 percent larger than currently thought. What effect would this finding have on the "standard" main sequence used in spectroscopic parallax?

17.7 Stellar Masses

LEARNING GOAL 7 What ultimately determines a star's position on the main sequence? The answer is its *mass* and its *composition*. Mass and composition are fundamental properties of any star. Together, they uniquely determine the star's internal structure, its external appearance, and even (as we will see in Chapter 20) its future evolution. The ability to measure these two key stellar properties is of the utmost importance if we are to understand how stars work. We have already seen how spectroscopy is used to determine composition. ∞ (Sec. 16.4) Now let's turn to the problem of finding a star's mass.

As with all other objects, we measure a star's mass by observing its gravitational influence on some nearby body—another star, perhaps, or a planet. If we know the distance between the two bodies, then we can use Newton's laws to calculate their masses. The extrasolar planetary systems that have recently been detected have not been studied well enough to provide independent stellar mass measurements, and we are a long way from placing our own spacecraft in orbit around other stars. ∞ (Sec. 15.5) Nevertheless, there are ways of determining stellar masses.

Binary Stars

Most stars are members of *multiple-star systems*—groups of two or more stars in orbit around one another. The majority of stars are found in **binary-star systems**, which consist of two stars in orbit about a common center of mass, held together by their mutual gravitational attraction. ∞ (Sec. 2.7) Other stars are members of triple, quadruple, or even more complex systems. The Sun is not part of a multiple-star system—if it has anything at all uncommon about it, it may be its lack of stellar companions.

Astronomers classify binary-star systems (or simply *binaries*) according to their appearance from Earth and the ease with which they can be observed. **Visual binaries** have widely separated members that are bright enough to be observed and monitored separately, as shown in Figure 17.19. The more common **spectroscopic binaries** are too distant to be resolved into separate stars, but they can be indirectly perceived by monitoring the back-and-forth Doppler shifts of their spectral lines as the stars orbit each

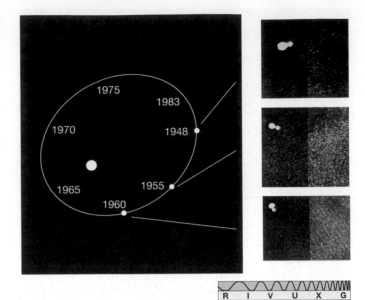

▲ **FIGURE 17.19 Visual Binary** The periods and separations of binary stars can be observed directly if each star is clearly seen. At the left is an orbital diagram for the double star Kruger 60; at the right are actual photographs taken in some of the years indicated. (*Harvard College Observatory*)

other. Recall that motion toward an observer shifts the lines toward the blue end of the electromagnetic spectrum and motion away from the observer shifts them toward the red end. ∞ (Sec. 3.5) In a *double-line* spectroscopic binary, two distinct sets of spectral lines—one for each star—shift back and forth as the stars move. Because we see particular lines alternately approaching and receding, we know that the objects emitting the lines are in orbit. In the more common *single-line* systems, such as that shown in Figure 17.20, one star is too faint for its spectrum to be distinguished, so only one set of lines is observed to shift back and forth. The shifting means that the star which is observed must be in orbit around another star, even though the companion cannot be observed directly. (If this idea sounds familiar, it should—all of the extrasolar planetary systems discovered to date are extreme examples of single-line spectroscopic binaries.) ∞ (Sec. 5.5)

In the much rarer **eclipsing binaries**, the orbital plane of the pair of stars is almost edge-on to our line of sight. In this situation, depicted in Figure 17.21, we observe a periodic decrease in starlight as one star passes in front of (transits) the other. By studying the variation in the light from the binary system—the binary's **light curve**—astronomers can derive detailed information not only about the stars' orbits and masses, but also about their radii. Thus, eclipsing binaries provide an alternative means of measuring stellar radii that is *independent* of either the direct or the indirect methods described in Section 17.4.

For example, in the sequence shown in Figure 17.21, the maximum brightness (frames 1, 3, and 5) represents

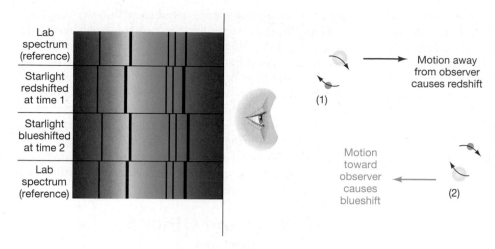

◀ **FIGURE 17.20 Spectroscopic Binary** Properties of binaries can be determined indirectly by measuring the periodic Doppler shift of one star relative to the other as they move in their orbits. The diagram shows a so-called single-line system, in which only one spectrum (from the brighter component) is visible. The observer is situated to the left of the diagram. In the upper middle frame, the visible component is moving away from the observer, so the spectrum is redshifted. In the lower middle frame, the star is moving toward the observer, so the spectrum is blueshifted. Reference lab spectra are shown at the top and bottom.

the combined brightnesses of the two stars, while the shallower minimum (frame 4) represents the brighter (larger) component only. These two pieces of information allow us to infer the individual brightnesses of the two stars. The deeper minima (frames 2 and 6) occur because the fainter red star partially blocks the light of the much brighter yellow star. The change in brightness tells us what fraction of the brighter star is obscured, and that in turn tells us the ratio of the *areas* of the two stars and hence (since area is proportional to radius squared) the ratio of their radii. If we also knew the components' orbital *speeds*—from Doppler measurements, say—then the widths of the minima and the time taken to go from minimum to maximum light would tell us the actual radii of the stars.

The preceding categories of binary-star systems are not mutually exclusive. For example, a single-line spectroscopic binary may also happen to be an eclipsing system. In that case, astronomers can use the eclipses to gain extra information about the fainter member of the pair. Occasionally, two unrelated stars just happen to lie close together in the sky, even though they are actually widely separated. These *optical doubles* are just chance superpositions and carry no useful information about stellar properties.

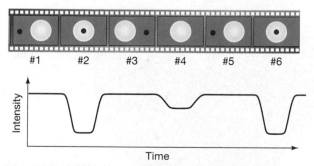

▲ **FIGURE 17.21 Eclipsing Binary** If the two stars in a binary-star system happen to eclipse one another, additional information on their radii and masses can be obtained by observing the periodic decrease in starlight as one star passes in front of the other.

Determination of Stellar Mass

By observing the actual orbits of the stars, the back-and-forth motion of the spectral lines, or the dips in the light curve—whatever information is available—astronomers can measure the binary's orbital period. Observed periods span a broad range—from hours to centuries. How much additional information can be extracted depends on the type of binary involved.

If the distance to a *visual* binary is known, the semimajor axis of its orbit can be determined directly, by simple geometry. Knowledge of the binary period and orbital semimajor axis is all we need to determine the combined mass of the component stars, using the modified form of Kepler's third law. ∞ (Sec. 2.7) Since the orbits of both stars can be separately tracked, it is also possible to determine each of the individual stars' masses. Recall from Section 2.7 that, in any system of orbiting objects, each object orbits the common center of mass. Measuring the distance from each star to the center of mass of a visual binary yields the ratio of the stellar masses. Knowing both the sum of the masses and their ratio, we can then find the mass of each star.

For *spectroscopic* binaries, it is not possible to determine the semimajor axis directly. Doppler-shift measurements give us information on the orbital velocities of the two stars, but only with regard to their radial components—that is, along the line of sight. As a result, we cannot determine the *inclination* of the orbit to our line of sight, and this imposes a limitation on how much information we can obtain—simply put, we cannot distinguish between a slow-moving binary seen edge-on and a fast-moving binary seen almost face-on (so that only a small component of the orbital motion is along the line of sight). We have already encountered this limitation in our study of extrasolar planets. ∞ (Sec. 15.5)

For a double-line spectroscopic system, individual radial velocities, and hence the ratio of the component masses, can be determined, but the uncertainty in the orbital inclination means that only lower limits on the individual masses can be obtained. For single-line systems, even less information is available, and only a fairly complicated relation between the component masses (known as the *mass function*) can be derived. However, if, as is often the case, the mass of the brighter star can be determined by other means (e.g., if the brighter star is recognized as a main-sequence star of a certain spectral class—see Figure 17.22), a lower limit can then be placed on the mass of the fainter, unseen star.

Finally, if a spectroscopic binary happens also to be an *eclipsing* system, then the uncertainty in the inclination is removed, as the binary is known to be edge-on or very nearly so. In that case, both masses can be determined for a double-line binary. For a single-line system, the mass function is simplified to the point where the mass of the unseen star is known if the mass of the brighter star can be found by other means (e.g., by recognizing it as a main-sequence star of known spectral type).

Despite all these qualifications and difficulties, the masses of individual component stars have been obtained for many nearby binary systems. Virtually all we know

about the masses of stars is based on such observations. As a simple example, consider the nearby visual binary-star system made up of the bright star Sirius A and its faint companion Sirius B. Their orbital period is observed to be 50 years, and their orbital semimajor axis is 20 A.U.—7.6″ at a distance of 2.6 pc. ∞ (*More Precisely 1-3*) The modified form of Kepler's third law then implies that the sum of their masses is $20^3/50^2 = 3.2$ times the mass of the Sun. ∞ (Sec. 2.7) Further study of the orbital motion shows that Sirius A has roughly twice the mass of its companion. It then follows that the masses of Sirius A and Sirius B are 2.1 and 1.1 solar masses, respectively.

17.8 Mass and Other Stellar Properties

We end our introduction to the stars with a brief look at how mass is correlated with the other stellar properties discussed in this chapter. Table 17.5 summarizes the various observational and theoretical techniques used to measure the stars, listing the quantities that are assumed known (usually as a result of the application of other techniques listed in the table), those which are measured, and the theory that is applied to turn the observations into the desired result.

Figure 17.22 is a schematic H–R diagram showing how stellar mass varies along the main sequence. There is a clear progression from low-mass red dwarfs to high-mass blue giants. With few exceptions, main-sequence stars range in mass from about 0.1 to 20 times the mass of the

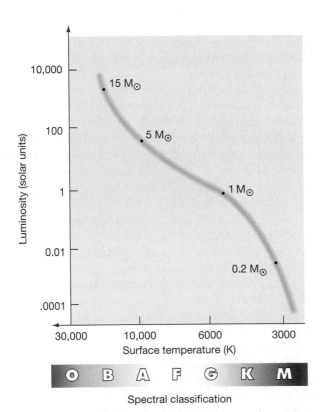

▲ **FIGURE 17.22 Stellar Masses** More than any other stellar property, mass determines a star's position on the main sequence. Low-mass stars are cool and faint, so they lie at the bottom of the main sequence. Very massive stars are hot and bright, so they lie at the top of the main sequence. (The symbol "M⊙" means "solar mass.")

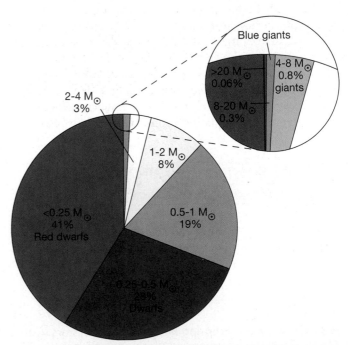

▲ **FIGURE 17.23 Stellar Mass Distribution** The range of masses of main-sequence stars, as determined from careful measurement of stars in the solar neighborhood.

Sun. The hot O- and B-type stars are generally about 10 to 20 times more massive than our Sun. The coolest K- and M-type stars contain only a few tenths of a solar mass. The mass of a star at the time of its formation determines the star's location on the main sequence. Based on observations of stars within a few hundred light-years of the Sun, Figure 17.23 illustrates how the masses of main-sequence stars are distributed. Notice the huge fraction of low-mass stars, as well as the tiny fraction contributed by stars of more than a few solar masses.

Figure 17.24 illustrates how a main-sequence star's radius and luminosity depend on its mass. The two plots shown, of the *mass–radius* and *mass–luminosity* relations, are based on observations of binary-star systems. Along the

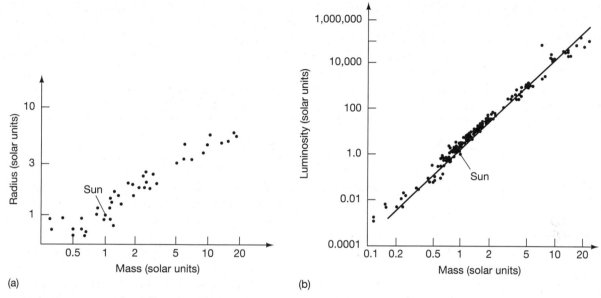

(a)   (b)

▲ **FIGURE 17.24 Stellar Radii and Luminosities** (a) Dependence of stellar radius on mass for main-sequence stars; actual measurements are plotted here. The radius increases roughly in proportion to the mass over much of the range. (b) Dependence of luminosity on mass. The luminosity increases roughly as the fourth power of the mass (indicated by the straight line).

TABLE 17.5 Measuring the Stars

Stellar Property	Measurement Technique	"Known" Quantity	Measured Quantity	Theory Applied	Section
Distance	stellar parallax spectroscopic parallax	astronomical unit main sequence	parallactic angle spectral type apparent magnitude	elementary geometry inverse-square law	17.1 17.6
Radial velocity		speed of light atomic spectra	spectral lines	Doppler effect	17.1
Transverse velocity	astrometry	distance	proper motion	elementary geometry	17.1
Luminosity		distance main sequence	apparent magnitude spectral type	inverse square law	17.2 17.6
Temperature	photometry spectroscopy		color spectral type	blackbody law atomic physics	17.3 17.3
Radius	direct indirect	distance	angular size luminosity temperature	elementary geometry radius–luminosity– temperature relationship	17.4 17.4
Composition	spectroscopy		spectrum	atomic physics	17.3
Mass	observations of binary stars	(distance)	binary period binary orbit orbital velocity	Newtonian gravity and dynamics	17.7

TABLE 17.6 Key Properties of Some Well-Known Main-Sequence Stars

Star	Spectral Type	Mass, *M* (Solar Masses)	Central Temperature (10^6 K)	Luminosity, *L* (Solar Luminosities)	Estimated Lifetime (*M/L*) (10^6 years)
Spica B*	B2V	6.8	25	800	90
Vega	A0V	2.6	21	50	500
Sirius	A1V	2.1	20	22	1000
Alpha Centauri	G2V	1.1	17	1.6	7000
Sun	G2V	1.0	15	1.0	10,000
Proxima Centauri	M5V	0.1	0.6	0.00006	16,000,000

The "star" Spica is, in fact, a binary system comprising a B1III giant primary (Spica A) and a B2V main-sequence secondary (Spica B).

main sequence, both radius and luminosity increase with mass. As an approximate rule of thumb, we can say that radius increases proportionally to stellar mass, while luminosity increases much faster—almost as the *fourth power* of the mass (as indicated by the line in Figure 17.24b). Thus, a 2-solar-mass main-sequence star has a radius about twice that of the Sun and a luminosity of 16 (2^4) solar luminosities; a 0.2-solar-mass main-sequence star has a radius of roughly 0.2 solar radii and a luminosity of around 0.0016 (0.2^4) solar luminosity.

Table 17.6 compares some key properties of several well-known main-sequence stars, arranged in order of decreasing mass. Notice that the central temperature (obtained from mathematical models similar to those discussed in Chapter 16) differs relatively little from one star to another, compared with the large spread in stellar luminosities. ∞ (Sec. 16.3) The rapid rate of nuclear burning deep inside a star releases vast amounts of energy per unit time. How long can the fire continue to burn? We can estimate a main-sequence star's *lifetime* simply by dividing the amount of fuel available (the mass of the star) by the rate at which the fuel is being consumed (the star's luminosity):

$$\text{stellar lifetime} \propto \frac{\text{stellar mass}}{\text{stellar luminosity}}.$$

The mass–luminosity relation tells us that a star's luminosity is roughly proportional to the fourth power of its mass, so we can rewrite this expression to obtain, approximately,

$$\text{stellar lifetime} \propto \frac{1}{(\text{stellar mass})^3}.$$

The final column in Table 17.6 lists estimated lifetimes, based on the above proportionality and noting that the lifetime of the Sun (see Chapter 20) is about 10 billion years.

For example, the lifetime of a 10-solar-mass main-sequence O-type star is roughly $10/10^4 = 1/1000$ of the lifetime of the Sun, or about 10 million years. The nuclear reactions in such a massive star proceed so rapidly that its fuel is quickly depleted, despite its large mass. We can be sure that all the O- and B-type stars we now observe are quite young—less than a few tens of millions of years old. Massive stars older than that have already exhausted their fuel and no longer emit large amounts of energy. They have, in effect, died.

At the opposite end of the main sequence, the cooler K- and M-type stars have less mass than our Sun has. With their low core densities and temperatures, their proton–proton reactions churn away rather sluggishly, much more slowly than those in the Sun's core. The small energy release per unit time leads to low luminosities for these stars, so they have very long lifetimes. Many of the K- and M-type stars we now see in the night sky will shine on for at least another trillion years. The evolution of stars—large and small—is the subject of Chapters 20 and 21.

CONCEPT CHECK

✔ How do we know the masses of stars that aren't components of binaries?

Chapter Review

SUMMARY

The distances to the nearest stars can be measured by trigonometric parallax. A star with a parallax of 1 arc second (1″) is 1 **parsec (p. 439)**—about 3.3 light-years—away from Earth. Stars have real motion through space as well as apparent motion as Earth orbits the Sun. A star's **proper motion (p. 439)**—its true motion across the sky—is a measure of the star's velocity perpendicular to our line of sight. The star's radial velocity—along the line of sight—is measured by the Doppler shift of the spectral lines emitted by the star.

The **apparent brightness (p. 442)** of a star is the rate at which energy from the star reaches a unit area of a detector. Apparent brightness falls off as the inverse square of the distance. Optical astronomers use the **magnitude scale (p. 443)** to express and compare stellar brightnesses. The greater the magnitude, the fainter is the star; a difference of five magnitudes corresponds to a factor of 100 in brightness. **Apparent magnitude (p. 443)** is a measure of apparent brightness. The **absolute magnitude (p. 444)** of a star is the apparent magnitude it would have if placed at a standard distance of 10 pc from the viewer. Absolute magnitude is a measure of the star's luminosity.

Astronomers often measure the temperatures of stars by measuring their brightnesses through two or more optical filters and then fitting a blackbody curve to the results. The measurement of the amount of starlight received through each member of a set of filters is called photometry. Astronomers classify stars according to the absorption lines in their spectra. The lines seen in the spectrum of a given star depend mainly on the temperature of the star, and spectroscopic observations of stars provide an accurate means of determining both stellar temperatures and stellar composition. The standard stellar **spectral classes (p. 449)**, in order of decreasing temperature, are O, B, A, F, G, K, and M.

Only a few stars are large enough and close enough that their radii can be measured directly. The sizes of most stars are estimated indirectly through the **radius–luminosity–temperature relationship (p. 450)**. Stars comparable in size to, or smaller than, the Sun are categorized as **dwarfs (p. 451)**, stars up to 100 times larger than the Sun are called **giants (p. 451)**, and stars more than 100 times larger than the Sun are known as **supergiants (p. 451)**. In addition to "normal" stars such as the Sun, two other important classes of star are **red giants (p. 451)**

(and **red supergiants**) **(p. 451)**, which are large, cool, and luminous, and **white dwarfs (p. 451)**, which are small, hot, and faint.

A plot of stellar luminosities versus stellar spectral classes (or temperatures) is called an **H–R diagram (p. 452)**, or a **color–magnitude diagram (p. 452)**. About 90 percent of all stars plotted on an H–R diagram lie on the **main sequence (p. 452)**, which stretches from hot, bright **blue supergiants (p. 453)** and **blue giants (p. 453)**, through intermediate stars such as the Sun, to cool, faint **red dwarfs (p. 453)**. Most main-sequence stars are red dwarfs; blue giants are quite rare. About 9 percent of stars are in the **white dwarf region (p. 455)**, and the remaining 1 percent are in the **red-giant region (p. 455)**.

By careful spectroscopic observations, astronomers can determine a star's **luminosity class (p. 457)**, allowing them to distinguish main-sequence stars from red giants or white dwarfs of the same spectral type (or color). Once a star is known to be on the main sequence, measurement of its spectral type allows its luminosity to be estimated and its distance to be measured. This method of determining distance, which is valid for stars up to several thousand parsecs from Earth, is called **spectroscopic parallax (p. 455)**.

Most stars are not isolated in space, but instead orbit other stars in **binary-star systems (p. 458)**. In a **visual binary (p. 458)**, both stars can be seen and their orbit charted. In a **spectroscopic binary (p. 458)**, the stars cannot be resolved, but their orbital motion can be detected spectroscopically. In an **eclipsing binary (p. 458)**, the orbit is oriented in such a way that one star periodically passes in front of the other as seen from Earth and dims the light we receive. The binary's **light curve (p. 458)** is a plot of the star's apparent brightness as a function of time.

Studies of binary-star systems often allow stellar masses to be measured. The mass of a star determines the star's size, temperature, and brightness. Fairly well defined mass–radius and mass–luminosity relations exist for main-sequence stars. Hot blue giants are much more massive than the Sun; cool red dwarfs are much less massive. The lifetime of a star can be estimated by dividing its mass by its luminosity. High-mass stars burn their fuel rapidly and have much shorter lifetimes than the Sun. Low-mass stars consume their fuel slowly and may remain on the main sequence for trillions of years.

REVIEW AND DISCUSSION

1. How is parallax used to measure the distances to stars?

2. What is a parsec? Compare it with the astronomical unit.

3. Explain two ways in which a star's real motion through space translates into motion that is observable from Earth.

4. How do astronomers go about measuring stellar luminosities?

5. Describe how astronomers measure stellar radii.

6. Describe some characteristics of red-giant and white-dwarf stars.

7. What is the difference between absolute and apparent brightness?

8. How do astronomers measure stellar temperatures?

9. Briefly describe how stars are classified according to their spectral characteristics.

10. Why do some stars have very few hydrogen lines in their spectra?

11. What information is needed to plot a star on the H–R diagram?

18

Interstellar space is the place both where stars are "born" and to which they return at "death." Rich in gas and dust, yet extraordinarily thinly distributed throughout the vast, dark regions among the stars, interstellar matter occasionally glows as nebulae or contracts to form new stars (and sometimes planets). Here, the spectacular nebula NGC604 houses more than 200 brilliant, young, blue stars spread across more than 1000 light-years. Those stars excite the surrounding gas, which then re-emits the characteristic red color of the hydrogen atoms that are ▶ most abundant in interstellar space. (STScI)

The Interstellar Medium

Gas and Dust among the Stars

Stars and planets are not the only inhabitants of our Galaxy. The space around us harbors invisible matter throughout the dark voids between the stars. The density of this matter is extremely low—approximately a trillion trillion times less dense than matter in either stars or planets, far more tenuous than the best vacuum attainable on Earth. Only because the volume of interstellar space is so vast does its mass amount to anything at all. So why bother to study this near-perfect vacuum? We do so for three important reasons. First, there is nearly as much mass in the "voids" among the stars as there is in the stars themselves. Second, interstellar space is the region out of which new stars are born. Third, interstellar space is also the region into which old stars expel their matter when they die. It is one of the most significant crossroads through which matter passes anywhere in our universe.

LEARNING GOALS

Studying this chapter will enable you to

1 Summarize the composition and physical properties of the interstellar medium.

2 Describe the characteristics of emission nebulae, and explain their significance in the life cycle of stars.

3 Discuss the properties of dark interstellar clouds.

4 Specify the radio techniques used to probe the nature of interstellar matter.

5 Discuss the nature and significance of interstellar molecules.

Visit astro.prenhall.com/chaisson for additional annotated images, animations, and links to related sites for this chapter.

18.1 Interstellar Matter

LEARNING GOAL 1 Figure 18.1 shows a large region of space, a much greater expanse of universal "real estate" than anything we have studied thus far. This optical image displays a 30-degree patch of sky spanning the densest part of the Milky Way Galaxy. The bright regions are congregations of innumerable stars, merging together into a continuous blur at the resolution of the telescope. However, the dark areas are not simply "holes" in the stellar distribution. They are regions of space where *interstellar matter* obscures (blocks) the light from stars beyond. Their very darkness means that they cannot easily be studied by the optical methods used to examine stellar matter. There is, quite simply, nothing to see!

From Figure 18.1 (see also Figure 18.5), it is evident that interstellar matter is distributed, very unevenly throughout space. In some directions, the dark obscuring matter is largely absent, allowing astronomers to study objects literally billions of parsecs from the Sun. In other directions, there are small amounts of interstellar matter, so the obscuration is moderate, preventing us from seeing objects more than a few thousand parsecs away, but still allowing us to study nearby stars. Still other regions are so heavily obscured that starlight from even relatively nearby stars is completely absorbed before reaching Earth.

Gas and Dust

The matter among the stars is collectively termed the **interstellar medium**. It is made up of two components—*gas* and *dust*—intermixed throughout all space. The gas is made up mainly of individual atoms, of average size 10^{-10} m (0.1 nm) or so, and small molecules, no larger than about 10^{-9} m across. Interstellar dust is more complex, consisting of clumps of atoms and molecules—not unlike chalk dust or the microscopic particles that make up smoke, soot, or fog.

Apart from numerous narrow atomic and molecular absorption lines, the gas alone does not block radiation to any great extent. The obscuration that is evident in Figure 18.1 is caused by the dust. Light from distant stars cannot penetrate the densest accumulations of interstellar dust any more than a car's headlights can illuminate the road ahead in a thick fog. As a rule of thumb, a beam of light can be absorbed or scattered only by particles having diameters comparable to or larger than the wavelength of the radiation involved. The amount of obscuration (that is, absorption or scattering) produced by particles of a given size increases with decreasing wavelength. The size of a typical interstellar dust particle—or **dust grain**—is about 10^{-7} m (0.1 μm), comparable in size to the wavelength of visible light. Consequently, dusty regions of interstellar space are transparent to long-wavelength radio and infrared radiation, but opaque to shorter wavelength optical and ultraviolet radiation. The overall dimming of starlight by interstellar matter is called **extinction**.

Because the interstellar medium is more opaque to short-wavelength radiation than to radiation of longer wavelengths, light from distant stars is preferentially robbed of its higher frequency ("blue") components. Hence, in addition to being generally diminished in brightness, stars also tend to appear redder than they really are. This effect, known as **reddening**, is conceptually

◀ **FIGURE 18.1** Milky Way Photo A wide-angle photograph of a great swath of space in the direction of the center of our Galaxy, showing regions of brightness (vast fields of stars) as well as regions of darkness (where interstellar matter obscures the light from more distant stars). The field of view is roughly 30° across. A few of the nebulae discussed later in the chapter are labeled. *(Palomar/Caltech)*

R I V U X G

similar to the process that produces spectacular red sunsets here on Earth. ∞ *(More Precisely 7-1)*

Reddening can be seen very clearly in Figure 18.2(a), which shows a type of compact, dusty interstellar cloud called a *globule*. (We will discuss such dark interstellar clouds in more detail in Section 18.3.) The center of this cloud, called Barnard 68, is opaque to all optical wavelengths, so starlight cannot pass through it. However, near the edges, where there is less intervening cloud matter, some light does make it through. Notice how stars seen through the cloud are both dimmed and reddened relative to those seen directly.

As illustrated in Figure 18.2(b), extinction and reddening change a star's apparent brightness and color. However, the patterns of absorption lines in the original stellar spectrum are still recognizable in the radiation reaching Earth, so the star's spectral class can be determined. Astronomers can use this fact to study the interstellar medium. From a main-sequence star's spectral and luminosity classes, astronomers learn the star's true luminosity and color. ∞ (Secs. 17.5, 17.6) They then measure the degree to which the starlight has been affected by extinction and reddening en route to Earth, and this, in turn, allows them to estimate both the numbers and the sizes of interstellar dust particles along the line of sight to the star. By repeating these measurements for stars in many different directions and at many different distances from Earth, astronomers have built up a picture of the distribution and overall properties of the interstellar medium in the solar neighborhood.

Overall Density

Gas and dust are found everywhere in interstellar space— no part of our Galaxy is truly devoid of matter. However, the density of the interstellar medium is extremely low.

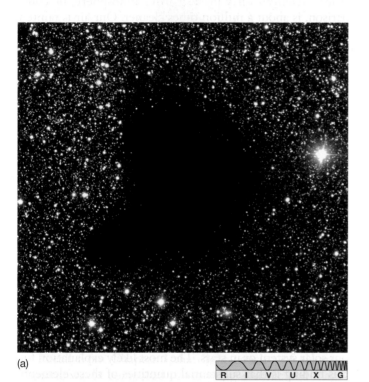

(a)

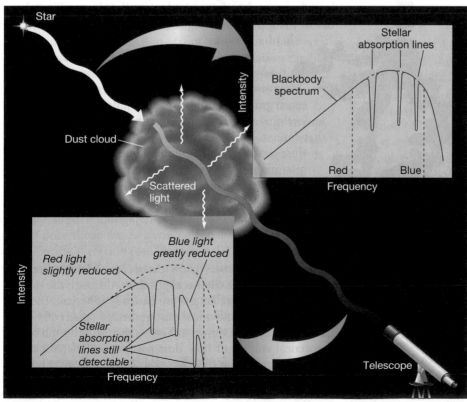

(b)

◀ **FIGURE 18.2 Reddening**
(a) This dusty interstellar cloud, called Barnard 68, is opaque to visible light, except near the edges, where some light from background stars can be seen. Because blue light is more easily scattered or absorbed by dust than is red light, stars seen through the cloud appear red. The cloud spans about 0.2 pc and lies about 160 pc away. (b) Starlight passing through a dusty region of space is both dimmed and reddened, but spectral lines are still recognizable in the light that reaches Earth. By recognizing stellar spectral features and inferring a star's intrinsic properties, astronomers can estimate the amount of obscuring dust along the line of sight. Note that this reddening has nothing to do with the Doppler effect—the frequencies of the lines are unchanged, although their intensities are reduced. *(ESO)*

▲ FIGURE 18.5 Milky Way Mosaic The Milky Way Galaxy, photographed almost from horizon to horizon, thus extending over nearly 180°. This band contains high concentrations of stars, as well as interstellar gas and dust. The field of view is several times wider than that of Figure 18.1, whose outline is superimposed on this image. *(Axel Mellinger)*

Today they are known as **emission nebulae**—glowing clouds of hot interstellar matter.

Observations of Emission Nebulae

Historically, astronomers have used the term **nebula** to refer to any "fuzzy" patch (bright or dark) on the sky—any region of space that was clearly distinguishable through a telescope, but not sharply defined, unlike a star or a planet. We now know that many (although not all) nebulae are clouds of interstellar dust and gas.

If a cloud happens to obscure stars lying behind it, we see it as a dark patch on a bright background, as in Figures 18.1, 18.2(a), and 18.5—a dark nebula. But if something within the cloud—a group of hot young stars, for example—causes it to glow, then we see a bright emission nebula instead. The method of spectroscopic parallax applied to stars that are visible within the emission nebulae shown in Figure 18.6 indicates that their distances from Earth range from 900 pc (M20) to 1800 pc (M16). ∞ (Sec. 17.6) Thus, all four nebulae are near the limit of visibility for any object embedded in the dusty Galactic plane. M16, at the top left, is approximately 1000 pc from M20, near the bottom.

We can gain a better appreciation of these nebulae by examining progressively smaller fields of view. Figure 18.7 is an enlargement of the region near the bottom of Figure 18.6, showing M20 at the top and M8 at the bottom, only a few degrees away. Figure 18.8 is an enlargement of the top of Figure 18.7, presenting a close-up of M20 and its immediate environment. The total area of the close-up view displayed measures some 10 pc across. Emission nebulae are among the most spectacular objects in the entire universe, yet they appear only as small, undistinguished patches of light when viewed in the larger context of the Milky Way, as in Figure 18.5. Perspective is crucial in astronomy.

Figure 18.9 shows enlargements of two of the nebulae visible in Figure 18.6. Notice the predominant red coloration of the emitted radiation and the hot, bright stars

embedded within the glowing nebular gas. The emission nebulae shown in Figures 18.7–18.9 are regions of glowing, ionized gas. At or near the center of each is at least one newly formed hot O- or B-type star producing huge amounts of ultraviolet light. As ultraviolet photons travel outward from the star, they ionize the surrounding gas. As electrons recombine with nuclei, they emit visible radiation, causing the gas to fluoresce, or glow. ∞ (Sec. 4.2)

The reddish hue of these nebulae—and, in fact, of all emission nebulae—results when hydrogen atoms emit light in the red part of the visible spectrum. Specifically, it is caused by the emission of radiation at 656.3 nm—the Hα line discussed in Chapter 4. ∞ (*More Precisely 4-1*) Other elements in the nebula also emit radiation as their electrons recombine, but because hydrogen is so plentiful, its emission usually dominates.

Woven through the glowing nebular gas, and plainly visible in Figures 18.7–18.9, are lanes of dark, obscuring dust. Recent studies have demonstrated that these **dust lanes** are part of the nebulae and are not just unrelated dust clouds that happen to lie along our line of sight. This relationship is particularly evident in Figures 18.9(b) and (d), where regions of gas and dust are simultaneously silhouetted against background nebular emission and illuminated by foreground nebular stars. Figure 18.10 sketches some of the key features of emission nebulae, illustrating the connection between the central stars, the nebula itself, and the surrounding interstellar medium.

The interaction between stars and gas is particularly striking in Figure 18.9(b). The three dark "pillars" visible in this spectacular *Hubble Space Telescope* image are part of the interstellar cloud from which the stars formed. The rest of the cloud in the vicinity of the new stars has already been heated and dispersed by their radiation, in a process known as *photoevaporation*. The fuzz around the edges of the pillars, especially at the top right and center, is the result of this ongoing process. As photoevaporation continues, it eats away the less dense material first, leaving

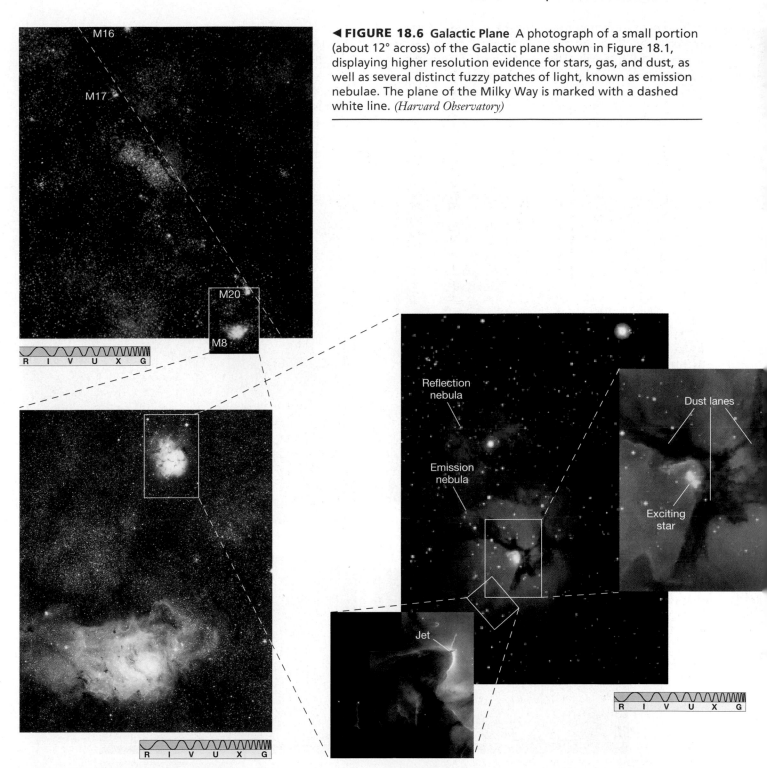

◀ **FIGURE 18.6 Galactic Plane** A photograph of a small portion (about 12° across) of the Galactic plane shown in Figure 18.1, displaying higher resolution evidence for stars, gas, and dust, as well as several distinct fuzzy patches of light, known as emission nebulae. The plane of the Milky Way is marked with a dashed white line. (*Harvard Observatory*)

▲ **FIGURE 18.7 M20–M8 Region** An enlargement of the bottom of Figure 18.6, showing M20 (top) and M8 (bottom) more clearly. The two nebulae are only a few degrees apart in the sky. (*Royal Observatory, Edinburgh*)

▲ **FIGURE 18.8 Trifid Nebula** Enlargements of the top of Figure 18.7, showing only M20 and its interstellar environment. The nebula itself (in red) is about 4 pc in diameter. It is often called the Trifid Nebula because of the dust lanes that trisect its midsection. (See inset at right.) The blue region is unrelated to the red emission nebula and is caused by starlight reflected from intervening dust particles. It is called a *reflection nebula*. The inset at the left shows a 0.5-pc-long jet, probably extending from a young star embedded in a pillar of gas and dust. (*AURA; D. Malin; NASA*)

behind delicate sculptures composed of the denser parts of the original cloud, just as wind and water create spectacular structures in Earth's deserts and shores by eroding away the softest rock. The process is a dynamic one: The pillars will eventually be destroyed, but probably not for another hundred thousand or so years.

The bluish region visible in Figures 18.7 and 18.8 immediately above M20 is another type of nebula unrelated to the red emission nebula itself. Called a *reflection nebula*, it is caused by starlight scattered from dust particles in interstellar clouds located just off the line of sight between Earth and the bright stars within M20. Reflection nebulae appear blue for much the same reason that Earth's daytime sky is blue: As indicated in Figure 18.10, short-wavelength blue light is more easily scattered by interstellar matter back toward Earth and into our detectors. ∞ (*More Precisely 7-1*)

Spectroscopists often refer to the *ionization state* of an atom by attaching a roman numeral to the chemical sym-

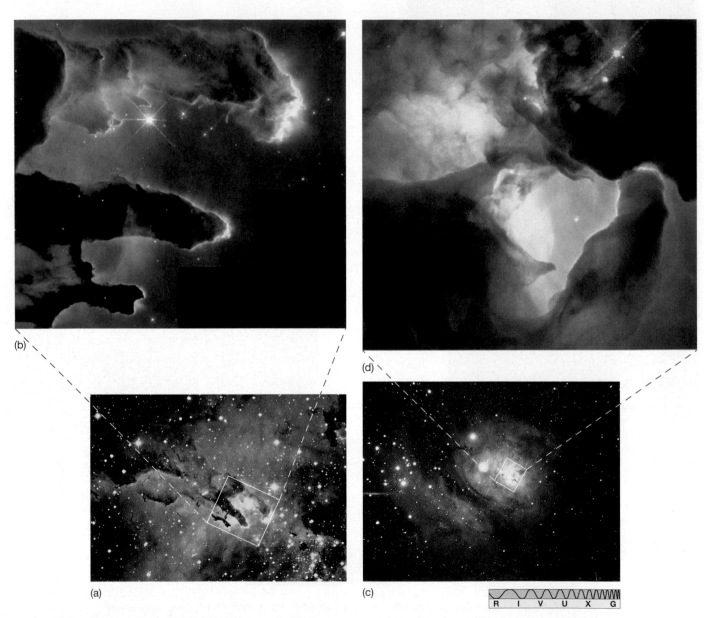

R I V U X G

▲ **FIGURE 18.9 Emission Nebulae** Enlargements of selected portions of Figure 18.6. (a) M16, the Eagle Nebula. (b) A *Hubble* image of huge pillars of cold gas and dust inside M16 shows delicate sculptures created by the action of stellar ultraviolet radiation on the original cloud. (c) M8, the Lagoon Nebula. (d) A high-resolution view of the core of M8, a region known as the Hourglass. Notice the irregular shape of the emitting regions, the characteristic red color of the light in the lower frames, the bright stars within the gas, and the patches of obscuring dust. The upper frames are not shown in true color. The various colors accentuate observations at different wavelengths: Green represents emission from hydrogen atoms, red emission from singly ionized sulfur, and blue emission from doubly ionized oxygen. (*AURA; NASA*)

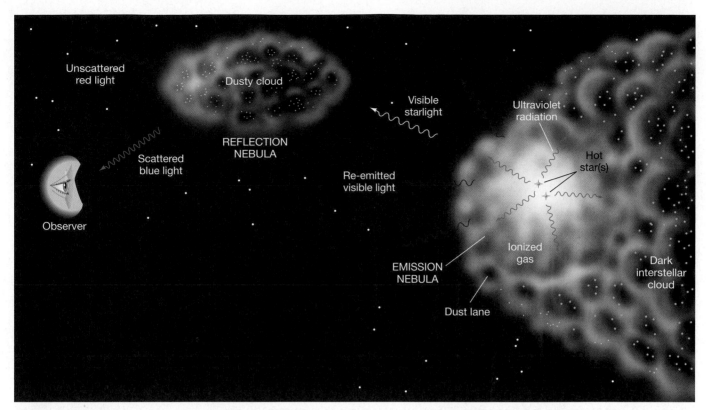

▲ FIGURE 18.10 Nebular Structure An emission nebula results when ultraviolet radiation from one or more hot stars ionizes part of an interstellar cloud. The nebula's reddish color is produced as electrons and protons recombine to form hydrogen atoms. Dust lanes may be seen if part of the parent cloud happens to obscure the emitting region. If some starlight chances to encounter another dusty cloud (or perhaps another part of the cloud harboring the emission nebula), some of the radiation, particularly at the shorter wavelength blue end of the spectrum, may be scattered back toward Earth, forming a reflection nebula.

bol for the atom—I for the neutral (that is, not ionized) atom, II for a singly ionized atom (an atom missing one electron), III for a doubly ionized atom (one missing two electrons), and so on. Because emission nebulae are composed mainly of ionized hydrogen, they are often referred to as *HII regions*. Regions of space containing primarily neutral (atomic) hydrogen are known as *HI regions*.

Nebular Spectra

Most of the photons emitted by the recombination of electrons with atomic nuclei escape from the emission nebulae. Unlike the ultraviolet photons originally emitted by the embedded stars, these reemitted photons do not have enough energy to ionize the nebular gas, so they pass through the nebula relatively unhindered. Some eventually reach Earth. By studying these lower energy photons, we can learn much about the detailed properties of emission nebulae.

Because at least one hot star resides near the center of every emission nebula, we might think that the combined spectrum of the star and the nebula would be hopelessly confused. In fact, they are not: We can easily distinguish nebular spectra from stellar spectra because the physical conditions in stars and emission nebulae differ so greatly. In particular, emission nebulae are made of hot, thin gas that, as we saw in Chapter 4, yields detectable *emission* lines. ∞ (Sec. 4.2) When our spectroscope is trained on a star, we see a familiar stellar spectrum, consisting of a blackbody-like continuous spectrum and absorption lines, together with superimposed emission lines from the nebular gas. When no star appears in the field of view, only the emission lines are seen.

Figure 18.11 shows a typical emission nebula, along with its spectrum, spanning part of the visible and near-ultraviolet range. Numerous emission lines can be seen, and information on the nebula can be extracted from all of them. Analyses of many nebular spectra show compositions close to those derived from observations of the Sun and other stars, and elsewhere in the interstellar medium: Hydrogen is about 90 percent abundant by number, followed by helium at about 9 percent; the heavier elements together make up the remaining 1 percent.

Unlike stars, nebulae are large enough for their actual sizes to be measurable by simple geometry. Coupling this

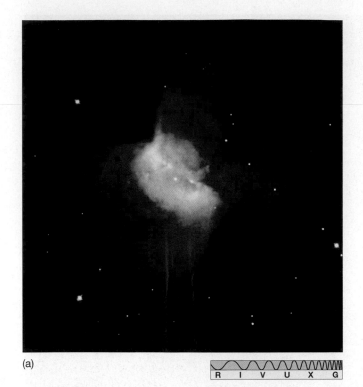

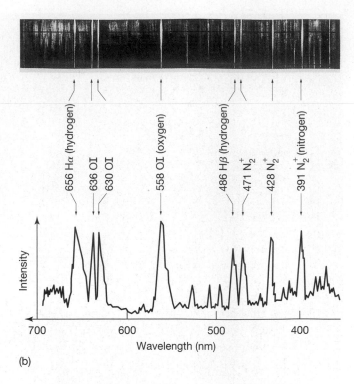

(a)

(b)

▲ **FIGURE 18.11 Emission Nebula Spectrum** (a) Nebula NGC 2346 is a glowing patch of gas about 0.2 pc across and residing some 700 pc away from our solar system. (b) The emission spectrum of NGC2346, showing light intensity over the entire visible portion of the electromagnetic spectrum, from red to deep violet. Some of the spectral lines, such as those of nitrogen, are from contaminants in Earth's atmosphere. (*NASA; Harvard Observatory*)

information on size with estimates of the amount of matter along our line of sight (as revealed by the nebula's total emission of light), we can find the nebula's density. Generally, emission nebulae have only a few hundred particles, mostly protons and electrons, in each cubic centimeter—a density some 10^{22} times lower than that of a typical planet. Spectral-line widths imply that the gas atoms and ions have temperatures around 8000 K. ∞ (Sec. 4.4) Table 18.1 lists some vital statistics for each of the nebulae shown in Figure 18.6.

"Forbidden" Lines

When astronomers first studied the spectra of emission nebulae, they found many lines that did not correspond to anything observed in terrestrial laboratories. For example, in addition to the dominant red coloration just discussed, many nebulae emit light with a characteristic green color. (See Figure 18.12). The greenish tint of portions of these nebulae puzzled astronomers in the early 20th century and defied explanation in terms of the properties of spectral lines known at the time, prompting speculation that the nebulae contained elements that were unknown on Earth. Some scientists even went so far as to invent the term "nebulium" for a supposed new element, much as the name helium came about when that element was first discovered in the Sun (recall also the fictitious element "coronium" from Chapter 16). ∞ (Sec. 16.3)

TABLE 18.1	Some Nebular Properties				
Object	Approximate Distance (pc)	Average Diameter (pc)	Density (10^6 particles/m³)	Mass (solar masses)	Temperature (K)
M8	1200	14	80	2600	7500
M16	1800	8	90	600	8000
M17	1500	7	120	500	8700
M20	900	4	100	150	8200

Later, with a fuller understanding of the workings of the atom, astronomers realized that these lines did in fact result from electron transitions within the atoms of familiar elements, but under unfamiliar conditions that were not reproducible in laboratories. Astronomers now understand that the greenish tint in Figure 18.12(b) and (c) is caused by a particular electron transition in doubly ionized oxygen. However, the structure of oxygen is such that an ion in the higher energy state for this transition tends to remain there for a very long time—many hours, in fact—before dropping back to the lower state and emitting a photon. Only if the ion is left undisturbed during this time, and not kicked into another energy state by a random interaction with another atom or molecule in the gas, will the transition actually occur and the photon be emitted.

In a terrestrial experiment, no atom or ion is left undisturbed for long. Even in a "low-density" laboratory gas, there are many trillions of particles per cubic meter, and each particle undergoes millions of collisions with other gas particles every second. The result is that an ion in the particular energy state that produces the peculiar green line in the nebular spectrum never has time to emit its photon in the lab—collisions kick it into some other state long before that occurs. For this reason, the line is usually called *forbidden*, even though it violates no law of physics; it simply occurs on Earth with such low probability that it is never seen.

In a typical emission nebula, the density is so low that collisions between particles are extremely rare. There is plenty of time for the excited ion to emit its photon, so the forbidden line is produced. Numerous forbidden lines are known in nebular spectra. These lines remind us once again that the environment in the interstellar medium is very different from conditions on Earth and warn us of the potential difficulties involved in extending our terrestrial experience from our laboratories to the study of interstellar space.

CONCEPT CHECK

✔ If emission nebulae are powered by ultraviolet radiation from very hot (blue-white) stars, why do they appear red?

1 pc

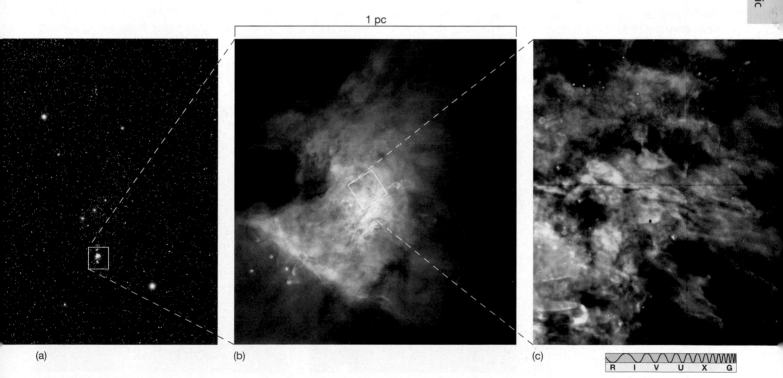

(a) (b) (c)

R I V U X G

▲ **FIGURE 18.12 Orion Nebula** (a) Lying some 450 pc from Earth, the Orion Nebula (M42) is visible to the naked eye as the fuzzy middle "star" of Orion's sword. (b) Like all emission nebulae, the Orion Nebula consists of hot, glowing gas powered by a group of bright stars in the center. In addition to exhibiting red Hα emission, parts of the nebula show a slight greenish tint, caused by a so-called forbidden transition in ionized oxygen. (c) A high-resolution, approximately true-color image shows rich detail in a region about 0.5 light-year across. Structural details are visible down to a level of 0.1", or six light-*hours*—a scale comparable to the dimensions of our solar system. (*NASA; ESO; NASA*)

18.3 Dark Dust Clouds

3 Emission nebulae are only one small component of the interstellar medium. Most of space—in fact, more than 99 percent of it—is devoid of such regions and contains no stars. It is simply dark. Look again at Figure 18.5, or just ponder the evening sky. The dark regions are by far the most representative of interstellar space.

The average temperature of a typical dark region of interstellar matter is about 100 K. Compare this with 273 K, at which water freezes, and 0 K, at which atomic and molecular motions cease. ∞ *(More Precisely 3-1)* As discussed further in *Discovery 18-1*, ultraviolet observations reveal that these cold regions are actually interspersed with hot "bubbles" of extremely low-density gas, apparently associated with the formation and explosive deaths of stars much more massive than the Sun (see Chapter 21). Somewhat like the Sun's faint corona, the dark regions are dark despite their high temperatures because the density of matter there is so low. ∞ (Sec. 16.4) However, apart from the superheated bubbles and the much smaller emission nebulae surrounding individual groups of stars, most of interstellar space is very cold.

Within these vast, dark interstellar voids lurks another type of astronomical object: the **dark dust cloud**. Dark dust clouds are even colder than their surroundings (with temperatures as low as a few tens of kelvins) and thousands or even millions of times denser. Along any given line of sight, their densities can range from 10^7 atoms/m^3 to more than 10^{12} atoms/m^3 (10^6 atoms/cm^3). Dark dust clouds are often called *dense* interstellar clouds by researchers, but we must recognize that even these densest interstellar regions are barely denser than the best vacuum achievable in terrestrial laboratories. Still, it is because their density is much larger than the average value of 10^6 atoms/m^3 in interstellar space that we can distinguish these clouds from the surrounding expanse of the interstellar medium.

Obscuration of Visible Light

Interstellar clouds bear little resemblance to terrestrial clouds. Most are much bigger than our solar system, and some are many parsecs across. (Yet even so, they make up no more than a few percent of the entire volume of interstellar space.) Despite their name, these clouds are made up primarily of gas, just like the rest of the interstellar medium. However, their ab- sorption of starlight is due almost entirely to the dust they contain.

Figure 18.13(a) shows a region called L977, in the constellation Cygnus. L977 is a classic example of a dark dust cloud. The dense globule Barnard 68, shown in Figure 18.2(a), is another. Some early (18th-century) observers thought that these dark patches on the sky were simply empty regions of space that happened to contain no bright stars. However, by the late 19th century, astronomers had discounted this idea. They realized that seeing clear spaces among the stars would be like seeing clear tunnels between the trees in a huge forest, and it was extremely unlikely that as many tunnels would lead away from Earth as would be required to explain the observed dark regions.

Despite this realization, before the advent of radio astronomy astronomers had no direct means of studying clouds like L977. Emitting no visible light, they are generally undetectable to the eye, except by the degree to which they dim starlight. However, as shown in Figure 18.13(b), the cloud's radio emission—in this case from carbon monoxide (CO) molecules contained within its volume—outlines the cloud clearly at radio wavelengths, providing an indispensible tool for the study of such objects. We will return to the subject of molecular emission from interstellar clouds in Section 18.5.

Figures 18.14(a) and (b) are optical photographs of another dark dust cloud. Taking its name from a neighboring star, Rho Ophiuchi, this dust cloud resides relatively nearby us—about 300 pc from the Sun. ∞ *(Discovery 17-1)* Pockets of intense blackness mark regions where the dust

(a)
R I V U X G

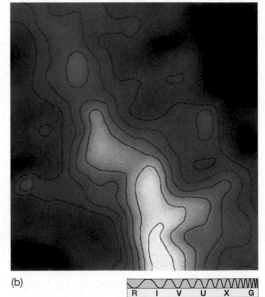
(b)
R I V U X G

▲ **FIGURE 18.13 Obscuration and Emission** (a) At optical wavelengths, this dark dust cloud (known as L977) can be seen only by its obscuration of background stars. (b) At radio wavelengths, it emits strongly in the CO molecular line, with the most intense radiation coming from the densest part of the cloud. *(C. and E. Lada)*

DISCOVERY 18-1

Ultraviolet Astronomy and the "Local Bubble"

The ultraviolet is that part of the electromagnetic spectrum where we expect to witness events and see objects involving temperatures in the hundreds of thousands—even millions—of kelvins—hot regions like the seething atmospheres or eruptive flares of stars, massive stars caught in the act of exploding as supernovae (Chapter 22), and active galaxies whose spinning hearts may harbor black holes (Chapter 25). Yet, ultraviolet astronomy has also contributed greatly to the study of the interstellar medium by allowing a unique mapping of our local cosmic neighborhood.

Short-wavelength ultraviolet radiation is effectively blocked by Earth's atmosphere. To study this region of the electromagnetic spectrum, instruments must be placed above our planet's atmospheric blanket. ∞ (Sec. 5.6) One of the most successful satellites yet launched—placed in orbit in 1978 and finally shut down in late 1996—was the *International Ultraviolet Explorer* (*IUE*). The accompanying photograph shows the *IUE* Science Operations Center at the University of California at Berkeley, where undergraduate students worked alongside scientists and engineers to command the spacecraft and acquire its data. More recently, the *Extreme Ultraviolet Explorer* (*EUVE*), which operated from 1992 until 2000, closed one of the few remaining gaps in the electromagnetic spectrum that was not well explored by other means.

Only 30 years ago, astronomers presumed that the gases filling the spaces among the stars would absorb virtually all short-wavelength UV radiation before it had a chance to reach Earth. But in 1975, during a historic linkup of the *Apollo* (U.S.) and *Soyuz* (USSR) space capsules, the onboard astronauts and cosmonauts performed a key experiment: They used a small telescope to detect extreme ultraviolet radiation from a few nearby very hot stars. Shortly thereafter, theoretical ideas changed as astronomers began to realize that the interstellar gas is not all uniformly distributed in space. Rather, it is spread very unevenly, with cool, dense clumps interspersed with regions of hot, low-density gas shaped like bubbles and tunnels.

Observations made by the *IUE* of weak spectral lines from highly excited atoms showed that some regions of interstellar space are much thinner (5000 atoms/m^3) and hotter (500,000 K) than previously expected. Part of the space among the dust clouds and the emission nebulae seems to contain extremely dilute, very high temperature gas, probably the result of the violent expansion of debris from stars that exploded long ago. Observations by the *EUVE* and other space-based instruments have found that these superheated interstellar "bubbles," making up the *intercloud medium*, may extend far into interstellar space beyond our local neighborhood and, conceivably, into the even vaster spaces among the galaxies.

The Sun seems to reside in one such low-density region—a huge cavity called the "*Local Bubble*," sketched in the second figure. Only because we live in this vast low-density bubble can we detect so many stars in the extreme UV—the hot, thin interstellar gas is virtually transparent to this radiation. The Local Bubble contains about 200,000 stars and extends for hundreds of trillions of kilometers, or nearly 100 pc. It was probably carved out by multiple supernova explosions that occurred several hundred thousand years ago in the Scorpius–Centaurus association, a rich cluster of bright young stars. (See Chapters 20 and 21.) Perhaps our hominid ancestors may have seen these ancient events—stellar catastrophies as bright as the full Moon—that now aid modern astronomers.

(UC/Berkeley)

(Rice University)

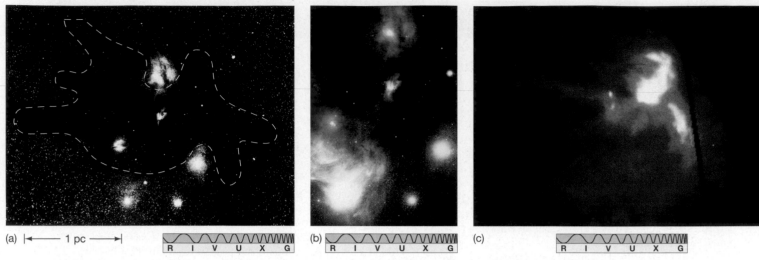

(a) |◄——— 1 pc ———►| R I V U X G   (b)   R I V U X G   (c)   R I V U X G

▲ **FIGURE 18.14 Dark Dust Cloud** (a) The dark dust cloud Rho Ophiuchi, is "visible" only because it blocks light coming from stars behind it. The dashed line indicates the cloud's approximate outline. (b) Another view of the region on a slightly enlarged scale, showing fainter foreground objects and more subtle colors. To orient (a) and (b), note the "pentagon" of bright objects that is clearly visible in each image. The bright star Antares is at the bottom. The bright star Antares is at the bottom left and to its right is a star cluster called M4. Rho Ophiuchi itself is the bright object near the top, surrounded by a blue reflection nebula. (c) An infrared map of the same region, to roughly the same scale as parts (a) and (b). The very bright source near the top of the cloud is a hot emission nebula, also visible in the optical images. The "streamers" at left are the dark dust lanes evident in parts (a) and (b) as well. (The black diagonal streak at the right is an instrumental artifact.) *(Harvard Observatory; D. Malin/AAT; NASA)*

and gas are especially concentrated and the light from background stars is completely obscured. Measuring several parsecs across, the Rho Ophiuchi cloud is only a tiny part of the grand mosaic shown in Figure 18.5. The cloud clearly is far from spherical. Indeed, most interstellar clouds are very irregularly shaped. Note especially the long "streamers" of (relatively) dense dust and gas in Figure 18.14(c).

Like all dark dust clouds, the Rho Ophiuchi cloud is too cold to emit any visible light. However, like L977 in Figure 18.13(a), it does radiate strongly at longer wavelengths. Figure 18.14(c) shows an infrared view of the same region, captured by sensitive detectors aboard the *Infrared Astronomy Satellite*. ∞ (Sec. 5.6) The bright patches within the dark region in Figure 18.14(a) are foreground objects—emission nebulae and groups of bright

▲ **FIGURE 18.15 Horsehead Nebula** Located in the constellation Orion, not far from the Orion Nebula, the Horsehead Nebula is a striking example of a dark dust cloud, silhouetted against the bright background of an emission nebula. The "neck" of the horse is about 0.25 pc across. This nebular region is roughly 1500 pc from Earth. *(Royal Observatory, Belgium; D. Malin/AAT)*

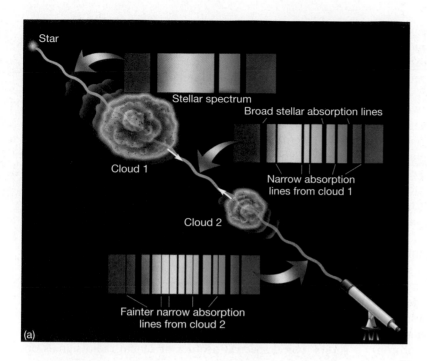

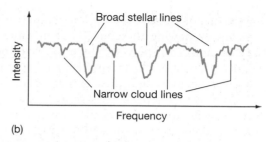

(b)

◀ **FIGURE 18.16 Absorption by Interstellar Clouds** (a) Simplified diagram of some interstellar clouds between a hot star and Earth. Optical observations might show an absorption spectrum like that traced in (b). The wide, intense lines are formed in the star's hot atmosphere; narrower, weaker lines arise from the cold interstellar clouds. The smaller the cloud, the weaker are the lines. The redshifts or blueshifts of the narrow absorption lines provide information on cloud velocities. The widths of all the spectral lines depicted here are greatly exaggerated for the sake of clarity.

stars. Some of them are part of the cloud itself, where newly formed stars near the surface have created a "hot spot" in the cold, dark gas. Others have no connection to the cloud and just happen to lie along our line of sight. The additional foreground stars in Figure 18.14(b) are too faint to be seen in Figure 18.14(a).

These dark and dusty interstellar clouds are sprinkled throughout our Galaxy. We can study them at optical wavelengths only if they happen to block the light emitted by more distant stars or nebulae. The dark outline of the Rho Ophiuchi cloud in Figure 18.14(a) and the dust lanes visible in Figures 18.8 and 18.9 are good examples of this obscuration. Figure 18.15 shows another striking example of a dark cloud: the Horsehead Nebula in Orion. This curiously shaped finger of gas and dust projects out from the much larger dark cloud in the bottom half of the image and stands out clearly against the red glow of a background emission nebula.

Absorption Spectra

Astronomers first became aware of the true extent of dark interstellar clouds in the 1930s, as they studied the optical spectra of distant stars. The gas in such a cloud absorbs some of the stellar radiation in a manner that depends on the cloud's own temperature, density, and elemental abundance. The absorption lines thus produced contain information about dark interstellar matter, just as stellar absorption lines reveal the properties of stars. ∞ (Sec. 4.2)

Because the interstellar absorption lines are produced by cold, low-density gas, astronomers can easily distinguish them from the much broader absorption lines formed in stars' hot lower atmospheres. ∞ (Sec. 4.4) Figure 18.16(a) illustrates how light from a star may pass through several interstellar clouds on its way to Earth. These clouds need

not be close to the star, and, indeed, they usually are not. Each absorbs some of the stellar radiation in a manner that depends on its own temperature, density, velocity, and elemental abundance. Figure 18.16(b) depicts part of a typical spectrum produced in this way.

The narrow absorption lines contain information about dark interstellar clouds, just as stellar absorption lines reveal the properties of stars and nebular emission lines tell us about conditions in hot nebulae. By studying these lines, astronomers can probe the cold depths of interstellar space. In most cases, the elemental abundances detected in interstellar clouds mirror those found in other astronomical objects—perhaps not surprising, since (as we will see in Chapter 19) interstellar clouds are the regions that spawn emission nebulae and stars.

CONCEPT CHECK

✔ How do astronomers use optical observations to probe the properties of dark dust clouds?

18.4 21-Centimeter Radiation

A basic difficulty with the optical technique just described is that we can examine interstellar clouds only along the line of sight to a distant star. To form an absorption line, a background source must provide radiation to absorb. The need to see stars through clouds also restricts this approach to relatively local regions, within a few thousand parsecs of Earth. Beyond that distance, stars are completely obscured, and optical observations are impossible. As we have seen, infrared observations provide a means of viewing the emission from some clouds, but they do not completely solve the problem, because only the denser, dustier clouds emit enough infrared radiation for astronomers to study them in that part of the spectrum.

To probe interstellar space more thoroughly, we need a more general, more versatile observational method—one that does not rely on conveniently located stars and nebulae. In short, we need a way to detect cold, neutral interstellar matter anywhere in space through its *own* radiation. This may sound impossible, but such an observational technique does in fact exist. The method relies on low-energy *radio* emissions produced by the interstellar gas itself.

Recall that a hydrogen atom has one electron orbiting a single-proton nucleus. Besides its orbital motion around the central proton, the electron also has some rotational motion—that is, *spin*—about its own axis. The proton also spins. This model is analogous to a planetary system in which, in addition to the orbital motion of a planet about a central star, both the planet (electron) and the star (proton) rotate about their own axes. But bear in mind the crucial difference between planetary and atomic systems: A planet orbiting the Sun is free to move in any orbit and spin at any rate, but within an atom, all physical quantities, such as energy, momentum, and angular momentum (spin), are quantized—they are permitted to take on only specific, distinct values. ∞ (Sec. 4.2)

The laws of physics dictate that there are exactly two possible spin configurations for a hydrogen atom in its ground state. The electron and proton can rotate in the same direction, with their spin axes parallel, or they can rotate with their axes antiparallel (i.e., parallel, but oppositely oriented). Figure 18.17 shows these two configura-

tions. The antiparallel configuration has slightly less energy than the parallel state.

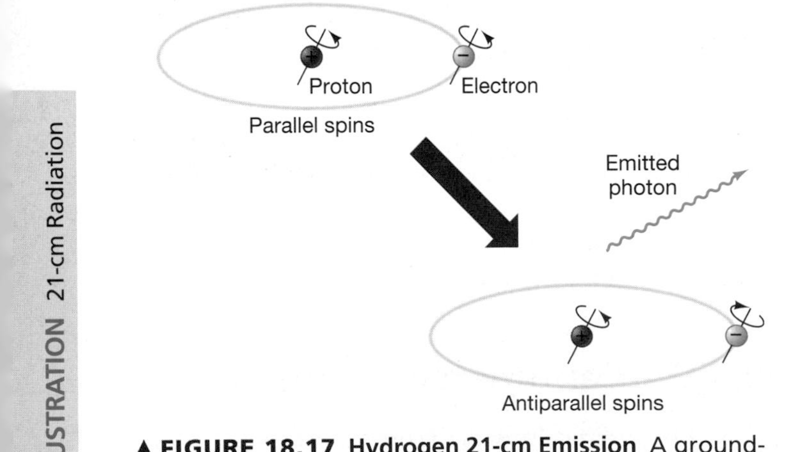

All matter in the universe tends to achieve its lowest possible energy state, and interstellar gas is no exception. A slightly excited hydrogen atom with the electron and proton spinning in the same direction eventually drops down to the less energetic, opposite-spin state as the electron suddenly and spontaneously reverses its spin. As with any other such change, the transition from a high-energy state to a low-energy state releases a photon with energy equal to the energy difference between the two levels.

Because that energy difference is very small, the energy of the emitted photon is very low. Consequently, the wavelength of the radiation is rather long—in fact, it is 21.1 cm, roughly the width of this book. That wavelength lies in the radio portion of the electromagnetic spectrum. Researchers refer to the spectral line that results from this hydrogen-spin-flip process as **21-centimeter radiation**. This spectral line provides a vital probe into any region of the universe containing atomic hydrogen gas. Figure 18.18 shows typical spectral profiles of 21-cm radio signals observed from several different regions of space. These tracings are the characteristic signatures of cold, atomic hydrogen in our Galaxy. Needing no visible starlight to help calibrate their signals, radio astronomers can observe *any* interstellar region that contains enough hydrogen gas to produce a detectable signal. Even the low-density regions between the dark clouds can be studied.

As can be seen in the figure, actual 21-cm lines are quite jagged and irregular, somewhat like nebular emission lines in appearance. The irregularities arise because there are usually numerous clumps of interstellar gas along any given line of sight, each with its own density, temperature, radial velocity, and internal motion. Thus, the intensity, width, and Doppler shift of the resultant 21-cm line vary from place to place. ∞ (Sec. 4.4) All these different lines are superimposed in the signal we eventually receive at Earth, and sophisticated computer analysis is generally required to disentangle them. The "average" figures quoted earlier for the temperatures (100 K) and densities (10^6 atoms/m^3) of the regions between the dark dust clouds are based on 21-cm measurements. Observations of the dark clouds themselves using 21-cm radiation yield densities and temperatures in good agreement with those obtained by optical spectroscopy.

All interstellar atomic hydrogen emits 21-cm radiation. But if all atoms eventually fall into their lowest-energy configuration, then why isn't all the hydrogen in the Galaxy in the lower energy state by now? Why do we see 21-cm radiation today? The answer is that the energy difference between the two states is comparable to the energy of a typical atom at a temperature of 100 K or so. As a result, atomic collisions in the interstellar medium are energetic enough to boost the electron into the higher energy configuration and so maintain comparable numbers of hydrogen atoms in either state. At any instant, any sample of

▲ FIGURE 18.17 Hydrogen 21-cm Emission A ground-state hydrogen atom changing from a higher energy state (electron and proton spinning in the same direction) to a lower energy state (the electron and proton spinning in opposite directions). The emitted photon carries away an energy equal to the energy difference between the two spin states.

18.5 Interstellar Molecules

LEARNING GOAL 4 **LEARNING GOAL 5** In some particularly cold (typically, 20-K) interstellar regions, densities can reach as high as 10^{12} particles/m³. Until the late 1970s, astronomers regarded these regions simply as abnormally dense interstellar clouds, but it is now recognized that they belong to an entirely new class of interstellar matter. The gas particles in these regions are not in atomic form at all; they are molecules. Because of the predominance of molecules in these dense interstellar regions, they are known as **molecular clouds**. They literally dwarf even the largest emission nebulae, which were previously thought to be the most massive residents of interstellar space.

Molecular Spectral Lines

As noted in Chapter 4, much like atoms, molecules can become excited through collisions or by absorbing radiation. ∞ (Sec. 4.3) Furthermore, again like atoms, molecules eventually return to their ground states, emitting radiation in the process. The energy states of molecules are much more complex than those of atoms, however. Once more like atoms, molecules can undergo internal electron transitions, but unlike atoms, they can also rotate and vibrate. They do so in specific ways, obeying the laws of quantum physics. Figure 18.19 depicts a simple molecule rotating rapidly—that is, a molecule in an excited rotational state. After a length of time that depends on its internal makeup, the molecule relaxes back to a slower rotational rate (a state of lower energy). This change causes a photon to be emitted, carrying an energy equal to the energy difference between the two rotational states involved. The energy differences between these states are generally very small, so the emitted radiation is usually in the radio range.

We are fortunate that molecules emit radio radiation, because they are invariably found in the densest and dustiest parts of interstellar space. These are regions where the

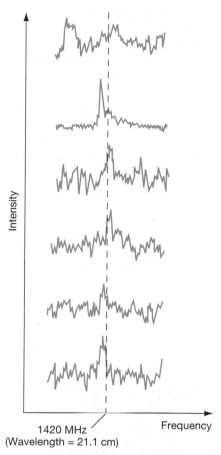

▲ **FIGURE 18.18 21-cm Lines** Typical 21-cm radio spectral lines observed from several different regions of interstellar space. The peaks do not all occur at a wavelength of exactly 21.1 cm, corresponding to a frequency of 1420 MHz, because the gas in our Galaxy is moving with respect to Earth.

(Graph y-axis label: Intensity; x-axis label: Frequency; marking: 1420 MHz (Wavelength = 21.1 cm))

interstellar hydrogen will contain many atoms in the upper level, so conditions will always be favorable for 21-cm radiation to be emitted.

Of great importance, the wavelength of this characteristic radiation is much larger than the typical size of interstellar dust particles. Accordingly, 21-cm radiation reaches Earth completely unscattered by interstellar debris. The opportunity to observe interstellar space well beyond a few thousand parsecs, and in directions lacking background stars, makes 21-cm observations among the most important and useful in all astronomy. We will see in Chapters 23 through 25 how such observations are indispensable in allowing astronomers to map out the large-scale structure of our own and other galaxies.

CONCEPT CHECK

✔ Why is 21-cm radiation so useful as a probe of galactic structure?

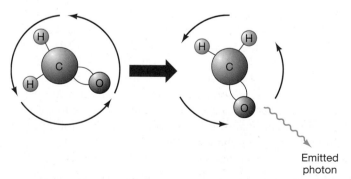

▲ **FIGURE 18.19 Molecular Emission** As a molecule changes from a rapid rotation (left) to a slower rotation (right), a photon is emitted that can be detected with a radio telescope. Depicted here is the formaldehyde molecule, H_2CO. The lengths of the curved arrows are proportional to the spin rate of the molecule.

absorption of shorter wavelength radiation is enough to prohibit the use of ultraviolet, optical, and most infrared techniques that might ordinarily detect changes in the energy states of the molecules. Only low-frequency radio radiation can escape.

Why are molecules found only in the densest and darkest of the interstellar clouds? One possible reason is that the dust serves to protect the fragile molecules from the normally harsh interstellar environment—the same absorption that prevents high-frequency radiation from getting out to our detectors also prevents it from getting in to destroy the molecules. Another possibility is that the dust acts as a catalyst that helps form the molecules. The grains provide both a place where atoms can stick together and react and a means of dissipating any heat associated with the reaction, which might otherwise destroy the newly formed molecules. Probably the dust plays both roles; the close association between dust grains and molecules in dense interstellar clouds argues strongly in favor of this view, although the details are still being debated.

Molecular Tracers

In mapping molecular clouds, radio astronomers are faced with a problem. Molecular hydrogen (H_2) is by far the most common constituent of these clouds, but unfortunately, despite its abundance, this molecule does not emit or absorb radio radiation. Rather, it emits only short-wavelength ultraviolet radiation, so it cannot easily be used as a probe of cloud structure. Nor are 21-cm observations helpful—they are sensitive only to *atomic* hydrogen, not to the *molecular* form of the gas. Theorists had expected H_2 to abound in these dense, cold pockets of interstellar space, but proof of its existence was hard to obtain. Only when spacecraft measured the ultraviolet spectra of a few stars located near the edges of some dense clouds was the presence of molecular hydrogen confirmed.

With hydrogen effectively ruled out as a probe of molecular clouds, astronomers must use observations of other molecules to study the dark interiors of these dusty regions. Molecules such as carbon monoxide (CO), hydrogen cyanide (HCN), ammonia (NH_3), water (H_2O), methyl alcohol (CH_3OH), formaldehyde (H_2CO), and about 120 others, some quite complex, are now known to exist in interstellar space.† These molecules are found only in very small quantities—they are generally 1 million to 1 billion times less abundant than H_2—but they are important as *tracers* of a cloud's structure and physical properties.

They are produced by chemical reactions within molecular clouds. When we observe them, we know that the regions under study must also contain high densities of molecular hydrogen, dust, and other important constituents.

The rotational properties of different molecules often make them suitable as probes of regions with different physical properties. Formaldehyde may provide the most useful information on one region, carbon monoxide on another, and water on yet another, depending on the densities and temperatures of the regions involved. The data obtained thereby equip astronomers with a sophisticated spectroscopic "toolbox" for studying the interstellar medium.

For example, Figure 18.20 shows some of the sites where formaldehyde molecules have been detected near M20. At practically every dark area sampled between M16 and M8, the formaldehyde molecule is present in surprisingly large abundance (although it is still far less common than H_2). Analyses of spectral lines at many locations along the 12°-wide swath shown in Figure 18.6 indicate that the temperature and density are much the same in all the molecular clouds studied (50 K and 10^{11} molecules/m^3, on average). Figure 18.21 shows a contour map of the

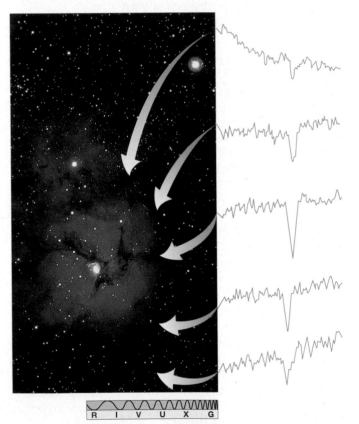

▲ **FIGURE 18.20 Molecular Absorption near M20** Spectra indicate that formaldehyde molecules exist around M20. (See arrows.) The lines, which are formed by the absorption of background radiation, are most intense both in the dark dust lanes trisecting the nebula and in the dark regions beyond the nebula as background radiation passes through. (*background image: AURA*)

---

†*Some remarkably complex organic molecules, including formaldehyde (H_2CO), ethyl alcohol (CH_3CH_2OH), methylamine (CH_3NH_2), and formic acid (H_2CO_2), have been found in the densest of the dark interstellar clouds. Their presence has fueled speculation about the origins of life, both on Earth and in the interstellar medium—especially since the report (still unconfirmed) by radio astronomers in the mid-1990s of evidence that glycine (NH_2CH_2COOH), one of the key amino acids that form the large protein molecules in living cells, may also be present in interstellar space.*

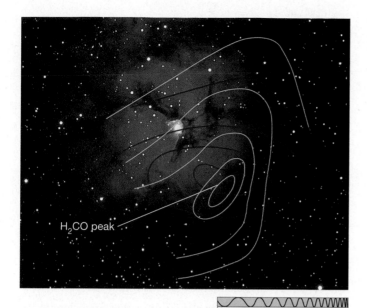

R I V U X G

◄ **FIGURE 18.21 M20 Radio Map** Contour map of the amount of formaldehyde near the M20 nebula, demonstrating how that gas is especially abundant in the darkest interstellar regions. Other kinds of molecules have been found to be similarly distributed. The contour values increase from the outside to the inside, so the maximum density of formaldehyde lies just to the bottom right of the visible nebula. The red and green contours outline the intensity of the formaldehyde lines at different rotational frequencies. The nebula itself is about 4 pc across. *(background image: AURA)*

distribution of formaldehyde molecules in the immediate vicinity of the M20 nebula. After radio spectral lines of formaldehyde were observed at various locations, contours connecting regions of similar abundance were drawn. Notice that the amount of formaldehyde (and, we assume, the amount of hydrogen) peaks in a dark region well away from the visible nebula.

Radio maps of interstellar gas and infrared maps of interstellar dust reveal that molecular clouds do not exist as distinct objects in space. Rather, they make up huge **molecular cloud complexes**, typically up to 50 pc across and containing enough gas to make a million stars like our Sun. About a thousand such giant complexes are currently known in our Galaxy.

The discovery of many interstellar molecules in the early 1970s forced astronomers to rethink and reobserve interstellar space. In doing so, they realized that this active and vital domain is far from the void suspected by theorists not long before. As we will see in Chapter 19, regions of space once thought to contain nothing more than galactic "garbage"—the cool, tenuous darkness among the stars—now play a critical role in our understanding of stars and the interstellar medium from which they are born.

CONCEPT CHECK

✔ In mapping molecular clouds, why do astronomers use observations of "minority" molecules such as carbon monoxide and formaldehyde when these molecules constitute only a tiny fraction of the total number of molecules in interstellar space?

Chapter Review

SUMMARY

The **interstellar medium (p. 468)** occupies the space among the stars. It is made up of cold (less than 100 K) gas, mostly atomic or molecular hydrogen and helium, and **dust grains (p. 468)**. Interstellar dust is highly effective at blocking our view of distant stars, even though the density of the interstellar medium is very low. The spatial distribution of interstellar matter is patchy. The general diminution of starlight by dust is called **extinction (p. 468)**. In addition, the dust preferentially absorbs short-wavelength radiation, leading to a distinct **reddening (p. 468)** of light passing through interstellar clouds. Interstellar dust is thought to be composed of silicates, graphite, iron, and "dirty ice." Interstellar dust particles are apparently elongated or rodlike. The **polarization (p. 470)** of starlight provides a means of studying these particles.

A **nebula (p. 472)** is a general term for any fuzzy bright or dark patch on the sky. **Emission nebulae (p. 472)** are extended clouds of hot, glowing interstellar gas. Associated with star formation, they result when hot O- and B-type stars heat and ionize their surroundings. Studies of the emission lines produced by excited nebular atoms allow astronomers to measure the properties of nebulae. Some excited atomic states take so long to emit a photon that the spectral lines associated with these transitions are never seen in terrestrial laboratories, where collisions always knock the atom into another energy state before it can emit any radiation. When these lines are seen in nebular spectra, they are called forbidden lines. Nebulae are often crossed by dark **dust lanes (p. 472)**, part of the larger cloud from which they formed.

Dark dust clouds (p. 478) are cold, irregularly shaped regions in the interstellar medium that diminish or completely obscure the light from background stars. Astronomers can learn about these clouds by studying the absorption lines they produce in starlight that passes through them. Another way to observe cold, dark regions of interstellar space is through **21-centimeter radiation (p. 482)**, atoms produced whenever the electron in an atom of hydrogen reverses its spin, changing its energy slightly in the process. This radio radiation is important because it is emitted by all cool atoms of hydrogen gas, even if the gas is undetectable by other means. In addition, 21-cm radiation is not appreciably absorbed by the interstellar medium, so radio astronomers making observations at this wavelength can "see" to great distances.

The interstellar medium also contains many cold, dark **molecular clouds (p. 483)**, which are observed mainly through the radio radiation emitted by the molecules they contain. Dust within these clouds probably both protects the molecules and acts as a catalyst to help them form. As with other interstellar clouds, hydrogen is by far the most common constituent, but molecular hydrogen happens to be very hard to observe. Astronomers usually study these clouds through observations of other "tracer" molecules that are less common, but much easier to detect. Often, several molecular clouds are found close to one another, forming an enormous **molecular cloud complex (p. 485)** millions of times more massive than the Sun.

REVIEW AND DISCUSSION

1. Give a brief description of the interstellar medium.
2. What is the composition of interstellar gas? What about interstellar dust?
3. Why is interstellar dust so much more effective than interstellar gas at absorbing starlight?
4. How dense, on average, is interstellar matter?
5. How is interstellar matter distributed throughout space?
6. What are some methods that astronomers use to study interstellar dust?
7. What is an emission nebula?
8. What is photoevaporation, and how does it change the structure and appearance of an emission nebula?
9. Why are some spectral lines that are observed in emission nebulae not normally seen in laboratories on Earth?
10. What is the local bubble? How did it form?
11. Describe some ways in which we can "see" a dark interstellar cloud.
12. Give a brief description of a dark dust cloud.
13. What is 21-cm radiation? With what element is it associated?
14. Why is 21-cm radiation useful to astronomers?
15. How does a molecular cloud differ from other interstellar matter?
16. Why can't astronomers use observations of hydrogen to explore the structure of molecular cloud complexes?
17. How do astronomers explore the structure of molecular cloud complexes?
18. If our Sun were surrounded by a cloud of gas, would this cloud be an emission nebula? Why or why not?
19. Compare the reddening of stars by interstellar dust with the reddening of the setting Sun.
20. Explain what it means for a star's light to be polarized. How does the polarization of starlight provide a means of studying the interstellar medium?

CONCEPTUAL SELF-TEST: TRUE OR FALSE/MULTIPLE CHOICE

1. Interstellar matter is quite evenly distributed throughout the Milky Way Galaxy.
2. To scatter a beam of radiation, a particle must be smaller in size than the wavelength of the radiation involved.
3. There is a lack of heavy elements in interstellar gas because these elements go into making interstellar dust.
4. The polarization of starlight tells us about the shape of interstellar dust particles.
5. Emission nebulae display spectra almost identical to those of the stars embedded in them.
6. Forbidden emission lines can occur in emission nebulae because the density of interstellar gas there is extremely low.
7. Because of the obscuration of visible light by interstellar dust, we can observe stars only within a few thousand parsecs of Earth in any direction.
8. Dark dust clouds radiate mainly in the ultraviolet part of the electromagnetic spectrum.
9. Twenty-one-centimeter radiation observations can be used to probe the interiors of molecular clouds.
10. Water, formaldehyde, carbon monoxide, and numerous organic molecules have all been detected in molecular clouds.
11. The chemical composition of the interstellar medium is basically similar to that of **(a)** the Sun; **(b)** Earth; **(c)** Venus; **(d)** Mars.
12. The density of atoms in the interstellar medium is most similar to **(a)** wildfire smoke; **(b)** dark rain clouds; **(c)** deep ocean water; **(d)** the interior of a TV tube.
13. Of the following objects, the one that shines most like an emission nebula shines is **(a)** a regular incandescent light bulb with a filament; **(b)** a red hot ember from a camp fire; **(c)** a glowing fluorescent light tube; **(d)** a star like the Sun.
14. Stars interact with emission nebulae by **(a)** exiting their atoms enough to emit light; **(b)** illuminating them like an advertising billboard; **(c)** causing them to contract; **(d)** heating them so they explode.
15. A dark interstellar globule is about the same size as **(a)** a cloud in Earth's atmosphere; **(b)** the entire planet Earth; **(c)** a star like the Sun; **(d)** the Oort cloud.
16. The Rho Ophiuchi cloud, shown in Figure 18.12(a) ("Dark Dust Cloud"), is dark because **(a)** there are no stars in this region; **(b)** the stars in this region are young and faint; **(c)** starlight from behind the cloud does not penetrate the cloud; **(d)** the region is too cold to sustain stellar fusion.

17. If a proton and an electron within a hydrogen atom initially have parallel spins, then change to have antiparallel spins, the atom must (a) absorb energy; (b) emit energy; (c) become hotter; (d) become larger.

18. Of the following telescopes, the one best suited to observing dark dust clouds is (a) an X-ray telescope; (b) a large visible-light telescope; (c) an orbiting ultraviolet telescope; (d) a radio telescope.

19. Of the following, the largest interstellar clouds are (a) molecular clouds; (b) dark dust clouds; (c) emission nebulae; (d) globules.

20. Molecular clouds are routinely studied using spectral lines from all but which of the following? (a) Molecular hydrogen; (b) Carbon monoxide; (c) Formaldehyde; (d) Water.

PROBLEMS

 Algorithmic versions of these questions are available in the Practice Problems module of the Companion Website at astro.prenhall.com/chaisson.

The number of squares preceding each problem indicates its approximate level of difficulty.

1. ■ The average density of interstellar gas within the "Local Bubble" is much lower than the value mentioned in the text—in fact, it is roughly 10^3 hydrogen atoms/m^3. Given that the mass of a hydrogen atom is 1.7×10^{-27} kg, calculate the total mass of interstellar matter contained within a Bubble volume equal in size to planet Earth.

2. ■ Assume the same average density as in the previous question, and calculate the total mass of interstellar hydrogen contained within a cylinder of cross-sectional area 1 m^2, extending from Earth to Alpha Centauri.

3. ■■ Given the average density of interstellar matter stated in Section 18.1, calculate how large a volume of space would have to be compressed to make a cubic meter of gas equal in density to air on Earth (1.2 kg/m^3).

4. ■ Assume a density of 3000 kg/m^3, and estimate the mass of the dust particle illustrated in Figure 18.3(a).

5. ■■ A beam of light shining through a dense molecular cloud is diminished in intensity by a factor of two for every 5 pc it travels. By how many magnitudes is the light from a background star dimmed if the total thickness of the cloud is 60 pc?

6. ■ Interstellar extinction is sometimes measured in magnitudes per kiloparsec (1 kpc = 1000 pc). Light from a star 1500 pc away is observed to be diminished in intensity by a factor of 20 over and above the effect of the inverse-square law. What is the average interstellar extinction, in mag/kpc, along the line of sight?

7. ■■ Spectroscopic observations of a certain star reveal it to be a B2II giant, with absolute magnitude −6. ∞ (Secs. 17.3, 17.6) The star's apparent magnitude is 14. Neglecting the effects of interstellar extinction, calculate the distance to the star. If the star's true distance is known (by other means) to be 5000 pc, calculate the average extinction, in mag/kpc, along the line of sight ∞ (More Precisely 17-1)

8. ■■ A star of apparent magnitude 10 lies 500 pc from Earth. If interstellar absorption results in an average extinction of 2 mag/kpc, calculate the star's absolute magnitude and luminosity.

9. ■■■ A star of known absolute magnitude −5 has apparent magnitude 10. If interstellar absorption results in an average extinction of 2 mag/kpc, calculate the star's distance. (*Note*: This problem does not have an algebraic solution. You will have to solve it by numerical means—essentially trial and error on a calculator.)

10. ■ To carry enough energy to ionize a hydrogen atom, a photon must have a wavelength of less than 9.12×10^{-8} m (91.2 nm). Using Wien's law, calculate the temperature a star must have for the peak wavelength of its blackbody curve to equal this value. ∞ (Sec. 3.4)

11. ■■ Estimate the escape speeds near the edges of the four emission nebulae listed in Table 18.1, and compare them with the average speeds of hydrogen nuclei in those nebulae. ∞ (*More Precisely 8-1*) Do you think it is possible that the nebulae are held together by their own gravity?

12. ■ What would the mass of M8 have to be in order for its escape speed to equal its average molecular speed?

13. ■■ If a group of interstellar clouds along the line of sight have radial velocities in the range 75 km/s (receding) to 50 km/s (approaching), calculate the range of frequencies and wavelengths over which the 21.1-cm (1420-MHz) line of hydrogen will be observed. ∞ (Sec. 3.5)

14. ■ Calculate the radius of a spherical molecular cloud whose total mass equals the mass of the Sun. Assume a cloud density of 10^{12} hydrogen atoms per cubic meter.

15. ■■■ A cloud of atomic hydrogen has a radius of 1 pc and an average density of 10^6 hydrogen atoms per cubic meter. Collisions between atoms ensure that, at any instant, $\frac{3}{4}$ of all atoms are in the upper (parallel-spin) state, as discussed in Section 18.4. The transition producing the 21-cm line is very unlikely—the probability that any given atom in the upper state will make the transition during any given second is about 3×10^{-15}. (Cf. with 10^8 for the Hα transition.) Use these figures, together with the Planck formula for the energy of the photon emitted in the transition, to estimate the total radio luminosity of the cloud. ∞ (Sec. 4.2)

 In addition to the Practice Problems module, the Companion Website at astro.prenhall.com/chaisson provides for each chapter a study guide module with multiple choice questions as well as additional annotated images, animations, and links to related Websites.

Few issues in astronomy are more basic than knowing how stars form. The trouble is that stars normally do so in the darkness in the hearts of dense interstellar clouds, so optical telescopes can grant only limited insight. Telescopes operating in the long-wavelength radio and infrared parts of the spectrum are essential to probe these otherwise invisible regions of space. Here, in this new infrared image downlinked from the Spitzer Space Telescope, the so-called Elephant Trunk Nebula (IC 1396) houses scores of young stars and protostars about 2500 light-years away. (NASA) ▶

Star Formation

A Traumatic Birth

We now move from the interstellar medium—the gas and dust among the stars—back to the stars themselves. The next four chapters discuss the formation and evolution of stars. We have already seen that stars must evolve as they consume their fuel supply, and we have extensive observational evidence of stars at many different evolutionary stages. With the help of these observations, astronomers have developed a good understanding of stellar evolution—the complex changes undergone by stars as they form, mature, grow old, and die. We begin by studying the process of star formation, through which interstellar clouds of gas and dust are transformed into the myriad stars we see in the night sky.

LEARNING GOALS

Studying this chapter will enable you to

1 Discuss the factors that compete against gravity in the process of star formation.

2 Summarize the sequence of events leading to the formation of a star like our Sun.

3 Explain how the process of star formation depends on stellar mass.

4 Describe some of the observational evidence supporting the modern theory of star formation.

5 Explain the nature of interstellar shock waves, and discuss their possible role in the formation of stars.

6 Distinguish between open and globular star clusters, and explain why the study of clusters is important to astronomers.

Visit astro.prenhall.com/chaisson for additional annotated images, animations, and links to related sites for this chapter.

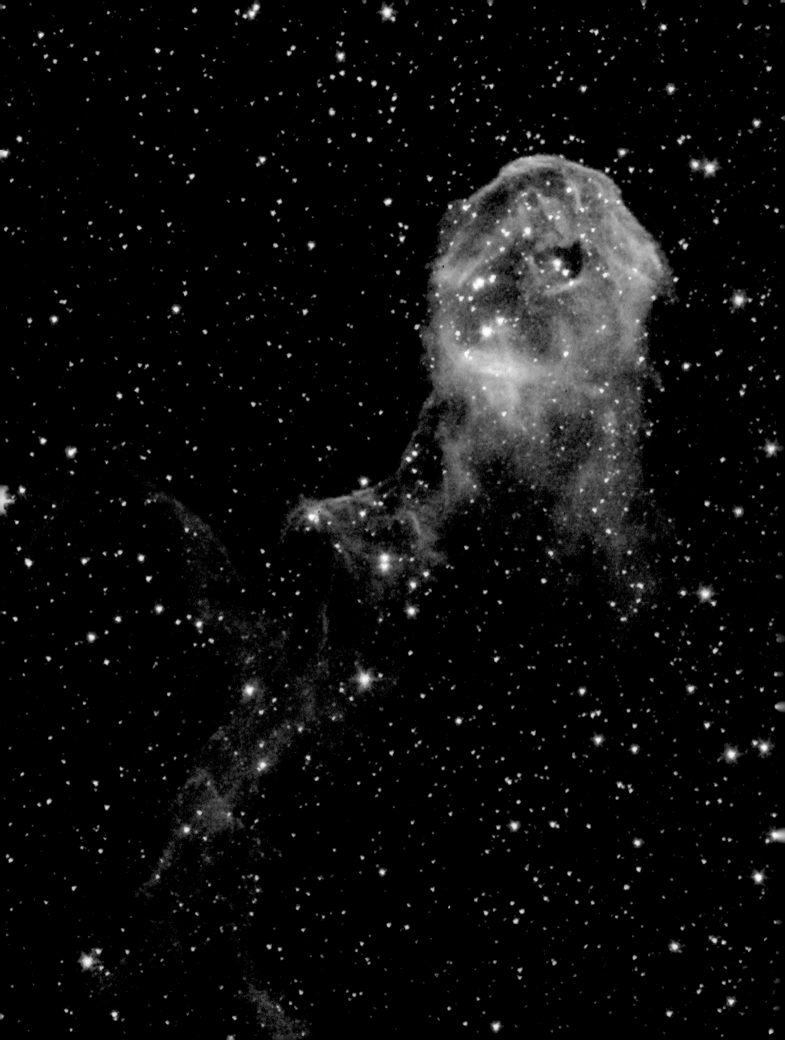

19.1 Star-Forming Regions

Our universe is constantly renewing itself. Literally billions of stars have been born, lived out their lives, and died since our Galaxy formed. We do not see this activity when we gaze at the nighttime sky, because the time scales on which stars play out this cosmic drama are enormously long by human standards. Even the shortest-lived O-type stars survive for millions of years. ∞ (Sec. 17.8) Nevertheless, we have plenty of evidence for ongoing stellar evolution throughout the cosmos.

Young Stars in the Universe

Our Sun, and probably most of the stars in our immediate cosmic neighborhood, formed billions of years ago. ∞ (Sec. 15.2) However, we know that many relatively nearby stars are much younger than this. The magnificent emission nebulae discussed in Chapter 18 and the ultraluminous, short-lived stars that power them are direct proof that star formation is a continuing process. ∞ (Sec. 18.2) The hottest stars in these regions must have formed less than a few million years ago—the blink of an eye, in cosmic terms—and there is no reason to suppose that Galactic star formation has recently and abruptly ceased! Stars are forming all across the Milky Way, even as you read this.

In fact, star-forming regions are observed in many regions of the universe far beyond our own Galaxy. Figure 19.1 shows one of the largest such regions discovered to date. It lies in a galaxy quite similar to our own, about 1 million parsecs away. Almost 500 pc across, this vast stellar nursery dwarfs any emission nebula known in our Galaxy. Conceivably, the Milky Way may contain similarly large nebulae. If they exist, they are obscured by so much interstellar matter in our Galactic plane that we cannot see them. But whatever their size—large or small—emission nebulae are the birthplaces of all the stars in our night sky.

How and where do stars form? What factors determine the masses, luminosities, and spatial distribution of stars in our Galaxy and beyond? The association of bright emission nebulae with much larger dark dust clouds provides the key. ∞ (Sec. 18.3) Simply put, star formation begins when part of the interstellar medium—one of those cold dark clouds—starts to collapse under its own weight. The cloud fragment heats up as it shrinks, and eventually its center becomes hot enough for nuclear fusion to begin. At that point, the contraction stops and a star is born. But what determines which interstellar clouds collapse? For that matter, since all clouds exert a gravitational pull, why didn't they all collapse long ago? To answer these questions and understand the processes leading to the stars we

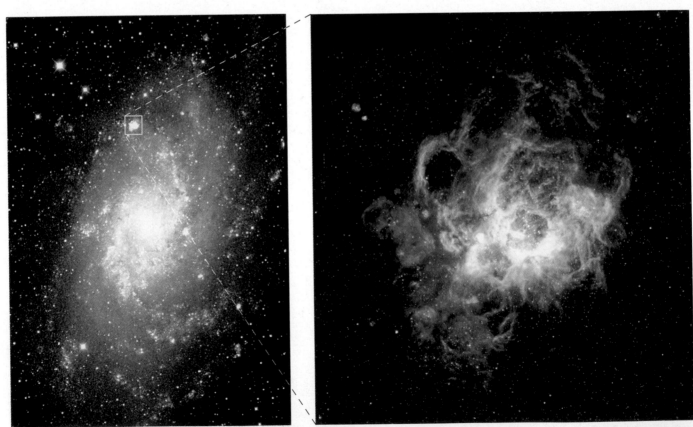

▲ **FIGURE 19.1 Extragalactic Star Formation** The giant star-forming region at the right, called NGC 604, is roughly 500 pc across. It is found in the nearby galaxy M33, displayed at the left on the much larger scale of 40,000 pc across. (*Palomar; NASA*)

R I V U X G

see, we must explore in a little more detail the factors that compete with gravity in determining a cloud's fate.

Gravity and Heat

Consider a small portion of a large cloud of interstellar gas. Concentrate first on just a few atoms, as shown in Figure 19.2. Even though the cloud's temperature is very low, each atom still has some random motion because of the cloud's heat. ∞ *(More Precisely 3-1)* Each atom is also influenced by the gravitational attraction of all its neighbors. The gravitational force is not large, however, because the mass of each atom is so small. When a few atoms accidentally cluster for an instant, as shown in Figure 19.2(b), their combined gravity is insufficient to bind them into a lasting, distinct clump of matter. This accidental cluster will disperse as quickly as it forms. The effect of heat—the random motion of the atoms—is much stronger than the effect of gravity.

Now consider a larger group of atoms. Imagine, for example, 50, 100, 1000—even a million—atoms, each gravitationally pulling on all the others. With increased mass, the force of gravity is now stronger than before. Will this many atoms exert a combined gravitational attraction strong enough to prevent the clump from dispersing again? The answer—at least under the conditions found in interstellar space—is still no. The gravitational attraction even of this mass of atoms is still far too weak to overcome the effect of heat.

We have already seen numerous instances of the competition between heat and gravity. ∞ *(More Precisely 8-1)* The temperature of a gas is simply a measure of the average speed of the atoms or molecules in it, so the higher the temperature, the greater the average speed of the molecules, and hence the higher the pressure of the gas. This is the main reason that the Sun and other stars don't collapse: The outward pressure of their heated gases exactly balances gravity's inward pull.

How many atoms must be accumulated in order for their collective pull of gravity to prevent them from dispersing back into interstellar space? The answer, even for a typical cool (100-K) cloud, is a truly huge number. Nearly 10^{57} atoms are required—much more than the 10^{25} grains of sand on all the beaches of the world and even more than the 10^{51} elementary particles that constitute all the atomic nuclei in our entire planet. There is simply nothing on Earth comparable to a star.

Some Complications

Heat is not the only factor that tends to oppose gravitational contraction: *Rotation*—that is, spin—can also compete with gravity's inward pull. As we saw in Chapter 15, a contracting cloud having even a small spin tends to develop a bulge around its midsection. ∞ (Sec. 6.7) As the cloud contracts, it must spin faster (to conserve its angular momentum), so the bulge grows and material on the edge tends to fly off into space. (Consider the analogy of mud flung from a rapidly rotating bicycle wheel.) Eventually, as in Figure 6.17, the cloud forms a flattened, rotating disk.

For material to remain part of the cloud and not be spun off into space, a force must be applied—in this case, the force of gravity. The more rapid the rotation, the greater is the tendency for the gas to escape, and the greater is the gravitational force needed to retain it. It is in this sense that we can regard rotation as opposing the inward pull of gravity. Should the rotation of a contracting gas cloud overpower gravity, the cloud would simply disperse. Thus, more mass is needed for a rapidly rotating interstellar cloud to contract to form a star than is needed for a cloud that does not rotate at all.

Magnetism can also hinder a cloud's contraction. Just as Earth, the outer planets, and the Sun all have some magnetism, magnetic fields permeate most interstellar clouds. As a cloud contracts, it heats up, and atomic encounters become violent enough to (partly) ionize the gas. As we noted in Chapter 7 in discussing Earth's Van Allen belts, and in Chapter 16 in discussing activity on the Sun, magnetic fields can exert electromagnetic control over charged particles. ∞ (Secs. 7.4, 16.5) In effect, the particles tend to become "tied" to the magnetic field—they are free to move *along* the field lines, but are inhibited from moving *perpendicular* to them.

As a result (Figure 19.3), interstellar clouds may contract in distorted ways. Because the charged particles and the magnetic field are linked, the field itself follows the contraction of the cloud. The charged particles literally pull the magnetic field toward the cloud's center in the direction perpendicular to the field lines. As the field lines are compressed, the magnetic field strength increases. In this way, the strength of the magnetism in a cloud can become much larger than that normally permeating general interstellar space. The primitive solar nebula may have contained a strong magnetic field created in just that manner.

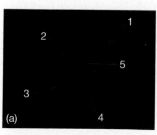
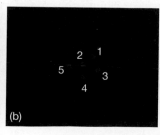
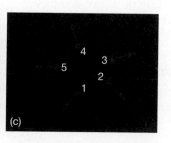

◀ FIGURE 19.2 Atomic Motion The motions of a few atoms within an interstellar cloud are influenced by gravity so slightly that the atoms' paths are hardly changed (a) before, (b) during, and (c) after an accidental, random encounter.

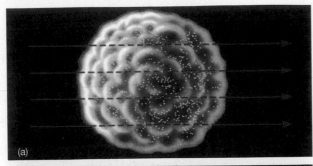

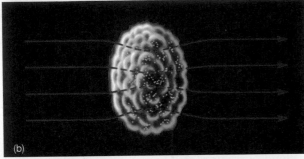

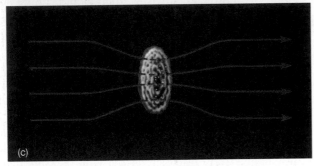

▲ **FIGURE 19.3 Interstellar Magnetic Field** Magnetism can hinder the contraction of a gas cloud, especially in directions perpendicular to the magnetic field (red lines). Frames (a), (b), and (c) trace the evolution of a slowly contracting interstellar cloud having some magnetism. The dashed lines within the cloud represent the regions where the field lines are distorted and compressed as the cloud shrinks.

Theory suggests that even small quantities of rotation or magnetism can compete quite effectively with gravity and can greatly alter the evolution of a typical gas cloud. Unfortunately, the interplay of these factors is not well understood—both can lead to highly complex behavior as a cloud contracts, and the combination of the two is extremely difficult to study theoretically. In this chapter, we will try to gain an appreciation for the broad outlines of the star-formation process by neglecting these two complicating factors. Bear in mind, however, that both are probably important in determining the details.

Modeling Star Formation

The next two sections describe the currently accepted theoretical view of star formation, derived in large part from numerical experiments performed on high-speed computers. The results are mathematical predictions of a multifaceted problem incorporating gravity, heat, rotation, magnetism, nuclear reaction rates, elemental abundances, and other physical conditions specifying the state of contracting interstellar clouds.

Scientific theories always develop in response to experimental or observational data, and theories of star formation are no exception. ∞ (Sec 1.2) The theory of star formation has evolved to explain innumerable observations of stars and star-forming regions. However, the phenomenology in this case is so complex and diverse that it is helpful to have a theoretical framework to "connect the dots" between phenomena that might otherwise appear unrelated. Accordingly, we present the theory first and then discuss how and where the observational data fit into and support the theoretical picture.

CONCEPT CHECK

✔ What basic competitive process controls star formation?

19.2 The Formation of Stars Like the Sun

LEARNING GOAL 2 Star formation begins when gravity begins to dominate over heat, causing a cloud to lose its equilibrium and start contracting. Only after the cloud has undergone radical changes in its internal structure is equilibrium finally restored.

Table 19.1 lists seven evolutionary stages that an interstellar cloud goes through in the process of becoming a main-sequence star like the Sun. The stages are characterized by varying central temperatures, surface temperatures, central densities, and radii of the prestellar object. They trace its progress from a cold, dark interstellar cloud to a hot, bright star. The numbers given in the table and in the following discussion are valid *only* for stars of approximately the same mass as that of the Sun. In the next section, we will relax this restriction and consider the formation of stars with masses different from that of the sun.

Stage 1: An Interstellar Cloud

The first stage in the star-formation process is a dense interstellar cloud—the core of a dark dust cloud or perhaps a molecular cloud. These clouds are truly vast, sometimes spanning tens of parsecs (10^{14}–10^{15} km) across. Typical temperatures are about 10 K throughout, with a density of perhaps 10^9 particles/m^3. Stage-1 clouds contain thousands of times the mass of the Sun, mainly in the form of cold atomic and molecular gas. (The dust in a stage-1

TABLE 19.1 Prestellar Evolution of a Solar-Type Star

Stage	Approximate Time to Next Stage (yr)	Central Temperature (K)	Surface Temperature (K)	Central Density (particles/m³)	Diameter* (km)	Object
1	2×10^6	10	10	10^9	10^{14}	Interstellar cloud
2	3×10^4	100	10	10^{12}	10^{12}	Cloud fragment
3	10^5	10,000	100	10^{18}	10^{10}	Cloud fragment/protostar
4	10^6	1,000,000	3000	10^{24}	10^8	Protostar
5	10^7	5,000,000	4000	10^{28}	10^7	Protostar
6	3×10^7	10,000,000	4500	10^{31}	2×10^6	Star
7	10^{10}	15,000,000	6000	10^{32}	1.5×10^6	Main-sequence star

*Round numbers; for comparison, recall that the diameter of the Sun is 1.4×10^6 km, while that of the solar system is roughly 1.5×10^{10} km.

cloud both cools the cloud as it contracts and plays a crucial role in planet formation, but it constitutes a negligible fraction of the total mass of the cloud.) ∞ (Sec. 6.7)

Despite their low internal temperatures, most observed dark interstellar clouds seem to have enough internal pressure to support themselves against the force of gravity. ∞ (*More Precisely 8-1*) However, if such a cloud is to be the birthplace of stars, it must become unstable, start to collapse under its own gravity, and eventually break up into smaller pieces. Most astronomers think that the process of star formation is triggered when some external event, such as the shock of a nearby stellar explosion or the pressure wave produced when a nearby O- or B-type star forms and ionizes its surroundings, squeezes a cloud beyond the point where pressure can resist gravity's inward pull. ∞ (Sec. 17.1) Or perhaps the cloud's supporting magnetic field leaks away as charged particles slowly drift across the confining field lines, leaving the gas unable to support its own weight.

Whatever the cause, theory suggests that once the collapse begins, fragmentation into smaller and smaller clumps of matter naturally follows, as gravitational instabilities continue to operate in the gas. As illustrated in Figure 19.4, a typical cloud can break up into tens, hundreds, or even thousands, of fragments, each imitating the shrinking behavior of the parent cloud and contracting ever faster. The whole process, from a single stable cloud to many collapsing fragments, takes a few million years.

In this way, depending on the precise conditions under which fragmentation takes place, an interstellar cloud can produce either a few dozen stars, each much larger than our Sun, or a whole cluster of hundreds of stars, each comparable to or smaller than our Sun. There is little evidence of stars born in isolation, one star from one cloud. Most stars—perhaps even all stars—appear to originate as members of multiple systems or large groups of stars. The Sun, which is now found alone and isolated in space, probably escaped from the larger system in which it formed, perhaps after an encounter with another star or some much larger object (such as a molecular cloud).

Stage 2: A Collapsing Cloud Fragment

The second stage in our evolutionary scenario represents the physical conditions in just one of the many fragments that develop in a typical interstellar cloud. A fragment destined to form a star like the Sun contains between one and two solar masses of material at this stage. Estimated to span a few hundredths of a parsec across, this fuzzy, gaseous blob is still about 100 times the size of our solar system. Its central density by this time is roughly 10^{12} particles/m³.

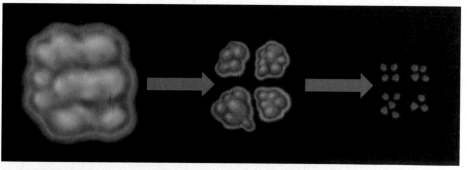

▲ FIGURE 19.4 Cloud Fragmentation As an interstellar cloud contracts, gravitational instabilities cause it to fragment into smaller pieces. The pieces themselves continue to fall inward and fragment, eventually forming many tens or hundreds of individual stars.

Even though the fragment has shrunk substantially, its average temperature is not much different from that of the original cloud. The reason is that the gas constantly radiates large amounts of energy into space. The material of the fragment is so thin that photons produced within it easily escape without being reabsorbed by the cloud, so virtually all the energy released in the collapse is radiated away and does not cause any significant increase in temperature. Only at the center, where the radiation must traverse the greatest amount of material to escape, is there any appreciable temperature rise. The gas there may be as warm as 100 K by this stage. For the most part, however, the fragment stays cold as it shrinks.

The process of continued fragmentation is eventually stopped by the increasing density within the shrinking cloud. As stage-2 fragments continue to contract, they eventually become so dense that radiation cannot get out of the cloud easily. The trapped radiation then causes the temperature to rise, the pressure to increase, and the fragmentation to cease.

Stage 3: Fragmentation Ceases

By the start of stage 3, several tens of thousands of years after it first began contracting, a typical stage-2 fragment has shrunk to roughly the size of our solar system (still 10,000 times the size of our Sun). The density in the inner regions has just become high enough that the gas is opaque to the radiation it emits, so the core of the fragment begins to heat up considerably, as noted in Table 19.1. The central temperature has reached about 10,000 K—hotter than the hottest steel furnace on Earth. However, the temperature in the fragment's outer parts has not increased much. The gas there is still able to radiate its energy into space and so remains cool. The density increases much faster in the center of the fragment than near the edge, so the outside is both cooler and thinner than the interior. By this time, the central density is approximately 10^{18} particles/m^3 (still only 10^{-9} kg/m^3 or so).

For the first time, our contracting cloud fragment is beginning to resemble a star. The dense, opaque region at the center is called a **protostar**—an embryonic object at the dawn of star birth. The protostar's mass grows as more and more material rains down on it from the surrounding, still shrinking fragment. However, the protostar's radius continues to decrease because pressure is still unable to overcome the relentless pull of gravity. After stage 3, we can distinguish a "surface" on the protostar—its *photosphere*. Inside the photosphere, the protostellar material is opaque to the radiation it emits.* From here on, the surface temperatures listed in Table 19.1 refer to the photosphere of the collapsing fragment, and not to its low-

density "periphery," where radiation can easily escape and the temperature remains low.

Stage 4: A Protostar

As the protostar evolves, it shrinks, its density grows, and its temperature rises, both in the core and at the photosphere. Some 100,000 years after the fragment began to form, it reaches stage 4, where its center seethes at about 1,000,000 K. Electrons and protons ripped from atoms whiz around at hundreds of kilometers per second, yet the temperature is still well short of the 10^7 K needed to ignite the proton–proton nuclear reactions that fuse hydrogen into helium. ∞ (Sec. 16.2) Still much larger than the Sun, our gassy heap is now about the size of Mercury's orbit. Heated by the material falling on it from above, it now has a surface temperature of a few thousand kelvins.

Knowing the protostar's radius and surface temperature, we can calculate its luminosity. Surprisingly, it turns out to be several thousand times the luminosity of the Sun.

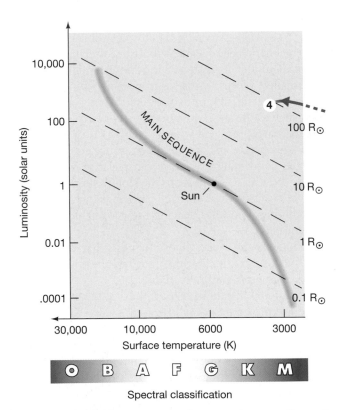

▲ **FIGURE 19.5 Protostar on the H–R Diagram** The red arrow indicates the approximate evolutionary track followed by an interstellar cloud fragment before it reaches the end of the Kelvin–Helmholtz contraction phase as a stage 4 protostar. The boldface numbers on this and subsequent H–R plots refer to the prestellar evolutionary stages listed in Table 19.1 and described in the text. Recall from Chapter 17 that "R$_\odot$" denotes the radius of the Sun.

---

*Note that this is the same definition of "surface" that we used for the Sun in Chapter 16. ∞ (Sec. 16.1)

Even though the protostar has a surface temperature only about half that of the Sun, it is hundreds of times larger, making its total luminosity very large indeed—in fact, much greater than the luminosity of most main-sequence stars. Because nuclear reactions have not yet begun, the protostar's luminosity is due entirely to the release of gravitational energy as the protostar continues to shrink and material from the surrounding fragment continues to fall onto its surface.

By the time stage 4 is reached, our protostar's physical properties can be plotted on the Hertzsprung–Russell (H–R) diagram, as shown in Figure 19.5. Recall that an H–R diagram is a plot of two key stellar properties: surface temperature (increasing to the left) and luminosity (increasing upward). ∞ (Sec. 17.6) The luminosity scale in the figure is expressed in terms of the solar luminosity (4×10^{26} W). Our G2-type Sun is plotted at a temperature of 6000 K and a luminosity of 1 unit. As before, the dashed diagonal lines in the H–R diagram represent an object's radius, allowing us to follow the changes in the protostar's size as it evolves. At each phase of the star's evolution, its surface temperature and luminosity can be represented by a point on the diagram. The motion of that point as the star evolves is known as the star's **evolutionary track**. It is a graphical representation of a star's life.

The red track in Figure 19.5 depicts the approximate path followed by our interstellar cloud fragment since it became a protostar at stage 3 (which itself lies off the right-hand edge of the figure). This early evolutionary track is known as the *Kelvin–Helmholtz contraction phase*, after the two European physicists (Lord Kelvin and Hermann von Helmholtz) who first studied the subject.

Figure 19.6 is an artist's sketch of an interstellar gas cloud proceeding along the evolutionary path outlined so far. As the stage-3 fragment contracts, it spins faster (to conserve angular momentum) and flattens into a rotating *protostellar disk* perhaps 100 A.U. in diameter, surrounding the central stage-4 protostar. ∞ (*More Precisely 6-2*) Recall that we first saw this process in Chapter 6, where we referred to the disk as the *solar nebula*. ∞ (Sec. 6.7) If the star is ultimately going to have a planetary system, by stage 4 that process is already well underway. ∞ (Sec. 15.2) However, regardless of whether planets actually form, astronomers think that protostellar disks are common—the vast majority of protostars (perhaps all) are accompanied by disks at this stage of their evolution.

Our protostar is still not in equilibrium. Even though its temperature is now so high that outward-directed pressure has become a powerful countervailing influence against gravity's continued inward pull, the balance is not yet perfect. The protostar's internal heat gradually diffuses out from the hot center to the cooler surface, where it is radiated away into space. As a result, the overall contraction slows, but it does not stop completely. From our perspective on Earth, this is quite fortunate: If the heated gas were somehow able to counteract gravity completely before the star reached the temperature and density needed to start nuclear burning in its core, the protostar would simply radiate away its heat and never become a true star. The night sky would be abundant in faint protostars, but completely lacking in the genuine article. Of course, there would be no Sun either, so it is unlikely that we, or any other intelligent life-form, would exist to appreciate these astronomical subtleties.

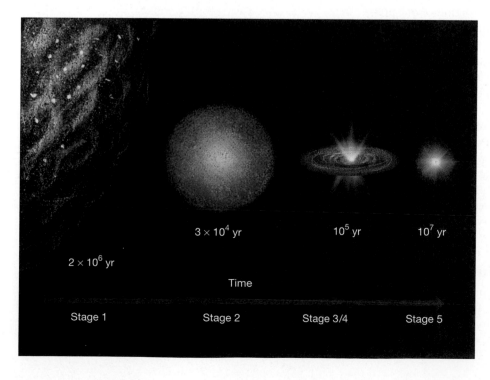

◀ **FIGURE 19.6 Interstellar Cloud Evolution** Artist's conception of the changes in an interstellar cloud during the early evolutionary stages outlined in Table 19.1. Shown are a stage 1 interstellar cloud; a stage 2 fragment; a smaller, hotter stage 4 fragment with jets; and a stage 5 protostar. (The pictures are not drawn to scale.) The duration of each stage, in years, is also indicated.

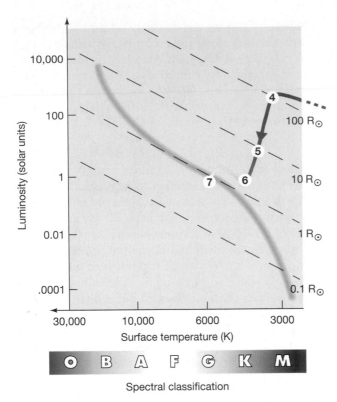

▲ **FIGURE 19.7 Newborn Star on the H–R Diagram** The changes in a protostar's observed properties are shown by the path of decreasing luminosity, from stage 4 to stage 6, often called the Hayashi track. At stage 7, the newborn star has arrived on the main sequence.

After stage 4, the protostar on the H–R diagram moves down (toward lower luminosity) and slightly to the left (toward higher temperature), as shown in Figure 19.7. Its surface temperature remains almost constant, and it becomes less luminous as it shrinks. This portion of our protostar's evolutionary path, running from point 4 to point 6 in Figure 19.7, is often called the *Hayashi track*, after C. Hayashi, a 20th-century Japanese astrophysicist whose groundbreaking work in the 1960s on the evolution of pre-main-sequence stars still provides the theoretical basis for all studies of star formation.

Protostars on the Hayashi track often exhibit violent surface activity during this phase of their evolution, resulting in extremely strong protostellar winds, much denser than the solar wind that flows from our own Sun. As mentioned previously, this portion of the protostar's evolution is often called the **T Tauri** phase, after T Tauri, the first "star" (actually protostar) to be observed in that stage of prestellar development. ∞ (Sec. 15.2)

Stage 5: Protostellar Evolution

By stage 5 on the Hayashi track, the protostar has shrunk to about 10 times the size of the Sun, its surface temperature is about 4000 K, and its luminosity has fallen to about 10 times the solar value. At this point, the central temperature has reached about 5,000,000 K. The gas is complete-

ly ionized by now, but the protons still do not have enough thermal energy to overcome their mutual electromagnetic repulsion and enter the realm of the nuclear binding force. ∞ (Sec. 16.2) The core is still too cool for nuclear fusion to begin.

Events proceed more slowly as the protostar approaches the main sequence. The initial contraction and fragmentation of the interstellar cloud occurred quite rapidly, but by stage 5, as the protostar nears the status of a full-fledged star, its evolution slows. The cause of this slowdown is heat: Even gravity must struggle to compress a hot object. The contraction is governed largely by the rate at which the protostar's internal energy can be radiated away into space. The greater this radiation of internal energy—that is, the more rapidly energy moves through the star to escape from its surface—the faster the contraction occurs. As the luminosity decreases, so, too, does the rate of contraction.

Stage 6: A Newborn Star

Some 10 million years after its first appearance, the protostar finally becomes a true star. By the bottom of the Hayashi track, at stage 6, when our roughly 1-solar-mass object has shrunk to a radius of about 1,000,000 km, the contraction has raised the central temperature to 10,000,000 K, enough to ignite nuclear burning. Protons begin fusing into helium nuclei in the core, and a star is born. As shown in Figure 19.7, the star's surface temperature at this point is about 4500 K, still a little cooler than the Sun. Even though the newly formed star is slightly larger in radius than our Sun, its lower temperature means that its luminosity is somewhat less than (actually, about two-thirds of) the solar value.

Stage 7: The Main Sequence at Last

Over the next 30 million years or so, the stage-6 star contracts a little more. In making this slight adjustment, the star's central density rises to about 10^{32} particles/m^3 (more conveniently expressed as 10^5 kg/m^3), the central temperature increases to 15,000,000 K, and the surface temperature reaches 6000 K. By stage 7, the star finally arrives at the main sequence, just about where our Sun now resides. Pressure and gravity are finally balanced, and the rate at which nuclear energy is generated in the core exactly matches the rate at which energy is radiated from the surface.

The evolutionary events just described occur over the course of some 40 to 50 million years. Although this is a long time by human standards, it is still less than 1 percent of the Sun's lifetime on the main sequence. Once an object begins fusing hydrogen in its core and establishes a "gravity-in, pressure-out" equilibrium, it is destined to burn steadily for a very long time. The star's location on the H–R diagram—that is, its surface temperature and luminosity—will remain virtually unchanged for the next 10 billion years.

CONCEPT CHECK

✔ What distinguishes a collapsing cloud from a protostar and a protostar from a star?

19.3 Stars of Other Masses

LEARNING GOAL 3 The numerical values and the evolutionary track just described are valid only for the case of a 1-solar-mass star. The temperatures, densities, and radii of prestellar objects of other masses exhibit similar trends, but the numbers and the tracks differ, in some cases considerably. Perhaps not surprisingly, the most massive fragments within interstellar clouds tend to produce the most massive protostars and, eventually, the most massive stars. Similarly, low-mass fragments give rise to low-mass stars.

The Zero-Age Main Sequence

Figure 19.8 compares the theoretical pre-main-sequence track taken by the Sun with the corresponding evolutionary tracks of a 0.3-solar-mass star and a 3-solar-mass star. All three tracks traverse the H–R diagram in the same general manner, but cloud fragments that eventually form stars more massive than the Sun approach the main sequence along a higher track on the diagram, while those

destined to form less massive stars take a lower track. The *time* required for an interstellar cloud to become a main-sequence star also depends strongly on its mass. The most massive cloud fragments heat up to the required 10 million K and become O-type stars in a mere million years, roughly $\frac{1}{50}$ the time taken by the Sun. The opposite is the case for prestellar objects having masses less than that of our Sun. A typical M-type star, for example, requires nearly a billion years to form.

Whatever the mass, the end point of the prestellar evolutionary track is the main sequence. A star is considered to have reached the main sequence when hydrogen burning begins in its core and the star's properties settle down to stable values. The main-sequence line thus predicted by theory is called the **zero-age main sequence** (or ZAMS, for short). The fact that the theoretically derived zero-age main sequence agrees very well with the actual main sequences observed for stars in the vicinity of the Sun and in more distant star clusters (see Section 19.6) provides strong support for the modern theory of star formation and stellar structure. ∞ (Sec. 1.2)

If all gas clouds contained precisely the same elements in exactly the same proportions, mass would be the sole determinant of a newborn star's location on the H–R diagram, and the zero-age main sequence would be a well-defined line rather than a broad band. However, the composition of a star affects its internal structure (mainly by changing the opacity of its outer layers), and this in turn affects both the star's temperature and its luminosity on the main sequence. Stars with more heavy elements tend to be cooler and slightly less luminous than stars that have the same mass, but contain fewer heavy elements. As a result, differences in composition between stars "blur" the zero-age main sequence into the broad band we observe.

It is important to realize that *the main sequence is itself not an evolutionary track—stars do not evolve along it.* Rather, it is just a "way station" on the H–R diagram where stars stop and spend most of their lives—low-mass stars at the bottom, high-mass stars at the top. Once on the main sequence, a star stays in essentially the same location on the H–R diagram during its whole time as a stage-7 object. (In other words, a star that arrives on the main sequence as, say, a G-type star can never "work its way up" to become a B- or an O-type main-sequence blue supergiant or move down to become an M-type red dwarf.) As we will see in Chapter 20, the next stage of stellar evolution occurs when a star moves away from the main sequence. A star leaving the main sequence and entering this next stage has pretty much the same surface temperature and luminosity it had when it arrived on the main sequence millions (or billions) of years earlier.

Failed Stars

Some cloud fragments are too small ever to become stars. The giant planet Jupiter is a good example. Jupiter formed in the Sun's protostellar disk (the solar nebula)

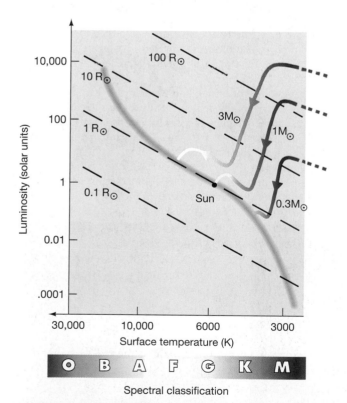

▲ **FIGURE 19.8 Prestellar Evolutionary Tracks** Some pre-main-sequence evolutionary paths for stars more massive and less massive than our Sun.

Observations of Brown Dwarfs

Cruelly put, brown dwarfs are "failed" stars—objects that formed through the contraction and fragmentation of an interstellar cloud, just as stars do, but somehow never reached the critical mass of about 0.08 solar mass (80 times the mass of Jupiter) needed to initiate hydrogen fusion in their cores. We mentioned in the text that interstellar space could contain large numbers of these small, dark, and hard-to-detect objects. Recent observations of star-forming regions have reinforced that idea by providing a plausible way in which the formation process could be stopped before a true star formed. Here we examine in a little more detail the observational evidence for these elusive objects.

Until 1994, this discussion would have been very short indeed, as there was no firm observational evidence for any brown dwarfs at all! Since then, however, continuing improvements in observational techniques have begun to bear fruit, and several dozen likely brown dwarfs are now known.

Detecting a brown dwarf is no easy task: Brown dwarfs are so faint that they are difficult—although not impossible—to detect directly. In searching for such objects in binary-star systems, astronomers use many of the same techniques they employ in the search for extrasolar planets, as discussed in Chapter 15. ∞ (Sec. 15.4) (Recall from Chapter 17 that most stars are binaries; the same may very well be true of brown dwarfs.) ∞ (Sec. 17.7) While numerous brown-dwarf candidates have been found in this way, the observational data seem to show a real deficit of brown dwarfs relative to planets—the same observations that have led to the discovery of many planets have turned up only a handful of brown dwarfs. It appears that brown dwarfs are not found as companions to solar-type stars, although recent observations of lower mass main-sequence stars suggest that the fraction of brown-dwarf companions may be higher.

Actually, the dividing line between brown dwarfs and Jupiter-like planets is not completely clear cut, especially given the properties of the extrasolar planets now known and the possibility that our solar system may not be the only, or even the most likely, type of planetary system to form. ∞ (Sec. 15.5) Most researchers adopt a division at about 12 times the mass of Jupiter. Above that mass (but below 80 Jupiter masses), although core temperatures never become high enough for hydrogen fusion to occur, a contracting fragment will experience a brief phase of *deuterium fusion*, as the core becomes hot enough for any deuterium nuclei that are present in the original cloud to combine. The phase ends once the deuterium is consumed, and the fragment's "nuclear" lifetime is over. Below 12 Jupiter masses, no nuclear fusion of any kind is expected.

The accompanying images show two binary-star systems containing brown-dwarf candidates (marked by arrows). The first (left) is an *HST* image of Gliese 623, which was originally identified as a binary system because of its variations in radial velocity. ∞ (Secs. 15.5, 17.7) From the binary's measured orbital separation and period, the mass of the faint companion appears to be approximately 0.1 solar mass—very close to the limit for a brown dwarf. ∞ (Sec. 2.7) Astronomers hope that continued observations of this system will allow the companion's mass to be measured with sufficient accuracy to determine whether or not it really is a brown dwarf. The "rings" in the image are instrumental artifacts.

The other two images show the binary star system Gliese 229, first identified as containing a possible brown dwarf by ground-based infrared observations (center) and subsequently imaged from space (right). The two objects are 7″ apart; the fainter "star" has a luminosity only a few millionths that of the Sun and an estimated mass about 50 times that of Jupiter. (The bright diagonal streak in the image is caused by a hardware problem in the CCD chip used to record it.)

Infrared and spectroscopic studies offer other ways of searching for brown dwarfs, especially those which are not binaries. The presence of molecules such as methane, which would be destroyed at the high temperatures found in a star, is a good indicator of a low-temperature object. Infrared observations are particularly effective because brown dwarfs emit most of their radiation in the infrared part of the spectrum, while true stars tend to be brightest at the near-infrared and optical ranges. (See Figure 19.21.)

The presence of lithium, an element that exists in interstellar gas, but is rapidly depleted in a star once nuclear fusion begins, is another telltale sign of a brown dwarf. Researchers using a spectrometer on the giant Keck telescope in Hawaii have identified several brown-dwarf candidates by this means. Recent infrared surveys of the sky have revealed about 100 low-temperature (less than 2000 K) objects and have so far identified lithium in about a quarter of them.

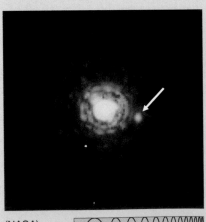

(NASA)

R I V U X G

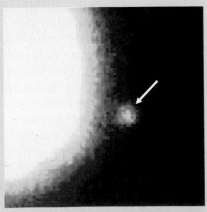

(Palomar)

R I V U X G

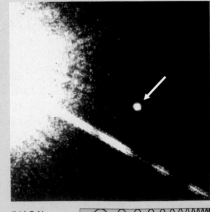

(NASA)

R I V U X G

and contracted under the influence of gravity. The resultant heat is still detectable, but the planet did not have enough mass for gravity to crush its matter to the point of nuclear ignition. ∞ (Sec. 11.3) Instead, Jupiter became stabilized by heat and rotation before the planet's central temperature became hot enough to fuse hydrogen. Thus, Jupiter never evolved beyond the protostar stage. If it, or any of the other jovian planets, had continued to accumulate gas from the solar nebula, it might have become a star (almost certainly to the detriment of life on Earth). However, that did not occur—virtually all the matter present during the formative stages of our solar system is now gone, swept away by the solar wind during the Sun's T Tauri phase. ∞ (Sec. 15.2)

Low-mass gas fragments simply lack the mass needed to initiate nuclear burning. Rather than turning into stars, they continue to cool, eventually becoming compact, dark "clinkers"—cold fragments of unburned matter—orbiting a star or moving alone through interstellar space. On the basis of theoretical modeling, astronomers think that the minimum mass of gas needed to generate core temperatures high enough to begin nuclear fusion is about 0.08 solar mass (80 times the mass of Jupiter). Our practical definition of a star requires that it shine via the energy released by nuclear fusion reactions in its core. Thus this mass of 0.08 times the mass of the Sun is a lower limit on the masses of all stars in the universe.

Vast numbers of "substellar" objects may well be scattered throughout the universe—fragments frozen in time somewhere along the Kelvin–Helmholtz contraction phase. Small, faint, and cool (and growing ever colder), they are known collectively as **brown dwarfs**. For theoretical reasons discussed in more detail in *Discovery 19-1*, researchers generally reserve the term *brown dwarf* to mean a low-mass prestellar fragment of more than about 12 Jupiter masses (so Jupiter itself is not a brown dwarf, by this definition). Anything smaller is simply called a *planet*.

Observationally, these faint, low-mass objects are difficult to study, be they planets or brown dwarfs associated with stars, or interstellar cloud fragments far from any star. We can detect stars by means of telescopes, and we can infer the presence of interstellar atoms and molecules by spectroscopic analysis, but astronomical objects of intermediate size outside our solar system remain hard to see. However, recent advances in observational hardware and image-processing techniques have begun to yield impressive results. (See *Discovery 19-1*.) Current observations suggest that up to 100 billion cold, dark substellar objects may lurk in the depths of interstellar space—a number comparable to the total number of "real" stars in our Galaxy.

CONCEPT CHECK

✔ Do stars evolve along the main sequence?

19.4 Observations of Cloud Fragments and Protostars

4 How can we verify the theoretical picture just outlined? The age of our entire civilization is much shorter than the time needed for a single interstellar cloud to contract and form a star. We can never observe individual objects proceed through the full panorama of star birth. However, we can do the next best thing: We can observe many different objects—interstellar clouds, protostars, and young stars approaching the main sequence—as they appear today at different stages of their evolutionary paths. Each observation is like part of a jigsaw puzzle. When properly oriented relative to all the others, the pieces can be used to build up a picture of the full life cycle of a star.

Evidence of Cloud Contraction

Prestellar objects at stages 1 and 2 are not yet hot enough to emit much infrared radiation, and certainly no optical radiation arises from their dark, cool interiors. The best way to study the early stages of cloud contraction and fragmentation is to observe the radio emission from interstellar molecules within those clouds. Consider again M20, the splendid emission nebula studied in Chapter 18. ∞ (Sec. 18.2) The brilliant region of glowing, ionized gas shown in Figure 18.8 is not our main interest here, however; instead, the youthful O- and B-type stars that energize the nebula alert us to the general environment in which stars are forming. Emission nebulae are indicators of star birth.

The region surrounding M20 contains galactic matter that seems to be contracting. The presence of (optically) invisible gas there was illustrated in Figure 18.20, which showed a contour map of the abundance of the formaldehyde (H_2CO) molecule. Formaldehyde and many other molecules are widespread in the vicinity of the nebula, especially throughout the dusty regions below and to the right of the emission nebula itself. Further analysis of the observations suggests that this region of greatest molecular abundance is also contracting and fragmenting, and is well on its way toward forming a star—or, more likely, a star cluster.

The interstellar clouds in and around M20 thus provide tentative evidence of three distinct phases of star formation, as shown in Figure 19.9. The huge, dark molecular cloud surrounding the visible nebula is the stage-1 cloud. Both its density and its temperature are low—about 10^8 particles/m^3 and 20 K, respectively. Greater densities and temperatures typify smaller regions within this large cloud. The totally obscured regions labeled A and B, where the molecular emission of radio energy is strongest, are such denser, warmer fragments. Here, the total gas density is observed to be at least 10^9 particles/m^3, and the temperature is about 100 K. The Doppler shifts of the radio lines observed in the vicinity of region B indicate that this portion of M20, labeled "contracting fragment" in the figure,

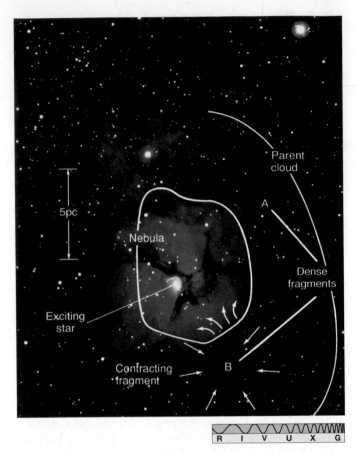

▲ **FIGURE 19.9 Star Formation Phases** The M20 region shows observational evidence for three broad phases in the birth of a star. The parent cloud is stage 1 of Table 19.1. The region labeled "contracting fragment" likely lies between stages 1 and 2. Finally, the emission nebula (M20 itself) results from the formation of one or more massive stars (stages 6 and 7). *(background photo: AURA)*

is contracting. Less than a light-year across, the region has a total mass over a thousand times the mass of the Sun—considerably more than the mass of M20 itself. The region lies somewhere between stages 1 and 2 of Table 19.1.

The third star-formation phase shown in Figure 19.9 is M20 itself. The glowing region of ionized gas results directly from a massive O-type star that formed there within the past million years or so. Because the central star is already fully formed, this final phase corresponds to stage 6 or 7 of our evolutionary scenario.

Evidence of Cloud Fragments

Other parts of our Milky Way Galaxy provide sketchy evidence for prestellar objects in stages 3 through 5. The Orion complex, shown in Figure 19.10, is one such region. Lit from within by several O-type stars, the bright Orion Nebula is partly surrounded by a vast molecular cloud that extends well beyond the roughly 2 × 3-pc region bounded by the photograph in Figure 19.10(b).

The Orion molecular cloud harbors several smaller sites of intense radiation emitted by molecules deep within the core of the cloud fragment. Their extent, shown in Figures 19.10(c) and (e) measures about 10^{10} km, or 1/1000 of a light-year, about the diameter of our solar system. Their density is about 10^{15} particles/m^3, much denser than the surrounding cloud. Although the temperature of these smaller regions cannot be estimated reliably, many researchers regard the regions as objects well on their way to stage 3. We cannot determine whether those regions will eventually form stars like the Sun, but it does seem certain that the intensely emitting objects in them are on the threshold of becoming protostars.

Evidence of Protostars

In the hunt for, and study of, objects at more advanced stages of star formation, radio techniques become less useful, because stages 4, 5, and 6 have increasingly higher temperatures. By Wien's law, their emission shifts toward shorter wavelengths, so these objects shine most strongly in the infrared. ∞ (Sec. 3.4) One particularly bright infrared emitter, known as the Becklin–Neugebauer object, was detected in the core of the Orion molecular cloud in the 1970s. Its luminosity is around a thousand times the luminosity of the Sun. Most astronomers agree that this warm, dense blob is a high-mass protostar, probably in or around stage 4.

Until the *Infrared Astronomy Satellite* (*IRAS*) was launched in the early 1980s, astronomers were aware of giant stars forming only in clouds far away. But *IRAS* showed that many such stars are forming much closer to home, and some of these protostars have masses comparable to that of our Sun. Figure 19.11 shows two examples of low-mass protostars, both spotted by *HST* in a rich star-forming region in Orion. Their infrared heat signatures are those expected of an object on the Hayashi track, at around stage 5.

The energy sources for some infrared objects seem to be luminous hot stars that are hidden from optical view by surrounding dark clouds. Apparently, these stars are already so hot that they emit large amounts of ultraviolet radiation, which is mostly absorbed by "cocoons" of dust surrounding them. The absorbed energy is then reemitted by the dust as infrared radiation. These bright infrared sources are known as *cocoon nebulae*. Two considerations support the idea that the hot stars heating the dust have only recently ignited: (1) The dust cocoons are predicted to disperse quite rapidly once their central stars form, and (2) they are invariably found in the dense cores of molecular clouds. The central stars probably lie near stage 6.

Protostellar Winds

Protostars often exhibit strong winds. Radio and infrared observations of hydrogen and carbon monoxide molecules in the Orion molecular cloud have revealed gas expanding outward at velocities approaching 100 km/s. High-resolution interferometric observations have disclosed expanding knots of water emission within the same star-forming

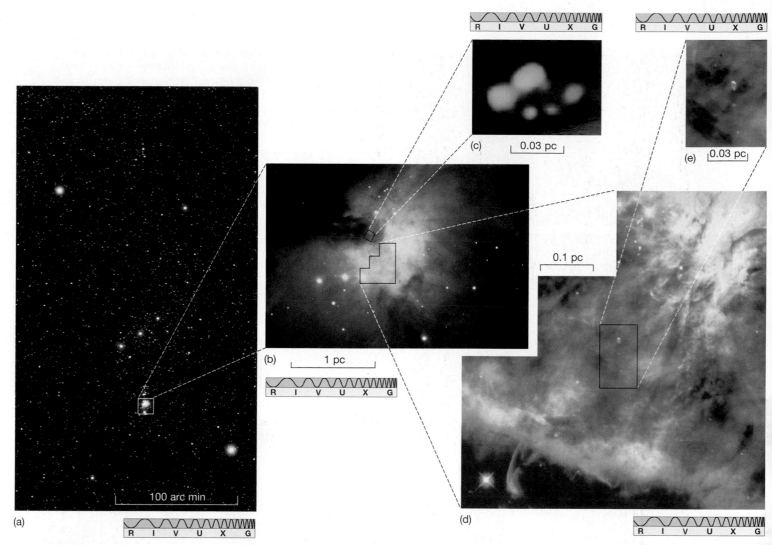

▲ FIGURE 19.10 Orion Nebula, Up Close (a) The constellation Orion, with the region around its famous emission nebula marked by a rectangle. The Orion Nebula is the middle "star" of Orion's sword. (b) Enlargement of the framed region in part (a), suggesting how the nebula is partly surrounded by a vast molecular cloud. Various parts of this cloud are probably fragmenting and contracting, with even smaller sites forming protostars. The three frames at the right show some of the evidence for those protostars: (c) a false-color radio image of some intensely emitting molecular sites, (d) a nearly real-color visible image of embedded nebular "knots" thought to harbor protostars, and (e) a high-resolution image of several young stars surrounded by disks of gas and dust out of which planets might ultimately form. *(Astrostock-Sanford; AURA; Harvard; NASA)*

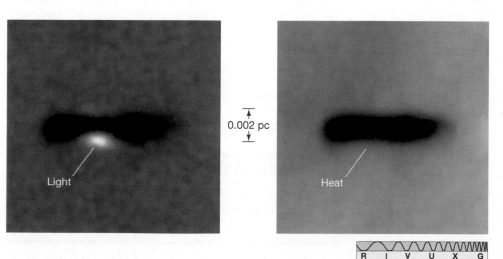

◀ FIGURE 19.11 Protostars Two infrared images of planetary-system-sized dusty disks in the Orion region, showing heat and light emerging from their centers. On the basis of their temperatures and luminosities, these unnamed sources appear to be low-mass protostars on the Hayashi track in the H–R diagram. *(NASA)*

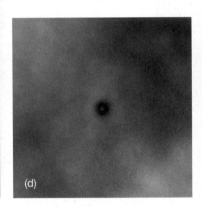

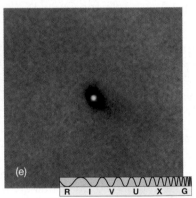

◄ **FIGURE 19.12 Protostellar Wind** (a) The nebular disk around a protostar can be the site of intense heating and strong outflows, forming a bipolar jet perpendicular to the disk. (b) As the disk is blown away by the wind, the jets fan out, eventually (c) merging into a spherical wind. Parts (d) and (e) show two actual circumstellar disks in the Orion region. (*NASA*)

region and have linked the strong winds to the protostars themselves. ∞ (Sec. 5.6) These winds may be related to the violent surface activity associated with many protostars.

As mentioned earlier, a young protostar may be embedded in an extensive protostellar disk of nebular material in which planets are forming. ∞ (Sec. 6.7) Strong heating within the turbulent disk and a powerful protostellar wind combine to produce a *bipolar flow*, expelling two "jets" of matter in the directions perpendicular to the disk, as illustrated in Figure 19.12. As the protostellar wind gradually destroys the disk, blowing it away into space, the outflow widens until, with the disk gone, the wind flows away from the star equally in all directions. Figure 19.13 shows the emission from an especially clear bipolar flow, along with an artist's conception of the system producing it.

These outflows can be very energetic. Figure 19.14 shows a portion of the Orion molecular cloud, south of the Orion Nebula, where a newborn star is seen still surrounded by a bright nebula, its turbulent wind spreading out into the interstellar medium. Below the star (enlarged in the inset) are twin jets known as HH1 and HH2. ("HH" stands for Herbig–Haro, the investigators who first cataloged such objects.) Formed in another (unseen) protostellar disk—the protostar itself is still hidden within the dusty cloud fragment from which it formed—these jets have traveled outward for almost half a light-year before colliding with interstellar matter. More Herbig–Haro objects can be seen in the upper right portion of the figure. One of them, the oddly shaped "waterfall," may be due to an earlier outflow from the same protostar responsible for the existence of HH1 and HH2.

CONCEPT CHECK

✔ How can a "snapshot" of the universe today test our theories of the evolution of individual objects?

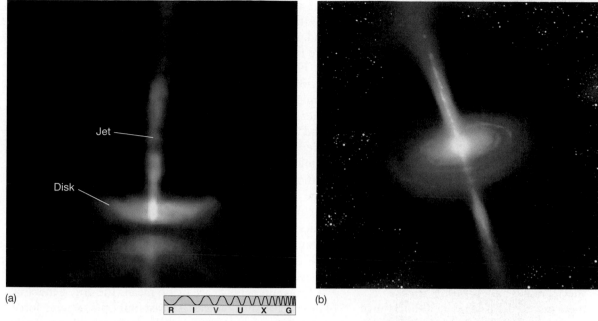

(a)

Jet

Disk

R I V U X G

(b)

▲ **FIGURE 19.13 Bipolar Jets** (a) This remarkable image shows two jets emanating from the young star system HH30, the result of infalling matter being expelled from an embryonic star near the center. (b) An idealized artist's conception of a young star system, showing two jets flowing perpendicular to the disk of gas and dust rotating around the star. *(NASA; D. Berry)*

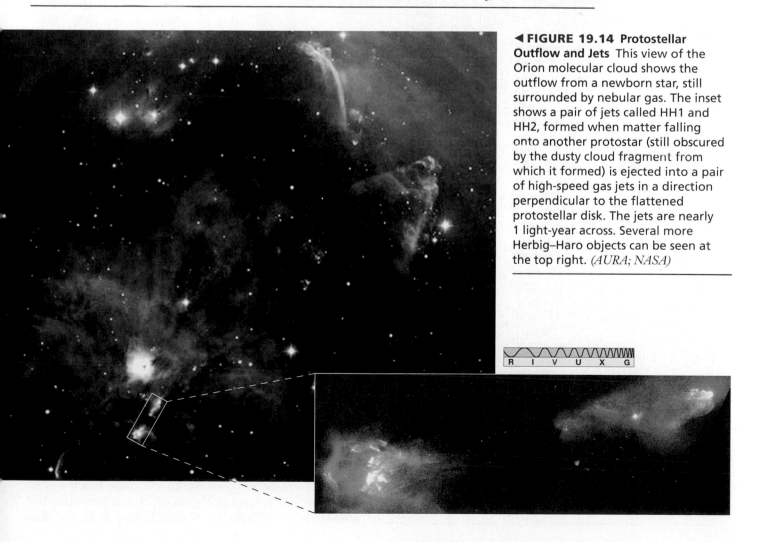

◄ **FIGURE 19.14 Protostellar Outflow and Jets** This view of the Orion molecular cloud shows the outflow from a newborn star, still surrounded by nebular gas. The inset shows a pair of jets called HH1 and HH2, formed when matter falling onto another protostar (still obscured by the dusty cloud fragment from which it formed) is ejected into a pair of high-speed gas jets in a direction perpendicular to the flattened protostellar disk. The jets are nearly 1 light-year across. Several more Herbig–Haro objects can be seen at the top right. *(AURA; NASA)*

R I V U X G

19.5 Shock Waves and Star Formation

The subject of star formation is really much more complicated than the preceding discussion suggests. Interstellar space is populated with many kinds of clouds, fragments, protostars, stars, and nebulae, all interacting in a complex fashion and each type of object affecting the behavior of all the other types. For example, the presence of an emission nebula in or near a molecular cloud probably influences the evolution of the entire region. We can easily imagine expanding waves of matter driven outward by the high temperatures and pressures in the nebula. As the waves crash into the surrounding molecular cloud, interstellar gas tends to pile up and become compressed. Such a shell of gas, rushing rapidly through space, known as a **shock wave**, can push ordinarily thin matter into dense sheets, just as a plow pushes snow.

Many astronomers regard the passage of a shock wave through interstellar matter as the triggering mechanism needed to initiate star formation in a galaxy. Calculations show that when a shock wave encounters an interstellar cloud, it races around the thinner exterior of the cloud more rapidly than it can penetrate the cloud's thicker interior. Thus, shock waves do not blast a cloud from only one direction, but effectively squeeze it from many directions. Atomic-bomb tests have experimentally demonstrated this squeezing: Shock waves created in the blast tend to surround buildings, causing them to be blown together (imploded) rather than apart (exploded). The "contracting fragment" in Figure 19.9 may well have been triggered by the shock wave from the M20 nebula. Note the correspondence between the shock-compressed region at the lower right and the high-density molecular gas revealed by radio studies (Figure 18.20). Once shock waves have begun compressing an interstellar cloud, natural gravitational instabilities take over, dividing the cloud into the fragments that eventually form stars.

Emission nebulae are by no means the only generators of interstellar shock waves. At least four other driving forces are available: the relatively gentle deaths of old stars in the form of planetary nebulae (to be discussed in Chapter 20); the much more violent ends of certain stars in supernova explosions (Chapter 21); the spiral-arm waves that plow through the Milky Way (Chapter 23); and interactions between galaxies (Chapter 24). Supernovae are by far the most energetic, and probably also the most efficient, means of piling up matter into dense clumps. However, they are relatively few and far between, so the other mechanisms may be more important overall in triggering star formation. Although the evidence is somewhat circumstantial, the presence of young (and thus fast-forming) O- and B-type stars in the vicinity of supernova remnants does suggest that the birth of stars is often initiated by the violent, explosive deaths of others.

This picture of shock-induced star formation is complicated by the fact that O- and B-type stars form quickly, live briefly, and die explosively. These massive stars, themselves perhaps born of a passing shock wave, may in turn create new shock waves, either through the expanding nebular gas produced by their births or through their explosive deaths. The new shock waves can produce "second-generation" stars, which in turn will explode and give rise to still more shock waves, and so on. As depicted in Figure 19.15, star formation resembles a chain reaction. Other, lighter stars are also

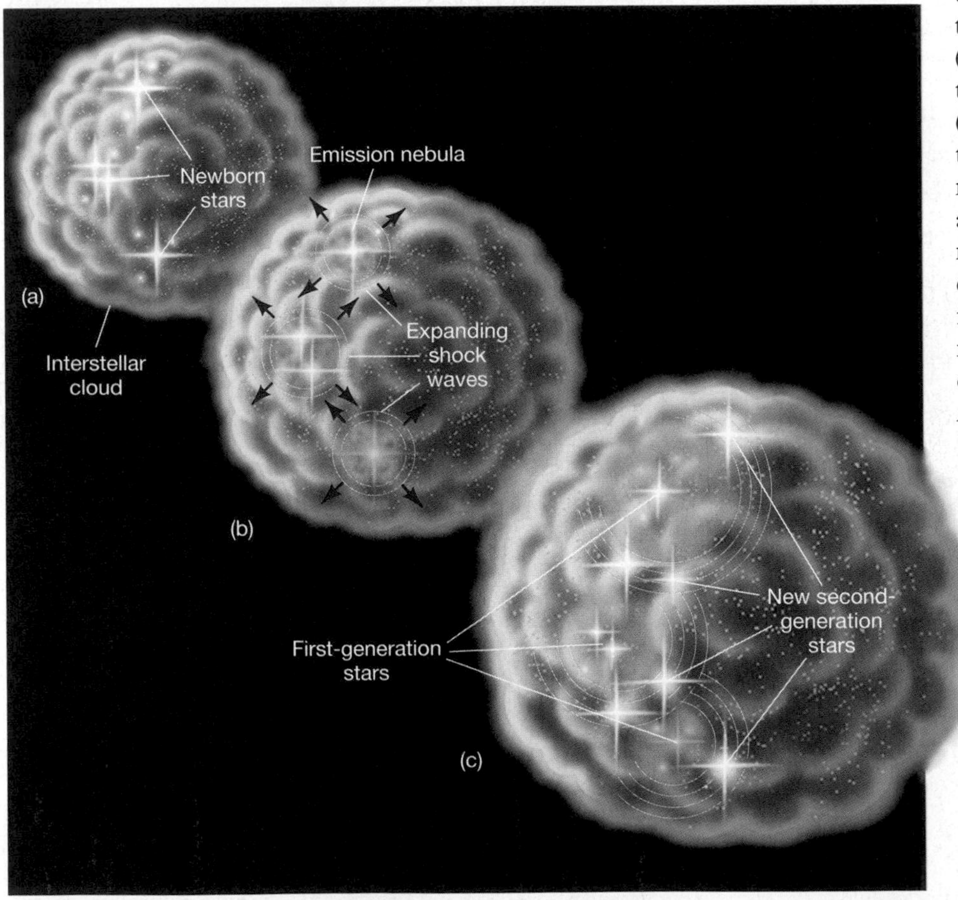

▲ **FIGURE 19.15 Generations of Star Formation** (a) Star birth and (b) shock waves lead to (c) more star births and more shock waves in a continuous cycle of star formation in many areas of our Galaxy. As in a chain reaction, old stars trigger the formation of new stars ever deeper into an interstellar cloud.

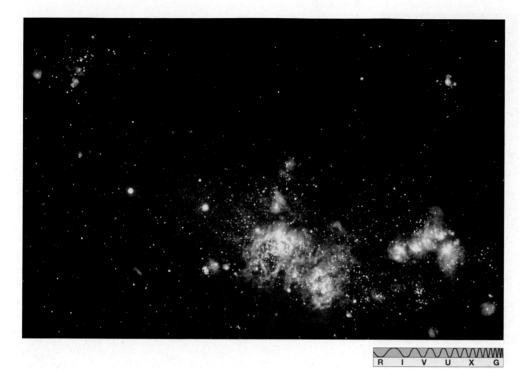

R I V U X G

formed in the process, of course, but they are largely "along for the ride." It is the O- and B-type stars that drive the star-formation wave through the cloud.

Observational evidence lends some support to this chain-reaction picture. Groups of stars nearest molecular clouds do indeed appear to be the youngest, whereas those farther away seem to be older. Figure 19.16 shows a spectacular *HST* image of a star-forming region in the galaxy NGC 4214, which lies some 13 million light-years from Earth. A series of bright emission nebulae, powered by hot young stars, can be seen, suggesting that a wave of star formation recently swept across the region, triggering the sequence seen here.

CONCEPT CHECK

✔ Why might we expect multiple episodes of star formation to occur in some locations?

19.6 Star Clusters

The end result of the collapse of a cloud is a group of stars, all formed from the same parent cloud and lying in the same region of space. Such a collection of stars is called a **star cluster**. Figure 19.17 shows a spectacular view of a newborn star cluster and (part of) the interstellar cloud from which it came. Because all the stars formed at the same time out of the same cloud of interstellar gas and under the same environmental conditions, clusters are near-ideal "laboratories" for stellar studies—not in the sense that astronomers can perform experiments on them, but because the properties of the stars are very tightly con-

strained. The only factor distinguishing one star from another in the same cluster is mass, so theoretical models of star formation and evolution can be compared with reality without the complications introduced by the broad spreads in age, chemical composition, and place of origin found when we consider all stars in our Galactic neighborhood.

Clusters and Associations

Figure 19.18(a) shows a small star cluster called the Pleiades, or Seven Sisters, a well-known naked-eye object in the constellation Taurus, lying about 120 pc from Earth. This type of loose, irregular cluster, found mainly in the plane of the Milky Way (see Figure 18.5), is called an **open cluster**. Open clusters typically contain from a few hundred to a few tens of thousands of stars and are a few parsecs across.

Figure 19.18(b) shows the H–R diagram of stars in the Pleiades. The cluster contains stars in almost all parts of the main sequence—only the very brightest main-sequence stars are missing. (The brightest six or seven stars in the diagram have just left the main sequence, as will be discussed in Chapter 20.) Thus, even though we have no direct evidence of the cluster's birth, we can estimate its age as less than about 100 million years, the lifetime of a main-sequence B-type star. ∞ (Sec. 17.8) If all the stars in the cluster formed at the same time, then the red stars must be young, too. The wisps of leftover gas evident in the photograph are further evidence of the cluster's relative youth. In addition, the system is abundant in heavy elements that (as we will see) could have been created only within the cores of many generations of ancient stars long since perished.

Less massive, but more extended, clusters are known as **associations**. These clusters typically contain no more than a few hundred bright stars, but may span many tens of parsecs. Associations tend to be rich in very young stars.

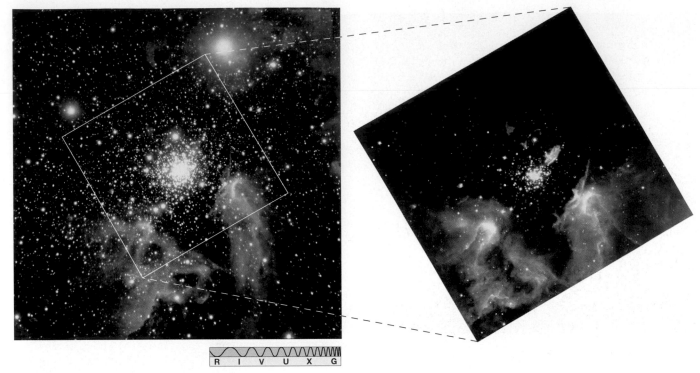

▲ **FIGURE 19.17 Newborn Cluster** The star cluster NGC 3603 and a portion of the larger molecular cloud in which it formed. The cluster contains about 2000 bright stars and lies some 6000 pc from Earth. The field of view shown here spans about 20 light-years. Radiation from the cluster has cleared a cavity in the cloud several light-years across. The inset shows the central area more clearly, including the most massive star in the region (called Sher 25, above and to the left of the cluster), which is already near the end of its lifetime, having ejected part of its outer layers and formed a ring of gas. Many low-mass stars, less massive than the Sun, can also be seen. *(ESO; NASA)*

Those containing many pre-main-sequence T Tauri stars are known as *T associations*, whereas those with prominent O- and B-type stars, such as the Trapezium in Orion (see Figure 19.21a below), are called *OB associations*. It is quite likely that the main difference between associations and open clusters is simply the efficiency (as measured by the fraction of gas that eventually ends up in stars) with which stars formed from the parent cloud.

Figure 19.19(a) shows a very different type of star cluster, called a **globular cluster**. All globular clusters are roughly spherical (which accounts for their name), are generally found away from the Milky Way plane, and contain hundreds of thousands, and sometimes millions, of stars spread out over about 50 pc. Figure 19.19(b) is an H–R diagram of the cluster shown, which is called Omega Centauri. Notice the many differences between this H–R diagram and that of Figure 19.18(b)—globular clusters present a stellar environment very different from that of open clusters like the Pleiades. The distance to Omega Centauri cluster has been determined by a variation on the method of spectroscopic parallax, applied to the entire cluster rather than to individual stars. ∞ (Sec. 17.6) It lies about 5000 pc from Earth.

The most outstanding spectroscopic feature of globular clusters is their lack of upper-main-sequence stars. As-

tronomers in the 1920s and 1930s, working with instruments incapable of detecting stars fainter than about 1 solar luminosity at the distances of globular clusters, and having no theory of stellar evolution to guide them, were puzzled by the H–R diagrams they saw when they looked at the globular clusters. Indeed, a comparison of just the top halves of the diagrams (so that the lower main sequences cannot be seen) reveals few similarities between Figures 19.18(b) and 19.19(b).

Most globular clusters contain no main-sequence stars with masses greater than about 0.8 times the mass of the Sun. The more massive O- through F-type stars have long since exhausted their nuclear fuel and disappeared from the main sequence (in fact becoming the red giants and other luminous stars above the main sequence, as we will see in Chapter 20). ∞ (Sec. 17.7) From the theory of stellar evolution (Chapter 20), the A-type stars in Figure 19.19b are now known to be stars at much later stages in their evolution that just happen to be passing through the location of the upper main sequence. On the basis of these and other observations, astronomers estimate that most globular clusters are at least 10 billion years old—they contain the oldest known stars in our Galaxy.

Other observations confirm the great ages of globular clusters. For example, their spectra show few heavy elements, implying that these stars formed in the distant past, when heavy elements were much less abundant than they are today (Chapter 21). Astronomers speculate that the 150 or so globular clusters observed today are just the survivors of a much larger population of clusters that formed long ago.

Clusters and Nebulae

How many stars form in a cluster, and of what type are they? How much gas is left over? What does the collapsed cloud look like once star formation has run its course? At present, although the main stages in the formation of individual stars (stages 3–7) are becoming clearer, the answers

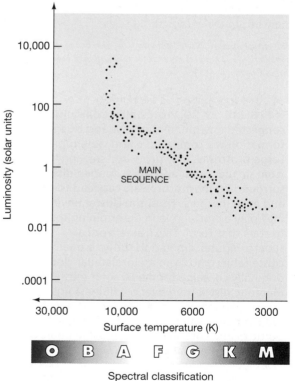

(a)

(a)

(b)

▲ **FIGURE 19.18 Open Cluster** (a) The Pleiades cluster (also known as the Seven Sisters or M45) lies about 120 pc from the Sun. The naked eye can see only six or seven of its brightest stars. (b) An H–R diagram of all the stars of this well-known open cluster. (*AURA*)

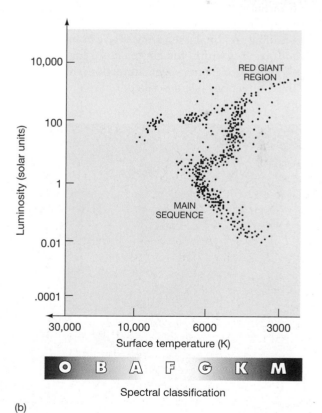

(b)

▲ **FIGURE 19.19 Globular Cluster** (a) The globular cluster Omega Centauri is approximately 5000 pc from Earth and spans some 40 pc in diameter. (b) An H–R diagram of some of its stars. (*Pat Seitzer*)

to these more general questions (involving stages 1 and 2) are still sketchy. They await a more thorough understanding of the star-formation process.

In general, the more massive the collapsing region, the more stars are likely to form there. In addition, we know from H–R diagrams of observed stars that low-mass stars are much more common than high-mass ones. ∞ (Sec. 17.8) For every O- or B-type giant, hundreds or even thousands of G-, K-, and M-type dwarfs may form. The precise number of stars of any given mass or spectral type likely depends in a complex (and poorly understood) way upon conditions within the parent cloud. The same is true of the *efficiency* of star formation—the fraction of the total mass that actually finds its way into stars—which determines the amount of leftover material. However, if, as is usually the case, one or more O- or B-type stars form, their intense radiation and winds will cause the surrounding gas to disperse, leaving behind a young star cluster.

In recent years, astronomers have come to realize that physical interactions—close encounters and even collisions—between protostars within a star cluster may be very important in determining the properties of the stars that eventually form. Detailed supercomputer simulations of star-forming clouds suggest that, while the seven stages presented earlier (and listed in Table 19.1) remain a good description of the overall formation process, the sequence of events leading to a main-sequence star can be strongly influenced by events within the cluster itself. Figure 19.20 presents frames from one such simulation, illustrating some of the interactions just described.

The simulations reveal that the strong gravitational fields of the most massive protostars give them a competitive advantage over their smaller rivals in attracting gas from the surrounding nebula, causing the giant protostars to grow even faster. Such encounters usually disrupt the smaller protostellar disks, terminating the growth of the central protostars and ejecting planets and low-mass brown dwarfs from the disk into intracluster space. In dense clusters these interactions may even lead to mergers and further growth of massive objects. Thus, even before the intense radiation of newborn O and B stars begins to disperse the cluster gas, the formation of a few large bodies can significantly inhibit the growth of smaller ones.*

All these considerations clearly illustrate the important role played by a future star's environment in the star-formation process and provide important insight into why low-mass stars are so much more common than high-mass stars. The first few massive bodies to form tend to prevent the formation of additional high-mass stars by stealing their "raw material" and ultimately disrupting the environment in which other stars are growing. This tendency also helps explain the existence of brown dwarfs, by providing at least two natural ways (disk destruction and gas dispersion) in which star formation can stop before nuclear fusion begins in a growing stellar core. *Discovery 19-2* describes another system in which the gas-dispersal process may be almost complete.

*Compare this picture of large objects dominating the accretion process at the expense of the smaller bodies around them with the standard view of planet formation presented in Chapter 6. ∞ (Sec. 6.7)

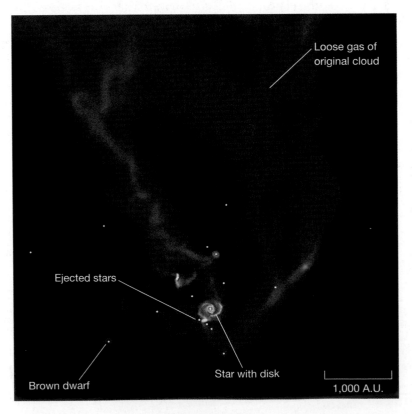

Loose gas of original cloud

Ejected stars

Brown dwarf

Star with disk

1,000 A.U.

◄ **FIGURE 19.20 Protostellar Collisions** In the congested environment of a young cluster, star formation is a competitive and violent process. Large protostars may grow by "stealing" gas from smaller ones, and the extended disks surrounding most protostars can lead to collisions and even mergers. Thus, the cluster environment may play a crucial role in determining the types of stars that form. This frame from a supercomputer simulation shows a small star cluster emerging from an interstellar cloud that originally contained some 50 solar masses of material, distributed over a volume 1 light year across. (Remnants of the cloud are shown here in red.) The simulation followed 3.5 million gas particles and consumed over 100,000 hours of computer time. *Matthew Bate (University of Exeter), Ian Bonnell (University of St. Andrews), and Volker Bromm (Harvard-Smithsonian Center for Astrophysics)*

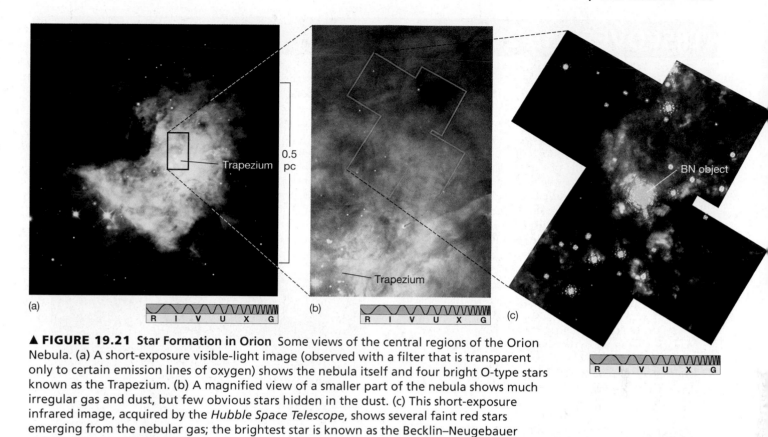

▲ **FIGURE 19.21** **Star Formation in Orion** Some views of the central regions of the Orion Nebula. (a) A short-exposure visible-light image (observed with a filter that is transparent only to certain emission lines of oxygen) shows the nebula itself and four bright O-type stars known as the Trapezium. (b) A magnified view of a smaller part of the nebula shows much irregular gas and dust, but few obvious stars hidden in the dust. (c) This short-exposure infrared image, acquired by the *Hubble Space Telescope*, shows several faint red stars emerging from the nebular gas; the brightest star is known as the Becklin–Neugebauer object. *(Lick Observatory; NASA)*

Until the 1990s, the existence of star clusters within emission nebulae was largely conjecture—the stars cannot be seen optically because they are obscured by dust. However, infrared observations have now clearly demonstrated that stars really are found within star-forming regions! Figure 19.21 compares optical and infrared views of the central regions of the Orion Nebula. The optical images in Figure 19.21(a) and (b) show the Trapezium, the group of four bright stars responsible for ionizing the nebula; the false-color infrared image in Figure 19.21(c) reveals an extensive cluster of stars within and behind the visible nebula. The Becklin–Neugebauer object (Section 19.4) can be seen as the central yellow spot within this region. It is thought to be a dust-shrouded B-type star just beginning to form its own emission nebula. These remarkable images show many stages of star formation.

Cluster Lifetimes

Eventually, star clusters dissolve into individual stars. In some cases, the ejection of unused gas reduces a cluster's mass so much that it becomes gravitationally unbound and dissolves rapidly. In clusters that survive the early gas-loss phase, stellar encounters tend to eject the lightest stars from the cluster, just as the gravitational slingshot effect can propel spacecraft around the solar system. ∞ *(Discovery 6-2)* At the same time, the tidal gravitational field of the Milky Way

Galaxy slowly strips outlying stars from the cluster. ∞ (Sec. 7.6) Occasional distant encounters with giant molecular clouds also tend to remove stars from a cluster. Even a near miss may disrupt the cluster entirely.

As a result of all these influences, most open clusters break up in a few hundred million years, although the actual lifetime depends on the cluster's mass. Loosely bound associations may survive for only a few tens of millions of years, whereas some very massive open clusters are known from their H–R diagrams to be almost 5 billion years old. In a sense, only when a star's parent cluster has completely dissolved is the star-formation process really complete. The road from a gas cloud to a single, isolated star like the Sun is long and tortuous indeed!

Take another look at the sky one clear, dark evening. Ponder all of the cosmic activity you have learned about as you peer upward at the stars. After studying this chapter, you may find that you have to modify your view of the night sky. Even the seemingly quiet nighttime darkness is dominated by continual change.

CONCEPT CHECK

✔ If stars in a cluster all form at the same time, how can some influence the formation of others?

DISCOVERY 19-2

Eta Carinae

At the heart of the Carina emission nebula (shown in the main figure) lies a remarkable object called Eta Carinae (right-hand images). With an estimated mass of around 100 times the mass of the Sun and a luminosity of 5 million times the solar value, Eta Carinae is one of the most massive stars known. Formed probably only a few hundred thousand years ago, this star has had an explosive, though brief, life, indeed. In the mid-19th century, Eta Carinae produced an outburst that made it one of the brightest stars in the southern sky (even though it lies some 3000 pc away from Earth, a very long way compared to most of the bright stars visible in our night sky). The star released as much visible energy as a supernova explosion (see Chapter 21), yet it somehow survived the event.

The bottom-right image, obtained by *HST* and carefully processed to reveal fine detail, is the highest-resolution view of the explosion obtained to date. Dust lanes, tiny condensations in the outflowing material, and dark radial streaks of unknown origin all appear with exquisite clarity. The star itself is the white dot at the center of the image. The two ends of the

"peanut" (at the top right and bottom left) are blobs of ejected material racing away from the star at hundreds of kilometers per second—perhaps enough to expel the surrounding nebular gas and convert the Carina Nebula into the Carina Cluster. Perpendicular to the line joining these two blobs is a thin disk of gas, also moving outward at high speed.

The details of the events leading to the Eta Carinae outburst are unclear. Quite possibly, such episodes of violent activity are the norm for supermassive stars. However, although a few comparable outbursts have been observed in other galaxies, they are so rare that astronomers still do not know what constitutes "typical" behavior for such exotic objects.

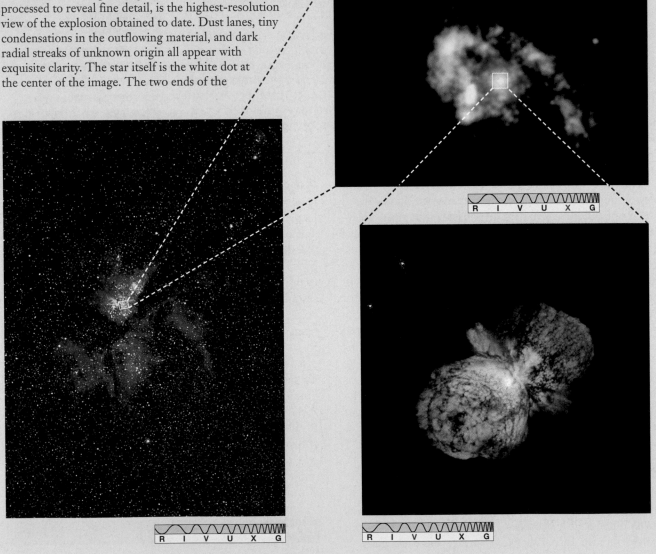

R I V U X G

Chapter Review

SUMMARY

Stars form when an interstellar cloud collapses under its own gravity and breaks up into pieces comparable in mass to our Sun. Heat, rotation, and magnetism all compete with gravity to influence the cloud's evolution. The evolution of the contracting cloud—the changes in its temperature and luminosity—can be conveniently represented as an **evolutionary track (p. 495)** on the Hertzsprung–Russell diagram. A cold interstellar cloud containing a few thousand solar masses of gas can fragment into tens or hundreds of smaller clumps of matter, from which stars eventually form.

As a collapsing prestellar fragment heats up and becomes denser, it eventually becomes a **protostar (p. 494)**—a warm, very luminous object that emits radiation mainly in the infrared portion of the electromagnetic spectrum. At this stage of its evolution, the protostar is also known as a **T Tauri (p. 496)** star, after the first object of that type discovered. Eventually, a protostar's central temperature becomes high enough for hydrogen fusion to begin, and the protostar becomes a star. For a star like the Sun, the whole formation process takes about 50 million years. Protostellar winds encounter less resistance in the directions perpendicular to the star's protostellar disk. Thus, they expel two jets of matter in the directions of the protostar's poles. As the protostellar winds gradually destroy the disk, the jets widen, until, with the disk gone, the wind flows away from the star equally in all directions.

More massive stars pass through similar stages, but much more rapidly. Stars less massive than the Sun take much longer to form. The **zero-age main sequence (p. 497)** is the region on the H–R diagram where stars lie when the formation process is over. Mass is the key property in determining a star's characteristics and life span. The most massive stars have the shortest formation times and main-sequence lifetimes. At the other extreme, some low-mass fragments never reach the point of nuclear ignition. The universe may be populated with a vast number of **brown dwarfs (p. 499)**—objects that are not massive enough to fuse hydrogen to helium in their interiors.

Many of the objects predicted by the theory of star formation have been observed in real astronomical objects. The dark interstellar regions near emission nebulae often provide evidence of cloud fragmentation and protostars. Radio telescopes are used in studying the early phases of cloud contraction and fragmentation; infrared observations allow us to see later stages of the process. Many well-known emission nebulae, lit by several O- and B-type stars, are partially engulfed by molecular clouds, portions of which are probably fragmenting and contracting, with smaller sites forming protostars. **Shock waves (p. 504)** can compress other interstellar clouds and trigger star formation. Star birth and the production of shock waves are thought to produce a chain reaction of star formation in molecular cloud complexes.

A single collapsing and fragmenting cloud can give rise to hundreds or thousands of stars—a **star cluster (p. 505)**. The formation of the most massive stars may play an important role in suppressing the further formation of stars from lower mass cluster members. **Open clusters (p. 505)**, with a few hundred to a few thousand stars, are found mostly in the plane of the Milky Way. They typically contain many bright blue stars, indicating that they formed relatively recently. **Globular clusters (p. 506)** are found mainly away from the Milky Way plane and may contain millions of stars. They include no main-sequence stars much more massive than the Sun, indicating that they formed long ago. Globular clusters are thought to date from the formation of our Galaxy. Loosely bound groups of newborn stars are called stellar **associations (p. 505)**. Infrared observations have revealed young star clusters or associations in several emission nebulae. Eventually, clusters break up into individual stars, although the entire process may take billions of years.

REVIEW AND DISCUSSION

1. Briefly describe the basic chain of events leading to the formation of a star like the Sun.
2. What is the role of heat in the process of stellar birth?
3. What is the role of rotation in the process of stellar birth?
4. What is the role of magnetism in the process of stellar birth?
5. What is an evolutionary track?
6. Why do stars tend to form in groups?
7. Why does the evolution of a protostar slow down as the star approaches the main sequence?
8. In what ways do the formative stages of high-mass stars differ from those of stars like the Sun?
9. What are brown dwarfs?
10. What are T Tauri stars?
11. Stars live much longer than we do, so how do astronomers test the accuracy of theories of star formation?
12. At what evolutionary stages must astronomers use radio and infrared radiation to study prestellar objects? Why can't they use visible light?
13. Why has it been difficult until recently to demonstrate that stars and protostars actually exist within star-forming regions?
14. What is a shock wave? Of what significance are shock waves in star formation?
15. Explain the usefulness of the H–R diagram in studying the evolution of stars. Why can't evolutionary stages 1–3 be plotted on the diagram?

16. Compare the times necessary for the various stages in the formation of a star like the Sun. Why are some so short and others so long?

17. What do star clusters and associations have to do with star formation?

18. Compare and contrast the observed properties of open star clusters and globular star clusters.

19. How can we tell whether a star cluster is young or old?

20. In the formation of a star cluster with a wide range of stellar masses, is it possible for some stars to die out before others have finished forming? Do you think this will have any effect on the cluster's formation?

CONCEPTUAL SELF-TEST: TRUE OR FALSE/MULTIPLE CHOICE

1. Given the typical temperatures found in interstellar space, a cloud containing as few as a million atoms has sufficient gravity for it to begin to collapse.

2. Star clusters form when an interstellar cloud explodes into smaller pieces.

3. Most stars form as members of groups or clusters.

4. An evolutionary track charts a star or protostar's changing position within its parent cluster.

5. More massive stars form more rapidly.

6. Brown dwarfs take a long time to form, but will eventually become visible as stars on the lower main sequence.

7. Stages 1 and 2 of star formation can be observed with optical telescopes.

8. Shock waves produced by emission nebulae can initiate star formation in nearby molecular clouds.

9. The formation of the first high-mass stars in a collapsing cloud tends to inhibit further star formation within that cloud.

10. The gas in an emission nebula eventually dissipates into space, leaving behind a star cluster.

11. If a newly forming star has an excess of heat, then it will likely have **(a)** more gravity; **(b)** less gravity; **(c)** a slower contraction rate; **(d)** a rapid contraction rate.

12. The gravitational contraction of an interstellar cloud is primarily the result of its **(a)** mass; **(b)** composition; **(c)** diameter; **(d)** pressure.

13. The interstellar cloud from which our Sun formed was **(a)** slightly larger than the Sun; **(b)** comparable in size to Saturn's orbit; **(c)** comparable in mass to the solar system; **(d)** thousands of times more massive than the Sun.

14. A protostar that will eventually turn into a star like the Sun is significantly **(a)** smaller; **(b)** more luminous; **(c)** fainter; **(d)** less massive than the Sun.

15. Prestellar objects in which nuclear fusion never starts are referred to as **(a)** terrestrial planets; **(b)** brown dwarfs; **(c)** protostars; **(d)** globules.

16. The current theory of star formation is based upon **(a)** amassing evidence from many different regions of the Galaxy; **(b)** carefully studying the births of a few stars; **(c)** systematically measuring the masses and rotation rates of interstellar clouds; **(d)** observations made primarily at short wavelengths.

17. If the initial interstellar cloud in Figure 19.15 ("Generations of Star Formation") were much more massive, the result would be **(a)** the formation of more stars; **(b)** contraction of the cloud due to stronger gravitational attraction; **(c)** stars forming closer together; **(d)** stronger shock waves.

18. One of the primary differences between the Pleiades cluster, shown in Figure 19.18(a), and Omega Centauri, shown in Figure 19.19(a), is that the Pleiades cluster is much **(a)** larger; **(b)** younger; **(c)** farther away; **(d)** denser.

19. If the HR Diagram shown in Figure 19.19(b) ("Globular Cluster") were redrawn to illustrate a much younger cluster, the main-sequence turnoff would shift to **(a)** higher temperature; **(b)** higher pressure; **(c)** higher frequency; **(d)** a spectral classification of K or M.

20. A typical open cluster will dissolve in about the same amount of time as the time since **(a)** North America was first visited by Europeans; **(b)** dinosaurs walked on Earth; **(c)** Earth was formed; **(d)** the universe formed.

PROBLEMS

The number of squares preceding each problem indicates its approximate level of difficulty.

1. ■■ In order for an interstellar gas cloud to contract, the average speed of its constituent particles must be less than half the cloud's escape speed. ∞ (*More Precisely 8-1*) Will a (spherical) molecular hydrogen cloud with a mass of 1000 solar masses, a radius of 10 pc, and a temperature of 10 K begin to collapse? Why or why not?

2. ■■ Under the same assumptions as in Problem 1, estimate the minimum mass needed to cause a 1000-K, 1-pc cloud to collapse.

3. ■■ Use the radius–luminosity–temperature relation to explain how a protostar's luminosity changes as it moves from stage 4 (temperature 3000 K, radius 2×10^8 km) to stage 6 (temperature 4500 K, radius 10^6 km). What is the change in absolute magnitude? ∞ (*Sec. 17.2*)

4. ■ A protostar on the Hayashi track evolves from a temperature of 3500 K and a luminosity 5000 times that of the Sun to a temperature of 5000 K and a luminosity of 3 solar units. What is the protostar's radius (a) at the start and (b) at the end of the evolution?

5. ■ What is the (approximate) absolute magnitude of a stage-5 protostar? (See Figure 19.7.)

6. ■■ Use the H–R diagrams in this chapter to estimate by what factor a 1000-solar-luminosity, 3000-K protostar is larger than a main-sequence star of the same luminosity.

7. ■ By how many magnitudes does a 3-solar-mass star decrease in brightness as it evolves from stage 4 to stage 6? (See Figure 19.8.)

8. ■■ As a simple model of the final stage of star formation, imagine that, between stages 6 and 7, a star's surface temperature increases with time at a constant rate, while the luminosity remains constant at the stage-7 level. The stage-7

radius is equal to the solar value. Using the temperatures given in Table 19.1, calculate the star's radius at a time exactly halfway between these two stages.

9. ■ What is the luminosity, in solar units, of a brown dwarf whose radius is 0.1 solar radius and whose surface temperature is 600 K (0.1 times that of the Sun)?

10. ■■ What is the maximum distance at which the brown dwarf in the previous problem could be observed by a telescope of limiting apparent magnitude (a) 18, (b) 30?

11. ■ A shock wave from a supernova explosion moves at a speed of about 5000 km/s. How long will such a disturbance take to cross a molecular cloud 20 pc in diameter?

12. ■■ The luminosity of a hypothetical star-forming region is dominated by five bright O-type stars, each of absolute magnitude −8. What is the net absolute magnitude of the region?

13. ■■ If the star-forming region in Problem 12 is in a galaxy 10 Mpc (1 megaparsec = 1,000,000 pc) from Earth, calculate the apparent magnitude of the region.

14. ■■■ An open cluster has a diameter of 5 pc and a mass 1000 times the mass of the Sun. (a) Estimate the typical speed of its component stars. (b) On the basis of the speed you calculated in (a), estimate the number of times a typical star orbits the center of the cluster in the 500 million years it takes for the cluster to dissolve in the galactic tidal field.

15. ■■■ Approximating the gravitational field of our Galaxy as a mass of 10^{11} solar masses at a distance of 8000 pc (see Chapter 23), estimate the "tidal radius" of a 20,000-solar-mass open-star cluster—that is, the distance from the cluster's center outside of which the galactic tidal force overwhelms the cluster's own gravity.

Surprisingly, given that no one has ever seen a single star move through all its evolutionary paces, the theory of stellar evolution is rather well understood. Like archaeologists who study bones and artifacts of different ages, astronomers observe stars of differing ages and then try to assemble the pieces into a consistent picture of how stars evolve over billions of years. In this striking mosaic of images telemetered to Earth by the Advanced Camera for Surveys aboard the Hubble Space Telescope, we see the Helix Nebula, whose aged central star is shedding its outer ▶ envelope. (STScI)

Stellar Evolution

The Life and Death of a Star

After reaching the main sequence, a newborn star changes little in outward appearance for more than 90 percent of its lifetime. However, at the end of that period, as the star begins to run out of fuel and die, its properties once again change greatly. Aging stars travel along evolutionary tracks that take them far from the main sequence as they end their lives. In this and the next two chapters, we will study the evolution of stars during and after their main-sequence burning stages. We will find that the ultimate fate of a star depends primarily on its mass—although interactions with other stars can also play a decisive role—and that the final states of stars can be strange, indeed. By continually comparing theoretical calculations with detailed observations of stars of all types, astronomers have refined the theory of stellar evolution into a precise and powerful tool for understanding the universe.

LEARNING GOALS

Studying this chapter will enable you to

1 Explain why stars evolve off the main sequence.

2 Outline the events that occur after a Sun-like star exhausts the supply of hydrogen in its core.

3 Summarize the stages in the death of a typical low-mass star, and describe the resulting remnant.

4 Contrast the evolutionary histories of high-mass and low-mass stars.

5 Discuss the observations that help verify the theory of stellar evolution.

6 Explain how the evolution of stars in binary systems may differ from that of isolated stars.

Visit astro.prenhall.com/chaisson for additional annotated images, animations, and links to related sites for this chapter.

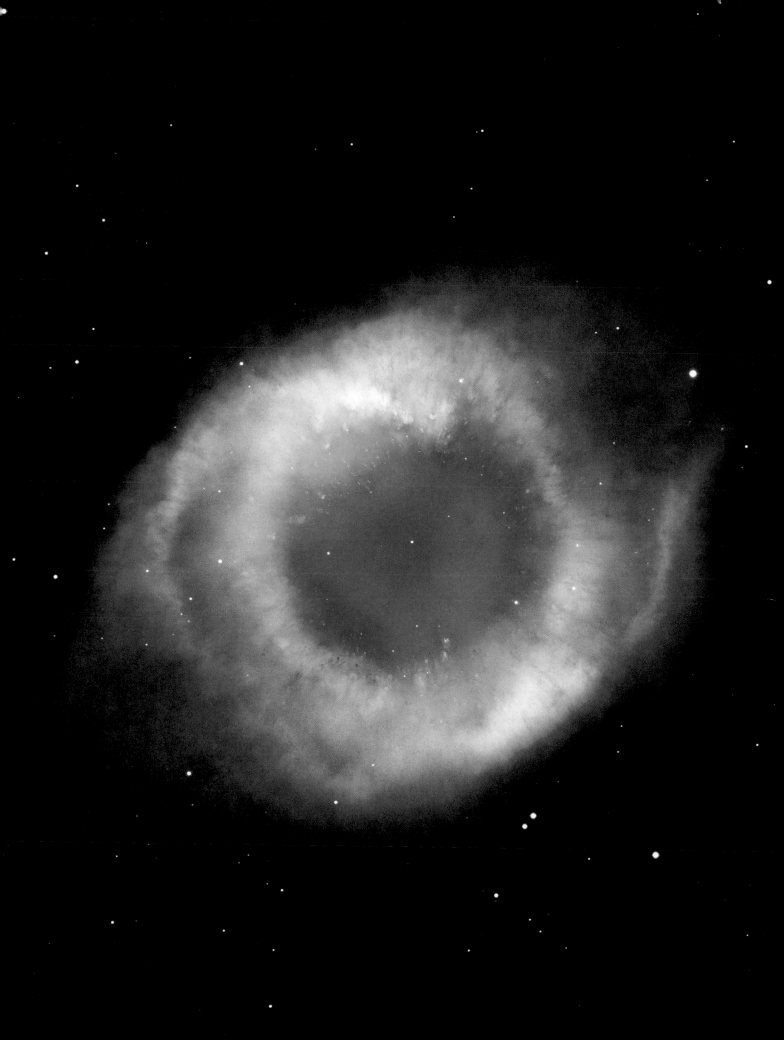

20.1 Leaving the Main Sequence

Most stars spend most of their lives on the main sequence. A star like the Sun, for example, after spending a few tens of millions of years in its formative stages (1–6 in Chapter 19), resides on or near the main sequence (stage 7) for 10 billion years before turning into something else. ∞ (Sec. 19.2) That "something else" is the main topic of this chapter.

Stars and the Scientific Method

Before we embark on our study of how and why stars evolve, let's take a moment to reflect on both the scope and the scientific basis of the theory we are about to present.

Realize first that no one has ever witnessed (or ever will witness) the complete lifetime of any star, from birth to death. Stars take an enormously long time—millions, billions, and even trillions of years—to evolve. ∞ (Sec. 17.8) Virtually all the low-mass stars that have ever formed still exist as stars. The coolest M-type stars—red dwarfs—consume their nuclear fuel so slowly that not one of them has yet left the main sequence. Some of them will shine steadily for a trillion years or more. By contrast, the most massive O- and B-type stars exhaust their fuel and leave the main sequence after only a few million years. Most of the high-mass stars that have ever existed perished long ago. Yet, despite the fact that it is physically impossible for any one human being to witness the lifetime of a star, over the past century astronomers have developed a detailed and comprehensive theory of stellar evolution, now arguably one of the best-tested theories in all of astronomy.

How can we can talk so confidently about what took place billions of years in the past and what will happen billions of years in the future? The answer is that, although we can never hope to follow the evolution of an individual star, we can observe billions of stars in the universe, and that is enough to see examples of every possible stage of stellar development, thereby allowing us to test and refine our theoretical ideas. Just as anthropologists might piece together a picture of the human life cycle by studying snapshots of all the residents of a large city, so astronomers can construct a picture of stellar evolution by studying the myriad stars we see in the night sky.

The modern theory of the lives and deaths of stars is one more excellent example of the scientific method in action. ∞ (Sec. 1.2) Faced with a huge volume of observational data, with little or no theory to organize or explain it, astronomers in the late 19th and early 20th centuries painstakingly classified and categorized the properties of the stars they observed. ∞ (Sec. 17.5) During the first half of the 20th century, as quantum mechanics began to yield detailed explanations of the behavior of light and matter on subatomic scales, theoretical explanations of many key stellar properties emerged. ∞ (Sec. 4.2) The individual pieces of the puzzle were being assembled, although the "big picture" was still unclear. Since the 1950s, a truly comprehensive theory has emerged, tying together the basic disciplines of atomic and nuclear physics, electromagnetism, thermodynamics, and gravitation into a coherent whole. Today, theory and observation proceed hand in hand, each refining and validating the details of the other as astronomers continue to hone their understanding of stellar evolution.

Finally, note that astronomers always use the term "evolution" in this context to mean change *during the lifetime of an individual star*. Contrast this usage with the meaning of the term in biology, where it refers to changes in the characteristics of a *population* of plants or animals over many generations. In fact, as we will see in Chapter 21, populations of stars do evolve in the latter "biological" sense, as the overall composition of the interstellar medium (and hence of each new stellar generation) changes slowly over time due to nuclear fusion in stars. However, in astronomical parlance, "stellar evolution" always refers to changes that occur during a single stellar lifetime.

Structural Change

On the main sequence, a star slowly fuses hydrogen into helium in its core. This process of nuclear fusion is called **core hydrogen burning**. In Chapter 16, we saw how the proton–proton fusion chain powers the Sun. ∞ (Sec. 16.2) *More Precisely 20-1* describes another sequence of nuclear reactions—one that is of great importance in stars more massive than the Sun. This sequence accomplishes the same basic result as the proton–proton chain, but in a very different way. Here is another instance where astronomers use a fairly familiar term in a quite unfamiliar way: To astronomers, "burning" always means nuclear fusion in a star's core, and not the chemical reaction (such as the combustion of wood or gasoline in air) we would normally think of in everyday speech. Chemical burning does not directly affect atomic nuclei.

As illustrated schematically in Figure 20.1, a main-sequence star is in a state of *hydrostatic equilibrium*, in which pressure's outward push exactly counteracts gravity's inward pull. This is a stable balance between gravity and pressure in which a small change in one always results in a small compensating change in the other. For example, as shown in the figure, a small increase in the star's central temperature leads to an increase in pressure, causing the star to expand and cool, thus recovering its equilibrium. Conversely, a small decrease in temperature leads to a slight decrease in pressure, and gravity causes the star to contract and heat up; again, equilibrium is restored. You

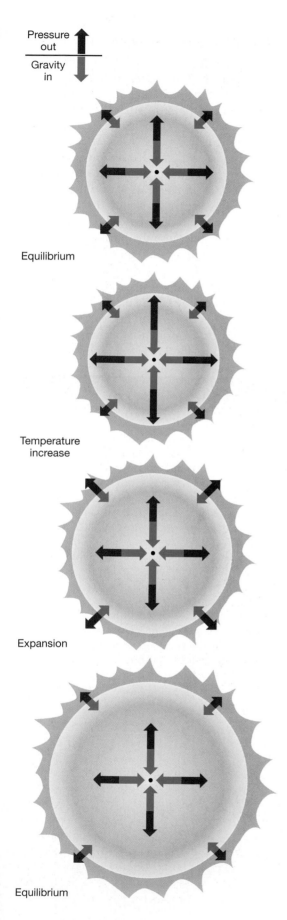

should keep Figure 20.1 in mind as you study the various stages of stellar evolution described next. Much of a star's complex behavior can be understood in these simple terms.

As the main-sequence star ages, its core temperature rises, and both its luminosity and radius increase. These changes are very slow, though—only a factor of three or four in luminosity over the Sun's entire 10-billion-year main-sequence lifetime, for example. As a result, the star's location on the H–R diagram remains almost unchanged during this phase (which is why we see the main sequence when we plot an H–R diagram for any reasonably large group of stars). ∞ (Sec. 17.5) Eventually, as the hydrogen in the core is consumed, the star's internal balance starts to shift, and both its internal structure and its outward appearance begin to change more rapidly: The star leaves the main sequence.

Once a star begins to move away from the main sequence, its days are numbered. The post-main-sequence stages of stellar evolution—the end of a star's life—depend critically on the star's mass. As a rule of thumb, we can say that low-mass stars die gently, whereas high-mass stars die catastrophically. The dividing line between these two very different outcomes lies around eight times the mass of the Sun, and in this chapter we will refer to stars of more than 8 solar masses as "high-mass" stars. Within both the "high-mass" and the "low-mass" (i.e., less than 8 solar masses) categories, there are substantial variations, some of which we will point out as we proceed.

Rather than dwelling on the many details, we will concentrate on a few representative evolutionary sequences. We begin by considering the evolution of a fairly low mass star like the Sun. The stages described in the next few sections pertain to the Sun as it nears the end of its fusion cycle 5 billion years from now. In fact, most of the qualitative features of the discussion apply to any low-mass star, although the exact numbers vary considerably. Later, we will broaden our discussion to include all stars, large and small.

CONCEPT CHECK

✔ In what sense is the Sun stable?

◀ **FIGURE 20.1 Hydrostatic Equilibrium** In a steadily burning star on the main sequence, the outward pressure of hot gas exactly balances the inward pull of gravity. This is true at every point within the star, guaranteeing its stability. From top to bottom, the frames illustrate how the star's size and internal structure change when the temperature in the interior increases. Whatever the change, the star adjusts, and equilibrium is restored.

20.2 Evolution of a Sun-like Star

The surface of a main-sequence star like the Sun occasionally erupts in flares and spots, but for the most part the star does not exhibit any sudden, large-scale changes in its properties. Its average surface temperature remains fairly constant, while its luminosity increases very slowly with time. The Sun has roughly the same surface temperature as it had when it formed nearly 5 billion years ago and is some 30 percent brighter than it was at that time.

This stable state of affairs cannot continue indefinitely, however. Eventually, drastic changes occur in the star's interior structure. After approximately 10 billion years of steady core hydrogen burning, a Sun-like star begins to run out of fuel. The situation is a little like that of an automobile cruising effortlessly along a highway at a constant speed for many hours, only to have the engine suddenly cough and sputter as the gas gauge reaches empty. Unlike automobiles, though, stars are not easy to refuel.

Stage 8: The Subgiant Branch

As nuclear fusion proceeds, the composition of the star's interior changes. Figure 20.2 illustrates the increase in helium abundance and the corresponding decrease in hydrogen abundance that take place in the stellar core as the star ages. Three cases are shown: (a) the chemical composition of the original core, (b) the composition after 5 billion years, and (c) the composition after 10 billion years. Case (b) represents approximately the present state of our Sun.

The star's helium content increases fastest at the center, where temperatures are highest and the burning is fastest. The helium content also increases near the edge of the core, but more slowly because the burning rate is less rapid there. The inner, helium-rich region becomes larger and more deficient in hydrogen as the star continues to shine. Eventually, about 10 billion years after the star arrived on the main sequence (Figure 20.2c), hydrogen becomes depleted at the center, the nuclear fires there subside, and the location of principal burning moves to

---

MORE PRECISELY 20-1

The CNO Cycle

The proton–proton chain is not the only nuclear process operating in the Sun and other late-generation stars. Another fusion mechanism capable of converting hydrogen into helium, starting from carbon-12 (^{12}C), proceeds according to the following six steps, with nitrogen (N) and oxygen (O) nuclei created as intermediate products:

$$^{12}\text{C} + {}^{1}\text{H} \rightarrow {}^{13}\text{N} + \text{energy}$$
$$^{13}\text{N} \rightarrow {}^{13}\text{C} + \text{positron} + \text{neutrino}$$
$$^{13}\text{C} + {}^{1}\text{H} \rightarrow {}^{14}\text{N} + \text{energy}$$
$$^{14}\text{N} + {}^{1}\text{H} \rightarrow {}^{15}\text{O} + \text{energy}$$
$$^{15}\text{O} \rightarrow {}^{15}\text{N} + \text{positron} + \text{neutrino}$$
$$^{15}\text{N} + {}^{1}\text{H} \rightarrow {}^{12}\text{C} + {}^{4}\text{He}$$

These six steps are termed the *CNO cycle*. Aside from the radiation and neutrinos produced, notice that the sum total of these six reactions is

$$^{12}\text{C} + 4({}^{1}\text{H}) \rightarrow {}^{12}\text{C} + {}^{4}\text{He}.$$

In other words, the net result is the fusion of four protons into a single helium-4 nucleus, just as in the proton–proton chain. The carbon-12 acts merely as a *catalyst*—an agent of change that is not itself consumed in the reaction.

The electromagnetic forces of repulsion operating in the CNO cycle are greater than in the proton–proton chain because the charges on the heavy-element nuclei are larger (six times more for carbon, seven times for nitrogen, and eight for oxygen). Accordingly, higher temperatures are required to propel the heavy nuclei into the realm of the strong nuclear force and to ignite fusion. The accompanying figure presents

a numerical estimate of the energy released in a star of solar composition by the proton–proton chain and by the CNO cycle, each as a function of gas temperature. The proton–proton chain dominates at lower temperatures, up to about 16 million K. Above that temperature, the CNO cycle is the more important fusion process in stars of solar composition. Both processes contribute to core hydrogen burning described in the text.

According to our theoretical models of the Sun, the temperature of the solar core is 15 million K, so the curves shown indicate that the proton–proton cycle is (barely) the dominant source of solar energy. (Notice that each step on the vertical scale corresponds to a *factor* of 100 in energy generation.) The CNO cycle contributes no more than 10 percent of the observed solar radiation. However, stars more massive than our Sun often have core temperatures much higher than 20 million K, making the CNO cycle the dominant energy-production mechanism in them.

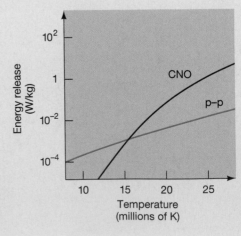

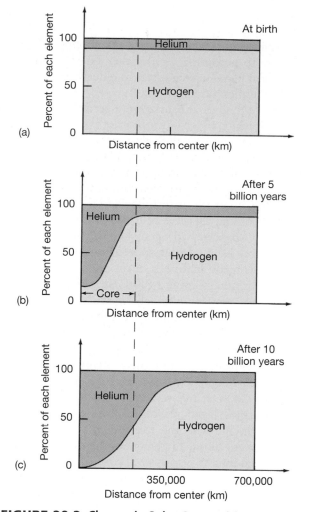

▲ FIGURE 20.2 Change in Solar Composition
Theoretical estimates of the changes in a Sun-like star's composition. Hydrogen (yellow) and helium (orange) abundances are shown (a) at birth, on the zero-age main sequence; (b) after 5 billion years; and (c) after 10 billion years. At stage (b), only about 5 percent of the star's total mass has been converted from hydrogen into helium. This change speeds up as the nuclear burning rate increases with time.

higher layers in the core. An inner core of nonburning pure helium starts to grow.

Without nuclear burning to maintain it, the outward-pushing gas pressure weakens in the helium inner core. However, the inward pull of gravity does not. Once the outward push against gravity is relaxed—even a little—structural changes in the star become inevitable. As the hydrogen is consumed, the inner core begins to contract. When all the hydrogen at the center is gone, the process accelerates.

If more heat could be generated, then the core might regain its equilibrium. For example, if helium in the core were to begin fusing into some heavier element, then energy would be created as a by-product of helium burning, and the necessary gas pressure would be reestablished. But the helium at the center cannot burn—not yet, anyway. Despite its high temperature, the core is far too cold to fuse helium into anything heavier.

Recall from Chapter 16 that a minimum temperature of about 10^7 K is needed to fuse hydrogen into helium. Only above that temperature do colliding hydrogen nuclei (i.e., protons) have enough speed to overwhelm the repulsive electromagnetic force between them. ∞ (Sec. 16.2) Because helium nuclei, with two protons each, carry a greater positive charge, their electromagnetic repulsion is larger, and even higher temperatures are needed to cause them to fuse—at least 10^8 K. A core composed of helium at 10^7 K thus cannot generate energy through fusion.

The shrinkage of the helium core releases gravitational energy, driving up the central temperature and heating the overlying burning layers. The higher temperatures—now well over 10^7 K (but still less than 10^8 K)—cause hydrogen nuclei to fuse even more rapidly than before. Figure 20.3 depicts this situation, in which hydrogen is burning at a furious rate in a shell surrounding the non-burning inner core of helium "ash" in the center. This phase is known as the **hydrogen-shell-burning** stage. The hydrogen shell generates energy faster than did the original main-sequence star's hydrogen-burning core, and the shell's energy production continues to increase as the helium core continues to shrink. Strange as it may seem, the star's response to the disappearance of the fire at its center is to get brighter!

Table 20.1 summarizes the key stages through which a solar-mass star evolves. The table is a continuation of Table 19.1, except that the density units have been changed from particles per cubic meter to the more convenient kilograms per cubic meter, and sizes are expressed as radii rather than diameters. The numbers in the "Stage" column refer to the evolutionary stages noted in the figures and discussed in the text.

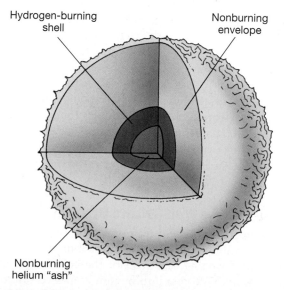

▲ FIGURE 20.3 Hydrogen-Shell Burning As a star's core converts more and more of its hydrogen into helium, the hydrogen in the shell surrounding the nonburning helium "ash" burns ever more violently.

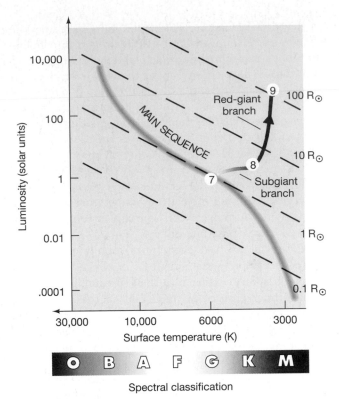

▲ FIGURE 20.4 Red Giant on the H–R Diagram As its helium core shrinks and its intermediate stellar layers expand, the star leaves the main sequence (stage 7). At stage 8, the star is well on its way to becoming a red giant. The star continues to brighten and grow as it ascends the red-giant branch at stage 9. As noted in Chapter 19, the dashed diagonals are lines of constant radius, allowing us to gauge the changes in the size of the star.

After a lengthy stay on the main sequence, the star's temperature and luminosity are once again beginning to change, and we can trace these changes via the star's evolutionary track on the H–R diagram. ⚭ (Sec. 19.2) Figure 20.4 shows the star's path away from the main sequence, labeled as stage 7. The star first evolves to the right on the diagram, its surface temperature dropping while its luminosity increases only slightly. The star's roughly horizontal path from its main-sequence location (stage 7) to stage 8 on the figure is called the **subgiant branch**. By stage 8, the star's radius has increased to about three times the radius of the Sun.

Stage 9: The Red-Giant Branch

Our aging star is now far from the main sequence and no longer in stable equilibrium. The helium core is unbalanced and shrinking. The rest of the core is also unbalanced, fusing hydrogen into helium at an ever-increasing rate. The gas pressure produced by this enhanced hydrogen burning causes the star's nonburning outer layers to increase in radius. Not even gravity can stop this inexorable change. While the core is shrinking and heating up,

the overlying layers are expanding and cooling. The star is on its way to becoming a red giant. The transformation from normal main-sequence star to elderly red giant takes about 100 million years.

By stage 8, the star's surface temperature has fallen to the point at which much of the interior is opaque to the radiation from within. Beyond this point, convection carries the core's enormous energy output to the surface. One consequence of that convection is that the star's surface temperature remains nearly constant between stages 8 and 9. The almost vertical path followed by the star between those stages is known as the **red-giant branch** of the H–R diagram. By stage 9, hydrogen-shell burning in the still-shrinking core is proceeding so ferociously that the giant's luminosity is many hundreds of times the solar value. Its radius by this time is around 100 solar radii.

The red giant is huge—about the size of Mercury's orbit. In contrast, its helium core is surprisingly small—only about a thousandth the size of the entire star, making the core just a few times larger than Earth. The central density is enormous: Continued shrinkage of the red giant's core has compacted its helium gas to approximately 10^8 kg/m^3. Contrast this value with the 10^{-3} kg/m^3 in the giant's outermost layers, with the 5000 kg/m^3 average density of Earth, and with the 150,000 kg/m^3 in the present core of the Sun. About 25 percent of the mass of the entire star is packed into its planet-sized core.

A familiar example of a low-mass star in the red-giant phase is the KIII giant Arcturus (see Figure 17.12), one of the brightest stars in the sky. Its mass is about 1.5 times that of the Sun. Currently in the hydrogen-shell-burning stage and ascending the red-giant branch, Arcturus has a radius some 21 times that of the Sun and emits about 160 times more energy than the Sun, much of it in the infrared part of the spectrum.

Stage 10: Helium Fusion

Should the unbalanced state of a red-giant star continue, the core would eventually collapse, and the rest of the star would slowly drift into space. The forces and pressures at work inside a red giant would literally pull it apart. In fact, for stars less than about one-quarter the mass of the Sun, that is precisely what will eventually happen (in a few hundred billion years—see Section 20.3). However, for a star like the Sun, this simultaneous shrinking and expanding does not continue indefinitely. A few hundred million years after a solar-mass star leaves the main sequence, something else happens: Helium begins to burn in the core. By the time the central density has risen to about 10^8 kg/m^3 (at stage 9), the temperature has reached the 10^8 K needed for helium to fuse into carbon, and the central fires reignite.

The reaction that transforms helium into carbon occurs in two steps. First, two helium nuclei come together to form a nucleus of beryllium-8 (^{8}Be), a highly unstable isotope that would normally break up into two helium nuclei in about 10^{-12} s. However, at the high densities found

TABLE 20.1   Evolution of a Sun-like Star

Stage	Approximate Time to Next Stage (Yr)	Central Temperature (10^6 K)	Surface Temperature (K)	Central Density (kg/m^3)	Radius (km)	Radius (solar radii)	Object
7	10^{10}	15	6000	10^5	7×10^5	1	Main-sequence star
8	10^8	50	4000	10^7	2×10^6	3	Subgiant branch
9	10^5	100	4000	10^8	7×10^7	100	Red-giant branch
10	5×10^7	200	5000	10^7	7×10^6	10	Horizontal branch
11	10^4	250	4000	10^8	4×10^8	500	Asymptotic-giant branch
12	10^5	300	100,000	10^{10}	10^4	0.01	Carbon core
			3000	10^{-17}	7×10^8	1000	Planetary nebula*
13	—	100	50,000	10^{10}	10^4	0.01	White dwarf
14	—	Close to 0	Close to 0	10^{10}	10^4	0.01	Black dwarf

Values refer to the envelope.

in the core of a red giant, it is possible that the beryllium-8 nucleus will encounter another helium nucleus before breakup occurs, fusing with the helium nucleus to form carbon-12 (^{12}C). This is the second step of the helium-burning reaction. In part, it is because of the electrostatic repulsion between beryllium-8 (containing four protons) and helium-4 (containing two) that the temperature must reach 10^8 K before that step can take place.

Symbolically, we can represent this next stage of stellar fusion as follows:

$$^4\text{He} + {}^4\text{He} \rightarrow {}^8\text{Be} + \text{energy},$$
$$^8\text{Be} + {}^4\text{He} \rightarrow {}^{12}\text{C} + \text{energy}.$$

Helium-4 nuclei are traditionally known as *alpha particles*. The term dates from the early days of nuclear physics, when the true nature of these particles, emitted by many radioactive materials, was unknown. Because three alpha particles are required to get from helium-4 to carbon-12, the foregoing reaction is usually called the **triple-alpha process**.

The Helium Flash

For stars comparable in mass to the Sun, a complication arises when helium fusion begins. At the high densities found in the core, the gas has entered a new state of matter whose properties are governed by the laws of quantum mechanics (the branch of physics describing the behavior of matter on subatomic scales), rather than by those of classical physics. ∞ (Sec. 4.2) Up to now, we have been concerned primarily with the nuclei—protons, alpha particles, and so on—that make up virtually all the star's mass and that participate in the reactions which generate its energy. However, the star contains another important constituent: a vast sea of electrons stripped from their parent nuclei by the ferocious heat in the stellar interior. At this

stage in our story, these electrons play an important role in determining the star's evolution.

Under the conditions found in the stage-9 red-giant core, a rule of quantum mechanics known as the *Pauli exclusion principle* (after Wolfgang Pauli, one of the founding fathers of quantum physics) prohibits the electrons in the core from being squeezed too close together. In effect, the exclusion principle tells us that we can think of the electrons as tiny rigid spheres which can be squeezed relatively easily up to the point of contact, but which become virtually incompressible thereafter. In the language of quantum mechanics, this condition is known as *electron degeneracy*; the pressure associated with the contact of the tiny electron spheres is called **electron degeneracy pressure**.* It has nothing to do with the thermal pressure (due to the star's heat) that we have been studying up to now. In fact, in our red-giant core, the pressure resisting the force of gravity is supplied almost entirely by degenerate electrons. Hardly any of the core's support results from "normal" thermal pressure, and this fact has dramatic consequences once the helium begins to burn.

Under normal ("nondegenerate") circumstances, the core could react to, and accommodate, the onset of helium burning, but in the core's degenerate state, the burning becomes unstable, with literally explosive consequences. In a star supported by thermal pressure, the increase in temperature produced by the onset of helium fusion would lead to an increase in pressure. The gas would then expand and cool, reducing the burning rate and reestablishing equilibrium, just as discussed earlier. In the electron-supported core of a solar-mass red giant, however, the pressure is largely *independent* of the temperature. When burning starts and

*The term refers to an idealized condition in which all particles (electrons in this case) are in their lowest possible energy states. As a result, the star cannot be compressed into a more compact configuration.

the temperature increases, there is no corresponding rise in pressure, no expansion of the gas, no drop in the temperature, and no stabilization of the core. Instead, the core is unable to respond to the rapidly changing conditions within it. The pressure remains more or less unchanged as the nuclear reaction rates increase, and the temperature rises rapidly in a runaway condition called the **helium flash**.

For a few hours, the helium burns ferociously. Eventually, the flood of energy released by this period of runaway fusion heats the core to the point at which normal thermal pressure once again dominates. Finally able to react to the energy dumped into it by helium burning, the core expands, its density drops, and equilibrium is restored as the inward pull of gravity and the outward push of gas pressure come back into balance. The core, now stable, begins to fuse helium into carbon at temperatures well above 10^8 K.

The helium flash terminates the giant star's ascent of the red-giant branch of the H–R diagram. Yet, despite the violent ignition of helium in the core, the flash does *not* increase the star's luminosity. On the contrary, the energy released in the helium flash expands and cools the core and ultimately results in a *reduction* in the energy output. On the H–R diagram, the star jumps from stage 9 to stage 10, a stable state with steady helium burning in the core. As indicated in Figure 20.5, the surface temperature is now higher than it was on the red-giant branch, but the luminosity is considerably less than at the helium flash. This adjustment in the star's properties occurs quite quickly—in about 100,000 years.

At stage 10, our star is now stably burning helium in its core and fusing hydrogen in a shell surrounding it. The star resides in a well-defined region of the H–R diagram known as the **horizontal branch**, where core-helium-burning stars remain for a time before resuming their journey around the H–R diagram. The star's specific position within this region is determined mostly by its mass—not its original mass, but whatever mass remains after its ascent of the red-giant branch. The two masses differ because, during the red-giant stage, strong stellar winds eject large amounts of matter from a star's surface. As much as 20 to 30 percent of the original stellar mass may escape during that period. It so happens that more massive stars have lower surface temperatures at this stage, but all stars have roughly the same luminosity after the helium flash. As a result, stage-10 stars tend to lie along a horizontal line on the H–R diagram, with more massive stars to the right and less massive ones to the left.

Stage 11: Back to the Giant Branch

The nuclear reactions in our star's helium core burn on, but not for long. Whatever helium exists in the core is rapidly consumed. The triple-alpha helium-to-carbon fusion reaction—like the proton–proton and CNO-cycle hydrogen-to-helium reactions before it—proceeds at a rate that increases rapidly with temperature. At the extremely high temperatures found in the horizontal-branch core, the helium fuel doesn't last long—no more than a few tens of millions of years after the initial flash.

As helium fuses to carbon, a new carbon-rich inner core begins to form, and phenomena similar to those which took place during the earlier buildup of helium recur. Helium becomes depleted at the center of the star, and eventually fusion ceases there. The nonburning carbon core shrinks in size—even as its mass increases due to helium fusion—and heats up as gravity pulls it inward, causing the hydrogen- and helium-burning rates in the overlying layers of the core to increase. The star now contains a contracting carbon core surrounded by a helium-burning shell, which is in turn surrounded by a hydrogen-burning shell. The outer envelope of the star—the nonburning layers surrounding the core—expands, much as it did earlier during the first red-giant stage. By the time it reaches stage 11 in Figure 20.6, the star has become a swollen red giant for the second time. Figure 20.7 depicts the star's interior structure during this period.

To distinguish the second ascent of the giant branch from the first, the star's track during the second phase is often referred to as the **asymptotic-giant branch**.* The

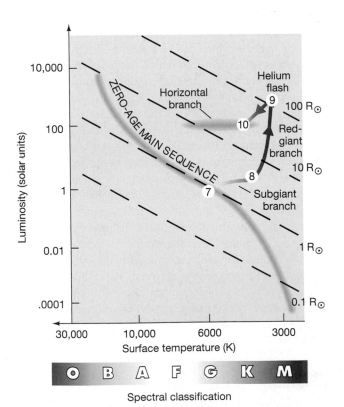

▲ **FIGURE 20.5 Horizontal Branch** A large increase in luminosity occurs as a star ascends the red-giant branch, ending in the helium flash. The star then settles down into another equilibrium state at stage 10, on the horizontal branch.

*The term is borrowed from mathematics. An asymptote to a curve is a second curve that approaches ever closer to the first as the two are extended to infinity. Theoretically, if the star remained intact, the asymptotic-giant branch would approach the red-giant branch from the left as the luminosity increased and would effectively merge with the red-giant branch near the top of Figure 20.7. However, as we will see in Section 20.3, a Sun-like star will not live long enough for that to occur.

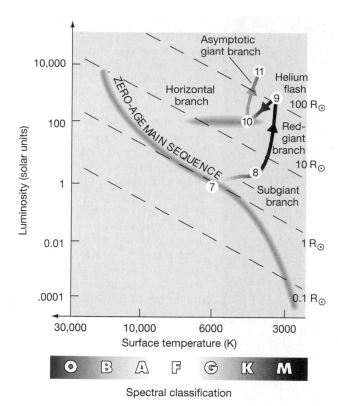

▲ FIGURE 20.6 Red-Giant Branch Revisited A carbon-core star reascends the giant branch of the H–R diagram—this time on a track called the *asymptotic-giant branch*—for the same reason it took that path the first time around: The lack of nuclear fusion at the core causes the core to contract and the overlying layers to expand.

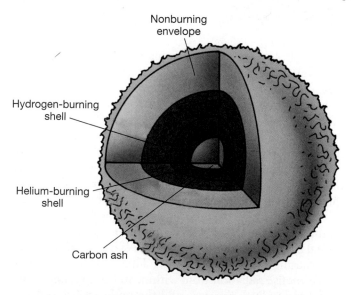

▲ FIGURE 20.7 Helium-Shell Burning Within a few million years after the onset of helium burning, carbon ash accumulates in the star's inner core, above which hydrogen and helium are still burning in concentric shells.

burning rates in the shells around the carbon core are much fiercer this time around, and the star's radius and luminosity increase to values even greater than those reached at the helium flash on the first ascent. The carbon core grows in mass as more and more carbon is produced in the helium-burning shell above it, but continues to shrink in radius, driving the hydrogen-burning and helium-burning shells to higher and higher temperatures and luminosities.

CONCEPT CHECK

✔ Why does a star get brighter as it runs out of fuel in its core?

20.3 The Death of a Low-Mass Star

Figure 20.8 illustrates the stages through which a G-type star like the Sun will pass over the course of its evolution. As our star moves from stage 10 (the horizontal branch) to stage 11 (the asymptotic-giant branch), its envelope swells, while its inner carbon core, too cool for further nuclear burning, continues to contract. If the central temperature could become high enough for carbon fusion to occur, still heavier products could be synthesized, and the newly generated energy might again support the star, restoring for a time the equilibrium between gravity and heat. For solar-mass stars, however, this does not occur. The temperature never reaches the 600 million K needed for a new round of nuclear reactions to occur. The red giant is very close to the end of its nuclear-burning lifetime.

The Fires Go Out

Before the carbon core can attain the incredibly high temperatures needed for carbon ignition, its density reaches a point beyond which it cannot be compressed further. At about 10^{10} kg/m^3, the electrons in the core once again become degenerate, the contraction of the core ceases, and the core's temperature stops rising. This stage (stage 12 in Table 20.1) represents the maximum compression that the star can achieve—there is simply not enough matter in the overlying layers to bear down any harder.

The core density at this stage is extraordinarily high. A single cubic centimeter of core matter would weigh 1000 kg on Earth—a ton of matter compressed into a volume about the size of a grape! Yet, despite the extreme compression of the core, the central temperature is "only" about 300 million K. Some oxygen is formed via reactions between carbon and helium at the inner edge of the helium-burning shell—that is,

$$^{12}\text{C} + {}^{4}\text{He} \rightarrow {}^{16}\text{O} + \text{energy}.$$

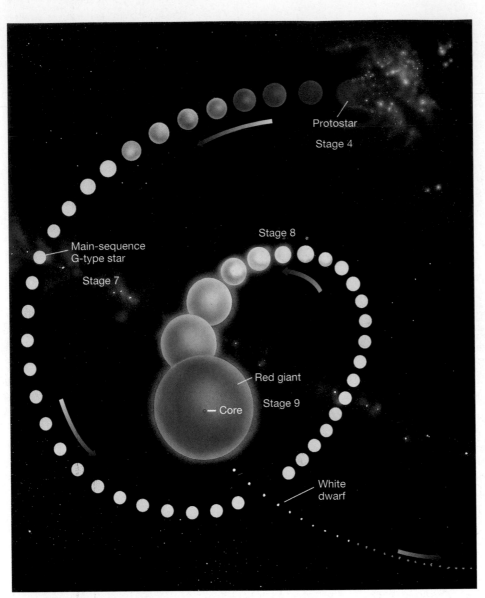

◀ **FIGURE 20.8 G-Type Star Evolution** Artist's conception of the relative sizes and changing colors of a normal G-type star (such as our Sun) during its formative stages, on the main sequence, and while passing through the red-giant and white-dwarf stages. At maximum swelling, the red giant is approximately 70 times the size of its main-sequence parent; the core of the giant is about one-fifteenth the main-sequence size and would be barely discernible if the figure were drawn exactly to scale. The duration of time spent in the various stages—protostar, main-sequence star, red giant, and white dwarf—is roughly proportional to the length of this imaginary trek through space. The star's brief stay on the horizontal and asymptotic-giant branches are not shown here.

However, collisions among nuclei are neither frequent nor violent enough to create any heavier elements. For all practical purposes, the central fires go out once carbon has formed.

Stage 12: A Planetary Nebula

Our aged stage-12 star is now in quite a predicament. Its inner carbon core no longer generates energy. The outer-core shells continue to burn hydrogen and helium, and as more and more of the inner core reaches its final, high-density state, the nuclear burning increases in intensity. Meanwhile, the envelope continues to expand and cool, reaching a maximum radius of about 300 times that of the Sun—big enough to engulf the planet Mars.

Around this time, the burning becomes quite unstable. The helium-burning shell is subject to a series of explosive *helium-shell flashes*, caused by the enormous pressure in the helium-burning shell and the extreme sensitivity of the triple-alpha burning rate to small changes in temperature.

The flashes produce large fluctuations in the intensity of the radiation reaching the star's outermost layers, causing those layers to pulsate violently as the envelope repeatedly is heated, expands, cools, and then contracts (Figure 20.9). The amplitude of the pulsations grows as the temperature of the core continues to increase and the nuclear burning intensifies in the surrounding shells.

Compounding the star's problems is the increasing instability of its surface layers. Around the peak of each pulsation, the surface temperature drops below the point at which electrons can recombine with nuclei to form atoms. ∞ (Sec. 4.2) Each recombination produces additional photons, giving the gas a little extra outward "push" and causing some of it to escape. Thus, driven by increasingly intense radiation from within, and accelerated by instabilities in both the core and the outer layers, virtually all of the star's envelope is ejected into space in less than a few million years at a speed of a few tens of kilometers per second.

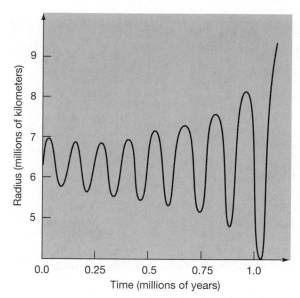

▲ FIGURE 20.9 Red-Giant Instability Buffeted by helium-shell flashes from within, and subject to the destabilizing influence of recombination, the outer layers of a red giant become unstable and enter into a series of growing pulsations. Eventually, the envelope is ejected and forms a planetary nebula.

In time, a rather unusual-looking object results. The "star" now consists of two distinct parts, both of which constitute stage 12 of Table 20.1. At the center is a small, well-defined core of mostly carbon ash. Hot, dense, and still very luminous, only the outermost layers of this core still fuse helium into carbon and oxygen. Well beyond the core lies an expanding cloud of dust and cool gas—the ejected envelope of the giant—spread over a volume roughly the size of our solar system.

As the core exhausts its last remaining fuel, it contracts and heats up, moving to the left in the H–R diagram. Eventually, it becomes so hot that its ultraviolet radiation ionizes the inner parts of the surrounding cloud, producing a spectacular display called a **planetary nebula**. Some well-known examples are shown in Figures 20.10 and 20.11. In all, more than 1500 planetary nebulae are known in our Galaxy. The word *planetary* here is misleading, for these objects have no association with planets. The name originated in the 18th century, when, viewed at poor resolution through small telescopes, these shells of glowing gas looked to some astronomers like the circular disks of planets in our solar system.

Note that the mechanism by which planetary nebulae shine is basically the same as that powering the emission nebulae we studied earlier: ionizing radiation from a hot star embedded in a cool gas cloud. ∞ (Sec. 18.2) However, recognize that these two classes of object have very different origins and represent completely separate phases of stellar evolution. The emission nebulae discussed in Chapter 18 are signposts of recent stellar birth. Planetary nebulae, by contrast, indicate impending stellar death.

Astronomers once thought that the escaping giant envelope would be more or less spherical in shape, completely surrounding the core in three dimensions, just as it had while still part of the star. Figure 20.10(a) shows an example where this may well in fact be the case. The "ring" of this planetary nebula is in reality a three-dimensional shell of glowing gas—its halo-shaped appearance is only an illusion. As illustrated in Figure 20.10(b), the nebula looks brighter near the edges simply because there is more emitting gas along the line of sight there, creating the illusion of a bright ring.

However, such cases now seem to be in the minority. There is growing evidence that, for reasons not yet fully understood, the final stages of red-giant mass loss are often decidedly *non*spherical. For example, the famous Ring Nebula shown in Figure 20.10(c) may well actually *be* a ring, and not just our view of a glowing spherical shell as was once believed, and many planetary nebulae are much more complex even than that. As illustrated in Figure 20.11, planetary nebulae may exhibit jetlike and other irregular structures. Apparently, both the details of the gas-ejection process and the star's environment (such as whether a binary companion is present) play important roles in determining the nebula's shape and appearance.

The central star fades and cools, and the expanding gas cloud becomes more and more diffuse, eventually dispersing into interstellar space. After just a few tens of thousands of years, the glowing planetary nebula disappears from view. As the cloud rejoins the interstellar medium, it plays a vital role in the evolution of our Galaxy. During the final stages of the red giant's life, nuclear reactions between carbon and unburned helium in the core create oxygen and, in some cases, even heavier elements, such as neon and magnesium. Some of these reactions also release neutrons, which, carrying no electrical charge, have no electrostatic barrier to overcome and hence can interact with existing nuclei to form still heavier elements (see Chapter 21). All of these elements—helium, carbon, oxygen, and heavier ones—are "dredged up" from the depths of the core into the envelope by convection during the star's final years, to enrich the interstellar medium when the giant envelope escapes. The evolution of low-mass stars is the source of virtually all the carbon-rich dust observed throughout the plane of our own and other galaxies. ∞ (Sec. 18.1)

Stages 13 and 14: White and Black Dwarfs

The carbon core—the stellar remnant at the center of the planetary nebula—continues to evolve. Formerly concealed by the atmosphere of the red-giant star, the core becomes visible as the envelope recedes. Several tens of thousands of years are needed for the core to appear from behind the veil of expanding gas. The core is very small. By the time the envelope is ejected as a planetary nebula,

(a)

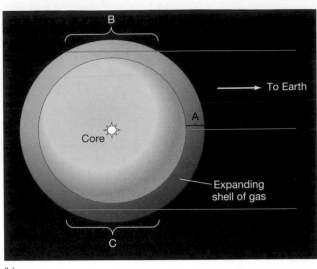

(b)

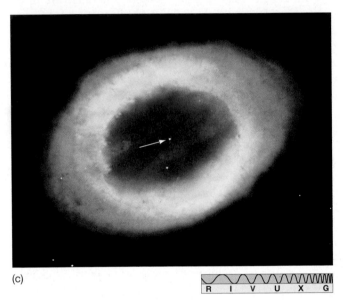

(c)

R I V U X G

◄ **FIGURE 20.10 Ejected Envelope** A planetary nebula is an extended region of glowing gas surrounding an intensely hot central star (marked with an arrow in each case shown here). The star is the core of a former red giant. The gas is what remains of the giant's envelope, now ejected into space. (a) This "classic" planetary nebula, called Abell 39, seems to be an almost spherical shell of gas some 2100 pc away and about 1.5 pc across. (b) The appearance of Abell 39 can be explained once we realize that the shell of glowing gas around the central core is actually quite thin. There is very little gas along the line of sight between the observer and the central star (path A), so that part of the shell is invisible. Near the edge of the shell, however, there is more gas along the line of sight (paths B and C), so the observer sees a glowing ring. (c) Perhaps the most famous of all planetary nebula, the Ring Nebula, is roughly 1500 pc away, 0.5 pc across, and too small and dim to be seen with the naked eye. Astronomers once thought its appearance could be explained in much the same way as that of Abell 39. However, it now seems that the Ring really is ring shaped! Researchers are still unsure as to why a spherical star should eject a ring of material during its final days. *(AURA; NASA)*

the core has shrunk to about the size of Earth. (In some cases, it may be even smaller than our planet.) Its mass is about half the mass of the Sun. Shining only by stored heat, not by nuclear reactions, this small "star" has a white-hot surface when it first becomes visible, although it appears dim because of its small size. The core's temperature and size give rise to its new name: *white dwarf*. This is stage 13 of Table 20.1. The approximate path followed by the star on the H–R diagram as it evolves from stage-11 red giant to stage-13 white dwarf is shown in Figure 20.12.

Not all white-dwarf stars are found as the cores of planetary nebulae: Several hundred have been discovered "naked" in our Galaxy, their envelopes expelled to invisibility (or perhaps stripped away by a binary companion—to be discussed shortly) long ago. Figure 20.13 shows an example of a white dwarf, Sirius B, that happens to lie particularly close to Earth; it is the faint binary companion of the much brighter and better known Sirius A. ∞ (Sec. 17.4) Some properties of Sirius B are listed in Table 20.2. With more

TABLE 20.2 Sirius B, a Nearby White Dwarf	
Mass	1.1 solar masses
Radius	0.0073 solar radius (5100 km)
Luminosity (total)	0.025 solar luminosity (9.8×10^{24} W)
Surface temperature	27,000 K
Average density	3.9×10^9 kg/m³

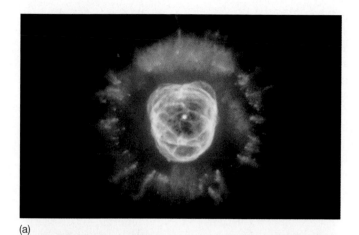

(a)

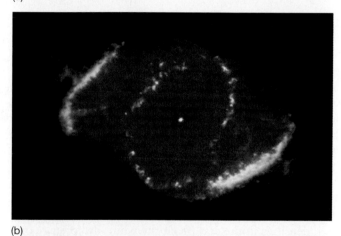

(b)

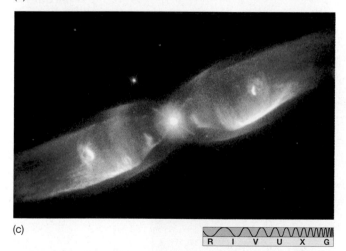

(c)

R I V U X G

▲ **FIGURE 20.11 Planetary Nebulae** (a) The Eskimo Nebula clearly shows several "bubbles" (or shells) of material being blown into space from this planetary nebula, which resides some 1500 pc away in the constellation Gemini. (b) The Cat's Eye Nebula, about 1000 pc away and 0.1 pc across, is an example of a much more complex planetary nebula, possibly produced by a pair of binary stars (unresolved at center) that have both shed envelopes. (c) M2-9, some 600 pc away and 0.5 pc end to end, shows surprising twin lobes (or jets) of glowing gas emanating from a central dying star and racing out at speeds of about 300 km/s. (*AURA; NASA*)

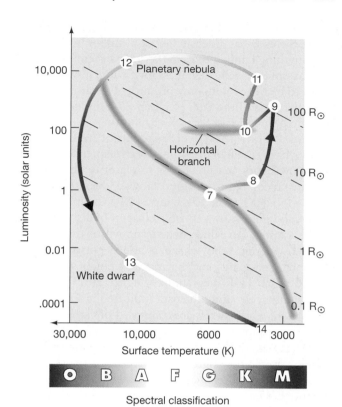

▲ **FIGURE 20.12 White Dwarf on the H–R Diagram** A star's passage from the horizontal branch (stage 10) to the white-dwarf stage (stage 13) by way of the asymptotic-giant branch creates an evolutionary path that cuts across the entire H–R diagram.

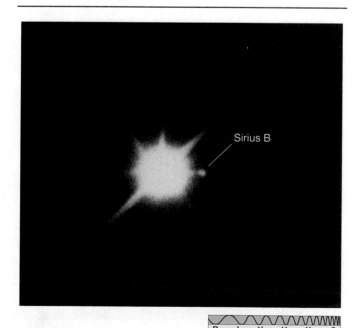

R I V U X G

▲ **FIGURE 20.13 Sirius Binary System** Sirius B (the speck of light to the right of the much larger and brighter star Sirius A) is a white dwarf star, a companion to Sirius A. The "spikes" on the image of Sirius A are not real; they are caused by the support struts of the telescope. (*Palomar Observatory*)

than the mass of the Sun packed into a volume smaller than Earth, Sirius B has a density about a million times greater than anything familiar to us in the solar system. In fact, Sirius B has an unusually high mass for a white dwarf—it is believed to be the evolutionary product of a star roughly four times the mass of the Sun. *Discovery 20-1* discusses another possible peculiarity of Sirius B's evolution.

Hubble Space Telescope observations of nearby globular clusters have revealed the white-dwarf sequences long predicted by theory, but previously too faint to detect at such large distances. Figure 20.14(a) shows a ground-based view of the globular cluster M4, lying 2100 pc from Earth. Part (b) of the figure shows an *HST* closeup of a small portion of the cluster, revealing dozens of white dwarfs (some marked) among the cluster's much brighter main-sequence, red-giant, and horizontal-branch stars. When plotted on an H–R diagram (see Figure 20.15), the white dwarfs fall nicely along the path indicated in Figure 20.12.

Not all white dwarfs are composed of carbon and oxygen. As mentioned earlier, theory predicts that very low mass stars (less than about one-quarter the mass of the Sun) will never reach the point of helium fusion. Instead, the core of such a star will become supported by electron degeneracy pressure before its central temperature reaches the 100 million K needed to start the triple-alpha process. Eventually, the star's envelope will be ejected in a manner similar to that of a more massive star, forming a *helium white dwarf*.

The time needed for this kind of transformation to occur is very long—hundreds of billions of years—so no helium white dwarfs have ever actually formed in this way. ∞ (Sec. 17.8) However, if a solar-mass star is a member of a binary system, it is possible for its envelope to be stripped away during the red-giant stage by the gravitational pull of its companion (see Section 20.6), exposing the helium core and terminating the star's evolution before helium fusion can begin. Several such low-mass helium white dwarfs have in fact been detected in binary systems.

Finally, in stars much more massive than the Sun (close to the 8-solar-mass limit on "low-mass" stars at the time the carbon core forms), temperatures in the core may become high enough that an additional reaction,

$$^{16}O + {}^4He \rightarrow {}^{20}Ne + energy,$$

can occur, ultimately leading to the formation of a rare *neon–oxygen white dwarf*.

Once an isolated star becomes a white dwarf, its evolution is over. (As we will see in Chapter 21, white dwarfs in binary systems may undergo further activity.) The isolated white dwarf continues to cool and dim with time, following the white–yellow–red track near the bottom of the H–R diagram of Figure 20.12 and eventually becoming a *black dwarf*—a cold, dense, burned-out ember in space. This is stage 14 of Table 20.1, the graveyard of stars.

The cooling dwarf does not shrink much as it fades away. Even though its heat is leaking away into space, gravity does not compress it further. At the enormously high densities in the star (from the white-dwarf stage on),

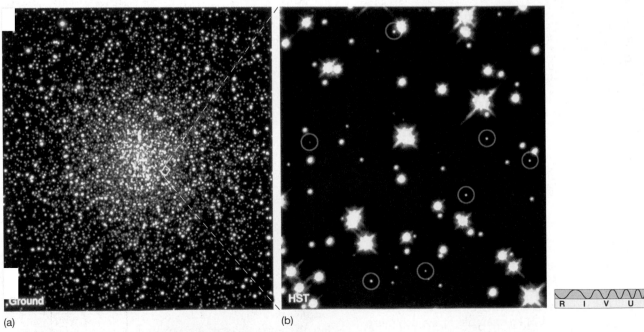

(a)   (b)

▲ **FIGURE 20.14 Distant White Dwarfs** (a) The globular cluster M4, as seen through a large ground-based telescope at Kitt Peak National Observatory in Arizona. At 2100 pc away, M4 is the closest globular cluster to us; it spans some 16 pc. (b) A peek at M4's "suburbs" by the *Hubble Space Telescope* shows nearly a hundred white dwarfs within a small 0.2 square-parsec region. Some of the brightest ones are circled in blue. (*AURA; NASA*)

(a)

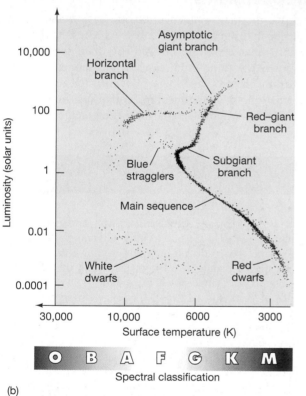

(b)

▲ FIGURE 20.15 Globular Cluster H–R Diagram (a) The globular cluster M80, some 8 kpc away. (b) Combined H–R diagram, based on ground- and space-based observations, for several globular clusters similar in overall composition to M80. The various evolutionary stages predicted by theory and depicted schematically in Figure 20.12 are clearly visible. Note also the blue stragglers—main-sequence stars that appear to have been "left behind" as other stars evolved into giants. They are probably the result of merging binary systems or actual collisions between stars of lower mass in this remarkably dense stellar system. (See also Figure 20.20.) *(NASA; data courtesy W.E. Harris)*

the resistance of electrons to being squeezed together—the same electron degeneracy that prevailed in the red-giant core around the time of the helium flash—supports the star, even as its temperature drops almost to absolute zero. As the dwarf cools, it remains about the size of Earth.

Comparing Theory with Reality

All the H–R diagrams and evolutionary tracks presented so far are theoretical constructs based largely on computer models of the interior workings of stars. Before continuing our study of stellar evolution, let's take a moment to compare our models with actual observations. Figure 20.15(a) shows the beautiful globular cluster M80, which lies about 8000 pc from Earth. Figure 20.15(b) shows a composite H–R diagram recently constructed by using the stars of a number of other globular clusters of roughly the same age and composition as M80. The diagram spans the entire range of stellar luminosities, from bright red giants to faint red and white dwarfs. Fitting theoretical models of the main-sequence, giant, and horizontal branches (see Section 20.5) implies an age of about 12 billion years, making these clusters among the oldest-known objects in the Milky Way Galaxy and, as such, key indicators of conditions in the early universe.

The great age of this cluster means that stars more massive than about 0.8 solar mass have already evolved beyond the red-giant stage, becoming mainly white dwarfs. The H–R diagram for this cluster can therefore be compared directly with Figure 20.12, as the red-giant, horizontal-branch, and asymptotic-giant-branch stars are all of roughly 1 solar mass. The similarity between theory and observation is striking: Stars in each of the evolutionary stages 7–13 can be seen, in numbers consistent with the theoretical models. (See also the acetate inset in this chapter.) Astronomers place great confidence in the theory of stellar evolution precisely because its predictions are so often found to be in excellent agreement with plots of real stars.

Note that the points in Figure 20.15(b) are actually shifted a little to the left relative to Figure 20.12. This is because of differences in composition between stars such as the Sun and stars in globular clusters. For reasons to be discussed more fully in Chapter 21, the old globular cluster stars contain much lower concentrations of "heavy" elements (astronomical jargon for anything more massive than helium). One result of this relative paucity of these elements is that the interiors and atmospheres of those stars tend to be slightly more transparent to radiation from within, allowing the energy to escape more easily and making the stars slightly smaller and hotter than solar-type stars of the same mass.

The objects labeled as *blue stragglers* in Figure 20.15(b) appear at first sight to contradict the theory just described. They lie on the main sequence, but in locations suggesting that they should have evolved into white dwarfs long ago, given the cluster's age of 12 billion years. Blue stragglers are observed in many star clusters. They are main-sequence stars, but they did not form when the cluster did. Instead, they formed much more recently, through *mergers* of lower mass stars—so recently, in fact, that they have not yet had time to evolve into giants.

How did these stars merge? In some cases, such mergers are probably the result of stellar evolution in binary systems, as the component stars evolve, grow, and come into contact. (See Section 20.6.) In others, the mergers are thought to be the result of actual *collisions* between stars. The core of M80 contains a huge number of stars packed into a relatively small volume. For example, a sphere of radius 2 pc centered on the Sun contains exactly four stars, including the Sun itself. ∞ (Sec. 17.1) At the center of M80, the same 2-pc sphere would contain more than 10 *million* stars—our night sky would be ablaze with thousands of objects brighter than Venus! The dense central cores of globular clusters are among the few places in the entire universe where stellar collisions are likely to occur.

CONCEPT CHECK

✔ Why does fusion cease in the core of a low-mass star?

20.4 Evolution of Stars More Massive than the Sun

High-mass stars evolve much faster than their low-mass counterparts. The more massive a star, the more ravenous is its fuel consumption and the shorter is its main-sequence lifetime. The Sun will spend a total of some 10 billion years on the main sequence, but a 5-solar-mass B-type star will remain there for only a hundred million years. A 10-solar-mass O-type star will depart in just 20 million years or so. This trend toward much faster evolution for more massive stars continues even after the star leaves the main sequence. All evolutionary changes happen much more rapidly for high-mass stars because their larger mass and stronger gravity generate more heat, speeding up *all* phases of stellar evolution.

Red Supergiants

Stars leave the main sequence for one basic reason: They run out of hydrogen in their cores. As a result, the early stages of stellar evolution beyond the main sequence are qualitatively the same in all cases: Main-sequence hydrogen burning in the core (stage 7) eventually gives way to the formation of a nonburning, collapsing helium core surrounded by a hydrogen-burning shell (stages 8 and 9).

DISCOVERY 20-1

Learning Astronomy from History

Sirius A, the brighter of the two objects shown in Figure 20.13, appears twice as luminous as any other visible star, excluding the Sun. Its absolute brightness is not very great, but because it is not very far from us (less than 3 pc away), its apparent brightness is very large. ∞ (Sec. 17.2) Sirius has been prominent in the nighttime sky since the beginning of recorded history. Cuneiform texts of the ancient Babylonians refer to the star as far back as 1000 B.C., and historians know that the star strongly influenced the agriculture and religion of the Egyptians of 3000 B.C.

Even though a star's evolution takes such a long time, we might have a chance to detect a slight change in Sirius because the recorded observations of this star go back several thousand years. The chances for success are improved in this case because Sirius A is so bright that even the naked-eye observations of the ancients should be reasonably accurate. Interestingly, recorded history does suggest that Sirius A has changed in appearance, but the observations are confusing. Every piece of information about Sirius recorded between the years 100 B.C. and A.D. 200 claims that this star was *red*. (No earlier records of its color are known.) In contrast, modern observations now show it to be white or bluish white—definitely *not* red.

If these reports are accurate, then Sirius has apparently changed from red to blue white in the intervening years. According to the theory of stellar evolution, however, no star should be able to change its color in this way in that short a time. Such a color change should take at least several tens of thousands of years and perhaps a lot longer. It should also leave some evidence of its occurrence.

Astronomers have offered several explanations for the rather sudden change in Sirius A, including the suggestions that (1) some ancient observers were wrong and other scribes copied their mistaken writings; (2) a galactic dust cloud passed between Sirius A and Earth some 2000 years ago, reddening the star much as Earth's dusty atmosphere often reddens our Sun at dusk; and (3) the companion to Sirius A, Sirius B, was a red giant and the dominant star of this double-star system 2000 years ago, but has since expelled its planetary nebular shell to reveal the white-dwarf star that we now observe.

Each of these explanations presents problems. How could the color of the sky's brightest star have been incorrectly recorded for hundreds of years? Where is the intervening galactic cloud now? Where is the shell of the former red giant? We are left with the uneasy feeling that the sky's brightest star doesn't fit particularly well into the currently accepted scenario of stellar evolution.

A high-mass star leaves the main sequence on its journey toward the red-giant region with an internal structure quite similar to that of its low-mass cousin. Thereafter, however, their evolutionary tracks diverge.

Figure 20.16 compares the post-main-sequence evolution of three stars respectively having masses 1, 4, and 15 times the mass of the Sun. Note that, whereas stars like the Sun ascend the red-giant branch almost vertically, stars of higher mass move nearly horizontally across the H–R diagram after leaving the upper main sequence. Their luminosities stay roughly constant as their radii increase and their surface temperatures drop.

In stars having more than about 2.5 times the mass of the Sun, helium burning begins smoothly and stably, *not* explosively—there is no helium flash. Calculations indicate that the more massive a star, the lower is its core density when the temperature reaches the 10^8 K necessary for helium ignition, and the smaller is the contribution to the pressure from degenerate electrons. As a result, above 2.5

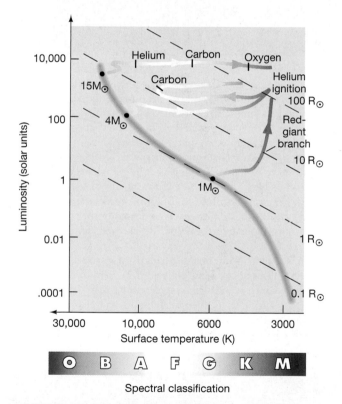

▲ **FIGURE 20.16 High-Mass Evolutionary Tracks** Evolutionary tracks for stars of 1, 4, and 15 solar masses (shown only up to the point of the helium flash in the low-mass cases). Stars with masses comparable to that of the Sun ascend the giant branch almost vertically, whereas stars with higher mass move roughly horizontally across the H–R diagram from the main sequence into the red-giant region. The most massive stars make smooth transitions into each new burning stage. No helium flash occurs for stars more massive than about 2.5 solar masses. Some points are labeled with the element that has just started to fuse in the inner core.

solar masses, the unstable core conditions described earlier do not occur. The 4-solar-mass red giant in Figure 20.16 remains a red giant as helium starts to fuse into carbon. There is no sudden jump to the horizontal branch and no subsequent reascent of the giant branch. Instead, the star loops smoothly back and forth near the top of the H–R diagram.

A much more important divergence occurs at approximately 8 solar masses—the dividing line between high and low mass mentioned in Section 20.1. A low-mass star never achieves the 600 million K needed to fuse carbon nuclei, so it ends its life as a carbon–oxygen (or possibly neon–oxygen) white dwarf. A high-mass star, however, can fuse not only hydrogen and helium, but also carbon, oxygen, and even heavier elements as its inner core continues to contract and its central temperature continues to rise. The rate of burning accelerates as the core evolves.

Evolution proceeds so rapidly in the 15-solar-mass star of Figure 20.16 that the star doesn't even reach the red-giant region before helium fusion begins. The star achieves a central temperature of 10^8 K while it is still quite close to the main sequence. As each element is burned to depletion at the center, the core contracts and heats up, and fusion starts again. A new inner core forms, contracts again, heats again, and so on. The star's evolutionary track continues smoothly across the supergiant region of the H–R diagram, seemingly unaffected by each new phase of burning. The star's radius increases as its surface temperature drops, so the star swells to become a **red supergiant.** ∞ (Sec. 17.4)

With heavier and heavier elements forming at an ever-increasing rate, the high-mass star shown in Figure 20.16 is very close to the end of its life. We will discuss the evolution and ultimate fate of such a star in more detail in the next chapter, but suffice it to say here that it is destined to die in a violent supernova—a catastrophic explosion releasing energy that will most likely literally blow the star to pieces—soon after carbon and oxygen begin to fuse in its core. High-mass stars evolve so rapidly that, for most practical observational purposes, they explode and die shortly after leaving the main sequence.

A good example of a post-main-sequence blue supergiant is the bright star Rigel in the constellation Orion. With a radius some 70 times that of the Sun and a total luminosity of more than 60,000 solar luminosities, Rigel is thought to have had an original mass about 17 times that of the Sun, although a strong stellar wind has probably carried away a significant fraction of its mass since it formed. Although still near the main sequence, Rigel is probably already fusing helium into carbon in its core.

Perhaps the best-known red supergiant is Betelgeuse (shown in Figures 17.8 and 17.11), also in Orion and Rigel's rival for the title of brightest star in the constellation. Its luminosity is roughly 10^4 times that of the Sun in visible light and perhaps four times that in the infrared. Astronomers believe that Betelgeuse is currently fusing helium into

DISCOVERY 20-2

Mass Loss from Giant Stars

Astronomers now know that stars of all spectral types are active and have stellar winds. Consider the highly luminous, hot, blue O- and B-type stars, which have by far the strongest winds. Satellite and rocket observations have shown that their wind speeds may reach 3000 km/s. The result is a yearly mass loss sometimes exceeding 10^{-6} solar mass per year. Over the relatively short span of 1 million years, these stars blow a tenth of their total mass—more than an entire solar mass of material—into space. The powerful stellar winds, driven directly by the pressure of the intense ultraviolet radiation emitted by the stars themselves, hollow out vast cavities in the interstellar gas.

The black-and-white photograph here is a *Hubble Space Telescope* image of the supergiant star AG Carinae—50 times more massive than the Sun and a million times brighter—shedding its outer atmosphere. The star is shown puffing out vast clouds of gas and dust. (The star, at the center, is intentionally obscured to show the surrounding faint nebula more clearly; the bright vertical line is also an artifact—an effect of the optical system used to hide the star.)

The four-part accompanying *Hubble* image captures another stellar outburst in the second half of the year 2002, during which a star brightened more than a half-million times our Sun's luminosity. This star, with the tongue-twisting name V838 Monocerotis, is a highly variable (and poorly understood) red supergiant about 20,000 light-years distant. Actually, what we are seeing here is not matter being expelled outward as fast as the images imply; rather, a burst of light—often called a "light echo"—is illuminating shells of gas and dust now surrounding the star, but that had been shed long ago. For scale, the rightmost image is about 7 light-years across.

Observations made with radio, infrared, and optical telescopes have shown that luminous cool stars (e.g., K- and M-type red giants) also lose mass at rates comparable to those at which luminous hot stars lose mass. Red-giant wind velocities, however, are much lower, averaging merely 30 km/s. They carry roughly as much mass into space as do O-type stellar winds, because their densities are generally much greater.

(NASA)

R I V U X G

Also, because luminous red stars are inherently cool objects (with surface temperatures of only about 3000 K), they emit virtually no ultraviolet radiation, so the mechanism driving the winds must differ from that driving the winds of luminous hot stars. We can only surmise that gas turbulence, magnetic fields, or both in the atmospheres of the red giants are somehow responsible. The surface conditions in red giants are in some ways similar to those in T Tauri protostars, which are also known to exhibit strong winds. Possibly the same basic mechanism—violent surface activity—is responsible for both kinds of winds.

Unlike winds from hot stars, winds from these cool stars are rich in dust particles and molecules. Nearly all stars eventually evolve into red giants, so such winds are a major source of new gas and dust to interstellar space and also provide a vital link between the cycle of star formation and the evolution of the interstellar medium.

 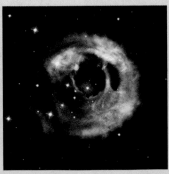 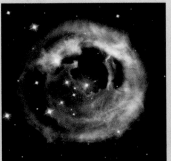 

(NASA)

R I V U X G

TABLE 20.3   End Points of Evolution for Stars of Different Masses

Initial Mass (Solar Masses)	Final State
less than 0.08	(hydrogen) brown dwarf
0.08–0.25	helium white dwarf
0.25–8	carbon–oxygen white dwarf
8–12 (approx.)[*]	neon–oxygen white dwarf
greater than 12[*]	supernova (Chapter 21)

[*]*Precise numbers depend on the (poorly known) amount of mass lost while the star is on, and after it leaves, the main sequence.*

carbon and oxygen in its core, but its eventual fate is uncertain. As best we can tell, the star's mass at formation was between 12 and 17 times the mass of the Sun. However, like Rigel and many other supergiants, Betelgeuse has a strong stellar wind and is known to be surrounded by a huge shell of dust of its own making (see *Discovery 20-2*). It also pulsates, varying in radius by about 60 percent. The pulsations and strong wind may be related to the huge spots observed on the star's surface (Figure 17.11). Together, they suggest that Betelgeuse has lost a lot of mass since it formed, but just how much remains uncertain.

The End of the Road

Protostars and stars evolve because gravity always tends to cause a nonburning stellar core to contract and heat up. The contraction continues until it is halted either by electron degeneracy pressure or by the onset of a new round of nuclear fusion. In the latter case, a new nonburning core builds up, and the process repeats. The more massive the star, the more repetitions occur before the star finally dies. Table 20.3 lists some possible outcomes of stellar evolution for stars of different masses. For completeness, brown dwarfs—the end product of low-mass protostars unable even to fuse hydrogen in their cores—are included in the list ∞ (Sec. 19.3)

Note that our earlier dividing line of 8 solar masses between "low mass" and "high mass" really refers to the mass at the time the carbon core forms. Since very luminous stars often have strong stellar winds (*Discovery 20-2*), main-sequence stars as massive as 10 to 12 times the mass of the Sun may still manage to avoid going supernova. Unfortunately, we do not know exactly how much mass either Rigel or Betelgeuse has lost, so we cannot yet tell whether they are above or below the threshold for becoming a supernova. Either might explode or instead become a neon–oxygen white dwarf, but for now we can't say which. We may just have to wait and see—in a million years or so we will know for sure!

CONCEPT CHECK

✔ What is the essential evolutionary difference between high-mass and low-mass stars?

20.5 Observing Stellar Evolution in Star Clusters

5 Star clusters provide excellent test sites for the theory of stellar evolution. Every star in a given cluster formed at the same time, from the same interstellar cloud, and with virtually the same composition. Only the mass varies from one star to another, thus allowing us to check the accuracy of our theoretical models in a very straightforward way. Having studied the evolutionary tracks of individual stars in some detail, let's now consider how their collective appearance changes in time.

In Chapter 19, we saw how astronomers estimate the ages of star clusters by determining which of their stars have already left the main sequence. ∞ (See. 19.6) In fact, the main-sequence lifetimes that go into those age measurements represent only a tiny fraction of the data obtained from theoretical models of stellar evolution. Starting from the zero-age main sequence, astronomers can predict exactly how a newborn cluster should look at any subsequent time. Although we cannot see into the interiors of stars to test our models, we can compare stars' outward appearances with theoretical predictions. The agreement—in detail—between theory and observation is remarkably good.

We begin our study shortly after the cluster's formation, with the upper main sequence already fully formed and burning steadily, and with stars of lower mass just beginning to arrive on the main sequence, as shown in Figure 20.17(a). The appearance of the cluster at this early stage is dominated by its most massive stars: the bright blue supergiants. Now let's follow the cluster forward in time and see how it evolves via an H–R diagram.

Figure 20.17(b) shows the appearance of our cluster's H–R diagram after 10 million years. The most massive O-type stars have left the main sequence. Most have already exploded and vanished, as just discussed, but one or two may still be visible as red supergiants. The remaining stars in the cluster are largely unchanged in appearance—their evolution is slow enough that little happens to them in such a relatively short period. The cluster's H–R diagram shows the main sequence slightly cut off, along with a rather poorly defined red-giant region. Figure 20.18 shows the twin open clusters h and chi Persei, along with their combined H–R diagram. Comparing Figure 20.18(b) with such diagrams as those in Figure 20.17, astronomers estimate the age of this pair of clusters to be about 10 million years.

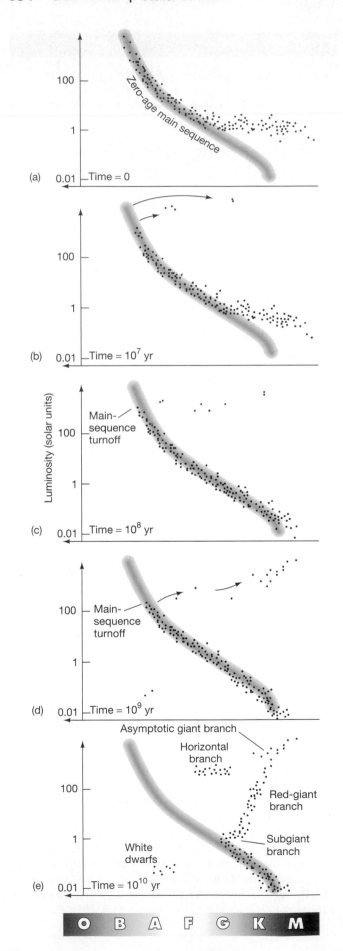

(a) Time = 0

(b) Time = 10⁷ yr

(c) Time = 10⁸ yr

(d) Time = 10⁹ yr

(e) Time = 10¹⁰ yr

Zero-age main sequence

Main-sequence turnoff

Main-sequence turnoff

Asymptotic giant branch

Horizontal branch

Red-giant branch

Subgiant branch

White dwarfs

Luminosity (solar units)

O B A F G K M

◀ **FIGURE 20.17 Cluster Evolution on the H–R Diagram** The changing H–R diagram of a hypothetical star cluster. (a) Initially, stars on the upper main sequence are already burning steadily while the lower main sequence is still forming. (b) At 10^7 years, O-type stars have already left the main sequence, and a few red giants are visible. (c) By 10^8 years, stars of spectral type B have left the main sequence. More red giants are visible, and the lower main sequence is almost fully formed. (d) At 10^9 years, the main sequence is cut off at about spectral type A. The subgiant and red-giant branches are just becoming evident, and the formation of the lower main sequence is complete. A few white dwarfs may be present. (e) At 10^{10} years, only stars less massive than the Sun still remain on the main sequence. The cluster's subgiant, red-giant, horizontal, and asymptotic-giant branches are all discernible. Many white dwarfs have now formed.

After 100 million years (Figure 20.17c), stars brighter than type B5 or so (about 4–5 solar masses) have left the main sequence, and a few more red supergiants are visible. By this time, most of the cluster's low-mass stars have finally arrived on the main sequence, although the dimmest M-type stars may still be in their contraction phase. The appearance of the cluster is now dominated by bright B-type main-sequence stars and brighter red supergiants.

At any time during the cluster's evolution, the original main sequence is intact up to some well-defined stellar mass, corresponding to the stars that are just leaving the main sequence at that instant. We can imagine the main sequence being "peeled away" from the top down, with fainter and fainter stars turning off and heading for the giant branch as time goes on. Astronomers refer to the high-luminosity end of the observed main sequence as the **main-sequence turnoff**. The mass of a star that is just evolving off the main sequence at any moment is known as the *turnoff mass*.

At 1 billion years, the main-sequence turnoff mass is around 2 solar masses, corresponding roughly to spectral type A2. The subgiant and giant branches associated with the evolution of low-mass stars are just becoming visible, as indicated in Figure 20.17(d). The formation of the lower main sequence is now complete. In addition, the first white dwarfs have just appeared, although they are often too faint to be observed at the distances of most clusters. Figure 20.19 shows the Hyades open cluster and its H–R diagram, which appears to lie between Figures 20.17(c) and 20.17(d), suggesting that the cluster's age is about 600 million years.

At 10 billion years, the turnoff point has reached solar-mass stars of spectral type G2. The subgiant and giant branches are now clearly discernible (see Figure 20.17e), and the horizontal and asymptotic-giant branches appear as distinct regions in the H–R diagram. Many white dwarfs are also present in the cluster. Although stars in all these evolutionary stages are also present in the 1-billion-year-old cluster shown in Figure 20.17(d), they are few in number then—typically only a few percent of the total number of stars in the cluster. Also, because they evolve so rapidly, these high-mass stars spend very little time in the various regions. Low-mass stars are much more numerous and evolve more slowly, so more of them spend more time in any given region of the H–R diagram, allowing their evolutionary tracks to be more easily discerned.

Figure 20.20 shows the globular cluster 47 Tucanae. By carefully adjusting their theoretical models until the

cluster's main sequence, subgiant, red-giant, and horizontal branches are all well matched, astronomers have determined the age of 47 Tucanae to be between 10 and 12 billion years, a little older than our hypothetical cluster in

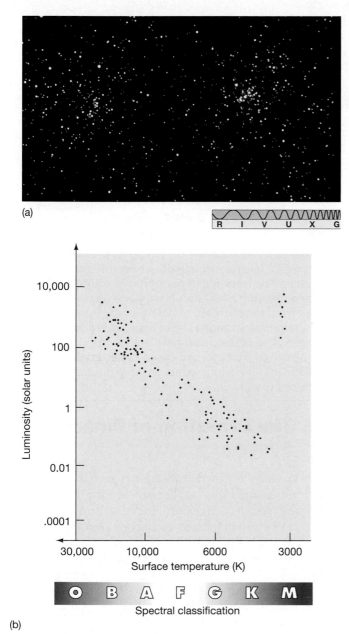

(a)

(b)

▲ **FIGURE 20.18 Newborn-Cluster H–R Diagram** (a) The "double cluster" h and chi Persei, two open clusters that apparently formed at the same time, possibly even orbiting one another. (b) The H–R diagram of the pair indicates that the stars are very young—probably only about 10 million years old. Even so, the most massive stars have already left the main sequence. (*AURA*)

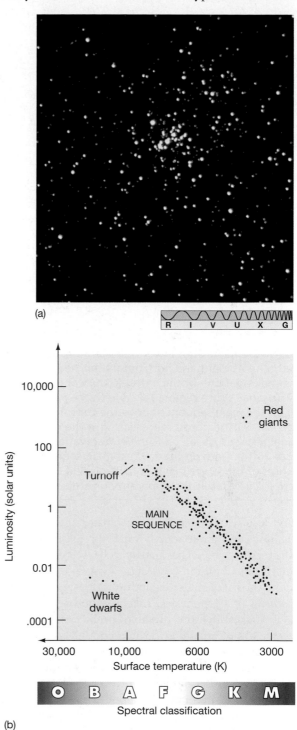

(a)

(b)

▲ **FIGURE 20.19 Young-Cluster H–R Diagram** (a) The Hyades cluster, a relatively young group of stars visible to the naked eye. The cluster lies 46 pc away in the constellation Taurus. (b) The H–R diagram for this cluster is cut off at about spectral type A, implying an age of about 600 million years. A few massive stars have already become white dwarfs. (*AURA*)

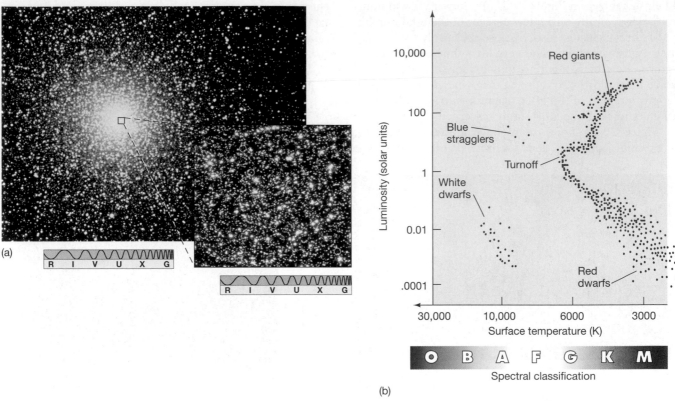

▲ FIGURE 20.20 Old-Cluster H–R Diagram (a) The southern globular cluster 47 Tucanae. (b) Fitting its main-sequence turnoff and its giant and horizontal branches to theoretical models gives 47 Tucanae an age of between 12 and 14 billion years, making it one of the oldest-known objects in the Milky Way Galaxy. The inset is a high-resolution ultraviolet image of 47 Tucanae's core region, taken with the *Hubble Space Telescope* and showing many blue stragglers—massive stars lying on the main sequence above the turnoff point, resulting perhaps from the merging of binary-star systems. (See also Figure 20.15.) The points representing white dwarfs, some red dwarfs, and blue stragglers have been added to the original data set, based on *Hubble* observations of this and other clusters. The white-dwarf data are for the cluster M4 (Figure 20.14). Data on the faintest main-sequence stars shown were obtained from ground-based observations. The thickness of the lower main sequence is due almost entirely to observational limitations, which make it difficult to determine accurately the apparent brightnesses and colors of low-luminosity stars. (*ESO; NASA*)

Figure 20.17(e). In fact, globular-cluster ages determined in this way show a remarkably small spread: All the globular clusters in our Galaxy appear to have formed between about 10 and 12 billion years ago.

The theory of stellar evolution is one of the great success stories of astrophysics. Like all good scientific theories, it makes definite testable predictions about the universe, while remaining flexible enough to incorporate new discoveries as they occur. Theory and observation have advanced hand in hand. At the start of the 20th century, many scientists despaired of ever knowing even the compositions of the stars, let alone why they shine and how they change. Today, the theory of stellar evolution is a cornerstone of modern astronomy.

CONCEPT CHECK

✔ Why are observations of star clusters so important to the theory of stellar evolution?

20.6 The Evolution of Binary-Star Systems

We have noted that most stars in our Galaxy are not isolated objects, but are actually members of binary-star systems. However, our discussion of stellar evolution has so far focused exclusively on isolated stars. This narrow focus prompts us to ask how membership in a binary-star system changes the evolutionary tracks we have just described. Indeed, because nuclear burning occurs deep in a star's core, does the presence of a stellar companion have any significant effect at all? Perhaps not surprisingly, the answer depends on the distance between the two stars in question.

For a binary system whose component stars are very widely separated—that is, the distance between the stars is greater than perhaps a thousand stellar radii—the two stars evolve more or less independently of one another, each following the track appropriate to an isolated star of its particular mass. However, if the two stars are closer, then the gravitational pull of one may strongly influence the envelope of the other. In that case, the physical properties of both may deviate greatly from those calculated for isolated single stars.

As an example, consider the star Algol (Beta Persei, the second-brightest star in the constellation Perseus). By studying its spectrum and the variation in its light intensity, astronomers have determined that Algol is actually a binary (in fact, an eclipsing double-lined spectroscopic binary, as described in Chapter 17), and they have measured its properties very accurately. ∞ (Sec. 17.7) Algol consists of a 3.7-solar-mass main-sequence star of spectral type B8 (a blue giant) with a 0.8-solar-mass red-subgiant companion moving in a nearly circular orbit around it. The stars are 4 million km apart and have an orbital period of about 3 days.

A moment's thought reveals that there is something odd about these findings. On the basis of our earlier discussion, the more massive main-sequence star should have evolved *faster* than the less massive component. If the two stars formed at the same time (as is assumed to be the case), there should be no way that the 0.8-solar-mass star could be approaching the giant stage first. Either our theory of stellar evolution is seriously in error, or something has modified the evolution of the Algol system. Fortunately for theorists, the latter is the case.

As sketched in Figure 20.21, each star in a binary system is surrounded by its own teardrop-shaped "zone of influence," inside of which its gravitational pull dominates the effects of both the other star and the overall rotation of the binary. Any matter within that region "belongs" to the star and cannot easily flow onto the other component or out of the system. Outside the two regions, it is possible for gas to flow toward either star relatively easily. The two teardrop-shaped regions are called **Roche lobes**, after Edouard Roche, the French mathematician who first studied the binary-system problem in the 19th century and whose work we have already encountered in the context of planetary rings. ∞ (Sec. 12.4) The Roche lobes of the two stars meet at a point on the line joining them—the inner Lagrangian point (L_1), which we saw in Chapter 14 in discussing asteroid motions in the solar system. ∞ (Sec. 14.1) This Lagrangian point is a place where the gravitational pulls of the two stars exactly balance the rotation of the binary system. The greater the mass of one component, the larger is its Roche lobe and the farther from its center (and the closer to the other star) is the Lagrangian point.

Normally, both stars lie well within their respective Roche lobes, and such a binary system is said to be

detached, as in Figure 20.22(a). However, as a star leaves the main sequence and moves toward the giant branch, it is possible for its radius to become so large that the star overflows its Roche lobe. Its gas begins to flow onto the companion through the Lagrangian point. The binary in this case is said to be *semidetached* (Figure 20.22b). Because matter flows from one star onto the other, semidetached binaries are also known as **mass-transfer binaries**. If, for some reason, the other star also overflows its Roche lobe (either because of stellar evolution or because so much extra material is dumped onto it), the surfaces of the two stars merge. The binary system then consists of two nuclear-burning stellar cores surrounded by a single continuous common envelope—a **contact binary**, shown in Figure 20.22(c).

In a binary system in which the two stars are very close together, neither star has to turn far off the main sequence before it overflows its Roche lobe and mass transfer begins. In a wide binary, however, both stars may evolve all the way up the giant branch without either surface ever reaching the Lagrangian point, and they evolve just as though they were isolated. Depending on the stars involved and their orbital separations, many different possibilities arise for the eventual outcome of their evolution. Binary evolution is a complex subject, but we can make these ideas more definite by returning to the question of how the binary star Algol may have reached its present state.

Astronomers believe that Algol started off as a detached binary. For reference, let us label the component that is now the 0.8-solar-mass subgiant as star 1 and the

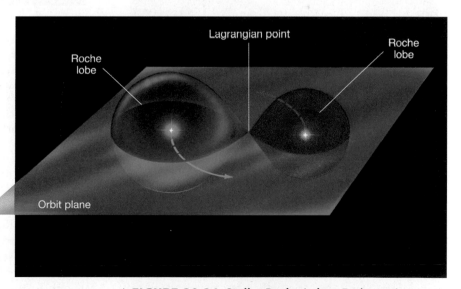

▲ **FIGURE 20.21 Stellar Roche Lobes** Each star in a binary system can be pictured as being surrounded by a "zone of influence," or Roche lobe, inside of which matter may be thought of as being "part" of that star. The two teardrop-shaped Roche lobes meet at the Lagrangian point between the two stars. Outside the Roche lobes, matter may flow onto either star with relative ease.

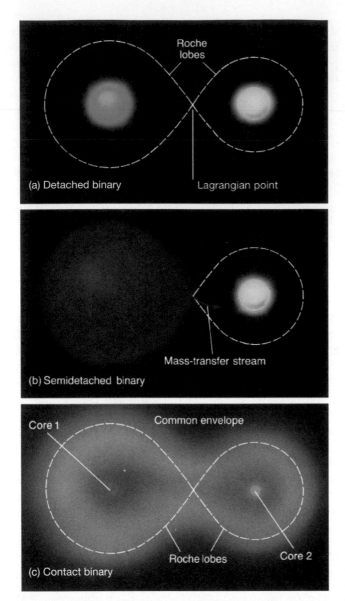

(a) Detached binary   Lagrangian point

(b) Semidetached binary

Mass-transfer stream

(c) Contact binary

▲ **FIGURE 20.22 Close Binary-Star Systems** (a) In a detached binary, each star lies within its respective Roche lobe. (b) In a semidetached binary, one of the stars fills its Roche lobe and transfers matter onto the other star, which still lies within its own Roche lobe. (c) In a contact, or common-envelope, binary, both stars have overflowed their Roche lobes, and a single star with two distinct nuclear-burning cores results.

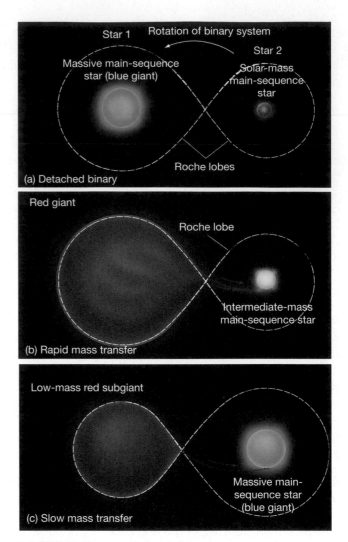

(a) Detached binary

(b) Rapid mass transfer

(c) Slow mass transfer

▲ **FIGURE 20.23 Algol Evolution** The evolution of the binary star Algol. (a) Initially, Algol was probably a detached binary made up of two main-sequence stars: a relatively massive blue giant and a less massive companion similar to the Sun. (b) As the more massive component (star 1) left the main sequence, it expanded to fill, and eventually overflow, its Roche lobe, transferring large amounts of matter onto its smaller companion (star 2). (c) Today, star 2 is the more massive of the two, but it is on the main sequence. Star 1 is still in the subgiant phase and fills its Roche lobe, causing a steady stream of matter to pour onto its companion.

3.7-solar-mass main-sequence star as star 2. Initially, star 1 was the more massive of the two, having perhaps three times the mass of the Sun. It thus turned off the main sequence first. Star 2 was originally a less massive star, perhaps comparable in mass to the Sun. As star 1 ascended the giant branch, it overflowed its Roche lobe and gas began to flow onto star 2. This transfer of matter had the effect of reducing the mass of star 1 and increasing that of star 2, which in turn caused the Roche lobe of star 1 to shrink as its gravity decreased. As a result, the rate at which star 1

overflowed its Roche lobe increased, and a period of unstable *rapid mass transfer* ensued, transporting most of star 1's envelope onto star 2. Eventually, the mass of star 1 became less than that of star 2. Detailed calculations show that the rate of mass transfer dropped sharply at that point, and the stars entered the relatively stable state we see today. These changes in Algol's components are illustrated in Figure 20.23.

Being part of a binary system has radically altered the evolution of both stars in the Algol system. The original high-mass star 1 is now a low-mass red subgiant, while the roughly solar mass star 2 is now a massive blue-giant main-sequence star. The removal of mass from the envelope of star 1 may prevent it from ever reaching the helium flash. Instead, its naked core may eventually be left behind as a *helium white dwarf*. In a few tens of millions of years, star 2 will itself begin to ascend the giant branch and fill its own Roche lobe. If star 1 is still a subgiant or a giant at that time, a contact binary system will result. If, instead, star 1 has by then become a white dwarf, a new mass-transfer period—with matter streaming from star 2 back onto star 1—will begin. In that case (as we will see in Chapter 21), Algol may have a very active and violent future in store for it.

Just as molecules exhibit few of the physical or chemical properties of their constituent atoms, binaries can display types of behavior that are quite different from the behavior of either of their component stars. The Algol system is a fairly simple example of binary evolution, yet it gives us an idea of the sorts of complications that can arise when two stars evolve interdependently. A significant fraction of all the binary stars in the Galaxy will pass through some sort of mass-transfer or common-envelope phase. In this chapter, we have seen one result of mass transfer involving main-sequence stars. We will return to the subject in the next two chapters, when we continue our discussion of stellar evolution and the strange states of matter that may ensue.

CONCEPT CHECK

✔ Why is it important to understand the evolution of binary stars?

Chapter Review

SUMMARY

Stars spend most of their lives on the main sequence, in the **core-hydrogen-burning (p. 516)** phase of stellar evolution, stably fusing hydrogen into helium at their centers. Stars leave the main sequence when the hydrogen in their cores is exhausted. The Sun, which is about halfway through its main-sequence lifetime, will reach this stage about 5 billion years from now. Low-mass stars evolve much more slowly than the Sun, and high-mass stars evolve much faster.

When the central nuclear fires in the interior of a solar-mass star cease, the helium in the star's core is still too cool to fuse into anything heavier. With no internal energy source, the helium core is unable to support itself against its own gravity and begins to shrink. At this stage, the star is in the **hydrogen-shell-burning (p. 519)** phase, in which the nonburning helium at the center is surrounded by a layer of burning hydrogen. The energy released by the contracting helium core heats the hydrogen-burning shell, greatly increasing the nuclear reaction rates there. As a result, the star becomes much brighter, while the envelope expands and cools. A low-mass star like the Sun moves off the main sequence on the H–R diagram first along the **subgiant branch (p. 520)** and then almost vertically up the **red-giant branch (p. 520)**.

As the helium core contracts, it heats up. Eventually, the core of a star more than 0.25 times the mass of the Sun reaches the point at which helium begins to fuse into carbon. The net effect of the fusion reactions is that three helium nuclei (or alpha particles) combine to form a nucleus of carbon in the **triple-alpha process (p. 521)**. In a star like the Sun, conditions at the onset of helium burning are such that the electrons in the core have become degenerate—they can be thought of as tiny, hard spheres that, once brought into contact, present stiff resistance to being compressed any further. This **electron degeneracy pressure (p. 521)** makes the core unable to "react" to the new energy source, and helium burning begins violently in a **helium flash (p. 522)**. The flash expands the core and reduces the star's luminosity, sending the star onto the **horizontal branch (p. 522)** of the H–R diagram. The star now has a core of burning helium surrounded by a shell of burning hydrogen.

As helium burns in the core, it forms an inner core of nonburning carbon. The carbon core shrinks and heats the overlying burning layers, and the star once again becomes a red giant, even more luminous than before. It reenters the red-giant region of the H–R diagram along the **asymptotic-giant branch (p. 522)**. The core of a low-mass star never becomes hot enough to fuse carbon. Such a star continues to ascend the asymptotic-giant branch until its envelope is ejected into space as a **planetary nebula (p. 525)**. At that point, the core becomes visible as a hot, faint, and extremely dense white dwarf, while the planetary nebula diffuses into space, carrying helium and some carbon into the interstellar medium. The white dwarf cools and fades, eventually becoming a cold black dwarf. Most white dwarfs are composed of carbon and oxygen, although stars in binary systems may give rise to helium white dwarfs, while more massive stars may become neon–oxygen white dwarfs.

High-mass stars evolve more rapidly than low-mass stars because larger mass results in higher central temperature. High-mass stars never initiate a helium flash, and they attain central temperatures high enough to fuse carbon. These stars become **red supergiants**, forming heavier and heavier elements in their cores at an increasingly rapid pace, and eventually die explosively.

The theory of stellar evolution can be tested by observing star clusters, all of whose stars formed at the same time. As time goes by, the most massive stars leave the main sequence first, then the intermediate-mass stars, and so on. At any instant, no stars with masses above the cluster's **main-sequence turnoff (p. 534)** mass remain on the main sequence. Stars below this mass have not yet evolved into giants and so still lie on the main sequence. By comparing a particular cluster's main-sequence turnoff mass with theoretical predictions, astronomers can measure the age of the cluster.

Stars in binary systems can evolve quite differently from isolated stars because of interactions with their companions. Each star is surrounded by a teardrop-shaped **Roche lobe (p. 537)**, which defines the region of space within which matter "belongs" to the star. As a binary star evolves into the giant phase, it may overflow its Roche lobe, forming a **mass-transfer binary** (p. 537), with gas flowing from the giant onto its companion. If both stars overflow their Roche lobes, a **contact binary (p. 537)** results. Stellar evolution in binaries can produce states that are not achievable in single stars. In a sufficiently wide binary, both stars evolve as though they were isolated.

REVIEW AND DISCUSSION

1. Why don't stars live forever? Which types of stars live the longest?

2. What is hydrostatic equilibrium?

3. How long can a star like the Sun keep burning hydrogen in its core?

4. Why is the depletion of hydrogen in the core of a star such an important event?

5. What makes an ordinary star become a red giant?

6. Roughly how big (in A.U.) will the Sun become when it enters the red-giant phase?

7. How long does it take for a star like the Sun to evolve from the main sequence to the top of the red-giant branch?

8. Do all stars eventually fuse helium in their cores?

9. What is a helium flash?

10. Describe an important way in which winds from red-giant stars are linked to the interstellar medium.

11. How do the late evolutionary stages of high-mass stars differ from those of low-mass stars?

12. What is the internal structure of a star on the asymptotic-giant branch?

13. What is a planetary nebula? Why do many planetary nebulae appear as rings?

14. What are white dwarfs? What is their ultimate fate?

15. Can you think of a way in which a helium white dwarf might exist today?

16. Why are white dwarfs hard to observe?

17. Do many black dwarfs exist in the Galaxy?

18. How can astronomers measure the age of a star cluster?

19. What are the Roche lobes of a binary system?

20. Why is it odd that the binary system Algol consists of a low-mass red giant orbiting a high-mass main-sequence star? How did Algol come to be in this configuration?

CONCEPTUAL SELF-TEST: TRUE OR FALSE/MULTIPLE CHOICE

1. While on the main sequence, a star slowly fuses helium into hydrogen in its core.

2. All of the single red-dwarf stars that ever formed are still on the main sequence today.

3. The Sun will get brighter as it begins to run out of fuel in its core.

4. As a star evolves away from the main sequence, it gets larger.

5. As a star evolves away from the main sequence, it gets hotter.

6. A planetary nebula is the disk of matter around a star that will eventually form a planetary system.

7. When helium fuses, it produces oxygen and releases neutrinos.

8. Stellar evolution in binary stars has much the same outcome as in isolated stars.

9. As a red giant, the Sun will eventually become about 10 times its present size.

10. The various stages of stellar evolution predicted by theory can best be tested via observations of stars in clusters.

11. A star will evolve "off the main sequence" when it uses up **(a)** all of its hydrogen; **(b)** half of its hydrogen; **(c)** most of the hydrogen in the core; **(d)** all of its gas.

12. On the main sequence, massive stars **(a)** conserve their hydrogen fuel by burning helium; **(b)** burn their hydrogen fuel more rapidly than the Sun; **(c)** burn their fuel more slowly than the Sun; **(d)** evolve into stars like the Sun.

13. Compared to other stars on the H-R diagram, red giant stars are so named because they are **(a)** cooler; **(b)** fainter; **(c)** denser; **(d)** younger.

14. When the Sun is on the red giant branch, it will be found at the **(a)** upper left; **(b)** upper right; **(c)** lower right; **(d)** lower left of the H-R diagram.

15. After the core of a Sun-like star starts to fuse helium on the horizontal branch, the core becomes **(a)** hotter; **(b)** cooler; **(c)** larger; **(d)** dimmer with time.

16. If the evolutionary track in *Overlay 3*, showing a Sun-like star, were instead illustrating a significantly more massive star, its starting point (stage 7) would be **(a)** up and to the right; **(b)** down and to the left; **(c)** up and to the left; **(d)** down and to the right.

17. A white dwarf is supported by the pressure of tightly packed **(a)** electrons; **(b)** protons; **(c)** neutrons; **(d)** photons.

18. When the Sun leaves the main sequence, it will become **(a)** hotter; **(b)** brighter; **(c)** more massive; **(d)** younger.

19. A star like the Sun will end up as a **(a)** blue giant; **(b)** white dwarf; **(c)** binary star; **(d)** red dwarf.

20. Compared to the Sun, stars plotted near the bottom left of the HR diagram are much **(a)** younger; **(b)** more massive; **(c)** brighter; **(d)** denser.

PROBLEMS

 Algorithmic versions of these questions are available in the Practice Problems module of the Companion Website at astro.prenhall.com/chaisson.

The number of squares preceding each problem indicates its approximate level of difficulty.

1. ■ The Sun will leave the main sequence when roughly 10 percent of its hydrogen has been fused into helium. Using the data given in Section 16.5 and Table 16.2, calculate the total amount of mass destroyed (i.e., converted into energy) and the total energy released by the fusion of that amount of matter.

2. ■ Use the radius–luminosity–temperature relation to calculate the radius of a red supergiant with temperature 3000 K (half the solar value) and total luminosity 10,000 times that of the Sun. ∞ (Sec. 17.3) How many planets of our solar system would this star engulf?

3. ■ What would be the luminosity of the Sun if its surface temperature were 3000 K and its radius were (a) 1 A.U., (b) 5 A.U.?

4. ■ Use the radius–luminosity–temperature relation to calculate the radius of a 12,000-K (twice the temperature of the Sun), 0.0004-solar-luminosity white dwarf.

5. ■ Use the graph in *More Precisely 20-1* to estimate the factor by which CNO energy production outstrips proton–proton energy production in a 10-solar-mass star with a central temperature of 25 million K. What do you think the factor would be if the abundances of C, N, and O were just one-tenth the solar value?

6. ■■ A main-sequence star at a distance of 20 pc is barely visible through a certain telescope. The star subsequently ascends the giant branch, during which time its temperature drops by a factor of three and its radius increases a hundredfold. What is the new maximum distance at which the star would still be visible in the same telescope?

7. ■ A Sun-like star goes through a rapid change in luminosity between stages 8 and 9, when its luminosity increases by about a factor of 100 in 10^5 years. On average, how rapidly does the star's absolute magnitude change, in magnitudes per year? Do you think this change would be noticeable in a distant star within a human lifetime?

8. ■ Calculate the average density of a red-giant core of 0.25 solar mass and radius 15,000 km. Compare your answer with the average density of the giant's envelope, if it has a 0.5 solar mass and its radius is 0.5 A.U. Compare each of the two densities with the central density of the Sun. ∞ (Sec. 16.2)

9. ■ How long will it take the Sun's planetary nebula, expanding at a speed of 50 km/s, to reach the orbit of Neptune? How long to reach the nearest star?

10. ■ What are the escape speed (in km/s) and surface gravity (relative to Earth's gravity) of Sirius B? (See Table 20.2.)

11. ■ A 15-solar-mass blue supergiant with a surface temperature of 20,000 K becomes a red supergiant with the same total luminosity and a temperature of 4000 K. By what factor does its radius change?

12. ■■ The radius of Betelgeuse varies by about 60 percent within a period of 3 years. If the star's surface temperature remains constant, by how much does its absolute magnitude change during this time?

13. ■■ The Sun will reside on the main sequence for 10^{10} years. If the luminosity of a main-sequence star is proportional to the fourth power of the star's mass, what is the mass of a star that is just now leaving the main sequence in a cluster that formed (a) 400 million years ago, (b) 2 billion years ago?

14. ■■■ In roughly 5 billion years, the Sun will eject its envelope as a planetary nebula. Suppose that, before then, the Sun loses 20 percent of its mass on the giant branch. (a) If Jupiter stays in a circular orbit while this mass is being lost and the angular momentum of Jupiter stays constant, what will be the planet's orbital radius and period when the Sun's mass has fallen to 0.8 solar mass? (b) When the planetary nebula is formed, the Sun loses a further 0.3 solar mass rapidly enough that we can regard the loss as immediate. Jupiter's instantaneous velocity is unchanged, but the planet's orbit is no longer circular, due to the Sun's smaller mass. Jupiter's location at that moment becomes the perihelion of the new orbit, and the orbital semimajor axis increases (in this case) by a factor of 2.5. What are the eccentricity and period of the new orbit?

15. ■■■ From the discussions presented in *More Precisely 2-3* and *More Precisely 6-2*, it may be shown that the angular momentum of a circular binary system of separation r and component masses m_1 and m_2 is proportional to $m_1 \times m_2 \times \sqrt{r}$ (if the total mass is constant). Such a binary has component masses one and two times the mass of the Sun, respectively, and an orbital period of 2 years. (a) What is its orbital separation r? (b) Mass transfer moves 0.2 solar mass of material from the more massive to the less massive star, keeping the total mass of the system fixed and conserving angular momentum. Assuming that the binary remains circular, calculate its new separation and orbital period.

 In addition to the Practice Problems module, the Companion Website at astro.prenhall.com/chaisson provides for each chapter a study guide module with multiple choice questions as well as additional annotated images, animations, and links to related Websites.

Elements heavier than iron were synthesized during the explosive deaths of massive stars—in supernovae, which astronomers have observed in many locations across the sky. There is something philosophically pleasing about the idea that the deaths of some stars cause the births of others, at the same time creating many of the elements that comprise our own world. Here, the Chandra X-Ray Observatory has imaged X rays of different energies (colors) from glowing debris of the supernova remnant N132D, the remains of a formerly massive star about 180,000 light-years away. (CXO/SAO) ▶

Stellar Explosions

Novae, Supernovae, and the Formation of the Elements

What fate awaits a star when it runs out of fuel? For a low-mass star, the white-dwarf stage is not necessarily the end of the road: The potential exists for further violent activity if a binary companion can provide additional fuel. High-mass stars—whether they are or are not members of binaries—are also destined to die explosively, releasing vast amounts of energy, creating many heavy elements, and scattering the debris throughout interstellar space. These cataclysmic explosions may trigger the formation of new stars, continuing the cycle of stellar birth and death. In this chapter, we will study in more detail both the processes responsible for the explosions and the mechanisms that create the elements from which we ourselves are made.

LEARNING GOALS

Studying this chapter will enable you to

1. Explain how white dwarfs in binary-star systems can become explosive.

2. Summarize the sequence of events leading to the violent death of a massive star.

3. Describe the two types of supernovae, and explain how each is produced.

4. Describe the observational evidence for the occurrence of supernovae in our Galaxy.

5. Explain the origin of elements heavier than helium, and discuss the significance of these elements for the study of stellar evolution.

6. Outline how the universe continually recycles matter through stars and the interstellar medium.

Visit astro.prenhall.com/chaisson for additional annotated images, animations, and links to related sites for this chapter.

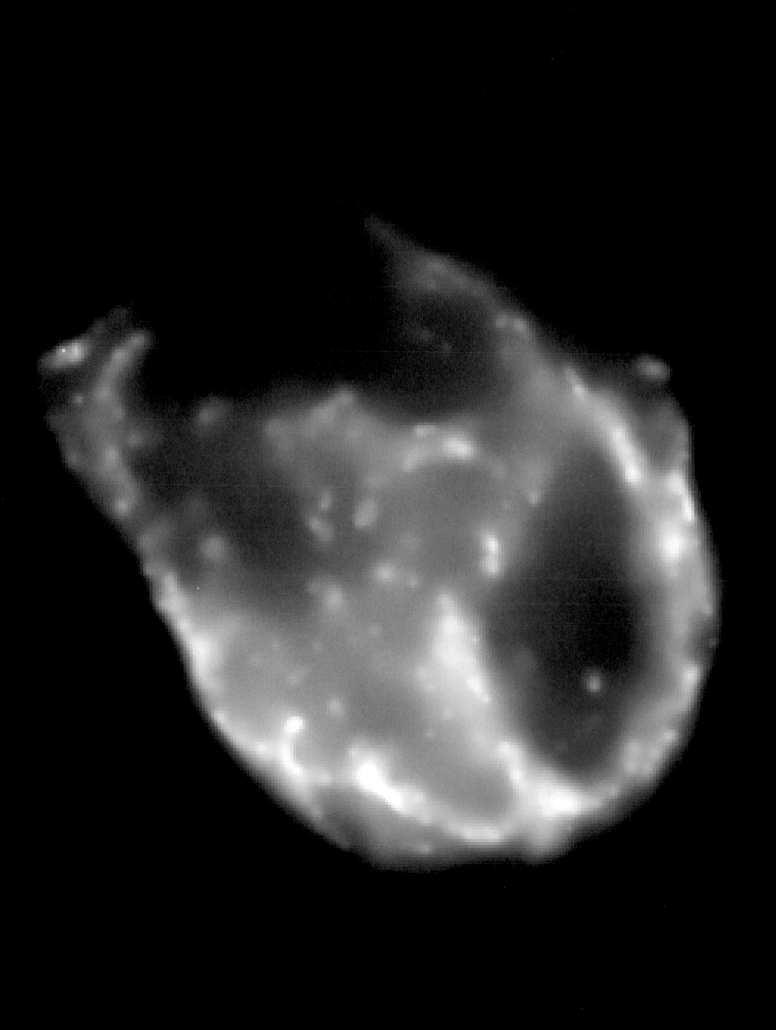

21.1 Life after Death for White Dwarfs

Although most stars shine steadily day after day and year after year, some change dramatically in brightness over very short periods of time. One type of star, called a **nova** (plural: *novae*), may increase enormously in brightness—by a factor of 10,000 or more—in a matter of days and then slowly return to its initial luminosity over a period of weeks or months.

The word *nova* means "new" in Latin, and to early observers these stars did indeed seem new, because they appeared suddenly in the night sky. Astronomers now recognize that a nova is not a new star at all. It is instead a white dwarf—a normally very faint star—undergoing an explosion on its surface that results in a rapid, temporary increase in the star's luminosity. Figures 21.1(a) and (b) illustrate the brightening of a typical nova. Figure 21.1(c) shows a nova light curve, demonstrating how the luminosity rises dramatically in a matter of days and then fades slowly back to normal over the course of several months. On average, two or three novae are observed each year. Astronomers also know of many *recurrent novae*—stars that have been observed to "go nova" several times over the course of a few decades.

What could cause such an explosion on a faint, dead star? The energy involved is far too great to be explained by flares or other surface activity, and as we saw in the previous chapter, there is no nuclear activity in the dwarf's interior. ∞ (Sec. 20.3) To understand what happens, we must consider again the fate of a low-mass star after it enters the white-dwarf phase.

We noted in Chapter 20 that the white-dwarf stage represents the end point of a star's evolution. Subsequently, the star simply cools, eventually becoming a black dwarf—a burned-out ember in interstellar space. This scenario is quite correct for an *isolated* star, such as our Sun. However, should the star be part of a *binary* system, an important new possibility exists. If the distance between the two stars is small enough, then the dwarf's tidal gravitational field can pull matter—primarily hydrogen and helium—away from

the surface of its main-sequence or giant companion. ∞ (Sec. 7.6) The system then becomes a mass-transferring binary, similar to those discussed in Chapter 20. A stream of gas leaves the companion through the inner (L_1) Lagrangian point and flows onto the dwarf. ∞ (Sec. 20.6)

Because of the binary's rotation and the white dwarf's small size, material leaving the companion does not fall directly onto the dwarf, as indicated in Figure 20.23. Instead, such material "misses" the compact star, loops around behind it, and goes into orbit around it, forming a swirling, flattened disk of matter called an **accretion disk** (shown in Figure 21.2). Due to the effects of viscosity (i.e., friction) within the gas, the orbiting matter in the disk drifts gradually inward, its temperature increasing steadily as it spirals

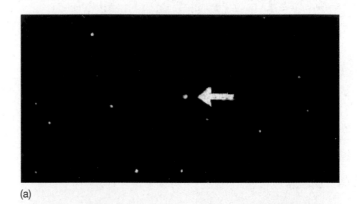
(a)

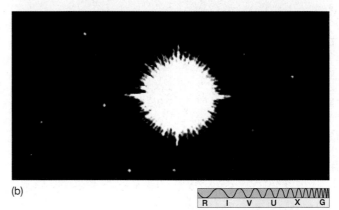
(b)

R I V U X G

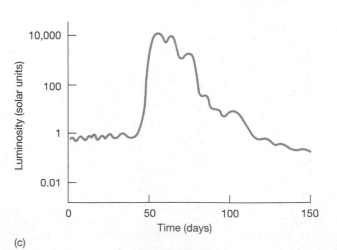
(c)

▶ **FIGURE 21.1 Nova** A nova is a star that suddenly increases enormously in brightness and then slowly fades back to its original luminosity. Novae are the result of explosions on the surfaces of faint white dwarf stars, caused by matter falling onto their surfaces from the atmospheres of larger binary companions. Shown is Nova Herculis 1934 in (a) March 1935 and (b) May 1935, after brightening by a factor of 60,000. (c) The light curve of a typical nova. The rapid rise and slow decline in the light received from the star, as well as the maximum brightness attained, are in good agreement with the explanation of a nova as a nuclear flash on a white dwarf's surface. *(UC/Lick Observatory)*

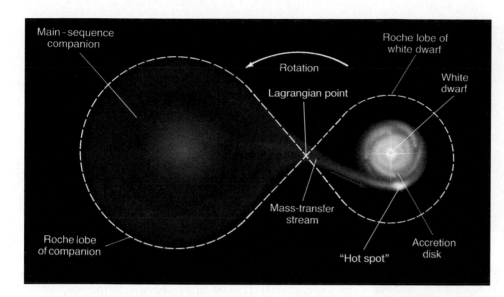

◀ **FIGURE 21.2 Close Binary System** If a white dwarf in a semidetached binary system is close enough to its companion (in this case, a main-sequence star), its gravitational field can tear matter from the companion's surface. (Cf. Figure 20.23.) Notice that, unlike the scenario shown in the earlier figure, the matter does not fall directly onto the white dwarf's surface. Instead, it forms an accretion disk of gas spiraling down onto the dwarf.

down onto the dwarf's surface. The inner part of the accretion disk becomes so hot that it radiates strongly in the visible, the ultraviolet, and even the X-ray portions of the electromagnetic spectrum. In many systems, the disk outshines the white dwarf itself and is the main source of the light emitted between nova outbursts. X rays from the hot disk are routinely observed in many galactic novae. The point at which the infalling stream of matter strikes the accretion disk often forms a turbulent "hot spot," causing detectable fluctuations in the light emitted by the binary system.

The "stolen" gas becomes hotter and denser as it builds up on the white dwarf's surface. Eventually, its temperature exceeds 10^7 K, and the hydrogen ignites, fusing into helium at a furious rate. (Figures 21.3a–d illustrate the sequence of events.) This surface-burning stage is as brief as it is violent: The star suddenly flares up in luminosity and then fades away as some of the fuel is exhausted and the remainder is blown off into space. If the event happens to be visible from Earth, we see a nova. Figure 21.4 shows

two novae apparently caught in the act of expelling mass from their surfaces. A nova's decline in brightness results from the expansion and cooling of the white dwarf's surface layers as they are blown into space. Studies of the details of the brightness curve associated with a nova provide astronomers with a wealth of information about both the dwarf and its binary companion.

A nova represents one way in which a star in a binary system can extend its "active lifetime" well into the white-dwarf stage. Recurrent novae can, in principle, repeat their violent outbursts many dozens, if not hundreds, of times. But even more extreme possibilities exist at the end of stellar evolution. Vastly more energetic events may be in store, given the right circumstances.

CONCEPT CHECK

✔ Will the Sun ever become a nova?

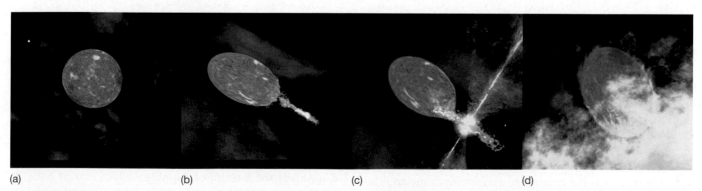

(a) (b) (c) (d)

▲ **FIGURE 21.3 Nova Explosion** In this artist's conception, a white-dwarf star (upper left) orbits a cool red giant (a); material accumulates on the white dwarf's surface after being accreted from the companion star (b and c) and then ignites in hydrogen fusion as a nova outburst (d). *(D. Berry)*

(a)

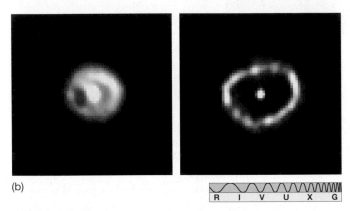

(b)

R I V U X G

▲ **FIGURE 21.4 Nova Matter Ejection** The ejection of material from a star's surface can clearly be seen in these images, which correspond approximately to Figure 21.3(d). (a) Nova Persei, taken some 50 years after it suddenly brightened by a factor of 40,000 in 1901. (b) Nova Cygni, imaged here with a European camera on the *Hubble Space Telescope*, erupted in 1992. At left, more than a year after the blast, a rapidly billowing bubble is seen; at right, seven months after that, the shell continues to expand and distort. The image is fuzzy because the object is more than 10,000 light-years away. *(Palomar Observatory; ESA)*

21.2 The End of a High-Mass Star

LEARNING GOAL 2 A low-mass star—a star with a mass of less than about 8 solar masses—never becomes hot enough to burn carbon in its core. It ends its life as a carbon–oxygen (or possibly neon–oxygen) white dwarf. ∞ (Sec. 20.3) A high-mass star, however, can fuse not just hydrogen and helium, but also carbon, oxygen, and even heavier elements as its inner core continues to contract and its central temperature continues to rise. ∞ (Sec. 20.4) The burning rate accelerates as the core evolves. Can anything stop this runaway process? Is there a stable "white-dwarf-like" state at the end of the evolution of a high-mass star? What is the ultimate fate of such a star? To answer these questions, we must look more carefully at fusion in massive stars.

Fusion of Heavy Elements

Figure 21.5 is a cutaway diagram of the interior of a highly evolved star of large mass. Note the numerous layers in which various nuclei burn. As the temperature increases with depth, the ash of each burning stage becomes the fuel for the next stage. At the relatively cool periphery of the core, hydrogen fuses into helium. In the intermediate layers, shells of helium, carbon, and oxygen burn to form heavier nuclei. Deeper down reside neon, magnesium, silicon, and other heavy nuclei, all produced by nuclear fusion in the layers overlying the core. (Recall that, to astronomers, a "heavy" element is anything more massive than helium.) The core itself is composed of iron. We will study the key reactions in this burning chain in more detail later in the chapter.

As each element is burned to depletion at the center, the core contracts, heats up, and starts to fuse the ash of the previous burning stage. A new inner core forms, contracts again, heats again, and so on. Through each period of stability and instability, the star's central temperature increases, the nuclear reactions speed up, and the newly released energy supports the star for ever-shorter periods of time. For example, in round numbers, a star 20 times more massive than the Sun burns hydrogen for 10 million years, helium for 1 million years, carbon for a thousand years, oxygen for a year, and silicon for a week. Its iron core grows for less than a day.

Collapse of the Iron Core

Once the inner core begins to change into iron, our high-mass star is in trouble. As illustrated in Figure 21.6, iron is

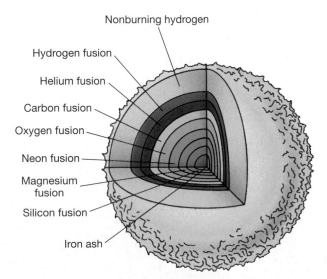

Nonburning hydrogen

Hydrogen fusion

Helium fusion

Carbon fusion

Oxygen fusion

Neon fusion

Magnesium fusion

Silicon fusion

Iron ash

▲ **FIGURE 21.5 Heavy-Element Fusion** Cutaway diagram of the interior of a highly evolved star of mass greater than 8 solar masses (not to scale). The interior resembles the layers of an onion, with shells of progressively heavier elements burning at smaller and smaller radii and at higher and higher temperatures.

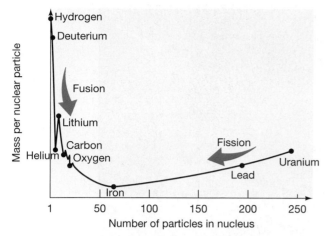

▲ FIGURE 21.6 Nuclear Masses This graph shows how the masses (per nuclear particle—proton or neutron) of most known nuclei vary with nuclear mass. It contains an enormous amount of information about nuclear structure and stability. When light nuclei fuse (left side of the figure), the mass per particle decreases and energy is released. ∞ (Sec. 16.2) Similarly, when heavy nuclei split apart (right side), the total mass again decreases and energy is again released. The nucleus with the smallest mass per nuclear particle—the most stable element—is iron. It can be neither fused nor split to release energy. The difference in mass between a given nucleus and the value for hydrogen represents the amount of mass lost (or energy released) if free neutrons and protons were combined to form that nucleus—or equivalently, the amount of energy that must be provided to split the nucleus into its component particles. ∞ (*More Precisely 4-1*)

the *most stable* element there is. To understand the figure, imagine fusing four protons to form helium-4. According to the figure, the mass per particle of a helium-4 nucleus is less than the mass of a proton, so mass is lost and (in accordance with the law of conservation of mass and energy) energy is released. ∞ (Sec. 16.2) Similarly, combining three helium-4 nuclei to form carbon results in a net loss of mass, again releasing energy. In other words, the left side of the figure shows how light elements can fuse to release energy. The right side of the figure shows the opposite process, known as *fission*. Here, combining nuclei will increase the total mass per particle and hence absorb energy, so fusion can't occur. However, splitting a heavy nucleus (such as uranium, or plutonium, which lies just off the right edge of the figure) into lighter nuclei does release energy—this is how nuclear reactors and atomic bombs work.

Iron lies at the dividing line between these two types of behavior—at the lowest point of the curve in the figure. Iron nuclei are so compact that energy cannot be extracted either by combining them into heavier elements or by splitting them into lighter ones. In effect, iron plays the role of a fire extinguisher, damping the inferno in the stel-

lar core. With the appearance of substantial quantities of iron, the central fires cease for the last time, and the star's internal support begins to dwindle. The star's foundation is destroyed, and its equilibrium is gone forever. Even though the temperature in the iron core has reached several billion kelvins by this stage, the enormous inward gravitational pull of matter ensures catastrophe in the very near future. Gravity overwhelms the pressure of the hot gas, and the star implodes, falling in on itself.

The core temperature rises to nearly 10 billion K. According to Wien's law, at that temperature individual photons have tremendously high energies—enough to split iron into lighter nuclei and then to break those lighter nuclei apart until only protons and neutrons remain. ∞ (Sec. 3.4) This process is known as *photodisintegration* of the heavy elements in the core. In less than a second, the collapsing core undoes all the effects of nuclear fusion that occurred during the previous 10 million years! But to split iron and lighter nuclei into smaller pieces requires a lot of energy (Figure 21.6, moving from iron to the left). After all, this splitting is just the opposite of the fusion reactions that generated the star's energy during earlier times. Photodisintegration *absorbs* some of the core's thermal energy—in other words, it cools the core and thus reduces the pressure there. As nuclei are destroyed, the core of the star becomes even less able to support itself against its own gravity. The collapse accelerates.

Now the core consists entirely of simple elementary particles—electrons, protons, neutrons, and photons—at enormously high densities, and it is still shrinking. As the density of the core continues to rise, the protons and electrons are crushed together, forming neutrons and neutrinos:

$$p + e \rightarrow n + \text{neutrino}.$$

This process is sometimes called the *neutronization* of the core. Recall from our discussion in Chapter 16 that the neutrino is an extremely elusive particle that hardly interacts at all with matter. ∞ (Sec. 16.2) Even though the central density by this time may have reached 10^{12} kg/m³ or more, most of the neutrinos produced by neutronization pass through the core as if it weren't there. They escape into space, carrying away energy as they go, further reducing the core's pressure support.

The disappearance of the electrons and the escape of the neutrinos make matters even worse for the core's stability. There is now nothing to prevent it from collapsing all the way to the point at which the neutrons come into contact with one another, at the incredible density of about 10^{15} kg/m³. At this point, the neutrons in the shrinking core play a role similar in many ways to that of the electrons in a white dwarf. Far apart, they offer little resistance to compression, but brought into contact, they produce enormous pressures that strongly oppose further

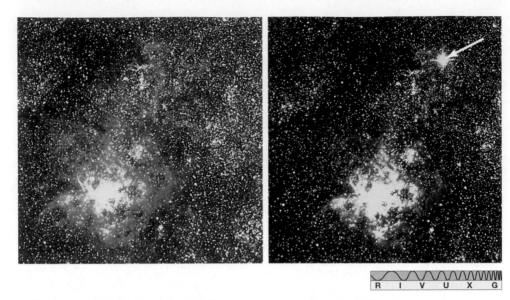

▲ **FIGURE 21.7 Supernova 1987A** A supernova called SN 1987A (arrow) was exploding near this nebula (called 30 Doradus) at the moment the photograph on the right was taken. The photograph on the left shows the star field prior to the supernova. (See *Discovery 21-1.*) *(AURA)*

gravitational collapse. This *neutron degeneracy pressure*, akin to the electron degeneracy pressure that operates in red giants and white dwarfs, finally begins to slow the star's collapse. ∞ (Sec. 20.2) By the time the collapse is actually halted, however, the core has overshot its point of equilibrium, and may reach densities as high as 10^{17} or 10^{18} kg/m³ before turning around and beginning to reexpand. Like a fast-moving ball hitting a brick wall and bouncing back, the core becomes compressed, stops, and then rebounds—with a vengeance!

The events just described do not take long. Only about a second elapses from the start of the collapse to the "bounce" at nuclear densities. At that point, the core rebounds. An enormously energetic shock wave sweeps through the star at high speed, blasting all the overlying layers—including all the heavy elements just formed outside the iron inner core—into space. Although computer models are still somewhat inconclusive, and the details of how the shock reaches the surface and destroys the star remain uncertain, the end result is not: In one of the most energetic events known in the universe the star explodes. (See Figure 21.7.) For a period of a few days, the exploding star may rival in brightness the entire galaxy in which it resides. This spectacular death rattle of a high-mass star is known as a **core-collapse supernova**.

CONCEPT CHECK

✔ Why does the iron core of a high-mass star collapse?

21.3 Supernovae

Novae and Supernovae

Let's compare a supernova with a nova. Like a nova, a **supernova** is a star that suddenly increases dramatically in brightness and then slowly dims again, eventually fading from view. In its unexploded state, a star that will become a supernova is known as the supernova's *progenitor*. In some cases, supernovae light curves can appear quite similar to those of novae, and a distant supernova can look a lot like a nearby nova—so much so, in fact, that the difference between the two was not fully appreciated until the 1920s. But novae and supernovae are now known to be quite different phenomena. Supernovae are much more energetic events, driven by very different underlying physical processes.*

Well before they understood the causes of either novae or supernovae, astronomers knew of clear observational differences between them. The most important of these differences is that a supernova is more than a million times brighter than a nova. A supernova produces a burst of light billions of times brighter than the Sun, reaching that level of brightness within just a few hours after the start of the outburst. The total amount of electromagnetic energy radiated by a supernova during the few months it

---

*Note that, in discussing novae and supernovae, astronomers tend to blur the distinction between the observed event (the sudden appearance and brightening of an object in the sky), the process responsible for the event (a violent explosion in or on a star), and the object per se (the star itself is called a nova or a supernova, as the case may be). The two terms can have any of the three meanings, depending on the context.

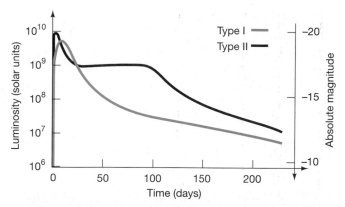

▲ FIGURE 21.8 Supernova Light Curves Light curves of typical Type I and Type II supernovae both show the maximum brightness or intensity reaching nearly 10 billion solar luminosities, but there are characteristic differences in the falloff of the luminosity after the initial peak. Type I light curves resemble those of novae somewhat (Figure 21.1). Type II curves have a characteristic plateau in the declining phase.

takes to brighten and fade away is roughly 10^{43} J—nearly as much energy as the Sun will radiate during its *entire* 10^{10}-year lifetime! (Enormous as this energy is, however, it pales in comparison with the energy emitted in the form of neutrinos, which may be 100 times greater.)

A second important difference is that the same star may become a nova many times, but a star can become a supernova only once. This fact was unexplained before astronomers knew the precise nature of novae and supernovae, but it is easily understood now that we understand how and why these explosions occur. The nova accretion–explosion cycle described earlier can take place over and over again, but a supernova destroys the star involved, with no possibility of a repeat performance.

In addition to the distinction between novae and supernovae, there are also important observational differences *among* supernovae. Some supernovae contain very little hydrogen, according to their spectra, whereas others contain a lot. Also, the light curves of the hydrogen-poor supernovae are qualitatively different from those of the hydrogen-rich ones. On the basis of these observations, astronomers divide supernovae into two classes, known simply as Type I and Type II. **Type I supernovae**, the hydrogen-poor kind, have a light curve somewhat similar in shape to that of typical novae; **Type II supernovae**, whose spectra show lots of hydrogen, usually have a characteristic "plateau" in the light curve a few months after the maximum. (See Figure 21.8.) Observed supernovae are divided roughly equally between these two categories.

Carbon-Detonation Supernovae

What is responsible for these differences among supernovae? Is there more than one way in which a supernova explosion can occur? The answer is yes. To understand the alternative supernova mechanism, we must return to the processes that cause novae and consider the long-term consequences of their accretion–explosion cycle.

Novae eject matter from a white dwarf's surface, but they do not necessarily expel or burn all the material that has accumulated since the last outburst. In other words, there is a tendency for the dwarf's mass to increase slowly with each new nova cycle. As its mass grows and the internal pressure required to support its weight rises, the white dwarf can enter into a new period of instability—with disastrous consequences.

Recall that a white dwarf is held up not by thermal pressure (heat), but by the degeneracy pressure of electrons that have been squeezed so close together that they have effectively come into contact with one another. ∞ (Sec. 20.3) However, there is a limit to the pressure that these electrons can exert. Consequently, there is a limit to the mass of a white dwarf, above which electrons cannot provide the pressure needed to support the star. Detailed calculations show that the maximum mass of a white dwarf is about 1.4 solar masses, a mass often called the *Chandrasekhar mass*, after the Indian-American astronomer Subramanyan Chandrasekhar, whose work in theoretical astrophysics earned him a Nobel prize in physics in 1983.

If an accreting white dwarf exceeds the Chandrasekhar mass, the pressure of the degenerate electrons in its interior becomes unable to withstand the pull of gravity, and the star immediately starts to collapse. Its internal temperature rapidly rises to the point at which carbon can fuse into heavier elements. Carbon fusion begins everywhere throughout the white dwarf almost simultaneously, and the entire star explodes in another type of supernova—a so-called **carbon-detonation supernova**—comparable in violence to the "implosion" supernova associated with the death of a high-mass star, but born of a very different cause. In an alternative and (many astronomers think) possibly more common scenario, two white dwarfs in a binary system may collide and merge to form a massive, unstable star. The end result is the same: a carbon-detonation supernova.

We can now understand the differences between Type I and Type II supernovae. The explosion resulting from the detonation of a carbon white dwarf, the descendant of a low-mass star, is a supernova of Type I. Because this conflagration stems from a system containing virtually no hydrogen, we can readily see why the spectrum of a Type I supernova shows little evidence of that element. The appearance of the light curve (as we will soon see) results almost entirely from the radioactive decay of unstable heavy elements produced in the explosion itself.

The implosion–explosion of the core of a massive star, described earlier, produces a Type II supernova. Detailed computer models indicate that the characteristic shape of the Type II light curve is just what would be expected from the expansion and cooling of the star's outer envelope as it is blown into space by the shock wave sweeping up from below. The expanding material consists mainly of unburned gas—hydrogen and helium—so it is not surprising that those elements are strongly represented in the supernova's observed spectrum. (See *Discovery 21-1* for an account of a well-studied Type II supernova that confirmed many basic theoretical predictions, while also forcing astronomers to revise the details of their models.)

Figure 21.9 summarizes the processes responsible for the two different types of supernovae. We emphasize that, despite the similarity in the total amounts of energy involved, Type I and Type II supernovae are unrelated to one another. They occur in stars of very different types, under very different circumstances. All high-mass stars become Type II (core-collapse) supernovae, but only a tiny fraction of low-mass stars evolve into white dwarfs that ultimately explode as Type I (carbon-detonation) supernovae. How-

DISCOVERY 21-1

Supernova 1987A

In 1987, astronomers were treated to a spectacular supernova in the Large Magellanic Cloud (LMC), a small satellite galaxy orbiting our own. (See Section 24.2.) Observers in Chile first saw the explosion on February 24, and within a few hours, nearly all Southern Hemisphere telescopes and every available orbiting spacecraft were focused on the object. It was officially named SN 1987A. (The SN stands for "supernova," 1987 gives the year, and A identifies the supernova as the first seen that year.) This was one of the most dramatic changes observed in the universe in nearly 400 years. A 15-solar-mass B-type supergiant star with the catalog name SK-69°202 exploded and outshone all the other stars in the LMC combined for a few weeks, as shown in the "before" and "after" images of Figure 21.7.

Because the LMC is relatively close to Earth and because the explosion was detected so soon after it occurred, SN 1987A has provided astronomers with a wealth of detailed information on supernovae, allowing them to make key comparisons between theoretical models and observational reality. By and large, the theory of stellar evolution described in the text has held up very well. Still, SN 1987A did hold some surprises.

According to its hydrogen-rich spectrum, the supernova was of Type II—the core-collapse type—as expected for a high-mass parent star such as SK-69°202. But according to Figure 20.16 (which was computed for stars in our own Galaxy), the parent star should have been a red supergiant at the time of the explosion—not a blue supergiant, as actually observed. This unexpected finding caused theorists to scramble in search of an explanation. It now seems that, relative to young stars in the Milky Way, the parent star's envelope was deficient in heavy elements. This deficiency had little effect on the evolution of the core and on the supernova explosion, but it did change the star's evolutionary track on the H–R diagram. Unlike a Milky Way star with the same mass, SK-69°202 shrank and looped back toward the main sequence once helium ignited in its core. Following the ignition of carbon, the star, with a surface temperature of around 20,000 K, had just begun to return to the right on the H–R diagram when the rapid chain of events leading to the supernova occurred.

The shape of the light curve of SN 1987A, shown in the first figure, also differed somewhat from the "standard" Type II shape. (See Figure 21.8.) The peak brightness was less than

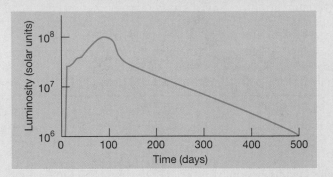

the expected value. For a few days after its initial detection, the supernova faded as it expanded and cooled rapidly. After about a week, the surface temperature had dropped to about 5000 K, at which point electrons and protons near the expanding surface recombined into atomic hydrogen, making the surface layers less opaque and allowing more radiation from the interior to leak out. As a result, the supernova brightened rapidly as it grew. The temperature of the expanding layers reached a peak in late May, by which point the radius of the expanding photosphere was about 2×10^{10} km—a little larger than our solar system. Subsequently, the photosphere cooled as it expanded, and the luminosity dropped as the internal supply of heat from the explosion dissipated into space.

Much of the preceding description would apply equally well to a Type II supernova in our own Galaxy. The differences between the SN 1987A light curve shown here and the Type II light curve in Figure 21.8 are mainly the result of the (relatively) small size of SN 1987A's parent star. The peak luminosity of SN 1987A was less than that of a "normal" Type II supernova because SK-69°202 was small and quite tightly bound by gravity. A lot of the energy emitted in the form of visible radiation (and evident in Figure 21.8) was used up in expanding SN 1987A's stellar envelope, so far less was left over to be radiated into space. Thus, SN 1987A's luminosity during the first few months was lower than expected, and the early peak evident in the figure did not occur. The peak in the SN 1987A light curve at about 80 to 100 days actually corresponds to the plateau in the Type II light curve in the figure.

About 20 hours before the supernova was detected optically, a brief (13-second) burst of neutrinos was simultaneously recorded by underground detectors in Japan and the United States. ∞ (Sec. 16.6) As discussed in the text, the neu-

ever, there are far more low-mass stars than high-mass stars, so, by a remarkable coincidence, the two types of supernova occur at roughly the same rate.

Supernova Remnants

We have plenty of evidence that supernovae have occurred in our Galaxy. Occasionally, the explosions themselves are visible from Earth. In many other cases, we can detect their glowing remains, or **supernova remnants**. One of the best-studied supernova remnants is known as the Crab Nebula, shown in Figure 21.10. The crab has greatly dimmed now, but the original explosion in the year A.D. 1054 was so brilliant that manuscripts of ancient Chinese and Middle Eastern astronomers claim that its brightness greatly exceeded that of Venus and—according to some (possibly exaggerated) accounts—even rivaled that of the Moon. For nearly a month, this exploded star reportedly could be seen in

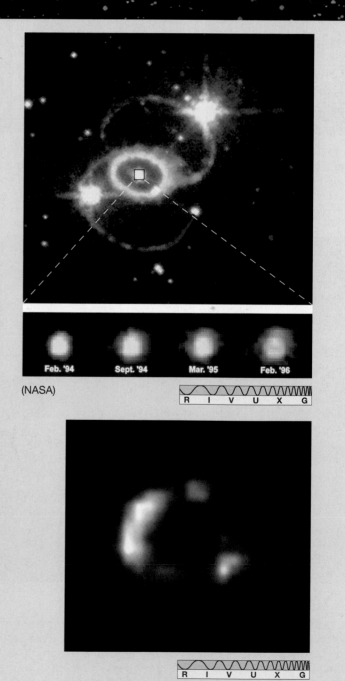

(NASA)

trinos are predicted to arise when electrons and protons in the star's collapsing core merge to form neutrons. The neutrinos preceded the light because they escaped during the collapse, whereas the first light of the explosion was emitted only after the supernova shock had plowed through the body of the star to the surface. In fact, theoretical models consistent with these observations suggest that vastly more energy was emitted in the form of neutrinos than in any other form. The supernova's neutrino luminosity was many tens of thousands of times greater than its optical energy output.

Despite some unresolved details in SN 1987A's behavior, the detection of the neutrino pulse is considered to be a brilliant confirmation of theory. This singular event—the detection of neutrinos—may well herald a new age of astronomy. For the first time, astronomers have received information from a specific body beyond the solar system by radiation outside the electromagnetic spectrum.

Theory predicts that the expanding remnant of SN 1987A is now on the verge of being resolvable by optical telescopes. The accompanying photographs show the barely resolved remnant (at the right) surrounded by a much larger shell of glowing gas (in yellow). Scientists reason that the progenitor of the supernova expelled this shell during its red-giant phase, some 40,000 years before the explosion. The image we see results from the initial flash of ultraviolet light from the supernova hitting the ring and causing it to glow brightly. As the debris from the explosion itself strikes the ring, the ring has become a temporary, but intense, source of X rays. As shown in the *Chandra* image at the bottom right, taken in 2000, the fastest-moving ejecta have already reached the ring, forming the small (1000 A.U. in diameter) glowing regions on its left side.

These images also show core debris moving outward toward the ring. The four insets at the bottom (as well as the sickle-shaped region in the bottom right image) show material expanding at nearly 3000 km/s. The main image (at the top) also revealed, to everyone's surprise, two additional faint rings that might be caused by radiation sweeping across an hour-glass-shaped bubble of gas. Why the gas should exhibit this odd structure remains unclear.

Buoyed by the success of stellar-evolution theory and armed with firm theoretical predictions of what should happen next, astronomers eagerly await future developments in the story of this remarkable object.

ANIMATION Shockwaves Hit the Ring Around Supernova 1987A/Composition and Structure of the Ring Around Supernova 1987A

(a) Type I Supernova

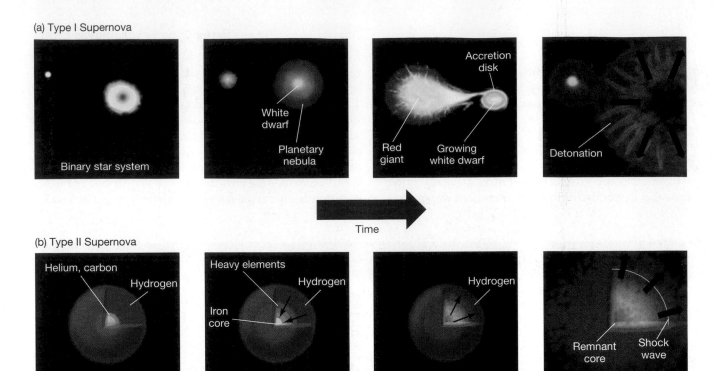

(b) Type II Supernova

▲ **FIGURE 21.9 Two Types of Supernova** Type I and Type II supernovae have different causes. These sequences depict the evolutionary history of each type. (a) A Type I supernova usually results when a carbon-rich white dwarf pulls matter onto itself from a nearby red-giant or main-sequence companion. (b) A Type II supernova occurs when the core of a high-mass star collapses and then rebounds in a catastrophic explosion.

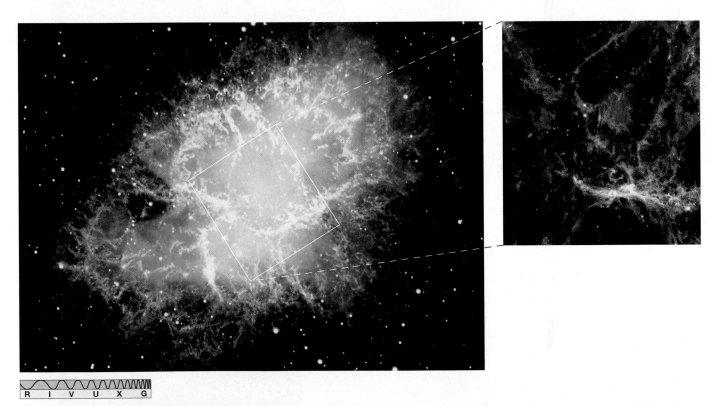

R I V U X G

▲ **FIGURE 21.10 Crab Supernova Remnant** This remnant of an ancient Type II supernova is called the Crab Nebula (or M1 in the Messier catalog). It resides about 1800 pc from Earth and has an angular diameter about one-fifth that of the full Moon. Because its debris is scattered over a region of "only" 2 pc, the Crab is considered to be a young supernova remnant. In A.D. 1054, Chinese astronomers observed this supernova. The main image was taken with the Very Large Telescope of the European Southern Observatory in Chile, the inset by the *Hubble* telescope in orbit. *(ESO; NASA)*

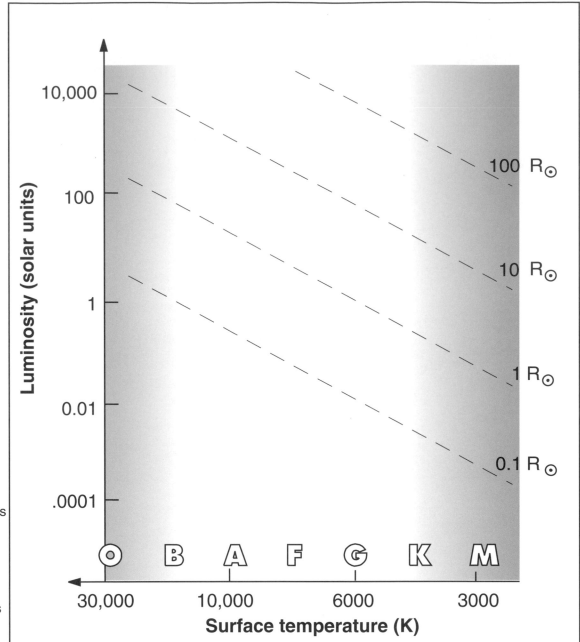

The H-R diagram plots stars by luminosity (vertical axis) and temperature, or spectral class (horizontal axis). The dashed diagonal lines are lines of constant radius.

broad daylight. Native Americans also left engravings of the event in the rocks of what is now the southwestern United States.

The Crab Nebula certainly has the appearance of exploded debris. Even today, the knots and filaments give a strong indication of past violence (and continuing activity—see *Discovery 21-2*). In fact, astronomers have proof that this matter was ejected from some central explosion. Doppler-shifted spectral lines indicate that the nebula—the envelope of the high-mass star that exploded to create this Type II supernova—is expanding into space at several thousand kilometers per second. A vivid illustration of the phenomenon is provided by Figure 21.11, which was made by superimposing a positive image of the Crab Nebula taken in 1960 and a negative image taken in 1974. If the gas were not in motion, the positive and negative images would overlap perfectly, but they do not. The gas moved outward in the intervening 14 years. Tracing the motion backward in time, astronomers have found that the explosion must have occurred about nine centuries ago, consistent with the Chinese observations.

The nighttime sky harbors many relics of stars that blew up long ago. Figure 21.12 is another example. It shows the Vela supernova remnant, whose expansion velocities imply that its central star exploded around 9000 B.C. The remnant lies only 500 pc away from Earth. Given its proximity, the Vela supernova may have been as bright as the Moon for several months. We can only speculate what impact such a bright supernova might have had on the myths, religions, and cultures of Stone Age humans when it first appeared in the sky.

Although hundreds of supernovae have been observed in other galaxies during the 20th century, no astronomer using modern equipment has ever observed a supernova in our own Galaxy. A viewable Milky Way star has not exploded since Galileo first turned his telescope to the heavens almost four centuries ago. The last supernovae observed in our Galaxy, by Tycho in 1572 and Kepler (and others) in 1604, caused a worldwide sensation in Renaissance times. The sudden appearance and subsequent fading of these very bright objects helped shatter the Aristotelian idea of an unchanging universe.

On the basis of stellar evolutionary theory, astronomers calculate that an observable supernova ought to occur in our Galaxy every 100 years or so. Even at a distance of several kiloparsecs, a supernova would (temporarily) outshine Venus, the brightest planet in our sky, so it seems unlikely that astronomers could have missed any since the last one nearly four centuries ago. Our part of the Milky Way seems long overdue for a supernova. However, a truly nearby supernova—within a few hundred parsecs, say—would be a very rare event, occurring only every 100,000 years or so. Humanity may be destined to see all supernovae from a distance.

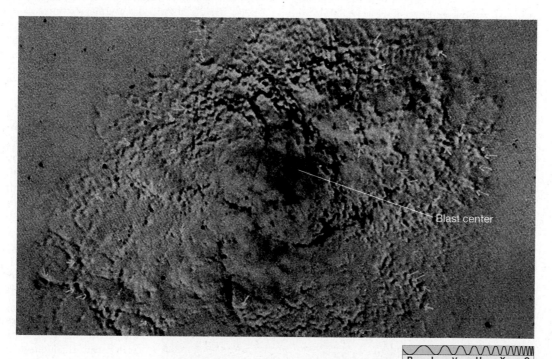

R I V U X G

▲ **FIGURE 21.11 The Crab in Motion** Positive and negative photographs of the Crab Nebula taken 14 years apart do not superimpose exactly, indicating that the gaseous filaments are still moving away from the site of the explosion. The positive image in glowing white was taken first, and then the black (negative) filaments were overlaid later—hence the reason the black (but still glowing) debris is farther from the center of the blast. The scale is roughly the same as in Figure 21.10. (*Harvard College Observatory*)

DISCOVERY 21-2

The Crab Nebula in Motion

The Crab Nebula is one of the most fascinating and instructive cosmic objects anywhere. Within this remarkable region of space we find a wide variety of physical processes at work, all the result of a supernova explosion that occurred almost a thousand years ago. A famous physics professor at MIT once taught a course exclusively on this object, claiming that if you understand all that transpires within the Crab, then you have mastered much of modern physics. Within this remarkable gas cloud, extending for a couple of parsecs and residing about 1800 pc away, we find applications of particle physics, nuclear physics, electromagnetism, thermodynamics, condensed-matter physics, plasma physics, and gravitational physics, to name but a few.

The accompanying figures present observations suggesting that the Crab Nebula is even more dynamic than had previously been thought. Conditions are changing rapidly among the debris, right before our eyes, giving astronomers a rare chance to study evolutionary changes that actually occur during a single human lifetime.

The first figure consists of two frames. On the left is a true-color, visible-light photo taken with the 5-m Hale telescope on Mount Palomar in southern California. This image is similar to that shown in the main frame of Figure 21.10. The visible light arises in two distinct ways. The outermost filaments seen in red, yellow, and green are literally glowing from the heat and violence of the explosion long ago. However, the faint bluish light toward the center of the nebula arises from a nonthermal process that releases energy as rapidly moving electrons spiral around lines of the object's magnetic field. (This process is not studied in the text until Chapter 25, but you can skip ahead and look at Figure 25.20 to get the gist of it.) The filamentary structure of the wispy debris—the re-

December 29,1995

(NASA)

February 1,1996

April 16,1996

Pulsar

Ripple

Polar jet
direction

R I V U X G

sult of turbulent gases running at high speed into the surrounding interstellar medium—leaves little doubt that the Crab Nebula is the remnant of an exploded star.

The right-hand frame is a recent image acquired by the *Hubble Space Telescope*, covering the portion of the first image contained within the white box. The image was taken at a wavelength of around 550 nm; the red color was added artificially during computer enhancement.

The central "engine" at the heart of the Crab Nebula is a pulsar—a rapidly spinning, compact starlike object which managed to survive the supernova explosion that created the nebula itself. (We will discuss these strange objects in more detail in Chapter 22.) The pulsar can be seen in this image as the left member of the pair of stars near the center of the frame. The pulsar powers the Crab today by accelerating elementary particles into the nebular debris, causing knots and wisps of energetic matter to stream away from the core.

The second set of images (above) allows us to see in action the tiny pulsar powering the Crab Nebula (again, the left star of the pair at the upper center). These three images were taken sequentially over the course of a few months. The field of view is even smaller than in the right frame of the previous figure—we are now virtually inside the Crab, witnessing the minute changes among the nebular gases near the site of the ancient explosion. Thanks to these and other *Hubble* observations, astronomers can now watch the changes in the shock-driven features that brighten and fade, over and over, like ripples in a pond. The light-year-long ripples in the Crab, seen throughout the bottom half of these frames, are moving outward from the pulsar at about half the speed of light. The final frame indicates the direction of the jet from the pulsar's pole (see Chapter 22) that may be the cause of the ripples we see.

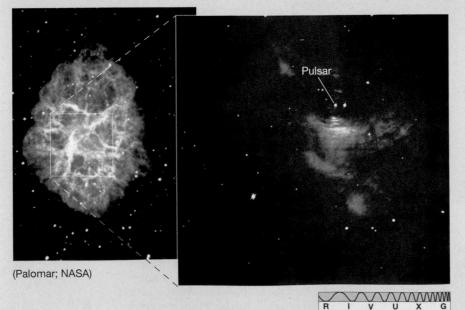

Pulsar

(Palomar; NASA)

R I V U X G

▲ **FIGURE 21.12 Vela Supernova Remnant** The glowing gases of the Vela supernova remnant are spread across a large 6° of the sky. The inset shows more clearly some of the details of the nebula's filamentary structure. (The long diagonal streak was caused by the passage of an Earth-orbiting satellite while the photo was being exposed.) *(D. Malin/AAT)*

CONCEPT CHECK

✔ How did astronomers know, even before the mechanisms were understood, that there were at least two distinct physical processes at work in creating supernovae?

21.4 The Formation of the Elements

Up to now, we have studied nuclear reactions mainly for their role in stellar energy generation. Now let's consider them again, but this time as the processes responsible for creating much of the world in which we live. The evolution of the elements, combining nuclear physics with astronomy, is a complex subject and a very important problem in modern astronomy.

Types of Matter

We currently know of 113 different elements, ranging from the simplest—hydrogen, containing one proton—to the most complex, discovered in 1998 and known for now as ununquadium, which has 114 protons and 185 neutrons in its nucleus. (See Appendix 3, Table 2. Element 113, while believed to exist, has not yet been created experimentally. In 1999, researchers claimed the discovery of elements 116 and 118, but the experimental findings have never been replicated, and these elements are not "officially" recognized.) All elements exist in different *isotopes*, each having the same number of protons, but a different number of neutrons. We often think of the most common or stable isotope as being the "normal" form of an element. Some elements, and many isotopes, are radioactively unstable, meaning that they eventually decay into other, more stable, nuclei.

The 81 stable elements found on Earth make up the overwhelming bulk of matter in the universe. In addition, 10 radioactive elements—including radon and uranium—also occur naturally on our planet. Even though the half-lives (the time required for half the nuclei to decay into something else) of these elements are very long (typically, millions or even billions of years), their slow, but steady, decay over the 4.5 billion years since the solar system formed means that they are scarce on Earth, in meteorites, and in lunar samples. ∞ (Sec. 6.7, *More Precisely 7-2*) They are not observed in stars—there is just too little of them to produce detectable spectral lines.

Besides these 10 naturally occurring radioactive elements, 19 more radioactive elements have been artificially produced under special conditions in nuclear laboratories on Earth. The debris collected after nuclear weapons tests also contains traces of some of these elements. Unlike the naturally occurring radioactive elements, the artificial ones decay into other elements quite quickly (in much less than a million years). Consequently, they, too, are extremely rare in nature. Two other elements round out our list: Promethium is a stable element that is found on our planet only as a by-product of nuclear laboratory experiments; technetium is an unstable element that is found in stars, but does not exist on Earth—any technetium that existed in our planet at its formation decayed long ago.

Abundance of Matter

How and where did all these elements form? Were they always present in the universe, or were they created after the universe formed? Since the 1950s, astronomers have come to realize that the hydrogen and most of the helium in the universe are *primordial*—that is, these elements date from the very earliest times. (See Chapter 27.) All other elements in our universe result from **stellar nucleosynthesis**—that is, they were formed by nuclear fusion in the hearts of stars.

To test this idea, we must consider not just the different kinds of elements and isotopes, but also their observed abundances, graphed in Figure 21.13. The curve shown is derived largely from spectroscopic studies of stars, including the Sun. The essence of the figure is summarized in Table 21.1, which combines all the known elements into eight groups based on the total numbers of nuclear particles (protons and neutrons) that they contain. (All isotopes of all elements are included in both the table and the figure, although only a few elements are marked by dots and labeled in the figure.) Any theory proposed for the creation of the elements must reproduce these observed abundances. The most obvious feature is that heavy elements are generally much less abundant than lighter elements. However, the many peaks and troughs evident in the figure also represent important constraints.

Hydrogen and Helium Burning

Let's begin by reviewing the reactions leading to the production of heavy elements at various stages of stellar evolution. Look again at Figure 21.6 as we discuss the reactions involved. Stellar nucleosynthesis begins with the proton–proton chain studied in Chapter 16. ∞ (Sec. 16.2) Provided that the temperature is high enough—at least 10 million K—a series of nuclear reactions occurs, ultimately forming a nucleus of ordinary helium (^{4}He) from four protons (^{1}H):

$$4(^1\text{H}) \rightarrow {}^4\text{He} + 2 \text{ positrons} + 2 \text{ neutrinos} + \text{energy}.$$

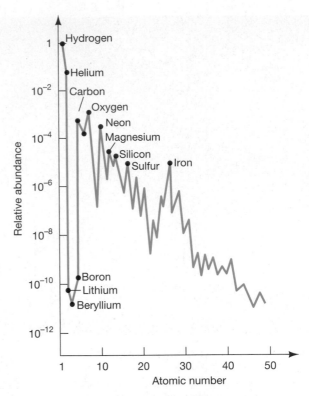

▲ **FIGURE 21.13 Elemental Abundance** A summary of the cosmic abundances of the elements and their isotopes, expressed relative to the abundance of hydrogen. The horizontal axis shows each of the listed elements' atomic number—the number of protons in the nucleus. Notice how many common terrestrial elements are found on "peaks" of the distribution, surrounded by elements that are tens or hundreds of times less abundant. Notice also the large peak around the element iron. The reasons for the peaks are discussed in the text.

TABLE 21.1 Cosmic Abundances of the Elements

Elemental Group of Particles	Percent Abundance by Number[*]
Hydrogen (1 nuclear particle)	90
Helium (4 nuclear particles)	9
Lithium group (7–11 nuclear particles)	0.000001
Carbon group (12–20 nuclear particles)	0.2
Silicon group (23–48 nuclear particles)	0.01
Iron group (50–62 nuclear particles)	0.01
Middle-weight group (63–100 nuclear particles)	0.00000001
Heaviest-weight group (over 100 nuclear particles)	0.000000001

[*]*The total does not equal 100 percent, because of uncertainties in the abundance of helium. All isotopes of all elements are included.*

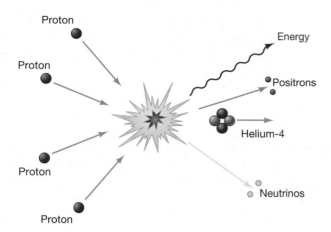

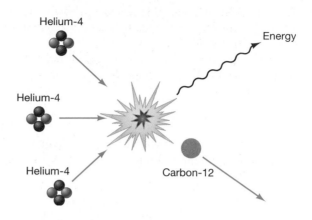

▲ **FIGURE 21.14 Proton Fusion** Diagram of the basic proton–proton hydrogen-burning reaction. Four protons combine to form a nucleus of helium-4, releasing energy in the process.

▲ **FIGURE 21.15 Helium Fusion** Diagram of the basic triple-alpha helium-burning reaction occurring in post-main-sequence stars. Three helium-4 nuclei combine to form carbon-12.

Recall that the positrons immediately interact with nearby free electrons, producing high-energy gamma rays through matter–antimatter annihilation. The neutrinos rapidly escape, carrying energy with them, but playing no direct role in nucleosynthesis. The existence of these reactions has been directly confirmed in nuclear experiments conducted in laboratories around the world during recent decades. In massive stars, the CNO cycle may greatly accelerate the hydrogen-burning process, but the basic four-protons-to-one-helium-nucleus reaction, illustrated in Figure 21.14, is unchanged. ∞ *(More Precisely 20-1)*

As helium builds up in the core of a star, the burning ceases, and the core contracts and heats up. When the temperature exceeds about 100 million K, helium nuclei can overcome their mutual electrical repulsion, leading to the *triple-alpha reaction*, which we discussed in Chapter 20: ∞ (Sec. 20.2)

$$3(^4\text{He}) \rightarrow {}^{12}\text{C} + \text{energy}.$$

The net result of this reaction is that three helium-4 nuclei are combined into one carbon-12 nucleus (Figure 21.15), releasing energy in the process.

Carbon Burning and Helium Capture

At higher and higher temperatures, heavier and heavier nuclei can gain enough energy to overcome the electrical repulsion between them. At about 600 million K (reached only in the cores of stars much more massive than the Sun), carbon nuclei can fuse to form magnesium, as depicted in Figure 21.16(a):

$$^{12}\text{C} + {}^{12}\text{C} \rightarrow {}^{24}\text{Mg} + \text{energy}.$$

However, because of the rapidly mounting nuclear charges—that is, the increasing number of protons in the nuclei—fusion reactions between any nuclei larger than

carbon require such high temperatures that they are actually quite uncommon in stars. The formation of most heavier elements occurs by way of an easier path. For example, the repulsive force between two carbon nuclei is three times greater than the force between a nucleus of carbon and one of helium. Thus, carbon–helium fusion occurs at a lower temperature than that at which carbon–carbon fusion occurs. As we saw in Section 20.3, at temperatures above 200 million K, a carbon-12 nucleus colliding with a helium-4 nucleus can produce oxygen-16:

$$^{12}\text{C} + {}^4\text{He} \rightarrow {}^{16}\text{O} + \text{energy}.$$

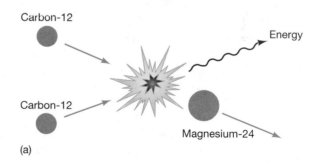

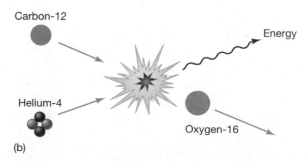

▲ **FIGURE 21.16 Carbon Fusion** Carbon can form heavier elements (a) by fusion with other carbon nuclei or, more commonly, (b) by fusion with a helium nucleus.

If any helium-4 is present, this reaction, shown in Figure 21.16(b), is much more likely to occur than the carbon–carbon reaction.

Similarly, the oxygen-16 thus produced may fuse with other oxygen-16 nuclei at a temperature of about 1 billion K to form sulfur-32:

$$^{16}O + {}^{16}O \rightarrow {}^{32}S + \text{energy.}$$

However, it is much more probable that an oxygen-16 nucleus will capture a helium-4 nucleus (if one is available) to form neon-20:

$$^{16}O + {}^{4}He \rightarrow {}^{20}Ne + \text{energy.}$$

The second reaction is more likely because it occurs at a lower temperature than that necessary for oxygen–oxygen fusion.

Thus, as the star evolves, heavier elements tend to form through **helium capture** rather than by fusion of like nuclei. As a result, elements with nuclear masses of 4 units (i.e., helium itself), 12 units (carbon), 16 units (oxygen), 20 units (neon), 24 units (magnesium), and 28 units (silicon) stand out as prominent peaks in Figure 21.13, our chart of cosmic abundances. Each element is built by combining the preceding element and a helium-4 nucleus as the star evolves.

Helium capture is by no means the only type of nuclear reaction occurring in evolved stars. As nuclei of many different kinds accumulate, a great variety of reactions become possible. In some, protons and neutrons are freed from their parent nuclei and are absorbed by others, forming new nuclei with masses intermediate between those formed by helium capture. Laboratory studies confirm that common nuclei, such as fluorine-19, sodium-23, phosphorus-31, and many others, are created in this way. However, their abundances are not as great as those produced directly by helium capture, simply because the helium-capture reactions are much more common in stars. For this reason, many of these elements (those with masses not divisible by four, the mass of a helium nucleus) are found in the troughs of Figure 21.13.

Iron Formation

Around the time silicon-28 appears in the core of a star, a competitive struggle begins between the continued capture of helium to produce even heavier nuclei and the tendency of more complex nuclei to break down into simpler ones. The cause of this breakdown is heat. By now, the star's core temperature has reached the unimaginably large value of 3 billion K, and the gamma rays associated with that temperature have enough energy to break a nucleus apart, as illustrated in Figure 21.17(a). This is the same process of photodisintegration that will ultimately accelerate the star's iron core in its final collapse toward a Type II supernova.

Under the intense heat, some silicon-28 nuclei break apart into seven helium-4 nuclei. Other nearby nuclei that have not yet photodisintegrated may capture some or all of these helium-4 nuclei, leading to the formation of still heavier elements (Figure 21.17b). The process of photodisintegration provides raw material that allows helium capture to proceed to greater masses. Photodisintegration continues, with some heavy nuclei being destroyed and

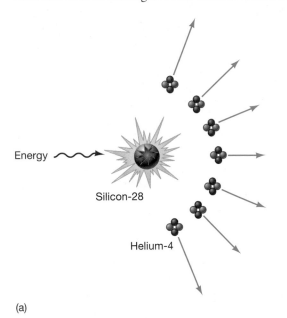

(a)

◀ **FIGURE 21.17 Alpha Process** (a) At high temperatures, heavy nuclei (such as silicon, shown here) can be broken apart into helium nuclei by high-energy photons. (b) Other nuclei can capture the helium nuclei—or alpha particles—thus produced, forming heavier elements by the so-called alpha process. This process continues all the way to the formation of nickel-56 (in the iron group).

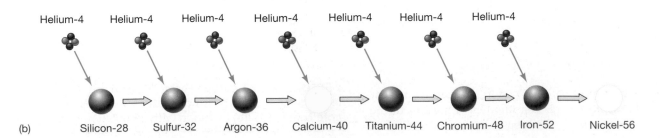

(b)   Silicon-28   Sulfur-32   Argon-36   Calcium-40   Titanium-44   Chromium-48   Iron-52   Nickel-56

others increasing in mass. In succession, the star forms sulfur-32, argon-36, calcium-40, titanium-44, chromium-48, iron-52, and nickel-56. The chain of reactions building from silicon-28 up to nickel-56 is

$$^{28}\text{Si} + 7(^4\text{He}) \rightarrow\ ^{56}\text{Ni} + \text{energy}.$$

This two-step process—photodisintegration followed by the direct capture of some or all of the resulting helium-4 nuclei (or alpha particles)—is often called the *alpha process.*

Nickel-56 is unstable, decaying rapidly first into cobalt-56 and then into a stable iron-56 nucleus. Any unstable nucleus will continue to decay until stability is achieved, and iron-56 is the most stable of all nuclei (Figure 21.6). Thus, the alpha process leads inevitably to the buildup of iron in the stellar core.

Another way of describing Figure 21.6 is to say that iron's 26 protons and 30 neutrons are bound together more strongly than the particles in any other nucleus. Iron is said to have the greatest *nuclear binding energy* of any element—more energy per particle is required to break up (unbind) an iron-56 nucleus than the nucleus of any other element. This enhanced stability of iron explains why some of the heavier nuclei in the iron group are more abundant than many lighter nuclei (see Table 21.1 and Figure 21.13): Nuclei tend to "accumulate" near iron as stars evolve.

Making Elements Beyond Iron

If the alpha process stops at iron, how did heavier elements, such as copper, zinc, and gold, form? To form them, some nuclear process other than helium capture must have been involved. That other process is **neutron capture**: the formation of heavier nuclei by the absorption of neutrons.

Deep in the interiors of highly evolved stars, conditions are ripe for neutron capture to occur. Neutrons are produced as "by-products" of many nuclear reactions, so there are many of them present to interact with iron and other nuclei. Neutrons have no charge, so there is no repulsive barrier for them to overcome in combining with positively charged nuclei. As more and more neutrons join a nucleus, its mass continues to grow.

Adding neutrons to a nucleus—iron, for example—does not change the element. Rather, a more massive isotope of the same element is produced. Eventually, however, so many neutrons have been added to the nucleus that it becomes unstable and then decays radioactively to form a stable nucleus of some other element. The neutron-capture process then continues. For example, an iron-56 nucleus can capture a single neutron to form a relatively stable isotope, iron-57:

$$^{56}\text{Fe} + \text{n} \rightarrow\ ^{57}\text{Fe}.$$

This reaction may be followed by another neutron capture:

$$^{57}\text{Fe} + \text{n} \rightarrow\ ^{58}\text{Fe}.$$

Thus, another relatively stable isotope, iron-58, is produced, and this isotope can capture yet another neutron to produce an even heavier isotope of iron:

$$^{58}\text{Fe} + \text{n} \rightarrow\ ^{59}\text{Fe}.$$

Iron-59 is known from laboratory experiments to be radioactively unstable. It decays in about a month into cobalt-59, which is stable. The neutron-capture process then resumes: Cobalt-59 captures a neutron to form the unstable cobalt-60, which in turn decays to nickel-60, and so on.

Each successive capture of a neutron by a nucleus typically takes about a year, so most unstable nuclei have plenty of time to decay before the next neutron comes along. Researchers usually refer to this "slow" neutron-capture mechanism as the *s-process.* It is the origin of the copper and silver in the coins in our pockets, the lead in our car batteries, and the gold (and the zirconium) in the rings on our fingers. As mentioned earlier, similar slow neutron capture processes involving nuclei of lower mass are responsible for many of the elements intermediate between those formed by helium capture. These reactions are thought to be particularly important during the late (asymptotic-giant branch) stages of low-mass stars. ∞ (Sec. 20.3)

Making the Heaviest Elements

The s-process explains the synthesis of stable nuclei up to, and including, bismuth-209, the heaviest-known nonradioactive nucleus, but it cannot account for the heaviest nuclei, such as thorium-232, uranium-238, or plutonium-242. Any attempt to form elements heavier than bismuth-209 by slow neutron capture fails because the new nuclei decay back to bismuth as fast as they form. Accordingly, there must be yet another nuclear mechanism that produces the very heaviest nuclei. This process is called the *r-process* (where r stands for "rapid," in contrast to the "slow" s-process just described). The r-process operates very quickly, occurring (we think) literally during the supernova explosion that signals the death of a massive star.

During the first 15 minutes of the supernova blast, the number of free neutrons increases dramatically as heavy nuclei are broken apart by the violence of the explosion. Unlike the s-process, which stops when it runs out of stable nuclei, the neutron-capture rate during the supernova is so great that even unstable nuclei can capture many neutrons before they have time to decay. Jamming neutrons into light- and middleweight nuclei, the r-process is responsible for the creation of the heaviest-known elements. The heaviest of the heavy elements, then, are actually born *after* their parent stars have died. However, because the time available for synthesizing these heaviest nuclei is so brief, they never become very abundant. Elements heavier than iron (see Table 21.1) are a billion times less common than hydrogen and helium.

Observational Evidence for Stellar Nucleosynthesis

The modern picture of the formation of the elements involves many different types of nuclear reactions occurring at many different stages of stellar evolution, from main-sequence stars all the way to supernovae. Elements of the periodic table from hydrogen to iron are built first by fusion and then by alpha capture, with proton and neutron capture filling in the gaps. Elements beyond iron form by neutron capture and radioactive decay. Ultimately, these elements are ejected into interstellar space as the stars in which they form reach the ends of their lives.

Scientific theories must continually be tested and validated by experiment and observation, and the theory of stellar nucleosynthesis is no exception. ∞ (Sec. 1.2) Yet almost all of the nuclear processes just described take place deep in the hearts of stars, hidden from our view, and the stars responsible for the heavy elements we see today are all long gone. How, then, can we be sure that the sequences of events presented here actually occurred (and are still occurring today)? The answer is that the theory of stellar nucleosynthesis makes many detailed predictions about the numbers and types of elements formed in stars, affording astronomers ample opportunity to observe and test its consequences. We are reassured of the theory's basic soundness by three particularly convincing pieces of evidence.

First, the rates at which various nuclei are captured and the rates at which they decay are known from laboratory experiments. When these rates are incorporated into detailed computer models of the nuclear processes occurring in stars and supernovae, the resulting elemental abundances agree extremely well, point by point, with the observational data presented in Figure 21.13 and Table 21.1. The match is remarkably good for elements up through iron and is still fairly close for heavier nuclei. Although the reasoning is indirect, the agreement between theory and observation is so striking that most astronomers regard it as very strong evidence in support of the entire theory of stellar evolution and nucleosynthesis.

Second, the presence of one particular nucleus—technetium-99—provides direct evidence that heavy elements really do form in the cores of stars. Laboratory measurements show that the technetium nucleus has a radioactive half-life of about 200,000 years, a very short time, astronomically speaking. No one has ever found even traces of naturally occurring technetium on Earth, because it all decayed long ago. The observed presence of technetium in the spectra of many red-giant stars implies that it must have been synthesized in their cores through neutron capture—the only known way in which technetium can form—within the past few hundred thousand years and then transported by convection to the surface. Otherwise, we would not observe it. Many astronomers consider the spectroscopic evidence for technetium as proof that the s-process really does operate in evolved stars.

Third, the study of typical light curves from Type I supernovae indicates that radioactive nuclei form as a result of the explosion. Figure 21.18(a) (see also Figure 21.8) displays the dramatic rise in luminosity at the moment of explosion and the characteristic slower decrease in brightness. Depending on the initial mass of the exploded star, the luminosity takes from several months to many years to decrease to its original value, but the *shape* of the decay curve is nearly the same for all exploded stars. These curves have two distinct features: After the initial peak, the luminosity declines rapidly; then it decreases at a slower rate. This abrupt change in the rate of luminosity decay invariably occurs about two months after the explosion, regardless of the intensity of the outburst.

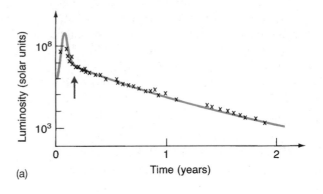

(a)

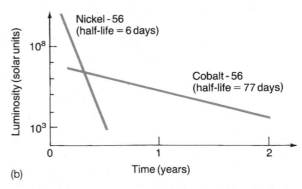

(b)

▲ **FIGURE 21.18 Supernova Energy Emission** (a) The light curve of a Type I supernova, showing not only the dramatic increase and slow decrease in luminosity, but also the characteristic change in the rate of decay about two months after the explosion (after the time indicated by the arrow). This particular supernova occurred in the faraway galaxy IC 4182 in 1938. The crosses are the actual observations of the supernova's light. (b) Theoretical calculations of the light emitted by the radioactive decay of nickel-56 and cobalt-56 produce a light curve similar to those actually observed in real supernova explosions, lending strong support to the theory of stellar nucleosynthesis.

We can explain the two-stage decline of the luminosity curve in Figure 21.18(a) in terms of the radioactive decay of unstable nuclei, notably nickel-56 and its decay product cobalt-56, produced in abundance during the early moments of the supernova. From theoretical models of the explosion, we can calculate the amounts of these elements expected to form, and we know their half-lives from laboratory experiments. Because each radioactive decay produces a known amount of energy, we can then determine how the light emitted by these unstable elements should vary in time. The result is in very good agreement with the observed light curve in Figure 21.18(b)—the luminosity of a Type I supernova is entirely consistent with the decay of about 0.6 solar mass of nickel-56. More direct evidence for the presence of these unstable nuclei was first obtained in the 1970s, when a gamma-ray spectral feature of decaying cobalt-56 was identified in a supernova observed in a distant galaxy.

CONCEPT CHECK

✔ Why are the elements carbon, oxygen, neon, and magnesium, whose masses are multiples of four, as well as the element iron, so common on Earth?

21.5 The Cycle of Stellar Evolution

The theory of stellar nucleosynthesis can naturally account for the observed differences in the abundances of heavy elements between the old globular-cluster stars and stars now forming in our Galaxy. ∞ (Sec. 20.5) Even though an evolved star continuously creates new heavy elements in its interior, changes in the star's composition are confined largely to the core, and the star's spectrum gives little indication of events within its core. Convection may carry some reaction products (such as the technetium observed in many red giants) from the core into the envelope, but the outer layers largely retain the star's original composition. Only at the end of the star's life are its newly created elements released and scattered into space.

Thus, the spectra of the *youngest* stars show the *most* heavy elements, because each new generation of stars increases the concentration of these elements in the interstellar clouds from which the next generation forms. Accordingly, the photosphere of a recently formed star contains a much greater abundance of heavy elements than that of a star which formed long ago. Knowledge of stellar evolution allows astronomers to estimate the ages of stars from purely spectroscopic studies, even when the stars are isolated and are not members of any cluster. ∞ (Sec. 20.5) In the last three chapters, we have seen all the ingredients that make up the complete cycle of star formation and evolution in our Galaxy. Let's briefly summarize that process, which is illustrated in Figure 21.19:

1. Stars form when part of an interstellar cloud is compressed beyond the point at which it can support itself against its own gravity. The cloud collapses and fragments, forming a cluster of stars. The hottest stars heat and ionize the surrounding gas, sending shock waves through the surrounding cloud, modifying the formation of lower mass stars, and possibly triggering new rounds of star formation. ∞ (Sec. 19.6)

2. Within the cluster, stars evolve. The most massive stars evolve fastest, creating the heaviest elements in their cores and spewing them forth into the interstellar medium in supernovae. Lower mass stars take longer to evolve, but they, too, can create heavy elements and contribute significantly to the "seeding" of interstellar space when they shed their envelopes as planetary nebulae. Roughly speaking, low-mass stars are responsible for most of the carbon, nitrogen, and oxygen that make life on Earth possible. High-mass stars produced the iron and silicon that make up Earth itself, as well as the heavier elements on which much of our technology is based.

3. The creation and explosive dispersal of newly formed elements are accompanied by further shock waves, whose passage through the interstellar medium simultaneously enriches the medium and compresses it into further star formation. Each generation of stars increases the concentration of heavy elements in the interstellar clouds from which the next generation forms. As a result, recently formed stars contain a much greater abundance of heavy elements than do stars that formed long ago.

In this way, although some material is used up in each cycle—turned into energy or locked up in low-mass stars—the Galaxy continuously recycles its matter. Each new round of formation creates stars with more heavy elements than the preceding generation had. From the old globular clusters, which are observed to be deficient in heavy elements relative to the Sun, to the young open clusters, containing much larger amounts of these elements, we observe this enrichment process in action. Our Sun is the product of many such cycles. We ourselves are another. Without the elements synthesized in the hearts of stars, neither Earth nor the life it harbors would exist.

CONCEPT CHECK

✔ Why is stellar evolution important to life on Earth?

◄ **FIGURE 21.19 Stellar Recycling** The cycle of star formation and evolution continuously replenishes the Galaxy with new heavy elements and provides the driving force for the creation of new generations of stars. Clockwise from the top are an interstellar cloud (Barnard 68), a star-forming region in our Galaxy (RCW 38), a massive star ejecting a "bubble" and about to explode (NGC 7635), and a supernova remnant and its heavy-element debris (N49). *(ESO; NASA)*

Chapter Review

SUMMARY

A **nova (p. 544)** is a star that suddenly increases greatly in brightness and then slowly fades back to its normal appearance over a period of months. A nova results when a white dwarf in a binary system draws hydrogen-rich material from its companion. The gas builds up on the white dwarf's surface, eventually becoming hot and dense enough for the hydrogen to burn explosively, temporarily causing a large increase in the dwarf's luminosity. The matter flowing from the companion star does not fall directly onto the surface of the dwarf. Instead, it goes into orbit around it, forming an **accretion disk (p. 544)**. Friction within the disk causes the gas to spiral slowly inward, heating up and glowing brightly as it nears the dwarf's surface.

Stars more massive than about 8 solar masses are able to attain high enough central temperatures to burn carbon and heavier nuclei. As they burn, their cores form a layered structure consisting of burning shells of successively heavier elements. A nonburning core of iron builds up at the center. Iron is special in that its nuclei can neither be fused together nor split apart to produce energy. As a result, stellar nuclear burning stops at iron. As a star's iron core grows in mass, it eventually becomes unable to

support itself against gravity and begins to collapse. At the enormous densities and temperatures produced during the collapse, iron nuclei are broken down into their constituent particles: protons and neutrons. The protons combine with electrons to form more neutrons. Eventually, when the core has become so dense that the neutrons are effectively brought into physical contact with one another, their resistance to further squeezing stops the collapse and the core rebounds, sending a violent shock wave out through the rest of the star. The star is blown to pieces in a **core-collapse supernova (p. 548)**.

Astronomers classify **supernovae (p. 548)** into two broad categories: Type I and Type II. These classes differ by their light curves and their composition. **Type I supernovae (p. 549)** are hydrogen poor and have a light curve similar in shape to that of a nova. **Type II supernovae (p. 549)** are hydrogen rich and have a characteristic plateau in the light curve a few months after maximum. A Type II supernova is a core-collapse supernova. A Type I supernova occurs when a carbon–oxygen white dwarf in a binary system exceeds about 1.4 solar masses (the Chandrasekhar mass)—the maximum mass that can be supported against gravity by electron degeneracy pressure. The star collapses and explodes as its carbon ignites. This type of supernova is called a **carbon-detonation supernova (p. 549)**.

Theory predicts that a supernova visible from Earth should occur within our Galaxy about once a century, although none has been observed in the last 400 years. We can see evidence of a past supernova in the form of a **supernova remnant (p. 551)**—a shell of exploded debris surrounding the site of the explosion and expanding into space at a speed of thousands of kilometers per second.

All elements heavier than helium are formed by **stellar nucleosynthesis (p. 556)**—the production of new elements by nuclear reactions in the cores of evolved stars. Elements heavier than carbon tend to form by **helium capture (p. 558)**, rather than by the fusion of more massive nuclei. Therefore, nuclei whose masses are a multiple of the mass of a helium nucleus tend to be more

common than others. At high enough core temperatures, photodisintegration breaks apart some heavy nuclei, providing helium-4 nuclei for the synthesis of even more massive elements, leading to a buildup of iron-56 in the core. Elements beyond iron form by **neutron capture (p. 559)** in the cores of evolved stars. With no repulsive electromagnetic barrier to overcome, neutrons can easily combine with nuclei. During a supernova, rapid neutron capture occurs, producing the heaviest nuclei of all. Comparisons between theoretical predictions of element production and observations of element abundances in stars and supernovae provide strong support for the theory of stellar nucleosynthesis.

The processes of star formation, evolution, and explosion form a cycle that constantly enriches the interstellar medium with heavy elements and sows the seeds of new generations of stars. Without the elements produced in supernovae, life on Earth would be impossible.

REVIEW AND DISCUSSION

1. Under what circumstances will a binary star produce a nova?

2. What is an accretion disk, and how does one form?

3. What is a light curve? How can it be used to identify a nova or a supernova?

4. What occurs in a massive star to cause it to explode?

5. How do photodisintegration and neutronization contribute to the demise of a massive star?

6. What is neutron degeneracy pressure?

7. What are the observational differences between Type I and Type II supernovae?

8. What is the Chandrasekhar mass, and what does it have to do with supernovae?

9. How do the mechanisms responsible for Type I and Type II supernovae explain their observed differences?

10. Roughly how often would we expect a supernova to occur in our own Galaxy? How often would we expect to *see* a galactic supernova?

11. What evidence is there that many supernovae have occurred in our Galaxy?

12. How can astronomers estimate the age of an isolated star?

13. What proof do astronomers have that heavy elements are formed in stars?

14. As a star evolves, why do heavier elements tend to form by helium capture rather than by fusion of like nuclei?

15. Why do the cores of massive stars evolve into iron and not heavier elements?

16. How are nuclei heavier than iron formed?

17. What is the r-process? When and where does it occur?

18. Why was supernova 1987A so important?

19. Why are neutrino detectors important to the study of supernovae?

20. Describe the role played by supernovae in "recycling" galactic matter.

CONCEPTUAL SELF-TEST: TRUE OR FALSE/MULTIPLE CHOICE

1. A nova is a sudden outburst of light coming from an old main-sequence star.

2. Novae occur in binary-star systems.

3. It takes less and less time to fuse heavier and heavier elements inside a high-mass star.

4. In a core-collapse supernova, the outer part of the core rebounds from the inner, high-density core, destroying the entire outer part of the star.

5. A supernova is the same as a nova, but it appears much brighter because it occurs closer to us.

6. Stellar nucleosynthesis can account for the existence of all elements except hydrogen and helium.

7. When a proton and an electron are forced together, they destroy one another and release gamma rays.

8. The first detection of supernova 1987A came in the form of a burst of neutrinos.

9. Neutron capture is responsible for the formation of all elements heavier than iron.

10. Because of stellar nucleosynthesis, the spectra of old stars show more heavy elements than those of young stars.

11. A white dwarf can dramatically increase in brightness only if it (a) has another star nearby; (b) can avoid nuclear fusion in its core; (c) is spinning very rapidly; (d) is descended from a very massive star.

12. A nova differs from a supernova in that the nova (a) can occur only once; (b) is much more luminous; (c) involves only high-mass stars; (d) is much less luminous.

13. Which of the following stars will become hot enough to form elements heavier than oxygen? (a) A star that is half the mass of the Sun. (b) A star having the same mass as the Sun. (c) A star that is twice as massive as the Sun. (d) A star that is eight times more massive than the Sun.

14. A massive star becomes a supernova when it **(a)** collides with a stellar companion; **(b)** forms iron in its core; **(c)** suddenly increases in surface temperature; **(d)** suddenly increases in mass.

15. Figure 21.8 ("Supernova Light Curves") indicates that a supernova whose luminosity declines steadily in time is most likely associated with a star that is **(a)** without a binary companion; **(b)** more than eight times the mass of the Sun; **(c)** on the main sequence; **(d)** comparable in mass to the Sun.

16. An observable supernova should occur in our Galaxy about once every **(a)** year; **(b)** decade; **(c)** century; **(d)** millennium.

17. Which one of the following *does not* provide evidence that supernovae have occurred in our Galaxy? **(a)** The rapid expansion and filamentary structure of the Crab nebula.

(b) Historical records from China and Europe. **(c)** The existence of binary stars in our Galaxy. **(d)** The existence of iron on Earth.

18. Nuclear fusion in the Sun will **(a)** never create elements heavier than helium; **(b)** create elements up to and including oxygen; **(c)** create all elements up to and including iron; **(d)** create some elements heavier than iron.

19. Most of the carbon in our bodies originated in **(a)** the core of the Sun; **(b)** the core of a red-giant star; **(c)** a supernova; **(d)** a nearby galaxy.

20. The silver atoms found in jewelry originated in **(a)** the core of the Sun; **(b)** the core of a red-giant star; **(c)** a supernova; **(d)** a nearby galaxy.

PROBLEMS

 Algorithmic versions of these questions are available in the **Practice Problems** module of the **Companion Website** at astro.prenhall.com/chaisson.

The number of squares preceding each problem indicates its approximate level of difficulty.

1. ■■■ Estimate how close a 0.5-solar-mass white dwarf must come to the center of a 2-solar-mass subgiant with radius 10 times that of the Sun in order for the white dwarf's tidal field to strip matter from the companion's surface.

2. ■ Calculate the orbital speed of matter in an accretion disk just above the surface of a 0.6-solar-mass, 15,000-km-diameter white dwarf.

3. ■ A certain telescope can just detect the Sun at a distance of 10,000 pc. What is the apparent magnitude of the Sun at this distance? (For convenience, take the Sun's absolute magnitude to be 5.) What is the maximum distance at which the telescope can detect a nova having a peak luminosity of 10^5 solar luminosities?

4. ■ Repeat the previous calculation for a supernova having a peak luminosity 10^{10} times that of the Sun. What would be the apparent magnitude of the explosion if it occurred at a distance of 10,000 Mpc? Would it be detectable by any existing telescope?

5. ■■ At what distance would a supernova of absolute magnitude -20 look as bright as the Sun? As the Moon? Would you expect a supernova to occur that close to us?

6. ■ A (hypothetical) supernova at a distance of 150 pc has an absolute magnitude of -20. Compare its apparent magnitude with that of (a) the full Moon and (b) Venus at its brightest. (See Figure 17.7.) Would you expect a supernova to occur this close to us?

7. ■ A supernova's energy is often compared to the total energy output of the Sun over its lifetime. Using the Sun's current

energy output, calculate its total energy output, assuming that the sun has a 10^{10} year main-sequence lifetime. How does this compare with the energy released by a supernova?

8. ■■ The *Hubble Space Telescope* is observing a distant Type I supernova with peak apparent magnitude 24. Using the light curve in Figure 21.8, estimate how long after the peak brightness the supernova will become too faint to be seen.

9. ■ The Crab Nebula is now about 1 pc in radius. If it was observed to explode in A.D. 1054, roughly how fast is it expanding? (Assume a constant expansion rate. Is that a reasonable assumption?)

10. ■■ Suppose that stars form in our Galaxy at an average rate of 10 per year. Suppose also that all stars greater than 8 solar masses explode as supernovae. Use Figure 17.23 to estimate the rate of Type II supernovae in our Galaxy.

11. ■■■ Assuming an interstellar extinction of 2 mag/kpc, calculate the maximum distance at which we could see (with the naked eye, with a limiting magnitude of 6) a Galactic supernova of absolute magnitude −19. (See Chapter 18, problem 9, for more on how to go about solving the equation you obtain here.)

12. ■■■ Repeat the previous question, but for a survey telescope with a limiting magnitude of 18.

13. ■■ As we will see in Chapter 23, the star-forming portion of our Galaxy consists of a highly flattened circular disk about 30 kpc (30,000 pc) in diameter. Interstellar extinction limits our view to within a radius of about 5 kpc of the Sun. If supernovae occur in the Galaxy roughly once every 30 years, on average, and are uniformly spread throughout the disk, calculate how often we should expect to see a supernova.

14. ■■ Assuming the data in the previous question, taking all supernovae, for simplicity, to have absolute magnitude −20, and ignoring interstellar extinction for such nearby events, calculate how often we should expect to observe a supernova brighter than the full Moon (apparent magnitude −12.5).

15. ■■ Based on the data in Table 21.1, estimate the fraction by mass of "iron-group" elements and the total mass of all elements in the Sun. Compare your answer with Earth's mass.

In addition to the Practice Problems module, the Companion Website at astro.prenhall.com/chaisson provides for each chapter a study guide module with multiple choice questions as well as additional annotated images, animations, and links to related Websites.

22

Neutron stars and black holes are among the more exotic members of the vast population of stars throughout the universe. These objects represent the end states of stellar systems and often exhibit accretion disks and high-speed jets, at least for a while after "death." Remarkably, they all fit within our models of stellar evolution; indeed they were predicted by theory long before they were discovered in space. This magnificent piece of art captures, in cutaway painting, a disk-shaped region of rapidly whirling hot gas and thick dust about to be consumed by the ▶ hole whose accretion disk sends jets of matter spewing forth. (Dana Berry)

Neutron Stars and Black Holes

Strange States of Matter

Our study of stellar evolution has led us to some very unusual and unexpected objects. Red giants, white dwarfs, and supernovae surely represent extreme states of matter completely unfamiliar to us here on Earth. Yet stellar evolution can have even more bizarre consequences. The strangest states of all result from the catastrophic implosion–explosion of stars much more massive than our Sun. The almost unimaginable violence of a supernova may bring into being objects so extreme in their behavior that they require us to reconsider some of our most hallowed laws of physics. They open up a science-fiction writer's dream of fantastic phenomena. They may even one day force scientists to construct a whole new theory of the universe.

LEARNING GOALS

Studying this chapter will enable you to

1 Describe the properties of neutron stars, and explain how these strange objects are formed.

2 Explain the nature and origin of pulsars, and account for their characteristic radiation.

3 List and explain some of the observable properties of neutron-star binary systems.

4 Discuss the basic characteristics of gamma-ray bursts and some theoretical attempts to explain them.

5 Describe how black holes are formed, and discuss their effects on matter and radiation in their vicinity.

6 Relate the phenomena that occur near black holes due to the warping of space around them.

7 Discuss the difficulties that arise in observing black holes, and explain some of the ways in which the presence of a black hole might be detected.

 Visit astro.prenhall.com/chaisson for additional annotated images, animations, and links to related sites for this chapter.

22.1 Neutron Stars

LEARNING GOAL 1 What remains after a supernova? Is the entire progenitor (parent) star blown to bits and dispersed throughout interstellar space, or does some portion of it survive? For a Type I (carbon-detonation) supernova, most astronomers regard it as quite unlikely that any central remnant is left after the explosion. The entire star is shattered by the blast. However, for a Type II supernova, involving the implosion and subsequent rebound of a massive star's iron core, theoretical calculations indicate that part of the star may survive. ∞ (Sec. 21.2) The explosion destroys the parent star, but it may leave a tiny ultracompressed remnant* at its center. Even by the high-density standards of a white dwarf, however, the matter within this severely compacted core is in a very strange state, unlike anything we are ever likely to find (or create) on Earth.

Recall from Chapter 21 that during the moment of implosion of a massive star—just prior to the supernova itself—the electrons in the core violently smash into the protons there, forming neutrons and neutrinos. ∞ (Sec. 21.2) The neutrinos leave the scene at (or nearly at) the speed of light, accelerating the collapse of the neutron core, which continues to contract until its particles come into contact. At that point, neutron degeneracy pressure causes the central portion of the core to rebound, creating a powerful shock wave that races outward through the star, expelling matter violently into space.

The key point here is that the shock wave does not start at the very center of the collapsing core. The innermost part of the core—the region that "bounces"— remains intact as the shock wave it causes destroys the rest of the star. After the violence of the supernova has subsided, this ball of neutrons is all that is left. Researchers colloquially call this core remnant a **neutron star,** although it is not a star in any true sense of the word, because all of its nuclear reactions have ceased forever.

Neutron stars are extremely small and very massive. Composed purely of neutrons packed together in a tight ball about 20 km across, a typical neutron star is not much bigger than a small asteroid or a terrestrial city (see Figure 22.1), yet its mass is greater than that of the Sun. With so much mass squeezed into such a small volume, neutron stars are incredibly dense. Their average density can reach 10^{17} or even 10^{18} kg/m^3, nearly a billion times denser than a white dwarf. (For comparison, the density of a normal atomic nucleus is about 3×10^{17} kg/m^3.) A single thimbleful of neutron-star material would weigh 100 million tons—about as much as a good-sized terrestrial mountain.

*Astronomers commonly use the term remnant to mean "whatever remains of a star's inner core after the star's evolution has ended." Such remnants are small and compact—no larger than Earth in the case of a white dwarf and far smaller still for a neutron star. They should not be confused with supernova remnants: glowing clouds of debris scattered across many parsecs of interstellar space. ∞ (Sec. 21.3)

In a sense, we can think of a neutron star as a single enormous nucleus, with an atomic mass of around 10^{57}!

Neutron stars are solid objects. Provided that a sufficiently cool one could be found, you might even imagine standing on it. However, doing so would not be easy, as a neutron star's gravity is extremely powerful. A 70-kg (150-pound) human would weigh the Earth equivalent of about 10 trillion kg (10 billion tons). The severe pull of a neutron star's gravity would flatten you much thinner than this piece of paper!

In addition to large mass and small size, newly formed neutron stars have two other very important properties. First, they *rotate* extremely rapidly, with periods measured in fractions of a second. This is a direct result of the law of conservation of angular momentum (Chapter 15), which tells us that any rotating body must spin faster as it shrinks. ∞ (*More Precisely 6-2*) Even if the core of the progenitor star were initially rotating quite slowly (once every couple of weeks, say, as is observed in many upper main-sequence stars), it would be spinning a few times per second by the time it had reached a diameter of 20 km. Second, newborn neutron stars have very strong *magnetic fields*. The original field of the progenitor star is amplified by the collapse of the core because the contracting material squeezes the magnetic field lines closer together, creating a magnetic field trillions of times stronger than Earth's.

In time, theory indicates, our neutron star will spin more and more slowly as it radiates its energy into space, and its magnetic field will diminish. However, for a few million years after its birth, these two properties combine to provide the primary means by which this strange object can be detected and studied.

▲ **FIGURE 22.1 Neutron Star** Neutron stars are not much larger than many of Earth's major cities. In this fanciful comparison, a typical neutron star sits alongside Manhattan Island. *(NASA)*

✔ Are all supernovae expected to lead to neutron stars?

22.2 Pulsars

Can we be sure that objects as strange as neutron stars really exist? The answer is a confident yes. The first observation of a neutron star occurred in 1967, when Jocelyn Bell, a graduate student at Cambridge University, made a surprising discovery. She observed an astronomical object emitting radio radiation in the form of rapid *pulses*. Each pulse consisted of a 0.01-second (s) burst of radiation, after which there was nothing. Then, 1.34 s later, another pulse would arrive. The interval between the pulses was astonishingly uniform—so accurate, in fact, that the repeated emissions could be used as a precise clock. Figure 22.2 is a recording of part of the radio radiation from the pulsating object Bell discovered.

More than 1500 of these pulsating objects are now known in the Milky Way Galaxy. They are called **pulsars.** Each has its own characteristic pulse period and duration. The pulse periods of some pulsars are so stable that they are by far the most accurate natural clocks known in the universe—more accurate even than the best atomic clocks on Earth. In some cases, the period is predicted to change by only a few seconds in a million years.

When Bell made her discovery in 1967, she did not know what she was looking at. Indeed, no one at the time knew what a pulsar was. The explanation of pulsars as spinning neutron stars won Bell's thesis advisor, Antony Hewish, a share of the 1974 Nobel prize in physics. Hewish reasoned that the only physical mechanism consistent with such precisely timed pulsations is a small rotating source of radiation. Only rotation can cause the high degree of regularity of the observed pulses, and only a small object can account for the sharpness of each pulse. Radiation emitted from different regions of an object larger than a few tens of kilometers across would arrive at Earth at slightly different times, blurring the pulse profile. The best current model

describes a pulsar as a compact, spinning neutron star that periodically flashes radiation toward Earth.

Figure 22.3 outlines the important features of this pulsar model. Two "hot spots" on the surface of a neutron star, or in the magnetosphere just above the surface, continuously emit radiation in a narrow "searchlight" pattern. These spots are most likely localized regions near the neutron star's magnetic poles, where charged particles, accelerated to extremely high energies by the star's rotating magnetic field, emit radiation along the star's magnetic axis. The hot spots radiate more or less steadily, and the resulting beams sweep through space, like a revolving lighthouse beacon, as the neutron star rotates. Indeed, this pulsar model is often known as the **lighthouse model.** If the neutron star happens to be oriented such that the beam sweeps across Earth, we see the star as a pulsar. The beams are observed as a series of rapid pulses—each time one of the beams flashes past Earth, a pulse is seen. The period of the pulses is the star's rotation period.

A few pulsars are clearly associated with supernova remnants, although not all such remnants have a detectable pulsar within them. Figure 22.4(a) shows a pair of optical photographs of the Crab pulsar, at the center of the Crab supernova remnant. ∞ (Sec. 21.3) In the left frame, the pulsar is off; in the right frame, it is on. The rapid variation in the pulsar's light, with a pulse period of about 33 milliseconds, is shown in Figure 22.4(b). The Crab also pulses in the radio and X-ray parts of the spectrum; Figure 22.4(c) is a pair of *ROSAT* images showing the nebula's X-ray variation. By observing the speed and direction of the Crab's ejected matter, astronomers can work backward to pinpoint the location in space at which the explosion must have occurred and where the supernova core remnant should be located. It turns out that is precisely the region of the Crab Nebula from which the pulsating signals arise. The Crab pulsar is evidently all that remains of the once-massive star whose supernova was observed in A.D. 1054.

As indicated in Figure 22.3, the neutron star's strong magnetic field and rapid rotation channel high-energy particles from near the star's surface into the surrounding nebula. (Cf. the expanding envelope of the 1054 supernova.) The result is an energetic *pulsar wind* that flows outward at

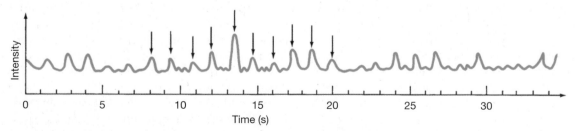

▲ **FIGURE 22.2 Pulsar Radiation** Pulsars emit periodic bursts of radiation. This recording shows the regular change in the intensity of the radio radiation emitted by the first such object known, discovered in 1967. Some of the object's pulses are marked by arrows.

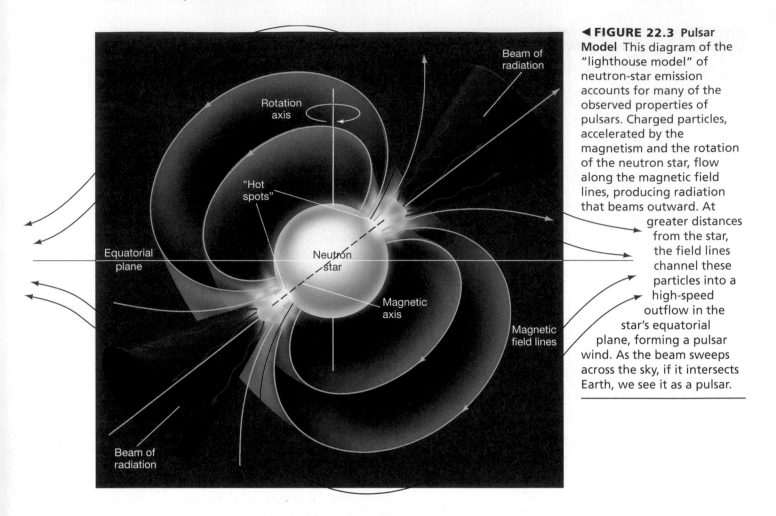

◀ **FIGURE 22.3** Pulsar
Model This diagram of the
"lighthouse model" of
neutron-star emission
accounts for many of the
observed properties of
pulsars. Charged particles,
accelerated by the
magnetism and the rotation
of the neutron star, flow
along the magnetic field
lines, producing radiation
that beams outward. At
greater distances
from the star,
the field lines
channel these
particles into a
high-speed
outflow in the
star's equatorial
plane, forming a pulsar
wind. As the beam sweeps
across the sky, if it intersects
Earth, we see it as a pulsar.

almost the speed of light, primarily in the star's equatorial plane. As it slams into the nebula, the wind heats the gas to very high temperatures. Figure 22.5 shows this process in action—the combined *Hubble/Chandra* image reveals rings of hot X-ray-emitting gas moving rapidly away from the pulsar. Also visible in the image is a jet of hot gas (*not* the beam of radiation from the pulsar) escaping perpendicular to the equatorial plane. Eventually, the energy from the pulsar wind is deposited into the Crab nebula, where it is radiated into space by the nebular gas, powering the spectacular display we see from Earth. ∞ (Fig. 21.10) *Discovery 21-2* presents more detail on the complex interaction between the pulsar and the nebula in which it resides.

Most pulsars emit pulses in the form of radio radiation, but some (like the Crab) have been observed to pulse in the visible, X-ray, and gamma-ray parts of the spectrum as well. Figure 22.6 shows the Crab and the nearby Geminga pulsar in gamma rays. Geminga is unusual in that, while it pulsates strongly in gamma rays, it is barely detectable in visible light and not at all at radio wavelengths. Whatever types of radiation are produced, these electromagnetic flashes at different frequencies all occur at regular, repeated intervals, as we would expect, since they arise from the same object. However, pulses at different frequencies do not necessarily all occur at the same instant

in the pulse cycle. The periods of most pulsars are quite short, ranging from about 0.03 s to 0.3 s (that is, flashing between 3 and 30 times per second). The human eye is insensitive to such rapid flashes, making it impossible to observe the flickering of a pulsar by eye, even with a large telescope. Fortunately, instruments can record pulsations of light that the human eye cannot detect.

Most known pulsars are observed (usually by Doppler measurements) to have high speeds—much greater than the typical speeds of stars in the Galaxy. The most likely explanation for these anomalously high speeds is that neutron stars may receive substantial "kicks" due to asymmetries in the supernovae in which they formed. Such asymmetries, which are predicted by theory, are generally not very pronounced, but if the supernova's enormous energy is channeled even slightly in one direction, the newborn neutron star can recoil in the opposite direction with a speed of many tens or even hundreds of kilometers per second. Thus, observations of pulsar velocities give theorists additional insight into the detailed physics of supernovae.

All pulsars are neutron stars, but not all neutron stars are observed as pulsars, for two reasons. First, the two ingredients that make the neutron star pulse—rapid rotation and a strong magnetic field—both diminish with time, so the pulses gradually weaken and become

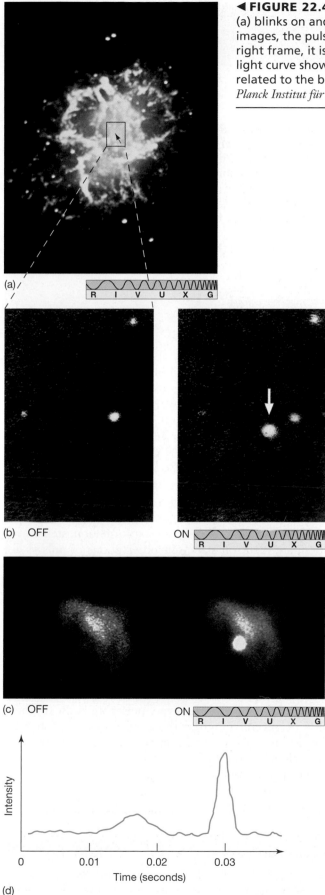

◀ **FIGURE 22.4 Crab Pulsar** The pulsar (arrow) in the core of the Crab Nebula (a) blinks on and off about 30 times each second. (b) In this pair of optical images, the pulsing can be clearly seen. In the left frame, the pulsar is off; in the right frame, it is on. (c) The same phenomenon is also detected in X rays. (d) The light curve shows the main pulse and its precursor. (The latter is thought to be related to the beam that is directed away from us). *(UC/Lick Observatory; Max Planck Institut für Extraterrestrische Physik)*

less frequent. Theory indicates that, within a few tens of millions of years, the beam weakens and the pulses all but stop. Second, even a young, bright neutron star is not necessarily detectable as a pulsar from our vantage point on Earth. The pulsar beam depicted in Figure 22.3 is relatively narrow—perhaps as little as a few degrees across in some cases. Only if the neutron star happens to be oriented in just the right way do we actually see pulses. When we see those pulses from Earth, we call the body a pulsar. Note that we are using the term "pulsar" here to mean the pulsing object we observe if the beam crosses Earth. However, many astronomers use the word more generically to mean *any* young neutron star producing beams of radiation as in Figure 22.3. Such an object will be a pulsar as seen from some directions—just not necessarily ours!

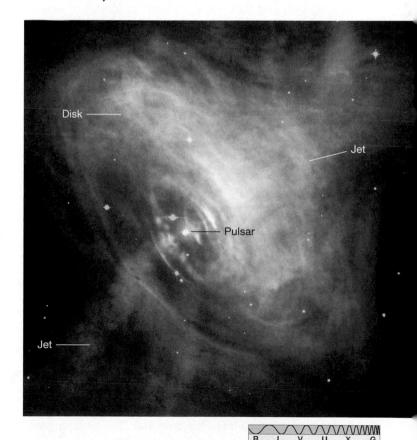

▲ **FIGURE 22.5 Pulsar Wind** A recent *Chandra* X-ray image of the Crab, superimposed on a *Hubble* optical image, shows more clearly the central pulsar, as well as the disk and the jets of outflowing material. *(NASA)*

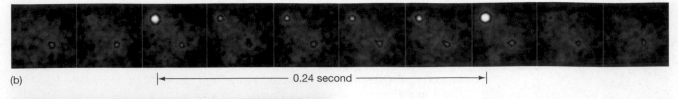

(b) |←————— 0.24 second —————→|

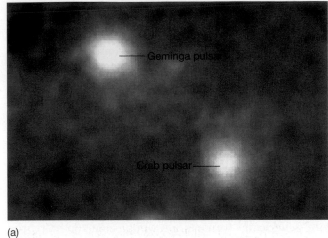

(a)

◄ **FIGURE 22.6 Gamma-Ray Pulsars** (a) The Crab and Geminga pulsars, which happen to lie fairly close to one another (about 4.5 degrees apart) in the sky. Unlike the Crab, Geminga is barely visible at optical wavelengths and undetectable in the radio region of the spectrum. (b) Sequence of *Compton Gamma-Ray Observatory* images showing Geminga's 0.24-s pulse period. (The Crab's 33-ms period is too rapid to be resolved by the detector.) *(NASA)*

R I V U X G

Figure 22.7 shows three superimposed *Hubble* images of a neutron star that is not a pulsar (at least, not as seen from Earth). First identified by *ROSAT* and *EUVE* observations of its soft X-ray emission, this object is just 30 km across and has a surface temperature of about 700,000 K. Despite this enormous temperature, its small size means that the star is very faint—just 25th magnitude in visible light. ∞ (Sec. 17.4) The figure shows the neutron star's proper motion over a period of almost three years. Small wobbles in the motion due to Earth's parallax can also (barely) be discerned. The star is thought to be about 1 million years old, lies about 60 pc away, and is moving at a speed of 110 km/s across our line of sight.

Such faint objects are very difficult to detect, and a detection of a "bare" neutron star like this is a rare event. Most cataloged neutron stars have been detected either as pulsars or via their interaction with a "normal" stellar companion in a binary system. (See Section 22.3.) However, given our current knowledge of star formation, stellar evolution, and neutron stars, our observations of pulsars are consistent with the ideas that (1) *every* high-mass star dies in a supernova explosion, (2) most supernovae leave a neutron star behind (a few result in black holes, as discussed in a moment), and (3) *all* young neutron stars emit beams of radiation, just like the pulsars we actually detect. A few pulsars are definitely associated with supernova remnants, clearly establishing those pulsars' explosive origin.

On the basis of estimates of the rate at which massive stars have formed over the lifetime of the Milky Way, astronomers reason that, for every pulsar we know of, there must be several hundred thousand more neutron stars moving unseen somewhere in the Galaxy. Some formed relatively recently—less than a few million years ago—and simply happen not to be beaming their energy toward Earth. However, the vast majority are old, their youthful pulsar phase long past.

R I V U X G

▲ **FIGURE 22.7 Isolated Neutron Star** This lone neutron star was first detected by its X-ray emission and subsequently imaged by *Hubble*. It lies about 60 pc from Earth and is thought to be about 1 million years old. This triple exposure shows the star streaking across the sky at more than 100 km/s. *(NASA)*

CONCEPT CHECK

✔ Why don't we see all neutron stars as pulsars?

22.3 Neutron-Star Binaries

3 In Chapter 17, we noted that most stars are not single, but instead are members of binary systems. ∞ (Sec. 17.7) Although many pulsars are known to be isolated (i.e., not part of any binary), at least some do have binary companions, and the same is true of neutron stars in general (even the ones not seen as pulsars). One important consequence of this pairing is that the masses of some neutron stars have been determined quite accurately. All the measured masses are fairly close to 1.4 times the mass of the Sun—the Chandrasekhar mass of the stellar core that collapsed to form the neutron-star remnant.

X-Ray Sources

The late 1970s saw several important discoveries about neutron stars in binary-star systems. Numerous X-ray sources were found near the central regions of our Galaxy and also near the centers of a few rich star clusters. Some of these sources, known as **X-ray bursters,** emit much of their energy in violent eruptions, each thousands of times more luminous than our Sun, but lasting only a few seconds. A typical burst is shown in Figure 22.8.

This X-ray emission is thought to arise on or near neutron stars that are members of binary systems. Matter torn from the surface of the (main-sequence or giant) companion by the neutron star's strong gravitational pull accumulates on the neutron star's surface. As in the case of white-dwarf accretion (see Chapter 21), the material does not fall directly onto the surface. Instead, as illustrated in Figure 22.9(a), it forms an accretion disk. (Cf. with Figure 21.2, which depicts the white-dwarf equivalent.) ∞ (Sec. 21.1) The gas goes into a tight orbit around the neutron star and then spirals slowly inward. The inner portions of the accretion disk become extremely hot, releasing a steady stream of X rays.

As gas builds up on the neutron star's surface, its temperature rises due to the pressure of overlying material. Soon the temperature becomes hot enough to fuse hydrogen. The result is a sudden period of rapid nuclear burning that releases a huge amount of energy in a brief, but intense, flash of X rays—an *X-ray burst*. After several hours of renewed accumulation, a fresh layer of matter produces the next burst. Thus, an X-ray burst is much like a nova on a white dwarf, but occurring on a far more violent scale because of the neutron star's much stronger gravity. ∞ (Sec. 21.1)

Not all the infalling gas makes it onto the neutron star's surface, however; in at least one case—an object known as SS 433*—we have direct observational evidence that some material is instead shot completely out of the system at enormously high speeds. SS 433 expels more than one Earth mass of material every year in the form of two oppositely directed narrow jets moving roughly perpendicular to the disk. Observations of the Doppler shifts of optical emission lines produced within the jets themselves imply speeds of almost 80,000 km/s—more than 25 percent of the speed of light! As the jets interact with the interstellar medium, they emit radio radiation, as shown in Figure 22.9(b).

Jets of this sort are apparently quite common in astronomical systems in which an accretion disk surrounds a compact object (such as a neutron star or a black hole). They are thought to be produced by the intense radiation and magnetic fields near the inner edge of the disk, although the details of their formation are still uncertain. (Once again, note that these jets are *not* the "lighthouse" beams of radiation from the neutron star itself, shown in Figure 22.3, that can result in a pulsar, nor are they associated with a pulsar wind, as in Figure 22.5.) Although SS 433 is the only stellar-mass compact object currently known to produce jets, we will see examples of similar phenomena on much larger scales in later chapters. One of the most important aspects of SS 433 is that we can actually

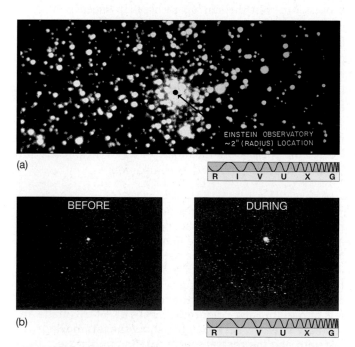

(a)

R I V U X G

BEFORE   DURING

(b)

R I V U X G

▲ **FIGURE 22.8 X-Ray Burster** An X-ray burster produces a sudden, intense flash of X rays, followed by a period of relative inactivity lasting as long as several hours. Then another burst occurs. The bursts are thought to be caused by explosive nuclear burning on the surface of an accreting neutron star, similar to the explosions on a white dwarf that give rise to novae. (a) An optical photograph of the globular star cluster Terzan 2, showing a 2″ dot at the center where the X-ray bursts originate. (b) X-ray images taken before and during the outburst. The most intense X rays correspond to the position of the dot shown in frame (a). *(SAO)*

The name simply identifies the object as the 433rd entry in a particular catalog of stars with strong optical emission lines.

study both the disk and the jets, instead of simply having to assume their existence, as we do in more distant cosmic objects.

Millisecond Pulsars

In the mid-1980s an important new category of pulsars was found: a class of very rapidly rotating objects called **millisecond pulsars.** More than 100 are currently known in the Milky Way Galaxy. These objects spin hundreds of times per second (i.e., their pulse period is a few milliseconds). This speed is about as fast as a typical neutron star can spin without flying apart. In some cases, the star's equator is moving at more than 20 percent of the speed of light, a speed that suggests a phenomenon bordering on the incredible: a cosmic object of kilometer dimensions, more massive than our Sun, spinning almost at breakup speed and making nearly a thousand complete revolutions *every second*! Yet the observations and their interpretation leave little room for doubt.

The story of these remarkable objects is further complicated because many of them are found in globular clusters. This is odd, since globular clusters are known to be very old—10 billion years, at least. Yet, Type II supernovae (the kind that create neutron stars) are associated with massive stars that explode within a few tens of *millions* of years after their formation, and no stars have formed in any globular cluster since the cluster itself came into being. Thus, no new neutron star has been produced in a globular cluster in a very long time. But, as mentioned earlier, the pulsar produced by a supernova is expected to slow down in only a few million years, and after 10 billion years its rotation should have all but ceased. Accordingly, the rapid rotation of the pulsars found in globular clusters cannot be a relic of their birth. Instead, these objects must have been "spun up"— that is, had their rotation rates increased—by some other, much more recent, mechanism.

The most likely explanation for the high rotation rate of pulsars is that the neutron star has been spun up by drawing in matter from a companion star. As matter spirals down onto the star's surface in an accretion disk, it provides the "push" needed to make the neutron star spin faster (see Figure 22.10). Theoretical calculations indicate that this process can spin the star up to breakup speed in about a hundred million years. Subsequently, an encounter with another star may eject the neutron star from the binary, or the pulsar's radiation may destroy its companion, so an isolated millisecond pulsar results. This general picture is supported by the finding that, of the 80 or so millisecond pulsars seen in globular clusters, roughly half are known to be members of binary systems. The remaining solo millisecond pulsars were probably formed when an encounter with another star ejected the pulsar from the binary or when the pulsar's own intense radiation destroyed its companion.

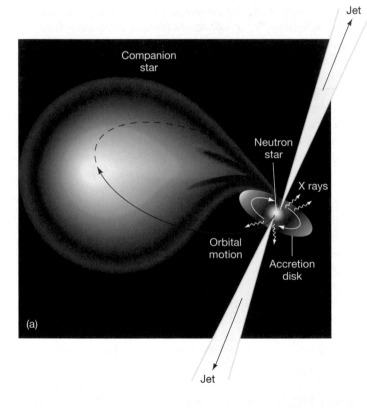

(a)

◀ **FIGURE 22.9 X-Ray Emission** (a) Matter flows from a normal star toward a compact neutron-star companion and falls toward the surface in an accretion disk. As the gas spirals inward under the neutron star's intense gravity, it heats up, becoming so hot that it emits X rays. In at least one instance—the peculiar object SS 433—some material may be ejected in the form of two high-speed jets of gas. (b) False-color radiographs of SS 433, made at monthly intervals (left to right), show the jets moving outward and the central source rotating under the gravitational influence of the companion star. *(NRAO)*

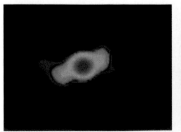
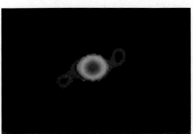
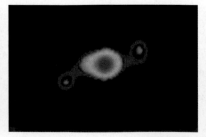
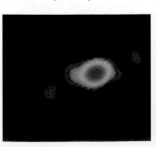

(b)

R I V U X G

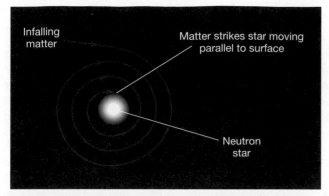

▲ **FIGURE 22.10 Millisecond Pulsar** Gas from a companion star spirals down onto the surface of a neutron star. As the infalling matter strikes the star, it moves almost parallel to the surface, so it tends to make the star spin faster. Eventually, this process can result in a millisecond pulsar—a neutron star spinning at the incredible rate of hundreds of revolutions per second.

Thus, although a pulsar like the Crab is the direct result of a supernova, millisecond pulsars are the product of a two-stage process. The neutron star was formed in an ancient supernova, billions of years ago. Only relatively recently, through its interaction with a binary companion, has the neutron star achieved the rapid spin that we observe today. Once again, we see how members of a binary system can evolve in ways quite different from the manner in which single stars evolve. Notice that the scenario of accretion onto a neutron star from a binary companion is the same scenario that we just used to explain the existence of X-ray bursters. In fact, the two phenomena are closely linked. Many X-ray bursters may be on their way to becoming millisecond pulsars, and many millisecond pulsars are X-ray sources, powered by the trickle of material still falling onto them from their binary companions.

Figure 22.11 shows the globular cluster 47 Tucanae, together with a *Chandra* image of its core showing no fewer than 108 X-ray sources—about 10 times the number that had been known in the cluster prior to *Chandra*'s launch. Roughly half of these sources are millisecond pulsars; the cluster also contains two or three "conventional" neutron-star binaries. Most of the remaining sources are white-dwarf binaries, similar to those discussed in Chapter 21. ∞ (Sec. 21.1)

The way in which a neutron star can become a member of a binary system is the subject of active research, because the violence of a supernova explosion would be expected to blow the binary apart in many cases. Only if the supernova progenitor lost a lot of mass before the explosion would the binary system be likely to survive. Alternatively, by interacting with an existing binary and displacing one of its components, a neutron star may become part of a binary system *after* it is formed, as depicted in Figure 22.12. Astronomers are eagerly searching the skies for more millisecond pulsars to test their ideas.

Pulsar Planets

Radio astronomers can capitalize on the precision with which pulsar signals repeat themselves to make extremely accurate measurements of pulsar motion. In January 1992, radio astronomers at the Arecibo Observatory found that the pulse period of a recently discovered millisecond pulsar lying some 500 pc from Earth varied in an unexpected, but quite regular, way. Careful analysis of the data

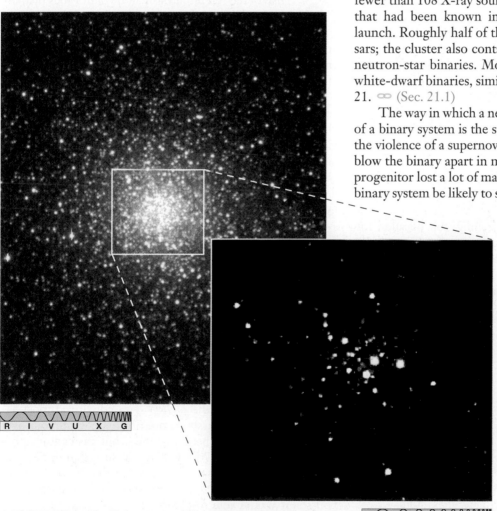

R I V U X G

▲ **FIGURE 22.11 Cluster X-Ray Binaries** The dense core of the old globular cluster 47 Tucanae (left) harbors more than 100 separate X-ray sources (shown in the *Chandra* image at the right). Roughly half of these are thought to be binary millisecond pulsars, still accreting small amounts of gas from their companions after an earlier period of mass transfer spun them up to millisecond speeds. *(ESO; NASA)*

has revealed that the period fluctuates on two distinct time scales—one of 67 days, the other of 98 days. The changes in the pulse period are small—less than one part in 10^7—but repeated observations have confirmed their reality.

The leading explanation for these fluctuations holds that they are caused by the Doppler effect as the pulsar wobbles back and forth in space. But what causes the wobble? The Arecibo group believes that it is the result of the combined gravitational pulls of not one, but *two*, planets, each about three times the mass of Earth! One orbits the pulsar at a distance of 0.4 A.U. and the other at a distance of 0.5 A.U. Their orbital periods are 67 and 98 days, respectively, matching the timing of the fluctuations. In April 1994, the group announced further observations that not only confirmed their earlier findings, but also revealed the presence of a *third* body, with mass comparable to Earth's Moon, orbiting only 0.2 A.U. from the pulsar.

These remarkable results constituted the first definite evidence of planet-sized bodies outside our solar system. A few other millisecond pulsars have since been found with similar behavior. However, it is unlikely that any of these planets formed in the same way as our own. Any planetary system orbiting the pulsar's progenitor star was almost certainly destroyed in the supernova explosion that created the pulsar. As a result, scientists are still unsure about how these planets came into being. One possibility involves the binary companion that provided the matter necessary to spin the pulsar up to millisecond speeds. Possibly, the pulsar's intense radiation and strong gravity destroyed the companion and then spread its matter out into a disk (a little like the solar nebula) in whose cool outer regions the planets might have condensed.

Astronomers have been searching for decades for planets orbiting main-sequence stars like our Sun, on the assumption that planets are a natural by-product of star formation. ∞ (Sec. 15.2) As we have seen, these searches have now identified many extrasolar planets, although the planetary systems discovered to date don't look much like our own, and nothing comparable in mass to Earth has yet been detected. ∞ (Sec. 15.5) It is ironic that the first and only Earth-sized planets to be found outside the solar system orbit a dead star and have little or nothing in common with our own world!

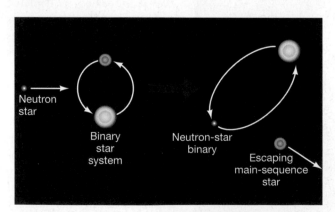

▲ **FIGURE 22.12 Binary Exchange** A neutron star can encounter a binary made up of two low-mass stars, ejecting one of them and taking its place. This mechanism provides a means of forming a binary system with a neutron-star component (which may later evolve into a millisecond pulsar) without having to explain how the binary survived the supernova explosion that formed the neutron star.

CONCEPT CHECK

✔ What is the connection between X-ray sources and millisecond pulsars?

22.4 Gamma-Ray Bursts

Discovered serendipitously in the late 1960s by military satellites looking for violators of the Nuclear Test Ban Treaty, and first made public in the 1970s, **gamma-ray bursts** have developed into one of the deepest mysteries in astronomy today. The bursts consist of bright, irregular flashes of gamma rays typically lasting only a few seconds (Figure 22.13a). Until the 1990s, it was thought that gamma-ray bursts were basically "scaled-up" versions of X-ray bursters in which matter accreted from the binary companion was subjected to even more violent nuclear burning, accompanied by the release of the more energetic gamma rays. However, it now appears that this is not the case.

Distances and Luminosities

Figure 22.13(b) shows an all-sky plot of the positions of 2704 bursts detected by the *Compton Gamma-Ray Observatory* (*CGRO*) during its nine-year operational lifetime. ∞ (Sec. 5.6) On average, *CGRO* detected gamma-ray bursts at the rate of about one a day. Note that the bursts are distributed uniformly across the sky (their distribution is said to be "isotropic"), rather than being confined to the relatively narrow band of the Milky Way. (Cf. Figure 5.40.) The bursts seemingly never repeat at the same location, show no obvious clustering, and appear unaligned with any known large-scale structure, near or far.

The isotropy of the *CGRO* data convinced most astronomers that the bursts do not originate within our own Galaxy, as had once been assumed, but instead are produced far beyond the Milky Way—at so-called *cosmological distances*, comparable to the scale of the universe itself. However, while *CGRO* detected thousands of gamma-ray bursts, it was unable to determine the distance to any of them. The reason for this brings us back to a fundamental issue in astronomy: the difficulties involved in measuring distances in the universe. ∞ (Sec. 17.6)

Gamma-ray observations in and of themselves do not provide enough information to tell us how far away a burst is. Rather, in order to determine the distance, astronomers

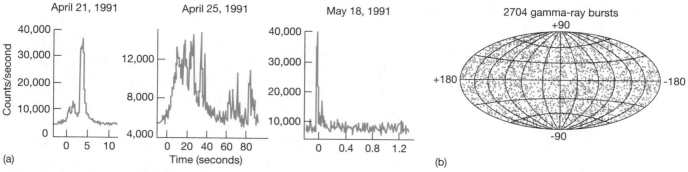

April 21, 1991   April 25, 1991   May 18, 1991

2704 gamma-ray bursts

(a)

(b)

▲ **FIGURE 22.13 Gamma-Ray Bursts** (a) Plots of intensity versus time (in seconds) for three gamma-ray bursts. Note the substantial differences between the plots, indicating that some bursts are irregular and spiky, whereas others are much more smoothly varying. Whether this wide variation in the appearance of the bursts means that more than one physical process is at work is unknown. (b) Positions on the sky of all the gamma-ray bursts detected by *Compton Observatory* during its nearly nine-year operating lifetime. The bursts appear to be distributed isotropically (uniformly) across the entire sky. The plane of the Milky Way Galaxy runs horizontally across the center of the map, which is also the direction to the center of our Galaxy. *(NASA)*

must somehow associate the burst with some other object in the sky whose distance can be measured by other means. Such objects are referred to as burst *counterparts*, and the techniques for studying them usually involve observations in the optical or X-ray parts of the electromagnetic spectrum. Unfortunately, as we saw in Chapter 5, gamma rays are far too penetrating to be focused by conventional optics. ∞ (Sec. 5.6) As a result, *CGRO's* burst positions were uncertain by several degrees, so optical or X-ray telescopes had to scan very large regions of the sky as they looked for the counterpart. Compounding this problem, the "afterglows" of a burst at X-ray or optical wavelengths generally fade quite rapidly, severely limiting the time available to complete the search.

The first direct measurement of the distance to a gamma-ray burst was made on May 8, 1997, thanks to a combination of gamma-ray, X-ray, and optical observations of the gamma-ray burst GRB 970508. (The number simply indicates the date of the burst's detection.) The Italian–Dutch *BeppoSAX* satellite recorded the burst in both the gamma- and X-ray regions of the spectrum. The importance of the X-ray observations is that they allowed a

much more accurate determination of the burst's location in the sky—in fact, to within a few arc minutes. That was enough for ground-based optical astronomers to look for and find GRB 970508's optical afterglow. For the first time, astronomers had observed an optical counterpart to a gamma-ray burst.

The counterpart's optical spectrum, obtained with the Keck telescope, revealed a very important piece of information. Several absorption lines of iron and magnesium were identified, but they were redshifted by almost a factor of two in wavelength. Such redshifts, as we will see in Chapter 24, are the result of the expansion of the universe, and they are clear proof—a "smoking gun," if you will—that at least this gamma-ray burst, and presumably all others, really did occur at cosmological distances. The events responsible for GRB 970508 occurred more than 2 *billion* parsecs from Earth.

Figure 22.14 shows optical images of another gamma-ray burst, GRB 971214, along with the "host" galaxy in which it resided. According to the redshift of lines observed in the galaxy's spectrum, this burst was almost 5 billion pc away. By the time the *Hubble* image (b) was taken (about two months after the Keck image and

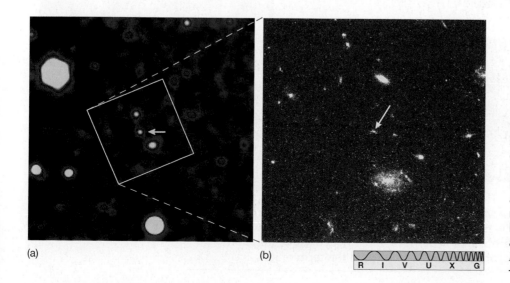

(a)   (b)

R I V U X G

◀ **FIGURE 22.14 Gamma-Ray Burst Counterparts** Two optical images of the gamma-ray burst GRB 971214. Image (a) shows the visible afterglow of the gamma-ray source (arrow) to be quite bright, comparable to two other prominent sources in the overlaid box. A Keck spectrum of the afterglow showed it to be highly redshifted, placing it near the limits of the observable universe, almost 5 billion parsecs away. *(Keck; NASA)*

four months after the initial burst of gamma rays), the afterglow had faded, but a faint image of a host galaxy remained. In all, nearly 100 optical or X-ray afterglows of gamma-ray bursts have been detected, and about two dozen distances are known. All are very large, implying that the bursts must be extremely energetic, since otherwise they wouldn't be detectable by our equipment. If we assume that the gamma rays are emitted equally in all directions (a big assumption!), then we can easily calculate the total energy emitted from the fraction we see, using the inverse-square law. ∞ (Sec. 17.2) We find that each burst apparently generates more energy—and in some cases hundreds of times more energy—than a typical supernova explosion, all in a matter of seconds! According to this calculation, GRB 971214 was the most violent explosion ever observed in the universe.

As we have just seen, finding the counterpart to a gamma-ray burst requires an accurate measurement of the burst's location (to limit the region of the sky that must be scanned) and fast communication (so that other telescopes can start to look before the afterglow fades). In October 2000, NASA launched the *High Energy Transient Explorer-2 (HETE-2)* satellite to continue the work begun by *CGRO*. With an accuracy of 10′ in gamma rays and 10″ in X rays, *HETE-2* instantly relays accurate burst positions to other instruments in space and on the ground, allowing

rapid searches for counterparts to be carried out. Since its launch, the satellite has almost doubled the number of known burst counterparts and has played a pivotal role in advancing our understanding of these violent phenomena.

What Causes the Bursts?

What are gamma-ray bursts? Frankly, we still don't know! Before the first distance measurements, some theorists had attempted to explain gamma-ray bursts in terms of nearby—and hence much less energetic—events, perhaps in the outer parts of our own Galaxy (see Chapter 23) or even in the Oort cloud surrounding the Sun. ∞ (Sec. 14.2) However, it now seems clear that gamma-ray bursts are very distant, extremely energetic events.

What's more, the millisecond flickering in the light curves recorded by *Compton* imply that, whatever the origin of gamma-ray bursts, all their energy must come from an extremely small volume—in fact, no larger than a *few hundred kilometers* across. The reasoning is as follows: If the emitting region were, say, 300,000 km—1 light-second—across, even an instantaneous change in intensity at the source would be smeared out over a time interval of 1 s as seen from Earth, because light from the far side of the object would take 1 s longer to reach us than light from the near side. For the gamma-ray variation not to be blurred

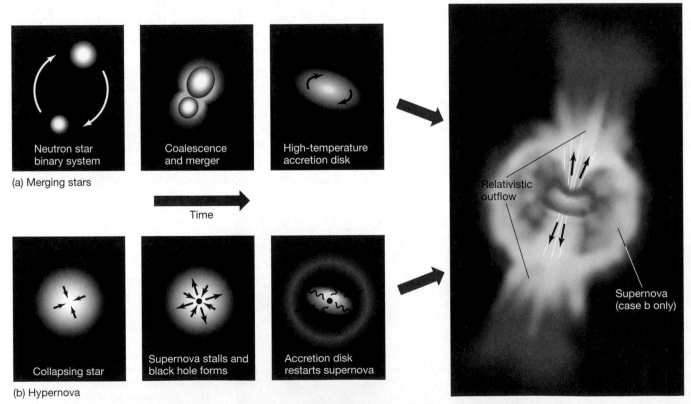

▲ **FIGURE 22.15 Gamma-Ray Burst Models** Two models have been proposed to explain gamma-ray bursts. Part (a) depicts the merger of two neutron stars; part (b) shows the core collapse, implosion, and stalled supernova, or *hypernova*, of a single massive star. Both models predict relativistic fireballs, probably releasing energy in the form of jets.

by the light-travel time, the source cannot be more than 1 light-millisecond, or 300 km, in diameter.

Most theoretical attempts to explain gamma-ray bursts picture the burst as a *relativistic fireball*—an expanding region, probably a jet, of superhot gas radiating furiously in the gamma-ray part of the spectrum. (The term "relativistic" here means that particles are moving at nearly the speed of light and that Einstein's theory of relativity is needed to describe them—see Section 22.6.) The complex burst structure and afterglows are produced as the fireball expands, cools, and interacts with its surroundings.

If the gamma-ray burst is emitted as a jet, then its total energy may be reduced to more "manageable" levels, as the bright flash we see is representative of only a small fraction of the sky. This may make the huge luminosity of the December 1997 burst somewhat easier to explain. As an analogy, consider a handheld laser pointer of the sort commonly used in talks and lectures. It radiates only a few milliwatts of power, much less than a household lightbulb, but it appears enormously bright if you happen to look directly into the beam. (*Don't* do this, by the way!) The laser beam is bright because all of its energy is concentrated in almost a single direction, instead of being radiated isotropically into space.

Two leading models for the energy source have emerged, as sketched in Figure 22.15. The first (Figure 22.15a) is the "true" end point of a binary-star system. Suppose that both members of the binary evolve to become neutron stars. As the system continues to evolve, gravitational radiation (see *Discovery 22-1*) is released, and the two ultradense stars spiral in toward each other. Once they are within a few kilometers of one another, coalescence is inevitable. Such a merger will likely produce a violent explosion comparable in energy to that generated by a supernova and perhaps energetic enough to explain the flashes of gamma rays we observe. The overall rotation of the binary system may channel the energy into a high-speed, high-temperature jet.

The second model (Figure 22.15b), sometimes called a *hypernova*, is a "failed" supernova—but what a failure! In this picture, a very massive star undergoes core collapse much as described earlier for a Type II supernova, but instead of forming a neutron star, the core collapses to a black hole (see Section 22.5). ∞ (Sec. 21.2) At the same time, the blast wave racing outward through the star stalls. Instead of being blown to pieces, the inner part of the star begins to implode, forming an accretion disk around the black hole and creating a relativistic jet. The jet punches its way out of the star, producing a gamma-ray burst as it slams into the surrounding shells of gas expelled from the star during the final stages of its nuclear-burning lifetime. ∞ (*Discovery 20-2*) At the same time, intense radiation from the accretion disk may restart the stalled supernova, blasting what remains of the star into space.

The idea of a relativistic fireball has become widely accepted among workers in this fast-changing branch of astrophysics. Many researchers regard it as likely that the energy is released as a jet and that the most intense gamma-ray bursts occur when the jet is directed toward us ("looking into the laser"). There is still no consensus among experts on which (if either) of the two models just described is correct, although the apparent association of some recent supernovae with earlier gamma-ray bursts argues in favor of the hypernova scenario. The strongest evidence of this connection was obtained in 2003, when *HETE-2* detected a particularly bright burst known as GRB 030329 (Figure 22.16). Within 24 hours, the 8.2-m *VLT* in Chile had observed the counterpart and measured its spectrum, and continued to monitor the afterglow as it faded over the next month. ∞ (Sec. 5.3) Both the spectrum and the light curve of the counterpart are precisely

April 3, 2003

GRB030329

(a)

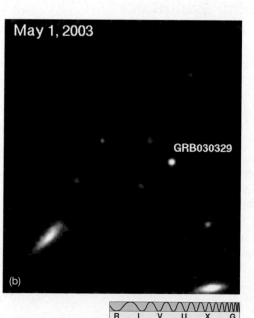

May 1, 2003

GRB030329

(b)

R I V U X G

◀ **FIGURE 22.16**

Hypernova? The gamma-ray burst GRB 030329 may prove crucial to theorists' understanding of the physical processes underlying these violent phenomena. Observed at radio, optical, and X-ray wavelengths, the burst counterpart has all the hallmarks of a high-mass supernova, lending strong support to the hypernova model. Here, the counterpart is shown as seen through the 8.2-m European VLT telescope in Chile, (a) near the moment of the burst and (b) fading a month later. *(ESO)*

what astronomers would expect from the supernova produced by a very massive (25-solar-mass) star.

The hypernova model predicts bursts of relatively long duration and is now the leading explanation of the "long" gamma-ray bursts, conventionally defined as those lasting more than about two seconds, such as the second event shown in Figure 22.13a. However, the model has difficulty explaining the many "short" bursts that are also observed (e.g., the other two bursts shown in that figure). Detailed calculations indicate that the neutron-star merger model naturally produces short bursts. Thus, it remains possible that there are in fact two distinct classes of gamma-ray bursts and that both of the models depicted in Figure 22.15 contribute to the total. Clearly, the final word on these enigmatic objects has yet to be written.

CONCEPT CHECK

✔ What are gamma-ray bursts, and why do they pose such a challenge to current theory?

22.5 Black Holes

Neutron stars are supported by the resistance of tightly packed neutrons to further compression. Squeezed together, the neutrons form a hard ball of matter that not even gravity can compress further. Or can it? Is it possible that, given enough matter packed into a small enough volume, the collective pull of gravity can eventually crush any opposing pressure? Can gravity continue to compress a massive star into an object the size of a planet, a city, a pinhead—even smaller? The answer, apparently, is yes.

The Final Stage of Stellar Evolution

Although the precise figure is uncertain, mainly because the behavior of matter at very high densities is not well enough understood, most researchers concur that the mass of a neutron star cannot exceed about 3 solar masses. That is the neutron-star equivalent of the white dwarf's Chandrasekhar mass limit discussed in the previous chapter. ∞ (Sec. 21.3) Above this limit, not even tightly packed neutrons can withstand the star's gravitational pull. In fact, we know of *no* force that can counteract gravity once neutron degeneracy pressure is overwhelmed. If enough material is left behind after a supernova such that the central core exceeds the 3-solar-mass limit, gravity wins the battle with pressure once and for all, and the star's central core collapses forever. Stellar evolution theory indicates that this is the fate of any star whose main-sequence mass exceeds about 25 times the mass of the Sun.

The limit of 3 solar masses is uncertain, in part because it ignores the effects of magnetism and rotation, both of which are surely present in the cores of evolved

stars. Because these effects can compete with gravity, they influence stellar evolution. ∞ (Sec. 19.1) In addition, we do not know precisely how the basic laws of physics might change in regions of very dense matter that is both rapidly spinning and strongly magnetized. Generally speaking, theorists expect that the neutron-star mass limit increases when magnetism and rotation are included, because even larger amounts of mass will then be needed for gravity to compress stellar cores into neutron stars or black holes, but the amount of the increase is not currently known.

As the stellar core shrinks, the gravitational pull in its vicinity eventually becomes so great that even light itself is unable to escape. The resultant object therefore emits no light, no radiation, and no information whatsoever. Astronomers call this bizarre end point of stellar evolution, in which a massive core remnant collapses in on itself and vanishes forever, a **black hole.**

Escape Speed

Newtonian mechanics—up to now our reliable and indispensable tool in understanding the universe—cannot adequately describe conditions in or near black holes. ∞ (Sec. 2.7) To comprehend these collapsed objects, we must turn instead to the modern theory of gravity: Einstein's *general theory of relativity*, discussed in Section 22.6. Still, we can usefully discuss some aspects of these strange bodies in more or less Newtonian terms. Let's consider again the familiar Newtonian concept of escape speed—the speed needed for one object to escape from the gravitational pull of another—supplemented by two key facts from relativity: (1) Nothing can travel faster than the speed of light, and (2) all things, *including light*, are attracted by gravity.

A body's escape speed is proportional to the square root of the body's mass divided by the square root of its radius. ∞ (Sec. 2.7) Earth's radius is 6400 km, and the escape speed from Earth's surface is just over 11 km/s. Now consider a hypothetical experiment in which Earth is squeezed on all sides by a gigantic vise. As our planet shrinks under the pressure, its mass remains the same, but its escape speed increases because the planet's radius is decreasing. For example, suppose Earth were compressed to one-fourth its present size. Then the proportionality mentioned in the first sentence of this paragraph predicts that our planet's escape speed would double. Consequently, any object escaping from this hypothetically compressed Earth would need a speed of at least 22 km/s to do so.

Imagine compressing Earth some more. Squeeze it by an additional factor of, say, a thousand, making its radius hardly more than a kilometer. Now a speed of about 630 km/s would be needed to escape from the planet's gravitational pull. Compress Earth still further, and the escape speed continues to rise. If our hypothetical vise were to squeeze Earth hard enough to crush its radius to about a centimeter, then the speed needed to escape the planet's surface would reach 300,000 km/s. But this is no ordinary speed—it is the speed of light, the fastest speed allowed by the laws of physics as we currently know them.

Thus, if, by some fantastic means, the entire planet Earth could be compressed to less than the size of a grape, the escape speed would exceed the speed of light. However, because nothing can in fact exceed that speed, the compelling conclusion is that *nothing—absolutely nothing—could escape from the surface of such a compressed body.* Even radiation—radio waves, visible light, X rays—indeed, photons of all wavelengths—would be unable to escape the intense gravity of our reshaped Earth. With no photons leaving, our planet would be invisible and uncommunicative—no signal of any sort could be sent to the universe beyond. The origin of the term *black hole* now becomes clear: For all practical purposes, such a supercompact Earth could be said to have disappeared from the universe! Only its gravitational field would remain behind, betraying the presence of its mass, now shrunk to a point.*

The Event Horizon

Astronomers have a special name for the critical radius at which the escape speed from an object would equal the speed of light and within which the object could no longer be seen. It is the **Schwarzschild radius,** after Karl Schwarzschild, the German scientist who first studied its properties. The Schwarzschild radius of any object is simply proportional to the object's mass. For Earth, the Schwarzschild radius is 1 cm; for Jupiter, which is about 300 Earth masses, it is approximately 3 m; for the Sun, with a mass of 300,000 Earth masses, it is 3 km. For a 3-solar-mass stellar core remnant, the Schwarzschild radius is about 9 km. As a convenient rule of thumb, the Schwarzschild radius of an object is simply 3 km, multiplied by the object's mass, measured in solar masses. Every object has a Schwarzschild radius; it is the radius to which the object would have to be compressed for it to become a black hole. Put another way, a black hole is an object that happens to lie within its own Schwarzschild radius.

The surface of an imaginary sphere with radius equal to the Schwarzschild radius and centered on a collapsing star is called the **event horizon.** It defines the region within which no event can ever be seen, heard, or known by anyone outside. Even though there is no matter of any sort associated with it, we can think of the event horizon as the "surface" of a black hole.

A 1.4-solar-mass neutron star has a radius of about 10 km and a Schwarzschild radius of 4.2 km. If we were to

keep increasing the star's mass, the star's Schwarzschild radius would grow, although its actual physical radius would not. In fact, the radius of a neutron star *decreases* slightly with increasing mass. By the time the neutron star's mass exceeded about 3 solar masses, it would lie just within its own event horizon, and it would collapse of its own accord. It would not stop shrinking at the Schwarzschild radius: The event horizon is not a physical boundary of any kind—just a communications barrier. The remnant would shrink right past the Schwarzschild radius to ever-diminishing size on its way to being crushed to a point.

Thus, provided that at least 3 solar masses of material remain behind after a supernova explosion, the remnant core will collapse catastrophically, diving below the event horizon in less than a second. The core simply "winks out," disappearing and becoming a small dark region from which nothing can escape—a literal black hole in space. Theory indicates that this is the likely fate of stars having more than about 20 to 25 times the mass of the Sun.

22.6 Einstein's Theories of Relativity

By the latter part of the 19th century, physicists were well aware of the special status of the speed of light, c. It was, they knew, the speed at which all electromagnetic waves traveled, and, as best they could tell, it represented an upper limit on the speeds of *all* known particles. Scientists struggled without success to construct a theory of mechanics and radiation in which c was a natural speed limit.

Special Relativity

In 1887, a fundamental experiment carried out by the American physicists A. A. Michelson and E. W. Morley compounded theorists' problems further by demonstrating another important and unique aspect of light: The measured speed of a beam of light is *independent* of the motion of either the observer or the source (see *More Precisely 22-1*). No matter what our motion may be relative to the source of the radiation, we always measure precisely the same value for c: 299,792.458 km/s.

A moment's thought tells us that this is a decidedly nonintuitive statement. For example, if we were traveling in a car moving at 100 km/h and we fired a bullet forward with a speed of 1000 km/h relative to the car, an observer standing at the side of the road would see the bullet pass by at $100 + 1000 = 1100$ km/h, as illustrated in Figure 22.17(a). However, the Michelson–Morley experiment tells us that if we were traveling in a rocket ship at one-tenth the speed of light, $0.1c$, and we shone a searchlight beam ahead of us (Figure 22.17b), an outside observer would measure the speed of the beam not as $1.1c$, as the example of the bullet would suggest, but as c. The rules that apply to particles moving at or near the speed of light are different from those we are used to in everyday life.

In fact, we now know that, regardless of the composition or condition of the object that formed the hole, only three physical properties can be measured from the outside: the hole's mass, charge, and angular momentum. All other information is lost once matter enters the hole. Thus, three numbers alone are required to completely describe a black hole's outward appearance. In this chapter, we will consider only black holes that formed from nonrotating, electrically neutral matter. Such objects are completely specified once their masses are known.

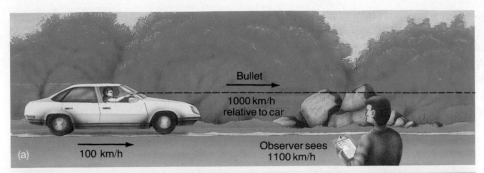

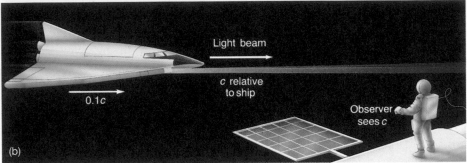

◀ **FIGURE 22.17 Speed of Light**
(a) A bullet fired from a speeding car would be measured by an outside observer to have a speed equal to the sum of the speeds of the car and of the bullet. (b) A beam of light shining forward from a high-speed spacecraft would still be observed to have speed c, regardless of the speed of the spacecraft. The speed of light is thus independent of the speed of the source or of the observer.

The *special theory of relativity* (or just *special relativity*) was proposed by Einstein in 1905 to deal with the preferred status of the speed of light. The theory is the mathematical framework that allows us to extend the familiar laws of physics from low speeds (i.e., speeds much less than c, which are often referred to as *nonrelativistic*) to very high (or *relativistic*) speeds, comparable to c.

The essential features of special relativity are as follows:

1. The speed of light, c, is the maximum possible speed in the universe, and all observers measure the same value for c, regardless of their motion. Einstein broadened this statement into the *principle of relativity*: The basic laws of physics are *the same* to all unaccelerated observers.

2. There is no absolute frame of reference in the universe; that is, there is no "preferred" observer relative to whom all other velocities can be measured. Instead, only relative velocities between observers matter (hence the term "relativity").

3. Neither space nor time can be considered independently of one another. Rather, they are each components of a single entity: *spacetime*. There is no absolute, universal time—observers' clocks tick at different rates, depending on the observers' motions relative to one another.

Special relativity is equivalent to Newtonian mechanics in describing objects that move much more slowly than the speed of light, but it differs greatly in its predictions at relativistic velocities. (See *More Precisely 22-1*.) Yet, despite their often nonintuitive nature, all of the theory's predictions have been repeatedly verified to a high degree of accuracy. Today, special relativity lies at the heart of modern science. No scientist seriously doubts its validity.

General Relativity

Einstein's special theory of relativity is cast in terms of frames of reference ("observers") moving at *constant* speeds with respect to one another. In constructing his theory, Einstein rewrote the laws of motion expounded by Newton more than two centuries previously. ∞ (Sec. 2.7) But Newton's other great legacy—the theory of gravitation—does not deal with observers moving at constant relative velocities. Rather, gravity causes observers to *accelerate* relative to one another, making for a much more complex mathematical problem. Fitting gravity into special relativity took Einstein another decade. The result once again overturned scientists' conception of the universe.

In 1915, Einstein illustrated the connection between special relativity and gravity with the following famous "thought experiment." Imagine that you are enclosed in an elevator with no windows, so that you cannot directly observe the outside world, and the elevator is floating in space. You are weightless. Now suppose that you begin to feel the floor press up against your feet. Weight has apparently returned. There are two possible explanations for this, as shown in Figure 22.18. A large mass could have come nearby, and you are feeling its downward gravitational attraction (Figure 22.18a), *or* the elevator has begun to accelerate upward, and the force you feel is that exerted by the elevator as it accelerates you at the same rate (Figure 22.18b). The crux of Einstein's argument is this: There is *no* experiment that you can perform within the elevator (without looking outside) that will let you distinguish between these two possibilities.

Thus, Einstein reasoned, there is no way to tell the difference between a gravitational field and an accelerated frame of reference (such as the rising elevator in the thought experiment). This statement is known more for-

PHYSLET® ILLUSTRATION Equivalence of Gravitational Field and Acceleration

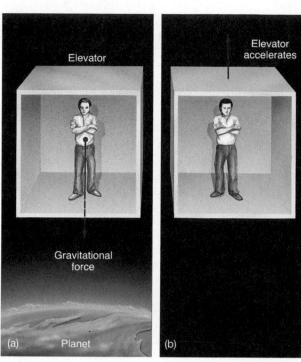

▲ FIGURE 22.18 Einstein's Elevator A passenger in a closed elevator floating in space feels weight return—he feels a force exerted on his feet by the elevator floor. Einstein reasoned that no experiment conducted entirely within the elevator can tell the passenger whether the force is (a) due to the gravity of a nearby massive object or (b) caused by the acceleration of the elevator itself.

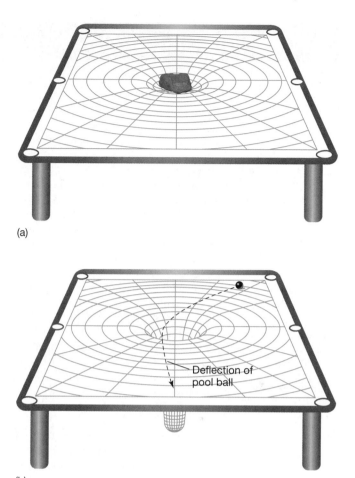

▲ FIGURE 22.19 Curved Space (a) A pool table made of a thin rubber sheet sags when a weight is placed on it. Likewise, space is bent, or warped, in the vicinity of any massive object. (b) As the weight increases, so does the warping of space. A ball shown rolling across the table is deflected by the curvature of the surface, in much the same way as a planet's curved orbit is determined by the curvature of spacetime produced by the Sun.

TUTORIAL SuperSpaceship-Voyage to the Sun

mally as the *equivalence principle*. Using it, Einstein set about incorporating gravity into special relativity as a general acceleration of all particles. However, he found that another major modification to the theory of special relativity had to be made. As we have just seen, a central concept in relativity is the notion that space and time are not separate quantities, but instead must be treated as a single entity known as *spacetime*. To incorporate the effects of gravity, the mathematics forced Einstein to the unavoidable conclusion that spacetime had to be *curved*. The resulting theory, the result of including gravity within the framework of special relativity, is called *general relativity*.

The central concept of general relativity is this: Matter—all matter—tends to "warp" or curve space in its vicinity. Objects such as planets and stars react to this warping by changing their paths. In the Newtonian view of gravity, particles move on curved trajectories because they are acted upon by a gravitational force. ∞ (Sec. 2.7) In Einsteinian relativity, those same particles move on curved trajectories because they are falling freely through space, following the curvature of spacetime produced by some nearby massive object. The more the mass, the greater is the warping. Thus, in general relativity, there is no such thing as a "gravitational force" in the Newtonian sense. Objects move as they do because they follow the curvature of spacetime, which is determined by the amount of matter

present. Stated more loosely, "Spacetime tells matter how to move, and matter tells spacetime how to curve."*

Some props may help you visualize these ideas. Bear in mind, however, that these props are not real, but only tools to help you grasp some exceedingly strange concepts. Imagine a pool table with the tabletop made of a thin rubber sheet rather than the usual hard felt. As Figure 22.19 suggests, such a rubber sheet becomes distorted when a heavy weight (e.g., a rock) is placed on it. The heavier the rock (Figure 22.19b), the larger is the distortion.

Trying to play pool on this table, you would quickly find that balls passing near the rock were deflected by the curvature of the tabletop (Figure 22.19b). The pool balls are not attracted to the rock in any way; rather, they respond to the curvature of the sheet produced by the rock's

*This characterization of gravity is due to the renowned relativist John Archibald Wheeler.

MORE PRECISELY 22-1

Special Relativity

The early 20th century was a tumultuous time in science. In a period of less than 20 years, the conventional views of matter and radiation were overturned by new experimental data and the emerging theory of quantum mechanics. ∞ (Sec. 4.2) Albert Einstein was a key player in this revolution. ∞ (*Discovery 4-1*) But despite his having won a Nobel prize for his pivotal work on the photoelectric effect, Einstein is probably best known for his **theories of relativity**—the successors to Newtonian mechanics that, together with quantum theory, form the foundation of modern physics.

As discussed in the text, Einstein's theory of special relativity solved the puzzle of the 1887 Michelson–Morley experiment, which had demonstrated that the observed speed of light is *independent* of the observer's motion through space. By measuring the speed of light at different times of the day (so that the orientation of their equipment would change as Earth rotated) and on different days of the year (so that Earth's velocity would vary as our planet orbited the Sun), Michelson and Morley attempted to determine Earth's motion relative to the "absolute" space through which light supposedly moved. But the experiment—combined with Einstein's theory explaining its results—had just the opposite effect.

As illustrated in the accompanying figure, Michelson and Morley expected to measure a faster moving beam of light when it was aligned with Earth's motion (to the left in the figure) and a slower beam when it opposed Earth's motion (to the right). But they did not; in fact, they measured precisely the same velocity of light for any orientation of their appara-

tus. This meant either that Earth was not moving through space—which conflicts with the fact that we detect stellar parallax—or that Newtonian thinking and human intuition go awry when light is involved. The upshot was that, instead of measuring the properties of absolute space, the Michelson–Morley experiment ultimately demolished that concept and, with it, the 19th-century view of the universe.

With the special theory of relativity, Einstein elevated the speed of light to the status of a constant of nature, rewrote the laws of mechanics to reflect that new fact, and opened the door to a flood of new physics and a much deeper understanding of the universe. But many commonsense ideas had to be abandoned in the process and replaced with some decidedly less intuitive concepts. We describe here some of those odd consequences of Einstein's theory.

Imagine that you are an observer watching a rocket ship fly past at relative velocity v and that the craft is close enough for you to make detailed observations inside its cabin. If v is much less than the speed of light, c, you would see nothing out of the ordinary—special relativity is consistent with familiar Newtonian mechanics at low velocities. However, if v is comparable to c, unexpected things start to happen.

As the ship's velocity increases, you begin to notice that the ship appears to *contract* in the direction in which it is moving. A meterstick on board, identical at launch to the one you keep in your laboratory, is now shorter than its twin. This is called *Lorentz contraction* (or sometimes Lorentz-Fitzgerald contraction). The graph shows the stick's measured length aboard the moving ship: At low velocities (bottom) the meterstick measures a meter, or close to it; at high velocities (top), the stick is shortened to considerably less than a meter. Because of Lorentz contraction, a meterstick moving at 90 percent of the velocity of light would shrink to a little less than half a meter. (This is not an optical illusion.)

At the same time, the ship's clock, synchronized prior to launch with your own, now ticks *more slowly*. This phenomenon, known as *time dilation*, has been observed many times in laboratory experiments in which fast-moving radioactive particles are observed to decay more slowly than if they were at rest in the lab (see Example 2 on the facing page). Their internal clocks—their half-lives—are slowed by their rapid motion. ∞ (*More Precisely 7-2*) Although no material particle can actually reach the speed of light, Einstein's theory implies that, as v approaches c, the measured length of the meterstick will shrink to nearly zero and the clock will slow to a virtual stop.

Of course, from the point of view of an astronaut on board the spaceship, *you* are the one moving rapidly. Hence, as seen from the ship, *you* appear to be compressed in the direction of motion, and *your* clock runs slowly! How can both observations be correct? After all, you both know that nothing has happened to your metersticks or clocks. You feel no force squeezing your bodies. What is going on?

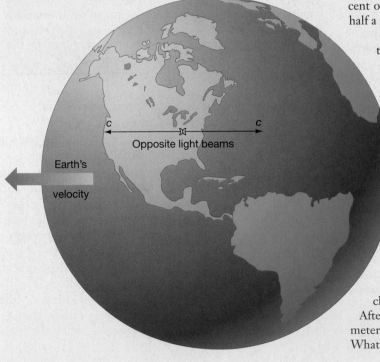

Earth's velocity

c ⋈ c

Opposite light beams

When measuring the length of the moving meterstick, you do so by noting the positions of the two ends *at the same time, according to your clock*. But those two events—the two measurements you make—do *not* occur at the same time as seen by the astronaut on the spaceship. In relativity, time is relative, and simultaneity (the idea that two events happen "at the same time") is no longer a-well defined concept. From the astronaut's viewpoint, your measurement of the leading end of the meterstick occurs *before* your measurement of the trailing end. This time discrepancy ultimately results in the Lorentz contraction you observe. A similar argument applies to measurements of time, such as the period between two clock ticks. Time dilation occurs because the measurements occur at the same location and different times in one frame, but at *different places and times* in the other.

Einstein's ideas were revolutionary at the time, requiring physicists to abandon some long-held, cherished, and "obvious" facts about the universe. Initially, they encountered opposition among his more conservative colleagues. Reading the above explanation, you can perhaps begin to understand why special relativity initially encountered such resistance among some scientists (and why it still confuses nonscientists today!). Nevertheless, the gain in scientific understanding ultimately overcame the price in unfamiliarity. Within just a few years, special relativity had become almost universally accepted (and Einstein was well on his way to becoming the best-known scientist on the planet).

Additional careful experimentation would reveal that the *mass* of the rocket ship—the quantity *m* that appears in Newton's second law of motion—also rises as the ship's relative velocity increases and theoretically becomes nearly infinitely large as the speed of the ship approaches that of light. ∞ (Sec. 2.7) Thus, objects become harder to accelerate as their speeds increase. Finally, perhaps the best-known prediction of special relativity is that the rocket ship's *energy* and mass are directly proportional to one another, connected via the famous equation $E = mc^2$. This is one way to see why objects with any mass can never actually reach the speed of light—an infinite amount of energy would be required to get there!

Again, Einstein's formula is consistent with Newtonian mechanics for velocities much less than that of light, but diverges radically at speeds near *c*. It also has the important consequence that, even at zero velocity, a particle has *rest-mass energy* by virtue of its mass alone. As we have seen, the equivalence of mass and energy is the key to nuclear energy and, hence, to our understanding of how stars live and die. ∞ (Sec. 16.2) (In less productive form, as the energy source for hydrogen bombs, it may yet have similar relevance to the fate of the human race!)

EXAMPLE 1: The effects of Lorentz contraction, time dilation, and mass increase are all governed by the same quantity, usually called the *Lorentz factor* and denoted by the Greek letter γ (gamma). For a particle moving with velocity *v*, this factor is given simply by

$$\gamma = \frac{1}{\sqrt{1 - v^2/c^2}}$$

As *v* increases, lengths are reduced, clocks slow down, and masses are increased by the factor γ (which is always greater than or equal to unity). For example, Earth, orbiting the Sun at 30 km/s, has $v = 0.0001c$, and $\gamma = 1/\sqrt{0.99999999} = 1.000000005$, so relativistic effects are negligibly small. However, for a charged particle in a pulsar wind (Section 22.2) traveling at 90 percent of the speed of light, $v/c = 0.9$, and now $\gamma = 1/\sqrt{0.19} = 2.3$. Note that speeds close to that of light are needed before the effect becomes large; even at one-tenth the speed of light, 30,000 km/s, γ is still only 1.005, so relativistic effects are barely noticeable.

EXAMPLE 2: A muon is an elementary particle closely related to the electron (see More Precisely 27-1). Muons are unstable, and are known to decay (in fact, into an electron and two neutrinos) in 2.0 μs, as measured in their "rest frame" (that is, by an observer relative to whom they are at rest). However, muons produced in experiments using high-energy particle accelerators, or when energetic particles called cosmic rays strike Earth's upper atmosphere, are usually traveling at relativistic speeds. As a result their lifetimes, measured in the terrestrial (laboratory) frame of reference, can be greatly increased by the effects of relativity. The amount of time dilation is determined by the factor γ given above. A striking example of this occurs in the case of muons created by cosmic rays. Without time dilation, they would travel only a few hundred meters from their formation site high in the atmosphere. However, moving at speeds of more than 99.9 percent the speed of light, some of these particles have γ factors of $1/\sqrt{1 - 0.999^2} = 22$ or more, increasing their lifetimes to 2.0 μs $\times$ 22 = 44 μs, long enough for them to reach Earth's surface and be detected before they decay.

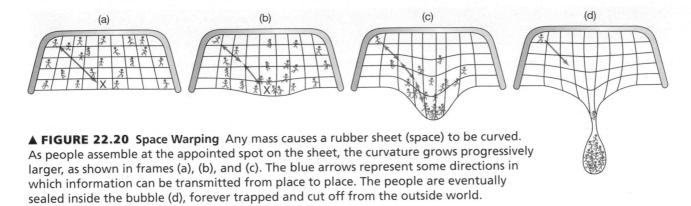

▲ **FIGURE 22.20 Space Warping** Any mass causes a rubber sheet (space) to be curved. As people assemble at the appointed spot on the sheet, the curvature grows progressively larger, as shown in frames (a), (b), and (c). The blue arrows represent some directions in which information can be transmitted from place to place. The people are eventually sealed inside the bubble (d), forever trapped and cut off from the outside world.

presence. In much the same way, anything that moves through space—matter *or radiation*—is deflected by the curvature of spacetime near a star. For example, Earth's orbital path is the trajectory that results as our planet falls freely in the relatively gentle curvature of space created by our Sun. When the curvature is small (i.e., gravity is weak), both Einstein and Newton predict the same orbit—the one we observe. However, as the gravitating mass increases, the two theories begin to diverge.

Curved Space and Black Holes

Modern notions about black holes rest squarely on the general theory of relativity. Although white dwarfs and (to a lesser extent) neutron stars can be adequately described by the classical Newtonian theory of gravity, only the modern Einsteinian theory of relativity can properly account for the bizarre physical properties of black holes.

In Figure 22.19, we saw how the distortion of space (the rubber sheet in our analogy) increases as the mass of the object causing the distortion increases. In these terms, a *black hole* is a region of space where the gravitational field becomes overwhelming and the curvature of space extreme. At the event horizon itself, the curvature is so great that space "folds over" on itself, causing objects within to become trapped and disappear.

Let's consider another analogy. Imagine a large extended family of people living on a huge rubber sheet—a sort of gigantic trampoline. Deciding to hold a reunion, they converge on a given place at a given time. As shown in Figure 22.20, one person remains behind, not wishing to attend. She keeps in touch with her relatives by means of "message balls" rolled out to her (and back from her) along the surface of the sheet. These message balls are the analog of radiation carrying information through space.

As the people converge, the rubber sheet sags more and more. Their accumulating mass creates an increasing amount of space curvature. The message balls can still reach the lone person far away in nearly flat space, but they arrive less frequently as the sheet becomes more and more warped and stretched—as shown in Figures 22.20(b) and (c)— and the balls have to climb out of a deeper and deep-

er well. Finally, when enough people have arrived at the appointed spot, the mass becomes too great for the rubber to support them. As illustrated in Figure 22.20(d), the sheet pinches off into a "bubble," compressing the people into oblivion and severing their communications with the lone survivor outside. This final stage represents the formation of an event horizon around the reunion party.

Right up to the end—the pinching off of the bubble— two-way communication is possible. Message balls can reach the outside from within (but at a slower and slower rate as the rubber stretches), and messages from outside can get in without difficulty. However, once the event horizon (the bubble) forms, balls from the outside can still fall in, but they can no longer be sent back out to the person left behind, no matter how fast they are rolled. They cannot make it past the "lip" of the bubble in Figure 22.20(d). This analogy (very) roughly depicts how a black hole warps space completely around on itself, isolating its interior from the rest of the universe. The essential ideas—the slowing down and eventual cessation of outward-going signals and the one-way nature of the event horizon once it forms—all have parallels in the case of stellar black holes.

CONCEPT CHECK

✔ How do Newton's and Einstein's theories differ in their descriptions of gravity?

22.7 Space Travel Near Black Holes

Black holes are *not* cosmic vacuum cleaners. They don't cruise around interstellar space, sucking up everything in sight. The orbit of an object near a black hole is basically the same as its orbit near a star of the same mass. Only if the object happens to pass within a few Schwarzschild radii (perhaps 50 or 100 km for a typical 5–10-solar-mass black hole formed in a supernova) of the event horizon is there any significant difference between its actual orbit and the one predicted by Newtonian gravity and described by Kepler's laws. Of course, if some matter does happen to fall into a black hole—if the object's orbit happens to take it too close to the event horizon—it will be unable to get out. Black holes are like turnstiles, permitting matter to flow in only one direction: inward.

Because a black hole will accrete at least a little material from its surroundings, its mass, and hence also the radius of its event horizon, tends to increase slowly over time.

Tidal Forces

Matter flowing into a black hole is subject to great tidal stress. An unfortunate person falling feet first into a solar-mass black hole would find himself stretched enormously in height and squeezed unmercifully laterally. He would be torn apart even before he reached the event horizon, for the pull of gravity would be much stronger at his feet (which are closer to the hole) than at his head. The tidal forces at work in and near a black hole are the same basic phenomenon that is responsible for ocean tides on Earth and the spectacular volcanoes on Io. The only difference is that the tidal forces near a black hole are far stronger than any other force we know in the solar system.

As illustrated (with some artistic license) in Figure 22.21, a similar fate awaits any kind of matter falling into a black hole. Whatever falls in—gas, people, space probes—is vertically stretched and horizontally squeezed—and accelerated to high speeds in the process. The net result of

▲ **FIGURE 22.21 Black-Hole Heating** Any matter falling into the clutches of a black hole will become severely distorted and heated. This sketch shows an imaginary planet being pulled apart by a black hole's gravitational tides.

all this stretching and squeezing is numerous and violent collisions among the torn-up debris, causing a great deal of frictional heating of the infalling matter. Material is simultaneously torn apart and heated to high temperatures as it plunges into the hole.

So efficient is the heating, that, before reaching the hole's event horizon, matter falling into the hole emits radiation of its own accord. For a black hole of mass comparable to the Sun, the energy is expected to be emitted in the form of X rays. In effect, the gravitational energy of matter outside the black hole is converted into heat as that matter falls toward the hole. Thus, contrary to what we might expect from an object whose defining property is that nothing can escape from it, the region surrounding a black hole is expected to be a *source* of energy. Of course, once the hot matter falls below the event horizon, its radiation is no longer detectable—it never leaves the hole.

Approaching the Event Horizon

One safe way to study a black hole would be to go into orbit around it well beyond the disruptive influence of the hole's strong tidal forces. After all, Earth and the other planets of our solar system orbit the Sun without falling into it and without being torn apart. The gravity field around a black hole is basically no different; however, even from a stable circular orbit, a close investigation of the hole would be unsafe for humans. Endurance tests conducted on astronauts of the United States and the former Soviet Union indicate that the human body cannot withstand stress greater than about 10–20 times the pull of gravity on Earth's surface. This breaking point would occur about 3000 km from a 10-solar-mass black hole (which, recall, would have a 30-km event horizon). Closer than that, the tidal effect of the hole would tear a human body apart.

Let's instead send an imaginary indestructible astronaut—a mechanical robot, say—in a probe toward the center of the hole. Watching from a safe distance in our orbiting spacecraft, we can then examine the nature of space and time near the hole. Our robot will be a useful explorer of theoretical ideas, at least down to the event horizon. After that boundary is crossed, there is no way for the robot to return any information about its findings.

Suppose, for example, our robot has an accurate clock and a light source of known frequency mounted on it. From our safe vantage point far outside the event horizon, we could use telescopes to read the clock and measure the frequency of the light we receive. What might we discover? We would find that the light from the robot would become more and more redshifted as the robot neared the event horizon. Even if the robot used rocket engines to remain motionless, a redshift would still be detected. The redshift is not caused by motion of the light source, nor is it the result of the Doppler effect arising as the robot falls into the hole. Rather, it is a redshift induced by the black hole's gravitational field, predicted by Einstein's general theory of relativity and known as **gravitational redshift.**

We can explain gravitational redshift as follows: According to general relativity, photons are attracted by gravity. As a result, in order to escape from a source of gravity, photons must expend some energy. They have to do work to get out of the gravitational field. They don't slow down at all—photons always move at the speed of light—they just lose energy. Because a photon's energy is proportional to the frequency of its radiation, light that loses energy must have its frequency reduced (or, equivalently, its wavelength lengthened). In other words, as illustrated in Figure 22.22, radiation coming from the vicinity of a massive object will be redshifted to a degree depending on the strength of the object's gravitational field.

As photons traveled from the robot's light source to the orbiting spacecraft, they would become gravitationally redshifted. From our standpoint on the orbiting spacecraft, a green light, say, would become yellow and then red as the robot astronaut neared the black hole. From the robot's perspective, the light would remain green. As the

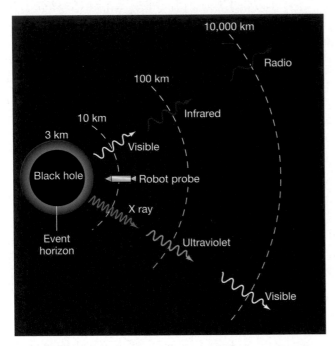

▲ **FIGURE 22.22 Gravitational Redshift** As a photon escapes from the strong gravitational field close to a black hole, it must expend energy to overcome the hole's gravity. This energy does not come from a change in the speed at which the photon travels. (That speed is always 300,000 km/s, even under these extreme conditions.) Rather, the photon "gives up" energy by increasing its wavelength. Thus, the photon's frequency changes, and the photon is (red)shifted into a less energetic region of the spectrum. This figure shows the effect on two beams of radiation, one of visible light and one of X rays, emitted from a space probe as it nears the event horizon of a 1-solar-mass black hole. (Note that the light is *not* coming from within the black hole itself.) The beams are shifted to longer and longer wavelengths as they move farther from the event horizon.

robot got closer to the event horizon, the radiation from its light source would become undetectable with optical telescopes. The radiation reaching us in the orbiting spacecraft would by then be lengthened so much that infrared and then radio telescopes would be needed to detect it. When the robot probe got closer still to the event horizon, the radiation it emitted as visible light would be shifted to wavelengths even longer than conventional radio waves by the time it reached us.

Light emitted *from the event horizon itself* would be gravitationally redshifted to infinitely long wavelengths. In other words, each photon would use all its energy trying to escape from the edge of the hole. What was once light (on the robot) would have no energy left upon its arrival at the safely orbiting spacecraft. Theoretically, this radiation would reach us—still moving at the speed of light—but with zero energy. Thus, the light radiation originally emitted would be redshifted beyond our perception.

Now, what about the robot's clock? Assuming that we could read it, what time would it tell? Would there be any observable change in the rate at which the clock ticked as it moved deeper into the hole's gravitational field? From the safely orbiting spacecraft, we would find that any clock close to the hole would appear to tick more *slowly* than an equivalent clock on board the spacecraft. The closer the clock came to the hole, the slower it would appear to run. On reaching the event horizon, the clock would seem to stop altogether. It would be as if the robot astronaut had found immortality! All action would become virtually frozen in time. Consequently, an external observer would never actually witness an infalling astronaut sink below the event horizon. Such a process would appear to take forever.

This apparent slowing down of the robot's clock is known as **time dilation.** It is another clear prediction of general relativity and in fact is closely related to the gravitational redshift. To see this connection, imagine that we use our light source as a clock, with the passage of (say) a wave crest constituting a "tick." The clock thus ticks at the frequency of the radiation. As the wave is redshifted, the frequency drops, and fewer wave crests pass the distant observer each second—the clock appears to slow down. This thought experiment demonstrates that the redshift of the radiation and the slowing of the clock are one and the same.

From the point of view of the indestructible robot, however, relativity theory predicts no strange effects at all. To the infalling robot, the light source hasn't reddened, and the clock keeps perfect time. In the robot's frame of reference, everything is normal. Nothing prohibits it from coming within the Schwarzschild radius of the hole. No law of physics constrains an object from passing through an event horizon. There is no barrier at the event horizon and no sudden lurch as it is crossed; it is only an imaginary boundary in space. Travelers passing through the event horizon of a sufficiently massive hole (such as might lurk in the heart of our own Galaxy, as we will see) might not even know it—at least until they tried to get out!

The gravitational fields of most astronomical objects are far too weak to produce any significant gravitational redshift, although in many cases the effect can still be measured. Delicate laboratory experiments on Earth and on satellites in near-Earth orbits have succeeded in detecting the tiny gravitational redshift produced by even our own planet's weak gravity. Sunlight is redshifted by only about a thousandth of a nanometer. A few white-dwarf stars do show some significant gravitational reddening of their emitted light, however. Their radii are much smaller than that of our Sun, so their surface gravity is very much stronger than the Sun's. Neutron stars should show a substantial shift in their radiation, but it is difficult to disentangle the effects of gravity, magnetism, and environment on the signals we observe.

Deep Down Inside

No doubt you are wondering what lies within the event horizon of a black hole. The answer is simple: No one really knows. However, the question is of great interest to theorists, as it raises some fundamental issues that lie at the forefront of modern physics.

Can an entire star simply shrink to a point and vanish? General relativity predicts that, without some agent to compete with gravity, the core remnant of a high-mass star will collapse all the way to a point at which both its density and its gravitational field become infinite. Such a point is called a **singularity.** We should not take this prediction of infinite density too literally, however. Singularities are not physical—rather, they always signal the breakdown of the theory producing them. In other words, the present laws of physics are simply inadequate to describe the final moments of a star's collapse.

As it stands today, the theory of gravity is incomplete, because it does not incorporate a proper (i.e., a quantum mechanical) description of matter on very small scales. As our collapsing stellar core shrinks to smaller and smaller radii, we eventually lose our ability even to describe, let alone predict, its behavior. Perhaps matter trapped in a black hole never actually reaches a singularity. Perhaps it just approaches this bizarre state in a manner that we will someday understand as the subject of *quantum gravity*— the merger of general relativity with quantum mechanics—develops.

Having said that, we can at least estimate how small the core can get before current theory fails. It turns out that by the time that stage is reached, the core is already much smaller than any elementary particle. Thus, although a complete description of the end point of stellar collapse may well require a major overhaul of the laws of physics, for all practical purposes the prediction of collapse to a point is valid. Even if a new theory somehow succeeds

in doing away with the central singularity, it is unlikely that the external appearance of the hole or the existence of its event horizon will change. Any modifications to general relativity are expected to occur only on submicroscopic scales, not on the macroscopic (kilometer-sized) scale of the Schwarzschild radius.

Singularities are places where the rules break down, and some very strange things may occur near them. Many possibilities have been envisaged—gateways into other universes, time travel, the creation of new states of matter—but none of them has been proved, and certainly none of them has ever been observed. Because these regions are places where science fails, their presence causes serious problems for many of our cherished laws of physics, from causality (the idea that cause should precede effect, which runs into immediate and severe problems if time travel is possible) to energy conservation (which is violated if material can hop from one universe to another through a black hole). It is currently unclear whether the removal of the central singularity by some future all-encompassing theory would necessarily also eliminate all of these problematic side effects.

Disturbed by the possibility of such chaos in science, some researchers have even proposed a "principle of cosmic censorship": Nature always hides *any* singularity, such as that found at the center of a black hole, inside an event horizon. In that case, even though physics fails, its breakdown cannot affect us outside, so we are safely insulated from any effects the singularity may have. What would happen if we one day found a so-called *naked singularity* somewhere—a singularity uncloaked by an event horizon? Would relativity theory still hold there? For now, we just don't know.

What sense are we to make of black holes? Do black holes and all the strange phenomena that occur in and around them really exist? The basis for understanding these weird objects is the relativistic concept that mass warps spacetime—which has already been found to be a good representation of reality, at least for the weak gravitational fields produced by stars and planets (see *More Precisely 22-2* and *Discovery 22-1*). The larger the concentration of mass, the greater is the spacetime warping, and, apparently, the stranger are the observational consequences. These consequences are part and parcel of general relativity, and black holes are one of its most striking predictions. As long as general relativity stands as the correct theory of gravity in the universe, black holes are real.

CONCEPT CHECK

✔ Why would you never actually witness an infalling object crossing the event horizon of a black hole?

22.8 Observational Evidence for Black Holes

So, theoretical ideas aside, is there any observational evidence for black holes? Can we prove that these strange invisible objects really do exist?

Stellar Transits?

One way in which we might think we would detect a black hole is if we observed it transiting (passing in front of) a star. Unfortunately, such an event would be extremely hard to see. The 12,000-km-diameter planet Venus is barely noticeable when it transits the Sun, so a 10-km-wide object moving across the image of a faraway star would be completely invisible with either current equipment or any equipment available in the foreseeable future.

Actually, we are even worse off than the previous paragraph suggests. Suppose we were close enough to the star to resolve the disk of the transiting black hole. Then the observable effect would not be a black dot superimposed on a bright background. Instead, the background starlight would be deflected as it passed the black hole on its way to Earth, as indicated in Figure 22.23. The effect is the same as the bending of distant starlight around the edge of the Sun, a phenomenon that has been repeatedly measured during solar eclipses throughout the last several decades (see *More Precisely 22-2*). With a black hole, much larger deflections would occur. As a result, the image of a black hole in front of a bright companion star would show, not a

---

MORE PRECISELY 22-2

Tests of General Relativity

Special relativity is the most thoroughly tested and most accurately verified theory in the history of science. General relativity, however, is on somewhat less firm experimental ground.

The problem with verifying general relativity is that its effects on Earth and in the solar system—the places where we can most easily perform tests—are very small. Just as special relativity produces major departures from Newtonian mechanics only when velocities approach the speed of light, general relativity predicts large departures from Newtonian gravity only when extremely *strong* gravitational fields are involved—in effect, when orbital speeds and escape velocities become relativistic.

We will encounter other experimental and observational tests of general relativity elsewhere in this chapter. (See *Discovery 22-1*.) Here, we consider just two "classical" tests of the theory. These tests are solar-system observations that helped ensure the acceptance of Einstein's theory. Later, more accurate measurements confirmed and strengthened the test results. Bear in mind, however, that there are no known tests of general relativity in the "strong-field" regime—that part of the theory which predicts black holes, for example—so the full theory has never been experimentally tested.

At the heart of general relativity is the premise that everything, including light, is affected by gravity because of the curvature of spacetime. Shortly after he published his theory in 1915, Einstein noted that light from a star should be deflected by a measurable amount as it passes the Sun. According to the theory, the closer to the Sun the light comes, the more it is deflected. Thus, the maximum deflection should occur for a ray that just grazes the solar surface. Einstein calculated that the deflection angle should be 1.75″— a small, but detectable, amount. Of course, it is normally impossible to see stars close to the Sun. However, during a solar eclipse, when the Moon blocks the Sun's light, the observation becomes possible, as illustrated in the first (highly exaggerated) figure.

In 1919, a team of observers led by the British astronomer Sir Arthur Eddington succeeded in measuring the deflection of starlight during an eclipse. The results were in

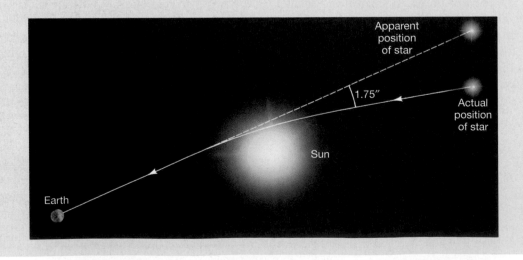

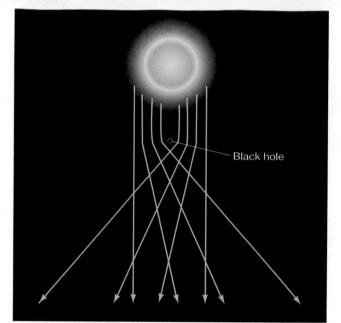

Black hole

◀ **FIGURE 22.23 Gravitational Light Deflection** The gravitational bending of light around the edges of a small, massive black hole makes it impossible to observe the hole as a black dot superimposed against the bright background of its stellar companion.

neat, well-defined black dot, but rather, a fuzzy image virtually impossible to resolve, even from nearby.

Black Holes in Binary Systems

A much better way to find black holes is to look for their effects on other objects. Our Galaxy harbors many binary-star systems in which only one object can be seen. Recall from our study of binary-star systems in Section 17.9 that

excellent agreement with the prediction of general relativity. Virtually overnight Einstein became world famous. His previous major accomplishments notwithstanding, this single prediction assured him a permanent position as the best-known scientist on Earth! Recently, the high-precision *Hipparcos* satellite ∞ (*Discovery 17-1*) has observed shifts in the apparent positions of many stars, even those whose line of sight is far from the Sun. The shifts are exactly as predicted by Einstein's theory.

EXAMPLE: According to general relativity, a beam of light passing an object of mass M at distance R is deflected through an angle (in radians) of $4GM/Rc^2$, where $G = 6.67 \times 10^{-11}$ N m^2/kg^2 is the gravitational constant and $c = 3.00 \times 10^8$ m/s is the speed of light. Putting in the numbers for the Sun, we obtain $M = 1.99 \times 10^{30}$ kg and $R = 696,000$ km, and remembering that 1 radian = 57.3°, we can calculate the deflection to be $(4 \times 6.67 \times 10^{-11} \times 1.99 \times 10^{30})/(6.96 \times 10^8 \times [3.00 \times 10^8]^2) \times 57.3$ (degrees per radian) $\times$ 3600 (arc seconds per degree) = 1.75″, as previously stated. ∞ (*More Precisely 1-3*) In more convenient units, we can write

$$\text{deflection (arc seconds)} = 1.75 \frac{M(\text{solar masses})}{R(\text{solar radii})}$$

Note that the deflection is proportional to the mass M and inversely proportional to the distance R. Thus, Earth, with mass $M = 3.0 \times 10^{-6}$ and radius $R = 9.2 \times 10^{-3}$ solar units, would produce a deflection of just 0.57 milliarcsecond (thousandths of an arcsecond), while a white dwarf such as Sirius B, with $M = 1.1$ and $R = 0.0073$ in the same units would deflect the beam by 4.4 arc minutes. ∞ (See. 20.3) (Neutron stars and black holes produce even greater effects, but the preceding simple formula is valid only when the deflection is small—less than a few degrees.)

Another prediction of general relativity is that planetary orbits deviate slightly from the perfect ellipses of Kepler's

laws. Again, the effect is greatest where gravity is strongest—that is, closest to the Sun. Thus, the largest relativistic effects are found in the orbit of Mercury. Relativity predicts that Mercury's orbit is not a closed ellipse. Instead, its orbit should rotate slowly, as shown in the second (again exaggerated) diagram. The amount of rotation is very small—only 43″ per century—but Mercury's orbit is so well charted that even this tiny effect is measurable.

In fact, the observed rotation rate is 574″ per century, much greater than that predicted by relativity. However, when other (nonrelativistic) gravitational influences—primarily the perturbations due to the other planets—are taken into account, the rotation is in complete agreement with the foregoing prediction.

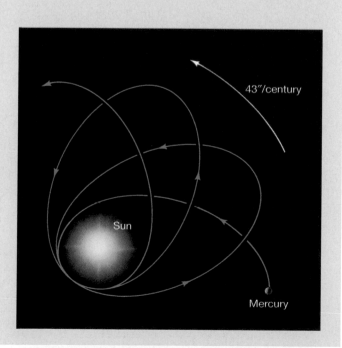

43″/century

Sun

Mercury

we need to observe the motion of only one star to infer the existence of an unseen companion and measure some of its properties. In the majority of cases, the invisible companion is simply small and dim, nothing more than an M-type star hidden in the glare of an O- or B-type partner or perhaps shrouded by dust or other debris, making it invisible to even the best available equipment. In either case, the invisible object is *not* a black hole.

A few close binary systems, however, have peculiarities suggesting that one of their members may be a black hole. Some of the most interesting observations, made during the 1970s and 1980s by Earth-orbiting satellites, revealed binary systems in which the invisible member emits large amounts of X rays. The mass of the emitting object is measured as several solar masses, so we know that it is not simply a small, dim star. Nor is it likely that visible radiation from

the X-ray source is obscured by dusty circumstellar debris—in the cases of interest, intense radiation from the binary components would have dispersed the debris into interstellar space long ago.

One particular binary system drawing much attention lies in the constellation Cygnus. Figure 22.24(a) shows the area of the sky in Cygnus where astronomers have reasonably good evidence for a black hole. The rectangle outlines the celestial system of interest, some 2000 pc from Earth. The black-hole candidate is an X-ray source called Cygnus X-1, studied in detail by the *Uhuru* satellite in the early 1970s. The main observational features of this binary system are as follows:

1. The visible companion of the X-ray source—a blue B-type supergiant with the catalog name HDE 226868—

DISCOVERY 22-1

Gravity Waves: A New Window on the Universe

Electromagnetic waves are common, everyday phenomena. Whether they are radio, infrared, visible, ultraviolet, X-ray, or gamma-ray radiation, all electromagnetic waves involve periodic changes in the strengths of electric and magnetic fields. ∞ (Sec. 3.2) Electromagnetic waves move through space and transport energy. Any accelerating charged particle, such as an electron in a broadcasting antenna or on the surface of a star, generates electromagnetic waves.

The modern theory of gravity—Einstein's theory of relativity (see *More Precisely 22-2*)—also predicts waves that move through space. A *gravity wave* is the gravitational counterpart of an electromagnetic wave. *Gravitational radiation* results from changes in the strength of a gravitational field. In principle, any time an object of any mass accelerates, a gravity wave should be emitted at the speed of light. The passage of a grav-

ity wave should produce small distortions in the space through which it passes. Gravity is an exceedingly weak force compared with electromagnetism, so these distortions are expected to be very small—in fact, much smaller than the diameter of an atomic nucleus for waves that may be produced by any known astrophysical source. Yet many researchers think that these tiny distortions are measurable. No one has yet succeeded in detecting gravity waves, but their detection would provide such strong support for the theory of relativity that scientists are eager to search for them.

Theorists are still debating which kinds of astronomical objects should produce gravity waves detectable on Earth. Leading candidates include (1) close binary systems containing black holes, neutron stars, or white dwarfs, (2) the collapse of a star into a black hole, and (3) supernovae and neutron-star oscillations. Each of these possibilities involves the acceleration of huge masses, resulting in rapidly changing gravitational fields.

Of the candidates just mentioned, the first probably presents the best chance to detect gravity waves, at least for the present. Binary-star systems should emit gravitational radiation as the component stars orbit one another. As energy escapes in the form of gravity waves, the two stars slowly spiral toward one another, orbiting more rapidly and emitting even more gravitational radiation. This runaway situation can lead to the decay and eventual merger of close binary systems in a relatively short time (which, in this case, means tens or hundreds of millions of years, although most of the radiation is emitted during the last few seconds). As we saw in the text (Section 22.3), the merger of such a system could be the origin of gamma-ray bursts, so gravitational radiation may provide an alternative means of studying these violent and mysterious phenomena.

As a matter of fact, a slow but steady decay in the orbit of a binary system has in fact been detected. In 1974, radio astronomer Joseph Taylor and his student Russell Hulse at the University of Massachusetts discovered an unusual binary system. Both components are neutron stars, and one is ob-

Companion star · 250 Million years · Now · Diameter of the Sun

was identified a few years after Cygnus X-1 was discovered. Assuming that the companion lies on the main sequence, we know that its mass must be around 25 times the mass of the Sun.

2. Spectroscopic observations indicate that the binary system has an orbital period of 5.6 days. Combining this information with further spectroscopic measurements of the visible component's orbital speed, astronomers estimate the total mass of the system to be around 35 solar masses, implying that Cygnus X-1 has a mass about 10 times the mass of the Sun. ∞ (Sec. 17.7)

3. Other detailed studies of Doppler-shifted spectral lines suggest that hot gas is flowing from the bright star toward an unseen companion. ∞ (Sec. 4.4)

4. X-ray radiation emitted from the immediate neighborhood of Cygnus X-1 implies the presence of very high temperature gas, perhaps as hot as several million kelvins (see Figure 22.24b).

5. Rapid time variations of this X-ray radiation imply that the size of the X-ray-emitting region of Cygnus X-1 must be extremely small—in fact, less than a few hundred kilometers across. The reasoning is basically the same as in the discussion of gamma-ray bursts in Section 22.4: X rays from Cygnus X-1 have been observed to vary in intensity on time scales as short as a millisecond. For this variation not to be blurred by the travel time of light across the source, Cygnus X-1 cannot be more than 1 light-millisecond, or 300 km, in diameter.

servable from Earth as a pulsar. This system has become known as the *binary pulsar*. Measurements of the periodic Doppler shift of the pulsar's radiation prove that its orbit is slowly shrinking. Furthermore, the rate at which the orbit is shrinking is exactly what would be predicted by relativity theory if the energy were being carried off by gravity waves. The two neutron stars should merge in an energetic burst of gravitational radiation and gamma rays in less than 300 million years. Even though the waves themselves have not been detected, the binary pulsar is regarded by most astronomers as a very strong piece of evidence in favor of general relativity (Sec. 22.6). Taylor and Hulse received the 1993 Nobel prize in physics for their discovery. The first accompanying figure illustrates the scale of, and predicted orbital changes in, the binary pulsar's orbit.

In January 2004, radio astronomers announced the discovery of a *double-pulsar* binary system with an even shorter period than the binary pulsar, implying stronger relativistic effects and a shorter merger time—less than 100 million years. Because both components are pulsars, and since the system, by pure luck, also happens to be seen almost exactly edge on by observers on Earth, leading to eclipses, we can expect this system to provide a wealth of information on both neutron stars and gravitational physics in coming years.

Gravity waves should contain a great deal of information about physical events in some of the most exotic regions of space. In 1992, funding was approved for an ambitious gravity-wave observatory called LIGO—short for Laser Interferometric Gravity-wave Observatory. Twin detectors, one in Hanford, Washington (shown in the second figure) and the other in Livingston, Louisiana, use laser beams to measure the extremely small distortions of space produced by gravitational radiation. The beams are designed to detect the tiny changes (less than one-thousandth the diameter of an atomic nucleus) produced in the lengths of the 4-km-long arms should

a gravity wave pass by. The instrument is (in theory) capable of detecting gravity waves from many Galactic and extragalactic sources. The combination of two detectors allows scientists to determine the *direction* of incoming waves—in other words, LIGO is designed to operate as a rudimentary gravity-wave *telescope*.

The detectors were completed in 1999 and began taking astronomical data in January 2003. So far, they have not detected any gravity waves. However, the system is scheduled to be upgraded and its sensitivity greatly increased in 2007, and much more sensitive space-based instruments literally a million times larger than LIGO are planned to come on-line around 2010. If any of these instruments is successful, the discovery of gravity waves will herald a new age in astronomy, in much the same way that invisible electromagnetic waves, virtually unknown a century ago, revolutionized classical astronomy and led to the field of modern astrophysics.

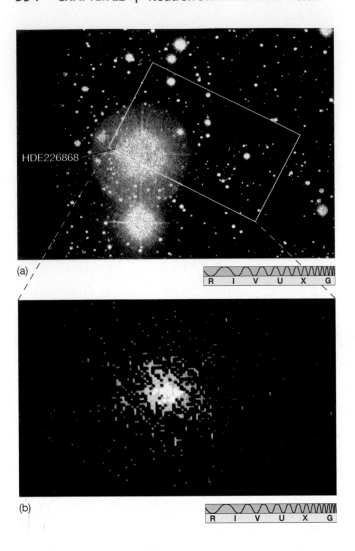

(a)

R I V U X G

(b)

R I V U X G

◀ **FIGURE 22.24 Cygnus X-1** (a) The brightest star in this photograph is a member of a binary system whose unseen companion, called Cygnus X-1, is a leading candidate for a black hole. (b) X rays emitted by the Cygnus X-1 source were analyzed by changing them into electronic signals that were then viewed on a video screen, from which this picture was taken. (The field of view here is outlined by the rectangle in part a.) *(Harvard-Smithsonian Center for Astrophysics)*

These properties suggest that the invisible X-ray-emitting companion could be a black hole. The X-ray-emitting region is likely an accretion disk formed as matter drawn from the visible star spirals down onto the unseen component. The rapid variability of the X-ray emission indicates that the unseen component must be compact—a neutron star or a black hole. The mass limit of the dark component argues for the latter, for a neutron star's mass cannot exceed about 3 solar masses. Figure 22.25 is an artist's conception of this intriguing object. Note that most of the gas drawn from the visible star ends up in a doughnut-shaped accretion disk of matter. As the gas flows toward the black hole, it becomes superheated and emits the X rays we observe just before they are trapped forever below the event horizon.

A few other black-hole candidates are known. For example, the third X-ray source ever discovered in the Large Magellanic Cloud—called LMC X-3—is an invisible object that, like Cygnus X-1, orbits a bright companion star. LMC X-3's visible companion seems to be distorted into the shape of an egg by the unseen object's intense gravitational pull. Reasoning similar to that applied to Cygnus X-1 leads to the conclusion that the compact object LMC X-3 has a mass nearly 10 times that of the Sun, making it too massive to be anything but a black hole. Similarly, the X-ray binary system A0620-00 has been found to contain an invisible compact object of mass 3.8 times the mass of

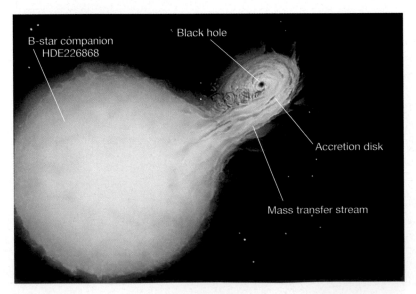

B-star companion
HDE226868

Black hole

Accretion disk

Mass transfer stream

◀ **FIGURE 22.25 Black-Hole Binary** Artist's conception of a binary system containing a large, bright, visible star and an invisible, X-ray-emitting black hole. This painting is based on data obtained from detailed observations of Cygnus X-1. *(L. Chaisson)*

ANIMATION Black Hole Geometry

the Sun. In total, there are perhaps two dozen known objects in or near our Galaxy that may turn out to be black holes, although Cygnus X-1, LMC X-3, and A0620-00 probably have the strongest claims.

Black Holes in Galaxies

Perhaps the strongest evidence for black holes comes not from binary systems in our own Galaxy, but from observations of the centers of many galaxies, including our own. Using high-resolution observations at wavelengths ranging from radio to ultraviolet, astronomers have found that

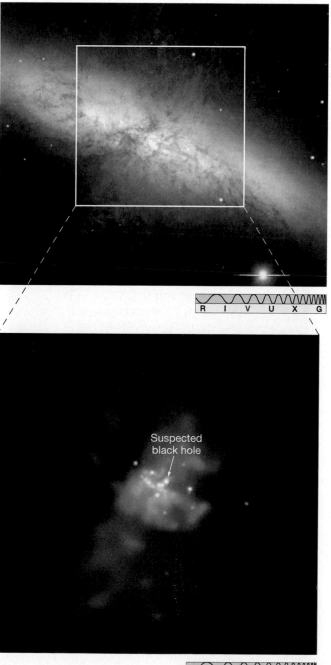

stars and gas near the centers of many galaxies are moving extremely rapidly, orbiting some very massive, unseen object. Masses inferred from Newton's laws range from millions to billions of times the mass of the Sun. ∞ *(More Precisely 2-3)* The intense energy emission from the centers of these galaxies and the short-timescale fluctuations in that emission suggest the presence of massive, compact objects. The leading (and at present, the only) explanation is that these objects are black holes. We will return to the observations, and the question of how such *supermassive black holes* might have formed, in Chapters 23 and 25.

In 2000, X-ray astronomers reported evidence for a long-sought, but elusive, missing link between "stellar-mass" black holes in binaries and the supermassive black holes in the hearts of galaxies. Figure 22.26 shows an unusual looking galaxy called M82, currently the site of an intense and widespread burst of star formation (see Chapter 24.) The inset shows a *Chandra* image of the innermost few thousand parsecs of M82, revealing a number of bright X-ray sources close to—but *not* at—the center of the galaxy. Their spectra and X-ray luminosities strongly suggest that they are accreting compact objects with masses ranging from 100 to almost 1000 times the mass of the Sun. If confirmed, they will be the first *intermediate-mass black holes* ever observed.

Too large to be remnants of normal stars and too small to warrant the "supermassive" label, these objects present a puzzle to astronomers. Supermassive black holes are thought to have formed long ago, and to have played a role in galaxy formation, but these intermediate-mass black holes appear to have formed recently. Where did they come from? One possible origin is suggested by follow-up infrared observations with the Subaru and Keck telescopes on Mauna Kea. These observations reveal that more than half of the X-ray sources are apparently associated with dense, young star clusters. ∞ (Secs. 5.3, 19.6) Theorists speculate that collisions between high-mass main-sequence stars in the congested cores of such clusters could lead to the runaway growth of extremely massive and highly unstable stars, which could then collapse to form intermediate-mass black holes. However, both the observations and this explanation remain controversial, and further systematic studies of many other galaxies will be needed to resolve the issue.

◄ **FIGURE 22.26 Intermediate-Mass Black Holes?** X-ray observations (inset) of the center of the starburst galaxy M82 reveal a collection of bright sources thought to be the result of matter accreting onto intermediate-mass black holes. The black holes are probably young, have masses between 100 and 1000 times the mass of the Sun, and lie relatively far from the center of M82. The brightest (and possibly most massive) intermediate-mass black hole candidate is marked by an arrow. *(Subaru; NASA)*

Do Black Holes Exist?

You may have noticed that the identification of an object as a black hole really proceeds by elimination. Loosely stated, the argument goes as follows: "Object X is compact and very massive. We don't know of anything else that can be that small and that massive. Therefore, object X is a black hole." For the very massive compact objects observed (or inferred to be) in the centers of galaxies, the absence of viable alternatives means that the black-hole hypothesis has become widely accepted among astronomers. However, Cygnus X-1 and the other suspected stellar-mass black holes in binary systems all have masses relatively close to the dividing line separating neutron stars from black holes. Given the present uncertainties in both observation and theory, might they conceivably be merely dim, dense neutron stars and not black holes at all?

Most astronomers do not regard this as a likely possibility, but it highlights a problem. It is difficult to unambiguously distinguish a 10-solar-mass black hole from, say, a 10-solar-mass neutron star (if one could somehow exist). Both objects would affect a companion star's orbit in the same way; both would tear mass from its surface, and both would form an accretion disk around themselves that would emit intense X rays (although some researchers think that the accretion disks may differ sufficiently in their detailed properties that the nature of the central object might be identifiable from observations).

We have stressed throughout this text that scientific theories unsupported by observational or experimental evidence are destined not to survive. ∞ (Sec. 1.2) Earlier, we noted that black holes are a clear prediction of Einstein's general theory of relativity, which is widely regarded as the correct description of gravity in the presence of strong fields and orbital speeds comparable to the speed of light. But we have also seen that general relativity has been tested the most thoroughly in situations where gravity is weak and velocities are relatively low, and not at all under the extreme conditions likely to be encountered near a black hole. So we can legitimately ask, "Do we have any unambiguous evidence that the massive, compact objects just described really are black holes?"

The short answer—at least, if measurements of mass and size alone are insufficient to convince you of a black hole's reality—is no. Detailed measurements of black-hole properties are hard to make and even harder to interpret. However, in 2001, two groups of researchers independently reported evidence for one of the true defining properties of black holes: an event horizon.

First, using instruments on board *Chandra*, X-ray astronomers found that X-ray novae (mass-transferring binary systems whose X-ray emission suddenly and unpredictably flares up, probably due to instabilities in the inner parts of the accretion disk) suspected of harboring black holes emitted far less radiation between outbursts than those in systems thought to contain neutron stars. The explanation of these astronomers is that the accreting

gas in the latter case glows brightly as it collides violently with the surface of the neutron star. In the case of a black hole, the gas simply crosses the event horizon and vanishes.

Second, researchers using the *Hubble Space Telescope* observed rapidly fading pulses of ultraviolet light from another X-ray binary in our Galaxy. They interpret their observations as blobs of hot gas spiraling down from the inner edge of an accretion disk onto the black hole at the center, as sketched in Figure 22.27. In this view, the pulses mark the decaying orbit of a blob, and the absence of a flash at the end of the pulse indicates that, at the end of the plunge, the gas simply vanished rather than hitting a solid surface.

Neither observation is conclusive, and neither finding has been confirmed, but these reports at least illustrate the sorts of measurements needed to identify a compact object as a black hole. As technology continues to improve, we can expect many more such observations, with increasing precision, as astronomers strive to test a key prediction of Einstein's theory.

So have black holes really been discovered? Despite the uncertainties, the answer is probably yes. Skepticism is healthy in science, but only the most stubborn astronomers (and some do exist!) would take serious issue with the many lines of theoretical reasoning that support the case for black holes. The crucial role played by black holes in the theories of stellar evolution, gamma-ray

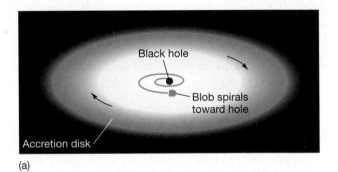

(a)

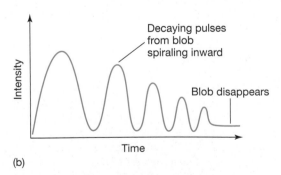

(b)

▲ **FIGURE 22.27 Infalling Matter Observed?** (a) As a blob of gas spirals inward from the inner edge of an accretion disk onto a black hole, the blob's luminosity varies in a characteristic way as its decaying orbit loops around the hole. (b) To an outside observer, the result is a series of pulses of radiation, cut off after just a few orbits as the blob crosses the event horizon and is lost forever.

bursts, and (as we will see in Chapters 24 and 25) the structure and evolution of galaxies is a clear indication of their widespread acceptance in astronomy.

Can we guarantee that future modifications to the theory of compact objects will not invalidate some or all of our arguments? No, but similar statements could be made in many areas of astronomy—indeed, about any theory in any area of science. We conclude that, strange as they are, black holes have been detected, both in our Galaxy and beyond. Perhaps someday, future generations of space travelers will visit Cygnus X-1 or the center of our Galaxy and (carefully!) test these conclusions firsthand. Until then, we will have to continue to rely on improving theoretical models and observational techniques to guide our discussions of the mysterious objects known as black holes.

CONCEPT CHECK

✔ How do astronomers "see" black holes?

Chapter Review

SUMMARY

A core-collapse supernova may leave behind a remnant—an ultracompressed ball of material called a **neutron star (p. 568)**. The processes that form neutron stars ensure that these stars are rapidly rotating and strongly magnetized at birth. **Pulsars (p. 569)** are objects that appear to emit regular bursts of electromagnetic energy. The accepted explanation for our observations of pulsars is the **lighthouse model (p. 569)**, in which a rotating neutron star sends a beam of energy into space. If the beam sweeps past Earth, we see a pulsar. The pulse period is the rotation period of the neutron star. Because the pulse energy is beamed into space and because neutron stars slow down as they radiate energy into space, not all neutron stars are seen as pulsars.

A neutron star in a close binary system can draw matter from its companion, forming an accretion disk. The material in the disk heats up even before it reaches the neutron star, and the disk is usually a strong source of X rays. As gas builds up on the star's surface, the star eventually becomes hot enough to fuse hydrogen. As with a nova explosion on a white dwarf, when hydrogen burning starts on a neutron star, it does so explosively. An **X-ray burster (p. 573)** results. The rapid rotation of the inner part of the accretion disk causes the neutron star to spin faster as new gas arrives on its surface. The eventual result is a very rapidly rotating neutron star—a **millisecond pulsar (p. 574)**. Many millisecond pulsars are found in the hearts of old globular clusters. They cannot have formed recently, so they must have been spun up by interactions with other stars. Careful analysis of the radiation received has shown that some pulsars are orbited by planet-sized objects. The origin of these "pulsar planets" is still uncertain.

Gamma-ray bursts (p. 576) are very energetic flashes of gamma rays observed about once per day, distributed uniformly over the entire sky. In some cases, their distances have been measured, placing them far away from us and implying that they are extremely luminous. The leading theoretical models for these explosions postulate the violent merger of neutron stars in a distant binary system, or the recollapse and subsequent violent explosion following a "failed" supernova in a very massive star.

The upper limit on the mass of a neutron star is about 3 solar masses. Beyond that mass, the star can no longer support itself against its own gravity, and it must collapse. No known force can prevent the material from collapsing all the way to a pointlike **singularity (p. 589)**—a region of extremely high density where the known laws of physics break down. Surrounding the singularity, at a distance of a few kilometers for a solar-mass object, is a region of space from which even light cannot escape: a **black hole (p. 580)**. Astronomers think that the most massive stars form black holes, rather than neutron stars, after they explode in a supernova.

Conditions in and near black holes cannot be described by Newtonian mechanics. A proper description involves the **theories of relativity (p. 584)** developed by Albert Einstein early in the 20th century. Even relativity theory fails right at the singularity, however. Relativity describes gravity in terms of a warping, or bending, of space by the presence of mass. The more mass, the greater is the warping. All particles—including photons—respond to that warping by moving along curved paths. A black hole is a region where the warping is so great that space folds back on itself, cutting off the interior of the hole from the rest of the universe.

The "surface" of a black hole is the **event horizon (p. 581)**; its distance from a singularity is called the **Schwarzschild radius (p. 581)**. At the event horizon, the escape speed equals the speed of light. Within this distance, nothing can escape. Photons passing too close to a black hole are deflected onto paths that cross the event horizon and become trapped. To a distant observer, the clock on a spaceship falling into a black hole would show **time dilation (p. 588)**—the clock would appear to slow down as the ship approached the event horizon. The observer would never see the ship reach the surface of the hole. At the same time, light leaving the ship would be subject to **gravitational redshift (p. 587)** as it climbed out of the hole's intense gravitational field. Light emitted just at the event horizon would be redshifted to infinite wavelength. Both phenomena are predictions of the theory of relativity. The gravitational redshifts due to both Earth and the Sun are very small, but have been detected experimentally.

Once matter falls into a black hole, it can no longer communicate with the outside. However, on its way in, it can form an accretion disk and emit X rays, just as in the case of a neutron star. The best candidates for black holes are binary systems in which one component is a compact X-ray source. Cygnus X-1, a well-studied X-ray source in the constellation Cygnus, is a long-standing black-hole candidate. There is also substantial evidence for more massive black holes residing in or near the centers of many galaxies, including our own.

REVIEW AND DISCUSSION

1. How does the way in which a neutron star forms determine some of its most basic properties?

2. What would happen to a person standing on the surface of a neutron star?

3. Why aren't all neutron stars seen as pulsars?

4. Why do many neutron stars move at high speeds relative to their neighbors?

5. What are X-ray bursters?

6. What is the favored explanation for the rapid spin rates of millisecond matter?

7. Why do you think astronomers were surprised to find a pulsar with a planetary system?

8. Why do astronomers think that gamma-ray bursts are very distant and very energetic?

9. Describe two leading models for gamma-ray bursts.

10. What does it mean to say that the measured speed of a light beam is independent of the motion of the observer?

11. Use your knowledge of escape speed to explain why black holes are said to be "black."

12. According to special relativity, what is special about the speed of light?

13. Why is it so difficult to test the predictions of general relativity? Describe two tests of the theory.

14. What would happen to someone falling into a black hole?

15. What is an event horizon?

16. What is the principle of cosmic censorship? Do you think it is a sound scientific principle?

17. What makes Cygnus X-1 a good black-hole candidate?

18. What evidence is there for black holes much more massive than the Sun?

19. Imagine that you had the ability to travel at will through the Galaxy. Explain why you would discover many more neutron stars than those known to observers on Earth. Where would you be most likely to find these objects?

20. Do you think that planet-sized objects discovered in orbit around a pulsar should be called planets? Why or why not?

CONCEPTUAL SELF-TEST: TRUE OR FALSE/MULTIPLE CHOICE

1. The density of a neutron star is comparable to the density of an atomic nucleus.

2. All millisecond pulsars are now, or once were, members of binary-star systems.

3. The fact that gamma-ray bursts are so distant means that they must be very energetic events.

4. Nothing can travel faster than the speed of light.

5. All things, except light, are attracted by gravity.

6. According to general relativity, space is warped, or curved, by matter.

7. Although visible light cannot escape from a black hole, high-energy radiation, like gamma rays, can escape.

8. If you could touch it, the surface of a black hole—the event horizon—would be very hard.

9. The present laws of physics break down near the center of a black hole.

10. Thousands of black holes have now been identified in our Galaxy.

11. A neutron star is about the same size as (a) a school bus; (b) a U.S. city; (c) the Moon; (d) Earth.

12. A neutron's star immense gravitational attraction is due primarily to its small radius and (a) rapid rotation rate; (b) strong magnetic field; (c) large mass; (d) high temperature .

13. The most rapidly "blinking" pulsars are those which (a) spin fastest; (b) are oldest; (c) are most massive; (d) are hottest.

14. The X-ray emission from a neutron star in a binary system comes mainly from (a) the hot surface of the neutron star itself; (b) heated material in an accretion disk around the neutron star; (c) the neutron star's magnetic field; (d) the surface of the companion star.

15. Gamma-ray bursts are observed to occur (a) mainly near the Sun; (b) throughout the Milky Way galaxy; (c) approximately uniformly over the entire sky; (d) near pulsars.

16. Black holes result from stars having initial masses (a) less than the mass of the Sun; (b) between 1 and 2 times the mass of the Sun; (c) up to 8 times the mass of the Sun; (d) more than 25 times the mass of the Sun.

17. If the Sun were magically to turn into a black hole of the same mass, (a) Earth would start to spiral inward; (b) Earth's orbit would remain unchanged; (c) Earth would fly off into space; (d) Earth would be torn apart.

18. Radio signals sent from space probes near the planet Mercury are redshifted by the time they reach Earth because (a) the probe is moving away from Earth; (b) radio photons lose energy as they escape from the intense gravitational field of the Sun; (c) the waves are partly absorbed by the solar corona; (d) the probe is heated up by the Sun's radiation.

19. The best place to search for black holes is in a region of space that (a) is dark and empty; (b) has recently lost some stars; (c) has strong X-ray emission; (d) is cooler than its surroundings.

20. The best evidence for supermassive black holes in the centers of galaxies is (a) the absence of stars there; (b) rapid gas motion and intense energy emission; (c) gravitational redshift of radiation emitted from near the center; (d) unknown visible and X-ray spectral lines.

PROBLEMS

 Algorithmic versions of these questions are available in the Practice Problems module of the Companion Website at astro.prenhall.com/chaisson.

The number of squares preceding each problem indicates its approximate level of difficulty.

1. ■ The angular momentum of a solid body is proportional to the angular velocity of the body times the square of its radius. ∞ (*More Precisely 6-2*) Using the law of conservation of angular momentum, estimate how fast a collapsed stellar core would spin if its initial spin rate was 1 revolution per day and its radius decreased from 10,000 km to 10 km.

2. ■ What would your mass be if you were composed entirely of neutron-star material of density 3×10^{17} kg/m³? (Assume that your average density is 1000 kg/m³.) Compare your answer with the mass of a typical 10-km-diameter rocky asteroid.

3. ■ Calculate the surface gravitational acceleration and escape speed of a 1.4-solar-mass neutron star with a radius of 10 km. What would be the escape speed from a neutron star of the same mass and radius 4 km?

4. ■■ Use the radius–luminosity–temperature relation to calculate the luminosity of a 10-km-radius neutron star for temperatures of 10^5 K, 10^7 K, and 10^9 K. At what wavelengths does the star radiate most strongly in each case? What do you conclude about the detectability of neutron stars? Could the brightest of them be plotted on an H–R diagram?

5. ■■ A gamma-ray detector of area 0.5 m² in the midst of a gamma-ray burst records photons having total energy 10^{-8} joule. If the burst occurred 1000 Mpc away, calculate the total amount of energy it released (assuming that the energy was emitted isotropically), and compare your answer with the Sun's total energy output on the main sequence. How would this figure change if the burst occurred 10,000 pc away instead, in the halo of our Galaxy? What if it occurred within the Oort cloud of our own solar system, at a distance of 50,000 A.U.?

6. ■■ A gamma-ray burst 5000 Mpc away releases 10^{45} joules of energy isotropically in the form of gamma rays, each of energy 250 keV. ∞ (*More Precisely 4-1*) Some of the rays are detected by an instrument in Earth orbit with an effective collecting area of 0.75 m². How many gamma-ray photons strike the detector?

7. ■ A 10-km-radius neutron star is spinning 1000 times per second. Calculate the speed of a point on its equator, and compare your answer with the speed of light. (Take the equator to be the circumference of a circle, and recall that the circumference is equal to $2\pi r$, where r is the diameter of the star.) Also, calculate the orbital speed of a particle in a circular orbit just above the surface if the neutron star's mass is 1.4 times the mass of the Sun.

8. ■ Supermassive black holes are thought to exist in the centers of many galaxies. What would be the Schwarzschild radii of black holes of 1 million and 1 billion solar masses, respectively? How does the 1-million-solar-mass black hole compare in size with the Sun? How does the 1-billion-solar-mass black hole compare in size with the solar system?

9. ■■ Use the information presented in *More Precisely 22-2* to estimate the deflection of a beam of light that just grazes the surface of (a) the Moon, (b) Jupiter, and (c) Sirius B. (d) A future generation of space astrometry missions will be able to measure angles as small as 10^{-6} arcsec accurately. At what distance from the Sun would this deflection occur?

10. ■■ Gravitational redshift is not confined to the neighborhoods of black holes. It can be shown that the fractional decrease in frequency due to the gravitational redshift of a light ray emitted at Earth's surface and detected at a height h above the surface is approximately gh/c^2, where $g = 9.80$ m/s² is the acceleration due to gravity at Earth's surface and c is the speed of light. Calculate the change in frequency of a 10-GHz (10^{10}-Hz) radio signal received by a satellite directly overhead at an altitude of 500 km.

11. ■■ Calculate the tidal acceleration on a 2-m-tall human falling feet-first into a 1-solar-mass black hole; that is, compute the difference in the accelerations (forces per unit mass) on their head and their feet just as the feet cross the event horizon. ∞ (*Sec. 7.6*) Repeat the calculation for a 1-million-solar-mass black hole and for a 1-billion-solar-mass black hole (see question 8). Compare these accelerations with the acceleration due to gravity on Earth ($g = 9.8$ m/s²).

12. ■■■ Endurance tests suggest that the human body cannot withstand stress greater than about 10 times the acceleration due to gravity on Earth's surface. At what distance from a 1-solar-mass black hole would the human in the previous question be torn apart?

13. ■■■ Using the data from the previous question, calculate the minimum mass of a black hole for which an infalling human could just reach the event horizon intact.

14. ■■ An intermediate-mass black hole in the galaxy M82 has a mass 100 times that of the Sun. What is the radius of the black hole's event horizon? Estimate how close to the hole a star like the Sun could approach before being torn apart by tidal forces.

15. ■■ Using the data given in the text, calculate the orbital separation of Cyg X-1. If the companion star's radius is 20 million km, verify (approximately) that the black hole's tidal field is sufficient to draw matter from the companion's surface.

 In addition to the Practice Problems module, the Companion Website at astro.prenhall.com/chaisson provides for each chapter a study guide module with multiple choice questions as well as additional annotated images, animations, and links to related Websites.

PART | 4
Galaxies and Cosmology

Andromeda Galaxy as observed in 1890 (*I. Roberts*)

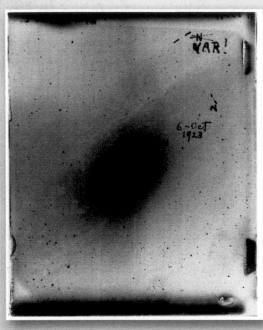

Edwin Hubble's discovery of variable stars in Andromeda (*Carnegie Institute*)

It is hard to imagine now, but less than a hundred years ago the Sun was considered the center of the universe. Earlier studies by Copernicus, Kepler, Galileo and others had dethroned Earth from a central position, but the Sun itself, at the center of our Solar System, was still assumed also to be the center of the Milky Way—which a century ago, was identical to the universe.

Enter the American astronomer Harlow Shapley (1885-1972), who, by studying variable stars in globular clusters, was able to deduce the size and scale of the Milky Way Galaxy, as well as our position in it. His results, announced in 1918, showed not only that our extended home in space was immensely larger than had previously been realized—about 100,000 light-years across—but also that Earth resided in what he called the "galactic suburbs," now known to be about 25,000 light-years from the center of the Galaxy. Shapley demonstrated that our Sun is not central, unique, or special in any way. His work was a milestone in our understanding of our place in the universe, certainly one of the most important astronomical discoveries of the 20th century.

Ironically, Shapley's dramatic discovery of the increased size and scale of the Milky Way led him astray regarding another, even more profound, advance in our knowledge at that time: the realization that our Galaxy is only one of many galaxies in the universe. The sheer size of the Milky Way implied by his observations caused him to oppose the idea of a vastly larger universe—he found it hard to believe that there could be other, distant galaxies as huge as our own. Even among eminent scientists, personal biases can sometimes cloud scientific judgment.

The stage was set for a "Great Debate" that occurred at the National Academy of Sciences in Washington in 1920. At issue were the fuzzy "spiral nebulae" (which today we call galaxies): Were they close enough to be part of our own Milky Way, or were they sufficiently distant to be whole galaxies unto themselves? Shapley held that, given that his research had revised upward the size of the Milky Way, the spiral nebulae must be part of our own Galaxy. His opponent, Heber Curtis of California's Lick Observatory, while incorrectly rejecting the great size of our Galaxy, correctly argued that the spirals were remote aggregates of stars similar to the Milky Way. Both men presented other scientific arguments supporting their views (see Section 23.2), but both also let personal feelings affect their comprehension of our home Galaxy. With no objective measurements of the true distances to the nebulae, the debate ended in a draw.

Harlow Shapley working at his rotating octagonal desk (*Harvard*)

Andromeda Galaxy as observed today (*R. Gendler*)

Shapley's rival, the Caltech astronomer Edwin Hubble (1884-1953), broke the stalemate just a few years later by using the premier optical telescope of the day, the 2.5-m (100-inch) reflector atop Mt. Wilson. He first resolved Andromeda into individual stars and then carefully measured its variable stars, thereby proving that Andromeda was a genuine galaxy millions of light-years distant, well outside our Milky Way. Ironically, Hubble used the same basic technique that Shapley and his Harvard colleagues had pioneered. It was yet another milestone along the road extending the Copernican principle: Neither Earth nor the Sun is special in any way, and even the Galaxy in which we live is just one of myriad galaxies in a much, much larger cosmos.

oday, astronomers have extensively mapped the distribution of variable stars in Andromeda. Curiously, we are still struggling to determine that galaxy's distance to better than 10 percent accuracy. Even in the decade-long lifetime of this textbook, Andromeda's quoted distance has fluctuated from about 2.2 to as much as 2.9 million light-years; in this edition, we have averaged the most recent measurements to arrive at a value of 2.5 million light-years. The correct value is important, for upon it rests a key rung in the so-called distance ladder. This cosmic yardstick is used to measure the ranges to billions of other, more distant galaxies and hence to gauge the vastly larger realm of the universe itself.

The Shapley-Curtis debate of yesteryear, together with our current struggles to pin down accurate distances to the truly faraway galaxies, constitute good case studies of how the scientific method actually works. Science is practiced by human beings, and scientists are no different from others who have strong emotions and personal values. Yet, over the course of time, and through much criticism and debate, scientific issues eventually gain a measure of objectivity. By demanding tests and proven facts, the scientific community gradually damps the subjectivity of individuals and arrives at a more objective view among a community of critical thinkers. Reasoned skepticism and repeated testing are hallmarks of the modern scientific method.

Hubble Ultra Deep Field (*STScI*)

23

Stars are not distributed randomly in space. Rather, they cluster into gargantuan assemblages called galaxies, of which our Milky Way Galaxy is just one among more than a hundred billion others. Our Sun sits in the suburbs of the Milky Way, nearly 30,000 light-years from its central regions, which appear to be a good deal more violent than the part of the Galaxy in which we live. Here displayed is the magnificent spiral galaxy, NGC 4526, which resides nearly 100 million light-years away, and whose size, shape, and mass resemble those of our own Galaxy. (STScI) ▶

The Milky Way Galaxy

A Spiral in Space

Looking up on a dark, clear night, we are struck by two aspects of the night sky. The first is that the individual stars we see are roughly uniformly distributed in all directions. They all lie relatively close to us, mapping out the local Galactic neighborhood within a few hundred parsecs of the Sun. But this is only a local impression. Ours is a rather provincial view. Beyond those nearby stars, the second thing we notice is a fuzzy band of light—the Milky Way—stretching across the heavens. From the Northern Hemisphere, this band is most easily visible in the summertime, arcing high above the horizon. Its full extent forms a great circle that encompasses the entire celestial sphere. This is the insider's view of the galaxy in which we live—the blended light of countless distant stars. As we consider much larger volumes of space, on scales far, far greater than the distances between neighboring stars, a new level of organization becomes apparent as the large-scale structure of the Milky Way Galaxy is revealed.

LEARNING GOALS

Studying this chapter will enable you to

1 Describe the overall structure of the Milky Way Galaxy, and specify how the various regions differ from one another.

2 Explain the importance of variable stars in determining the size and shape of our Galaxy.

3 Describe the orbital paths of stars in different regions of the Galaxy, and explain how these motions are accounted for by our understanding of how the Galaxy formed.

4 Discuss some possible explanations for the existence of the spiral arms observed in our own and many other galaxies.

5 Explain what studies of Galactic rotation reveal about the size and mass of our Galaxy, and discuss the possible nature of dark matter.

6 Describe some of the phenomena observed at the center of our Galaxy.

 Visit astro.prenhall.com/chaisson for additional annotated images, animations, and links to related sites for this chapter.

23.1 Our Parent Galaxy

LEARNING GOAL 1 A **galaxy** is a gargantuan collection of stellar and interstellar matter—stars, gas, dust, neutron stars, black holes—isolated in space and held together by its own gravity. Astronomers are aware of literally billions of galaxies beyond our own. The particular galaxy we happen to inhabit is known as the *Milky Way Galaxy*, or just *the Galaxy*, with a capital *G*.

Our Sun lies in a part of the Galaxy known as the **Galactic disk**—an immense, circular, flattened region containing most of our Galaxy's luminous stars and interstellar matter (and virtually everything we have studied so far in this book). Figure 23.1 illustrates how, viewed from within, the Galactic disk appears as a band of light stretching across our night sky, a band known as the *Milky Way*. As indicated in the figure, if we look in a direction away from the Galactic disk (red arrows), we see relatively few stars in our field of view. However, if our line of sight happens to lie within the disk (white and blue arrows), we see so many stars that their light merges into a continuous blur.

Paradoxically, although we can study individual stars and interstellar clouds that lie near the Sun in great detail, our location within the Galactic disk makes deciphering our Galaxy's large-scale structure from Earth a very diffi-cult task—a little like trying to unravel the layout of paths, bushes, and trees in a city park without being able to leave one particular park bench. In some directions, the interpretation of what we see is ambiguous and inconclusive. In others, foreground objects completely obscure our view of what lies beyond, but we cannot move around them to get a better look. As a result, astronomers who study the Milky Way Galaxy are often guided in their efforts by comparisons with more distant, but much more easily observable, systems.

Figures 23.2 and 23.3 show three galaxies thought to resemble our own in overall structure. Figure 23.2 is the Andromeda Galaxy, the nearest major galaxy to the Milky Way Galaxy, lying nearly 800 kpc (roughly 2.5 million light years) away. Andromeda's apparent elongated shape is a consequence of the angle at which we happen to view it. In fact, our Galaxy, like this one, consists of a circular galactic disk of matter that fattens to a **Galactic bulge** at the center. The disk and bulge are embedded in a roughly spherical ball of faint old stars known as the **Galactic halo**. These three basic galactic regions are indicated on the figure. (The halo stars are so faint that they cannot be discerned here.) Figures 23.3(a) and (b) show views of two other galaxies—one seen face-on, the other edge-on—that illustrate these points more clearly.

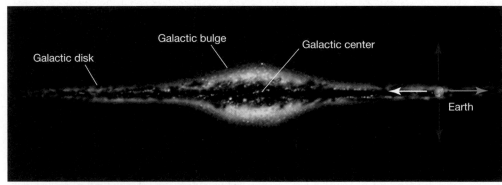

Galactic disk · Galactic bulge · Galactic center · Earth

(a) Artist's view of Milky Way from afar

(b) Real image of Milky Way from inside

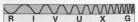

R I V U X G

◀ **FIGURE 23.1 Galactic Plane** (a) This "edge-on" artist's conception shows our position in the Milky Way Galaxy. Looking toward the Galactic center (white arrow), we see myriad stars stacked up within the thin band of the Milky Way. Looking in the opposite direction (blue arrow), we still see the Milky Way band, but now it is fainter because our position is far from the Galactic center, so we see more stars when looking toward the center than in the opposite direction. Perpendicular to the disk (red arrows), we see even fewer stars. (b) This is a real image of what we see looking from Earth toward the Galactic center—in the direction of the white arrow in part (a). The white band dominating the view is the disk of our Milky Way Galaxy—dimly visible to the naked eye from very dark locations on Earth. (*Lund Observatory*)

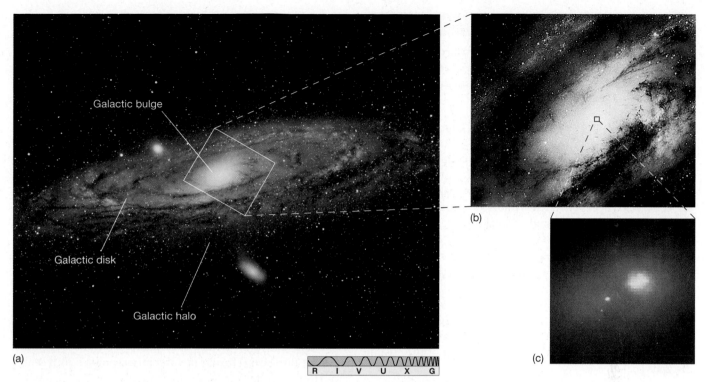

(a)

(b)

(c)

R I V U X G

▲ **FIGURE 23.2 Andromeda Structure** (a) The Andromeda Galaxy probably resembles the overall layout of our own Milky Way Galaxy fairly closely. The disk and bulge are clearly visible in this image, which is about 30,000 pc across. The faint stars of the halo, completely surrounding the disk and bulge, cannot be seen here. The white stars sprinkled all across this image are not part of Andromeda's halo; they are foreground stars in our own Galaxy, lying in the same part of the sky as Andromeda, but about a thousand times closer. (b) More detail within the inner parts of the galaxy. (c) The galaxy's peculiar—and still unexplained—double core; this inset covers a region only 15 pc across. *(R. Gendler; Palomar/Caltech; NASA)*

(a)

(b)

R I V U X G

▲ **FIGURE 23.3 Spiral Galaxies** (a) This galaxy, catalogued as M101 and seen nearly face-on, is somewhat similar in its overall structure to our own Milky Way Galaxy and Andromeda. (b) The galaxy NGC 4565 happens to be oriented in such a way that we see it edge-on, allowing us to make out its disk and central bulge. *(R. Gendler)*

23.2 Measuring the Milky Way

Before the 20th century, astronomers' conception of the cosmos differed markedly from the modern view. The fact that we live in just one of many enormous "islands" of matter separated by even larger tracts of apparently empty space was completely unknown, and the clear distinction between "our Galaxy" and "the universe" did not exist. The twin ideas that (1) the Sun is not at the center of the Galaxy and (2) the Galaxy is not at the center of the universe required both time and hard observational evidence before they gained widespread acceptance. The growth in our knowledge of our Galaxy, as well as the realization that there are many other distant galaxies similar to our own, has gone hand in hand with the development of the cosmic distance scale.

Star Counts

In the late 18th century, long before the distances to any stars were known, the English astronomer William Herschel tried to estimate the size and shape of our Galaxy simply by counting how many stars he could see in different directions in the sky. Assuming that all stars were of about equal brightness, he concluded that the Galaxy was a somewhat flattened, roughly disk-shaped collection of stars lying in the plane of the Milky Way, with the Sun at its *center* (Figure 23.4). Subsequent refinements to this approach led to much the same picture. Early in the 20th century, some astronomers went so far as to estimate the dimensions of this "Galaxy" as about 10 kpc in diameter by 2 kpc thick.

Today the Milky Way Galaxy is known to be several tens of kiloparsecs across, and the Sun is known to lie far from the center. How could the older picture have been so flawed? The answer is that the earlier observations were made in the visible part of the electromagnetic spectrum, and astronomers failed to take into account the (then-unknown) absorption of visible light by interstellar gas and dust. ∞ (Sec. 18.1) Only in the 1930s did astronomers begin to realize the true extent and importance of the interstellar medium.

Any objects in the Galactic disk more than a few kiloparsecs away from us are hidden from our view (in visible light) by the effects of interstellar dust. The apparent falloff in the density of stars with distance in the plane of the Milky Way is thus not a real thinning of their numbers in space, but simply a consequence of the murky environment in the Galactic disk. The long "fingers" in Herschel's map are directions in which the obscuration happens to be a little less severe than in others. However, because some obscuration occurs in all directions in the disk, the falloff is roughly similar no matter which way we look, so the Sun appears to be more or less at the center. The horizontal extent of Figure 23.4 corresponds approximately to the span of the blue and white arrows in Figure 23.1.

Radiation coming to us from above or below the plane of the Galaxy, where there is less gas and dust along the line of sight, arrives on Earth relatively unscathed. There is still some patchy obscuration, but the Sun happens to be located where the view out of the disk is largely unimpeded by nearby interstellar clouds.

Spiral Nebulae and Globular Clusters

We have just seen how astronomers' attempts to probe the Galactic disk by optical means are frustrated by the effects of the interstellar medium. However, looking in other directions, out of the Milky Way plane, we can see to much greater distances. During the first quarter of the 20th century, studies of the large-scale structure of our Galaxy focused on two particularly important classes of objects, both found mainly away from the Milky Way. The first is *globular clusters*, those tightly bound swarms of old, reddish stars we initially met in Chapter 19. ∞ (Sec. 19.6) About 150 are now known in our own Galaxy. The second class consisted of objects that were known at the time as *spiral nebulae*. Examples are shown in Figures 23.2(a) and 23.3(a). We know them today as **spiral galaxies**, comparable in size to our own.

Early 20th-century astronomers had no means of determining the distances to any of these objects. They are too far away to have any observable parallax, and with the technology of the day, main-sequence stars (after the discovery of the main sequence in 1911) could not be clearly identified and measured. For these reasons, neither of the techniques discussed in Chapter 17 was applicable. ∞ (Secs. 17.1, 17.7) As a result, even the most basic properties—size, mass, and stellar and interstellar content—of globular clusters and spiral nebulae were unknown. It was

▲ **FIGURE 23.4 Herschel's Galaxy Model** Eighteenth-century astronomer William Herschel constructed this "map" of the Galaxy by counting the numbers of stars he saw in different directions in the sky. Our Sun (marked by the large yellow dot) appears to lie near the center of the distribution. The long axis of the diagram lies in the plane of the Galactic disk.

assumed that the globular clusters lay within our own Galaxy, which was thought at the time to be relatively small (using the size estimates just mentioned). The locations of the spiral nebulae were much less clear.

Knowing the distance to an object is vitally important to understanding its true nature. As an example, consider again the Andromeda "nebula" (Figure 23.2). In the late 19th century, when improved telescopes and photographic techniques allowed astronomers to obtain images showing detail comparable to that in Figure 23.2(a), the newly released photographs caused great excitement among astronomers, who thought they were seeing the formation of a star from a swirling gaseous disk! Comparing Figure 23.2(a) with the figures in Chapter 15 (especially Figure 15.1b), we can perhaps understand how such a mistake could be made—*if* we believed that we were looking at a relatively close, star-sized object. Far from demonstrating that Andromeda was distant and large, the new observations seemed to confirm that it was just a small part of our own Galaxy.

Further observations soon made it clear that Andromeda was not a star-forming region. Andromeda's parallax is too small to measure, indicating that it must be at least several hundred parsecs from Earth, and, even at 100 pc—which we now know is vastly less than Andromeda's true distance—an object the size of the solar nebula would be impossible to resolve and simply would not look like Figure 23.2(a). (See Section 22.4 for another, more recent example of how distance measurements directly affect our theoretical understanding of observational data.)

During the first quarter of the 20th century, both the size of our Galaxy and the distances to the spiral nebulae were hotly debated in astronomical circles (see below, and also the discussion on p. 600). One school of thought maintained that the spiral nebulae were relatively small systems contained within our Galaxy. Other astronomers held that the spirals were much larger objects, lying far outside the Milky Way Galaxy and comparable to it in size. However, with no firm distance information, both arguments were inconclusive. Only with the discovery of a new distance-measurement technique—which we discuss next—was the issue finally settled in favor of the latter view. However, in the process, astronomers' conception of our own Galaxy changed radically and forever.

A New Yardstick

An important by-product of the laborious effort to catalog stars around the turn of the 20th century was the systematic study of *variable stars*—stars whose luminosity changes with time, some quite erratically, others more regularly. Only a small fraction of stars fall into this category, but those which do are of great astronomical significance.

We encountered several examples of variable stars in earlier chapters. Often, the variability is the result of membership in a binary system. In an eclipsing binary, for example, the total brightness varies because one star periodically blocks the light of the other. ∞ Sec. 17.7 In novae and supernovae, binary membership has more violent consequences. ∞ Sec. 21.3 These latter objects are called *cataclysmic variables*, because of their sudden, large changes in brightness.

In other instances, however, the variability is a basic trait of a star and is not dependent on its being a part of a binary system. We call such a star an *intrinsic variable*. A particularly important class of intrinsic variables is the **pulsating variable stars**, which vary cyclically in luminosity in very characteristic ways. Two types of pulsating variable stars that have played central roles in revealing both the true extent of our Galaxy and the distances to our galactic neighbors are the **RR Lyrae** and **Cepheid** variables. Following long-standing astronomical practice, the names come from the first star of each class to be discovered—in this case, the variable star labeled RR in the constellation Lyra and the variable star Delta Cephei, the fourth brightest star in the constellation Cepheus. ∞ (*Discovery 17-1*) Note, by the way, that these stars have *nothing* whatsoever to do with the pulsars discussed in the previous chapter! Pulsars are rapidly rotating neutron stars beaming energy into space as they spin; as we will see in a moment, pulsating variable stars are "normal" stars undergoing a temporary period of instability as they evolve. ∞ (Sec. 22.2)

RR Lyrae and Cepheid variable stars are recognizable by the characteristic shapes of their light curves. RR Lyrae stars all pulsate similarly (Figure 23.5a), with only small differences in period between them. Observed periods range from about 0.5 to 1 day. Cepheid variables also pulsate in distinctive ways (the regular "sawtooth" pattern in Figure 23.5b), but different Cepheids can have quite different pulsation periods, ranging from about 1 to 100 days. The period of any given RR Lyrae or Cepheid variable is, to a high degree of accuracy, the same from one cycle to the next. The key point is that pulsating variable stars can be recognized and identified *just by observing the variations in the light they emit*.

Why do Cepheids and RR Lyrae variables pulsate? The basic mechanism was first suggested by the British astrophysicist Sir Arthur Eddington in 1941. The structure of any star is determined in large part by how easily radiation can travel from the core to the photosphere—that is, by the *opacity* of the interior, the degree to which the gas hinders the passage of light through it. If the opacity rises, the radiation becomes trapped, the internal pressure increases, and the star "puffs up." If the opacity falls, radiation can escape more easily, and the star shrinks. According to theory, under certain circumstances a star can become unbalanced and enter a state in which the flow of radiation causes the opacity first to rise—making the star expand, cool, and diminish in luminosity—and then to fall, leading to the pulsations we observe.

The conditions necessary to cause pulsations are not found in main-sequence stars. Rather, they occur in evolved post-main-sequence stars as they pass through a region of the H–R diagram known as the *instability strip* (Figure 23.6). When a star's temperature and luminosity place it in this strip, the star becomes internally unstable. Both its temperature and its radius vary in a regular way, causing the pulsations we observe: For the reasons just described, as the star brightens, its surface becomes hotter

and its radius shrinks; as its luminosity decreases, the star expands and cools. As we learned in Chapter 20, high-mass stars evolve across the upper part of the H–R diagram. When their evolutionary tracks take them into the instability strip, they become Cepheid variables. ∞ (Sec. 20.4) RR Lyrae variables are low-mass horizontal-branch stars that lie within the lower portion of the instability strip. ∞ (Sec. 20.2) Thus, pulsating variables are normal stars passing through a brief—roughly million-year—phase of instability as a natural part of stellar evolution.

The importance of these stars to Galactic astronomy lies in the fact that once we recognize a star as being of the RR Lyrae or Cepheid type, we can infer its luminosity, and that in turn allows us to measure its distance. The distance calculation is precisely the same as that presented in Chapter 17 during our discussion of spectroscopic parallax. ∞ (Sec. 17.6) Comparing the star's (known) luminosity with its (observed) apparent brightness yields an estimate of its distance, by the inverse-square law: ∞ (Sec. 17.2)

$$\text{apparent brightness} \propto \frac{\text{luminosity}}{\text{distance}^2}.$$

In this way, astronomers can use pulsating variables as a means of determining distances, both within our own Galaxy and far beyond.

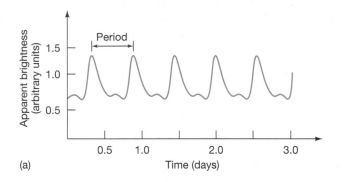

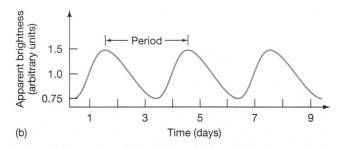

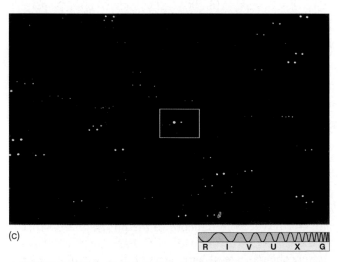

▲ **FIGURE 23.5 Variable Stars** (a) Light curve of the pulsating variable star RR Lyrae. All RR Lyrae-type variables have similar light curves, with periods of less than a day. (b) The light curve of a Cepheid variable star called WW Cygni, having a period of several days. (c) This Cepheid is shown here (boxed) on successive nights, near its maximum and minimum brightness; two photos, one from each night were superimposed and then slightly displaced. (*Harvard College Observatory*)

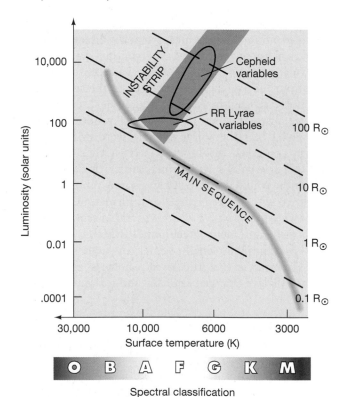

▲ **FIGURE 23.6 Variable Stars on the H–R Diagram** Pulsating variable stars are found in the instability strip of the H–R diagram. As a high-mass star evolves through the strip, it becomes a Cepheid variable. Low-mass horizontal-branch stars in the instability strip are RR Lyrae variables.

How do we infer a variable star's luminosity? For RR Lyrae stars, doing so is simple. As we saw in Chapter 20, all such stars have basically the same luminosity (averaged over a complete pulsation cycle)—about 100 times that of the Sun. ∞ (Sec. 20.2) Thus, once a variable star is recognized as being of the RR Lyrae type, its luminosity is immediately known. For Cepheids, we make use of a close correlation between average luminosity and pulsation period, discovered in 1908 by Henrietta Leavitt of Harvard University (see *Discovery 23-1*) and known simply as the **period–luminosity relationship**. Cepheids that vary slowly—that is, that have long periods—have large luminosities; conversely, short-period Cepheids have low luminosities.

Figure 23.7 illustrates the period–luminosity relationship for Cepheids found within a thousand parsecs or so of Earth. Astronomers can plot such a diagram for relatively nearby stars because they can measure their distances by using stellar or spectroscopic parallax. Once the distances are known, the luminosities of those stars can be calculated. We know of no exceptions to the period–luminosity relationship, and it is consistent with theoretical calculations of pulsations in evolved stars. Consequently, we assume that it holds for all Cepheids, near and far. Thus, a simple measurement of a Cepheid variable's pulsation period immediately tells us its luminosity—we just read it off the plot in Figure 23.7. (The roughly constant luminosities of the RR Lyrae variables are also indicated in the figure.)

This distance-measurement technique works well, provided that the variable star can be clearly identified and its pulsation period measured. With Cepheids, the method allows astronomers to estimate distances out to about 25 million parsecs, enough to take us all the way to the nearest galaxies. The less luminous RR Lyrae stars are not so easily seen as Cepheids, so their useful range is not as great. However, they are much more common, so, within their limited range, they are actually more useful than Cepheids.

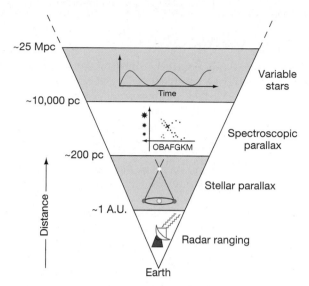

▲ **FIGURE 23.8 Variable Stars on Distance Ladder** Applying the period–luminosity relationship of Cepheid variable stars allows us to determine distances out to about 25 Mpc with reasonable accuracy.

Figure 23.8 extends our cosmic distance ladder, begun in Chapter 2 with radar ranging in the solar system and expanded in Chapter 17 to include stellar and spectroscopic parallax, by adding variable stars as a fourth method of determining distance. Note that, because the period–luminosity relationship is calibrated by using nearby stars, this latest rung inherits any and all uncertainties and errors present in the lower levels. Uncertainties also arise from the "scatter" shown in Figure 23.7: Although the overall connection between period and luminosity is unmistakable, the individual data points do not quite lie on a straight line; instead, a range of possible luminosities corresponds to any measured period.

The Size and Shape of our Galaxy

Many RR Lyrae variables are found in globular clusters. Early in the 20th century, the American astronomer Harlow Shapley used observations of RR Lyrae stars to make two very important discoveries about the Galactic globular cluster system. First, he showed that most globular clusters reside at great distances—many thousands of parsecs—from the Sun. Second, by measuring the direction and distance of each cluster, he was able to determine the three-dimensional distribution of the clusters in space (Figure 23.9). In this way, Shapley demonstrated that the globular clusters map out a truly gigantic, and roughly *spherical*, volume of space, about 30 kpc across.* However, the center of the distribution lies nowhere near our Sun;

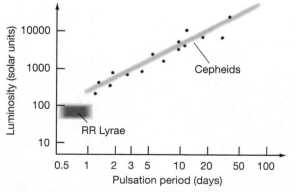

▲ **FIGURE 23.7 Period–Luminosity Plot** A plot of pulsation period versus average absolute brightness (i.e., luminosity) for a group of Cepheid variable stars. The two properties are quite tightly correlated. The pulsation periods of some RR Lyrae variables are also shown.

*The Galactic globular cluster system and the Galactic halo, of which it is a part, are somewhat flattened in the direction perpendicular to the disk, but the degree of flattening is quite uncertain. The halo is certainly much less flattened than the disk, however.

Early Computers

A large portion of the early research in observational astronomy focused on monitoring stellar luminosities and analyzing stellar spectra. Much of this pioneering work was done using photographic methods. What is not so well known is that most of the labor was accomplished by women. Around the turn of the twentieth century, a few dozen dedicated women—assistants at the Harvard College Observatory—created an enormous database by observing, sorting, measuring, and cataloging photographic information that helped form the foundation of modern astronomy. Some of them went far beyond their duties in the lab to make several of the basic astronomical discoveries often taken for granted today.

The first photograph below, taken in 1910, shows several of those women carefully examining star images and measuring variations in luminosity or wavelengths of spectral lines. In the cramped quarters of the Harvard Observatory these women inspected image after image to collect a vast body of data on hundreds of thousands of stars. Note the plot of stellar luminosity changes pasted on the wall at the left. The cyclical pattern is so regular that it likely belongs to a Cepheid variable. Known as "computers" (there were no electronic devices then), these women were paid 25 cents an hour.

The second photograph, taken in 1913, shows a more formal portrait of another group of staff members, along with their director, E.C. Pickering. Though looking rather stern here, Pickering was often described as a true Victorian gentleman who championed a policy, unique at the time, of admitting women to the staff. Also prominent here (and symmetrically positioned to Pickering's left) is Annie Cannon, perhaps the most accomplished of the early group of women who, beginning in 1880, undertook a survey of the skies that lasted for more than half a century—work that netted Cannon the first Oxford honorary degree awarded to a woman.

More than anything else, these women were master cataloguers, tabulating millions of measurements of stars in scores of exquisitely crafted log books that exist to this day. Their first major accomplishment was a record of the brightnesses and spectra of tens of thousands of stars, published in 1890 under the direction of Williamina Fleming (seen standing in her supervisory role in the first photograph). On the basis of this compilation, several of these women made fundamental contributions to astronomy. In 1897 Antonia Maury (who is also pictured in the first photo at left rear) undertook the most detailed study of stellar spectra to that time, enabling Hertzsprung and Russell independently to develop what is now called the H–R diagram. In 1898 Annie Cannon proposed the spectral classification system (describe in Chapter 17) that is now the international standard for categorizing stars. ∞ (Sec. 17.5) And in 1908 Henrietta Leavitt discovered the period–luminosity relationship for Cepheid variable stars, which later allowed Pickering's successor as director, Harlow Shapley (see the introductory essay for Part IV), to recognize our Sun's true position in the universe.

All was not work, however, as socializing was common among this generation of astronomers. The third photograph shows a 1920s scene from a humorous play portraying life at the Observatory, starring (at center) the then youngest of the "lady computers," Cecilia Payne, who would go on to become one of the foremost astronomers of the 20th century (see the introductory essay for Part III).

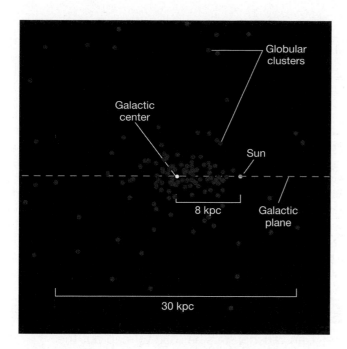

◄ **FIGURE 23.9 Globular Cluster Distribution** Our Sun does not coincide with the center of the very large collection of globular clusters. Instead, more globular clusters are found in one direction than in any other. The Sun resides closer to the edge of the collection, which measures roughly 30 kpc across. We now know that the globular clusters outline the true distribution of stars in the Galactic halo.

the halo. Since Shapley's time, astronomers have identified many individual stars—that is, stars not belonging to any globular cluster—within the Galactic halo.

Shapley's bold interpretation of the globular clusters as defining the overall distribution of stars in our Galaxy was an enormous step forward in human understanding of our place in the universe. Five hundred years ago, Earth was considered the center of all things. Copernicus argued otherwise, demoting our planet to an undistinguished location removed from the center of the solar system. In Shapley's time, as we have just seen, the prevailing view was that our Sun was the center not only of the Galaxy, but also of the universe. Shapley showed otherwise. With his observations of globular clusters, he simultaneously increased the size of our Galaxy by almost a factor of 10 over earlier estimates and banished our parent Sun to its periphery, virtually overnight!

The Shapley–Curtis Debate

Curiously, Shapley's dramatic revision of the size of the Milky Way Galaxy and our place in it only strengthened his erroneous opinion that the spiral nebulae were part of our Galaxy and that our Galaxy was essentially the entire

rather, it is located nearly 8 kpc away from us, in the direction of the constellation Sagittarius.

In a brilliant intellectual leap, Shapley realized that the distribution of globular clusters maps out the true extent of stars in the Milky Way Galaxy—the region that we now call the Galactic halo. The hub of this vast collection of matter, 8 kpc from the Sun, is the **Galactic center**. Figure 23.9 shows the distribution, based on modern data, of the 138 globular clusters lying within 20 kpc of the center. As illustrated in Figure 23.10, we live in the "suburbs" of this huge ensemble—in the Galactic disk, the thin sheet of young stars, gas, and dust that cuts through the center of

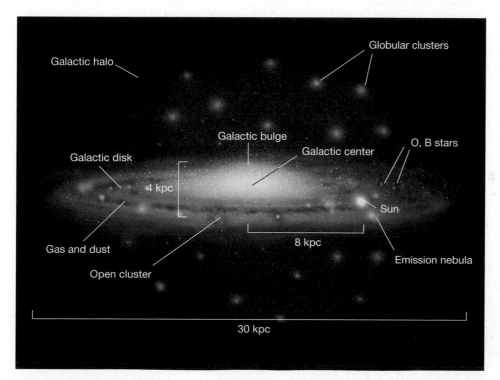

◄ **FIGURE 23.10** Stellar Populations in Our Galaxy Artist's conception of a (nearly) edge-on view of the Milky Way Galaxy, showing the distributions of young blue stars, open clusters, old red stars, and globular clusters. (The brightness and size of the Sun are greatly exaggerated for clarity.)

universe. He regarded as beyond belief the idea that there could be other structures as large as our Galaxy. The scientific issues involved in understanding the nature of the spiral nebulae were clearly drawn in a famous 1920 debate between Shapley and Lick Observatory astronomer Heber Curtis. (See also p. 600.) We list here some key elements of the debate:

1. *Size of the Milky Way.* Shapley correctly asserted that the diameter of the Milky Way Galaxy was much larger than the "conventional" scale based on star counts, but then incorrectly concluded that similar-sized galaxies beyond our own could not exist. Curtis incorrectly accepted the smaller size for our Galaxy, but correctly argued that similar galaxies might exist beyond our own.

2. *Distribution of the nebulae on the sky.* Curtis noted that the observed spiral nebulae were generally found away from the plane of the Galaxy, and he suggested that our Galaxy had a "ring" of occulting material in its plane, like those observed in many edge-on spirals, preventing us from seeing the nebulae in the plane. Shapley simply had to accept the notion that spiral nebulae were, for some unknown reason, not found in the Galactic plane. Curtis was almost correct in this point. Note, however, that the effects of absorption by interstellar dust were completely unknown at the time. ∞ (Sec. 18.1)

3. *Observations of novae.* Shapley argued (correctly) that the observed apparent brightnesses of some "novae" seen in spiral nebulae implied enormous luminosities if the nebulae lay at large distances. ∞ (Secs. 17.2, 21.1) Curtis suggested (also correctly) that these anomalous events might be members of a second, much brighter, class of nova—today we call them *supernovae.* ∞ (Sec. 21.3)

4. *Brightness and spectra of the nebulae.* Shapley pointed out that the measured brightnesses and colors of spiral nebulae differed from what he would have expected to see if our Galaxy were observed from afar, suggesting that the nebulae were somehow fundamentally different from the Milky Way. Curtis had no answer. We know today that these differences exist because of interstellar absorption and reddening, which prevent astronomers from getting a comparable view of our own Galaxy, but all this was unknown at the time. ∞ (Sec. 18.1) Curtis did correctly note that spectral lines seen in spiral nebulae were generally consistent with the nebulae's being assemblages of large numbers of stars, supporting his argument that they were stellar systems comparable to our own Galaxy. ∞ (Sec. 4.2)

5. *Rotation of the nebulae.* Shapley cited published measurements of the angular rotation speeds of some spiral nebulae, which implied that the nebulae would have to be spinning faster than the speed of light if they were very distant and hence very large. ∞ (*More Precisely 1-3*) Curtis simply responded that the observations were in error. Curtis was right, but he couldn't prove it at the time.

We see that both men made some correct and some incorrect statements (or conclusions) about the problem. However, with the observations of the day, their disagreements could not be resolved, and the debate was inconclusive. But technology marches on, and just a few years later, in 1925, American astronomer Edwin Hubble reported that he had observed Cepheids in the Andromeda Galaxy and finally succeeded in measuring its distance. His work firmly established Andromeda as a separate galaxy lying far beyond our own, finally extending the Copernican principle to the Galaxy itself.

CONCEPT CHECK

✔ Can variable stars be used to map out the structure of the Galactic disk?

23.3 Galactic Structure

Based on optical, infrared, and radio observations of stars, gas, and dust, Figure 23.10 illustrates the different spatial distributions of the disk, bulge, and halo of the Milky Way Galaxy. The extent of the halo is based largely on optical observations of globular clusters and other halo stars. However, as we have seen, optical techniques can cover only a small portion of the dusty Galactic disk. Much of our knowledge of the structure of the disk on larger scales is based on radio observations, particularly of the 21-cm radio emission line produced by atomic hydrogen. ∞ (Sec. 18.4)

The center of the gas distribution coincides roughly with the center of the globular cluster system, lying about 8 kpc from the Sun. In fact, the location of the Galactic center is determined most accurately from radio observations of Galactic gas. The densities of both stars and gas in the disk decline quite rapidly beyond about 15 kpc from the Galactic center (although some radio-emitting gas has been observed out to at least 50 kpc).

The Spatial Distribution of Stars

Perpendicular to the Galactic plane, the disk in the vicinity of the Sun is relatively thin—"only" 300 pc thick, or about one hundredth of the 30-kpc Galactic diameter. Don't be fooled, though: Even if you could travel at the speed of light, it would take you a thousand years to traverse the thickness of the Galactic disk. The disk may be thin compared with the Galactic diameter, but it is huge by human standards.

Actually, the thickness of the Galactic disk depends on the kinds of objects measured. Young stars and interstellar gas are more tightly confined to the plane than are stars like the Sun, and solar-type stars in turn are more tightly confined than are older K- and M-type dwarfs. The reason for these differences is that stars form in interstellar clouds close to the plane of the disk, but then tend to drift out of the disk over time, mainly due to their interactions with other stars and molecular clouds. Thus, as stars age, their abundance above and below the plane of the disk slowly increases. Note that these considerations do not apply to the Galactic halo, whose ancient stars and globular clusters extend far above and below the Galactic plane. As we will see in a moment, the halo is a remnant of an early stage of our Galaxy's evolution and predates the formation of the disk.

Recently, improved observational techniques have revealed an intermediate category of Galactic stars, midway between the old halo and the younger disk, both in age and in spatial distribution. Consisting of stars with estimated ages in the range of 7–10 billion years, this *thick-disk* component of the Milky Way Galaxy measures some 2–3 kpc from top to bottom. Its thickness is too great to be explained by the slow drift just described. Like the halo, it appears to be a vestige of our Galaxy's distant past.

Also shown in Figure 23.10 is our Galaxy's central bulge, measuring roughly 6 kpc across in the plane of the Galactic disk by 4 kpc perpendicular to the plane. Obscuration by interstellar dust makes it difficult to study the detailed structure of the Galactic bulge in optical images of the Milky Way. (See, for example, Figure 18.5, which would clearly show a large portion of the bulge were it not for interstellar absorption.) However, at longer wavelengths, which are less affected by interstellar matter, a much clearer picture emerges (Figure 23.11; cf. Figure 23.3b). Detailed measurements of the motion of gas and stars in and near the bulge imply that it is actually football shaped, about half as wide as it is long, with the long axis of the "football" lying in the Galactic plane. On the basis of these observations, astronomers speculate that the central part of our Galaxy may have a distinctly elongated, or barlike, appearance and that we may live in a galaxy of the "barred-spiral" type, as discussed further in Chapter 24.

Stellar Populations

Aside from their distributions in space, the three components of the Galaxy—disk, bulge, and halo—have several other properties that distinguish them from one another. First, the halo contains almost *no* gas or dust—just the opposite of the disk and bulge, in which interstellar matter is common. Second, there are clear differences in both *appearance* and *composition* among disk, bulge, and halo stars: Stars in the Galactic bulge and halo are found to

be distinctly *redder* than stars found in the disk. Observations of other spiral galaxies also show this trend—the blue-white tint of the disk and the yellowish coloration of the bulge are evident in Figures 23.2(a) and 23.3(a).

All the bright, blue stars visible in our sky are part of the Galactic disk, as are the young, open star clusters and star-forming regions. In contrast, the cooler, redder stars—including those found in the old globular clusters—are more uniformly distributed throughout the disk, bulge, and halo. Galactic disks appear bluish because main-sequence O- and B- type blue supergiants are very much brighter than G-, K-, and M-type dwarfs, even though the dwarfs are present in far greater numbers.

The explanation for the marked difference in stellar content between disk and halo is this: Whereas the gas-rich Galactic disk is the site of ongoing star formation and so contains stars of all ages, all the stars in the Galactic halo are *old*. The absence of dust and gas in the halo means that no new stars are forming there, and star formation apparently ceased long ago—at least 10 billion years in the past, judging from the types of halo stars we now observe. (Recall from Chapter 20 that most globular clusters are thought to be between 10 and 12 billion years old.) ∞ (Sec. 20.5) The gas density is very high in the inner part of the Galactic bulge, making that region the site of vigorous ongoing star formation, and both very old and very young stars mingle there. The bulge's gas-poor outer regions have properties more similar to those of the halo.

Support for this picture comes from studies of the spectra of halo stars, which indicate that these stars are far less abundant in heavy elements (i.e., elements heavier than helium) than are nearby stars in the disk. In Chapter 21, we saw how each successive cycle of star formation and evolution enriches the interstellar medium with the products of

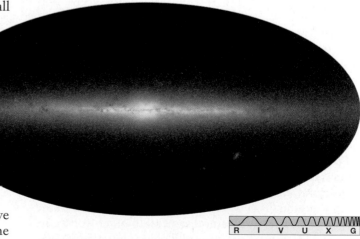

▲ **FIGURE 23.11 Infrared View of the Milky Way** A wide-angle infrared image of the disk and bulge of the Milky Way Galaxy, as observed by the Two Micron All Sky Survey. [Cf. Figure 23.3(b).] *(UMass/Caltech)*

stellar nucleosynthesis, leading to a steady increase in heavy elements with time. ∞ (Sec. 21.5) Thus, the scarcity of these elements in halo stars is consistent with the view that the halo formed long ago.

Astronomers often refer to young disk stars as *Population I* stars and to old halo stars as *Population II* stars. The idea of two stellar "populations" dates from the 1940s, when the differences between disk and halo stars first became clear. The names are something of an oversimplification, as there is actually a continuous variation in stellar ages throughout the Milky Way Galaxy, not a simple division of stars into two distinct "young" and "old" categories. Nevertheless, the terminology is widely used.

Orbital Motion

Now let's turn our attention to the *dynamics* of the Milky Way Galaxy—that is, to the motion of the stars, dust, and gas it contains. Are the internal motions of our Galaxy's members chaotic and random, or are they part of some gigantic "traffic pattern"? The answer depends on our perspective. The motion of stars and clouds we see on small scales (within a few tens of parsecs from the Sun) seems random, but on larger scales (hundreds or thousands of parsecs) the motion is much more orderly.

As we look around the Galactic disk in different directions, a clear pattern of motion emerges (Figure 23.12). Radiation received from stars and interstellar gas clouds in the upper right and lower left quadrants of the figure is generally *blueshifted*. At the same time, radiation from stars and gas sampled in the upper left and lower right quadrants tends to be *redshifted*. In other words, some regions of the Galaxy (those in the blueshifted directions) are approaching the Sun, while others (the redshifted ones) are receding from us.

Careful study of the positions and velocities of stars and gas clouds near the Sun leads us to two important conclusions about the motion of the Galactic disk. First, the entire disk is *rotating*—stars, gas, and dust all move in roughly circular paths around the Galactic center, their orbits governed by the Galaxy's gravitational pull. The orbital speed in the vicinity of the Sun is about 220 km/s. Thus, at the Sun's distance of 8 kpc from the Galactic center, material takes about 225 million years (an interval of time sometimes called 1 *Galactic year*) to complete one circuit.

Second, the Galactic rotation period depends on distance from the Galactic center, being shorter closer to the center and longer at greater distances. In other words, the Galactic disk rotates not as a solid object, but *differentially*. Accurate measurements, made by the *Hipparcos* satellite, of stars within a few hundred parsecs of the Sun have proved particularly valuable in measuring these important Galactic properties. ∞ (*Discovery 17-2*) Similar differential rota-

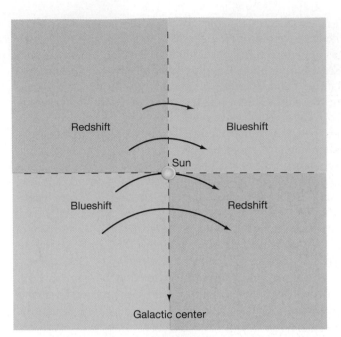

▲ **FIGURE 23.12 Orbital Motion in the Galactic Disk** Stars and interstellar clouds in the neighborhood of the Sun show systematic Doppler motions, implying that the disk of the Galaxy is spinning in a well-ordered way. These four Galactic quadrants are drawn to intersect not at the Galactic center, but at the Sun, the location from which our observations are made. The curved arrows represent the angular speed of the disk material. Because the Sun orbits faster than stars and gas at larger radii, we are moving away from material at the top left and gaining on that at the top right, resulting in the Doppler shifts indicated. Similarly, stars and gas in the bottom left quadrant are gaining on us, while material at the bottom right is pulling away.

tion is observed in Andromeda and many other spiral galaxies.

This picture of orderly circular motion about the Galactic center applies only to the disk: Stars in the Galactic halo and bulge are not so well behaved. The old globular clusters in the halo and the faint, reddish individual stars in both the halo and the bulge do *not* share the disk's well-defined rotation. Instead, their orbital orientations are largely random.* Although these objects do orbit the Galactic center, they move in all directions, their paths filling an entire three-dimensional volume rather than a nearly two-dimensional disk.

Figure 23.13 contrasts the motion of bulge and halo stars with the much more regular orbits of stars in the

Halo stars do, in fact, have some net rotation about the Galactic center, but the rotational component of their motion is overwhelmed by the larger random component. The motion of bulge stars also has a rotational component, larger than that of the halo, but still smaller than the random component of stellar motion in the bulge.

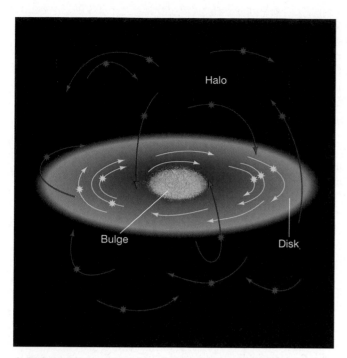

▲ **FIGURE 23.13 Stellar Orbits in Our Galaxy** Stars in the Galactic disk move in orderly, circular orbits about the Galactic center. In contrast, halo stars have orbits with largely random orientations and eccentricities. The orbit of a typical halo star takes it high above the Galactic disk, then down through the disk's plane, and then out the other side and far below the disk. The orbital properties of bulge stars are intermediate between those of disk stars and those of halo stars.

Galactic disk. At any given distance from the Galactic center, bulge or halo stars move at speeds comparable to the disk's rotation speed at that radius, but in *all* directions, not just one. Their orbits carry these stars repeatedly through the plane of the disk and out the other side. (They don't collide with stars in the disk because interstellar distances are huge compared with the diameters of individual stars—a star or even an entire star cluster passes through the disk almost as though it weren't there—see *Discovery 24-1*.) Some well-known stars in the vicinity of the Sun—the bright giant Arcturus, for example—are actually halo stars that are "just passing through" the disk on orbits that take them far above and below the Galactic plane.

Recently, astronomers have detected numerous *tidal streams* in the Galactic halo—groups of stars thought to be the remnants of globular clusters and even small satellite galaxies (see Section 24.1) torn apart by our Galaxy's tidal field. Just as micrometeoroid swarms in our solar system follow the orbit of their disrupted parent comet long after the comet itself is gone, stars in a tidal stream are now spread out around the entire original orbit of their parent cluster or galaxy. ∞ (Sec. 14.3)

Table 23.1 compares some key properties of the three basic components of the Galaxy.

CONCEPT CHECK

✔ Why do astronomers regard the disk and the halo as different components of our Galaxy?

TABLE 23.1 Overall Properties of the Galactic Disk, Halo, and Bulge

Galactic Disk	Galactic Halo	Galactic Bulge
highly flattened	roughly spherical—mildly flattened	somewhat flattened and elongated in the plane of the disk ("football shaped")
contains both young and old stars	contains old stars only	contains both young and old stars; more old stars at greater distances from the center
contains gas and dust	contains no gas and dust	contains gas and dust, especially in the inner regions
site of ongoing star formation	no star formation during the last 10 billion years	ongoing star formation in the inner regions
gas and stars move in circular orbits in the Galactic plane	stars have random orbits in three dimensions	stars have largely random orbits, but with some net rotation about the Galactic center
spiral arms	no obvious substructure	ring of gas and dust near center; central galactic nucleus
overall white coloration, with blue spiral arms	reddish in color	yellow-white

23.4 The Formation of the Milky Way

Is there some evolutionary scenario that can naturally account for the Galactic structure we see today? The answer is that there is, and it takes us all the way back to the birth of our Galaxy, more than 10 billion years ago. ∞ (Sec. 20.5) Not all the details are agreed upon by all astronomers, but the overall picture is now fairly widely accepted. For simplicity, we confine our discussion here to the Galactic disk and halo; in many ways, the bulge is intermediate in its properties between these two extremes.

Figure 23.14 illustrates the current view of our Galaxy's evolution, starting (not unlike the star-formation scenario outlined in Chapter 19) from a contracting cloud of pregalactic gas. ∞ (Sec. 19.1) When the first Galactic stars and globular clusters formed, the gas in our Galaxy had not yet accumulated into a thin disk. Instead, it was spread out over an irregular and quite extended region of space, spanning many tens of kiloparsecs in all directions. When the first stars formed, they were distributed throughout that volume. Their distribution today (the Galactic halo) reflects that fact—it is an imprint of their

birth. Many astronomers believe that the very first stars formed even earlier, in smaller systems that later merged to create our Galaxy (Figure 23.14a). Probably, many more stars were born during the mergers themselves, as interstellar gas clouds collided and began to collapse. ∞ (Sec.19.5) Whatever the details, the present-day halo would look much the same in either case.

Since those early times, rotation has flattened the gas in our Galaxy into a relatively thin disk. Physically, the process is similar to the flattening of the solar nebula during the formation of the solar system, as described in Chapter 15, except on a vastly larger scale. ∞ (Sec. 6.7) Star formation in the halo ceased billions of years ago when the raw materials fell to the Galactic plane. Ongoing star formation in the disk gives it its bluish tint, but the halo's short-lived blue stars have long since burned out, leaving only the long-lived red stars that give the halo its characteristic pinkish glow. The Galactic halo is ancient, whereas the disk is full of youthful activity. The thick disk, with its intermediate-age stars, may represent an intermediate stage of star formation that occurred while the gas was still flattening into the plane.

The chaotic orbits of the halo stars are also explained by this theory. When the halo developed, the ir-

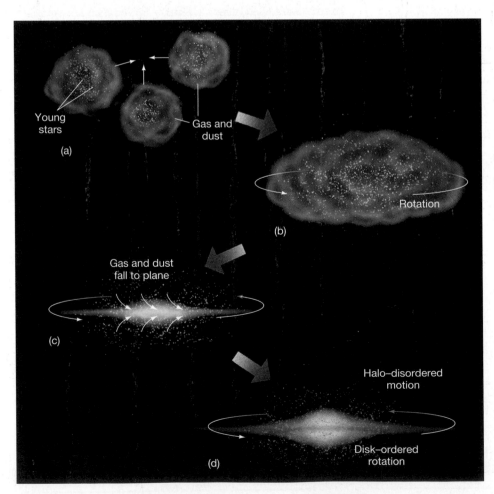

Young stars

Gas and dust

(a)

Rotation

(b)

Gas and dust fall to plane

(c)

Halo–disordered motion

Disk–ordered rotation

(d)

◀ **FIGURE 23.14 Milky Way Formation** (a) The Milky Way Galaxy may have formed through the merger of several smaller systems. (b) Astronomers reason that, early on, our Galaxy was irregularly shaped, with gas distributed throughout its volume. When stars formed during this stage, there was no preferred direction in which they moved and no preferred location in which they were found. (c) In time, the gas and dust fell to the Galactic plane and formed a spinning disk. The stars that had already formed were left behind in the halo. (d) New stars forming in the disk inherit its overall rotation and so orbit the Galactic center on ordered, circular orbits.

regularly shaped Galaxy was rotating only very slowly, so there was no strongly preferred direction in which matter tended to move. As a result, halo stars were free to travel along nearly any path once they formed (or when their parent systems merged), leading to the random halo orbits we observe today. As the Galactic disk formed, however, conservation of angular momentum caused it to spin more rapidly. Stars that formed from the gas and dust of the disk inherited its rotational motion and so move on well-defined, circular orbits. Again, the thick disk's orbital properties are consistent with the idea that it formed while gas was still sinking to the Galaxy's midplane.

In principle, the structure of our Galaxy bears witness to the conditions that created it. In practice, however, the interpretation of the observations is made difficult by the sheer complexity of the system we inhabit and by the many competing physical processes that have modified its appearance since it formed. As a result, the early stages of the Milky Way are still quite poorly understood. We will return to the subject of galaxy formation in Chapters 24 and 25.

CONCEPT CHECK

✔ Why are there no young halo stars?

23.5 Galactic Spiral Arms

If we want to look beyond our immediate neighborhood and study the full extent of the Galactic disk, we cannot rely on optical observations, as interstellar absorption severely limits our vision. In the 1950s, astronomers developed a very important tool for exploring the distribution of gas in our Galaxy: spectroscopic radio astronomy.

Radio Maps of the Milky Way

The keys to observing Galactic interstellar gas are the 21-cm radio emission line produced by atomic hydrogen and the many radio molecular lines formed in molecular cloud complexes. ∞ (Sec. 18.4) Long-wavelength radio waves are largely unaffected by interstellar dust, so they travel more or less unimpeded through the Galactic disk, allowing us to "see" to great distances. Because hydrogen is by far the most abundant element in interstellar space, the 21-cm signals are strong enough that a large portion of the disk can be observed in this way. As noted in Chapter 18, observations of spectral lines from "tracer" molecules, such as carbon monoxide, allow us to study the distribution of the densest interstellar clouds. ∞ (Sec. 18.5)

Earlier, we noted that observations of stars within several hundred parsecs of the Sun have allowed astronomers to measure the rotation rate of the Galaxy in the solar neighborhood. As indicated in Figure 23.15, in order to probe to greater distances, astronomers often turn to radio observations (illustrated here with 21-cm radiation), because long-wavelength radio waves are largely unaffected by interstellar dust, allowing astronomers to study virtually the entire Galactic disk. ∞ (Sec. 18.4)

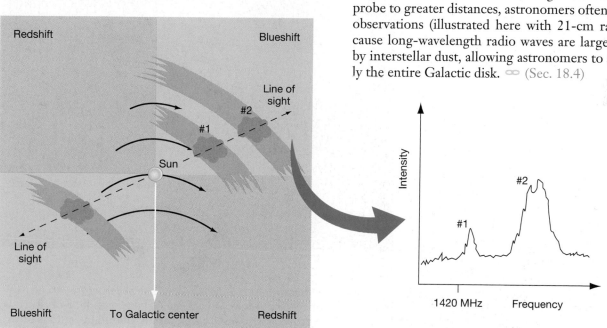

▲ **FIGURE 23.15 Gas in the Galactic Disk** Because the disk of our Galaxy is rotating differentially, 21-cm radio signals from different clumps of hydrogen matter along any given line of sight are Doppler shifted by different amounts, allowing both the densities and the rotation rates of those clumps to be measured. As in Figure 23.12, the black arrows represent the angular speed at which gas orbits the Galactic center. The graph at the right is a typical radio spectrum, showing how one clump that is part of an outer spiral arm (upper right quadrant) is blueshifted more than another clump (in an inner spiral arm). By repeating these observations in many different directions, astronomers map out the distribution of gas in our Galaxy.

However, the distances to the clouds emitting the radio radiation are often poorly known. How, then, can we determine just where in the disk a cloud lies? Astronomers accomplish this by using all available data, coupled with our knowledge of Newtonian mechanics, to construct a *mathematical model* of the rotation of stars and gas throughout the Galactic disk. ∞ (Sec. 2.7) Assuming circular orbits, the model allows us to turn a measured radial velocity into a distance along the line of sight. As in so many areas of astronomy, theory and observations complement one another: The data refine the theoretical model, while the model in turn provides the framework needed to interpret further observations. ∞ (Sec. 1.2)

Radio astronomers couple their observations with this Galactic model to turn their measurements into detailed information about the distribution of gas along the line of sight. Because of differential rotation, the measured velocity of a cloud depends on its *distance* from the Sun (Figure 23.15), and the Galactic model provides the connection between the two. Furthermore, the strength of the signal is a measure of the *density* of gas in the cloud—denser clouds contain more gas and emit more radiation. Thus, knowing direction, distance, and density, astronomers can use observations along different lines of sight to map out the radio-emitting gas in our Galaxy.

Spiral Structure

Interstellar gas in the Galactic disk exhibits an organized pattern on a grand scale. Near the center, the gas in the disk fattens markedly in the Galactic bulge. Radio-emitting gas has been observed out to at least 50 kpc from the Galactic center. Over much of the inner 20 kpc or so of the disk, the gas is confined within about 100 pc of the Galactic plane. Beyond that distance, the gas distribution spreads out somewhat, to a thickness of several kiloparsecs, and shows definite signs of being "warped," possibly because of the gravitational influence of a pair of nearby galaxies (to be discussed in Chapter 24; see also Figure 23.16).

Radio studies provide perhaps the best direct evidence that we live in a spiral galaxy. Figure 23.16 is an artist's conception (based on observational data) of the appearance of our Galaxy as seen from far above the disk. The figure clearly shows our Galaxy's **spiral arms**, pinwheellike structures originating close to the Galactic bulge and extending outward throughout much of the Galactic disk. Our Sun lies near the edge of one of these arms, which wraps around a large part of the disk. Notice, incidentally, the scale markers on Figures 23.9, 23.10, and 23.15: The Galactic globular-cluster distribution (Figure 23.9), the luminous stellar component of the disk (Figure 23.10), and the known spiral structure (Figure 23.16) all have roughly the *same* diameter—about 30 kpc. This scale is fairly typical of spiral galaxies observed elsewhere in the universe.

Persistence of the Spiral Arms

The spiral arms in our Galaxy are made up of much more than just interstellar gas and dust. Studies of the Galactic disk within a kiloparsec or so of the Sun indicate that young stellar and prestellar objects—emission nebulae, O- and B-type stars, and recently formed open clusters—are also distributed in a spiral pattern that closely follows the distribution of interstellar clouds. The obvious conclusion is that the spiral arms are the part of the Galactic disk where star formation takes place. The brightness of the young stellar objects just listed is the main reason that the spiral arms of other galaxies are easily seen from afar (e.g., Figure 23.3a).

A central problem facing astronomers trying to understand spiral structure is how that structure persists over long periods. The basic issue is simple: Differential rotation makes it impossible for any large-scale structure "tied" to the disk material to survive. Figure 23.17 shows how a spiral pattern consisting always of the same group of stars

30 kpc

▲ **FIGURE 23.16 Milky Way Spiral Structure** An artist's conception of our Milky Way Galaxy, seen face-on. This illustration is based on data accumulated by legions of astronomers during the past few decades, including radio maps of gas density in the Galactic disk. Painted from the perspective of an observer 100 kpc above the Galactic plane, the spiral arms are at their best-determined positions. All the features are drawn to scale (except for the oversized yellow dot near the top, which represents our Sun). The two small blotches to the left are irregular dwarf galaxies, called the Magellanic Clouds. We will study them in Chapter 24. (*L. Chaisson*)

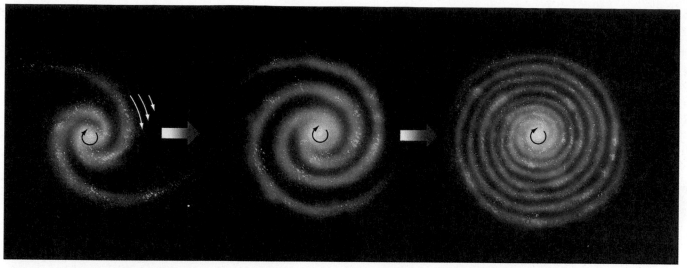

▲ **FIGURE 23.17 Differential Galactic Rotation** The disk of our Galaxy rotates differentially—stars close to the center take less time to orbit the Galactic center than those farther out. (The red arrows represent angular speed.) If spiral arms were somehow tied to the material of the Galactic disk, this differential rotation would cause the spiral pattern to disappear in a few hundred million years. Spiral arms would be too short lived to be consistent with the numbers of spiral galaxies we observe today.

and gas clouds would necessarily disappear within a few hundred million years. How, then, do the Galaxy's spiral arms retain their structure over long periods in spite of differential rotation?

A leading explanation for the existence of spiral arms holds that they are **spiral density waves**—coiled waves of gas compression that move through the Galactic disk, squeezing clouds of interstellar gas and triggering the process of star formation as they go. ∞ (Sec. 19.5) The spiral arms we observe are defined by the denser-than-normal clouds of gas the density waves create and by the new stars formed as a result of the spiral waves' passage.

This explanation of spiral structure avoids the problem of differential rotation, because the wave pattern is not tied to any particular piece of the Galactic disk. The spirals we see are merely patterns moving through the disk, not great masses of matter being transported from place to place. The density wave moves through the collection of stars and gas making up the disk just as a sound wave moves through air or an ocean wave passes through water, compressing different parts of the disk at different times. Even though the rotation rate of the disk material varies with distance from the Galactic center, the wave itself remains intact, defining the Galaxy's spiral arms.

In fact, over much of the visible portion of the Galactic disk (within about 15 kpc of the center), the spiral wave pattern is predicted to rotate *more slowly* than the stars and gas. Thus, as shown in Figure 23.18, Galactic material catches up with the wave, is temporarily slowed down and

R I V U X G

▲ **FIGURE 23.18 Density Wave Theory** This theory holds that the spiral arms seen in our own and many other galaxies are waves of gas compression moving through the material of the Galactic disk and forming stars as they go. In the painting at the right, gas motion is indicated by red arrows and arm motion by white arrows. Gas enters an arm from behind, is compressed, and forms stars. The spiral pattern is delineated by dust lanes, regions of high gas density, and newly formed O- and B-type stars. The inset shows the spiral galaxy NGC 1566, which displays many of the features just described. *(AURA)*

compressed as it passes through, and then continues on its way. (For a more down-to-earth example of an analogous process, see *Discovery 23-2*.)

As material enters the density wave from behind, the gas is compressed and forms stars. Dust lanes mark the regions of highest-density gas. The most prominent stars—the bright O- and B-type blue giants—live for only a short time, so young stellar associations, emission nebulae, and open clusters with long main sequences are found only within the arms, near their birth sites, just ahead of the dust lanes. The brightness of these young systems emphasizes the spiral structure. Further downstream, ahead of the spiral arms, we see mostly older stars and star clusters. These objects have had enough time since their formation to out-

DISCOVERY 23-2

Density Waves

In the late 1960s, American astrophysicists C. C. Lin and Frank Shu proposed a way in which spiral arms in the Galaxy could persist for many Galactic rotations. They argued that the arms themselves contain no "permanent" matter. They should thus not be viewed as assemblages of stars, gas, and dust moving intact through the disk—those would quickly be destroyed by differential rotation. Instead, a spiral arm should be envisaged as a density wave—a wave of compression and expansion sweeping through the Galaxy.

A wave in water builds up material temporarily in some places (crests) and lets it down in others (troughs). Similarly, as the spiral density wave encounters galactic matter, the gas is compressed to form a region of slightly higher than normal density. Galactic material enters the wave, is temporarily slowed down and compressed as it passes through, and then continues on its way. The compression triggers the formation of new stars and nebulae. In this way, the spiral arms are formed and re-formed repeatedly, without disappearing completely. Lin and Shu showed that the process can in fact maintain a spiral pattern for very long periods.

The accompanying figure illustrates the formation of a density wave in a much more familiar context: a traffic jam on a highway, triggered by the presence of a repair crew moving slowly down the road. As cars approach the crew, they slow down temporarily. Then they speed up again as they pass the work site and continue on their way. The result, as might be reported by a high-flying traffic helicopter, is a region of high traffic density, concentrated around the location of the work crew and moving with it. An observer on the side of the road, however, sees that the jam never contains the same cars for very long. Cars constantly catch up to the bottleneck, move slowly through it, and then speed up again, only to be replaced by more cars arriving from behind.

The traffic jam is analogous to the region of high stellar density in a Galactic spiral arm. Just as the traffic density wave is not tied to any particular group of cars, the spiral arms are not attached to any particular piece of disk material. Stars and gas enter a spiral arm, slow down for a while, then continue on their orbits around the Galactic center. The result is a moving region of high stellar and gas density, involving different parts of the disk at different times. Notice also that, just as in our Galaxy, the wave moves more slowly than, and independently of, the overall traffic flow.

We can extend our traffic analogy a little further. Most drivers are well aware that the effects of such a tie-up can persist long after the road crew responsible for it has stopped work and gone home for the night. Similarly, spiral density waves can continue to move through the disk even after the disturbance that originally produced them has long since subsided. According to spiral density wave theory, that is precisely what has happened in the Milky Way. Some disturbance in the past produced the wave, which has been moving through the Galactic disk ever since.

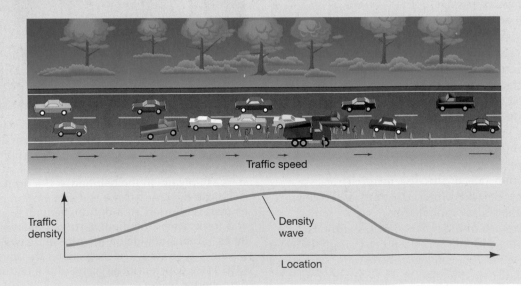

distance the wave and pull away from it. Over millions of years, their random individual motions, superimposed on their overall rotation around the Galactic center, distort and eventually destroy their original spiral configuration, and they become part of the general disk population.

Note, incidentally, that, although the spirals shown in Figure 23.18 have two arms each, astronomers are not certain how many arms make up the spiral structure in our own Galaxy (see Figure 23.16). The theory makes no strong predictions on this point.

An alternative possibility is that the formation of stars drives the waves, instead of the other way around. Imagine a row of newly formed massive stars somewhere in the disk. The emission nebula created when these stars form, and the supernovae when they die, send shock waves through the surrounding gas, triggering new star formation. ∞ (Sec. 21.5) Thus, as illustrated in Figure 23.19(a), the formation of one group of stars provides the mechanism for the creation of others. Computer simulations suggest that it is possible for the "wave" of star formation created in this manner to take on the form of a partial spiral and for the pattern to persist for some time. However, the process, sometimes known as **self-propagating star formation**, can produce only pieces of spirals, as are seen in some galaxies (Figure 23.19b). It apparently cannot produce the galaxywide spiral arms seen in other galaxies and present in our own. It may well be that there is more than one process at work in the spectacular spirals we see.

Origin of Spiral Structure

An important question (but one that unfortunately is not answered by either of the two theories just described) is Where do these spirals come from? What was responsible for generating the density wave in the first place or for creating the line of newborn stars whose evolution drives the advancing spiral arm? Scientists speculate that (1) the gravitational effects of our satellite galaxies (the Magellanic Clouds, to be discussed in Chapter 24), (2) instabilities in the gas near the Galactic bulge, or (3) the possible barlike asymmetry within the bulge itself may have had a big enough influence on the disk to get the process going.

The first possibility is supported by growing evidence that many other spiral galaxies seem to have been affected by gravitational interactions with neighboring systems in the relatively recent past (see Chapter 24). However, many astronomers still regard the other two possibilities as equally likely. For example, they point to *isolated* spirals, whose structure clearly cannot be the result of an external interaction. The fact is that we still don't know for sure how galaxies—including our own—acquire such beautiful spiral arms.

CONCEPT CHECK

✔ Why can't spiral arms simply be clouds of gas and young stars orbiting the Galactic center?

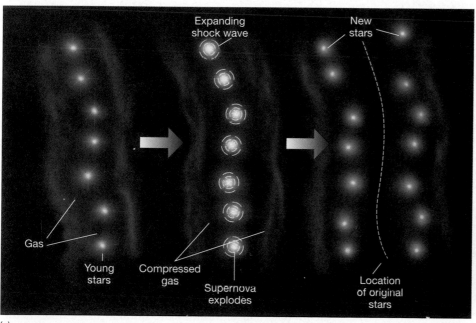

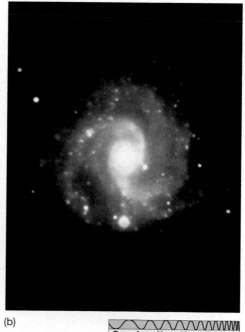

(a)   (b)

▲ **FIGURE 23.19 Self-Propagating Star Formation** (a) In this theory of the formation of spiral arms, the shock waves produced by the formation and later evolution of a group of stars provide the trigger for new rounds of star formation. We have used supernova explosions to illustrate the point here, but the formation of emission nebulae and planetary nebulae is also important. (b) This process may well be responsible for the partial spiral arms seen in some galaxies, such as NGC 3184, shown here in true color. The distinct blue appearance of this galaxy derives from the vast numbers of young stars that pepper its ill-defined spiral arms. *(R. Gendler)*

23.6 The Mass of the Milky Way Galaxy

5 We can measure our Galaxy's mass by studying the motions of gas clouds and stars in the Galactic disk. Recall from Chapter 2 that Kepler's third law (as modified by Newton) connects the period, orbital size, and masses of any two objects in orbit around each other: ∞ (Sec. 2.7)

$$\text{total mass (solar masses)} = \frac{\text{orbital size (A.U.)}^3}{\text{orbital period (years)}^2}.$$

As we saw earlier, the distance from the Sun to the Galactic center is about 8 kpc, and the Sun's orbital period is 225 million years. Substituting these numbers into the preceding equation, we obtain a mass of $(8,000 \times 206,000)^3/(225,000,000)^2$, or almost 9×10^{10} solar masses—90 *billion* times the mass of our Sun! The Milky Way Galaxy is truly enormous, in mass as well as in size.

But what mass have we just measured? When we performed the analogous calculation in the case of a planet orbiting the Sun, there was no ambiguity: The result of our calculation was the mass of the Sun. ∞ (Sec. 2.7) However, the Galaxy's matter is not concentrated at the Galactic center (as the Sun's mass is concentrated at the center of the solar system); instead, Galactic matter is distributed over a large volume of space. Some of it lies inside the Sun's orbit (i.e., within 8 kpc of the Galactic center), and some lies outside, at large distances from both the Sun and the center of the Galaxy. What portion of the Galaxy's mass controls the Sun's orbit? Isaac Newton answered this question three centuries ago: The Sun's orbital period is determined by the portion of the Galaxy that lies *within the orbit of the Sun* (Figure 23.20). This is the mass computed in the foregoing equation.

Dark Matter

The Sun's motion around the Galactic center tells us that the total Galactic mass within the Sun's orbit is about 90 billion solar masses, but it says nothing about the mass

lying outside that orbit—that is, more than 8 kpc from the center. To determine the mass of the Galaxy on larger scales, we must measure the orbital motion of stars and gas at greater distances from the Galactic center. Astronomers have found that the most effective way to do this is to make radio observations of gas in the Galactic disk, because radio waves are relatively unaffected by interstellar absorption and allow us to probe to great distances, far beyond the Sun's orbit. On the basis of these studies, radio astronomers have determined our Galaxy's rotation rate at various distances from the Galactic center. The resultant plot of rotation speed versus distance from the center (Figure 23.21) is called the Galactic **rotation curve**.

Knowing the Galactic rotation curve, we can now repeat our earlier calculation to compute the total mass that lies within any given distance from the Galactic center. We find, for example, that the mass within about 15 kpc from the center—the volume defined by the globular clusters and the known spiral structure—is roughly 2×10^{11} solar masses, about twice the mass contained within the Sun's orbit. Does the distribution of matter in the Galaxy "cut off" at this point, where the luminosity drops off sharply? Surprisingly, it does not.

Newton's laws of motion predict that if all of the mass of the Galaxy were contained within the edge of the visible structure, then the orbital speed of stars and gas beyond 15 kpc would decrease with increasing distance from the Galactic center, just as the orbital speeds of the planets diminish as we move outward from the Sun. The dashed line in Figure 23.21 indicates what the rotation curve would look like in that case. However, the true rotation curve is quite different: Far from falling off at larger distances, it *rises* slightly, out to the limits of our measurement capabilities. This slight rise implies that the amount of mass contained within successively larger radii continues to grow beyond the orbit of the Sun, apparently out to a distance of at least 40 or 50 kpc.

According to the equation presented at the beginning of this section, the amount of mass within 40 kpc is approximately 6×10^{11} solar masses. Since 2×10^{11} solar masses lie within 15 kpc of the Galactic center, we have to conclude that at least twice as much mass lies *outside* the luminous part of our Galaxy—the part made up of stars, star clusters, and spiral arms—as lies inside!

On the basis of these observations of the Galactic rotation curve, astronomers now believe that the luminous portion of the Milky Way Galaxy—the region outlined by the globular clusters and by the spiral arms—is merely the "tip of the Galactic iceberg." Our Galaxy is in reality very much larger. The luminous region is surrounded by an extensive, invisible **dark halo**, which dwarfs the inner halo of stars and globular clusters and extends well beyond the 15-kpc radius once thought to represent the limit of our Galaxy. But what is the composition of this dark halo? We do not detect enough stars or interstellar matter to account

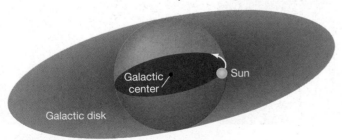

▲ **FIGURE 23.20 Weighing the Galaxy** The orbital speed of a star or gas cloud moving around the Galactic center is determined only by the mass of the Galaxy lying inside the orbit (within the grey-shaded sphere). Thus, to measure the Galaxy's total mass, we must look for objects orbiting at large distances from the center.

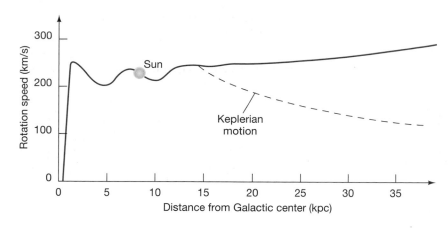

◄ FIGURE 23.21 Galaxy Rotation Curve
The rotation curve for the Milky Way Galaxy plots rotation speed versus distance from the Galactic center. We can use this curve to compute the mass of the Galaxy that lies within any given radius. The dashed curve is the rotation curve expected if the Galaxy "ended" abruptly at a radius of 15 kpc, the limit of most of the known spiral structure and the globular cluster distribution. The fact that the red curve does not follow this dashed line, but instead stays well above it, indicates that additional unseen matter must be beyond that radius.

for the mass that our computations tell us must be there. We are inescapably drawn to the conclusion that most of the mass in our Galaxy exists in the form of invisible **dark matter**, which we presently do not understand.

The term *dark* here does not refer just to matter that is undetectable in visible light: The material has (so far) escaped detection at *all* wavelengths, from radio to gamma rays. Only by its gravitational pull do we know of its existence. Dark matter is not hydrogen gas (atomic or molecular), nor is it made up of ordinary stars. Given the amount of matter that must be accounted for, we would have been able to detect it with present-day instruments if it were in either of those forms. Its nature and its consequences for the evolution of galaxies and the universe are among the most important questions in astronomy today.

Many candidates have been suggested for this dark matter, although none is proven. Stellar-mass black holes may supply some of the unseen mass, but given that they are the evolutionary products of (relatively rare) massive stars, it is unlikely that there could be enough of them to

hide large amounts of Galactic matter. ∞ (Sec. 22.8) Currently among the strongest "stellar" contenders are brown dwarfs—low-mass prestellar objects that never reached the point of core nuclear burning—white dwarfs, and faint, low-mass red dwarfs. ∞ (Secs. 19.3, 20.3) In the jargon of the field, these objects are collectively known as *MAssive Compact Halo Objects*, or MACHOs for short. In principle, they could exist in great numbers throughout the Galaxy, yet would be exceedingly hard to see because they are so faint.

Hubble Space Telescope observations of globular clusters seem to argue against at least the last of the three possibilities listed for MACHOs. Figure 23.22 shows a *Hubble* image of a relatively nearby globular cluster—one close enough that very faint red dwarfs could have been detected if any existed. The *Hubble* data suggest that there is a cutoff at about 0.2 solar mass, below which stars form much less frequently than had previously been supposed. As a result, stars with very low mass may be unexpectedly rare, at least in the Galactic halo.

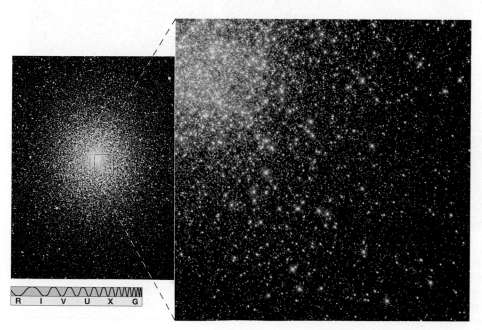

◄ FIGURE 23.22 Missing Red Dwarfs Sensitive visible-light observations with the *Hubble Space Telescope* have apparently ruled out faint red-dwarf stars as candidates for dark matter. The object shown here, the globular cluster 47 Tucanae, is one of many regions searched in the Milky Way. The inset, 0.4 pc on a side, is a high-resolution *Hubble* image of part of the cluster. The red dwarfs that would be expected if they existed in sufficient numbers to account for the dark matter in the Galaxy are not found. (The red stars that are seen are giants.) *(AAT; NASA)*

A radically different alternative is that the dark matter is made up of exotic *subatomic particles* that pervade the entire universe. In order to account for the properties of dark matter, these particles must have mass (to produce the observed gravitational effects), but also must interact hardly at all with "normal" matter (because otherwise we would be able to see them). One class of candidate particles satisfying these requirements has been dubbed *W*eakly *I*nteracting *M*assive *P*articles, or WIMPs. Many theoretical astrophysicists believe that such "dark-matter particles" could have been produced in abundance during the very earliest moments of our universe. If they survived to the present day, there might be enough of them to account for all the dark matter we believe must be out there. We will discuss this possibility and its far-reaching implications in more detail in Chapter 27. These ideas are hard to test, however, because these particles would necessarily be very difficult to detect. Several detection experiments on Earth have been attempted, so far without success.

The Search for Stellar Dark Matter

Recently, researchers have obtained insight into the distribution of stellar dark matter by using a key element of Albert Einstein's theory of general relativity (see Chapter 22): the prediction that a beam of light can be deflected by a gravitational field, which has already been verified in the case of starlight that passes close to the Sun. ∞ (*More Precisely 22-2*) The effect is small in the case of light grazing the Sun, but it has the potential for making distant and otherwise invisible stellar objects observable from Earth. Here's how:

Imagine looking at a distant star as a faint foreground object (a MACHO, such as a brown or white dwarf) happens to cross your line of sight. As illustrated in Figure 23.23, the intervening object deflects a little more starlight than usual toward you, resulting in a temporary, but quite substantial, *brightening* of the distant star. In some ways, the effect is like the focusing of light by a lens, so the process is known as **gravitational lensing**. The fore-

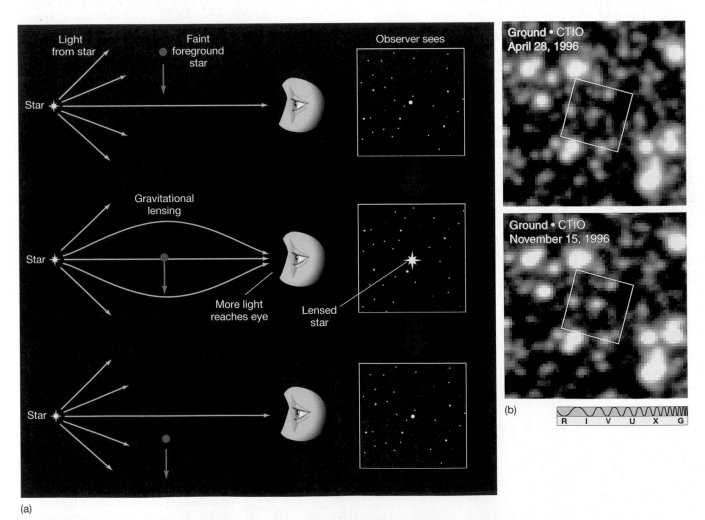

(a)

(b)

R I V U X G

▲ **FIGURE 23.23 Gravitational Lensing** (a) Gravitational lensing by a faint foreground object (such as a brown dwarf) can temporarily cause a background star to brighten significantly, providing a means of detecting otherwise invisible stellar dark matter. (b) The brightening of a star during a lensing event, this one implying that a massive, but unseen, object passed in front of the unnamed star at the centers of the two boxes imaged six months apart. (*AURA*)

TUTORIAL Gravitational Lensing

ground object is referred to as a *gravitational lens*. The amount of brightening and the duration of the effect depend on the mass, distance, and speed of the lensing object. Typically, the apparent brightness of the background star increases by a factor of two to five for a period of several weeks. Thus, even though the foreground object cannot be seen directly, its effect on the light of the background star makes it detectable. (In Chapter 25, we will encounter other instances of gravitational lensing in the universe, but on very much larger scales.)

Of course, stars are very small compared with the distance scale of the Galaxy, and the probability that one star will pass almost directly in front of another, as seen from Earth, is extremely low. But by observing millions of stars every few days over a period of years (using automated telescopes and high-speed computers to reduce the burden of coping with so much data), astronomers have been able to see enough of these events to let them estimate the amount of stellar dark matter in the Galactic halo. The technique represents an exciting new means of probing the structure of our Galaxy. The first lensing events were reported in late 1993. Subsequent observations are consistent with lensing by low-mass white dwarfs and suggest that such stars could account for at least half—but, it now seems, apparently not all—of the dark matter inferred from dynamical studies.

Bear in mind, though, that the identity of the dark matter is not necessarily an all-or-nothing proposition. It is perfectly conceivable—and, in fact, most astronomers think it likely—that more than one type of dark matter exists. For example, it is quite possible that most of the dark matter in the inner (visible) parts of galaxies is in the form of brown dwarfs and very low mass stars, while the dark matter farther out may be primarily in the form of exotic particles. We will return to this perplexing problem in later chapters, when we discuss some theories of how galaxies form and evolve, and how matter in the universe may have come into being.

CONCEPT CHECK

✔ In what sense is dark matter "dark"?

23.7 The Galactic Center

Theory predicts that the Galactic bulge, and especially the region close to the Galactic center, should be densely populated with billions of stars. However, we are unable to see this region of our Galaxy—the interstellar medium in the Galactic disk shrouds what otherwise would be a stunning view. Figure 23.24 shows the (optical) view we do have of the region of the Milky Way toward the Galactic center, in the general direction of the constellation Sagittarius.

With the help of infrared and radio techniques, we can peer more deeply into the central regions of our Galaxy than we can by optical means. Infrared observations (Figure 23.25a) indicate that the heart of our Galaxy harbors roughly 50,000 stars per cubic parsec—a stellar density about a million times greater than that in our solar neighborhood and high enough that stars must experience frequent close encounters and even collisions. Infrared radiation has also been detected from what appear to be huge clouds rich in dust. In addition, radio observations indicate a ring of molecular gas nearly 400 pc across, containing some 30,000 solar masses of material and rotating around the Galactic center at about 100 km/s. The origin of this ring is unclear, although researchers suspect that the gravitational influence of our Galaxy's elongated, rotating bulge may well be involved.

High-resolution radio observations show more structure on smaller scales. Figure 23.25(b) shows a region called Sagittarius A. (The name simply means that it is the brightest radio source in the constellation Sagittarius.) It lies at the center of the boxed region in Figure 23.22 and Figure 23.25(a)—and, we think, at the center of our Galaxy. On a scale of about 100 pc, extended filaments can be seen. Their presence suggests to many astronomers that strong magnetic fields operate in the vicinity of the center, creating structures similar in appearance to (but much larger than) those observed on the active Sun. On even smaller scales (Figure 23.25c), the observations indicate a rotating ring or disk of matter only a few parsecs across—and within that, as revealed by the *Chandra X–Ray Observatory* (Figure 23.25d), lies a bright X-ray source apparently embedded within a supernova remnant.

What could cause all this activity? An important clue comes from the Doppler broadening of infrared spectral lines emitted from the central swirling whirlpool of gas. The extent of the broadening indicates that the gas is moving very rapidly. In order to keep this gas in orbit, whatever is at the center must be extremely massive—more than a million solar masses. Given the twin requirements of large mass and small size, a leading contender is a supermassive black hole.

The hole itself is not the source of the energy, of course. Instead, the vast accretion disk of matter being drawn toward the hole by the enormous gravity emits the energy as it falls in, just as we saw (on a much smaller scale) in Chapter 22 when we discussed X-ray emission from neutron stars and stellar-mass black holes. ∞ (Secs. 22.3 and 22.8) The strong magnetic fields, thought to be generated within the accretion disk as matter spirals inward, may act as "particle accelerators," creating extremely high energy particles detected on Earth as *cosmic rays*. In the late 1990s, the *Compton Gamma Ray Observatory* found indirect evidence for a fountain of high-energy particles, possibly produced by violent processes close to the event horizon, gushing outward from the hole into the halo more than a thousand parsecs beyond the Galactic center.

1 pc

◄ **FIGURE 23.24 Galactic Center in Visible Light** A photograph of stellar and interstellar matter in the direction of the Galactic center. Because of heavy obscuration, even the largest optical telescopes can see no farther than one-tenth the distance to the center. The M8 nebula (arrow) can be seen at the extreme top center. The field is roughly 20° across and is a continuation of the bottom part of Figure 18.6. (Cf. Figure 18.1.) The overlaid box outlines the location of the center of our Galaxy, and the inset shows the best optical view in that direction, heavily confused by stars along the line of sight. (The arrow points to the dynamical core of the Milky Way.) *(AURA; ESO)*

Astronomers have reason to suspect that similar events are occurring at the centers of many other galaxies.

At the very center of our Galaxy, at the heart of Sagittarius A, is a remarkable object with the odd-sounding name Sgr A* (pronounced "saj ay star"). By the standards of the active galaxies to be studied in Chapter 24, this compact *Galactic nucleus* is not particularly energetic. Still, radio observations made during the past two decades, along with more recent X- and gamma-ray observations, suggest that it is nevertheless a pretty violent place. Its total energy output (at all wavelengths) is estimated to be 10^{33} W, which is more than a million times that of the Sun.

VLBI observations using radio telescopes arrayed from Hawaii to Massachusetts imply that Sgr A* cannot be much larger than 10 A.U., and it is probably a good deal smaller than that. ∞ (Sec. 5.6) This size is consistent with the view that the energy source is a massive black hole. Figure 23.26 is perhaps the strongest evidence to date supporting the black-hole picture. It shows a high-resolution infrared image of the innermost 0.05 pc (or 10,000 A.U. across) near the Galactic center, centered on Sgr A*. Using advanced adaptive-optics techniques on the Keck telescopes and the VLT, U.S. and European researchers have created the first-ever diffraction-limited (0.05″ resolution) images of the region. ∞ (Secs. 5.4, 17.5)

Remarkably, the quality of the image is good enough that the *proper motions* of several of the stars—their orbits around the Galactic center—can clearly be seen. The inset shows a series of observations of one of the brightest stars—called S2—over a 10 year period. The motion is consistent with an orbit around a massive object at the location of Sgr A*, in accordance with Newton's laws of motion. ∞ (Sec. 2.7) The solid curve on the figure shows the elliptical orbit that best fits the observations: a 15-year orbit with a semimajor axis of 950 A.U., corresponding (from Kepler's third law, as modified by Newton) to a central mass of 3.7 million solar masses.

Other observations, using adaptive-optics infrared imaging techniques, have revealed a bright source very close to Sgr A* that seems to vary with a 10-minute period. ∞ (Sec. 5.4) The source could be a hot spot on the accretion disk that circles the purported hole. Note that, even with the large mass just mentioned, if Sgr A* is a genuine black hole, the size of its event horizon is still only 0.02 A.U. ∞ (Sec. 22.5) Such a small region, 8 kpc away, is currently unresolvable with any telescope now in existence.

Figure 23.27 places these findings into a simplified perspective. Each frame is centered on the Galaxy's core, and each increases in resolution by a factor of 10. Frame (a) renders the Galaxy's overall shape, as painted in Figure 23.16. The scale of this frame measures about 100 kpc from top to bottom. Frame (b) spans a distance of 10 kpc from top to bottom and is nearly filled by the great circular sweep of the innermost spiral arm. Moving inward to a 1-kpc span, frame (c) depicts the 400-pc ring of matter mentioned earlier. The dark blobs represent giant molecu-lar clouds, the small red-white patches emission nebulae associated with star formation within those clouds.

In frame (d), at 100 pc, a pinkish (thin, warm) region of ionized gas surrounds the reddish (thicker, warmer) heart of the Galaxy, corresponding to the images shown in Figure 23.25(b) and (c). The source of energy producing this vast ionized cloud is the activity in the Galactic center. Frame (e), spanning 10 pc, depicts, up close, the tilted, spinning whirlpool of hot (10^4 K) gas and myriad stars (many more than depicted here) seen in Figure 23.25(c).

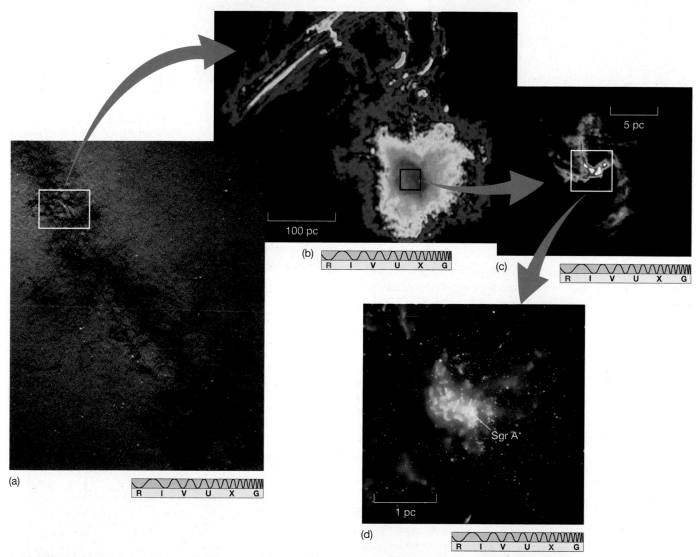

▲ **FIGURE 23.25 Galactic Center Close-up** (a) An infrared image of the region around the center of our Galaxy shows many bright stars packed into a relatively small volume. The average density of matter in this region is estimated to be about a million times that in the solar neighborhood. (b) The central portion of our Galaxy, as seen in the radio part of the spectrum. This image shows a region about 200 pc across surrounding the Galactic center (which lies within the bright blob at the bottom right). The long-wavelength radio emission cuts through the Galaxy's dust, providing a view of matter in the immediate vicinity of the Galaxy's center. (c) The spiral pattern of radio emission arising from Sagittarius A, the center of the Galaxy. The data suggest a rotating ring of matter only 5 pc across. (d) A recent *Chandra* image showing the relation of a hot supernova remnant (red) and Sgr A*, the suspected black hole at the very center of our Galaxy. All the images are false color, since they lie outside the visible spectrum. *(UMass/Caltech; NRAO; NASA)*

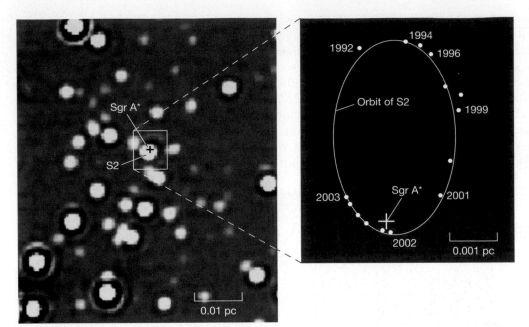

◄ **FIGURE 23.26 Orbits Near the Galactic Center** This extremely close-up map of the Galactic center (left) was obtained by infrared adaptive optics, resulting in a high-resolution image of the innermost 0.1 pc of the Milky Way. The resolution is high enough that the orbits of individual stars can be tracked with confidence around a still-unseen object, called Sgr A* and marked with a cross. The inset shows the orbit of the central star in the frame, labeled S2, between 1992 and 2003. The solid line shows the best-fitting orbit for S2 around a black hole of 3.7 million solar masses, located at Sgr A*.

The final painted frame (f) depicts a swiftly spinning horde of stars and hot, million-kelvin gas nearly engulfing a massive black hole too small in size to be pictured (even as a minute dot) on this scale.

The last decade has seen an explosion in our knowledge of the innermost few parsecs of our Galaxy, and astronomers are working hard to decipher the clues hidden within its invisible radiation. Still, we are only now beginning to appreciate the full magnitude of this strange new realm deep in the heart of the Milky Way.

CONCEPT CHECK

✔ What is the most likely explanation of the energetic events observed at the Galactic center?

▼ **FIGURE 23.27 Galactic Center Zoom** Six artist's conceptions, centered on the Galactic center and increasing in resolution by successive factors of 10. Frame (a) shows the same scene as Figure 23.16. Frame (f) is a rendition of a vast whirlpool within the innermost parsec of our Galaxy. The data imaged in Figure 23.25 are displayed at specific wavelengths (so that, e.g., only cooler gas is seen in the radio and mainly hotter stars in X rays). Thus, these artistic renderings present a coherent (though highly simplified) view at many wavelengths.

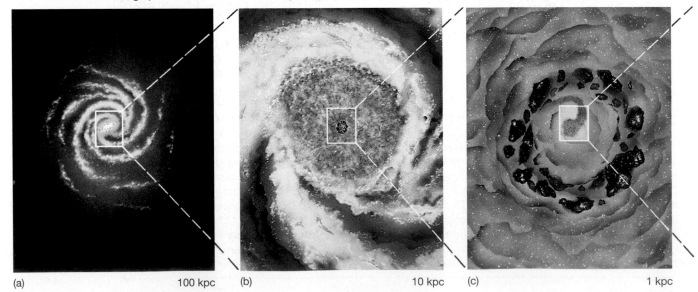

(a)   100 kpc   (b)   10 kpc   (c)   1 kpc

Chapter Review

SUMMARY

A **galaxy (p. 604)** is a huge collection of stellar and interstellar matter isolated in space and bound together by its own gravity. Because we live within it, the **Galactic disk (p. 604)** of our own Milky Way Galaxy appears as a broad band of light across the sky, a band called the Milky Way. Near the center, the disk thickens into the **Galactic bulge (p. 604)**. The disk is surrounded by a roughly spherical **Galactic halo (p. 604)** of old stars and star clusters. Like many others visible in the sky, our Galaxy is a **spiral galaxy (p. 606)**.

Distant regions of our Galaxy and others can be studied by examining variable stars, whose luminosity changes with time. **Pulsating variable stars (p. 607)** vary in brightness in a repetitive and predictable way. **RR Lyrae variables (p. 607)** and **Cepheid variables (p. 607)** are pulsating variables with characteristic light curves that make them easily recognizable. All RR Lyrae stars have roughly the same luminosity. The **period–luminosity relationship (p. 608)** is a simple correlation between Cepheid period and absolute brightness. The brightest Cepheids can be seen at distances of millions of parsecs. RR Lyrae stars are fainter, but much more numerous, making them useful within the Milky Way. In the early 20th century, Harlow Shapley used RR Lyrae stars to determine the distances to many of the Galaxy's globular clusters. He found that the clusters have a roughly spherical distribution in space, but the center of the sphere lies far from the Sun. The globular clusters map out the true extent of the luminous portion of the Milky Way Galaxy. The center of their distribution is close to the **Galactic center (p. 611)**, which lies about 8 kpc from the Sun. The luminous portion of our Galaxy has a diameter of about 30 kpc.

The Galactic halo lacks gas and dust, so no new stars are forming there. All halo stars are old. The gas-rich disk is the site of current star formation and contains many young stars. Stars in the halo and bulge move on largely random three-dimensional orbits that pass repeatedly through the plane of the disk, but have no preferred orientation. Stars and gas in the disk move on roughly circular orbits around the Galactic center. Halo stars appeared early on, before the Galactic disk took shape, when there was no preferred orientation for their orbits. After the disk formed, stars born there inherited its overall spin and so move on circular orbits in the Galactic plane.

Astronomers use radio observations to explore the Galactic disk because radio waves are largely unaffected by interstellar dust. These observations clearly reveal the extent of our Galaxy's **spiral arms (p. 618)**. The spiral arms in spiral galaxies are regions of the densest interstellar gas and are the places where star formation is taking place. The spirals cannot be "tied" to the disk material, as the disk's differential rotation would have made them disappear long ago. Instead, they may be **spiral density waves (p. 619)** that move through the disk, triggering star formation as they pass by. Alternatively, the spirals may arise from **self-propagating star formation (p. 621)**, whereby shock waves produced by the formation and evolution of one generation of stars triggers the formation of the next.

The Galactic **rotation curve (p. 622)** plots the orbital speed of matter in the disk versus distance from the Galactic center. By applying Newton's laws of motion, astronomers can determine the mass of the Galaxy. They find that the Galactic mass continues to increase beyond the radius defined by the globular clusters and the spiral structure we observe. Our Galaxy, like many others, has an invisible **dark halo (p. 622)** containing far more mass than can be accounted for in the form of luminous matter. The **dark matter (p. 623)** making up these dark halos is of unknown composition. Leading candidates include low-mass stars and exotic subatomic particles. Recent attempts to detect stellar dark matter have used the fact that a faint foreground object can occasionally pass in front of a more distant star, deflecting the star's

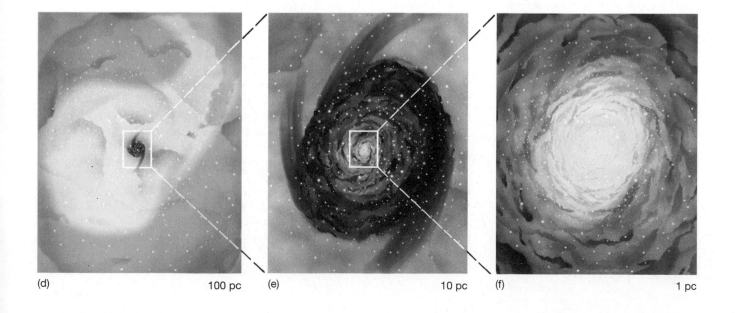

(d)   100 pc   (e)   10 pc   (f)   1 pc

light and causing its apparent brightness to increase temporarily. This deflection is called **gravitational lensing (p. 624)**.

Astronomers working at infrared and radio wavelengths have uncovered evidence of energetic activity within a few par-

secs of the Galactic center. The leading explanation is that a black hole 3–4 million times more massive than the Sun resides at the heart of our Galaxy.

REVIEW AND DISCUSSION

1. What evidence do we have that we live in a disk galaxy?

2. Why is it difficult to map out our Galaxy from our vantage point on Earth?

3. What are spiral nebulae? How did they get that name?

4. In what region of the Galaxy are globular clusters found?

5. How are Cepheid variables used in determining distances?

6. Roughly how far out into space (looking out of the Galactic disk) can we use Cepheids to measure distance?

7. What important discoveries were made early in this century by using RR Lyrae variables?

8. Of what use is radio astronomy in the study of Galactic structure?

9. Contrast the motions of disk and halo stars.

10. How do we know that the Milky Way Galaxy has spiral arms?

11. Explain why galactic spiral arms are believed to be regions of recent and ongoing star formation.

12. Describe what happens to interstellar gas as it passes through a spiral density wave.

13. What is self-propagating star formation?

14. What do the red stars in the Galactic halo tell us about the history of the Milky Way?

15. What does the rotation curve of our Galaxy tell us about the Galaxy's total mass?

16. What evidence is there for dark matter in the Galaxy?

17. Describe some candidates for Galactic dark matter.

18. What is gravitational lensing, and can astronomers use it to search for dark matter?

19. Why can't optical astronomers easily study the center of our Galaxy?

20. Describe some ways in which astronomers can observe the Galactic center.

CONCEPTUAL SELF-TEST: TRUE OR FALSE/MULTIPLE CHOICE

1. Herschel's attempt to map the Milky Way by counting stars led to an inaccurate estimate of the Galaxy's size because he was unaware of absorption by interstellar dust.

2. Cepheids and RR Lyrae variables lie in the red-giant region of the H–R diagram.

3. Globular clusters trace out the large-scale structure of the Galactic disk.

4. The Galactic halo contains only old stars.

5. Stars and gas in the Galactic disk move in roughly circular orbits around the Galactic center.

6. Halo stars are often involved in collisions as they pass through the Galactic disk.

7. Astronomers use 21-cm radiation to study Galactic molecular clouds.

8. Rotational velocities in the outer part of the Galaxy are smaller than would be expected on the basis of observed stars and gas, indicating the presence of dark matter.

9. The Galactic center has been studied extensively at optical wavelengths.

10. The most likely explanation of the high-speed motion of stars and gas near the Galactic center is that the stars and gas are orbiting a supermassive black hole.

11. Most of the bright stars in our Galaxy are located in the Galactic **(a)** center; **(b)** bulge; **(c)** halo; **(d)** disk.

12. A variable star's pulsation period is most directly related to the star's **(a)** rotation rate; **(b)** age; **(c)** central temperature; **(d)** luminosity.

13. Globular clusters are found mainly **(a)** in the Galactic center; **(b)** in the Galactic disk; **(c)** in spiral arms; **(d)** in the Galactic halo.

14. Shapley measured the distances to globular clusters by using **(a)** trigonometric parallax; **(b)** a comparison of the absolute and apparent magnitudes of variable stars; **(c)** spectroscopic parallax; **(d)** radar ranging.

15. In the Milky Way Galaxy, our Sun is located **(a)** near the Galactic center; **(b)** about halfway out from the center; **(c)** at the outer edge; **(d)** in the halo.

16. A telescope searching for newly formed stars would make the most discoveries if it were pointed **(a)** directly away from the Galactic center; **(b)** perpendicular to the Galactic disk; **(c)** within a spiral arm; **(d)** between spiral arms.

17. The first stars that formed in the Milky Way now **(a)** have chaotic orbits; **(b)** orbit in the Galactic plane; **(c)** orbit closest to the Galactic center; **(d)** orbit in the same direction the Milky Way spins.

18. Stars in the outermost regions of the Milky Way Galaxy **(a)** are youngest; **(b)** orbit faster than astronomers would expect on the basis of the Galactic mass we can see; **(c)** are more likely to explode as supernovae; **(d)** are more luminous than other stars.

19. Most of the mass of the Milky Way exists in the form of **(a)** stars; **(b)** gas; **(c)** dust; **(d)** dark matter.

20. A black hole probably exists at the Galactic center because **(a)** stars near the center of the Milky Way are disappearing;

(b) no stars can be seen in the vicinity of the Galactic center; **(c)** stars near the center of the Milky Way have been observed orbiting some unseen object; **(d)** the Galaxy rotates faster than astronomers would expect.

PROBLEMS

 Algorithmic versions of these questions are available in the Practice Problems module of the Companion Website at astro.prenhall.com/chaisson.

The number of squares preceding each problem indicates its approximate level of difficulty.

1. ■ Calculate the angular diameter of a prestellar nebula of radius 100 A.U. lying 100 pc from Earth. Compare this with the roughly 6° diameter of the Andromeda Galaxy (Figure 23.2a).

2. ■■ How close would the nebula in the previous question have to be in order to have the same angular diameter as Andromeda? Calculate the apparent magnitude of the central star if it had a luminosity 10 times that of the Sun.

3. ■ What is the greatest distance at which an RR Lyrae star of absolute magnitude 0 could be seen by a telescope capable of detecting objects as faint as 20th magnitude?

4. ■ A typical Cepheid variable is 100 times brighter than a typical RR Lyrae star. How much farther away than RR Lyrae stars can Cepheids be used as distance-measuring tools?

5. ■■ An astronomer looking through the *Hubble Space Telescope* can see a star with solar luminosity at a distance of 100,000 pc. The brightest Cepheids have luminosities 30,000 times greater than that of the Sun. Taking the Sun's absolute magnitude to be 5, calculate the absolute magnitudes of these bright Cepheids. Neglecting interstellar absorption, how far away can *HST* see them?

6. ■■■ What is the maximum distance at which *Hubble* could see the Cepheid in the previous question if it lay in the Galactic disk, with an average interstellar extinction of 2.5 magnitudes per kiloparsec?

7. ■■ Calculate the proper motion (in arc seconds per year) of a globular cluster with a transverse velocity (relative to the Sun) of 200 km/s and a distance of 3 kpc. Do you think that this motion is measurable?

8. ■■ Calculate the total mass of the Galaxy lying within 20 kpc of the Galactic center if the rotation speed at that radius is 240 km/s.

9. ■■ Using the data presented in Figure 23.21, estimate the distance from the Galactic center at which matter takes **(a)** 100 million years and **(b)** 500 million years to complete one orbit.

10. ■■ Consider the motion of the Sun and another star lying 100 pc farther from the Galactic center. Both stars move in circular orbits in the Galactic plane. Initially, the stars lie along the same radial line through the center (so that they are 100 pc apart). How far apart will they be after the Sun has completed exactly one orbit?

11. ■■■ Using the data presented in Figure 23.21, calculate how long it takes the Sun to "lap" stars orbiting 15 kpc from the Galactic center. How long does matter at 5 kpc take to lap us?

12. ■ A density wave made up of two spiral arms is moving through the Galactic disk. At the 8-kpc radius of the Sun's orbit around the Galactic center, the wave's speed is 120 km/s and the Galactic rotation speed is 220 km/s. Calculate how many times the Sun has passed through a spiral arm since forming 4.6 billion years ago.

13. ■ Given the data in the previous question and the fact that O-type stars live at most 10 million years before exploding as supernovae, calculate the maximum distance at which an O-type star (orbiting at the Sun's distance from the Galactic center) can be found from the density wave in which it formed.

14. ■■ Material at an angular distance of 0.2″ from the Galactic center is observed to have an orbital speed of 1200 km/s. If the Sun's distance to the Galactic center is 8 kpc, and the material's orbit is circular and is seen edge-on, calculate the radius of the orbit and the mass of the object around which the material is orbiting.

15. ■■■ The best-fitting orbit shown in Figure 23.26 has an eccentricity of 0.87. Use the information presented in Section 23.7 of the text to calculate the distance of closest approach between the star S2 and the central black hole. Was the star in any danger of being tidally disrupted by the black hole's gravity at that distance?

 In addition to the Practice Problems module, the Companion Website at astro.prenhall.com/chaisson provides for each chapter a study guide module with multiple choice questions as well as additional annotated images, animations, and links to related Websites.

24

Galaxies are candidates for the grandest, most beautiful objects in the universe. They are colossal collections of typically a hundred billion stars, held together by gravity within one loose structure. Despite their ubiquity in space, astronomers do not fully understand how galaxies originated. Here, amid a backdrop of faraway galaxies, the majestic spiral galaxy NGC3370 looms in the foreground, yet still ▶ about 100 million light-years away. (STScI)

Normal and Active Galaxies

Building Blocks of the Universe

As our field of view expands to truly cosmic scales, the focus of our studies shifts dramatically. Planets become inconsequential, stars themselves mere points of hydrogen consumption. Now entire galaxies become the "atoms" from which the universe is built—distant realms completely unknown to scientists just a century ago. We know of literally millions of galaxies beyond our own. Most are smaller than the Milky Way, some comparable in size, a few much larger. Many are sites of explosive events far more energetic than anything ever witnessed in our own Galaxy. All are vast, gravitationally bound assemblages of stars, gas, dust, dark matter, and radiation separated from us by almost incomprehensibly large distances. The light we receive tonight from the most distant galaxies was emitted long before Earth existed. By studying the properties of galaxies and the violence that ensues when they collide, we gain insight into the history of our Galaxy and the universe in which we live.

LEARNING GOALS

Studying this chapter will enable you to

1 Describe the basic properties of the main types of normal galaxies.

2 Discuss the distance-measurement techniques that enable astronomers to map the universe beyond the Milky Way.

3 Describe how galaxies clump into clusters.

4 State Hubble's law and explain how it is used to derive distances to the most remote objects in the observable universe.

5 Specify the basic differences between active and normal galaxies.

6 Describe the important features of active galaxies.

7 Explain what drives the central engine thought to power all active galaxies.

 Visit astro.prenhall.com/chaisson for additional annotated images, animations, and links to related sites for this chapter.

24.1 Hubble's Galaxy Classification

Figure 24.1 shows a vast expanse of space lying about 100 million pc from Earth. Almost every patch or point of light in this figure is a separate galaxy—hundreds can be seen in just this one photograph. Over the years, astronomers have accumulated similar images of many millions of galaxies. We begin our study of these enormous accumulations of matter simply by considering their appearance on the sky.

Seen through even a small telescope, images of galaxies look distinctly nonstellar. They have fuzzy edges, and many are quite elongated—not at all like the sharp, point-like images normally associated with stars. Although it is difficult to tell from the photograph, some of the blobs of light in Figure 24.1 are spiral galaxies like the Milky Way Galaxy and Andromeda. Others, however, are definitely not spirals—no disks or spiral arms can be seen. Even when we take into account their different orientations in space, galaxies do *not* all look the same.

The American astronomer Edwin Hubble was the first to categorize galaxies in a comprehensive way. Working with the then recently completed 2.5-m optical telescope on Mount Wilson in California in 1924, he classified the galaxies he saw into four basic types—*spirals*, *barred spirals*, *ellipticals*, and *irregulars*—solely on the basis of their visual appearance. Many modifications and refinements have been incorporated over the years, but the basic **Hubble classification scheme** is still widely used today.

Spirals

We saw several examples of **spiral galaxies** in Chapter 23—for example, our own Milky Way Galaxy and our neighbor Andromeda. ∞ (Sec. 23.1) All galaxies of this type contain a flattened galactic disk in which spiral arms are found, a central galactic bulge with a dense nucleus, and an extended halo of faint, old stars. ∞ (Sec. 23.3) The stellar density (i.e., the number of stars per unit volume) is greatest in the *galactic nucleus*, at the center of the bulge. However, within this general description, spiral galaxies exhibit a wide variety of shapes, as illustrated in Figure 24.2.

In Hubble's scheme, a spiral galaxy is denoted by the letter S and classified as type a, b, or c according to the size of its central bulge. Type Sa galaxies have the largest bulges, Type Sc the smallest. The tightness of the spiral pattern is quite well correlated with the size of the bulge (although the correspondence is not perfect). Type Sa spiral galaxies tend to have tightly wrapped, almost circular,

(a)

R I V U X G

▲ **FIGURE 24.1 Coma Cluster** (a) A collection of many galaxies, each consisting of hundreds of billions of stars. Called the Coma Cluster, this group of galaxies lies more than 100 million pc from Earth. (The blue spiked object at the top right is a nearby star; virtually every other object visible is a galaxy.) (b) A recent *Hubble Space Telescope* image of part of the cluster. (*AURA; NASA*)

(b)

(a) M81   Type Sa   (b) M51   Type Sb   (c) NGC 2997   Type Sc

▲ **FIGURE 24.2 Spiral Galaxy Shapes** Variation in shape among spiral galaxies. As we progress from Type Sa to Sb to Sc, the bulges become smaller while the spiral arms tend to become less tightly wound. *(R. Gendler; NOAO; D. Malin/AAT)*

R I V U X G

spiral arms, Type Sb galaxies typically have more open spiral arms, and Type Sc spirals often have a loose, poorly defined spiral structure. The arms also tend to become more "knotty," or clumped, in appearance as the spiral pattern becomes more open.

The bulges and halos of spiral galaxies contain large numbers of reddish old stars and globular clusters, similar to those observed in our own Galaxy and in Andromeda. Most of the light from spirals, however, comes from A-through G-type stars in the galactic disk, giving these galaxies an overall whitish glow. We assume that thick disks exist, too, but their faintness makes this assumption hard to confirm—the thick disk in the Milky Way contributes only a percent or so of our Galaxy's total light. ∞ (Sec. 23.3)

Like the disk of the Milky Way, the flat disks of typical spiral galaxies are rich in gas and dust. The 21-cm radio radiation emitted by spirals betrays the presence of interstellar gas, and obscuring dust lanes are clearly visible in many systems (see Figures 24.2b and c). Stars are forming within the spiral arms, which contain numerous emission nebulae and newly formed O- and B-type stars. ∞ (Secs. 18.2, 23.5) The arms appear bluish because of the presence of bright blue O- and B-type stars there. Type Sc galaxies contain the most interstellar gas and dust, Sa galaxies the least. The photo of the Sc galaxy NGC 2997 shown in Figure 24.2(c) clearly reveals the preponderance of interstellar gas, dust, and young blue stars tracing the spiral pat-

tern. Spirals are not necessarily young galaxies, however: Like our own Galaxy, they are simply rich enough in interstellar gas to provide for continued stellar birth.

Most spirals are not seen face-on, as they are shown in Figure 24.2. Many are tilted with respect to our line of sight, making their spiral structure hard to discern. However, we do not need to see spiral arms to classify a galaxy as a spiral. The presence of the disk, with its gas, dust, and newborn stars, is sufficient. For example, the galaxy shown in Figure 24.3 is classified as a spiral because of the clear line of obscuring dust seen along its midplane. (Incidentally, this relatively nearby galaxy was another of the "nebulae" figuring in the Shapley–Curtis debate discussed in Chapter 23. ∞ (Sec. 23.2) The visible dust lane was interpreted by Curtis as an obscuring "ring" of material, leading him to suggest that our Galactic plane might contain a similar feature.)

A variation of the spiral category in Hubble's classification scheme is the **barred-spiral galaxy**. Barred spirals differ from ordinary spirals mainly by the presence of an elongated "bar" of stellar and interstellar matter passing through the center and extending beyond the bulge, into the disk. The spiral arms project from near the ends of the bar rather than from the bulge (their origin in normal spirals). Barred spirals are designated by the letters SB and are subdivided, like the ordinary spirals, into categories SBa, SBb, and SBc, depending on the size of the bulge. Again like ordinary spirals, the

◀ **FIGURE 24.3 Sombrero Galaxy** The Sombrero Galaxy, a spiral system seen edge-on. Officially cataloged as M104, this galaxy has a dark band composed of interstellar gas and dust. The large size of the galaxy's central bulge marks it as Type Sa, even though its spiral arms cannot easily be seen. *(NASA)*

R I V U X G

tightness of the spiral pattern is correlated with the size of the bulge. Figure 24.4 shows the variation among barred-spiral galaxies. In the case of the SBc category, it is often hard to tell where the bar ends and the spiral arms begin.

Frequently, astronomers cannot distinguish between spirals and barred spirals, especially when a galaxy happens to be oriented with its galactic plane nearly edge-on toward Earth, as in Figure 24.3. Because of the physical and chemical similarities of spiral and barred-spiral galaxies, some researchers do not even bother to distinguish between them. Others, however, regard the differences in their structures as very important, arguing that these differences suggest basic dissimilarities in the ways the two types of galaxies formed and evolved.

The discovery that the bulge of our own Galaxy is elongated suggests that the Milky Way may be a barred spiral, of type SBb or SBc. However, the full extent of the bar remains uncertain. Some astronomers place our Galaxy in an intermediate category lying between Hubble's spiral and barred spiral types. ∞ (Sec. 23.3)

Ellipticals

Unlike the spirals, **elliptical galaxies** have no spiral arms and, in most cases, no obvious galactic disk—in fact, other

than possessing a dense central nucleus, they often exhibit little internal structure of any kind. As with spirals, the stellar density increases sharply in the central nucleus. Denoted by the letter E, these systems are subdivided according to how elliptical they appear on the sky. The most circular are designated E0, slightly flattened systems are labeled E1, and so on, all the way to the most elongated ellipticals, of type E7 (Figure 24.5).

Notice, by the way, that the Hubble type of an elliptical galaxy depends both on the galaxy's intrinsic three-dimensional shape *and* on its orientation relative to the line of sight. Consider, for example, a spherical galaxy, a cigar-shaped galaxy seen end on, and a lozenge-shaped galaxy seen face on; all would appear circular in the sky. As a result, it is often difficult to decipher a galaxy's true shape from its visual appearance.

There is a large range in both the size and the number of stars contained in elliptical galaxies. The largest elliptical galaxies are much larger than our own Milky Way Galaxy. These *giant ellipticals* can range up to a few megaparsecs across and contain trillions of stars. At the other extreme, *dwarf ellipticals* may be as small as 1 kpc in diameter and contain fewer than a million stars. The significant observational differences between giant and dwarf ellipticals have led many astronomers to conclude that these galaxies are members of separate classes, with quite differ-

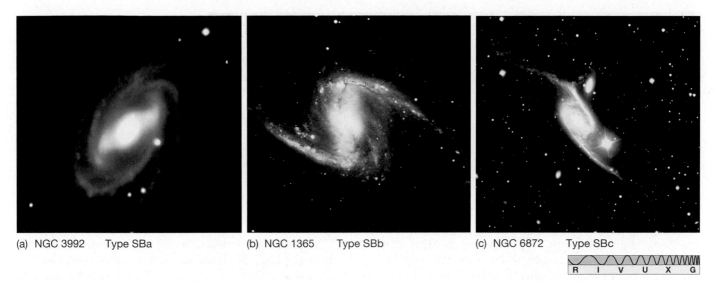

(a) NGC 3992   Type SBa

(b) NGC 1365   Type SBb

(c) NGC 6872   Type SBc

▲ **FIGURE 24.4 Barred-spiral Galaxy Shapes** Variation in shape among barred-spiral galaxies. The variation from SBa to SBc is similar to that for the spirals in Figure 24.2, except that now the spiral arms begin at either end of a bar through the galactic center. In frame (c), the bright star is a foreground object in our own Galaxy; the object at the top center is another galaxy that is probably interacting with NGC 6872. *(NOAO; AAT; ESO)*

ent histories of formation and stellar content. The dwarfs are by far the most common type of ellipticals, outnumbering their brighter counterparts by about 10 to 1. However, most of the *mass* that exists in the form of elliptical galaxies is contained in the larger systems.

The absence of spiral arms is not the only difference between spirals and ellipticals: Most ellipticals also contain little or no cool gas and dust. The 21-cm radio emission from neutral hydrogen gas is, with few exceptions, completely absent, and no obscuring dust lanes are seen. In most cases, there is no evidence of young stars or ongoing

star formation. Like the halo of our own Galaxy, ellipticals are made up mostly of old, reddish, low-mass stars. Also, as in the halo of our Galaxy, the orbits of stars in ellipticals are disordered, exhibiting little or no overall rotation; objects move in all directions, not in regular, circular paths as in our Galaxy's disk. Ellipticals differ from our Galaxy's halo in at least one important respect, however: X-ray observations reveal large amounts of very *hot* (several million kelvins) interstellar gas distributed throughout their interiors, often extending well beyond the visible portions of the galaxies (Figure 24.5c).

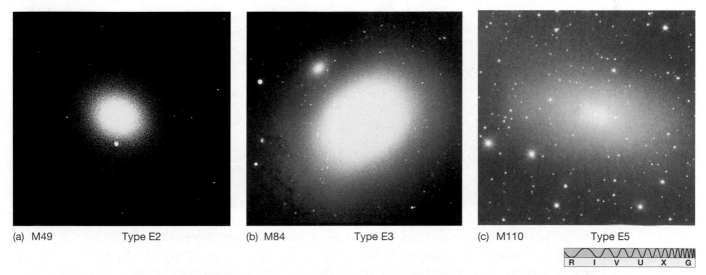

(a) M49   Type E2

(b) M84   Type E3

(c) M110   Type E5

▲ **FIGURE 24.5 Elliptical Galaxy Shapes** Variation in shape among elliptical galaxies. (a) The E1 galaxy M49 is nearly circular in appearance. (b) M84 is a slightly more elongated elliptical galaxy, classified as E3. Both galaxies lack spiral structure, and neither shows evidence of cool interstellar dust or gas, although each has an extensive X-ray halo of hot gas that extends far beyond the visible portion of the galaxy. (c) M110 is a dwarf elliptical companion to the much larger Andromeda Galaxy. *(AURA; SAO; R. Gendler)*

◄ **FIGURE 24.6 S0 Galaxies** (a) S0 (or lenticular) galaxies contain a disk and a bulge, but no interstellar gas and no spiral arms. They are in many respects intermediate between E7 ellipticals and Sa spirals in their properties. (b) SB0 galaxies are similar to S0 galaxies, except for a bar of stellar material extending beyond the central bulge. *(Palomar/Caltech)*

(a) NGC 1201   Type S0   (b) NGC 2859   Type SB0

R I V U X G

Some giant ellipticals are exceptions to many of the foregoing general statements about elliptical galaxies, as they have been found to contain disks of gas and dust in which stars are forming. Astronomers think that these systems may be the results of collisions among gas-rich galaxies (see *Discovery 25-1*). Indeed, galactic collisions may have played an important role in determining the appearance of many of the systems we observe today.

Intermediate between the E7 ellipticals and the Sa spirals in the Hubble classification is a class of galaxies that show evidence of a thin disk and a flattened bulge, but that contain no gas and no spiral arms. Two such objects are shown in Figure 24.6. These galaxies are known as **S0 galaxies** if no bar is evident and **SB0 galaxies** if a bar is present. They are also known as *lenticular* galaxies, because of their lens-shaped appearance. They look a little like spirals whose dust and gas have been stripped away, leaving behind just a stellar disk. Observations in recent years have shown that many normal elliptical galaxies have faint disks within them, like the S0 galaxies. As with the S0s, the origin of these disks is uncertain, but some researchers suspect that S0s and ellipticals may be closely related.

Irregulars

The final class of galaxies identified by Hubble is a catch-all category—**irregular galaxies**—so named because their visual appearance does not allow us to place them into any

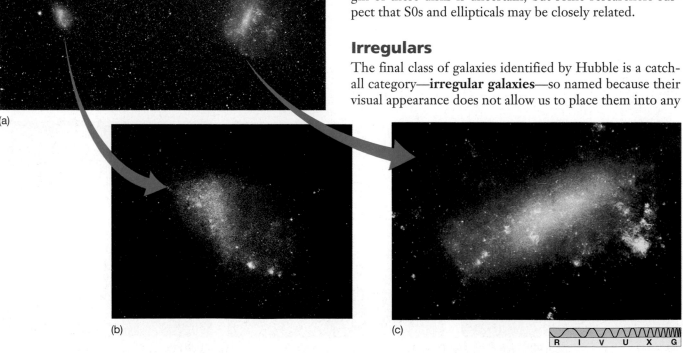

▲ **FIGURE 24.7 Magellanic Clouds** The Magellanic Clouds are prominent features of the night sky in the Southern Hemisphere. Named for the 16th-century Portuguese explorer Ferdinand Magellan, whose around-the-world expedition first brought word of these fuzzy patches of light to Europe, they are dwarf irregular (Irr I) galaxies, gravitationally bound to our own Milky Way Galaxy. They orbit our Galaxy and accompany it on its trek through the cosmos. (a) The Clouds' relationship to one another in the southern sky. Both the Small (b) and the Large (c) Magellanic Cloud have distorted, irregular shapes, although some observers claim they can discern a single spiral arm in the Large Cloud. *(F. Espenak; Harvard Observatory)*

of the other categories just discussed. Irregulars tend to be rich in interstellar matter and young, blue stars, but they lack any regular structure, such as well-defined spiral arms or central bulges. They are divided into two subclasses: Irr I galaxies and Irr II galaxies. The Irr I galaxies often look like misshapen spirals.

Irregular galaxies tend to be smaller than spirals, but somewhat larger than dwarf ellipticals. They typically contain between 10^8 and 10^{10} stars. The smallest such galaxies are called *dwarf irregulars*. As with elliptical galaxies, the dwarf type is the most common irregular. Dwarf ellipticals and dwarf irregulars occur in approximately equal numbers and together make up the vast majority of galaxies in the universe. They are often found close to a larger "parent" galaxy.

Figure 24.7 shows the **Magellanic Clouds**, a famous pair of Irr I galaxies that orbit the Milky Way Galaxy. They are shown to proper scale in Figure 23.16. Studies of Cepheid variables within the Clouds show them to be approximately 50 kpc from the center of our Galaxy. ∞ (Sec. 23.2) The Large Cloud contains about 6 billion solar masses of material and is a few kiloparsecs across. Both Clouds contain lots of gas, dust, and blue stars (and the recent, well-documented supernova discussed in *Discovery 21-1*), indicating ongoing star formation. Both also contain many old stars and several old globular clusters, so we know that star formation has been going on in them for a very long time.

Radio studies hint at a possible bridge of hydrogen gas connecting the Milky Way to the Magellanic Clouds, although more observational data are still needed to establish this link beyond doubt. It is possible that the tidal force of the Milky Way tore a stream of gas from the Clouds the last time their orbits brought them close to our Galaxy. Of course, gravity works both ways, and many researchers reason that the forces exerted by the Clouds may in turn be responsible for distorting our Galaxy, warping and thickening the outer parts of the Galactic disk. ∞ (Sec. 23.5)

The much rarer Irr II galaxies (Figure 24.8), in addition to their irregular shape, have other peculiarities, often exhibiting a distinctly explosive or filamentary appearance. Their appearance once led astronomers to suspect that violent events had occurred within them. However, it now seems more likely that, in some (but probably not all) cases, we are seeing the result of a close encounter or collision between two previously "normal" systems.

The Hubble Sequence

Table 24.1 summarizes the basic characteristics of the various types of galaxies. When he first developed his classification scheme, Hubble arranged the galaxies into the "tuning fork" diagram shown in Figure 24.9. The variation in types across the diagram, from ellipticals to spirals to irregulars, is often referred to as the *Hubble sequence*.

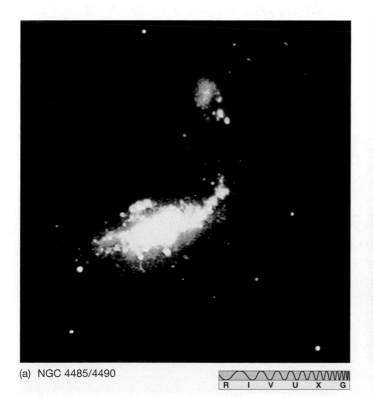

(a) NGC 4485/4490

(b) M82

▲ **FIGURE 24.8 Irregular Galaxy Shapes** Some irregular (Irr II) galaxies. (a) The oddly shaped galaxies NGC 4485 and NGC 4490 may be close to one another and interacting gravitationally. (b) The galaxy M82 seems to show an explosive appearance and has experienced a recent galaxywide burst of star formation. (*AURA; Subaru*)

TABLE 24.1   Galaxy Properties by Type

	Spiral/Barred Spiral (S/SB)	Elliptical[1] (E)	Irregular (Irr)
Shape and structural properties	Highly flattened disk of stars and gas, containing spiral arms and thickening central bulge. Sa and SBa galaxies have the largest bulges, the least obvious spiral structure, and roughly spherical stellar halos. SB galaxies have an elongated central "bar" of stars and gas.	No disk. Stars smoothly distributed through an ellipsoidal volume ranging from nearly spherical (E0) to very flattened (E7) in shape. No obvious substructure other than a dense central nucleus.	No obvious structure. Irr II galaxies often have "explosive" appearances.
Stellar content	Disks contain both young and old stars; halos consist of old stars only.	Contain old stars only.	Contain both young and old stars.
Gas and dust	Disks contain substantial amounts of gas and dust; halos contain little of either.	Contain hot X-ray emitting gas, little or no cool gas and dust.	Very abundant in gas and dust.
Star formation	Ongoing star formation in spiral arms.	No significant star formation during the last 10 billion years.	Vigorous ongoing star formation.
Stellar motion	Gas and stars in disk move in circular orbits around the galactic center; halo stars have random orbits in three dimensions.	Stars have random orbits in three dimensions.	Stars and gas have highly irregular orbits.

[1] *As noted in the text, some giant ellipticals appear to be the result of collisions between gas-rich galaxies and are exceptions to many of the statements listed here.*

Hubble's primary aim in creating this diagram was to indicate similarities in appearance among galaxies. However, he also regarded the tuning fork as an evolutionary sequence from left to right, with E0 ellipticals evolving into flatter ellipticals and S0 systems and ultimately forming disks and spiral arms. Indeed, Hubble's terminology referring to ellipticals as "early-type" and spirals as "late-type" galaxies is still widely used today. However, as far as

modern astronomers can tell, there is no direct evolutionary connection of this sort along the Hubble sequence. Isolated normal galaxies do *not* evolve from one type to another. Spirals are not ellipticals that have grown arms, nor are ellipticals spirals that have somehow expelled their star-forming disks. Some astronomers do suspect that bars may be transient features and that barred-spiral galaxies may therefore evolve into ordinary spirals, but, in general,

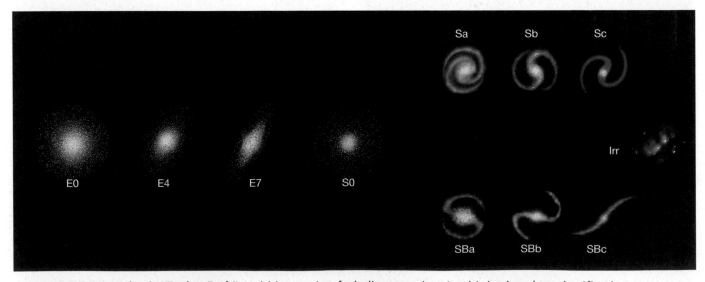

▲ **FIGURE 24.9 Galactic "Tuning Fork"** Hubble's tuning fork diagram, showing his basic galaxy classification scheme. The placement of the four basic types of galaxies—ellipticals, spirals, barred spirals, and irregulars—in the diagram is suggestive, but the tuning fork has no physical meaning.

astronomers know of no simple parent–child relationship among Hubble types.

However, the key word in the previous paragraph is *isolated*. As described in Section 25.2, there is now strong observational evidence that collisions and tidal interactions *between* galaxies are commonplace and that these encounters are the main physical processes driving the evolution of galaxies. We will return to this important subject in Chapter 25.

CONCEPT CHECK

✔ In what ways are large spirals like the Milky Way and Andromeda *not* representative of galaxies as a whole?

24.2 The Distribution of Galaxies in Space

Now that we have seen some of their basic properties, let us ask how galaxies are spread through the expanse of the universe beyond the Milky Way. Galaxies are not distributed uniformly in space. Rather, they tend to clump into still larger agglomerations of matter. As we will see, this uneven distribution is crucial in determining both their appearance and their evolution. As always in astronomy, our understanding hinges on our ability to tell how far away an object lies. We therefore begin by looking more closely at the means used by astronomers to measure distances to galaxies.

Extending the Distance Scale

Astronomers estimate that some 40 billion galaxies exist in the observable universe. Some reside close enough for the Cepheid variable technique to work—astronomers have detected and measured the periods of Cepheids in galaxies as far away as 25 Mpc. (See Figure 24.10.) ∞ (Sec. 23.2) However, some galaxies contain no Cepheid stars (can you think of some reasons that this might be?), and, in any case, most known galaxies lie much farther away than 25 Mpc. Cepheid variables in very distant galaxies simply cannot be observed well enough, even through the world's most sensitive telescopes, to allow us to measure their apparent brightnesses and periods. To extend our distance-measurement ladder, therefore, we must find some new class of object to study. What individual objects are bright enough for us to observe at great distances?

One way in which researchers have tackled this problem is through observations of **standard candles**—easily recognizable astronomical objects whose luminosities are confidently known. The basic idea is very simple. Once an object is identified as a standard candle—by its appearance or by the shape of its light curve, say—its luminosity can be estimated. Comparison of the luminosity with the apparent brightness then gives the object's distance and,

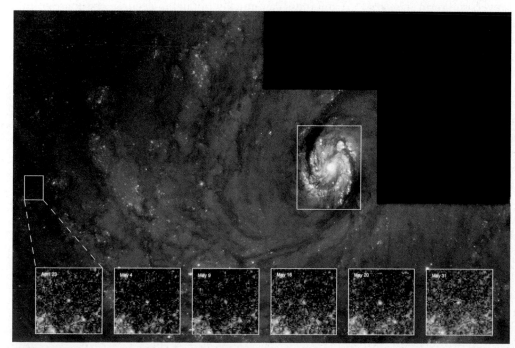

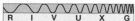

▲ **FIGURE 24.10 Cepheid in Virgo** This sequence of six snapshots chronicles the periodic changes in a Cepheid variable star in the spiral galaxy M100, a member of the Virgo Cluster of galaxies. The Cepheid appears at the center of each inset, taken at the different times indicated during 1994. The star looks like a square because of the high magnification of the digital CCD camera—we are seeing individual pixels of the image. The 24th-magnitude star varies by about a factor of two in brightness every seven weeks. (Cf. image of M100 in *Discovery 5-1*.) *(NASA)*

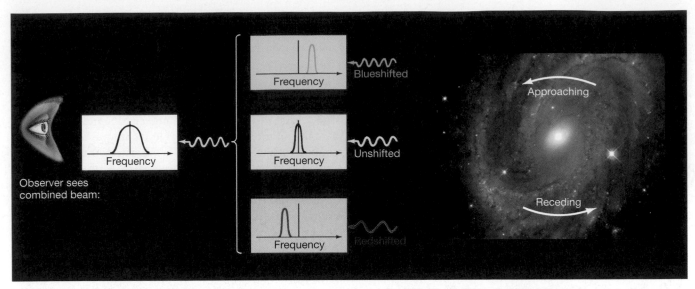

▲ FIGURE 24.11 Galaxy Rotation A galaxy's rotation causes some of the radiation it emits to be blueshifted and some to be redshifted (relative to what the emission would be from an unmoving source). From a distance, when the radiation from the galaxy is combined into a single beam and analyzed spectroscopically, the redshifted and blueshifted components combine to produce a broadening of the galaxy's spectral lines. The amount of broadening is a direct measure of the rotation speed of the galaxy, such as the one at the right, NGC 4603, about 100 million light-years away. *(NASA)*

hence, the distance to the galaxy in which it resides. ∞ (Sec. 17.2) Note that, apart from the way in which the luminosity is determined, the Cepheid variable technique relies on identical reasoning. However, the term *standard candle* tends to be applied only to very bright objects.

To be most useful, a standard candle must (1) have a narrowly defined luminosity, so that the uncertainty in estimating its brightness is small, and (2) be bright enough to be seen at large distances. Over the years, astronomers have explored the use of many types of objects as standard candles—novae, emission nebulae, planetary nebulae, globular clusters, Type I (carbon-detonation) supernovae, and even entire galaxies have been employed. Not all have been equally useful, however: Some have larger intrinsic spreads in their luminosities than others, making them less reliable for measuring distances.

In recent years, planetary nebulae and Type I supernovae have proved particularly reliable as standard candles. ∞ (Secs. 20.3, 21.3) The latter have remarkably consistent peak luminosities and are very bright, allowing them to be identified and measured out to distances of many hundreds of megaparsecs. The small luminosity spread of Type I supernovae is a direct consequence of the circumstances in which these violent events occur. As discussed in Chapter 21, an accreting white dwarf explodes when it reaches the well-defined critical mass at which carbon fusion begins. ∞ (Sec. 21.3) The magnitude of the explosion is relatively insensitive to the details of how the white dwarf formed or how it subsequently reached critical mass, with the result that all such super-

novae have quite similar properties.* Thus, when a Type I supernova is observed in a distant galaxy (we assume that it occurs *in the galaxy*, not in the foreground), astronomers can quickly obtain an accurate estimate of the galaxy's distance.

An important alternative to standard candles was discovered in the 1970s, when astronomers found a close correlation between the rotational speeds and the luminosities of spiral galaxies within a few tens of megaparsecs of the Milky Way Galaxy. Rotation speed is a measure of a spiral galaxy's total mass, so it is perhaps not surprising that this property should be related to luminosity. ∞ (Sec. 23.5) What *is* surprising, though, is how tight the correlation is. The **Tully–Fisher relation**, as it is now known (after its discoverers), allows us to obtain a remarkably accurate estimate of a spiral galaxy's luminosity simply by observing how fast the galaxy rotates. As usual, comparing the galaxy's (true) luminosity with its (observed) apparent brightness yields its distance.

To see how the method is used, imagine that we are looking edge-on at a distant spiral galaxy and observing one particular emission line, as illustrated in Figure 24.11. Radiation from the side of the galaxy where matter is generally approaching us is blueshifted by the

---

Recall from Chapter 21 that a Type II supernova also occurs when a growing stellar core—this time at the center of a massive star—reaches a critical mass. ∞ (Sec. 21.2) However, the outward appearance of the explosion can be significantly modified by the amount of stellar material through which the blast wave must travel before it reaches the star's surface, resulting in a greater spread in observed luminosities.

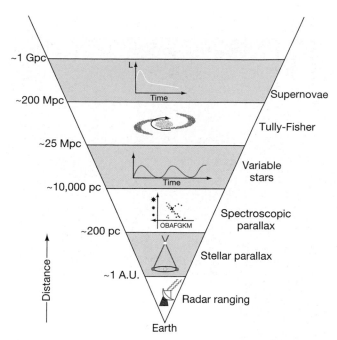

▲ FIGURE 24.12 Extragalactic Distance Ladder An inverted pyramid summarizes the distance techniques used to study different realms of the universe. The techniques shown in the bottom four layers—radar ranging, stellar parallax, spectroscopic parallax, and variable stars—take us as far as the nearest galaxies. To go farther, we must use new techniques—the Tully–Fisher relation and the use of standard candles—based on distances determined by the four lowest techniques.

Doppler effect. Radiation from the other side, which is receding from us, is redshifted by a similar amount. The overall effect is that line radiation from the galaxy is "smeared out," or broadened, by the galaxy's rotation. The faster the rotation, the greater the amount of broadening (see Figure 4.18 for the stellar equivalent). By measuring the amount of broadening, we can therefore determine the galaxy's rotation speed. Once we know that, the Tully–Fisher relation tells us the galaxy's luminosity.

The particular line normally used in these studies actually lies in the radio part of the spectrum. It is the 21-cm line of cold, neutral hydrogen in the galactic disk. ∞ (Sec. 18.4) This line is used in preference to optical lines because (1) optical radiation is strongly absorbed by dust in the disk under study and (2) the 21-cm line is normally very narrow, making the broadening easier to observe. In addition, astronomers often use *infrared*, rather than optical, luminosities, to avoid absorption problems caused by dust, both in our own Galaxy and in others.

The Tully–Fisher relation can be used to measure distances to spiral galaxies out to about 200 Mpc, beyond which the line broadening becomes increasingly difficult to measure accurately. A somewhat similar connection, relating line broadening to a galaxy's *diameter*, exists for ellipti-

cal galaxies. Once the galaxy's diameter and angular size are known, its distance can be computed from elementary geometry. ∞ (*More Precisely 1-3*) These methods bypass many of the standard candles often used by astronomers and so provide independent means of determining distances to faraway objects.

As indicated in Figure 24.12, standard candles and the Tully–Fisher relation form the fifth and sixth rungs of our cosmic distance ladder, introduced in Chapter 1 and expanded in Chapters 17 and 23. ∞ (Secs. 1.7, 17.1, 17.6, 23.2) In fact, they stand for perhaps a dozen or so related, but separate, techniques that astronomers have employed in their quest to map out the universe on large scales. Just as with the lower rungs, we calibrate the properties of these new techniques by using distances measured by more local means. In this way, the distance-measurement process "bootstraps" itself to greater and greater distances. However, at the same time, the errors and uncertainties in each step accumulate, so the distances to the farthest objects are the least well known.

Clusters of Galaxies

Figure 24.13 sketches the locations of all the known major astronomical objects within about 1 Mpc of the Milky Way. Our Galaxy appears with its dozen or so satellite galaxies—including the two Magellanic Clouds discussed earlier and a recently discovered companion (labeled "Sagittarius dwarf" in the figure) lying almost within our own Galactic plane. The Andromeda Galaxy, lying 800 kpc from us, is also shown, surrounded by satellites of its own. Two of Andromeda's galactic neighbors are shown in insets. M33 is a spiral, while M32 is a dwarf elliptical, easily seen in Figure 23.2(a) below and to the right of Andromeda's central bulge.

All told, some 45 galaxies are known to populate our Galaxy's neighborhood. Three of them (the Milky Way, Andromeda, and M33) are spirals; the remainder are dwarf irregulars and dwarf ellipticals. Together, these galaxies form the **Local Group**—a new level of structure in the universe above the scale of our Galaxy. As indicated in Figure 24.13, the Local Group's diameter is a little over 1 Mpc. The Milky Way Galaxy and Andromeda are by far its largest members, and most of the smaller galaxies are gravitationally bound to one or the other of them. The combined gravity of the galaxies in the Local Group binds them together, like stars in a star cluster, but on a millionfold larger scale. More generally, a group of galaxies held together by their mutual gravitational attraction is called a **galaxy cluster**.

Moving beyond the Local Group, the next large concentration of galaxies we come to is the *Virgo Cluster* (Figure 24.14), named after the constellation in which it is found. Lying some 18 Mpc from the Milky Way, the Virgo Cluster does not contain a mere 45 galaxies, however. Rather, it houses more than 2500 galaxies, bound by gravity into a tightly knit group about 3 Mpc across.

Wherever we look in the universe, we find galaxies, and the majority of galaxies are members of galaxy clusters.

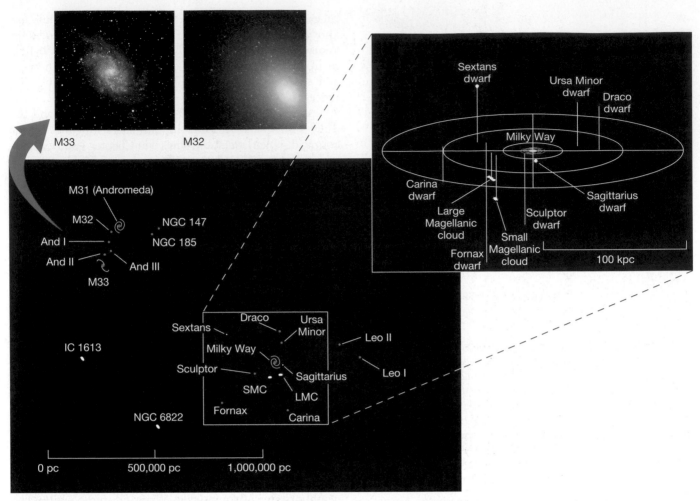

▲ **FIGURE 24.13 Local Group** The Local Group is made up of some 45 galaxies within approximately 1 Mpc of our Milky Way Galaxy. Only a few are spirals; most of the rest are dwarf-elliptical or irregular galaxies, only some of which are shown here. Spirals are colored blue, ellipticals pink, and irregulars white. The inset map at the right shows the Milky Way in relation to some of its satellite galaxies. The photographic insets (top) show two well-known neighbors of the Andromeda Galaxy (M31): the spiral galaxy M33 and the dwarf elliptical galaxy M32 (also visible in Figure 23.2a, a larger scale view of the Andromeda system). (*M. BenDaniel; NASA*)

Small clusters, such as the Local Group, contain only a few galaxies and are quite irregular in shape. Large, "rich" clusters like Virgo contain thousands of individual galaxies distributed fairly smoothly in space. The Coma cluster, shown in Figure 24.1 and lying approximately 100 Mpc away, is another example of a rich cluster. Figure 24.15 is a long-exposure photograph of a much more distant rich cluster, lying almost 1 billion parsecs from Earth. A sizeable minority of galaxies (perhaps 20 to 30 percent) are not members of any cluster, but are apparently isolated systems, moving alone through intercluster space.

CONCEPT CHECK

✔ What are some of the problems astronomers encounter in measuring the distances to faraway galaxies?

24.3 Hubble's Law

Now that we have seen some basic properties of galaxies throughout the universe, let's turn our attention to the large-scale *motions* of galaxies and galaxy clusters. Within a galaxy cluster, individual galaxies move more or less randomly. You might expect that, on even larger scales, the clusters themselves would also have random, disordered motion—some clusters moving this way, some that. In fact, that is not the case: On the largest scales, galaxies and galaxy clusters alike move in a very *ordered* way.

Universal Recession

In 1912, the American astronomer Vesto M. Slipher, working under the direction of Percival Lowell, discovered that virtually every spiral galaxy he observed had a redshifted spectrum—it was *receding* from our Galaxy.

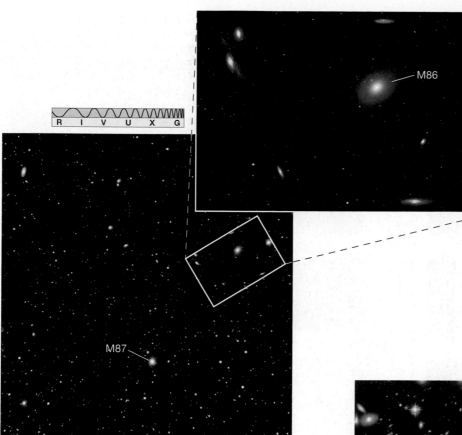

◄ **FIGURE 24.14** Virgo Cluster The central region of the Virgo Cluster of galaxies, about 18 Mpc from Earth. Many large spiral and elliptical galaxies can be seen. The inset shows several galaxies surrounding the giant elliptical known as M86. An even bigger elliptical galaxy, M87, noted at the bottom, will be discussed later in the chapter. (*M. BenDaniel; AURA*)

∞ (Sec. 3.5) It is now known that, except for a few nearby systems, *every* galaxy takes part in a general motion away from us in all directions. Individual galaxies that are not part of galaxy clusters are steadily receding. Galaxy clusters, too, have an overall recessional motion, although their individual member galaxies move randomly with respect to one another. (Consider a jar full of fireflies that has been thrown into the air. The fireflies within the jar, like the galaxies within the cluster, have random motions due to their individual whims, but the jar as a whole, like the galaxy cluster, has some directed motion as well.)

Figure 24.16 shows the optical spectra of several galaxies, arranged in order of increasing distance from the Milky Way Galaxy. The spectra are redshifted, indicating that the associated galaxies are receding. Furthermore, the extent of the redshift increases progressively from top to bottom in the figure. There is a connection between Doppler shift and distance: The greater the distance, the greater the redshift. This trend holds for nearly all galaxies in the universe. (Two galaxies within our Local Group, including Andromeda, and a few galaxies in the Virgo Cluster display blueshifts and so are moving toward us, but this results from their local motions within their parent clusters—recall the fireflies in the jar.)

Figure 24.17(a) shows recessional velocity plotted against distance for the galaxies of Figure 24.16. Figure

▲ **FIGURE 24.15** Distant Galaxy Cluster The galaxy cluster Abell 1689 contains huge numbers of galaxies and resides roughly 2 billion parsecs from Earth. Virtually every patch of light in this photograph is a separate galaxy. Thanks to the high resolution of the optics on board the *Hubble Space Telescope*, we can now discern, even at this great distance, spiral structure in some of the galaxies. We also see many galaxies colliding—some tearing matter from one another, others merging into single systems. (*NASA*)

ANIMATION Cluster Merger

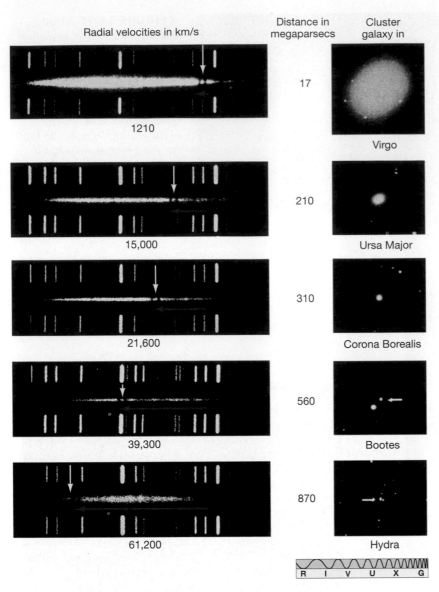

Radial velocities in km/s

Distance in megaparsecs

Cluster galaxy in

1210 — 17 — Virgo

15,000 — 210 — Ursa Major

21,600 — 310 — Corona Borealis

39,300 — 560 — Bootes

61,200 — 870 — Hydra

R I V U X G

◀ **FIGURE 24.16 Galaxy Spectra** Optical spectra, shown at left, of several galaxies named on the right. Both the extent of the redshift (denoted by the horizontal red arrows) and the distance from the Milky Way Galaxy to each galaxy (numbers in center column) increase from top to bottom. The vertical yellow arrow in each spectrum highlights a particular spectral feature (a pair of dark absorption lines). The horizontal red arrows indicate how this feature shifts to longer wavelengths in spectra of more distant galaxies. The white lines at the top and bottom of each spectrum are laboratory references. *(Palomar/Caltech)*

24.17(b) is a similar plot for some more galaxies within about 1 billion parsecs of Earth. Plots like these were first made by Edwin Hubble in the 1920s and now bear his name: *Hubble diagrams*. The data points generally fall close to a straight line, indicating that the rate at which a galaxy recedes is *directly proportional* to its distance from us. This rule is called **Hubble's law**. We can construct such a diagram for any group of galaxies, provided that we can determine their distances and velocities. The universal recession described by the Hubble diagram is sometimes called the *Hubble flow*.

The recessional motions of the galaxies prove that the cosmos is neither steady nor unchanging on the largest scales. The universe (actually, *space itself*—see Section 26.2) is expanding! However, let's be clear on just *what* is expanding and what is not. Hubble's law does not mean that humans, Earth, the solar system, or even individual galaxies and galaxy clusters are physically increasing in size. These groups of atoms, rocks, planets, stars, and galaxies are held together by their own internal forces and are not

themselves getting bigger. Only the largest framework of the universe—the vast distances separating the galaxy clusters—is expanding.

To distinguish recessional redshift from redshifts caused by motion *within* an object—for example, galactic orbits within a cluster or explosive events in a galactic nucleus—the redshift resulting from the Hubble flow is called the **cosmological redshift**. Objects that lie so far away that they exhibit a large cosmological redshift are said to be at *cosmological distances*—distances comparable to the scale of the universe itself.

Hubble's law has some fairly dramatic implications. If nearly all galaxies show recessional velocity according to Hubble's law, then doesn't that mean that they all started their journey from a single point? If we could run time backward, wouldn't all the galaxies fly back to this one point, perhaps the site of some explosion in the remote past? The answer is yes—but not in the way you might expect! In Chapters 26 and 27, we will explore the ramifications of the Hubble flow for the past and future evolution

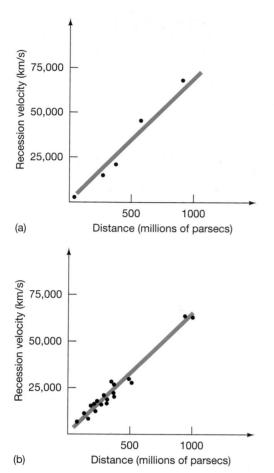

(a)

(b)

▲ **FIGURE 24.17 Hubble's Law** Plots of recessional velocity against distance (a) for the galaxies shown in Figure 24.16 and (b) for numerous other galaxies within about 1 billion pc of Earth.

of our universe. For now, however, we set aside its cosmic implications and use Hubble's law simply as a convenient distance-measuring tool.

Hubble's Constant

The constant of proportionality between recessional velocity and distance in Hubble's law is known as **Hubble's constant**, denoted by the symbol H_0. The data shown in Figure 24.17 then obey the equation

$$\text{recessional velocity} = H_0 \times \text{distance}.$$

The value of Hubble's constant is the slope of the straight line—recessional velocity divided by distance—in Figure 24.17(b). Reading the numbers off the graph, we get roughly 70,000 km/s divided by 1000 Mpc, or 70 km/s/Mpc (kilometers per second per megaparsec, the most commonly used unit for H_0). Astronomers continually strive to refine the accuracy of the Hubble diagram and the resulting estimate of H_0, because Hubble's constant is one of the most fundamental quantities of nature; it specifies the rate of expansion of the entire cosmos.

The precise value of Hubble's constant is the subject of considerable debate. The most recent measurements, made by many different research groups using different sets of galaxies and a wide variety of distance-measurement techniques, give results mainly between 50 and 80 km/s/Mpc. Most astronomers would be quite surprised if the true value of H_0 turned out to lie outside this range. However, the width of the quoted range is not the result of measurement uncertainties in any one method; rather, there remain real, and as yet unresolved, inconsistencies between the different techniques currently in use. Infrared Tully–Fisher measurements and studies of Cepheid variables, the latter now extended to include the Virgo cluster by researchers using the *Hubble Space Telescope*, generally produce results at the high end of the range, 70–80 km/s/Mpc. ∞ (Secs. 23.2, 24.2) However, visible-light Tully–Fisher studies and techniques using standard candles, including Type I supernovae, tend to return lower values, in the range 50–65 km/s/Mpc. ∞ (Sec. 24.2) Other methods give results scattered between 50 and 80 km/s/Mpc.

For now, astronomers must simply live with this uncertainty. For the remainder of the text, we will adopt $H_0 = 70$ km/s/Mpc (roughly the median of all recent results and also a value consistent with some precise measurements to be discussed in Chapter 27) as the best current estimate of Hubble's constant. Bear in mind, though, that there is considerable ambiguity—and dispute—among experts as to the true value of this very important number.

The Top of the Distance Ladder

Using Hubble's law, we can derive the distance to a remote object simply by measuring the object's recessional velocity and dividing by Hubble's constant. Hubble's law thus tops our inverted pyramid of distance-measurement techniques (Figure 24.18). This seventh method simply assumes that Hubble's law holds. If that assumption is correct, Hubble's law enables us to measure great distances in the universe—so long as we can obtain an object's spectrum, we can determine how far away it is. Notice, however, that the uncertainty in Hubble's constant translates directly into a similar uncertainty in all distances determined from Hubble's law.

Many redshifted objects have recessional motions that are a substantial fraction of the speed of light. The most distant objects thus far observed in the universe—some young galaxies and quasars (see Section 24.4)—have redshifts (fractional increases in wavelength) of more than 6, meaning that their radiation has been stretched in wavelength not by just a few percent, as with most of the objects we have discussed, but *sevenfold*. Their ultraviolet spectral lines are shifted all the way into the infrared part of the spectrum! *More Precisely 24-1* discusses in more detail the meaning and interpretation of such large redshifts, apparently implying recessional velocities comparable to the

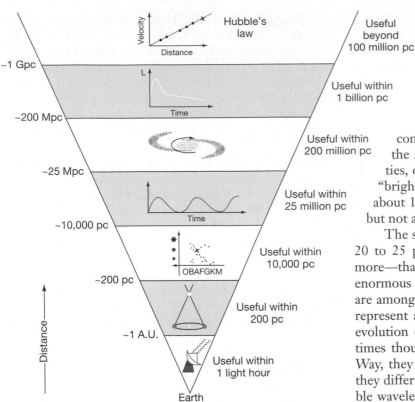

◄ FIGURE 24.18 Cosmic Distance Ladder
Hubble's law tops the hierarchy of distance-measurement techniques. It is used to find the distances of astronomical objects all the way out to the limits of the observable universe.

comparison, in round numbers, the luminosity of the Milky Way Galaxy is 2×10^{10} solar luminosities, or roughly 10^{37} W. For our purposes, the term "bright" will be taken to mean anything more than about 10^{10} times the solar value. Our Galaxy is bright, but not abnormally so.

The substantial minority of bright galaxies—perhaps 20 to 25 percent, although some researchers would say more—that don't fit well into the Hubble scheme are of enormous interest to astronomers. Some of these galaxies are among the most energetic objects known, and all may represent an important, if intermittent, phase of galactic evolution (see Section 25.3). Having luminosities sometimes thousands of times greater than that of the Milky Way, they are known collectively as **active galaxies**, and they differ significantly from their normal cousins. At visible wavelengths, they often *look* like normal galaxies—familiar components such as disks, bulges, stars, gas, and dust can be identified. At other wavelengths, however, their unusual properties are much more apparent.

speed of light. According to Hubble's law, the objects that exhibit these redshifts lie almost 9000 Mpc away from us, as close to the limits of the observable universe as astronomers have yet been able to probe.

The speed of light is finite. It takes time for light—or, for that matter, any kind of radiation—to travel from one point in space to another. The radiation that we now see from these most distant objects originated long ago. Incredibly, that radiation was emitted almost 13 billion years ago (see Table 24.1), well before our planet, our Sun, and perhaps even our Galaxy came into being!

CONCEPT CHECK

✔ How does the use of Hubble's law differ from the other extragalactic distance-measurement techniques we have seen in this text?

24.4 Active Galactic Nuclei

The galaxies described in the previous sections—those falling into the various Hubble classes—are generally referred to as *normal* galaxies. Probably more than 75 percent of "bright" galaxies (and a larger fraction of all galaxies) fall into this broad category. The luminosities of normal galaxies range from a million or so times that of the Sun for dwarf ellipticals and irregulars to more than a trillion solar luminosities for the largest giant ellipticals. For

Galactic Radiation

Active galaxies differ fundamentally from normal galaxies both in their overall luminosities and in the *character* of the radiation they emit. Most of a normal galaxy's energy is emitted in or near the visible portion of the electromagnetic spectrum, much like the radiation from stars. Indeed, to a large extent, the light we see from a normal galaxy *is* just the accumulated light of its many component stars (once the effects of interstellar dust are taken into account). By contrast, as illustrated schematically in Figure 24.19, the radiation from active galaxies does *not* peak in the visible range. Most active galaxies do emit substantial amounts of visible radiation, but far more of their energy is emitted at invisible wavelengths, both longer and shorter than those in the visible range. Put another way, the radiation from active galaxies is *inconsistent* with what we would expect if it were the combined radiation of myriad stars. Their radiation is said to be *nonstellar*.

Many luminous galaxies with nonstellar emission are known to be *starburst galaxies*—previously normal systems currently characterized by widespread episodes of star formation, most likely as a result of interactions with a neighbor. The irregular galaxy M82 shown in Figure 24.8 is a prime example. We will study these important systems and their role in galaxy evolution in Chapter 25. For purposes of this text, however, we will use the term "active galaxy"

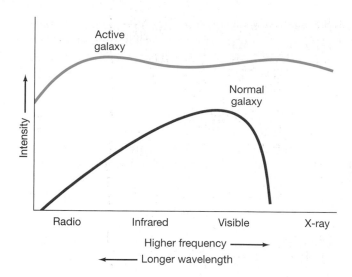

◀ **FIGURE 24.19 Galaxy Energy Spectra** The energy emitted by a normal galaxy differs significantly from that emitted by an active galaxy. This plot illustrates the general run of intensity for all galaxies of a particular type and does not represent any one individual galaxy.

to mean a system whose abnormal activity is related to violent events occurring in or near the galactic nucleus. Such systems are also known as **active galactic nuclei**.

Even with this restriction, there is still considerable variation in the properties of galaxies, and astronomers have identified and cataloged a bewildering array of systems falling into the "active" category. For example, Figure 24.20 shows an active galaxy exhibiting both nuclear activity and widespread star formation, with a blue-tinted ring of newborn stars surrounding an extended 1-kpc-wide core of intense emission. Rather than attempting to describe the entire "zoo" of active galaxies, we will instead discuss three basic species: the energetic *Seyfert galaxies* and *radio galaxies* and the even more luminous *quasars*. Although these objects all lie toward the "high-luminosity" end of the active range and represent perhaps only a few percent of the total number of active galaxies, their properties will allow us to identify and discuss features common to active galaxies in general.

Astronomers once distinguished between active galaxies and quasars on the basis of their appearance, spectra, and distance from us. Quasars are generally so far away that little structure can be discerned, giving most quasars a "starlike" appearance. But improving observations now reveal "galactic" components in many quasars, so the distinction between quasars and active galaxies is not clear cut. Most astronomers think that quasars are simply an early stage of galactic evolution and that the same basic processes power all active objects.

Seyfert Galaxies

In 1943, Carl Seyfert, an American optical astronomer studying spiral galaxies from Mount Wilson Observatory, discovered the type of active galaxy that now bears his name. **Seyfert galaxies** are a class of astronomical objects whose properties lie between those of normal galaxies and those of the most energetic active galaxies known.

Superficially, Seyferts resemble normal spiral galaxies (Figure 24.21). Indeed, the stars in a Seyfert's galactic disk and spiral arms produce about the same amount of visible radiation as do the stars in a normal spiral galaxy. However, most of a Seyfert's energy is emitted from a small central region known as the **galactic nucleus**—the center of the overexposed white patch in the figure. The nucleus of a Seyfert galaxy is some 10,000 times brighter than the center of our own Galaxy. In fact, the brightest Seyfert nuclei are 10 times more energetic than the *entire* Milky Way.

Some Seyferts produce radiation spanning a broad range in wavelengths, from the infrared all the way through ultraviolet and even X rays. However, the majority (about 75 percent) emit most of their energy in the infrared. Scientists think that much of the high-energy radiation in these Seyferts is absorbed by dust in or near the nucleus and then reemitted as infrared radiation.

Seyfert spectral lines have many similarities to those observed toward the center of our own Galaxy. ∞ (Sec. 23.6) Some of the lines are very broad, most likely indicating rapid (5000 km/s or more) internal motion within the nuclei. ∞ (Sec. 4.4) However, not all of the lines are broad, and

R I V U X G

▲ **FIGURE 24.20 Active Galaxy** This image of the galaxy NGC 7742 resembles a fried egg, with a ring of blue star-forming regions surrounding a very bright yellow core that spans about 1 kpc. An active galaxy, NGC 7742 combines star formation with intense emission from its central nucleus and lies roughly 24 Mpc away. *(NASA)*

◄ FIGURE 24.21 Seyfert Galaxy The Circinus galaxy, a Seyfert with a bright compact core, lies some 4 Mpc away. It is one of the closest active galaxies. *(NASA)*

R I V U X G

some Seyferts show no broad lines at all. In addition, their energy emission often varies in time (Figure 24.22). A Seyfert's luminosity can double or halve within a fraction of a year. These rapid fluctuations in luminosity lead us to conclude that the source of energy emissions in Seyfert galaxies must be quite compact—simply put, as we saw in Chapter 22, an object cannot "flicker" in less time than radiation takes to cross it. ∞ (Sec. 22.4) The emitting region must therefore be less than one light-year across—an extraordinarily small region, considering the amount of energy emanating from it.

MORE PRECISELY 24-1

Relativistic Redshifts and Look-Back Time

In discussing very distant objects, astronomers usually talk about their redshifts rather than their distances. Indeed, it is common for researchers to speak of an event occurring "at" a certain redshift—meaning that the light received today from that event is redshifted by the specified amount. Of course, because of Hubble's law, redshift and distance are equivalent to one another. However, redshift is the preferred quantity because it is a directly observable property of an object, whereas distance is derived from redshift with the use of Hubble's constant, whose value is not accurately known. (In Chapter 26 we will see another, much more fundamental, reason why astronomers favor the use of redshift in studies of the cosmos.)

The redshift of a beam of light is, by definition, the *fractional* increase in the wavelength of the light resulting from the recessional motion of the source. ∞ (Sec. 3.5) Thus, a redshift of 1 corresponds to a *doubling* of the wavelength. From the formula for the Doppler shift given previously, the redshift of radiation received from a source moving away from us with speed v is given by

$$\text{redshift} = \frac{\text{observed wavelength} - \text{true wavelength}}{\text{true wavelength}}$$

$$= \frac{\text{recessional velocity } v}{\text{speed of light, } c}.$$

Let's illustrate this relationship with two examples, rounding the speed of light, c, to 300,000 km/s. A galaxy at a distance of 100 Mpc has a recessional speed (by Hubble's law) of 70 km/s/Mpc × 100 Mpc = 7,000 km/s. Its redshift is therefore 7,000 km/s ÷ 300,000 km/s = 0.023. Conversely, an object that has a redshift of 0.05 has a recessional velocity of 0.05 × 300,000 km/s = 15,000 km/s and hence a distance of 15,000 km/s ÷ 70 km/s/Mpc = 210 Mpc.

Unfortunately, while the foregoing equation is correct for low speeds, it does not take into account the effects of relativity. As we saw in Chapter 22, the rules of everyday physics have to be modified when speeds begin to approach the speed of light. ∞ (*More Precisely 22-1*) The formula for the Doppler shift is no exception. In particular, while the formula is valid for speeds much less than the speed of light, when $v = c$ the redshift is not unity, as the equation suggests, but is in fact *infinite*. That is, radiation received from an object moving away from us at nearly the speed of light is redshifted to almost infinite wavelength.

Thus, do not be alarmed to find that many galaxies and quasars have redshifts greater than unity. This does not mean that they are receding faster than light! It simply means that the preceding simple formula is not applicable. In fact, the real connection between redshift and distance is quite complex, requiring us to make key assumptions about the past history of the universe (see Chapter 26). In place of a formula, we can use Table 24.2, which presents a conversion chart relating redshift and distance. All of the values shown are based on reasonable assumptions and are usable even for large redshifts. We take Hubble's constant to be 70 km/s/Mpc and assume a flat universe (see Section 26.5) in which matter (mostly dark) contributes just over one-quarter of the total density. The conversions in the table are used consistently throughout this text. The column headed "v/c" gives equivalent recessional velocities based on the Doppler effect, taking relativity properly into account. Even though this is *not* the correct interpretation of the redshift (see Section 26.2), we include it here for comparison, simply because it is so often quoted in the popular media.

Because the universe is expanding, the "distance" to a galaxy is not very well defined. Do we mean the distance to the galaxy when it emitted the light we see today, the present distance to the galaxy (as presented in the table, even though we do not see the galaxy as it is today), or some other, more appropriate measure? Largely because of this ambiguity, as-

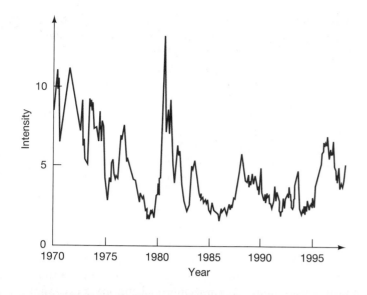

◀ **FIGURE 24.22 Seyfert Time Variability** This graph illustrates the irregular variations in a Seyfert galaxy's luminosity over two decades. Because this Seyfert, called 3C 84, emits strongly in the radio part of the electromagnetic spectrum, these observations were made with large radio telescopes. The optical and X-ray luminosities vary as well. *(NRAO)*

Together, the rapid time variability and large radio and infrared luminosities observed in Seyferts imply violent nonstellar activity in their nuclei. This activity may well be similar in *nature* to processes occurring at the center of our own Galaxy, but its *magnitude* is thousands of times greater than the comparatively mild events within our own Galaxy's heart. ∞ (Sec. 23.7)

tronomers prefer to work in terms of a quantity known as the *look-back time* (shown in the last column of Table 24.2), which is simply how long ago an object emitted the radiation we see today. While astronomers talk frequently about redshifts and sometimes about look-back times, they hardly ever talk of distances to high-redshift objects (and *never* about recession velocities, despite what you hear on the news!). Bear in mind, however, that redshift is the only unambiguously measured quantity in this discussion. Statements about "derived" quantities, such as distances and look-back times, all require that we make specific assumptions about how the universe has evolved with time.

For nearby sources, the look-back time is numerically equal to the distance in light-years: The light we receive tonight from a galaxy at a distance of 100 million light-years was emitted 100 million years ago. However, for more distant objects, the look-back time and the present distance in light-years differ because of the expansion of the universe, and the divergence increases dramatically with increasing redshift.

As a simple analogy, imagine an ant crawling across the surface of an expanding balloon at a constant speed of 1 cm/s relative to the balloon's surface. After 10 seconds, the ant may think it has traveled a distance of 10 cm, but an outside observer with a tape measure will find that it is actually more than 10 cm from its starting point (measured along the surface of the balloon) because of the balloon's expansion. In exactly the same way, the present distance to a galaxy with a given redshift depends on how the universe expanded in the past. For example, a galaxy now located 15 billion light-years from Earth was much closer to us when it emitted the light we now see. Consequently, its light has taken considerably less than 15 billion years—in fact, about 10 billion years—to reach us.

TABLE 24.2 Redshift, Distance, and Look-Back Time

Redshift	V/C	Present Distance (Mpc)	Present Distance (10^6 light-years)	Look-Back Time (millions of years)
0.000	0.000	0	0	0
0.010	0.010	42	137	137
0.025	0.025	105	343	338
0.050	0.049	209	682	665
0.100	0.095	413	1350	1290
0.200	0.180	809	2640	2410
0.250	0.220	999	3260	2920
0.500	0.385	1880	6140	5020
0.750	0.508	2650	8640	6570
1.000	0.600	3320	10,800	7730
1.500	0.724	4400	14,400	9320
2.000	0.800	5250	17,100	10,300
3.000	0.882	6460	21,100	11,500
4.000	0.923	7310	23,800	12,100
5.000	0.946	7940	25,900	12,500
6.000	0.960	8420	27,500	12,700
10.000	0.984	9660	31,500	13,200
50.000	0.999	12,300	40,100	13,600
100.000	1.000	12,900	42,200	13,700
∞	1.000	14,600	47,500	13,700

Radio Galaxies

As the name suggests, **radio galaxies** are active galaxies that emit large amounts of energy in the radio portion of the electromagnetic spectrum. They differ from Seyferts not only in the wavelengths at which they radiate, but also in both the appearance and the extent of their emitting regions.

Figure 24.23 shows the radio galaxy Centaurus A, which lies about 4 Mpc from Earth. Almost none of this galaxy's radio emission comes from a compact nucleus. Instead, the energy is released from two huge extended regions called **radio lobes**—roundish clouds of gas spanning about half a megaparsec and lying well beyond the visible galaxy.* Undetectable in visible light, the radio lobes of radio galaxies are truly enormous. From end to end, they typically span more than 10 times the size of the Milky Way Galaxy, comparable in scale to the entire Local Group.

Figure 24.24 shows the relationship between the galaxy's visible, radio, and X-ray emissions. In visible light, Centaurus A is apparently a large E2 galaxy some 500 kpc in diameter, bisected by an irregular band of dust. Centaurus A is a member of a small cluster of galaxies, and numerical simulations suggest that this peculiar galaxy is probably the result of a collision between an elliptical galaxy and a smaller spiral galaxy about 500 million years ago. In the crowded confines of a cluster, such collisions may be commonplace (Section 25.2). The radio lobes are roughly sym-

The term "visible galaxy" is commonly used to refer to those components of an active galaxy that emit visible "stellar" radiation, as opposed to the nonstellar and invisible "active" component of the galaxy's emission.

metrically placed, jutting out from the center of the visible galaxy and roughly perpendicular to the dust lane, suggesting that they consist of material ejected in opposite directions from the galactic nucleus. This conclusion is strengthened by the presence of a pair of smaller secondary lobes closer to the visible galaxy and by the presence of a roughly 1-kpc-long jet of matter in the galactic center, all aligned with the main lobes (and marked in the figure).

If the material was ejected from the nucleus at close to the speed of light and has subsequently slowed, then Centaurus A's outer lobes were created a few hundred million years ago, quite possibly around the time of the collision thought to be responsible for the galaxy's odd optical appearance. The secondary lobes were expelled more recently. Apparently, some violent process at the center of Centaurus A—most probably triggered by the collision—started up around that time and has been intermittently firing jets of matter out into intergalactic space ever since.

Centaurus A is a relatively low luminosity source that happens to lie very close to us, astronomically speaking, making it particularly easy to study. Figure 24.25 shows a much more powerful emitter, called Cygnus A, lying roughly 250 Mpc from Earth. The high-resolution radio map in Figure 24.25(b) clearly shows two narrow, high-speed jets joining the radio lobes to the center of the visible galaxy (the dot at the center of the radio image). Notice that, as with Centaurus A, Cygnus A is a member of a small group of galaxies, and the optical image (Figure 24.25a) appears to show two galaxies colliding.

The radio lobes of the brightest radio galaxies (such as Cygnus A) emit roughly 10 times more energy than the

(a)

R I V U X G

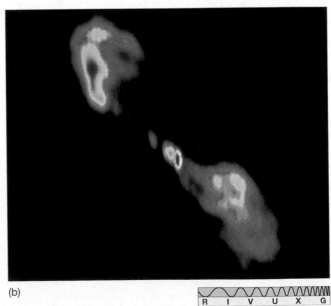

(b)

R I V U X G

▲ **FIGURE 24.23 Centaurus A Radio Lobes** Radio galaxies, such as Centaurus A, shown here optically in (a), often have giant radio-emitting lobes (b) extending a million parsecs or more beyond the central galaxy. The lobes cannot be imaged in visible light and are observable only with radio telescopes. The lobes in part (b) are shown in false color, with decreasing intensity from red to yellow to green to blue. *(ESO; NRAO)*

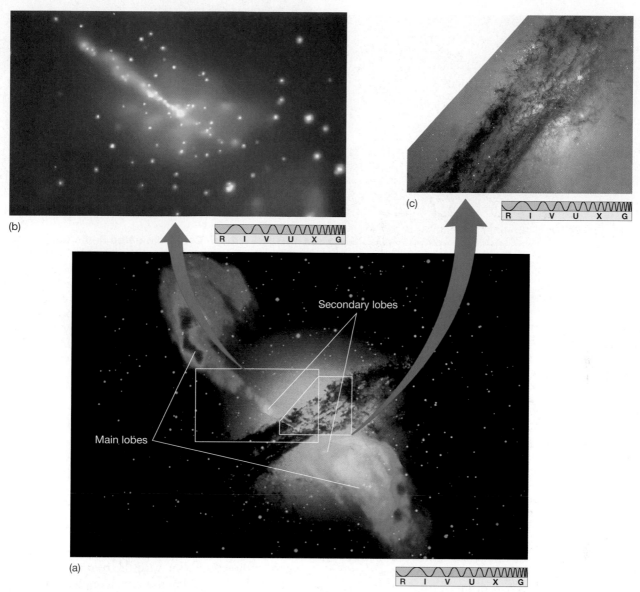

▲ **FIGURE 24.24 Centaurus A, Close Up** The main image (a) shows an optical photograph of Centaurus A, one of the most massive and peculiar galaxies known. Centaurus A is thought to be the result of a collision between two galaxies that took place 500 million years ago. The pastel false colors mark the radio emission shown in Figure 24.23; the data here were more recently acquired and are of higher resolution. (b) Although the radio jets emit no visible light, they do emit X rays, as shown in this *Chandra* image. (c) Increasingly high resolution optical views of the galaxy's core region, taken by the *Hubble Space Telescope*. (*NASA; SAO; J. Burns*)

Milky Way Galaxy does at all wavelengths, coincidentally about the same amount of energy emitted by the most luminous Seyfert nuclei. However, despite their names, radio galaxies actually radiate far more energy at shorter wavelengths. Their total energy output can be a hundred (or more) times greater than their radio emission. Most of this energy comes from the nucleus of the visible galaxy. With total luminosities up to a thousand times that of the Milky Way, bright radio galaxies are among the most energetic objects known in the universe. Their radio emission lets us study in detail the connection between the small-scale nucleus and the large-scale radio lobes.

Not all radio galaxies have obvious radio lobes. Figure 24.26 shows a *core-dominated* radio galaxy, most of whose energy is emitted from a small central nucleus (which radio astronomers refer to as the *core*) less than 1 pc across. Weaker radio emission comes from an extended region surrounding the nucleus. It is likely that all radio galaxies have jets and lobes, but what we observe depends on our perspective. As illustrated in Figure 24.27, when a radio galaxy is viewed from the side, we see the jets and lobes. However, if we view the jet almost head-on—in other words, looking *through* the lobe—we see a core-dominated system.

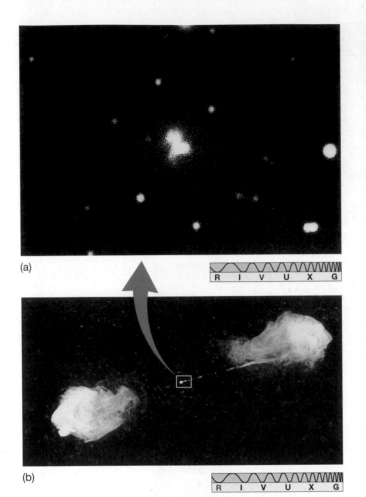

(a)

(b)

▲ **FIGURE 24.25 Cygnus A** (a) Like Centaurus A, Cygnus A appears to be two galaxies colliding. (b) On a much larger scale, Cygnus A displays radio-emitting lobes on either side of the optical image. The optical galaxy in (a) is about the size of the small dot at the center of (b). Note the thin line of radio-emitting material joining the right lobe to the central galaxy. The distance from one lobe to the other is approximately a million light-years. *(NOAO; NRAO)*

Common Features of Active Galaxies

The basic properties just described for Seyfert galaxies and radio galaxies are fairly representative of active galaxies in general. In all cases, a huge amount of energy, spanning much of the electromagnetic spectrum, is generated within—and usually also emitted from—a central, compact nucleus.

Jets are a common feature of active galaxies. Figure 24.28 presents several images of the giant elliptical galaxy M87, a prominent member of the Virgo Cluster (Figure 24.14). A long time exposure (Figure 24.28a) shows a large, fuzzy ball of light—a fairly normal-looking E1 galaxy about 100 kpc across. A shorter exposure of M87 (Figure 24.28b), capturing only the galaxy's bright inner regions, reveals a long (2 kpc) thin jet of matter ejected from the galactic center at nearly the speed of light. Computer en-

▲ **FIGURE 24.26 Core-Dominated Radio Galaxy** On this radio contour map of the radio galaxy M86, we can see that the radio emission comes from a bright central nucleus, which is surrounded by an extended, less intense radio halo. The radio map is superimposed on an optical image of the galaxy and some of its neighbors, a wider field version of which was shown previously in Figure 24.14. *(Harvard-Smithsonian Center for Astrophysics)*

hancement shows that the jet is made up of a series of distinct "blobs" more or less evenly spaced along its length, suggesting that the material was ejected during bursts of activity. The jet has also been imaged in the radio, infrared (Figure 24.28c), and X-ray regions of the spectrum.

Our location with respect to a jet also affects the *type* of radiation we see. The theory of relativity tells us that radiation emitted by particles moving close to the speed of light is strongly concentrated, or beamed, in the direction of motion. ∞ *(More Precisely 22-1)* As a result, if we happen to be directly in line with the beam, the radiation we receive is both very intense and Doppler shifted toward short wavelengths. ∞ *(Sec. 3.5)* The resulting object is called a **blazar** (Figure 5.39). Much of the luminosity of the hundred or so known blazars is received in the form of X or gamma rays.

Finally, note that all the active galaxies described so far show signs of *interactions* with other galaxies. We have already discussed this in the context of Centaurus A and Cygnus A, and M87 lies near the center of the Virgo cluster, having probably achieved its present size via mergers with other, smaller galaxies. The galaxies shown in Figures 24.20 and 24.21 also show evidence for bursts of star formation within the past few million years—a clear signature of a recent galactic encounter.

CONCEPT CHECK

✔ The energy emission from an active galactic nucleus does not resemble a blackbody curve. Why is this important?

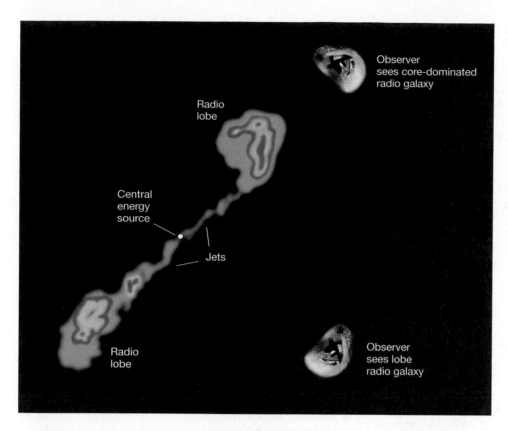

◄ **FIGURE 24.27 Radio Galaxy**
A central energy source produces high-speed jets of matter that interact with intergalactic gas to form radio lobes. The system may appear to us as either radio lobes or a core-dominated radio galaxy, depending on our location with respect to the jets and lobes.

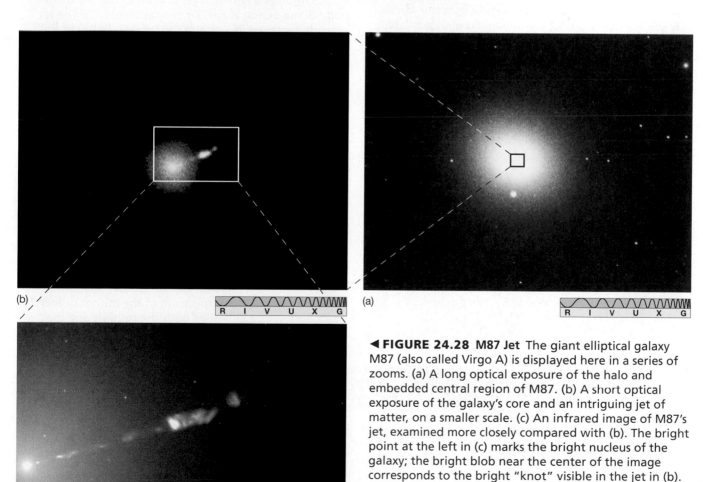

◄ **FIGURE 24.28 M87 Jet** The giant elliptical galaxy M87 (also called Virgo A) is displayed here in a series of zooms. (a) A long optical exposure of the halo and embedded central region of M87. (b) A short optical exposure of the galaxy's core and an intriguing jet of matter, on a smaller scale. (c) An infrared image of M87's jet, examined more closely compared with (b). The bright point at the left in (c) marks the bright nucleus of the galaxy; the bright blob near the center of the image corresponds to the bright "knot" visible in the jet in (b). *(NOAO; NASA)*

Quasars

In the early days of radio astronomy, many radio sources were detected for which no corresponding visible object was known. By 1960, several hundred such sources were listed in the *Third Cambridge Catalog*, and astronomers were scanning the skies in search of visible counterparts to these radio sources. Their job was made difficult both by the low resolution of the radio observations (which meant that the observers did not know exactly where to look) and by the faintness of the objects at visible wavelengths.

In 1960, astronomers detected what appeared to be a faint blue star at the location of the radio source 3C 48 (the 48th object on the third *Cambridge* list) and obtained its spectrum. Containing many unknown and unusually broad emission lines, the object's peculiar spectrum defied interpretation. 3C 48 remained a unique curiosity until 1962, when another similar-looking—and similarly mysterious—faint blue object with "odd" spectral lines was discovered and identified with the radio source 3C 273 (Figure 24.29).

The following year saw a breakthrough when astronomers realized that the strongest unknown lines in 3C 273's spectrum were simply familiar spectral lines of hydrogen redshifted by a very unfamiliar amount—about 16 percent, corresponding to a recession velocity of 48,000 km/s! Figure 24.30 shows the spectrum of 3C 273. Some prominent emission lines and the extent of their redshift are marked on the diagram. Once the nature of the strange spectral lines was known, astronomers quickly found a similar explanation for the spectrum of 3C 48, whose 37 percent redshift implied that it was receding from Earth at the astonishing rate of almost one-third the speed of light!

Their huge speeds mean that neither of these two objects can be members of our Galaxy. In fact, their large redshifts indicate that they are very far away indeed. Applying Hubble's law (with our adopted value of the Hub-

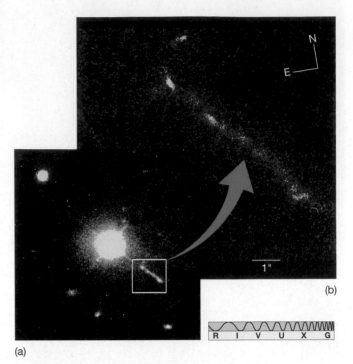

(a)

(b)

▲ **FIGURE 24.29** Quasar 3C 273 (a) The bright quasar 3C 273 displays a luminous jet of matter, but the main body of the quasar is starlike in appearance. (b) The jet extends for about 30 kpc and can be seen better in this high-resolution image. *(AURA)*

ble constant, $H_0 = 70$ km/s/Mpc), we obtain distances of 650 Mpc for 3C 273 and 1400 Mpc for 3C 48. (See again *More Precisely 24-1* for more information of how these distances are determined and what the large redshifts really mean.)

However, this explanation of the unusual spectra created an even deeper mystery. A simple calculation using the inverse-square law reveals that, despite their unimpressive optical appearance (see Figure 24.31), these faint "stars" are in fact among the brightest-known objects in

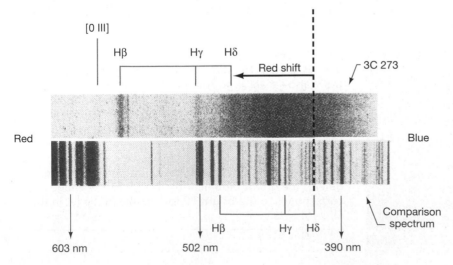

◀ **FIGURE 24.30** Quasar Spectrum Optical spectrum of the distant quasar 3C 273. Notice both the redshift and the widths of the three hydrogen spectral lines marked as Hβ, Hγ, and Hδ. The redshift indicates the quasar's enormous distance. The width of the lines implies rapid internal motion within the quasar. *(Adapted from Palomar/Caltech)*

▲ FIGURE 24.31 Typical Quasar Although quasars are the most luminous objects in the universe, they are often unimpressive in appearance. In this optical image, a distant quasar (marked by an arrow) is seen close (in the sky) to a nearby spiral galaxy. The quasar's much greater distance makes it appear much fainter than the galaxy. *(NOAO)*

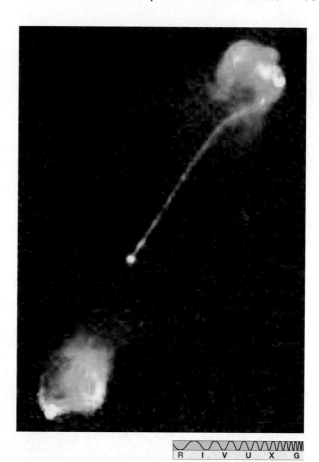

▲ FIGURE 24.32 Quasar Jets Radio image of the quasar 3C 175, showing radio jets feeding faint radio lobes. The bright (white) central object is the quasar, some 3000 Mpc away. The lobes themselves span approximately a million light-years. *(NRAO)*

the universe! 3C 273, for example, has a luminosity of about 10^{40} W, comparable to 20 trillion Suns or a thousand Milky Way Galaxies. More generally, quasars range in luminosity from around 10^{38} W—about the same as the brightest Seyferts—up to nearly 10^{42} W. A value of 10^{40} W (comparable to the luminosity of a bright radio galaxy) is fairly typical.

Clearly not stars (because of their enormous luminosities), these objects became known as *quasi-stellar radio sources* ("quasi-stellar" means "starlike"), or **quasars**. (The name persists even though we now know that not all such highly redshifted, starlike objects are strong radio sources.) More than 30,000 quasars are now known, and the numbers are increasing rapidly as large-scale surveys probe deeper and deeper into space (see *Discovery 25-1*). The distance to the *closest* quasar is 240 Mpc, and the farthest lies more than 9000 Mpc away. Most quasars lie well over 1000 Mpc from Earth. Since light travels at a finite speed, these faraway objects represent the universe as it was in the distant past. The implication is that most quasars date back to much earlier periods of galaxy formation and evolution, rather than more recent times. The prevalence of these energetic objects at great distances tells us that the universe was once a much more violent place than it is today.

Quasars share many properties with Seyferts and radio galaxies. Their radiation is nonstellar and may vary irregularly in brightness over periods of months, weeks, days, or

(in some cases) even hours, and some quasars show evidence of jets and extended emission features. Note the jet of luminous matter in 3C 273 (Figure 24.29), reminiscent of the jet in M87 and extending nearly 30 kpc from the center of the quasar. Figure 24.32 shows a quasar with radio lobes similar to those seen in Cygnus A (Figure 24.25b). Quasars have been observed in all parts of the electromagnetic spectrum, although many emit most of their energy in the infrared. For all these reasons, as mentioned earlier, most astronomers think that quasars *are* in fact the intensely bright cores of distant active galaxies lying too far away for the galaxies themselves to be seen. (Figure 25.16 presents *Hubble Space Telescope* observations of several quasars in which the surrounding galaxies can clearly be seen.)

CONCEPT CHECK

✔ Why did astronomers initially have difficulty recognizing quasars as highly luminous, very distant objects?

24.5 The Central Engine of an Active Galaxy

7 **6** The present consensus among astronomers is that, despite their differences in appearance and luminosity, Seyferts, radio galaxies, and quasars share a common energy-generation mechanism.

As a class, active galactic nuclei have some or all of the following properties:

1. They have *high luminosities*, generally greater than the 10^{37} W characteristic of a bright normal galaxy.

2. Their energy emission is mostly *nonstellar*—it cannot be explained as the combined radiation of even trillions of stars.

3. Their energy output can be highly *variable*, implying that their energy is emitted from a small central nucleus much less than a parsec across.

4. They often exhibit *jets* and other signs of explosive activity.

5. Their optical spectra may show broad emission lines, indicating *rapid internal motion* within the energy-producing region.

The principal questions, then, are "How can such vast quantities of energy arise from these relatively small regions of space?" "Why is the radiation nonstellar?" and "What is the origin of the jets and extended radio-emitting lobes?" We first consider how the energy is *produced* and then turn to the question of how it is actually *emitted* into intergalactic space.

Energy Production

To develop a feeling for the enormous emissions of active galaxies, consider for a moment an object having a luminosity of 10^{38} W. In and of itself, this energy output is not inconceivably large. The brightest giant ellipticals are comparably powerful. Thus, some 10^{12} stars—a few normal galaxies' worth of material—could *equivalently* power a typical active galaxy. However, in an active galaxy, this energy production is packed into a region much less than a parsec in diameter!

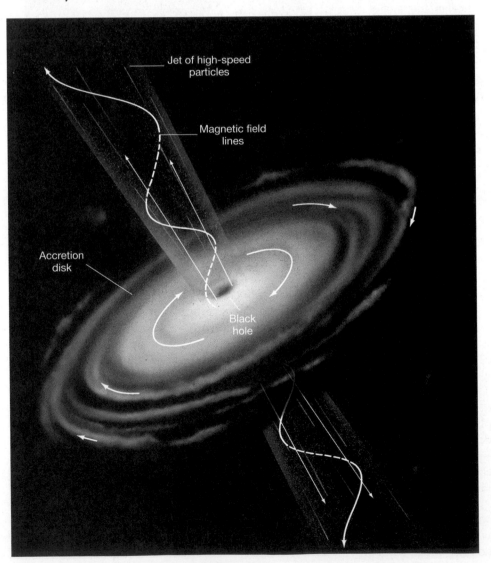

Jet of high-speed particles

Magnetic field lines

Accretion disk

Black hole

◀ **FIGURE 24.33 Active Galactic Nucleus** The leading theory for the energy source in active galactic nuclei holds that these objects are powered by material accreting onto a supermassive black hole. As matter spirals toward the hole, the matter heats up, producing large amounts of energy. At the same time, high-speed jets of gas may be ejected perpendicular to the accretion disk, forming the jets and lobes seen in many active objects. Magnetic fields generated in the disk are carried by the jets out to the radio lobes, where they play a crucial role in producing the observed radiation.

ANIMATION Active Galaxy

The twin requirements of large energy generation and small physical size bring to mind our discussion of X-ray sources in Chapter 13. ∞ (Secs. 22.3, 22.8) The presence of the jets in M87 and 3C 273 and the radio lobes in Centaurus A and Cygnus A strengthen the connection, as similar phenomena have also been observed in some stellar X-ray-emitting systems. Recall that the best current explanation for those "small-scale" phenomena involves the accretion of material onto a compact object—a neutron star or a black hole. Large amounts of energy are produced as matter spirals down onto the central object. In Chapter 23, we suggested that a similar mechanism, involving a *supermassive black hole*—having a mass of around 3–4 million suns—may also be responsible for the energetic radio and infrared emission observed at the center of our own Galaxy. ∞ (Sec. 23.6)

As illustrated in Figure 24.33, the leading model for the central engine of active galaxies is basically a scaled-up version of the same accretion process taking place in "normal" galaxies—only now the black holes involved are millions or even billions of times more massive than the Sun. As with this model's smaller scale counterparts, infalling gas forms an accretion disk and spirals down toward the black hole, heating up to high temperatures by friction within the disk and emitting large amounts of radiation as a result. In the case of an active galaxy, however, the origin of the accreted gas is not a binary companion, as it is in stellar X-ray sources, but entire stars and clouds of interstellar gas—most likely diverted into the galactic center by an encounter with another galaxy—that come too close to the hole and are torn apart by its strong gravity.

Accretion is extremely efficient at converting infalling mass (in the form of gas) into energy (in the form of electromagnetic radiation). Detailed calculations indicate that as much as 10 or 20 percent of the total mass–energy of the infalling matter can be radiated away before it crosses the

hole's event horizon and is lost forever. ∞ (Sec. 22.5) Since the total mass–energy of a star like the Sun—the mass times the speed of light squared—is about 2×10^{47} J, it follows that the 10^{38}-W luminosity of a bright active galaxy can be accounted for by the consumption of "only" 1 solar mass of gas per decade by a billion-solar-mass black hole. More or less luminous active galaxies would require correspondingly more or less fuel. Thus, to power a 10^{40}-W quasar, which is 100 times brighter, the black hole simply consumes 100 times more fuel, or 10 stars per year. The central black hole of a 10^{36}-W Seyfert galaxy would devour only one Sun's worth of material every thousand years.

The small size of the emitting region is a direct consequence of the compact central black hole. Even a billion-solar-mass black hole has a radius of only 3×10^9 km, or 10^{-4} pc—about 20 A.U.—and theory suggests that the part of the accretion disk responsible for most of the emission would be much less than 1 pc across. ∞ (Sec. 22.5) Instabilities in the accretion disk can cause fluctuations in the energy released, leading to the variability observed in many objects. The broadening of the spectral lines seen in the nuclei of many active galaxies may result from the rapid orbital motion of the gas in the black hole's intense gravity.

The jets consist of material (mainly protons and electrons) blasted out into space—and completely out of the visible portion of the galaxy—from the inner regions of the disk. The details of how jets form remain uncertain, but there is a growing consensus among theorists that jets are a common feature of accretion flows, large and small. They are most likely formed by strong magnetic fields produced within the accretion disk itself. These fields accelerate charged particles to nearly the speed of light and eject them parallel to the disk's rotation axis. Figure 24.34 shows a *Hubble Space Telescope* image of a disk of gas and dust at the core of the radio galaxy NGC 4261 in the Virgo Cluster.

ANIMATION Cosmic Jets

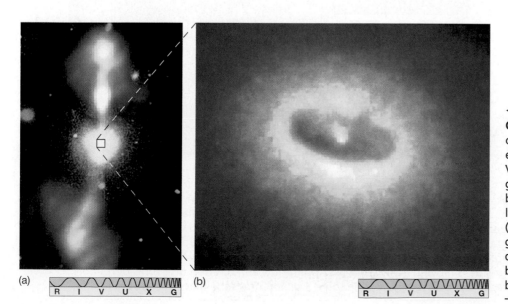

(a)

R I V U X G

(b)

R I V U X G

◀ **FIGURE 24.34 Giant Elliptical Galaxy** (a) A combined optical–radio image of the giant elliptical galaxy NGC 4261, in the Virgo Cluster, shows a white visible galaxy at the center, from which blue-orange (false-color) radio lobes extend for about 60 kpc. (b) A close-up photograph of the galaxy's nucleus reveals a 100-pc-diameter disk surrounding a bright hub thought to harbor a black hole. (*NRAO; NASA*)

Consistent with the theory just described, the disk is perpendicular to the huge jets emanating from the galaxy's center.

Figure 24.35 shows imaging and spectroscopic data from the center of M87, suggesting a rapidly rotating disk of matter orbiting the galaxy's center, again perpendicular to the jet. Measurements of the gas velocity on opposite sides of the disk indicate that the mass within a few parsecs of the center is approximately 3×10^9 solar masses; we assume that this is the mass of the central black hole. At M87's distance, *HST*'s resolution of 0.05 arc second corresponds to a scale of about 5 pc, so we are still far from seeing the (solar-system-sized) central black hole itself, but the improved "circumstantial" evidence has convinced many astronomers of the basic correctness of the theory.

Energy Emission

Theory suggests that the radiation emitted by the hot accretion disk surrounding a supermassive black hole would span a broad range of wavelengths, from infrared through X rays, and would account for the observed spectra of at least some active galactic nuclei. However, as mentioned earlier, it appears that in many cases the high-energy radiation emitted from the accretion disk itself is "reprocessed"—that is, absorbed and reemitted at longer, particularly infrared, wavelengths—by dust surrounding the nucleus before eventually reaching our detectors.

Researchers suspect that the most likely site of this reprocessing is a rather fat, donut-shaped ring of gas and dust surrounding the inner accretion disk where the energy is actually produced. As illustrated in Figure 24.36, if our line of sight to the black hole does not intersect the dusty donut, then we see the "bare" energy source, emitting large amounts of high-energy radiation (with broad emission lines, since we can see the rapidly moving gas near the black hole). ∞ (Sec. 4.4) If the donut intervenes, we see instead large amounts of infrared radiation reradiated from the dust (and only narrow emission lines, from gas farther from the center).

A different reprocessing mechanism operates in many jets and radio lobes. This mechanism involves the *magnetic fields* possibly produced within the accretion disk and transported by the jets into intergalactic space (Figure 24.33). As sketched in Figure 24.37(a), whenever a charged particle (here an electron) encounters a magnetic field, the particle tends to spiral around the magnetic field lines. We have already encountered this idea in the discussions of Earth's magnetosphere and solar activity. ∞ (Secs. 7.5, 16.5)

As the particles whirl around, they emit electromagnetic radiation. ∞ (Sec. 3.2) The radiation produced in this way—called **synchrotron radiation**, after the type of particle accelerator in which it was first observed—is *nonthermal* in nature, meaning that there is no link between the emission and the temperature of the radiating object. Hence,

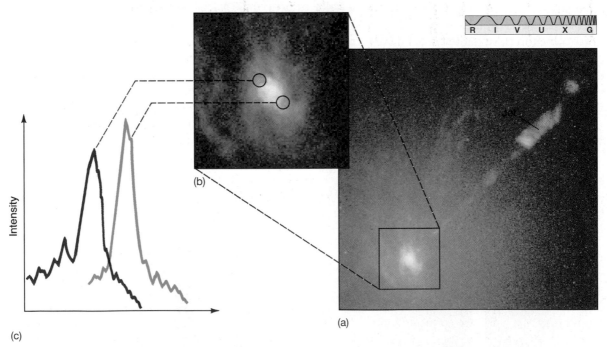

▲ FIGURE 24.35 M87 Disk Recent images and spectra of M87 support the idea of a rapidly whirling accretion disk at the galaxy's heart. (a) An image of the central region of M87, similar to that shown in Figure 24.28(c), shows the galaxy's bright nucleus and jet (marked). (b) A magnified view of the nucleus suggests a spiral swarm of stars, gas, and dust. (c) Spectral-line features observed on opposite sides of the nucleus show opposite Doppler shifts, implying that material on one side of the nucleus is coming toward us and material on the other side is moving away from us. Apparently, an accretion disk spins perpendicular to the jet. At the center of the disk is a black hole some 3 billion times the mass of the Sun. (*NASA*)

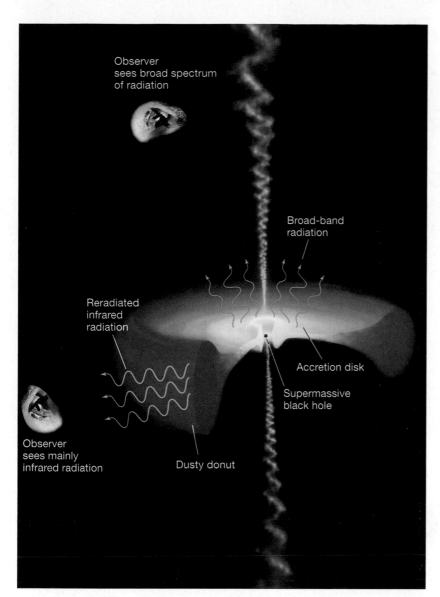

◀ **FIGURE 24.36 Dusty Donut** The accretion disk surrounding a massive black hole, drawn here with some artistic licence, consists of hot gas at many different temperatures (hottest nearest the center). When viewed from above or below, the disk is seen to radiate a broad spectrum of electromagnetic energy extending into the X-ray band. However, the dusty infalling gas that ultimately powers the system is thought to form a rather fat, donut-shaped region outside the accretion disk (shown here in dull red). The donut-shaped region effectively absorbs much of the high-energy radiation reaching it and re-emits it mainly in the form of cooler, infrared radiation. Thus, when the accretion disk viewed from the side, strong infrared emission is observed. The appearance of the jets, radiating mostly radio waves and X rays, also depends on the viewing angle. (See Figure 24.27.) *(D. Berry)*

the radiation is not described by a blackbody curve. Instead, its intensity decreases with increasing frequency, as shown in Figure 24.37(b). This is just what is needed to explain the overall spectrum of radiation from active galaxies. (Cf. Figure 24.37b with Figure 24.19.) Observations of the radiation received from the jets and radio lobes of active galaxies are completely consistent with synchrotron radiation.

Eventually, the jet is slowed and stopped by the intergalactic medium, the flow becomes turbulent, and the magnetic field grows tangled. The result is a gigantic radio lobe emitting virtually all of its energy in the form of synchrotron radiation. Thus, even though the radio *emission* comes from an enormously extended volume of space that dwarfs the visible galaxy, the *source* of the energy is still the accretion disk—a billion billion times smaller in volume than the radio lobe—lying at the galactic center. The jets serve merely as a conduit that transports energy from the nucleus, where it is generated, into the lobes, where it is finally radiated into space.

The existence of the inner lobes of Centaurus A and the blobs in M87's jet imply that the formation of a jet may be an intermittent process (or, as in the case of the Seyferts

discussed earlier, may not occur at all), and, as we have seen, there is also evidence to indicate that much, if not all, of the activity observed in nearby active galaxies has been sparked by recent interaction with a neighbor. Many nearby active galaxies (e.g., Centaurus A) appear to have been "caught in the act" of interacting with another galaxy, suggesting that the fuel supply can be "turned on" by a companion. The tidal forces involved divert gas and stars into the galactic nucleus, triggering an outburst that may last for many millions of years.

What do active galaxies look like between active outbursts? What is their connection with the normal galaxies we see? To answer these important questions, we must delve more deeply into the subject of *galaxy evolution*, to which we turn in Chapter 25.

CONCEPT CHECK

✔ How does accretion onto a supermassive black hole power the energy emission from the extended radio lobes of a radio galaxy?

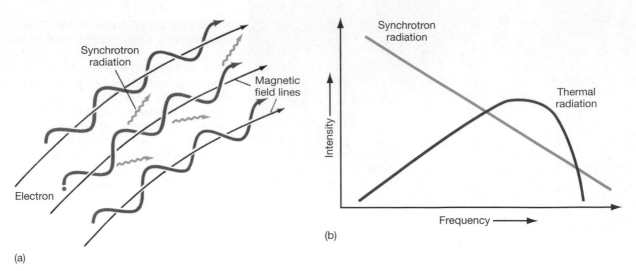

(a)

(b)

▲ **FIGURE 24.37 Nonthermal Radiation** (a) Charged particles, especially fast-moving electrons (red), emit synchrotron radiation (blue) while spiraling in a magnetic field (black). This process is not confined to active galaxies; it occurs as well, though on smaller scales, when charged particles interact with magnetism in Earth's Van Allen belts ∞ (Sec. 7.5), when charged matter arches above sunspots on the Sun ∞ (Sec. 16.5), in the vicinity of neutron stars ∞ (Sec. 22.2), and at the center of our own Galaxy ∞ (Sec. 23.7). (b) Variation of the intensity of thermal and synchrotron (nonthermal) radiation with frequency. Thermal radiation, described by a blackbody curve, peaks at some frequency that depends on the temperature of the source. Nonthermal synchrotron radiation, by contrast, is more intense at low frequencies and is independent of the temperature of the emitting object. (Cf. with Figure 24.19.)

Chapter Review

SUMMARY

The **Hubble classification scheme (p. 634)** divides galaxies into several classes, depending on their appearance. **Spiral galaxies (p. 634)** have flattened disks, central bulges, and spiral arms. Their halos consist of old stars, whereas the gas-rich disks are the sites of ongoing star formation. **Barred-spiral galaxies (p. 635)** contain an extended "bar" of material projecting beyond the central bulge. **Elliptical galaxies (p. 636)** have no disk and contain little or no cool gas or dust, although very hot interstellar gas is observed within them. In most cases, they consist entirely of old stars. Elliptical galaxies range in size from dwarf ellipticals, which are much less massive than the Milky Way Galaxy, to giant ellipticals, which may contain trillions of stars.

S0 and **SB0 galaxies (p. 638)** are intermediate in their properties between ellipticals and spirals. **Irregular galaxies (p. 638)** are galaxies that are neither spiral nor elliptical. Some may be the result of collisions or close encounters with other galaxies. Many irregulars are rich in gas and dust and are the sites of vigorous star formation. The **Magellanic Clouds (p. 639)**, two small systems that orbit the Milky Way Galaxy, are examples of this type of galaxy.

Astronomers often use **standard candles (p. 641)** as distance-measuring tools. These are objects that are easily identifiable and whose luminosities lie within some reasonably well defined range. Comparing luminosity and apparent brightness, astronomers determine distance with the use of the inverse-square law. An alternative approach is the **Tully–Fisher relation (p. 642)**, an empirical correlation between rotational velocity and luminosity in spiral galaxies.

The Milky Way, Andromeda, and several other smaller galaxies form the **Local Group (p. 643)**, a small **galaxy cluster (p. 643)**. Galaxy clusters consist of a collection of galaxies orbiting one another, bound together by their own gravity. The nearest large galaxy cluster to the Local Group is the Virgo Cluster.

Distant galaxies are observed to be receding from the Milky Way at speeds proportional to their distances from us. This relationship is called **Hubble's law (p. 646)**. The constant of proportionality in the law is **Hubble's constant (p. 647)**. Its value is thought to be around 70 km/s/Mpc. Astronomers use Hubble's law to determine distances to the most remote objects in the universe. The redshift associated with the Hubble expansion is called the **cosmological redshift (p. 646)**.

Active galaxies (p. 648) are much more luminous than normal galaxies and have nonstellar spectra, emitting most of their energy outside the visible part of the electromagnetic spectrum. A significant minority of galaxies shows some sort of activity in the form of **active galactic nuclei (p. 649)**. A **Seyfert galaxy**

(p. 649) looks like a normal spiral, but has an extremely bright central **galactic nucleus (p. 649)**. Spectral lines from Seyfert nuclei are very broad, indicating rapid internal motion, and the rapid variability in the luminosity of Seyferts implies that the source of the radiation is much less than one light-year across. **Radio galaxics (p. 652)** emit large amounts of energy in the radio part of the spectrum. The corresponding visible galaxy is usually elliptical. Often, the energy comes from enormous **radio lobes (p. 652)** that lie far beyond the visible portion of the galaxy. **Quasars (p. 657)** are the most luminous objects known. In visible light they appear starlike, and their spectra are usually substantially redshifted. All quasars are very distant, indicating that we see them as they were in the remote past.

Many active galaxies have high-speed, narrow jets of matter shooting out from their central nuclei. Astronomers think that, in radio galaxies, the jets transport energy from the nucleus, where it is generated, to the lobes, where it is radiated into space. The jets often appear to be made up of distinct "blobs" of gas, sug-gesting that the process which generates the energy is intermit-tent. If the jet happens to be directed toward us, we see an intense **blazar (p. 654)**.

The generally accepted explanation for the observed proper-ties of all active galaxies is that their energy is generated by the accretion of galactic gas onto a supermassive (million- to billion-solar-mass) black hole lying in the galactic center. The small size of the accretion disk explains the compact extent of the emitting region, and the high-speed orbit of gas in the black hole's intense gravity accounts for the rapid motion that is observed. Typical lu-minosities of active galaxies require the consumption of about 1 solar mass of material every few years. Some of the infalling mat-ter is blasted out into space, producing magnetized jets that cre-ate and feed the galaxy's radio lobes. Charged particles spiraling around the magnetic field lines produce **synchrotron radiation (p. 660)**, whose spectrum is consistent with the nonstellar radia-tion observed in radio galaxies and jets.

REVIEW AND DISCUSSION

1. What distinguishes one type of spiral galaxy from another?

2. Describe some similarities and differences between elliptical galaxies and the halo of our own Galaxy.

3. Describe the four rungs in the distance-measurement ladder used to determine the distance to a galaxy lying 5 Mpc away.

4. Describe the contents of the Local Group. How much space does it occupy compared with the volume of the Milky Way?

5. What are standard candles, and why are they important to astronomy?

6. How is the Tully–Fisher relation used to measure distances to galaxies?

7. What is the Virgo Cluster?

8. What is Hubble's law?

9. How is Hubble's law used by astronomers to measure dis-tances to galaxies?

10. What is the most likely range of values for Hubble's con-stant? Why is the exact value uncertain?

11. Name two basic differences between normal galaxies and ac-tive galaxies.

12. Are there any "nearby" active galaxies—within 50 Mpc of Earth, say?

13. Describe some of the basic properties of Seyfert galaxies.

14. What is the evidence that the radio lobes of some active galaxies consist of material ejected from the galaxy's cen-ter?

15. How do we know that the energy-emitting regions of many active galaxies must be very small?

16. What was it about the spectra of quasars that was so unex-pected and surprising?

17. Why do astronomers prefer to speak in terms of redshifts rather than distances to faraway objects?

18. How do we know that quasars are extremely luminous?

19. Briefly describe the leading model for the central engine of an active galaxy.

20. How is the process of synchrotron emission related to obser-vations of active galaxies?

CONCEPTUAL SELF-TEST: TRUE OR FALSE/MULTIPLE CHOICE

1. Most elliptical galaxies contain only young stars.

2. Most galaxies are spirals like the Milky Way.

3. Irregular galaxies, although small, have lots of star formation taking place in them.

4. Every galaxy is a member of some galaxy cluster.

5. Most galaxies are receding from the Milky Way Galaxy.

6. Hubble's law can be used to determine distances to the far-thest objects in the universe.

7. The spectrum of an active galaxy is well described by a blackbody curve.

8. Radio galaxies emit large amounts of energy from regions much larger in size than the visible galaxy.

9. Astronomers began to understand quasar spectra when it was discovered that their radiation is redshifted by an unexpect-edly large amount.

10. For all types of active galaxy, the actual source of the tremen-dous energy emitted is accretion onto a black hole in the galactic nucleus.

11. Stars in a galactic disk are **(a)** evenly distributed within and between spiral arms; **(b)** mostly found in the space between

spiral arms; **(c)** mostly found in the spiral arms; **(d)** older than stars in the halo.

12. Astronomers classify elliptical galaxies by **(a)** the number of stars they contain; **(b)** their colors; **(c)** how flattened they appear; **(d)** their diameters.

13. Using the method of standard candles, we can, in principle, find the distance to a campfire if we know **(a)** the number of logs used; **(b)** the fire's temperature; **(c)** the length of time the fire has been burning; **(d)** the type of wood used in the fire.

14. If the galaxy in Figure 24.11 ("Galaxy Rotation") were smaller and spinning more slowly, then, in order to represent it correctly, the figure should be redrawn to show **(a)** a greater blueshift; **(b)** a greater redshift; **(c)** a narrower combined line; **(d)** a larger combined amplitude.

15. Within 30 Mpc of the Sun, there are about **(a)** 3 galaxies; **(b)** 30 galaxies; **(c)** a few thousand galaxies; **(d)** a few million galaxies.

16. Hubble's law states that **(a)** more distant galaxies are younger; **(b)** the greater the distance to a galaxy, the greater is the galaxy's redshift; **(c)** most galaxies are found in clusters; **(d)** the greater the distance to a galaxy, the fainter the galaxy appears.

17. Compared with a normal galaxy, an active galaxy **(a)** is much larger; **(b)** emits more energy at long wavelengths; **(c)** is blueshifted; **(d)** is brighter at visible wavelengths.

18. If the light from a galaxy fluctuates in brightness very rapidly, the region producing the radiation must be **(a)** very large; **(b)** very small; **(c)** very hot; **(d)** rotating very rapidly.

19. Quasar spectra **(a)** are strongly redshifted; **(b)** show no spectral lines; **(c)** look like the spectra of stars; **(d)** contain emission lines from unknown elements.

20. Active galaxies are very luminous because they **(a)** are hot; **(b)** contain black holes in their cores; **(c)** are surrounded by hot gas; **(d)** emit jets.

PROBLEMS

 Algorithmic versions of these questions are available in the practice problems module of the companion Website at astro.prenhall.com/chaisson.

The number of squares preceding each problem indicates its approximate level of difficulty.

1. ■ A supernova of luminosity 1 billion times the luminosity of the Sun is used as a standard candle to measure the distance to a faraway galaxy. From Earth, the supernova appears as bright as the Sun would appear from a distance of 10 kpc. What is the distance to the galaxy?

2. ■■ A Cepheid variable star in the Virgo cluster has an absolute magnitude of −5 and is observed to have an apparent magnitude of 26.3. Use these figures to calculate the distance to the Virgo cluster.

3. ■ According to Hubble's law, with $H_0 = 70$ km/s/Mpc, what is the recessional velocity of a galaxy at a distance of 200 Mpc? How far away is a galaxy whose recessional velocity is

4000 km/s? How do these answers change if $H_0 = 60$ km/s/Mpc? If $H_0 = 80$ km/s/Mpc?

4. ■■ According to Hubble's law, with $H_0 = 70$ km/s/Mpc, how long will it take for the distance from the Milky Way Galaxy to the Virgo Cluster to double?

5. ■■■ Assuming Hubble's law with $H_0 = 70$ km/s/Mpc, what would be the angular diameter of an E0 galaxy of actual diameter 80 kpc if its 656.3-nm Hα line is observed at 700 nm?

6. ■■ A certain quasar has a redshift of 0.25 and an apparent magnitude of 13. Using the data from Table 24.1, calculate the quasar's absolute magnitude and hence its luminosity. Compare the apparent brightness of the quasar, viewed from a distance of 10 pc, with that of the Sun as seen from Earth.

7. ■ What are the absolute magnitude and luminosity of a quasar with a redshift of 5 and an apparent magnitude of 22?

8. ■ On the basis of the data in Table 24.1, estimate the apparent magnitude of a quasar with absolute magnitude −24 and redshift 1.

9. ■■ Spectral lines from a Seyfert galaxy are observed to be redshifted by 0.5 percent and to have broadened emission lines indicating an orbital speed of 250 km/s at an angular distance of 0.1″ from its center. Assuming circular orbits, use Kepler's laws to estimate the mass within this 0.1″ radius. ∞ (Sec. 23.6)

10. ■ On the basis of the data presented in the text, calculate the orbital speed of material orbiting at a distance of 0.5 pc from the center of M87.

11. ■ Centaurus A—from one radio lobe to the other—spans about 1 Mpc and lies at a distance of 4 Mpc from Earth. What is the angular size of Centaurus A? Compare your answer with the angular diameter of the Moon.

12. ■■ Calculate the energy flux—that is, the energy received per unit area per unit time—that would be observed at Earth from a 10^{37}-W Seyfert nucleus located at the Galactic center, 8 kpc away, neglecting the effects of interstellar extinction. Using the data presented in Appendix 3, Table 5, compare this energy flux with that received from Sirius A, the brightest star in the night sky. From what you know about active-galaxy energy emission, is it reasonable to ignore interstellar extinction?

13. ■ Assuming a jet speed of $0.75c$, calculate the time taken for material in Cygnus A's jet to cover the 500 kpc between the galaxy's nucleus and its radio-emitting lobes.

14. ■ Assuming the upper end of the efficiency range indicated in the text, calculate how much energy an active galaxy would generate if it consumed 1 Earth mass of material every day. Compare this value with the luminosity of the Sun.

15. ■ A quasar consumes 1 solar mass of material per year, converting 15 percent of it directly into energy. What is the quasar's luminosity, in solar units?

 In addition to the Practice Problems module, the Companion Website at astro.prenhall.com/chaisson provides for each chapter a study guide module with multiple choice questions as well as additional annotated images, animations, and links to related Websites.

25

Active galaxies are much more energetic than the normal galaxy—the Milky Way—in which we live. The "central engines" powering them are thought to be supermassive black holes. Many active galaxies are found within huge clusters of galaxies—clusters that contain numerous other galaxies amidst much loose, extremely hot gas. Here, a radio image (red and yellow) is superposed on an optical photo (blue), showing its core at the center and huge lobes of radiation to either ▶ side. The extent of the lobes is nearly a million light-years, end to end. (NRAO)

Galaxies and Dark Matter

The Large-Scale Structure of the Cosmos

On scales much larger than even the largest galaxy clusters, the dynamics of the universe itself becomes apparent, new levels of structure are revealed, and a humbling new reality emerges. We may be star stuff, the product of countless cycles of stellar evolution, but we are not the stuff of the cosmos. The universe in the large is composed of matter fundamentally different from the familiar atoms and molecules that make up our bodies, our planet, our star and galaxy, and all the luminous matter we observe in the heavens. Only its gravity announces the presence of this strange kind of matter, providing the backdrop against which galaxies form and evolve. By comparing and classifying the properties of galaxies near and far, astronomers have begun to understand their complex dynamics and evolution. By mapping out the distribution of those galaxies in space, astronomers trace out the immense realms of the universe. Points of light in the uncharted darkness, they remind us that our position in the universe is no more special than that of a boat adrift at sea.

Visit astro.prenhall.com/chaisson for additional images, animations, and links to related sites for this chapter.

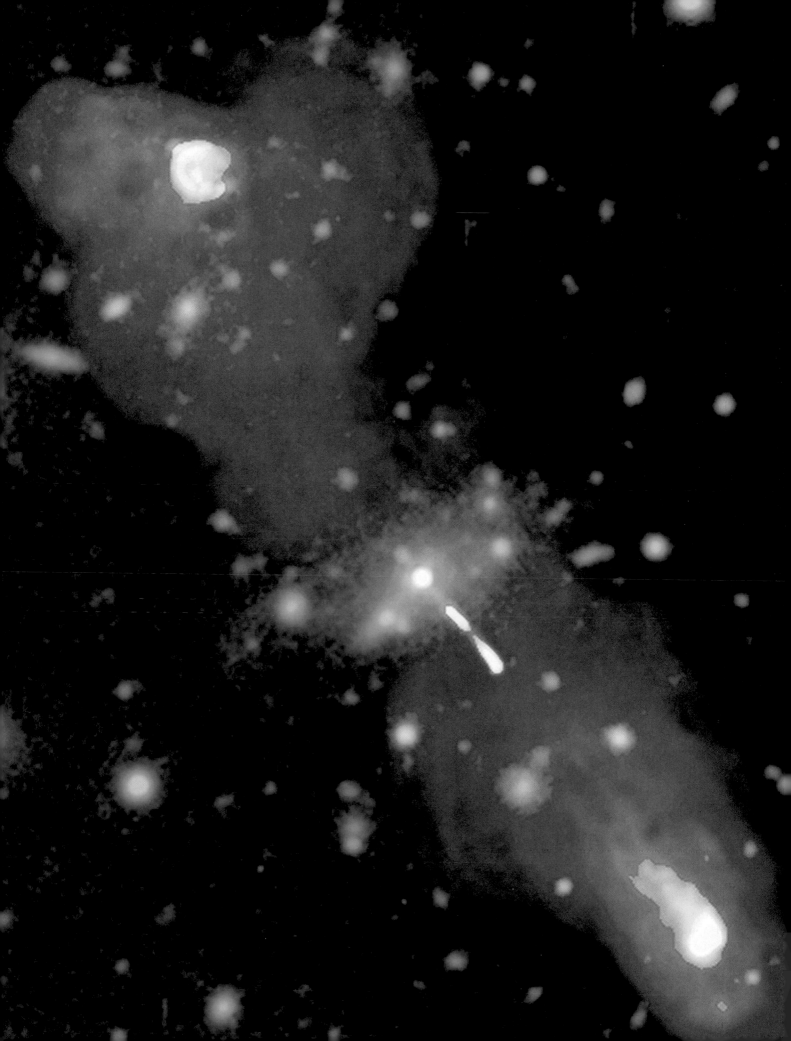

25.1 Dark Matter in the Universe

In Chapter 23, we saw how measurements of the orbital velocities of stars and gas in our own Galaxy reveal the presence of an extensive *dark-matter halo* surrounding the galaxy we see. ∞ (Sec. 23.6) Do other galaxies have similar dark halos? And what evidence do we have for dark matter on larger scales? To answer these questions, we need a way to calculate the masses of galaxies and galaxy clusters.

How can we measure the masses of such large systems? Surely, we can neither count all their stars nor estimate their interstellar content very well: Galaxies are just too complex for us to take a direct inventory of their material makeup. Instead, we must rely on indirect techniques. Despite their enormous sizes, galaxies and galaxy clusters obey the same physical laws as do the planets in our own solar system. To calculate the masses of galaxies, we turn as usual to Newton's law of gravity.

Galaxy Masses

Astronomers can calculate the masses of some spiral galaxies by determining their *rotation curves*, which plot rotation speed versus distance from the galactic center, as illustrated in Figure 25.1. Rotation curves for a few nearby spirals are shown in Figure 25.1(b). The mass within any given radius then follows directly from Newton's laws. ∞ (Sec. 2.7) The rotation curves shown imply masses ranging from about 10^{11} to 5×10^{11} solar masses within about 25 kpc of the center—comparable to the results obtained for our own Galaxy with the use of the same technique. ∞ (Sec. 23.6)

Distant galaxies are generally too far away for such detailed curves to be drawn. Nevertheless, by observing the broadening of their spectral lines—as discussed in Chapter 24 in the context of the Tully–Fisher relation—we can still measure the overall rotation speed of these galaxies. ∞ (Sec. 24.2) Estimating a galaxy's size then leads to an estimate of its mass. Similar techniques have been applied to ellipticals and irregulars. In general, the approach is useful for measuring the mass lying within about 50 kpc of a galaxy's center—the extent of the electromagnetic emission from stellar and interstellar material.

To probe farther from the centers of galaxies, astronomers turn to *binary* galaxy systems (Figure 25.2a), whose components may lie hundreds of kiloparsecs apart. The orbital period of such a system is typically billions of years, far too long for the orbit to be accurately measured. However, by estimating the period and semimajor axis from the available information—the line-of-sight velocities and the angular separation of the components—an approximate total mass can be derived. ∞ (*More Precisely 2-3*)

Galaxy masses obtained in this way are fairly uncertain. However, by combining many such measurements,

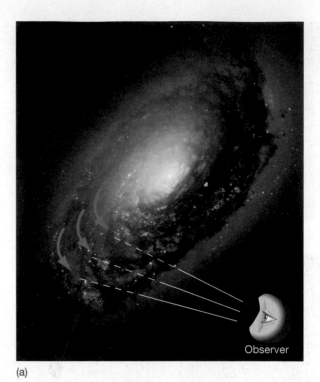

(a)

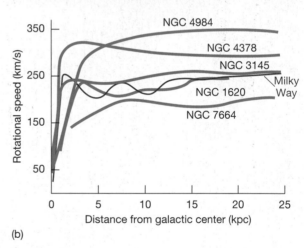

(b)

▲ **FIGURE 25.1 Galaxy Rotation Curves** (a) By observing orbital velocities at many different distances from the center of a disk galaxy, astronomers can plot a rotation curve for the galaxy. This is M64, the "Evil Eye" galaxy, some 5 Mpc distant. (b) Rotation curves for some nearby spiral galaxies indicate masses of a few hundred billion times the mass of the Sun. The corresponding curve for our own Galaxy (from Figure 23.21) is marked in red for comparison (*NASA*)

astronomers can obtain quite reliable *statistical* information about galaxy masses. Most normal spirals (the Milky Way Galaxy included) and large ellipticals contain between 10^{11} and 10^{12} solar masses of material. Irregular galaxies often contain less mass, about 10^8 to 10^{10} times that of the Sun. Dwarf ellipticals and dwarf irregulars can contain as little as 10^6 or 10^7 solar masses of material.

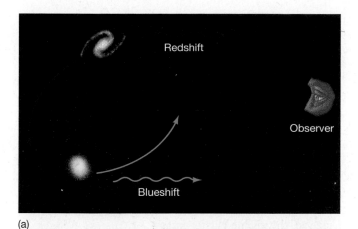

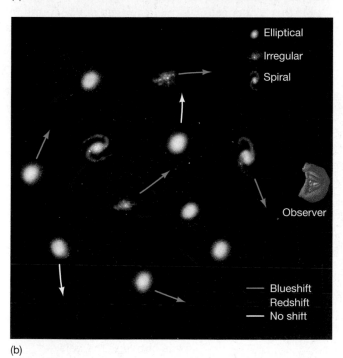

(b)

▲ **FIGURE 25.2 Galaxy Masses** (a) In a binary galaxy system, galaxy masses may be estimated by observing the orbit of one galaxy about the other. (b) The mass of a galaxy cluster may be estimated by observing the motion of many galaxies in the cluster and then estimating how much mass is needed to prevent the cluster from flying apart.

We can use another statistical technique to derive the combined mass of all the galaxies within a galaxy cluster. As depicted in Figure 25.2(b), each galaxy within a cluster moves relative to all other members of the cluster, and we can estimate the cluster's mass simply by asking how massive it must be in order to bind its galaxies gravitationally. For example, if we find that galaxies in a cluster are moving with an average speed of 1000 km/s and the cluster radius is 3 Mpc (both typical values), it follows from Newton's laws—*assuming* that the cluster is gravitationally bound—that the mass of the cluster must be around

$(3 \text{ Mpc}) \times (1000 \text{ km/s})^2/G \approx 7 \times 10^{14}$ solar masses. ∞ (*More Precisely 2-3*) Cluster masses obtained in this way generally lie in the range of 10^{14}–10^{15} solar masses. Notice that this calculation gives us no information whatsoever about the masses of individual galaxies. It tells us only about the *total* mass of the cluster.

Visible Matter and Dark Halos

The rotation curves of the spiral galaxies shown in Figure 25.1 remain flat (i.e., do not decline and may even rise slightly) far beyond the visible images of the galaxies themselves, implying that these galaxies—and perhaps all spiral galaxies—have invisible dark halos similar to that surrounding the Milky Way. ∞ (Sec. 23.6) Spiral galaxies seem to contain from 3 to 10 times more mass than can be accounted for in the form of visible matter. Some studies of elliptical galaxies suggest similarly large dark halos surrounding these galaxies, too.

Curiously, some of the least luminous galaxies seem to have the largest fractions of dark matter, raising the intriguing possibility of *dark galaxies*—systems composed almost entirely of dark matter, emitting virtually no visible light. Direct observational evidence for such objects is naturally hard to obtain, although some astronomers have suggested that the peculiar appearance of the dwarf elliptical galaxy UGC 10214 (Figure 25.3), which has a stream of matter flowing out of it, apparently toward nothing, may be due to a dark companion.

Astronomers find even greater discrepancies between visible light and total mass when they study galaxy clusters. Calculated cluster masses range from 10 to nearly 100

R I V U X G

▲ **FIGURE 25.3 Dark Galaxy?** Astronomers speculate that some galaxies may be composed almost entirely of dark matter, emitting virtually no visible light. Has the long stream of gas leaving this galaxy, called UGC 10214, been torn out by a close encounter with a dark companion at the right?

times the mass suggested by the light emitted by individual cluster galaxies. Put another way, a lot more mass is needed to bind galaxy clusters than we can see. Thus, the problem of dark matter exists not just in our own Galaxy, but also in other galaxies and, to an even greater degree, in galaxy clusters as well. In that case, we are compelled to accept the fact that *upwards of 90 percent of the matter in the universe is dark*—and not just in the visible portion of the spectrum, but it goes undetected at *any* electromagnetic wavelength.

As discussed in Chapter 23, many possible explanations for the dark matter have been suggested, ranging from stellar remnants of various sorts to exotic subatomic particles. ∞ (Sec. 23.6) Whatever its nature, the dark matter in clusters apparently cannot simply be the accumulation of dark matter within individual galaxies. Even including the galaxies' dark halos, we still cannot account for all the dark matter in galaxy clusters. As we look on larger and larger scales, we find that a larger and larger fraction of the matter in the universe is dark.

Intracluster Gas

In addition to the luminous matter observed within the cluster galaxies themselves, astronomers also have evidence for large amounts of *intracluster gas*—superhot (more than 10 million K), diffuse intergalactic matter filling the space among the galaxies. Satellites orbiting above Earth's atmosphere have detected substantial amounts of X-ray radiation from many clusters. Figure 25.4 shows false-color X-ray images of two such systems. In each case, the X-ray-emitting region is centered on, and comparable in size to, the visible cluster image.

Further evidence for intracluster gas can be found in the appearance of the radio lobes of some active galaxies. ∞ (Sec. 24.4) In some systems, known as *head–tail* radio

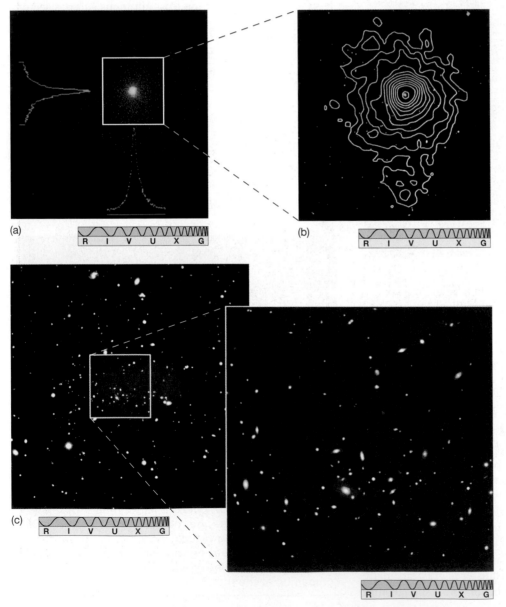

(a)

R I V U X G

(b)

R I V U X G

(c)

R I V U X G

R I V U X G

◀ **FIGURE 25.4 Galaxy Cluster X-Ray Emission** (a) X-ray image of Abell 85, an old, distant cluster of galaxies, taken by the *Einstein X-ray Observatory*. The cluster's X-ray emission is shown in orange. The green graphs display a smooth, peaked intensity profile centered on the cluster, but not associated with individual galaxies. (b) The contour map of X rays is superimposed on an optical photo, showing its X rays peaked on Abell 85's central supergiant galaxy. Images like these demonstrate that the space between the galaxies within galaxy clusters is filled with superheated gas. (c) Superposition of infrared and X-ray radiation from another distant galaxy cluster. The X rays are shown as a fuzzy, bluish cloud of hot gas filling the intracluster spaces among the galaxies. The inset is a longer infrared exposure of the central region, showing the richness of this cluster, which spans about a million parsecs. *(NASA; AURA; ESA)*

galaxies, the lobes seem to form a "tail" behind the main part of the galaxy. For example, the lobes of radio galaxy NGC 1265, shown in Figure 25.5, appear to be "swept back" by some onrushing wind, and, indeed, this is the most likely explanation for the galaxy's appearance. If NGC 1265 were at rest, it would be just another double-lobe source, perhaps quite similar to Centaurus A (Figure 24.23). However, the galaxy is traveling through the intergalactic medium of its parent galaxy cluster (known as the Perseus Cluster), and the outflowing matter forming the lobes tends to be left behind as NGC 1265 moves.

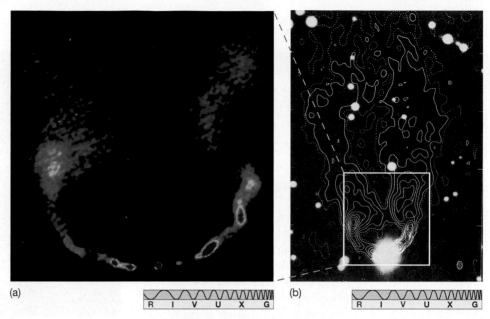

(a)

(b)

▲ **FIGURE 25.5** Head–Tail Radio Galaxy (a) Radiograph, in false color, of the head–tail radio galaxy NGC 1265. (b) The same radio data, in contour form, superposed on the optical image of the galaxy. Astronomers reason that this object is moving rapidly through space, trailing a "tail" behind as it goes. (*NRAO; Palomar/Caltech*)

These observations reveal that at least as much matter—and, in most cases, significantly *more*—exists within clusters in the form of hot gas as is visible in the form of stars. Where did the gas come from? There is so much of it that it could not have been expelled from the galaxies themselves. Instead, astronomers think that it is mainly *primordial*—gas that has been around since the universe formed and that never became part of a galaxy. However, the intracluster gas does contain some heavy elements—carbon, nitrogen, and so on—implying that at least some of it is material ejected from galaxies after enrichment by stellar evolution. ∞ (Sec. 21.5) Why is the gas so hot? Simply because its particles are bound by gravity and hence are moving at speeds comparable to those of the galaxies in the cluster—1000 km/s or so. Since temperature is just a measure of the speed at which the gas particles move, this speed translates (for protons) directly into a temperature of 40 million K. ∞ (*More Precisely 8-1*)

The amount of mass in the form of hot gas in clusters is substantial, but not enough to impact the dark-matter problem. To account for the total masses of galaxy clusters implied by dynamical studies, we would have to find from 10 to 100 times more mass in gas than in stars. In the mid-1990s, *ROSAT* observations of gas within the Virgo Cluster actually compounded the problem (Figure 25.6). The gas was so hot that the amount of dark matter needed to bind it to the cluster and prevent it from dispersing into intercluster space was far greater than had previously been suspected to exist!

▶ **FIGURE 25.6** Virgo in X Rays This combined optical and X-ray (*ROSAT*) image shows hot (more than 10^7 K) X-ray-emitting gas (falsely colored blue, increasing to yellow and red) within the rich Virgo Cluster of galaxies. (See also Figure 24.14.) The amount of mass required to bind the gas (especially that surrounding M87, shown in blue) to the cluster greatly exceeds the total mass of the galaxies themselves. (*ESA*)

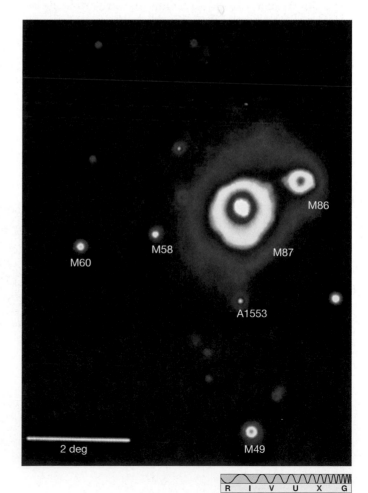

CONCEPT CHECK

✔ What assumptions are we making when we infer the mass of a galaxy cluster from observations of the spectra of its constituent galaxies?

25.2 Galaxy Collisions

LEARNING GOAL 3 Contemplating the congested confines of a rich galaxy cluster (such as Virgo or Coma, with thousands of member galaxies orbiting within a few megaparsecs), we might expect that collisions among galaxies would be common. ∞ (Sec. 24.2) Gas particles collide in our atmosphere, and hockey players collide in the rink. So, do galaxies in clusters collide, too? The answer is yes.

Figure 25.7 apparently shows the aftermath of a bull's-eye collision between a small galaxy (perhaps one of the two at the right, although that is by no means certain) and the larger galaxy at the left. The result is the "Cartwheel" galaxy, about 150 Mpc from Earth, its halo of young stars resembling a vast ripple in a pond. The ripple is most likely a density wave created by the passage of the smaller galaxy through the disk of the larger one. ∞ (Sec. 23.5) The disturbance is now spreading outward from the region of impact, creating new stars as it goes.

Figure 25.8 shows an example of a close encounter that hasn't (yet) led to an actual collision. Two spiral galaxies are apparently passing each other like majestic ships in the night. The larger and more massive galaxy on the left is called NGC 2207; the smaller one on the right is IC 2163. Analysis of this image suggests that IC 2163 is now swinging past NGC 2207 in a counterclockwise direction, having made a close approach some 40 million years ago. The two galaxies seem destined to undergo further close

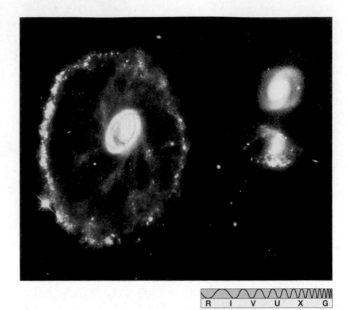

▲ **FIGURE 25.7 Cosmic Cartwheel** The "Cartwheel" galaxy (left) appears to be the result of a collision (possibly with one of the smaller galaxies at the right) that has led to an expanding ring of star formation moving outward through the galactic disk. *(NASA)*

encounters, as IC 2163 apparently does not have enough energy to escape the gravitational pull of NGC 2207. Each time the two galaxies experience a close encounter, bursts of star formation erupt in both as their interstellar clouds of gas and dust are pushed, shoved, and shocked. In roughly a billion years, these two galaxies will probably merge into a single, massive galaxy.

No human will ever witness an entire galaxy collision, for it lasts many millions of years. However, modern computers can follow the event in a matter of hours. Large-scale simulations modeling in detail the gravitational

◀ **FIGURE 25.8 Galaxy Encounter** This encounter between two spirals, NGC 2207 (left) and IC 2163, has already led to bursts of star formation in each. Eventually the two will merge, but probably not for a billion years or so. *(NASA)*

ANIMATION Collision of Two Spiral Galaxies

interactions among stars and gas, and incorporating the best available models of gas dynamics, allow astronomers to better understand the effects of a collision on the galaxies involved and even estimate the eventual outcome of the interaction.

The particular calculation shown in Figure 25.9(b) began with two colliding spiral galaxies, not so different from those shown in Figure 22.8, but the details of the original structure have been largely obliterated by the collision. Notice the similarity to the real image of NGC 4038/4039 (Figure 22.9a), the so-called Antennae galaxies, which show extended tails, as well as double galactic centers only a few hundred parsecs across. Star formation induced by the collision is clearly traced by the blue light from thousands of young, hot stars. The simulations indicate that, as with the galaxies in Figure 22.8, ultimately the two galaxies will merge into one.

Galaxies in clusters apparently collide quite often. Many collisions and near misses similar to those shown in the previous figures have been observed (see also Section 24.4), and a straightforward calculation reveals that, given the crowded conditions in even a modest cluster, close encounters are the norm rather than the exception. The reason is simple: The distance between adjacent galaxies in a cluster averages a few hundred thousand parsecs, which is not much greater (certainly less than five times more) than the size of a typical galaxy, including its extended dark halo. Galaxies simply do not have that much room to roam around without bumping into one another. Many re-

searchers think that most galaxies in most clusters have been strongly influenced by collisions, in some cases in the relatively recent past.

In the smaller groups, the galaxies' speeds are low enough that interacting galaxies tend to "stick together," and mergers, as shown in the computer simulation, are the most common outcome. In larger groups, galaxies move faster and tend to pass through one another without sticking. Either way, the encounters have substantial effects on the galaxies involved (see Section 25.3). If we wait long enough, we will have an opportunity to see for ourselves what a galaxy collision is really like: Our nearest large neighbor, the Andromeda galaxy (Fig. 23.2), is currently approaching us at a velocity of 120 km/s. In a few billion years it will collide with the Milky Way, and we will then be able to test astronomers' theories firsthand!

Curiously, although a collision may wreak havoc on the large-scale structure of the galaxies involved, it has essentially no effect on the individual stars they contain. The stars within each galaxy just glide past one another. In contrast to galaxies in the cluster, the stars in a galaxy are so small compared with the distances between them, that when two galaxies collide, the star population merely doubles for a time, and the stars continue to have so much space that they do not run into each other. Collisions can rearrange the stellar and interstellar contents of each galaxy, often producing a spectacular burst of star formation that may be visible to enormous distances, but from the point of view of the stars, it's clear sailing.

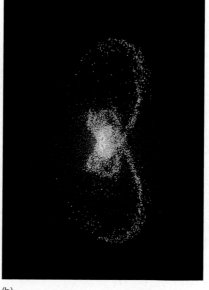

(a)

R I V U X G

(b)

▲ **FIGURE 25.9 Galaxy Collision** (a) The "Antennae" galaxies collided a few tens of millions of years ago. The long tidal "tails" (black and white image at the left) mark their final plunge. Strings of young, bright "super star clusters" (color *HST* image at the right) are the result of violent shock waves produced in the gas disks of the two colliding galaxies. (b) A simulation of the encounter shows many of the same features as the real thing, strengthening the case that we really are seeing a collision in progress. *(AURA; NASA; J. Barnes)*

25.3 Galaxy Formation and Evolution

6 With Hubble's law as our guide to distances in the universe, and armed now with knowledge of the distribution of dark matter on galactic and larger scales, let's turn to the question of how galaxies came to be the way they are. Can we explain the different types of galaxy we see? Astronomers know of no simple evolutionary connections among the various categories in the Hubble classification scheme. ∞ (Sec. 24.1) To answer the question, then, we must understand how galaxies formed.

Unfortunately, compared with the theories of star formation and stellar evolution, the theory of galaxy formation is still very much in its infancy. Galaxies are much more complex than stars, they are harder to observe, and the observations are harder to interpret. In addition, we have only a partial understanding of conditions in the universe immediately preceding galaxy formation, quite unlike the corresponding situation for stars. ∞ (Sec. 18.3) Finally, whereas stars almost never collide with one another, with the result that most single stars and binaries evolve in isolation, galaxies may suffer numerous collisions during their lives, making it much harder to decipher their pasts. Nevertheless,

some general ideas have begun to gain widespread acceptance, and we can offer some insights into the processes responsible for the galaxies we see.

Mergers and Acquisitions

3 The seeds of galaxy formation were sown in the very early universe, when small density fluctuations in the primordial matter began to grow (see Section 27.5). Our discussion here begins with these "pregalactic" blobs of gas already formed. The masses of the various fragments were quite small—only a few million solar masses, comparable to the masses of the smallest present-day dwarf galaxies, which may in fact be remnants of that early time. Most astronomers think that galaxies grew by repeated *merging* of smaller objects, as illustrated in Figure 25.10(a). Contrast this with the process of star formation, in which a large cloud fragments into smaller pieces that eventually become stars. ∞ (Sec. 19.2)

Theoretical evidence for this picture is provided by computer simulations of the early universe, which clearly show merging taking place. Further strong support comes from observations that galaxies at large redshifts (meaning that they are very distant and the light we see was emitted long ago) are distinctly smaller and more irregular than

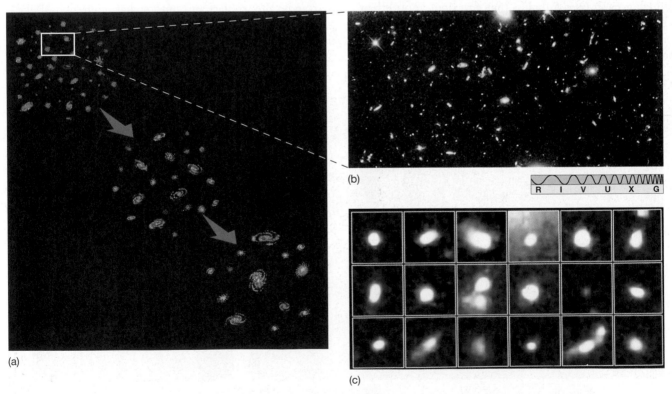

(a)

(b)

(c)

R I V U X G

▲ **FIGURE 25.10 Galaxy Formation** (a) The present view of galaxy formation holds that large systems were built up from smaller ones through collisions and mergers, as shown schematically in this drawing. (b) This photograph, one of the deepest ever taken of the universe (i.e., looking at the faintest, most distant objects), provides "fossil evidence" for hundreds of galaxy shards and fragments, up to 5000 Mpc distant. (c) Enlargements of selected portions of (b) reveal rich (billion-star) "star clusters," all lying within a relatively small volume of space (about 1 Mpc across). Their proximity to one another suggests that we may be seeing a group of pregalactic fragments about to merge to form a galaxy. The events pictured took place about 10 billion years ago. *(NASA)*

those found nearby. Figures 25.10(b) and 25.11 show some of these images, which include objects up to 5 *billion* parsecs away. The vague bluish patches are separate small galaxies, each containing only a few percent of the mass of the Milky Way Galaxy. Their irregular shape is thought to be the result of galaxy mergers; the bluish coloration comes from young stars that formed during the merging process.

Figure 25.10(c) shows more detailed views of some of the objects in Figure 25.10(b), all lying in the same region of space, about 1 Mpc across and almost 5000 Mpc from Earth. Each blob seems to contain several billion stars

R I V U X G

▲ **FIGURE 25.11 Hubble Deep Field** Numerous small, irregularly shaped young galaxies can be seen in this very deep optical image. Known as the Hubble Deep Field–North, this region of the sky, imaged with an exposure of approximately 100 hours, captured objects as faint as 30th magnitude. (As in Figure 25.10, "deep" in this context means "faint," indicating that we are looking at objects far away and as they were long ago.) ∞ (Sec. 17.2) Redshift measurements (see *More Precisely 24-1*) indicate that some of these galaxies lie well over 1000 Mpc from Earth. Their size, color, and irregular appearance support the theory that galaxies grew by mergers and were smaller and less regular in the past. The field of view is about 2 arc minutes across. (*NASA*)

spread throughout a distorted spheroid about a kiloparsec across. Their decidedly bluish tint suggests that active star formation is already under way. We see them as they were nearly 10 billion years ago, a group of young galaxies possibly poised to merge into one or more larger objects.

Interactions and Evolution

Left alone, a galaxy will evolve slowly and fairly steadily as interstellar clouds of gas and dust are turned into new generations of stars and main sequence stars evolve into giants and, ultimately, into compact remnants—white dwarfs, neutron stars, and black holes. ∞ (Sec. 19.2) The galaxy's overall color, composition, and appearance change in a more or less predictable way as the cycle of stellar evolution recycles and enriches the galaxy's interstellar matter. ∞ (Sec. 21.5)

But many—perhaps most—galaxies are not alone. They reside in small groups and clusters, and, as we have just seen, their orderly "internal" evolution may be significantly complicated by external events—interactions with other galaxies over extended periods. As described in Section 25.2, these interactions can rearrange a galaxy's internal structure and trigger a sudden, intense burst of star formation. The result is called a **starburst galaxy**; some examples are shown in Figure 25.12. Encounters may also divert fuel to a central black hole, powering violent activity in some galactic nuclei. ∞ (Sec. 24.4) Thus, starbursts and nuclear activity are key indicators of interactions and mergers between galaxies.

Careful studies of starburst galaxies and active galactic nuclei indicate that most galactic encounters probably took place long ago—at redshifts greater than about 1, when the clusters were more compact and galaxy collisions were correspondingly more frequent. ∞ (*More Precisely 24-1*) We see the majority of these violent events as they unfolded roughly 10 billion years ago. Nevertheless, the galaxy interactions observed locally are extensions of this same basic process into the present day. We have ample evidence that galaxies evolved, and are still evolving, in response to external factors long after the first pregalactic fragments formed and merged.

Computer simulations performed over the past decade have shown that the

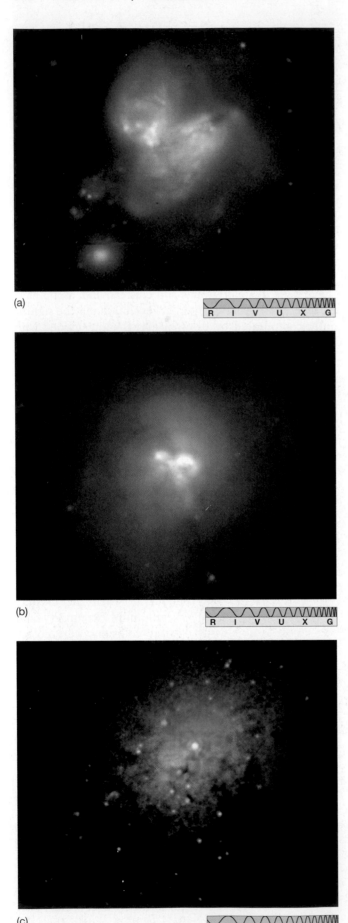

(a)

R I V U X G

(b)

R I V U X G

(c)

R I V U X G

◄ **FIGURE 25.12 Starburst Galaxies** (a) This interacting galaxy pair (IC 694, at the left, and NGC 3690) shows starbursts now under way in both galaxies—hence the bluish tint. Such intense, short-lived bursts probably last for no more than a few tens of millions of years—a small fraction of a typical galaxy's lifetime. (b) This infrared image of a starburst galaxy (Arp 220) shows a 300-light-year disk, which may be fueling a supermassive black hole within it, and rapid star formation outside it. (c) The peculiar (Irr II) galaxy NGC 1275 contains a system of long filaments that seem to be exploding outward into space. Its blue blobs, as revealed by the *Hubble Space Telescope*, are probably young globular clusters formed by the collision of two galaxies. *(W. Keel; NASA)*

extensive dark-matter halos surrounding most, if not all, galaxies are crucial to galaxy interactions. Consider first two galaxies orbiting one another—a binary galaxy system. As they orbit, the galaxies interact with each other's dark halos, one galaxy stripping halo material from the other by tidal forces. The freed matter is redistributed between the galaxies or is entirely lost from the binary system. In either case, the interaction changes the orbits of the galaxies, causing them to spiral toward one another and eventually to merge.

If one galaxy of the pair happens to have a much lower mass than the other, the process is colloquially termed *galactic cannibalism*. Such cannibalism might explain why supermassive galaxies are often found at the cores of rich galaxy clusters. Having "dined" on their companions, they now lie at the center of the cluster, waiting for more "food" to arrive. Figure 25.13 is a remarkable combination of images that has apparently captured this process at work. Closer to home, the Magellanic Clouds (Figure 24.8) will eventually suffer a similar fate at the center of the Milky Way.

Now consider two interacting disk galaxies, one a little smaller than the other, but each having a mass comparable to the Milky Way Galaxy. As shown in the computer-generated frames of Figure 25.14, the smaller galaxy can distort the larger one substantially, causing spiral arms to appear where none existed before. The entire event requires several hundred million years—a span of evolution that a supercomputer can model in minutes. The final frame of the figure looks remarkably similar to the double galaxy shown in Figure 24.2(b), and in fact, the simulation was constructed to mimic the sizes, shapes, and velocities in that binary galaxy system. The magnificent spiral galaxy is M51, popularly known as the Whirlpool Galaxy, about 10 Mpc from Earth. Its smaller companion is an irregular galaxy that may have drifted past M51 millions of years ago.

Did this smaller galaxy cause the spiral structure we see in M51? Does the model mirror reality? Perhaps. We need more evidence from other galaxies to confirm the accuracy of these and similar simulations. Still, the computer simulation does demonstrate a plausible way in which two

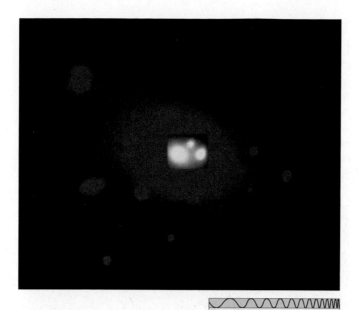

R I V U X G

◀ **FIGURE 25.13 Galactic Cannibalism** This computer-enhanced, false-color composite optical photograph of the galaxy cluster Abell 2199 is thought to show an example of galactic cannibalism. The large central galaxy of the cluster (itself 120 kpc along its long axis) is displayed with a superimposed "window." (The image results from a shorter time exposure, which shows only the brightest objects that fall within the frame.) Within the core of the large galaxy are several smaller galaxies (the three bright yellow images at the center) apparently already "eaten" and now being "digested" (i.e., being torn apart and becoming part of the larger system). Other small galaxies swarm on the outskirts of the swelling galaxy, almost certainly to be "eaten," too. *(SAO)*

galaxies may have interacted millions of years ago and how spiral arms may be created or enhanced as a result.

Making the Hubble Sequence

If galaxies form and evolve by repeated mergers, can we account for the Hubble sequence and, specifically, differences between spirals and ellipticals? ∞ (Sec. 24.1) The details are still far from certain, but, remarkably, the answer now seems to be a qualified yes. Collisions and close encounters are random events and do not represent a "genuine" evolutionary sequence linking all galaxies. However, computer simulations suggest a plausible way in which the observed Hubble types might have arisen, starting from a universe populated only by irregular, gas-rich galaxy fragments.

The simulations reveal that a "major" merger—a collision between galaxies of comparable size—can destroy a spiral galaxy's disk, creating a galaxywide starburst episode. The violence of the merger and the effects of subsequent supernovae eject much of the remaining gas into inter-

galactic space, creating the hot intracluster gas noted in Section 24.1. Once the burst of star formation has subsided, the resulting object looks very much like an elliptical galaxy. The elliptical's hot X-ray halo is the last vestige of the original spiral's disk. The Irr II galaxy shown in Figure 25.12(c) may be an example of this phenomenon in progress. The blue blobs are thought to be young star clusters formed during the starburst, and the explosive appearance suggests that we are witnessing the gas and dust being ejected.

The simulations also indicate that "minor" mergers, in which a small galaxy interacts with, and ultimately is absorbed by, a larger one, generally leave the larger galaxy intact, with more or less the same Hubble type as it had before the merger. This is the most likely way for large spirals to grow—in particular, our own Galaxy probably formed in such a manner. The "streams" of halo stars in the Milky Way Galaxy, all with similar orbits and composition, may well be the stellar remnants of such mergers in the past. ∞ (Sec. 23.3) The small Sagittarius dwarf galaxy (Figure 24.12) and the Magellanic clouds (Figure 24.8) will probably be the Milky Way's next victims.

Supporting evidence for this general picture comes from observations that spiral galaxies are relatively rare in regions of high galaxy density, such as the central regions

Time ━━▶

▲ **FIGURE 25.14 Galaxy Interaction** Galaxies can change their shapes long after their formation. In this computer-generated sequence, two galaxies closely interact over several hundred million years. The smaller galaxy, in red, has gravitationally disrupted the larger galaxy, in blue, changing it into a spiral galaxy. Compare the result of this supercomputer simulation with Figure 24.2b, a photograph of M51 and its small companion. *(J. Barnes & L. Hernquist)*

of rich galaxy clusters. These observations are consistent with the view that the fragile disks of spiral galaxies are easily destroyed by collisions, which are more common in dense galactic environments. Spirals also seem to be more common at larger redshifts (i.e., in the past), implying that their numbers are decreasing with time, presumably also as the result of collisions. However, nothing in this area of astronomy is clear cut, and astronomers know of numerous isolated elliptical galaxies in low-density regions of the universe that are hard to explain as the result of mergers.

In principle, the starbursts associated with galaxy mergers leave their imprint on the star-formation history of the universe in a way that can be correlated with the properties of galaxies. As a result, studies of star formation in distant galaxies have become a very important way of testing and quantifying the details of the entire merger scenario.

CONCEPT CHECK

✔ Other than scale, in what important ways does galaxy evolution differ from that of stars?

25.4 Black Holes and Active Galaxies

LEARNING GOAL 4 Now let's ask how quasars and active galaxies fit into the framework of galaxy evolution just described. The fact that quasars are more common at great distances from us demonstrates that they were much more prevalent in the past than they are today. ∞ (Sec. 24.4) Quasars have been observed with redshifts of up to 6.4 (the current record, as of early 2004), so the process must have started at least 13 billion years ago (see Table 24.1). However, most quasars have redshifts between 2 and 3, corresponding to an epoch some 2 billion years later. Most astronomers agree that quasars represent an early stage of galaxy evolution—an "adolescent" phase of development, prone to frequent flare-ups and "rebellion" before settling into more steady "adulthood." This view is reinforced by the fact that the same black-hole energy-generation mechanism can account for the luminosities of quasars, active galaxies, and the central regions of normal galaxies like our own.

▶ **FIGURE 25.15 Binary Black Hole** These (a) optical (*Hubble*) and (b) X-ray (*Chandra*) images of the starburst galaxy NGC 6420 show two supermassive black holes (the blue-white objects near the center of the X-ray image) orbiting about 1 kpc apart. According to theoretical estimates, they will merge in about 400 million years, releasing an intense burst of gravitational radiation in the process. The colors of the optical image are real; the false colors in the X-ray image indicate a range of energies in the X-ray band. ∞ (*Discovery 22-1*) (*NASA*)

The Quasar Epoch

In Chapter 24 we presented the standard model of active galactic nuclei accepted by most astronomers—accretion of gas onto a supermassive black hole. ∞ (Sec. 24.5) But where did these supermassive black holes come from? To be honest, the answer is not yet known. The processes whereby the first billion-solar-mass black holes formed early in the history of the universe are not fully understood. However, the accretion responsible for the energy emission also naturally accounts for the mass of the black

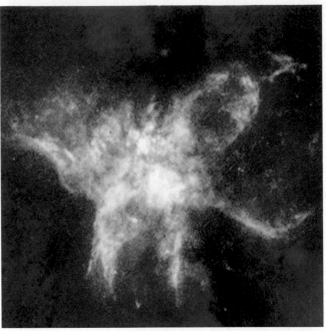

(a)

R I V U X G

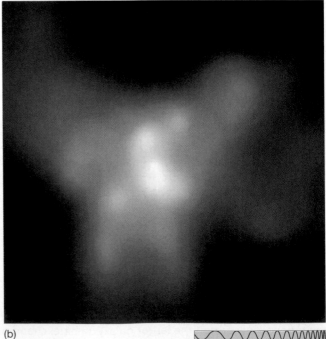

(b)

R I V U X G

hole, and simple estimates suggest that the accretion rates needed to power the quasars are generally consistent with black-hole masses inferred by other means.

Since the brightest known quasars devour about a thousand solar masses of material every year, it is unlikely that they could maintain their luminosity for very long—even a million years would require a billion solar masses, enough to account for the most massive black holes known. ∞ (Sec. 24.4) This suggests that a typical quasar spends only a fairly short amount of time in its highly luminous phase—perhaps only a few million years in some cases—before running out of fuel. Thus, most quasars were relatively brief events that occurred long ago.

To make a quasar, we need a black hole and enough fuel to power it. While fuel was abundant at early times in the universe's history in the form of gas and newly formed stars, black holes were not. They had yet to form, most probably by the same basic stellar evolutionary processes we saw in Chapter 21, although the details are not well known. ∞ (Sec. 21.2) The building blocks of the supermassive black holes that would ultimately power the quasars may well have been relatively small black holes having masses tens or perhaps a few hundreds of times the mass of the Sun. These small black holes sank to the center of their still-forming parent galaxy and merged to form a single, more massive black hole.

As galaxies merged, so, too, did their central black holes, and eventually supermassive (1-million- to 1-billion-solar-mass) black holes existed in the centers of many young galaxies. Some supermassive black holes may have formed directly by the gravitational collapse of the dense central regions of a protogalactic fragment or perhaps by accretion or a rapid series of mergers in a particularly dense region of the universe. These events resulted in the

earliest (redshift-6) quasars known, shining brightly 13 billion years ago. However, in most cases, all those mergers took time—roughly another 2 billion years. By then (at redshifts between 2 and 3, roughly 11 billion years ago), many supermassive black holes had formed, and there was still plenty of merger-driven fuel available to power them. This was the height of the "quasar epoch" in the universe.

Until recently, astronomers were confident that black holes would merge when their parent galaxies collided, but they had no direct evidence of the process—no image of two black holes "caught in the act." In 2002, the *Chandra* X-ray observatory discovered a binary black hole—two supermassive objects, each having a mass a few tens of millions of times that of the Sun—in the center of the ultraluminous starburst galaxy NGC 6240, itself the product of a galaxy merger some 30 million years ago. Figure 25.15 shows optical and X-ray views of the system. The black holes are the two blue-white objects near the center of the (false-color) X-ray image. Orbiting just 1000 pc apart, they are losing energy via interactions with stars and gas and are predicted to merge in about 400 million years. NGC 6240 lies just 120 Mpc from Earth, so we are far from seeing a quasar merger in the early universe. Nevertheless, astronomers think that events similar to this must have occurred countless times billions of years ago, as galaxies collided and quasars blazed.

Distant galaxies are generally much fainter than their bright quasar cores. As a result, until quite recently, astronomers were hard pressed to discern any galactic structure in quasar images. Since the mid-1990s, several groups of astronomers have used the *Hubble Space Telescope* to search for the "host" galaxies of moderately distant quasars. After removing the bright quasar core from the *HST* images and carefully analyzing the remnant light, the

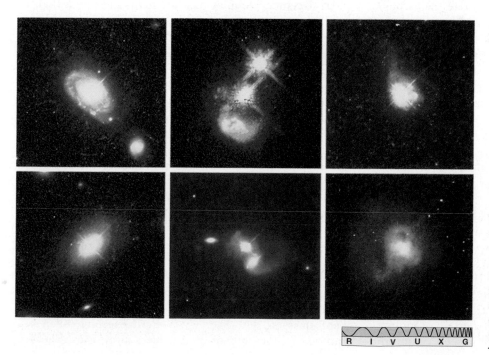

R I V U X G

◀ **FIGURE 25.16 Quasar Host Galaxies** These long-exposure *Hubble Space Telescope* images of distant quasars clearly show the young host galaxies in which the quasars reside, lending strong support to the view that quasars represent an early, highly luminous phase of galactic evolution. The quasar at the top left is the best example, having the catalog name PG0052 + 251 and residing roughly 690 Mpc from Earth. Note that several of the quasars appear to be associated with interacting galaxies, consistent with current theories of galaxy evolution. (*NASA*)

researchers have reported that, in every case studied—several dozen quasars so far—a host galaxy can be seen enveloping the quasar. Figure 25.16 shows some of the longest quasar exposures ever taken. Even without sophisticated computer processing, the hosts are clearly visible.

As we saw in Chapter 24, the connection between active galaxies and galaxy clusters is well established, and many relatively nearby quasars are also known to be members of clusters. ∞ (Sec. 24.4) The link is less clear cut for the most distant quasars, however, simply because they are so far away that other cluster members are very faint and extremely hard to see. However, as the number of known quasars continues to increase, evidence for quasar clustering (and presumably, therefore, for quasar membership in young galaxy clusters) mounts. Thus, as best we can tell, quasar activity—and, in fact, galaxy activity of *all* sorts—is intimately related to interactions and collisions in galaxy clusters.

Active and Normal Galaxies

Early on, frequent mergers may have replenished the quasar's fuel supply, extending its luminous lifetime. However, as the merger rate declined, these systems spent less and less of their time in the "bright" phase. The rapid decline in the number of bright quasars roughly 10 billion years ago marks the end of the quasar epoch. Today, the number of quasars has dropped virtually to zero (recall that the nearest lies hundreds of megaparsecs away). ∞ (Sec. 24.4)

Large black holes do not simply vanish. If a galaxy contained a bright quasar 10 billion years ago, the black hole responsible for all that youthful activity must still be present in the center of the galaxy today. We see some of these black holes as active galaxies, but since the difference between a normal galaxy and an active one seems to be mainly a matter of fuel supply, the implication is that many—perhaps most—normal galaxies should also contain black holes. Do we have any evidence that this in fact the case? The answer appears to be yes. In recent years, astronomers have amassed evidence that many bright normal galaxies contain black holes at their centers. Our own Galaxy is a case in point. ∞ (Sec. 23.7) The 3–4-million-solar-mass black hole at the center of the Milky Way is not currently active, but if fresh fuel were supplied (say, by a star coming too close to the hole's intense gravitational field), it might well become a (relatively weak) active galactic nucleus.

Perhaps the most compelling evidence for a supermassive black hole at the center of a normal galaxy comes from a radio study of NGC 4258, a spiral galaxy about 6 Mpc away. Using the Very Long Baseline Array, a continentwide interferometer comprising 10 radio telescopes, a U.S.-Japanese team has achieved an angular resolution hundreds of times better than that attainable with *HST*. ∞ (Sec. 5.6) The observations reveal a group of molecular clouds swirling in an organized fashion about the

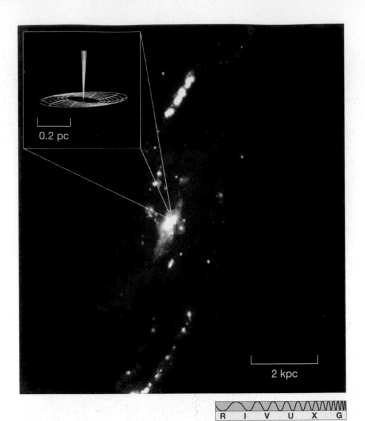

▲ **FIGURE 25.17 Galactic Black Hole** A network of radio telescopes has probed the core of the spiral galaxy NGC 4258, shown here in the light of mostly hydrogen emission. Within the innermost 0.2 pc (inset), observations of Doppler-shifted molecular clouds (designated by red, green, and blue dots) show that they obey Kepler's third law perfectly. The observations also have revealed a slightly warped disk of rotating gas (shown in the inset as an artist's conception). At the center of the disk presumably lurks a huge black hole. (*J. Moran*)

galaxy's center. Doppler measurements indicate a slightly warped, spinning disk centered precisely on the galaxy's heart. (See Figure 25.17.) The rotation speeds imply the presence of more than 40 million solar masses packed into a region less than 0.2 pc across.

Discoveries such as these blur the distinction between normal and active galaxies. When the fuel runs out and a quasar shuts down, its central black hole remains behind, its energy output reduced to a relative trickle. But the black holes at the hearts of normal galaxies are simply quiescent, awaiting another interaction to trigger a new active outburst. Occasionally, two nearby galaxies may interact with each other, causing a flood of new fuel to be directed toward the central black hole of one or both. The engine starts up for a while, giving rise to the nearby active galaxies—radio galaxies, Seyferts, and others—we observe.

Should this general picture be correct, it follows that many relatively nearby galaxies (but probably *not* our own Milky Way, whose central black hole is even now only a paltry 3–4 million solar masses) must once have been brilliant quasars. ∞ (Sec. 23.7) Perhaps some alien astronomer, thousands of megaparsecs away, is at this very moment observing the progenitor of M87 in the Virgo cluster—seeing it as it was billions of years ago—and is commenting on its enormous luminosity, nonstellar spectrum, and high-speed jets and wondering what exotic physical process can account for its violent activity! ∞ (Sec. 24.4)

Active-Galaxy Evolution

The galaxy merger scenario, combined with local observations of supermassive black holes, may also provide some insight into the connection between actvity and galaxy type. Some possible (but unproven) evolutionary connections among quasars, active, and normal galaxies are illustrated in Figure 25.18.

Astronomers have found that the largest black holes tend to be found in the most massive galaxies (Figure 25.19). Furthermore, it is likely that the most massive black holes power the brightest active galactic nuclei. In

that case, we would expect that the most luminous nuclei should reside in the largest galaxies, which probably came into being via "major" mergers of other large galaxies. Since the products of such mergers are elliptical galaxies, we have a plausible explanation of why the brightest active galaxies—the radio galaxies—should be associated with large ellipticals. ∞ (Sec. 24.4) Furthermore, the path to spiral galaxies would necessarily have entailed a series of mergers involving smaller galaxies, resulting in the less violent Seyferts.

When active galactic nuclei—and especially quasars—were first discovered, their extreme properties defied conventional explanation. Initially, the idea of supermassive (million- to billion-solar-mass) black holes in galaxies was just one of several competing, and very different, hypotheses advanced to account for the enormous luminosities and small sizes of those baffling objects. However, as observational evidence mounted, the other hypotheses were abandoned one by one, and massive black holes in galactic

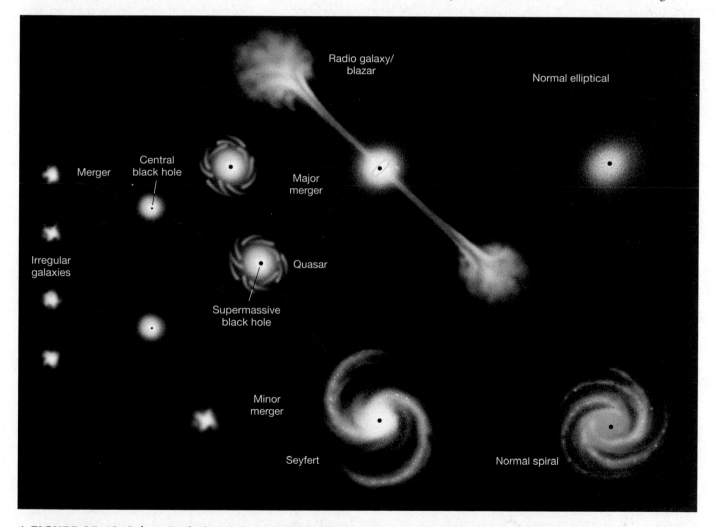

▲ **FIGURE 25.18 Galaxy Evolution** Some possible evolutionary sequences for galaxies, beginning with galaxy mergers leading to the highly luminous quasars, decreasing in violence through the radio and Seyfert galaxies, and ending with normal ellipticals and spirals. The central black holes that powered the early activity are still there at later times; they simply run out of fuel as time goes on.

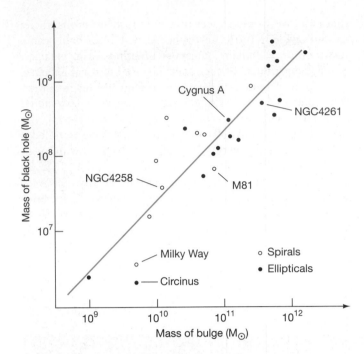

◀ **FIGURE 25.19** **Black-Hole Masses** Careful observations of nearby normal and active galaxies reveal that the mass of the central black hole is well correlated with the mass of the galactic bulge. In this diagram, each point represents a different galaxy. The straight line is the best fit to the data, implying a black-hole mass of 1/200 the mass of the bulge. (*Data courtesy L. Ferrarese*)

nuclei became first the leading, and eventually the standard, theory of active galaxies.

As often happens in science, a theory once itself considered extreme is now the accepted explanation for these phenomena. Far from threatening the laws of physics, as some astronomers once feared, active galaxies are now an integral part of our understanding of how galaxies form and evolve. The synthesis of studies of normal and active galaxies, galaxy formation, and large-scale structure is one of the great triumphs of extragalactic astronomy.

CONCEPT CHECK

✔ Does *every* galaxy have the potential for activity?

25.5 The Universe on Large Scales

5 Many galaxies, including our own, are members of galaxy clusters—megaparsec-sized structures held together by their own gravity. ∞ (Sec. 24.2) Our own small cluster is called the Local Group. Figure 25.20 shows the locations of the Virgo cluster, the closest "large" large cluster, and of several other well-defined clusters in our cosmic neighborhood. The region displayed is about 70 Mpc across. Each point in the figure represents an entire galaxy whose distance has been determined by one of the methods described in Chapter 24.

Clusters of Clusters

Does the universe have even greater groupings of matter, or do galaxy clusters top the cosmic hierarchy? Most astronomers have concluded that the galaxy clusters are

themselves clustered, forming titanic agglomerations of matter known as **superclusters**.

Together, the galaxies and clusters shown in Figure 25.20 form the *Local Supercluster*, also known as the Virgo Supercluster. Aside from the Virgo Cluster itself, it contains the Local Group and numerous other clusters lying within about 20–30 Mpc of Virgo. Most of the galaxies depicted in the figure are fairly large spirals and ellipticals; the fainter irregulars and dwarfs are not included in the diagram. The galaxies are false colored according to the local galaxy density, with white and yellow indicating the most congested regions, green less dense regions, and blue the least dense. The white, yellow, and green galaxies trace out approximately the supercluster's extent. Figure 25.21 shows a three-dimensional rendering of an even wider view, illustrating the Virgo supercluster (near the center) relative to other "nearby" galaxy superclusters

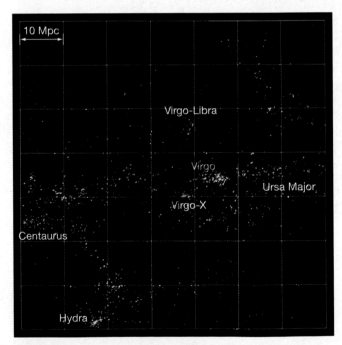

▲ **FIGURE 25.20** **Local Supercluster** The locations of numerous galaxies and galaxy clusters in the neighborhood of the Virgo cluster. More than 4500 galaxies are represented here, and several prominent galaxy clusters are labeled. The diagram shows the Virgo Supercluster roughly as it appears from the direction of our own Galaxy, which is located approximately 20 Mpc (two grid squares) above the page. Notice the supercluster's irregular, elongated shape. (*Data courtesy B. Tully, U. Hawaii; visualization by S. Levy, NCSA*)

within a vast imaginary rectangle roughly 100 Mpc on its short side.

All told, the Local Supercluster is about 40–50 Mpc across, contains some 10^{15} solar masses of material (several tens of thousands of galaxies), and is very irregular in shape. The Local Supercluster is significantly elongated perpendicular to the line joining the Milky Way to Virgo, with its center lying near the Virgo Cluster. By now it should perhaps come as no surprise that the Local Group is *not* found at the heart of the Local Supercluster—we live far off in the periphery, about 18 Mpc from the center.

Redshift Surveys

The farther we peer into deep space, the more galaxies, clusters of galaxies, and superclusters we see. Is there structure on scales even larger than superclusters? To answer these questions, astronomers use Hubble's law to map out the distribution of galaxies in the universe.

Figure 25.22 shows part of an early survey of the universe performed by astronomers at Harvard University in the 1980s. Using Hubble's law as a distance indicator, the team systematically mapped out the locations of galaxies within about 200 Mpc of the Milky Way in a series of wedge-shaped "slices," each 6° thick, starting in the northern sky. The first slice (shown in the figure) covered a region of the sky containing the Coma Cluster (see Figure 24.1), which happens to lie in a direction almost perpendicular to our Galaxy's plane. Because redshift is used as the primary distance indicator, these studies are known as *redshift surveys*.

The most striking feature of maps such as that of Figure 25.22 is that the distribution of galaxies on very large scales is decidedly nonrandom. The galaxies appear to be arranged in a network of strings, or filaments, surrounding large, relatively unpopulated regions of space known as

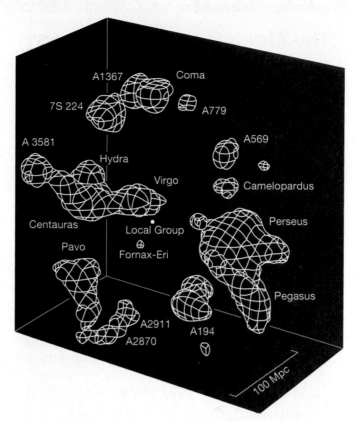

▲ **FIGURE 25.21 Virgo Supercluster in 3-D** The structure of the Virgo supercluster (left center) is mapped relative to other neighboring galaxy superclusters within about 100 Mpc. Individual galaxies are not shown; rather, smoothed contour plots outline galaxy clusters, each one named or numbered by its most prominent member.

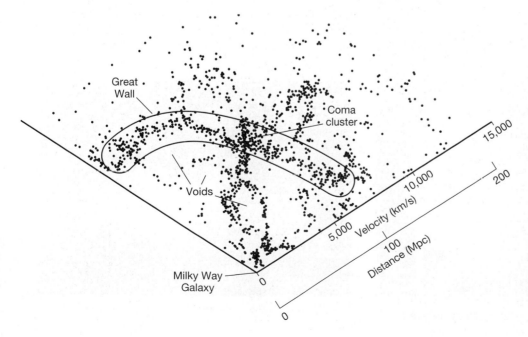

◀ **FIGURE 25.22 Galaxy Survey** A slice of a survey of the universe, covering 1732 galaxies out to an approximate distance of 200 Mpc, shows clearly that galaxies and clusters are not randomly distributed on large scales. Instead, they appear to have a filamentary structure, surrounding vast, nearly empty voids. We assume that $H_0 = 70$ km/s/Mpc. The slice covers 6° of the sky in the direction out of the plane of the paper. (*Harvard-Smithsonian Center for Astrophysics*)

The Sloan Digital Sky Survey

Many of the photographs used in this book—not to mention most of the headline-grabbing imagery found in the popular media—come from large, high-profile, and usually very expensive, instruments such as NASA's *Hubble Space Telescope* and the European Southern Observatory's *Very Large Telescope* in Chile. ∞ (Secs. 5.3, 5.4) Their spectacular views of deep space have revolutionized our view of the universe. Yet a less well known, considerably cheaper, but no less ambitious, project currently underway may, in the long run, have every bit as great an impact on astronomy and our understanding of the cosmos.

The Sloan Digital Sky Survey (SDSS) is a five year project designed to systematically map out a quarter of the entire sky on a scale and at a level of precision never before attempted. By the time the project is completed in 2005, it will have catalogued more than 100 million celestial objects, recording their apparent brightnesses at 5 different colors (wavelength ranges) spread across the optical and near-infrared part of the spectrum. In addition, spectroscopic follow-up observations will determine redshifts and hence distances to 1 million galaxies and 100,000 quasars. These data will be used to construct even more detailed redshift surveys than those described in the text, and to probe the structure of the universe on very large scales. The sensitivity of the survey is such that it can detect bright galaxies like our own out to distances of more than 1 billion parsecs. Very bright objects, such as quasars and young starburst galaxies, are detectable almost throughout the entire observable universe.

The first figure shows the Sloan Survey telescope, a special-purpose 2.5-m instrument sited in Apache Point Observatory, near Sunspot, New Mexico. This reflecting telescope (whose box-like structure protects it from the wind) is not space-based, does not employ active or adaptive optics, and cannot probe as deep (*i.e.* far) into space as larger instruments. How can it possibly compete with these other systems? The answer is that, unlike most other large telescopes in current use, where hundreds or even thousands of observers share the instrument and compete for its time, the SDSS telescope was designed specifically for the purpose of the survey. It has a wide field of view and is dedicated to the task, carrying out observations of the sky on *every* clear night during the 5-year duration of the survey project.

The use of a single instrument night after night, combined with tight quality controls on which nights' data are actually incorporated into the survey (nights with poor seeing or other problematic conditions are discarded and the observations repeated) mean that the end-product is a database of exceptionally high quality and uniformity spanning an enormous volume of space—a monumental achievement and an indispensible tool for cosmology. The survey field of view covers much of the sky away from the Galactic plane in the north, together with three broad "wedges" in the south.

Archiving images and spectra on millions of galaxies produces a lot of data. The full survey will consist of roughly 15 *trillion* bytes of information—comparable to the entire Library of Congress! As of mid 2004, data for roughly one third of the survey area, comprising 88 million objects (360,000 with measured spectra), has been released to the public. The second figure shows the distant galaxy NGC 5792 and a bright red star much closer to us, in fact in our own Milky Way Galaxy, just one of hundreds of thousands of images that will make up the full dataset. Among recent highlights, SDSS has detected the largest known structure in the universe, observed the most distant known galaxies and quasars, and has been instrumental in pinning down the key observational parameters describing our universe (see Chapter 26).

SDSS will impact astronomy in areas as diverse as the large-scale structure of the universe, the origin and evolution of galaxies, the nature of dark matter, the structure of the Milky Way, and the properties and distribution of interstellar matter. Its uniform, accurate, and detailed database is likely to be used by generations of scientists for decades to come.

voids. The biggest voids measure some 100 Mpc across. The most likely explanation for the voids and the filamentary structure shown in the figure is that the galaxies and galaxy clusters are spread across the surfaces of vast "bubbles" in space. The voids are the interiors of these gigantic bubbles. The galaxies seem to be distributed like beads on strings only because of the way our slice of the universe cuts through the bubbles. Like suds on soapy water, these bubbles fill the entire universe. The densest clusters and superclusters lie in regions where several bubbles meet. The elongated shape of the Virgo Supercluster (Figure 25.20) is a local example of this same filamentary structure.

Most theorists think that this "frothy" distribution of galaxies, and in fact all structure on scales larger than a few megaparsecs, traces its origin directly to conditions in the very earliest stages of the universe (Chapter 27). Consequently, studies of large-scale structure are vital to our efforts to understand the origin and nature of the cosmos itself.

The idea that the filaments are the intersection of a survey slice with much larger structures (the bubble surfaces) was confirmed when the next three slices of the survey, lying above and below the first, were completed. The region of Figure 25.22 indicated by the red outline was found to continue through both the other slices. This extended sheet of galaxies, which has come to be known as the *Great Wall*, measures at least 70 Mpc (out of the plane

of the page) by 200 Mpc (across the page). It is one of the largest known structures in the universe.

Figure 25.23 shows a more recent redshift survey, considerably larger than the one presented in Figure 25.22. This survey includes nearly 24,000 galaxies within about 750 Mpc of the Milky Way. Numerous voids and "Great Wall-like" filaments can be seen (some are marked), but, apart from the general falloff in numbers of galaxies at large distances—basically because the more distant galaxies are harder to see due to the inverse-square law—there is no obvious evidence for any structures on scales *larger* than about 200 Mpc. Careful statistical analysis confirms this impression. Apparently, voids and walls represent the largest structures in the universe. We will return to the far-reaching implications of this fact in Chapter 26.

Quasar Absorption Lines

How can we probe the structure of the universe on very large scales? As we have seen, much of the matter is dark, and even the "luminous" component is so faint that it is hard to detect at large distances. One way to study large-scale structure is to take advantage of the great distances, pointlike appearance, and large luminosities of quasars. Since quasars are so far away, light traveling from a quasar to Earth has a pretty good chance of passing through or near "something interesting" en route. By analyzing quasar images and spectra, it is possible to piece together a partial picture of the intervening space.

The quasar approach is reminiscent of the use of bright stars to probe the interstellar medium near the Sun, and it suffers from the same basic drawback: We can study only regions of the sky where quasars happen to be found. ∞ (Sec. 18.1) However, this problem will diminish in time as ongoing and planned large-scale surveys scan the sky for fainter and fainter objects. Foremost among these surveys is the Sloan Digital Sky Survey *(Discovery 25-1)*, an ambitious project, currently in progress, that, by 2005, will construct a map of much of the northern sky, including several million galaxies and over 100,000 quasars.

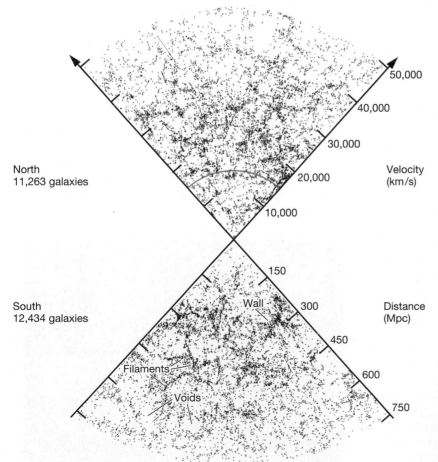

North
11,263 galaxies

South
12,434 galaxies

Velocity (km/s)

Distance (Mpc)

◀ **FIGURE 25.23 The Universe on Larger Scales** This large-scale galaxy survey, carried out at the Las Campanas Observatory in Chile, consists of 23,697 galaxies within about 1000 Mpc, in two 80° × 4.5° wedges of the sky. Many voids and "walls" on scales of up to 100–200 Mpc can be seen, but no larger structures are evident. For scale, the extent of the survey shown in Figure 25.22 is marked as a thin blue arc in the northern sky.

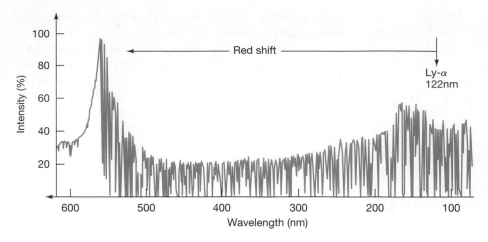

◀ **FIGURE 25.24** Lyman-Alpha
"Forest" When light from a distant quasar passes through a foreground cloud of atomic hydrogen, the absorption lines produced bear the cloud's redshift. The huge number of absorption lines in the spectrum of this quasar (called QSO 1422 + 2309) are the ultraviolet Lyman-alpha lines from hundreds of clouds of foreground hydrogen gas, each redshifted by a slightly different amount (but less than the quasar itself). The peak marks the Lyman-alpha emission line from the quasar, emitted at 122 nm, but redshifted here to a wavelength of 564 nm, in the visible range.

In addition to exhibiting their own strongly redshifted spectra, many quasars show additional absorption features that are redshifted by substantially *less* than the lines from the quasar itself. For example, the quasar PHL 938 has an emission-line redshift of 1.954, placing the quasar at a distance of some 5700 Mpc, but it also shows absorption lines having redshifts of just 0.613. These lines with lesser redshifts are interpreted as arising from intervening gas that is much closer to us (only about 2400 Mpc away) than the quasar itself. Most probably, this gas is part of an otherwise invisible galaxy lying along the line of sight. Quasar spectra, then, afford astronomers a means of probing previously undetected parts of the universe.

The absorption lines of atomic hydrogen are of particular interest, since hydrogen makes up so much of all matter in the cosmos. Specifically, hydrogen's ultraviolet (122-nm) "Lyman-alpha" line, associated with transitions between the ground and first excited states, is often used in this context. ∞ (Sec. 4.3) As illustrated in Figure 25.24, when astronomers observe the spectrum of a high-redshift quasar, they typically see a "forest" of absorption lines, starting at the (redshifted) wavelength of the quasar's own Lyman-alpha emission line and extending to shorter wavelengths. These lines are interpreted as Lyman-alpha absorption features produced by gas clouds in foreground structures—galaxies, clusters, and so on—giving astronomers crucial information about the distribution of matter along the line of sight.

The quasar light thus explores an otherwise invisible component of cosmic gas. In principle, every intervening cloud of atomic hydrogen leaves its own characteristic imprint on the quasar's spectrum, in a form that lets us explore the distribution of matter in the universe. By comparing these *Lyman-alpha forests* with the results of simulations, astronomers hope to refine many key elements of the theories of galaxy formation and the evolution of large-scale structure.

Quasar "Mirages"

In 1979, astronomers were surprised to discover what appeared to be a binary quasar—two quasars with exactly the same redshift and similar spectra, separated by only a few arc seconds on the sky. Remarkable as the discovery of such a binary would have been, the truth about this pair of quasars turned out to be even more amazing: Closer study of the quasars' radio emission revealed that they were *not* two distinct objects; instead, they were two separate images of the *same* quasar! Optical views of such a *twin quasar* are shown in Figure 25.25.

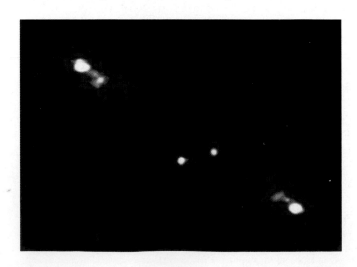

▶ **FIGURE 25.25 Twin Quasar** This "double" quasar (designated AC114 and located about 2 billion parsecs away) is not two separate objects at all. Instead, the two large "blobs" (at upper left and lower right) are images of the same object, created by a gravitational lens. The lensing galaxy itself is probably not visible in this image—the two objects near the center of the frame are thought to be unrelated galaxies in a foreground cluster. (*NASA*)

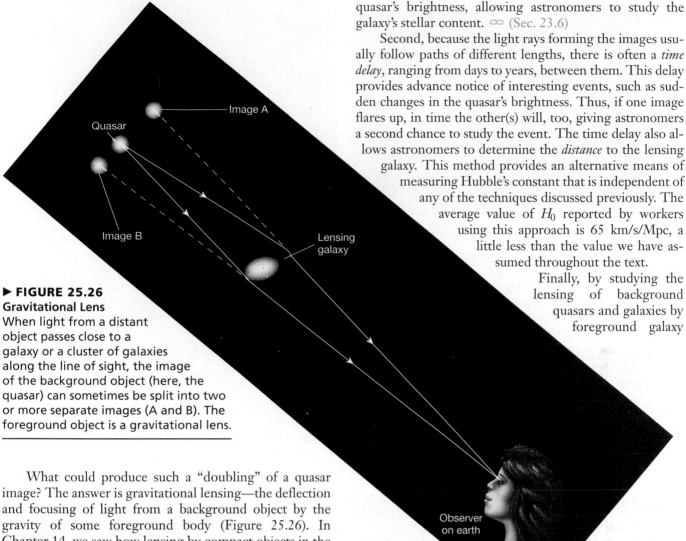

► FIGURE 25.26
Gravitational Lens
When light from a distant object passes close to a galaxy or a cluster of galaxies along the line of sight, the image of the background object (here, the quasar) can sometimes be split into two or more separate images (A and B). The foreground object is a gravitational lens.

quasar's brightness, allowing astronomers to study the galaxy's stellar content. ∞ (Sec. 23.6)

Second, because the light rays forming the images usually follow paths of different lengths, there is often a *time delay*, ranging from days to years, between them. This delay provides advance notice of interesting events, such as sudden changes in the quasar's brightness. Thus, if one image flares up, in time the other(s) will, too, giving astronomers a second chance to study the event. The time delay also allows astronomers to determine the *distance* to the lensing galaxy. This method provides an alternative means of measuring Hubble's constant that is independent of any of the techniques discussed previously. The average value of H_0 reported by workers using this approach is 65 km/s/Mpc, a little less than the value we have assumed throughout the text.

Finally, by studying the lensing of background quasars and galaxies by foreground galaxy

What could produce such a "doubling" of a quasar image? The answer is gravitational lensing—the deflection and focusing of light from a background object by the gravity of some foreground body (Figure 25.26). In Chapter 14, we saw how lensing by compact objects in the halo of the Milky Way Galaxy may amplify the light from a distant star, allowing astronomers to detect otherwise invisible stellar dark matter. ∞ (Sec. 23.6) In the case of quasars, the idea is the same, except that the foreground lensing object is an entire galaxy or galaxy cluster, and the deflection of the light is so great (a few arc seconds) that several separate images of the quasar may be formed, as shown in Figure 25.27.* About two dozen such gravitational lenses are known. As telescopes probe the universe with greater and greater sensitivity, astronomers are beginning to realize that gravitational lenses are relatively common features of the cosmos.

The existence of these multiple images provides astronomers with a number of useful observational tools. First, lensing by a foreground galaxy tends to amplify the light of the quasar, as just mentioned, making it easier to observe. At the same time, *microlensing* by individual stars within the galaxy may cause large fluctuations in the

clusters, astronomers can obtain a better understanding of the distribution of dark matter in those clusters, an issue that has great bearing on the large-scale structure of the cosmos.

Mapping Dark Matter

Astronomers have extended the ideas first learned from studies of quasars to the lensing of any distant object in order to better probe the universe. Distant, faint irregular galaxies—the raw material of the universe if current theories are correct (see Section 16.4)—are of particular interest here, as they are far more common than quasars, so they provide much better coverage of the sky. By studying the lensing of background quasars and galaxies by foreground galaxy clusters,

*In fact, much of the theory of gravitational lensing was worked out after the first lensed quasar observations and subsequently applied to dark-matter searches in our Galaxy.

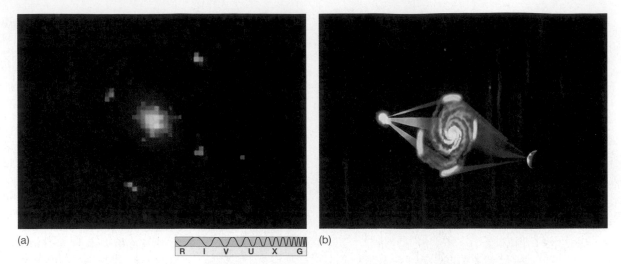

(a)

R I V U X G

(b)

▲ **FIGURE 25.27 Einstein Cross** (a) The "Einstein Cross," a multiply imaged quasar. In this *Hubble* view, spanning only a couple of arc seconds, four separate images of the same quasar have been produced by the galaxy at the center. (b) A simplified artist's conception of what might be occurring here, with Earth at the right and the distant quasar at the left. *(NASA; D. Berry)*

astronomers can obtain a better understanding of the distribution of dark matter on large scales.

Figures 25.28(a) and (b) show how the images of faint background galaxies are bent into arcs by the gravity of a foreground galaxy cluster. The degree of bending allows the total mass of the cluster (*including* the mass of the dark matter) to be measured. The (mostly blue) loop- and arc-

shaped features visible in Figure 25.28(b) are thought to be multiple images of a single distant (unseen) spiral or ring-shaped galaxy, lensed by the foreground galaxy cluster (the yellow-red blobs in the image). Astronomers are now trying to delicately "reassemble" an undistorted image of the distant object by retracing the light rays back to a common point.

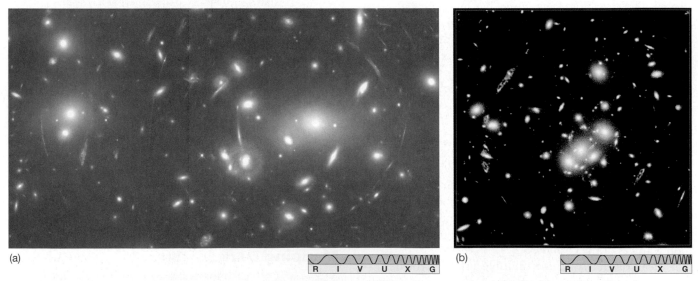

(a)

R I V U X G

(b)

R I V U X G

▲ **FIGURE 25.28 Galaxy Cluster Lensing** (a) This spectacular example of gravitational lensing shows more than a hundred faint arcs from very distant galaxies. The wispy pattern spread across the intervening galaxy cluster (A 2218, about a billion parsecs distant) resembles a spider's web, but it is really an illusion caused by the gravitational field of the lensing cluster, which deflects the light from background galaxies and distorts their appearance. By measuring the extent of this distortion, astronomers can estimate the mass of the intervening cluster. (b) An approximately true-color image of the galaxy cluster known only by its catalog name, 0024 + 1654, residing some 1.5 billion pc away. The reddish-yellow blobs are mostly normal elliptical galaxy members of the cluster, concentrated toward the center of the image. The bluish looplike features are thought to be images of a single background galaxy. *(NASA)*

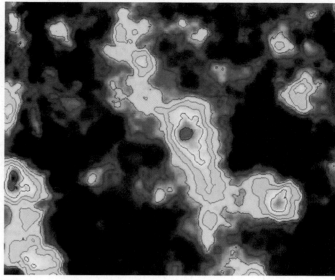

(a)

R I V U X G

(b)

▲ **FIGURE 25.29 Dark-Matter Map** (a) An optical view of a region of the sky containing a small galaxy cluster, too diffuse to be easily seen in this image, but including the clump of yellowish galaxies near the center of the frame. The fuzzy blue specks scattered across the frame are the much more distant background galaxies whose distortions are used to estimate the cluster's dark-matter content. (b) The distribution of dark matter in and near the visible cluster, on the same scale as (a), obtained by analyzing the images of the background galaxies. *(Data courtesy J. A. Tyson, Bell Labs; NSF/NOAO)*

It is even possible to reconstruct the foreground dark-matter distribution by carefully analyzing the distortions of the background objects, thereby providing a means of tracing out the distribution of mass on scales far larger than have previously been possible. Figure 25.29(a) is an optical image of a (hard-to-see) foreground galaxy cluster, set against a background of much fainter distant galaxies. Figure 25.29(b) is the reconstructed dark-matter image, revealing the presence of dark mass many megaparsecs from the center of the cluster. Notice the elongated struc-ture of the dark-matter distribution, reminiscent of the Virgo Supercluster and filamentary structure seen in large-scale galaxy surveys.

CONCEPT CHECK

✔ How do observations of distant quasars tell us about the structure of the universe closer to home?

Chapter Review

SUMMARY

The masses of nearby spiral galaxies can be determined by studying their rotation curves. Astronomers also use studies of binary galaxies and galaxy clusters to obtain statistical estimates of the masses of the galaxies involved. The measurements reveal the presence of large amounts of dark matter, with the fraction of dark matter growing as the scale under consideration increases. Large amounts of hot X-ray-emitting gas have been detected among the galaxies in many clusters, but not enough to account for the dark matter inferred from dynamical studies.

Researchers know of no simple evolutionary sequence that links spiral, elliptical, and irregular galaxies. Most astronomers think that large galaxies formed by the merger of smaller ones and that collisions and mergers among galaxies play very important roles in galactic evolution. A **starburst galaxy (p. 675)** may result when a galaxy has a close encounter or a collision with a neighbor. The strong tidal distortions caused by the encounter compress galactic gas, resulting in a widespread burst of star formation. Mergers between spirals most likely result in elliptical galaxies.

Quasars, active galaxies, and normal galaxies may represent an evolutionary sequence. When galaxies began to form and merge, conditions may have been suitable for the formation of large black holes at their centers, and a highly luminous quasar could have been the result. The brightest quasars consume so much fuel that their energy-emitting lifetimes must be quite short. As the fuel supply of such a quasar diminished, the quasar dimmed, and the galaxy in which it was embedded became intermittently visible as an active galaxy. At even later times, the nucleus became virtually inactive, and a normal galaxy was all that remained. Many normal galaxies have been found to contain massive central black holes, suggesting that most galaxies in clusters have the capacity for activity should they interact with a neighbor.

Galaxy clusters themselves tend to clump together into **superclusters (p. 682)**. The Virgo Cluster, the Local Group, and several other nearby clusters form the Local Supercluster. On even larger scales, galaxies and galaxy clusters are arranged on the surfaces of enormous "bubbles" of matter surrounding vast low-density regions called **voids (p. 685)**. The origin of this structure is thought to be closely related to conditions in the very earliest epochs of the universe.

Quasar spectra can be used as probes of the universe along the observer's line of sight. Some quasars have been observed to have double or multiple images, caused by gravitational lensing, in which the gravitational field of a foreground galaxy or galaxy cluster bends and focuses the light from the more distant quasar. Analysis of the images of distant galaxies, distorted by the gravitational effect of a foreground cluster, provides a means of determining the masses of galaxy clusters—including the dark matter within them—far beyond the information that the optical images of the galaxies themselves afford.

REVIEW AND DISCUSSION

1. Describe two techniques for measuring the mass of a galaxy.

2. Why do astronomers believe that galaxy clusters contain more mass than we can see?

3. Why are galaxies at great distances from us generally smaller and bluer than nearby galaxies?

4. What evidence do we have that galaxies collide with one another?

5. Describe the role of collisions in the formation and evolution of galaxies.

6. Give an example of how mergers can transform one type of galaxy into another.

7. Do you think that collisions between galaxies constitute "evolution" in the same sense as the evolution of stars?

8. Do we have any evidence that our own Galaxy has collided with other galaxies in the past?

9. What are starburst galaxies, and what do they have to do with galaxy evolution?

10. What conditions result in a head–tail radio galaxy?

11. What evidence do astronomers have for supermassive black holes in galactic nuclei?

12. Why do astronomers think that quasars represent an early stage of galaxy evolution?

13. Why do astronomers think that quasars are relatively short-lived phenomena?

14. What happened to the energy source at the center of a quasar?

15. Why does the theory of galaxy evolution suggest that there should be supermassive black holes at the centers of many normal galaxies?

16. What is a redshift survey?

17. What are voids?

18. Describe the distribution of galactic matter on very large (more than 100-Mpc) scales.

19. How can observations of distant quasars be used to probe the space between them and Earth?

20. How do astronomers "see" dark matter?

CONCEPTUAL SELF-TEST: TRUE OR FALSE/MULTIPLE CHOICE

1. More than 90 percent of matter in the universe is dark.

2. Intergalactic gas in galaxy clusters emits large amounts of energy in the form of radio waves.

3. Distant galaxies appear to be much larger than those nearby.

4. Many nearby normal galaxies may become active in the future.

5. Collisions between galaxies are rare and have little or no effect on the stars and interstellar gas in the galaxies involved.

6. The quasar stage of a galaxy ends because the central black hole swallows up all the matter around it.

7. Elliptical galaxies may be formed by mergers between spirals.

8. On the largest scales, galaxies in the universe appear to be arranged on huge sheets surrounding nearly empty voids.

9. The fact that a typical quasar would consume an entire galaxy's worth of mass in 10 billion years suggests that quasar lifetimes are relatively long.

10. The image of a distant quasar can be split into several images by the gravitational field of a foreground cluster.

11. The more massive a galaxy is, **(a)** the more distant it is; **(b)** the faster star formation in it occurs; **(c)** the larger the proportion of old stars it contains; **(d)** the faster it rotates.

12. A galaxy containing substantial amounts of dark matter will **(a)** appear darker; **(b)** spin faster; **(c)** repel other galaxies; **(d)** have more tightly wound arms.

13. According to X-ray observations, the space between galaxies in a galactic cluster is **(a)** completely devoid of matter; **(b)** very cold; **(c)** very hot; **(d)** filled with faint stars.

14. Relative to luminous stellar matter, the fraction of dark matter in clusters is **(a)** greater than the fraction in galaxies; **(b)** less than the fraction in galaxies; **(c)** the same as the fraction in galaxies; **(d)** unknown.

15. The "Hubble Deep Field" (Figure 25.11) shows a patch of sky that has the same angular size as **(a)** the thickness of a piece of string; **(b)** a dime; **(c)** a clenched fist; **(d)** a basketball held at arm's length.

16. Galaxies evolve by (a) fragmenting into smaller galaxies; (b) merging to form larger galaxies; (c) ejecting their gas and dust into intergalactic space; (d) using up all their gas and eventually becoming ellipticals.

17. According to current theories of galactic evolution, quasars occur (a) early in the evolutionary sequence; (b) near the Milky Way; (c) when elliptical galaxies merge; (d) late in the evolutionary sequence.

18. Many nearby galaxies (a) will become black holes; (b) contain quasars; (c) have radio lobes; (d) were more active in the past.

19. If light from a distant quasar did not pass through any intervening atomic hydrogen clouds, then Figure 25.24 ("Lyman-Alpha 'Forest' ") would have to be redrawn to show (a) more absorption features; (b) few absorption features; (c) a single large absorption feature; (d) more features at short wavelengths, but fewer at long wavelengths.

20. If Figure 25.26 ("Gravitational Lens") showed a more massive lensing galaxy, the quasar images would be (a) farther apart; (b) closer together; (c) fainter; (d) redder.

PROBLEMS

 Algorithmic versions of these questions are available in the Practice Problems module of the Companion Website at astro.prenhall.com/chaisson.

The number of squares preceding each problem indicates its approximate level of difficulty.

1. ■ The Andromeda Galaxy is approaching our Galaxy with a radial velocity of 120 km/s. Given the galaxies' present separation of 800 kpc, and neglecting *both* the transverse component of the velocity *and* the effect of gravity in accelerating the motion, estimate when the two galaxies will collide.

2. ■■ Based on the data in Figure 25.1, estimate the mass of the galaxy NGC 4984 inside 20 kpc.

3. ■■ Based on the data in Figure 25.1, estimate the amount of line broadening (maximum minus minimum wavelength) of the 656.3-nm Hα line observed in NGC 4984.

4. ■■ Two galaxies are orbiting each other at a distance of 500 kpc. Their orbital period is estimated to be 30 billion years. Use Kepler's law (as stated in Section 23.6) to find the total mass of the pair.

5. ■■ Use Kepler's third law (Section 23.6) to estimate the mass required to keep a galaxy moving at 750 km/s in a circular orbit of radius 2 Mpc around the center of a galaxy cluster. Given the approximations involved in calculating this mass, do you think it is a good estimate of the cluster's true mass?

6. ■■■ Assuming that the average speed of the protons in the (ionized) X-ray-emitting gas in a galaxy cluster is the same as the mean orbital speed of the galaxies in the cluster, estimate the temperature of the X-ray emitting intracluster gas in a galaxy cluster of mass 10^{15} solar masses and radius 3 Mpc. ∞ (*More Precisely 8-1*)

7. ■■ Calculate the average speed of hydrogen nuclei (protons) in a gas of temperature 20 million K. Compare your answer with the speed of a galaxy moving in a circular orbit of radius 1 Mpc around a galaxy cluster of mass 10^{14} solar masses.

8. ■ In a galaxy collision, two similar-sized galaxies pass through each other with a combined relative velocity of 1500 km/s. If each galaxy is 100 kpc across, how long does the event last?

9. ■■■ A small satellite galaxy is moving in a circular orbit around a much more massive parent and just happens to be moving exactly parallel to the line of sight as seen from Earth. The recession velocities of the satellite and the parent galaxy are measured to be 6450 km/s and 6500 km/s, respectively, and the two galaxies are separated by an angle of 0.1° in the sky. Assuming that $H_0 = 70$ km/s/Mpc, calculate the mass of the parent galaxy.

10. ■■■ Galaxies in a distant galaxy cluster are observed to have recession velocities ranging from 12,500 km/s to 13,500 km/s. The cluster's angular diameter is 55'. Use these data to estimate the cluster's mass. What assumptions do you have to make in order to obtain this estimate? Take $H_0 = 70$ km/s/Mpc.

11. ■ Assuming an energy-generation efficiency (i.e., the ratio of energy released to total mass–energy available) of 10 percent, calculate how much mass a 10^{41}-W quasar would consume if it shone for 10 billion years.

12. ■■ A quasar has a luminosity of 10^{40} W and 10^8 solar masses of fuel available. Assuming constant luminosity and 20 percent efficiency, estimate the quasar's lifetime.

13. ■ The spectrum of a quasar with a redshift of 0.20 contains two sets of absorption lines, redshifted by 0.15 and 0.155, respectively. If $H_0 = 70$ km/s/Mpc, estimate the distance between the intervening galaxies responsible for the two sets of lines.

14. ■■■ Light from a distant quasar is deflected through an angle of 3″ by an intervening lensing galaxy and is subsequently detected on Earth (Figure 25.16). If Earth, the galaxy, and the quasar are all aligned, the quasar's redshift is 3.0, and if the galaxy lies midway between Earth and the quasar, calculate the minimum distance between the light ray and the center of the galaxy.

15. ■■ Light from a distant star is deflected by 1.75″ as it grazes the Sun. ∞ (*More Precisely 22-2*) Given that the deflection angle is proportional to the mass of the gravitating body and inversely proportional to the minimum distance between the light ray and the body, estimate the mass of the galaxy described in the previous question.

 In addition to the Practice Problems module, the Companion Website at astro.prenhall.com/chaisson provides for each chapter a study guide module with multiple choice questions as well as additional annotated images, animations, and links to related Websites.

26

The Universe began in a fiery expansion some 14 billion years ago, and out of this maelstrom emerged all the energy that would later form galaxies, stars, and planets. The story of the origin and fate of all these systems—and especially of the entire universe—comprises the subject of cosmology. This image—called the Ultra Deep Field—was taken with the Advanced Camera for Surveys aboard the Hubble Space Telescope. More than a thousand galaxies are crowded into this one image. In all, astronomers estimate that the observable universe contains about 40 billion ▶ such galaxies. (NASA/ESA)

Cosmology

The Big Bang and the Fate of the Universe

Our field of view now extends for billions of parsecs into space and billions of years back in time. We have asked and answered many questions about the structure and evolution of planets, stars, and galaxies. At last we are in a position to address the central issues of the biggest puzzle of all: How big is the universe? How long has it been around, and how long will it last? What was its origin, and what will be its fate? Is the universe a one-time event, or does it recur and renew itself, in a grand cycle of birth, death, and rebirth? How and when did matter, atoms, and our Galaxy form? These are basic questions, but they are hard questions. Many cultures have asked them, in one form or another, and have developed their own cosmologies—theories about the nature, origin, and destiny of the universe—to answer them. In this and the next chapter, we will see how modern scientific cosmology addresses these important issues and what it has to tell us about the universe we inhabit. After more than 10,000 years of civilization, science may be ready to provide some insight regarding the origin of all things.

LEARNING GOALS

Studying this chapter will enable you to

1 State the cosmological principle, and explain both its significance and its observational underpinnings.

2 Explain how the age of the universe is determined, and discuss the uncertainties involved.

3 Summarize the leading evolutionary models of the universe.

4 Explain the relationship between the future of the universe and the overall geometry of space.

5 Discuss the factors that determine whether the universe will expand forever.

6 Explain what observations of the distant universe reveal about the composition of the universe.

7 Describe the cosmic microwave background radiation, and explain its importance to our understanding of cosmology.

Visit astro.prenhall.com/chaisson for additional annotated images, animations, and links to related sites for this chapter.

26.1 The Universe on the Largest Scales

The universe shows structure on every scale we have examined so far. Subatomic particles form nuclei and atoms. Atoms form planets and stars. Stars form star clusters and galaxies. Galaxies form galaxy clusters, superclusters, and even larger structures—voids, filaments, and sheets that stretch across the sky. ∞ (Sec. 25.5) From the protons in a nucleus to the galaxies in the Great Wall, we can trace a hierarchy of "clustering" of matter from the very smallest to the very largest scales. It is natural to ask, "Does the clustering ever end? Is there some scale on which the universe can be regarded as more or less smooth and featureless?" Perhaps surprisingly, given the trend we have just described, most astronomers think the answer is yes.

The End of Structure

We saw in Chapter 25 how astronomers use *redshift surveys* to explore the large-scale distribution of galaxies in space, combining position on the sky with distances inferred from Hubble's law to construct three-dimensional maps of the universe on truly "cosmic" scales. ∞ (Sec. 25.5) Figure 26.1 is a map similar in concept to those shown in Chapter 25, but based on data from the most extensive redshift survey to date: the Sloan Digital Sky Survey. ∞ (*Discovery 25-1*) The figure plots the positions of some 67,000 galaxies lying in a 120° wide "slice" of the sky within a few degrees of the celestial equator. The distances shown assume a Hubble constant of $H_0 = 70$ km/s/Mpc. ∞ (Sec. 24.3)

The region of the sky shown in Figure 26.1 extends out to a distance of approximately 1000 Mpc. This reach is comparable to the extent of Figure 25.23, but because the Sloan survey includes much fainter galaxies, the map in Figure 26.1 contains nearly three times more points than the earlier figure, making structure somewhat easier to discern, particularly at large distances. The extended "filament" of galaxies near the center of the wedge, some 300 Mpc from Earth, is called the *Sloan Great Wall*. Discovered in 2003 and measuring some 250 Mpc long by 50 Mpc thick, it is currently the largest known structure in the universe.

Plots such as this contain huge amounts of information about the structure and evolution of the universe. Yet, although they cover wide areas of the sky and enormous volumes of space, the studies on which they are based are still relatively "local," in the sense that they span only about 10 percent of the distance to the farthest quasars (which lie over 8000 Mpc from Earth). ∞ (Sec. 25.3) The main obstacle to extending these wide-angle surveys to much greater distances is the sheer observational effort involved in measuring the redshifts of all the galaxies within larger and larger volumes of space.

An alternative approach is to narrow the field of view to only a few small patches of the sky, but to study extremely faint (and hence very distant) galaxies within those patches. The volume surveyed then becomes a long, thin "pencil beam" extending deep into space rather than a wide swath through the local universe. *Discovery 26-1* discusses some very deep surveys that have generated a great deal of excitement among astronomers. As illustrated in

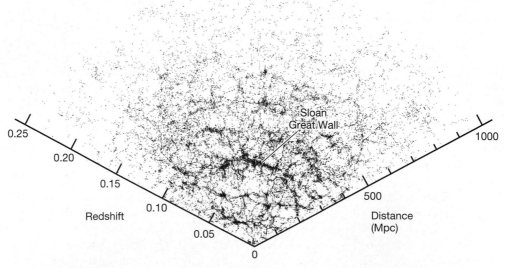

▲ **FIGURE 26.1 Galaxy Survey** This map of the universe is drawn using data from the Sloan Digital Sky Survey. ∞ (*Discovery 25-1*) The map shows the locations of 66,976 galaxies lying within 12 degrees of the celestial equator and extends out to a distance of almost 1000 Mpc (or a redshift of about 0.25). The largest known structure in the universe, the "Sloan Great Wall," is marked, stretching nearly 300 Mpc across the center of the frame. (*Astrophysical Research Consortium/SDSS Collaboration*)

Figure 26.2, along the line of sight clusters and walls show up as "spikes" in the distribution—groups of galaxies with similar redshifts, separated by broad empty regions of space (the voids).

The data from both kinds of survey seem to agree that the largest known structures in the local universe are "only" 200–300 Mpc across. Rich superclusters measure tens of megaparsecs across, while the largest voids are perhaps 100 Mpc in diameter. Most walls and filaments are less than 100 Mpc in length, and even the largest structures—the Great Walls mentioned previously—can be explained statistically as chance superpositions of smaller structures. No larger voids, superclusters, or walls of galaxies are seen. Studies of Lyman-alpha forests in quasar spectra lead to generally similar conclusions. ∞ (Sec. 25.5) In short, there is *no* evidence for structure in the universe on scales greater than about 300 Mpc.

We will turn to the *origin* of large-scale cosmic structure in Chapter 27. In the current chapter, however, we focus on the *absence* of structure on the very largest scales to frame our discussion of the future of the universe.

The Cosmological Principle

LEARNING GOAL 1 The results of the large-scale studies just mentioned strongly suggest that the universe is **homogeneous** (the same everywhere) on scales greater than a few hundred megaparsecs. In other words, if we took a huge cube—300 Mpc on a side, say—and placed it anywhere in the universe, its overall contents would look much the same no matter where it was centered. Some of the galaxies it contained would be clustered and clumped into fairly large structures and some would not, and we would see numerous walls and voids, but the total numbers of these objects would not vary much as the cube was moved from place to place. In this sense, the universe appears *smooth* on the largest scales.

The universe also appears to be **isotropic** (the same in all directions) on these large scales. Excluding directions that are obscured by our Galaxy, we count roughly the same number of galaxies per square degree in any patch of the sky we choose to observe, provided that we look deep (far) enough that local inhomogeneities don't distort our sample. In other words, any deep pencil-beam survey of the sky should count about the same number of galaxies, regardless of which patch of the sky is chosen.

In the science of **cosmology**—the study of the structure and evolution of the entire universe—researchers generally assume that the universe is homogeneous and isotropic on sufficiently large scales. No one knows whether these assumptions are precisely correct, but we can at least say that they are consistent with current observations, and they provide helpful guidance to our studies of the cosmos. In this chapter, we simply assume that they hold. The twin assumptions of homogeneity and isotropy are known as the **cosmological principle**. Note that this principle also includes the important assumption made

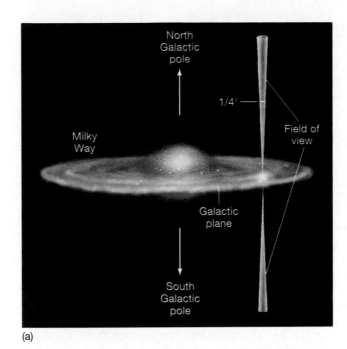

(a)

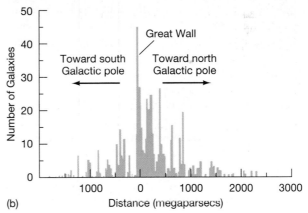

(b)

▲ **FIGURE 26.2 Pencil-Beam Survey** The results of a deep "pencil-beam" survey of two small portions of the sky in opposite directions from Earth, perpendicular to the Galactic plane (a), are plotted in (b). The graph shows the number of galaxies found at different distances from us, out to about 2000 Mpc. Wherever we look on the sky, the distinctive "picket fence" pattern highlights voids and sheets of galaxies on scales of 100 to 200 Mpc, but gives no indication of any larger structures.

throughout this book (and indeed throughout astronomy) that the laws of physics are the same everywhere.

The cosmological principle has far-reaching implications. For example, it implies that there can be no edge to the universe, because that would violate the assumption of homogeneity. Furthermore, it implies that there is no *center*, because that would mean that the universe would not be the same in all directions from any noncentral point, a violation of the assumption of isotropy. This is the familiar Copernican principle expanded to truly cosmic proportions—not only that are we not central to the universe, but that *no one* can be central, because *the universe has no center!* ∞ (Sec. 2.3)

26.2 The Expanding Universe

Every time you go outside at night and notice that the sky is dark, you are making a profound cosmological observation. Here's why.

Olbers's Paradox

Let's assume that, in addition to being homogeneous and isotropic, the universe is infinite in spatial extent and unchanging in time—precisely the view of the universe that prevailed until the early part of the 20th century. On average, then, the universe is uniformly populated with galaxies filled with stars. In that case, when you look up at the night sky, your line of sight must *eventually* encounter a star, as illustrated in Figure 26.3. The star may lie at an enormous distance in some remote galaxy, but the laws of probability dictate that, in an infinite universe, sooner or later any line drawn outward from Earth will run into a bright stellar surface.

Of course, faraway stars appear fainter than those nearby, because of the inverse-square law. ∞ (Sec. 17.3) However, they are also much more numerous, because the number of stars we see at any given distance in fact increases as the *square* of the distance. (Just consider the area of a sphere of increasing radius.) Thus, the diminishing brightnesses of distant stars are exactly balanced by their increasing numbers, and stars at all distances contribute equally to the total amount of light received on Earth. This fact has a dramatic implication: No matter where you look, the sky should be as bright as the surface of a star; in other words, the entire night sky should be as brilliant as the surface of the Sun! The obvious difference between this prediction and the actual appearance of the night sky is known as **Olbers's paradox**, after the 19th-century German astronomer Heinrich Olbers, who popularized the idea.

So why is it dark at night? Given that the universe appears to be homogeneous and isotropic, one (or both) of the other two assumptions must be false. Either the universe is finite in extent, or it evolves in time. In fact, the answer involves a little of each and is intimately tied to the behavior of the universe on the largest scales.

The Birth of the Universe

In Chapter 24, we saw that all the galaxies in the universe are rushing away from us in a manner described by Hubble's law,

recession velocity = H_0 × distance,

where we take Hubble's constant H_0 to be 70 km/s/Mpc. ∞ (Sec. 24.3) Up to now, we have used this relation as a

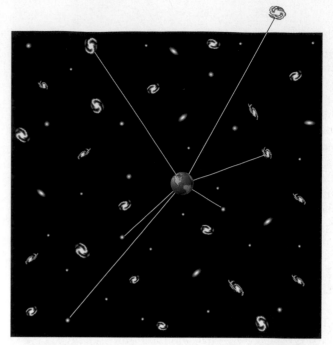

▲ **FIGURE 26.3 Olbers's Paradox** If the universe were homogeneous, isotropic, infinite in extent, and unchanging, then any line of sight from Earth should eventually run into a star, and the entire night sky should be bright. This obvious contradiction of the facts is known as Olbers's paradox.

convenient means of determining the distances to galaxies and quasars, but it is much more than that.

Assuming for the moment that all velocities have remained constant in time, we can ask a simple question: How long has it taken for any given galaxy to reach its present distance from us? The answer follows from Hubble's law. The time taken is simply the distance traveled divided by the velocity, so

$$\text{time} = \frac{\text{distance}}{\text{velocity}}$$
$$= \frac{\text{distance}}{H_0 \times \text{distance}} \quad \text{(using Hubble's law for the velocity)}$$
$$= \frac{1}{H_0}.$$

For H_0 = 70 km/s/Mpc, this time is about 14 billion years. Notice that it is *independent* of the distance: Galaxies twice as far away are moving twice as fast, so the time they took to cross the intervening distance is the same.

Hubble's law therefore implies that, at some time in the past—14 billion years ago, according to the foregoing simple calculation—*all* the galaxies in the universe lay right on top of one another. In fact, astronomers believe that *everything* in the universe—matter and radiation alike—was confined to a single point at that instant. Then the point expanded, its newly acquired volume flying apart at high speeds. The present locations and velocities of the galaxies are a direct consequence of that primordial blast.

This gargantuan explosion, involving everything in the universe, is known as the **Big Bang**. It marked the beginning of the universe.

Thus, by measuring Hubble's constant, we can estimate the age of the universe to be $1/H_0 \approx 14$ billion years. The range of possible error in this age is considerable, both because Hubble's constant is not known precisely and because the assumption that galaxies moved at constant speed in the past is not a good one. We will refine our estimate in a moment, but regardless of the details, the critical fact here is that the age of the universe is *finite*.

The Big Bang provides the resolution of Olbers's paradox. Whether the universe is actually finite or infinite in extent is irrelevant, at least as far as the appearance of the night sky is concerned. We see only a finite part of it—the region lying within roughly 14 billion light-years of us. What lies beyond is unknown—its light has not yet had time to reach us.

Note that, even though it appears to place us at the center of the expansion, Hubble's law does *not* violate the cosmological principle in any way. To see this, consider Figure 26.4, which shows how observers in five hypothetical galaxies perceive the motion of their neighbors. For simplicity, the galaxies are taken to be equally spaced, 100 Mpc apart, and they are separating in accordance with Hubble's law with $H_0 = 70$ km/s/Mpc, as seen by an observer in the middle galaxy, number 3. The first pair of numbers beneath each galaxy represents its distance and recessional velocity as measured by that observer. For definiteness, let's take galaxy number 3 to be the Milky Way and the observer to be an astronomer on Earth.

Now consider how the expansion looks from the point of view of the observer in galaxy 2. Galaxy 4, for example, is moving with velocity 7000 km/s to the right relative to galaxy 3, and galaxy 3 in turn is moving at 7000 km/s to the right as seen by observer 2. Therefore, galaxy 4 is moving at a velocity of 14,000 km/s to the right as seen by the observer in galaxy 2. But the distance between the two galaxies is 200 Mpc, so the Hubble constant measured by the observer on galaxy 2 is 14,000 km/s / 200 Mpc = 70 km/s/Mpc, the same as the Hubble constant measured by the observer on galaxy 3. The distances and velocities that would be measured by observer 2 are noted in the second row. You can verify for yourself that the ratio of recession velocity to distance is the same for all galaxies.

Similarly, the measurements made by an observer on galaxy 1 are noted in the third row. Again, the ratio of velocity to distance is the same. The conclusion is clear: *Each observer sees an overall expansion described by Hubble's law, and the constant of proportionality—Hubble's constant—is the same in all cases.* Far from singling out any one observer as central, Hubble's law is in fact the *only* expansion law possible if the cosmological principle holds.

Where Was the Big Bang?

Now we know *when* the Big Bang occurred. Is there any way of telling *where*? We think that the universe is the same everywhere, yet we have just seen that the observed recession of the galaxies described by Hubble's law implies that all the galaxies expanded from a point at some time in the past. Wasn't that point, then, different from the rest of

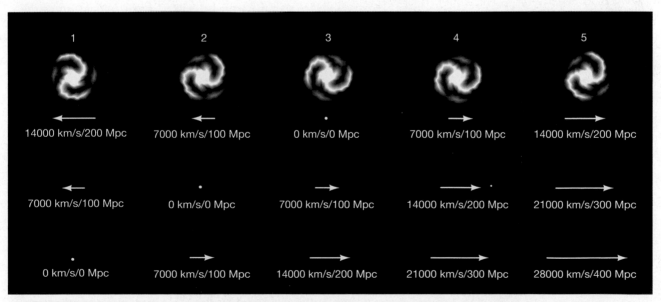

▲ **FIGURE 26.4 Hubble Expansion** Hubble's law is the same, regardless of who makes the measurements. The top numbers are the distances and recessional velocities as seen by an observer on the middle of five galaxies, galaxy 3. The bottom two sets of numbers are from the points of view of observers on galaxies 2 and 1, respectively. In all cases, Hubble's law holds: The ratio of the observed recession velocity to the distance is the same.

the universe, violating the assumption of homogeneity expressed in the cosmological principle? The answer is a definite *no*!

To understand why there is no "center" to the expansion, we must make a great leap in our perception of the universe. If we were to imagine the Big Bang as simply an enormous explosion that spewed matter out into space, ultimately to form the galaxies we see, then the foregoing reasoning would be quite correct—there would be a center and an edge, and the cosmological principle would not apply. But the Big Bang was *not* an explosion in an otherwise featureless, empty universe. The only way that we can have Hubble's law hold *and* retain the cosmological princi-

ple is to realize that the Big Bang involved the entire universe—not just the matter and radiation within it, but the universe *itself*. In other words, the galaxies are not flying apart into the rest of the universe. The universe itself is expanding. Like raisins in a loaf of raisin bread that move apart as the bread expands in an oven, the galaxies are just along for the ride.

Let's consider again some of our earlier statements in light of this new perspective. We now recognize that Hubble's law describes the expansion of the universe itself. Although galaxies have some small-scale, individual random motions, on average they are *not* moving with respect to the fabric of space—any such overall motion would pick

DISCOVERY 26-1

Stunning Views of Deep Space

As we probe deep into space, we are looking far back in time. Telescopes are time machines, and astronomers are historians. Now, thanks to NASA's orbiting observatories, we have deep,

(STScI)

R I V U X G

detailed views of extragalactic space. The first image shows the *Hubble Deep Field* (*HDF*) in approximately true color, that is, approximately as the human eye would see. ∞ (Sec. 26.1) It was assembled from nearly 300 exposures taken during more than 100 orbits over the course of 10 consecutive days, for a total exposure time of about 100 hours. All the data were combined to make this one spectacular image. (Figure 25.11 shows a smaller view.) The field is in a region of the sky near the Big Dipper that happens to lie in a direction perpendicular to the congested plane of the Milky Way. The superimposed numbers are redshifts of various galaxies as measured by the Keck Observatory in Hawaii. Note the enormous spread in distances (see Table 24.1 and Figure 26.2) evident in even the small sample of redshifts shown.

Virtually every blob of light in this image is a galaxy. (A few clear exceptions, such as the bright star with the artificial "spikes" at the bottom left of the image, are local stars in our own Milky Way.) In all, approximately 1600 galaxies of all types can be counted in this one image. Hundreds of these galaxies are so faint that they have never been seen before. The image extends to 30th magnitude, making the HDF among the most sensitive photographs ever taken. ∞ (Sec. 17.2) The image is "deep" in the sense that it allows us to see very faint objects. However, the word *deep* should not be construed to mean that this image goes beyond the realm of ground-based telescopes. *Hubble* does not see significantly farther into the universe than any other large telescope, but it does see more sensitively; hence, it is able to pick up light from very dim objects and

out a "special" direction in space and violate the assumption of isotropy. On the contrary, the portion of the galaxies' motion that makes up the Hubble flow is really an expansion of space itself. The expanding universe remains homogeneous at all times. There is no "empty space" beyond the galaxies into which they rush. At the time of the Big Bang, the galaxies did not reside at a point located at some well-defined place within the universe. Rather, the *entire universe* was a point. That point was in no way different from the rest of the universe; that point *was* the universe. Therefore, there was no one point where the Big Bang "happened"—because the Big Bang involved the entire universe, it happened *everywhere* at once.

To illustrate these ideas, imagine an ordinary balloon with coins taped to its surface, as shown in Figure 26.5. (Better yet, do the experiment yourself!) The coins represent galaxies, and the two-dimensional surface of the balloon represents the "fabric" of our three-dimensional universe. The cosmological principle applies to the balloon because every point on the balloon looks pretty much the same as every other. Imagine yourself as a resident of one of the three dark-colored coin "galaxies" in the leftmost frame, and note your position relative to your neighbors. As the balloon inflates (i.e., as the universe expands), the other galaxies recede from you; more distant galaxies recede more rapidly. (Notice, incidentally, that the coins

often to resolve them better than ground-based instruments can.

Much good science has been mined from this single, very long exposure, and many new projects have been triggered by it. Indeed, as soon as *HDF* was unveiled, pressure mounted to take longer, deeper exposures elsewhere in the sky. The second image is known as the *Hubble Ultra Deep Field*, or *UDF* (see also the chapter-opening image). Like *HDF*, it shows a selection of cosmic objects distributed along a narrow corridor of space approximately 3000 Mpc, or 10 billion light-years, in length. It represents the deepest visible-light portrait of the universe ever made, showing in this one image nearly 10,000 galaxies. In all, data were acquired during a million-second (300 hour) exposure over the course of 11 days of observing during late 2003 when the orbiting telescope was pointed toward the constellation Fornax, just south of Orion.

Both images show many galaxies of many ages. Disentangling young, distant galaxies from old, nearby ones is tricky, but, as a general rule of thumb (and perhaps a little counterintuitively, given that we have gone to some lengths to stress that distant objects are characterized by large redshifts), the blue objects are probably the farthest away (since we see their most luminous stars in their blazing ultraviolet youth), whereas red objects tend to be closer (we see them by the light of their old, reddened stars). Bright spiral galaxies are most easily seen, while ellipticals appear mostly as reddish blobs. Other galaxies are irregularly shaped, probably the result of collisions or close encounters with neighboring galaxies, seen in this image when the universe was younger and denser.

Of particular interest are the small, faint smudges of blue light in the images, the faraway protogalaxies that gave rise to the galaxies we see today. Some of the faintest galaxies in these images existed when the universe was a mere billion years old, eons before the birth of our solar system. However, only spectroscopic observations can tell if these dim smudges are truly distant. The big ground-based telescopes in Hawaii, in Chile, and on Kitt Peak are far better suited than *Hubble* is for identifying redshifts, and this work is underway.

These deep fields cover only tiny parts of the entire sky, as the images shown here are just a few arc minutes across. To map the entire sky to this depth would take *Hubble* nearly a

million years! One way to put these images into a larger perspective is to think of them as core samples, much like those a geologist might take on Earth. By examining the dirt and rocks in their samples, geologists try to reconstruct the history of events that occurred as Earth evolved. Likewise, by studying the galaxies in these deep fields, astronomers seek to understand better the history of the universe.

Despite their narrow fields of view, astronomers consider these deep fields to be representative of the typical spread of galaxies in space. Statistically, the two images seem virtually identical. This is another way of stating the cosmological principle—the universe on the largest scales looks pretty much the same in all directions. Extrapolating the contents of these images over the entire sky, astronomers then estimate that the total number of galaxies in the observable universe is approximately 40 billion.

(STScI)

R I V U X G

themselves do *not* expand along with the balloon, any more than people, planets, stars, or galaxies—all of which are held together by their own internal forces—expand along with the universe.) ∞ (Sec. 24.5)

Regardless of which galaxy you chose to consider, you would see all the other galaxies receding from you. Nothing is special or peculiar about the fact that all the galaxies are receding from you. Such is the cosmological principle: No observer anywhere in the universe has a privileged position. There is no center to the expansion and no position that can be identified as the location from which the universal expansion began. Everyone sees an overall expansion described by Hubble's law, with the same value of Hubble's constant in all cases.

Now imagine letting the balloon deflate. This corresponds to running the universe backward from the present time to the Big Bang. *All* the galaxies (coins) would arrive at the same place at the same time—at the instant the balloon reached zero size. But there is no one point on the balloon that could be said to be *the* place where that occurred. The entire balloon expanded from a point, just as the Big Bang encompassed the entire universe and expanded from a point.

This analogy has its shortcomings. The main difficulty with it is that we see the balloon, which, in our illustration, we imagined as two dimensional, expanding into the third dimension of space. This might suggest that the three-dimensional universe is expanding "into" some fourth spatial dimension. It is not, so far as we know. At the very least, if higher spatial dimensions are involved, they are not relevant to our theory of the universe.

The Cosmological Redshift

This view of the expanding universe requires us to reinterpret the cosmological redshift. ∞ (Sec. 24.5) Previously, we explained the redshift of galaxies as a Doppler shift—a consequence of their motion relative to us. However, we have just argued that the galaxies are *not* in fact moving with respect to the universe, in which case the Doppler in-

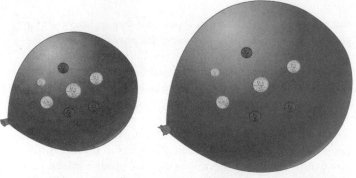

▲ **FIGURE 26.5 Receding Galaxies** Coins taped to the surface of a spherical balloon recede from one another as the balloon inflates (left to right). Similarly, galaxies recede from one another as the universe expands. As the coins recede, the distance between any two of them increases, and the rate of increase of this distance is proportional to the distance between the two coins. Thus, the balloon expands according to Hubble's law.

terpretation is incorrect. The true explanation is that, as a photon moves through space, its wavelength is influenced by the expansion of the universe. In a sense, we can think of the photon as being attached to the expanding fabric of space, so its wavelength expands along with the universe, as illustrated in Figure 26.6. Although it is common practice in astronomy to refer to the cosmological redshift in terms of recessional velocity, bear in mind that, strictly speaking, that is not the right thing to do. The cosmological redshift is a consequence of the changing size of the universe—it is *not* related to velocity at all.

The redshift of a photon measures the amount by which the universe has expanded since that photon was emitted. For example, when we measure the light from a quasar and find that it has a redshift of 5, it means that the observed wavelength is 6 times (1 plus the redshift) greater than the wavelength at the time of emission, and this in turn means that the light was emitted at a time when the universe was just one-sixth its present size (and we are observing the quasar as it was at that time). ∞ (*More Precisely 24-1*) In general, the larger a photon's redshift, the smaller the universe was at the time the photon was emitted, so the longer ago that emission occurred. Because the

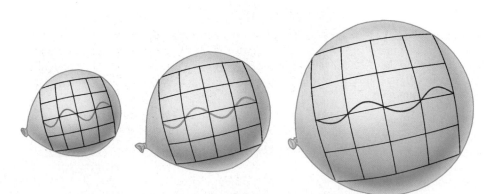

◄ **FIGURE 26.6 Cosmological Redshift** As the universe expands, photons of radiation are stretched in wavelength, giving rise to the cosmological redshift. In this case, as the baseline in the diagram stretches, the radiation shifts from the short-wavelength blue region of the spectrum to the longer wavelength red region. ∞ (Sec. 3.1)

universe expands with time and redshift is related to that expansion, cosmologists routinely use redshift as a convenient means of expressing time.

Relativity and the Universe

These concepts are difficult to grasp. The notion of the entire universe expanding from a point—with *nothing*, not even space and time, outside—takes some getting used to. Nevertheless, this picture of the universe lies at the heart of modern cosmology. The description of the universe itself (not just its contents) as a dynamic, evolving object is far beyond the capabilities of Newtonian mechanics, which we have used almost everywhere throughout the book. ∞ (Sec. 2.7) Instead, the more powerful techniques of Einstein's *general relativity*, with its built-in notions of warped space and dynamic spacetime, are needed.

We encountered general relativity in Chapter 22 when we discussed the strange properties of black holes. ∞ (Sec. 22.6) We can loosely summarize its description of the universe by saying that the presence of matter or energy causes a curvature of space (more correctly, *spacetime*) and that the curved trajectories of freely falling particles within warped space are what Newton thought of as orbits under the influence of gravity. The amount of curvature depends on the amount of matter present, and the orbits of particles in turn depend on the curvature. In a homogeneous universe, the overall curvature of space must be uniform.

Must we keep this difficult notion of a warped, expanding, homogeneous, universe constantly in the forefront of our thoughts if we are to comprehend the evolution of the cosmos? Perhaps surprisingly, given that relativity is the only theory that properly describes the large-scale behavior of the universe, the answer is no. Although general relativity predicts some curious consequences for the overall *geometry* of space on large scales, as we will see in Section 26.5, much of the *dynamics* of the universe can be understood with concepts that would have been thoroughly familiar to Newton.

Consider two neighboring points A and B in the expanding universe, as pictured in Figure 26.7. From the perspective of point A, every other point, including B, is rushing away from it, in accordance with Hubble's law. How does the distance between A and B change in time? Does it just keep growing, or does its rate of increase eventually slow down and stop? From a Newtonian perspective, we might expect that the overall gravitational pull of the universe would tend to slow the expansion, just as Earth's gravity tends to slow the upward motion of an object projected from the surface. Even though we have just argued that Newtonian physics is inadequate to describe the large-scale dynamics of the universe, a surprising prediction of general relativity is that the Newtonian description of the universe's motion—with some important adjustments, to be discussed shortly—can still be used! We can, in fact, discuss the expansion of the universe in more or less Newtonian terms, although we need general relativity to justify our doing so.

CONCEPT CHECK

✔ Why does Hubble's law imply a Big Bang?

26.3 The Fate of the Cosmos

LEARNING GOAL 3 Will the expansion of the universe continue forever? This fundamental question about the fate of the universe has been at the heart of cosmology since Hubble's law was first discovered. Until the late 1990s, the prevailing view among cosmologists was that the answer would most likely be found by determining the extent to which gravity would slow, and perhaps ultimately reverse, the current expansion. However, it now appears that the answer is more subtle—and perhaps a lot more profound in its implications—than was hitherto thought.

Critical Density

LEARNING GOAL 3 Let's begin with an analogy. Assume for the moment that gravity is the only force affecting large-scale motion in the universe, and consider a rocket ship launched from the surface of a planet. Until relatively recently, this scenario was much more than just an analogy—this basic picture and its implications represented the conventional wisdom among cosmologists. However, as we will see in Section 26.5, new observations have forced fundamental changes in astronomers' view of the universe. Nevertheless, the simplified view we now present is a convenient starting point, as it allows us to define some basic ideas and terminology.

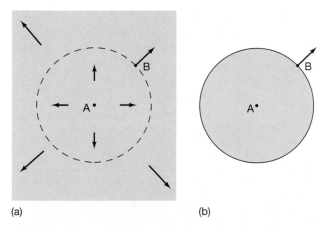

(a)   (b)

▲ **FIGURE 26.7 Expanding Universe** The relative motion of any two points in the expanding universe (a) can be addressed as a problem in Newtonian mechanics, even though Einstein's theory of general relativity is needed to explain why it is correct to do so. If the rest of the universe is ignored—as in part (b)—then the Newtonian calculation of B's motion relative to A gives the same answer as a calculation based on general relativity.

What are the likely outcomes of the rocket ship's motion? According to Newtonian mechanics, there are just two basic possibilities, depending on the launch speed of the ship relative to the escape speed of the planet. ∞ (Sec. 2.7) If the launch speed is high enough, it will exceed the planet's escape speed, and the ship will never return to the surface. The speed will diminish because of the planet's gravitational pull, but it will never reach zero. The spacecraft leaves the planet on an unbound trajectory, as illustrated in Figure 26.8(a). Alternatively, if the launch speed is lower than the escape speed, the ship will reach a maximum distance from the planet and then fall back to the surface. Its bound trajectory is shown in Figure 26.8(b).

Similar reasoning applies to the expansion of the universe. Consider Figure 26.7 again, but now imagine that A and B are galaxies at some known distance from each other, with their present relative velocity given by Hubble's law. Since we know we can apply familiar Newtonian concepts to the problem, the same two basic possibilities exist for these galaxies as for our spacecraft: The distance between them can increase forever, or it can increase for a while and then start to decrease. What's more, the cosmological principle says that, whatever the outcome for A and B, it must be the same for *any* two galaxies—in other words, the same statement applies to the universe *as a whole*. Thus, as illustrated in Figure 26.9, the universe has only two options: It can continue to expand forever, or the present expansion will someday stop and turn around into a contraction. The two curves in the figure are drawn so that they pass through the same point at the present time. Both are possible descriptions of the universe, given its current size and expansion rate.

What determines which of the two possibilities will actually occur? In the case of a rocket ship of fixed launch speed (analogous to a universe with a given expansion rate), the *mass* of the planet (for given radius) determines whether or not escape will occur—a more massive planet has a higher escape speed, making it less likely that the rocket can escape. For the universe, the corresponding factor is the *density* of the cosmos. A high-density universe contains enough mass to stop the expansion and eventually cause a collapse. A low-density universe, conversely, will expand forever.

The dividing line between these outcomes—the density corresponding to a universe in which *gravity acting alone* would be just sufficient to halt the present expansion—is called the universe's **critical density**. For $H_0 = 70$ km/s/Mpc, the critical density is about 9×10^{-27} kg/m^3. That's an extraordinarily low density—just five hydrogen atoms per cubic meter, a volume the size of a typical household closet. In more "cosmological" terms, it corresponds to about 0.1 Milky Way Galaxy (including the dark matter) per cubic megaparsec.

Two Futures

The two possibilities just presented represent radically different futures for our universe. If the cosmos emerged from the Big Bang with sufficiently high density, then it

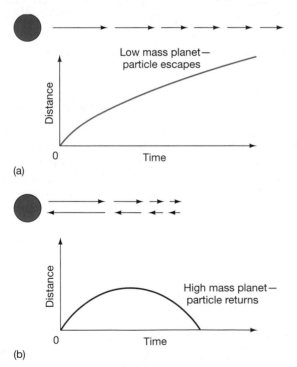

(a)

(b)

▲ **FIGURE 26.8 Escape Speed** (a) A spacecraft (arrow) leaving a planet (blue ball) with a speed greater than the planet's escape speed follows an unbound trajectory. The graph shows the distance between the ship and the planet as a function of time. (b) If the launch speed is less than the escape speed, the ship eventually drops back to the planet. Its distance from the planet first rises and then falls.

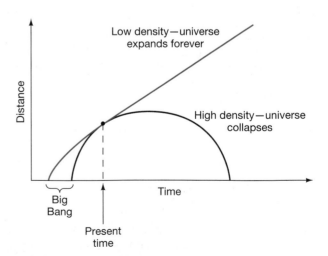

▲ **FIGURE 26.9 Model Universes** Distance between two galaxies as a function of time in each of the two basic universes discussed in the text: a low-density universe that expands forever and a high-density cosmos that collapses. The point where the two curves touch represents the present time.

contains enough matter to halt its own expansion, and the recession of the galaxies will eventually stop. At some time in the future, astronomers everywhere—on any planet within any galaxy—will announce that the radiation received from nearby galaxies is no longer redshifted. (The light from *distant* galaxies will still be redshifted, however, because we will see them as they were in the past, at a time when the universe was still expanding.) The bulk motion of the universe, and of the galaxies within, will be stilled—at least momentarily.

The expansion may stop, but the pull of gravity will not. The universe will begin to contract. Nearby galaxies will begin to show blueshifts, and both the density and the temperature of the universe will start to rise as matter collapses back onto itself. As illustrated in Figure 26.10(a), the universe will collapse to a point, requiring just as much time to fall back as it took to rise. First galaxies and then stars will collide with increasing frequency and violence as the available space diminishes and the entire universe shrinks toward a superdense, superhot singularity much like the one from which it originated. The cosmos will ultimately—billions of years from now—experience a "heat death," in which all matter and life are destined to be incinerated. Some astronomers call the final collapse of this high-density universe the "Big Crunch." Cosmologists do not know what will happen to the universe if it ever reaches the point of collapse. The laws of physics as we presently understand them are simply inadequate to describe those extreme conditions.

A quite different fate awaits a low-density universe whose gravity is too weak to halt the present expansion. As illustrated in Figure 26.10(b), such a universe will expand forever, the galaxies continually receding, their radiation steadily weakening with increasing distance. In time, an observer on Earth will see no galaxies in the sky beyond the Local Group (which is not itself expanding). Even with the most powerful telescope, the rest of the observable universe will appear dark, the distant galaxies too faint to be seen. Eventually, the Milky Way and the Local Group, too, will peter out as their fuel supply is consumed. This universe will ultimately experience a "cold death": All radiation, matter, and life are eventually destined to freeze.

How long might the "cold death" of the universe take? Astronomers estimate that our Galaxy probably contains enough gas to keep forming stars for several tens of billions of years, and the majority of stars (the low-mass red dwarfs) can shine for hundreds of billions of years or more. ∞ (Sec. 17.8) Thus, we can expect our Galaxy (and our neighbor Andromeda) to shine on—albeit feebly—for another trillion years or so.

We will see in a moment that the separation between never-ending expansion and cosmic collapse is not quite as straightforward as the foregoing simple reasoning would suggest. Several independent lines of evidence now indicate that gravity is *not* the only influence on the dynamics of the universe on large scales (Section 26.5). As a result, while the "futures" just described are still the only two possibilities for the long-term evolution of the universe, the distinction between them turns out to be more than just a matter of density alone. Nevertheless, the density of the universe—or, more precisely, the *ratio of the total density to the critical value*—is a vitally important quantity in cosmology.

CONCEPT CHECK

✔ What are the two basic possibilities for the future expansion of the universe?

26.4 The Geometry of Space

LEARNING GOAL 4 We have reverted to the familiar notions of Newtonian mechanics and gravity, and moved away from the more correct concept of warped spacetime, because speaking in Newtonian terms makes our discussion of the evolution of the universe easier to understand. However, general relativity makes some predictions about the universe that do not have a simple description within Newton's theory.

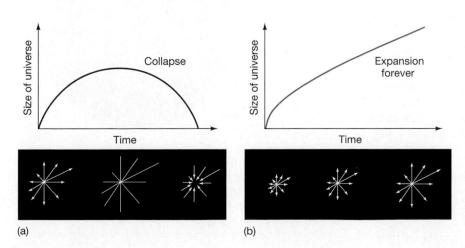

(a)   (b)

◀ **FIGURE 26.10 Two Futures** (a) A high-density universe has a beginning, an end, and a finite lifetime. The lower frames illustrate its evolution, from explosion, to maximum size, to collapse. (b) A low-density universe expands forever from its explosive beginning, with galaxies getting farther and farther apart as time passes.

Foremost among these non-Newtonian predictions is the fact that space (again, actually spacetime) is *curved*, and the degree of curvature is determined by the *total density* of the cosmos. ∞ (Sec. 22.6) But there is an important clarification to be made here. In our earlier discussion, we did not specify exactly what we meant by "density." Our discussion implicitly focused on the density of matter, but according to the theory of relativity, mass (m) and energy (E) are equivalent, connected by Einstein's famous relation $E = mc^2$. ∞ (*More Precisely 22-1*) What, then, actually contributes to the warping of spacetime? General relativity's answer is clear: Both matter *and* energy must be taken into account, with energy properly "converted" into matter units via division by the square of the speed of light. [That is, an energy of 1 joule is counted as its mass equivalent of $1 \text{ J}/(3 \times 10^8 \text{ m/s})^2 = 1.1 \times 10^{-17}$ kg—not much, but it adds up!] In this way, the density of the universe includes not just the atoms and molecules that make up the familiar "normal" matter around us, but also the invisible dark matter that dominates the masses of galaxies and galaxy clusters, as well as *everything* that carries energy—photons, relativistic neutrinos, gravity waves, and anything else we can think of.

In a homogeneous universe, the curvature (on sufficiently large scales) must be the same everywhere, so there are really only three distinct possibilities for the large-scale geometry of space. (For more information on the different types of geometry involved, see *More Precisely 26-1*.) General relativity tells us that the geometry of the universe depends only on the ratio of the density of the universe to the critical density (defined in the previous section). As just noted, for $H_0 = 70$ km/s/Mpc, the critical density is 9×10^{-27} kg/m³. Cosmologists conventionally call the ratio of the universe's actual density to the critical value the *cosmic density parameter* and denote it by the symbol Ω_0 ("omega nought"). In terms of this quantity, then, a universe with density equal to the critical value has $\Omega_0 = 1$, a "low-density" cosmos has Ω_0 less than 1, and a "high-density" universe has Ω_0 greater than 1.

In a high-density universe (Ω_0 greater than 1), space is curved so much that it bends back on itself and "closes off," making this universe *finite* in size. Such a universe is known as a **closed universe**. It is difficult to visualize a three-dimensional volume uniformly arching back on itself in this way, but the two-dimensional version is well known: It is just the surface of a sphere, like that of the balloon we discussed earlier. Figure 26.5, then, is the two-dimensional likeness of a three-dimensional closed universe. Like the surface of a sphere, a closed universe has no boundary, yet is finite in extent.* One remarkable property of a closed universe is illustrated in Figure 26.11: Just as a traveler on the surface of a sphere can keep moving forward in a straight line and eventually return to her starting point, a flashlight beam shone in some direction in space might eventually traverse the entire universe and return from the opposite direction!

The surface of a sphere curves, loosely speaking, "in the same direction," no matter which way we move from a given point. A sphere is said to have *positive curvature*. However, if the average density of the universe is below the critical value, the surface curves like a saddle, in which case it has *negative curvature*. Most people have a good idea of what a saddle looks like—it curves "up" in one direction and "down" in another—but no one has ever seen a uniformly negatively curved surface, for the simple reason that it cannot be constructed in three-dimensional Euclidean space! It is just "too big" to fit. A low-density, saddle-curved universe is infinite in extent and is usually called an **open universe**.

The intermediate case, in which the density is precisely equal to the critical density (i.e., $\Omega_0 = 1$), is the easiest to visualize. This universe, called a **critical universe**, has no curvature. It is said to be "flat" and is infinite in extent. In this case, and *only* in this case, the geometry of space on large scales is precisely the familiar Euclidean geometry taught in high schools. Apart from its overall expansion, this is basically the universe that Newton knew.

Euclidean geometry—the geometry of flat space—is familiar to most of us because it is a good description of space in the vicinity of Earth. It is the geometry of every-

---

*Notice that, for the sphere analogy to work, we must imagine ourselves as two-dimensional "flatlanders" who cannot visualize or experience in any way the third dimension perpendicular to the sphere's surface. Flatlanders and their light rays are confined to the sphere's surface, just as we are confined to the three-dimensional volume of our universe.

▲ **FIGURE 26.11 Einstein's Curve Ball** In a closed universe, a beam of light launched in one direction might return someday from the opposite direction after circling the universe, just as motion in a "straight line" on Earth's surface will eventually encircle the globe.

day experience. Does this mean that the universe is flat, which would in turn mean that it has exactly the critical density? Not necessarily: Just as a flat street map is a good representation of a city, even though we know Earth is really a sphere, Euclidean geometry is a good description of space within the solar system, or even the Galaxy, because the curvature of the universe is negligible on scales smaller than about 1000 Mpc. Only on the very largest scales would the geometric effects we have just discussed become evident.

MORE PRECISELY 26-1

Curved Space

Euclidean geometry is the geometry of flat space—the geometry taught in high schools everywhere. Set forth by one of the most famous of the ancient Greek mathematicians, Euclid, who lived around 300 B.C., it is the geometry of everyday experience. Houses are usually built with flat floors. Writing tablets and blackboards are also flat. We work easily with flat, straight objects, because the straight line is the shortest distance between any two points.

When we construct houses or any other straight-walled buildings on the surface of Earth, the other basic axioms of Euclid's geometry also apply: Parallel lines never meet, even when extended to infinity; the angles of any triangle always sum to 180°; the circumference of a circle equals π times the diameter of the circle. (See the accompanying figure.) If these axioms did not hold, walls and roof would never meet to form a house!

In reality, though, the geometry of Earth's surface is not really flat; it is curved. We live on the surface of a sphere, and on that *surface*, Euclidean geometry breaks down. Instead, the rules for the surface of a sphere are those of *Riemannian geometry*, named after the 19th-century German mathematician Georg Friedrich Riemann. There are no parallel "straight" lines on a sphere. The analog of a straight line on a sphere's surface is a "great circle"—the arc formed when a plane passing through the center of the sphere intersects the surface. Any two such lines must eventually intersect. The sum of a triangle's angles, when drawn on the surface of a sphere, exceeds 180°—in the 90°–90°–90° triangle shown in the accompanying figure, the sum is actually 270°—and the circumference of a circle is less than π times the circle's diameter.

We see that the curved surface of a sphere, governed by the spherical geometry of Riemann, differs greatly from the flat-space geometry of Euclid. The two are approximately the same only if we confine ourselves to a small patch on the surface. If the patch is small enough compared with the sphere's radius, the surface looks "flat" nearby, and Euclidean geometry is approximately valid. This is why we can draw a usable map of our home, our city, and even our state, on a flat sheet of paper, but an accurate map of the entire Earth must be drawn on a globe.

When we work with larger parts of Earth, we must abandon Euclidean geometry. World navigators are fully aware of this. Aircraft do not fly along what might appear on most maps as a straight-line path from one point to another. Instead, they follow a great circle on Earth's surface. On the curved surface of a sphere, such a path is always the shortest distance between two points. For example, as illustrated in the figure, a flight from Los Angeles to London does not proceed directly across the United States and the Atlantic Ocean, as you might expect from looking at a flat map. Instead, it goes far to the north, over Canada and Greenland, above the Arctic Circle, finally coming in over Scotland for a landing at London. This is the great-circle route—the shortest path between the two cities, as you can easily see if you inspect a globe.

The "positively curved" space of Riemann is not the only possible departure from flat space. Another is the "negatively curved" space first studied by Nikolai Ivanovich Lobachevsky, a 19th-century Russian mathematician. In this geometry, there are an *infinite* number of lines through any given point that are parallel to another line, the sum of the angles of a triangle is *less* than 180° (see the first figure), and the circumference of a circle is *greater* than π times its diameter. This type of space is described by the surface of a curved saddle, rather than a flat plane or a curved sphere. It is a hard geometry to visualize!

Most of the local realm of the *three*-dimensional universe (including the solar system, the neighboring stars, and even our Milky Way Galaxy) is correctly described by Euclidean geometry. If the currently favored cosmology described in the text turns out to be correct, then the whole universe is, too!

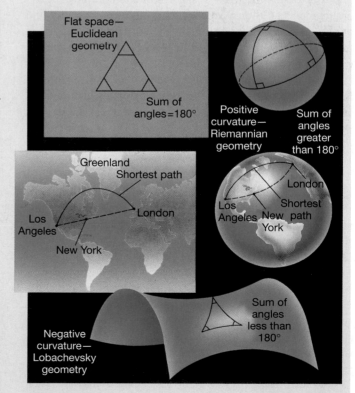

CONCEPT CHECK

✔ How is the curvature of space related to the density of the universe?

26.5 Will the Universe Expand Forever?

Is there any way for us to determine which of the futures we have described actually applies to our universe (i.e., apart from waiting to find out)? Will the universe end as a small, dense point much like that from which it began? Or will it expand forever? And can we hope to measure the geometry of the vast cosmos we inhabit? Finding answers to these questions has been the dream of astronomers for decades. We are fortunate to live at a time when astronomers can subject these questions to intensive observational tests and come up with definite answers—even though they aren't what most cosmologists expected! Let's begin by looking at the density of the universe (or, equivalently, the cosmic density parameter Ω_0).

The Density of the Universe

How might we determine the density of the universe? On the face of it, it would seem simple: Just measure the total mass of the galaxies residing within some large parcel of space, calculate the volume of that space, and then divide the mass by the volume to compute the average density. When astronomers do this, they usually find a little less than 10^{-28} kg/m^3 in the form of luminous matter. Largely independent of whether the chosen region contains many scattered galaxies or only a few rich galaxy clusters, the resulting density is about the same, within a factor of two or three. Galaxy counts thus yield a value of Ω_0 of only a few percent. If that measure were correct, and galaxies were all that existed, then we would live in a low-density open universe destined to expand forever.

But there is a catch. We have noted (Chapters 23 and 25) that most of the matter in the universe is *dark*—it exists in the form of invisible material that has been detected only through its gravitational effect in galaxies and galaxy clusters. ∞ (Secs. 23.6, 25.1) Currently, we do not know what the dark matter is, but we *do* know that it is there. Galaxies may contain as much as 10 times more dark matter than luminous material, and the figure for galaxy clusters is even higher—perhaps as much as 95 percent of the total mass in clusters is invisible. Even though we cannot see it, dark matter contributes to the density of the universe and plays its part in opposing the cosmic expansion. Including all the dark matter that is known to exist in galaxies and galaxy clusters increases the value of Ω_0 to 0.2 or 0.3.

Unfortunately, the distribution of dark matter on larger scales is not very well known. We can infer its presence in galaxies and galaxy clusters, but we are largely ignorant of its extent in superclusters, voids, or other larger structures. Still, there are indications that dark matter accounts for an even greater fraction of the mass on large scales than it does in galaxy clusters. Observations of gravitational lensing by galaxy clusters suggest that dark matter may be considerably more extensive than is indicated by the motions of galaxies within the clusters. ∞ (Sec. 25.5) One of the few mass estimates for an object much larger than a supercluster comes from optical and infrared observations of the overall motion of galaxies (including the Local Group) within the Local Supercluster. The measured velocities suggest the presence of a nearby huge accumulation of mass known as the *Great Attractor*, with a total mass of about 10^{17} solar masses and a size of 100–150 Mpc. The average density of this structure may be quite close to the critical value.

However, it seems that, when averaged over the entire universe, even the densities of objects like the Great Attractor don't raise the overall cosmic density by much. In short, there doesn't seem to be much additional dark matter "tucked away" on very large scales. Most cosmologists agree that the density of matter (luminous plus dark) in the universe is not much greater than about 30 percent of the critical value—*not* enough to halt the universe's current expansion.

Cosmic Acceleration

Determining the mass density of the universe is an example of a *local* measurement that provides an estimate of Ω_0. But the result we obtain depends on just how local our measurement is—as we have just seen, there are many uncertainties in the result, especially on large scales. In an attempt to get around this problem, astronomers have devised alternative methods that rely instead on *global* measurements, covering much larger regions of the observable universe. In principle, such global tests should indicate the universe's overall density, not just its value in our cosmic neighborhood.

One such global method is based on observations of Type I (carbon-detonation) supernovae. ∞ (Sec. 21.3) Recall that these objects are very bright and have a remarkably narrow spread in luminosities, making them particularly useful as standard candles. ∞ (Sec. 24.2) They can be used as probes of the universe because, by measuring their distances (*without* using Hubble's law) and their redshifts, we can determine the rate of cosmic expansion in the distant past. Here's how the method works:

Suppose the universe is decelerating, as we would expect if gravity were slowing its expansion. Then, because the expansion rate is decreasing, objects at great distances—that is, objects that emitted their radiation long ago—should appear to be receding *faster* than Hubble's law predicts. Figure 26.12(a) illustrates this concept. If the universal expansion were constant in time, recessional velocity and distance would be related by the black line. (The line is not quite straight, because it takes the expansion of the universe properly into account in computing the distance.) ∞ (*More Precisely 24-1*) In a decelerating universe, the velocities of distant objects should lie *above* the black curve, and the deviation from that curve should be greater for a denser universe, in which gravity has been more effective at slowing the expansion.

How does theory compare with reality? In the late 1990s, two groups of astronomers announced the results of independent, systematic surveys of distant supernovae. Some of these supernovae are shown in Figure 26.12(b); the data are marked on Figure 26.12(a). Far from clarifying the picture of cosmic deceleration, these findings indicate that the expansion of the universe is not slowing, but actually accelerating! According to the supernova data, galaxies at large distances are receding *less* rapidly than Hubble's law would predict. The deviations from the decelerating curves appear small in the figure, but they are statistically very significant, and both groups report similar findings.

These observations are *inconsistent* with the standard Big Bang model just described and have sparked a major revision of our view of the cosmos. However, the measurements are difficult, and the results depend quite sensitively on just how "standard" the supernovae luminosities really are; thus, some astronomers initially questioned the accuracy of the method. In particular, if supernovae at great distances (i.e., long ago) were for some reason slightly less luminous than those nearby, then we would think that these distant supernovae were farther away than they actually are, and the error would appear as a deviation to the right of the black curve in Figure 26.12(a)—in other words, as an acceleration in the cosmic expansion rate.

Not surprisingly, since so much hangs on this measurement, the reliability of the supernova measurement technique has been the subject of intense scrutiny by cosmologists. However, no convincing argument against the

method has yet been put forward, so there is no reason to believe that we are somehow being "fooled" by nature. As far as we can tell, the measurements are good, and the acceleration is real.

What could cause an overall acceleration of the universe? Frankly, cosmologists don't know, although several possibilities have been suggested. Whatever it is, the mysterious cosmic field causing the universe to accelerate is neither matter nor radiation. Although it carries energy, it exerts an overall *repulsive* effect on the universe, causing empty space to expand. It has come to be known simply as **dark energy**, and it is perhaps *the* leading puzzle in astronomy today.

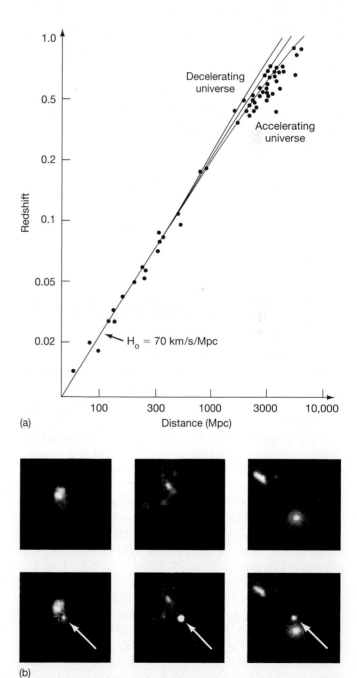

▶ **FIGURE 26.12 Accelerating Universe** (a) Observations of distant supernovae allow astronomers to measure changes in the rate of expansion of the universe. In a decelerating universe (purple and red curves), redshifts of distant objects are greater than would be predicted from Hubble's law (black curve). The reverse is true for an accelerating universe. The points showing observations of some 50 supernovae strongly suggest that the cosmic expansion is accelerating. The vertical scale shows redshift; for small velocities, redshift is just velocity divided by the speed of light. ∞ (*More Precisely 24-1*) (b) The bottom frames show three supernovae (marked by arrows) that exploded in distant galaxies when the Universe was nearly half its current age. The top frames are wider views of areas around these supernovae, which were originally discovered in 1997 during a ground-based survey with the Canada–France–Hawaii Telescope on Mauna Kea. (*P. Garnavich/Harvard-Smithsonian Center for Astrophysics/NASA*)

One leading dark-energy candidate is an additional "vacuum pressure" force associated with empty space and effective only on very large scales. Known simply as the **cosmological constant**, it has a long and checkered history. It was first proposed by Einstein as a way to force his new theory of general relativity into "predicting" a static universe, but was subsequently dropped from Einstein's equations following Hubble's discovery that the universe is not static, but instead is expanding (see *Discovery 26-2*). Since the 1990s, the cosmological constant has arisen again, to become a staple of astronomers' models of the universe. Note, however, that although models that take this force into account can fit the observational data, as described in the next section, astronomers have *no* clear physical interpretation of what the force actually is. It is neither required nor explained by any known law of physics.

The repulsive effect of the cosmological constant is proportional to the size of the universe, so it increases as the universe expands. Thus, it was negligible at early times, but today, given the magnitude of the observed acceleration, it is the major factor controlling the cosmic expansion. Furthermore, since the effect of gravity *weakens* as the expansion proceeds, it follows that, once the cosmological constant begins to dominate, gravity can never catch up, and the universe will continue to accelerate at an ever-increasing pace. Thus, despite the considerable uncertainty as to the nature of dark energy, we can at least make one definite statement: By opposing the attractive force of gravity, the repulsive effect of the dark energy strengthens our earlier conclusion that the universe will expand forever.

CONCEPT CHECK

✔ Why do astronomers think the universe will expand forever?

26.6 Dark Energy and Cosmology

As we proceed through the remainder of this chapter and the next, it is worth bearing in mind that the Big Bang *is* a scientific theory and, like any other, must continually be challenged and scrutinized. ∞ (Sec. 1.2) The Big Bang theory makes detailed, testable predictions about the state and history of the cosmos and must change—or be replaced—if these predictions are found to be at odds with observations. The supernova observations just described are a case in point.

Even though the supernova observations and their interpretation have so far withstood intense scrutiny, the idea of an accelerating universe driven by some completely unknown field called dark energy probably would not have gained such rapid and widespread acceptance among cosmologists were it not strongly supported by several other pieces of evidence. In this section, we discuss how the exis-

tence of dark matter fits in with various observations of the universe and even helps to resolve some other long-standing riddles. Every independent piece of evidence, and every old puzzle solved, provides further support not just for the idea of dark energy, but also for the entire Big Bang theory of the universe.

Cosmic Composition

In addition to measuring density and acceleration, astronomers have several other means of estimating the "cosmological parameters" that describe the large-scale properties of our universe.

Theoretical studies of the early universe (to be discussed in more detail in Chapter 27) strongly suggest that the geometry of the universe should be precisely flat—that is, that the total density of the cosmos should exactly equal the critical value. This idea first became widespread in the 1980s, and for many years there seemed to be a major discrepancy between it and observations that clearly showed a cosmic matter density of less than 30 percent of the critical value, even taking the dark matter into account. Dark energy resolves that conflict by providing another form in which the "extra" density can exist, although not all astronomers are happy at the price of this resolution, which has introduced yet another unknown component into the cosmic mix!

Recent detailed measurements of the radiation field known to fill the entire cosmos (see Section 26.7 and Chapter 27) strongly support the theoretical prediction that $\Omega_0 = 1$ and are also consistent with the dark energy inferred from the supernova studies. Further independent corroboration comes from careful analyses of galaxy surveys such as those discussed in Section 26.1, which allow astronomers to measure the growth of large-scale structure in the universe. Simply put, the more mass there is in the universe, the easier it is for clusters, superclusters, walls, and voids to grow as gravity gathers matter into larger and larger clumps. Higher density implies more rapid formation of structure—or, equivalently, *less* structure in the past, given the structure we see around us today. Thus, structure measurements constrain the value of Ω_0.

Remarkably, all the approaches just described yield consistent results! As of mid-2004, the consensus among cosmologists is that the universe is of precisely critical density, $\Omega_0 = 1$, but that this density is made up of both matter (mostly dark) *and* dark energy (converted into mass units as discussed earlier in Section 26.4). Radiation contributes negligibly to the total (see Section 27.1). Thus, the current best estimate, based on all the available data, is that matter accounts for 27 percent of the total and dark energy for the remaining 73 percent. Furthermore, the structure observations appear to favor the cosmological constant over the other candidates for dark energy. These combined assumptions underlie Table 24.1 and are used consistently throughout this book.

DISCOVERY 26-2

Einstein and the Cosmological Constant

Even the greatest minds are fallible. The first scientist to apply general relativity to the universe was, not surprisingly, the theory's inventor, Albert Einstein. When he derived and solved the equations describing the behavior of the universe, Einstein discovered that they predicted a universe that evolved in time. But in 1917, neither he nor anyone else knew about the expansion of the universe, as described by Hubble's law, which would not be discovered for another 10 years. ∞ (Sec. 24.3) At the time, Einstein, like most scientists, believed that the universe was static—that is, unchanging and everlasting. The discovery that there was no static solution of his equations seemed to Einstein to be a near-fatal flaw in his new theory.

To bring the theory into line with the beliefs, Einstein tinkered with the equations, introducing a "fudge factor" describing a hypothetical repulsive force operating on large scales in the universe. This factor is now known as the cosmological constant. As illustrated in the accompanying figure, which shows the effect of introducing such a factor into the equations describing the expansion of a critical-density universe, the cosmological constant allows many solutions of Einstein's equations. One of these solutions describes a "coasting" universe, whose radius remains constant for an indefinite period. Einstein took this to be the static universe he expected.

Instead of predicting an evolving cosmos, which would have been one of general relativity's greatest triumphs, Einstein yielded to a preconceived notion of the way the universe "should be," unsupported by observational evidence. Later, when the expansion of the universe was discovered and Einstein's equations—without the fudge factor—were found to describe it perfectly, he declared that the cosmological constant was the biggest blunder of his scientific career.

Scientists are reluctant to introduce unknown quantities into their equations purely to make the results "come out right." Einstein introduced the cosmological constant to fix what he thought was a serious problem with his equations, but he discarded it immediately once he realized that no problem actually existed. As a result, the cosmological constant fell out of favor among astronomers for many years.

In the 1980s, the concept made something of a comeback with the realization by physicists that the very early universe may have gone through a phase when its evolution was determined by a "cosmological constant" of sorts (see Section 27.4), and this idea is now firmly entrenched in many cosmologists' models of the universe. Today, as discussed in the text, the cosmological constant has apparently been completely rehabilitated and identified as a leading candidate for the "dark energy" whose existence is inferred from studies of the universe on very large scales. The green curves in the figure show how the inclu-

sion of a suitable dark-energy term in Einstein's equations can cause the expansion of the universe to accelerate, instead of slowing down as it would if only gravity were involved.

For many researchers—Einstein included—the main problem with the cosmological constant was (and still is) the fact that we have no clear explanation for either its existence or its present value. The leading theories of the structure of matter do in fact predict repulsive forces of this sort, but these forces generally operate only under extreme conditions, and, in any case, their "natural" scale is vastly greater (by something like a factor of 10^{120}!) than anything consistent with cosmological observations.

An additional problem is that the present value of the repulsive force is comparable to the attractive force of gravity opposing further expansion. Why is that a problem? Simply because, when we calculate the evolution of a universe containing a cosmological constant consistent with current observations (Figure 26.14), we find that this state of affairs was not true in the early universe (when galaxies were forming, say), nor will it be true in 10 or 20 billion years' time. In other words, presupposing such a cosmological constant seems to suggest that we live at a special time in the history of the universe—a conclusion viewed with considerable suspicion by astronomers who grew up with the Copernican principle as their guide.

Attempts are underway to construct theories of dark-energy fields that preserve the character of the cosmological constant, yet also account for its value in some natural way, perhaps somehow coupling it to the density of matter. Before we make too many sweeping statements about the role of the cosmological constant in cosmology, we should probably remember the experience of its inventor and bear in mind that—at least for now—its physical meaning remains completely unknown.

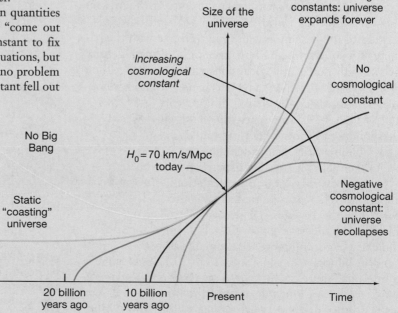

▲ **FIGURE 26.13 Geometry of the Universe** As best we can tell, the universe on the largest scales is geometrically flat—governed by the same familiar Euclidean geometry taught in high schools.

Note that such a universe will expand forever and, the heavy machinery of general relativity and curved space-time notwithstanding, is perfectly flat (Figure 26.13)—an irony that would no doubt have amused Newton!

The Age of the Universe

We have at least one other independent, noncosmological, way of testing the preceding important conclusion. In Section 26.2, when we estimated the age of the universe from the accepted value of Hubble's constant, we made the assumption that the expansion speed of the cosmos was constant in the past. However, as we have just seen, this is a considerable oversimplification. Gravity tends to slow the universe's expansion, while dark energy acts to accelerate it, and the actual expansion of the universe is the result of the competition between the two. In the absence of a cosmological constant, the universe would have expanded faster in the past than it does today, so the assumption of a constant expansion rate leads to an overestimate of the universe's age—such a universe is younger than the 14 billion years calculated earlier. Conversely, the repulsive effect of dark energy tends to increase the age of the cosmos.

Figure 26.14 illustrates these points. It is similar to Figure 26.9, except that we have added two extra lines, one corresponding to a constant expansion rate at the present value—a completely *empty* 14-billion-year-old universe—the other to the best-fit accelerating universe with the parameters just described. The age of a critical-density universe with no cosmological constant is about 9 billion years. A low-density open universe (again with no cosmological constant) is older than 9 billion years, but still less than 14 billion years old. The age corresponding to the accelerating universe is 13.7 billion years, coincidentally very close to the value for constant expansion.

How does this kind of calculation compare with an age estimated by other means? On the basis of the theory of stellar evolution, the oldest globular clusters formed about 12 billion years ago, and most are estimated to be between 10 and 12 billion years old. (Secs. 19.6, 20.5) This range is indicated in Figure 26.14. These ancient star clusters are thought to have formed at around the same time as our Galaxy, so they date the epoch of galaxy formation. More important, they can't be older than the universe! The fig-

ure shows that globular cluster ages are consistent with a 14-billion-year-old cosmos and even allow a couple of billion years for galaxies to form and grow, as discussed in Chapter 25. ∞ (Sec. 25.3) Note also that the cluster ages are *not* consistent with a critical-density universe without dark energy. This independent check of a key prediction is an important piece of evidence supporting the modern version of the Big Bang theory.

Thus, for $H_0 = 70$ km/s/Mpc, our current best guess of the history of the universe places the Big Bang at 14 billion years ago. The first quasars appeared about 13 billion years ago (at a redshift of 6), the peak quasar epoch (redshifts 2–3) occurred during the next 1 billion years, and the oldest known stars in our Galaxy formed during the 2 billion years after that. Even though astronomers do not presently understand the nature of dark energy, the good agreement between so many separate lines of reasoning has convinced many that the dark-matter, dark-energy Big Bang theory just described is the correct description of the universe. But astronomers aren't ready to relax just yet: The history of this subject suggests that there may be a few more unexpected twists and turns in the road before the details are finally resolved.

CONCEPT CHECK

✔ How does the age of the universe enter the debate over the existence of dark energy?

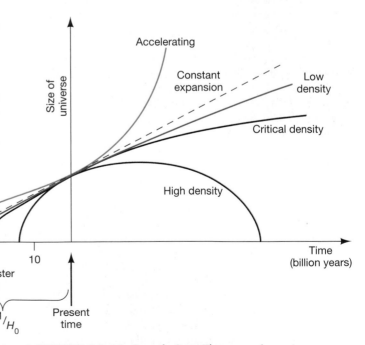

▲ **FIGURE 26.14 Cosmic Age** The age of a universe without a cosmological constant is always less than $1/H_0$ and decreases for larger values of the present-day density. The existence of a repulsive cosmological constant increases the age of the cosmos.

26.7 The Cosmic Microwave Background

LEARNING GOAL 7 Looking out into space is equivalent to looking back into time. ∞ *(More Precisely 24-1)* But how far back in time can we probe? Is there any way to study the universe beyond the most distant quasar? How close can we come to perceiving the edge of time—the very origin of the universe—directly?

A partial answer to these questions was discovered by accident in 1964, during an experiment designed to improve the U.S. telephone system. As part of a project to identify and eliminate interference in planned satellite communications, Arno Penzias and Robert Wilson, two scientists at Bell Telephone Laboratories in New Jersey, were studying the Milky Way's emission at microwave (radio) wavelengths, using the horn-shaped antenna shown in Figure 26.15. In their data, they noticed a bothersome background "hiss" that just would not go away—a little like the background static on an AM radio station. Regardless of where and when they pointed their antenna, the hiss persisted. Never diminishing or intensifying, the weak signal was detectable at any time of the day, any day of the year, apparently filling all space.

What was the source of this radio noise? And why did it appear to come uniformly from all directions, unchanging in time? Unaware that they had detected a signal of great cosmological significance, Penzias and Wilson sought many different origins for the excess emission, including atmospheric storms, interference from the ground, short circuits of equipment—even pigeon droppings inside the antenna! Eventually, after conversations with colleagues at Bell Labs

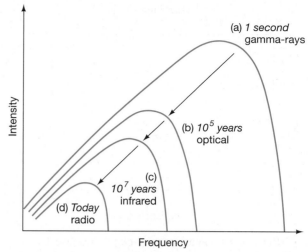

▲ **FIGURE 26.16** Cosmic Blackbody Curves Theoretically derived blackbody curves for the entire universe (a) 1 second after the Big Bang, (b) 100,000 years after the Big Bang, (c) 10 million years after the Big Bang, and (d) at present, approximately 10 billion years after the Big Bang

and theorists at nearby Princeton University, the two experimentalists realized that the origin of the mysterious static was nothing less than the fiery creation of the universe itself. The radio hiss that Penzias and Wilson detected is now known as the **cosmic microwave background**. Their discovery won them the 1978 Nobel prize in physics.

In fact, researchers had predicted the existence and general properties of the microwave background well before its discovery. As early as the 1940s, physicists had realized that, in addition to being extremely dense, the early universe must also have been very hot, and shortly after the Big Bang the universe must have been filled with extremely high-energy thermal radiation—gamma rays of very short wavelength. Researchers at Princeton had extended these ideas, reasoning that the frequency of this primordial radiation would have been redshifted (simply by cosmic expansion) from gamma ray, to X ray, to ultraviolet, and eventually all the way into the radio range of the electromagnetic spectrum as the universe expanded and cooled (Figure 26.16). ∞ (Sec. 3.4) By the present time, they argued, this redshifted "fossil remnant" of the primeval fireball should have a temperature of no more than a few tens of kelvins, peaking in the microwave part of the spectrum. The Princeton group was in the process of constructing a microwave antenna to search for this radiation when Penzias and Wilson announced their discovery.

The Princeton researchers confirmed the existence of the microwave background and estimated its temperature at about 3 K. However, because of atmospheric absorption, this part of the electromagnetic spectrum happens to be difficult to observe from the ground, and it was 25 years until astronomers could demonstrate conclusively that the radiation was described by a blackbody curve. In 1989, the *Cosmic Background Explorer*

▲ **FIGURE 26.15** Microwave Background Discoverers This "sugar-scoop" antenna was built to communicate with Earth-orbiting satellites, but was used by Robert Wilson (left) and Arno Penzias to discover the 2.7-K cosmic background radiation. *(Bell Labs)*

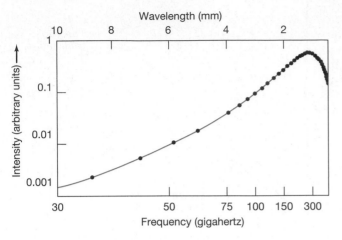

▲ **FIGURE 26.17 Microwave Background Spectrum** The intensity of the cosmic background radiation, as measured by the *COBE* satellite, agrees very well with theory. The curve is the best fit to the data, corresponding to a temperature of 2.735 K. The experimental errors in this remarkably accurate observation are smaller than the dots representing the data points.

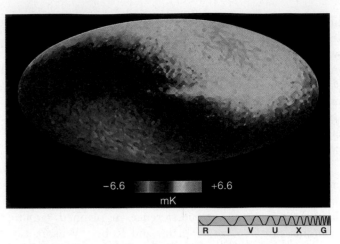

▲ **FIGURE 26.18 Microwave Sky** A *COBE* map of the microwave sky reveals that the microwave background appears a little hotter in the direction of the constellation Leo and a little cooler in the opposite direction. The maximum temperature deviation from the average is about 0.0034 K, corresponding to a velocity of 400 km/s in the direction of Leo. (*NASA*)

(*COBE*) satellite measured the intensity of the microwave background at wavelengths straddling the peak of the curve, from a half millimeter up to about 10 cm. The results are shown in Figure 26.17. The solid line is the blackbody curve that best fits the *COBE* data. The near-perfect fit corresponds to a universal temperature of about 2.7 K.

Figure 26.18 shows a *COBE* map of the microwave background temperature over the entire sky. The blue regions are hotter than average, by about 0.0034 K, the red regions cooler by the same amount. This temperature range is not an inherent property of the microwave background, however. Rather, it is a consequence of Earth's *motion* through space. If we were precisely at rest with respect to the universal expansion (like the coin taped to the surface of the expanding balloon in Figure 26.5), then we would see the microwave background as almost perfectly isotropic, as illustrated in Figure 26.19(a). However, if we are moving with respect to that frame of reference, as in Figure 26.19(b), then the radiation from in front of us should be slightly blueshifted by our motion, while that from behind should be redshifted. Thus, to a moving observer, the microwave background should appear a little hotter than average in front and slightly cooler behind.

The data indicate that Earth's velocity is about 380 km/s in the approximate direction of the constellation Leo. Once the effects of this motion are corrected for, the cosmic microwave background is found to be strikingly isotropic. Its intensity is virtually constant (in fact, to about one part in 10^5) from one direction on the sky to another, lending strong support to one of the key assumptions underlying the cosmological principle.

When we observe the microwave background, we are looking almost all the way to the very beginning of the

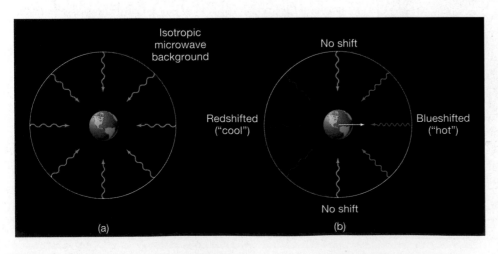

◀ **FIGURE 26.19 Earth's Motion through the Cosmos** (a) To an observer at rest with respect to the expanding universe, the microwave background appears isotropic. (b) A moving observer measures "hot" blueshifted radiation in one direction (the direction of motion) and "cool" redshifted radiation in the opposite direction.

universe. The photons that we receive as these radio waves today have not interacted with matter since the universe was a mere 400,000 years old, when, according to our models, it was less than a thousandth of its present size. To probe further, back to the Big Bang itself, requires that we enter the world of nuclear and particle physics. The Big Bang was the biggest and the most powerful particle accelerator of all! In the next chapter, we will see how studies of conditions in the primeval fireball aid us in understanding the present-day structure and future evolution of the universe in which we live.

CONCEPT CHECK

✔ When was the cosmic microwave background formed?

Chapter Review

SUMMARY

Redshift surveys reveal that, on scales larger than a few hundred megaparsecs, the universe appears roughly **homogeneous** (the same everywhere, **p. 695**) and **isotropic** (the same in all directions, **p. 695**). In **cosmology (p. 695)**—the study of the universe as a whole—researchers usually assume that the universe is homogeneous and isotropic. This assumption is known as the **cosmological principle (p. 695)** and implies that the universe cannot have a center or an edge. If the universe were homogeneous, isotropic, infinite, and unchanging, the night sky would be bright because any line of sight would eventually intercept a star. The fact that the night sky is dark is called **Olbers's paradox (p. 696)**. Its resolution lies in the fact that, regardless of whether the universe is finite or infinite, we see only a finite part of it from Earth—the region from which light has had time to reach us since the universe began.

Tracing the observed motions of galaxies back in time implies that some 14 billion years ago the universe consisted of a single point that then began to expand rapidly, at the time of the **Big Bang (p. 697)**. However, the galaxies are not flying apart into the rest of an otherwise empty universe; rather, space itself is expanding. The Big Bang did not happen at any particular location in space, because the entire universe was compressed to a point at that instant—the Big Bang happened everywhere at once. The cosmological redshift occurs as a photon's wavelength is "stretched" by cosmic expansion. The extent of the observed redshift is a direct measure of the expansion of the universe since the photon was emitted.

Although Einstein's theory of general relativity is needed for a full description of the cosmos, much of the dynamics of the universe can be understood in large part with simpler, Newtonian concepts. There are only two possible outcomes to the current expansion: Either the universe will expand forever, or it will recollapse to a point. The **critical density (p. 702)** is the density of matter needed to overcome the present expansion and cause the universe to collapse.

General relativity provides a description of the geometry of the universe on the largest scales. The curvature of spacetime is determined by the total density of the universe, including that of matter, radiation, and dark energy. The curvature in a high- (greater-than-critical) density universe is sufficiently large that the universe "bends back" on itself and is finite in extent, somewhat like the surface of a sphere. Such a universe is said to be a **closed universe (p. 704)**. A low-density **open universe (p. 704)** is infinite in extent and has a "saddle-shaped" geometry. The **critical universe (p. 704)** has a density precisely equal to the critical value and is spatially flat.

Luminous matter by itself contributes only a few percent of the critical density. When dark matter in galaxies and clusters is taken into account, the figure rises to 20 or 30 percent. The fraction of dark matter on larger scales is uncertain, but may be even greater than is found in galaxy clusters. Most astronomers believe that the present mass density of the universe is no more than about 30 percent of the critical value. Observations of distant supernovae indicate that the expansion of the universe is accelerating, possibly driven by the effects of **dark energy (p. 707)**. One candidate for this dark energy is the **cosmological constant (p. 708)**, a repulsive force that may exist throughout all space. Its physical nature is unknown. Other, independent observations are consistent with the idea that the universe is flat—that is, of exactly critical density—with (mostly dark) matter making up 27 percent of the total and dark energy making up the rest. Such a universe will expand forever.

For $H_0 = 70$ km/s/Mpc, the age of a critical-density universe without dark energy is about 9 billion years. This age estimate conflicts with the 10–12-billion-year ages of globular clusters derived from studies of stellar evolution. The inclusion of dark energy increases the age of the universe to 14 billion years, consistent with the cluster ages.

The **cosmic microwave background (p. 711)** is an isotropic blackbody radiation field that fills the entire universe. Its present temperature is about 3 K. The existence of the microwave background is direct evidence that the universe expanded from a hot, dense state. As the universe has expanded, the initially high energy radiation has been redshifted to lower and lower temperatures.

REVIEW AND DISCUSSION

1. What evidence do we have that there is no structure in the universe on very large scales? How large is "very large"?

2. What is the cosmological principle?

3. What is Olbers's paradox? How is it resolved?

4. Explain how an accurate measure of Hubble's constant can lead to an estimate of the age of the universe.

5. We appear to be at the center of the Hubble flow. Why doesn't this observation violate the cosmological principle?

6. Why isn't it correct to say that the expansion of the universe involves galaxies flying outward into empty space?

7. Where did the Big Bang occur?

8. How does the cosmological redshift relate to the expansion of the universe?

9. What properties of the universe determine whether it will or will not expand forever?

10. What will be the ultimate fate of the universe if it does expand forever?

11. Is there enough luminous matter to halt the current cosmic expansion?

12. Is there enough dark matter to halt the current cosmic expansion?

13. What do observations of distant supernovae tell us about the expansion of the universe?

14. What is the cosmological constant, and what does it have to do with the future of the universe?

15. Why are measurements of globular cluster ages important to cosmology?

16. What is the significance of the cosmic microwave background?

17. Why does the temperature of the microwave background fall as the universe expands?

18. How can we measure Earth's motion with respect to the universe?

19. Many cultures throughout history have developed their own cosmologies. Do you think the modern scientific cosmology is more likely to endure than any other? Why or why not?

20. Do you think it constitutes good science to explain the universe mainly in terms of dark matter and dark energy, neither of which is known or understood?

CONCEPTUAL SELF-TEST: TRUE OR FALSE/MULTIPLE CHOICE

1. Deep surveys of the universe indicate that the largest structures in space are no larger than about 50 Mpc in size.

2. If the universe had an edge, that fact would violate the assumption of isotropy in the cosmological principle.

3. Hubble's law implies that the universe will expand forever.

4. The cosmological redshift is a direct measure of cosmic expansion.

5. Recent observations of supernovae suggest that the universe is not expanding.

6. Current density estimates suggest that the universe will ultimately end in a "Big Crunch."

7. Gravity slows the expansion of the universe; dark matter accelerates it.

8. Astronomers think that we live in a flat universe.

9. The cosmic microwave background is the highly redshifted radiation of the early Big Bang.

10. The spectrum of the cosmic microwave background is that of a blackbody.

11. If observations made from the middle of a large city are isotropic, then (a) there are tall buildings in every direction; (b) all buildings are exactly the same height; (c) all buildings are the same color; (d) some buildings are taller than others.

12. The cosmological principle would be invalidated if we found that (a) the universe is not expanding; (b) galaxies are older than currently estimated; (c) the number of galaxies per square degree is the same in every direction; (d) the observed structure of the universe depends on the direction in which we look.

13. When we use Hubble's law to estimate the age of the universe, the answer we get (a) depends on which galaxies we choose; (b) is the same for all galaxies; (c) depends on the direction in the sky toward which we are looking; (d) proves that we are at the center of the universe.

14. Olbers's paradox is resolved by (a) the finite size of the universe; (b) the finite age of the universe; (c) light from distant galaxies being redshifted so we can't see it; (d) the fact that there is an edge to the universe.

15. The universe will collapse back on itself in a "Big Crunch" if (a) the Big Bang wasn't big enough; (b) the universe has sufficient mass; (c) the cosmological principle is violated; (d) the universe has less than the critical density.

16. The galactic distances used to measure the acceleration of the universe are determined by observations of (a) trigonometric parallax; (b) line broadening; (c) Cepheid variable stars; (d) exploding white dwarfs.

17. The observed acceleration of the universe means that (a) we understand the nature of dark energy; (b) the amount of dark energy is small compared with the luminous mass in galaxies; (c) the amount of dark energy exceeds the total mass–energy

of matter in the universe; **(d)** dark energy has a higher temperature than expected.

18. On the basis of our current best estimate of the present mass density of the universe, astronomers think that **(a)** the universe is finite in extent and will expand forever; **(b)** the universe is finite in extent and will eventually collapse; **(c)** the universe is infinite in extent and will expand forever; **(d)** the universe is infinite in extent and will eventually collapse.

19. The age of the universe is estimated to be **(a)** less than Earth's age; **(b)** the same as the age of the Sun; **(c)** the same as the age of the Milky Way galaxy; **(d)** greater than the age of the Milky Way galaxy.

20. The cosmic background radiation is observed to come from **(a)** the center of our Galaxy; **(b)** the center of the universe; **(c)** radio antennae in New Jersey; **(d)** all directions equally.

PROBLEMS

 Algorithmic versions of these questions are available in the Practice Problems module of the Companion Website at astro.prenhall.com/chaisson.

The number of squares preceding each problem indicates its approximate level of difficulty.

1. ■ What is the greatest distance at which a galaxy survey sensitive to objects as faint as 20th magnitude could detect a galaxy as bright as the Milky Way (absolute magnitude –20)?

2. ■■ From Table 24.1, estimate the redshift of the Milky Way at the distance calculated in the previous question.

3. ■ If the entire universe were filled with Milky Way-like galaxies, with an average density of 0.1 galaxy per cubic megaparsec, calculate the total number of galaxies observable by the survey in Problem 1 if it covered the entire sky.

4. ■■■ Assuming that the entire universe is uniformly filled with Sun-like stars with a density of 5 billion stars per cubic megaparsec (corresponding to 50 billion stars per galaxy and critical mass density), calculate how far out into space one would have to look, on average, before the line of sight intersects a star.

5. ■■ Eight galaxies are located at the corners of a cube. The present distance from each galaxy to its nearest neighbor is 10 Mpc, and the entire cube is expanding according to Hubble's law, with $H_0 = 70$ km/s/Mpc. Calculate the recession velocity of one corner of the cube relative to the opposite corner.

6. ■ According to the Big Bang theory described in this chapter, *without a cosmological constant*, what is the maximum possible age of the universe if $H_0 = 60$ km/s/Mpc? 70 km/s/Mpc? 80 km/s/Mpc?

7. ■ For a Hubble constant of 70 km/s/Mpc, the critical density is 9×10^{-27} kg/m^3. (a) How much mass does this correspond to within a volume of 1 cubic astronomical unit? (b) How large a cube would be required to enclose 1 Earth mass of material?

8. ■■ The Virgo Cluster is observed to have a recession velocity of 1200 km/s. Assuming that $H_0 = 70$ km/s/Mpc and assuming a critical-density universe, calculate the total mass contained within a sphere centered on Virgo and just enclosing the Milky Way Galaxy. What is the escape speed from the surface of this sphere?

9. ■■ Given the age range for star clusters stated in the text, calculate the *maximum possible* value of Hubble's constant in the cases of (a) a constant expansion velocity and (b) critical density without a cosmological constant.

10. ■■ Assuming critical density and using the distances presented in Table 24.1, estimate the total amount of matter in the universe out to a redshift of 6. Express your answer (a) in kilograms and (b) in solar masses.

11. ■■ The critical density is proportional to the square of Hubble's constant. If the critical density were equal to the known density of "normal" matter (not dark matter), namely, about 10^{-28} kg/m^3, what would be the corresponding value of Hubble's constant? Is this value within the currently accepted range of values?

12. ■ What was the temperature of the cosmic microwave background at the epoch of quasar formation (at a redshift of 6)?

13. ■■ (a) What is the present peak wavelength of the cosmic microwave background? Calculate the size of the universe relative to its present size when the radiation background peaked in (b) the infrared, at 10 μm, (c) in the ultraviolet, at 100 nm, and (d) in the gamma ray region of the spectrum, at 1 nm.

14. ■■ Using Wien's law and the Doppler effect, verify that an observer moving at the velocity quoted in the text should indeed see the stated temperature shift in the microwave background.

15. ■■■ Using the information given in Problem 8, calculate the distance from our location to the point that would one day become the center of the Virgo Cluster, at the time when the temperature of the microwave background was equal to the present surface temperature of the Sun.

 In addition to the Practice Problems module, the Companion Website at astro.prenhall.com/chaisson provides for each chapter a study guide module with multiple choice questions as well as additional annotated images, animations, and links to related Websites.

27

Direct astronomical observations of the early universe are not possible. However, particle accelerators on Earth can approximate, for fleeting moments, the incredibly hot and dense conditions that likely prevailed during the earliest epochs of the universe. Here, in this computer simulation, we see a typical event thought to have occurred about a trillionth of a second after the beginning of time. Two fast-moving protons (straight red lines, left and right) are shown colliding head on, the encounter ▶ producing a multitude of new particles depicted in the debris at center. (CERN)

The Early Universe

Toward the Beginning of Time

What were the conditions during the first few seconds of the universe, and how did those conditions change to give rise to the universe we see today? In studying the earliest moments of our universe, we enter a truly alien domain. As we move backward in time toward the Big Bang, our customary landmarks slip away one by one. Atoms vanish, then nuclei, and then even the elementary particles themselves. In the beginning, the universe consisted of pure energy at unimaginably high temperatures. As it expanded and cooled, the ancient energy gave rise to the particles that make up everything we see around us today. Modern physics has now arrived at the point where it can reach back almost to the instant of the Big Bang itself, allowing scientists to unravel some of the mysteries of our beginnings in time.

LEARNING GOALS

Studying this chapter will enable you to

1 Describe the characteristics of the universe immediately after its birth.

2 Explain how matter emerged from the primeval fireball.

3 List the epochs in the evolutionary history of the universe, and specify the major characteristics of each.

4 Explain how and when the simplest nuclei and atoms formed.

5 Summarize the horizon and flatness problems, and discuss the theory of cosmic inflation as a possible solution to these problems.

6 Explain the formation of large-scale structure in the cosmos, and discuss the observational evidence for our theories of the formation of structure in the universe.

 Visit astro.prenhall.com/chaisson for additional annotated images, animations, and links to related sites for this chapter.

27.1 Back to the Big Bang

(LEARNING GOAL 1) On the very largest scales, we can regard the universe as a roughly homogeneous mixture of matter, radiation, and dark energy. As best we can tell, we live in a geometrically "flat" universe in which the total mass–energy density exactly equals the critical value. ∞ (Secs. 26.3, 26.4) On the basis of the best available observational data, cosmologists have concluded that just under 30 percent of this density is in the form of matter (mostly dark); the rest is dark energy—the mysterious repulsive force field thought to fill even the apparent vacuum of intergalactic space. ∞ (Sec. 26.5) According to theoretical models, there is not enough matter in the cosmos for the attractive force of gravity to overcome the repulsion of dark energy and reverse the current expansion.

Thus, the future of the cosmos seems clear: The universe is destined to expand forever. In this chapter, we turn our attention to the past. To begin to understand the early universe, just after the Big Bang, we must look more closely at the respective roles played by matter and radiation in the cosmos.

Matter and Radiation

The matter in the universe consists of the familiar building blocks of atoms—protons, neutrons, and electrons—as well as dark matter, whose composition is still being debated by astronomers. For a Hubble constant $H_0 = 70$ km/s/Mpc, the critical density is 9×10^{-27} kg/m^3. ∞ (Sec. 24.3) Thus, on the basis of the above numbers, the matter density in the universe today is 30 percent of this figure, or roughly 3×10^{-27} kg/m^3.

Most of the radiation in the universe is in the form of the cosmic microwave background—the low-temperature (3 K) radiation field that fills all space. ∞ (Sec. 26.6) Surprisingly, although the microwave background radiation is very weak, it still contains more energy than has been emitted by all the stars and galaxies that have ever existed! The reason for this is that stars and galaxies, though very intense sources of radiation, occupy only a tiny fraction of space. When their energy is averaged out over the volume of the entire universe, it falls short of the energy of the microwave background by at least a factor of 10. For our current purposes, then, we can ignore most of the first 26 chapters of this book and regard the cosmic microwave background as the only significant form of radiation in the universe!

Is matter the dominant component of the universe, or does radiation also play an important role on large scales? In order to compare matter and radiation, we must first convert them to a "common currency"—either mass or energy. Let's choose to compare their masses. We can express the energy in the microwave background as an equivalent density by first calculating the number of photons in any cubic centimeter of space and then converting the total energy of these photons into a mass using the relation

$E = mc^2$. ∞ (Sec. 16.5) When we do this, we arrive at an equivalent density for the microwave background of about 5×10^{-31} kg/m^3. Thus, *at the present moment*, the density of matter in the universe far exceeds the density of radiation. In cosmological terminology (which predates the possibility of dark energy in the cosmos), we say that we live in a **matter-dominated** universe.

Was the universe always matter dominated? To answer this question, we must ask how the densities of both matter and radiation changed as the universe expanded. To this end, cosmologists construct *theoretical models* of the universe, taking into account the effects of Einstein's general relativity and incorporating the known properties of matter and radiation, as well as the assumed properties of dark energy. ∞ (Sec. 22.5) These models describe how cosmic quantities (such as the densities and temperatures of the matter and radiation fields) change as the universe evolves. They also make detailed *predictions*, which can be compared directly with observations. ∞ (Sec. 1.2) The outstanding agreement between models and reality (see Section 27.5) is the main reason that astronomers attach so much weight to the measurements of cosmic density, composition, and evolution described in the previous chapter.

As shown in Figure 27.1, the models indicate that, as the scale of the universe increases, the densities of matter

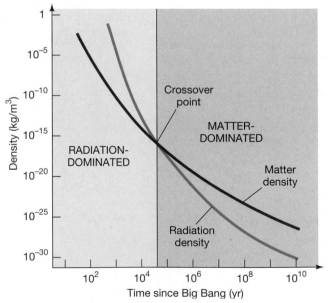

▲ **FIGURE 27.1 Radiation–Matter Dominance** As the universe expanded, the number of both matter particles and photons per unit volume decreased. However, the photons were also reduced in energy by the cosmological redshift, reducing their equivalent mass, and hence their density, still further. As a result, the density of radiation fell faster than the density of matter as the universe grew. Tracing the curves back from the densities observed today, we see that radiation dominated matter at early times, before the crossover point.

and radiation both decrease, with the expansion diluting the numbers of atoms and photons alike. But the radiation is also diminished in energy by the cosmological redshift, and its density falls *faster* than that of matter as the universe grows. Hence, as we look *back* in time, closer and closer to the Big Bang, the density of the radiation increases faster than that of matter. Accordingly, even though today the radiation density is much less than the matter density, there must have been a time in the past when they were equal. Before that time, radiation was the main constituent of the cosmos. The universe is said to have been **radiation dominated** then. Given our best estimates of the present densities, the crossover point—the time at which the densities of matter and radiation were equal—occurred about 50,000 years after the Big Bang, when the universe was about 6000 times smaller than it is today. The temperature of the background radiation at that time was about 16,000 K, so it peaked in the near-ultraviolet portion of the spectrum.

Throughout this book, we have been concerned exclusively with the history of the universe long after it became matter dominated—the formation and evolution of galaxies, stars, and planets as the universe thinned and cooled toward the state we see today. In this chapter, we consider some important events in the early, hot, radiation-dominated universe, long before any star or galaxy existed, that played no less a role in determining the present condition of the cosmos.

But what of dark energy? This component of the universe currently accounts for the largest fraction of the total cosmic density and is responsible for the acceleration of the Hubble expansion inferred from supernova studies. ∞ (Sec. 26.4) However, according to theory, and consistent with current observations, dark energy is a large-scale phenomenon, increasing in importance as the universe expands. In fact (as usual, converting energies and masses into the same units for purposes of comparison), the density associated with the dark energy remains *constant* as the universe expands, while, as we have just seen, the densities of matter and radiation both decrease. We therefore conclude that dark energy, at least as we currently understand it, *increases* in importance as the universe expands. Put another way, it was probably *unimportant* at early times, so we can neglect its influence in our discussion of conditions in the very early universe.

Particle Production in the Early Universe

2 The existence of the microwave background implies that the early universe was dominated by an intense radiation field whose temperature fell steadily as the cosmos expanded. The temperatures and densities prevailing at these times were far greater than anything we have encountered thus far, even in the hearts of supernovae. To understand conditions in the universe shortly after the Big Bang, we must delve a little more deeply

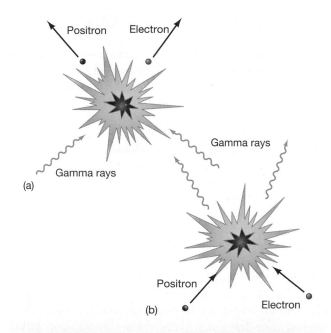

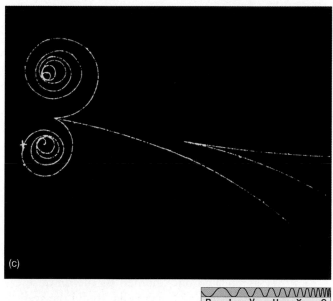

▲ **FIGURE 27.2 Pair Production** (a) Two photons can produce a particle–antiparticle pair—in this case an electron and a positron—if their total energy exceeds the mass energy of the particles produced. (b) The reverse process is particle–antiparticle annihilation, in which an electron and a positron destroy each other, vanishing in a flash of gamma rays. (c) A particle detector (an instrument designed to show the tracks of otherwise invisible subatomic particles) allows us to visualize pair production. Here, a gamma ray, whose path is invisible because it is electrically neutral, arrives from the left, dislodging an atomic electron and sending it flying (the longest path). At the same time, the gamma ray provides the energy to produce an electron–positron pair (the spiral paths, which curve in opposite directions in the detector's magnetic field because of their opposite electric charges). The V-shaped tracks at the right are the result of a separate particle interaction. (*Fermi National Laboratory*)

into the behavior of matter and radiation at very high temperatures.

The key to understanding events at very early times lies in a process called **pair production**, in which two photons give rise to a *particle–antiparticle* pair, as shown in Figure 27.2(a) for the particular case of electrons and positrons. Through pair production, matter is created directly from energy in the form of electromagnetic radiation. The reverse process can also occur: A particle and its antiparticle can *annihilate* each other to produce radiation, as depicted in Figure 27.2(b). In other words, energy in the form of radiation can be freely converted into matter in the form of particles and antiparticles, and particles and antiparticles can be freely converted back into radiation, subject only to the law of conservation of mass and energy.

The higher the temperature of a radiation field, the greater the frequency and hence the energy of the typical constituent photons, and the greater the masses of the particles that can be created by pair production. ∞ (Secs. 3.4, 4.2, 16.2) For any given particle, the critical temperature above which pair production is possible and below which it is not is called the particle's *threshold temperature*. The threshold temperature increases as the mass of the particle increases. For electrons, it is about 6×10^9 K. For protons, which are nearly 2000 times more massive, it is just over 10^{13} K.

As an example of how pair production affected the composition of the early universe, consider the production of electrons and positrons as the universe expanded and cooled. At high temperatures—above about 10^{10} K—most photons had enough energy to form an electron or a positron, and pair production was commonplace. As a result, space seethed with electrons and positrons, constantly created from the radiation field and annihilating one another to form photons again. Particles and radiation are said to have been in *thermal equilibrium*: New particle–antiparticle pairs were created by pair production at the same rate as they annihilated one another. As the universe expanded and the temperature decreased, so did the average photon energy. By the time the temperature had fallen below a billion or so kelvins, photons no longer had enough energy for pair production to occur, and only radiation remained. Figure 27.3 illustrates how this change took place.

Pair production in the very early universe was directly responsible for all the matter that exists in the universe today. *Everything we see around us was created out of radiation as the cosmos expanded and cooled.* Because we are here to ponder the subject and we ourselves are made of matter, we know that some matter must have survived those early violent moments. For some reason, there was a slight excess of matter over antimatter at early times—about one extra proton for every billion proton–antiproton pairs. That small residue of particles which outnumbered their antiparticles was left behind as the temperature dropped below the threshold for creating them. With no antiparticles left to annihilate them, the number of particles has re-

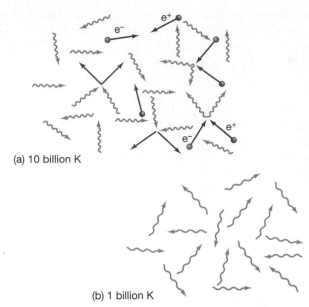

(a) 10 billion K

(b) 1 billion K

▲ **FIGURE 27.3 Thermal Equilibrium** (a) At 10 billion K, most photons have enough energy to create particle–antiparticle (electron–positron) pairs, so these particles exist in great numbers in equilibrium with the radiation. The label e⁻ refers to the electrons, e⁺ to positrons. (b) Below about 1 billion K, photons have too little energy for pair production to occur, so electrons and positrons are no longer in thermal equilibrium with the background radiation field.

mained constant ever since. These survivors are said to have *frozen out* of the radiation field as the universe expanded and cooled.

According to the models, the first hundred or so seconds of the universe's existence saw the creation of all of the basic "building blocks" of matter we know today. Protons and neutrons froze out when the temperature dropped below 10^{13} K, when the universe was only 0.0001 s old, and the lighter electrons froze out somewhat later, about a minute or so after the Big Bang, when the temperature fell below 10^9 K. This "matter-creation" phase of the universe's evolution ended when the electrons—the lightest known elementary particles—appeared out of the cooling primordial fireball. From that point on, matter has continued to evolve, clumping together into more and more complex structures, eventually forming the atoms, planets, stars, galaxies, and large-scale structure we see today, but for all practical purposes no new matter has been created since that early time. (Pair production still occurs today, but only in relatively rare, extreme environments, such as supernovae and experimental particle accelerators.)

CONCEPT CHECK

✔ What does it mean to say that the early universe was radiation dominated?

27.2 The Evolution of the Universe

For the first few thousand years after the Big Bang, the universe was small, dense, and dominated by radiation. We will refer to this period as the *Radiation Era*. Some matter existed during this time, but it was a mere contaminant in the blinding gamma-ray light of the primeval Big Bang fireball. Afterwards, in the *Matter Era*, matter came to dominate. Atoms, molecules, and galaxies formed as the universe cooled and thinned toward the state we see today.

Let's begin our study of the early universe by summarizing in broad terms the history of the cosmos, starting at the Big Bang. Table 27.1 presents the time, density, and temperature spans of eight major epochs in the

development of the universe, along with a brief description of the main physical events that dominated the universe during each epoch. The numbers in the table result from pushing the known laws of physics as far back in time as we can. In the next few sections, we will expand on some of these epochs in greater detail, but let's not lose sight of the big picture and the place of each epoch in it.

Before the Big Bang

The Big Bang was a *singularity* in space and time—an instant when the present laws of physics imply that the universe had zero size and infinite temperature and density. As we saw in Chapter 22, where we discussed the singularities at the center of black holes, these predictions should not be taken too literally. ∞ (Sec. 22.7) The presence of

---

TABLE 27.1   Major Epochs in the History of the Universe

Era	Epoch	Time (after big bang)	Density (kg/m³)	Temperature (K)	Main Events
Radiation Era					
		0 s	∞	∞	
	Planck				Unknown physics; quantum gravity
		10^{-43} s	10^{95}	10^{32}	
	GUT*				Strong, weak, and electromagnetic forces unified
		10^{-35} s	10^{75}	10^{27}	
	Quark				Strong force frozen out. Heavy and light particles all in thermal equilibrium. Electroweak force freezes out at 10^{15} K.
		10^{-4} s	10^{16}	10^{12}	
	Lepton				Only low-mass particles still in thermal equilibrium; neutrinos decouple at 10^{10} K.
		10^2 s	10^4	10^9	
	Nuclear				Deuterium and helium formed by fusion of protons and neutrons during first 1000 s.
		5×10^4 yr ($\approx 2 \times 10^{12}$ s)	6×10^{-16}	16,000	
Matter Era					
		5×10^4 yr ($\approx 2 \times 10^{12}$ s)	6×10^{-16}	16,000	
	Atomic				Matter begins to dominate; atoms form; electromagnetic radiation decouples.
		10^8 yr ($\approx 3 \times 10^{15}$ s)	10^{-22}	100	
	Galactic				Large-scale structure forms; the first stars and quasars shine.
		3×10^9 yr ($\approx 10^{17}$ s)	2×10^{-25}	10	
	Stellar				Galaxies merge and grow; stars continue to form.
		$>10^{10}$ yr (3×10^{17} s)	3×10^{-27}	3	

*Grand Unified Theory; see p. 724 of text for discussion.

singularities signals that, under extreme conditions, the theory—in this case, general relativity—making the predictions has broken down.

At present, no theory exists to let us penetrate the singularity at the start of the universe. We have no means of describing these earliest of times, so we have no way of answering the question "What came *before* the Big Bang?" Indeed, given the laws of physics as we currently know them, the question itself may be meaningless. The Big Bang represented the beginning of the entire universe—mass, energy, space, *and* time came into being at that instant. Without time, the notion of "before" does not exist. Consequently, some cosmologists maintain that asking what happened before the Big Bang is a little like asking what lies north of the North Pole! Others disagree, however, arguing that when the correct theory of quantum gravity—the "Theory of Everything" that unifies gravity and quantum mechanics—is constructed, it will remove the singularity and allow us to address the question of what came before.

The Radiation Era

The universe began with a rapid expansion from an incredibly hot and dense state. Precisely what state, we cannot say, and what "triggered" the Big Bang, we really don't know. To understand why the universe began expanding, or even more fundamentally, why the universe exists at all, is currently beyond science—there simply are no relevant data. Although ignorant of the moment of creation itself, theorists nevertheless believe that the physical conditions in the universe can be understood in terms of present-day physics back to an extraordinarily short time—a mere 10^{-43} s, in fact—after the Big Bang.

Why can't theorists push our knowledge back to the Big Bang itself? The answer is that we presently have no theory capable of describing the universe at these earliest of times. Under the extreme conditions of density and temperature within 10^{-43} s of the Big Bang, gravity and the other fundamental forces (electromagnetism, the strong force, and the weak force, as described in *More Precisely 27-1*) were indistinguishable from one another—a far cry from the radically different characteristics we see today. The four forces are said to have been *unified* at that early time—there was, in effect, only one force of nature.

The theory that combines quantum mechanics (the proper description of microscopic phenomena) with general relativity (which describes the universe on the largest scales) is generically known as quantum gravity. ∞ (Sec. 22.7) The period from the beginning to 10^{-43} s is often referred to as the *Planck epoch*, after Max Planck, one of the creators of quantum mechanics. Unfortunately, for now at least, there is no working theory of quantum gravity, so we simply cannot talk meaningfully about the universe during the Planck epoch.

By the end of the Planck epoch, the temperature was around 10^{32} K, and the universe was filled with radiation and a vast array of subatomic particles created by the mechanism of pair production. At around that time, gravity parted company with the other forces of nature—it became distinguishable from the "quantum" forces and has remained so ever since. The strong, weak, and electromagnetic forces were still unified. The present-day theories that describe this epoch are collectively known as **Grand Unified Theories**, or GUTs for short (see *More Precisely 27-1*). Accordingly, we refer to this period as the *GUT epoch*.

Theory indicates that at temperatures below 10^{28} K, the strong nuclear force becomes distinguishable from the electroweak force (the unified weak and electromagnetic force). Once the universe had cooled to that temperature, about 10^{-35} s after the Big Bang, the GUT epoch ended. According to many GUTs, one important legacy of that epoch may have been the appearance and subsequent freeze-out of a veritable "zoo" of very massive (and as yet unobserved) elementary particles that interact only very weakly with normal matter. These "exotic" particles are prime candidates for the dark matter of unknown composition thought to exist in abundance both within galaxies and in the unseen depths of intergalactic space. ∞ (Secs. 23.6, 24.3)

Our next major subdivision of the Radiation Era covers the period when all "heavy" elementary particles—that is, all the way down in mass to protons, neutrons, and their constituent quarks—were in thermal equilibrium with the radiation. We refer to this period as the *quark epoch*, since quarks are the fundamental components of all particles that interact via the strong force.

The universe continued to expand and cool. At a temperature of about 10^{15} K, 10^{-10} s after the Big Bang, the weak and the electromagnetic components of the electroweak force began to display their separate characters. By about 0.1 millisecond (10^{-4} s) after the Big Bang, the temperature had dropped well below the 10^{13} K threshold for the creation of protons and neutrons (the lightest stable particles composed of quarks), and the quark epoch ended. The main constituents of the universe were now lightweight particles—muons (see *More Precisely 27-1*), electrons, neutrinos, and their antiparticles—all still in thermal equilibrium with the radiation. Compared with the numbers of these lighter particles, only very few protons and neutrons remained at this stage, because most had been annihilated.

Electrons, muons, and neutrinos are collectively known as *leptons*, after the Greek word meaning "light" (i.e., not heavy). Accordingly, this period in the history of the universe is known as the *lepton epoch*. During that epoch, at a temperature of about 3×10^{10} K—approximately 1 second after the Big Bang—the rapidly thinning universe became transparent to neutrinos, and these ghostly particles have been streaming freely through space ever since. (Indeed, most neutrinos have not interacted with any other particle since the universe was a few seconds old!) The lepton epoch ended

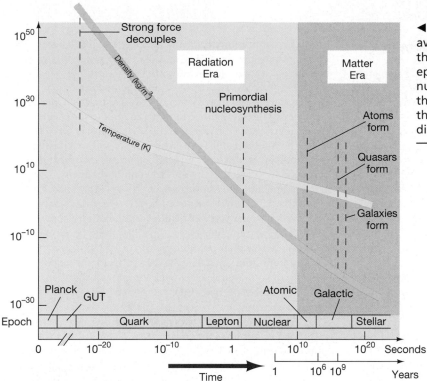

when the universe was about 100 s old and the temperature fell to about 1 billion K—too low for electron–positron pair production to occur. The density of the universe by this time was about 10 times the density of water.

The final significant event in the Radiation Era occurred when protons and neutrons began to fuse into heavier nuclei. At the start of this period, which we will call the *nuclear epoch*, the temperature was a few hundred million kelvins, and fusion occurred very rapidly, forming deuterium and helium in quick succession before conditions became too cool for further reactions to occur. By the time the universe was about 15 minutes old, much of the helium we observe today had been formed.

Figure 27.4 illustrates graphically how both the temperature and the density of the universe dropped rapidly during the Radiation and Matter Eras. Notice how the time scale on the horizontal axis increases from tiny fractions of a second to thousands of years as we move fom left to right. The rate of change of the cosmos slowed radically as the universe expanded.

The Matter Era

Time passed, the universe continued to expand and cool, and radiation gave way to matter as the dominant constituent of the universe. Our next major epoch extends in time from 50,000 years (the end of the Radiation Era) to about 100 million years after the Big Bang. As the primeval fireball diminished in intensity, a crucial change occurred—perhaps the most important change in the history of the universe. At the end of the nuclear epoch, radiation still overwhelmed matter. As fast as protons and electrons

combined, radiation broke them apart again, preventing the formation of even simple atoms or molecules. However, as the universe expanded and cooled, the early dominance of radiation eventually ended. Once formed, atoms remained intact. We will call this period the *atomic epoch*. It ended about 100 million years after the Big Bang, when the first stars formed and their intense radiation *reionized* the universe.

The last two epochs together bring us to the current age of the universe. During these late stages, change happened at a much more sedate pace. By the time the universe was about 3 billion years old, galaxies and large-scale structure had formed. For the first time, the visible universe departed from homogeneity on macroscopic scales. The largely uniform universe of the Radiation Era became a universe containing large agglomerations of matter. We call the period from 100 million to 3 billion years after the Big Bang the *galactic epoch*. At its end, large-scale structure and the first galaxies had formed, quasars were shining brightly, and new generations of stars were burning and exploding, helping to determine the future shape of their parent galaxies. Since then, galaxies have continued to merge and evolve, and stars, planets, and life have appeared in the universe. This final *stellar epoch* has been the subject of the first 25 chapters of this book.

CONCEPT CHECK

✔ Why did lighter and lighter particles "freeze out" of the universe as the cosmos expanded?

More on Fundamental Forces

In *More Precisely 16-1*, we noted that the behavior of all matter in the universe is ruled by just three *fundamental forces*: gravity, the electroweak force (the unification of the electromagnetic and weak forces), and the strong (nuclear) force. In terrestrial laboratories, these forces display properties that are very different from one another. Gravity and electromagnetism are long-range, inverse-square forces, whereas the strong and weak forces have very short ranges—10^{-15} and 10^{-17} m, respectively. Furthermore, the forces do not all affect the same particles. Gravity affects everything. The electromagnetic force affects only charged particles. The strong force operates between nuclear particles, such as protons and neutrons, but it does not affect electrons and neutrinos. The weak force shows up in certain nuclear reactions and radioactive decays. The strong force is 137 times stronger than the electromagnetic force, 100,000 times stronger than the weak force, and 10^{39} times stronger than gravity.

In fact, there is more structure below the level of the nucleus. Protons and neutrons are not truly "elementary" in nature, but are actually made of subparticles called *quarks*. (The name derives from a meaningless word coined by novelist James Joyce in his book *Finnegans Wake*.) According to current theory, there are six distinct types of quark in the universe (with the obscure names *up*, *down*, *charm*, *strange*, *top*, and *bottom*). The most massive, and most elusive, of them—the "top" quark—was discovered at the Fermi National Laboratory (Fermilab), Illinois, in 1994. What we call the strong nuclear force is actually a manifestation of the interactions that bind quarks to one another.

Table 27.2 lists some properties of the fundamental forces of nature and also indicates how physicists have sought to connect those properties with one another. On the face of it, one might not imagine that there could be any deep underlying connection between forces as dissimilar as those just described, yet there is strong evidence that they are really just different aspects of a single basic phenomenon. In the 1960s, theoretical physicists succeeded in explaining the electromagnetic and weak forces in terms of the electroweak force. Shortly thereafter, the first attempts were made at combining the strong and electroweak forces into a single all-encompassing "superforce." A central idea in the modern version of this superforce is that there is a one-to-one correspondence between the quarks, which interact via the strong force, and particles called *leptons*, which are affected only by the electroweak force. The six known types of quark are paired with six distinct types of lepton: the electron, two related "electronlike" particles (called muons and taus), and three types of neutrino.

Theories that combine the strong and electroweak forces into one are generically known as *Grand Unified Theories*, or GUTs for short. (Note that the term is plural—no one GUT has yet been proven to be "the" correct description of nature.) One general prediction of GUTs is that the three nongravitational forces are indistinguishable from one another only at enormously high energies, corresponding to temperatures in excess of 10^{28} K. Below that temperature, the superforce splits into two, displaying its separate strong and electroweak aspects. In particle-physics parlance, we say that there is a *symmetry* between the strong and the electroweak forces that is broken at temperatures below 10^{28} K, allowing the separate characters of the two forces to become apparent. At "low" temperatures—less than about 10^{15} K, a range that includes almost everything we know on Earth and in the stars—there is a second symmetry breaking, and the electroweak force splits to reveal its more familiar electromagnetic and weak natures.

The key predictions of the electroweak theory were experimentally verified in the 1970s, winning the theory's originators (Sheldon Glashow, Steven Weinberg, and Abdus Salam) the 1979 Nobel prize in physics. The GUTs have not yet been experimentally verified (or refuted), in large part because of the extremely high energies that must be reached in order to observe their predictions.

An important idea that has arisen from the realization that the strong and the electroweak forces can be unified is the notion of *supersymmetry*, which extends the idea of symmetry between fundamental forces to place all particles—those which are acted on by forces (such as protons and electrons) and those which transmit those forces (such as photons and gluons—see Section 27.4)—on an equal footing. One particularly important prediction of super symmetry is that all particles should have so-called *supersymmetric partners*—extra particles that must exist in order for the theory to remain self-consistent. None of these new particles has yet been detected, yet many physicists are convinced of the theory's essential correctness.

These new particles, if they exist, would have been produced in abundance in the Big Bang and should still be around today. They are also expected to be very massive—at least a thousand times heavier than a proton. So-called supersymmetric relics, the new particles are among the current leading candidates for the dark matter in the universe (see Section 27.5), although it must be admitted that recent experimental failures to detect them have dampened some astronomers' early enthusiasm.

Efforts to include gravity within this picture have so far been unsuccessful. Gravitation has not yet been incorporated into a single "SuperGUT," in which all the fundamental forces are united. Some theoretical efforts to merge gravity with the other forces have tried to fit gravity into the quantum world by postulating extra particles—called *gravitons*—that transmit the gravitational force. However, this is a different view of gravity from the geometric picture embodied in Einstein's general relativity, and combining the two into a consistent theory of quantum gravity has proved very difficult.

One theory that is currently under active investigation seeks to interpret all particles and forces in terms of particular modes of vibration of submicroscopic objects known as *strings*. As an analogy, consider a more familar string—one on a guitar or a violin. Many different musical notes can be created, or *excited*, in the string, depending on how it is struck and how tightly it is stretched. Similarly, in *string theory*, the many different elementary particles we see (and many more that we postulate, but haven't yet observed) can be thought of as "musical notes," or excited modes of elementary strings. As sketched in the first accompanying figure, one particular mode represents an electron, another a positron, another a photon, another a neutrino, etc. Particles and forces are unified because they are all particular aspects of a single fundamental object: the string. Also as shown in the figure, when two particles interact, their strings are temporarily replaced by a "composite" string, which subsequently splits into two other strings representing the results of the interaction.

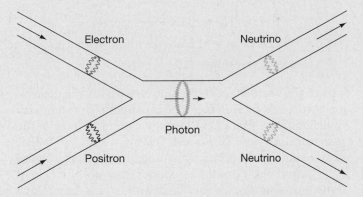

How big are these strings? They are truly tiny! The length scale on which string theory operates, called the *Planck length*, is just 4×10^{-35} m. This is the scale on which theory predicts that general relativity and quantum mechanics must be unified. (See Section 27.2.) Note that the Planck length is directly related to the time marking the end of the Planck epoch in Table 27.1—the *Planck time* (1.3×10^{-43} s)—by the simple fact that light moves a distance equal to the Planck length in the Planck time. Unfortunately, this scale effectively puts string theory far beyond hope of direct verification by any conceivable experiment. The energies needed to probe such fine structure exceed those attainable in even the largest particle accelerators by at least 15 orders of magnitude—a factor of 1000 trillion!

Nevertheless, string theory has much to recommend it. In part because it "smears" elementary particles out over finite (if tiny) regions of space, instead of treating them as mass points, this approach has so far avoided or eliminated many of the thorny technical problems that have derailed other promising theories. String theory is extremely complex in its mathematical

details, however. For example, in order for it to remain internally consistent, the strings must vibrate not in the four-dimensional spacetime we all know, but in no fewer than 11 spacetime dimensions—the familiar four, plus seven more! We don't see the extra seven dimensions because, in those directions, the universe is "rolled up" on very small scales (the Planck scale).

To understand this idea, consider as another analogy a drinking straw (second figure)—a tube much longer than it is wide. From a distance, we see only that the straw has some finite length—we might not even be able to tell that it has any width at all. This perspective is analogous to our macroscopic (large-scale) view of spacetime, where only the four "large" dimensions are evident. But if we look more closely—magnify the straw—we see that it has an extra "hidden" dimension, curving tightly back on itself, and that the surface of the straw is in reality a cylinder. This is the microscopic (close-up) view of the universe—only on scales close to the Planck scale do the extra, tightly curved dimensions of spacetime become apparent.

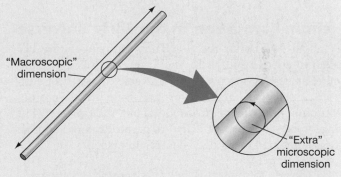

Despite these complexities, many theorists feel that string theory currently offers the greatest promise of unifying the forces of nature. Still, while that theory seems to be our "best bet," realize that, at present, no theory has yet succeeded in making any definite statement about conditions in the very early universe. A complete theory of quantum gravity continues to elude researchers.

TABLE 27.2 Fundamental Forces and Particles

Particles affected	Range (m)	Force	Unification (temperature)		
matter composed of quarks (protons, neutrons, etc.)	10^{-15}	strong		GUT/superforce (10^{28} K)	quantum gravity (10^{32} K)
charged particles (protons, electrons, etc.)	infinite	electromagnetic	electroweak (10^{15} K)		
leptons (electrons, muons, taus, neutrinos)	10^{-17}	weak			
everything	infinite	gravity			

27.3 The Formation of Nuclei and Atoms

4 We now have all the ingredients needed to complete our story of the creation of the elements, begun in Chapter 21, but never quite finished. ∞ (Sec. 21.4) The theory of stellar nucleosynthesis accounts very well for the observed abundances of heavy elements in the universe, but there are discrepancies between theory and observation when it comes to the abundances of the light elements, especially helium. Simply put, the total amount of helium in the universe today—about 25 percent by mass—is far too large to be explained by nuclear fusion in stars. The accepted explanation is that this base level of helium is *primordial*—that is, it was created during the early, hot epochs of the universe, before any stars had formed. The production of elements heavier than hydrogen by nuclear fusion shortly after the Big Bang is called **primordial nucleosynthesis**.

Helium Formation in the Early Universe

By about 100 s after the Big Bang, the temperature had fallen to about 1 billion K, and apart from "exotic" dark-matter particles, matter in the universe consisted of electrons, protons, and neutrons, with the protons outnumbering the neutrons by about five to one. The stage was set for nuclear fusion to occur. Protons and neutrons combined to produce deuterium nuclei:

$$^1\text{H (proton)} + \text{neutron} \rightarrow {}^2\text{H (deuteron)} + \text{energy}.$$

Although this reaction must have occurred frequently during the Lepton epoch, the temperature then was still so high that the deuterium nuclei were broken apart by high-energy gamma rays as soon as they formed. The universe had to wait until it became cool enough for the deuterium

to survive. This waiting period is sometimes called the *deuterium bottleneck*.

Only when the temperature of the universe fell below about 900 million K, roughly two minutes after the Big Bang, was deuterium at last able to form and endure. Once that occurred, the deuterium was quickly converted into heavier elements by numerous reactions, including:

$$^2\text{H} + {}^1\text{H} \rightarrow {}^3\text{He} + \text{energy},$$
$$^2\text{H} + {}^2\text{H} \rightarrow {}^3\text{He} + \text{neutron} + \text{energy},$$
$$^3\text{He} + \text{neutron} \rightarrow {}^4\text{He} + \text{energy}.$$

The result was that, once the universe passed the deuterium bottleneck, fusion proceeded rapidly and large amounts of helium were formed. In just a few minutes most of the free neutrons were consumed, leaving a universe whose matter content was primarily hydrogen and helium. Figure 27.5 illustrates some of the reactions responsible for helium formation. Contrast it with Figure 16.5, which depicts how helium is formed today in the cores of main-sequence stars such as the Sun.* ∞ (Sec. 16.2)

We might imagine that fusion could have continued to create heavier and heavier elements, just as it does in the cores of stars, but that did not occur. In stars, the density and the temperature both *increase* slowly with time, allowing more and more massive nuclei to form, but in the early universe exactly the opposite was true. The temperature and density were both *decreasing* rapidly, making conditions less and less favorable for fusion as time went on. Even before the supply of neutrons was completely used up, the nuclear reactions had effectively ceased. Reactions

The proton–proton chain that powers the Sun played no significant role in primordial helium formation. The proton–proton reaction that starts the chain is very slow compared with the proton–neutron reaction discussed here and is important in the Sun only because the solar interior contains no free neutrons to make the latter reaction possible.

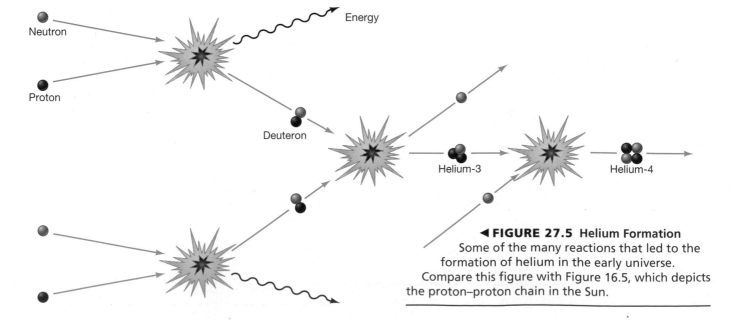

◄ **FIGURE 27.5 Helium Formation**
Some of the many reactions that led to the formation of helium in the early universe. Compare this figure with Figure 16.5, which depicts the proton–proton chain in the Sun.

Neutron

Proton

Energy

Deuteron

Helium-3

Helium-4

between helium nuclei and protons may also have formed trace amounts of lithium (the next element beyond helium) by this time, but for all practical purposes, the expansion of the universe caused fusion to stop at helium. The brief epoch of primordial nucleosynthesis was over about 15 minutes after it began.

By the end of the period of nucleosynthesis, some 1000 s after the Big Bang, the temperature of the universe was about 300 million K and the cosmic elemental abundances were set. Careful calculations indicate that about one helium nucleus had formed for every 12 protons remaining. Because a helium nucleus is four times more massive than a proton, helium accounted for about one-quarter of the total mass of matter in the universe:

$$\frac{1 \text{ helium nucleus}}{12 \text{ protons} + 1 \text{ helium nucleus}} = \frac{4 \text{ mass units}}{12 \text{ mass units} + 4 \text{ mass units}}$$

$$= \frac{4}{16} = \frac{1}{4}.$$

The remaining 75 percent of the matter in the universe was hydrogen. It would be almost a billion years before nucleosynthesis in stars would change these numbers. ∞ (Sec. 21.4)

The foregoing calculation implies that all stars and galaxies should contain *at least 25 percent* helium by mass. The figure for the Sun, for example, is about 28 percent. However, it is difficult to disentangle the contributions to the present-day helium abundance from primordial nucleosynthesis and later hydrogen burning in stars. Our best hope of determining the amount of primordial helium is to study the oldest stars known, since they formed early on, before stellar nucleosynthesis had had time to change the helium content of the universe significantly. Unfortunately, stars surviving from that early time are of low mass and hence quite cool, making the helium lines in their spectra very weak and hard to measure accurately. ∞ (Secs. 17.5, 17.8) Nevertheless, despite this uncertainty, the observations are generally consistent with the theory just described.

Bear in mind that while all this was going on, matter was just an insignificant "contaminant" in the radiation-dominated universe. Radiation outmassed matter by about a factor of 5000 at the time helium formed. The existence of helium is very important in determining the structure and appearance of stars today, but its creation was completely irrelevant to the evolution of the universe at the time.

Deuterium and the Density of the Cosmos

During the nuclear epoch, although most deuterium was quickly fused into helium as soon as it formed, a small amount was left over when the primordial nuclear reactions ceased. Observations of deuterium—especially those made by orbiting satellites able to capture deuterium's

strongest spectral feature, which happens to be emitted in the ultraviolet part of the spectrum—indicate a present-day abundance of about two deuterium nuclei for every 100,000 protons. However, unlike helium, deuterium is not produced to any significant degree in stars (in fact, deuterium tends to be destroyed in stars), so any deuterium we see today *must* be primordial.

This observation is of great importance to astronomers, because it provides them with a sensitive method—and one that is completely independent of the techniques discussed in previous chapters—of probing the present-day density of matter in the universe. According to theory, as illustrated in Figure 27.6, the denser the universe is today, the more particles there were at early times to react with deuterium as it formed, and the less deuterium was left over when nucleosynthesis ended. A comparison of the observed deuterium abundance (marked on the figure) with the theoretical results implies a present-day density of *at most* 3×10^{-28} kg/m^3—only a few percent of the critical density.

But before we jump to any far-reaching cosmic conclusions based on this number, we must make a very important qualification. As just described, primordial nucleosynthesis depends *only* on the presence of protons and neutrons in the early universe. Thus, measurements of the abundance of helium and deuterium tell us only about

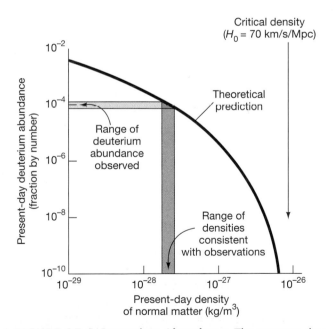

▲ FIGURE 27.6 Deuterium Abundance The present-day abundance of deuterium depends strongly on the amount of matter present at early times, and this, in turn, determines the present-day density of the universe. Thus, measuring the amount of deuterium in the universe gives us an estimate of the overall density of matter. The best deuterium measurements are marked; they imply that the density of matter in the universe is at most a few percent of the critical value.

the density of "normal" matter—matter made up of protons and neutrons—in the cosmos. This finding has a momentous implication for the overall composition of the universe. As we saw earlier, astronomers have concluded, for a variety of reasons, that the total density of matter is about the critical value. ∞ (Sec. 26.4) In that case, if the density of normal matter is only a few percent of the critical value, then we are forced to admit that not only is most of the matter in the universe dark, but most of the dark matter is *not* composed of protons and neutrons.

We will see in Section 27.5 that, because normal matter and dark matter interact differently with the background radiation field, studies of the cosmic microwave background allow us to distinguish between these two types of matter. Observations made by the *WMAP* satellite have found the density of normal matter to be just 4 percent of the critical density.

Thus, the bulk (about 90 percent) of the matter in the universe apparently exists in the form of elusive subatomic particles (for example, the WIMPs discussed as dark-matter candidates in Chapter 23) whose nature we do not fully understand and whose very existence has yet to be conclusively demonstrated in laboratory experiments. ∞ (Sec. 23.6) For the sake of brevity, from here on we will adopt the convention that the term "dark matter" refers only to these unknown particles and not to "stellar" dark matter, such as black holes and brown and white dwarfs (also discussed in Chapter 23), which are made of relatively well-understood normal matter.

The First Atoms

A few tens of thousands of years after the Big Bang, radiation ceased to be the dominant component of the universe. The Matter Era had begun. At the start of the atomic epoch, matter consisted of electrons, protons, helium nuclei (formed by primordial nucleosynthesis), and dark matter. The temperature was several tens of thousands of kelvins—far too hot for atoms of hydrogen to exist (although some helium ions may already have formed). During the next few hundred thousand years, a major change occurred: The universe expanded by another factor of 10, the temperature dropped to a few thousand kelvins, and electrons and nuclei combined to form neutral atoms. By the time the temperature had fallen to about 3000 K, the universe consisted of atoms, photons, and dark matter.

The period during which nuclei and electrons combined to form atoms is often called the epoch of **decoupling,*** for it was during this period that the radiation background parted company with normal matter. At early times, when matter was ionized, the universe was filled with large numbers of free electrons, which interacted frequently with electromagnetic radiation of all wavelengths. As a

result, a photon could not travel far before encountering an electron and scattering off it. In effect, the universe was opaque to radiation. Matter and radiation were strongly "tied," or *coupled*, to one another by these interactions.

After the electrons combined with nuclei to form atoms of hydrogen and helium, only certain wavelengths of radiation—the ones corresponding to the spectral lines of those atoms—could interact with matter. ∞ (Sec. 4.2) Radiation of other wavelengths could travel virtually forever without being absorbed. Thus, the universe became nearly transparent. From that time on, most photons passed generally unhindered through space. As the universe expanded, the radiation simply cooled, eventually becoming the microwave background we see today.

The microwave photons now detected on Earth have been traveling through the universe ever since they decoupled. According to the models that best fit the observational data, the last interaction these photons had with matter (at the epoch of decoupling) occurred when the universe was about 400,000 years old and roughly 1100 times smaller (and hotter) than it is today—that is, at a redshift of 1100. As illustrated in Figure 27.7, the epoch of atom formation created a kind of "photosphere" in the universe, completely surrounding Earth at a distance of approximately 14,000 Mpc, the distance at which photons

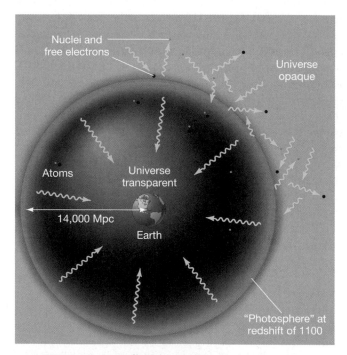

▲ **FIGURE 27.7 Radiation–Matter Decoupling** When atoms formed, the universe became virtually transparent to radiation. Thus, observations of the cosmic background radiation allow us to study conditions in the universe around a time when the redshift was 1100 and the temperature had dropped below about 4500 K. For an explanation of how we can see a region of space 14,000 Mpc (46 billion light-years) away when the universe is just 14 billion years old, see *More Precisely 24-1*.

**Some astronomers refer to this epoch as* recombination, *although the term is a misnomer, since protons and electrons had never previously combined to form atoms.*

last interacted before they decoupled. ∞ (*More Precisely 25-2*) On our side of the photosphere—that is, since decoupling—the universe is transparent. On the far side—before decoupling—it was opaque. Thus, by observing the microwave background, we are probing conditions in the universe almost all the way back in time to the Big Bang, in much the same way as studying sunlight tells us about the surface layers of the Sun.

CONCEPT CHECK

✔ How do we know that most of the dark matter in the universe is not of "normal" composition?

27.4 The Inflationary Universe

In the late 1970s, cosmologists trying to piece together the evolution of the universe were confronted with two nagging problems that had no easy explanation within the standard Big Bang model. The resolution of these problems has caused cosmologists to completely rethink their views of the very early universe.

The Horizon and Flatness Problems

The first problem is known as the **horizon problem**, and it concerns the remarkable isotropy of the cosmic microwave background. ∞ (Sec. 26.6) Recall that the temperature of this radiation is virtually constant, at about 2.7 K, in all directions. Imagine observing the microwave background in two opposite directions of the sky, as illustrated in Figure 27.8. As we have just seen, that radiation last interacted with matter in the universe at around a redshift of 1100. Thus, in observing these two distant regions of the universe, marked A and B on the figure, we are studying regions that were separated by several million parsecs at the time they emitted this radiation. The fact that the background radiation is isotropic to high accuracy means that regions A and B had similar densities and temperatures at the time the radiation we see left them. The problem is, according to the Big Bang theory as just described, there is *no* good reason why these regions should in fact be similar to each other.

To take an everyday example, we all know that heat flows from regions of high temperature to regions of low temperature, but it takes time for this to occur. If we light a fire in one corner of a room, we have to wait a while for the other corners to warm up. Eventually, the room reaches a more or less uniform temperature, but only after the heat from the fire—or, more generally, the *information* that the fire is there—has had time to spread. Similar reasoning applies to regions A and B in Figure 27.8. These regions are separated by many megaparsecs, and there has not been enough time for information, which can go no faster than the speed of light, to travel from one to the other. In cosmo-

logical parlance, the two regions are said to be outside each other's *horizon*. But if that is so, then how do they "know" that they are supposed to look the same? With no possibility of communication between them, the only alternative is that regions A and B simply started off looking alike—an assumption that cosmologists are reluctant to make.

The second problem with the standard Big Bang model is called the **flatness problem**. Whatever the exact value of Ω_0, it appears to be very close to unity—the total density of the universe is fairly near the critical value. In terms of spacetime curvature, we can assert that the universe is remarkably close to being flat. ∞ (Sec. 26.4) We say "remarkably" here because, again, there is no good reason that the universe should have formed with a density very close to the critical value. Why not a millionth of, or a million times, that value? Furthermore, as shown in Figure 27.9, a universe that starts off close to, but not exactly on, the critical curve soon deviates greatly from it, so if the universe is close to critical now, it must have been *extremely* close to critical in the past. For example, if $\Omega_0 = 0.3$ today (approximately the density of "known" dark matter), then the departure from critical density at the time of nucleosynthesis would have been only one part in 10^{15} (a thousand trillion)!

These observations constitute "problems" because cosmologists want to be able to *explain* the present condition of the universe, not just accept it "as is." They would

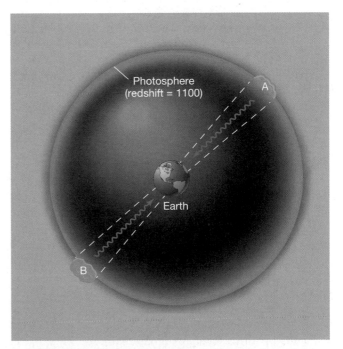

▲ **FIGURE 27.8 Horizon Problem** The isotropy of the microwave background indicates that regions A and B in the universe were similar to each other when the radiation we observe left them, but there has not been enough time since the Big Bang for them ever to have interacted physically with one another. Why, then, should they look the same?

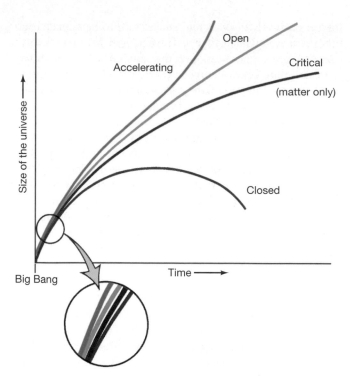

▲ **FIGURE 27.9 Flatness Problem** If the universe deviates even slightly from critical density, that deviation grows rapidly in time. For the universe to be as close to critical as it is today, it must have differed from the critical density in the past by only a tiny amount.

prefer to resolve the horizon and flatness problems in terms of physical processes that could have taken a universe with no special properties and caused it to evolve into the cosmos we now see. The resolution of both problems takes us back in time even earlier than nucleosynthesis or the formation of any of the elementary particles we know today—back, in fact, almost to the instant of the Big Bang itself.

Freeze-Out

Grand Unified Theories predict that three of the four basic forces of nature—electromagnetism and the strong and weak nuclear forces—are in reality aspects of a single, all-encompassing "superforce." (See *More Precisely 27–1*.) However, this unification is evident only at enormously high energies, corresponding to temperatures in excess of 10^{28} K. At lower temperatures, the superforce reveals its separate electromagnetic, strong, and weak characters.

A fundamental concept in quantum physics is the idea that forces between elementary particles are exerted, or mediated, by the exchange of another type of particle, generically called a *boson*. We might imagine the two particles as playing a rapid game of catch, using a boson as a ball, as illustrated in Figure 27.10. As the ball is thrown back and forth, the force is transmitted. In ordinary electromagnetism, the boson involved is the photon—a bundle of electromagnetic energy that always travels at the speed of light. The strong force is mediated by particles known

as *gluons*. The electroweak theory includes a total of four bosons: the massless photon and three other massive particles, called (for historical reasons) W^+, W^-, and Z^0, all of which have been observed in laboratory experiments. Gravity is (theoretically) mediated by *gravitons*. (See *More Precisely 27-1*.) Most of the particles we have encountered so far in this book—electrons, protons, neutrons, neutrinos—play "catch" with at least some of these "balls."

The fundamental forces we know today differ from one another because the particles that mediate these forces are different. However, in the very early universe, when the basic forces were unified, these particles were indistinguishable from one another. There is said to have been a *symmetry* between the forces of nature at those times. As the universe expanded and cooled, this symmetry was broken, and the separate characters of the forces were revealed.

Section 27.1 discussed how particles "froze out" of the universe as its temperature dropped below the threshold temperature for their creation by pair production. Now that we know that the basic forces of nature are also mediated by particles, we can understand—in general terms, at least—how the fundamental forces froze out, too, as the universe cooled. The W and Z particles responsible for the electroweak force have masses about 100 times the mass of a proton. The threshold temperature for their production—roughly 10^{15} K—is the temperature at which the weak and electromagnetic forces parted company near the end of the quark epoch. According to the GUTs, the particle that unifies the strong and electroweak forces is extremely massive—at least 10^{15} times the mass of the proton and possibly much more. It is because this particle is so massive that the unification of the strong and electroweak forces becomes evident only at extremely high temperatures.

The freezing out of the electroweak force at a temperature of 10^{15} K, some 10^{-10} s after the Big Bang, had little overall effect on the cosmic expansion. By contrast, the freezing out of the strong force, a mere 10^{-35} s after the Big Bang, may have produced one of the strangest events in the history of the cosmos.

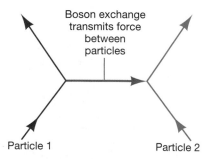

▲ **FIGURE 27.10 Fundamental Forces** Forces between elementary particles are transmitted through the exchange of other particles called bosons. As two particles interact, they exchange bosons, a little like playing catch with a submicroscopic ball.

Cosmic Inflation

As the universe expanded and the temperature fell below 10^{28} K, the strong force appeared for the first time as a separate entity—a little like a gas liquefying or water freezing as the temperature drops. Theory suggests that some parts of the universe briefly entered a very odd and unstable state that physicists call a "false vacuum." These regions are of direct interest to us because—if theorists are correct—we live in one!

In essence, due to random fluctuations at the quantum level, those regions found themselves in the "unified" condition a little too long, like water that has been cooled below freezing, but has not yet turned to ice. The temporary appearance of the false vacuum within such a region—ours, say—had dramatic consequences. For a short while, empty space acquired an enormous *pressure*, which overwhelmed the pull of gravity and caused the region to expand at a greatly accelerated rate. The pressure remained constant as the expansion proceeded, and the acceleration increased with time—in fact, the size of the region *doubled* every 10^{-34} s or so! This period of unchecked expansion, illustrated in Figure 27.11, is known as the **epoch of inflation**.

Eventually, the region reached its normal "true vacuum" state, and inflation stopped. The whole episode lasted a mere 10^{-32} s, but during that time the patch of the universe that had become unstable swelled in size by the incredible factor of about 10^{50}. After the inflationary phase,

the grand unified force was gone forever. In its place were the more familiar electroweak and strong forces that operate around us in the low-temperature universe of today. The universe once again resumed its (relatively) leisurely expansion. However, a number of important changes had occurred that would have far-reaching ramifications for the evolution of the cosmos.

Note that it is possible—even likely, according to many theorists—that not all of the universe underwent inflation. Only some regions became unstable, causing huge inflated "bubbles" to appear in the cosmos. We apparently live in one such bubble; the universe outside is probably unknowable to us. Henceforth, we will use the term "universe" to refer to just this bubble and its contents.

The theory just described, and the association of inflation with the end of the GUT epoch, was developed mainly in the 1980s. However, since then, researchers have realized that conditions suitable for inflation could have occurred under many different circumstances in the early universe. This generalization actually strengthens inflation as a theory by loosening the restrictions on when it might have occurred, although it casts some doubt on exactly when the inflationary epoch leading to "our" universe occurred. Nevertheless, the basic idea of a *quantum fluctuation* expanding to become the universe we know is now quite well established. Some theorists even speculate that we might be living in a sort of "self-creating universe" that erupted into existence spontaneously from inflation in one such random quantum fluctuation! This sort of "statistical" creation of the primal cosmic energy from absolutely nothing has been dubbed "the ultimate free lunch."

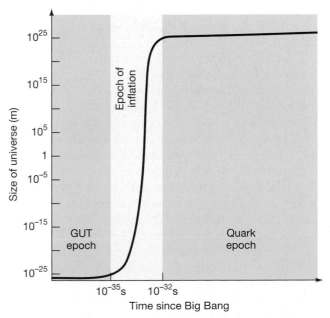

▲ **FIGURE 27.11 Cosmic Inflation** During the period of inflation at the end of the GUT epoch, the universe expanded enormously in a very short time. Afterward, it resumed its earlier "normal" expansion rate, except that then the size of the cosmos was about 10^{50} times bigger than it was before inflation.

Implications for the Universe

The inflationary epoch provides a natural solution to the horizon and flatness problems. The horizon problem is solved because inflation took regions of the universe that had already had time to communicate with one another—and so had established similar physical properties—and then dragged them far apart, well out of communication range of one another. For example, regions A and B in Figure 27.8 have been out of contact since 10^{-32} s after creation, but they were in contact before then. As illustrated in Figure 27.12, their properties are the same today because they were the same long ago, before inflation separated them.

Figure 27.12(a) shows a small piece of the universe just before the onset of inflation. The point that will one day become the site of the Milky Way Galaxy is at the center of the shaded region, which represents the portion of space "visible" to that point at that time—that is, there has been enough time since the Big Bang for light to have traveled from the edge of this region to its center. That entire region is more or less homogeneous, because different parts of it have been able to interact with one another, so any initial differences between the parts have largely been smoothed

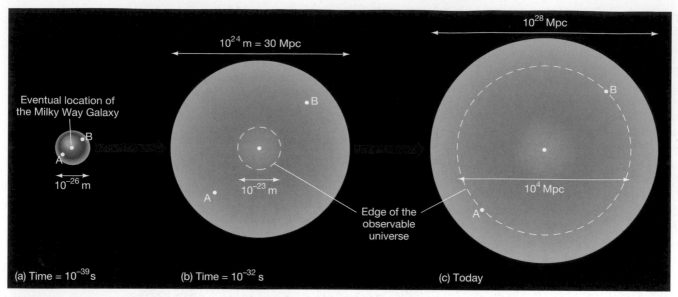

▲ **FIGURE 27.12 Inflation and the Horizon Problem** Inflation solves the horizon problem by taking a small region of the very early universe—whose parts had already had time to interact with one another and that had thus already become homogeneous—and expanding it to enormous size. In (a), points A and B are well within the (shaded) homogeneous region of the universe centered on the eventual site of the Milky Way Galaxy. In (b), after inflation, A and B are far outside the horizon (indicated by the dashed line), so they are no longer visible from our location. Subsequently, the horizon expands faster than the universe as a whole does, so that today (c) A and B are just reentering our field of view. They have similar properties now because they had similar properties before the inflationary epoch.

out. The points A and B of Figure 27.8 are also marked. They lie within the homogeneous patch, so they have very similar properties. The actual size of the shaded region is about 10^{-26} m—only a trillionth the size of a proton.

Immediately after inflation, as shown in Figure 27.12(b), the homogeneous region has expanded by 50 orders of magnitude, to a diameter of about 10^{24} m, or 30 Mpc—larger than the largest supercluster. By contrast, the visible portion of the universe, indicated by the dashed line, has grown only by a factor of a thousand and is still microscopic in size. In effect, the universe expanded much faster than the speed of light during the inflationary epoch, so what was once well within the horizon of the point that is to become the site of our Galaxy now lies far beyond it. In particular, points A and B are no longer visible, either to us or to each other, at this time. (Note that, while the theory of relativity restricts matter and energy to speeds less than the speed of light, it imposes no such limit on the universe as a whole.)

Since the end of inflation, the universe has expanded by a further factor of 10^{27}, so the size of the homogeneous region of space surrounding us is now about 10^{51} m (10^{28} Mpc)—10 trillion trillion times greater than the distance to the most distant quasar. As shown in Figure 27.12(c), the horizon has expanded faster than the universe, so points A and B are just now becoming visible again. As the portion of the universe that is now observable from Earth grows in time, it remains homogeneous be-

cause our cosmic field of view is simply reexpanding into a region of the universe that was within our horizon long ago. We will have to wait a very long time—at least 10^{35} years—before the edge of the homogeneous patch surrounding us comes back into view.

To see how inflation solves the flatness problem, let's return to our earlier balloon analogy. ∞ (Sec. 26.2) Imagine that you are a 1-mm-long ant sitting on the surface of the balloon as it expands, as illustrated in Figure 27.13. When the balloon is just a few centimeters across, you can easily perceive the surface to be curved—its circumference is only a few times your own size. When the balloon expands to, say, a few meters in diameter, the curvature of the surface is less pronounced, but perhaps still perceptible. However, by the time the balloon has expanded to a few kilometers across, an "ant-sized" patch of the surface will look quite flat, just as the surface of Earth looks flat to us.

Now imagine that the balloon expands 100 trillion trillion trillion trillion times, as the universe did during the period of inflation. Your local patch of the surface is now completely indistinguishable from a perfectly flat plane, deviating from flatness by no more than one part in 10^{50}. Exactly the same argument applies to the universe: Because it has expanded so much, for all practical purposes the universe is perfectly flat on all scales we can ever hope to observe.

Notice that this resolution of the flatness problem—the universe appears close to being flat because the uni-

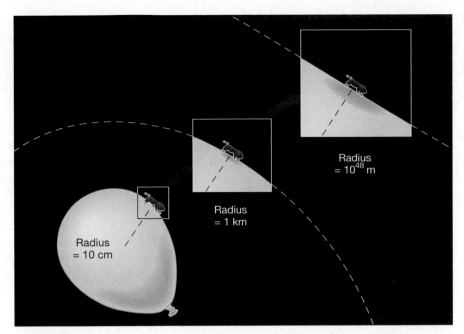

▲ FIGURE 27.13 Inflation and the Flatness Problem Inflation solves the flatness problem by taking a curved surface, here represented by the surface of the expanding balloon, and expanding it enormously in size. To an ant on the surface, the balloon looks virtually flat when the expansion is complete.

verse *is* in fact precisely flat, to very high accuracy—has a very important consequence: Because the universe is geometrically flat, relativity tells us that the total density must be exactly equal to the critical value of $\Omega_0 = 1$. ∞ (Sec. 26.4) This is the key result that led us to conclude in Chapter 26 that dark energy—whatever it is—must dominate the density of the universe. ∞ (Sec. 26.5)

Inflation as a Theory

Even though inflation solves the horizon and flatness problems in a quite convincing way, for nearly two decades after it was first proposed the theory was resisted by many astronomers. The main reason was that its prediction of $\Omega_0 = 1$ was clearly at odds with the growing evidence that the density of matter in the universe was no more than 30 percent or so of the critical value. Actually, many cosmologists *had* considered the possibility that a cosmological constant offered a way to account for the remaining 70 percent of the cosmic density, but without independent corroboration a conclusive case could not be made. That is why the supernova observations were so important: By providing empirical evidence for acceleration in the cosmic expansion rate, they established independent evidence for the effects of dark energy and, in doing so, reconciled inflation with the otherwise discrepant observations. ∞ (Sec. 26.3)

Thus, the combined weight of theory and observation force us to the conclusion that not only is most matter dark (Section 27.3), but *most of the cosmic density isn't made up of matter at all*. In a sense, this is the ultimate statement of the Copernican principle: Not only is Earth in no way central to the cosmos, but the stuff of which we are made is unrepresentative of matter in general, and matter itself is the minority constituent of the universe!

Physicists will probably never create in terrestrial laboratories conditions even remotely similar to those which existed in the universe during the inflationary epoch. The creation of a false vacuum is (safely) beyond our reach. Nevertheless, cosmic inflation seems to be a natural consequence of many Grand Unified Theories. It explains two otherwise intractable problems within the Big Bang theory—and, following the empirical observations of cosmic acceleration, it is now reconciled with measurements of the matter density of the universe via the inclusion of dark energy into the cosmic mix. For all these reasons, despite the absence of direct evidence for the process, inflation theory has become an integral part of modern cosmology. Inflation makes definite, testable predictions about the large-scale geometry and structure of the universe that are critically important to current theories of galaxy formation. As we will see in the next section, astronomers are now subjecting these predictions to rigorous scrutiny.

CONCEPT CHECK

✔ Why does the theory of inflation imply that much of the energy density of the universe may be neither matter nor radiation?

27.5 The Formation of Structure in the Universe

Just as stars formed from *inhomogeneities*—deviations from perfectly uniform density—in interstellar clouds, so galaxies, galaxy clusters, and larger structures are thought to have grown from small density fluctuations in the matter of the expanding universe. ∞ (Sec. 19.1) Given the conditions in the universe during the atomic and galactic epochs (Table 27.1), cosmologists calculate that regions of higher-than-average density which contained more than about a million times the mass of the Sun would have begun to contract. There was thus a natural tendency for million-solar-mass "pregalactic" objects to form. In Chapter 25, we learned how these pregalactic fragments might have interacted and merged to form galaxies. ∞ (Sec. 25.3) In the rest of this chapter, we concern ourselves mostly with the formation of structure on much larger scales.

The Growth of Inhomogeneities

By the early 1980s, cosmologists had come to realize that galaxies could not have formed from the contraction of inhomogeneities involving only *normal* matter. The following lines of reasoning led to this conclusion:

1. Calculations show that, before decoupling (which occurred at a redshift of 1100), the intense background radiation would have prevented clumps of normal matter from contracting. Matter and radiation were just too strongly coupled for structure to form. Thus, any such clumps would have had to wait until after decoupling before their densities could start to increase.

2. Because radiation was "tied" to normal matter up until decoupling, any variations in the matter density at that time would have led to temperature variations in the cosmic background radiation—denser regions would have been a little hotter than less dense ones. The high degree of isotropy observed in the microwave background indicates that any density variations from one region of space to another during the epoch of decoupling must have been small—at most a few parts in 10^5. ∞ (Sec. 26.7)

3. Galaxies—or, at least, quasars—are known to have formed at a redshift of 6, and some theorists believe that, in order to produce the densest galactic nuclei we see today, the formation process must have already been well established as long ago as a redshift of 10 or 20. ∞ (Sec. 25.3)

4. Theory shows that, because the contracting matter had to "fight" the general expansion of the universe, contracting pregalactic clumps could have increased in density by a factor of at most 50 to 100 in the time

available (even with fairly optimistic assumptions). As a result, the small inhomogeneities permitted by observations of the microwave background could not have grown into galaxies in the time available—the universe would still have been almost perfectly homogeneous at a time when we know galaxies had already formed.

Put another way, if galaxies had grown from density fluctuations in the normal-matter component of the early universe, then the fluctuations would have had to be so large as to leave a clearly observable imprint on the cosmic microwave background. That imprint is not observed.

Dark Matter

Fortunately for cosmology (and for life on Earth), much of the universe is made of *dark* matter, which has properties quite different from those of normal matter and which provides a natural explanation for the large-scale structure we see today. Whatever the nature of dark matter, its defining property is that it interacts only very weakly with normal matter and radiation, so its natural tendency to clump and contract under gravity was not hindered by the radiation background. Dark matter started clumping well before decoupling (redshift 1100)—in fact, density inhomogeneities in the dark-matter component of the universe probably began to grow as soon as matter first began to dominate the universe at a redshift of about 6000. Because the dark matter was not directly tied to the radiation, these inhomogeneities could have been quite large at the time of decoupling, without having a correspondingly large effect on the microwave background. In short, dark matter could clump to form large-scale structure in the universe without running into any the problems just described for normal matter.

Thus, as illustrated in Figure 27.14, dark matter determined the overall distribution of mass in the universe and clumped to form the observed large-scale structure without violating any observational constraints on the microwave background. Then, at later times, normal matter was drawn by gravity into the regions of highest density, eventually forming galaxies and galaxy clusters. This picture explains why so much dark matter is found outside the visible galaxies. The luminous material is strongly concentrated near the density peaks and dominates the dark matter there, but the rest of the universe is largely devoid of normal matter. Like foam on the crest of an ocean wave, the universe we can see is only a tiny fraction of the total.

Given that the nature of the dark matter is still unknown, theorists have considerable freedom in choosing its properties when they attempt to simulate the formation of structure in the universe. Cosmologists conventionally classify dark matter as either "hot" or "cold," on the basis of its temperature at the time when galaxies began to form. The two types predict quite different kinds of structure in the present-day universe.

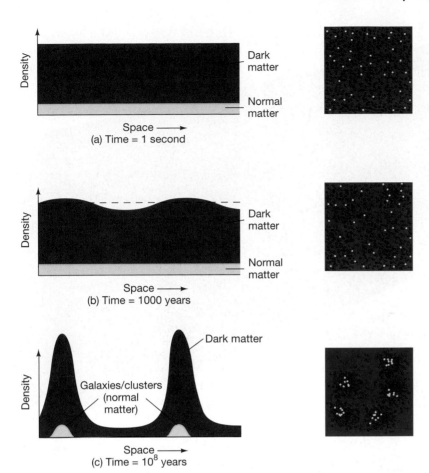

◄ **FIGURE 27.14 Structure Formation** The formation of structure in the cosmos depended crucially on the existence of dark matter. (a) The universe started out as a mixture of (mostly) dark and normal matter. (b) A few thousand years after the Big Bang, the dark matter began to clump. (c) Eventually, the dark matter formed large structures (represented here by the two high-density peaks) into which normal matter flowed, ultimately to form the galaxies we see today. The three frames at the right represent the densities of dark matter (red) and normal matter (yellow) in the respective graphs at the left.

Hot dark matter consists of lightweight particles—much less massive than the electron. Neutrinos, which appear to have small, but nonzero, masses, are leading candidates for hot dark-matter particles. ∞ (Sec. 16.6) However, simulations of a universe filled with hot dark matter indicate that, while large structures, such as superclusters and voids, form fairly naturally, structure on smaller scales does not. Small amounts of hot material tend to disperse, not clump together. As a result, most cosmologists have concluded that models based on hot dark matter are unable to explain the observed structure of the universe.

Cold dark matter consists of very massive particles, possibly formed during the GUT epoch or even before. Computer simulations modeling the universe with these particles as the dark matter easily produce small-scale structure. With the understanding that galaxies form preferentially in the densest regions—as is predicted particularly by models that include a cosmological constant—these models also predict large-scale structure that is in excellent agreement with what is actually observed.

Figure 27.15 shows the results of a recent supercomputer simulation of a "best-bet" universe consisting of 4 percent normal matter, 23 percent cold dark matter, and 73 percent dark energy (in the form of a cosmological constant). ∞ (Sec. 26.5) Yellow dots represent regions in each frame where significant star formation is occurring—

quasars at redshift 6 and bright interacting galaxies today. The similarities with real observations of cosmic structure, shown in Figures 25.21, 25.23, and 25.29, are striking, and more detailed statistical analysis confirms that these models agree extremely well with reality. Notice the large-scale extended filamentary structure evident in the last two frames, which may be compared to the observed structure presented earlier in the text. The visible galaxies are also surrounded by extensive dark-matter halos. Although calculations like this cannot prove that these models are the correct description of the universe, the agreement in detail between models and reality strongly favor the cosmological-constant–cold-dark-matter model of the cosmos.

The Microwave Background

Because dark matter does not interact directly with photons, its density variations do not cause large (and easily observable) temperature variations in the microwave background. However, that radiation is influenced slightly by the *gravity* of the growing dark clumps, which produces a slight gravitational redshift that varies from place to place, depending on the dark-matter density. As a result, cosmological models predict that there should be tiny "ripples" in the microwave background—temperature variations of only a few parts per million from place to place on the sky.

Until the late 1980s these ripples were too small to be measured accurately, although cosmologists were confident

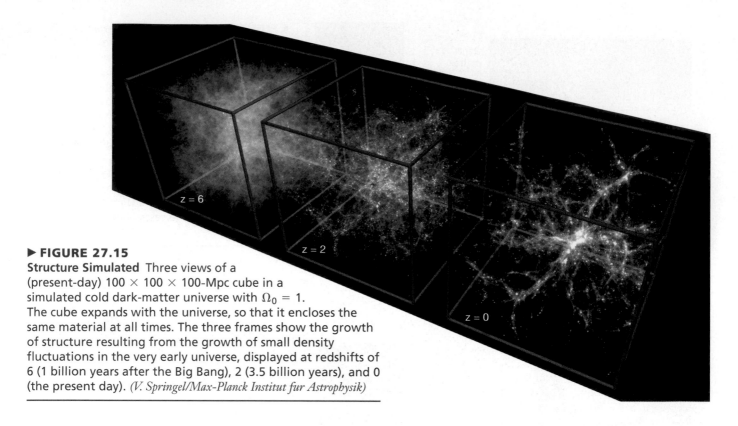

▶ **FIGURE 27.15**
Structure Simulated Three views of a (present-day) $100 \times 100 \times 100$-Mpc cube in a simulated cold dark-matter universe with $\Omega_0 = 1$. The cube expands with the universe, so that it encloses the same material at all times. The three frames show the growth of structure resulting from the growth of small density fluctuations in the very early universe, displayed at redshifts of 6 (1 billion years after the Big Bang), 2 (3.5 billion years), and 0 (the present day). (*V. Springel/Max-Planck Institut für Astrophysik*)

that they would be found. In 1992, after almost two years of careful observation, the *COBE* team announced that the expected ripples had indeed been detected. ∞ (Sec. 26.7) The temperature variations are tiny—only 30–40 millionths of a kelvin from place to place in the sky—but they are there. The *COBE* results are displayed as a temperature map of the microwave sky in Figure 27.16. The temperature variation due to Earth's motion (see Figure 26.18) has been subtracted out, as has the radio emission from the Milky Way, and temperature deviations from the average are displayed.

The ripples seen by *COBE*, combined with computer simulations such as that shown in Figure 27.15, predict present-day structure that is consistent with the superclusters, voids, filaments and Great Walls we see around us. Although the *COBE* data were limited by relatively low (roughly 7°) resolution, detailed analysis of the ripples also supports the key prediction of inflation theory—that the universe is of exactly critical density and hence spatially flat. For these reasons, the *COBE* observations rank alongside the discovery of the microwave background itself in terms of their importance to the field of cosmology.

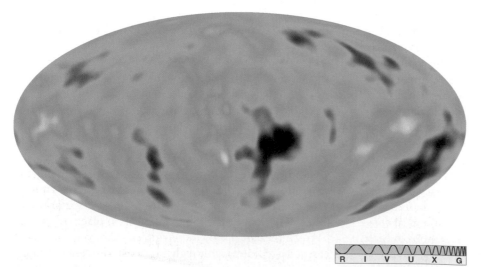

◀ **FIGURE 27.16** Cosmic Microwave Background Map A *COBE* map of temperature fluctuations in the cosmic microwave background over the entire sky. Hotter-than-average regions are in yellow, cooler-than-average regions in blue. The total temperature range shown is ±200 millionths of a kelvin. The temperature variation due to Earth's motion has been subtracted out, as has the radio emission from the Milky Way Galaxy, and temperature deviations from the average are displayed. (*NASA*)

R I V U X G

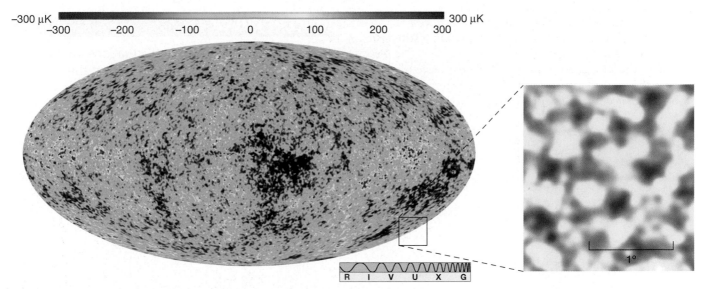

▲ **FIGURE 27.17** A Flat Universe The entire microwave sky, as seen by the *WMAP* spacecraft at frequencies between 22 and 90 GHz (wavelengths ranging from 14 to 3 mm). This image, taken in 2003, may be compared directly with the *COBE* map shown in Figure 27.16. The effective resolution of the *WMAP* multiwavelength map is about 20–30' (compared with 7° for *COBE*). The inset at the right shows a 7'-resolution map of a roughly 2° patch of the sky, obtained by the ground-based *Cosmic Background Imager* instrument at 30 GHz (1 cm wavelength). The bright blobs are slightly denser than average regions of the universe at an age of roughly 400,000 years; they will eventually collapse to form clusters of galaxies. *(NASA)*

The decade following the end of the *COBE* mission in late 1993 has seen dramatic improvements in both the resolution and sensitivity of microwave background measurements. Figure 27.17 shows results returned by two recent experiments. Both have radically improved our view of the microwave background, confirming and refining the *COBE* results.

The main image in Figure 27.17 shows an all-sky map made by the *Wilkinson Microwave Anisotropy Probe* (*WMAP*), launched by NASA in 2001. (See *Discovery 27-1*.) Permanently stationed some 1.5 million km outside Earth's orbit along the Sun–Earth line and always pointing away from the Sun to keep its delicate heat-sensitive detectors in shadow, *WMAP* completes a scan of the entire sky every six months. The instrument's angular resolution is roughly 20–30', some 20 times finer than that of *COBE* (cf. Figure 27.16), allowing extraordinarily detailed measurements of many cosmological parameters to be made. The inset shows a smaller scale (just 2° wide), but even higher resolution (7') image returned by *Cosmic Background Imager*, a ground-based microwave telescope located high in the Chilean Andes. (Recall from Chapter 2 that the microwave part of the spectrum is only partly transparent to electromagnetic radiation, so microwave detectors must be placed above as much of Earth's absorbing atmosphere as possible.) ∞ (Sec. 3.3)

Both maps show temperature fluctuations of a few hundred microkelvins, with a characteristic angular scale of about 1°. This temperature range is larger than the fluctuations seen by *COBE* because *COBE*'s low resolution ef-

fectively averaged the data over a large area of the sky, smearing out the peaks and troughs seen in the higher resolution observations.

The most important implications of the data are summarized in Figure 27.18, a complicated-looking plot with a very simple interpretation. The red dots (with error estimates indicated by vertical bars) show the amount of structure seen in the data on different angular scales (more technically, they show the correlation between different data points separated by those scales) and clearly show a peak at about 1 degree—pretty much what your eye sees in Figure 27.17. The data points are based on observations made by *WMAP* and several other instruments over the past few years. The blue curve is *not* a fit to the data. Rather, it shows a theoretical prediction for a universe with $\Omega_0 = 1$ (actually, with 27 percent matter and 73 percent dark energy, as quoted in Chapter 26 and earlier in the current chapter). The agreement is striking. In fact, the location of the large peak in the theoretical curve is a *direct* measure of the value of Ω_0. These data indicate that Ω_0 is very close to unity, in full agreement with the prediction from inflation, with a very small margin of error.

The other features—the smaller "bumps and wiggles"—on the curve depend on other cosmic parameters and provide a wealth of information on the history and composition of the universe. Indeed, the *WMAP* data are our primary source for the cosmological parameters used in *More Precisely 24-1* and throughout this text. Future ground- and space-based missions (such as ESA's *Planck* mission, scheduled for 2005) will further refine these observations, improving their accuracy and making still more detailed measurements of the microwave sky. The first decade of the 21st century may well see the basic parameters of the universe measured (even if not fully understood) to an accuracy only dreamed of just a few years ago.

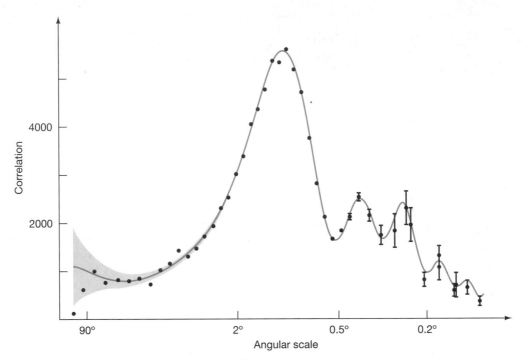

▲ **FIGURE 27.18 Cosmic Structure** The most obvious features on the high-resolution map of the microwave background shown in Figure 27.17 occur on a scale of about 1°—the approximate size of the "blobs" in the inset to that figure. The graph shown here quantifies this impression by plotting the correlation (vertical axis) between different points on the map separated by a given angle (horizontal axis). In effect, the height of the graph measures the "amount" of structure seen on different angular scales. The blue curve is a theoretical prediction for a flat universe with the cosmological parameters quoted in the text. The excellent agreement between theory and observations is regarded by most cosmologists as strong evidence that we live in a flat, critical-density universe.

Chapter Review

SUMMARY

At present, the density of matter in the universe greatly exceeds the equivalent mass density of radiation: The universe is **matter dominated (p. 718)**. The density of matter was much greater in the past, when the universe was smaller. However, because radiation is redshifted as the universe expands, the density of radiation was greater still. Thus, the early universe was **radiation dominated (p. 719)**. During the first few minutes after the Big Bang, matter was formed out of the primordial fireball by the process of **pair production (p. 720)**. In the early universe, matter and radiation were linked by this process. Particles "froze out" of the radiation background as the temperature fell below the threshold for creating them. The existence of matter today means that there must have been unequal amounts of matter and antimatter early on.

The physical state of the universe can be understood in terms of present-day physics back to about 10^{-43} s after the Big Bang. Before that, the four fundamental forces of nature—gravity, electromagnetism, the strong force, and the weak force—were all indistinguishable. There is presently no theory that can describe these extreme conditions. As the universe expanded and its temperature dropped, the forces became distinct from one another. First gravity, then the strong force, and then the weak and electromagnetic forces separated out.

Only a small fraction of the helium observed in the universe today was formed in stars. Most of it was created by **primordial nucleosynthesis (p. 726)** in the early universe. Some deuterium was also formed at these early times, and it provides a sensitive indicator of the present density of the universe in the form of "normal" (as opposed to dark) matter. Studies of deuterium indicate that normal matter can account for at most 3 or 4 percent of the critical density. The remaining mass inferred from studies of clusters must then be made of dark matter, in the form of unknown particles formed at some very early epoch. When the universe was about 1100 times smaller than it is today, the temperature became low enough for atoms to form. At that time, the (then-optical) background radiation **decoupled (p. 728)** from the matter. The universe became transparent. The photons that now make up the microwave background have been traveling freely through space ever since.

Why should regions of the universe that have not had time to "communicate" with one another look so similar? This is called the **horizon problem (p. 729)**. Inflation solves it by taking a small homogeneous patch of the early universe and expanding it enormously in size. The patch is still homogeneous, but it is now much larger than the portion of the universe we can see today. Why should the density of the universe be so near the critical value? This is called the **flatness problem (p. 729)**. According to modern **Grand Unified Theories (p. 722)**, the three nongravitational forces of nature began to display their separate characters about 10^{-35} s after the Big Bang. At that time, or possibly earlier, a brief period of rapid cosmic expansion called the **epoch of inflation (p. 731)** occurred, during which the size of the universe increased by a factor of about 10^{50}. Inflation implies that the total cosmic density is in fact exactly critical. Recent observations from the *WMAP* mission strongly support this conclusion. In that case, most of that density is in the form of dark energy, with matter (normal and dark) making up only about one-third of the total.

The large-scale structure observed in the universe could not have formed out of density fluctuations in normal gaseous matter—there simply has not been enough time, given the twin constraints of the smoothness of the microwave background and the epoch at which the first galaxies and quasars are known to have formed. Instead, dark matter clumped and grew to form the "skeleton" of the structure now observed. Normal matter then flowed into the densest regions of space, eventually forming the galaxies we now see. Cosmologists distinguish between hot dark matter and **cold dark matter (p. 735)**, depending on the temperature of dark matter at the end of the Radiation Era. In order to explain the observed large-scale structure in the universe, much of the dark matter must be cold. In 1992, the *COBE* satellite discovered the expected ripples in the cosmic microwave background associated with the clumping of dark matter at early times. Subsequently, the *WMAP* satellite and numerous ground-based instruments have made detailed observations of the microwave sky, determining the basic cosmological parameters to unprecedented accuracy.

REVIEW AND DISCUSSION

1. For how long was the universe dominated by radiation? How hot was the universe when the dominance of radiation ended?

2. What was the role of dark energy in the very early universe?

3. Why is our knowledge of the Planck epoch so limited?

4. When and how did the first atoms form?

5. Describe the universe at the end of the galactic epoch.

6. Why do all stars, regardless of their abundance of heavy elements, seem to contain at least one-quarter helium by mass?

7. Why didn't heavier and heavier elements form in the early universe, as they do in stars?

8. If large amounts of deuterium formed in the early universe, why do we see so little deuterium today?

9. How do measurements of the cosmic abundance of deuterium provide a reliable estimate of the density of normal matter?

10. How do we know that most matter in the universe is not "normal"?

11. When did the universe become transparent to radiation?

12. How can we observe the epoch at which the universe became transparent?

13. What is the epoch of inflation, and what happened to the early universe during that time?

14. How does inflation solve the horizon problem?

15. How does inflation solve the flatness problem?

16. What does inflation tell us about the total density of the universe?

17. What is the difference between hot and cold dark matter?

18. What is the connection between dark matter and the formation of large- and small-scale structures?

19. Why were the observations made by the *COBE* satellite so important to cosmology?

20. What key measurement was made by the *WMAP* experiment?

CONCEPTUAL SELF-TEST: TRUE OR FALSE/MULTIPLE CHOICE

1. The light emitted from all the stars currently in the universe far outshines the cosmic microwave background.

2. The present-day abundance of deuterium tells us that most of the matter in the universe is dark.

3. The microwave background radiation last interacted with matter around the time of decoupling.

4. The flatness problem is the fact that the observed density of matter is unexpectedly different from the critical density.

5. The universe grew in size by a factor of about 10,000 during the inflationary period.

6. The theory of inflation predicts that the density of the universe is exactly equal to the critical density.

7. Physicists have detected cold dark-matter particles in terrestrial laboratories.

8. Combining inflation with measurements of the density of dark matter leads to the conclusion that most of the universe is composed of neutrinos.

9. We know that density fluctuations in the normal-matter component of the early universe must have been very small because otherwise we would see their imprint on the microwave background.

10. The *WMAP* experiment found that the universe is spatially flat.

11. Immediately after its birth, the universe (a) was dominated by photons; (b) was made mostly of protons; (c) had equal amounts of matter and antimatter; (d) formed stars and galaxies.

12. Present-day Grand Unified Theories unite all of the fundamental forces except (a) the strong force; (b) the weak force; (c) the electromagnetic force; (d) the gravitational force.

13. About half a million years after the Big Bang, the universe had cooled to the point that (a) protons and electrons could combine to form atoms; (b) particle–antiparticle annihilation ceased; (c) gas could condense to form stars; (d) carbon condensed to make dust.

14. One of the problems with the standard Big Bang model is that (a) galaxies are redshifted; (b) the temperature is almost exactly the same everywhere; (c) the universe is hottest in the center; (d) the galaxy will expand forever.

15. According to our best estimates, the line that best describes the universe in Figure 27.9 ("Flatness Problem") is (a) accelerating; (b) open; (c) critical; (d) closed.

16. It is likely that the density of the universe is made up mostly of (a) hydrogen; (b) electromagnetic radiation; (c) dark energy; (d) cold dark matter.

17. The horizon problem in the standard Big Bang model is solved by having the universe (a) accelerate; (b) inflate rapidly early in its existence, (c) have tiny, but significant fluctuations in temperature; (d) be geometrically flat.

18. The structure we observe in the universe is the result of (a) dark matter clumping long ago; (b) galaxies colliding; (c) the freezing out of electrons; (d) radiation dominance in the early universe.

19. Elements more massive than lithium were not formed in the early universe because the temperature was (a) too high; (b) too low; (c) not related to density; (d) unstable.

20. Matter and energy clumping in the early universe result in (a) the formation of atoms; (b) rapid inflation; (c) small but observable red shifts (d) lower temperatures.

PROBLEMS

 Algorithmic versions of these questions are available in the Practice Problems module of the Companion Website at astro.prenhall.com/chaisson.

The number of squares preceding each problem indicates its approximate level of difficulty.

1. ■ What was the distance between the points that would someday become, respectively, the center of the Milky Way Galaxy and the center of the Virgo Cluster at the time of decoupling? (The present separation is 18 Mpc.)

2. ■■ What was the equivalent mass density of the cosmic radiation field when the universe was one-thousandth its present size? (*Hint*: Don't forget the cosmological redshift!)

3. ■ Assuming critical density today, what were the temperature and density of the universe when the first quasars formed?

4. ■■ Of matter and radiation, which dominated the universe, and by what factor in density (assuming critical density today), at the start of (a) decoupling, (b) nucleosynthesis?

5. ■■■ Estimate the temperature needed for electron–positron pair production. The mass of an electron is 9.1×10^{-31} kg. Use $E = mc^2$ to find the energy involved in the reaction

(*More Precisely 22-1*). Then use $E = hf$ and $\lambda f = c$ to find the wavelength λ of the two photons contributing to that energy (see Section 4.2 and Section 27.1). Finally, use Wien's law to find the temperature at which a blackbody spectrum peaks at that wavelength (Section 3.4). How does your answer compare with the threshold temperature given in the text?

6. ■■ Given that the threshold temperature for the production of electron–positron pairs is about 6×10^9 K and that a proton is 1800 times more massive than an electron, calculate the threshold temperature for proton–antiproton pair production.

7. ■ At what wavelength did the background radiation peak at the start of the epoch of nucleosynthesis? In what part of the electromagnetic spectrum does this wavelength lie?

8. ■■ By what factor did the volume of the universe increase during the epoch of primordial nucleosynthesis, from the time when deuterium could first survive until the time at which all nuclear reactions ceased? By what factor did the matter density of the universe decrease during that period?

9. ■ From Table 24.1, the "photosphere" of the universe corresponding to the epoch of decoupling presently lies some 14,000 Mpc from us. (See Figure 27.7.) How far away was a point on the photosphere when the background radiation we see today was emitted?

10. ■■ Calculate the mass, in kilograms, of the particle that unifies the strong and electroweak forces if it froze out at a temperature of 10^{28} K. Calculate the mass of a hypothetical particle that might unify gravity with the other forces, given that the particle froze out at the end of the Planck epoch (Table 27.1).

11. ■ How many times did the universe double in size during the inflationary period if it expanded by a factor of 10^{50}?

12. ■■ What would be the radius of a trillion-solar-mass patch of a homogeneous, critical-density universe (a) today? (b) at decoupling? (c) at the end of the nuclear epoch?

13. ■■ The blobs evident in the inset to Figure 27.17 are about $20'$ across. If those blobs represent clumps of matter around the time of decoupling (redshift = 1100), estimate the diameter of the clumps at the time of decoupling, assuming Euclidean geometry.

14. ■ If dark-matter particles are remnants of the GUT epoch, they must have decoupled from the rest of the matter and radiation in the universe at the end of that epoch. Repeat Problem 9, but for dark-matter particles instead of electromagnetic photons, assuming a present photosphere distance of 14,000 Mpc.

15. ■■■ Assuming critical density today, what size and mass scale correspond to the peak of the *WMAP* data (Figure 27.18), at an angular scale of $1°$ and a redshift of 1100?

 In addition to the Practice Problems module, the Companion Website at astro.prenhall.com/chaisson provides for each chapter a study guide module with multiple choice questions as well as additional annotated images, animations, and links to related Websites.

28

Is there intelligent life elsewhere in the universe? This fanciful painting by noted space artist Dana Berry suggests the view from an alien world well beyond our own solar system—a planetary body residing far out in the halo of our Galaxy. Despite blockbuster movies, science-fiction novels, and a host of claims for extraterrestrial contact, astronomers have found no unambiguous evidence of life, ▶ intelligent or otherwise, anywhere else in the universe.

Life in the Universe

Are We Alone?

Are we unique? Is life on our planet the only example of life in the universe? These are difficult questions, for the subject of extraterrestrial life is one on which we have no data, but they are important questions with profound implications for the human species. Earth is the only place in the universe where we know for certain that life exists. In this chapter, we take a look at how humans evolved on Earth and then consider whether those evolutionary steps might have happened elsewhere. Having done that, we will assess the likelihood of our having galactic neighbors and consider how we might learn about them if they exist.

LEARNING GOALS

Studying this chapter will enable you to

1 Summarize the process of cosmic evolution as it is currently understood.

2 Evaluate the chances of finding life elsewhere in the solar system.

3 Summarize the various probabilities used to estimate the number of advanced civilizations that might exist in the Galaxy.

4 Discuss some of the techniques we might use to search for extraterrestrials and to communicate with them.

WWW Visit astro.prenhall.com/chaisson for additional annotated images, animations, and links to related sites for this chapter.

742

28.1 Cosmic Evolution

Figure 28.1 identifies seven major phases in the history of the universe: *particulate, galactic, stellar, planetary, chemical, biological,* and *cultural* evolution. Together, these evolutionary stages make up the grand sweep of **cosmic evolution**—the continuous transformation of matter and energy that has led to the appearance of life and civilization on Earth. The first four phases represent, in reverse order, the contents of this book. We now briefly expand our field of view beyond astronomy to include the other three.

From the Big Bang, to the formation of galaxies, to the birth of the solar system, to the emergence of life, to the evolution of intelligence and culture, the universe has evolved from simplicity to complexity. We are the result of an incredibly complex chain of events that spanned billions of years. Were those events random, making us unique, or are they in some sense *natural*, so that technological civilization is inevitable? Put another way, are we alone in the universe, or are we just one among countless other intelligent life-forms in our Galaxy? In this chapter, we consider the development of life on Earth and try to assess the likelihood of finding intelligent life elsewhere in the cosmos.

Life In The Universe

Before embarking on our study, we need a working definition of *life*. Defining life, however, is not an easy task: The distinction between the living and the nonliving is not as obvious as we might at first think. Although most physicists would agree on the definitions of matter and energy, biologists have not arrived at a clear-cut definition of life. Generally speaking, scientists regard the following as characteristics of living organisms: (1) They can *react* to their environment and can often heal themselves when damaged; (2) they can *grow* by taking in nourishment from their surroundings and processing it into energy; (3) they can *reproduce*, passing along some of their own characteristics to their offspring; and (4) they have the capacity for genetic change and can therefore *evolve* from generation to generation and adapt to a changing environment.

These rules are not strict, and there is great leeway in interpreting them. Stars, for example, react to the gravity of their neighbors, grow by accretion, generate energy, and "reproduce" by triggering the formation of new stars, but no one would suggest that they are alive. By contrast, a virus (see *Discovery 28-1*) is inert when isolated from living organisms, but once inside a living system, it exhibits all the properties of life, seizing control of a living cell and using the cell's own genetic machinery to grow and reproduce. Most researchers now believe that the distinction between living and nonliving matter is more one of structure and complexity than a simple checklist of rules.

The general case in favor of extraterrestrial life is summed up in what are sometimes called the *assumptions of mediocrity*: (1) Because life on Earth depends on just a few basic molecules, and (2) because the elements that make up these molecules are (to a greater or lesser extent) common to all stars, and (3) if the laws of science we know apply to the entire universe, as we have supposed throughout this book, then—given sufficient time—life must have originated elsewhere in the cosmos. The opposing view maintains that intelligent life on Earth is the product of a series of extremely fortunate accidents—astronomical, geological, chemical, and biological events unlikely to have occurred anywhere else in the universe. The purpose of this chapter is to examine some of the arguments for each of these viewpoints.

▼ **FIGURE 28.1 Arrow of Time** Some highlights of cosmic history are noted along this arrow of time, from the beginning of the universe to the present. Along the bottom of the arrow are seven "windows" outlining the major phases of cosmic evolution: the evolution of primal energy into elementary particles; of atoms into galaxies and stars; of stars into heavy elements; of elements into solid, rocky planets; of those same elements into the molecular building blocks of life; of those molecules into life itself; and of advanced life-forms into intelligence, culture, and technological civilization. (*D. Berry*)

DISCOVERY 28-1

The Virus

The central idea of chemical evolution is that life evolved from nonliving molecules. But aside from insight based on biochemical knowledge and laboratory simulations of some key events on primordial Earth, do we have any direct evidence that life could have developed from nonliving molecules? The answer is yes. The smallest and simplest entity that sometimes appears to be alive is a virus. We say "sometimes" because viruses seem to have the attributes of both nonliving molecules and living cells. *Virus* is the Latin word for "poison," an appropriate name, since viruses often cause disease. Although they come in many sizes and shapes—a representative example is the polio virus, shown here magnified 300,000 times—all viruses are smaller than the size of a typical modern cell. Some are made of only a few thousand atoms. In terms of size, then, viruses seem to bridge the gap between cells that are living and molecules that are not.

Viruses contain some proteins and genetic information (in the form of DNA or the closely related molecule RNA, the two molecules responsible for transmitting genetic characteristics from one generation to the next), but not much else—none of the material by which living organisms normally grow and reproduce. How, then, can a virus be considered alive? Indeed, alone, it cannot; a virus is absolutely lifeless when it is isolated from living organisms. But when it is inside a living system, a virus has all the properties of life. Viruses come alive by transferring their genetic material into living cells. The genes of a virus seize control of a cell and establish themselves as the new master of chemical activity. Viruses grow and reproduce copies of themselves by using the genetic machinery of the invaded cell, often robbing the cell of its usual function. Some viruses multiply rapidly and wildly, spreading disease and—if unchecked—eventually killing the invaded organism. In a sense, then, viruses exist within the gray area between the living and the nonliving.

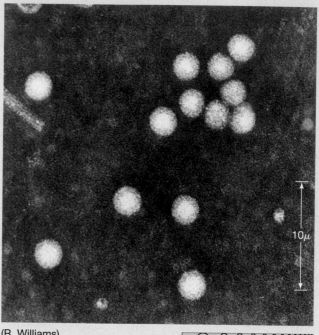

(R. Williams)

10μ

R I V U X G

Chemical Evolution

What information do we have about the earliest stages of planet Earth? Unfortunately, not very much. Geological hints about the first billion years or so were largely erased by violent surface activity, as volcanoes erupted and meteorites bombarded our planet; subsequent erosion by wind and water has seen to it that little evidence of that era has survived to the present. Scientists believe that the early Earth was barren, with shallow, lifeless seas washing upon grassless, treeless continents. Gases emanating from our planet's interior through volcanoes, fissures, and geysers produced an atmosphere rich in hydrogen, nitrogen, and carbon compounds and poor in free oxygen. As Earth cooled, ammonia, methane, carbon dioxide, and water formed. The stage was set for the appearance of life.

The surface of the young Earth was a very violent place. Natural radioactivity, lightning, volcanism, solar ultraviolet radiation, and meteoritic impacts all provided large amounts of energy that eventually shaped the ammonia, methane, carbon dioxide, and water on our planet into more complex molecules known as **amino acids** and **nucleotide bases**—organic (carbon-based) molecules that are the building blocks of life as we know it. Amino acids build *proteins*, and proteins control metabolism, the daily utilization of food and energy by means of which organisms stay alive and carry out their vital activities. Sequences of nucleotide bases form *genes*—parts of the DNA molecule—which direct the synthesis of proteins and thus determine the characteristics of the organism. These same genes, via the DNA contained within every cell in the organism, transfer hereditary characteristics from one generation to the next through reproduction. In all living creatures on Earth—from bacteria to amoebas to humans—genes mastermind life and proteins maintain it.

The idea that complex molecules could have evolved naturally from simpler ingredients found on the primitive Earth has been around since the 1920s. The first experimental verification was provided in 1953, when scientists Harold Urey and Stanley Miller, using laboratory equipment somewhat similar to that shown in Figure 28.2, took a mixture of the materials thought to be present on Earth long ago—a "primordial soup" of water, methane, carbon

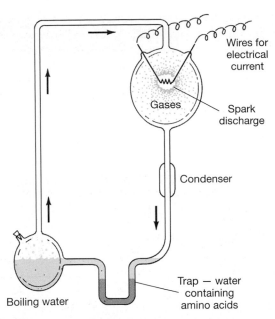

▲ **FIGURE 28.2 Urey–Miller Experiment** This chemical apparatus is designed to synthesize complex biochemical molecules by energizing a mixture of simple chemicals. A mixture of gases (ammonia, methane, carbon dioxide, and water vapor) is placed in the upper bulb to simulate the primordial Earth atmosphere and is then energized by spark-discharge electrodes (to simulate lightning). After about a week, amino acids and other complex molecules are found in the trap at the bottom, which simulates the primordial oceans into which heavy molecules produced in the overlying atmosphere would have fallen.

dioxide, and ammonia—and energized it by passing an electrical discharge ("lightning") through the gas. After a few days, they analyzed their mixture and found that it contained many of the same amino acids found in all living things on Earth. About a decade later, scientists succeeded in constructing nucleotide bases in a similar manner. These experiments have been repeated in many different forms, with more realistic mixtures of gases and a variety of energy sources, but always with the same basic outcomes.

Although none of these experiments has ever produced a living organism, or even a single strand of DNA, they do demonstrate conclusively that "biological" molecules—the molecules involved in the functioning of living organisms—can be synthesized by strictly *non*biological means, using raw materials available on the early Earth. More advanced experiments, in which amino acids are united under the influence of heat, have fashioned proteinlike blobs that behave to some extent like true biological cells. Such near-protein material resists dissolution in water (so it would remain intact when it fell from the primitive atmosphere into the ocean) and tends to cluster into small droplets called microspheres—a little like oil globules floating on the surface of water. Figure 28.3 shows some of these laboratory-made proteinlike microspheres, whose walls permit the inward passage of small molecules,

which then combine within the droplet to construct more complex molecules that are too large to pass back out through the walls. As the droplets "grow," they tend to "reproduce," forming smaller droplets.

Can we consider these proteinlike microspheres to be alive? Almost certainly not. Most biochemists would say that the microspheres are not life itself, but they contain many of the basic ingredients needed to form life. The microspheres lack the hereditary DNA molecule. However, as illustrated in Figure 28.4, they do have similarities to ancient cells found in the fossil record, which in turn have many similarities to modern organisms (such as blue-green algae). Thus, while no actual living cells have yet been created "from scratch" in any laboratory, many biochemists feel that the chain of events leading from simple nonbiological molecules almost to the point of life itself has been amply demonstrated.

An Interstellar Origin?

Recently, a dissenting view has emerged. Some scientists have argued that Earth's primitive atmosphere might *not* in fact have been a particularly suitable environment for the production of complex molecules. These scientists say that there may not have been sufficient energy available to power the necessary chemical reactions and the early atmosphere may not have contained enough raw material for

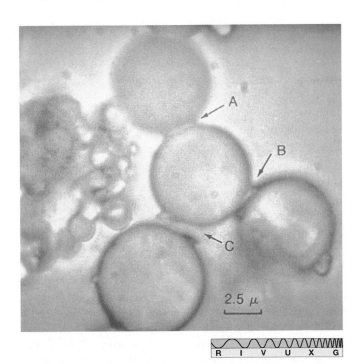

R I V U X G

▲ **FIGURE 28.3 Chemical Evolution** These carbon-rich, proteinlike droplets display the clustering of as many as a billion amino-acid molecules in a liquid. The droplets can "grow," and parts of them can separate from their "parent" droplet to become new individual droplets (as at A, B, and C). The scale of 2.5 microns noted here is 1/4000 of a centimeter. *(S. Fox)*

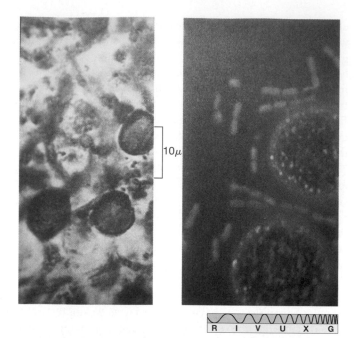

▲ **FIGURE 28.4 Primitive Cells** The photograph on the left, taken through a microscope, shows a fossilized organism found in sediments that radioactive dating indicates are 2 billion years old. This primitive system possesses concentric spheres or walls connected by smaller spheroids. The roundish fossils here measure about a thousandth of a centimeter in diameter. The photograph on the right, also taken through a microscope and on approximately the same scale, displays modern blue-green algae. (*E. Barghoorn*)

the reactions to have become important in any case. They suggest instead that much, if not all, of the organic (carbon-based) material that combined to form the first living cells was produced in *interstellar space* and subsequently arrived on Earth in the form of comets, interplanetary dust, and meteors that did not burn up during their descent through the atmosphere.

Several pieces of evidence support this idea. Interstellar molecular clouds are known to contain complex molecules—indeed, there have even been reports (still unconfirmed) of at least one amino acid (glycine) in interstellar space. ∞ (Sec. 18.5) To test the interstellar space hypothesis, NASA researchers have carried out their own version of the Urey–Miller experiment in which they exposed an icy mixture of water, methanol, ammonia, and carbon monoxide—representative of many interstellar grains—to ultraviolet radiation to simulate the energy from a nearby newborn star. As shown in Figure 28.5, when they later placed the irradiated ice in water and examined the results, they found that the ice had formed droplets surrounded by membranes and containing complex organic molecules. As with the droplets found in earlier experiments, no amino acids, proteins, or DNA were observed in the mix, but the results, repeated numerous times, clearly show that even the harsh, cold vacuum of in-

terstellar space can be a suitable medium in which complex molecules and primitive cellular structures can form.

As described in Chapter 15, these icy interstellar grains are believed to have formed the comets in our own solar system. ∞ (Sec. 15.3) Large amounts of organic material were detected on comet Halley by space probes when Halley last visited the inner solar system, and similarly complex molecules have been observed on many other well-studied comets, such as Hale–Bopp. ∞ (Sec. 14.2, *Discovery 14-2*) Also as discussed in Chapter 15, there is reason to believe that cometary impacts were responsible for most of Earth's water, and it is perhaps a small step to imagining that this water already contained the building blocks for life. ∞ (Sec. 15.3)

In addition, a small fraction of the meteorites that survive the plunge to Earth's surface—including perhaps the controversial "Martian meteorite" discussed in Chapter 10—contain organic compounds. ∞ (*Discovery 10-1*) The Murchison meteorite (Figure 28.6), which fell near Murchison, Australia, in 1969, is a particularly well studied example. Located soon after crashing to the ground, this meteorite has been shown to contain 12 of the amino acids normally found in living cells, although the detailed structures of these molecules indicate potentially important differences between those found in space and those found on Earth. At the very least, though, these discoveries argue that such molecules can form in an interplanetary or

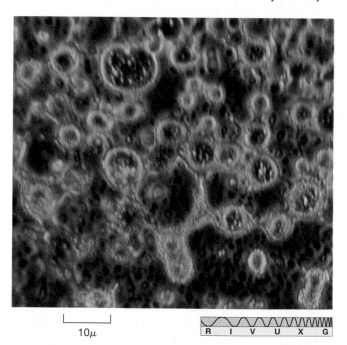

▲ **FIGURE 28.5 Interstellar Globules** These oily, hollow droplets rich in organic molecules were made by exposing a freezing mixture of primordial matter to harsh ultraviolet radiation. The larger ones span about 10 microns across and, when immersed in water, show cell-like membrane structure. Although they are not alive, they bolster the idea that life on Earth could have come from space. (*NASA*)

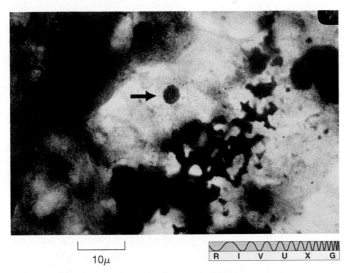

▲ FIGURE 28.6 Murchison Meteorite The Murchison meteorite contains relatively large amounts of amino acids and other organic material, indicating that chemical evolution of some sort has occurred beyond our own planet. In this magnified view of a meteorite fragment, the arrow points to a microscopic sphere of organic matter. *(NASA)*

interstellar environment and that they could have reached Earth's surface unscathed after their fiery descent.

Thus, the hypothesis that organic matter is constantly raining down on Earth from space in the form of interplanetary debris is quite plausible. However, whether this was the *primary* means by which complex molecules first appeared in Earth's oceans remains unclear.

Diversity and Culture

However the basic materials appeared on Earth, we know that life *did* appear. The fossil record chronicles how life on Earth became widespread and diversified over the course of time. The study of fossil remains shows the initial appearance of simple one-celled organisms such as blue-green algae more than 3.5 billion years ago. These were followed by more complex one-celled creatures, like the amoeba, about 2 billion years ago. Multicellular organisms such as sponges did not appear until about 1 billion years ago, after which there flourished a wide variety of increasingly complex organisms—insects, reptiles, and mammals.

The fossil record leaves no doubt that biological organisms have changed over time—all scientists accept the reality of *biological evolution*. As conditions on Earth shifted and Earth's surface evolved, those organisms which could best take advantage of their new surroundings succeeded and thrived—often at the expense of organisms that could not make the necessary adjustments and consequently became extinct. What led to these changes? Chance. An organism that happened to have a certain useful genetically determined trait—for example, the ability to run faster,

climb higher, or even hide more easily—would find itself with the upper hand in a particular environment. That organism was therefore more likely to reproduce successfully, and its advantageous characteristic would then be more likely to be passed on to the next generation. The evolution of the rich variety of life on our planet, including human beings, occurred as chance mutations—changes in genetic structure—led to changes in organisms over millions of years.

What about the development of intelligence? Many anthropologists believe that, like any other highly advantageous trait, intelligence *is* strongly favored by natural selection. As humans learned about fire, tools, and agriculture, the brain became more and more elaborate. The social cooperation that went with coordinated hunting efforts was another important competitive advantage that developed as brain size increased. Perhaps most important of all was the development of language. Indeed, some anthropologists have gone so far as to suggest that human intelligence *is* human language. Through language, individuals could signal one another while hunting for food or seeking protection. Even more importantly, now our ancestors could share ideas as well as food and shelter. Experience, stored in the brain as memory, could be passed down from generation to generation. A new kind of evolution had begun, namely, *cultural evolution*, the changes in the ideas and behavior of society. Within only the past 10,000 years or so, our more recent ancestors have created the entirety of human civilization.

To put all this into historical perspective, let's imagine the entire lifetime of Earth to be 46 years rather than 4.6 billion years. On this scale, we have no reliable record of the first decade of our planet's existence. Life originated at least 35 years ago, when Earth was about 10 years old. Our planet's middle age is largely a mystery, although we can be sure that life continued to evolve and that generations of mountain chains and oceanic trenches came and went. Not until about 6 years ago did abundant life flourish throughout Earth's oceans. Life came ashore about 4 years ago, and plants and animals mastered the land only about 2 years ago. Dinosaurs reached their peak about 1 year ago, only to die suddenly about four months later. ∞ *(Discovery 14-1)* Humanlike apes changed into apelike humans only last week, and the latest ice ages occurred only a few days ago. *Homo sapiens*—our species—did not emerge until about four hours ago. Agriculture was invented within the last hour, and the Renaissance—along with all of modern science—is just three minutes old!

CONCEPT CHECK

✔ Has chemical evolution been verified in the laboratory?

28.2 Life in the Solar System

2 Simple one-celled life-forms reigned supreme on Earth for most of our planet's history. It took time—a great deal of time—for life to emerge from the oceans, to evolve into simple plants, to continue to evolve into complex animals, and to develop intelligence, culture, and technology. Have those (or similar) events occurred elsewhere in the universe? Let's try to assess what little evidence we have on the subject.

Life as We Know It

"Life as we know it" is generally taken to mean carbon-based life that originated in a liquid water environment—in other words, life on Earth. Might such life exist elsewhere in our solar system?

The Moon and Mercury lack liquid water, protective atmospheres, and magnetic fields, so these two bodies are subjected to fierce bombardment by solar ultraviolet radiation, the solar wind, meteoroids, and cosmic rays. Simple molecules could not survive in such hostile environments. Venus, by contrast, has *far too much* protective atmosphere! Its dense, dry, scorchingly hot atmospheric blanket effectively rules it out as an abode for life, at least like us.

The jovian planets have no solid surfaces (although some researchers have suggested that life might have evolved in their atmospheres), and Pluto and most of the moons of the outer planets are too cold. However, the possibility of liquid water below Europa's icy surface has refueled speculation about the development of life there, making this moon of Jupiter a prime candidate for future exploration. ∞ (Sec. 11.5) For now at least, Europa is high on the priority list for missions by both NASA and the Eu-

ropean Space Agency. Saturn's moon Titan, with its atmosphere of methane, ammonia, and nitrogen, and possibly with some liquid on its surface, is conceivably a site where life might have arisen, although the results of the 1980 *Voyager 1* flyby suggest that Titan's frigid surface conditions are inhospitable to anything familiar to us. ∞ (Sec. 12.5)

The planet most likely to harbor life (or to have harbored it in the past) seems to be Mars. The Red Planet seems harsh by Earth standards: Liquid water is scarce, the atmosphere is thin, and the absence of magnetism and an ozone layer allows solar high-energy particles and ultraviolet radiation to reach the surface unabated. But the Martian atmosphere was thicker, and the surface probably warmer and much wetter, in the past. ∞ (Secs. 10.4, 10.5) Indeed, there is strong photographic evidence from orbiters such as *Viking* and *Mars Global Surveyor* for flowing and standing water on Mars in the distant (and perhaps even relatively recent) past. In 2004, the European *Mars Express* orbiter confirmed the long-hypothesized presence of water ice at the Martian poles, and NASA's *Opportunity* rover reported strong geological evidence that the region around its landing site was once "drenched" with water for an extended period.

All of these lines of reasoning strongly suggest that Mars—at least at some time in its past—harbored large amounts of liquid water. However, none of the Mars landers has detected anything that might be interpreted as the remains (fossilized or otherwise) of large plants or animals, and only the *Viking* landers carried equipment capable of performing the detailed biological analysis needed to detect bacterial life (or its fossil remnants). The *Viking* robots scooped up Martian soil (Figure 28.7) and tested it for the presence of life by conducting chemical experiments

◀ **FIGURE 28.7 Search for Martian Life** Several trenches were dug by the *Viking* robots, such as *Viking 2*, seen here on Mars's Utopia Planitia. Soil samples were scooped up and taken inside the robot, where instruments tested them for chemical composition and any signs of life. (*NASA*)

R   I   V   U   X   G

designed to detect the waste gases and other products of metabolic activity, but no unambiguous evidence of Martian life has emerged. ∞ *(Discovery 10-1)* Of course, we might argue that the landers touched down on the safest Martian terrain, not in the most interesting regions, such as near the moist polar caps.

Some scientists have suggested that a different type of biology may be operating on the Martian surface. They propose that Martian microbes capable of eating and digesting oxygen-rich compounds in the Martian soil could also explain the *Viking* results. This speculation would be greatly strengthened if recent announcements of fossilized bacteria in meteorites originating on Mars were confirmed (although it seems that the weight of scientific opinion is currently running against that interpretation of the data). ∞ *(Discovery 10-1)* The consensus among biologists and chemists today is that Mars does not house any life similar to that on Earth, but a solid verdict regarding past life on Mars will not be reached until we have thoroughly explored our intriguing neighbor.

In considering the emergence of life under adversity, however, we should perhaps not be too quick to rule out an environment solely on the basis of its extreme properties. Figure 28.8 shows a very hostile environment on Earth, one where life nonetheless thrives under conditions quite unlike anything on our planet's surface. These are sites of undersea volcanic activity where hydrothermal vents spill forth boiling-hot water from vertical tubes a few meters tall. Rich in sulfur and poor in oxygen, the upwelling water feeds "extremophilic" life by a process known as *chemosynthesis*, an analog of *photosynthesis*, whereby plants turn sunlight into energy. Chemosynthesis, however, operates in total darkness, creating the energy needed for life by purely chemical means. Such underground hot springs might conceivably exist on alien worlds, raising the possibility of life-forms with much greater diversity over a range of conditions much wider than those known to us on Earth.

▲ **FIGURE 28.8 Hydrothermal Vents** A small two-person submarine (the *Alvin*, partly seen at the bottom right) took this picture of a hot spring, or "black smoker"—one of many along the midocean ridge in the eastern Pacific Ocean. As hot water rich in sulfur pours out of the top of the vent's tube (near the center), black clouds billow forth, providing a strange environment for many life-forms thriving near the vent. *(WHOI)*

Alternative Biochemistries

Conceivably, some types of biology might be so different from life on Earth that we would not recognize them and would not know how to test for them. What might these other biologies be?

Some scientists have pointed out that the abundant element silicon has chemical properties somewhat similar to those of carbon and have suggested silicon as a possible alternative to carbon as the basis for living organisms. Ammonia (made of the common elements hydrogen and nitrogen) is sometimes put forward as a possible liquid medium in which life might develop, at least on a planet cold enough for ammonia to exist in the liquid state. Together or separately, these alternatives would surely give rise to organisms with biochemistries (the basic biological and chemical processes responsible for life) radically different from those we know on Earth. Conceivably, we might have difficulty even identifying these organisms as alive.

Although the possibility of such alien life-forms is a fascinating scientific problem, most biologists would argue that chemistry based on carbon and water is the one most likely to give rise to life. Carbon's flexible chemistry and water's wide liquid temperature range are just what are needed for life to develop and thrive. Silicon and ammonia seem unlikely to fare as well as bases for advanced life-forms. Silicon's chemical bonds are weaker than those of carbon and may not be able to form complex molecules—an apparently essential aspect of carbon-based life. Also, the colder the environment, the less energy there is to drive biological processes. The low temperatures necessary for ammonia to remain liquid might inhibit or even

completely prevent the chemical reactions leading to the equivalent of amino acids and nucleotide bases.

Still, we must admit that we know next to nothing about noncarbon, nonwater biochemistries, for the very good reason that there are no examples of them to study experimentally. We can speculate about alien life-forms and try to make general statements about their characteristics, but we can say little of substance about them.

CONCEPT CHECK

✔ Which solar-system bodies (other than Earth) are the leading candidates in the search for extraterrestrial life?

28.3 Intelligent Life in the Galaxy

LEARNING GOAL 3 With humans apparently the only intelligent life in the solar system, we must broaden our search for extraterrestrial intelligence to other stars and perhaps even other galaxies. At such distances, though, we have little hope of actually detecting life with current equipment. Instead, we must ask, "How likely is it that life in any form—carbon based, silicon based, water based, ammonia based, or something we cannot even dream of—exists? Let's look at some numbers to develop estimates of the probability of life elsewhere in the universe.

The Drake Equation

An early approach to this statistical problem is known as the **Drake equation**, after the U.S. astronomer who pioneered the analysis (see below).

Of course, several of the factors in this formula are largely a matter of opinion. We do not have nearly enough information to determine—even approximately—every factor in the equation, so the Drake equation cannot give us a hard-and-fast answer. Its real value is that it subdivides a large and difficult question into smaller pieces that we can attempt to answer separately. The equation provides the framework within which the problem can be addressed and parcels out the responsibility for the final solution among many different scientific disciplines. Figure 28.9 illustrates how, as our requirements become

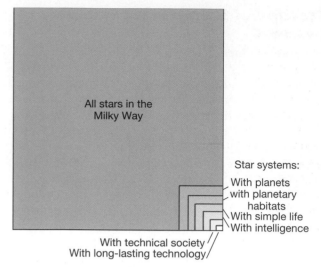

▲ **FIGURE 28.9 Drake Equation** Of all the star systems in our Milky Way (represented by the largest box), progressively fewer and fewer have each of the qualities typical of a long-lasting technological society (represented by the smallest box at the lower right corner).

more and more stringent, only a small fraction of star systems in the Milky Way is likely to generate the advanced qualities specified by the combination of factors on the right-hand side of the equation.

Let's examine the factors in the equation one by one and make some educated guesses about their values. Bear in mind, though, that if you ask two scientists for their best estimates of any given factor, you will likely get two very different answers!

Rate of Star Formation

We can estimate the average number of stars forming each year in the Galaxy simply by noting that at least 100 billion stars now shine in the Milky Way. Dividing this number by the 10-billion-year lifetime of the Galaxy, we obtain a formation rate of 10 stars per year. This rate may be an overestimate, because we think that fewer stars are forming now than formed at earlier epochs of the Galaxy, when more interstellar gas was available. However, we do know that stars are forming today, and our estimate does not include stars that formed in the past and have since died, so our value of 10 stars per year is probably reasonable when averaged over the lifetime of the Milky Way.

PHYSLET® ILLUSTRATION Are We Alone?

| number of technological, intelligent civilizations now present in the Galaxy | = | rate of star formation, averaged over the lifetime of the Galaxy | × | fraction of stars having planetary systems | × | average number of habitable planets within those planetary systems | × | fraction of those habitable planets on which life arises | × | fraction of those life-bearing planets on which intelligence evolves | × | fraction of those intelligent-life planets that develop technological society | × | average lifetime of a technologically competent civilization. |

Fraction of Stars Having Planetary Systems

Many astronomers believe that planet formation is a natural result of the star-formation process. If the condensation theory (Chapter 15) or some variant of it is correct, and if there is nothing special about our Sun, as we have argued throughout this book, then we would expect many stars to have at least one planet. ∞ (Sec. 15.2) Indeed, as we have seen, increasingly sophisticated observations indicate the presence of disks around young stars. Could these disks be protosolar systems? The condensation theory suggests that they are, and the short (theoretical) lifetimes of disks imply the existence of many planet-forming systems in the neighborhood of the Sun.

No planets like our own have yet been *seen* orbiting any other star.* The light reflected by an Earth-like planet circling even the closest star would be too faint to detect even with the best equipment. The light would be lost in the glare of the parent star. Large orbiting telescopes may soon be able to detect Jupiter-sized planets orbiting the nearest stars, but even those huge planets will be barely visible. However, as described in Chapter 15, there is now overwhelming evidence for planets orbiting other stars. ∞ (Sec. 15.5) The planets found so far are Jupiter sized rather than Earth sized and generally have rather eccentric orbits, but astronomers are confident that an Earth-like planet (in an Earth-like orbit) will one day be detected, and plans to build the necessary equipment are well underway.

Accepting the condensation theory and its consequences, and without being either too conservative or naïvely optimistic, we assign a value near unity to this factor—that is, we believe that nearly all stars form with planetary systems of some sort.

Number of Habitable Planets Per Planetary System

Temperature, more than any other single factor, determines the feasibility of life on a given planet. The surface temperature of a planet depends on two things: the planet's distance from its parent star and the thickness of the planet's atmosphere. Planets with a nearby parent star (but not too close) and some atmosphere (though not too thick) should be reasonably warm, like Earth or Mars. Planets far from the star and with no atmosphere, like Pluto, will surely be cold by our standards. And planets too close to the star and with a thick atmosphere, like Venus, will be very hot, indeed.

Figure 28.10 illustrates how a three-dimensional zone of "comfortable" temperatures—often called a *stellar habitable zone*—surrounds every star. The stellar habitable zone represents the range of distances within which a planet of

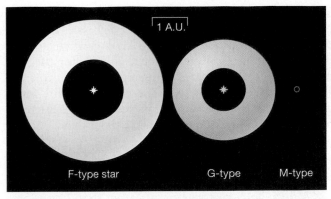

▲ **FIGURE 28.10 Stellar Habitable Zones** The extent of the habitable zone is much larger around a hot star than around a cool one. For a star like the Sun (a G-type star), the zone extends from about 0.85 to 2.0 A.U. For an F-type star, the range is 1.2 to 2.8 A.U. For a cool M-type star, only planets orbiting between about 0.02 and 0.06 A.U. would be habitable.

mass and composition similar to Earth's would have a surface temperature between the freezing and boiling points of water. (Our Earth-based bias is plainly evident here!) The hotter the star, the larger is this zone. For example, A- and F-type stars have a rather large habitable zone, but the size of the zone diminishes rapidly as we proceed through G-, K-, and M-type stars. O- and B-type stars are not considered in this regard because they are not expected to last long enough for life to develop, even if they do have planets. M-type stars, despite their large numbers, have such small habitable zones and, in any case, are thought to be prone to such violent surface activity, that they are not generally considered likely hosts to life-bearing planets. ∞ (Sec. 17.8)

Three planets—Venus, Earth, and Mars—reside within the habitable zone surrounding our Sun. However, Venus is too hot because of its exceptionally thick atmosphere and proximity to the Sun. ∞ (Sec. 9.5) Mars is a little too cold, because its atmosphere is much thinner than Earth's and it lies too far from the Sun. But if Venus had Mars's thin atmosphere, and if Mars had Venus's thick atmosphere, both of these nearby planets might conceivably have surface conditions resembling those on Earth. In that case, our solar system would have had three habitable planets instead of one.

A planet moving on a "habitable" orbit may still be rendered uninhabitable by external events. Many scientists think that the outer planets in our own solar system are critical to the habitability of the inner worlds, both by stabilizing their orbits and by protecting them from cometary impacts, deflecting would-be impactors away from the inner part of the solar system. The theory presented in Chapter 15 suggests that a star with inner terrestrial planets on stable orbits would probably also have the jovian worlds needed to safeguard their survival. ∞ (Sec. 15.3)

*Recall from Chapter 22 that Earth-mass planets have been observed orbiting some neutron stars. However, as we saw in that chapter, the formation of those planets was very different from that of Earth and the solar system ∞ (Sec. 22.3), so they are not thought to be likely candidates for the emergence of life.

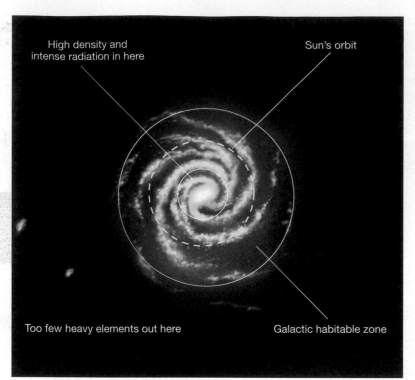

High density and
intense radiation in here

Sun's orbit

Too few heavy elements out here

Galactic habitable zone

◄ **FIGURE 28.11** Galactic Habitable Zone
Some regions of the Galaxy may be more
conducive to life than others. Too far from the
Galactic center, there may not be enough heavy
elements for terrestrial planets to form or
technological society to evolve. Too close, the
radiative or gravitational effects of nearby stars
may render life impossible. The result is a ring-
shaped habitable zone, colored here in green.
This zone presumably encloses the orbit of the
Sun, but its radial extent in both directions is
uncertain.

However, observations of extrasolar planets are not yet sufficiently refined to determine the fraction of stars having "outer planet" systems like our own. ∞ (Sec. 15.5)

Other external forces may also influence a planet's survival. Some researchers have suggested that there is a *galactic habitable zone* for stars in general, outside of which conditions are unfavorable for life (see Figure 28.11). Far from the Galactic center, the star formation rate is low and few cycles of star formation have occurred, so there are insufficient heavy elements to form terrestrial planets, or populate them with technological civilizations if any should form. ∞ (Sec. 21.5) Too close, and the radiation from bright stars and supernovae in the crowded inner part of the Galaxy might be detrimental to life. More importantly, the gravitational effects of nearby stars may send frequent showers of comets from the counterpart of the Oort cloud into the inner regions of a planetary system, striking the terrestrial planets and terminating any chain of evolution that might lead to intelligent life.

Thus, to estimate the number of habitable planets per planetary system, we must first take inventory of how many stars of each type shine in the Galactic habitable zone, then calculate the sizes of their stellar habitable zones, and, finally, estimate the number of planets likely to be found there. We must eliminate almost all of the 10 percent of surveyed stars around which planets have so far been observed, and, presumably, a similar fraction of stars in general, because the large jovian planets seen in most cases have eccentric orbits that would destabilize the motion of any inner terrestrial world, either ejecting it completely from the system or making conditions so extreme that the chances for the development of life are severely reduced. ∞ (Sec. 15.4)

Finally, we exclude the majority of binary-star systems, because a planet's orbit within the habitable zone of a binary would be unstable in many cases, as illustrated in Figure 28.12. Given the observed properties of binaries in our Galaxy, we expect that most habitable planetary orbits would be unstable, so there would not be time for life to develop.

The inner and outer radii of the Galactic habitable zone are not known with any certainty, and we still have insufficient data about most stars to make any definitive statement about their planetary systems. Taking these uncertainties into account as best we can, we assign a value of 1/10 to this factor in our equation. In other words, we believe that, on average, there is one potentially habitable planet for every 10 planetary systems that might exist in our Galaxy. Single F-, G-, and K-type stars are the best candidates.

Fraction of Habitable Planets on Which Life Actually Arises

The number of possible combinations of atoms is incredibly large. If the chemical reactions which led to the complex molecules that make up living organisms occurred completely at random, then it is extremely unlikely that those molecules could have formed at all. In that case, life is extraordinarily rare, this factor is close to zero, and we are probably alone in the Galaxy, perhaps even in the entire universe.

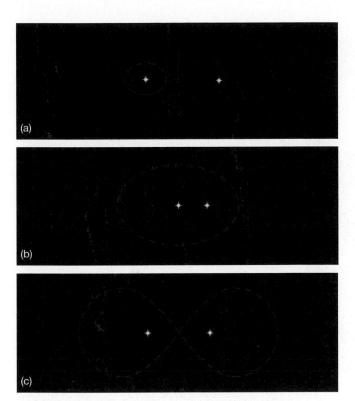

▲ **FIGURE 28.12 Binary-star Planets** In binary-star systems, planets are restricted to only a few kinds of orbits that are gravitationally stable. The orbit in case (a) is stable only if the planet lies very close to its "parent" star, so that the gravity of the other star is negligible. Case (b) shows a planet circulating at a great distance about both stars in an elliptical orbit. This orbit is stable only if it lies far from both stars. Another possible, but unstable, path, (c), interweaves between the two stars in a figure-eight pattern.

However, laboratory experiments (like the Urey–Miller experiment described earlier) seem to suggest that certain chemical combinations are strongly favored over others—that is, the reactions are *not* random. Of the billions upon billions of basic organic groupings that could occur on Earth from the random combination of all sorts of simple atoms and molecules, only about 1500 actually do occur. Furthermore, these 1500 organic groups of terrestrial biology are made from only about 50 simple "building blocks" (including the amino acids and nucleotide bases mentioned earlier). This suggests that molecules critical to life are not assembled by chance alone; apparently, additional factors are at work on the microscopic level. If a relatively small number of chemical "evolutionary tracks" are likely to exist, then the formation of complex molecules—and hence, we assume, life—becomes much more likely, given sufficient time.

To assign a very low value to this factor in the equation is to believe that life arises randomly and rarely. To assign a value close to unity is to believe that life is inevitable, given the proper ingredients, a suitable environment, and a long enough time. No simple experiment can distinguish be-

tween these extreme alternatives, and there is little or no middle ground. To many researchers, the discovery of life (past or present) on Mars, Europa, Titan, or some other object in our solar system would convert the appearance of life from an unlikely miracle to a virtual certainty throughout the Galaxy. We will take the optimistic view and adopt a value of unity.

Fraction of Life-Bearing Planets on Which Intelligence Arises

As with the evolution of life, the appearance of a well-developed brain is a highly unlikely event if only chance is involved. However, biological evolution through natural selection is a mechanism that generates apparently highly improbable results by singling out and refining useful characteristics. Organisms that profitably use adaptations can develop more complex behavior, and complex behavior provides organisms with the *variety* of choices needed for more advanced development.

One school of thought maintains that, given enough time, intelligence is inevitable. On this view, assuming that natural selection is a universal phenomenon, at least one organism on a planet will always rise to the level of "intelligent life." If this is correct, then the fifth factor in the Drake equation equals or nearly equals unity.

Others argue that there is only one known case of intelligence: human beings on Earth. For 2.5 billion years—from the start of life about 3.5 billion years ago to the first appearance of multicellular organisms about 1 billion years ago—life did not advance beyond the one-celled stage. Life remained simple and dumb, but it survived. If this latter view is correct, then the fifth factor in our equation is very small, and we are faced with the depressing prospect that humans may be the smartest form of life anywhere in the Galaxy. As with the previous factor, we will be optimistic and simply adopt a value of unity here.

Fraction of Planets on Which Intelligent Life Develops and Uses Technology

To evaluate the sixth factor of our equation, we need to estimate the probability that intelligent life eventually develops technological competence. Should the rise of technology be inevitable, this factor is close to unity, given a long enough time. If it is not inevitable—if intelligent life can somehow "avoid" developing technology—then this factor could be much less than unity. The latter scenario envisions a universe possibly teeming with intelligent civilizations, but very few among them ever becoming technologically competent. Perhaps only one managed it—ours.

Again, it is difficult to decide conclusively between these two views. We don't know how many prehistoric Earth cultures failed to develop simple technology or rejected its use. We do know that the roots of our present civilization arose independently at several different places

on Earth, including Mesopotamia, India, China, Egypt, Mexico, and Peru. Because so many of these ancient cultures originated at about the same time, it is tempting to conclude that the chances are good that some sort of technological society will inevitably develop, given some basic intelligence and enough time.

If technology is inevitable, then why haven't other life-forms on Earth also found it useful? Possibly the competitive edge given by intellectual and technological skills to humans, the first species to develop them, allowed us to dominate so rapidly that other species—gorillas and chimpanzees, for example—simply haven't had time to catch up. The fact that only one technological society exists on Earth does not imply that the sixth factor in our Drake equation must be very much less than unity. On the contrary, it is precisely because *some* species will probably always fill the niche of technological intelligence that we will take this factor to be close to unity.

Average Lifetime of a Technological Civilization

The reliability of the estimate of each factor in the Drake equation declines markedly from left to right. For example, our knowledge of astronomy enables us to make a reasonably good stab at the first factor, namely, the rate of star formation in our Galaxy, but it is much harder to evaluate some of the later factors, such as the fraction of life-bearing planets that eventually develop intelligence. The last factor on the right-hand side of the equation, the longevity of technological civilizations, is totally unknown. There is only one known example of such a civilization: humans on planet Earth. Our own civilization has survived in its "technological" state for only about 100 years, and how long we will be around before a natural or human-made catastrophe ends it all is impossible to tell. ∞ (*Discovery 14-1*)

One thing is certain: If the correct value for *any one factor* in the equation is very small, then few technological civilizations now exist in the Galaxy. In other words, the pessimistic view of the development of life or of intelligence is correct, then we are unique, and that is the end of our story. However, if both life and intelligence are inevitable consequences of chemical and biological evolution, as many scientists believe, and if intelligent life always becomes technological, then we can plug the higher, more optimistic values into the Drake equation. In that case, combining our estimates for the other six factors (and noting that $10 \times 1 \times 1/10 \times 1 \times 1 \times 1 = 1$), we have

number of technological, intelligent civilizations now present in the Milky Way Galaxy	=	average lifetime of a technologically competent civilization in years.

Thus, if civilizations typically survive for 1000 years, there should be 1000 of them currently in existence scattered throughout the Galaxy. If they live for a million years, on average, we would expect there to be a million advanced civilizations in the Milky Way, and so on.

Note that, even setting aside language and cultural issues, the sheer size of the Galaxy presents a significant hurdle to communication between technological civilizations. The minimum requirement for a two-way conversation is that we can send a signal and receive a reply in a time shorter than our own lifetime. If the lifetime is short, then civilizations are literally few and far between—small in number, according to the Drake equation, and scattered over the vastness of the Milky Way—and the distances between them (in light years) are much greater than their lifetimes (in years). In that case, two-way communication, even at the speed of light, will be impossible. However, as the lifetime increases, the distances get smaller as the Galaxy becomes more crowded, and the prospects improve.

Taking into account the size, shape, and distribution of stars in the Galactic disk (why do we exclude the halo?), and under the optimistic assumptions just made, we find that, unless the life expectancy of a civilization is *at least* a few thousand years, it is unlikely to have time to communicate with even its nearest neighbor.

CONCEPT CHECK

✔ How does the Drake equation assist astronomers in refining their search for extraterrestrial life?

28.4 The Search for Extraterrestrial Intelligence

Let us continue our optimistic assessment of the prospects for life and assume that civilizations enjoy a long stay on their parent planet once their initial technological "teething problems" are past. In that case, intelligent, technological, and perhaps also communicative cultures are likely to be plentiful in the Galaxy. How might we become aware of their existence?

Meeting Our Neighbors

For definiteness, let's assume that the average lifetime of a technological civilization is 1 million years—only 1 percent of the reign of the dinosaurs, but 100 times longer than human civilization has survived thus far. Given the size and shape of our Galaxy and the known distribution of stars in the Galactic disk, we can then estimate the average *distance* between these civilizations to be some 30 pc, or about 100 light-years. Thus, any two-way communication with our neighbors—using signals traveling at or below the speed of light—will take at least 200 years (100 years for the message to reach the planet and another 100 years for the reply to travel back to us).

One obvious way to search for extraterrestrial life would be to develop the capability to travel far outside our

solar system. However, that may never be a practical possibility. At a speed of 50 km/s, the speed of the fastest space probes operating today, the round-trip to even the nearest Sun-like star, Alpha Centauri, would take about 50,000 years. The journey to the nearest technological neighbor (assuming a distance of 30 pc) and back would take 600,000 years—almost the entire lifetime of our species! Interstellar travel at these speeds is clearly not feasible. Speeding up our ships to near the speed of light would reduce the travel time, but doing that is far beyond our present technology.

Actually, our civilization has already launched some interstellar probes, although they have no specific stellar destination. Figure 28.13 is a reproduction of a plaque mounted on board the *Pioneer 10* spacecraft launched in the mid-1970s and now well beyond the orbit of Pluto, on its way out of the solar system. Similar information was included aboard the *Voyager* probes launched in 1978. Although these spacecraft would be incapable of reporting back to Earth the news that they had encountered an alien culture, scientists hope that the civilization on the other end would be able to unravel most of its contents using the universal language of mathematics. The caption to Figure 28.13 notes how the aliens might discover from where and when the *Pioneer* and *Voyager* probes were launched.

Setting aside the many practical problems that arise in trying to establish direct contact with extraterrestrials,

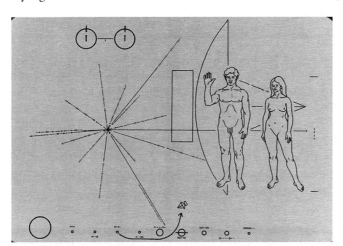

▲ **FIGURE 28.13** *Pioneer 10* **Plaque** A replica of a plaque mounted on board the *Pioneer 10* spacecraft. Included are scale drawings of the spacecraft, a man, and a woman; a diagram of the hydrogen atom undergoing a change in energy (top left); a starburst pattern representing various pulsars and the frequencies of their radio waves that can be used to estimate when the craft was launched (middle left); and a depiction of the solar system, showing that the spacecraft departed the third planet from the Sun and passed the fifth planet on its way to outer space (bottom). All the drawings have computer-coded (binary) markings from which actual sizes, distances, and times can be derived. *(NASA)*

some scientists have argued that it might not even be a particularly good idea. Our recent emergence as a technological civilization implies that we must be one of the least advanced technological intelligences in the entire Galaxy. Any other civilization that discovers us will almost surely be more advanced than us. Consequently, a healthy degree of caution is warranted. If extraterrestrials behave even remotely like human civilizations on Earth, then the most advanced aliens may naturally try to dominate all others. The behavior of the "advanced" European cultures toward the "primitive" races they encountered on their voyages of discovery in the seventeenth, eighteenth, and nineteenth centuries should serve as a clear warning of the possible undesirable consequences of contact. Of course, the aggressiveness of Earthlings may not apply to extraterrestrials, but given the history of the one intelligent species we know, the cautious approach may be in order.

Radio Communication

A cheaper and much more practical alternative to direct contact is to try to communicate with extraterrestrials by using only electromagnetic radiation, the fastest known means of transferring information from one place to another. Because light and other high-frequency radiation is heavily scattered while moving through dusty interstellar space, long-wavelength radio radiation seems to be the natural choice. We would not attempt to broadcast to *all* nearby candidate stars, however—that would be far too expensive and inefficient. Instead, radio telescopes on Earth would listen *passively* for radio signals emitted by other civilizations. Indeed, some preliminary searches of selected nearby stars are now underway, thus far without success.

In what direction should we aim our radio telescopes? The answer to this question, at least, is fairly easy: On the basis of our earlier reasoning, we should target all F-, G-, and K-type stars in our vicinity. But are extraterrestrials broadcasting radio signals? If they are not, this search technique will obviously fail. And even they are, how do we distinguish their artificially generated radio signals from signals naturally emitted by interstellar gas clouds? To what frequency should we tune our receivers? The answer to this question depends on whether the signals are produced deliberately or are simply "waste radiation" escaping from a planet.

Consider how Earth would look at radio wavelengths to extraterrestrials. Figure 28.14 shows the pattern of radio signals we emit into space. From the viewpoint of a distant observer, the spinning Earth emits a bright flash of radio radiation every few hours. In fact, Earth is now a more intense radio emitter than the Sun. The flashes result from the periodic rising and setting of hundreds of FM radio stations and television transmitters. Each station broadcasts mostly parallel to Earth's surface, sending a great "sheet" of electromagnetic radiation into interstellar

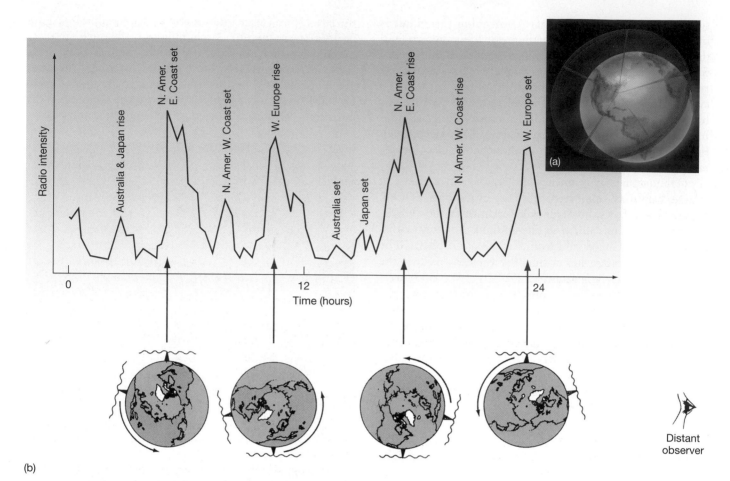

▲ **FIGURE 28.14 Earth's Radio Leakage** Radio radiation now leaks from Earth into space because of the daily activities of our technological civilization. (a) FM radio and television transmitters broadcast their energy parallel to Earth's surface, so they send a great "sheet" of electromagnetic radiation into interstellar space, producing the strongest signal in any given direction when they happen to lie on Earth's horizon, as seen from that direction. (The more common AM broadcasts are trapped below our ionosphere, so those signals never leave Earth.) (b) Because the great majority of transmitters are clustered in the eastern United States and western Europe, a distant observer would detect blasts of radiation from Earth as our planet rotates each day.

space, as illustrated in Figure 28.14(a). (The more common AM broadcasts are trapped below our ionosphere, so those signals never leave Earth.)

Because the great majority of these transmitters are clustered in the eastern United States and western Europe, a distant observer would detect periodic blasts of radiation from Earth as our planet rotates each day (Figure 28.14b). This radiation races out into space and has been doing so since the invention of these technologies more than seven decades ago. Another civilization at least as advanced as ours might have constructed devices capable of detecting these blasts of radiation. If any sufficiently advanced (and sufficiently interested) civilization resides on a planet orbiting any of the thousand or so stars within about 70 light-years (20 pc) of Earth, then we have already broadcast our presence to them.

Of course, it may very well be that, having discovered cable, most civilizations' indiscriminate transmissions cease after a few decades. In that case, radio silence becomes the hallmark of intelligence, and we must find an alternative means of locating our neighbors.

The Water Hole

Now let us suppose that a civilization has decided to assist searchers by actively broadcasting its presence to the rest of the Galaxy. At what frequency should we listen for such an extraterrestrial beacon? The electromagnetic spectrum is enormous; the radio domain alone is vast. To hope to detect a signal at some unknown radio frequency is like searching for a needle in a haystack. Are some frequencies more likely than others to carry alien transmissions?

Some basic arguments suggest that civilizations might communicate at a wavelength near 20 cm. As we saw in Chapter 18, the basic building blocks of the universe, namely, hydrogen atoms, radiate naturally at a wavelength of 21 cm. ∞ (Sec. 18.4) Also, one of the simplest molecules, hydroxyl (OH), radiates near 18 cm. Together, these

two substances form water (H_2O). Arguing that water is likely to be the interaction medium for life anywhere and that radio radiation travels through the disk of our Galaxy with the least absorption by interstellar gas and dust, some researchers have proposed that the interval between 18 and 21 cm is the best range of wavelengths for civilizations to transmit or monitor. Called the **water hole**, this radio interval might serve as an "oasis" where all advanced galactic civilizations would gather to conduct their electromagnetic business.

The water-hole frequency interval is only a guess, of course, but it is supported by other arguments as well. Figure 28.15 shows the water hole's location in the electromagnetic spectrum and plots the amount of natural emission from our Galaxy and from Earth's atmosphere. The 18- to 21-cm range lies within the quietest part of the spectrum, where the galactic "static" from stars and interstellar clouds happens to be minimized. Furthermore, the atmospheres of typical planets are also expected to interfere least at these wavelengths. Thus, the water hole seems like a good choice for the frequency of an interstellar beacon, although we cannot be sure of this reasoning until contact is actually achieved.

A few radio searches are now in progress at frequencies in and around the water hole. One of the most sensitive and comprehensive projects in the ongoing search for extraterrestrial intelligence (known to many by its acronym, SETI) was Project Phoenix, carried out during

the late 1990s. Large radio antennas, such as that in Figure 28.16(a), were used to search millions of channels simultaneously in the 1–3-GHz spectrum. Actually, in these searches, computers do most of the "listening;" humans get involved only if the signals look intriguing. Figure 28.16(b) shows what a typical narrowband, 1-Hz signal—a potential "signature" of an intelligent transmission—

(a)

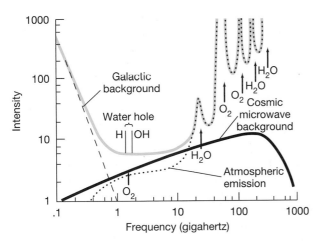

▲ FIGURE 28.15 "Water Hole" The "water hole" is bounded by the natural emission frequencies of the hydrogen (H) atom (21-cm wavelength) and the hydroxyl (OH) molecule (18-cm wavelength). ∞ (Secs. 18.4, 18.5) The topmost solid (blue) curve sums the natural emissions of our Galaxy (the dashed line on the left side of the diagram, labeled "Galactic background"), Earth's atmosphere (the dotted line on the right side of the diagram, denoted by various chemical symbols), and the cosmic microwave background. ∞ (Sec. 26.7) The sum is minimized near the water-hole frequencies. Perhaps all intelligent civilizations conduct their interstellar communications within this quiet "electromagnetic oasis."

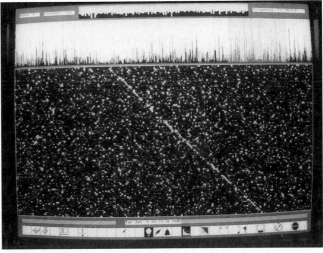

(b)

▲ FIGURE 28.16 Project Phoenix (a) This large radio telescope at the National Radio Astronomy Observatory in Green Bank, West Virginia, was used by Project Phoenix during the 1990s to search for extraterrestrial intelligent signals. (b) A typical recording of an alien signal—here, as a test, the Doppler-shifted broadcast from the *Pioneer 10* spacecraft, now well beyond the orbit of Pluto. The diagonal line across the computer monitor, in contrast to the random noise in the background, betrays the presence of an organized signal in the incoming data stream. *(NRAO; SETI Institute)*

would look like on a computer monitor. However, this observation was merely a test to detect the weak, redshifted radio signal emitted by the *Pioneer 10* robot, now receding into the outer realm of our solar system—a sign of intelligence, but one that we put there. Nothing resembling an extraterrestrial signal has yet been detected.

Right now, the space surrounding all of us could be flooded with radio signals from extraterrestrial civilizations. If only we knew the proper direction and frequency, we might be able to make one of the most startling discoveries of all time. The result would likely provide whole new opportunities to study the cosmic evolution of energy, matter, and life throughout the universe.

CONCEPT CHECK

✔ Why do many researchers regard the "water hole" as a likely place to search for extraterrestrial signals?

Chapter Review

SUMMARY

The history of the universe can be divided into seven phases: particulate, galactic, stellar, planetary, chemical, biological, and cultural. **Cosmic evolution (p. 744)** is the continuous process that has led to the appearance of galaxies, stars, planets, and life on Earth.

Living organisms may be characterized by their ability to react to their environment, to grow by taking in nutrition from their surroundings, and to reproduce, passing along some of their own characteristics to their offspring.

Powered by natural energy sources, reactions between simple molecules in the oceans of the primitive Earth are believed to have led to the formation of **amino acids (p. 745)** and **nucleotide bases (p. 745)**, the basic molecules of life. Alternatively, some complex molecules may have been formed in interstellar space and then delivered to Earth by meteors or comets.

Organisms that can best take advantage of their new surroundings succeed at the expense of those organisms which cannot make the necessary adjustments. Intelligence is strongly favored by natural selection.

The best hope for life beyond Earth in the solar system is the planet Mars, although no evidence for living organisms has been found there. Jupiter's moon Europa and Saturn's Titan may also be possibilities, but conditions on those bodies are harsh by terrestrial standards.

The **Drake equation (p. 751)** provides a means of estimating the probability of intelligent life in the Galaxy. The astronomical factors in the equation are the galactic star-formation rate, the likelihood of planets, and the number of habitable planets. Chemical and biological factors are the probability that life appears and the probability that it subsequently develops intelligence. Cultural and political factors are the probability that intelligence leads to technology and the lifetime of a civilization in the technological state. Taking an optimistic view of the development of life and intelligence leads to the conclusion that the total number of technologically competent civilizations in the Galaxy is approximately equal to the lifetime of a typical civilization, expressed in years.

Even with optimistic assumptions, the distance to our nearest intelligent neighbor is likely to be many hundreds of parsecs. Currently, space travel is not a feasible means of searching for intelligent life. Existing programs to discover extraterrestrial intelligence involve scanning the electromagnetic spectrum for signals. So far, no intelligible broadcasts have been received. A technological civilization would probably "announce" itself to the universe by the radio and television signals it emits into space. Observed from afar, our planet would appear as a radio source with a 24-hour period, as different regions of the planet rise and set.

The "**water hole**" (**p. 758**) is a region in the radio range of the electromagnetic spectrum, near the 21-cm line of hydrogen and the 18-cm line of hydroxyl, where natural emissions from the Galaxy happen to be minimized. Many researchers regard this region as the best part of the spectrum for communication purposes.

REVIEW AND DISCUSSION

1. Why is life difficult to define?

2. What is chemical evolution?

3. What is the Urey–Miller experiment? What important organic molecules were produced in that experiment?

4. What other experiments have attempted to produce organic molecules by inorganic means?

5. What are the basic ingredients from which biological molecules formed on Earth?

6. Why do some scientists think life might have originated in space?

7. How do we know anything at all about the early episodes of life on Earth?

8. What is the role of language in cultural evolution?

9. Where else, besides Earth, have organic molecules been found?

10. Where—besides Earth and the planet Mars—might we hope to find signs of life in our solar system?

11. Do we know whether Mars ever had life at any time during its past? What argues in favor of the position that it may once have harbored life?

12. What is generally meant by "life as we know it"? What other forms of life might be possible?

13. How many of the factors in the Drake equation are known with any degree of certainty? Which factor is least well known?

14. What factors determine the suitability of a star as the parent of a planet on which life might arise?

15. What is the relationship between the average lifetime of galactic civilizations and the possibility of our someday communicating with them?

16. How would Earth appear at radio wavelengths to extraterrestrial astronomers?

17. Do you think that advanced civilizations would continue to emit large amounts of radio energy as they evolved?

18. What are the advantages in using radio waves for communication over interstellar distances?

19. What is the "water hole"? What advantages does it offer for interstellar communication?

20. If you were designing a SETI experiment, what parts of the sky would you monitor?

CONCEPTUAL SELF-TEST: TRUE OR FALSE/MULTIPLE CHOICE

1. The definition of life requires only that, to be considered "alive," you must be able to reproduce.

2. Organic molecules exist only on Earth.

3. Laboratory experiments have created living cells from non-biological molecules.

4. Dinosaurs lived on Earth for a thousand times longer than human civilization has existed to date.

5. The *Viking* landers on Mars discovered microscopic evidence of life, but found no large fossil evidence.

6. We have no direct evidence for Earth-like planets orbiting other stars.

7. Our civilization has already launched probes into interstellar space and broadcast our presence to our neighbors.

8. The Drake equation estimates the number of Earth-like planets in the Milky Way Galaxy.

9. Planets in binary-star systems are not considered habitable, because the planetary orbits are usually unstable.

10. The development of life and intelligence on Earth are extremely unlikely if chance is the only evolutionary factor involved.

11. The "assumptions of mediocrity" suggest that **(a)** life should be common throughout the cosmos; **(b)** lower forms of life must evolve to higher forms; **(c)** lower forms of life have lower intelligence; **(d)** viruses are actually life-forms.

12. The chemical elements that form the basic molecules needed for life are found **(a)** in the cores of Sun-like stars; **(b)** commonly throughout the cosmos; **(c)** only on planets that have liquid water; **(d)** only on Earth.

13. Fossil records of early life-forms on Earth suggest that life began about **(a)** 6000 years ago; **(b)** 65 million years ago; **(c)** 3.5 billion years ago; **(d)** 14 billion years ago.

14. The discovery of bacteria on another planet would be an important discovery because bacteria **(a)** can easily survive in high temperatures; **(b)** are the only life-form to exist on Earth for most of the planet's history; **(c)** are the lowest form of life known to exist; **(d)** eventually evolve into intelligent beings.

15. The least-well-known factor in the Drake equation is **(a)** the rate of star formation; **(b)** the average number of habitable planets within planetary systems; **(c)** the average lifetime of a technologically competent civilization; **(d)** the diameter of the Milky Way Galaxy.

16. Although the habitable zone around a large B-class star is large, we don't often look for life on planets there because the star **(a)** has too much gravity; **(b)** is too short lived for life to evolve; **(c)** is at too low a temperature to sustain life; **(d)** would have only gas giant planets.

17. NASA's Space Shuttle orbits Earth at about 17,500 mph. If it traveled to the next Sun-like star at that speed, the trip would take at least **(a)** one week; **(b)** one decade; **(c)** one century; **(d)** 100 millennia.

18. If Figure 28.14 ("Earth's Radio Leakage") were to be redrawn for a planet spinning twice as fast, the new jagged line would be **(a)** unchanged; **(b)** taller; **(c)** stretched out horizontally; **(d)** compressed horizontally.

19. Radio telescopes cannot simply scan the skies looking for signals, because **(a)** astronomers don't know what frequencies alien civilizations might use; **(b)** many nonliving objects emit radio signals naturally; **(c)** Earth's radio communications drown out extraterrestrial signals; **(d)** inclement weather in the winter prevents the use of radio telescopes.

20. The strongest radio-wavelength emitter in the solar system is **(a)** human-made signals from Earth; **(b)** the Sun; **(c)** the Moon; **(d)** Jupiter.

PROBLEMS

 Algorithmic versions of these questions are available in the Practice Problems module of the Companion Website at astro.prenhall.com/chaisson.

The number of squares preceding each problem indicates its approximate level of difficulty.

1. ■ If Earth's 4.6-billion-year age were compressed to 46 years, as described in the text, what would be your age, in seconds? On that scale, how long ago was the end of World War II? The Declaration of Independence? Columbus's discovery of the New World? The extinction of the dinosaurs?

2. ■■■ According to the inverse-square law, a planet receives energy from its parent star at a rate proportional to the star's luminosity and inversely proportional to the square of the planet's distance from the star. ∞ (Sec. 17.2) According to Stefan's law, the rate at which the planet radiates energy into space is proportional to the fourth power of its surface temperature. ∞ (Sec. 3.4) In equilibrium, the two rates are equal. Based on this information, and given the fact that (taking into account the greenhouse effect) the Sun's habitable zone extends from 0.6 A.U. to 1.5 A.U., estimate the extent of the habitable zone surrounding a K-type main-sequence star of the luminosity of the Sun.

3. ■■ Using the data in the previous problem, how would the inner and outer radii of the Sun's habitable zone change if the solar luminosity increased by a factor of four?

4. ■■ The outer edge of the habitable zone corresponds roughly to the freezing point of water (273 K). Using the data in problem 2, how would this radius change if life could survive at temperatures as low as 150 K?

5. ■■■ Using the data in problem 2, what would be the orbital period of a planet orbiting at the outer edge of the habitable zone of an F-type main-sequence star of 1.5 solar masses and luminosity five times that of the Sun?

6. ■ Based on the numbers presented in the text, and assuming an average lifetime of 5 billion years for suitable stars, estimate the total number of habitable planets in the Galaxy.

7. ■■ A planet orbits one component of a binary-star system at a distance of 1 A.U. (See Figure 28.12a.) If both stars have the same mass and their orbit is circular, estimate the minimum distance between the stars for the tidal force due to the companion not to exceed a "safe" 0.01 percent of the gravitational force between the planet and its parent star.

8. ■ Suppose that each of the fractional factors in the Drake equation turns out to have a value of 1/10, that stars form at an average rate of 20 per year, and that each star has exactly one habitable planet orbiting it. Estimate the present number of technological civilizations in the Milky Way Galaxy if the average lifetime of a civilization is (a) 100 years (b) 10,000 years (c) 1 million years.

9. ■■■ If we adopt the estimate from the text that the number of technological civilizations in the Milky Way Galaxy is equal to the average lifetime of a civilization, it follows that the distance to our nearest neighbor decreases as the average lifetime increases. Assuming that civilizations are uniformly spread over a two-dimensional Galactic disk of radius 15 kpc and that all have the same lifetime, calculate the *minimum* lifetime for which two-way radio communication with our nearest neighbor would be possible before our civilization ends. Is the assumption that civilizations are uniformly spread across the disk a reasonable one?

10. ■■■ Repeat the calculation in the previous question for a round-trip personal visit, using current-technology spacecraft that travel at 50 km/s.

11. ■ How fast would a spacecraft have to travel in order to complete the trip from Earth to Alpha Centauri (a distance of 1.3 pc) and back in less than an average human lifetime (80 years, say)?

12. ■ Assuming that there are 10,000 FM radio stations on Earth, each transmitting at a power level of 50 kW, calculate the total radio luminosity of Earth in the FM band. Compare this value with the roughly 10^6 W radiated by the Sun in the same frequency range.

13. ■ Convert the water hole's wavelengths to frequencies. For practical reasons, any search of the water hole must be broken up into channels, much like those you find on a television, except that the water-hole's channels are very narrow in radio frequency, about 100 Hz wide. How many channels must astronomers search in the water hole?

14. ■ At what wavelength does the microwave background radiation peak, and what is the corresponding frequency? How does this frequency compare with the waterhole's frequency range?

15. ■ There are 20,000 stars within 100 light-years that are to be searched for radio communications. How long will the search take if one hour is spent looking at each star? What if one day is spent per star?

 In addition to the practice problems module, the Companion Website at astro.prenhall.com/chaisson provides for each chapter a study guide module with multiple choice questions as well as additional annotated images, animations, and links to related Websites.

APPENDIX 1

Scientific Notation

The objects studied by astronomers range in size from the smallest particles to the largest expanse of matter we know—the entire universe. Subatomic particles have sizes of about 0.000000000000001 meter, while galaxies (like that shown in Figure 1.3) typically measure some 1,000,000,000,000,000,000,000 meters across. The most distant known objects in the universe lie on the order of 100,000,000,000,000,000,000,000,000 meters from Earth.

Obviously, writing all those zeros is both cumbersome and inconvenient. More important, it is also very easy to make an error—write down one zero too many or too few and your calculations become hopelessly wrong! To avoid this, scientists always write large numbers using a short-hand notation in which the number of zeros following or preceding the decimal point is denoted by a superscript power, or *exponent*, of 10. The exponent is simply the number of places between the first significant (nonzero) digit in the number (reading from left to right) and the decimal point. Thus, 1 is 10^0, 10 is 10^1, 100 is 10^2, 1000 is 10^3, and so on. For numbers less than 1, with zeros between the decimal point and the first significant digit, the exponent is negative: 0.1 is 10^{-1}, 0.01 is 10^{-2}, 0.001 is 10^{-3}, and so on. Using this notation we can shorten the number describing subatomic particles to 10^{-15} meter, and write the number describing the size of a galaxy as 10^{21} meters.

More complicated numbers are expressed as a combination of a power of 10 and a multiplying factor. This factor is conventionally chosen to be a number between 1 and 10, starting with the first significant digit in the original number. For example, 150,000,000,000 meters (the distance from Earth to the Sun, in round numbers) can be more concisely written at 1.5×10^{11} meters, 0.000000025 meters as 2.5×10^{-8} meter, and so on. The exponent is simply the number of places the decimal point must be moved *to the left* to obtain the multiplying factor.

Some other examples of scientific notation are:

- the approximate distance to the Andromeda
  Galaxy = 2,500,000 light-years = 2.5×10^6 light-years
- the size of a hydrogen
  atom = 0.00000000005 meter = 5×10^{-11} meter

- the diameter of the Sun
  = 1,392,000 kilometers = 1.392×10^6 kilometers
- the U.S. national debt (as of June 1, 2004)
  = \$7,206,219,000,000.00 = \$7.206219 trillion
  = 7.206219×10^{12} dollars.

In addition to providing a simpler way of expressing very large or very small numbers, this notation also makes it easier to do basic arithmetic. The rule for multiplication of numbers expressed in this way is simple: Just multiply the factors and add the exponents. Similarly for division: Divide the factors and subtract the exponents. Thus, 3.5×10^{-2} multiplied by 2.0×10^3 is simply $(3.5 \times 2.0) \times 10^{-2+3} = 7.0 \times 10^1$— that is, 70. Again, 5×10^6 divided by 2×10^4 is just $(5/2) \times 10^{6-4}$, or 2.5×10^2 ($= 250$). Applying these rules to unit conversions, we find, for example, that 200,000 nanometers is $200,000 \times 10^{2-9}$ meter (since 1 nanometer $= 10^{-9}$ meter; see Appendix 2), or $2 \times 10^5 \times 10^{-9}$ meter, or $2 \times 10^{5-9} = 2 \times 10^{-4}$ meter $= 0.2$ mm. Verify these rules for yourself with a few examples of your own. The advantages of this notation when considering astronomical objects will soon become obvious.

Scientists often use "rounded-off" versions of numbers, both for simplicity and for ease of calculation. For example, we will usually write the diameter of the Sun as 1.4×10^6 kilometers, instead of the more precise number given earlier. Similarly, Earth's diameter is 12,756 kilometers, or 1.2756×10^4 kilometers, but for "ballpark" estimates we really don't need so many digits and the more approximate number 1.3×10^4 kilometers will suffice. Very often, we perform rough calculations using only the first one or two significant digits in a number, and that may be all that is necessary to make a particular point. For example, to support the statement, "The Sun is much larger than Earth," we need only say that the ratio of the two diameters is roughly 1.4×10^6 divided by 1.3×10^4. Since 1.4/1.3 is close to 1, the ratio is approximately $10^6/10^4 = 10^2$, or 100. The essential fact here is that the ratio is much larger than 1; calculating it to greater accuracy (to get 109.13) would give us no additional *useful* information. This technique of stripping away the arithmetic details to get to the essence of a calculation is very common in astronomy, and we use it frequently throughout this text.

APPENDIX 2

Astronomical Measurement

Astronomers use many different kinds of units in their work, simply because no single system of units will do. Rather than the *Système Internationale* (SI), or meter-kilogram-second (MKS), metric system used in most high school and college science classes, many professional astronomers still prefer the older centimeter-gram-second (CGS) system. However, astronomers also commonly introduce new units when convenient. For example, when discussing stars, the mass and radius of the Sun are often used as reference points. The solar mass, written as $M_\odot$, is equal to 2.0×10^{33} g, or 2.0×10^{30} kg (since 1 kg = 1000 g). The solar radius, $R_\odot$, is equal to 700,000 km, or 7.0×10^8 m (1 km = 1000 m). The subscript $\odot$ always stands for Sun. Similarly, the subscript $\oplus$ always stands for Earth. In this book, we try to use the units that astronomers commonly use in any given context, but we also give the "standard" SI equivalents where appropriate.

Of particular importance are the units of length astronomers use. On small scales, the *angstrom* (1 Å = 10^{-10} m = 10^{-8} cm), the *nanometer* (1 nm = 10^{-9} m = 10^{-7} cm), and the *micron* (1 μm = 10^{-6} m = 10^{-4} cm) are used. Distances within the solar system are usually expressed in terms of the *astronomical unit* (A.U.), the mean distance between Earth and the Sun. One A.U. is approximately equal to 150,000,000 km, or 1.5×10^{11} m On larger scales, the *light-year* (1 ly = 9.5×10^{15} m = 9.5×10^{12} km) and the *parsec* (1 pc = 3.1×10^{16} m = 3.1×10^{13} km = 3.3 ly) are commonly used. Still larger distances use the regular prefixes of the metric system: *kilo* for one thousand and *mega* for one million. Thus 1 kiloparsec (kpc) = 10^3 pc = 3.1×10^{19} m, 10 megaparsecs (Mpc) = 10^7 pc = 3.1×10^{23} m, and so on.

Astronomers use units that make sense within a context, and as contexts change, so do the units. For example, we might measure densities in grams per cubic centimeter (g/cm^3), in atoms per cubic meter (atoms/m^3), or even in solar masses per cubic megaparsec (M$_\odot$/Mpc^3), depending on the circumstances. The important thing to know is that once you understand the units, you can convert freely from one set to another. For example, the radius of the Sun could equally well be written as $R_\odot = 6.96 \times 10^8$ m, or 6.96×10^{10} cm, or 109 $R_\oplus$, or 4.65×10^{-3} A.U., or even 7.36×10^{-8} ly—whichever happens to be most useful. Some of the more common units used in astronomy, and the contexts in which they are most likely to be encountered, are listed below.

Length:		
1 angstrom (Å)	= 10^{-10} m	
1 nanometer (nm)	= 10^{-9} m	atomic physics, spectroscopy
1 micron (μm)	= 10^{-6} m	interstellar dust and gas
1 centimeter (cm)	= 0.01 m	
1 meter (m)	= 100 cm	in widespread use throughout all astronomy
1 kilometer (km)	= 1000 m = 10^5 cm	
Earth radius ($R_\oplus$)	= 6378 km	planetary astronomy
Solar radius ($R_\odot$)	= 6.96×10^8 m	
1 astronomical unit (A.U.)	= 1.496×10^{11} m	solar system, stellar evolution
1 light-year (ly)	= 9.46×10^{15} m = 63,200 A.U.	
1 parsec (pc)	= 3.09×10^{16} m = 206,000 A.U.	galactic astronomy, stars and star clusters
	= 3.26 ly	
1 kiloparsec (kpc)	= 1000 pc	
1 megaparsec (Mpc)	= 1000 kpc	galaxies, galaxy clusters, cosmology
Mass:		
1 gram (g)		
1 kilogram (kg)	= 1000 g	in widespread use in many different areas
Earth mass ($M_\oplus$)	= 5.98×10^{24} kg	planetary astronomy
Solar mass ($M_\odot$)	= 1.99×10^{30} kg	"standard" unit for all mass scales larger than Earth
Time:		
1 second (s)		in widespread use throughout astronomy
1 hour (h)	= 3600 s	
1 day (d)	= 86,400 s	planetary and stellar scales
1 year (yr)	= 3.16×10^7 s	virtually all processes occurring on scales larger than a star

APPENDIX 3

Tables

TABLE 1	Some Useful Constants and Physical Measurements[*]
astronomical unit	$1 \text{ A.U.} = 1.496 \times 10^8 \text{ km } (1.5 \times 10^8 \text{ km})$
light-year	$1 \text{ ly} = 9.46 \times 10^{12} \text{ km } (10^{13} \text{ km}; 6 \text{ trillion miles})$
parsec	$1 \text{ pc} = 3.09 \times 10^{13} \text{ km} = 206{,}000 \text{ A.U.} = 3.3 \text{ ly}$
speed of light	$c = 299{,}792.458 \text{ km/s } (3 \times 10^5 \text{ km/s})$
Stefan-Boltzmann constant	$\sigma = 5.67 \times 10^{-8} \text{ W/m}^2 \cdot \text{K}^4$
Planck's constant	$h = 6.63 \times 10^{-34} \text{ J s}$
gravitational constant	$G = 6.67 \times 10^{-11} \text{ N m}^2/\text{kg}^2$
mass of Earth	$M_\oplus = 5.98 \times 10^{24} \text{ kg } (6 \times 10^{24} \text{ kg, about 6000 billion billion tons})$
radius of Earth	$R_\oplus = 6378 \text{ km } (6500 \text{ km})$
mass of the Sun	$M_\odot = 1.99 \times 10^{30} \text{ kg } (2 \times 10^{30} \text{ kg})$
radius of the Sun	$R_\odot = 6.96 \times 10^5 \text{ km } (7 \times 10^5 \text{ km})$
luminosity of the Sun	$L_\odot = 3.90 \times 10^{26} \text{ W } (4 \times 10^{26} \text{ W})$
effective temperature of the Sun	$T_\odot = 5778 \text{ K } (5800 \text{ K})$
Hubble's constant	$H_0 = 70 \text{ km/s/Mpc}$
mass of an electron	$m_e = 9.11 \times 10^{-31} \text{ kg}$
mass of a proton	$m_p = 1.67 \times 10^{-27} \text{ kg}$

[*]*The rounded-off values used in the text are shown above in parentheses.*

Conversions Between Common English and Metric Units

English	Metric
1 inch	= 2.54 centimeters (cm)
1 foot (ft)	= 0.3048 meters (m)
1 mile	= 1.609 kilometers (km)
1 pound (lb)	= 453.6 grams (g) or 0.4536 kilograms (kg) [on Earth]

TABLE 2   Periodic Table of Elements

Legend:

2	Atomic number
He	Symbol of element
4.003	Atomic weight
Helium	Name of element

Group →	1	2	3	4	5	6	7	8	9	10	11	12	13	14	15	16	17	18
Period 1	1 **H** 1.0080 Hydrogen																	2 **He** 4.003 Helium
Period 2	3 **Li** 6.939 Lithium	4 **Be** 9.012 Beryllium											5 **B** 10.81 Boron	6 **C** 12.011 Carbon	7 **N** 14.007 Nitrogen	8 **O** 15.9994 Oxygen	9 **F** 18.998 Fluorine	10 **Ne** 20.183 Neon
Period 3	11 **Na** 22.990 Sodium	12 **Mg** 24.31 Magnesium											13 **Al** 26.98 Aluminum	14 **Si** 28.09 Silicon	15 **P** 30.974 Phosphorus	16 **S** 32.064 Sulfur	17 **Cl** 35.453 Chlorine	18 **Ar** 39.948 Argon
Period 4	19 **K** 39.102 Potassium	20 **Ca** 40.08 Calcium	21 **Sc** 44.96 Scandium	22 **Ti** 47.90 Titanium	23 **V** 50.94 Vanadium	24 **Cr** 52.00 Chromium	25 **Mn** 53.94 Manganese	26 **Fe** 55.85 Iron	27 **Co** 58.93 Cobalt	28 **Ni** 58.71 Nickel	29 **Cu** 63.54 Copper	30 **Zn** 65.37 Zinc	31 **Ga** 69.72 Gallium	32 **Ge** 72.59 Germanium	33 **As** 74.92 Arsenic	34 **Se** 78.96 Selenium	35 **Br** 79.909 Bromine	36 **Kr** 83.80 Krypton
Period 5	37 **Rb** 85.47 Rubidium	38 **Sr** 87.62 Strontium	39 **Y** 88.91 Yttrium	40 **Zr** 91.22 Zirconium	41 **Nb** 92.91 Niobium	42 **Mo** 95.94 Molybdenum	43 **Tc** (99) Technetium	44 **Ru** 101.1 Ruthenium	45 **Rh** 102.90 Rhodium	46 **Pd** 106.4 Palladium	47 **Ag** 107.87 Silver	48 **Cd** 112.40 Cadmium	49 **In** 114.82 Indium	50 **Sn** 118.69 Tin	51 **Sb** 121.75 Antimony	52 **Te** 127.60 Tellurium	53 **I** 126.9 Iodine	54 **Xe** 131.30 Xenon
Period 6	55 **Cs** 132.91 Cesium	56 **Ba** 137.34 Barium	71 **Lu** 174.97 Lutetium *	72 **Hf** 178.49 Hafnium	73 **Ta** 180.95 Tantalum	74 **W** 183.85 Tungsten	75 **Re** 186.2 Rhenium	76 **Os** 190.2 Osmium	77 **Ir** 192.2 Iridium	78 **Pt** 195.09 Platinum	79 **Au** 197.0 Gold	80 **Hg** 200.59 Mercury	81 **Tl** 204.37 Thallium	82 **Pb** 207.19 Lead	83 **Bi** 208.98 Bismuth	84 **Po** (210) Polonium	85 **At** (210) Astantine	86 **Rn** (222) Radon
Period 7	87 **Fr** (223) Francium	88 **Ra** 226.05 Radium	103 **Lw** (257) Lawrencium **	104 **Rf** (261) Rutherfordium	105 **Db** (262) Dubnium	106 **Sg** (263) Seaborgium	107 **Bh** (262) Bohrium	108 **Hs** (265) Hassium	109 **Mt** (266) Meitnerium	110 **Uun** (269) Ununnilium	111 **Uuu** (272) Unununium	112 **Uub** (277) Ununbium	113 **Uut** (undiscovered) Ununtrium	114 **Uuq** (285) Ununquadium				

Lanthanide series (*):

57 **La** 138.91 Lanthanum	58 **Ce** 140.12 Cerium	59 **Pr** 140.91 Praseodymium	60 **Nd** 144.24 Neodymium	61 **Pm** (147) Promethium	62 **Sm** 150.35 Samarium	63 **Eu** 151.96 Europium	64 **Gd** 157.25 Gadolinium	65 **Tb** 158.92 Terbium	66 **Dy** 162.50 Dysprosium	67 **Ho** 164.93 Holmium	68 **Er** 167.26 Erbium	69 **Tm** 168.93 Thulium	70 **Yb** 173.04 Ytterbium

Actinide series (**):

89 **AC** (227) Actinium	90 **Th** 232.04 Thorium	91 **Pa** (231) Protactinium	92 **U** 238.03 Uranium	93 **Np** (237) Neptunium	94 **Pu** (242) Plutonium	95 **Am** (243) Americium	96 **Cm** (247) Curium	97 **Bk** (249) Berkelium	98 **Cf** (251) Californium	99 **Es** (254) Einsteinium	100 **Fm** (253) Fermium	101 **Md** (256) Mendelevium	102 **No** (254) Nobelium

TABLE 3A   Planetary Orbital Data

Planet	Semi-Major Axis (A.U.)	(10^6 km)	Eccentricity (e)	Perihelion (A.U.)	(10^6 km)	Aphelion (A.U.)	(10^6 km)
Mercury	0.39	57.9	0.206	0.31	46.0	0.47	69.8
Venus	0.72	108.2	0.007	0.72	107.5	0.73	108.9
Earth	1.00	149.6	0.017	0.98	147.1	1.02	152.1
Mars	1.52	227.9	0.093	1.38	206.6	1.67	249.2
Jupiter	5.20	778.4	0.048	4.95	740.7	5.46	816
Saturn	9.54	1427	0.054	9.02	1349	10.1	1504
Uranus	19.19	2871	0.047	18.3	2736	20.1	3006
Neptune	30.07	4498	0.009	29.8	4460	30.3	4537
Pluto	39.48	5906	0.249	29.7	4437	49.3	7376

Planet	Mean Orbital Speed (km/s)	Sidereal Period (tropical years)	Synodic Period (days)	Inclination to the Ecliptic (degrees)	Greatest Angular Diameter as Seen from Earth (arc seconds)
Mercury	47.87	0.24	115.88	7.00	13
Venus	35.02	0.62	583.92	3.39	64
Earth	29.79	1.00	—	0.01	—
Mars	24.13	1.88	779.94	1.85	25
Jupiter	13.06	11.86	398.88	1.31	50
Saturn	9.65	29.42	378.09	2.49	21
Uranus	6.80	83.75	369.66	0.77	4.1
Neptune	5.43	163.7	367.49	1.77	2.4
Pluto	4.74	248.0	366.72	17.2	0.11

TABLE 3B   Planetary Physical Data

Planet	Equatorial Radius (km)	(Earth = 1)	Mass (kg)	(Earth = 1)	Mean Density (kg/m³)	Surface Gravity (Earth = 1)	Escape Speed (km/s)
Mercury	2440	0.38	3.30×10^{23}	0.055	5430	0.38	4.2
Venus	6052	0.95	4.87×10^{24}	0.82	5240	0.91	10.4
Earth	6378	1.00	5.97×10^{24}	1.00	5520	1.00	11.2
Mars	3394	0.53	6.42×10^{23}	0.11	3930	0.38	5.0
Jupiter	71,492	11.21	1.90×10^{27}	317.8	1330	2.53	60
Saturn	60,268	9.45	5.68×10^{26}	95.16	690	1.07	36
Uranus	25,559	4.01	8.68×10^{25}	14.54	1270	0.91	21
Neptune	24,766	3.88	1.02×10^{26}	17.15	1640	1.14	24
Pluto	1137	0.18	1.27×10^{22}	0.0021	2060	0.07	1.2

Planet	Sidereal Rotation Period (solar days)*	Axial Tilt (degrees)	Surface Magnetic Field (Earth = 1)	Magnetic Axis Tilt (degrees relative to rotation axis)	Albedo†	Surface Temperature‡ (K)	Number of Moons**
Mercury	58.6	0.0	0.011	<10	0.11	100–700	0
Venus	−243.0	177.4	<0.001		0.65	730	0
Earth	0.9973	23.45	1.0	11.5	0.37	290	1
Mars	1.026	23.98	0.001		0.15	180–270	2
Jupiter	0.41	3.08	13.89	9.6	0.52	124	16
Saturn	0.44	26.73	0.67	0.8	0.47	97	18
Uranus	−0.72	97.92	0.74	58.6	0.50	58	27
Neptune	0.67	29.6	0.43	46.0	0.5	59	13
Pluto	−6.39	118	?		0.6	40–60	1

*A negative sign indicates retrograde rotation.
†Fraction of sunlight reflected from surface.
‡Temperature is effective temperature for jovian planets.
**Moons more than 10 km in diameter.

TABLE 4 The Twenty Brightest Stars in Earth's Night Sky

Name	Star	Spectral Type* A	B	Parallax (arc seconds)	Distance (pc)	Apparent Visual Magnitude* A	B
Sirius	α CMa	A1V	wd[†]	0.379	2.6	−1.44	+8.4
Canopus	α Car	F0Ib–II		0.010	96	−0.62	
Arcturus	α Boo	K2III		0.089	11	−0.05	
Rigel Kentaurus (Alpha Centauri)	α Cen	G2V	K0V	0.742	1.3	−0.01	+1.4
Vega	α Lyr	A0V		0.129	7.8	+0.03	
Capella	α Aur	GIII	M1V	0.077	13	+0.08	+10.2
Rigel	β Ori	B8Ia	B9	0.0042	240	+0.18	+6.6
Procyon	α CMi	F5IV–V	wd[†]	0.286	3.5	+0.40	+10.7
Betelgeuse	α Ori	M2Iab		0.0076	130	+0.45	
Achernar	α Eri	B5V		0.023	44	+0.45	
Hadar	β Cen	B1III	?	0.0062	160	+0.61	+4
Altair	α Aq1	A7IV–V		0.194	5.1	+0.76	
Acrux	α Cru	B1IV	B3	0.010	98	+0.77	+1.9
Aldebaran	α Tau	K5III	M2V	0.050	20	+0.87	+13
Spica	α Vir	B1V	B2V	0.012	80	+0.98	2.1
Antares	α Sco	M1Ib	B4V	0.005	190	+1.06	+5.1
Pollux	β Gem	K0III		0.097	10	+1.16	
Formalhaut	α PsA	A3V	?	0.130	7.7	+1.17	+6.5
Deneb	α Cyg	A2Ia		0.0010	990	+1.25	
Mimosa	β Cru	B1IV		0.0093	110	+1.25	

Name	Visual Luminosity* (Sun = 1) A	B	Absolute Visual Magnitude[1] A	B	Proper Motion (arc seconds/yr)	Transverse Velocity (km/s)	Radial Velocity (km/s)
Sirius	22	0.0025	+1.5	+11.3	1.33	16.7	−7.6[‡]
Canopus	1.4×10^4		−5.5		0.02	9.1	20.5
Arcturus	110		−0.3		2.28	119	−5.2
Rigel Kentaurus	1.6	0.45	+4.3	+5.7	3.68	22.7	−24.6
Vega	50		+0.6		0.34	12.6	−13.9
Capella	130	0.01	−0.5	+9.6	0.44	27.1	30.2[‡]
Rigel	4.1×10^4	110	−6.7	−0.3	0.00	1.2	20.7[‡]
Procyon	7.2	0.0006	+2.7	+13.0	1.25	20.7	−3.2[‡]
Betelgeuse	9700		−5.1		0.03	18.5	21.0[‡]
Achernar	1100		−2.8		0.10	20.9	19
Hadar	1.3×10^4	560	−5.4	−2.0	0.04	30.3	−12[‡]
Altair	11		+2.2		0.66	16.3	−26.3
Acrux	4100	2200	−4.2	−3.5	0.04	22.8	−11.2
Aldebaran	150	0.002	−0.6	+11.5	0.20	19.0	54.1
Spica	2200	780	−3.5	−2.4	0.05	19.0	1.0[‡]
Antares	1.1×10^4	290	−5.3	−1.3	0.03	27.0	−3.2
Pollux	31		+1.1		0.62	29.4	3.3
Formalhaut	17	0.13	+1.7	+7.1	0.37	13.5	6.5
Deneb	2.6×10^5		−8.7		0.003	14.1	−4.6[‡]
Mimosa	3200		−3.9		0.05	26.1	—

*Energy output in the visible part of the spectrum; A and B columns identify individual components of binary-star systems.
[†] "wd" stands for "white dwarf."
[‡] Average value of variable velocity

TABLE 5 The Twenty Nearest Stars

Name	Spectral Type A	B	Parallax (arc seconds)	Distance (pc)	Apparent Visual Magnitude[*] A	B
Sun	G2V				−26.74	
Proxima Centauri	M5		0.772	1.30	+11.01	
Alpha Centauri	G2V	K1V	0.742	1.35	−0.01	+1.35
Barnard's Star	M5V		0.549	1.82	+9.54	
Wolf 359	M8V		0.421	2.38	+13.53	
Lalande 21185	M2V		0.397	2.52	+7.50	
UV Ceti	M6V	M6V	0.387	2.58	+12.52	+13.02
Sirius	A1V	wd[†]	0.379	2.64	−1.44	+8.4
Ross 154	M5V		0.345	2.90	+10.45	
Ross 248	M6V		0.314	3.18	+12.29	
ε Eridani	K2V		0.311	3.22	+3.72	
Ross 128	M5V		0.298	3.36	+11.10	
61 Cygni	K5V	K7V	0.294	3.40	+5.22	+6.03
ε Indi	K5V		0.291	3.44	+4.68	
Grm 34	M1V	M6V	0.290	3.45	+8.08	+11.06
Luyten 789-6	M6V		0.290	3.45	+12.18	
Procyon	F5IV–V	wd[†]	0.286	3.50	+0.40	+10.7
Σ 2398	M4V	M5V	0.285	3.55	+8.90	+9.69
Lacaille 9352	M2V		0.279	3.58	+7.35	
G51-15	MV		0.278	3.60	+14.81	

Name	Visual Luminosity[*] (Sun = 1) A	B	Absolute Visual Magnitude[*] A	B	Proper Motion (arc seconds/yr)	Transverse Velocity (km/s)	Radial Velocity (km/s)
Sun	1.0		+4.83				
Proxima Centauri	5.6×10^{-5}		+15.4		3.86	23.8	−16
Alpha Centauri	1.6	0.45	+4.3	+5.7	3.68	23.2	−22
Barnard's Star	4.3×10^{-4}		+13.2		10.34	89.7	−108
Wolf 359	1.8×10^{-5}		+16.7		4.70	53.0	+13
Lalande 21185	0.0055		+10.5		4.78	57.1	−84
UV Ceti	5.4×10^{-5}	0.00004	+15.5	+16.0	3.36	41.1	+30
Sirius	22	0.0025	+1.5	+11.3	1.33	16.7	−8
Ross 154	4.8×10^{-4}		+13.3		0.72	9.9	−4
Ross 248	1.1×10^{-4}		+14.8		1.58	23.8	−81
ε Eridani	0.29		+6.2		0.98	15.3	+16
Ross 128	3.6×10^{-4}		+13.5		1.37	21.8	−13
61 Cygni	0.082	0.039	+7.6	+8.4	5.22	84.1	−64
ε Indi	0.14		+7.0		4.69	76.5	−40
Grm 34	0.0061	0.00039	+10.4	+13.4	2.89	47.3	+17
Luyten 789-6	1.4×10^{-4}		+14.6		3.26	53.3	−60
Procyon	7.2	0.00055	+2.7	+13.0	1.25	2.8	−3
Σ 2398	0.0030	0.0015	+11.2	+11.9	2.28	38.4	+5
Lacaille 9352	0.013		+9.6		6.90	117	+10
G51-15	1.4×10^{-5}		+17.0		1.26	21.5	—

[*]*A and B columns identify individual components of binary-star systems.*
[†]*"wd" stands for "white dwarf."*

GLOSSARY

Key Terms which are boldface in the text are followed by a page reference in the Glossary.

A

A ring One of three Saturnian rings visible from Earth. The A ring is farthest from the planet and is separated from the B ring by the Cassini division. (p. 311)

aberration of starlight Small shift in the observed direction to a star, caused by Earth's motion perpendicular to the line of sight. (p. 43)

absolute brightness The apparent brightness a star would have if it were placed at a standard distance of 10 parsecs from Earth. (p. 441)

absolute magnitude The apparent magnitude a star would have if it were placed at a standard distance of 10 parsecs from Earth. (p. 444)

Absolute Zero The lowest possible temperature that can be obtained; all thermal motion ceases at this temperature.

absorption line Dark line in an otherwise continuous bright spectrum, where light within one narrow frequency range has been removed. (p. 85)

abundance Relative amounts of different elements in a gas.

acceleration The rate of change of velocity of a moving object. (p. 51)

accretion Gradual growth of bodies, such as planets, by the accumulation of other, smaller bodies. (p. 162)

accretion disk Flat disk of matter spiraling down onto the surface of a neutron star or black hole. Often, the matter originated on the surface of a companion star in a binary-star system. (p. 544)

active galactic nucleus Region of intense emission at the center of an active galaxy, responsible for virtually all of the galaxy's nonstellar luminosity. (p. 649)

active galaxies The most energetic galaxies, which can emit hundreds or thousands of times more energy per second than the Milky Way, mostly in the form of long-wavelength nonthermal radiation. (p. 648)

active optics Collection of techniques used to increase the resolution of ground-based telescopes. Minute modifications are made to the overall configuration of an instrument as its temperature and orientation change; used to maintain the best possible focus at all times. (p. 121)

active region Region of the photosphere of the Sun surrounding a sunspot group, which can erupt violently and unpredictably. During sunspot maximum, the number of active regions is also a maximum. (p. 426)

active Sun The unpredictable aspects of the Sun's behavior, such as sudden explosive outbursts of radiation in the form of prominences and flares. (p. 418)

adaptive optics Technique used to increase the resolution of a telescope by deforming the shape of the mirror's surface under computer control while a measurement is being taken; used to undo the effects of atmospheric turbulence. (p. 122)

aerosol Suspension of liquid or solid particles in air.

alpha particle A helium-4 nucleus.

alpha process Process occurring at high temperatures, in which high-energy photons split heavy nuclei to form helium nuclei.

ALSEP Acronym for Apollo Lunar Surface Experiments Package.

altimeter Instrument used to determine altitude.

amino acids Organic molecules which form the basis for building the proteins that direct metabolism in living creatures. (p. 745)

amor asteroid Asteroid that crosses only the orbit of Mars.

amplitude The maximum deviation of a wave above or below zero point. (p. 64)

angstrom Distance unit equal to 0.1 nanometers, or one ten-billionth of a meter.

angular diameter Angle made between the top (or one edge) of an object, the observer and the bottom (or opposite edge) of the object.

angular distance Angular separation between two objects as seen by some observer.

angular momentum Tendency of an object to keep rotating; proportional to the mass, radius, and rotation speed of the body.

angular resolution The ability of a telescope to distinguish between adjacent objects in the sky. (p. 117)

annular eclipse Solar eclipse occurring at a time when the Moon is far enough away from Earth that it fails to cover the disk of the Sun completely, leaving a ring of sunlight visible around its edge. (p. 21)

antiparallel Configuration of the electron and proton in a hydrogen (or other) atom when their spin axes are parallel but the two rotate in opposite directions.

antiparticle A particle of the same mass but opposite in all other respects (e.g., charge) to a given particle; when a particle and its antiparticle come into contact, they annihilate and release energy in the form of gamma rays.

aphelion The point on the elliptical path of an object in orbit about the Sun that is most distant from the Sun. (p. 46)

Apollo asteroid *See* Earth-crossing asteroid.

apparent brightness The brightness that a star appears to have, as measured by an observer on Earth. (p. 442)

apparent magnitude The apparent brightness of a star, expressed using the magnitude scale. (p. 443)

association Small grouping of (typically 100 or less) bright stars, spanning up to a few tens of parses across, usually rich in very young stars. (p. 505)

Assumption of Mediocrity Statements suggesting that the development of life on Earth did not require any unusual circumstances, suggesting that extraterrestrial life may be common.

asteroid One of thousands of very small members of the solar system orbiting the Sun between the orbits of Mars and Jupiter. Often referred to as "minor planets." (p. 356)

asteroid belt Region of the solar system, between the orbits of Mars and Jupiter, in which most asteroids are found. (p. 356)

asthenosphere Layer of Earth's interior, just below the lithosphere, over which the surface plates slide. (p. 184)

astrology Pseudoscience that purports to use the positions of the planets, sun, and moon to predict daily events and human destiny.

astronomical unit (A.U.) The average distance of Earth from the Sun. Precise radar measurements yield a value for the A.U. of 149,603,500 km. (p. 48)

astronomy Branch of science dedicated to the study of everything in the universe that lies above Earth's atmosphere. (p. 4)

asymptotic-giant branch Path on the Hertzsprung–Russell diagram corresponding to the changes that a star undergoes after helium burning ceases in the core. At this stage, the carbon core shrinks and drives the expansion of the envelope, and the star becomes a swollen red giant for a second time. (p. 522)

aten asteroid Earth-crossing asteroid with semimajor axis less than 1 A.U.

atmosphere Layer of gas confined close to a planet's surface by the force of gravity. (p. 170)

atom Building block of matter, composed of positively charged protons and neutral neutrons in the nucleus surrounded by negatively charged electrons. (p. 89)

atomic epoch Period after decoupling when the first simple atoms and molecules formed.

aurora Event which occurs when atmospheric molecules are excited by incoming charged particles from the solar wind, then emit energy as they fall back to their ground states. Aurorae generally occur at high latitudes, near the north and south magnetic poles. (p. 191)

autumnal equinox Date on which the Sun crosses the celestial equator moving southward, occurring on or near September 22. (p. 16)

B

B ring One of three Saturnian rings visible from Earth. The B ring is the brightest of the three, and lies just past the Cassini Division, closer to the planet than the A ring. (p. 311)

background noise Unwanted light in an image, from unresolved sources in the telescope's field of view, scattered light from the atmosphere, or instrumental "hiss" in the detector itself.

barred-spiral galaxy Spiral galaxy in which a bar of material passes through the center of the galaxy, with the spiral arms beginning near the ends of the bar. (p. 635)

basalt Solidified lava; an iron-magnesium-silicate mixture.

baseline The distance between two observing locations used for the purposes of triangulation measurements. The larger the baseline, the better the resolution attainable. (p. 25)

belt Dark, low-pressure region in the atmosphere of a jovian planet, where gas flows downward. (p. 278)

Big Bang Event that cosmologists consider the beginning of the universe, in which all matter and radiation in the entire universe came into being. (p. 697)

Big Crunch Point of final collapse of a bound universe.

binary asteroid Asteroids that have a partner in orbit around it.

binary pulsar Binary system in which both components are pulsars.

binary-star system A system which consists of two stars in orbit about their common center of mass, held together by their mutual gravitational attraction. Most stars are found in binary-star systems. (p. 458)

biological evolution Change in a population of biological organisms over time.

bipolar flow Jets of material expelled from a protostar perpendicular to the surrounding protostellar disk.

blackbody curve The characteristic way in which the intensity of radiation emitted by a hot object depends on frequency. The frequency at which the emitted intensity is highest is an indication of the temperature of the radiating object. Also referred to as the Planck curve. (p. 72)

black dwarf The endpoint of the evolution of an isolated, low-mass star. After the white-dwarf stage, the star cools to the point where it is a dark "clinker" in interstellar space. (p. 528)

black hole A region of space where the pull of gravity is so great that nothing—not even light—can escape. A possible outcome of the evolution of a very massive star. (p. 580)

blazar Particularly intense active galactic nucleus in which the observer's line of sight happens to lie directly along the axis of a high-speed jet of particles emitted from the active region. (p. 654)

blue giant Large, hot, bright star at the upper-left end of the main sequence on the Hertzsprung–Russell diagram. Its name comes from its color and size. (p. 453)

blueshift Motion-induced changes in the observed wavelength from a source that is moving toward us. Relative approaching motion between the object and the observer causes the wavelength to appear shorter (and hence bluer) than if there were no motion at all. (p. 78)

blue straggler Star found on the main sequence of the Hertzsprung–Russell diagram, but which should already have evolved off the main sequence, given its location on the diagram; thought to have formed from mergers of lower mass stars.

blue supergiant The very largest of the large, hot, bright stars at the uppermost-left end of the main sequence on the Hertzsprung–Russell diagram. (p. 453)

Bohr model First theory of the hydrogen atom to explain the observed spectral lines. This model rests on three ideas: that there is a state of lowest energy for the electron, that there is a maximum energy beyond which the electron is no longer bound to the nucleus, and that within these two energies the electron can only exist in certain energy levels. (p. 91)

bok globule Dense, compact cloud of interstellar dust and gas on its way to forming one or more stars.

boson Particle that exerts or mediates forces between elementary particles in quantum physics.

bound trajectory Path of an object with launch speed low enough that it cannot escape the gravitational pull of a planet.

brown dwarf Remnants of fragments of collapsing gas and dust that did not contain enough mass to initiate core nuclear fusion. Such objects are then frozen somewhere along their pre-main-sequence contraction phase, continually cooling into compact dark objects. Because of their small size and low temperature they are extremely difficult to detect observationally. (p. 499)

brown oval Feature of Jupiter's atmosphere that appears only at latitudes near 20 degrees N, this structure is a long-lived hole in the clouds that allows us to look down into Jupiter's lower atmosphere. (p. 283)

C

C ring One of three Saturnian rings visible from Earth. The C ring lies closest to the planet and is relatively thin compared to the A and B rings. (p. 311)

caldera Crater that forms at the summit of a volcano.

capture theory (Moon) Theory suggesting that the Moon formed far from Earth but was later captured by it.

carbonaceous asteroid The darkest, or least reflective, type of asteroid, containing large amounts of carbon.

carbon-based molecule Molecule containing atoms of carbon.

carbon-detonation supernova *See* Type I supernova.

cascade Process of deexcitation in which an excited electron moves down through energy states one at a time.

Cassegrain telescope A type of reflecting telescope in which incoming light hits the primary mirror and is then reflected upward toward the prime focus, where a secondary mirror reflects the light back down through a small hole in the main mirror into a detector or eyepiece. (p. 109)

Cassini Division A relatively empty gap in Saturn's ring system, discovered in 1675 by Giovanni Cassini. It is now known to contain a number of thin ringlets. (p. 311)

cataclysmic variable Collective name for novae and supernovae.

catalyst Something that causes or helps a reaction to occur, but is not itself consumed as part of the reaction.

catastrophic theory A theory that invokes statistically unlikely accidental events to account for observations. (p. 386)

celestial coordinates Pair of quantities—right ascension and declination—similar to longitude and latitude on Earth, used to pinpoint locations of objects on the celestial sphere. (p. 14)

celestial equator The projection of Earth's equator onto the celestial sphere. (p. 11)

celestial mechanics Study of the motions of bodies, such as planets and stars, that interact via gravity.

celestial pole Projection of Earth's north or south pole onto the celestial sphere. (p. 10)

celestial sphere Imaginary sphere surrounding Earth to which all objects in the sky were once considered to be attached. (p. 10)

Celsius Temperature scale in which the freezing point of water is 0 degrees and the boiling point of water is 100 degrees.

center of mass The "average" position in space of a collection of massive bodies, weighted by their masses. For an isolated system this point moves with constant velocity, according to Newtonian mechanics. (p. 55)

centigrade *See* Celsius.

centripetal force (literally "center seeking") Force directed toward the center of a body's orbit.

centroid Average position of the material in an object; in spectroscopy, the center of a spectral line.

Cepheid variable Star whose luminosity varies in a characteristic way, with a rapid rise in brightness followed by a slower decline. The period of a Cepheid variable star is related to its luminosity, so a determination of this period can be used to obtain an estimate of the star's distance. (p. 607)

chandrasekhar mass Maximum possible mass of a white dwarf.

chaotic rotation Unpredictable tumbling motion that nonspherical bodies in eccentric orbits, such as Saturn's satellite Hyperion, can exhibit. No amount of observation of an object rotating chaotically will ever show a well-defined period. (p. 322)

charge-coupled device (CCD) An electronic device used for data acquisition; composed of many tiny pixels, each of which records a buildup of charge to measure the amount of light striking it. (p. 111)

chemical bond Force holding atoms together to form a molecule.

chemosynthesis Analog of photosynthesis that operates in total darkness.

chromatic aberration The tendency for a lens to focus red and blue light differently, causing images to become blurred. (p. 110)

chromosphere The Sun's lower atmosphere, lying just above the visible atmosphere. (p. 407)

circumnavigation Traveling all the way around an object.

cirrus High-level clouds composed of ice or methane crystals.

closed universe Geometry that the universe as a whole would have if the density of matter is above the critical value. A closed universe is finite in extent and has no edge, like the surface of a sphere. It has enough mass to stop the present expansion and will eventually collapse. (p. 704)

CNO cycle Chain of reactions that converts hydrogen into helium using carbon, nitrogen, and oxygen as catalysts.

cocoon nebula Bright infrared source in which a surrounding cloud of gas and dust absorb ultraviolet radiation from a hot star and reemits it in the infrared.

cold dark matter Class of dark-matter candidates made up of very heavy particles, possibly formed in the very early universe. (p. 735)

collecting area The total area of a telescope capable of capturing incoming radiation. The larger the telescope, the greater its collecting area, and the fainter the objects it can detect. (p. 116)

collisional broadening Broadening of spectral lines due to collisions between atoms, most often seen in dense gases.

color index A convenient method of quantifying a star's color by comparing its apparent brightness as measured through different filters. If the star's radiation is well described by a blackbody spectrum, the ratio of its blue intensity (B) to its visual intensity (V) is a measure of the object's surface temperature. (p. 446)

color–magnitude diagram A way of plotting stellar properties, in which absolute magnitude is plotted against color index. (p. 452)

coma An effect occurring during the formation of an off-axis image in a telescope. Stars whose light enters the telescope at a large angle acquire cometlike tails on their images. The brightest part of a comet, often referred to as the "head." (p. 363)

comet A small body, composed mainly of ice and dust, in an elliptical orbit about the Sun. As it comes close to the Sun, some of its material is vaporized to form a gaseous head and extended tail. (p. 363)

common envelope Outer layer of gas in a contact binary.

comparative planetology The systematic study of the similarities and differences among the planets, with the goal of obtaining deeper insight into how the solar system formed and has evolved in time. (p. 145)

composition The mixture of atoms and molecules that make up an object.

condensation nuclei Dust grains in the interstellar medium which act as seeds around which other material can cluster. The presence of dust was very important in causing matter to clump during the formation of the solar system. (p. 161)

condensation theory Currently favored model of solar system formation which combines features of the old nebular theory with new information about interstellar dust grains, which acted as condensation nuclei. (p. 161)

conjunction Orbital configuration in which a planet lies in the same direction as the Sun, as seen from Earth.

conservation of mass and energy *See* law of conservation of mass and energy.

constellation A human grouping of stars in the night sky into a recognizable pattern. (p. 8)

constituents *See* composition.

contact binary A binary-star system in which both stars have expanded to fill their Roche lobes and the surfaces of the two stars merge. The binary system now consists of two nuclear burning stellar cores surrounded by a continuous common envelope. (p. 537)

continental drift The movement of the continents around Earth's surface.

continuous spectrum Spectrum in which the radiation is distributed over all frequencies, not just a few specific frequency ranges. A prime example is the blackbody radiation emitted by a hot, dense body. (p. 84)

convection Churning motion resulting from the constant upwelling of warm fluid and the concurrent downward flow of cooler material to take its place. (p. 171)

convection zone Region of the Sun's interior, lying just below the surface, where the material of the Sun is in constant convection motion. This region extends into the solar interior to a depth of about 20,000 km. (p. 407)

co-orbital satellites Satellites sharing the same orbit around a planet.

Copernican Principle The removal of Earth from any position of cosmic significance.

Copernican revolution The realization, toward the end of the sixteenth century, that Earth is not at the center of the universe. (p. 41)

core The central region of Earth, surrounded by the mantle. (p. 170); The central region of any planet or star. (p. 408)

core-accretion theory Theory that the jovian planets formed when icy protoplanetary cores became massive enough to capture gas directly from the solar nebula. *See* gravitational instability theory. (p. 388)

core-collapse supernova *See* Type II supernova.

core hydrogen burning The energy burning stage for main-sequence stars, in which the helium is produced by hydrogen fusion in the central region of the star. A typical star spends up to 90 percent of its lifetime in hydrostatic equilibrium brought about by the balance between gravity and the energy generated by core hydrogen burning. (p. 516)

cornea (eye) The curved transparent layer covering the front part of the eye.

corona One of numerous large, roughly circular regions on the surface of Venus, thought to have been caused by upwelling mantle material causing the planet's crust to bulge outward. (p. 239)

corona The tenuous outer atmosphere of the Sun, which lies just above the chromosphere and, at great distances, turns into the solar wind. (p. 407)

coronal hole Vast regions of the Sun's atmosphere where the density of matter is about 10 times lower than average. The gas there streams freely into space at high speeds, escaping the Sun completely. (p. 428)

coronal mass ejection Giant magnetic "bubble" of ionized gas that separates from the rest of the solar atmosphere and escapes into interplanetary space. (p. 427)

corpuscular theory Early particle theory of light.

cosmic density parameter Ratio of the universe's actual density to the critical value corresponding to zero curvature.

cosmic distance scale Collection of indirect distance-measurement techniques that astronomers use to measure distances in the universe. (p. 25)

cosmic evolution The collection of the seven major phases of the history of the universe; namely galactic, stellar, planetary, chemical, biological, cultural, and future evolution. (p. 744)

cosmic microwave background The almost perfectly isotropic radio signal that is the electromagnetic remnant of the Big Bang. (p. 711)

cosmic ray Very energetic subatomic particle arriving at Earth from elsewhere in the Galaxy.

cosmological constant Quantity originally introduced by Einstein into general relativity to make his equations describe a static universe. Now one of several candidates for the repulsive "dark energy" force responsible for the observed cosmic acceleration. (p. 708)

cosmological distance Distance comparable to the scale of the universe.

cosmological principle Two assumptions which make up the basis of cosmology, namely that the universe is homogeneous and isotropic on sufficiently large scales. (p. 695)

cosmological redshift The component of the redshift of an object which is due only to the Hubble flow of the universe. (p. 646)

cosmology The study of the structure and evolution of the entire universe. (p. 695)

cosmos The universe.

coudé focus Focus produced far from the telescope using a series of mirrors. Allows the use of heavy and/or finely tuned equipment to analyze the image.

crater Bowl-shaped depression on the surface of a planet or moon, resulting from a collision with interplanetary debris. (p. 203)

crescent Appearance of the Moon (or a planet) when less than half of the body's hemisphere is visible from Earth.

crest Maximum departure of a wave above its undisturbed state.

critical density The cosmic density corresponding to the dividing line between a universe that recollapses and one that expands forever. (p. 702)

critical universe Universe in which the density of matter is exactly equal to the critical density. The universe is infinite in extent and has zero curvature. The expansion will continue forever, but will approach an expansion speed of zero. (p. 704)

crust Layer of Earth which contains the solid continents and the seafloor. (p. 170)

C-type asteroid *See* carbonaceous asteroid.

cultural evolution Change in the ideas and behavior of a society over time.

current sheet Flat sheet on Jupiter's magnetic equator where most of the charged particles in the magnetosphere lie due to the planet's rapid rotation.

D

D ring Collection of very faint, thin rings, extending from the inner edge of the C ring down nearly to the cloudtops of Saturn. This region contains so few particles that it is completely invisible from Earth. (p. 314)

dark dust cloud A large cloud, often many parsecs across, which contains gas and dust in a ratio of about 1012 gas atoms for every dust particle. Typical densities are a few tens or hundreds of millions of particles per cubic meter. (p. 478)

dark energy Generic name given to the unknown cosmic force field thought to be responsible for the observed acceleration of the Hubble expansion. (p. 707)

dark halo Region of a galaxy beyond the visible halo where dark matter is believed to reside. (p. 622)

dark matter Term used to describe the mass in galaxies and clusters whose existence we infer from rotation curves and other techniques, but which has not been confirmed by observations at any electromagnetic wavelength. (p. 623)

dark matter particle Particle undetectable at any electromagnetic wavelength, but which can be inferred due to its gravitational influence.

Daughter/Fission theory Theory suggesting that the Moon originated out of Earth.

declination Celestial coordinate used to measure latitude above or below the celestial equator on the celestial sphere. (p. 14)

decoupling Event in the early universe when atoms first formed, after which photons could propagate freely through space. (p. 728)

deferent A construct of the geocentric model of the solar system which was needed to explain observed planetary motions. A deferent is a large circle encircling Earth, on which an epicycle moves. (p. 38)

degree Unit of angular measure. There are 360 degrees in one complete circle. (p. 11)

density A measure of the compactness of the matter within an object, computed by dividing the mass of the object by its volume. Units are kilograms per cubic meter (kg/m^3), or grams per cubic centimeter (g/cm^3). (p. 147)

detached binary Binary system where each star lies within its respective Roche lobe.

detector noise Readings produced by an instrument even when it is not observing anything; produced by the electronic components within the detector itself.

deuterium A form of hydrogen with an extra neutron in its nucleus.

deuterium bottleneck Period in the early universe between the start of deuterium production and the time when the universe was cool enough for deuterium to survive.

deuteron An isotope of hydrogen in which a neutron is bound to the proton in the nucleus. Often called "heavy hydrogen" because of the extra mass of the neutron. (p. 409)

differential rotation The tendency for a gaseous sphere, such as a jovian planet or the Sun, to rotate at a different rate at the equator than at the poles. More generally, a condition where the angular speed varies with location within an object. (p. 277)

differentiation Variation in the density and composition of a body, such as Earth, with low density material on the surface and higher density material in the core. (p. 180)

diffraction The ability of waves to bend around corners. The diffraction of light establishes its wave nature. (p. 66)

diffraction grating Sheet of transparent material with many closely spaced parallel lines ruled on it, designed to separate white light into a spectrum.

diffraction-limited resolution Theoretical resolution that a telescope can have due to diffraction of light at the telescope's aperture. Depends on the wavelength of radiation and the diameter of the telescope's mirror.

direct motion *See* prograde motion.

distance modulus Difference between the apparent and absolute magnitude of an object; equivalent to distance, by the inverse-square law.

diurnal motion Apparent daily motion of the stars, caused by Earth's rotation.

DNA Deoxyribonucleic acid, the molecule that carries genetic information and determine the characteristics of a living organism.

Doppler effect Any motion-induced change in the observed wavelength (or frequency) of a wave. (p. 77)

double-line spectroscopic binary Binary system in which spectral lines of both stars can be distinguished and seen to shift back and forth as the stars orbit one another.

double-star system System containing two stars in orbit around one another.

Drake equation Expression that gives an estimate of the probability that intelligence exists elsewhere in the galaxy, based on a number of supposedly necessary conditions for intelligent life to develop. (p. 751)

dust grain An interstellar dust particle, roughly 10^{-7} m in size, comparable to the wavelength of visible light. (p. 468)

dust lane A lane of dark, obscuring interstellar dust in an emission nebula or galaxy. (p. 472)

dust tail The component of a comet's tail that is composed of dust particles. (p. 364)

dwarf Any star with radius comparable to, or smaller than, that of the Sun (including the Sun itself). (p. 451)

dwarf elliptical Elliptical galaxy as small as 1 kiloparsec across, containing only a few million stars.

dwarf galaxy Small galaxy containing a few million stars.

dwarf irregular Small irregular galaxy containing only a few million stars.

dynamo theory Theory that explains planetary and stellar magnetic fields in terms of rotating, conducting material flowing in an object's interior. (p. 186)

E

E ring A faint ring, well outside the main ring system of Saturn, which was discovered by *Voyager* and is believed to be associated with volcanism on the moon Enceladus. (p. 314)

Earth-crossing asteroid An asteroid whose orbit crosses that of Earth. Earth-crossing asteroids are also called Apollo asteroids, after the first asteroid of this type discovered. (p. 359)

earthquake A sudden dislocation of rocky material near Earth's surface. (p. 171)

eccentricity A measure of the flatness of an ellipse, equal to the distance between the two foci divided by the length of the major axis. (p. 46)

eclipse Event during which one body passes in front of another, so that the light from the occulted body is blocked. (p. 15)

eclipse season Times of the year when the Moon lies in the same plane as Earth and Sun, so that eclipses are possible. (p. 23)

eclipse year Time interval between successive orbital configurations in which the line of nodes of the Moon's orbit points toward the Sun.

eclipsing binary Rare binary-star system that is aligned in such a way that from Earth we observe one star pass in front of the other, eclipsing the other star. (p. 458)

ecliptic The apparent path of the Sun, relative to the stars on the celestial sphere, over the course of a year. (p. 13)

effective temperature Temperature of a blackbody of the same radius and luminosity as a given star or planet.

ejecta (planetary) Material thrown outward by a meteoroid impact.

ejecta (stellar) Material thrown into space by a nova or supernova.

electric field A field extending outward in all directions from a charged particle, such as a proton or an electron. The electric field determines the electric force exerted by the particle on all other charged particles in the universe; the strength of the electric field decreases with increasing distance from the charge according to an inverse-square law. (p. 66)

electromagnetic energy Energy carried in the form of rapidly fluctuating electric and magnetic fields.

electromagnetic radiation Another term for light, electromagnetic radiation transfers energy and information from one place to another. (p. 62)

electromagnetic spectrum The complete range of electromagnetic radiation, from radio waves to gamma rays, including the visible spectrum. All types of electromagnetic radiation are basically the same phenomenon, differing only by wavelength, and all move at the speed of light. (p. 69)

electromagnetism The union of electricity and magnetism, which do not exist as independent quantities but are in reality two aspects of a single physical phenomenon. (p. 66)

electron An elementary particle with a negative electric charge; one of the components of the atom. (p. 65)

electron degeneracy pressure The pressure produced by the resistance of electrons to further compression once they are squeezed to the point of contact. (p. 521)

electrostatic force Force between electrically charged objects.

electroweak force Unification of the weak electromagnetic forces.

element Matter made up of one particular atom. The number of protons in the nucleus of the atom determines which element it represents. (p. 95)

elementary particle Technically, a particle that cannot be subdivided into component parts; however, the term is also often used to refer to particles such as protons and neutrons, which are themselves made up of quarks.

ellipse Geometric figure resembling an elongated circle. An ellipse is characterized by its degree of flatness, or eccentricity, and the length of its long axis. In general, bound orbits of objects moving under gravity are elliptical. (p. 46)

elliptical galaxy Category of galaxy in which the stars are distributed in an elliptical shape on the sky, ranging from highly elongated to nearly circular in appearance. (p. 636)

elongation Angular distance between a planet and the Sun.

emission line Bright line in a specific location of the spectrum of radiating material, corresponding to emission of light at a certain frequency. A heated gas in a glass container produces emission lines in its spectrum. (p. 84)

emission nebula A glowing cloud of hot interstellar gas. The gas glows as a result of one or more nearby young stars which ionize the gas. Since the gas is mostly hydrogen, the emitted radiation falls predominantly in the red region of the spectrum, because of the hydrogen-alpha emission line. (p. 472)

emission spectrum The pattern of spectral emission lines produced by an element. Each element has its own unique emission spectrum. (p. 85)

empirical Discovery based on observational evidence (rather than from theory).

Encke gap A small gap in Saturn's A ring. (p. 311)

energy flux Energy per unit area per unit time radiated by a star (or recorded by a detector).

epicycle A construct of the geocentric model of the solar system which was necessary to explain observed planetary motions. Each planet rides on a small epicycle whose center in turn rides on a larger circle (the deferent). (p. 38)

epoch of inflation Short period of unchecked cosmic expansion early in the history of the universe. During inflation, the universe swelled in size by a factor of about 10^{50}. (p. 731)

equinox *See* vernal equinox, autumnal equinox.

escape speed The speed necessary for one object to escape the gravitational pull of another. Anything that moves away from a gravitating body with more than the escape speed will never return. (p. 56)

euclidean geometry Geometry of flat space.

event horizon Imaginary spherical surface surrounding a collapsing star, with radius equal to the Schwarzschild radius, within which no event can be seen, heard, or known about by an outside observer. (p. 581)

evolutionary theory A theory which explains observations in a series of gradual steps, explainable in terms of well-established physical principles. (p. 386)

evolutionary track A graphical representation of a star's life as a path on the Hertzsprung–Russell diagram. (p. 495)

excited state State of an atom when one of its electrons is in a higher energy orbital than the ground state. Atoms can become excited by absorbing a photon of a specific energy, or by colliding with a nearby atom. (p. 89)

extinction The dimming of starlight as it passes through the interstellar medium. (p. 468)

extrasolar planet Planet orbiting a star other than the Sun. (p. 384)

extremophilic Adjective describing organisms that can survive in very harsh environments.

eyepiece Secondary lens through which an observer views an image. This lens is often chosen to magnify the image.

F

F ring Faint narrow outer ring of Saturn, discovered by *Pioneer 11* in 1979. The F ring lies just inside the Roche limit of Saturn, and was found by *Voyager 1* to be made up of several ring strands apparently braided together. (p. 315)

Fahrenheit Temperature scale in which the freezing point of water is 32 degrees and the boiling point of water is 212 degrees.

false vacuum Region of the universe that remained in the "unified" state after the strong and electroweak forces separated; one possible cause of cosmic inflation at very early times.

fault line Dislocation on a planet's surface, often indicating the boundary between two plates.

field line Imaginary line indicating the direction of an electric or magnetic field.

fireball Large meteor that burns up brightly and sometimes explosively in Earth's atmosphere.

firmament Old-fashioned term for the heavens (i.e., the sky).

flare Explosive event occurring in or near an active region on the Sun. (p. 426)

flatness problem One of two conceptual problems with the standard Big Bang model, which is that there is no natural way to explain why the density of the universe is so close to the critical density. (p. 729)

fluidized ejecta The ejecta blankets around some Martian craters, which apparently indicate that the ejected material was liquid at the time the crater formed. (p. 257)

fluorescence Phenomenon where an atom absorbs energy, then radiates photons of lower energy as it cascades back to the ground state; in astronomy, often produced as ultraviolet photons from a hot young star are absorbed by a neutral gas, causing some of the gas atoms to become excited and give off an optical (red) glow.

flyby Unbound trajectory of a spacecraft around a planet or other body.

focal length Distance from a mirror or the center of a lens to the focus.

focus One of two special points within an ellipse, whose separation from each other indicates the eccentricity. In a bound orbit, planets orbit in ellipses with the Sun at one focus. (p. 46)

forbidden line A spectral line seen in emission nebulae, but not seen in laboratory experiments because, under laboratory conditions, collisions kick the electron in question into some other state before emission can occur. (p. 478)

force Action on an object that causes its momentum to change. The rate at which the momentum changes is numerically equal to the force. (p. 51)

fragmentation The breaking up of a large object into many smaller pieces (for example, as the result of high-speed collisions between planetesimals and protoplanets in the early solar system). (p. 387)

Fraunhofer lines The collection of over 600 absorption lines in the spectrum of the Sun, first categorized by Joseph Fraunhofer in 1812. (p. 88)

frequency The number of wave crests passing any given point in a unit time. (p. 64)

full When the full hemisphere of the Moon or a planet can be seen from Earth.

full Moon Phase of the Moon in which it appears as a complete circular disk in the sky. (p. 15)

fusion *See* nuclear fusion.

G

G ring Faint, narrow ring of Saturn, discovered by Pioneer 11 and lying just outside the F ring. (p. 316)

galactic bulge Thick distribution of warm gas and stars around the galactic center. (p. 604)

galactic cannibalism A galaxy merger in which a larger galaxy consumes a smaller one. (p. 649)

galactic center The center of the Milky Way, or any other galaxy. The point about which the disk of a spiral galaxy rotates. (p. 611)

galactic disk Flattened region of gas and dust that bisects the galactic halo in a spiral galaxy. This is the region of active star formation. (p. 604)

galactic epoch Period from 100 million to 3 billion years after the Big Bang when large agglomerations of matter (galaxies and galaxy clusters) formed and grew.

galactic habitable zone Region of a galaxy in which conditions are conducive to the development of life.

galactic halo Region of a galaxy extending far above and below the galactic disk, where globular clusters and other old stars reside. (p. 604)

galactic nucleus Small, central, high-density region of a galaxy. Almost all the radiation from active galaxies is generated within the nucleus. (p. 649)

galactic rotation curve Plot of rotation speed versus distance from the center of a galaxy.

galactic year Time taken for objects at the distance of the Sun (about 8 kpc) to orbit the center of the Galaxy, roughly 225 million years.

galaxy Gravitationally bound collection of a large number of stars. The Sun is a star in the Milky Way Galaxy. (p. 604)

galaxy cluster A collection of galaxies held together by their mutual gravitational attraction. (p. 643)

Galilean Moon The four large moons of Jupiter discovered by Galileo Galilei.

Galilean satellites The four brightest and largest moons of Jupiter (Io, Europa, Ganymede, Callisto), named after Galileo Galilei, the seventeenth-century astronomer who first observed them. (p. 276)

gamma ray Region of the electromagnetic spectrum, far beyond the visible spectrum, corresponding to radiation of very high frequency and very short wavelength. (p. 62)

gamma-ray burst Object that radiates tremendous amounts of energy in the form of gamma rays, possibly due to the collision and merger of two neutron stars initially in orbit around one another. (p. 576)

gamma-ray spectrograph Spectrograph designed to work at gamma-ray wavelengths. Used to map the abundances of certain elements on the Moon and Mars.

gaseous Composed of gas.

gas-exchange experiment Experiment to look for life on Mars. A nutrient broth was offered to Martian soil specimens. If there were life in the soil, gases would be created as the broth was digested.

gene Sequence of nucleotide bases in the DNA molecule that determines the characteristics of a living organism.

general relativity Theory proposed by Einstein to incorporate gravity into the framework of special relativity.

geocentric model A model of the solar system which holds that Earth is at the center of the universe and all other bodies are in orbit around it. The earliest theories of the solar system were geocentric. (p. 38)

giant A star with a radius between 10 and 100 times that of the Sun. (p. 451)

giant elliptical Elliptical galaxy up to a few megaparsecs across, containing trillions of stars.

gibbous Appearance of the Moon (or a planet) when more than half (but not all) of the body's hemisphere is visible from Earth.

globular cluster Tightly bound, roughly spherical collection of hundreds of thousands, and sometimes millions, of stars spanning about 50 parsecs. Globular clusters are distributed in the halos around the Milky Way and other galaxies. (p. 506)

gluon Particle that exerts or mediates the strong force in quantum physics.

gradient Rate of change of some quantity (e.g., temperature or composition) with respect to location in space.

Grand Unified Theories Class of theories describing the behavior of the single force that results from unification of the strong, weak, and electromagnetic forces in the early universe. (p. 722)

granite Igneous rock, containing silicon and aluminum, that makes up most of Earth's crust.

granulation Mottled appearance of the solar surface, caused by rising (hot) and falling (cool) material in convective cells just below the photosphere. (p. 417)

gravitational force Force exerted on one body by another due to the effect of gravity. The force is directly proportional to the masses of both bodies involved, and inversely proportional to the square of the distance between them. (p. 52)

gravitational instability theory Theory that the jovian planets formed directly from the solar nebula via instabilities in the gas leading to gravitational contraction. *See* core-accretion theory. (p. 388)

gravitational lensing The effect induced on the image of a distant object by a massive foreground object. Light from the distant object is bent into two or more separate images. (p. 624)

gravitational radiation Radiation resulting from rapid changes in a body's gravitational field.

gravitational redshift A prediction of Einstein's general theory of relativity. Photons lose energy as they escape the gravitational field of a massive object. Because a photon's energy is proportional to its frequency, a photon that loses energy suffers a decrease in frequency, which corresponds to an increase, or redshift, in wavelength. (p. 587)

graviton Particle carrying the gravitational field in theories attempting to unify gravity and quantum mechanics.

gravity The attractive effect that any massive object has on all other massive objects. The greater the mass of the object, the stronger its gravitational pull. (p. 52)

gravity assist Using gravity to change the flight path of a satellite or spacecraft.

gravity wave Gravitational counterpart of an electromagnetic wave.

great attractor A huge accumulation of mass in the relatively nearby universe (within about 200 Mpc of the Milky Way).

Great Dark Spot Prominent storm system in the atmosphere of Neptune observed by *Voyager 2*, near the equator of the planet. The system was comparable in size to Earth. (p. 335)

Great Red Spot A large, high-pressure, long-lived storm system visible in the atmosphere of Jupiter. The Red Spot is roughly twice the size of Earth. (p. 278)

great wall Extended sheet of galaxies measuring at least 200 megaparsecs across; one of the largest known structures in the Universe.

greenhouse effect The partial trapping of solar radiation by a planetary atmosphere, similar to the trapping of heat in a greenhouse. (p. 173)

greenhouse gas Gas (such as carbon dioxide or water vapor) that efficiently absorbs infrared radiation.

ground state The lowest energy state that an electron can have within an atom. (p. 89)

GUT epoch Period when gravity separated from the other three forces of nature.

gyroscope System of rotating wheels that allows a spacecraft to maintain a fixed orientation in space.

H

habitable zone Three-dimensional zone of comfortable temperature (corresponding to liquid water) that surrounds every star.

half-life The amount of time it takes for half of the initial amount of a radioactive substance to decay into something else.

hayashi track Evolutionary track followed by a protostar during the final pre-main-sequence phase before nuclear fusion begins.

heat Thermal energy, the energy of an object due to the random motion of its component atoms or molecules.

heat death End point of a bound universe, in which all matter and life are destined to be incinerated.

heavy element In astronomical terms, any element heavier than hydrogen and helium.

heliocentric model A model of the solar system which is centered on the Sun, with Earth in motion about the Sun. (p. 39)

helioseismology The study of conditions far below the Sun's surface through the analysis of internal "sound" waves that repeatedly cross the solar interior. (p. 414)

helium-burning shell Shell of burning helium gas surrounding a non-burning stellar core of carbon ash.

helium capture The formation of heavy elements by the capture of a helium nucleus. For example, carbon can form heavier elements by fusion with other carbon nuclei, but it is much more likely to occur by helium capture, which requires less energy. (p. 558)

helium flash An explosive event in the post-main-sequence evolution of a low-mass star. When helium fusion begins in a dense stellar core, the burning is explosive in nature. It continues until the energy released is enough to expand the core, at which point the star achieves stable equilibrium again. (p. 522)

helium precipitation Mechanism responsible for the low abundance of helium of Saturn's atmosphere. Helium condenses in the upper layers to form a mist, which rains down toward Saturn's interior, just as water vapor forms into rain in the atmosphere of Earth. (p. 309)

helium shell flash Condition in which the helium-burning shell in the core of a star cannot respond to rapidly changing conditions within it, leading to a sudden temperature rise and a dramatic increase in nuclear reaction rates.

Hertzsprung–Russell diagram A plot of luminosity against temperature (or spectral class) for a group of stars. (p. 452)

high-energy astronomy Astronomy using X- or gamma-ray radiation rather than optical radiation.

high-energy telescope Telescope designed to detect X- and gamma radiation. (p. 133)

highlands Relatively light-colored regions on the surface of the Moon which are elevated several kilometers above the maria. Also called terrae. (p. 203)

high-mass star Star with a mass more than 8 times that of the Sun; progenitor of a neutron star or black hole.

HI region Region of space containing primarily neutral hydrogen.

HII region Region of space containing primarily ionized hydrogen.

homogeneity Assumed property of the universe such that the number of galaxies in an imaginary large cube of the universe is the same no matter where in the universe the cube is placed. (p. 695)

horizon problem One of two conceptual problems with the standard Big Bang model, which is that some regions of the universe which have very similar properties are too far apart to have exchanged information within the age of the universe. (p. 729)

horizontal branch Region of the Hertzsprung–Russell diagram where post-main-sequence stars again reach hydrostatic equilibrium. At this point, the star is burning helium in its core and fusing hydrogen in a shell surrounding the core. (p. 522)

hot dark matter A class of candidates for the dark matter in the universe, composed of lightweight particles such as neutrinos, much less massive than the electron. (p. 731)

hot Jupiter A massive, gaseous planet orbiting very close to its parent star. (p. 396)

hot longitudes (Mercury) Two opposite points on Mercury's equator where the Sun is directly overhead at perihelion.

Hubble classification scheme Method of classifying galaxies according to their appearance, developed by Edwin Hubble. (p. 634)

Hubble diagram Plot of galactic recession velocity versus distance; evidence for an expanding universe.

Hubble flow Universal recession described by the Hubble diagram and quantified by the Hubble Law.

Hubble's constant The constant of proportionality which gives the relation between recessional velocity and distance in Hubble's law. (p. 647)

Hubble's law Law that relates the observed velocity of recession of a galaxy to its distance from us. The velocity of recession of a galaxy is directly proportional to its distance away. (p. 646)

hydrocarbon Molecules consisting solely of hydrogen and carbon.

hydrogen envelope An invisible sheath of gas engulfing the coma of a comet, usually distorted by the solar wind and extending across millions of kilometers of space. (p. 364)

hydrogen shell burning Fusion of hydrogen in a shell that is driven by contraction and heating of the helium core. Once hydrogen is depleted in the core of a star, hydrogen burning stops and the core contracts due to gravity, causing the temperature to rise, heating the surrounding layers of hydrogen in the star, and increasing the burning rate there. (p. 519)

hydrosphere Layer of Earth which contains the liquid oceans and accounts for roughly 70 percent of Earth's total surface area. (p. 170)

hydrostatic equilibrium Condition in a star or other fluid body in which gravity's inward pull is exactly balanced by internal forces due to pressure. (p. 413)

hyperbola Curve formed when a plane intersects a cone at a small angle to the axis of the cone.

hypernova Explosion where a massive star undergoes core collapse and forms a black hole and a gamma-ray burst. *See* supernova.

I

igneous Rocks formed from molten material.

image The optical representation of an object produced when light from the object is reflected or refracted by a mirror or lens. (p. 106)

Impact Theory (Moon) Combination of the Capture and Daughter theories, suggesting that the Moon formed after an impact which dislodged some of Earth's mantle and placed it in orbit.

inertia The tendency of an object to continue moving at the same speed and in the same direction, unless acted upon by a force. (p. 51)

inferior conjunction Orbital configuration in which an inferior planet (Mercury or Venus) lies closest to Earth.

inflation *See* epoch of inflation. (p. 728)

infrared Region of the electromagnetic spectrum just outside the visible range, corresponding to light of a slightly longer wavelength than red light. (p. 62)

infrared telescope Telescope designed to detect infrared radiation. Many such telescopes are designed to be lightweight so that they can be carried above (most of) Earth's atmosphere by balloons, airplanes, or satellites. (p. 130)

inhomogeneity Deviation from perfectly uniform density; in cosmology, inhomogeneities in the universe are ultimately due to quantum fluctuations before inflation.

inner core The central part of Earth's core, believed to be solid, and composed mainly of nickel and iron. (p. 179)

instability strip Part of the Hertzsprung–Russell diagram where pulsating post-main-sequence stars are found.

intensity A basic property of electromagnetic radiation that specifies the amount or strength of the radiation. (p. 72)

intercloud medium Superheated bubbles of hot gas extending far into interstellar space.

intercrater plains Regions on the surface of Mercury that do not show extensive cratering but are relatively smooth. (p. 219)

interference The ability of two or more waves to interact in such a way that they either reinforce or cancel each other. (p. 67)

interferometer Collection of two or more telescopes working together as a team, observing the same object at the same time and at the same wavelength. The effective diameter of an interferometer is equal to the distance between its outermost telescopes. (p. 127)

interferometry Technique in widespread use to dramatically improve the resolution of radio and infrared maps. Several telescopes observe the object simultaneously, and a computer analyzes how the signals interfere with each other. (p. 127)

intermediate-mass black hole Black hole having a mass 100–1000 times greater than the mass of the Sun.

interplanetary matter Matter in the solar system that is not part of a planet or moon—cosmic "debris".

interstellar dust Microscopic dust grains that populate space between the stars, having their origins in the ejected matter of long-dead stars. (p. 386)

interstellar gas cloud A large cloud of gas found in the space among the stars.

interstellar medium The matter between stars, composed of two components, gas and dust, intermixed throughout all of space. (p. 468)

intrinsic variable Star that varies in appearance due to internal processes (rather than, say, interaction with another star).

inverse-square law The law that a field follows if its strength decreases with the square of the distance. Fields that follow the inverse-square law decrease rapidly in strength as the distance increases, but never quite reach zero. (p. 52)

Io plasma torus Doughnut-shaped region of energetic ionized particles, emitted by the volcanoes on Jupiter's moon Io and swept up by Jupiter's magnetic field. (p. 290)

ion An atom that has lost one or more of its electrons.

ionization state Term describing the number of electrons missing from an atom: I refers to a neutral atom, II refers to an atom missing one electron, and so on.

ion tail Thin stream of ionized gas that is pushed away from the head of a comet by the solar wind. It extends directly away from the Sun. Often referred to as a plasma tail. (p. 364)

ionosphere Layer in Earth's atmosphere above about 100 km where the atmosphere is significantly ionized and conducts electricity. (p. 167)

irregular galaxy A galaxy which does not fit into any of the other major categories in the Hubble classification scheme. (p. 638)

isotopes Nuclei containing the same number of protons but different numbers of neutrons. Most elements can exist in several isotopic forms. A common example of an isotope is deuterium, which differs from normal hydrogen by the presence of an extra neutron in the nucleus. (p. 410)

isotropy Assumed property of the universe such that the universe looks the same in every direction. (p. 695)

J

jet stream Relatively strong winds in the upper atmosphere channeled into a narrow stream by a planet's rotation. Normally refers to a horizontal, high-altitude wind.

joule The SI unit of energy.

jovian planet One of the four giant outer planets of the solar system, resembling Jupiter in physical and chemical composition. (p. 150)

K

Kelvin Scale Temperature scale in which absolute zero is at 0 K; a change of 1 Kelvin is the same as a change of 1 degree Celsius.

Kelvin-Helmholtz contraction phase Evolutionary track followed by a star during the protostar phase.

Kepler's laws of planetary motion Three laws, based on precise observations of the motions of the planets by Tycho Brahe, which summarize the motions of the planets about the Sun. (p. 44)

kinetic energy Energy of an object due to its motion.

Kirchhoff's laws Three rules governing the formation of different types of spectra. (p. 87)

Kirkwood gaps Gaps in the spacings of orbital semimajor axes of asteroids in the asteroid belt, produced by dynamical resonances with nearby planets, especially Jupiter. (p. 361)

Kuiper belt A region in the plane of the solar system outside the orbit of Neptune where most short-period comets are thought to originate. (p. 366)

Kuiper-belt object Small icy body orbiting in the Kuiper belt. (p. 371)

L

labeled-release experiment Experiment to look for life on Mars. Radioactive carbon compounds were added to Martian soil specimens. Scientists looked for indications that the carbon had been eaten or inhaled.

Lagrangian point One of five special points in the plane of two massive bodies orbiting one another, where a third body of negligible mass can remain in equilibrium. (p. 324)

lander Spacecraft that lands on the object it is studying.

laser ranging Method of determining the distance to an object by firing a laser beam at it and measuring the time taken for the light to return.

lava dome Volcanic formation formed when lava oozes out of fissures in a planet's surface, creating the dome, and then withdraws, causing the crust to crack and subside. (p. 239)

law of conservation of mass and energy A fundamental law of modern physics which states that the sum of mass and energy must always remain constant in any physical process. In fusion reactions, the lost mass is converted into energy, primarily in the form of electromagnetic radiation. (p. 409)

laws of planetary motion Three laws derived by Kepler describing the motion of the planets around the Sun.

leap year Year in which an additional day is inserted into the calendar in order to keep the calendar year synchronized with Earth's orbit around the Sun. (p. 18)

lens Optical instrument made of glass or some other transparent material, shaped so that, as a parallel beam of light passes through it, the rays are refracted so as to pass through a single focal point.

lens (eye) The part of the eye that refracts light onto the retina.

lepton Greek word for light, referring in particle physics to low-mass particles such as electrons, muons, and neutrinos that interact via the weak force.

lepton epoch Period when the light elementary particles (leptons) were in thermal equilibrium with the cosmic radiation field.

lidar Light Detection and Ranging—a device that uses laser-ranging to measure distance.

light See electromagnetic radiation.

light curve The variation in brightness of a star with time. (p. 458)

light element In astronomical terms, hydrogen and helium.

light-gathering power Amount of light a telescope can view and focus, proportional to the area of the primary mirror.

lighthouse model The leading explanation for pulsars. A small region of the neutron star, near one of the magnetic poles, emits a steady stream of radiation which sweeps past Earth each time the star rotates. The period of the pulses is the star's rotation period. (p. 569)

light-year The distance that light, moving at a constant speed of 300,000 km/s, travels in one year. One light-year is about 10 trillion kilometers. (p. 4)

limb Edge of a lunar, planetary, or solar disk.

linear momentum Tendency of an object to keep moving in a straight line with constant velocity; the product of the object's mass and velocity.

line of nodes Intersection of the plane of the Moon's orbit with Earth's orbital plane.

line of sight technique Method of probing interstellar clouds by observing their effects on the spectra of background stars.

lithosphere Earth's crust and a small portion of the upper mantle that make up Earth's plates. This layer of Earth undergoes tectonic activity. (p. 184)

Local Bubble The particular low-density intercloud region surrounding the Sun.

Local Group The small galaxy cluster that includes the Milky Way Galaxy. (p. 643)

local supercluster Collection of galaxies and clusters centered on the Virgo cluster. *See also* supercluster.

logarithm The power to which 10 must be raised to produce a given number.

logarithmic scale Scale using the logarithm of a number rather than the number itself; commonly used to compress a large range of data into more manageable form.

look-back time Time in the past when an object emitted the radiation we see today.

low-mass star Star with a mass less than 8 times that of the Sun; progenitor of a white dwarf.

luminosity One of the basic properties used to characterize stars, luminosity is defined as the total energy radiated by star each second, at all wavelengths. (p. 408)

luminosity class A classification scheme which groups stars according to the width of their spectral lines. For a group of stars with the same temperature, luminosity class differentiates between supergiants, giants, main-sequence stars, and subdwarfs. (p. 457)

lunar dust *See* regolith.

lunar eclipse Celestial event during which the moon passes through the shadow of Earth, temporarily darkening its surface. (p. 20)

lunar phase The appearance of the Moon at different points along its orbit. (p. 19)

Lyman-alpha forest Collection of lines in an object's spectra starting at the redshifted wavelength of the object's own Lyman-alpha emission line and extending to shorter wavelengths; produced by gas in galaxies along the line of sight.

M

macroscopic Large enough to be visible by the unaided eye.

Magellanic Clouds Two small irregular galaxies that are gravitationally bound to the Milky Way Galaxy. (p. 639)

magnetic field Field which accompanies any changing electric field and governs the influence of magnetized objects on one another. (p. 66)

magnetic poles Points on a planet where the planetary magnetic field lines intersect the planet's surface vertically.

magnetism The presence of a magnetic field.

magnetometer Instrument that measures magnetic field strength.

magnetopause Boundary between a planet's magnetosphere and the solar wind.

magnetosphere A zone of charged particles trapped by a planet's magnetic field, lying above the atmosphere. (p. 170)

magnitude scale A system of ranking stars by apparent brightness, developed by the Greek astronomer Hipparchus. Originally, the brightest stars in the sky were categorized as being of first magnitude, while the faintest stars visible to the naked eye were classified as sixth magnitude. The scheme has since been extended to cover stars and galaxies too faint to be seen by the unaided eye. Increasing magnitude means fainter stars, and a difference of five magnitudes corresponds to a factor of 100 in apparent brightness. (p. 443)

main sequence Well-defined band on the Hertzsprung–Russell diagram on which most stars are found, running from the top left of the diagram to the bottom right. (p. 452)

main-sequence turnoff Special point on the Hertzsprung–Russell diagram for a cluster, indicative of the cluster's age. If all the stars in the cluster are plotted, the lower mass stars will trace out the main sequence up to the point where stars begin to evolve off the main sequence toward the red giant branch. The point where stars are just beginning to evolve off is the main-sequence turnoff. (p. 534)

major axis The long axis of an ellipse.

mantle Layer of Earth just interior to the crust. (p. 170)

mare Relatively dark-colored and smooth region on the surface of the Moon (plural: *maria*). (p. 203)

marginally bound universe Universe that will expand forever but at an increasingly slow rate.

mass A measure of the total amount of matter contained within an object. (p. 51)

mass function Relation between the component masses of a single-line spectroscopic binary.

massive compact halo object (MACHO) Collective name for "stellar" candidates for dark matter, including brown dwarfs, white dwarfs, and low-mass red dwarfs.

mass–luminosity relation The dependence of the luminosity of a main-sequence star on its mass. The luminosity increases roughly as the mass raised to the third power. (p. 459)

mass–radius relation The dependence of the radius of a main-sequence star on its mass. The radius rises roughly in proportion to the mass. (p. 459)

mass transfer Process by which one star in a binary system transfers matter onto the other.

mass-transfer binary *See* semi-detached binary. (p. 537)

matter Anything having mass.

matter-antimatter annihilation Reaction in which matter and antimatter annihilate to produce high energy gamma rays. *See also* antiparticle.

matter-dominated universe A universe in which the density of matter exceeds the density of radiation. The present-day universe is matter dominated. (p. 718)

matter era (Current) era following the Radiation era when the universe is larger and cooler, and matter is the dominant constituent of the universe.

Maunder minimum Lengthy period of solar inactivity that extended from 1645 to 1715.

mean solar day Average length of time from one noon to the next, taken over the course of a year—24 hours.

medium Material through which a wave, such as sound, travels.

meridian An imaginary line on the celestial sphere through the north and south celestial poles, passing directly overhead at a given location. (p. 17)

mesosphere Region of Earth's atmosphere lying between the stratosphere and the ionosphere, 50–80 km above Earth's surface. (p. 167)

Messier object Member of a list of "fuzzy" objects compiled by astronomer Charles Messier in the 18th century.

metabolism The daily utilization of food and energy by which organisms stay alive.

metallic Composed of metal or metal compounds.

metamorphic Rocks created from existing rocks exposed to extremes of temperature or pressure.

meteor Bright streak in the sky, often referred to as a "shooting star," resulting from a small piece of interplanetary debris entering Earth's atmosphere and heating air molecules, which emit light as they return to their ground states. (p. 374)

meteor shower Event during which many meteors can be seen each hour, caused by the yearly passage of Earth through the debris spread along the orbit of a comet. (p. 375)

meteorite Any part of a meteoroid that survives passage through the atmosphere and lands on the surface of Earth. (p. 374)

meteoroid Chunk of interplanetary debris prior to encountering Earth's atmosphere. (p. 374)

meteoroid swarm Pebble-sized cometary fragments dislodged from the main body, moving in nearly the same orbit as the parent comet. (p. 375)

microlensing Gravitational lensing by individual stars in a galaxy.

micrometeoroid Relatively small chunks of interplanetary debris ranging from dust-particle size to pebble-sized fragments. (p. 375)

microsphere Small droplet of protein-like material that resists dissolution in water.

midocean ridge Place where two plates are moving apart, allowing fresh magma to well up.

Milky Way Galaxy The spiral galaxy in which the Sun resides. The disk of our Galaxy is visible in the night sky as the faint band of light known as the Milky Way. (p. 600)

millisecond pulsar A pulsar whose period indicates that the neutron star is rotating nearly 1000 times each second. The most likely explanation for these rapid rotators is that the neutron star has been spun up by drawing in matter from a companion star. (p. 574)

molecular cloud A cold, dense interstellar cloud which contains a high fraction of molecules. It is widely believed that the relatively high density of dust particles in these clouds plays an important role in the formation and preservation of the molecules. (p. 483)

molecular cloud complex Collection of molecular clouds that spans as much as 50 parsecs and may contain enough material to make millions of Sun-size stars. (p. 485)

molecule A tightly bound collection of atoms held together by the atoms' electromagnetic fields. Molecules, like atoms, emit and absorb photons at specific wavelengths. (p. 96)

molten In liquid form due to high temperatures.

moon A small body in orbit around a planet. (p. 144)

mosaic (photograph) Composite photograph made up of many smaller images.

M-type asteroid Asteroid containing large fractions of nickel and iron.

multiple star system Group of two or more stars in orbit around one another.

muon A type of lepton (along with the electron and tau).

N

naked singularity A singularity that is not hidden behind an event horizon.

nanobacterium Very small bacterium with diameter in the nanometer range.

nanometer One billionth of a meter.

neap tide The smallest tides, occurring when the Earth-Moon line is perpendicular to the Earth-Sun line at the first and third quarters.

nebula General term used for any "fuzzy" patch on the sky, either light or dark. (p. 472)

nebular theory One of the earliest models of solar system formation, dating back to Descartes, in which a large cloud of gas began to collapse under its own gravity to form the Sun and planets. (p. 160)

nebulosity "Fuzziness," usually in the context of an extended or gaseous astronomical object.

neon-oxygen white dwarf White dwarf formed from a low-mass star with a mass close to the "high-mass" limit, in which neon and oxygen form in the core.

neutrino Virtually massless and chargeless particle that is one of the products of fusion reactions in the Sun. Neutrinos move at close to the speed of light, and interact with matter hardly at all. (p. 410)

neutrino oscillations Possible solution to the solar neutrino problem, in which the neutrino has a very tiny mass. In this case, the correct number of neutrinos can be produced in the solar core, but on their way to Earth some can "oscillate," or become transformed into other particles, and thus go undetected. (p. 433)

neutron An elementary particle with roughly the same mass as a proton, but which is electrically neutral. Along with protons, neutrons form the nuclei of atoms. (p. 95)

neutron capture The primary mechanism by which very massive nuclei are formed in the violent aftermath of a supernova. Instead of fusion of like nuclei, heavy elements are created by the addition of more and more neutrons to existing nuclei. (p. 559)

neutron degeneracy pressure Pressure due to the Pauli exclusion principle, arising when neutrons are forced to come into close contact.

neutronization Process occurring at high densities, in which protons and electrons are crushed together to form neutrons and neutrinos.

neutron spectrometer Instrument designed to search for water ice by looking for hydrogen.

neutron star A dense ball of neutrons that remains at the core of a star after a supernova explosion has destroyed the rest of the star. Typical neutron stars are about 20 km across, and contain more mass than the Sun. (p. 568)

new Moon Phase of the moon during which none of the lunar disk is visible. (p. 15)

Newtonian mechanics The basic laws of motion, postulated by Newton, which are sufficient to explain and quantify virtually all of the complex dynamical behavior found on Earth and elsewhere in the universe. (p. 50)

Newtonian telescope A reflecting telescope in which incoming light is intercepted before it reaches the prime focus and is deflected into an eyepiece at the side of the instrument. (p. 108)

nodes Two points on the Moon's orbit when it crosses the ecliptic.

nonrelativistic Speeds that are much less than the speed of light.

nonthermal spectrum Continuous spectrum not well described by a blackbody.

north celestial pole Point on the celestial sphere directly above Earth's North Pole. (p. 8)

Northern and Southern lights Colorful display produced when atmospheric molecules, excited by charged particles from the Van Allen Belts, fall back to their ground state.

nova A star that suddenly increases in brightness, often by a factor of as much as 10,000, then slowly fades back to its original luminosity. A nova is the result of an explosion on the surface of a white-dwarf star, caused by matter falling onto its surface from the atmosphere of a binary companion. (p. 544)

nuclear binding energy Energy that must be supplied to split an atomic nucleus into neutrons and protons.

nuclear epoch Period when protons and neutrons fused to form heavier nuclei.

nuclear fusion Mechanism of energy generation in the core of the Sun, in which light nuclei are combined, or fused, into heavier ones, releasing energy in the process. (p. 409)

nuclear reaction Reaction in which two nuclei combine to form other nuclei, often releasing energy in the process. *See also* fusion.

nucleotide base An organic molecule, the building block of genes that pass on hereditary characteristics from one generation of living creatures to the next. (p. 745)

nucleus Dense, central region of an atom, containing both protons and neutrons, and orbited by one or more electrons. (p. 89); The solid region of ice and dust that composes the central region of the head of a comet. (p. 363)

nucleus The dense central core of a galaxy. (p. 634)

O

obscuration Blockage of light by pockets of interstellar dust and gas.

Olbers's paradox A thought experiment suggesting that if the universe were homogeneous, infinite, and unchanging, the entire night sky would be as bright as the surface of the Sun. (p. 696)

Oort cloud Spherical halo of material surrounding the solar system out to a distance of about 50,000 A.U., where most comets reside. (p. 366)

opacity A quantity that measures a material's ability to block electromagnetic radiation. Opacity is the opposite of transparency. (p. 70)

open cluster Loosely bound collection of tens to hundreds of stars, a few parsecs across, generally found in the plane of the Milky Way. (p. 505)

open universe Geometry that the universe would have if the density of matter were less than the critical value. In an open universe there is not enough matter to halt the expansion of the universe. An open universe is infinite in extent. (p. 704)

opposition Orbital configuration in which a planet lies in the opposite direction from the Sun, as seen from Earth.

optical double Chance superposition in which two stars appear to lie close together but are actually widely separated.

optical telescope Telescope designed to observe electromagnetic radiation at optical wavelengths.

orbital One of several energy states in which an electron can exist in an atom.

orbital period Time taken for a body to complete one full orbit around another.

orbiter Spacecraft that orbits an object to make observations.

organic compound Chemical compound (molecule) containing a significant fraction of carbon atoms; the basis of living organisms.

outer core The outermost part of Earth's core, believed to be liquid and composed mainly of nickel and iron. (p. 179)

outflow channel Surface feature on Mars, evidence that liquid water once existed there in great quantity; believed to be the relics of catastrophic flooding about 3 billion years ago. Found only in the equatorial regions of the planet. (p. 261)

outgassing Production of atmospheric gases (carbon dioxide, water vapor, methane, and sulphur dioxide) by volcanic activity.

ozone layer Layer of Earth's atmosphere at an altitude of 20–50 km where incoming ultraviolet solar radiation is absorbed by oxygen, ozone, and nitrogen in the atmosphere. (p. 172)

P

pair production Process in which two photons of electromagnetic radiation give rise to a particle-antiparticle pair. (p. 720)

parallax The apparent motion of a relatively close object with respect to a more distant background as the location of the observer changes. (p. 26)

parsec The distance at which a star must lie in order for its measured parallax to be exactly 1 arc second; 1 parsec equals 206,000 A.U. (p. 439)

partial eclipse Celestial event during which only a part of the occulted body is blocked from view. (p. 20)

particle A body with mass but of negligible dimension.

particle accelerator Device used to accelerate subatomic particles to relativistic speeds.

particle-antiparticle pair Pair of particles (e.g., an electron and a positron) produced by two photons of sufficiently high energy.

particle detector Experimental equipment that allows particles and antiparticles to be detected and identified.

Pauli exclusion principle A rule of quantum mechanics that prohibits electrons in dense gas from being squeezed too closely together.

penumbra Portion of the shadow cast by an eclipsing object in which the eclipse is seen as partial. (p. 21)

penumbra The outer region of a sunspot, surrounding the umbra, which is not as dark and not as cool as the central region. (p. 418)

perihelion The closest approach to the Sun of any object in orbit about it. (p. 46)

period The time needed for an orbiting body to complete one revolution about another body. (pp. 47, 63)

period–luminosity relation A relation between the pulsation period of a Cepheid variable and its absolute brightness. Measurement of the pulsation period allows the distance of the star to be determined. (p. 609)

permafrost Layer of permanently frozen water ice believed to lie just under the surface of Mars. (p. 259)

phase Appearance of the sunlit face of the Moon at different points along its orbit, as seen from Earth.

photodisintegration Process occurring at high temperatures, in which individual photons have enough energy to split a heavy nucleus (e.g., iron) into lighter nuclei.

photoelectric effect Emission of an electron from a surface when a photon of electromagnetic radiation is absorbed.

photoevaporation Process in which a cloud in the vicinity of a newborn hot star is dispersed by the star's radiation.

photometer A device that measures the total amount of light received in all or part of the image. (p. 103)

photometry Branch of observational astronomy in which the brightness of a source is measured through each of a set of standard filters. (p. 113)

photomicrograph Photograph taken through a microscope.

photon Individual packet of electromagnetic energy that makes up electromagnetic radiation. (p. 91)

photosphere The visible surface of the Sun, lying just above the uppermost layer of the Sun's interior, and just below the chromosphere. (p. 406)

photosynthesis Process by which plants manufacture carbohydrates and oxygen from carbon dioxide and water using chlorophyll and sunlight as the energy source.

pixel One of many tiny picture elements, organized into an array, making up a digital image. (p. 111)

Planck curve *See* blackbody curve.

Planck epoch Period from the beginning of the Universe to roughly 10^{-43} second, when the laws of physics are not understood.

Planck's constant Fundamental physical constant relating the energy of a photon to its radiation frequency (color).

planet One of nine major bodies that orbit the Sun, visible to us by reflected sunlight. (p. 144)

planetary nebula The ejected envelope of a red-giant star, spread over a volume roughly the size of our solar system. (p. 525)

planetary ring system Material organized into thin, flat rings encircling a giant planet, such as Saturn. (p. 309)

planetesimal Term given to objects in the early solar system that had reached the size of small moons, at which point their gravitational fields were strong enough to begin influencing their neighbors. (p. 385)

plasma A gas in which the constituent atoms are completely ionized.

plate tectonics The motions of regions of Earth's lithosphere, which drift with respect to one another. Also known as continental drift. (p. 182)

plutino Kuiper-belt object whose orbital period (like that of Pluto) is in a 3:2 resonance with the orbit of Neptune. (p. 392)

polarization The alignment of the electric fields of emitted photons, which are generally emitted with random orientations. (p. 470)

Population I and II stars Classification scheme for stars based on the abundance of heavy elements. Within the Milky Way, Population I refers to young disk stars and Population II refers to old halo stars.

positron Atomic particle with properties identical to those a negatively charged electron, except for its positive charge. This positron is the antiparticle of the electron. Positrons and electrons annihilate one another when they meet, producing pure energy in the form of gamma rays. (p. 409)

prebiotic compound Molecule that can combine with others to form the building blocks of life.

precession The slow change in the direction of the rotation axis of a spinning object, caused by some external gravitational influence. (p. 16)

primary atmosphere The chemical components that would have formed Earth's atmosphere. (p. 171)

primary mirror Mirror placed at the prime focus of a telescope (*see* prime focus).

prime focus The point in a reflecting telescope where the mirror focuses incoming light to a point. (p. 106)

prime-focus image Image formed at the prime focus of a telescope.

primordial matter Matter created during the early, hot epochs of the universe.

primordial nucleosynthesis The production of elements heavier than hydrogen by nuclear fusion in the high temperatures and densities which existed in the early universe. (p. 726)

principle of cosmic censorship A proposition to separate the unexplained physics near a singularity from the rest of the well-behaved universe. The principle states that nature always hides any singularity, such as a black hole, inside an event horizon, which insulates the rest of the universe from seeing it. (p. 589)

progenitor "Ancestor" star of a given object; e.g., the star that existed before a supernova explosion is the supernova's progenitor.

prograde motion Motion across the sky in the eastward direction.

prominence Loop or sheet of glowing gas ejected from an active region on the solar surface, which then moves through the inner parts of the corona under the influence of the Sun's magnetic field. (p. 426)

proper motion The angular movement of a star across the sky, as seen from Earth, measured in seconds of arc per year. This movement is a result of the star's actual motion through space. (p. 439)

protein Molecule made up of amino acids that controls metabolism.

proton An elementary particle carrying a positive electric charge, a component of all atomic nuclei. The number of protons in the nucleus of an atom dictates what type of atom it is. (p. 65)

proton–proton chain The chain of fusion reactions, leading from hydrogen to helium, that powers main-sequence stars. (p. 410)

protoplanet Clump of material, formed in the early stages of solar system formation, that was the forerunner of the planets we see today. (p. 162)

protostar Stage in star formation when the interior of a collapsing fragment of gas is sufficiently hot and dense that it becomes opaque to its own radiation. The protostar is the dense region at the center of the fragment. (p. 494)

protosun The central accumulation of material in the early stages of solar system formations, the forerunner of the present-day Sun. (p. 386)

Ptolemaic model Geocentric solar system model, developed by the second century astronomer Claudius Ptolemy. It predicted with great accuracy the positions of the then known planets. (p. 38)

pulsar Object that emits radiation in the form of rapid pulses with a characteristic pulse period and duration. Charged particles, accelerated by the magnetic field of a rapidly rotating neutron star, flow along the magnetic field lines, producing radiation that beams outward as the star spins on its axis. (p. 569)

pulsating variable star A star whose luminosity varies in a predictable, periodic way. (p. 607)

P-waves Pressure waves from an earthquake which travel rapidly through liquids and solids.

pyrolific-release experiment Experiment to look for life on Mars. Radioactively tagged carbon dioxide was added to Martian soil specimens. Scientists looked for indications that the radioactive material had been absorbed.

Q

quantization The fact that light and matter on small scales behave in a discontinuous manner, and manifest themselves in the form of tiny "packets" of energy, called quanta. (p. 89)

quantum fluctuation Temporary random change in the amount of energy at a point in space.

quantum gravity Theory combining general relativity with quantum mechanics.

quantum mechanics The laws of physics as they apply on atomic scales.

quark A fundamental matter particle that interacts via the strong force; basic constituent of protons and neutrons.

quark epoch Period when all heavy elementary particles (composed of quarks) were in thermal equilibrium with the cosmic radiation field.

quarter Moon Lunar phase in which the Moon appears as a half disk. (p. 15)

quasar Starlike radio source with an observed redshift that indicates an extremely large distance from Earth. The brightest nucleus of a distant active galaxy. (p. 657)

quasi-stellar object (QSO) *See* quasar.

quiescent prominence Prominence that persists for days or weeks, hovering high above the solar photosphere.

quiet Sun The underlying predictable elements of the Sun's behavior, such as its average photospheric temperature, which do not change in time. (p. 425)

R

radar Acronym for RAdio Detection And Ranging. Radio waves are bounced off an object, and the time taken for the echo to return indicates its distance. (p. 49)

radial motion Motion along a particular line of sight, which induces apparent changes in the wavelength (or frequency) of radiation received. (p. 78)

radial velocity Component of a star's velocity along the line of sight.

radian Angular measure equivalent to $180/\pi = 57.3$ degrees.

radiant Constellation from which a meteor shower appears to come.

radiation A way in which energy is transferred from place to place in the form of a wave. Light is a form of electromagnetic radiation. (p. 62)

radiation darkening The effect of chemical reactions that result when high-energy particles strike the icy surfaces of objects in the outer solar system. The reactions lead to a buildup of a dark layer of material. (p. 339)

radiation-dominated universe Early epoch in the universe, when the equivalent density of radiation in the cosmos exceeded the density of matter. (p. 719)

radiation era The first few thousand years after the Big Bang when the universe was small, dense, and dominated by radiation.

radiation zone Region of the Sun's interior where extremely high temperatures guarantee that the gas is completely ionized. Photons only occasionally interact with electrons, and travel through this region with relative ease. (p. 407)

radio Region of the electromagnetic spectrum corresponding to radiation of the longest wavelengths. (p. 62)

radio galaxy Type of active galaxy that emits most of its energy in the form of long-wavelength radiation. (p. 652)

radiograph Image made from observations at radio wavelengths.

radio lobe Roundish extended region of radio-emitting gas, lying well beyond the center of a radio galaxy. (p. 652)

radio telescope Large instrument designed to detect radiation from space at radio wavelengths. (p. 123)

radioactivity The release of energy by rare, heavy elements when their nuclei decay into lighter nuclei. (p. 180)

radius–luminosity–temperature relationship A mathematical proportionality, arising from Stefan's Law, which allows astronomers to indirectly determine the radius of a star once its luminosity and temperature are known. (p. 450)

rapid mass transfer Mass transfer in a binary system that proceeds at a rapid and unstable rate, transferring most of the mass of one star onto the other.

ray The path taken by a beam of radiation.

Rayleigh scattering Scattering of light by particles in the atmosphere.

recession velocity Rate at which two objects are separating from one another.

recurrent nova Star that "goes nova" several times over the course of a few decades.

red dwarf Small, cool faint star at the lower-right end of the main sequence on the Hertzsprung–Russell diagram. (p. 453)

red giant A giant star whose surface temperature is relatively low so that it glows red. (p. 451)

red-giant branch The section of the evolutionary track of a star corresponding to intense hydrogen shell burning, which drives a steady expansion and cooling of the outer envelope of the star. As the star gets larger in radius and its surface temperature cools, it becomes a red giant. (p. 520)

red-giant region The upper-right corner of the Hertzsprung–Russell diagram, where red-giant stars are found. (p. 455)

redshift Motion-induced change in the wavelength of light emitted from a source moving away from us. The relative recessional motion causes the wave to have an observed wavelength longer (and hence redder) than it would if it were not moving. (p. 79)

redshift survey Three-dimensional survey of galaxies, using redshift to determine distance.

red supergiant An extremely luminous red star. Often found on the asymptotic-giant branch of the Hertzsprung–Russell diagram. (p. 451)

reddening Dimming of starlight by interstellar matter, which tends to scatter higher-frequency (blue) components of the radiation more efficiently than the lower-frequency (red) components. (p. 468)

reflecting telescope A telescope which uses a mirror to gather and focus light from a distant object. (p. 106)

reflection nebula Bluish nebula caused by starlight scattering from dust particles in an interstellar cloud located just off the line of sight between Earth and a bright star.

refracting telescope A telescope which uses a lens to gather and focus light from a distant object. (p. 106)

refraction The tendency of a wave to bend as it passes from one transparent medium to another. (p. 106)

regolith Surface dust on the moon, several tens of meters thick in places, caused by billions of years of meteoritic bombardment.

relativistic Speeds comparable to the speed of light.

relativistic fireball Leading explanation of a gamma-ray burst in which an expanding region of superhot gas radiates in the gamma-ray part of the spectrum.

residual cap Portion of Martian polar ice caps that remains permanently frozen, undergoing no seasonal variations. (p. 264)

resonance Circumstance in which two characteristic times are related in some simple way, e.g., an asteroid with an orbital period exactly half that of Jupiter.

retina (eye) The back part of the eye onto which light is focused by the lens.

retrograde motion Backward, westward loop traced out by a planet with respect to the fixed stars. (p. 37)

revolution Orbital motion of one body about another, such as Earth about the Sun. (p. 13)

Riemannian geometry Geometry of positively curved space (such as the surface of a sphere).

right ascension Celestial coordinate used to measure longitude on the celestial sphere. The zero point is the position of the Sun at the vernal equinox. (p. 14)

rille A ditch on the surface of the Moon where molten lava flowed in the past. (p. 219)

ring *See* planetary ring system.

ringlet Narrow region in Saturn's planetary ring system where the density of ring particles is high. *Voyager* discovered that the rings visible from Earth are actually composed of tens of thousands of ringlets. (p. 313)

Roche limit Often called the tidal stability limit, the Roche limit gives the distance from a planet at which the tidal force (due to the planet) between adjacent objects exceeds their mutual attraction. Objects within this limit are unlikely to accumulate into larger objects. The rings of Saturn occupy the region within Saturn's Roche limit. (p. 312)

Roche lobe An imaginary surface around a star. Each star in a binary system can be pictured as being surrounded by a tear-drop shaped zone of gravitational influence, the Roche lobe. Any ma-

terial within the Roche lobe of a star can be considered to be part of that star. During evolution, one member of the binary system can expand so that it overflows its own Roche lobe and begins to transfer matter onto the other star. (p. 537)

rock Material made predominantly from compounds of silicon and oxygen.

rock cycle Process by which surface rock on Earth is continuously redistributed and transformed from one type into another. (p. 188)

rotation Spinning motion of a body about an axis. (p. 10)

rotation curve Plot of the orbital speed of disk material in a galaxy against its distance from the galactic center. Analysis of rotation curves of spiral galaxies indicates the existence of dark matter. (p. 622)

R-process "Rapid" process in which many neutrons are captured by a nucleus during a supernova explosion.

RR Lyrae star Variable star whose luminosity changes in a characteristic way. All RR Lyrae stars have more or less the same average luminosity. (p. 607)

runaway greenhouse effect A process in which the heating of a planet leads to an increase in its atmosphere's ability to retain heat and thus to further heating, causing extreme changes in the temperature of the surface and the composition of the atmosphere. (p. 245)

runoff channel Riverlike surface feature on Mars, evidence that liquid water once existed there in great quantities. They are found in the southern highlands, and are thought to have been formed by water that flowed nearly 4 billion years ago. (p. 261)

S

S0 galaxy Galaxy which shows evidence of a thin disk and a bulge, but which has no spiral arms and contains little or no gas. (p. 638)

Sagittarius A/Sgr A Strong radio source corresponding to the supermassive black hole at the center of the Milky Way.

Saros cycle Time interval between successive occurrences of the "same" solar eclipse, equal to 18 years, 11.3 days.

satellite A small body orbiting another larger body.

SB0 galaxy An S0-type galaxy whose disk shows evidence of a bar. (p. 638)

scarp Surface feature on Mercury believed to be the result of cooling and shrinking of the crust forming a wrinkle on the face of the planet. (p. 219)

Schmidt telescope Type of telescope having a very wide field of view, allowing large areas of the sky to be observed at once.

Schwarzschild radius The distance from the center of an object such that, if all the mass were compressed within that region, the escape speed would equal the speed of light. Once a stellar remnant collapses within this radius, light cannot escape and the object is no longer visible. (p. 581)

scientific method The set of rules used to guide science, based on the idea that scientific "laws" be continually tested, and modified or replaced if found inadequate. (p. 7)

scientific notation Expressing large and small numbers using power-of-10 notation.

seasonal cap Portion of Martian polar ice caps that is subject to seasonal variations, growing and shrinking once each Martian year. (p. 264)

seasons Changes in average temperature and length of day that result from the tilt of Earth's (or any planet's) axis with respect to the plane of its orbit. (p. 16)

secondary atmosphere The chemicals that composed Earth's atmosphere after the planet's formation, once volcanic activity outgassed chemicals from the interior. (p. 171)

sedimentary Rocks formed from the buildup of sediment.

seeing A term used to describe the ease with which good telescopic observations can be made from Earth's surface, given the blurring effects of atmospheric turbulence. (p. 119)

seeing disk Roughly circular region on a detector over which a star's pointlike images is spread, due to atmospheric turbulence. (p. 120)

seismic wave A wave that travels outward from the site of an earthquake through Earth. (p. 177)

seismology The study of earthquakes and the waves they produce in Earth's interior.

seismometer Equipment designed to detect and measure the strength of earthquakes (or quakes on any other planet).

selection effect Observational bias in which a measured property of a collection of objects is due to the way in which the measurement was made, rather than being intrinsic to the objects themselves. (p. 397)

self-propagating star formation Mode of star formation in which shock waves produced by the formation and evolution of one generation of stars triggers the formation of the next. (p. 621)

semi-detached Binary system where one star lies within its Roche lobe but the other fills its Roche lobe and is transferring matter onto the first star.

semimajor axis One-half of the major axis of an ellipse. The semimajor axis is the way in which the size of an ellipse is usually quantified. (p. 46)

SETI Acronym for Search for Extraterrestrial Intelligence.

Seyfert galaxy Type of active galaxy whose emission comes from a very small region within the nucleus of an otherwise normal-looking spiral system. (p. 649)

shepherd satellite Satellite whose gravitational effect on a ring helps preserve the ring's shape. Examples are two satellites of Saturn, Prometheus and Pandora, whose orbits lie on either side of the F ring. (p. 316)

shield volcano A volcano produced by repeated nonexplosive eruptions of lava, creating a gradually sloping, shield-shaped low dome. Often contains a caldera at its summit. (p. 239)

shock wave Wave of matter, which may be generated by a newborn star or supernova, which pushes material outward into the surrounding molecular cloud. The material tends to pile up, forming a rapidly moving shell of dense gas. (p. 504)

short-period comet Comet with orbital period less than 200 years.

SI Système International, the international system of metric units used to define mass, length, time, etc.

sidereal day The time needed between successive risings of a given star. (p. 13)

sidereal month Time required for the Moon to complete one trip around the celestial sphere. (p. 20)

sidereal year The time required for the constellations to complete one cycle around the sky and return to their starting points, as seen from a given point on Earth. Earth's orbital period around the Sun is one sidereal year (p. 16)

single-line spectroscopic binary Binary system in which one star is too faint for its spectrum to be distinguished, so only the spectrum of the brighter star can be seen to shift back and forth as the stars orbit one another.

singularity A point in the universe where the density of matter and the gravitational field are infinite, such as at the center of a black hole. (p. 589)

sister/coformation theory (Moon) Theory suggesting that the Moon formed as a separate object close to Earth.

soft landing Use of rockets, parachutes, or packaging to break the fall of a space probe as it lands on a planet.

solar constant The amount of solar energy reaching Earth per unit area per unit time, approximately 1400 W/m². (p. 408)

solar core The region at the center of the Sun, with a radius of nearly 200,000 km, where powerful nuclear reactions generate the Sun's energy output. (p. 407)

solar cycle The 22-year period that is needed for both the average number of spots and the Sun's magnetic polarity to repeat themselves. The Sun's polarity reverses on each new 11-year sunspot cycle. (p. 424)

solar day The period of time between the instant when the Sun is directly overhead (i.e., noon) to the next time it is directly overhead. (p. 13)

solar eclipse Celestial event during which the new Moon passes directly between Earth and the Sun, temporarily blocking the Sun's light. (p. 20)

solar interior The region of the Sun between the solar core and the photosphere. (p. 409)

solar maximum Point of the sunspot cycle during which many spots are seen. They are generally confined to regions in each hemisphere, between about 15 and 20 degrees latitude. (p. 421)

solar minimum Point of the sunspot cycle during which only a few spots are seen. They are generally confined to narrow regions in each hemisphere at about 25–30 degrees latitude. (p. 421)

solar nebula The swirling gas surrounding the early Sun during the epoch of solar system formation, also referred to as the primitive solar system. (p. 160)

solar neutrino problem The discrepancy between the theoretically predicted flux of neutrinos streaming from the Sun as a result of fusion reactions in the core, and the flux which is actually observed. The observed number of neutrinos is only about half the predicted number. (p. 431)

solar system The Sun and all the bodies that orbit it—Mercury, Venus, Earth, Mars, Jupiter, Saturn, Uranus, Neptune, Pluto, their moons, the asteroids, and the comets. (p. 144)

solar wind An outward flow of fast-moving charged particles from the Sun. (p. 421)

solstice *See* summer solstice, winter solstice.

south celestial pole Point on the celestial sphere directly above Earth's South Pole. (p. 8)

spacetime Single entity combining space and time in special and general relativity.

spatial resolution The dimension of the smallest detail that can be seen in an image.

special relativity Theory proposed by Einstein to deal with the preferred status of the speed of light.

speckle interferometry Technique whereby many short-exposure images of a star are combined to make a high-resolution map of the star's surface.

spectral class Classification scheme, based on the strength of stellar spectral lines, which is an indication of the temperature of a star. (p. 449)

spectral window Wavelength range in which Earth's atmosphere is transparent.

spectrograph Instrument used to produce detailed spectra of stars. Usually, a spectrograph records a spectrum on a CCD detector, for computer analysis. (p. 86)

spectrometer Instrument used to produce detailed spectra of stars. Usually, a spectrograph records a spectrum on a photographic plate, or more recently, in electronic form on a computer. (p. 115)

spectroscope Instrument used to view a light source so that it is split into its component colors. (p. 84)

spectroscopic binary A binary-star system which appears as a single star from Earth, but whose spectral lines show back-and-forth Doppler shifts as two stars orbit one another. (p. 458)

spectroscopic parallax Method of determining the distance to a star by measuring its temperature and then determining its absolute brightness by comparing with a standard Hertzsprung–Russell diagram. The absolute and apparent brightness of the star give the star's distance from Earth. (p. 455)

spectroscopy The study of the way in which atoms absorb and emit electromagnetic radiation. Spectroscopy allows astronomers to determine the chemical composition of stars. (p. 87)

spectrum The separation of light into its component colors.

speed Distance moved per unit time, independent of direction. *See* also velocity.

speed of light The fastest possible speed, according to the currently known laws of physics. Electromagnetic radiation exists in the form of waves or photons moving at the speed of light. (p. 67)

spicule Small solar storm that expels jets of hot matter into the Sun's lower atmosphere.

spin–orbit resonance State that a body is said to be in if its rotation period and its orbital period are related in some simple way. (p. 204)

spiral arm Distribution of material in a galaxy forming a pinwheel-shaped design, beginning near the galactic center. (p. 618)

spiral density wave A wave of matter formed in the plane of planetary rings, similar to ripples on the surface of a pond, which wrap around the rings forming spiral patterns similar to grooves in a record disk. Spiral density waves can lead to the appearance of ringlets. (p. 619)

spiral density wave A proposed explanation for the existence of galactic spiral arms, in which coiled waves of gas compression move through the galactic disk, triggering star formation. (p. 615)

spiral galaxy Galaxy composed of a flattened, star-forming disk component which may have spiral arms and a large central galactic bulge. (pp. 606, 634)

spiral nebula Historical name for spiral galaxies, describing their appearance.

spring tide The largest tides, occurring when the Sun, Moon, and Earth are aligned at new and full Moon.

S-process "Slow" process in which neutrons are captured by a nucleus; the rate is typically one neutron capture per year.

standard candle Any object with an easily recognizable appearance and known luminosity, which can be used in estimating distances. Supernovae, which all have the same peak luminosity (depending on type) are good examples of standard candles and are used to determine distances to other galaxies. (p. 641)

Standard Solar Model A self-consistent picture of the Sun, developed by incorporating the important physical processes that are believed to be important in determining the Sun's internal structure into a computer program. The results of the program are then compared with observations of the Sun and modifications are made to the model. The Standard Solar Model, which enjoys widespread acceptance, is the result of this process. (p. 412)

standard time System of dividing Earth's surface into 24 time zones, with all clocks in each zone keeping the same time. (p. 18)

star A glowing ball of gas held together by its own gravity and powered by nuclear fusion in its core. (p. 406)

star cluster A grouping of anywhere from a dozen to a million stars which formed at the same time from the same cloud of interstellar gas. Stars in clusters are useful to aid our understanding of stellar evolution because, within a given cluster, stars are all roughly the same age and chemical composition and lie at roughly the same distance from Earth. (p. 505)

starburst galaxy Galaxy in which a violent event, such as near-collision, has caused a sudden, intense burst of star formation in the recent past. (p. 675)

Stefan's law Relation that gives the total energy emitted per square centimeter of its surface per second by an object of a given temperature. Stefan's law shows that the energy emitted increases rapidly with an increase in temperature, proportional to the temperature raised to the fourth power. (p. 75)

stellar epoch Most recent period, when stars, planets, and life have appeared in the universe.

stellar nucleosynthesis The formation of heavy elements by the fusion of lighter nuclei in the hearts of stars. Except for hydrogen and helium, all other elements in our universe resulted from stellar nucleosynthesis. (p. 556)

stellar occultation The dimming of starlight produced when a solar system object such as a planet, moon, or ring passes directly in front of a star. (p. 344)

stratosphere The portion of Earth's atmosphere lying above the troposphere, extending up to an altitude of 40–50 km. (p. 167)

string theory Theory that interprets all particles and forces in terms of particular modes of vibration of submicroscopic strings.

strong nuclear force Short-range force responsible for binding atomic nuclei together. The strongest of the four fundamental forces of nature. (p. 409)

S-type asteroid Asteroid made up mainly of silicate or rocky material.

subatomic particle Particle smaller than the size of an atomic nucleus.

subduction zone Place where two plates meet and one slides under the other.

subgiant branch The section of the evolutionary track of a star that corresponds to changes that occur just after hydrogen is depleted in the core, and core hydrogen burning ceases. Shell hydrogen burning heats the outer layers of the star, which causes a general expansion of the stellar envelope. (p. 520)

sublimation Process by which element changes from the solid to the gaseous state, without becoming liquid.

summer solstice Point on the ecliptic where the Sun is at its northernmost point above the celestial equator, occurring on or near June 21. (p. 15)

sunspot An Earth-sized dark blemish found on the surface of the Sun. The dark color of the sunspot indicates that it is a region of lower temperature than its surroundings. (p. 422)

sunspot cycle The fairly regular pattern that the number and distribution of sunspots follows, in which the average number of spots reaches a maximum every 11 or so years, then fall off to almost zero. (p. 423)

supercluster Grouping of several clusters of galaxies into a larger, but not necessarily gravitationally bound, unit. (p. 682)

superforce An attempt to combine the strong and electroweak forces into one single force.

supergiant A star with a radius between 100 and 1000 times that of the Sun. (p. 451)

supergranulation Large-scale flow pattern on the surface of the Sun, consisting of cells measuring up to 30,000 km across, believed to be the imprint of large convective cells deep in the solar interior. (p. 413)

superior conjunction Orbital configuration in which an inferior planet (Mercury or Venus) lies farthest from Earth (on the opposite side of the Sun).

supermassive black hole Black hole having a mass a million to a billion times greater than the mass of the Sun; usually found in the central nucleus of a galaxy.

supernova Explosive death of a star, caused by the sudden onset of nuclear burning (Type I), or an enormously energetic shock wave (Type II). One of the most energetic events of the universe, a supernova may temporarily outshine the rest of the galaxy in which it resides. (p. 548)

supernova remnant The scattered glowing remains from a supernova that occurred in the past. The Crab Nebula is one of the best-studied supernova remnants. (p. 551)

supersymmetric relic Massive particles that should have been created in the Big Bang if supersymmetry is correct.

S-waves Shear waves from an earthquake, which can travel only through solid material and move more slowly than p-waves.

synchrotron radiation Type of nonthermal radiation produced by high-speed charged particles, such as electrons, as they are accelerated in a strong magnetic field. (p. 660)

synchronous orbit State of an object when its period of rotations is exactly equal to its average orbital period. The Moon is in a synchronous orbit, and so presents the same face toward Earth at all times. (p. 206)

synodic month Time required for the Moon to complete a full cycle of phases. (p. 20)

synodic period Time required for a body to return to the same apparent position relative to the Sun, taking Earth's own motion into account; (for a planet) the time between one closest approach to Earth and the next. (p. 215)

T

T Tauri star Protostar in the late stages of formation, often exhibiting violent surface activity. T Tauri stars have been observed to brighten noticeably in a short period of time, consistent with the idea of rapid evolution during this final phase of stellar formation. (p. 496)

tail Component of a comet that consists of material streaming away from the main body, sometimes spanning hundreds of millions of kilometers. May be composed of dust or ionized gases. (p. 363)

tau A type of lepton (along with the electron and muon).

tectonic fracture Cracks on a planet's surface, in particular on the surface of Mars, caused by internal geological activity.

telescope Instrument used to capture as many photons as possible from a given region of the sky and concentrate them into a focused beam for analysis. (p. 106)

temperature A measure of the amount of heat in an object, and an indication of the speed of the particles that comprise it. (p. 71)

tenuous Thin, having low density.

terminator The line separating night from day on the surface of the Moon or a planet.

terrae *See* highlands.

terrestrial planet One of the four innermost planets of the solar system, resembling Earth in general physical and chemical properties. (p. 150)

theories of relativity Einstein's theories, on which much of modern physics rests. Two essential facts of the theory are that nothing can travel faster than the speed of light, and that everything, including light, is affected by gravity. (p. 584)

theory A framework of ideas and assumptions used to explain some set of observations and make predictions about the real world. (p. 6)

thermal equilibrium Condition in which new particle-antiparticle pairs are created from photons at the same rate as pairs annihilate one another to produce new photons.

thick disk Region of a spiral galaxy where an intermediate population of stars resides, younger than the halo stars but older than stars in the disk. (p. 608)

threshold temperature Critical temperature above which pair production is possible, and below which pair production cannot occur.

tidal bulge Elongation of Earth caused by the difference between the gravitational force on the side nearest the Moon and the force on the side farthest from the Moon. The long axis of the tidal bulge points toward the Moon. More generally, the deformation of any body produced by the tidal effect of a nearby gravitating object. (p. 192)

tidal force The variation in one body's gravitational force from place to place across another body—for example, the variation of the Moon's gravity across Earth. (p. 192)

tidal locking Circumstance in which tidal forces have caused a moon to rotate at exactly the same rate at which it revolves around its parent planet, so that the moon always keeps the same face turned toward the planet.

tidal stability limit The minimum distance within which a moon can approach a planet before being torn apart by the planet's tidal force.

tides Rising and falling motion of terrestrial bodies of water, exhibiting daily, monthly and yearly cycles. Ocean tides on Earth are caused by the competing gravitational pull of the Moon and Sun on different parts of Earth. (p. 192)

time dilation A prediction of the theory of relativity, closely related to the gravitational reshift. To an outside observer, a clock lowered into a strong gravitational field will appear to run slow. (p. 588)

time zone Region on Earth in which all clocks keep the same time, regardless of the precise position of the Sun in the sky, for consistency in travel and communications. (p. 18)

total eclipse Celestial event during which one body is completely blocked from view by another. (p. 20)

transit Orbital configuration where a planet (i.e., Mercury or Venus) passes between Earth and the Sun.

transition zone The region of rapid temperature increases that separates the Sun's chromosphere from the corona. (p. 407)

transverse motion Motion perpendicular to a particular line of sight, which does not result in Doppler shift in radiation received. (p. 78)

transverse velocity Component of star's velocity perpendicular to the line of sight.

triangulation Method of determining distance based on the principles of geometry. A distant object is sighted from two well-separated locations. The distance between the two locations and the angle between the line joining them and the line to the distant object are all that are necessary to ascertain the object's distance. (p. 25)

triple-alpha process The creation of carbon-12 by the fusion of three helium-4 nuclei (alpha particles). Helium-burning stars occupy a region of the Hertzsprung–Russell diagram known as the horizontal branch. (p. 521)

triple star system Three stars that orbit one another, bound together by gravity.

Trojan asteroid One of two groups of asteroids which orbit at the same distance from the Sun as Jupiter, 60 degrees ahead of and behind the planet. (p. 360)

tropical year The time interval between one vernal equinox and the next. (p. 16)

troposphere The portion of Earth's atmosphere from the surface to about 15 km. (p. 167)

trough Maximum departure of a wave below its undisturbed state.

true space motion True motion of a star, taking into account both its transverse and radial motion according to the Pythagorean theorem.

Tully-Fisher relation A relation used to determine the absolute luminosity of a spiral galaxy. The rotational velocity, measured from the broadening of spectral lines, is related to the total mass, and hence the total luminosity. (p. 642)

turnoff mass The mass of a star that is just now evolving off the main sequence in a star cluster.

21-centimeter radiation Radio radiation emitted when an electron in the ground state of a hydrogen atom flips its spin to become parallel to the spin of the proton in the nucleus. (p. 482)

twin quasar Quasar that is seen twice at different locations in the sky due to gravitational lensing.

Type I supernova One possible explosive death of a star. A white dwarf in a binary-star system can accrete enough mass that it cannot support its own weight. The star collapses and temperatures become high enough for carbon fusion to occur. Fusion begins throughout the white dwarf almost simultaneously and an explosion results. (p. 549)

Type II supernova One possible explosive death of a star, in which the highly evolved stellar core rapidly implodes and then explodes, destroying the surrounding star. (p. 549)

U

ultraviolet Region of the electromagnetic spectrum, just beyond the visible range, corresponding to wavelengths slightly shorter than blue light. (p. 62)

ultraviolet telescope A telescope that is designed to collect radiation in the ultraviolet part of the spectrum. Earth's atmosphere is partially opaque to these wavelengths, so ultraviolet telescopes are put on rockets, balloons, and satellites to get high above most or all of the atmosphere. (p. 132)

umbra Central region of the shadow cast by an eclipsing body. (p. 21)

umbra The central region of a sunspot, which is its darkest and coolest part. (p. 418)

unbound An orbit which does not stay in a specific region of space, but where an object escapes the gravitational field of another. Typical unbound orbits are hyperbolic in shape. (p. 56)

unbound trajectory Path of an object with launch speed high enough that it can escape the gravitational pull of a planet.

uncompressed density The density a body would have in the absence of any compression due to its own gravity.

universal time Mean solar time at the Greenwich meridian. (p. 18)

universe The totality of all space, time, matter, and energy. (p. 4)

unstable nucleus Nucleus that cannot exist indefinitely, but rather must eventually decay into other particles or nuclei.

upwelling Upward motion of material having temperature higher than the surrounding medium.

V

Van Allen belts At least two doughnut-shaped regions of magnetically trapped, charged particles high above Earth's atmosphere. (p. 189)

variable star A star whose luminosity changes with time. (p. 603)

velocity Displacement (distance plus direction) per unit time. *See also* speed.

vernal equinox Date on which the Sun crosses the celestial equator moving northward, occurring on or near March 21. (p. 16)

visible light The small range of the electromagnetic spectrum that human eyes perceive as light. The visible spectrum ranges from about 400–700 nm, corresponding to blue through red light. (p. 62)

visible spectrum The small range of the electromagnetic spectrum that human eyes perceive as light. The visible spectrum ranges from about 4000–7000 angstroms, corresponding to blue through red light. (p. 64)

visual binary A binary-star system in which both members are resolvable from Earth. (p. 458)

void Large, relatively empty region of the universe around which superclusters and "walls" of galaxies are organized. (p. 685)

volcano Upwelling of hot lava from below Earth's crust to the planet's surface. (p. 173)

W

wane (referring to the Moon or a planet) To shrink. The Moon appears to wane, or shrink in size, for two weeks after full Moon.

warm longitudes (Mercury) Two opposite points on Mercury's equator where the Sun is directly overhead at aphelion. Cooler than the hot longitudes by 150 degrees.

water hole The radio interval between 18 cm and 21 cm, the respective wavelengths at which hydroxyl (OH) and hydrogen (H) radiate, in which intelligent civilizations might conceivably send their communication signals. (p. 758)

water volcano Volcano that ejects water (molten ice) rather than lava (molten rock) under cold conditions.

watt/kilowatt Unit of power: 1 watt (W) is the emission of 1 joule (J) per second; 1 kilowatt (kW) is 1000 watts.

wave A pattern that repeats itself cyclically in both time and space. Waves are characterized by the speed at which they move, their frequency, and their wavelength. (p. 63)

wavelength The distance from one wave crest to the next, at a given instant in time. (p. 63)

wave period The amount of time required for a wave to repeat itself at a specific point in space. (p. 64)

wave theory of radiation Description of light as a continuous wave phenomenon, rather than as a stream of individual particles. (p. 67)

wax (referring to the Moon or a planet) To grow. The Moon appears to wax, or grow in size, for two weeks after new Moon.

weakly interacting massive particle (WIMP) Class of subatomic particles that might have been produced early in the history of the universe; dark-matter candidates.

weak nuclear force Short-range force, weaker than both electromagnetism and the strong force, but much stronger than gravity; responsible for certain nuclear reactions and radioactive decays. (p. 410)

weight The gravitational force exerted on you by Earth (or the planet on which you happen to be standing). (p. 51)

weird terrain A region on the surface of Mercury with oddly rippled features. This feature is thought to be the result of a strong impact which occurred on the other side of the planet, and sent seismic waves traveling around the planet, converging in the weird region. (p. 215)

white dwarf A dwarf star with sufficiently high surface temperature that it glows white. (p. 451)

white-dwarf region The bottom-left corner of the Hertzsprung–Russell diagram, where white-dwarf stars are found. (p. 455)

white oval Light-colored region near the Great Red Spot in Jupiter's atmosphere. Like the red spot, such regions are apparently rotating storm systems. (p. 283)

Wien's law Relation between the wavelength at which a blackbody curve peaks and the temperature of the emitter. The peak wavelength is inversely proportional to the temperature, so the hotter the object, the bluer its radiation. (p. 73)

winter solstice Point on the ecliptic where the Sun is at its southernmost point below the celestial equator, occurring on or near December 21. (p. 15)

wispy terrain Prominent light-colored streaks on Rhea, one of Saturn's moons.

X

X ray Region of the electromagnetic spectrum corresponding to radiation of high frequency and short wavelength, far beyond the visible spectrum. (p. 62)

X-ray burster X-ray source that radiates thousands of times more energy than our Sun in short bursts lasting only a few seconds. A neutron star in a binary system accretes matter onto its surface until temperatures reach the level needed for hydrogen fusion to occur. The result is a sudden period of rapid nuclear burning and release of energy. (p. 573)

X-ray nova Nova explosion detected at X-ray wavelengths.

Z

Zeeman effect Broadening or splitting of spectral lines due to the presence of a magnetic field.

zero-age main sequence The region on the Hertzsprung–Russell diagram, as predicted by theoretical models, where stars are located at the onset of nuclear burning in their cores. (p. 497)

zodiac The twelve constellations on the celestial sphere through which the Sun appears to pass during the course of a year. (p. 13)

zonal flow Alternating regions of westward and eastward flow, roughly symmetrical about the equator of Jupiter, associated with the belts and zones in the planet's atmosphere. (p. 279)

zone Bright, high-pressure region in the atmosphere of a jovian planet, where gas flows upward. (p. 278)

ANSWERS TO CONCEPT CHECK QUESTIONS

Chapter 1

1.1 (*p. 8*) A theory can never become proven "fact," because it can always be invalidated, or forced to change, by a single contradictory observation. However, once a theory's predictions have been repeatedly verified by experiments over many years, it is often widely regarded as "true." **1.2** (*p. 13*) (1) Because the celestial sphere provides a natural means of specifying the locations of stars on the sky. Celestial coordinates are directly related to Earth's orientation in space, but are independent of Earth's rotation. (2) Distance information is lost. **1.3** (*p. 17*) (1) In the Northern Hemisphere, summer occurs when the Sun is near its highest (northernmost) point on the celestial sphere, or, equivalently, when Earth's North Pole is "tipped" toward the Sun, the days are longest and the Sun is highest in the sky. Winter occurs when the Sun is lowest in the sky (near its southernmost point on the celestial sphere) and the days are shortest. (2) We see different constellations because Earth has moved halfway around its orbit between these seasons and the darkened hemisphere faces an entirely different group of stars. **1.4** (*p. 18*) No, because the time shown by your watch (if it is accurate) is the standard time in your time zone, and noon is not directly related to the time when the Sun passes overhead at your location. **1.5** (*p. 25*) (1) The angular size of the Moon would remain the same, but that of the Sun would be halved, making it easier for the Moon to eclipse the Sun. We would expect to see total or partial eclipses, but no annular ones. (2) If the distance is halved, the angular size of the Sun would double and total eclipses would never be seen, only partial or annular ones. **1.6** (*p. 29*) Because astronomical objects are too distant for direct ("measuring tape") measurements, so we must rely on indirect means and mathematical reasoning.

Chapter 2

2.1 (*p. 41*) In the geocentric view, retrograde motion is the real backward motion of a planet as it moves on its epicycle. In the heliocentric view, the backward motion is only apparent, caused by Earth "overtaking" the planet in its orbit. **2.2** (*p. 44*) The discovery of the phases of Venus could not be reconciled with the geocentric model, and observations of Jupiter's moons proved that some objects in the universe did not orbit Earth. **2.3** (*p. 48*) Because the laws were derived using only the orbits of the planets Mercury through Saturn, before the outermost planets were known. **2.4** (*p. 50*) Because Kepler determined the overall geometry of the solar system by triangulation using Earth's orbit as a baseline, so all distances were known only relative to the scale of Earth's orbit—the astronomical unit. **2.5** (*p. 56*) In the absence of any force, a planet would move in a straight line with constant velocity (Newton's first law) and therefore tends to move along the tangent to its orbital path. The Sun's gravity causes the planet to accelerate toward the Sun (Newton's second law), bending its trajectory into the orbit we observe.

Chapter 3

3.1 (*p. 68*) Light is an electromagnetic wave produced by accelerating charged particles. All waves on the list are characterized by their wave periods, frequencies, and wavelengths, and all carry energy and information from one location to another. Unlike waves in water or air, however, light waves require no physical medium in which to propagate. **3.2** (*p. 71*) All are electromagnetic radiation and travel at the speed of light. In physical terms, they differ only in frequency (or wavelength), although their effects on our bodies (or our detectors) differ greatly. **3.3** (*p. 77*) As the switch is turned and the temperature of the filament rises, the bulb's brightness increases rapidly, by Stefan's law, and its color shifts, by Wien's law, from invisible infrared to red to yellow to white. **3.4** (*p. 79*) According to *More Precisely 2-3*, measuring masses in astronomy usually entails measuring the orbital speed of one object—a companion star, or a planet, perhaps—around another. In most cases, the Doppler effect is an astronomer's only way of making such measurements.

Chapter 4

4.1 (*p. 88*) They are characteristic frequencies (wavelengths) at which matter absorbs or emits photons of electromagnetic radiation. They are unique to each atom or molecule, and thus provide a means of identifying the gas producing them. **4.2** (*p. 90*) Electron orbits can occur only at certain specific energies and there is a ground state which has the lowest possible energy. Planetary orbits can have any energy. Planets can stay in any orbit indefinitely; in an atom, the electron must eventually fall to the ground state, emitting electromagnetic radiation in the process. Planets may reasonably be thought of as having specific trajectories around the Sun—there is no ambiguity as to "where" a planet is. Electrons in an atom are smeared out into an electron cloud, and we can only talk of the electron's location in probabilistic terms. **4.3** (*p. 96*) Spectral lines correspond to transitions between specific orbitals within an atom. The structure of an atom determines the energies of these orbitals, hence the possible transitions, and hence the energies (colors) of the photons involved. **4.4** (*p. 97*) In addition to changes involving electron energies, changes involving a molecule's vibration or rotation can also result in emission or absorption of radiation. **4.5** (*p. 101*) Because with few exceptions, spectral analysis is the only way we have of determining the physical conditions—composition, temperature, density, velocity, etc.,—in a distant object. Without spectral analysis, astronomers would know next to nothing about the properties of stars and galaxies.

Chapter 5

5.1 (*p. 110*) Because reflecting telescopes are easier to design, build, and maintain than refracting instruments. **5.2** (*p. 116*) First, photographic plates are inefficient; CCDs are almost always used. Second, because they want to make detailed measurements of brightness, variability, and spectra, which require the use of other, non-imaging detectors. **5.3** (*p. 119*) The need to gather as much light as possible, and the need to achieve the highest possible angular resolution. **5.4** (*p. 123*) To reduce or overcome the effects of atmospheric absorption, instruments are placed on high mountains or in space. To compensate for atmospheric turbulence, adaptive optics techniques probe the air above the observing site and adjust the mirror surface accordingly to try to recover the undistorted image. **5.5** (*p. 127*) Radio observations allow us to see objects whose visible light is obscured by intervening matter, or which simply do not emit most of their energy in the visible portion of the spectrum. **5.6** (*p. 129*) The long wavelength of radio radiation. Astronomers use the largest radio telescopes possible and interferometry, which combines the signals from two or more separate telescopes to create the effect of a single instrument of much larger diameter. **5.7** (*p. 136*) Benefits: they are above the atmosphere, so they are unaffected by seeing or absorption; they can also make round-the-clock observations. Drawbacks: cost, smaller size, inaccessibility, vulnerability to damage by radiation and cosmic rays.

Chapter 6

6.1 (*p. 145*) They provide new tests of our theories of solar-system formation, and new examples of the sorts of planetary systems that are possible. **6.2** (*p. 148*) By applying the laws of geometry and Newtonian mechanics to observations made from Earth and (more recently) by visiting spacecraft. **6.3** (*p. 149*) The planets orbit in very nearly the same plane—the ecliptic. Viewed from outside the system, only the orbits of Mercury and Pluto would deviate noticeably from this plane. **6.4** (*p. 151*) Because the two classes of planet differ in almost every physical property—orbit, mass, radius, composition, existence of rings, and number of moons. **6.5** (*p. 152*) Because it is thought to be much less evolved than material now found in planets, and hence a better indicator of conditions in the early solar system. **6.6** (*p. 165*) Because dust grains formed condensation nuclei that began the process leading to the formation of planetesimals and eventually planets.

Chapter 7

7.1 (*p. 175*) It raises Earth's average surface temperature above the freezing point of water, which was critical for the development of life on our planet. Should the greenhouse effect continue to strengthen, however, it may conceivably cause catastrophic climate changes on Earth. **7.2** (*p. 180*) We would have far less detailed direct (from volcanoes) or indirect (from seismic studies following earthquakes) information on our planet's interior.   **7.3** (*p. 189*) Convection currents in the upper mantle cause portions of Earth's crust—plates—to slide around on the surface. As the plates move and interact, they are responsible for volcanism, earthquakes, the formation of mountain ranges and ocean trenches, and the creation and destruction of oceans and continents.   **7.4** (*p. 191*) It tells us that the planet has a conducting, liquid core in which the magnetic field is continuously generated.   **7.5** (*p. 194*) A tidal force is the *variation* in one body's gravitational force from place to place across another. Tidal forces tend to deform a body, rather than causing an overall acceleration, and they decrease proportional to the inverse cube, rather than the inverse square, of the distance.

Chapter 8

8.1 (*p. 203*) Both bodies have substantially lower escape speeds than Earth, and any atmosphere they may have once had has escaped into space long ago.   **8.2** (*p. 205*) The maria are younger, denser, and much less heavily cratered than the highlands.   **8.3** (*p. 212*) In the case of the Moon, Earth's tidal force has slowed the spin to the point where the rotation rate is now exactly synchronized with the Moon's orbital period around Earth. For Mercury, the Sun's tidal force has caused the rotation period to become exactly 2/3 of the orbital period, and the rotation axis to be exactly perpendicular to the planet's orbit plane. A synchronous orbit is not possible because of Mercury's eccentric orbit around the Sun.   **8.4** (*p. 219*) Heavy bombardment long ago created the basins which later filled with lava to form the maria. Subsequent impacts created virtually all the lunar craters, large and small, we see today. Meteoritic bombardment is the main agent of lunar erosion, although the rate is much smaller than the erosion rate on Earth.   **8.5** (*p. 220*) They are thought to have formed when the planet's core cooled and contracted, causing the crust to crumple. Faults on Earth are the result of tectonic activity.   **8.6** (*p. 222*) Generation of a magnetic field is thought to require a rapidly rotating body with a conducting liquid core. Neither the Moon nor Mercury rotates rapidly and, while Mercury's core may still be partly liquid, the Moon's probably is not.   **8.7** (*p. 223*) The impact theory holds that a collision created the Moon essentially from Earth's mantle, accounting for the composition similarities. If the collision occurred after Earth had already differentiated and the dense material had formed a core, relatively little of this material would have found its way into the newborn Moon.

Chapter 9

9.1 (*p. 232*) It is very slow and retrograde. The reason is unknown, but may simply be a matter of chance.   **9.2** (*p. 233*) Because they were observing in the optical and could not see the surface. Their measurements pertained to the upper atmosphere, above the planet's reflective cloud layers.   **9.3** (*p. 242*) No—they are mostly shield volcanoes, where lava upwells through a "hot spot" in the crust. There appears to be no plate tectonic activity on Venus.   **9.4** (*p. 246*) Given Venus's other bulk similarities to Earth, the planet might well have had an Earth-like climate.   **9.5** (*p. 246*) The dynamo model of planetary magnetism implies that both a conducting liquid core and rapid rotation are needed to generate a magnetic field. Venus's rotation is the slowest of any planet in the solar system.

Chapter 10

10.1 (*p. 253*) Because they occur when Martian opposition happens to coincide approximately with Martian perihelion. Such an alignment happens every 7 Martian synodic years.   **10.2** (*p. 255*) Mars does have seasons, since its rotation axis is inclined to its orbit plane in much the same way as Earth's is. However, the Martian seasons are affected by the planet's eccentric orbit. The appearance of the planet changes seasonally, although the changes have nothing to do with growing cycles, as was once thought.   **10.3** (*p. 259*) The lowlands are much less heavily cratered, implying that they have been resurfaced by volcanism (or smoothed by erosion) since the highlands formed.   **10.4** (*p. 266*) Some of it may have escaped into space. Of the rest, some exists in the form of permafrost (or perhaps liquid water) below the surface. The rest is contained in the Martian polar caps.   **10.5** (*p. 270*) Some of it was lost due to the planet's weak gravity, aided by violent meteoritic impacts. The rest became part of the planet's surface rocks, the polar caps, or the subsurface permafrost.   **10.6** (*p. 270*) The planet's small size, which allowed the planet's internal heat to escape, effectively shutting down the processes that drive mantle convection, volcanism, and plate tectonics.   **10.7** (*p. 271*) They are much smaller, and orbit much closer to the parent planet. In addition, they seem to have a different formation history—their composition differences from Mars suggest that they were captured by the planet long after they (and Mars) formed; for Earth's Moon, this scenario does not seem to work.

Chapter 11

11.1 (*p. 278*) The magnetic field is generated by the motion of electrically conducting liquid in the deep interior, and therefore presumably shares the rotation of that region of the planet.   **11.2** (*p. 283*) Like weather systems on Earth, the belts and zones are regions of high and low pressure and are associated with convective motion. However, unlike storms on Earth, they wrap all the way around the planet because of Jupiter's rapid rotation. In addition, the clouds are arranged in three distinct layers and the bright colors are the result of cloud chemistry unlike anything operating in Earth's atmosphere. Jupiter's spots are somewhat similar to hurricanes on Earth, but they are far larger and longer-lived.   **11.3** (*p. 285*) By constructing theoretical models contrained by detailed spacecraft observations of the planet's bulk properties.   **11.4** (*p. 286*) Because Jupiter rotates more rapidly than Earth and because the volume of conducting fluid responsible for the field is much greater.   **11.5** (*p. 298*) (1) Jupiter's gravitational field, via its tidal effect on the moons. (2) Because liquid water is thought to have played a critical role in the appearance of life on Earth, and is presumed to be similarly important elsewhere in the solar system.

Chapter 12

12.1 (*p. 305*) Because Earth crossed Saturn's ring plane in 1995. The images from before then see the rings from above, those from later see the rings from below.   **12.2** (*p. 309*) Saturn's cloud and haze layers are thicker than those on Jupiter because of Saturn's lower gravity, so we usually see only the upper level of the atmosphere even though the same basic features and cloud layers are there.   **12.3** (*p. 310*) Roughly half of it has precipitated into the planet's interior, reducing the helium fraction in the outer layers (and releasing gravitational energy as it fell).   **12.4** (*p. 317*) The Roche limit is the radius inside of which a moon will be torn apart by tidal forces. Planetary rings are found inside the Roche limit, (most) moons outside. Orbital resonances between ring particles and moons are responsible for much of the fine structure in the rings. They may also maintain the orbits of shepherd satellites that keep some ring features sharp.   **12.5** (*p. 320*) It has a thick atmosphere (denser than Earth's), unlike any other moon in the solar system.   **12.6** (*p. 324*) Because they are tidally locked by Saturn's gravity into synchronous orbits, and therefore all have permanently leading and trailing faces that interact differently with the environment around the planet.

Chapter 13

13.1 (*p. 331*) Uranus's orbit was observed to deviate from a perfect ellipse, leading astronomers to try to compute the mass and location of the body responsible for those discrepancies. That body was Neptune.   **13.2** (*p. 333*) For unknown reasons, Uranus rotates "on its side"—with its rotation axis almost in the plane of the ecliptic.   **13.3** (*p. 336*) Neptune lies far from the Sun and is very cold, so the source of the energy for these atmospheric phenomena is not known. Uranus, closer to the Sun but with a similar atmospheric temperature, shows much less activity.   **13.4** (*p. 338*) In each case, the field is significantly offset from the planet's center and inclined to the rotation axis.   **13.5** (*p. 343*) Triton shows evidence for surface activity—nitrogen geysers and water volcanoes—that seems to have erased most of its impact craters.   **13.6** (*p. 347*) Both are thin rings that require shepherd satellites to prevent them from spreading out and dispersing.   **13.7** (*p. 350*) They have similar masses, radii, composition, and perhaps also similar origins in the Kuiper belt.

Chapter 14

14.1 (*p. 358*) Similarities: all orbit in the inner solar system, and are solid bodies of generally "terrestrial" composition. Differences: asteroids are much smaller than the terrestrial planets, and their orbits are much less regular.   **14.2** (*p. 362*) Because most Earth-crossing asteroids will eventually come very close to or even collide with our planet, with potentially catastrophic results.   **14.3** (*p. 366*) Most comets never come close enough to the Sun for us to see them.   **14.4** (*p. 374*) (1) Asteroids are generally rocky. Comets are predominantly made of ice, with some dust and other debris mixed in. (2) Because of its similarities to the known Kuiper belt objects, as well as to the icy moons of the outer planets, it is now more commonly regarded as the largest member of the Kuiper belt, rather than as the smallest planet.   **14.5** (*p. 378*) They are fragments of comets or asteroids orbiting the Sun in interplanetary space. They are important to planetary scientists because they generally consist of ancient material, and contain vital information on conditions in the early solar system.

Chapter 15

15.1 (*p. 385*) Because it must explain certain general features of solar system architecture, while accommodating the fact that there are exceptions to many of them. **15.2** (*p. 389*) Because the planets could start to form only after the solar nebula had formed, but their formation was terminated when the Sun reached the T Tauri phase and dispersed the disk. **15.3** (*p. 393*) Yes—if a star formed with a disk of matter around it then the basic processes of condensation and accretion would probably have occurred there too, even if it doesn't have planets like Earth. **15.4** (*p. 393*) The fact that larger objects were built up through collisions between smaller objects. The general trends are predictable, but the details, at the level of individual collisions, are not. **15.5** (*p. 397*) (1) Only massive planets have so far been found; (2) The planets' orbits are often quite eccentric; (3) Many of the observed planets orbit very close to their parent star. **15.6** (*p. 399*) Because the search techniques are most sensitive to massive planets orbiting close to their parent stars, which are exactly what have been observed.

Chapter 16

16.1 (*p. 408*) When we simply multiply the solar constant by the total area to obtain the solar luminosity, we are implicitly assuming that the same amount of energy reaches every square meter of the large sphere in Figure 16.3. **16.2** (*p. 411*) Because the Sun shines by nuclear fusion, which converts mass into energy as hydrogen turns into helium. **16.3** (*p. 417*) The energy may be carried in the form of (1) radiation, where energy travels in the form of light, and (2) convection, where energy is carried by physical motion of upwelling solar gas. **16.4** (*p. 421*) The spectrum shows (1) emission lines of (2) highly ionized elements, implying high temperature. **16.5** (*p. 431*) There is a strong field, with a well-defined east–west organization, just below the surface. The field direction in the southern hemisphere is opposite that in the north. However, the details of the fields are very complex. **16.6** (*p. 433*) Almost all of the neutrinos produced in the solar core reach Earth, with very little chance of interaction with intervening matter. Electromagnetic energy, produced in the form of gamma radiation, interacts frequently with solar matter, taking possibly thousands of years to reach the surface.

Chapter 17

17.1 (*p. 442*) (1) Because stars are so far away that their parallaxes relative to any baseline on Earth are too small to measure accurately. (2) Because the transverse component must be determined by measuring the star's proper motion, which decreases as the star's distance increases, and is too small to measure for most distant stars. **17.2** (*p. 444*) Nothing—we need to know their distances before the luminosities (or absolute magnitudes) can be determined. **17.3** (*p. 449*) Because temperature controls which excited states the star's atoms and ions are in, and hence which atomic transitions are possible. **17.4** (*p. 451*) Yes, using the radius–luminosity–temperature relationship, but only if we can find a method of determining the luminosity that doesn't depend on the inverse-square law (Sec. 17.3). **17.5** (*p. 455*) Because giants are intrinsically very luminous, and can be seen to much greater distances than the more common main-sequence stars or white dwarfs. **17.6** (*p. 458*) All stars would be further away, but their measured spectral types and apparent brightnesses would be unchanged, so their luminosities would be greater than previously thought. The main sequence would therefore move vertically upward in the H–R diagram. (We would then use larger luminosities in the method of spectroscopic parallax, so distances inferred by that method would also increase.) **17.7** (*p. 462*) We don't—we assume that their masses are the same as similar stars found in binaries.

Chapter 18

18.1 (*p. 471*) Because the scale of interstellar space is so large that even very low densities can add up to a large amount of obscuring matter along the line of sight to a distant star. **18.2** (*p. 477*) Because the UV light is absorbed by hydrogen gas in the surrounding cloud, ionizing it to form the emission nebulae. The red light is Hα radiation; part of the visible hydrogen spectrum emitted as electrons and protons recombine to form hydrogen atoms. **18.3** (*p. 481*) By studying absorption lines caused by atoms and molecules in the clouds, and the general extinction

due to dust, it is possible to map out a cloud's properties—so long as enough stars are conveniently located behind it. **18.4** (*p. 483*) Because most of the interstellar matter in the Galactic disk is made up of atomic hydrogen, and 21-centimeter radiation provides a probe of that gas largely unaffected by interstellar absorption. **18.5** (*p. 485*) Because the main constituent, hydrogen, is very hard to observe, so astronomers must use other molecules as tracers of the cloud's properties.

Chapter 19

19.1 (*p. 492*) The competing effects of gravity, which tends to make an interstellar cloud collapse, and heat (pressure), which opposes that collapse. **19.2** (*p. 497*) (1) The existence of a photosphere, meaning that the inner part of the cloud becomes opaque to its own radiation, signaling the slowing of the collapse phase. (2) Nuclear fusion in the core and equilibrium between pressure and gravity. **19.3** (*p. 499*) No—different parts of the main sequence correspond to stars of different masses. A typical star stays at roughly the same location on the main sequence of most of its lifetime. **19.4** (*p. 502*) We assume that we observe objects at many different evolutionary stages, and that the snapshot therefore provides a representative sample of the evolutionary stages that stars go through. **19.5** (*p. 505*) Star formation may be triggered by some external event, which might cause several interstellar clouds to start contracting at once. Alternatively, the shock wave produced when an emission nebula forms may be sufficient to send another nearby part of the same cloud into collapse. **19.6** (*p. 509*) Because stars form at different rates—high-mass stars reach the main sequence and start disrupting the parent cloud long before lower-mass stars have finished forming.

Chapter 20

20.1 (*p. 517*) Hydrostatic equilibrium means that, if some property of the Sun changes a little, the star's structure adjusts to compensate. Small changes in the Sun's internal temperature or pressure will not lead to large changes in its radius or luminosity. **20.2** (*p. 523*) Because the nonburning inner core, unsupported by fusion, begins to shrink, releasing gravitational energy, heating the overlying layers and causing them to burn more vigorously, thus increasing the luminosity. **20.3** (*p. 530*) Because the core's contraction is halted by the pressure of degenerate (tightly packed) electrons before it reaches a temperature high enough for the next stage of fusion to begin. In fact, this statement is true whether the "next round" is hydrogen fusion (brown dwarf), helium fusion (helium white dwarf), or carbon fusion (carbon-oxygen white dwarf). **20.4** (*p. 533*) Fusion in high-mass stars is not halted by electron degeneracy pressure. Temperatures are always high enough that each new burning stage can start before degeneracy becomes important. Such stars continue to fuse more and more massive nuclei, faster and faster, eventually exploding in a supernova. **20.5** (*p. 536*) Because a star cluster gives us a "snapshot" of stars of many different masses, but of the same age and initial composition, allowing us to directly test the predictions of the theory. **20.6** (*p. 539*) Because many, if not most, stars are found in binaries, and stars in binaries can follow evolutionary paths quite different from those they would follow if single.

Chapter 21

21.1 (*p. 545*) No, because it is of low mass and not a member of a binary-star system. **21.2** (*p. 548*) Because iron cannot fuse to produce energy. As a result, no further nuclear reactions are possible, and the core's equilibrium cannot be restored. **21.3** (*p. 555*) Because the two types of supernova differ in their spectra and their light curves, making it impossible to explain them in terms of a single phenomenon. **21.4** (*p. 561*) Because they are readily formed by helium capture, a process common in evolved stars. Other elements (with masses not multiples of four) had to form via less common reactions involving proton and neutron capture. **21.5** (*p. 561*) Because it is responsible for creating and dispersing all the heavy elements out of which we are made. In addition, it may also have played an important role in triggering the collapse of the interstellar cloud from which our solar system formed.

Chapter 22

22.1 (*p. 569*) No—only Type II supernova. According to theory, the rebounding central core of the original star becomes a neutron star. **22.2** (*p. 572*) Because (1) not all supernovae form neutron stars, (2) the pulses are beamed, so not all pulsing neutron stars are visible from Earth, and (3) pulsars spin down and become too faint to observe after a few tens of millions of years. **22.3** (*p. 576*) Some X-ray sources are binaries containing accreting neutron stars, which may be in the process of being spun up to form millisecond pulsars. **22.4** (*p. 580*) They are energetic bursts of gamma rays, roughly isotropically distributed on the sky, occurring about once per day. They pose a challenge because they are very distant, and hence extremely luminous, but their energy originates in a region less than a few hundred kilometers across. **22.5** (*p. 586*) Newton's theory describes gravity as a force produced by a massive object that influences all other massive objects. Einstein's relativity describes gravity as a curvature of space-time produced by a massive object; that curvature then determines the trajectories of all particles—matter or radiation—in the universe. **22.6** (*p. 589*) Because the object would appear to take infinitely long to reach the event horizon, and its light would be infinitely redshifted by the time it got there. **22.7** (*p. 597*) By observing their gravitational effects on other objects, and from the X-rays emitted when matter falls into them.

Chapter 23

23.1 (*p. 606*) The Milky Way is the thin plane of the Galactic disk, seen from within. When our line of sight lies in the plane of the Galaxy, we see many stars blurring into a continuous band. In other directions, we see darkness. **23.2** (*p. 612*) No, because even the brightest Cepheids are unobservable at distances of more than a kiloparsec or so through the obscuration of interstellar dust. **23.3** (*p. 615*) Because the compositions, ages, and orbits of the two classes of stars are quite different from one another. **23.4** (*p. 617*) The halo formed early in our Galaxy's history, before gas and dust had formed a spinning, flattened disk. Disk stars are still forming today. Consequently, halo stars are old and move in more or less random three-dimensional orbits, while the disk contains stars of all ages, all moving in roughly circular orbits around the Galactic center. **23.5** (*p. 621*) Because differential rotation would destroy the spiral structure within a few hundred million years. **23.6** (*p. 625*) Its emission has not been observed at any electromagnetic wavelength. Its presence is inferred from its gravitational effect on stars and gas orbiting the Galactic center. **23.7** (*p. 628*) Observations of rapidly moving stars and gas, and the variability of the radiation emitted suggest the presence of a 2–3 million-solar-mass black hole.

Chapter 24

24.1 (*p. 641*) Most galaxies are not large spirals—the most common galaxy types are dwarf ellipticals and dwarf irregulars. **24.2** (*p. 644*) Because distance-measurement techniques ultimately rely upon the existence of very bright objects whose luminosities can be inferred by other means. Such objects become increasingly hard to find and calibrate the farther we look out into intergalactic space. **24.3** (*p. 648*) It doesn't use the inverse-square law. The other methods all provide a way of determining the luminosity of a distant object, which then is converted to a distance using the inverse-square law. Hubble's law gives a direct connection between redshift and distance. **24.4** (*p. 654*) It means that the energy source cannot simply be the summed energy of a huge number of stars—some other mechanism must be at work. **24.5** (*p. 657*) They often appear starlike, but it was at first not realized that their unusual spectra were actually highly redshifted. **24.6** (*p. 661*) The energy is generated in an accretion disk in the central nucleus of the visible galaxy, then transported by jets out of the galaxy and into the lobes, where it is eventually emitted by the synchrotron process in the form of radio waves.

Chapter 25

25.1 (*p. 672*) First, that the galaxies are gravitationally bound to the cluster. Second, and more fundamentally, that the laws of physics as we know them in the solar system—gravity, atomic structure, the Doppler effect—all apply on very large scales, and to systems possibly containing a lot of dark matter. **25.2** (*p. 678*) Stars form by collapse and fragmentation of a large interstellar cloud and subsequently evolve largely in isolation. Galaxies form by mergers of smaller objects, and interactions with other galaxies play a major role in their evolution. **25.3** (*p. 682*) Probably not. There may well be galaxies which do not harbor supermassive central black holes, and in any case, only galaxies in clusters are likely to experience the encounters that trigger activity. **25.4** (*p. 689*) Light traveling from these objects to Earth is influenced by

cosmic structure all along the line of sight. Light rays are deflected by the gravitational field of intervening concentrations of mass, and gas along the line of sight produces absorption lines whose redshifts tell us the distance at which each feature formed. The light received on Earth thus gives astronomers a "core sample" through the universe, from which detailed information can be extracted.

Chapter 26

26.1 (*p. 696*) On very large scales—more than 300 Mpc—the distribution of galaxies seems to be roughly the same everywhere and in all directions. **26.2** (*p. 701*) Because, tracing the motion backwards in time, it implies that all galaxies, and in fact everything in the entire universe, were located at a single point at the same instant in the past. **26.3** (*p. 703*) The universe can expand forever, in which case we die a cold death in which all activity gradually fades away, or the expansion can stop and the universe will recollapse to a fiery Big Crunch. **26.4** (*p. 706*) A low-density universe has negative curvature, a critical density universe is spatially flat (Euclidean), and a high-density universe has positive curvature (and is finite in extent). **26.5** (*p. 708*) There doesn't seem to be enough matter to halt the collapse and, in addition, the observed cosmic acceleration suggests the existence of a large-scale repulsive force in the cosmos that also opposes recollapse. **26.6** (*p. 710*) Dark energy tends to accelerate the expansion of the universe, competing with the effect of gravity, which tends to slow it. The age of the universe that we infer based on present-day measurements depends on the outcome of that competition. **26.7** (*p. 713*) At the time of the Big Bang. It is the electromagnetic remnant of the primeval fireball.

Chapter 27

27.1 (*p. 720*) It means that the total mass-energy density of the universe, which today is made up almost entirely of matter and dark energy, was once comprised almost entirely of radiation. We know this because, going back in time toward the Big Bang, the dark energy density stays constant as the universe contracts, the matter density increases because the volume shrinks, but the radiation density increases even faster because of the cosmological redshift. Thus, at sufficiently early times, radiation was the dominant component of the cosmos. **27.2** (*p. 723*) Once the temperature of the expanding, cooling universe dropped below the point where particle–antiparticle pairs could no longer be created from the radiation background, the particles separated out of the radiation field. Particles and antiparticles annihilated one another, and any leftover "frozen out" matter has survived to the present day. **27.3** (*p. 729*) Because the amount of deuterium observed in the universe today implies that the present density of normal matter is at most a few percent of the critical value—much less than the density of dark matter inferred from dynamical studies. **27.4** (*p. 733*) Inflation implies that the universe is flat, and hence that the total cosmic density equals the critical value. However, the matter density seems to be only about one third of the critical value, and the density of electromagnetic radiation (the microwave background) is a tiny fraction of the critical density. The remaining density may be in the form of the "dark energy" thought to be powering the accelerating cosmic expansion (Section 27.3). **27.5** (*p. 738*) They allow us to measure the value of W_0—and in fact imply that it is very close to 1.

Chapter 28

28.1 (*p. 748*) The formation of complex molecules from simple ingredients by nonbiological processes has been repeatedly demonstrated, but no living cell or self-replicating molecule has ever been created. **28.2** (*p. 751*) Mars remains the most likely site, although Europa and Titan also have properties that might have been conducive to the emergence of living organisms. **28.3** (*p. 755*) It breaks a complex problem up into simpler "astronomical," "biochemical," "anthropological," and "cultural" pieces, which can be analyzed separately. It also identifies the types of stars where a search might be most fruitful. **28.4** (*p. 759*) It is in the radio part of the spectrum, where Galactic absorption is least, at a region where natural Galactic background "static" is minimized, and in a portion of the spectrum characterized by lines of hydrogen and hydroxyl, both of which would likely have significance to a technological civilization.

ANSWERS TO SELF-TEST QUESTIONS

Chapter 1

True-False 1.1 T, 1.2 F, 1.3 T, 1.4 F, 1.5 T, 1.6 F, 1.7 T, 1.8 F, 1.9 T, 1.10 F
Multiple Choice 1.11 b, 1.12 b, 1.13 d, 1.14 a, 1.15 c, 1.16 a, 1.17 c, 1.18 c, 1.19 a, 1.20 d
Odd-Numbered Problems 1.1 (c) the Moon (384,000 km) 1.3 It would decrease by roughly eight minutes. 1.5 (a) 7.14 solar days; (b) infinite 1.7 1.02 km/s; 1.9 (a) 57,300 km; (b) 3.44×10^6 km; (c) 2.06×10^8 km 1.11 6.5×10^{-5} arc sec 1.13 391 1.15 $0°$

Chapter 2

True-False 2.1 F, 2.2 T, 2.3 F, 2.4 F, 2.5 F, 2.6 F, 2.7 F, 2.8 T, 2.9 F, 2.10 T
Multiple Choice 2.11 a, 2.12 d, 2.13 b, 2.14 c, 2.15 c, 2.16 a, 2.17 c, 2.18 b, 2.19 c, 2.20 a
Odd-Numbered Problems 2.1 (a) 110 km; (b) 44,000 km; (c) 370,000 km 2.3 Pluto perihelion = 29.65 A.U.; Neptune perihelion = 29.80 A.U. 2.5 143 days 2.7 8.1″, if Mercury is at aphelion (0.47 A.U.) and Earth is at perihelion (0.98 A.U.) at the point of closest approach 2.9 9.42×10^{-4} solar = 1.88×10^{27} kg 2.11 10,000 A.U., 16 million years 2.13 7.8 km/s, 7.3 km/s, 4.9 km/s; equal in all cases 2.15 1.7 km/s; 2.4 km/s

Chapter 3

True-False 3.1 T, 3.2 F, 3.3 F, 3.4 F, 3.5 F, 3.6 T, 3.7 T, 3.8 T, 3.9 T, 3.10 T
Multiple Choice 3.11 a, 3.12 c, 3.13 b, 3.14 b, 3.15 d, 3.16 a, 3.17 b, 3.18 d, 3.19 a, 3.20 b
Odd-Numbered Problems 3.1 1480 m/s 3.3 23 Hz; radio 3.5 9.4 cm; radio 3.7 310; 9.4 microns; infrared 3.9 2.9 microns 3.11 6.4×10^7 W/m²; 3.9×10^{26} W 3.13 300 km/s away 3.15 1.94×10^{27} kg

Chapter 4

True-False 4.1 F, 4.2 T, 4.3 F, 4.4 T, 4.5 F, 4.6 F, 4.7 F, 4.8 T, 4.9 T, 4.10 T
Multiple Choice 4.11 c, 4.12 c, 4.13 d, 4.14 c, 4.15 b, 4.16 b, 4.17 b, 4.18 b, 4.19 b, 4.20 d
Odd-Numbered Problems 4.1 2.8, 6.2 4.3 620 nm, 12,400 nm, 0.25 nm 4.5 2.4×10^8 4.7 39 microns, 7.7×10^{12} Hz, infrared; 4.5 cm, 6.7 GHz, radio; 46 m, 6.6 MHz, radio 4.9 the first six Balmer lines, ranging in wavelength from 656 nm (Hα) to 410 nm (Hζ) 4.11 137 km/s, approaching 4.13 0.05 nm, to one significant figure 4.15 0.3 rev/day, to one significant figure

Chapter 5

True-False 5.1 F, 5.2 F, 5.3 F, 5.4 F, 5.5 T, 5.6 T, 5.7 T, 5.8 T, 5.9 F, 5.10 F
Multiple Choice 5.11 c, 5.12 d, 5.13 d, 5.14 b, 5.15 c, 5.16 c, 5.17 a, 5.18 d, 5.19 b, 5.20 c
Odd-Numbered Problems 5.1 0.3 arc seconds; 6.8 pixels 5.3 6.7 minutes; 1.7 minutes 5.5 (a) 0.022″, (b) 0.062″ 5.7 160 light-years, using the formula in the text to compute the resolution 5.9 5.5 km, 92 1.8 m. 5.11 roughly 8 times 5.13 (a) 0.003 arc seconds, (b) 0.005 arc seconds 5.15 No, it is not. The wavelength of a 1-keV photon is 1.2 nm, which would give an angular resolution of about 0.000026″ for the quoted diameter. The effective resolution is complicated by the X-ray mirror arrangement, and is actually determined by other design factors.

Chapter 6

True-False 6.1 T, 6.2 F, 6.3 T, 6.4 T, 6.5 F, 6.6 T, 6.7 T, 6.8 F, 6.9 F, 6.10 T
Multiple Choice 6.11 a, 6.12 d, 6.13 a, 6.14 b, 6.15 c, 6.16 a, 6.17 a, 6.18 b, 6.19 b, 6.20 d
Odd-Numbered Problems 6.1 19″, 16 days 6.3 Mercury: 0.31 A.U., 0.47 A.U., Mars: 1.38 A.U., 1.67 A.U., Pluto: 29.7 A.U., 49.3 A.U. 6.5 7×10^{20} kg, 0.01 percent of Earth's mass 6.7 68 A.U., beyond Pluto 6.9 0.21, 1.26 A.U, 0.71 years 6.11 — 6.13 (a) 630 years, (b) 91 days 6.15 increases proportional to the radius; surface gravity doubles, escape speed doubles

Chapter 7

True-False 7.1 T, 7.2 T, 7.3 F, 7.4 F, 7.5 F, 7.6 F, 7.7 F, 7.8 T, 7.9 F, 7.10 F
Multiple Choice 7.11 b, 7.12 d, 7.13 a, 7.14 c, 7.15 b, 7.16 b, 7.17 c, 7.18 b, 7.19 d, 7.20 b
Odd-Numbered Problems 7.1 — 7.3 5.0×10^{18} kg = 8.4×10^{-7} times Earth's mass 7.5 21 m 7.7 (a) 8.4×10^{-3}, (b) 0.16, (c) 0.84, (d) 0.014 7.9 1.9 billion years 7.11 3.1×10^{-6}, 3.1×10^{-7} of surface gravity 7.13 1.6×10^{-7} 7.15 Smaller by a factor of 6.0×10^{-6}. No!

Chapter 8

True-False 8.1 T, 8.2 F, 8.3 T, 8.4 F, 8.5 F, 8.6 F, 8.7 T, 8.8 T, 8.9 T, 8.10 T
Multiple Choice 8.11 a, 8.12 b, 8.13 b, 8.14 c, 8.15 b, 8.16 a, 8.17 b, 8.18 d, 8.19 a, 8.20 a
Odd-Numbered Problems 8.1 8.8 minutes 8.3 57 kg equivalent 8.5 1.74°, 1.15° 8.7 1.81 hours 8.9 (a) 60.4 km/s, or 6.5°/day, (b) 39.7 km/s, or 2.8°/day 8.11 4.2×10^{23} kg, or roughly 26 percent greater than it is now 8.13 need at least 4.8×10^5 such craters, requiring at least 4.8 trillion years; rate would have to increase by a factor of about 1000 8.15 (a) 4 million years for a 2 cm deep bootprint, (b) 4 million years, accepting literally the factor of 10,000 given in the text (c) 1.6 trillion years, assuming Reinhold is 8 km deep

Chapter 9

True-False 9.1 F, 9.2 F, 9.3 F, 9.4 F, 9.5 T, 9.6 F, 9.7 T, 9.8 T, 9.9 F, 9.10 T
Multiple Choice 9.11 a, 9.12 c, 9.13 c, 9.14 b, 9.15 b, 9.16 c, 9.17 b, 9.18 c, 9.19 a, 9.20 a
Odd-Numbered Problems 9.1 (a) 35.4″, (b) 23.3″, (c) 9.7 9.3 470 seconds, 280 seconds (for a separation of 0.28 A.U.) 9.5 4.1°/day; 1.6°/day; 145 (Earth) days 9.7 tidal acceleration due to Earth is 7.3×10^{-12} times Venus's surface gravity; tidal acceleration due to the Sun is 1.4×10^{-8} times Venus's surface gravity; unlikely that Earth's tidal influence would cause the resonance, as it is much smaller than the tidal force due to the Sun 9.9 yes—the smallest detectable feature would be about 20 km across, much smaller than the largest impact features 9.11 400 km/h, 250 mph 9.13 35 times 9.15 197 minutes, 94 minutes

Chapter 10

True-False 10.1 T, 10.2 F, 10.3 F, 10.4 F, 10.5 F, 10.6 F, 10.7 F, 10.8 T, 10.9 F, 10.10 T
Multiple Choice 10.11 b, 10.12 c, 10.13 b, 10.14 c, 10.15 b, 10.16 d, 10.17 a, 10.18 c, 10.19 b, 10.20 a
Odd-Numbered Problems 10.1 780 Earth days 10.3 41° 10.5 2.2 minutes longer 10.7 28 kg equivalent, assuming you weigh 70 kg on Earth 10.9 3.2×10^{16} kg, taking Earth's atmospheric mass to be 5.0×10^{18} kg; 2.8 times the seasonal polar cap mass of 1.1×10^{16} kg 10.11 1.3×10^{20} kg; 3900 times the current atmospheric mass 10.13 2.9×10^{17} kg; 4.5×10^{-7} times the mass of Mars 10.15 16.1′, 2.7′; no

Chapter 11

True-False 11.1 F, 11.2 T, 11.3 F, 11.4 T, 11.5 T, 11.6 T, 11.7 F, 11.8 F, 11.9 T, 11.10 T
Multiple Choice 11.11 c, 11.12 b, 11.13 a, 11.14 c, 11.15 a, 11.16 d, 11.17 b, 11.18 a, 11.19 a, 11.20 c
Odd-Numbered Problems 11.1 2.5 times greater 11.3 61 days 11.5 1.5 percent of the actual mass 11.7 4800 times greater; ratio for Earth–Moon is 81 11.9 — 11.11 3.8×10^{-4}, or 0.038 percent 11.13 3.5 days (retrograde); 7.3×10^{-5} 11.15 Io: 36′, Europa: 18′, Ganymede: 18, Callisto: 9′, Sun: 6′. Yes.

Chapter 12

True-False 12.1 F, 12.2 F, 12.3 T, 12.4 F, 12.5 T, 12.6 T, 12.7 T, 12.8 T, 12.9 F, 12.10 F
Multiple Choice 12.11 b, 12.12 c, 12.13 d, 12.14 c, 12.15 d, 12.16 d, 12.17 b, 12.18 c, 12.19 a, 12.20 c

Odd-Numbered Problems 12.1 47″   12.3 7.3 × 10^{22} kg; 1.3 × 10^{-4} times the actual mass, 1.2 percent of Earth's mass   12.5 74 K   12.7 1.1 × 10^{18} 12.9 1.36 m/s^2 (Earth: 9./80 m/s^2); 2.6 km/s   12.11 142,000 km; 117,000 km; 142,000 km; 237,000 km   12.13 Saturn's tidal acceleration is 7.9 × 10^{-5} times Titan's surface gravity; no—the tidal acceleration is small, comparable to Jupiter's tidal force on Callisto and, regardless of the tidal effect, there are no nearby large moons to perturb Titan's orbit, so the conditions leading to the heating of the inner Galilean moons do not arise   12.15 approximately 50 km

Chapter 13

True-False 13.1 T,   13.2 T,   13.3 T,   13.4 F,   13.5 T,   13.6 T,   13.7 F,   13.8 T,   13.9 T,   13.10 F
Multiple Choice 13.11 b,   13.12 b,   13.13 c,   13.14 d,   13.15 c,   13.16 d,   13.17 d,   13.18 a,   13.19 a,   13.20 d
Odd-Numbered Problems 13.1 (a) 171 years (1.05 Neptune years), (b) 489 years (3 Neptune years)   13.3 1.7′, 13.2′; yes, but the next one is not expected until 2007 (see Figure 13.6)!   13.5 10,000 km/h—it moves halfway around the planet in just over eight hours; from the Data Box, rotation speed at the equator is about 10,000 km/h.   13.7 0.44 Earth Moon masses; 2.4 Pluto masses   13.9 Yes, barely—at 37 K, an escape speed of 1.3 km/s would be needed; Triton's escape speed is 1.5 km/s   13.11 2.1′, assuming that Pluto is close to perihelion   13.13 7.2 times farther out   13.15 11.1 hours, 20,000 km

Chapter 14

True-False 14.1 T,   14.2 F,   14.3 F,   14.4 T,   14.5 T,   14.6 F,   14.7 F,   14.8 T,   14.9 T,   14.10 T
Multiple Choice 14.11 c,   14.12 d,   14.13 c,   14.14 c,   14.15 a,   14.16 d,   14.17 b,   14.18 d,   14.19 d,   14.20 d
Odd-Numbered Problems 14.1 (a) 3.2 kg equivalent,   (b) 0.41 km/s   14.3 230 km diameter   14.5 1.1 A.U., 2.0 A.U.; no—the orbital inclination differs from Earth's   14.7 21 hours, taking the middle of the stated mass range and assuming a circular orbit   14.9 (a) 4.0 million years,   (b) 49 A.U.   14.11 3 × 10^{12} kg; 0.06 percent   14.13 5 trillion   14.15 1.9 billion times (2 crossings per orbit)

Chapter 15

True-False 15.1 F,   15.2 T,   15.3 F,   15.4 F,   15.5 T,   15.6 T,   15.7 F,   15.8 T,   15.9 F,   15.10 F
Multiple Choice 15.11 a,   15.12 d,   15.13 a,   15.14 a,   15.15 c,   15.16 a,   15.17 c,   15.18 c,   15.19 d,   15.20 b
Odd-Numbered Problems 15.1 (a) 1.9 × 10^{43}, 7.8 × 10^{42}, and 2.7 × 10^{40} kg·m^2/s,   (b) 1.0 × 10^{31} kg·m^2/s   15.3 8   15.5 3.0 × 10^7 newtons; tidal force = 8.1 × 10^9 newtons, 270 times larger   15.7 2 × 10^8; once every 2.5 years   15.9 188 years   15.11 0.21 A.U.   15.13 1900 K   15.15 0.56

Chapter 16

True-False 16.1 T,   16.2 T,   16.3 F,   16.4 T,   16.5 F,   16.6 F,   16.7 F,   16.8 F,   16.9 T,   16.10 F
Multiple Choice 16.11 c,   16.12 b,   16.13 c,   16.14 a,   16.15 a,   16.16 b,   16.17 b,   16.18 c,   16.19 c,   16.20 b
Odd-Numbered Problems 16.1 14,600 W/m^2; 52 W/m^2 16.3 310,000 years   16.5 4 × 10^{28} 16.7 (a) 3000 km,   (b) 1500,   (c) 1/33 of the 167-minute orbital period   16.9 36 percent (64 percent less)   16.11 5.2 solar radii, assuming a coronal temperature of 3 million K   16.13 95 million years   16.15 radiation mass loss = 4.3 million tons/s ≈ twice the mass loss due to the solar wind

Chapter 17

True-False 17.1 F,   17.2 F,   17.3 T,   17.4 F,   17.5 T,   17.6 F,   17.7 T,   17.8 F,   17.9 T,   17.10 T
Multiple Choice 17.11 a,   17.12 d,   17.13 d,   17.14 a,   17.15 c,   17.16 b,   17.17 c,   17.18 b,   17.19 c,   17.20 d
Odd-Numbered Problems 17.1 83 pc; 0.36^{11} 17.3 80 solar luminosities   17.5 B is three times farther away   17.7 3.3 × 10^{-10} W/m^2; 2.3 × 10^{-13} times the solar constant   17.9 −1.0   17.11 (a) 100 pc, (b) 4000 pc,   (c) 160,000 pc,   (d) 1,000,000 pc   17.13 3.5, 2.3 solar masses   17.15 (a) 3.4 solar masses,   (b) 0.22 solar masses

Chapter 18

True-False 18.1 F,   18.2 T,   18.3 T,   18.4 T,   18.5 F,   18.6 T,   18.7 F,   18.8 F,   18.9 F,   18.10 T

Multiple Choice 18.11 a,   18.12 d,   18.13 c,   18.14 a,   18.15 d,   18.16 c,   18.17 b,   18.18 d,   18.19 a,   18.20 a
Odd-Numbered Problems 18.1 1.9 grams   18.3 7.1 × 10^{20} m^3, or a cube of side 8900 km   18.5 9   18.7 100,000 pc, 2.6   18.9 1.8 kpc   18.11 escape speeds (km/s): 1.8, 1.1, 1.1, 0.80; average molecular speeds: 13.6, 14.0, 14.6, 14.2; No   18.13 frequency: 1419.65–1420.24 MHz, wavelength: 21.1053–21.0965 cm   18.15 photon energy = 9.4 × 10^{-25} J; 2.6 × 10^{17} W

Chapter 19

True-False 19.1 F,   19.2 F,   19.3 T,   19.4 F,   19.5 T,   19.6 F,   19.7 F,   19.8 T,   19.9 T,   19.10 T
Multiple Choice 19.11 c,   19.12 a,   19.13 d,   19.14 b,   19.15 b,   19.16 a,   19.17 a,   19.18 b,   19.19 a,   19.20 d
Odd-Numbered Problems 19.1 yes, barely: the escape speed is 0.93 km/s, molecular speed is 0.35 km/s   19.3 luminosity decreases by a factor of 7900; absolute magnitude increases by 9.7   19.5 3.2   19.7 8.4   19.9 10^{-6} solar luminosities   19.11 4000 years   19.13 20.25   19.15 37 pc

Chapter 20

True-False 20.1 F,   20.2 T,   20.3 T,   20.4 T,   20.5 F,   20.6 F,   20.7 F,   20.8 F,   20.9 F,   20.10 T
Multiple Choice 20.11 c,   20.12 b,   20.13 a,   20.14 b,   20.15 a,   20.16 c,   20.17 a,   20.18 b,   20.19 b,   20.20 d
Odd-Numbered Problems 20.1 9.9 × 10^{26} kg destroyed, 8.9 × 10^{43} J emitted   20.3 (a) 3300,   (b) 82,000 solar luminosities   20.5 roughly a factor of 80; factor of 10 smaller   20.7 5 × 10^{-5}; not noticeable to the naked eye, but (just) measurable with a telescope and photometer   20.9 2.9 years, 26,000 years   20.11 increases by a factor of 25   20.13 (a) 2.9 solar masses,   (b) 1.7 solar masses   20.15 (a) 2.29 A.U.   (b) 1.96 A.U., 1.59 yr

Chapter 21

True-False 21.1 F,   21.2 T,   21.3 T,   21.4 T,   21.5 F,   21.6 T,   21.7 F,   21.8 T,   21.9 T,   21.10 F
Multiple Choice 21.11 a,   21.12 d,   21.13 d,   21.14 b,   21.15 d,   21.16 c,   21.17 c,   21.18 b,   21.19 b,   21.20 c
Odd-Numbered Problems 21.1 14.7 solar radii   21.3 20, 3.2 Mpc   21.5 0.46 pc; no—there are no O or B stars (in fact no stars at all) within that distance of us. For the Moon: 320 pc; yes, but rarely—see problem 14   21.7 1.2 × 10^{44} J; about 12 times greater than a supernova's electromagnetic output, and about 1/8 the output in the form of neutrinos   21.9 roughly 1000 km/s; not too bad an assumption—the nebula is moving too fast to be affected much by gravity, although it is probably slowing down as it runs into the interstellar medium   21.11 5.6 kpc   21.13 one per 270 years   21.15 0.43 percent, 8.6 × 10^{27} kg, or 1400 Earth masses

Chapter 22

True-False 22.1 T,   22.2 T,   22.3 T,   22.4 T,   22.5 F,   22.6 T,   22.7 F,   22.8 F,   22.9 T,   22.10 F
Multiple Choice 22.11 b,   22.12 c,   22.13 a,   22.14 b,   22.15 c,   22.16 d,   22.17 b,   22.18 b,   22.19 c,   22.20 b
Odd-Numbered Problems 22.1 1 million revolutions per day, or 11.6 revolutions per second   22.3 1.9 × 10^{12} m/s^2, or 190 billion Earth gravities; 190,000 km/s, or 64 percent of the speed of light; 306,000 km/s   22.5 2.4 × 10^{44} J, roughly twice the Sun's total energy output; 2.4 × 10^{34} J; 1.4 × 10^{25} J   22.7 63,000 km/s = 21 percent of the speed of light; 137,000 km/s   22.9 (a) 26 microarc sec,   (b) 0.016 arc sec,   (c) 4.1 arc min,   (d) 8100 A.U. = 0.039 pc   22.11 2 × 10^{10} m/s^2 = 2 × 10^9 g; 2 × 10^{-3} g; 2 × 10^{-9} g   22.13 14,000 solar masses   22.15 3 × 10^7 km = 0.2 A.U., tidal force due to the hole is 1.4 times greater than the star's gravity at a radius of 20 million km

Chapter 23

True-False 23.1 T,   23.2 F,   23.3 F,   23.4 T,   23.5 T,   23.6 F,   23.7 F,   23.8 F,   23.9 F,   23.10 T
Multiple Choice 23.11 d,   23.12 d,   23.13 d,   23.14 b,   23.15 b,   23.16 c,   23.17 a,   23.18 b,   23.19 d,   23.20 c
Odd-Numbered Problems 23.1 2.0″, much less than the angular diameter of Andromeda   23.3 100 kpc   23.5 −6.2; 17 Mpc   23.7 0.014″/yr; proper motion has in fact been measured for several globular clusters   23.9 (a) 3.5 kpc,   (b) 19 kpc   23.11 570 million years (assuming a rotation speed of 240 km/s at 15 kpc); 490 million years (assuming a rotation speed of 200 km/s at 5 kpc)   23.13 1 kpc   23.15 124 A.U.; no

Chapter 24

True-False 24.1 F, 24.2 F, 24.3 T, 24.4 F, 24.5 T, 24.6 T, 24.7 F, 24.8 F, 24.9 T, 24.10 T
Multiple Choice 24.11 a, 24.12 c, 24.13 a, 24.14 c, 24.15 c, 24.16 b, 24.17 b, 24.18 b, 24.19 a, 24.20 b
Odd-Numbered Problems 24.1 320 Mpc 24.3 14,000 km/s, 57 Mpc; 12,000 km/s, 67 Mpc; 16,000 km/s, 50 Mpc 24.5 58 arc seconds 24.7 -22.5; 8.7×10^{10} solar luminosities 24.9 5100 km/s 24.11 1.3×10^{-5} W/m^2; 1.0×10^{-7} W/m^2; yes, since much of the emission from the nucleus is in the infrared 24.13 2.2 million years 24.15 2.2×10^{12}

Chapter 25

True-False 25.1 T, 25.2 F, 25.3 F, 25.4 T, 25.5 F, 25.6 T, 25.7 T, 25.8 T, 25.9 F, 25.10 T
Multiple Choice 25.11 d, 25.12 b, 25.13 c, 25.14 a, 25.15 a, 25.16 b, 25.17 a, 25.18 d, 25.19 b, 25.20 a
Odd-Numbered Problems 25.1 6.5 billion years 25.3 1.5 nm, taking the rotation speed to be 350 km/s 25.5 2.6×10^{14} solar masses; reasonable to within a factor of perhaps 2–3 25.7 700 km/s; 660 km/s—comparable 25.9 9.4×10^{10} solar masses 25.11 1.8×10^{12} solar masses 25.13 21 Mpc 25.15 3.6×10^{12} solar masses

Chapter 26

True-False 26.1 F, 26.2 F, 26.3 F, 26.4 T, 26.5 F, 26.6 F, 26.7 F, 26.8 T, 26.9 T, 26.10 T
Multiple Choice 26.11 a, 26.12 d, 26.13 b, 26.14 b, 26.15 b, 26.16 d, 26.17 c, 26.18 c, 26.19 d, 26.20 d
Odd-Numbered Problems 26.1 1000 Mpc 26.3 4×10^8 26.5 1200 km/s 26.7 (a) 30,000 tons, (b) 2.8 pc 26.9 (a) 82, (b) 55 km/s/Mpc 26.11 7.4 km/s/Mpc; no 26.13 (a) 1.1 mm, (b) 0.0093, (c) 9.3×10^{-5}, (d) 9.3×10^{-7} 26.15 8.0 kpc

Chapter 27

True-False 27.1 F, 27.2 T, 27.3 T, 27.4 F, 27.5 F, 27.6 T, 27.7 F, 27.8 F, 27.9 T, 27.10 T
Multiple Choice 27.11 a, 27.12 d, 27.13 a, 27.14 b, 27.15 a, 27.16 c, 27.17 b, 27.18 a, 27.19 b, 27.20 c
Odd-Numbered Problems 27.1 6 kpc 27.3 20 K (redshift = 6.5); 3×10^{-24} kg/m^3 27.5 1.2×10^9 K; 1/5 the temperature in the text 27.7 3×10^{-12} m; hard X ray/gamma ray 27.9 12.7 Mpc 27.11 166 27.13 74 kpc 27.15 220 kpc, 2×10^{18} solar masses in a cube of that size

Chapter 28

True-False 28.1 F, 28.2 F, 28.3 F, 28.4 T, 28.5 F, 28.6 T, 28.7 T, 28.8 F, 28.9 T, 28.10 T
Multiple Choice 28.11 a, 28.12 b, 28.13 c, 28.14 c, 28.15 c, 28.16 b, 28.17 d, 28.18 d, 28.19 b, 28.20 a
Odd-Numbered Problems 28.1 6.3 seconds (for a 20 year old reader); 19 seconds (in 2004); 72 seconds; 162 seconds; 237 days 28.3 They would both increase by a factor of two 28.5 five years 28.7 27 A.U. 28.9 3400 years; no, civilizations are more likely to be found within the Galactic habitable zone 28.11 32,000 km/s 28.13 1.43×10^9 to 1.67×10^9 Hz; 2.4×10^6 channels 28.15 2.3 years; 55 years

INDEX

Star Charts

Have you ever become lost in an unfamiliar city or state? Chances are you used two things to get around: a map and some signposts. In much the same way, these two items can help you find your way around the night sky in any season. Fortunately, in addition to the seasonal Star Charts on the following pages, the sky provides us with two major signposts. Each seasonal description will talk about the Big Dipper—a group of seven bright stars that dominates the constellation Ursa Major the Great Bear. Meanwhile, the constellation Orion the Hunter plays a key role in finding your way around the sky from late autumn until early spring.

Each chart portrays the sky as seen from near 35° north latitude at the times shown at the top of the page. Located just outside the chart are the four directions: north, south, east, and west. To find stars above your horizon, hold the map overhead and orient it so a direction label matches the direction you're facing. The stars above the map's horizon now match what's in the sky.

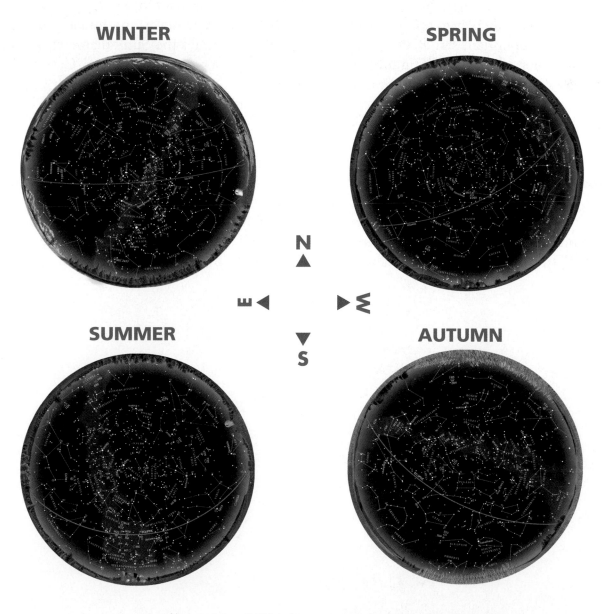

WINTER   SPRING

N ▲

E ◀   ► W

SUMMER   AUTUMN

▼ S

Exploring the Winter Sky

Winter finds the Big Dipper climbing the northeastern sky, with the three stars of its handle pointing toward the horizon and the four stars of its bowl standing highest. The entire sky rotates around a point near Polaris, a 2nd-magnitude star found by extending a line from the uppermost pair of stars in the bowl across the sky to the left of the Dipper. Polaris also performs two other valuable functions: The altitude of the star above the horizon equals your latitude north of the equator, and a straight line dropped from the star to the horizon.

Turn around with your back to the Dipper and you'll be facing the diamond-studded winter sky. The second great signpost in the sky, Orion the Hunter, is central to the brilliant scene. Three closely spaced, 2nd-magnitude stars form a straight line that represents the unmistakable belt of Orion. Extending the imaginary line joining these stars to the upper right leads to Taurus the Bull and its orangish 1st-magnitude star, Aldebaran. Reverse the direction of your gaze to the belt's lower left and you cannot miss Sirius the Dog Star—brightest in all the heavens at magnitude −1.5.

Now move perpendicular to the belt from its westernmost star, Mintaka, and find at the upper left of Orion the red supergiant star Betelgeuse. Nearly a thousand times the Sun's diameter, Betelgeuse marks one shoulder of Orion. Continuing this line brings you to a pair of bright stars, Castor and Pollux. Two lines of fainter stars extend from this pair back toward Orion—these represent Gemini the Twins. At the northeastern corner of this constellation lies the beautiful open star cluster M35. Head south of the belt instead and your gaze will fall on the blue supergiant star Rigel, Orion's other luminary.

Above Orion and nearly overhead on winter evenings is brilliant Capella, in Auriga the Charioteer. Extending a line through the shoulders of Orion to the east leads you to Procyon in Canis Minor the Little Dog. Once you have these principal stars mastered, using the chart to discover the fainter constellations will be a whole lot easier. Take your time, and enjoy the journey. Before leaving Orion, however, aim your binoculars at the line of stars below the belt. The fuzzy "star" in the middle is actually the glorious Orion Nebula (M42), a stellar nursery illuminated by bright, newly formed stars.

WINTER

2 a.m. on December 1; midnight on January 1; 10 p.m. on February 1

N

E

W

S

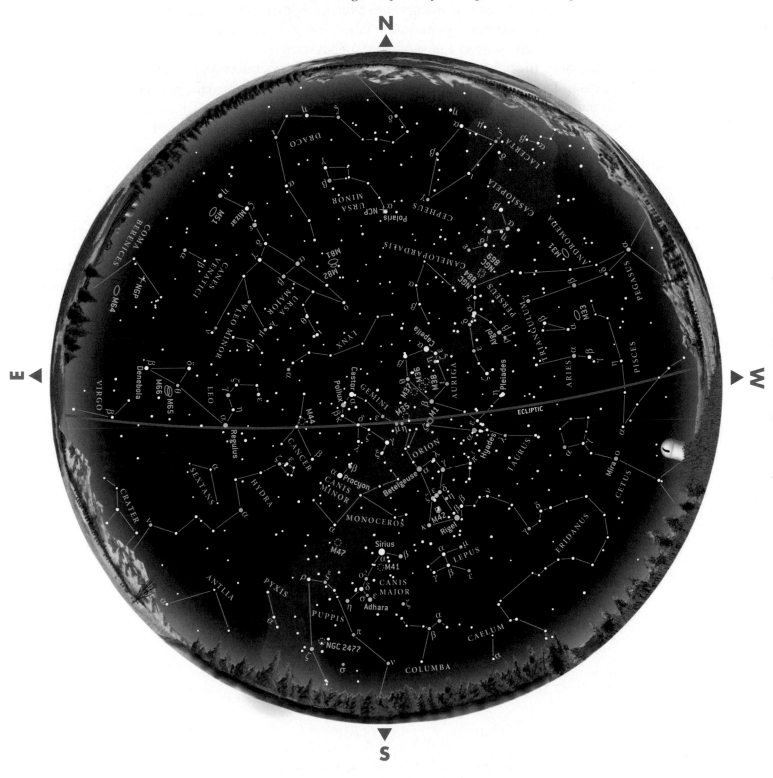

Exploring the Spring Sky

The Big Dipper, our signpost in the sky, swings high overhead during the spring and lies just north of the center of the chart. This season of rejuvenation encourages us to move outdoors with the milder temperatures, and with the new season a new set of stars beckons us.

Follow the arc of stars outlining the handle of the Dipper away from the bowl and you will land on brilliant Arcturus. This orangish star dominates the spring sky in the kite-shaped constellation Boötes the Herdsman. Well to the west of Boötes lies Leo the Lion. You can find its brightest star, Regulus, by using the pointers of the Dipper in reverse. Regulus lies at the base of a group of stars shaped like a sickle or backward question mark, which represents the head of the lion.

Midway between Regulus and Pollux in Gemini, which is now sinking in the west, is the diminutive group Cancer the Crab. Centered in this group is a hazy patch of light that binoculars reveal as the Beehive star cluster (M44).

To the southeast of Leo lies the realm of the galaxies and the constellation Virgo the Maiden. Virgo's brightest star, Spica, shines at magnitude 1.0.

During springtime, the Milky Way lies level with the horizon, and it's easy to visualize that we are looking out of the plane of our galaxy. In the direction of Virgo, Leo, Coma Berenices, and Ursa Major lie thousands of galaxies whose light is unhindered by intervening dust in our own galaxy. However, all these galaxies are elusive to the untrained eye and require binoculars or a telescope to be seen.

Boötes lies on the eastern border of this galaxy haven. Midway between Arcturus and Vega, the bright "summer" star rising in the northeast, is a region where no star shines brighter than 2nd magnitude. A semicircle of stars represents Corona Borealis the Northern Crown, and adjacent to it is a large region that houses Hercules the Strongman, the fifth-biggest constellation in the sky. It is here we can find the northern sky's brightest globular star cluster, M13. A naked-eye object from a dark site, it looks spectacular when viewed through a telescope.

Returning to Ursa Major, check the second-to-last star in the Dipper's handle. Most people will see it as double, while binoculars show this easily. The pair is called Mizar and Alcor, and they lie just 0.2° apart. A telescope reveals Mizar itself to be a double. Its companion star shines at magnitude 4.0 and lies 14 arcseconds away.

SPRING

1 a.m. on March 1; 11 p.m. on April 1; 9 p.m. on May 1. Add one hour for daylight-saving time

N

E

W

S

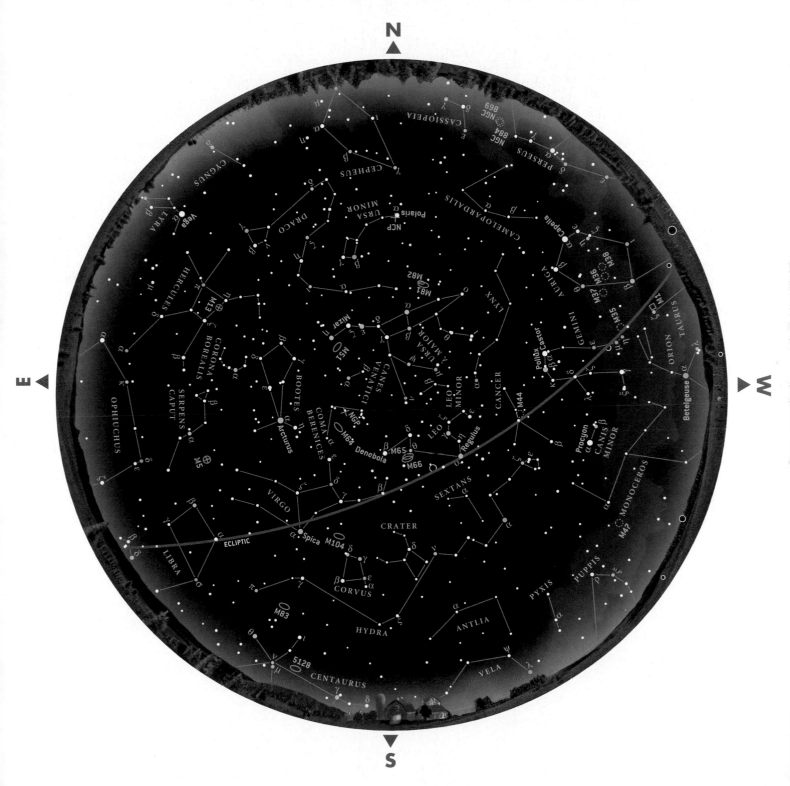

Reproduced by permission. © 2004, Astronomy magazine, Kalmbach Publishing Co.

Exploring the Summer Sky

The richness of the summer sky is exemplified by the splendor of the Milky Way. Stretching from the northern horizon in Perseus, through the cross-shaped constellation Cygnus overhead, and down to Sagittarius in the south, the Milky Way is packed with riches. These riches include star clusters, nebulae, double stars, and variable stars.

Let's start with the Big Dipper, our perennial signpost, which now lies in the northwest with its handle still pointing toward Arcturus. High overhead, and the first star to appear after sunset, is Vega in Lyra the Harp. Vega forms one corner of the summer triangle, a conspicuous asterism of three stars. Near Vega lies the famous double-double, Epsilon (ε) Lyrae. Two 5th-magnitude stars lie just over 3 arcminutes apart and can be split when viewed through binoculars. Each of these two stars is also double, but you need a telescope to split them.

To the east of Vega lies the triangle's second star: Deneb in Cygnus the Swan (some see a cross in this pattern). Deneb marks the tail of this graceful bird, the cross represents its outstretched wings, and the base of the cross denotes its head, which is marked by the incomparable double star Albireo. Albireo matches a 3rd-magnitude yellow star and a 5th-magnitude blue star and offers the finest color contrast anywhere in the sky. Deneb is a supergiant star that pumps out enough light to equal 60,000 Suns. Also notice that the Milky Way splits into two parts in Cygnus, a giant rift caused by interstellar dust blocking starlight from beyond.

Altair, the third star of the summer triangle and the one farthest south, is the second brightest of the three. Lying 17 light-years away, it's the brightest star in the constellation Aquila the Eagle.

Frequently overlooked to the north of Deneb lies the constellation Cepheus the King. Shaped rather like a bishop's hat, the southern corner of Cepheus is marked by a compact triangle of stars that includes Delta (Δ) Cephei. This famous star is the prototype of the Cepheid variable stars used to determine the distances to some of the nearer galaxies. It varies regularly from magnitude 3.6 to 4.3 and back again with a 5.37-day period.

Hugging the southern horizon, the constellations Sagittarius the Archer and Scorpius the Scorpion lie in the thickest part of the Milky Way. Scorpius's brightest star, Antares, is a red supergiant star whose name means "rival of Mars" and derives from its similarity to the planet in both color and brightness.

SUMMER

1 a.m. on June 1; 11 p.m. on July 1; 9 p.m. on August 1. Add one hour for daylight-saving time

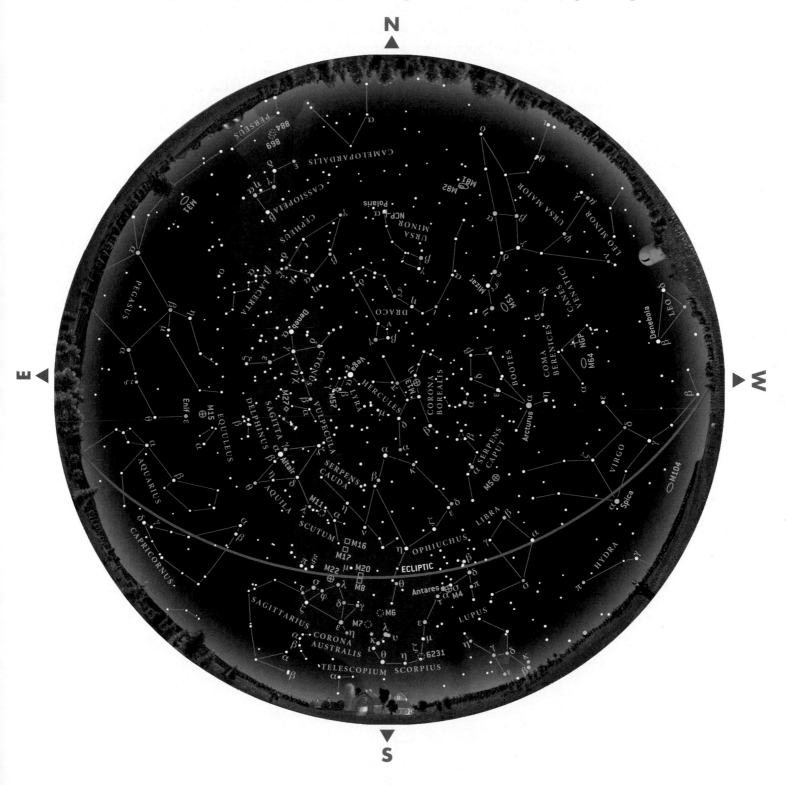

Exploring the Autumn Sky

The cool nights of autumn are here to remind us the chill of winter is not far off. Along with the cool air, the brilliant stars of the summer triangle descend in the west to be replaced with a rather bland-looking region of sky. But don't let initial appearances deceive you. Hidden in the fall sky are gems equal to summertime.

The Big Dipper swings low this season, and for parts of the Southern United States it actually sets. Cassiopeia the Queen, a group of five bright stars in the shape of a "W" or "M," reaches its highest point overhead, the same spot the Big Dipper reached six months ago. To the east of Cassiopeia, Perseus the Hero rises high. Nestled between these two groups is the wondrous Double Cluster—NGC 869 and NGC 884—a fantastic sight in binoculars or a low-power telescope.

Our view to the south of the Milky Way is a window out of the plane of our galaxy in the opposite direction to that visible in spring. This allows us to look at the Local Group of galaxies. Due south of Cassiopeia is the Andromeda Galaxy (M31), a 4th-magnitude smudge of light that passes directly overhead around 9 p.m. in mid-November. Farther south, between Andromeda and Triangulum, lies M33, a sprawling face-on spiral galaxy best seen in binoculars or a rich-field telescope.

The Great Square of Pegasus passes just south of the zenith. Four 2nd- and 3rd-magnitude stars form the square, but few stars can be seen inside of it. If you draw a line between the two stars on the west side of the square and extend it southward, you'll find 1st-magnitude Fomalhaut in Piscis Austrinus the Southern Fish. Fomalhaut is the solitary bright star low in the south. Using the eastern side of the square as a pointer to the south brings you to Diphda in the large, faint constellation of Cetus the Whale.

To the east of the Square lies the Pleiades star cluster (M45) in Taurus, which reminds us of the forthcoming winter. By late evening in October and early evening in December, Taurus and Orion have both cleared the horizon and Gemini is rising in the northeast. In concert with the reappearance of winter constellations, the view to the northwest finds summertime's Cygnus and Lyra about to set. The autumn season is a great transition period, both on Earth and in the sky, and a fine time to experience the subtleties of these constellations.

AUTUMN

1 a.m. on September 1; 11 p.m. on October 1; 9 p.m. on November 1. Add one hour for daylight-saving time

N
▲

E ◀   ▶ **W**

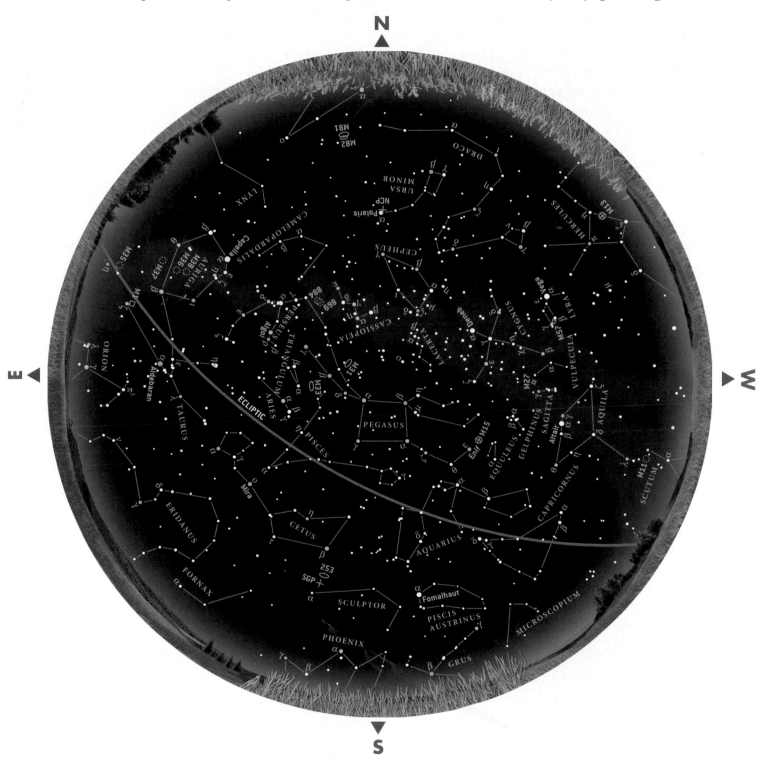

▼
S

Reproduced by permission. © 2004, Astronomy magazine, Kalmbach Publishing Co.

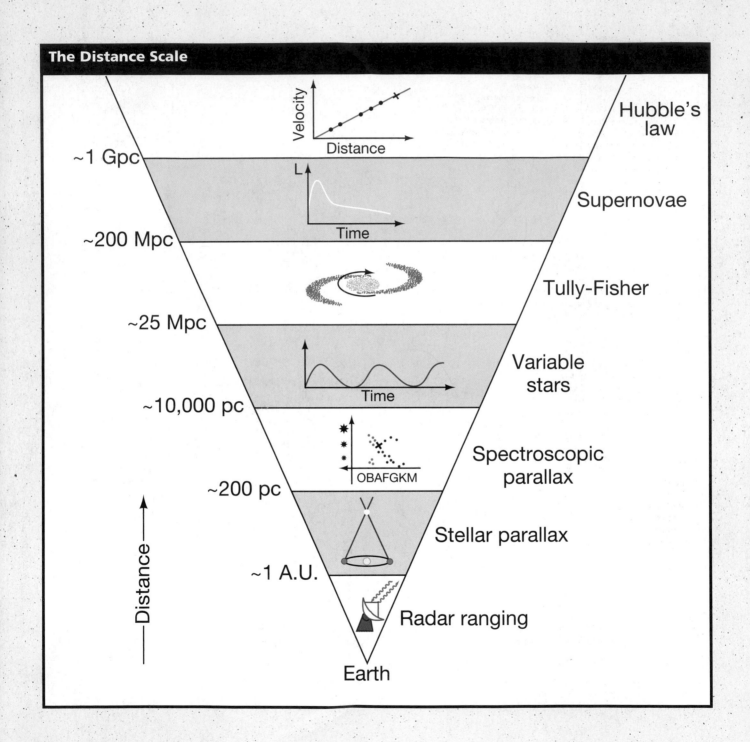

The Distance Scale

Hubble's law

~1 Gpc

Supernovae

~200 Mpc

Tully-Fisher

~25 Mpc

Variable stars

~10,000 pc

Spectroscopic parallax

~200 pc

Stellar parallax

~1 A.U.

Radar ranging

Earth

Distance